Physiology and Function of Glial Cells in Health and Disease

Bilal El-Mansoury
Faculty of Sciences, Chouaib Doukkali University, Morocco

Omar El Hiba
Chouaib Doukkali University, Morocco

Arumugam Radhakrishnan Jayakumar
University of Miami, USA

A volume in the Advances in Bioinformatics and Biomedical Engineering (ABBE) Book Series

IGI Global
PUBLISHER of TIMELY KNOWLEDGE

Published in the United States of America by
IGI Global
Engineering Science Reference (an imprint of IGI Global)
701 E. Chocolate Avenue
Hershey PA, USA 17033
Tel: 717-533-8845
Fax: 717-533-8661
E-mail: cust@igi-global.com
Web site: http://www.igi-global.com

Library of Congress Cataloging-in-Publication Data

Names: El-Mansoury, Bilal, 1995- editor. | El Hiba, Omar, 1985- editor. | Jayakumar, Arumugam, 1965- editor.
Title: Physiology and function of glial cells in health and disease / edited by Bilal El-Mansoury, Omar El Hiba, Arumugam Jayakumar.
Description: Hershey, PA : Engineering Science Reference, [2024] | Includes bibliographical references and index. | Summary: "This book leads to a deep understanding of neuroglia functions and their interaction with neurons and give exciting insights into new therapeutic approaches to various brain abnormalities such as neurodegenerative diseases, neurological disorders, gliomas, and metabolic diseases by targeting glial cells"-- Provided by publisher.
Identifiers: LCCN 2023040604 (print) | LCCN 2023040605 (ebook) | ISBN 9781668496756 (hardcover) | ISBN 9781668496763 (ebook)
Subjects: LCSH: Neuroglia. | Nervous system--Abnormalities--Etiology.
Classification: LCC QP363.2 .P497 2024 (print) | LCC QP363.2 (ebook) | DDC 612.8/1046--dc23/eng/20230927
LC record available at https://lccn.loc.gov/2023040604
LC ebook record available at https://lccn.loc.gov/2023040605

This book is published in the IGI Global book series Advances in Bioinformatics and Biomedical Engineering (ABBE) (ISSN: 2327-7033; eISSN: 2327-7041)

British Cataloguing in Publication Data
A Cataloguing in Publication record for this book is available from the British Library.

Advances in Bioinformatics and Biomedical Engineering (ABBE) Book Series

Ahmad Taher Azar
College of Computer & Information Sciences, Prince Sultan University, Riyadh, Saudi Arabia & Faculty of Computers and Artificial Intelligence, Benha University, Benha, Egypt

ISSN:2327-7033
EISSN:2327-7041

Mission

The fields of biology and medicine are constantly changing as research evolves and novel engineering applications and methods of data analysis are developed. Continued research in the areas of bioinformatics and biomedical engineering is essential to continuing to advance the available knowledge and tools available to medical and healthcare professionals.

The **Advances in Bioinformatics and Biomedical Engineering (ABBE) Book Series** publishes research on all areas of bioinformatics and bioengineering including the development and testing of new computational methods, the management and analysis of biological data, and the implementation of novel engineering applications in all areas of medicine and biology. Through showcasing the latest in bioinformatics and biomedical engineering research, ABBE aims to be an essential resource for healthcare and medical professionals.

Coverage

- Biostatistics
- DNA Sequencing
- Databases
- Neural Engineering
- Robotics and Medicine
- Genetic Engineering
- Biomedical Sensors
- Applied Biosystems
- Molecular Simulations
- Dental Engineering

IGI Global is currently accepting manuscripts for publication within this series. To submit a proposal for a volume in this series, please contact our Acquisition Editors at Acquisitions@igi-global.com or visit: http://www.igi-global.com/publish/.

The Advances in Bioinformatics and Biomedical Engineering (ABBE) Book Series (ISSN 2327-7033) is published by IGI Global, 701 E. Chocolate Avenue, Hershey, PA 17033-1240, USA, www.igi-global.com. This series is composed of titles available for purchase individually; each title is edited to be contextually exclusive from any other title within the series. For pricing and ordering information please visit http://www.igi-global.com/book-series/advances-bioinformatics-biomedical-engineering/73671. Postmaster: Send all address changes to above address.

Titles in this Series

For a list of additional titles in this series, please visit: www.igi-global.com/book-series

Applications of Synthetic Biology in Health, Energy, and Environment
Muhammad Arshad (International Islamic University, Islamabad, Pakistan)
Medical Information Science Reference • © 2023 • 454pp • H/C (ISBN: 9781668465776) • US $325.00

Isolation, Characterization, and Therapeutic Applications of Natural Bioactive Compounds
Ajeet Singh (ICAR-Indian Institute of Wheat and Barley Research, India)
Medical Information Science Reference • © 2022 • 401pp • H/C (ISBN: 9781668473375) • US $345.00

AI-Enabled Smart Healthcare Using Biomedical Signals
Rahul Kumar Chaurasiya (Maulana Azad National Institute of Technology, Bhopal, India) Dheeraj Agrawal (Maulana Azad National Institute of Technology, Bhopal, India) and Ram Bilas Pachori (Indian Institute of Technology, Indore, India)
Medical Information Science Reference • © 2022 • 322pp • H/C (ISBN: 9781668439470) • US $435.00

Advancements in Controlled Drug Delivery Systems
Shekhar Verma (University College of Pharmacy, Pandit Deendayal Upadhyay Memorial Health Science, India & Ayush University of Chattisgarh, India) and Santosh Kumar Verma (Yulin University, China)
Medical Information Science Reference • © 2022 • 395pp • H/C (ISBN: 9781799889083) • US $360.00

Polymer-Based Functional Materials for Biomedical Applications
Nithin Kundachira Subramani (The National Institute of Engineering, India) and D.V. Gowda (JSS College of Pharmacy, India)
Engineering Science Reference • © 2022 • 320pp • H/C (ISBN: 9781799871989) • US $265.00

Emerging Developments and Applications of Low Temperature Plasma
Aamir Shahzad (Government College University, Faisalabad, Pakistan) and Maogang He (Xi'an Jiaotong University, China)
Engineering Science Reference • © 2022 • 239pp • H/C (ISBN: 9781799883982) • US $250.00

Biomedical Computing for Breast Cancer Detection and Diagnosis
Wellington Pinheiro dos Santos (Universidade Federal de Pernambuco, Brazil) Washington Wagner Azevedo da Silva (Universidade Federal de Pernambuco, Brazil) and Maira Araujo de Santana (Universidade Federal de Pernambuco, Brazil)
Medical Information Science Reference • © 2021 • 357pp • H/C (ISBN: 9781799834564) • US $325.00

IGI Global
PUBLISHER of TIMELY KNOWLEDGE

701 East Chocolate Avenue, Hershey, PA 17033, USA
Tel: 717-533-8845 x100 • Fax: 717-533-8661
E-Mail: cust@igi-global.com • www.igi-global.com

To the memory of my father, Ahmed
To my mother
To my wonderful wife, Maryam
To my son, Ahmed
To my brother, Mohammed
To all my family and friends
Omar El Hiba

To my father, Lahcen
To my mother, Fatiha
To my uncle, Ahmed
To all my family and friends
Bilal El-Mansoury

Editorial Advisory Board

Table of Contents

Detailed Table of Contents

Glial cells were once thought of as simple support players in the central nervous system (CNS). However, the latest studies of glial cells have shown that they are actually of significance and play a variety of functions. Once thought to be passive supportive cells, astrocytes are now important in maintaining neurotransmitter balance, controlling synaptic activity, and regulating blood flow in the brain. Oligodendrocytes and Schwann cells, besides making myelin, are closely involved in controlling how fast nerve signals travel and maintaining the health of axons. Microglia had been thought to be only immune observers, but they also help to control the immune response and control synapses. Ependymal cells, which are sometimes ignored, are important in regulating cerebrospinal fluid circulation, directing the location of neural stem cells, and enabling cell communication. This chapter explores distinct glial cell types -astrocytes, oligodendrocytes, Schwann cells, microglia, and ependymal cells- highlighting their newfound, intricate functionalities and interactions with neurons.

Astrocytes are the most proliferate glial cells in the central nervous system (CNS); they are considered as the supporting cells in the CNS, and play a big role in behavioral, circuit, and synaptic functions. Astrocytes are also important for neuronal repair, neurogenesis, and survival. Astrocytes play a main role in brain protection, by maintaining a regulated ion balance and blood flow, and preserving an antioxidant functions in brain. In this chapter, the authors elaborated an overview of the role of astrocytes on the developing brain.

Astrocytes, once considered passive support cells in the central nervous system, are now recognized as dynamic contributors to neuronal processes. They play pivotal roles in regulating synaptic transmission, modulating excitability, and influencing synapse formation. These non-neuronal cells release gliotransmitters like glutamate, affecting synaptic activity. Dysfunctions in astrocytes are linked to neurodegenerative and psychiatric disorders. In neurodegenerative disorders like Alzheimer's and Parkinson's, astrocytic dysfunction plays distinct roles. While astrocytes may not significantly contribute to Alzheimer's progression, they are involved in neuroinflammation, Aβ metabolism, and calcium regulation. Conversely, in Parkinson's, astrocytes contribute to mitochondrial dysfunction, impacting dopaminergic neurons. This comprehensive exploration sheds light on the intricate and multifaceted roles of astrocytes in cognition and their potential implications for therapeutic interventions in neurological and psychiatric conditions.

Microglia are the resident macrophages of the central nervous system (CNS). These cells of mesodermal/mesenchymal origin migrate into all regions of the CNS. Recent studies indicate that even in the normal brain, microglia have highly motile processes by which they scan their territorial domains. By a large number of signaling pathways, they can communicate with macroglial cells (e.g. astrocytes) and neurons and with cells of the immune system. Under normal physiological conditions, microglia constantly monitor their microenvironment and survey neurons. Microglia have other functions including the participation in the formation of new blood vessels or angiogenesis, cognitive function, the regulation of synaptic plasticity, and neurogenesis and they play a crucial role in the CNS through communication with other brain cells. This chapter will provide an overview of the functions of microglial cells within the CNS.

Brain aging is a complicated and diversified process that includes an ongoing decline in cognitive performance as well as various cellular and molecular changes in the brain. During an individual's life, glial cells such as oligodendrocytes, microglia, and astrocytes are crucial for maintaining healthy brain function. Astrocytes support neurons but may become less efficient with age, contributing to cognitive decline. Microglia, the brain's immune cells, can turn overactive with age, causing inflammation which can lead to neuronal damage. Oligodendrocytes, responsible for myelin sheath maintenance, also undergo changes, affecting neural communication. With advancing age, changes in glial cell function may lead to neuroinflammation, oxidative damage, and abnormalities in neural circuitry, all of which contribute to age related cognitive decline and neurodegenerative disorders. This chapter delves into the intricate dynamics of brain aging, focusing on the profound influence of age-related changes in glial cells.

Himel Mondal, All India Institute of Medical Sciences, Deoghar, India
Shaikat Mondal, Raiganj Governmnet Medical College, West Bengal, India

Microglia, specialized immune cells within the central nervous system (CNS), are central to maintaining CNS health and protecting against damage and infections. However, their improper activation can contribute to the development and progression of various neurological diseases, including Alzheimer's, Parkinson's, multiple sclerosis, autism, and traumatic brain injury. Neuroinflammation, a complex process involving microglia and other glial cells, is considered a key factor in the pathogenesis of many neurological disorders. To comprehensively address this, this chapter introduces microglia's essential role in CNS homeostasis and sets the stage for understanding the concept of neuroinflammation, range of neurological and psychiatric disorders, cutting-edge imaging and genetic tools, the potential therapeutic strategies for modulating microglial function and targeting neuroinflammation.

Hafida El Ghachi, Faculty of Sciences Semlalia, Cadi Ayyad University, Morocco
Asmaa Haj-khlifa, Faculty of Sciences Semlalia, Cadi Ayyad University, Morocco
Mjid Oukhrib, Faculty of Sciences Semlalia, Cadi Ayyad University, Morocco
Lahcen Tamegart, Faculty of Sciences, Abdelmalek Essaadi University, Morocco
Youssef Ait Hamdan, Higher Normal School, Cadi Ayyad University, Morocco
Abdelali Ben Maloui, Faculty of Sciences Semlalia, Cadi Ayyad University, Morocco
Najib Kissani, Department of Neurology, Medical Research Center, University Hospital Mohamed VI, Morocco
Halima Gamrani, Faculty of Sciences Semlalia, Cadi Ayyad University, Morocco

Chronic pain is a pathological health condition caused by the imbalance between inhibitory and excitatory signaling pain pathways. This clinical problem affects about 30% of adults worldwide, and current treatments are insufficient with side effects. The mechanisms underlying this disease were thought to be a result of neuronal dysfunction, however, several animal studies clearly showed that non-neuronal cells (glial cells), primarily microglia and astrocytes, and their interaction with neurons, had a key role in pain persistence. A long-lasting state of activated glia turns to a pathological element causing a painful sensation. This chapter aims to elucidate glial dysfunction by shedding light on peripheral and central glial cells involved in the pathogenesis of chronic pain.

Brain injury is damage to the brain caused by external forces. They are the leading cause of death and ill health in young men, with a high incidence worldwide. Glial cells make up the majority of central nervous system cells and play a key role in the brain's response to brain injury. Astrocytes play a crucial role in maintaining the brain's ionic and water balance, energy metabolism, regulation of the blood-brain barrier and immune response. In response to brain injury, astrocytes undergo morphological and behavioral changes that modify their protein expression. To sustain astrocytic responses in certain severe situations, therapeutic strategies that make use of biomaterials represent a significant advance in the field of regenerative medicine, as they create a microenvironment more conducive to nerve tissue regeneration. This chapter examines the behavior and responses of astrocytes to brain injury in the context of traumatic brain injury (TBI), as well as looking at the implication of biomaterials in the treatment of TBI.

It is well known that the decline in the viability of dopamine neurons slowly leads to the appearance of various symptoms characteristic of Parkinson's disease (PD). These symptoms depend mainly on dysfunction of nigrostriatal dopaminergic denervation. The cause of neuronal death has not yet been elucidated, although there are several hypotheses suggesting different factors that may trigger it. One possible mechanism of neurodegeneration is the establishment of chronic inflammation in the central nervous system, where glial cells are key regulators of inflammatory responses. They also play a phagocytic role, engulfing synapses, apoptotic cells, cellular debris and released toxic proteins. An imbalance in the activation of these cells can lead to an overproduction of cytotoxic factors, which contribute to the death of dopamine neurons. As PD involves not only the loss of dopamine neurons, but also the dysfunction of glial cells, whose loss or excessive activation can contribute to neuronal death, there is a need to better understand the role of these cells in PD in order to develop effective therapies.

Amyotrophic Lateral Sclerosis (ALS) is a devastating neurodegenerative disease, causing death within two to five years following diagnosis. ALS is a disease characterized by motor neuron degeneration and destruction of the neuromuscular junction (NMJ). To date, numerous genetic mutations have been associated with ALS, but the mutation in the gene coding for superoxide dismutase (SOD1) has been the most extensively studied. Nevertheless, studies carried out on rodent models and ALS patients have shown that damage to glial cells contributes directly to the development and progression of the disease. However, few studies have focused on the properties of glial cells in ALS. The accumulation of knowledge on the active role and pathological mechanisms of each glial type in the disease must be carefully applied to better understand the ALS pathophysiology, but also for the development of targeted therapy for glial cells. Therefore, in this chapter the authors aimed to compile the most recent information on how each type of glial cell contributes to the development of ALS.

Multiple sclerosis (MS) is a complex autoimmune disease of the central nervous system (CNS), which is characterized by demyelination and neurodegeneration. Oligodendrocytes have an essential role in maintaining the integrity of myelin, the protective sheath that surrounds nerve fibers and is indispensable for efficient signal transmission. Nevertheless, in multiple sclerosis, oligodendrocytes become dysfunctional, leading to myelin deterioration and axonal degeneration. Recent research suggests that metabolic changes, including mitochondrial dysfunction and alterations in glucose and lipid metabolism, contribute significantly to the pathogenesis of MS. Mitochondrial dysfunction is observed in both immune cells and CNS oligodendrocytes of MS patients. Impaired mitochondrial function leads to energy deficits, affecting crucial processes such as impulse transmission and axonal transport, ultimately contributing to neurodegeneration. Understanding the complex relationship between these mechanisms is crucial to the development of an effective treatment for MS.

and the tumor's infiltrative nature restrict their effectiveness, frequently leading to tumor recurrence and illness progression. To create new therapeutic approaches and enhance patient outcomes, it is crucial to comprehend the physiopathology of glioblastoma and its associated consequences. To improve treatment and quality of life for patients with glioblastoma, further research is required to clarify molecular causes, discover therapeutic targets, and address the difficulties provided by comorbidities.

Although the pathophysiology of Alzheimer's disease (AD) is exceedingly complex and poorly understood, the illness is nonetheless of great interest to the scientific community. Recent advances in AD research have allowed for the possibility that further treatment advantages might be found, which would assist patients all around the world. However, recent studies suggest that glial cells, such as microglia, astrocytes, oligodendrocytes, and oligodendrocyte progenitor cells (NG2 glia), are linked to the pathogenesis of AD and may offer several potential therapeutic targets against AD. Previous research on AD has focused primarily on neurons. Glial cells are essential to the structural integrity of neurons and are necessary for regulating homeostasis (concentration of ions, neurotransmitters) within the central nervous system. This chapter investigates the following topics: (i) the function of glial cells in the pathogenesis of AD; (ii) the intricate functions of the constituent parts; and (iii) prospective therapeutic targets that may one day improve the quality of life for AD patients.

Neurodegenerative disorders (NDD) are chronic conditions that lead to nerve cells degeneration in the central nervous system. The most common NDD are Alzheimer's disease (AD), Parkinson's disease (PD), amyotrophic lateral sclerosis (ALS), multiple sclerosis (MS), and Huntington's disease (HD). Parkinson's disease (PD) is recognized as the second most common neurodegenerative disease after Alzheimer's disease (AD) and Huntington disease (HD). Astrocytes reprogramming has been recently proposed as one of the most important treatments to stop the progression of these neurological diseases. By targeting the microRNA, astrocytes in the animal brain could be reprogrammed into various functional neural systems. Activation of endogenous genes could be a revolutionary therapeutic approach for treating neurological diseases. Reprogrammed astrocytes could also be used to generate various types of neurons lost in case of Huntington's disease, Parkinson's disease, and Alzheimer's disease. The ability to reprogram astrocytes into functional neurons was classified as a new advancement for PD, AD, and HD treatments.

Depression is a complex neuropsychiatric illness affecting millions worldwide. Furthermore, its exact cause remains uncertain. Theories include the monoamine hypothesis, changes in hormonal systems, inflammation, immune alterations, neurogenesis issues, and environmental factors. Recent research has highlighted the role of glial cells, particularly microglia and astrocytes, in depression. This chapter explores the implications of glial cells in depression, shedding light on their involvement in neuroinflammation, synaptic plasticity, and the regulation of mood-related circuits. An intricate interplay between glial cells, proinflammatory cytokines, and neuronal processes appears to underlie the etiology and progression of depression. Understanding the dynamic interactions between glial cells and neurons in the context of depression offers promising avenues for novel therapeutic interventions targeting this debilitating disorder's neuroinflammatory components.

Lysosomal storage disorders (LSDs) are a group of about 70 life-threatening conditions caused by genetic defects affecting lysosomal components. The underscoring molecular deficiency leads to widespread cellular dysfunction through most tissues in the body, including peripheral organs and the central nervous system (CNS). Clinical and experimental strategies encompassing enzyme replacement, gene and cell therapies, substrate reduction, and chemical chaperones are showing considerable potential in attenuating the peripheral pathology. Microglial and astrocyte activation is a hallmark of many LSDs that affect the CNS, which often precedes and predicts regions where eventual neuron loss will occur. However, the timing, intensity, and duration of neuroinflammation may ultimately dictate the impact on CNS homeostasis. For example, a transient inflammatory response following CNS insult/injury can be neuroprotective, as glial cells attempt to remove the insult and provide trophic support to neurons.

Glial cells, although often overlooked in comparison with neurons, play a crucial role in the structure and function of the nervous system. Their loss can lead to impairment of functions such as learning and memory, contributing to neurodegenerative diseases. To combat these diseases effectively, therapeutic strategies must restore the function of neuronal and glial cells. Natural resources offer promising therapeutic molecules for the treatment of neurodegenerative diseases. Chitosan (CTS), chitooligosaccharide (COS) and their derivatives from arthropod exoskeletons have attracted the attention of neuroscientists because of their neuroprotective effects and their ability to transport therapeutic substances across the blood-brain barrier (BBB). This chapter explores the role of natural resources in the fight against neuronal diseases.

Neuroinflammation is a complex, well-orchestrated process involving various groups of CNS glial cells and immune cells. Glial cells play a remarkable structural and functional role in the nervous system compared to neurons. However, neuroinflammation at the central level is a key player in various neurological disorders, including neurodegenerative diseases and CNS lesions. Therapeutic approaches to combat human neurodegenerative diseases must therefore restore neuronal and glial cell function. Natural resources are a source of potential therapeutic molecules for the treatment of neurodegenerative diseases. Currently, chitosan and its derivatives from arthropod exoskeletons are endowed with powerful anti-neuroinflammatory properties, as well as the ability to transport therapeutic substances across the BBB. This chapter discusses possible therapeutic options, the mechanism and role of chitosan in alleviating neuroinflammation at the central level, and the resulting diseases, in particular glial cell disorders.

Glial cells play a remarkable structural and functional role in the nervous system compared to neurons. However, damage to glial cells leads to degeneration and the onset of numerous neurodegenerative diseases. It is well known that the central nervous system (CNS) has a limited capacity for tissue regeneration. The use of chitosan-based biomaterials has emerged as a potential alternative solution to overcome these limitations. Interest in these biomaterials for CNS biomedical implementation has increased due to their ability to cross the BBB, their mucoadherence, and especially their hydrogel-forming capacity. In addition, their ability to form porous scaffolds and to carry cells and biomolecules has offered a means of achieving glial cell regeneration. Thus, this chapter aims to bring together recent work that highlights the potential of chitosan-based biomaterials suitable for glial cell tissue engineering.

Preface

Glial cells, including microglia, astrocytes, oligodendrocytes, and their progenitors NG2-glia, serve as key players in maintaining structural integrity and complex brain homeostasis. They actively participate in neurotransmission, energy metabolism, synaptic plasticity, neurogenesis, ion balance, immune defense, and the clearance of neuronal debris. However, the physiological functions of glial cells are often compromised in aging, neurodegenerative diseases such as Alzheimer's disease, Parkinson's disease, amyotrophic lateral sclerosis, and multiple sclerosis, as well as in gliomas, which are brain tumors that demand specialized understanding for effective therapeutic interventions.

In our book *Physiology and Function of Glial Cells in Health and Disease,* we provided a comprehensive exploration of the vital role played by glial cells in maintaining neural homeostasis within the central nervous system (CNS). This book delves into the intricate interaction between glial cells and neurons, shedding light on their essential contributions to neural function and overall brain health.

Readers will gain a deep understanding of the intricate relationship between glial cells and neurons, paving the way for novel therapeutic approaches to address neurodegenerative diseases, neurological disorders, gliomas, and metabolic conditions. The book also highlights emerging research on astrocyte reprogramming for the management of neurodegenerative diseases, offering a glimpse into potential future therapies. This book serves as an essential resource for researchers, clinicians, and students in the field of neuroscience. With in-depth discussions and cutting-edge insights, this research publication is a valuable reference for anyone seeking a thorough understanding of the role of glial cells in both health and disease.

The book is a collection of a total of 21 chapters.

Through a total of 21 chapters, the current book:

Addresses specific aspects of glial cell function and their interaction with neurons, unveiling fascinating insights into the mechanisms underlying various brain abnormalities. From exploring the classification and functions of glial cells in the mature CNS to investigating the role of astrocytes, microglia, and oligodendrocytes in neuroinflammation, neurodegeneration, and cognitive processes, this book offers a comprehensive overview in the field.

In the first chapter "Glial Cells in the Mature Central Nervous System: Classification and Functions" the authors explore distinct glial cell types -astrocytes, oligodendrocytes, Schwann cells, microglia, and ependymal cells- highlighting their newfound, intricate functionalities and interactions with neurons, while chapter 2 "Astrocytes and the Developing Brain" elaborates an overview of the role of astrocytes on the developing brain. In chapter 3 "Cognitive Function Involving Glial Cells: The Surprising Role of Astrocytes" sheds light on the intricate and multifaceted roles of astrocytes in cognition and their potential implications for therapeutic interventions in neurological and psychiatric conditions. Chapter

4 "Microglial Cells Function in the Central Nervous System: Beyond the Immune Function" however, provides an overview of the functions of microglial cells within the CNS. In chapter 5 "Brain Aging and Glial Cells: What is the Link?" the authors describe the intricate dynamics of brain aging, focusing on the profound influence of age-related changes in glial cells. While in chapter 6 "Microglia and Neuroinflammation", introduces microglia's essential role in CNS homeostasis and sets the stage for understanding the concept of neuroinflammation, range of neurological and psychiatric disorders, cutting-edge imaging and genetic tools, the potential therapeutic strategies for modulating microglial function and targeting neuroinflammation. Chapter 7 "Glial Cells Dysfunction and Chronic Pain" elucidate glial dysfunction by shedding light on peripheral and central glial cells involved in the pathogenesis of chronic pain. While Chapter 8 "Astrocytes functions and their involvement in brain injury: A focus on the biomaterials role in mitigating traumatic brain injury" examines the behavior and responses of astrocytes to brain injury in the context of traumatic brain injury (TBI), as well as looking at the implication of biomaterials in the treatment of TBI. In chapter 9 "Parkinson's Disease Involving Glial Cells Dysfunction" the authors try to understand the role of glial cells in Parkinson's disease leading to the development of effective therapies. Whereas, chapter 10 "Amyotrophic Lateral Sclerosis Involving Gliopathy: Insights into the Underlying Mechanisms" summarizes the most recent data on how each type of glial cell contributes to the development of Amyotrophic Lateral Sclerosis Involving. In chapter 11 "Oligodendrogliopathy in Multiple Sclerosis" we describe the role of oligodendrocytes and there pathology contribution to the neuropathology of Multiple Sclerosis. Chapter 12 "Hepatic Encephalopathy as a Gliopathy: The Mechanisms of Alzheimer Type II Astrocytosis" however, provides a brief overview of hepatic encephalopathy with a focus on the mechanisms of Alzheimer type II astrocytosis. Regarding chapter 13 "Astroglia and Acute Hepatic Encephalopathy: The Mechanisms of Cytotoxic Brain Edema" focusing on the same pathology (hepatic encephalopathy), the authors describe the pathomechanisms underlying astrocyte swelling in AHE, with the eventual evolution into macroscopic brain edema and the deathly herniation in humans as well as animal models of acute hepatic encephalopathy. In chapter 14 "Glioblastoma: Physiopathology and Complications" we try to well understand the physiopathology of glioblastoma involving glial cells and its associated consequences. While chapter 15 "Glial Cell Biology and Their Multifaceted Functions in Alzheimer's Disease: New Therapeutic Prospects" investigates the function of glial cells in the pathogenesis of in Alzheimer's disease and the intricate functions of the constituent parts along with the prospective therapeutic targets that may one day improve the quality of life for these patients. Chapter 16 "Astrocytes Reprogramming for Neurodegenerative Disease Management" we describe the ability to reprogram astrocytes into functional neurons as a new advancement for neurodegenerative diseases treatments. In chapter 17 "Involvement of Glial Cells in the Pathophysiology and Treatment of Depression" the authors aim to understand the dynamic interactions between glial cells and neurons in the context of depression with special focus on the therapeutic approaches. Chapter 18 "Therapeutic Strategies for Lysosomal Storage Diseases by Targeting Glial Cells" describes the possible therapeutic strategy aiming to target glial cells in the case of Lysosomal Storage Diseases. In chapter 19 "Investigation of potential neuroprotective role of chitosan-based biomaterials and their derivatives by targeting glial cells" explores the role of natural resources in the fight against neuronal diseases. While chapter 20 "Chitosan polysaccharides: modulation of neuroinflammation" aims to bring together recent work that highlights the potential of chitosan-based biomaterials suitable for glial cell tissue engineering. Finally chapter 21 "Chitosan-based biomaterials for tissue engineering of glial cells" aims to bring together recent work that highlights the potential of chitosan-based biomaterials suitable for glial cell tissue engineering.

Bilal El-Mansoury
Chouaib Doukkali University, Morocco

Omar El Hiba
Chouaib Doukkali University, Morocco

Arumugam Radhakrishnan Jayakumar
University of Miami, US

Chapter 1
Glial Cells in the Mature Central Nervous System: Classification and Functions

Kamal Smimih
Faculty of Sciences and Techniques, Sultan Moulay Slimane University, Morocco

Bilal El-Mansoury
https://orcid.org/0000-0002-2583-7515
Faculty of Sciences, University Chouaib Doukkali, Morocco

Chaima Azzouhri
Faculty of Sciences and Techniques, Sultan Moulay Slimane University, Morocco

Youssef Ait Hamdan
https://orcid.org/0009-0005-8616-8852
Higher Normal School, Cadi Ayyad University, Morocco

Fatima Ez-zahraa Saad
Faculty of Sciences, University Chouaib Doukkali, Morocco

Merzouki Mohamed
https://orcid.org/0000-0002-9837-4312
Faculty of Sciences and Techniques, Sultan Moulay Slimane University, Morocco

ABSTRACT

Glial cells were once thought of as simple support players in the central nervous system (CNS). However, the latest studies of glial cells have shown that they are actually of significance and play a variety of functions. Once thought to be passive supportive cells, astrocytes are now important in maintaining neurotransmitter balance, controlling synaptic activity, and regulating blood flow in the brain. Oligodendrocytes and Schwann cells, besides making myelin, are closely involved in controlling how fast nerve signals travel and maintaining the health of axons. Microglia had been thought to be only immune observers, but they also help to control the immune response and control synapses. Ependymal cells, which are sometimes ignored, are important in regulating cerebrospinal fluid circulation, directing the location of neural stem cells, and enabling cell communication. This chapter explores distinct glial cell types -astrocytes, oligodendrocytes, Schwann cells, microglia, and ependymal cells- highlighting their newfound, intricate functionalities and interactions with neurons.

DOI: 10.4018/978-1-6684-9675-6.ch001

INTRODUCTION

The intricate functioning of the nervous system is no longer attributed solely to neurons; the once-overlooked glial cells have taken center stage due to their vital contributions. Glial cells, which are often called "supporting cast" of neurons, have important roles to play in ensuring neuronal health, promoting communication, and influencing many aspects of brain function. Glial cells, or neuroglia, constitute a wide array of non-neuronal cells that make up the CNS and peripheral nervous system (PNS). Historically perceived as passive structural elements, glial cells are now recognized as active participants in neural circuits.

While neurons are renowned for their ability to transmit electrical signals, glial cells play a multifaceted role. Astrocytes, a prominent type of glial cell in the CNS, contribute to neurotransmitter regulation, maintenance of ion balance, and provision of metabolic support to neurons (Verkhratsky and Nedergaard 2018). Oligodendrocytes, on the other hand, specialize in myelination, crucial for rapid signal propagation (Saab et al. 2016). Microglia, the immune cells of the CNS, serve as sentinels, detecting and responding to pathogens or tissue damage (Kettenmann et al. 2011). Ependymal cells have been shown to actively participate in the circulation of cerebrospinal fluid and the maintenance of brain balance (Del Bigio 2010). These diverse roles differentiate glial cells from neurons and establish their importance in the intricate network of the nervous system.

The vital role that glial cells play in the neurological system is obscured by the typical focus on neurons. Neurotransmission, synaptic plasticity, and general brain homeostasis are all actively influenced by glial cells. Astrocytes regulate extracellular ion concentrations, optimizing synaptic transmission and preventing excitotoxicity (Khakh and Sofroniew 2015). Oligodendrocytes ensure rapid conduction through axonal myelination, enhancing the efficiency of neural communication (Fields 2015). Microglia, once thought solely to combat infections, are now understood to sculpt neural circuits, contributing to learning and memory processes (Parkhurst et al. 2013).

This chapter introduces glial cells and sheds insight on their recently discovered roles and functions as key players in the nervous system. It also makes the way for an exploration of the complex interactions that underlie neural function by looking at their differences from neurons and emphasizing their multifaceted activities.

TYPES OF GLIAL CELLS

Astrocytes Structure

The CNS contains astrocytes, a subclass of glial cells that support neural function and maintain neural homeostasis in many of ways. Morphologically, astrocytes typically exhibit a star-like shape, with a central cell body containing a nucleus and numerous radiating processes (Oberheim, Goldman, and Nedergaard 2012). These processes can be categorized into two major types: fibrous and protoplasmic (Bushong, Martone, and Ellisman 2004) (Figure 1). Fibrous astrocytes, predominantly found in white matter, have fewer, longer processes that align with axons, while protoplasmic astrocytes, located in gray matter, possess shorter, highly branched processes that interact extensively with neurons and synapses (Halassa et al. 2007).

Figure 1. Illustration depicting fibrous and protoplasmic astrocytes within the CNS (Kim, Park, and Choi 2019)

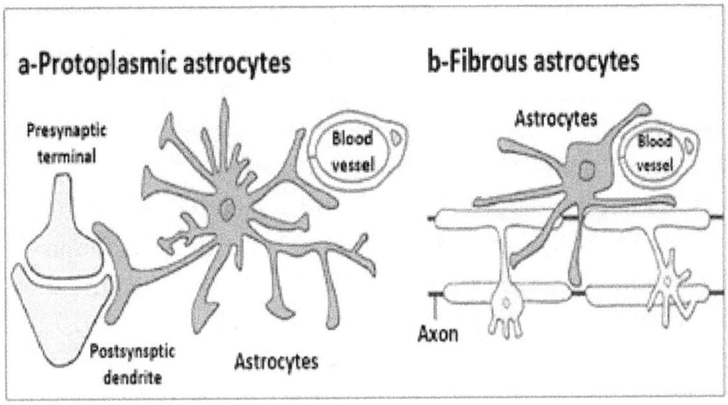

These intricate processes enable astrocytes to establish close associations with various components of the neural tissue, including blood vessels, neurons, and other glial cells, forming a network known as the "tripartite synapse" (Allen and Eroglu 2017). The structural arrangement of astrocytic processes allows them to cover a wide spatial domain, providing comprehensive coverage of neuronal synapses and regulating neurotransmission, ion homeostasis, and the extracellular environment (Nedergaard and Verkhratsky 2012). Astrocytic processes are enriched with various structural proteins, including aquaporin-4 (AQP4) water channels, which are vital for maintaining water balance and facilitating waste clearance in the brain (Nagelhus and Ottersen 2013).

The structural complexity of astrocytes is further exemplified by the formation of specialized structures called endfeet, which envelop blood vessels and contribute to the blood-brain barrier (BBB) function (Abbott, Rönnbäck, and Hansson 2006). These endfeet possess specific transporters and channels that regulate the exchange of nutrients, ions, and waste products between the blood and brain parenchyma. Additionally, astrocytes exhibit intricate heterogeneity in their morphological characteristics across brain regions and during development, suggesting their capacity to adapt and perform specialized functions in diverse microenvironments (Khakh and Sofroniew 2015).

Roles and Functions of Astrocytes

Structural support

Astrocytes play a fundamental role in providing structural support to neurons and maintaining the overall integrity of the neural tissue. Their complex and highly branched processes allow them to establish physical connections with neurons, blood vessels, and other glial cells, forming an intricate network that contributes to the structural organization of the CNS (Oberheim, Goldman, and Nedergaard 2012). Astrocytes contribute to the formation and maintenance of the BBB, where their endfeet ensheath blood vessels and regulate the exchange of ions, nutrients, and metabolites between the blood and brain parenchyma (Abbott, Rönnbäck, and Hansson 2006). Furthermore, their strategic positioning around synapses and neuronal processes provides critical structural support to neuronal connections, influencing synapse formation, stability, and plasticity (Allen and Eroglu 2017).

The astrocytic processes are enriched with cytoskeletal elements, including glial fibrillary acidic protein (GFAP) and actin filaments, which help maintain the structural integrity of these cells (Smith et al. 2014). These cytoskeletal components enable astrocytes to create a scaffold that guides the growth and organization of neurons during development and contributes to the maintenance of neuronal circuits in the mature brain. Additionally, astrocytes are involved in the formation of specialized microdomains known as "perisynaptic astroglial sheaths," which encapsulate synapses and provide physical support to synaptic contacts (Bernardinelli, Muller, and Nikonenko 2014).

Through their structural interactions, astrocytes facilitate neuronal migration, axonal guidance, and the establishment of neural networks during development. In conditions of injury or disease, astrocytes respond by undergoing structural changes, including hypertrophy and the formation of glial scars, which limit the spread of damage and aid in tissue repair (Sofroniew and Vinters 2010). The structural support provided by astrocytes extends to their role in maintaining extracellular ion homeostasis, ensuring proper neural function and preventing excessive excitability that could lead to neuronal damage (Kimelberg and Nedergaard 2010).

Regulation of Ion Balance

Astrocytes carefully regulate the ion balance in the surroundings of neurons. In order to maintain proper neuronal activity and prevent excitotoxicity in the extracellular environment, they help in maintaining the precise ion concentrations required. Astrocytes play an active role in controlling the potassium (K+) levels in the brain. Through a process known as "spatial buffering," astrocytes uptake excess K+ ions that accumulate in the synaptic cleft during neuronal activity. This uptake helps prevent excessive neuronal excitability and stabilizes the resting membrane potential. Astrocytes achieve this by using potassium inward rectifier channels (Kir4.1) and sodium-potassium pumps that help to redistribute and clear excess K+ ions (Kofuji and Newman 2004).

Astrocytes contribute to the regulation of calcium (Ca2+) levels in the CNS. They express various channels and transporters that influence Ca2+ signaling, impacting neuronal excitability, synaptic transmission, and plasticity. The astrocytic endfeet that surround blood vessels also help in the regulation of calcium ion concentrations by controlling the local microenvironment and influencing vasodilation and vasoconstriction through the modulation of endothelial cells (Filosa and Iddings 2013).

The regulation of ion balance by astrocytes extends to sodium (Na+), hydrogen (H+), and chloride (Cl-) ions as well. Astrocytes play a role in the uptake and redistribution of these ions, maintaining proper pH levels, water homeostasis, and cellular osmolarity in the brain (Toth et al. 2008). Dysregulation of ion balance by astrocytes can have detrimental effects on neuronal function, contributing to disorders such as epilepsy and ischemia (Seifert and Steinhäuser 2013).

Energy Metabolism of Neurons

Astrocytes play an important role in ensuring a steady supply of energy substrates for neuronal activity, hence supporting the energy metabolism of neurons in the brain. Through a process known as the astrocyte-neuron lactate shuttle, astrocytes take up glucose from the bloodstream and convert it into lactate through glycolysis, even in the presence of oxygen (aerobic glycolysis) (Dutrillaux, Xie, and Dutrillaux 2007). This lactate is then released by astrocytes and taken up by neurons, serving as an important energy source to fuel their high energy demands during synaptic transmission and other processes(Dienel 2013).

Moreover, astrocytes contribute to the generation of energy-rich molecules such as ATP, which is crucial for neuronal function. Astrocytes store glycogen, a polysaccharide composed of glucose units, and can break it down to provide glucose when energy demand increases. This glycogenolysis process is essential for providing neurons with a rapid and localized source of glucose during increased activity (Sciotti and Van Wylen 1993).

Astrocytes are also involved in neurotransmitter recycling, a process that consumes a significant amount of energy. They take up released neurotransmitters, such as glutamate, from the synaptic cleft and convert them into glutamine, which is then transported back to neurons for resynthesis of neurotransmitters. This process prevents excitotoxicity and ensures proper neurotransmission (Danbolt 2001).

Regulation of Blood-Brain Barrier

Astrocytes play a crucial role in the upkeep and maintenance of the BBB, which is a highly selective interface that links the CNS to the circulation. Astrocytes help to create and maintain the BBB through intricate interactions with endothelial cells, pericytes, and neurons, which is crucial for maintaining the right environment for brain function. Astrocytic endfeet envelop blood vessels and create a physical and functional interface that modulates the exchange of ions, nutrients, and waste products between the blood and the brain parenchyma (Abbott, Rönnbäck, and Hansson 2006).

Astrocytes influence the tight junctions between endothelial cells, which form the physical barrier of the BBB. They release various signaling molecules, such as transforming growth factor-β (TGF-β), that help regulate the formation and maintenance of tight junctions, thereby controlling the permeability of the barrier (Alvarez et al. 2011). Additionally, astrocytes release factors like sonic hedgehog that influence the expression of transporters, enzymes, and receptors in endothelial cells, contributing to BBB integrity (Daneman et al. 2010).

Furthermore, astrocytes are involved in the regulation of cerebral blood flow (CBF), which is closely linked to BBB function. Through mechanisms involving the release of vasoactive substances such as arachidonic acid metabolites and potassium ions, astrocytes modulate CBF and help match blood supply to local metabolic demands of neurons (Iadecola and Nedergaard 2007).

The role of astrocytes in BBB regulation extends to their responses during injury and inflammation. They actively participate in immune responses, contributing to the maintenance of BBB integrity during various pathological conditions. Astrocytes release cytokines, chemokines, and other signaling molecules that can either protect or compromise the BBB depending on the context (Sofroniew and Vinters 2010).

Involvement in Synaptic Plasticity

Astrocytes provide an important part in the modulation of synaptic plasticity, a fundamental process underlying learning and memory in the CNS. Beyond their conventional supportive role, astrocytes actively participate in bidirectional communication with neurons at the synapse, contributing to both long-term potentiation (LTP) and long-term depression (LTD) – two key forms of synaptic plasticity. Astrocytes interact with synapses through specialized structures known as perisynaptic astroglial processes, which envelop synapses and establish close associations with pre- and postsynaptic elements (Allen and Eroglu 2017).

Astrocytes release various signaling molecules, termed gliotransmitters, including glutamate, ATP, and D-serine, that modulate synaptic strength and plasticity. For instance, astrocyte-derived D-serine is an essential co-agonist for the N-methyl-D-aspartate (NMDA) receptor, a critical component of synaptic plasticity (Henneberger et al. 2020). Astrocytes contribute to the clearance of neurotransmitters like glutamate from the synaptic cleft, preventing excessive activation of receptors and promoting the termination of synaptic transmission, which is crucial for proper plasticity induction and maintenance (Araque et al. 1999).

Astrocytes also influence the structural changes associated with synaptic plasticity. They participate in the formation of dendritic spines – small protrusions on neurons that are crucial for synaptic connections. Astrocytes release factors that promote spine formation, and their physical interactions with spines provide mechanical support that contributes to the stability of synaptic contacts (Christopherson et al. 2005).

Astrocytes are involved in homeostatic plasticity, a process that maintains overall network stability. During chronic changes in neuronal activity, astrocytes can adjust the strength of synapses to counterbalance the perturbations, thereby maintaining an optimal level of excitability in neuronal networks (Turrigiano 2011).

OLIGODENDROCYTES AND SCHWANN CELLS

Structure and Differences

Oligodendrocytes and Schwann cells, the key players in myelination, exhibit notable structural differences that reflect their distinct roles in the CNS and peripheral nervous system (PNS), respectively (Nave and Werner 2014) (Kristjan R Jessen and Mirsky 1999). Oligodendrocytes, found in the CNS, are characterized by their complex morphology. A single oligodendrocyte can extend multiple processes, each of which interacts with and myelinates segments of several nearby axons. This multiterminal myelination enables oligodendrocytes to efficiently support and myelinate multiple axons, optimizing neural circuitry within the CNS (Butt, Papanikolaou, and Rivera 2019). In contrast, Schwann cells, situated in the PNS, are primarily associated with uniterminal myelination, where each Schwann cell ensheathes a single axon. A Schwann cell wraps around an axonal segment to form a myelin sheath, and nodes of Ranvier, gaps between adjacent segments, facilitate rapid saltatory conduction. This distinct pattern of myelination is crucial for the proper functioning of the peripheral nerves (Sherman and Brophy 2005) (Kristjan R Jessen and Arthur-Farraj 2019) (Figure 2).

Structurally, oligodendrocytes possess a larger cell body with fewer, extensively branching processes that enable them to interact with multiple axons in their vicinity. Schwann cells, on the other hand, have smaller cell bodies with numerous delicate processes. These processes wrap around the axon in a one-to-one relationship, contributing to the formation of individual myelin segments (Salzer 2015).

The differences in myelination patterns and cellular structures of these glial cells are deeply tied to the functional requirements of the CNS and PNS. Oligodendrocytes optimize neural connections over larger areas in the CNS, while Schwann cells ensure precise myelination of individual axons in the PNS. These adaptations highlight the remarkable diversity and specialization of glial cells in supporting and regulating neural communication.

Figure 2. Cell types involved in myelination: CNS Oligodendroglial cells vs. PNS Schwann cells (Carlson 2019)

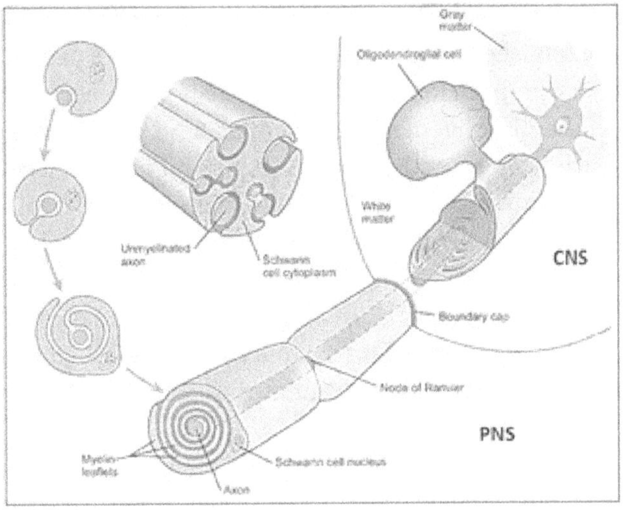

Role In Myelin Formation and Maintenance

Oligodendrocytes in the CNS and Schwann cells in the PNS play crucial roles in the formation and maintenance of myelin, the insulating sheath around axons that facilitates rapid and efficient nerve conduction. Oligodendrocytes extend their processes to ensheath multiple axons, forming myelin segments that cover the axons within the CNS. In contrast, Schwann cells individually encase single axons, generating myelin segments in the PNS. Myelin formation involves a series of intricate steps, including membrane extension, wrapping around the axon, and compaction of the myelin layers, guided by specialized proteins like myelin basic protein (MBP) and proteolipid protein (PLP) (Nave and Werner 2014) (Braun and Brostoff 1977).

Beyond the initial myelination, both types of glial cells are essential for myelin maintenance and plasticity. Oligodendrocytes continually monitor the integrity of myelin and axons, and they are involved in the dynamic process of myelin remodeling. They can form new myelin segments in response to changes in neural activity or damage. Similarly, Schwann cells contribute to myelin maintenance and repair by aiding in the clearance of debris, phagocytosis, and guiding regenerating axons to their original targets (Fields 2008) (K. R. Jessen and Mirsky 2016).

Disruptions in myelin formation or maintenance have significant consequences for neural function. Diseases such as multiple sclerosis (MS) involve demyelination due to autoimmune attacks on oligodendrocytes, leading to impaired nerve signaling. Inherited disorders like Charcot-Marie-Tooth disease affect Schwann cells, causing defects in myelin structure and nerve conduction (Franklin and ffrench-Constant 2008) (Ma et al. 2023).

The intricate interplay between oligodendrocytes, Schwann cells, and axons during myelination and myelin maintenance ensures proper nervous system function. Their combined efforts provide the structural and functional foundation for efficient neural communication.

Importance of Myelin in Nerve Conduction

Myelin is of paramount importance in facilitating efficient nerve conduction within the nervous system. Myelin, the lipid-rich insulating sheath that surrounds axons, serves as an electrical insulator, allowing for the rapid and saltatory propagation of action potentials along nerve fibers. The myelin sheath is formed by specialized glial cells, oligodendrocytes in the CNS and Schwann cells in the peripheral nervous system (PNS), which wrap around the axons in a spiral fashion. The myelin sheath acts as a physical barrier that prevents the dissipation of electrical signals, thereby greatly increasing the speed of nerve impulse transmission (Nave and Werner 2014) (Simons and Nave 2016). Furthermore, the nodes of Ranvier, the small gaps between adjacent myelin segments, are enriched with ion channels, enabling the efficient propagation of action potentials by allowing for rapid saltatory conduction, where the impulse "jumps" from one node to the next. This saltatory conduction significantly reduces the energy expenditure required for maintaining the action potential, thus optimizing the overall efficiency of nerve conduction (Salzer 2015) (Fields 2008).

Myelin's importance is underscored by the debilitating effects of demyelinating diseases, such as multiple sclerosis (MS), where the immune system attacks and damages the myelin sheath in the CNS, leading to disrupted nerve conduction and neurological impairments (Franklin and ffrench-Constant 2008). Additionally, genetic disorders affecting myelin, like Charcot-Marie-Tooth disease, further emphasize the critical role of myelin in maintaining proper neural communication (Brosnan and Raine 2013). Furthermore, myelin dysfunction has been implicated in a range of neurological disorders, including neuropathic pain, cognitive impairment, and neurodegenerative diseases (Trapp and Nave 2008).

MICROGLIA

Structure and Origin

Microglia, the CNS's resident immune cells, have a particular shape and origin, which contribute to their critical functions in maintaining brain homeostasis and reacting to pathological situations. These cells grow through a distinct developmental mechanism that distinguishes them from other immune cells. Microglia originate from primitive myeloid precursor cells that emerge from the yolk sac during embryonic development (Ginhoux and Prinz 2015) (Figure 3). These precursor cells migrate to the developing CNS, where they differentiate into microglia and colonize the brain and spinal cord prior to birth (Prinz and Priller 2017). This distinct origin endows microglia with a specialized ability to adapt to the unique environment of the CNS, actively participating in immune surveillance, synaptic pruning, and neuronal support (Wake and Fields 2011) (Colonna and Butovsky 2017).

Structurally, microglia possess a highly ramified morphology characterized by multiple fine processes that constantly survey their surroundings. These processes enable microglia to interact with neurons, synapses, and other glial cells, allowing them to monitor neural health and respond to disturbances in the microenvironment. In the resting state, microglia exhibit a "surveying" phenotype, which involves constant monitoring of their surroundings without inducing an immune response (Nimmerjahn, Kirchhoff, and Helmchen 2005) (Schafer, Lehrman, and Stevens 2013). However, when activated by injury, infection, or inflammation, microglia undergo a process known as "activation," where they transform into an amoeboid shape and become more mobile. This activation allows them to migrate toward sites of injury or inflammation, phagocytose debris, and release immune modulators to restore tissue integrity (Ransohoff and Cardona 2010).

Figure 3. Microglia development: Yolk sac origin in embryonic development
(Jurga, Paleczna, and Kuter 2020)

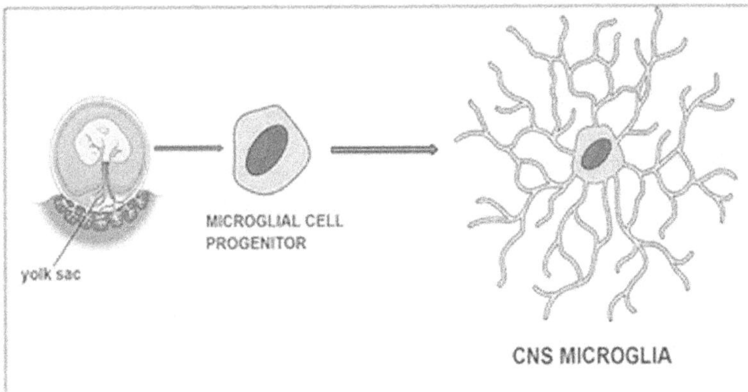

Immune and Inflammatory Functions

Microglia play pivotal roles in immune surveillance and inflammation, contributing to both neuroprotection and neuropathology. These versatile cells are equipped with an intricate array of immune and inflammatory functions that help maintain CNS homeostasis and respond to insults. In their surveilling state, microglia constantly monitor the neural microenvironment and promptly respond to changes by phagocytosing cellular debris, apoptotic cells, and protein aggregates (Ransohoff and Cardona 2010) (Prinz and Priller 2017). Additionally, they contribute to synaptic pruning during development, refining neural circuits and optimizing synaptic connectivity (Schafer, Lehrman, and Stevens 2013). However, microglia's immune and inflammatory functions extend beyond homeostatic maintenance.

Activated microglia mount immune responses upon detecting signs of injury, infection, or pathological conditions. They release proinflammatory cytokines, such as interleukin-1β (IL-1β) and tumor necrosis factor-alpha (TNF-α), as well as reactive oxygen species and nitric oxide, which are essential for defense against invading pathogens (Tang and Le 2016). While these responses are protective, excessive activation and release of proinflammatory factors can lead to neuroinflammation and contribute to neural damage in conditions like neurodegenerative diseases (Colonna and Butovsky 2017). Conversely, microglia also display anti-inflammatory functions by producing anti-inflammatory cytokines like interleukin-10 (IL-10) and promoting tissue repair through phagocytosis of apoptotic cells (Qin et al. 2007).

The balance between immune defense and inflammation moderation is crucial, as dysregulated microglial responses can contribute to various neurological disorders. Conditions such as Alzheimer's disease, Parkinson's disease, and multiple sclerosis showcase the dual nature of microglial functions—wherein chronic inflammation may exacerbate disease progression (Saijo and Glass 2011).

Role In Phagocytosis and Response to Injury

Microglia are important in phagocytosis and in reacting to damage in the neuronal microenvironment. Their remarkable ability to phagocytose cellular debris, apoptotic cells, and protein aggregates is essential for maintaining CNS homeostasis (Ransohoff and Cardona 2010). In response to injury, microglia undergo a phenotypic transformation from their surveilling state to an activated state, allowing them to efficiently remove cellular waste and contribute to tissue repair (Hanisch and Kettenmann 2007). Microglia recognize injury-associated signals through pattern recognition receptors (PRRs), including Toll-like receptors (TLRs), and initiate the phagocytic process (Neumann, Kotter, and Franklin 2008). This response includes the engulfment of damaged cells and the subsequent secretion of proinflammatory cytokines such as interleukin-1β (IL-1β) and IL-6, which aid in immune defense and tissue restoration (Huizinga et al. 2012).

The phagocytic activity of microglia is instrumental in clearing cellular debris resulting from neural injury, thereby preventing the accumulation of harmful components that could trigger inflammation and impair neuronal function (Li and Barres 2018). This activity also plays a pivotal role in shaping neural circuits during development through the elimination of excess synapses—a process essential for fine-tuning connectivity and optimizing brain function (Schafer, Lehrman, and Stevens 2013). However, uncontrolled or chronic microglial activation can lead to neuroinflammation and exacerbate neural damage, contributing to various neurodegenerative disorders (Tay et al. 2017).

EPENDYMAL CELLS

Structure

Ependymal cells are specialized epithelial cells that line the cerebral ventricles and central canal of the spinal cord and are essential for the generation and circulation of cerebrospinal fluid (CSF) throughout the CNS (MacDonald et al. 2021) . These cells form a barrier between the CSF and the neural tissue while also contributing to the maintenance of the microenvironment within the ventricular system (Del Bigio 2010). Ependymal cells are characterized by a columnar or cuboidal morphology with cilia on their apical surface, which actively move CSF to facilitate its circulation and distribution (Sawamoto et al. 2006). Their apical surfaces also possess microvilli that further enhance their absorptive capacity. Tight junctions between ependymal cells contribute to the formation of the blood-cerebrospinal fluid barrier, regulating the exchange of substances between the CSF and the brain parenchyma (Oshio et al. 2005) (Figure 4).

Ependymal cells originate from radial glial cells during embryonic development (Luskin, Pearlman, and Sanes 1988). Radial glial cells, critical for neural migration, differentiate into ependymal cells after they withdraw from the cell cycle (Malatesta et al. 2003). The transformation involves changes in gene expression and cytoskeletal organization, resulting in the acquisition of ependymal cell characteristics. In regions like the lateral ventricles, ependymal cells can also harbor neural stem cell properties and contribute to neurogenesis under specific conditions (Mirzadeh et al. 2008).

Figure 4. Structure and location of Ependymal cells
(Sheng et al. 2022)

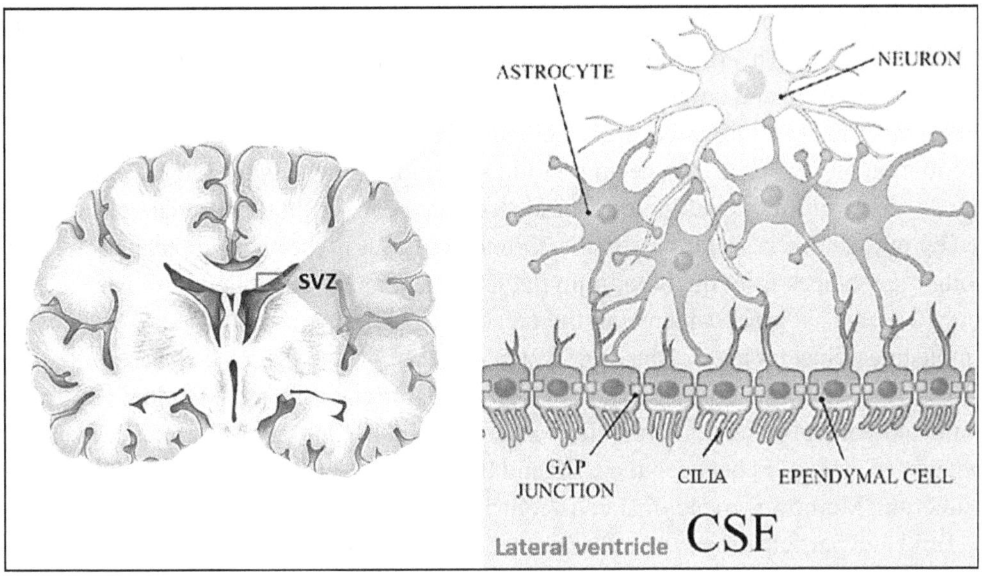

Localisation and Role in The Brain's Ventricles

Ependymal cells, strategically localized within the brain's ventricular system, fulfill crucial roles in maintaining the cerebral environment and facilitating the CSF (Del Bigio 2010). These specialized epithelial cells line the ventricles and central canal of the spinal cord, creating a physical barrier between the CSF and neural tissue (Johanson et al. 2008). Their distribution varies across the ventricular system, with ependymal cells in different regions exhibiting distinct morphological features and functional properties (Lim and Alvarez-Buylla 2016).

Ependymal cells are primarily responsible for regulating the composition of CSF and ensuring its proper circulation. Their apical surfaces are adorned with cilia, which beat rhythmically to generate directional fluid flow within the ventricles. This ciliary movement propels CSF along the ventricular system, facilitating the distribution of nutrients, removal of waste products, and transport of signaling molecules (Mirzadeh et al. 2010). Furthermore, microvilli on their apical surfaces enhance the absorptive and secretory capabilities of ependymal cells, contributing to the maintenance of CSF homeostasis (Oshio et al. 2005).

Beyond their role in fluid dynamics, ependymal cells also contribute to the formation and maintenance of neural stem cell niches within the brain (Sawamoto et al. 2006). In regions such as the lateral ventricles, ependymal cells are associated with specialized niches where neurogenesis occurs throughout adulthood (Marzesco et al. 2005). These cells create a microenvironment that supports the proliferation and differentiation of neural stem cells, further highlighting their importance in brain plasticity and regeneration (Gritti et al. 1996).

Production and Circulation of Cerebrospinal Fluid

Ependymal cells are equipped with cilia on their apical surfaces, which beat in a coordinated manner to create fluid movement. This rhythmic ciliary motion propels CSF throughout the ventricular system, facilitating its distribution and contributing to the removal of waste products and the transportation of essential nutrients and signaling molecules (Del Bigio 2010).

The production of CSF primarily occurs within structures known as the choroid plexuses; specialized tissues located within the ventricles. These structures consist of ependymal cells and capillaries ensheathed by pia mater, a vascular membrane. Ependymal cells in the choroid plexus actively transport ions and other substances from the blood into the CSF, generating an osmotic gradient that drives the movement of water. This process results in the creation of CSF that is rich in ions, proteins, and other essential molecules (Spector and Johanson 1989).

The CSF generated by the choroid plexuses fills the ventricles and subarachnoid spaces, providing buoyancy and mechanical protection to the CNS. Ependymal cells lining the ventricles actively participate in the exchange of molecules between the CSF and the neural tissue, ensuring a stable environment for neuronal function. Moreover, ependymal cells contribute to the regulation of CSF volume by reabsorbing excess fluid and participating in waste clearance (Oshio et al. 2005).

The intricate balance between ependymal cells, choroid plexuses, and the dynamics of CSF circulation is essential for CNS homeostasis. Disruptions in these processes can lead to conditions such as hydrocephalus, characterized by abnormal accumulation of CSF within the ventricles due to impaired fluid circulation (Lim and Alvarez-Buylla 2016).

INTERACTION BETWEEN GLIAL CELLS AND NEURONS

The interaction between glial cells and neurons, known as neuron-glia communication, constitutes a complex network profoundly shaping neural function, synaptic transmission, and brain homeostasis (Araque et al. 1999). Glial cells, including astrocytes, oligodendrocytes, and microglia, contribute to this intricate interplay through multifaceted mechanisms that extend beyond their traditional roles.

Astrocytes, pivotal in neuron-glia communication, actively engage in neurotransmitter uptake, particularly glutamate, from the synaptic cleft, thus preventing excitotoxicity and maintaining synaptic strength (Kimelberg and Nedergaard 2010). In response to neuronal activity, astrocytes release gliotransmitters such as ATP, D-serine, and glutamate, influencing nearby neurons and modulating synaptic plasticity. This astrocytic signaling influences synaptic strength through mechanisms like LTP and LTD (Santello, Calì, and Bezzi 2012). The release of gliotransmitters, including ATP, triggers purinergic signaling that can modulate synaptic transmission (Verkhratsky and Nedergaard 2018).

Oligodendrocytes, originally associated with myelination, now emerge as contributors to neuronal function. They express neurotransmitter receptors and release signaling molecules, including glutamate, which impact axonal activity and neuronal communication (Saab et al. 2016). Recent research has highlighted the role of oligodendrocyte-derived extracellular vesicles in modulating neural circuits (Xiao et al. 2016).

Microglia, CNS immune cells, release proinflammatory cytokines and chemokines upon neuronal activity or injury, shaping the neural microenvironment. As well, microglia contribute to synaptic pruning, eliminating weak or unnecessary synapses during brain development and in response to experience, refining neural circuits (Parkhurst et al. 2013). The influence of microglia on synaptic transmission involves the release of signaling molecules such as brain-derived neurotrophic factor (BDNF) (Lenz et al. 2013).

These intricate interactions have substantial implications for brain function and neurological disorders. Neuron-glia communication dysfunction has been linked to neurological diseases, mood disorders, and epilepsy.

CONCLUSION

Glial cell research has evolved, showing a complex environment in which, these previously neglected "supporting actors" have evolved into critical players in the inner functioning of brain physiology. Several kinds of dynamic functions and roles that glial cells, such as astrocytes, oligodendrocytes, microglia, and ependymal cells, carry out in the nervous system have recently come into focus in research.

Glial cells have proven to be much more than passive support structures. Astrocytes, with their strategic spatial arrangement, actively regulate neurotransmitter balance, ion homeostasis, and energy metabolism, crucially influencing synaptic transmission and plasticity. Oligodendrocytes excel in myelination, ensuring rapid and efficient signal propagation while also participating in axonal maintenance and energy supply. Microglia, once seen as mere immune sentinels, orchestrate immune responses, synaptic pruning, and tissue maintenance, thus shaping neural circuits and influencing cognitive processes. Ependymal cells have recently come to light as active contributors to cerebrospinal fluid circulation and the preservation of brain balance.

As we explore new frontiers, it is evident that glial cells, once in the shadows, are now shining beacons of promise for uncovering the secrets of the brain and opening the way for breakthrough therapeutic options that hold the potential to change lives.

REFERENCES

Abbott, N. J., Rönnbäck, L., & Hansson, E. (2006). 'Astrocyte–Endothelial Interactions at the Blood–Brain Barrier'. *Nature Reviews. Neuroscience*, *7*(1), 41–53. doi:10.1038/nrn1824 PMID:16371949

Allen, N. J., & Eroglu, C. (2017). 'Cell Biology of Astrocyte-Synapse Interactions'. *Neuron*, *96*(3), 697–708. doi:10.1016/j.neuron.2017.09.056 PMID:29096081

Alvarez, J. I., Dodelet-Devillers, A., Kebir, H., Ifergan, I., Fabre, P. J., Terouz, S., Sabbagh, M., Wosik, K., Bourbonnière, L., Bernard, M., van Horssen, J., de Vries, H. E., Charron, F., & Prat, A. (2011). 'The Hedgehog Pathway Promotes Blood-Brain Barrier Integrity and CNS Immune Quiescence'. *Science*, *334*(6063), 1727–1731. doi:10.1126cience.1206936 PMID:22144466

Araque, A., Parpura, V., Sanzgiri, R. P., & Haydon, P. G. (1999). 'Tripartite Synapses: Glia, the Unacknowledged Partner'. *Trends in Neurosciences*, *22*(5), 208–215. doi:10.1016/S0166-2236(98)01349-6 PMID:10322493

Bernardinelli, Y., Muller, D., & Nikonenko, I. (2014). 'Astrocyte-Synapse Structural Plasticity'. *Neural Plasticity*, *2014*, 1–13. doi:10.1155/2014/232105 PMID:24511394

Braun, P. E., & Brostoff, S. W. (1977). Proteins of Myelin. In P. Morell (Ed.), *Myelin* (pp. 201–231). Springer US. doi:10.1007/978-1-4615-7514-6_6

Brosnan, C. F., & Raine, C. S. (2013). 'The Astrocyte in Multiple Sclerosis Revisited'. *Glia, 61*(4), 453–465. doi:10.1002/glia.22443 PMID:23322421

Bushong, E. A., Martone, M. E., & Ellisman, M. H. (2004). 'Maturation of Astrocyte Morphology and the Establishment of Astrocyte Domains during Postnatal Hippocampal Development'. *International Journal of Developmental Neuroscience, 22*(2), 73–86. doi:10.1016/j.ijdevneu.2003.12.008 PMID:15036382

Butt, A. M., Papanikolaou, M., & Rivera, A. 2019. 'Physiology of Oligodendroglia'. In Neuroglia in Neurodegenerative Diseases. Springer Singapore. doi:10.1007/978-981-13-9913-8_5

Carlson, B. M. (2019). Tissues. In *The Human Body* (pp. 27–63). Elsevier. doi:10.1016/B978-0-12-804254-0.00002-8

Christopherson, K. S., Ullian, E. M., Stokes, C. C. A., Mullowney, C. E., Hell, J. W., Agah, A., Lawler, J., Mosher, D. F., Bornstein, P., & Barres, B. A. (2005). 'Thrombospondins Are Astrocyte-Secreted Proteins That Promote CNS Synaptogenesis'. *Cell, 120*(3), 421–433. doi:10.1016/j.cell.2004.12.020 PMID:15707899

Colonna, M., & Butovsky, O. (2017). 'Microglia Function in the Central Nervous System During Health and Neurodegeneration'. *Annual Review of Immunology, 35*(1), 441–468. doi:10.1146/annurev-immunol-051116-052358 PMID:28226226

Danbolt, N. C. (2001). 'Glutamate Uptake'. *Progress in Neurobiology, 65*(1), 1–105. doi:10.1016/S0301-0082(00)00067-8 PMID:11369436

Daneman, R., Zhou, L., Kebede, A. A., & Barres, B. A. (2010). 'Pericytes Are Required for Blood–Brain Barrier Integrity during Embryogenesis'. *Nature, 468*(7323), 562–566. doi:10.1038/nature09513 PMID:20944625

Del Bigio, M. R. (2010). 'Ependymal Cells: Biology and Pathology'. *Acta Neuropathologica, 119*(1), 55–73. doi:10.100700401-009-0624-y PMID:20024659

Dienel, G. A. (2013). Astrocytic Energetics during Excitatory Neurotransmission: What Are Contributions of Glutamate Oxidation and Glycolysis? *Neurochemistry International, 63*(4), 244–258. doi:10.1016/j.neuint.2013.06.015 PMID:23838211

Dutrillaux, A. M., Xie, H., & Dutrillaux, B. (2007). 'High Chromosomal Polymorphism and Heterozygosity in *Cyclocephala Tridentata* from Guadeloupe: Chromosome Comparison with Some Other Species of Dynastinae (Coleoptera: Scarabaeidae)'. *Cytogenetic and Genome Research, 119*(3–4), 248–254. doi:10.1159/000112070 PMID:18253038

Fields, R. D. (2008). 'White Matter in Learning, Cognition and Psychiatric Disorders'. *Trends in Neurosciences, 31*(7), 361–370. doi:10.1016/j.tins.2008.04.001 PMID:18538868

Fields, R. D. (2015). 'A New Mechanism of Nervous System Plasticity: Activity-Dependent Myelination'. *Nature Reviews. Neuroscience, 16*(12), 756–767. doi:10.1038/nrn4023 PMID:26585800

Filosa, J. A., & Iddings, J. A. (2013). 'Astrocyte Regulation of Cerebral Vascular Tone'. *American Journal of Physiology. Heart and Circulatory Physiology*, *305*(5), H609–H619. doi:10.1152/ajpheart.00359.2013 PMID:23792684

Franklin, R. J. M., & ffrench-Constant, C. (2008). 'Remyelination in the CNS: From Biology to Therapy'. *Nature Reviews. Neuroscience*, *9*(11), 839–855. doi:10.1038/nrn2480 PMID:18931697

Ginhoux, F., & Prinz, M. (2015). 'Origin of Microglia: Current Concepts and Past Controversies'. *Cold Spring Harbor Perspectives in Biology*, *7*(8), a020537. doi:10.1101/cshperspect.a020537 PMID:26134003

Gritti, A., Parati, E., Cova, L., Frolichsthal, P., Galli, R., Wanke, E., Faravelli, L., Morassutti, D. J., Roisen, F., Nickel, D. D., & Vescovi, A. L. (1996). 'Multipotential Stem Cells from the Adult Mouse Brain Proliferate and Self-Renew in Response to Basic Fibroblast Growth Factor'. *The Journal of Neuroscience : The Official Journal of the Society for Neuroscience*, *16*(3), 1091–1100. doi:10.1523/JNEUROSCI.16-03-01091.1996 PMID:8558238

Halassa, M. M., Fellin, T., Takano, H., Dong, J.-H., & Haydon, P. G. (2007). 'Synaptic Islands Defined by the Territory of a Single Astrocyte'. *The Journal of Neuroscience : The Official Journal of the Society for Neuroscience*, *27*(24), 6473–6477. doi:10.1523/JNEUROSCI.1419-07.2007 PMID:17567808

Hanisch, U.-K., & Kettenmann, H. (2007). 'Microglia: Active Sensor and Versatile Effector Cells in the Normal and Pathologic Brain'. *Nature Neuroscience*, *10*(11), 1387–1394. doi:10.1038/nn1997 PMID:17965659

Henneberger, C., Bard, L., Panatier, A., Reynolds, J. P., Kopach, O., Medvedev, N. I., Minge, D., Herde, M. K., Anders, S., Kraev, I., Heller, J. P., Rama, S., Zheng, K., Jensen, T. P., Sanchez-Romero, I., Jackson, C. J., Janovjak, H., Ottersen, O. P., Nagelhus, E. A., & Rusakov, D. A. (2020). 'LTP Induction Boosts Glutamate Spillover by Driving Withdrawal of Perisynaptic Astroglia'. *Neuron*, *108*(5), 919–936.e11. doi:10.1016/j.neuron.2020.08.030 PMID:32976770

Huizinga, R., & Van, B. J. (2012). 'Phagocytosis of Neuronal Debris by Microglia Is Associated with Neuronal Damage in Multiple Sclerosis'. *Glia*, *60*(3), 422–431. doi:10.1002/glia.22276 PMID:22161990

Iadecola, C., & Nedergaard, M. (2007). 'Glial Regulation of the Cerebral Microvasculature'. *Nature Neuroscience*, *10*(11), 1369–1376. doi:10.1038/nn2003 PMID:17965657

Jessen, K. R., & Arthur-Farraj, P. (2019). 'Repair Schwann Cell Update: Adaptive Reprogramming, EMT, and Stemness in Regenerating Nerves'. *Glia*, *67*(3), 421–437. doi:10.1002/glia.23532 PMID:30632639

Jessen, K. R., & Mirsky, R. (1999). 'Schwann Cells and Their Precursors Emerge as Major Regulators of Nerve Development'. *Trends in Neurosciences*, *22*(9), 402–410. doi:10.1016/S0166-2236(98)01391-5 PMID:10441301

Jessen, K. R., & Mirsky, R. (2016). 'The Repair Schwann Cell and Its Function in Regenerating Nerves'. *The Journal of Physiology*, *594*(13), 3521–3531. doi:10.1113/JP270874 PMID:26864683

Johanson, C. E., Duncan, J. A. III, Klinge, P. M., Brinker, T., Stopa, E. G., & Silverberg, G. D. (2008). 'Multiplicity of Cerebrospinal Fluid Functions: New Challenges in Health and Disease'. *Cerebrospinal Fluid Research*, *5*(1), 10. doi:10.1186/1743-8454-5-10 PMID:18479516

Jurga, A. M., Paleczna, M., & Kuter, K. Z. (2020). 'Overview of General and Discriminating Markers of Differential Microglia Phenotypes'. *Frontiers in Cellular Neuroscience*, *14*, 198. https://www.frontiersin.org/articles/10.3389/fncel.2020.00198. doi:10.3389/fncel.2020.00198 PMID:32848611

Kettenmann, H., Hanisch, U.-K., Noda, M., & Verkhratsky, A. (2011). 'Physiology of Microglia'. *Physiological Reviews*, *91*(2), 461–553. doi:10.1152/physrev.00011.2010 PMID:21527731

Khakh, B. S., & Sofroniew, M. V. (2015). 'Diversity of Astrocyte Functions and Phenotypes in Neural Circuits'. *Nature Neuroscience*, *18*(7), 942–952. doi:10.1038/nn.4043 PMID:26108722

Kim, Y., Park, J., & Choi, Y. K. (2019). 'The Role of Astrocytes in the Central Nervous System Focused on BK Channel and Heme Oxygenase Metabolites: A Review'. *Antioxidants*, *8*(5), 121. doi:10.3390/antiox8050121 PMID:31060341

Kimelberg, H. K., & Nedergaard, M. (2010). 'Functions of Astrocytes and Their Potential as Therapeutic Targets'. *Neurotherapeutics; the Journal of the American Society for Experimental NeuroTherapeutics*, *7*(4), 338–353. doi:10.1016/j.nurt.2010.07.006 PMID:20880499

Kofuji, P., & Newman, E. A. (2004). 'Potassium Buffering in the Central Nervous System'. *Neuroscience*, *129*(4), 1043–1054. doi:10.1016/j.neuroscience.2004.06.008 PMID:15561419

Lenz, K. M., Nugent, B. M., Haliyur, R., & McCarthy, M. M. (2013). 'Microglia Are Essential to Masculinization of Brain and Behavior'. *The Journal of Neuroscience : The Official Journal of the Society for Neuroscience*, *33*(7), 2761–2772. doi:10.1523/JNEUROSCI.1268-12.2013 PMID:23407936

Li, Q., & Barres, B. A. (2018). 'Microglia and Macrophages in Brain Homeostasis and Disease'. *Nature Reviews. Immunology*, *18*(4), 225–242. doi:10.1038/nri.2017.125 PMID:29151590

Lim, D. A., & Alvarez-Buylla, A. (2016). 'The Adult Ventricular–Subventricular Zone (V-SVZ) and Olfactory Bulb (OB) Neurogenesis'. *Cold Spring Harbor Perspectives in Biology*, *8*(5), a018820. doi:10.1101/cshperspect.a018820 PMID:27048191

Luskin, M. B., Pearlman, A. L., & Sanes, J. R. (1988). 'Cell Lineage in the Cerebral Cortex of the Mouse Studied in Vivo and in Vitro with a Recombinant Retrovirus'. *Neuron*, *1*(8), 635–647. doi:10.1016/0896-6273(88)90163-8 PMID:3272182

Ma, M., Li, Y., Dai, S., Chu, M., Sun, L., Liu, L., & Zhou, J.-C. (2023). 'A Meta-Analysis on the Prevalence of Charcot–Marie–Tooth Disease and Related Inherited Peripheral Neuropathies'. *Journal of Neurology*, *270*(5), 2468–2482. doi:10.100700415-023-11559-8 PMID:36631678

MacDonald, A., Lu, B., Caron, M., Caporicci-Dinucci, N., Hatrock, D., Petrecca, K., Bourque, G., & Stratton, J. A. (2021). 'Single Cell Transcriptomics of Ependymal Cells Across Age, Region and Species Reveals Cilia-Related and Metal Ion Regulatory Roles as Major Conserved Ependymal Cell Functions'. *Frontiers in Cellular Neuroscience*, *15*, 703951. https://www.frontiersin.org/articles/10.3389/fncel.2021.703951. doi:10.3389/fncel.2021.703951 PMID:34335193

Malatesta, P., Hack, M. A., Hartfuss, E., Kettenmann, H., Klinkert, W., Kirchhoff, F., & Götz, M. (2003). 'Neuronal or Glial Progeny: Regional Differences in Radial Glia Fate'. *Neuron*, *37*(5), 751–764. doi:10.1016/S0896-6273(03)00116-8 PMID:12628166

Marzesco, A.-M., Janich, P., Wilsch-Bräuninger, M., Dubreuil, V., Langenfeld, K., Corbeil, D., & Huttner, W. B. (2005). 'Release of Extracellular Membrane Particles Carrying the Stem Cell Marker Prominin-1 (CD133) from Neural Progenitors and Other Epithelial Cells'. *Journal of Cell Science, 118*(13), 2849–2858. doi:10.1242/jcs.02439 PMID:15976444

Mirzadeh, Z., Doetsch, F., Sawamoto, K., Wichterle, H., & Alvarez-Buylla, A. (2010, May). 'The Subventricular Zone En-Face: Wholemount Staining and Ependymal Flow'. *Journal of Visualized Experiments, 1938*(39). doi:10.3791/1938-v PMID:20461052

Mirzadeh, Z., Merkle, F. T., Soriano-Navarro, M., Garcia-Verdugo, J. M., & Alvarez-Buylla, A. (2008). 'Neural Stem Cells Confer Unique Pinwheel Architecture to the Ventricular Surface in Neurogenic Regions of the Adult Brain'. *Cell Stem Cell, 3*(3), 265–278. doi:10.1016/j.stem.2008.07.004 PMID:18786414

Nagelhus, E. A., & Ottersen, O. P. (2013). 'Physiological Roles of Aquaporin-4 in Brain'. *Physiological Reviews, 93*(4), 1543–1562. doi:10.1152/physrev.00011.2013 PMID:24137016

Nave, K.-A., & Werner, H. B. (2014). 'Myelination of the Nervous System: Mechanisms and Functions'. *Annual Review of Cell and Developmental Biology, 30*(1), 503–533. doi:10.1146/annurev-cellbio-100913-013101 PMID:25288117

Nedergaard, M., & Verkhratsky, A. (2012). 'Artifact versus Reality-How Astrocytes Contribute to Synaptic Events'. *Glia, 60*(7), 1013–1023. doi:10.1002/glia.22288 PMID:22228580

Neumann, H., Kotter, M. R., & Franklin, R. J. M. (2008). 'Debris Clearance by Microglia: An Essential Link between Degeneration and Regeneration'. *Brain, 132*(2), 288–295. doi:10.1093/brain/awn109 PMID:18567623

Nimmerjahn, A., Kirchhoff, F., & Helmchen, F. (2005). 'Resting Microglial Cells Are Highly Dynamic Surveillants of Brain Parenchyma in Vivo'. *Science, 308*(5726), 1314–1318. doi:10.1126cience.1110647 PMID:15831717

Oberheim, N. A., Goldman, S. A., & Nedergaard, M. 2012. 'Heterogeneity of Astrocytic Form and Function'. In Astrocytes. Totowa, NJ: Humana Press. doi:10.1007/978-1-61779-452-0_3

Oshio, K., Binder, D. K., Liang, Y., Bollen, A., Feuerstein, B., Berger, M. S., & Manley, G. T. (2005). 'Expression of the Aquaporin-1 Water Channel in Human Glial Tumors'. *Neurosurgery, 56*(2), 375–381. doi:10.1227/01.NEU.0000148904.57841.6B PMID:15670385

Parkhurst, C. N., Yang, G., Ninan, I., Savas, J. N., Yates, J. R. III, Lafaille, J. J., Hempstead, B. L., Littman, D. R., & Gan, W.-B. (2013). 'Microglia Promote Learning-Dependent Synapse Formation through Brain-Derived Neurotrophic Factor'. *Cell, 155*(7), 1596–1609. doi:10.1016/j.cell.2013.11.030 PMID:24360280

Prinz, M., & Priller, J. (2017). 'The Role of Peripheral Immune Cells in the CNS in Steady State and Disease'. *Nature Neuroscience, 20*(2), 136–144. doi:10.1038/nn.4475 PMID:28092660

Qin, L., Wu, X., Block, M. L., Liu, Y., Breese, G. R., Hong, J.-S., Knapp, D. J., & Crews, F. T. (2007). 'Systemic LPS Causes Chronic Neuroinflammation and Progressive Neurodegeneration'. *Glia, 55*(5), 453–462. doi:10.1002/glia.20467 PMID:17203472

Ransohoff, R. M., & Cardona, A. E. (2010). 'The Myeloid Cells of the Central Nervous System Parenchyma'. *Nature*, *468*(7321), 253–262. doi:10.1038/nature09615 PMID:21068834

Saab, A. S., Tzvetavona, I. D., Trevisiol, A., Baltan, S., Dibaj, P., Kusch, K., Möbius, W., Goetze, B., Jahn, H. M., Huang, W., Steffens, H., Schomburg, E. D., Pérez-Samartín, A., Pérez-Cerdá, F., Bakhtiari, D., Matute, C., Löwel, S., Griesinger, C., Hirrlinger, J., & Nave, K.-A. (2016). 'Oligodendroglial NMDA Receptors Regulate Glucose Import and Axonal Energy Metabolism'. *Neuron*, *91*(1), 119–132. doi:10.1016/j.neuron.2016.05.016 PMID:27292539

Saijo, K., & Glass, C. K. (2011). 'Microglial Cell Origin and Phenotypes in Health and Disease'. *Nature Reviews. Immunology*, *11*(11), 775–787. doi:10.1038/nri3086 PMID:22025055

Salzer, J. L. (2015). 'Schwann Cell Myelination'. *Cold Spring Harbor Perspectives in Biology*, *7*(8), a020529. doi:10.1101/cshperspect.a020529 PMID:26054742

Santello, M., Calì, C., & Bezzi, P. (2012). 'Gliotransmission and the Tripartite Synapse'. In Synaptic Plasticity. Springer Vienna. doi:10.1007/978-3-7091-0932-8_14

Sawamoto, K., Wichterle, H., Gonzalez-Perez, O., Cholfin, J. A., Yamada, M., Spassky, N., Murcia, N. S., Garcia-Verdugo, J. M., Marin, O., Rubenstein, J. L. R., Tessier-Lavigne, M., Okano, H., & Alvarez-Buylla, A. (2006). 'New Neurons Follow the Flow of Cerebrospinal Fluid in the Adult Brain'. *Science*, *311*(5761), 629–632. doi:10.1126cience.1119133 PMID:16410488

Schafer, D. P., Lehrman, E. K., & Stevens, B. (2013). 'The "Quad-partite" Synapse: Microglia-synapse Interactions in the Developing and Mature CNS'. *Glia*, *61*(1), 24–36. doi:10.1002/glia.22389 PMID:22829357

Sciotti, V. M., & Van Wylen, D. G. L. (1993). 'Increases in Interstitial Adenosine and Cerebral Blood Flow with Inhibition of Adenosine Kinase and Adenosine Deaminase'. *Journal of Cerebral Blood Flow and Metabolism*, *13*(2), 201–207. doi:10.1038/jcbfm.1993.24 PMID:8436611

Seifert, G., & Steinhäuser, C. (2013). 'Neuron–Astrocyte Signaling and Epilepsy'. *Experimental Neurology*, *244*(June), 4–10. doi:10.1016/j.expneurol.2011.08.024 PMID:21925173

Sheng, J., Li, Q., Liu, T., & Wang, X. (2022). 'Cerebrospinal Fluid Dynamics along the Optic Nerve'. *Frontiers in Neurology*, *13*(August), 931523. doi:10.3389/fneur.2022.931523 PMID:36046631

Sherman, D. L., & Brophy, P. J. (2005). 'Mechanisms of Axon Ensheathment and Myelin Growth'. *Nature Reviews. Neuroscience*, *6*(9), 683–690. doi:10.1038/nrn1743 PMID:16136172

Simons, M., & Nave, K.-A. (2016). 'Oligodendrocytes: Myelination and Axonal Support'. *Cold Spring Harbor Perspectives in Biology*, *8*(1), a020479. doi:10.1101/cshperspect.a020479 PMID:26101081

Smith, M. J., Wallace, A. J., Bennett, C., Hasselblatt, M., Elert-Dobkowska, E., Evans, L. T., Hickey, W. F., van Hoff, J., Bauer, D., Lee, A., Hevner, R. F., Beetz, C., du Plessis, D., Kilday, J.-P., Newman, W. G., & Evans, D. G. (2014). 'Germline *SMARCE1* Mutations Predispose to Both Spinal and Cranial Clear Cell Meningiomas'. *The Journal of Pathology*, *234*(4), 436–440. doi:10.1002/path.4427 PMID:25143307

Sofroniew, M. V., & Vinters, H. V. (2010). 'Astrocytes: Biology and Pathology'. *Acta Neuropathologica*, *119*(1), 7–35. doi:10.100700401-009-0619-8 PMID:20012068

Spector, R., & Johanson, C. E. (1989). 'The Mammalian Choroid Plexus'. *Scientific American, 261*(5), 68–74. doi:10.1038cientificamerican1189-68 PMID:2479097

Tang, Yu., & Le, W. (2016). 'Differential Roles of M1 and M2 Microglia in Neurodegenerative Diseases'. *Molecular Neurobiology, 53*(2), 1181–1194. doi:10.100712035-014-9070-5 PMID:25598354

Tay, T. L., Savage, J. C., Chin, W. H., Bisht, K., & Tremblay, M.-È. (2017). 'Microglia across the Lifespan: From Origin to Function in Brain Development, Plasticity and Cognition: Microglia across the Lifespan'. *The Journal of Physiology, 595*(6), 1929–1945. doi:10.1113/JP272134 PMID:27104646

Toth, E., Gersner, R., Wilf-Yarkoni, A., Raizel, H., Dar, D. E., Richter-Levin, G., Levit, O., & Zangen, A. (2008). 'Age-Dependent Effects of Chronic Stress on Brain Plasticity and Depressive Behavior: Coping with Chronic Stress'. *Journal of Neurochemistry, 107*(2), 522–532. doi:10.1111/j.1471-4159.2008.05642.x PMID:18752645

Trapp, B. D., & Nave, K.-A. (2008). Multiple Sclerosis: An Immune or Neurodegenerative Disorder? *Annual Review of Neuroscience, 31*(1), 247–269. doi:10.1146/annurev.neuro.30.051606.094313 PMID:18558855

Turrigiano, G. (2011). 'Too Many Cooks? Intrinsic and Synaptic Homeostatic Mechanisms in Cortical Circuit Refinement'. *Annual Review of Neuroscience, 34*(1), 89–103. doi:10.1146/annurev-neuro-060909-153238 PMID:21438687

Verkhratsky, A., & Nedergaard, M. (2018). 'Physiology of Astroglia'. *Physiological Reviews, 98*(1), 239–389. doi:10.1152/physrev.00042.2016 PMID:29351512

Wake, H., & Douglas Fields, R. (2011). 'Physiological Function of Microglia'. *Neuron Glia Biology, 7*(1), 1–3. doi:10.1017/S1740925X12000166 PMID:22857736

Xiao, L., Ohayon, D., McKenzie, I. A., Sinclair-Wilson, A., Wright, J. L., Fudge, A. D., Emery, B., Li, H., & Richardson, W. D. (2016). 'Rapid Production of New Oligodendrocytes Is Required in the Earliest Stages of Motor-Skill Learning'. *Nature Neuroscience, 19*(9), 1210–1217. doi:10.1038/nn.4351 PMID:27455109

Chapter 2
Astrocytes and the Developing Brain

Fatima Ez-Zahraa Saad
Faculty of Sciences, Chouaib Doukkali University, Morocco

Kamal Smimih
Faculty of Sciences and Techniques, Sultan Moulay Slimane University, Mococco

Abdelali Bitar
Faculty of Sciences, Chouaib Doukkali University, Morocco

Manal Khanouchi
https://orcid.org/0000-0002-7935-3049
Faculty of Sciences, Chouaib Doukkali University, Morocco

ABSTRACT

Astrocytes are the most proliferate glial cells in the central nervous system (CNS); they are considered as the supporting cells in the CNS, and play a big role in behavioral, circuit, and synaptic functions. Astrocytes are also important for neuronal repair, neurogenesis, and survival. Astrocytes play a main role in brain protection, by maintaining a regulated ion balance and blood flow, and preserving an antioxidant functions in brain. In this chapter, the authors elaborated an overview of the role of astrocytes on the developing brain.

INTRODUCTION

The nervous system is made up of two main cell types, neurons and glial cells. Glial cells are more numerous than neurons; representing 90% of the human brain (He and Sun, 2007), and the two types of cell cover a similar proportion of the area within the nervous tissue. The principal types of glial cells found in the central nervous system include astrocytes and oligodendrocytes, while in the peripheral nervous system they include Schwann cells, enteric glial cells and satellite cells (Jessen, 2004). In this chapter, we concentrate on the astrocyte, which has numerous clearly defined functions; astrocytes regulate ion

DOI: 10.4018/978-1-6684-9675-6.ch002

homeostasis, absorb glutamate at synapses and give metabolic support to nearby neurons (Ullian et al., 2004; He and Sun, 2007), and most importantly synapse formation. Protoplasmic astrocytes in grey matter develop widely branched stalks that terminate in delicately ramified leaf-like formations called perisynaptic astrocytic processes (PAPs), enabling single astrocytes to establish connections with hundreds or even millions of synapses, according to the species (Zhou et al., 2019; Torres-Ceja and Olsen, 2022). The development of astrocyte morphological complexity in the mouse cortex is timed to coincide precisely with a phase of explosive synaptogenesis, indicating that the two processes are linked (Stogsdill et al., 2017). Astrocytes have a key role in the development of synapses via the production of synaptogenic factors. This chapter will focus on some of these factors and how they mediate synapse formation during brain development, after highlighting the role of astrocytes in the adult brain.

Astrocytes Functions in The Brain

Astrocytes are the most proliferate cells in the Central Nervous System (CNS) (Markiewicz and Lukomska, 2006), in mammalian brains, astrocytes account for around 20-40% of the overall number of neuronal cells (Herculano-Houzel, 2014).Following the initial finding that astroglial functions go way deeper than passive structural support, they have established their position as critical contributors to neuronal functioning. Also they are the key for neural repair and neurogenesis. By now, astrocytes exhibit a wide range of cellular-level functions, like synapse maturation, formation and elimination, neurotransmitter clearance, ion homeostasis, extracellular space volume regulation and modulation of both synaptic plasticity and activity (Araque et al., 2014; Dallérac and Rouach, 2016). Furthermore, they have been shown to be implicated in rhythm production as well as neural networks (Fellin, 2009; Lee et al., 2014; Poskanzer and Yuste, 2016).

By studying post-mortem human tissues, Oberheim and colleagues subsequently presented a more detailed study of human astrocyte categories. Four types of astrocytes, characterised by a high expression of the glial fibrillary acidic protein (GFAP), that rises with aging (Nichols et al., 1993); have been found in the human brain: interlaminar, located in layers I and II of the cortex; protoplasmic in layers III and IV of the cortex; varicose projections in layers V and VI of the cortex; and fibrous astroglia in the white matter (Figure 1).

Interlaminar Astroglia

Typical interlaminar astrocytes in human cortex are resident in layer I and extend in long threads through several laminae, ending in layer III/IV (Colombo and Reisin, 2004; Falcone et al., 2019).Interlaminar astrocytes in human brains are different to those in primate. In contrast to the oblong somata found in primates, interlaminar astrocytes in humans are more abundant and display smaller, rounder cell bodies. Yet a further noticeable distinction is the presence, in humans, of shorter processes that spread out in all directions and contribute to the boundaries of the pial glia via a network of GFAP fibers. While the functional importance of interlaminarastroglia remains to be defined, it has been suggested by Oberheim and associates that its properties point to a potential role in long-distance intra-cortical coordination and communication (Vasile et al., 2017).

Figure 1. The different types of astrocytes in human brain and their location in the distinct cortex layers (Vasile et al., 2017)

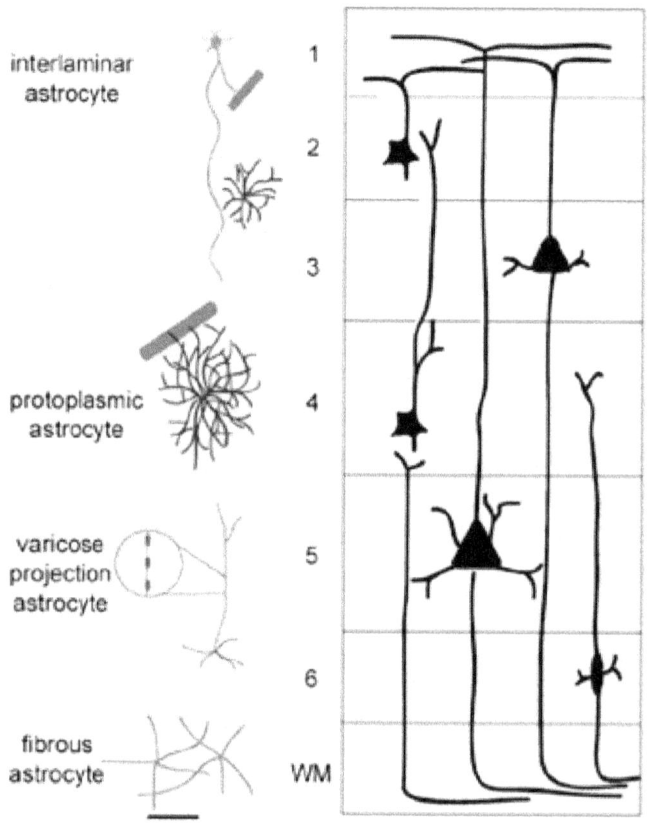

Protoplasmic Astroglia

Protoplasmic astroglia is a major astroglial type found in human, rodent, and primate brain. They are situated in layers II to VI of the brain cortex, and even though the human protoplasmic astrocytes' cell body size is equal to that of rodents, they have significantly longer processes (97.9 μm vs. 37.2 μm in rodents) (Oberheim et al., 2009). Human protoplasmic astrocytes are ~2.5 times larger in diameter, ~16.5 times bigger in volume (Oberheim et al., 2009), and have a 10-fold higher number of GFAP-positive processes, both symmetrically and radially in all directions from the soma, compared with rodent astrocytes (Oberheim et al., 2009). Regarding their role, protoplasmic astrocytes are in intimate connection with the pre- and postsynaptic compartments and contribute strongly to synaptic function and development in the majority of mammalian brains (Farhy-Tselnicker and Allen, 2018).

Varicose Projections Astroglia

The Varicose Projection Astroglia is the 3rd class of human brain GFAP+ astrocytes defined by Oberheim et al. (2009). Unique to humans and high-grade primates like chimpanzees, but much smaller and less complex in chimpanzees than in human (Vasile et al., 2017).They are located in layers V and VI of the cortex (Vasile et al., 2017), and also localized in the superficial white matter of the cortex (Falcone et al., 2022); and are distinguished with its long fibers with varicosities distributed around 10 μm from each other. Varicose projections astroglia make less synaptic contacts comparing to protoplasmic astroglia, because the principal GFAP+ processes of these astrocytes are straighter and have fewer branches (Vasile et al., 2017).

In terms of their functions, Oberheim and partners found that varicosities have been proposed to provide compartmentalisation of subcellular areas alongside the process, while extended processes could provide long-distance communication between the different cortical layers, like interlaminar astrocytes, and as well as between the grey and white matter (Oberheim et al., 2009). Also, a new study by Falcone and associates, suggests that VPAs may form a cellular reaction to certain brain conditions, notably inflammation or stress (Falcone et al., 2022).

Fibrous Astroglia

Fibrous astroglia are located further down in the cerebral cortex, in the white matter (Vasile et al., 2017). They differ from protoplasmic astrocytes in being larger, having fewer straight, ramified processes and less fine processes (Oberheim et al., 2009). With regard to their functions, Vasile et al confirm that a role for fibrous astrocytes in metabolic maintenance is evident, because most of them are in close connection with the vascular system, although in view of the lack of synapses in the white matter, they are unlikely to be responsible for modulating neuronal activity (Vasile et al., 2017).

The roles of astrocytes in brain are wide. Any disruption or malfunction in these functions can lead to serious trouble in SNC and brain capabilities. Yet, these important informations encourage us to search and write about the role of astrocytes in the developing brain, focusing on the dysfunction of these astrocytes and how can it harm neonatal and adult brain.

ROLE OF ASTROCYTES IN SYNAPSE FORMATION IN THE DEVELOPING BRAIN

Astrocytes have a number of clearly identified functions, such as the regulation of ion homeostasis, the glutamate uptake at the synapse and the support of the metabolic function of surrounding nerve cells. But their role in synapse formation and dysfunction is broader and more important (He and Sun, 2007; Ullian et al., 2004).

Throughout the development of the brain, cortical pathways are formed by a complicated process of synaptogenesis, maturation as well as synaptic pruning. During each of these stages, astrocytes fulfil a number of essential functions to assist the creation of these complex networks (Bosworth and Allen, 2017).

During development, the presence of astrocytes affects synapse formation. In the cerebral cortex of rodents, the astrocytes differentiate and mature at the exact same time as the synapses form and mature. Extensive synapse formation and maturation occurs in the second and third postnatal weeks (Risher et al., 2014).

Astrocytes Liberate Soluble Factors That Affect Synapse Creation

Many factors released by astrocytes (Figure 1), link to neuronal receptors to adjust the creation, maintenance and maturation of synapses and promote the formation of neural circuits (Allen and Eroglu, 2017). Yet, astrocytes' capacity to use secreted factors to stimulate synapse creation was discovered for the first time in purified primary cultures of neurons from the retinal ganglion cells (RGCs) (Farizatto and Baldwin, 2023). It has been shown that a direct connection with astrocytes or astrocyte conditioned medium (ACM) significantly increases the number of synapses and synaptic activity (Ullian et al., 2001; Nägler et al., 2001). The finding that ACM alone is enough to produce structural and functional synapses points to the identity of soluble factors liberated by astrocytes that have the ability to promote synapses. In fact, this is what happened when these factors were identified (Barker and Ullian, 2008). Cholesterol has been demonstrated to work in a presynaptic manner to increase the quantum contents and effectiveness of the transmitter liberation (Mauch et al., 2001).

Figure 2. Astrocytes liberate soluble factors that affect synapse formation and maturation

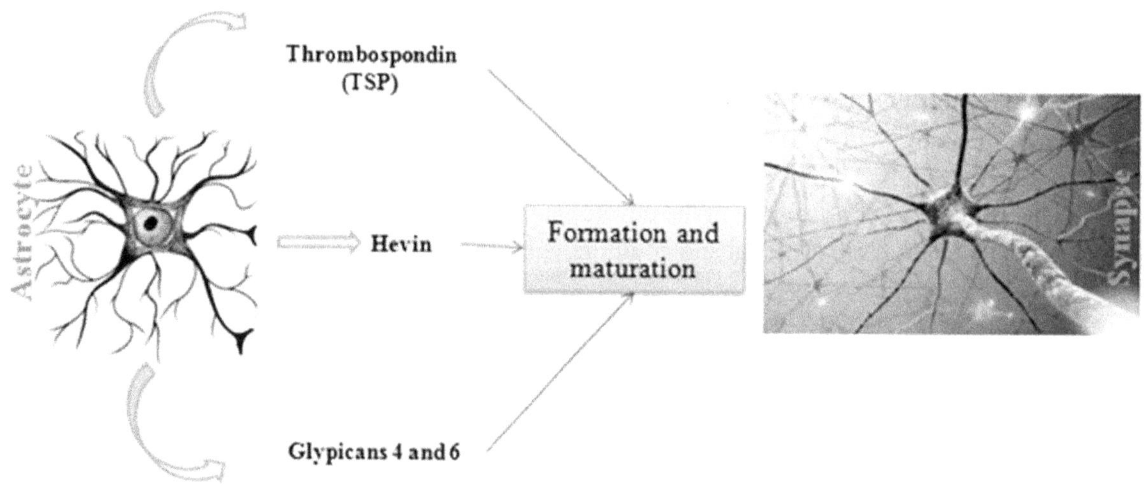

Thrombospondins

It has been shown that proteins released from astrocytes, thrombospondin (TSP) 1 and 2, induce an expansion in the number of synapse in cultured RGCs (Christopherson et al., 2005). Although the TSP-induced synapses showed a normal ultrastructural state, they still missed functional post-synaptic activity, which highlights the complex involvement of astrocytes in the synapse formation process (Barker and Ullian, 2008). Also, for Farizatto and Baldwin TSP1 and TSP2, stimulate the formation of excitatory intracortical synapses through association with a2d-1, their neuronal receptor.In 2021,Mazur et al showed that TSP2's synaptogenic impact on cortical neurons is gender-specific, it promotes strong synapse creation in male, but not female, primary cultures of cortical neurons. This is probably due to differential oestrogen secretion between male and female neurons, since suppression of 17b-estradiol

(E2) production cancels out this sex difference (Mazur et al., 2021). Yet, Fossati et al., 2019 showed that astrocytes, both express and release, pentraxin 3 (PTX3) throughout early post-natal development, which improves synaptic excitatory transmission by enhancing the integration of the AMPA synaptic receptor (AMPAR). This is achieved by PTX3 re-modelling the extracellular matrix around excitatory synapses and involving b1 integrin signalling. Excitingly, TSP1 directly blocks PTX3, indicating that there is a complicated molecular crosstalk controlling the balancing act between synaptic formation and maturation (Fossati et al., 2019).Recently, Farizatto and Baldwin (2023), confirmed that in the lack of glia, RGC nerve cells would persist, but would only form limited numbers of synapses, confirming the importance of astrocytes in synapses formation (Farizatto and Baldwin, 2023).

In addition to TSP 1 and 2, a certain number of factors are involved in the development and structural and functional maturation of synapses, in particular, Hevin, Glypicans 4 and 6, TNFα, Chordin-like 1, Pentraxin-3 and TBGb (Figure 2) (Farizatto and Baldwin, 2023).

Hevin

Hevin protein produced by astrocytes causesynaptogenic action in the primary visual cortex (Singh et al., 2016). The hevin protein is essential for the formation of thalamocorticalsynapses throughout development. It promotes the creation of thalamocortical synapses by regulating the trans-synaptic connection between presynaptic neurexin isoforms and non-compatible postsynaptic neuroligin isoforms, thus promoting the structuring of presynaptic and postsynaptic proteins (Bosworth and Allen, 2017).A quarter of postsynaptic dendritic spines within juvenile visual cortex accept stimulatory input from multiple presynaptic contacts (SMEC), and these developmental transitional structures are extensively suppressed in adults. In Hevin KO animals, the SMEC remain in adult life and the number of immature synapses is higher directly on the neuron's dendritic stalk than on the spines, suggesting an alteration in synaptic maturation. A population of axons may initially be competing for a cluster of similar postsynaptic contacts and that hevin is required for the resolving of these SMECs into single-contact mature spines (Risher et al., 2014). Thus,besides its synaptogenic effects, hevin is therefore involved in the establishment of mature, correctly localised synapses (Bosworth and Allen, 2017).

Glypicans 4 and 6

The heparin-sulphate proteoglycans Glypican 4 (Gpc4) and Glypican 6 (Gpc6) stimulate the development of functional synapses by stimulating the post-synaptic recruitment and assembly of glutamate receptors (Allen et al., 2012). According to one study, Gpc6 markedly boosts surface expression of NMDA glutamate receptor sub-units in neurons derived from human induced pluripotent stem cells and increases cell responsiveness to exogenous glutamate (Sato et al., 2016). On the other hand the Gpc4 protein attaches to the presynaptic receptor for protein tyrosine phosphatase d (RPTPd), triggering the secretion of neuronal pentraxin 1 (NP1), an AMPAR assembly factor. NP1 promotes postsynaptic binding of GluA1 AMPARs, causing the production of functional synapses (Farhy-Tselnicker et al., 2017). In mice with a deficiency in Gpc4, excitatory synapses and synapses containing GluA1 are fewer in number (Farizatto and Baldwin, 2023). In humans, reduced Gpc4 production is linked to autism spectrum disorders. Consequently, loss of Gpc4 leads to hyperactivity in juvenile mice and a deterioration in social interaction in adult mice (Dowling and Allen, 2018). As the brain develops, synapses that are both newly formed and functional release AMPARs that contain GluA1, which are not permeable to calcium. On maturation, these are substituted by GluA2-containing calcium-permeable AMPARs. This crucial stage in synapse maturation is controlled by the Chordin-like 1 (Chrdl1) protein produced by astrocytes (Blanco-Suarez et al., 2018). Whereas Chrdl1 is required to promote synaptic maturation in RGC culture, its expression in mouse brain is developmentally limited to visual cortex, with a peak in expression at the moment of this developmental change. Thus, Chrdl1 upregulates the presence of GluA2 AMPARs at excitatory intracortical and thalamocortical synapses and limits experimentally-dependent visual plasticity (Farizatto and Baldwin, 2023). Yet, Chrdl1 only regulates the production of thalamocortical, but not intracortical, excitatory synapses, suggesting that Chrdl1 works in a region- and cell-type-specific way to mediate synapse creation and maturation (Blanco-Suarez et al., 2018).

SPARC

In addition, a neighbouring homolog of Hevin/Sparcl1, called SPARC (Secreted Protein Acidic Rich in Cysteine), is released by astrocytes. Indeed, when purified neuronal cell cultures were treated with ACM from which SPARC had been removed, a significant increase in the number of excitatory synapses formed in RGCs was observed (Kucukdereli et al., 2011).

In vivo inactivation (KO) of SPARC increases the development of thalamo-cortical synapses within the superior colliculus by P14 (Tan et al., 2021). Likewise, it has been suggested that SPARC participates in the induction of GluA1 and GluA2 AMPAR sub-units at developmental synapses, since hippocampal neurons that grow on SPARC KO astrocyte culture exhibit wider GluA1 and GluA2 areal display (Jones et al., 2011).

Remarkably, SPARC also negatively controls the maturity and subsequent development of cholinergic-type synapses, in parallel with thalamocortical synapses. For example, treatment of neonatal rat superior cervical ganglion neuron cultures with SPARC reduced the amount of cholinergic synapses produced, providing further evidence that SPARC contributes to limiting synapse formation (Tan et al., 2021). SPARC administration can also cause the reduction and deletion of existing synapses, indicating that SPARC plays a role in regulating synaptic connection following synaptogenesis (Lo'pez-Murcia et al., 2015).

These findings indicate that SPARC produced by astrocytes inhibits the formation and maturity of pre-synaptic excitatory and cholinergic endings.

TNFα

TNFα, is a pro-inflammatory cytokine, responsible for controlling the production of AMPAR sub-units at synaptic surfaces, and it is also produced by astrocytes (Tan et al., 2021). It was found that TNFα elevates the frequency of mEPSCs and concomitantly the synaptic expression of GluA1, but causes a decline in the mIPSC frequency and GABAaR representation, indicating that TNFα is able to participate in the maturation of excitation synapses and the downregulation of inhibition synapses (Stellwagen et al., 2005).

Tan et al., 2021, confirmed that once the activity at the synapse is reduced, TNFα boosts this activity by augmenting AMPAR surface expression in order to balance the global activity decline. As a result, TNFα released by astrocytes becomes relevant for GluA1 recruitment and expression at excitatory synapses, as well as for the control of synaptic potency following modifications in synaptic activity after the full development of synaptic pathways (Tan et al., 2021).

Pentraxin-3

Another recent example is pentraxin-3 (PTX3), a protein released by astrocytes that promotes the formation and proper development of excitatory synapses by targeting AMPAR subunits (Fossati et al., 2019). This soluble molecule of innate immunity is synthesised in particular by astrocytes towards the end of the embryonic stage and the beginning of the post-natal stage (Tan et al., 2021). Also, Tan et al., 2021, confirmed that PTX3 treatment of cultured hippocampal neurons enhanced the amplitude as well as the frequency of mEPSCs without impacting the amount of dendritic spines or glutamatergic synapses. The quantity of AMPAR subunits at the synapses was likewise increased by PTX3 therapy. Neurons treated with PTX3 did not exhibit alterations in NP1 or Gpc4 mRNA amounts, indicating that PTX3 induces a rise in postsynaptic AMPARs due to an approach different from that of glypicans and NP-1 (Tan et al., 2021). Through changing the extracellular matrix, which is the structure around excitatory synapses and boosting b1-integrin signaling, PTX3 achieves this (Farizatto and Baldwin, 2023). Remarkably, TSP1, which has been shown to bind to the extracellular matrix and enhance excitatory synaptogenesis, acts as an antagonist of PTX3, preventing PTX3 from upregulating the expression of the AMPAR component (Christopherson et al., 2005; Fossati et al., 2019). This relationship implies that chemicals released by astrocytes can influence one another in both directions in addition to acting on neuronal pathways (Tan et al., 2021).

Chordin-like 1

Astrocytes not only control the insertion of GluA1 AMPAR subunits into growing synapses but are also essential for the maturational change, which occurs when GluA1 subunits are switched out for calcium-impermeable components of GluA2. However, glypicans do not actually govern this changeover, which normally occurs in the mouse brain during P7 and P16 (Tan et al., 2021). Rather, when given to glia-free RGC cells, Chordin-Like 1 (Chrdl-1) was observed to greatly improve the surface cluster of GluA2 subunits on RGC dendrites (Blanco- Suarez et al., 2018). Also, Tan et al., 2021, confirmed that there was no change in the quantity of VGlut1+ intracortical synapses in the primary visual cortex of Chrdl-1 KO mice, while there was a decrease in the overall number of VGlut2+ thalamocortical synapses, indicating a critical function for Chrdl-1 in the development of thalamocortical synapses (Tan et al., 2021). Amazingly, greater synapse reorganization following visual loss indicates that Chrdl-1 KO mice have a considerably longer critical time after monocular enucleation (Blanco-Suarez et al., 2018). Because of a lack in GluA2 subunit surface expression, synapses produced in the absence of astrocytic Chrdl-1 may be relatively immature, which might account for this greater plasticity. thereby, astrocytes have a role in important phase control in addition to supporting the real development and maturation of synapses (Tan et al., 2021).

ASTROCYTE-NEURON CONTACT REGULATING SYNAPSE FORMATION AND MATURATION

The identification of novel contact-mediated mechanisms that control synapse and astrocyte development represents a significant recent breakthrough in our understanding of astrocyte-synapse interaction. In this section we go across a number of new, significant research studies.

Many molecules that bind to cells, such as members of the Neurexin (Nrxn) and Neuroligin (NL) families, are expressed by astrocytes and are essential for nerve cells over synaptic and circuit assembly (Farizatto and Baldwin, 2023). A research investigation showed that astrocyte connecting complexity can be created in vitro by direct contact with neurons using an innovative astrocyte-neuron interaction system. Nrnx1 in neurons and NL1, NL2, and NL3 in astrocytes mediate this contact mechanism (Stogsdill et al., 2017). In the developing mouse cortex, astrocyte complexity in morphology is substantially decreased by a lack of NL1, 2, or 3, as well as excitatory synapse quantity and perform decrease by removal of NL2 from astrocytes. Without affecting the quantity of inhibitory synapses, astrocyte-specific NL2 elimination also improves inhibitory synaptic activity (Stogsdill et al., 2017).

CONCLUSION

Astrocytes play a key role in synapse formation and maturation through the release of their soluble factors such as TSP 1 and 2, Hevin, Glypicans 4 and 6, TNFα, Chordin-like 1 and Pentraxin-3. In this chapter, we have tried to highlight the mechanisms of action of these factors and how they stimulate synapse formation and maturation, but there is still work to be done to clarify the pathways involved.

REFERENCES

Allen, N. J., Bennett, M. L., Foo, L. C., Wang, G. X., Chakraborty, C., Smith, S. J., & Barres, B. A. (2012). Astrocyte glypicans 4 and 6 promote formation of excitatory synapses via GluA1 AMPA receptors. *Nature*, *486*(7403), 410–414. doi:10.1038/nature11059 PMID:22722203

Allen, N. J., & Eroglu, C. (2017). Cell biology of astrocyte-synapse interactions. *Neuron*, *96*(3), 697–708. doi:10.1016/j.neuron.2017.09.056 PMID:29096081

Araque, A., Carmignoto, G., Haydon, P. G., Oliet, S. H., Robitaille, R., & Volterra, A. (2014). Gliotransmitters travel in time and space. *Neuron*, *81*(4), 728–739. doi:10.1016/j.neuron.2014.02.007 PMID:24559669

Barker, A. J., & Ullian, E. M. (2008). New roles for astrocytes in developing synaptic circuits. *Communicative & Integrative Biology*, *1*(2), 207–211. doi:10.4161/cib.1.2.7284 PMID:19513261

Blanco-Suarez, E., Liu, T. F., Kopelevich, A., & Allen, N. J. (2018). Astrocyte-secreted chordin-like 1 drives synapse maturation and limits plasticity by increasing synaptic GluA2 AMPA receptors. *Neuron*, *100*(5), 1116–1132. doi:10.1016/j.neuron.2018.09.043 PMID:30344043

Bosworth, A. P., & Allen, N. J. (2017). The diverse actions of astrocytes during synaptic development. *Current Opinion in Neurobiology*, *47*, 38–43. doi:10.1016/j.conb.2017.08.017 PMID:28938161

Christopherson, K. S., Ullian, E. M., Stokes, C. C., Mullowney, C. E., Hell, J. W., Agah, A., Lawler, J., Mosher, D. F., Bornstein, P., & Barres, B. A. (2005). Thrombospondins are astrocyte-secreted proteins that promote CNS synaptogenesis. *Cell*, *120*(3), 421–433. doi:10.1016/j.cell.2004.12.020 PMID:15707899

Colombo, J. A., & Reisin, H. D. (2004). Interlaminarastroglia of the cerebral cortex: A marker of the primate brain. *Brain Research*, *1006*(1), 126–131. doi:10.1016/j.brainres.2004.02.003 PMID:15047031

Dallérac, G., & Rouach, N. (2016). Astrocytes as new targets to improve cognitive functions. *Progress in Neurobiology*, *144*, 48–67. doi:10.1016/j.pneurobio.2016.01.003 PMID:26969413

Dowling, C., & Allen, N. J. (2018). Mice lacking glypican 4 display juvenile hyperactivity and adult social interaction deficits. *Brain Plasticity*, *4*(2), 197–209. doi:10.3233/BPL-180079 PMID:30598870

Falcone, C., McBride, E. L., Hopkins, W. D., Hof, P. R., Manger, P. R., Sherwood, C. C., Noctor, S. C., & Martínez-Cerdeño, V. (2022). Redefining varicose projection astrocytes in primates. *Glia*, *70*(1), 145–154. doi:10.1002/glia.24093 PMID:34533866

Falcone, C., Wolf-Ochoa, M., Amina, S., Hong, T., Vakilzadeh, G., Hopkins, W. D., Hof, P. R., Sherwood, C. C., Manger, P. R., Noctor, S. C., & Martínez-Cerdeño, V. (2019). Cortical interlaminar astrocytes across the therian mammal radiation. *The Journal of Comparative Neurology*, *527*(10), 1654–1674. doi:10.1002/cne.24605 PMID:30552685

Farhy-Tselnicker, I., & Allen, N. J. (2018). Astrocytes, neurons, synapses: A tripartite view on cortical circuit development. *Neural Development*, *13*(1), 1–12. doi:10.118613064-018-0104-y PMID:29712572

Farhy-Tselnicker, I., van Casteren, A. C., Lee, A., Chang, V. T., Aricescu, A. R., & Allen, N. J. (2017). Astrocyte-secreted glypican 4 regulates release of neuronal pentraxin 1 from axons to induce functional synapse formation. *Neuron*, *96*(2), 428–445. doi:10.1016/j.neuron.2017.09.053 PMID:29024665

Farizatto, K. L., & Baldwin, K. T. (2023). Astrocyte-synapse interactions during brain development. *Current Opinion in Neurobiology*, *80*, 102704. doi:10.1016/j.conb.2023.102704 PMID:36913751

Fellin, T. (2009). Communication between neurons and astrocytes: Relevance to the modulation of synaptic and network activity. *Journal of Neurochemistry*, *108*(3), 533–544. doi:10.1111/j.1471-4159.2008.05830.x PMID:19187090

Fossati, G., Pozzi, D., Canzi, A., Mirabella, F., Valentino, S., Morini, R., Ghirardini, E., Filipello, F., Moretti, M., Gotti, C., Annis, D. S., Mosher, D. F., Garlanda, C., Bottazzi, B., Taraboletti, G., Mantovani, A., Matteoli, M., & Menna, E. (2019). Pentraxin 3 regulates synaptic function by inducing AMPA receptor clustering via ECM remodeling and β1-integrin. *The EMBO Journal*, *38*(1), e99529. doi:10.15252/embj.201899529 PMID:30396995

He, F., & Sun, Y. E. (2007). Glial cells more than support cells? *The International Journal of Biochemistry & Cell Biology*, *39*(4), 661–665. doi:10.1016/j.biocel.2006.10.022 PMID:17141551

Herculano-Houzel, S. (2014). The glia/neuron ratio: How it varies uniformly across brain structures and species and what that means for brain physiology and evolution. *Glia*, *62*(9), 1377–1391. doi:10.1002/glia.22683 PMID:24807023

Jessen, K. R. (2004). Glial cells. *The International Journal of Biochemistry & Cell Biology*, *36*(10), 1861–1867. doi:10.1016/j.biocel.2004.02.023 PMID:15203098

Jones, E. V., Bernardinelli, Y., Tse, Y. C., Chierzi, S., Wong, T. P., & Murai, K. K. (2011). Astrocytes control glutamate receptor levels at developing synapses through SPARC–β-integrin interactions. *The Journal of Neuroscience : The Official Journal of the Society for Neuroscience*, *31*(11), 4154–4165. doi:10.1523/JNEUROSCI.4757-10.2011 PMID:21411656

Kucukdereli, H., Allen, N. J., Lee, A. T., Feng, A., Ozlu, M. I., Conatser, L. M., Chakraborty, C., Workman, G., Weaver, M., Sage, E. H., Barres, B. A., & Eroglu, C. (2011). Control of excitatory CNS synaptogenesis by astrocyte-secreted proteins Hevin and SPARC. *Proceedings of the National Academy of Sciences of the United States of America, 108*(32), E440–E449. doi:10.1073/pnas.1104977108 PMID:21788491

Lee, H. S., Ghetti, A., Pinto-Duarte, A., Wang, X., Dziewczapolski, G., Galimi, F., Huitron-Resendiz, S., Piña-Crespo, J. C., Roberts, A. J., Verma, I. M., Sejnowski, T. J., & Heinemann, S. F. (2014). Astrocytes contribute to gamma oscillations and recognition memory. *Proceedings of the National Academy of Sciences of the United States of America, 111*(32), E3343–E3352. doi:10.1073/pnas.1410893111 PMID:25071179

López-Murcia, F. J., Terni, B., & Llobet, A. (2015). SPARC triggers a cell-autonomous program of synapse elimination. *Proceedings of the National Academy of Sciences of the United States of America, 112*(43), 13366–13371. doi:10.1073/pnas.1512202112 PMID:26420865

Markiewicz, I., & Lukomska, B. (2006). The role of astrocytes in the physiology and pathology of the central nervous system. *Acta Neurobiologiae Experimentalis, 66*(4), 343–358. PMID:17265695

Mauch, D. H., Nagler, K., Schumacher, S., Goritz, C., Müller, E. C., Otto, A., & Pfrieger, F. W. (2001). CNS synaptogenesis promoted by glia-derived cholesterol. *Science, 294*(5545), 1354–1357. doi:10.1126cience.294.5545.1354 PMID:11701931

Mazur, A., Bills, E. H., DeSchepper, K. M., Williamson, J. C., Henderson, B. J., & Risher, W. C. (2021). Astrocyte-derived thrombospondin induces cortical synaptogenesis in a sex-specific manner. *eNeuro, 8*(4), ENEURO.0014-21.2021. doi:10.1523/ENEURO.0014-21.2021 PMID:34266964

Nägler, K., Mauch, D. H., & Pfrieger, F. W. (2001). Glia-derived signals induce synapse formation in neurones of the rat central nervous system. *The Journal of Physiology, 533*(Pt 3), 665–679. doi:10.1111/j.1469-7793.2001.00665.x PMID:11410625

Nichols, N. R., Day, J. R., Laping, N. J., Johnson, S. A., & Finch, C. E. (1993). GFAP mRNA increases with age in rat and human brain. *Neurobiology of Aging, 14*(5), 421–429. doi:10.1016/0197-4580(93)90100-P PMID:8247224

Oberheim, N. A., Takano, T., Han, X., He, W., Lin, J. H., Wang, F., Xu, Q., Wyatt, J. D., Pilcher, W., Ojemann, J. G., Ransom, B. R., Goldman, S. A., & Nedergaard, M. (2009). Uniquely hominid features of adult human astrocytes. *The Journal of Neuroscience : The Official Journal of the Society for Neuroscience, 29*(10), 3276–3287. doi:10.1523/JNEUROSCI.4707-08.2009 PMID:19279265

Oberheim, N. A., Wang, X., Goldman, S., & Nedergaard, M. (2006). Astrocytic complexity distinguishes the human brain. *Trends in Neurosciences, 29*(10), 547–553. doi:10.1016/j.tins.2006.08.004 PMID:16938356

Poskanzer, K. E., & Yuste, R. (2016). Astrocytes regulate cortical state switching in vivo. *Proceedings of the National Academy of Sciences of the United States of America, 113*(19), E2675–E2684. doi:10.1073/pnas.1520759113 PMID:27122314

Risher, W. C., Patel, S., Kim, I. H., Uezu, A., Bhagat, S., Wilton, D. K., Pilaz, L.-J., Singh Alvarado, J., Calhan, O. Y., Silver, D. L., Stevens, B., Calakos, N., Soderling, S. H., & Eroglu, C. (2014). Astrocytes refine cortical connectivity at dendritic spines. *eLife, 3*, e04047. doi:10.7554/eLife.04047 PMID:25517933

Sato, K., Takahashi, K., Shigemoto-Mogami, Y., Chujo, K., & Sekino, Y. (2016). Glypican 6 enhances N-methyl-D-aspartate receptor function in human-induced pluripotent stem cell-derived neurons. *Frontiers in Cellular Neuroscience, 10*, 259. doi:10.3389/fncel.2016.00259 PMID:27895553

Singh, S. K., Stogsdill, J. A., Pulimood, N. S., Dingsdale, H., Kim, Y. H., Pilaz, L. J., Kim, I. H., Manhaes, A. C., Rodrigues, W. S. Jr, Pamukcu, A., Enustun, E., Ertuz, Z., Scheiffele, P., Soderling, S. H., Silver, D. L., Ji, R.-R., Medina, A. E., & Eroglu, C. (2016). Astrocytes assemble thalamocortical synapses by bridging NRX1α and NL1 via hevin. *Cell, 164*(1), 183–196. doi:10.1016/j.cell.2015.11.034 PMID:26771491

Stellwagen, D., Beattie, E. C., Seo, J. Y., & Malenka, R. C. (2005). Differential regulation of AMPA receptor and GABA receptor trafficking by tumor necrosis factor-α. *The Journal of Neuroscience : The Official Journal of the Society for Neuroscience, 25*(12), 3219–3228. doi:10.1523/JNEUROSCI.4486-04.2005 PMID:15788779

Stogsdill, J. A., Ramirez, J., Liu, D., Kim, Y. H., Baldwin, K. T., Enustun, E., Ejikeme, T., Ji, R.-R., & Eroglu, C. (2017). Astrocytic neuroligins control astrocyte morphogenesis and synaptogenesis. *Nature, 551*(7679), 192–197. doi:10.1038/nature24638 PMID:29120426

Tan, C. X., Lane, C. J. B., & Eroglu, C. (2021). Role of astrocytes in synapse formation and maturation. *Current Topics in Developmental Biology, 142*, 371–407. doi:10.1016/bs.ctdb.2020.12.010 PMID:33706922

Torres-Ceja, B., & Olsen, M. L. (2022). A closer look at astrocyte morphology: Development, heterogeneity, and plasticity at astrocyte leaflets. *Current Opinion in Neurobiology, 74*, 102550. doi:10.1016/j.conb.2022.102550 PMID:35544965

Ullian, E. M., Christopherson, K. S., & Barres, B. A. (2004). Role for glia in synaptogenesis. *Glia, 47*(3), 209–216. doi:10.1002/glia.20082 PMID:15252809

Ullian, E. M., Sapperstein, S. K., Christopherson, K. S., & Barres, B. A. (2001). Control of synapse number by glia. *science, 291*(5504), 657-661.

Vasile, F., Dossi, E., & Rouach, N. (2017). Human astrocytes: Structure and functions in the healthy brain. *Brain Structure & Function, 222*(5), 2017–2029. doi:10.100700429-017-1383-5 PMID:28280934

Zhou, B., Zuo, Y. X., & Jiang, R. T. (2019). Astrocyte morphology: Diversity, plasticity, and role in neurological diseases. *CNS Neuroscience & Therapeutics, 25*(6), 665–673. doi:10.1111/cns.13123 PMID:30929313

Chapter 3
Cognitive Function Involving Glial Cells:
The Surprising Role of Astrocytes

Samira Boulbaroud
Polydisciplinary Faculty of Beni Mellal, Morocco

Hanane Khalki
Polydisciplinary Faculty of Beni Mellal, Morocco

Fatima Zahra Azzaoui
Laboratory of Biology and Health, Faculty of Sciences, Ibn Tofail University, Kenitra, Morocco

ABSTRACT

Astrocytes, once considered passive support cells in the central nervous system, are now recognized as dynamic contributors to neuronal processes. They play pivotal roles in regulating synaptic transmission, modulating excitability, and influencing synapse formation. These non-neuronal cells release gliotransmitters like glutamate, affecting synaptic activity. Dysfunctions in astrocytes are linked to neurodegenerative and psychiatric disorders. In neurodegenerative disorders like Alzheimer's and Parkinson's, astrocytic dysfunction plays distinct roles. While astrocytes may not significantly contribute to Alzheimer's progression, they are involved in neuroinflammation, Aβ metabolism, and calcium regulation. Conversely, in Parkinson's, astrocytes contribute to mitochondrial dysfunction, impacting dopaminergic neurons. This comprehensive exploration sheds light on the intricate and multifaceted roles of astrocytes in cognition and their potential implications for therapeutic interventions in neurological and psychiatric conditions.

INTRODUCTION

Glial cells, a heterogeneous group of non-neuronal cells, are crucial for modulating and supporting neuronal function in the central nervous system (CNS). Glial cells, including astrocytes, microglia, and oligodendrocytes, constitute the majority of cells in the CNS (Jakel & Dimou., 2017). While neurons

DOI: 10.4018/978-1-6684-9675-6.ch003

have received more attention traditionally, glial cells are increasingly acknowledged for their essential roles in maintaining brain homeostasis, synaptic function, neuronal survival, and information processing. Glial cells provide metabolic support, regulate extracellular ion concentrations, recycle neurotransmitters, contribute to the blood-brain barrier, and actively participate in immune responses and synaptic plasticity (Barker et al., 2008).

Astrocytes, the most abundant glial cell type in the brain, have been traditionally associated with functions such as metabolic support, regulation of extracellular ion concentrations, and maintenance of the blood-brain barrier (Herculano-houzel., 2014; Vasile et al., 2017). However, recent studies have unveiled their unexpected involvement in cognitive functions, challenging the conventional view of astrocytes as passive support cells., Research has shown that astrocytes actively participate in various cognitive processes, including learning, memory, and synaptic plasticity (Dossi et al., 2018). They regulate synaptic transmission, modulate neuronal excitability, and influence the formation and elimination of synapses. Additionally, astrocytes release gliotransmitters, such as glutamate, ATP, and Adenosine, which can directly modulate synaptic activity and contribute to information processing in the brain (Park et al., 2015; Chen et al., 2013; Lee et al., 2013; Fujii et al., 2014; Sibille et al., 2014, 2015). Astrocyte dysfunctions have been implicated in several neurodegenerative and psychiatric disorders, further highlighting their importance in cognitive functions. Dysregulated astrocyte calcium signaling, impaired glutamate uptake, and disrupted astrocyte-neuron interactions have been observed in conditions such as Alzheimer's disease, Parkinson's disease, and schizophrenia. Overall, the coexistence and intertwining of protoplasmic astrocytes and varicose projection astrocytes in the human cortex suggest that they are distinct subtypes of cells with specialized functions in regulating synaptic activity and supporting the communication between different brain regions. Astrocytes, previously known as supportive cells, are now recognized as active contributors to cognitive processes. The multifaceted nature of their roles, encompassing metabolic assistance, modulatory control over synaptic activity, and involvement in neural plasticity, underscores the complex interplay between neurons and astrocytes within the CNS (khakh & Sofroniew., 2015). Gaining insight into the functions of astrocytes in cognition may offer novel avenues for therapeutic interventions aimed at addressing neurodegenerative and psychiatric conditions (Liddelow & Barre., 2017; Dallerac & Rouach., 2016).

THE MULTIFACETED ROLES OF ASTROCYTES IN COGNITIVE FUNCTION: INSIGHTS AND IMPLICATIONS

Heterogeneity of Astrocytes and the Multiplicity of Their Origin

Astrocytes, a type of glial cells in the CNS, perform various functions, including metabolic and structural support for neurons, regulation of the extracellular environment, and contribution to neural development and signaling (Chen and Swanson, 2003). The existence of two basic subtypes of astrocytes in rodents, the protoplasmic and fibrous astrocytes, has been established (Miller & Raff, 1984).

Protoplasmic astrocytes exhibit bushy and highly branched morphology, with numerous fine processes emanating from the cell body. They are primarily located in the gray matter of the CNS, particularly in the cerebral cortex, where they are closely associated with neurons and synaptic structures. Protoplasmic astrocytes play a vital role in maintaining microenvironment surrounding neurons, providing metabolic support, regulating neurotransmitter levels, and modulating synaptic activity (Tabata, 2015).

On the other hand, Fibrous astrocytes are commonly situated in the white matter regions of the brain and spinal cord. They possess elongated and slender processes with numerous delicate branches that extend in diverse directions, forming a dense network. These processes serve as a structural foundation and contribute to the creation of the blood-brain barrier, which aids in regulating the exchange of substances between the blood and the brain tissue. In this cell type, the expression level of glial fibrillary acidic protein (GFAP), an intermediate filament protein, is higher than that in the protoplasmic astrocyte, in which the GFAP protein is sometimes found only in the end feet on the blood vessels (Oberheim et al., 2009). Apart from their structural function, fibrous astrocytes are involved in various other functions. They maintain ion balance, supply nutrients to neurons, and remove waste products, thereby regulating the extracellular environment. Fibrous astrocytes also play a pivotal role in the process of repair and scar formation after CNS injury, as they can proliferate and form a glial scar that helps safeguard the injured area (*Figure 2*).

Figure 1. Role of astrocytes in a micro-environment-dependent mode: (A) In normal, healthy conditions, astrocytes perform functions that support the overall homeostasis of the nervous tissue, and (B) Reactive astrocytosis is a complex phenomenon with debated effects

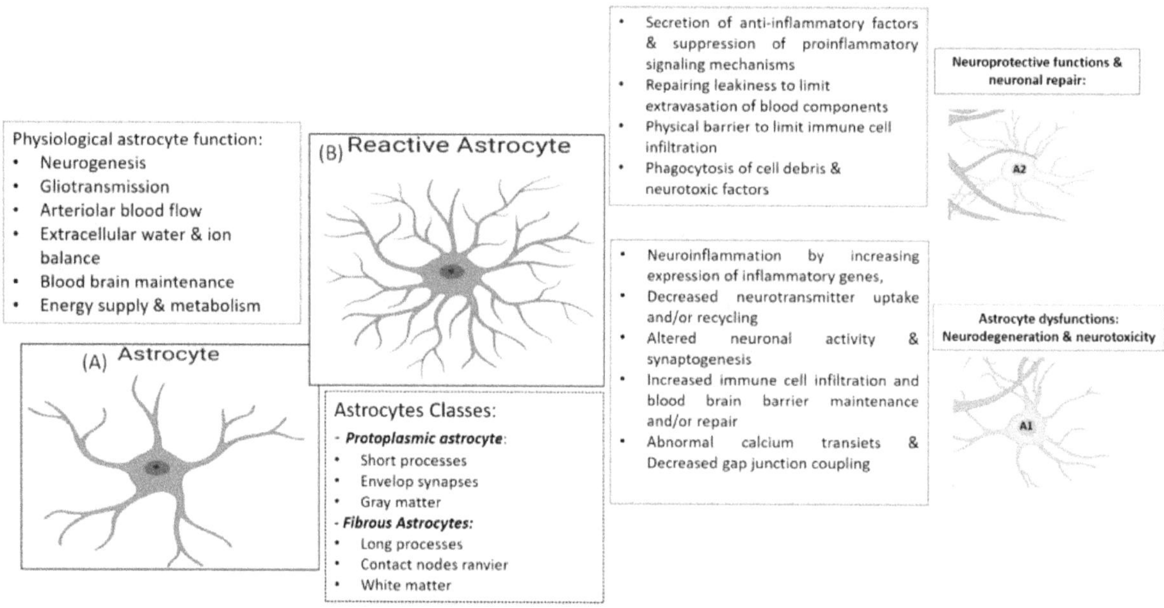

It appears to have a dual role, potentially contributing to both cell death and neuroprotection, and the outcome may depend on the specific context in which it occurs.

In humans, both protoplasmic and fibrous astrocytes display a unique structural composition. It has been documented that their diameters are significantly larger in the human cortex, measuring approximately 2~2.5-fold greater than in mice (Oberheim et al., 2009). In the human cortex, astrocytes are morphologically categorized into four subtypes; interlaminar, protoplasmic, varicose projections, and fibrous astroglia, located in the layers I and II, III and IV, V and VI and in the white matter, respectively (Vasile et al., 2017). Moreover, human protoplasmic astrocytes establish exclusive domains similar to those in rodents, with a single domain encompassing roughly 2,000,000 synapses. In the deeper layers of the human cortex, protoplasmic astrocytes and varicose projection astrocytes coexist, and their processes are intertwined, implying that they are distinct subtypes of cells that serve varying functions. The overall coexistence and intertwining of protoplasmic astrocytes and varicose projection astrocytes within the human cortex implies that these are discrete cell subtypes that possess specialized functions in the regulation of synaptic activity and in supporting communication between diverse brain regions. Astrocytes are regionally and subregionally heterogeneous both morphologically and functionally. RNA sequencing analysis conducted by Zeisel et al. (2018) has revealed the existence of seven distinct types of astrocytes, which exhibit molecular differences in their expression of neurotransmission systems. Moreover, these astrocytic subtypes show regional specificity within the CNS. For example, Bergmann glia is found in the cerebellum and plays a crucial role in cerebellar development and function (De Zeeuw & Hoogland., 2015). Ependymal glia line the ventricles of the brain and contribute to the production and circulation of cerebrospinal fluid. Marginal glia is located in the marginal zone of the neocortex and is thought to be involved in neuronal migration and synapse formation. Perivascular glia, as the name suggests, are situated around blood vessels and are implicated in regulating blood-brain barrier integrity and nutrient exchange. Radial glia, present during development, serves as a scaffold for migrating neurons and is crucial for proper neuronal positioning. Tanycytes are specialized astrocytes located in the hypothalamus and are involved in regulating hormone release and energy homeostasis (Prevot et al., 2018). Velate glia are unique to the olfactory system and are believed to participate in odor processing. The identification and classification of these distinct astrocytic subtypes highlight the complexity and heterogeneity of astrocytes in the CNS. These differences in morphology, location, and function are likely influenced by both genetic factors and the specific microenvironment of each brain region (Lee et al., 2022, Makarava et al., 2023). Further research is still ongoing to fully understand the functional significance and molecular mechanisms underlying the diversity of astrocytes and their contributions to brain function.

Examination of Astroglial Structural Plasticity

Throughout evolution, astrocytes undergo a series of changes in their structural complexity, reaching their highest level in the human brain. In the human brain, astrocytes exhibit a remarkable degree of branching and intricate overall structure, surpassing that observed in other species (Oberheim et al., 2009, 2006, 2012). Human astrocytes also demonstrate more effective and rapid induction of intracellular Ca2+ signaling in comparison to their rodent counterparts. Previous research studies that focused on the number and morphology of astrocytes in rodents and humans primarily relied on glial fibrillary acidic protein (GFAP) as a marker. Nevertheless, this approach has limitations as GFAP expression is not present in every astrocyte and, even when present, it only represents a small fraction of the total cell volume (approximately 15%). Consequently, relying on GFAP staining provides an insufficient understanding of the intricate membranous structure of astrocytes and underestimates their presence at the tripartite synapse. Regarding astrocytic branching patterns, two models are still a subject of intense debate. One model proposes that astrocytes follow a dendritic architecture similar to a tree (reminiscent of most neural cells), while others favor a spongiform architecture that is established through internal connectivity via branch fusions or reflexive gap junctions (Khakh & Sofroniew, 2015a; Rusakov, 2015).

Tripartite Synapse: Neuron–Astrocyte Interactions

For many years, astrocytes were primarily regarded as passive support cells for neurons, providing metabolic support, neurotransmitter precursors, and ion buffering. Studies have shown that removing astrocytes in vivo or culturing neurons without astrocytes leads to neuronal degeneration and death (reviewed in (Allen.,2014; Jäkel & Dimou., 2017). The essential role of astrocytes in supporting neuronal survival has hindered a comprehensive understanding of their involvement in other aspects of neuronal function, such as synapse formation activity, until more recent investigations. It is now recognized that astrocytes play a crucial role in regulating synapses through both direct physical contact (Stogsdill et al., 2017., Nishida & Okabe., 2007) and the secretion of soluble factors that target pre- and postsynaptic sites, thereby influencing the structure and function of both excitatory and inhibitory synapses (Ullian et al., 2001; Christopherson et al., 2005). This has led to the concept of the "tripartite synapse," where a synapse is considered a functional unit composed of two neurons and an astrocyte.

In the tripartite conceptualization of the synapse, perisynaptic astrocytes are present along with the standard presynaptic and postsynaptic neurons (Bushong et al., 2004; Halassa et al., 2007; Santello & Volterra, 2009; Bernardinelli et al., 2014; Filosa et al., 2015). Contact made by perisynaptic astrocytes with the synaptic cleft depends on the type and location of synapses (Bernardinelli et al., 2014; Halassa et al., 2007; Genoud et al., 2006). Moreover, Research focusing on the interaction between astrocytes and excitatory synapses has revealed that the extent of astrocyte contact with synapses can vary across different brain regions. A single astrocyte in the rat hippocampus is estimated to occupy a territory of 66,000 μm3 of neuropil and make contact with over 140,000 synapses (Khakh & Sofroniew., 2015; Allen & Eroglu., 2017; Goenaga et al., 2023). This positioning allows a single astrocyte to effectively modulate the activity of a vast neuronal network. In comparison to rodents, human astrocytes have a volume approximately 27 times larger with a 2.5-fold increased diameter, and each astrocyte is estimated to make contact with up to approximately 2,000,000 synapses (Nanclares et al., 2021 (see review), Oberheim et al., 2009). In the cerebellum, astrocytes contact approximately 60-90% of synapses, while in the barrel cortex, they contact around 90% of synapses. In the hippocampus, this contact ranges from 50% to 90% of synapses, and in the striatum, approximately 80% of synapses are contacted by astrocytes (Farhy-Tselnicker& Allen., 2018; Zhou et al., 2019). These findings indicate that astrocytes have a significant presence and influence on synaptic function in these brain regions (Ventura & Harris., 1999). Additionally, these astrocyte-synapse contacts are not static but can be dynamically altered by neuronal activity. Increased neuronal activity has been shown to result in an expansion of astrocyte coverage of dendritic spines and an increased number of spines that are contacted by astrocytic processes (Bernardinelli et al., 2014; Chai et al., 2017). These dynamic changes in astrocyte-synapse interactions suggest that astrocytes play an active role in modulating synaptic activity and may contribute to activity-dependent synaptic plasticity (Bernardinelli et al., 2014; Chai et al., 2017). In the hippocampus, for instance, 64% of synapses are contacted by perisynaptic astrocytes at the synaptic cleft (Bushong et al., 2004). The intricate arborization and ramifications of astrocytes allow them to tightly enwrap the synaptic terminal to modulate synaptic processes (Bernardinelli et al., 2014; Halassa et al., 2007). Previous studies suggest that astrocytes respond to neurotransmitter release by increasing their intracellular calcium levels and controlling neuronal excitability through the release of gliotransmitters (Araque et al., 1999; Parpura & Verkhratsky, 2012). Based on findings from studies by Araque's group (Araque et al., 1999; Halassa et al., 2007) and others (Bushong et al., 2004; Genoud et al., 2006), which explain the functioning of the tripartite synapse, more attention has been given to the potential role of how astrocytes aid memory. In areas known for synaptic plasticity, such as the hippocampus, astroglial membranes appear to surround the majority of larger axo-dendritic synapses and around 60% of all synapses in the hippocampus (Genoud et al., 2006). In astrocytes that are part of a tripartite synapse, calcium peaks, which correspond to calcium oscillations tuned to neuronal activity, cause the proximal and distal release of glutamate from the astrocyte to neighboring neurons (Ota et al., 2013).

Astrocytes Are Interconnected via Gap Junction Channels

In addition to contacting neurons, astrocytes are interconnected with each other by gap junctions, specialized channels that allow nutrients and ions to diffuse between networks of astrocytes, expanding further the range and magnitude of synaptic regulation of neurons by astrocytes (Pannasch & Rouach., 2013). Astrocytes are interconnected via gap junction channels composed of connexins (Cxs), the gap junction channel–forming proteins. the astroglial network formed by these Cx-mediated gap junction channels plays a crucial role in regulating the efficient clearance of extracellular potassium (K+) and glutamate at synapses (Pannasch et al., 2012; Hardy et al., 2023). Additionally, it facilitates the long-distance transport of energy metabolites to actively functioning synapses (Rouach et al., 2008). It is important to note that Cxs not only form gap junction channels with other astrocytes but can also function as hemichannels, allowing for direct exchange with the extracellular space (Chever et al., 2014; Roux et al., 2015). In the mouse brain, astrocytes express two main types of connexins (Cxs): Cx43 and Cx30. Cx43 is known to play a critical role in brain development and various physiological processes (Ghézali et al., 2020). On the other hand, Cx30 expression begins after the second week of postnatal development (Ghézali et al., 2018), and its levels are influenced by neuronal activity. The deletion of Cx30 and Cx43 impairs spatial learning and memory in adult mice (Hösli et al., 2022; Walrave et al., 2016). The loss of astrocytic connexins and coupling activates astrocytes and microglia, which may contribute to the observed impairments in synaptic plasticity and spatial learning. A recent study conducted by Hardy et al. (2023) investigated the effects of selectively increasing Cx30 expression in astrocytes in the CA1 region of the hippocampus. Their findings revealed that mice with heightened levels of Cx30 exhibited an enlarged gap junction network but experienced reduced synaptic transmission at the CA1 Schaffer collateral synapses (Hösli et al., 2022). The increased Cx30 expression also led to decreased neuronal excitability, impairing synaptic plasticity and learning processes. These results highlight the intricate role of astrocytic connexins in regulating synaptic function and plasticity. Disruptions in astrocytic connexins can impact the astroglial network, synaptic transmission, and learning processes. Further research in this area may provide insights into the mechanisms underlying brain function and potential therapeutic targets for cognitive disorders.

Maintenance of the Blood-Brain Barrier and Support of Neuronal Synapses

The blood-brain barrier (BBB) is a highly selective barrier that separates the blood circulation from the brain tissue, controlling the movement of substances into and out of the brain (Zhao et al., 2015; Montagne et al., 2017). In the mammalian brain, the supply of blood to CNS circuits is facilitated by cerebral arteries, arterioles, and capillaries. In response to neuronal stimuli, these blood vessels increase the rate of cerebral blood flow (CBF) and oxygen delivery, a process known as neurovascular coupling ((NVC) (Kisler et al., 2017; Iadecola., 2017). The NVC involves the coordinated activity of various cell types within the neurovascular unit (NVU), notably vascular cells, such as endothelial cells that line the blood vessels, smooth muscle cells in the vessel walls, and pericytes that surround the capillaries, glia (e.g., astrocytes, microglia), and neurons. These vascular cells play crucial roles in regulating blood flow and maintaining the integrity of the blood-brain barrier (BBB) (Iadecola & Nedergaard., 2007; Zlokovic., 2011). Among the neurovascular unit, the astrocytes seem to play a crucial role in the regulation of the blood-brain barrier (BBB) and neurovascular coupling (NVC) (Kaplan, et al., 2020). These simultaneous direct interactions with neurons and the rest of the NVU make astrocytes attractive candidates for relaying signals between the parenchyma and the vasculature (Kaplan, et al., 2020). Astrocytes are essential for the formation and maintenance of the BBB by providing secreted factors that lead to the adequate association between the cells of the BBB and the formation of strong tight junctions (Luissint et al., 2012). Astrocytic endfeet, which make contact with the brain vasculature, endothelial cells (ECs), and pericytes, serve a crucial role in modulating both neuronal activity and cerebral blood flow (Cabezas et al., 2014, Hattori., 2022). This modulation is facilitated by the elevation of intracellular calcium levels in the endfeet (Zonta et al., 2003). Notably, astrocytic endfeet express specialized molecules such as Kir4.1 potassium channels and aquaporin 4, which regulate ionic concentrations at the blood-brain barrier (BBB), as well as protein transporters like glucose transporter-1 and P-glycoprotein (Bataveljić et al., 2012; Wagner et al., 2022; Olsen & Sontheimer.,2008; Abbott, 2002). This suggests the significance of endfeet in astrocyte polarization and the regulation of BBB integrity (O'Brown et al., 2018). Furthermore, astrocytes communicate with each other through gap junctions, forming a functional syncytium that exhibits well-coordinated responses. It has been proposed that inter-astroglial gap junctions facilitate the transmission of astrocytic mechanisms involved in vasodilation and vasoconstriction (Alvarez et al., 2013). Astrocytes also play a crucial role in the development and maintenance of BBB characteristics in ECs by releasing growth factors such as vascular endothelial growth factor (VEGF), glial cell line-derived neurotrophic factor (GDNF), basic fibroblast growth factor (bFGF), and angiopoietin-1 (ANG-1) (Igarashi., 1999; Lécuyer et al., 2016). These growth factors promote the formation of tight junctions, enhance enzymatic systems, and contribute to transporter polarization (Alvarez et al., 2013; Wong et al., 2013). In summary, astrocytic endfeet is critical in mediating the interaction between neurons, the BBB, and cerebral blood flow regulation. Their specialized molecules, gap junction communication, and secretion of growth factors contribute to the modulation of neuronal activity, maintenance of BBB integrity, and support of neuronal health in various physiological and pathological contexts (Figure 2).

Astrocytic endfeet play a crucial dual role in shaping neuronal activity and cerebral blood flow. Their strategic contact with the neurovascular unit (NVU) components, including brain vasculature, endothelial cells, and pericytes, positions them as key influencers in orchestrating the delicate balance between neural function and blood supply.

Regulation of Neurotransmitter Uptake and Release

Astrocytes play a crucial role in neurotransmitter uptake and release, making it one of their most important functions. Due to their strategic positioning around synapses, astrocytes efficiently clear excess neurotransmitters, optimizing neurotransmission. Astrocytes express a wide range of transporters and receptors for chemical transmitters, including both metabotropic and ionotropic receptors. Activation of these receptors, such as metabotropic glutamate receptor 5 (mGluR5) and purinergic receptors (P2Y1), leads to the generation of inositol trisphosphate (IP3) through phospholipase C. IP3 then binds to IP3-regulated channels present in intracellular calcium stores, increasing cytosolic calcium levels. Subsequently, activated astrocytes release gliotransmitters, including glutamate, GABA, and ATP/adenosine,

Figure 2. Astrocytic endfeet: Architectural maestros of neural-vascular harmony

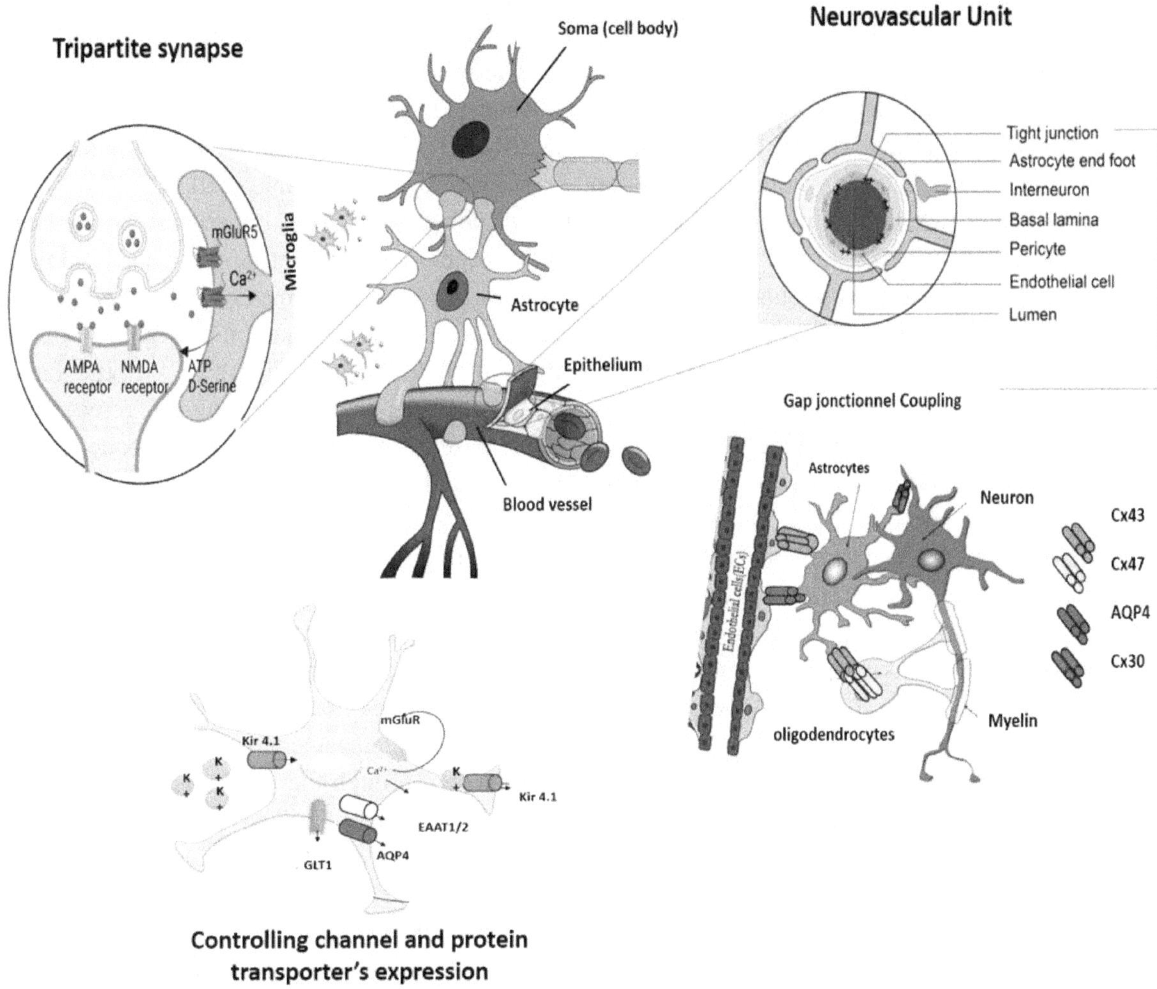

which serve as neuroactive substances. These gliotransmitters can modulate synaptic transmission and plasticity by activating GPCRs present in neurons. Such signaling mechanisms play a critical role in regulating synaptic transmission and plasticity.

Neurotransmitter Glutamate: Glutamate serves as the primary excitatory neurotransmitter in the CNS and exerts its effects on astrocytes in various brain regions. While there have been some reports proposing the presence of ionotropic glutamate receptors in astrocytes, the prevailing consensus is that glutamatergic signaling in these glial cells predominantly occurs through metabotropic glutamate receptors (mGluRs). Extensive evidence supports the notion that mGluR5 is the most prominent glutamate receptor expressed in astrocytes both in cultured cells and in situ environments (Schools & Kimelberg., 1999; Bernardinelli et al., 2014; Umpierre et al., 2019). The loss of IP3-dependent calcium activity, which serves as a critical mediator of astrocyte mGluR5 signaling, has been shown to decrease the formation of tripartite synapses in the hippocampus (Petravicz et al., 2008). Consequently, this reduction in synapse formation affects the kinetics of astrocyte glutamate transport, as reported by Tanaka et al. in

2013. Furthermore, the activation of mGluR5 signaling in astrocytes induces activity-dependent changes in the expression of GLT-1 (EAAT2), the primary glutamate transporter found in the mature hippocampus (Kellneret al., 2021; Umpierre et al., 2019). The removal of glutamate from the synaptic cleft by astrocytes is essential for proper synaptic transmission, underscoring the active role of astrocytes as key participants rather than mere supporters of synapses (Kinney and Spain, 2002). Additionally, astrocytes contribute to synaptic modulation by releasing neurotransmitters and modulators such as glutamate, purines (ATP, adenosine, and guanine), GABA, and D-serine (Covelo & Araque., 2018).

Neurotransmitter Adenosine triphosphate (ATP): Purine receptors can be broadly categorized into two groups: adenosine receptors (P1 receptors) and ATP receptors (P2 receptors). Various glial cells including astrocytes have been found to express both groups of receptors (James & Butt., 2002; Liu et al., 2019, Vincenzi et al., 2023). Adenosine receptors, which are G-protein coupled, primarily play a role in glial development and pathological responses (Butt., 2006; Sheth et al., 2014). On the other hand, P2 receptors can be further classified into ionotropic P2X receptors and metabotropic P2Y receptors. Activation of both P2X and P2Y receptors leads to an increase in intracellular calcium concentration ([Ca2+]i) and has been associated with important functions in glial cell physiology and pathology (James & Butt., 2002; Glaser et al., 2013; Puchałowicz et al., 2014).

Neurotransmitter GAB: Rapid uptake of gamma-aminobutyric acid (GABA) by astrocytes prevents the inhibitory signal from spreading to neighboring regions, ensuring precise signaling (Kinney & Spain, 2002). In studies conducted on hippocampal slices, it has been observed that the activation of GABAergic interneurons leads to the generation of calcium signals in astrocytes. This astrocytic response is effectively inhibited by the $GABA_B$ receptor antagonist CGP55845A (Kang et al., 1998; Perea et al., 2016). Furthermore, direct stimulation of astrocytes results in the enhancement of miniature inhibitory postsynaptic currents (mIPSCs) in pyramidal neurons, achieved through the activation of kainate receptors present in inhibitory terminals (Kang et al., 1998). The selective removal of $GABA_B$ receptors specifically from astrocytes through conditional knockout experiments resulted in notable changes in hippocampal oscillatory activity during in vivo studies. These findings provide compelling evidence for the direct involvement of $GABA_B$ receptor signaling in astrocytes in the processing of neuronal information (Perea et al., 2016). Hence, it is plausible for an inhibitory signal mediated by gamma-aminobutyric acid (GABA) to transform into an excitatory signal due to the activation of astrocytes. Moreover, various other elements of synaptic modulation facilitated by astrocytes have been associated with $GABA_B$ receptors in astrocytes (Anderson et al., 2007; Nagai et al., 2019). Interestingly, the activation of interneurons and the $GABA_B$ receptor signaling in astrocytes also can potentiate glutamatergic transmission in pyramidal neurons (Cuellar-Santoyo et al., 2023).

Neurotransmitter Dopamine: astrocytes also express receptors for various other chemical transmitters, and there is some evidence suggesting the presence of dopamine receptors as well. Recent studies have provided evidence that the activation of astrocytic dopamine D2 receptors (DRD2) exerts suppressive effects on neuroinflammation in conditions such as intracerebral hemorrhage and neurotoxicity induced by 1-methyl-4-phenyl-1,2,3,6-tetrahydropyridine (MPTP) (Shao et al., 2013; Zhang et al., 2015). Furthermore, αB-crystallin (CRYAB), a heat-shock protein, has demonstrated anti-inflammatory properties and has been found to regulate the astrocytic DRD2-mediated anti-inflammatory pathway in both neurodegenerative disorders and models of cerebral injury (Qiu et al., 2016; Corkrum et al., 2020). Furthermore, using electron microscopy, Corkrum et al., 2020 have provided evidence for the existence of D1 receptors (D1Rs) in astrocytes located in the nucleus accumbens and demonstrated that the activation of these receptors, both in vivo and in situ, by dopamine released synaptically, leads to the elevation of intracellular calcium levels in astrocytes. This process occurs through the G protein-coupled receptor (GPCR) signaling cascade involving IP3 receptor type 2 (IP3R2) and the mobilization of intracellular calcium. These data suggest the potential for dopamine to have a broader impact on drug abuse by influencing astrocytic functions (Sobolczyk & Boczek., 2022).

ASTROCYTIC DYSFUNCTION IN NEURODEGENERATIVE DISEASES: UNRAVELING THEIR ROLES IN NEURODEGENERATION

Astrocytes play a crucial role in maintaining the health and function of synapses, which are essential for cognitive processes. Reactive astrocytes, due to injury, disease, or aging, may lead to synapse loss Synapse loss is observed in neurodegenerative diseases like Alzheimer's (Henstridge et al., 2019; Lawrence et al., 2023).). Astrocytes interact with microglia to respond to brain injury and neuroinflammation, facilitating the clearance of cellular debris through phagocytosis. However, dysfunctional astrocytes can contribute to neurodegenerative diseases, causing vascular dysfunction, excessive glutamate release, and loss of neural stem cells. Understanding the role of astrocyte dysfunction in dementia and cognitive decline can lead to therapeutic interventions for preserving synaptic integrity and brain health. It is increasingly recognized that addressing astrocyte dysfunction and related mechanisms may be equally important as targeting amyloid-β plaques in Alzheimer's treatment. Astrocytes play a crucial role in maintaining the health and function of synapses, which are essential for cognitive processes. Reactive astrocytes, due to injury, disease, or aging, may lead to synapse loss. Synapse loss is observed in neurodegenerative diseases like Alzheimer's. Astrocytes interact with microglia to respond to brain injury and neuroinflammation, facilitating the clearance of cellular debris through phagocytosis (Kempuraj et al., 2020; Piccioni et al., 2021). However, dysfunctional astrocytes can contribute to neurodegenerative diseases, causing vascular dysfunction, excessive glutamate release, and loss of neural stem cells. Understanding the role of astrocyte dysfunction in dementia and cognitive decline can lead to therapeutic interventions for preserving synaptic integrity and brain health

Astrocytic Dysfunction in The Pathophysiology of Alzheimer's Disease

Alzheimer's disease (AD) represents the most prevalent form of dementia in the elderly population. It is distinguished by a range of pathological features, including the accumulation of extracellular amyloid plaques, the presence of intraneuronal neurofibrillary tangles, synaptic dysfunction, and neurodegeneration (Stancu et al., 2014). One significant facet of AD pathology is the presence of neuroinflammation within affected brains. This neuroinflammatory response is characterized by the activation of astrocytes (reactive astrogliosis), microglia (microgliosis), and an increased release of inflammatory cytokines (Heneka et al., 2015; Orsini et al., 2014; Gonzalez-Reyes RE et al., 2017). Reactive astrogliosis in AD is particularly evident in regions surrounding amyloid plaques and throughout affected brain areas. Activated astrocytes release various molecules, including cytokines and chemokines, which contribute to the inflammatory response and impact neuronal function (Heneka et al., 2015)

- **The Complex Role of Astrocytes in Alzheimer's Disease**

In AD, astrocytes serve as multifaceted players with profound implications for disease progression. These cells experience significant disruptions in energy metabolism and are exposed to oxidative stress, factors that are believed to render them potentially detrimental to neurons (Allaman et al., 2011). These alterations in astrocytic function contribute significantly to the complex pathophysiology of AD.

- **Astrocytes and Aβ Metabolism**

Furthermore, astrocytes play a pivotal role in the metabolism of Aβ (amyloid-beta), a hallmark feature of AD pathology. Upon exposure to Aβ, either through direct contact or proximity, astrocytes exhibit a marked reduction in neuronal viability in vitro studies (Allaman et al., 2011). This observation raises concerns regarding the potential role of astrocytes in the neurotoxicity associated with Aβ accumulation, given their abundance in the brain.

- **Astrocytic Aβ Production**

Intriguingly, astrocytes can produce Aβ themselves, particularly under specific inflammatory conditions. For example, the presence of TGF-β1 can act as a catalyst, prompting astrocytes to generate Aβ40/42, ultimately leading to elevated Aβ levels in TGF-β1 transgenic mice (Lesné et al., 2003; Zhang et al.,2016). Even more remarkably, when neurons coexist with astrocytes in a shared environment, the detrimental impact on neuronal survival is further amplified upon exposure to Aβ. Crucially, this injurious effect on neurons does not necessitate direct physical contact between neurons and astrocytes, suggesting that astrocytes can exert influence on neuronal health through secreted factors or intricate signaling mechanisms.

- **Astrocytes in Alzheimer's Pathogenesis**

This dual role of astrocytes, both in responding to Aβ and potentially exacerbating its neurotoxic effects, underscores their significance in the pathogenesis of AD. The complex interactions unfolding between astrocytes, Aβ, and neurons constitute a pivotal realm of investigation in the ongoing quest to fathom and target the mechanisms underpinning AD.

- **Astrocytic Contributions to Astrocytes in Alzheimer's: Calcium and Connectivity**

Moreover, studies have revealed disruptions in calcium homeostasis in reactive astrocytes near amyloid-beta (Aβ) plaques in mouse AD brains (Kuchibhotla et al., 2009). Alterations in the immunoreactivity pattern of connexin 43 (Cx43), a hemichannel protein involved in astrocytic gap junctions, suggest changes in connectivity between astrocytes in AD (Nagy et al., 1996; Mei et al., 2010). Elevated levels of Cx43 have been associated with increased release of toxic glutamate and adenosine triphosphate (ATP) in vitro (Orellana et al., 2011a, b).

- **Modulation of Hemichannels**

While glutamate release may have limited relevance in vivo, ATP release can significantly affect synaptic transmission (Pascual et al., 2005; Barres, 2008). Notably, blocking hemichannels in vivo has demonstrated enhanced memory function without impacting Aβ plaque deposition and a delay in disease progression, as observed in amyotrophic lateral sclerosis (ALS) (Takeuchi et al., *2011*).

- **Versatile Roles of Astrocytes**

On another front, within the context of AD, astrocytes reveal versatile roles (Monterey et al.,2021), each carrying significant implications for disease progression:

- **S100β and Inflammation:** The S100 β protein, known for its neurotrophic role during fetal development, possesses the ability to induce astrocytic reactivity. In AD, a notable concentration of S100-positive cells is observed in proximity to neuritic plaques, particularly in severely affected brain regions like the hippocampus. Interestingly, the antiprotozoal medication pentamidine, which inhibits S100 β activity, exhibits promising results by reducing astrocytic activation and mitigating the neuroinflammatory response (GFAP, S100β, RAGE) in AD mouse models (Cirillo et al., 2015).
- **Cholesterol and Lipid Processing:** Astrocytes play a central role in cholesterol and lipid metabolism in the brain, essential for normal functioning. Dysregulated cholesterol processing has been implicated in AD development. Excessive cholesterol production in neurons, triggered by elevated levels, promotes increased β-amyloid deposition by suppressing the expression of ATP-binding cassette transporter 1 (ABCA1). Moreover, high-fat diets have been linked to β-amyloid accumulation and plaque formation. Ceramide, a metabolite derived from fatty acids, is elevated in AD brains and drives astrocytic production of inflammatory cytokines, subsequently enhancing β-amyloid production in neurons (Loera-Valencia et al., 2019; Osborn et al., 2016).
- **Glutamate Regulation:** Astrocytes play a pivotal role in recycling glutamate and GABA into glutamine through the enzyme glutamine synthetase, thereby providing nutrients and protection against oxidative stress. However, in AD, a reduction in mRNA levels of glutamate transporters GLT-1 and GLAST is observed, resulting in disrupted glutamate homeostasis. This disturbance leads to impaired clearance of excitatory neurotransmitters and increased β-amyloid and tau levels originating from astrocytes (Jacob et al., 2007; Mookherjee et al., 2011; Matos et al., 2008).
- **GABA and Senescence:** Astrocytes have been implicated in the GABAergic system, with elevated GABA and MAOB content observed in AD patients and models. Dysfunctional GABA signaling within astrocytes can influence AD symptomology, particularly memory loss (Kilb & Kirischuk, 2022; Monterey et al., 2021). Furthermore, astrocytes can undergo stress-induced premature senescence, resulting in altered secretion profiles (senescence-associated secretory phenotype or SASP) that contribute to AD pathology. Senescent astrocytes share similarities with reactive astrocytes and may play a role in neurodegenerative processes (Lazic et al., 2022).

While these roles of astrocytes in AD are intricate and multifaceted, their involvement in inflammation, lipid processing, neurotransmitter regulation, and potential senescence underscores their significance in the progression of this neurodegenerative disease. Further research is warranted to fully elucidate the molecular mechanisms underlying astrocyte-related contributions to AD **(Table I).**

Table 1. Astrocytic dysfunction in the pathophysiology of Alzheimer's disease

Astrocytic changes	Models	Brain structures	Astrocytic reactivity consequences	REFERENCES
Astrocytic cytoskeletal changes within the mPFC and hippocampus	PDAPP-J20 mouse 3 × Tg-AD mice	Hippocampus (CA3) Amygdala Entorhinal cortex Prefrontal cortex	Decreased number and cell complexity of GFAP+ astrocytes. Decreased in the surface area and volume of GFAP-positive expression.	(Beauquis et al., 2014) (Kulijewicz-Nawrot et al., 2012)
Astroglial glutamine synthetase (GS) changes.	3 × Tg-AD mice	Hippocampus (Dentate gyrus, hippocampal CA1)	Decreased GFAP in the entorhinal cortex (EC). The presence of GFAP does not significantly impact glutamate homeostasis in the (EC). Unchanged numerical, surface, and volume of cells containing GS-IR and GS/GFAP-IR.	(Yeh et al., 2013).
Astrocytic EAAT2 variations and their interactions with amyloid-β (Aβ)	Tg2576 AD mouse model for Aβ	Hippocampal CA1	Aβ-induced calcium elevations in astrocytes can lead to the induction of NMDAR-mediated slow inward currents (SICs) in CA1 neurons. Amyloid-β (Aβ) Interaction with α7 Nicotinic Acetylcholine Receptors (nAChRs) which are highly permeable to calcium ions.	(Pirttimaki et al., 2013)
Changes in glutamate dynamics and GLT-1 expression	APP/PS1 transgenic mice	Somatosensory cortex Cortical samples	Anomalous glutamate fluctuations /reduction of GLT-1 expression around Aβ plaques. Understand the Timing of Pathological Processes of GLT-1 dysfunction in Amyloid-β (Aβ) Pathology	(Hefendehl et al., 2016; Mookherjee et al., 2011)
D-serine levels	AβPPswe/ PS1ΔE9 mice Human	Hippocampal and cortical samples cerebrospinal fluid (CSF)	D-serine levels elevation associated with AD-like pathology. Alterations in D-serine levels in both post-mortem brain tissue and cerebrospinal fluid (CSF) samples of individuals with AD.	(Madeira et al., 2015)
Abnormal release of GABA from reactive astrocytes	APP/PS1 mice	Dentate gyrus	Inhibitory gliotransmitter GABA by monoamine oxidase-B (MaoB) and abnormally release GABA through the bestrophin 1 channel. the abnormal release of GABA from reactive astrocytes affects neuronal function.	(Jo et al., 2014)

Astrocytic Dysfunction in the Pathophysiology of Parkinson's Disease

Parkinson's disease (PD) is a complex neurodegenerative disorder that affects both motor and non-motor functions. It is characterized by the progressive degeneration of dopaminergic neurons in a specific brain region called the substantia nigra pars compacta (SNpc). Additionally, the presence of abnormal protein aggregates known as Lewy bodies and neurites, composed mainly of α-synuclein, is a hallmark of PD. Recent research has shed light on the role of a particular type of glial cell called astrocytes in the development and progression of PD (Kordower., 2013) (Figure 3).

Figure 3. Astrocytes in Parkinson's disease

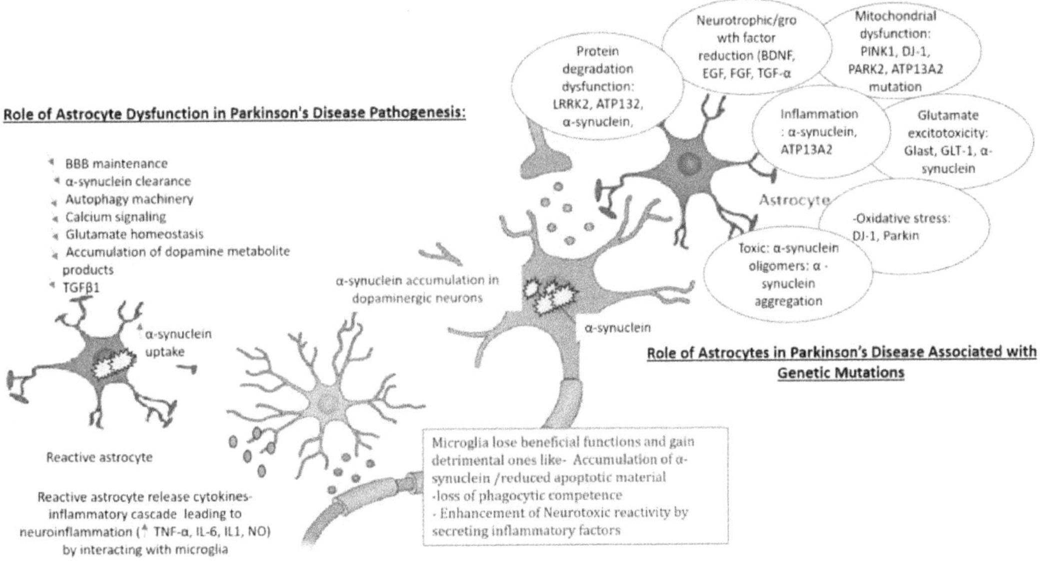

Toxic α-synuclein convert resting astrocytes into reactive astrocyte, which interact with neurons, microglia, and other CNS cells by releasing feed-forward signals and contributing to the vicious cycle leading to neurodegeneration. Astrocyte involvement in Parkinson's disease and genetic mutations; gene factors have been suggested to impact various aspects of astrocytic functions.

- **Astrocytes and PD pathology**

Emerging evidence indicates that astrocytes, a type of glial cell, are intimately involved in PD pathogenesis. These star-shaped cells express several genes linked to PD pathology. Notably, genes such as SNCA, DJ-1, PLA2G6, ATP13A2, LRRK2, and Parkin are expressed in astrocytes and are associated with astrocyte activation in response to inflammation (Booth et al., 2017). PARK7, which encodes the DJ-1 protein, is one such gene. Postmortem brain samples have revealed higher levels of PARK7 expression in astrocytes compared to neurons, suggesting its significant role in astrocyte biology. Furthermore, upregulated PARK7 expression has been observed in reactive astrocytes in PD patients, indicating its involvement in the disease process (Bonifati et al., 2003; Pang et al., 2019)

- **Mitochondria Dysfunction in Astrocytes**

Studies using animal models of PD have provided insights into the impact of astrocyte dysfunction. Astrocytes expressing mutant α-synuclein or lacking the Parkin protein exhibit functional and morphological impairments related to mitochondria, the cellular energy-producing organelles. Mitochondrial dysfunction within astrocytes may disrupt normal brain metabolism and contribute to the degeneration of dopaminergic neurons. Moreover, these impaired astrocytes are less effective in supporting neuronal differentiation, hindering the maturation and functionality of immature neurons. Introduction of the PD-linked A53T mutant α-synuclein within astrocytes has been associated with the downregulation of glutamate transporters and the widespread induction of gliosis (Teravskis et al., 2018; Banerjee et al., 2023).

- **Dopamine Regulation and Neurotrophic Factors**

Astrocytes play a pivotal role in the regulation of dopamine (DA) levels within the brain. They express various dopamine receptors, including D1R and D2R, which, when activated by DA, induce calcium (Ca++) transients within astrocytes (Hisahara & Shimohama, 2011; Asanuma et al., 2014; Oda & Funato, 2023). This interaction leads to transient increases in calcium ion concentrations within astrocytic cells. Furthermore, astrocytes are involved in the production and release of neurotrophic factors, including glial-cell-line-derived neurotrophic factor (GDNF) (Prunell & Olivera-Bravo, 2022; Zhang et al., 2022). GDNF is crucial for the survival and maturation of dopaminergic neurons. Additionally, astrocytes secrete fibroblast growth factors (FGF), such as FGF20, in a site-specific manner. This localized release may influence FGF receptors in the substantia nigra pars compacta (SNpc) dopaminergic neurons (Prunell & Olivera-Bravo, 2022).

To sum up, the accumulating evidence strongly suggests that dysfunction in astrocytes plays a substantial role in the underlying mechanisms of Parkinson's disease. The expression of PD-associated genes in astrocytes, coupled with their impairments affecting mitochondrial function, neuronal differentiation, and neurotransmission, underscores the critical role of astrocytes in the disease process. However, the precise mechanisms underlying astrocytic dysfunction in PD require further investigation. Future research efforts should aim to elucidate these mechanisms and explore potential therapeutic interventions targeting astrocytes.

CONCLUSION

In summary, astrocytes, a diverse group of glial cells in the CNS, play a multifaceted role in supporting and modulating neuronal function. They are not merely passive support cells but actively participate in various critical processes, including the regulation of synaptic activity, maintenance of the blood-brain barrier, and neurotransmitter uptake and release. Astrocytes are regionally and functionally heterogeneous, with different subtypes specialized for specific functions. Their interactions with neurons and other glial cells are vital for brain homeostasis and cognitive function. Understanding the complexity of astrocyte function offers potential insights into therapeutic interventions for neurodegenerative and psychiatric disorders. Further research is needed to fully unravel the intricate mechanisms underlying astrocyte-related contributions to some illnesses such as AD and PD.

REFERENCES

Abbott, N. J. (2002). Astrocyte-endothelial interactions and blood-brain barrier permeability. *Journal of Anatomy*, *200*(6), 629–638. doi:10.1046/j.1469-7580.2002.00064.x PMID:12162730

Allaman, I., Bélanger, M., & Magistretti, P. J. (2011). Astrocyte-neuron metabolic relationships: For better and for worse. *Trends in Neurosciences*, *34*(2), 76–87. doi:10.1016/j.tins.2010.12.001 PMID:21236501

Allen, N.J. (2014). Astrocyte regulation of synaptic behavior. *Annual Review of Cell and Developmental Biology, 30*(1), 439–63. https://doi.org/ bio-100913-013053 doi:10.1146/annual-cell

Allen, N. J., & Eroglu, C. (2017). Cell Biology of Astrocyte-Synapse Interactions. *Neuron*, *96*(3), 697–708. doi:10.1016/j.neuron.2017.09.056 PMID:29096081

Alvarez, J. I., Katayama, T., & Prat, A. (2013). Glial influence on the blood-brain barrier. *Glia*, *61*(12), 1939–1958. doi:10.1002/glia.22575 PMID:24123158

Andersson, M., Blomstrand, F., & Hanse, E. (2007). Astrocytes play a critical role in transient heterosynaptic depression in the rat hippocampal CA1 region. *The Journal of Physiology*, *585*(3), 843–852. doi:10.1113/jphysiol.2007.142737 PMID:17962333

Araque, A., Parpura, V., Sanzgiri, R. P., & Haydon, P. G. (1999). Tripartite synapses: Glia, the unacknowledged partner. *Trends in Neurosciences*, *22*(5), 208–215. doi:10.1016/S0166-2236(98)01349-6 PMID:10322493

Asanuma, M., Miyazaki, I., Murakami, S., Diaz-Corrales, F. J., & Ogawa, N. (2014). Striatal Astrocytes Act as a Reservoir for L-DOPA. *PLoS One*, *9*(9), e106362. doi:10.1371/journal.pone.0106362 PMID:25188235

Banerjee, R., Raj, A., Potdar, C., Pal, P. K., Yadav, R., Kamble, N., Holla, V., & Datta, I. (2023). Astrocytes Differentiated from LRRK2-I1371V Parkinson's-Disease-Induced Pluripotent Stem Cells Exhibit Similar Yield but Cell-Intrinsic Dysfunction in Glutamate Uptake and Metabolism, ATP Generation, and Nrf2-Mediated Glutathione Machinery. *Cells*, *12*(12), 1592. doi:10.3390/cells12121592 PMID:37371062

Barker, A. J., Koch, S. M., Reed, J., Barres, B. A., & Ullian, E. M. (2008). Developmental Control of Synaptic Receptivity. *The Journal of Neuroscience : The Official Journal of the Society for Neuroscience*, *28*(33), 8150–8160. doi:10.1523/JNEUROSCI.1744-08.2008 PMID:18701677

Barres, B. A. (2008). The mystery and magic of glia: A perspective on their roles in health and disease. *Neuron*, *60*(3), 430–440. doi:10.1016/j.neuron.2008.10.013 PMID:18995817

Bataveljić, D., Nikolić, L., Milosević, M., Todorović, N., & Andjus, P. R. (2012). Changes in the astrocytic aquaporin-4 and inwardly rectifying potassium channel expression in the brain of the amyotrophic lateral sclerosis SOD1(G93A) rat model. *Glia*, *60*(12), 1991–2003. doi:10.1002/glia.22414 PMID:22987392

Beauquis, J., Vinuesa, A., Pomilio, C., Pavía, P., Saravia, F. (2014). Alteraciones hipocampales y cambios cognitivos preceden al depósito de placas amiloides en un modelo murino de la enfermedad de Alzheimer [Hippocampal and cognitive alterations precede amyloid deposition in a mouse model for Alzheimer's disease]. *Medicina (B Aires), 74*(4), 282-6.

Bernardinelli, Y., Muller, D., & Nikonenko, I. (2014). Astrocyte-synapse structural plasticity. *Neural Plasticity*, *2014*, 232105. doi:10.1155/2014/232105 PMID:24511394

Bernardinelli, Y., Randall, J., Janett, E., Nikonenko, I., König, S., Jones, E. V., Flores, C. E., Murai, K. K., Bochet, C. G., Holtmaat, A., & Muller, D. (2014). Activity-dependent structural plasticity of perisynaptic astrocytic domains promotes excitatory synapse stability. *Current Biology*, *24*(15), 79–88. doi:10.1016/j.cub.2014.06.025 PMID:25042585

Bernardinelli, Y., Randall, J., Janett, E., Nikonenko, I., König, S., Jones, E. V., Flores, C. E., Murai, K. K., Bochet, C. G., Holtmaat, A., & Muller, D. (2014). Activity-dependent structural plasticity of perisynaptic astrocytic domains promotes excitatory synapse stability. *Current Biology*, *24*(15), 1679–1688. doi:10.1016/j.cub.2014.06.025 PMID:25042585

Bonifati, V., Rizzu, P., van Baren, M. J., Schaap, O., Breedveld, G. J., Krieger, E., Dekker, M. C., Squitieri, F., Ibanez, P., Joosse, M., van Dongen, J. W., Vanacore, N., van Swieten, J. C., Brice, A., Meco, G., van Duijn, C. M., Oostra, B. A., & Heutink, P. (2003). Mutations in the DJ-1 gene associated with autosomal recessive early-onset parkinsonism. *Science*, *299*(5604), 256–259. doi:10.1126cience.1077209 PMID:12446870

Booth, H. D. E., Hirst, W. D., & Wade-Martins, R. (2017). The Role of astrocyte dysfunction in Parkinson's disease pathogenesis. Trends in Neurosciences, 40, 358–370. https://doi.org/. tins.2017.04.001 doi:10.1016/j

Bushong, E. A., Martone, M. E., & Ellisman, M. H. (2004). Maturation of astrocyte morphology and the establishment of astrocyte domains during postnatal hippocampal development. *International Journal of Developmental Neuroscience*, *22*(2), 73–86. doi:10.1016/j.ijdevneu.2003.12.008 PMID:15036382

Butt, A. M. (2006). Neurotransmitter-mediated calcium signaling in oligodendrocyte physiology and pathology. *Glia*, *54*(7), 666–675. doi:10.1002/glia.20424 PMID:17006895

Cabezas, R., Avila, M., Gonzalez, J., El-Bachá, R. S., Báez, E., García-Segura, L. M., Jurado Coronel, J. C., Capani, F., Cardona-Gomez, G. P., & Barreto, G. E. (2014). Astrocytic modulation of the blood-brain barrier: Perspectives on Parkinson's disease. *Frontiers in Cellular Neuroscience*, *8*, 211. doi:10.3389/fncel.2014.00211 PMID:25136294

Chai, H., Diaz-Castro, B., Shigetomi, E., Monte, E., Octeau, J. C., Yu, X., Cohn, W., Rajendran, P. S., Vondriska, T. M., Whitelegge, J. P., Coppola, G., & Khakh, B. S. (2017). Neural Circuit-Specialized Astrocytes: Transcriptomic, Proteomic, Morphological, and Functional Evidence. *Neuron*, *95*(3), 531–549. e9. doi:10.1016/j.neuron.2017.06.029 PMID:28712653

Chen, J., Tan, Z., Zeng, L., Zhang, X., He, Y., Gao, W., Wu, X., Li, Y., Bu, B., Wang, W., & Duan, S. (2013). Heterosynaptic long-term depression mediated by ATP released from astrocytes. *Glia*, *61*(2), 178–191. doi:10.1002/glia.22425 PMID:23044720

Chen, Y., & Swanson, R. A. (2003). Astrocytes and brain injury. *Journal of Cerebral Blood Flow and Metabolism*, *23*(2), 137–149. doi:10.1097/01.WCB.0000044631.80210.3C PMID:12571445

Chever, O., Lee, C. Y., & Rouach, N. (2014). Astroglial connexin43 hemichannels tune basal excitatory synaptic transmission. *The Journal of Neuroscience : The Official Journal of the Society for Neuroscience*, *34*(34), 11228–11232. doi:10.1523/JNEUROSCI.0015-14.2014 PMID:25143604

Christopherson, K. S., Ullian, E. M., Stokes, C. C. A., Mullowney, C. E., Hell, J. W., Agah, A., Lawler, J., Mosher, D. F., Bornstein, P., & Barres, B. A. (2005). Thrombospondins are astrocyte-secreted proteins that promote CNS synaptogenesis. *Cell*, *120*(3), 421–433. doi:10.1016/j.cell.2004.12.020 PMID:15707899

Cirillo, C., Capoccia, E., Iuvone, T., Cuomo, R., Sarnelli, G., Steardo, L., & Esposito, G. (2015). S100B inhibitor pentamidine attenuates reactive gliosis and reduces neuronal loss in a mouse model of Alzheimer's disease. *BioMed Research International*, *2015*, 508342. doi:10.1155/2015/508342 PMID:26295040

Corkrum, M., Covelo, A., Lines, J., Bellocchio, L., Pisansky, M., Loke, K., Quintana, R., Rothwell, P. E., Lujan, R., Marsicano, G., Martin, E. D., Thomas, M. J., Kofuji, P., & Araque, A. (2020, March 18). Dopamine-Evoked Synaptic Regulation in the Nucleus Accumbens Requires Astrocyte Activity. *Neuron*, *105*(6), 1036–1047.e5. doi:10.1016/j.neuron.2019.12.026 PMID:31954621

Covelo, A., & Araque, A. (2018). Neuronal activity determines distinct gliotransmitter release from a single astrocyte. *eLife*, *7*, e32237. doi:10.7554/eLife.32237 PMID:29380725

Cuellar-Santoyo, A. O., Ruiz-Rodríguez, V. M., Mares-Barbosa, T. B., Patrón-Soberano, A., Howe, A. G., Portales-Pérez, D. P., Miquelajáuregui Graf, A., & Estrada-Sánchez, A. M. (2023). Revealing the contribution of astrocytes to glutamatergic neuronal transmission. *Frontiers in Cellular Neuroscience*, *16*, 1037641. doi:10.3389/fncel.2022.1037641 PMID:36744061

Dallérac, G., & Rouach, N. (2016). Astrocytes as new targets to improve cognitive functions. *Progress in Neurobiology*, *144*, 48–67. doi:10.1016/j.pneurobio.2016.01.003 PMID:26969413

De Zeeuw, C. I., & Hoogland, T. M. (2015). Reappraisal of Bergmann glial cells as modulators of cerebellar circuit function. *Frontiers in Cellular Neuroscience*, *9*, 246. doi:10.3389/fncel.2015.00246 PMID:26190972

Dossi, E., Vasile, F., & Rouach, N. (2018). Human astrocytes in the diseased brain. *Brain Research Bulletin*, *136*, 139–156. doi:10.1016/j.brainresbull.2017.02.001 PMID:28212850

Farhy-Tselnicker, I., & Allen, N. J. (2018). Astrocytes, neurons, synapses: A tripartite view on cortical circuit development. *Neural Development*, *13*(1), 7. doi:10.118613064-018-0104-y PMID:29712572

Filosa, J. A., Morrison, H. W., Iddings, J. A., Du, W., & Kim, K. J. (2015). Beyond neurovascular coupling, the role of astrocytes in the regulation of vascular tone. *Neuroscience*, *323*, 96–109. doi:10.1016/j.neuroscience.2015.03.064 PMID:25843438

Fujii, S., Tanaka, K. F., Ikenaka, K., & Yamazaki, Y. (2014). Increased adenosine levels in mice expressing mutant glial fibrillary acidic protein in astrocytes fail induction of LTP reversal (depotentiation) in hippocampal CA1 neurons. *Brain Research*, *1578*, 1–13. doi:10.1016/j.brainres.2014.07.005 PMID:25017946

Genoud, C., Quairiaux, C., Steiner, P., Hirling, H., Welker, E., & Knott, G. W. (2006). Plasticity of astrocytic coverage and glutamate transporter expression in adult mouse cortex. *PLoS Biology*, *4*(11), e343. doi:10.1371/journal.pbio.0040343 PMID:17048987

Ghézali, G., Calvo, C., Pillet, L., Llense, F., Ezan, P., Pannasch, U., Bemelmans, A., Etienne Manneville, S., & Rouach, N. (2018). Connexin 30 controls astroglial polarization during postnatal brain development. *Development (Cambridge, England), 145*(4), dev155275. doi:10.1242/dev.155275 PMID:29475972

Ghézali, G., Vasile, F., Curry, N., Fantham, M., Cheung, G., Ezan, P., Cohen-Salmon, M., Kaminski, C., & Rouach, N. (2020). Neuronal Activity Drives Astroglial Connexin 30 in Perisynaptic Processes and Shapes Its Functions. *Cerebral Cortex (New York, N.Y.), 30*(2), 753–766. doi:10.1093/cercor/bhz123 PMID:31271200

Glaser, T., Resende, R. R., & Ulrich, H. (2013). Implications of purinergic receptor-mediated intracellular calcium transients in neural differentiation. *Cell Communication and Signaling, 11*(1), 12. doi:10.1186/1478-811X-11-12 PMID:23414261

Goenaga, J., Araque, A., Kofuji, P., & Herrera Moro Chao, D. (2023). Calcium signaling in astrocytes and gliotransmitter release. *Frontiers in Synaptic Neuroscience, 15*, 1138577. doi:10.3389/fnsyn.2023.1138577 PMID:36937570

González-Reyes, R. E., Nava-Mesa, M. O., Vargas-Sánchez, K., Ariza-Salamanca, D., & Mora-Muñoz, L. (2017). Involvement of Astrocytes in Alzheimer's Disease from a Neuroinflammatory and Oxidative Stress Perspective. *Frontiers in Molecular Neuroscience, 10*, 427. doi:10.3389/fnmol.2017.00427 PMID:29311817

Halassa, M. M., Fellin, T., & Haydon, P. G. (2007). The tripartite synapse: Roles for gliotransmission in health and disease. *Trends in Molecular Medicine, 13*(2), 54–63. doi:10.1016/j.molmed.2006.12.005 PMID:17207662

Hardy, E., Moulard, J., Walter, A., Ezan, P., Bemelmans, A., Mouthon, F., Charvériat, M., Rouach, N., & Rancillac, A. (2023). Upregulation of astroglial connexin 30 impairs hippocampal synaptic activity and recognition memory. *PLoS Biology, 21*(4), e3002075. doi:10.1371/journal.pbio.3002075 PMID:37040348

Hattori, Y. (2022). The Multiple Roles of Pericytes in Vascular Formation and Microglial Functions in the Brain. *Life (Basel, Switzerland), 12*(11), 1835. doi:10.3390/life12111835 PMID:36362989

Hefendehl, J., LeDue, J., Ko, R. W. Y., Mahler, J., Murphy, T. H., & MacVicar, B. A. (2016). Mapping synaptic glutamate transporter dysfunction in vivo to regions surrounding Aβ plaques by iGluSnFR two-photon imaging. *Nature Communications, 7*(1), 13441. doi:10.1038/ncomms13441 PMID:27834383

Heneka, M. T., Carson, M. J., El Khoury, J., Landreth, G. E., Brosseron, F., & Feinstein, D. L. (2015). Neuroinflammation in Alzheimer's disease. *Lancet Neurology, 14*(4), 388–405. doi:10.1016/S1474-4422(15)70016-5 PMID:25792098

Henstridge, C. M., Tzioras, M., & Paolicelli, R. C. (2019). Glial contribution to excitatory and inhibitory synapse Loss in Neurodegeneration. *Frontiers in Neuroscience, 13*, 63. doi:10.3389/fncel.2019.00063 PMID:30863284

Herculano-Houzel, S. (2014). The glia/neuron ratio: How it varies uniformly across brain structures and species and what that means for brain physiology and evolution. *Glia, 62*(9), 1377–1391. doi:10.1002/glia.22683 PMID:24807023

Hisahara, S., & Shimohama, S. (2011). Dopamine receptors and Parkinson's disease. *International Journal of Medicinal Chemistry*, *2011*, 403039. doi:10.1155/2011/403039 PMID:25954517

Hösli, L., Binini, N., Ferrari, K. D., Thieren, L., Looser, Z. J., Zuend, M., Zanker, H. S., Berry, S., Holub, M., Möbius, W., Ruhwedel, T., Nave, K.-A., Giaume, C., Weber, B., & Saab, A. S. (2022). Decoupling astrocytes in adult mice impairs synaptic plasticity and spatial learning. *Cell Reports*, *38*(10), 110484. doi:10.1016/j.celrep.2022.110484 PMID:35263595

Iadecola, C. (2017). The Neurovascular Unit Coming of Age: A Journey through Neurovascular Coupling in Health and Disease. *Neuron*, *96*(1), 17–42. doi:10.1016/j.neuron.2017.07.030 PMID:28957666

Iadecola, C., & Nedergaard, M. (2007). Glial regulation of the cerebral microvasculature. *Nature Neuroscience*, *10*(11), 1369–1376. doi:10.1038/nn2003 PMID:17965657

Igarashi, Y., Utsumi, H., Chiba, H., Yamada-Sasamori, Y., Tobioka, H., Kamimura, Y., Furuuchi, K., Kokai, Y., Nakagawa, T., Mori, M., & Sawada, N. (1999). Glial cell line-derived neurotrophic factor induces barrier function of endothelial cells forming the blood-brain barrier. *Biochemical and Biophysical Research Communications*, *261*(1), 108–112. doi:10.1006/bbrc.1999.0992 PMID:10405331

Jacob, C. P., Koutsilieri, E., Bartl, J., Neuen-Jacob, E., Arzberger, T., Zander, N., Ravid, R., Roggendorf, W., Riederer, P., & Grünblatt, E. (2007). Alterations in expression of glutamatergic transporters and receptors in sporadic Alzheimer's disease. *Journal of Alzheimer's Disease*, *11*(1), 97–116. doi:10.3233/JAD-2007-11113 PMID:17361039

Jäkel, S., & Dimou, L. (2017). Glial Cells and Their Function in the Adult Brain: A Journey through the History of Their Ablation. *Frontiers in Cellular Neuroscience*, *11*, 24. doi:10.3389/fncel.2017.00024 PMID:28243193

James, G., & Butt, A. M. (2002). P2Y and P2X purinoceptor mediated Ca2+ signaling in glial cell pathology in the central nervous system. *European Journal of Pharmacology*, *447*(2-3), 247–260. doi:10.1016/S0014-2999(02)01756-9 PMID:12151016

Jo, S., Yarishkin, O., Hwang, Y. J., Chun, Y. E., Park, M., Woo, D. H., Bae, J. Y., Kim, T., Lee, J., Chun, H., Park, H. J., Lee, D. Y., Hong, J., Kim, H. Y., Oh, S. J., Park, S. J., Lee, H., Yoon, B. E., Kim, Y., ... Lee, C. J. (2014). GABA from reactive astrocytes impairs memory in mouse models of Alzheimer's disease. *Nature Medicine*, *20*(8), 886–896. doi:10.1038/nm.3639 PMID:24973918

Kang, J., Jiang, L., Goldman, S. A., & Nedergaard, M. (1998). Astrocyte-mediated potentiation of inhibitory synaptic transmission. *Nature Neuroscience*, *1*(8), 683–692. doi:10.1038/3684 PMID:10196584

Kaplan, L., Chow, B. W., & Gu, C. (2020). Neuronal regulation of the blood-brain barrier and neurovascular coupling. *Nature Reviews. Neuroscience*, *21*(8), 416–432. doi:10.103841583-020-0322-2 PMID:32636528

Kellner, V., Kersbergen, C. J., Li, S., Babola, T. A., Saher, G., & Bergles, D. E. (2021). Dual metabotropic glutamate receptor signaling enables the coordination of astrocyte and neuron activity in developing sensory domains. *Neuron*, *109*(16), 2545–2555.e7. doi:10.1016/j.neuron.2021.06.010 PMID:34245686

Kempuraj, D., Ahmed, M., Selvakumar, G., Thangavel, R., Dhaliwal, A., Dubova, I., Mentor, S., Premkumar, K., Saeed, D., Zahoor, H., Raikwar, S., Zaheer, S., Iyer, S., & Zaheer, A. (2020). Brain injury-Mediated Neuroinflammatory response and Alzheimer's disease. *The Neuroscientist*, *26*(2), 134–155. doi:10.1177/1073858419848293 PMID:31092147

Khakh, B. S., & Sofroniew, M. V. (2015). Diversity of astrocyte functions and phenotypes in neural circuits. *Nature Neuroscience*, *18*(7), 942–952. doi:10.1038/nn.4043 PMID:26108722

Kilb, W., & Kirischuk, S. (2022). GABA Release from Astrocytes in Health and Disease. *International Journal of Molecular Sciences*, *23*(24), 15859. doi:10.3390/ijms232415859 PMID:36555501

Kinney, G. A., & Spain, W. J. (2002). Synaptically evoked GABA transporter currents in neocortical glia. *Journal of Neurophysiology*, *88*(6), 2899–2908. doi:10.1152/jn.00037.2002 PMID:12466416

Kisler, K., Nelson, A. R., Rege, S. V., Ramanathan, A., Wang, Y., Ahuja, A., Lazic, D., Tsai, P. S., Zhao, Z., Zhou, Y., Boas, D. A., Sakadžić, S., & Zlokovic, B. V. (2017). Pericyte degeneration leads to neurovascular uncoupling and limits oxygen supply to the brain. *Nature Neuroscience*, *20*(3), 406–416. doi:10.1038/nn.4489 PMID:28135240

Kordower, J. H., Olanow, C. W., Dodiya, H. B., Chu, Y., Beach, T. G., Adler, C. H., Halliday, G. M., & Bartus, R. T. (2013). Disease duration and the integrity of the nigrostriatal system in Parkinson's disease. *Brain*, *136*(8), 2419–2431. doi:10.1093/brain/awt192 PMID:23884810

Kuchibhotla, K. V., Lattarulo, C. R., Hyman, B. T., & Bacskai, B. J. (2009). Synchronous hyperactivity and intercellular calcium waves in astrocytes in Alzheimer's mice. *Science*, *323*(5918), 1211–1215. doi:10.1126cience.1169096 PMID:19251629

Kulijewicz-Nawrot, M., Verkhratsky, A., Chvátal, A., Syková, E., Rodríguez, J.J. (2012). Astrocytic cytoskeletal atrophy in the medial prefrontal cortex of a triple transgenic mouse model of Alzheimer's disease. *Journal of Anatomy, 221*(3), 252-62. . doi:10.1111/j.1469-7580.2012.01536.x

Lawrence, J., Schardien, K., Wigdahl, B., & Nonnemacher, M. (2023). Roles of neuropathology-associated reactive astrocytes: A systematic review. *Acta Neuropathologica Communications*, *11*(1), 42. doi:10.118640478-023-01526-9 PMID:36915214

Lazic, A., Balint, V., Stanisavljevic Ninkovic, D., Peric, M., & Stevanovic, M. (2022). Reactive and Senescent Astroglial Phenotypes as Hallmarks of Brain Pathologies. *International Journal of Molecular Sciences*, *23*(9), 4995. doi:10.3390/ijms23094995 PMID:35563385

Lécuyer, M. A., Kebir, H., & Prat, A. (2016). Glial influences on BBB functions and molecular players in immune cell trafficking. *Biochimica et Biophysica Acta*, *1862*(3), 472–482. doi:10.1016/j.bbadis.2015.10.004 PMID:26454208

Lee, H. U., Yamazaki, Y., Tanaka, K. F., Tanaka, K. F., Furuya, K., Sokabe, M., Hida, H., Takao, K., Miyakawa, T., Miyakawa, T., Fujii, S., & Ikenaka, K. (2013). Increased astrocytic ATP release results in enhanced excitability of the hippocampus. *Glia*, *61*(2), 210–224. doi:10.1002/glia.22427 PMID:23018918

Lee, J., Kim, S. W., & Kim, K. T. (2022). Region-Specific Characteristics of Astrocytes and Microglia: A Possible Involvement in Aging and Diseases. *Cells*, *11*(12), 1902. doi:10.3390/cells11121902 PMID:35741031

Lesné, S., Docagne, F., Gabriel, C., Liot, G., Lahiri, D.K., Buée, L., Plawinski, L., Delacourte, A., MacKenzie, E.T., Buisson, A., & Vivien, D. (2003). Transforming growth factor-beta 1 potentiates amyloid-beta generation in astrocytes and in transgenic mice. *JBC, 278*(20), 18408-18. . doi:10.1074/jbc.M300819200

Liddelow, S. A., & Barres, B. A. (2017). Reactive Astrocytes: Production, Function, and Therapeutic Potential. *Immunity, 46*(6), 957-967. doi:10.1016/j.immuni.2017.06.006

Liu, Y. J., Chen, J., Li, X., Zhou, X., Hu, Y. M., Chu, S. F., Peng, Y., & Chen, N. H. (2019). Research progress on adenosine in central nervous system diseases. *CNS Neuroscience & Therapeutics*, *25*(9), 899–910. doi:10.1111/cns.13190 PMID:31334608

Loera-Valencia, R., Goikolea, J., Parrado-Fernandez, C., Merino-Serrais, P., & Maioli, S. (2019). Alterations in cholesterol metabolism as a risk factor for developing Alzheimer's disease: Potential novel targets for treatment. *The Journal of Steroid Biochemistry and Molecular Biology*, *190*, 104–114. doi:10.1016/j.jsbmb.2019.03.003 PMID:30878503

Luissint, A. C., Artus, C., Glacial, F., Ganeshamoorthy, K., & Couraud, P. O. (2012). Tight junctions at the blood-brain barrier: Physiological architecture and disease-associated dysregulation. *Fluids and Barriers of the CNS*, *9*(1), 23. doi:10.1186/2045-8118-9-23 PMID:23140302

Madeira, C., Lourenco, M. V., Vargas-Lopes, C., Suemoto, C. K., Brandão, C. O., Reis, T., Leite, R. E. P., Laks, J., Jacob-Filho, W., Pasqualucci, C. A., Grinberg, L. T., Ferreira, S. T., & Panizzutti, R. (2015). D-serine levels in Alzheimer's disease: Implications for novel biomarker development. *Translational Psychiatry*, *5*(5), e561. doi:10.1038/tp.2015.52 PMID:25942042

Makarava, N., Mychko, O., Molesworth, K., Chang, J. C.-Y., Henry, R. J., Tsymbalyuk, N., Gerzanich, V., Simard, J. M., Loane, D. J., & Baskakov, I. V. (2023). Region-Specific Homeostatic Identity of Astrocytes Is Essential for Defining Their Response to Pathological Insults. *Cells*, *12*(17), 2172. doi:10.3390/cells12172172 PMID:37681904

Matos, M., Augusto, E., Oliveira, C. R., & Agostinho, P. (2008). Amyloid-beta peptide decreases glutamate uptake in cultured astrocytes: Involvement of oxidative stress and mitogen-activated protein kinase cascades. *Neuroscience*, *156*(4), 898–910. doi:10.1016/j.neuroscience.2008.08.022 PMID:18790019

Mei, X., Ezan, P., Giaume, C., & Koulakoff, A. (2010). Astroglial connexin immunoreactivity is specifically altered at β-amyloid plaques in β-amyloid precursor protein/presenilin1 mice. *Neuroscience Letters*, *474*(2), 71–75. doi:10.1016/j.neuroscience.2010.08.001 PMID:20227462

Miller, R., & Raff, M. (1984). Fibrous and protoplasmic astrocytes are biochemically and developmentally distinct. *The Journal of Neuroscience : The Official Journal of the Society for Neuroscience*, *4*(2), 585–592. doi:10.1523/JNEUROSCI.04-02-00585.1984 PMID:6366155

Montagne, A., Zhao, Z., & Zlokovic, B. V. (2017). Alzheimer's disease: A matter of blood-brain barrier dysfunction? *The Journal of Experimental Medicine*, *214*(11), 3151–3169. doi:10.1084/jem.20171406 PMID:29061693

Monterey, M. D., Wei, H., Wu, X., & Wu, J. Q. (2021). The Many Faces of Astrocytes in Alzheimer's Disease. *Frontiers in Neurology*, *12*, 619626. doi:10.3389/fneur.2021.619626 PMID:34531807

Mookherjee, P., Green, P. S., Watson, G. S., Marques, M. A., Tanaka, K., Meeker, K. D., Meabon, J. S., Li, N., Zhu, P., Olson, V. G., & Cook, D. G. (2011). GLT-1 loss accelerates cognitive deficit onset in an Alzheimer's disease animal model. *Journal of Alzheimer's Disease*, *26*(3), 447–455. doi:10.3233/JAD-2011-110503 PMID:21677376

Nagai, J., Rajbhandari, A. K., Gangwani, M. R., Hachisuka, A., Coppola, G., Masmanidis, S. C., Fanselow, M. S., & Khakh, B. S. (2019). Hyperactivity with Disrupted Attention by Activation of an Astrocyte Synaptogenic Cue. *Cell*, *177*(5), 1280–1292 e1220. doi:10.1016/j.cell.2019.03.019 PMID:31031006

Nagy, J. I., Patel, D., Ochalski, P. A., & Stelmack, G. L. (1996). Connexin30 in rodent, cat and human brain: Selective expression in gray matter astrocytes, co-localization with connexin43 at gap junctions and late developmental appearance. *Neuroscience*, *76*(4), 931–953. doi:10.1016/S0306-4522(98)00191-2 PMID:10197766

Nanclares, C., Baraibar, A. M., Araque, A., & Kofuji, P. (2021). egulation of astrocytes-neuronal communication in Alzheimer's disease. *International Journal of Molecular Sciences*, *22*(15), 7887. doi:10.3390/ijms22157887 PMID:34360652

Nishida, H., & Okabe, S. (2007). Direct astrocytic contacts regulate the local maturation of dendritic spines. *The Journal of Neuroscience : The Official Journal of the Society for Neuroscience*, *27*(2), 331–340. doi:10.1523/JNEUROSCI.4466-06.2007 PMID:17215394

O'Brown, N. M., Pfau, S. J., & Gu, C. (2018). Bridging barriers: A comparative look at the blood-brain barrier across organisms. *Genes & Development*, *32*(7-8), 466–478. doi:10.1101/gad.309823.117 PMID:29692355

Oberheim, N. A., Goldman, S. A., & Nedergaard, M. (2012). Heterogeneity of astrocytic form and function. *Methods in Molecular Biology (Clifton, N.J.)*, *814*, 23–45. doi:10.1007/978-1-61779-452-0_3 PMID:22144298

Oberheim, N. A., Takano, T., Han, X., He, W., Lin, J. H., Wang, F., Xu, Q., Wyatt, J. D., Pilcher, W., Ojemann, J. G., Ransom, B. R., Goldman, S. A., & Nedergaard, M. (2009). Uniquely hominid features of adult human astrocytes. *The Journal of Neuroscience : The Official Journal of the Society for Neuroscience*, *9*(10), 3276–3287. doi:10.1523/JNEUROSCI.4707-08.2009 PMID:19279265

Oberheim, N. A., Wang, X., Goldman, S., & Nedergaard, M. (2006). Astrocytic complexity distinguishes the human brain. *Trends in Neurosciences*, *29*(10), 547–553. doi:10.1016/j.tins.2006.08.004 PMID:16938356

Oda, S., & Funato, H. (2023). D1- and D2-type dopamine receptors are immunolocalized in pial and layer I astrocytes in the rat cerebral cortex. *Frontiers in Neuroanatomy*, *17*, 1111008. doi:10.3389/fnana.2023.1111008 PMID:36865631

Olsen, M. L., & Sontheimer, H. (2008). Functional implications for Kir4.1 channels in glial biology: From K+ buffering to cell differentiation. *Journal of Neurochemistry*, *107*(3), 589–601. doi:10.1111/j.1471-4159.2008.05615.x PMID:18691387

Orellana, J. A., Froger, N., Ezan, P., Jiang, J. X., Bennett, M. V. L., Naus, C. C. G., & Sáez, J. C. (2011a). ATP and glutamate released via astroglial connexin 43 hemichannels mediate neuronal death through activation of pannexin 1 hemichannels. *The Journal of Neuroscience : The Official Journal of the Society for Neuroscience*, *31*(49), 17690–17701. doi:10.1111/j.1471-4159.2011.07210.x PMID:21294731

Orellana, J. A., Sáez, P. J., Shoji, K. F., Schalper, K. A., Palacios-Prado, N., Velarde, V., & Sáez, J. C. (2011b). Modulation of brain hemichannels and gap junction channels by pro-inflammatory agents and their possible role in neurodegeneration. *Antioxidants & Redox Signaling*, *11*(2), 369–399. doi:10.1089/ars.2008.2130 PMID:18816186

Orsini, F., De Blasio, D., Zangari, R., Zanier, E. R., & De Simoni, M. G. (2014). The versatility of the complement system in neuroinflammation, neurodegeneration, and brain homeostasis. *Frontiers in Cellular Neuroscience*, *8*, 380. doi:10.3389/fncel.2014.00380 PMID:25426028

Osborn, L. M., Kamphuis, W., Wadman, W. J., & Hol, E. M. (2016). Astrogliosis: An integral player in the pathogenesis of Alzheimer's disease. *Progress in Neurobiology*, *144*, 121–141. doi:10.1016/j.pneurobio.2016.01.001 PMID:26797041

Ota, Y., Zanetti, A. T., & Hallock, R. M. (2013). The role of astrocytes in the regulation of synaptic plasticity and memory formation. *Neural Plasticity*, *2013*, 185463. doi:10.1155/2013/185463 PMID:24369508

Pang, S., Ho, P., Liu, H., Leung, C., Li, L., Chang, E., Ramsden, D., & Ho, S. (2019). The interplay of aging, genetics, and environmental factors in the pathogenesis of Parkinson's disease. *Translational Neurodegeneration*, *8*(1), 23. doi:10.118640035-019-0165-9 PMID:31428316

Pannasch, U., Derangeon, M., Chever, O., & Rouach, N. (2012). Astroglial gap junctions shape neuronal network activity. *Communicative & Integrative Biology*, *5*(3), 248–254. doi:10.4161/cib.19410 PMID:22896785

Pannasch, U., & Rouach, N. (2013). An emerging role for astroglial networks in information processing: From synapse to behavior. *Trends in Neurosciences*, *36*(7), 405–417. doi:10.1016/j.tins.2013.04.004 PMID:23659852

Park, H., Han, K. S., Seo, J., Lee, J., Dravid, S. M., Woo, J., Chun, H., Cho, S., Bae, J. Y., An, H., An, H., Koh, W., Yoon, B.-E., Yoon, B.-E., Berlinguer-Palmini, R., Mannaioni, G., Traynelis, S. F., Bae, Y. C., Choi, S.-Y., Lee, C. J. (2015). Channel-mediated astrocytic glutamate modulates hippocampal synaptic plasticity by activating postsynaptic NMDA receptors. *Molecular Brain*, *8*(1), pp. 8:7. doi:10.1186/s13041-015-0097-y

Parpura, V., & Verkhratsky, A. (2012). The astrocyte excitability brief: From receptors to gliotransmission. *Neurochemistry International*, *61*(4), 610–621. doi:10.1016/j.neuint.2011.12.001 PMID:22178457

Pascual, O., Casper, K. B., Kubera, C., Zhang, J., Revilla-Sanchez, R., Sul, J. Y., & Haydon, P. G. (2005). Astrocytic purinergic signaling coordinates synaptic networks. *Science*, *310*(5745), 113–116. doi:10.1126cience.1116916 PMID:16210541

Perea, G., Gomez, R., Mederos, S., Covelo, A., Ballesteros, J. J., Schlosser, L., Hernández-Vivanco, A., Martín-Fernández, M., Quintana, R., Rayan, A., Díez, A., Fuenzalida, M., Agarwal, A., Bergles, D. E., Bettler, B., Manahan-Vaughan, D., Martín, E. D., Kirchhoff, F., & Araque, A. (2016). Activity-dependent switch of GABAergic inhibition into glutamatergic excitation in astrocyte-neuron networks. *eLife*, *5*, e20362. doi:10.7554/eLife.20362 PMID:28012274

Petravicz, J., Fiacco, T. A., & McCarthy, K. D. (2008, May 7). Loss of IP3 receptor-dependent Ca2+ increases in hippocampal astrocytes does not affect baseline CA1 pyramidal neuron synaptic activity. *The Journal of Neuroscience : The Official Journal of the Society for Neuroscience*, *28*(19), 4967–4973. doi:10.1523/JNEUROSCI.5572-07.2008 PMID:18463250

Piccioni, G., Mango, D., Saidi, A., Corbo, M., & Nisticò, R. (2021). Targeting Microglia-synapse interaction in Alzheimer's disease. *International Journal of Molecular Sciences*, *22*(5), 2342. doi:10.3390/ijms22052342 PMID:33652870

Pirttimaki, T.M., Codadu, N.K., Awni, A., Pratik, P., Nagel, D.A., Hill, E.J., Dineley, K.T., Parri, H.R.(2013). α7 Nicotinic receptor-mediated astrocytic gliotransmitter release: Aβ effects in a preclinical Alzheimer's mouse model. *PLoS One, 28*(11), e81828. . doi:10.1371/journal.pone.0081828

Prevot, V., Dehouck, B., Sharif, A., Ciofi, P., Giacobini, P., & Clasadonte, J. (2018). The versatile tanycyte: A hypothalamic integrator of reproduction and energy metabolism. *Endocrine Reviews*, *39*(3), 333–368. doi:10.1210/er.2017-00235 PMID:29351662

Prunell, G., & Olivera-Bravo, S. (2022). A Focus on Astrocyte Contribution to Parkinson's Disease Etiology. *Biomolecules*, *12*(12), 1745. doi:10.3390/biom12121745 PMID:36551173

Puchałowicz, K., Tarnowski, M., Baranowska-Bosiacka, I., Chlubek, D., & Dziedziejko, V. (2014). P2X and P2Y receptors—Role in the pathophysiology of the nervous system. *International Journal of Molecular Sciences*, *15*(12), 23672–23704. doi:10.3390/ijms151223672 PMID:25530618

Qiu, J., Yan, Z., Tao, K., Li, Y., Li, J., Dong, Y., Feng, D., & Chen, H. (2016). Sinomenine activates astrocytic dopamine D2 receptors and alleviates neuroinflammatory injury via the CRYAB/STAT3 pathway after ischemic stroke in mice. *Journal of Neuroinflammation*, *13*(1), 263. doi:10.118612974-016-0739-8 PMID:27724964

Roux, L., Madar, A., Lacroix, M. M., Yi, C., Benchenane, K., & Giaume, C. (2015). Astroglial Connexin 43 Hemichannels Modulate Olfactory Bulb Slow Oscillations. *The Journal of Neuroscience : The Official Journal of the Society for Neuroscience*, *35*(46), 15339–15352. doi:10.1523/JNEUROSCI.0861-15.2015 PMID:26586821

Rusakov, D. A. (2015). Disentangling calcium-driven astrocyte physiology. *Nature Reviews. Neuroscience*, *16*(4), 226–233. doi:10.1038/nrn3878 PMID:25757560

Santello, M., & Volterra, A. (2009). Astrocytes as major integrators in neurovascular coupling. *The Neuroscientist*, *15*(3), 262–273.

Schools, G. P., & Kimelberg, H. K. (1999). mGluR3 and mGluR5 are the predominant metabotropic glutamate receptor mRNAs expressed in hippocampal astrocytes acutely isolated from young rats. *Journal of Neuroscience Research*, *58*(4), 533–543. doi:10.1002/(SICI)1097-4547(19991115)58:4<533::AID-JNR6>3.0.CO;2-G PMID:10533045

Shao, W., Zhang, S.Z., Tang, M., Zhang, X.H., Zhou, Z., Yin, Y.Q., Zhou, Q.B., Huang, Y.Y., Liu, Y.J., Wawrousek, E., Chen, T., Li, S.B., Xu, M., Zhou, J.N., Hu, G., Zhou, J.W. (2013). Suppression of neuroinflammation by astrocytic dopamine D2 receptors via αB-crystallin. *Nature, 7*(7435), 90-4. . doi:10.1038/nature11748

Sheth, S., Brito, R., Mukherjee, D., Rybak, L. P., & Ramkumar, V. (2014). Adenosine receptors: Expression, function and regulation. *International Journal of Molecular Sciences*, *15*(2), 2024–2052. doi:10.3390/ijms15022024 PMID:24477263

Sibille, J., Sibille, J., Pannasch, U., & Rouach, N. (2014). Astroglial potassium clearance contributes to the short-term plasticity of synaptically evoked currents at the tripartite synapse. *Journal of Physiology, 592*(1), pp. 87-102. doi:10.1113/jphysiol.2013.261735

Sibille, J., Zapata, J., Teillon, J., & Rouach, N. (2015). Astroglial calcium signaling displays short-term plasticity and adjusts synaptic efficacy. *Frontiers in Cellular Neuroscience*, *9*, 189. doi:10.3389/fncel.2015.00189 PMID:26074766

Sobolczyk, M., & Boczek, T. (2022). Astrocytic Calcium and cAMP in Neurodegenerative Diseases. *Frontiers in Cellular Neuroscience*, *16*, 889939. doi:10.3389/fncel.2022.889939 PMID:35663426

Stancu, I. C., Vasconcelos, B., Terwel, D., & Dewachter, I. (2014). Models of β-amyloid induced Tau-pathology: The long and "folded" road to understand the mechanism. *Molecular Neurodegeneration*, *9*(1), 51. doi:10.1186/1750-1326-9-51 PMID:25407337

Stogsdill, J. A., Ramirez, J., Liu, D., Kim, Y. H., Baldwin, K. T., Enustun, E., Ejikeme, T., Ji, R. R., & Eroglu, C. (2017). Astrocytic neuroligins control astrocyte morphogenesis and synaptogenesis. *Nature*, *551*(7679), 192–197. doi:10.1038/nature24638 PMID:29120426

Tabata, H. (2015). Diverse subtypes of astrocytes and their development during corticogenesis. *Frontiers in Neuroscience*, *9*, 114. doi:10.3389/fnins.2015.00114 PMID:25904839

Takeuchi, H., Mizoguchi, H., Doi, Y., Jin, S., Noda, M., Liang, J., & Suzumura, A. (2011). Blockade of gap junction hemichannel suppresses disease progression in mouse models of amyotrophic lateral sclerosis and Alzheimer's disease. *PLoS One*, *6*(6), e21108. doi:10.1371/journal.pone.0021108 PMID:21712989

Tanaka, M., Shih, P. Y., Gomi, H., Yoshida, T., Nakai, J., Ando, R., Furuichi, T., Mikoshiba, K., Semyanov, A., & Itohara, S. (2013). Astrocytic Ca^{2+} signals are required for the functional integrity of tripartite synapses. *Molecular Brain*, *6*(1), 6. doi:10.1186/1756-6606-6-6 PMID:23356992

Teravskis, P., Covelo, A., Miller, E., Singh, B., Martell-Martínez, H., Benneyworth, M., Gallardo, C., Oxnard, B., Araque, A., Lee, M., & Liao, D. (2018). A53T Mutant Alpha-Synuclein Induces Tau-Dependent Postsynaptic Impairment Independently of Neurodegenerative Changes. *The Journal of Neuroscience : The Official Journal of the Society for Neuroscience*, *38*(45), 9754–9767. doi:10.1523/JNEUROSCI.0344-18.2018 PMID:30249789

Ullian, E.M., Sapperstein, S.K., Christopherson, K.S., Barres, B.A. (2001). Control of synapse number by glia. *Science, 291*(5504), 657–61. doi:10.1126/science.291.5504.657

Umpierre, A. D., West, P. J., White, J. A., & Wilcox, K. S. (2019). Conditional Knock-out of mGluR5 from Astrocytes during Epilepsy Development Impairs High-Frequency Glutamate Uptake. *The Journal of Neuroscience : The Official Journal of the Society for Neuroscience, 39*(4), 727–742. doi:10.1523/JNEUROSCI.1148-18.2018 PMID:30504280

Vasile, F., Dossi, E., & Rouach, N. (2017). Human astrocytes: Structure and functions in the healthy brain. *Brain Structure & Function, 222*(5), 2017–2029. doi:10.100700429-017-1383-5 PMID:28280934

Ventura, R., & Harris, K. M. (1999). Three-dimensional relationships between hippocampal synapses and astrocytes. *The Journal of Neuroscience : The Official Journal of the Society for Neuroscience, 19*(16), 6897–6906. doi:10.1523/JNEUROSCI.19-16-06897.1999 PMID:10436047

Vincenzi, F., Pasquini, S., Contri, C., Cappello, M., Nigro, M., Travagli, A., Merighi, S., Gessi, S., Borea, P. A., & Varani, K. (2023). Pharmacology of Adenosine Receptors: Recent Advancements. *Biomolecules, 13*(9), 1387. doi:10.3390/biom13091387 PMID:37759787

Wagner, K., Unger, L., Salman, M. M., Kitchen, P., Bill, R. M., & Yool, A. J. (2022). Signaling Mechanisms and Pharmacological Modulators Governing Diverse Aquaporin Functions in Human Health and Disease. *International Journal of Molecular Sciences, 23*(3), 1388. doi:10.3390/ijms23031388 PMID:35163313

Walrave, L., Vinken, M., Albertini, G., De Bundel, D., Leybaert, L., & Smolders, I. J. (2016). Inhibition of Connexin43 Hemichannels Impairs Spatial Short-Term Memory without Affecting Spatial Working Memory. *Frontiers in Cellular Neuroscience, 10*, 288. doi:10.3389/fncel.2016.00288 PMID:28066184

Wong, A. D., Ye, M., Levy, A. F., Rothstein, J. D., Bergles, D. E., & Searson, P. C. (2013). The blood-brain barrier: An engineering perspective. *Frontiers in Neuroengineering, 6*, 7. doi:10.3389/fneng.2013.00007 PMID:24009582

Yeh, C. Y., Verkhratsky, A., Terzieva, S., & Rodríguez, J. J. (2013). Glutamine synthetase in astrocytes from entorhinal cortex of the triple transgenic animal model of Alzheimer's disease is not affected by pathological progression. *Biogerontology, 14*(6), 777–787. doi:10.100710522-013-9456-1 PMID:23990215

Zeisel, A., Hochgerner, H., Lönnerberg, P., Johnsson, A., Memic, F., van der Zwan, J., Häring, M., Braun, E., Borm, L. E., La Manno, G., Codeluppi, S., Furlan, A., Lee, K., Skene, N., Harris, K. D., Hjerling-Leffler, J., Arenas, E., Ernfors, P., Marklund, U., & Linnarsson, S. (2018). Molecular Architecture of the Mouse Nervous System. *Cells, 174*(4), 999–1014.e22. doi:10.1016/j.cell.2018.06.021 PMID:30096314

Zhang, X., Huang, W. J., & Chen, W. W. (2016). TGF-β1 factor in the cerebrovascular diseases of Alzheimer's disease. *European Review for Medical and Pharmacological Sciences, 20*(24), 5178–5185. PMID:28051272

Zhang, Y., Chen, Y., Wu, J., Manaenko, A., Yang, P., Tang, J., Fu, W., Zhang, J.H. (2015). Activation of Dopamine D2 Receptor Suppresses Neuroinflammation Through αB-Crystalline by Inhibition of NF-κB Nuclear Translocation in Experimental ICH Mice Model. *Stroke, 46*(9), 2637-46. . doi:10.1161/STROKEAHA.115.009792

Zhang, Z., Zhang, N., & Ding, S. (2022). Glial cell line-derived neurotrophic factor in brain repair after focal ischemic stroke. *Neural Regeneration Research*, *17*(8), 1735–1736. doi:10.4103/1673-5374.332141 PMID:35017424

Zhao, Z., Nelson, A. R., Betsholtz, C., & Zlokovic, B. V. (2015). Establishment and Dysfunction of the Blood-Brain Barrier. *Cell*, *163*(5), 1064–1078. doi:10.1016/j.cell.2015.10.067 PMID:26590417

Zhou, B., Zuo, Y. X., & Jiang, R. T. (2019). Astrocyte morphology: Diversity, plasticity, and role in neurological diseases. *CNS Neuroscience & Therapeutics*, *25*(6), 665–673. doi:10.1111/cns.13123 PMID:30929313

Zlokovic, B. V. (2011). Neurovascular pathways to neurodegeneration in Alzheimer's disease and other disorders. *Nature Reviews. Neuroscience*, *12*(12), 723–738. doi:10.1038/nrn3114 PMID:22048062

Zonta, M., Sebelin, A., Gobbo, S., Fellin, T., Pozzan, T., & Carmignoto, G. (2003). Glutamate-mediated cytosolic calcium oscillations regulate a pulsatile prostaglandin release from cultured rat astrocytes. *The Journal of Physiology*, *553*(Pt 2), 407–414. doi:10.1113/jphysiol.2003.046706 PMID:14500777

Chapter 4
Microglial Cells Function in the Central Nervous System:
Beyond the Immune Function

Bilal El-Mansoury
https://orcid.org/0000-0002-2583-7515
Faculty of Sciences, Chouaib Doukkali University, Morocco

Ahmed Draoui
https://orcid.org/0000-0002-7700-059X
Faculty of Science Semlalia, Cadi Ayyad University, Morocco

Kamal Smimih
Faculty of Sciences and Techniques, Sultan Moulay Slimane University, Morocco

Samira Boulbaroud
Polydisciplinary Faculty, Sultan Moulay Slimane University, Morocco

Youssef Ait Hamdan
https://orcid.org/0009-0005-8616-8852
Higher Normal School, Cadi Ayyad University, Morocco

Arumugam Radhakrishnan Jayakumar
Miller School of Medicine, University of Miami, USA

ABSTRACT

Microglia are the resident macrophages of the central nervous system (CNS). These cells of mesodermal/mesenchymal origin migrate into all regions of the CNS. Recent studies indicate that even in the normal brain, microglia have highly motile processes by which they scan their territorial domains. By a large number of signaling pathways, they can communicate with macroglial cells (e.g. astrocytes) and neurons and with cells of the immune system. Under normal physiological conditions, microglia constantly monitor their microenvironment and survey neurons. Microglia have other functions including the participation in the formation of new blood vessels or angiogenesis, cognitive function, the regulation of synaptic plasticity, and neurogenesis and they play a crucial role in the CNS through communication with other brain cells. This chapter will provide an overview of the functions of microglial cells within the CNS.

DOI: 10.4018/978-1-6684-9675-6.ch004

INTRODUCTION

The term "microglia" was introduced more than a century ago by Pio del Rio-Hortega in a published a paper in 1919. He distinguished between oligodendrocytes and microglia using an enhanced silver-staining technique (Matejuk & Ransohoff, 2020). For a very long time, it was believed that the main function of glial cells in the brain is to support and provide an ideal environment for the neurons. Neuroscientists have just lately concentrated on the distinct functions of these non-neuronal cells in maintaining brain homeostasis (Kettenmann & Verkhratsky, 2008). Therefore, it is evident that glial cells actively contribute to neuronal function and dysfunction rather of just filling "the space not occupied by neurons," as Virchow argued in the late nineteenth century (Peferoen et al., 2014a).

Microglial cells take their origin from yolk sac erythro-myeloid progenitors and colonize the developing brain before the blood-brain-barrier (BBB) closure (Ginhoux et al., 2010). After migrating as microglia progenitors to the developing CNS, these cells rapidly multiply and produce a reservoir of residual cells and have the capacity to renew themselves apart from the hematopoietic system. Microglia mature completely by the end of the second postnatal week and express adult gene signatures (Li & Barres, 2018).

It is well known that microglia play a crucial role in immune regulation of the CNS (Peferoen et al., 2014a). The role of microglia in the CNS's generation and maintenance of inflammatory responses has long been understood, but it is now obvious that they ensure other functions as well, especially in the healthy brain. Indeed, microglia, which were once thought to be "silent" in the healthy brain, have now been shown to play an active role in a number of physiological processes related to the brain, such as adult hippocampal neurogenesis, cognitive and behavioral function, biochemical homeostasis maintenance, neuronal circuit maturation during development, and experience-dependent remodeling of adult brain neuronal circuits. Furthermore, owing to their expression of a wide range of receptors and molecules on their cell surface, microglia can interact in bidirectional manner with the other brain cells including neurons, astrocytes, and oligodendrocytes. While microglia activation and the subsequent neuroinflammation has been recently observed in several conditions including Alzheimer disease (AD) (W. Chen et al., 2016), Parkinson disease (PD) (El-Mansoury et al., 2023), multiple sclerosis (MS) (El-Mansoury et al., 2023), Huntington disease (HD), amyotrophic lateral sclerosis (ALS) (W. Chen et al., 2016), and hepatic encephalopathy (HE) (B El-Mansoury et al., 2023), among others. This chapter will discuss the roles of microglial cells beyond immune function with focus on synaptic function, cognitive functions, angiogenesis and neurogenesis as well as their cross talk with the other CNS' cells including astrocytes, neurons, and oligodendrocytes.

THE DISCOVERY AND DEFINITION OF MICROGLIA

Microglial cells are well known by their role in regulating immune system function in the CNS (Peferoen et al., 2014a). In 1932, Pio del Rio-Hortega wrote a book chapter titled "Microglia" for the influential book Cytology and Cellular Pathology of the Nervous System, edited by Wilder Penfield. This chapter popularized the idea of microglia as a recognized cellular component of the CNS. Del Rio-Hortega made the following assertions: 1) during early development, microglia penetrate the developing brain. 2) The mesodermal origin of these invasive cells is indicated by their amoeboid shape. 3) They migrate throughout all parts of the brain using arteries and white matter tracts as guidance structures. 4) In the more developed brain, they change into a branching, ramified morphological pattern that is now referred to be the resting microglia. 6) Each cell appears to have a specific area to it. 7) Following a pathogenic occurrence, these cells transform. 8) Transformed cells take on an amoeboid shape that resembles that seen in the early stages of development. 9) These cells have the ability to move, multiply, and phagocytize (Helmut et al., 2011).

FUNCTIONAL STATES OF MICROGLIA: RESTING AND ACTIVATED MICROGLIA ARE HIGHLY DYNAMIC

In the brain, microglia are highly ramified and have small cell bodies (Kettenmann et al., 2013), and were thought to be functionally quiescent (Colonna & Butovsky, 2017). Ramified microglia are often functionally inactive, and are hence, referred to as "quiescent" or "resting" in normal physiological circumstances. Two articles that identified the fine processes of ramified microglia as highly dynamic surveyors of the healthy brain posed a significant challenge to this argument (Davalos et al., 2005; Nimmerjahn et al., 2005), giving them the new name of "surveillant microglia," and that the description of these cells as "resting" is no longer necessary, presented a significant challenge to this claim (Petersen & Dailey, 2004; Streit et al., 1999).

In response to CNS injury and pathogenic stimuli, microglia (Thomas, 1992), undergo morphological changes, from the highly ramified ''surveillant'' state to a globular ''amoeboid'' activated form (Kitamura et al., 1978; Stence et al., 2001), migrating to sites of damage (Nolte et al., 1996; Stence et al., 2001). It is also recognized that these activated cells can provide neuroprotective functions (Streit, 2005), can direct the encroaching vasculature, stimulate neurogenesis, eliminate apoptotic cells, affect synaptogenesis, and regulate developmental apoptosis (Bessis et al., 2007; Ransohoff & Cardona, 2010; Saijo & Glass, 2011). According to descriptions used for peripheral macrophage activation, microglia activation states are classified as the classical M1 and alternative M2 (Morris et al., 2013).

M1 activation is a proinflammatory and neurotoxic state producing a variety of proinflammatory cytokines and chemokines, such as IL-6, TNF-α, IL-12, IL-1β, and CCL2. They also generates ROS because they express the NADPH oxidase, in addition to inducible nitric oxidase, which converts arginase into nitric oxide (NO). Eventually, NO promotes the toxic effect of glutamate, thus potentiating N-methyl-D-aspartate (NMDA) receptor–mediated neurotoxicity. Therefore, M1 microglia state is responsible for inflammatory and neurotoxic effects (Colonna & Butovsky, 2017).

Microglia's inflammatory-suppressive and reparative functions are described by M2 activation. It can be brought on by IL-4, IL-13, IL-10, in addition to neurotrophic substances like brain-derived neurotrophic factor (BDNF), nerve growth factor (NGF), and glial cell-derived neurotrophic factor (GDNF). M2 microglia also secrete growth factors including insulin-like growth factor I (IGF-I) and fibroblast growth factor (FGF) (Colonna & Butovsky, 2017). However, the relevance of these confined criteria is not yet evident because activation state is probably more complex than this (Morris et al., 2013). Indeed, in spite of the fact that the M1 and M2 categories have been useful for conceptualizing microglia activities in *in vitro* studies, it becomes evident that the M1/M2 paradigm is insufficient to represent microglia and macrophage activation *in vivo*.

MICROGLIA FUNCTIONS

Microglia ensure several functions in the CNS beyond the immune regulatory roles. Indeed, this cell type is of great importance in the brain owing to its multiple roles including synaptic plasticity and neurogenesis, learning and memory, promoting angiogenesis, preventing neurotoxicity and maintaining brain homeostasis as depicted in figure 1.

Figure 1. A summary of microglial cells functions in the brain

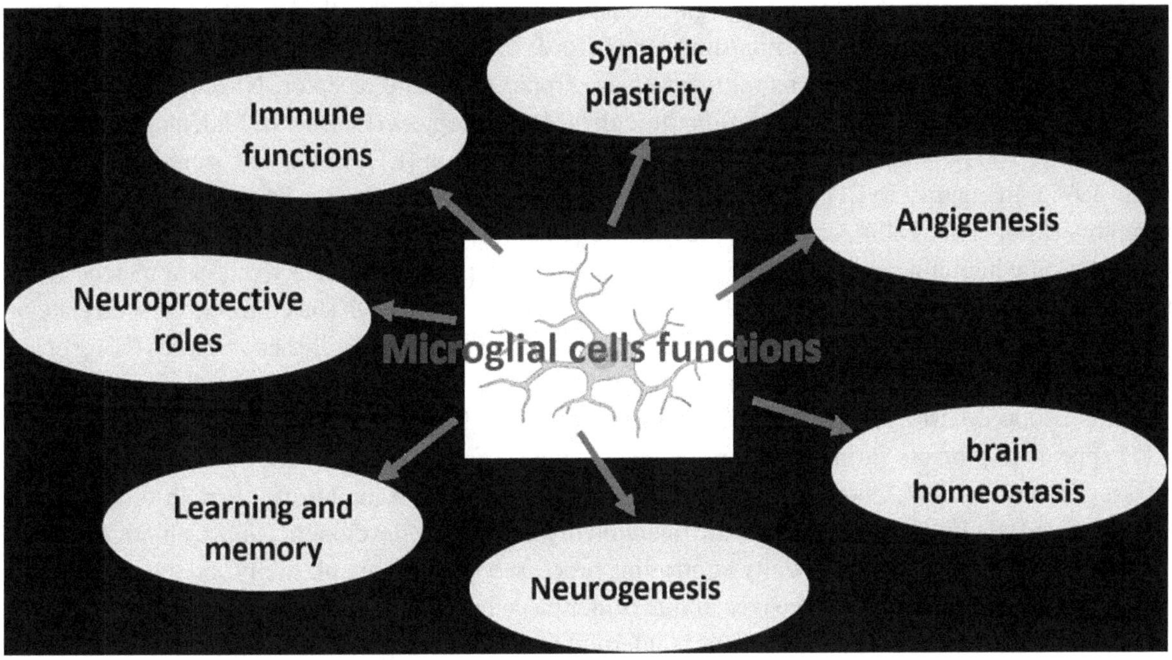

Microglia Cells Involvement in Neurogenesis Synaptic Plasticity

Microglia and Synaptic Plasticity

Synaptic Plasticity

The term synaptic plasticity is referred to as the ability of synapses to change the strength of their connections. A synapse's "strength" is defined as the reaction brought on by pre- and post-synapse activity (Atwood & Karunanithi, 2002). The response elicited in a target cell by stimulation of a pre-synaptic terminal can be altered by changes in both the chemical expression of synaptic molecules and changes in the structural makeup of synapses. These changes can take place in both pre- and post-synaptic dendritic spines (Atwood & Karunanithi, 2002). Indeed, synaptic plasticity can involve structural changes through the alteration, development, and removal of existing synapses, a process known as "structural plasticity", in addition to chemical changes in existing synapses (Morris et al., 2013). Additionally, synaptic plasticity can occur quickly, utilizing a range of chemical pathways to modify, delete, and add new synaptic connections as well as change the strength of existing ones (Bear, 1999; Holtmaat et al., 2005; Lüscher & Malenka, 2012). Synaptic scaling, a type of homeostasis that involves uniform modifications in the strength of all synapses in a cell, is another way that synapses might improve their connection. This happens as a result of sustained changes in a cell's electrical activity, which enhances or suppress synaptic firing in order to stabilize neural connections (Turrigiano, 2008). Therefore, synaptic plasticity is a key component for changing brain networks (Morris et al., 2013).

In 1949, the influential psychologist, Donald Hebb, introduced his famous theory that can be summarized with the common phrase ''cells that fire together, wire together'' (Hebb, 1949). Hebb's law was initially supported by the finding of a phenomenon called long-term potentiation (LTP) (Bliss & Lømo, 1973). LTP is induced by brief trains of high-frequency stimulation within the excitatory synapses, leading a long lasting increase in neurotransmission (Bliss & Collingridge, 1993; Cooke & Bliss, 2006). It boosts the synaptic membrane's recruitment of ionotropic glutamate receptor; NMDA receptors, and this LTP typically depends on Ca2+ influx through NMDA receptors (Lüscher & Malenka, 2012; Matsuzaki et al., 2004). A related phenomenon, long-term depression (LTD), was discovered later (Lynch et al., 1977). In contrast to LTP, which strengthens synapses, LTD causes the internalization of AMPA receptors (Man et al., 2000), activating presynaptic NMDA receptors and lowering intracellular Ca2+ levels, thereby inhibiting synaptic activity. LTD can also be dependent on the activation of metabotropic type glutamate receptor (mGluR) (Snyder et al., 2005; Xiao et al., 2001). The conventional correlate of plasticity, LTP, seems to trigger global increases in the number of synaptically active spines (Lang et al., 2004; Maletic-Savatic et al., 1999), transforming mushroom-shaped tiny spines into thicker ones (Kopec et al., 2006; Matsuzaki et al., 2004), forming new spines (Engert & Bonhoeffer, 1999). In contrast, LTD appears to promote shrinkage and retraction of spines (Y. Chen et al., 2004; S. H. Lee et al., 2002; Massey & Bashir, 2007; Zhou et al., 2004). LTP and LTD are two examples of the traditional theories of synaptic plasticity, but these theories are undoubtedly continually developing (Mozzachiodi & Byrne, 2010). Accordingly, synaptic plasticity-controlling processes, such as metaplasticity and the synaptic tag and capture theory in particular, are now taking center place in behavioral neuroscience and in a number of well-known hypotheses about the neural underpinnings of learning and memory (Abraham & Bear, 1996; Finnie & Nader, 2012; Redondo & Morris, 2011).

Microglial Cells Involvement in Synaptic Plasticity

Synaptic connections in the adult CNS are constantly changing, leading to changes in strength and connectivity. Microglia are highly involved in synaptic plasticity and microglial processes have been shown to be highly mobile, constantly expanding, retracting, and interacting with synapses in revolutionary in vivo imaging investigations (Davalos et al., 2005; Nimmerjahn et al., 2005; Tremblay et al., 2010; Wake et al., 2009). It is believed that this dynamic process (motility), enables microglia to evaluate or "survey" nearby synapses and react appropriately. It has been reported that 94% of microglial processes in the visual cortex make contact with synaptic components at any given time, and these connections have been linked to both the formation and disappearance of tiny spines (Tremblay et al., 2010). Additionally, during dark adaptation, the visual cortex receives less synaptic input, which results in a reduction in microglia process motility. This effect is reversed after being exposed to light and visual stimulation again (Tremblay et al., 2010).

It is well known that in the adult brain, synapse production and elimination are reduced, and spines are more stable (Grutzendler et al., 2002; Majewska et al., 2006). Research on microglia's function in eliminating damaged synapses dates back to 1968 (Blinzinger & Kreutzberg, 1968). Although the purpose of this was not recognized at the time, activated microglia were shown to physically separate pre- and post-synaptic boutons. The precise mechanism of synaptic stripping and the microglial phenotype that causes it are unknown (Kalla et al., 2001). Complement cascade proteins have recently been linked to microglial-controlled synaptic pruning during development. The mouse retinogeniculate system, a frequently researched area for studying developmental synaptic elimination, has been shown to contain extra synaptic connections in animals defective in the complement cascade protein C1q or a downstream complement protein C3 (Stevens et al., 2007). Additionally, the segregation of retinal ganglion cell axonal projections into the lateral geniculate nucleus is compromised in these mice lacking CR3 or C3 because they are unable to efficiently prune retinogeniculate synapses (Colonna & Butovsky, 2017b).

The Role of Microglia in Promoting Neurogenesis

Neurogenesis refers to the generation of new neurons in the brain. It is a lifelong process in the adult brain of mammals, including humans (Eriksson et al., 1998; Roy et al., 2000; Congmin Wang et al., 2011). In rodents, neurogenesis predominately takes place in the subventricular zone (SVZ) of the lateral ventricle and the subgranular zone (SGZ) of the dentate gyrus in the hippocampal formation (Gage, 2000). In the SGZ, there is a population of proliferating radial and non-radial precursor cells that are the source of adult hippocampal neurogenesis. These cells give rise to neural progenitor cells (NPCs), which in turn produce neuroblasts (Ehninger & Kempermann, 2008). Only a very small percentage of the neurons produced by NPCs growth survive for a prolonged period of time (Gemma & Bachstetter, 2013). During normal neurogenesis in the healthy brain, a considerable number of neural progenitors undergo apoptosis within their first days in life. Scavenging abilities of microglia may play a significant regulatory function in the early phases of neurogenesis by rapidly eliminating cell debris (Sierra et al., 2010). It is important to note that ramified microglia phagocytizes apoptotic cells during neurogenesis and the high proportion of apoptotic cells are ingested by ramified microglia.

Numerous studies have demonstrated the crucial involvement of microglia in neurogenesis, a role that goes beyond just removing cells. Microglia play a crucial role in the neurogenic niches of the SVZ and SGZ of the dentate gyrus in the adult brain, where they produce neurons that integrate into the olfactory bulb and the hippocampus, respectively (Colonna & Butovsky, 2017). According to available data, microglia release substances that may affect a newborn cell's ability to proliferate, differentiate into neurons or glia, and survive (Gemma & Bachstetter, 2013). Microglia can influence proliferation, survival and neuronal differentiation (Gemma & Bachstetter, 2013). In vitro studies on neural stem cell cultures provide some of the earliest evidence for the positive effects of microglia or the mediators they release, showing that co-culturing with microglia or microglia's conditioned media can prevent the production of neuroblasts (Aarum et al., 2003; Walton et al., 2006). This finding was confirmed by another study showing that adding microglia-conditioned media in SVZ-derived culture leads to increased neuroblast production (Walton et al., 2006). Since then, a number of investigations from the M. Schwartz research have provided more evidence for this "pro-neurogenic" activity of microglia. They first showed that microglia stimulated by T helper cell-related cytokines increased neurogenesis in vitro (Butovsky et al., 2006). Moreover, a greater number of Iba1+ microglia co-labeled with the anti-inflammatory proteins IGF-1 and MHC class II, as well as an increase in T cell recruitment and neurogenesis, were found in the hippocampus SGZ of rodents living in an enriched environment (Ekdahl, 2012). According to Bachstetter et al. (2011), who investigated the neuronal transmembrane chemokine fractalkine (FKN)/CX3CL1 and its microglia-expressed receptor, CX3CR1-pathway, microglia play a protective role for newly generated neurons. In young adult rodents, disruption of this route reduced hippocampal neural progenitor cell survival and proliferation (Bachstetter et al., 2011). Additionally, the data reveals that microglia can produce growth factors that support neurogenesis, such as BDNF and IGF-I, and that in proliferating cultures, the loss of inducible neurogenesis was accompanied with microglia depletion (Ziv et al., 2006). All this compelling evidence sustains the potential role of microglial cells in promoting neurogenesis.

MICROGLIAL CELLS INTERACTION WITH OTHER BRAIN CELLS

Microglia-Neuron Interactions

In the healthy brain, the 'surveying' phenotype of microglia is dependent on a dynamic crosstalk between these cells (microglia) and neurons (Szepesi et al., 2018). Indeed, several neuronal signals can regulate microglia activation (Biber et al., 2007), and can also transmit signals to microglia to keep them in their surveillance state, pointing to a neuron-microglia interaction. Numerous neuron-mediated signals have an anti-inflammatory effect at the level of the microglia under physiological circumstances. For instance, all of the neuro-immunoregulatory proteins Cluster of differentiation (CD) 200, CD47, CD55, and CX3CL1 (fractalkine receptor) are constitutively produced in healthy neurons and have a corresponding receptor on microglia (fig.2) (Kierdorf & Prinz, 2013). One of the best-characterized neuronal signals that controls microglia function is fractalkine, and the anatomical expression of fractalkine on neurons and CXRCR1 receptor on microglia led to the hypothesis of a novel signaling in which neurons may keep microglia in a surveillant/ramified state through a repressive fractalkine signal (Gemma & Bachstetter, 2013). Additionally, recent research has shown that microglial process motility is not passive but controlled by neuronal activity (Schafer et al., 2012; Tremblay et al., 2010). Importantly, microglia cells use ATP as a primary means of communication with neurons and as a primary trigger to migrate toward ATP sources.

On the other hand, microglia express most major classes and subtypes of ion channels and excitatory and inhibitory neurotransmitter receptors that are normally found at neuronal synapses (Morris et al., 2013). Numerous studies have evaluated how these receptors and ion channels affect the way that microglia react to inflammatory stimuli. By inhibiting NMDA receptors, for instance, microglial cells are less likely to become inflammatory in response to insults like lipopolysaccharide (LPS) (Glezer et al., 2003), and in vitro, activation of the NMDA receptor in microglia leads to an increase in oxidative stress, cytokines, and chemokines (A M Kaindl et al., 2008; Angela M Kaindl et al., 2012).

While, the reciprocal neuron–microglia communication is mediated thanks to numerous soluble factors, extracellular vesicles as well as contact-dependent mechanisms (Fig.2) (Szepesi et al., 2018). Intercellular communication is further facilitated by indirect interactions through soluble substances such as neuromodulators, growth factors, and cytokines. While, the most efficient, fast, and dynamic communication between the two cell types can be achieved through direct microglia-neuron connections via tight membrane-membrane contacts (Cserép et al., 2021).

Cross Talk Between Microglia and Astrocytes

Recently, microglia-astrocyte interaction has taken center place in glial research. By producing a range of signaling molecules following a CNS insult or damage, microglia and astrocytes both establish autocrine feedback and their bidirectional communication for a tight reciprocal modulation (Jha et al., 2018). Additionally, interactions between microglia and astrocytes during brain development start as soon as these cells settle in the developing brain (Jha et al., 2018). There is growing evidence that microglia and astrocytes communicate with each other in both directions through molecules that they release (Clarke et al., 2018; Jha et al., 2016; Liddelow et al., 2017; Vainchtein et al., 2018). Different chemicals, like as chemokines, growth factors, and cytokines, are released to facilitate this intercellular communication (Jha et al., 2018), in addition to neurotransmitters, gliotransmitters, and tissue damage molecules (e.g., ATP) (Matejuk & Ransohoff, 2020). Another pathway for astrocyte-microglia communication has been identified in mice. This pathway involves the gut-brain axis, where commensal flora-controlled metabolites of dietary tryptophan directly affect CNS' microglia and regulate astrocyte pathogenic activities during inflammation and neurodegeneration (Rothhammer et al., 2018).

Microglial cells serve as the primary immune cells of the CNS and also regulate the innate immune functions of astrocytes (Jha et al., 2018). Microglia promote astrocyte activation and influence the fate of astrocytes since they often respond to pathogenic stimuli more quickly than astrocytes (Jha et al., 2018). Indeed, microglia are the first cell type to respond to CNS insult, and this microglial reaction is followed by reactive astrocytosis (Jha et al., 2018). In order to connect with astrocytes and control neuronal activity and synaptic transmission, microglia express a variety of membrane receptors for neurotransmitters. For example, in the context of neuroinflammation, microglia release ATP, which stimulates astrocytes to release glutamate (Jha et al., 2018), suggesting that microglial cells can have a non-negligible role in the astrocyte-mediated modulation of excitatory neurotransmission (Pascual et al., 2012). Moreover, proinflammatory mediators produced by microglia reduce the ability of astrocytes to metabolize glutamate, highlighting how important microglia-astrocyte interactions are for the CNS homeostasis (fig. 2) (Jha et al., 2018).

In contrast, astrocytes control the phenotypes and activities of microglia, including their motility and phagocytosis, through the chemicals they release (Jha et al., 2018). It has been suggested that astrocytes triggers the activation of microglia and regulate their biological processes (Jha et al., 2018). Indeed, astrocytes influence the phenotypes and activities of microglia through a variety of astrocyte-derived substances, including cytokines, chemokines, Ca2+, complement proteins, and other inflammatory mediators (fig.2). Astrocytes experience changes in the production of numerous soluble molecules that enable this cell type to communicate with microglia after pathological insults or in the presence of inflammatory stimuli. These astrocytic substances that are produced modify the activation, migration, and phagocytosis of microglial cells, which in turn influences astrocyte-microglia interactions that are critical in the development of several neuropathologies, including neuroinflammation (Jha et al., 2018). It has been revealed that astrocytes produce innate immunological pattern recognition receptors (PRRs), including complement receptors, mannose receptors, scavenger receptors, and Toll-like receptors (TLR) (fig.2) (Farina et al., 2007; Tian et al., 2012). These astrocytic PRRs' activation creates a crosstalk with the nearby microglia. Astrocytes are important providers of inflammatory mediators in the CNS, including numerous complement components, cytokines, and chemokines, upon activation by TLR or NLR ligands (Ransohoff & Brown, 2012), this may lead to the removal of harmful or infectious agents, the restoration of integrity, or the creation of scars (Jha et al., 2018). Interestingly, astrocytic responses to activation of TLR2, TLR3, and TLR4 is suggested to be totally dependent on the presence of microglia (Holm et al., 2012). Otherwise, astrocytes reduce inflammatory responses from microglia by releasing GABA, which prevents microglia from becoming activated (Lee et al., 2011), and by inducing the antioxidant protein heme oxygenase 1 via the microglial transcription factor Nrf2 (Min et al., 2006).

Cross Talk Between Microglia and Oligodendrocytes

Although oligodendrocytes are well recognized for their function as CNS myelinating cells that facilitate rapid and effective transfer of neuronal communication through the myelination of axons, they are also important for preserving axonal integrity, supporting axonal metabolism, and promoting neuronal survival (Bankston et al., 2013; Fünfschilling et al., 2012). Moreover, through creating immune regulatory factors and expressing receptors to connect with microglia, they have been demonstrated to play a crucial role in immune-mediated processes (Peferoen et al., 2014b). Oligodendrocytes are increasingly recognized as critical players in the etiology of neurological conditions such as AD, spinal cord injury, MS, PD, and ALS due to their crucial functions in safeguarding neurons (Peferoen et al., 2014a).

Typically, microglia activation has been reported in many oligodendropathies (Amor et al., 2010; Kato & Kanba, 2013). Due to their high metabolic activity and energy requirements, oligodendrocytes are particularly vulnerable to microglia-derived substances (proinflammatory cytokines). As response, oligodendrocytes frequently produce low-quality myelin, which might help explain the pathology seen in neurological disorders. Studies employing the endotoxin lipopolysaccharide (LPS), a significant part of the cell wall of Gram-negative bacteria that triggers a potent immune response, serve as an illustration of this. For instance, intracerebral injection of LPS in young rats activates microglia and astrocytes, prompting a variety of pro-inflammatory cytokines to be produced as a result, which ultimately causes hypomyelination (Chew et al., 2013; Pang et al., 2003). In vitro, microglia activated with LPS arrest oligodendrocyte progenitor cell (OPC) proliferation and induce OPC death (Li et al., 2008; Sherwin & Fern, 2005). Furthermore, microglial activity and oligodendrocyte damage in MS might be related because a variety of pro-inflammatory cytokines (fig.2), such as interleukins (IL) 1, 2, 3, interferons (IFNs) a, b, and c, TNF-a, and lymphotoxin, have been found in demyelinating MS lesions (Benveniste, 1997). Therefore, several byproducts of microglia activation might be harmful to oligodendrocytes. Hence, understanding the processes that underlie this interaction and result in oligodendrocyte destruction may offer fresh insights into the pathophysiology of a wide range of neurological disorders (Peferoen et al., 2014a). Conversely, oligodendrocytes can in turn regulate the activity of microglia by producing chemokines, cytokines, and chaperokines (fig.2) (Peferoen et al., 2014a).

On the other hand, microglia might also be extremely important for the CNS's regeneration, repair, and wound healing processes. The CNS can regenerate myelin by attracting, growing, and maturing OPCs into a demyelinated area. Microglia appear to be essential in this process, according to a number of lines of evidence. Indeed, studies using animal models have demonstrated that the removal or inhibition of microglia negatively impacts healing, indicating a crucial interaction between microglia and oligodendrocytes during remyelination (Li et al., 2005; Tanaka et al., 2013). Although TNF-a is typically thought of as an inflammatory cytokine that promotes oligodendrocyte destruction, there is mounting evidence that it also has strong immunosuppressive effects and takes role in the healing process (Kassiotis & Kollias, 2001). In experimental models, TNF-a suppression or deletion reduces the number of proliferating OPCs and delays remyelination (Arnett et al., 2001). This suggests that encouraging an environment that is conducive to repair promotes remyelination and may represent a promising therapeutic strategy for demyelinating conditions (Peferoen et al., 2014a).

This bidirectional interaction is essential for maintaining brain homeostasis, but also in promoting the development of several conditions.

Figure 2. Microglial cells interaction with the other brain cells

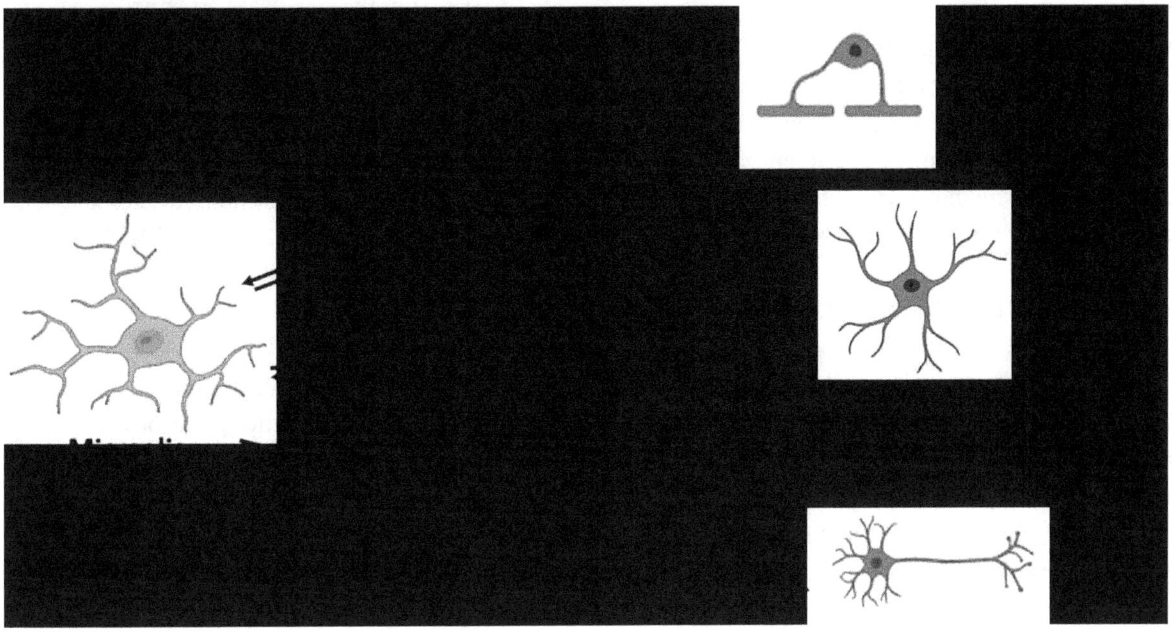

OTHER FUNCTIONS OF MICROGLIAL CELLS IN THE CNS

Microglia and Learning and Memory

The majority of the research that have looked at the function of microglia in learning and memory have found that in inflammatory and sick conditions, microglia have a negative impact on these processes (Cornell et al., 2022). However, new research has shown that microglial cells play a positive role in learning and memory. Indeed, as discussed before, microglia play a role in a variety of physiological synaptic-related processes, including hippocampus neurogenesis (Diaz-Aparicio et al., 2020), synaptic plasticity (Wu et al., 2015), and synaptic modulation, as well as learning and memory processes (Blank & Prinz, 2012). Additionally, microglia control the development and maintenance of synapses, which promotes neurogenesis in memory and learning (Bennett & Molofsky, 2019; Chao Wang et al., 2020). Following neurogenesis, newborn neurons play a role in learning and memory that is dependent on the hippocampal region (Diaz-Aparicio et al., 2020). Otherwise, microglia mediate forgetting through phagocytosis (Diaz-Aparicio et al., 2020; Chao Wang et al., 2020). Interestingly, microglia-mediated forgetting can be reverse through the inhibition of complement pathways, indicating that complement-mediated synaptic elimination is the primary mechanism by which microglia regulate forgetting of distant memories (Cornell et al., 2022).

It has been shown that the repopulation of microglia following its removal has been shown to improve short-term memory and learning tasks (De Luca et al., 2020) as well as improved spine efficacy (De Luca et al., 2020; Gipson & Olive, 2017). Additionally, Parkhurst and al. (2013) revealed that the physiologic role microglial cells in learning and memory could be as result of promoting learning-related synapse formation through BDNF signaling (Parkhurst et al., 2013). Otherwise, it is thought that microglia regulate memory quality through the interleukin 33 (IL-33) signaling (Nguyen et al., 2020). The loss of

IL-33 receptors on microglia or neuronal IL-33 inhibits the integration of newborn neurons and affects the precision of fear memory as well as synaptic plasticity. In the adult hippocampus, IL-33 is primarily produced in neurons in a manner that is experience-dependent. IL-33 KO mice exhibit intact fear memory and no difference in anxiety as compared to controls, however; 14 or 28 days after fear conditioning training, these mice exhibit greater freezing in an unconditioned setting (Nguyen et al., 2020), suggesting that microglia can regulate contextual memory quality through the IL-33 signaling (Cornell et al., 2022). Hence, memory impairment can arise or progress due to changes in the number and/or shape of microglial cells in the brain regions responsible for cognitive functions (Blank & Prinz, 2012).

Microglia Prevent Excitotoxicity

Excessive neurotransmitter release causes excitotoxicity (excessive release of glutamate, leads to increased activation of NMDA receptors), which results in prolonged depolarization, neurotoxicity, and axonal swelling. Recent studies of organotypic hippocampal slice cultures have reported the crucial role of microglia in protecting against NMDA-induced toxicity (Masuch et al., 2016). Additionally, Kato et al. (2016a), using cortical brain slices from mice, they used two-photon imaging and patch clamping to investigate the effects of repetitive supramaximal stimulation on axonal morphology, neuronal membrane potential, and microglial migration simultaneously (Kato et al., 2016a). They showed that in situations of neuronal excitation, microglia have an immediate and highly localized neuroprotective effect. A significant, prolonged depolarization of the soma membrane potential was observed in conjunction with the swelling of axons but not dendrites when recurrent action potentials were evoked in individual layer 2/3 pyramidal neurons (Kato et al., 2016a). Additionally, they have demonstrated that ATP and glutamate release via volume-activated anion channels are involved in the migration of microglial processes to these enlarged axons. Following this movement, the impacted axons underwent extensive microglial wrapping, and in certain cases, axonal debris were removed. This caused the soma membrane to rapidly repolarize back to resting potentials (Kato et al., 2016a). Interestingly, activity-induced depolarization persisted until cell death when microglial migration was pharmacologically inhibited, showing that the microglia-axon interaction maintained neuronal survival by preventing aberrant depolarization of the soma (Kato et al., 2016a).

Microglia Promote Angiogenesis

The recognition of the functional significance of the physical interaction between microglia and the neurovasculature has received increasing attention (Arnold & Betsholtz, 2013; Dudvarski Stankovic et al., 2016). Numerous studies have documented that in the developing brain, the meticulous development of the cerebral blood artery network comes before the appearance of microglia (Fantin et al., 2010). In line with these findings, it has also been demonstrated that macrophages generated from the yolk sac that grow into microglia invading the CNS before and independently of neurovascular development (Cuadros et al., 1993; Kurz & Christ, 1998). Even though they do not emerge simultaneously in the developing CNS, microglia and the neurovasculature frequently interact physically. Early research showed that microglial cells were situated near the vasculature in the developing mouse brain (Ashwell et al., 1989). This has been verified multiple times, indicating that microglia and the vasculature communicate pertinently—possibly in both directions. Indeed, real-time confocal imaging also showed "juxtavascular" microglia migrating along the walls of cerebral blood arteries in brain slices (Grossmann et al., 2002). Moreover, microglia maintain close physical contact with components of the neurovasculature, as evidenced by in vivo two photon imaging in adult mice (Matsuo et al., 2013).

Microglia have a role in angiogenesis, the process of creating new blood vessels, which is one of the increasingly recognized roles of these cells. This assumes prior mechanistic commonalities in early development and has been shown most frequently in diseased circumstances. Some of the genetic and pharmacological data that has been offered to support this; the genetic depletion of macrophage colony stimulating factor, often known as mCSF, first offered evidence in favor of the involvement of microglia in embryonic angiogenesis. This gene regulates the growth and survival of monocytic cells, including brain microglia (Cecchini et al., 1994). The pharmacological suppression of mCSF with neutralizing anti-mCSF antibodies also resulted in a reduction in blood vessel complexity, supporting a role for mCSF (Kubota et al., 2009). A disadvantage of using this pharmaceutical approach is the potential for mCSF to target cells other than microglial cells that contain mCSF receptors. Second, reduced vascularization was seen in the developing mouse retina after clodronate liposomes, which kill microglia after they ingest the liposomes, were used to target microglia. (Checchin et al., 2006). To a certain extent, these results consistently suggest that microglia promote developmental angiogenesis (Matsuo et al., 2013). Similar to mCSF, the PU.1 gene regulates hematopoietic cell differentiation, and genetic disruption of its activity limits the generation of microglia (McKercher et al., 1996; Scott et al., 1994). Interestingly, PU.1 mutant mice also exhibit less elaborate blood vessel complexity (Fantin et al., 2010). All this evidence supports the potential role of microglia in promoting angiogenesis.

CONCLUSION

Microglial cells are well known by their roles in regulation the immune functions in the CNS, however; recently several emerging roles have been attributed to these cells making them a crucial cell type in the brain. These functions include learning and memory, neurogenesis and synaptic plasticity and promoting angiogenesis among others. Microglia also interact in complex pathways with the other brain cells owing to their expression of multiple receptors and molecules.

REFERENCES

Aarum, J., Sandberg, K., Haeberlein, S. L. B., & Persson, M. A. A. (2003). Migration and differentiation of neural precursor cells can be directed by microglia. *Proceedings of the National Academy of Sciences of the United States of America, 100*(26), 15983–15988. doi:10.1073/pnas.2237050100 PMID:14668448

Abraham, W. C., & Bear, M. F. (1996). Metaplasticity : Plasticity of synaptic. *Trends in Neurosciences, 19*(4), 126–130. doi:10.1016/S0166-2236(96)80018-X PMID:8658594

Amor, S., Puentes, F., Baker, D., & Van Der Valk, P. (2010). Inflammation in neurodegenerative diseases. *Immunology, 129*(2), 154–169. doi:10.1111/j.1365-2567.2009.03225.x PMID:20561356

Arnett, H. A., Mason, J., Marino, M., Suzuki, K., Matsushima, G. K., & Ting, J. P.-Y. (2001). TNFα promotes proliferation of oligodendrocyte progenitors and remyelination. *Nature Neuroscience, 4*(11), 1116–1122. doi:10.1038/nn738 PMID:11600888

Arnold, T., & Betsholtz, C. (2013). Erratum to: The importance of microglia in the development of the vasculature in the central nervous system. *Vascular Cell*, *5*, 1–7.

Ashwell, K. W. S., Holländer, H., Streit, W., & Stone, J. (1989). The appearance and distribution of microglia in the developing retina of the rat. *Visual Neuroscience*, *2*(5), 437–448. doi:10.1017/S0952523800012335 PMID:2487081

Atwood, H. L., & Karunanithi, S. (2002). Diversification of synaptic strength: Presynaptic elements. *Nature Reviews. Neuroscience*, *3*(7), 497–516. doi:10.1038/nrn876 PMID:12094207

Bachstetter, A. D., Morganti, J. M., Jernberg, J., Schlunk, A., Mitchell, S. H., Brewster, K. W., Hudson, C. E., Cole, M. J., Harrison, J. K., Bickford, P. C., & Gemma, C. (2011). Fractalkine and CX3CR1 regulate hippocampal neurogenesis in adult and aged rats. *Neurobiology of Aging*, *32*(11), 2030–2044. doi:10.1016/j.neurobiolaging.2009.11.022 PMID:20018408

Bankston, A. N., Mandler, M. D., & Feng, Y. (2013). Oligodendroglia and neurotrophic factors in neurodegeneration. *Neuroscience Bulletin*, *29*(2), 216–228. doi:10.100712264-013-1321-3 PMID:23558590

Bear, M. F. (1999). Homosynaptic long-term depression: A mechanism for memory? *Proceedings of the National Academy of Sciences of the United States of America*, *96*(17), 9457–9458. doi:10.1073/pnas.96.17.9457 PMID:10449713

Bennett, F. C., & Molofsky, A. V. (2019). The immune system and psychiatric disease: A basic science perspective. *Clinical and Experimental Immunology*, *197*(3), 294–307. doi:10.1111/cei.13334 PMID:31125426

Benveniste, E. N. (1997). Role of macrophages/microglia in multiple sclerosis and experimental allergic encephalomyelitis. *Journal of Molecular Medicine*, *75*(3), 165–173. doi:10.1007001090050101 PMID:9106073

Bessis, A., Béchade, C., Bernard, D., & Roumier, A. (2007). Microglial control of neuronal death and synaptic properties. *Glia*, *55*(3), 233–238. doi:10.1002/glia.20459 PMID:17106878

Biber, K., Neumann, H., Inoue, K., & Boddeke, H. W. G. M. (2007). Neuronal 'On' and 'Off' signals control microglia. *Trends in Neurosciences*, *30*(11), 596–602. doi:10.1016/j.tins.2007.08.007 PMID:17950926

Blank, T., & Prinz, M. (2012). *Microglia as Modulators of Cognition and Neuropsychiatric Disorders.*, *000*, 1–9. doi:10.1002/glia.22372 PMID:22740320

Blinzinger, K., & Kreutzberg, G. (1968). Displacement of synaptic terminals from regenerating motoneurons by microglial cells. *Zeitschrift für Zellforschung und Mikroskopische Anatomie*, *85*(2), 145–157. doi:10.1007/BF00325030 PMID:5706753

Bliss, T. V. P., & Collingridge, G. L. (1993). A synaptic model of memory: Long-term potentiation in the hippocampus. *Nature*, *361*(6407), 31–39. doi:10.1038/361031a0 PMID:8421494

Bliss, T. V. P., & Lømo, T. (1973). Long-lasting potentiation of synaptic transmission in the dentate area of the anaesthetized rabbit following stimulation of the perforant path. *The Journal of Physiology*, *232*(2), 331–356. doi:10.1113/jphysiol.1973.sp010273 PMID:4727084

Butovsky, O., Ziv, Y., Schwartz, A., Landa, G., Talpalar, A. E., Pluchino, S., Martino, G., & Schwartz, M. (2006). Microglia activated by IL-4 or IFN-γ differentially induce neurogenesis and oligodendrogenesis from adult stem/progenitor cells. *Molecular and Cellular Neurosciences*, *31*(1), 149–160. doi:10.1016/j.mcn.2005.10.006 PMID:16297637

Cecchini, M. G., Dominguez, M. G., Mocci, S., Wetterwald, A., Felix, R., Fleisch, H., Chisholm, O., Hofstetter, W., Pollard, J. W., & Stanley, E. R. (1994). Role of colony stimulating factor-1 in the establishment and regulation of tissue macrophages during postnatal development of the mouse. *Development (Cambridge, England)*, *120*(6), 1357–1372. doi:10.1242/dev.120.6.1357 PMID:8050349

Checchin, D., Sennlaub, F., Levavasseur, E., Leduc, M., & Chemtob, S. (2006). Potential role of microglia in retinal blood vessel formation. *Investigative Ophthalmology & Visual Science*, *47*(8), 3595–3602. doi:10.1167/iovs.05-1522 PMID:16877434

Chen, W., Zhang, X. I. A., & Huang, W. (2016). Role of neuroinflammation in neurodegenerative diseases. *Molecular Medicine Reports*, *13*(4), 3391–3396. doi:10.3892/mmr.2016.4948 PMID:26935478

Chen, Y., Bourne, J., Pieribone, V. A., & Fitzsimonds, R. M. (2004). The role of actin in the regulation of dendritic spine morphology and bidirectional synaptic plasticity. *Neuroreport*, *15*(5), 829–832. doi:10.1097/00001756-200404090-00018 PMID:15073524

Chew, L.-J., Fusar-Poli, P., & Schmitz, T. (2013). Oligodendroglial alterations and the role of microglia in white matter injury: Relevance to schizophrenia. *Developmental Neuroscience*, *35*(2–3), 102–129. doi:10.1159/000346157 PMID:23446060

Clarke, L. E., Liddelow, S. A., Chakraborty, C., Münch, A. E., Heiman, M., & Barres, B. A. (2018). Normal aging induces A1-like astrocyte reactivity. *Proceedings of the National Academy of Sciences of the United States of America*, *115*(8), E1896–E1905. doi:10.1073/pnas.1800165115 PMID:29437957

Colonna, M., & Butovsky, O. (2017a). *Microglia Function in the Central Nervous System During Health and Neurodegeneration.*

Cooke, S. F., & Bliss, T. V. P. (2006). Plasticity in the human central nervous system. *Brain*, *129*(7), 1659–1673. doi:10.1093/brain/awl082 PMID:16672292

Cornell, J., Salinas, S., Huang, H., & Zhou, M. (2022). *Microglia regulation of synaptic plasticity and learning and memory, 17*(4), 705–716.

Cserép, C., Pósfai, B., & Dénes, Á. (2021). Shaping Neuronal Fate: Functional Heterogeneity of Direct Microglia-Neuron Interactions. *Neuron*, *109*(2), 222–240. doi:10.1016/j.neuron.2020.11.007 PMID:33271068

Cuadros, M. A., Martin, C., Coltey, P., Almendros, A., & Navascués, J. (1993). First appearance, distribution, and origin of macrophages in the early development of the avian central nervous system. *The Journal of Comparative Neurology*, *330*(1), 113–129. doi:10.1002/cne.903300110 PMID:8468399

Davalos, D., Grutzendler, J., Yang, G., Kim, J. V., Zuo, Y., Jung, S., Littman, D. R., Dustin, M. L., & Gan, W.-B. (2005). ATP mediates rapid microglial response to local brain injury in vivo. *Nature Neuroscience*, *8*(6), 752–758. doi:10.1038/nn1472 PMID:15895084

De Luca, S. N., Soch, A., Sominsky, L., Nguyen, T.-X., Bosakhar, A., & Spencer, S. J. (2020). Glial remodeling enhances short-term memory performance in Wistar rats. *Journal of Neuroinflammation, 17*(1), 1–18. doi:10.118612974-020-1729-4 PMID:32028971

Diaz-Aparicio, I., Paris, I., Sierra-Torre, V., Plaza-Zabala, A., Rodríguez-Iglesias, N., Márquez-Ropero, M., Beccari, S., Huguet, P., Abiega, O., Alberdi, E., Matute, C., Bernales, I., Schulz, A., Otrokocsi, L., Sperlagh, B., Happonen, K. E., Lemke, G., Maletic-Savatic, M., Valero, J., & Sierra, A. (2020). Microglia actively remodel adult hippocampal neurogenesis through the phagocytosis secretome. *The Journal of Neuroscience : The Official Journal of the Society for Neuroscience, 40*(7), 1453–1482. doi:10.1523/JNEUROSCI.0993-19.2019 PMID:31896673

Do, H. (1949). *The organization of behavior.*

Dudvarski Stankovic, N., Teodorczyk, M., Ploen, R., Zipp, F., & Schmidt, M. H. H. (2016). Microglia–blood vessel interactions: A double-edged sword in brain pathologies. *Acta Neuropathologica, 131*(3), 347–363. doi:10.100700401-015-1524-y PMID:26711460

Ehninger, D., & Kempermann, G. (2008). Neurogenesis in the adult hippocampus. *Cell and Tissue Research, 331*(1), 243–250. doi:10.100700441-007-0478-3 PMID:17938969

Ekdahl, C. T. (2012). Microglial activation-tuning and pruning adult neurogenesis. *Frontiers in Pharmacology, 3 MAR*(March), 1–9. doi:10.3389/fphar.2012.00041

El-Mansoury, B., Laadraoui, J., El Khiat, A., El Hiba, O., Jayakumar, A. R., Ferssiwi, A., & Bitar, A. (2023). *Ataxia in Multiple Sclerosis: From Current Understanding to Therapy.*

El-Mansoury, B. El Hiba, O., Agnaou, M., Jayakumar, A. R., El Khiat, A., Kahim, K., Boulbaroud, S., Zouhairi, N., Tibssirte, O., & Bouighajd, H. (2023). Neuropathology of Parkinson's Disease: An Update of the Recent Findings. In Experimental and Clinical Evidence of the Neuropathology of Parkinson's Disease (pp. 82–101). IGI Global.

El-Mansoury, B, Smimih, K., Aimrane, A., Draoui, A., Ferssiwi, A., Bitar, A., Jayakumar, A. R., & El Hiba, O. (2023). P8 Cognitive Impairment Associated With Cortical and Hippocampal Microglia Activation and Altered Synaptic Function in a Rat Model of Chronic Hepatic Encephalopathy. *Official Journal of the American College of Gastroenterology| ACG, 118*(9S), S7.

Engert, F., & Bonhoeffer, T. (1999). Dendritic spine changes associated with hippocampal long-term synaptic plasticity. *Nature, 399*(6731), 66–70. doi:10.1038/19978 PMID:10331391

Eriksson, P. S., Perfilieva, E., Björk-Eriksson, T., Alborn, A.-M., Nordborg, C., Peterson, D. A., & Gage, F. H. (1998). Neurogenesis in the adult human hippocampus. *Nature Medicine, 4*(11), 1313–1317. doi:10.1038/3305 PMID:9809557

Fantin, A., Vieira, J. M., Gestri, G., Denti, L., Schwarz, Q., Prykhozhij, S., Peri, F., Wilson, S. W., & Ruhrberg, C. (2010). Tissue macrophages act as cellular chaperones for vascular anastomosis downstream of VEGF-mediated endothelial tip cell induction. *Blood, 116*(5), 829–840. doi:10.1182/blood-2009-12-257832 PMID:20404134

Farina, C., Aloisi, F., & Meinl, E. (2007). Astrocytes are active players in cerebral innate immunity. *Trends in Immunology*, *28*(3), 138–145. doi:10.1016/j.it.2007.01.005 PMID:17276138

Finnie, P. S. B., & Nader, K. (2012). The role of metaplasticity mechanisms in regulating memory destabilization and reconsolidation. *Neuroscience and Biobehavioral Reviews*, *36*(7), 1667–1707. doi:10.1016/j.neubiorev.2012.03.008 PMID:22484475

Fünfschilling, U., Supplie, L. M., Mahad, D., Boretius, S., Saab, A. S., Edgar, J., Brinkmann, B. G., Kassmann, C. M., Tzvetanova, I. D., Möbius, W., Diaz, F., Meijer, D., Suter, U., Hamprecht, B., Sereda, M. W., Moraes, C. T., Frahm, J., Goebbels, S., & Nave, K.-A. (2012). Glycolytic oligodendrocytes maintain myelin and long-term axonal integrity. *Nature*, *485*(7399), 517–521. doi:10.1038/nature11007 PMID:22622581

Gage, F. H. (2000). Mammalian neural stem cells. *Science*, *287*(5457), 1433–1438. doi:10.1126cience.287.5457.1433 PMID:10688783

Gemma, C., & Bachstetter, A. D. (2013). The role of microglia in adult hippocampal neurogenesis. *Frontiers in Cellular Neuroscience*, *7*(NOV), 1–5. doi:10.3389/fncel.2013.00229 PMID:24319411

Ginhoux, F., Greter, M., Leboeuf, M., Nandi, S., See, P., Gokhan, S., Mehler, M. F., Conway, S. J., Ng, L. G., Stanley, E. R., Samokhvalov, I. M., & Merad, M. (2010). Fate mapping analysis reveals that adult microglia derive from primitive macrophages. *Science*, *330*(6005), 841–845. doi:10.1126cience.1194637 PMID:20966214

Gipson, C. D., & Olive, M. F. (2017). Structural and functional plasticity of dendritic spines–root or result of behavior? *Genes Brain & Behavior*, *16*(1), 101–117. doi:10.1111/gbb.12324 PMID:27561549

Glezer, I., Zekki, H., Scavone, C., & Rivest, S. (2003). Modulation of the innate immune response by NMDA receptors has neuropathological consequences. *The Journal of Neuroscience : The Official Journal of the Society for Neuroscience*, *23*(35), 11094–11103. doi:10.1523/JNEUROSCI.23-35-11094.2003 PMID:14657167

Grossmann, R., Stence, N., Carr, J., Fuller, L., Waite, M., & Dailey, M. E. (2002). Juxtavascular microglia migrate along brain microvessels following activation during early postnatal development. *Glia*, *37*(3), 229–240. doi:10.1002/glia.10031 PMID:11857681

Grutzendler, J., Kasthuri, N., & Gan, W.-B. (2002). Long-term dendritic spine stability in the adult cortex. *Nature*, *420*(6917), 812–816. doi:10.1038/nature01276 PMID:12490949

Helmut, K., Hanisch, U. K., Noda, M., & Verkhratsky, A. (2011). Physiology of microglia. *Physiological Reviews*, *91*(2), 461–553. doi:10.1152/physrev.00011.2010 PMID:21527731

Holm, T. H., Draeby, D., & Owens, T. (2012). Microglia are required for astroglial Toll-like receptor 4 response and for optimal TLR2 and TLR3 response. *Glia*, *60*(4), 630–638. doi:10.1002/glia.22296 PMID:22271465

Holtmaat, A. J. G. D., Trachtenberg, J. T., Wilbrecht, L., Shepherd, G. M., Zhang, X., Knott, G. W., & Svoboda, K. (2005). Transient and persistent dendritic spines in the neocortex in vivo. *Neuron*, *45*(2), 279–291. doi:10.1016/j.neuron.2005.01.003 PMID:15664179

Jha, M. K., Jo, M., Kim, J., & Suk, K. (2018). Microglia-Astrocyte Crosstalk : An Intimate Molecular Conversation. *The Neuroscientist.* Advance online publication. doi:10.1177/1073858418783959 PMID:29931997

Jha, M. K., Lee, W.-H., & Suk, K. (2016). Functional polarization of neuroglia: Implications in neuroinflammation and neurological disorders. *Biochemical Pharmacology, 103*, 1–16. doi:10.1016/j.bcp.2015.11.003 PMID:26556658

Kaindl, A. M., Degos, V., Peineau, S., Gouadon, E., Chhor, V., Loron, G., Le Charpentier, T., Josserand, J., Ali, C., Vivien, D., Collingridge, G. L., Lombet, A., Issa, L., Rene, F., Loeffler, J.-P., Kavelaars, A., Verney, C., Mantz, J., & Gressens, P. (2012). Activation of microglial N-methyl-D-aspartate receptors triggers inflammation and neuronal cell death in the developing and mature brain. *Annals of Neurology, 72*(4), 536–549. doi:10.1002/ana.23626 PMID:23109148

Kaindl, A. M., Degos, V., Peineau, S., Gouadon, E., Loron, G., Lombet, A., Verney, C., Mantz, J., & Gressens, P. (2008). Microglia express functional NMDA receptors: A novel finding and a promise for innovative treatment of excitotoxic and inflammatory brain disease. *Neuropediatrics, 39*(05), V23. doi:10.1055-0029-1215736

Kalla, R., Liu, Z., Xu, S., Koppius, A., Imai, Y., Kloss, C. U. A., Kohsaka, S., Gschwendtner, A., Möller, J. C., Werner, A., & Raivich, G. (2001). Microglia and the early phase of immune surveillance in the axotomized facial motor nucleus: Impaired microglial activation and lymphocyte recruitment but no effect on neuronal survival or axonal regeneration in macrophage-colony stimulating factor-defici. *The Journal of Comparative Neurology, 436*(2), 182–201. doi:10.1002/cne.1060 PMID:11438923

Kassiotis, G., & Kollias, G. (2001). Uncoupling the proinflammatory from the immunosuppressive properties of tumor necrosis factor (TNF) at the p55 TNF receptor level: Implications for pathogenesis and therapy of autoimmune demyelination. *The Journal of Experimental Medicine, 193*(4), 427–434. doi:10.1084/jem.193.4.427 PMID:11181695

Kato, G., Inada, H., Wake, H., Akiyoshi, R., Miyamoto, A., Eto, K., Ishikawa, T., Moorhouse, A. J., Strassman, A. M., & Nabekura, J. (2016). Microglial Contact Prevents Excess Depolarization. *eNeuro, 3*(June), 1–9. PMID:27390772

Kato, T. A., & Kanba, S. (2013). Are microglia minding us? Digging up the unconscious mind-brain relationship from a neuropsychoanalytic approach. *Frontiers in Psychology, 7*, 13. doi:10.3389/fnhum.2013.00013 PMID:23443737

Kettenmann, H., Kirchhoff, F., & Verkhratsky, A. (2013). Microglia: New roles for the synaptic stripper. *Neuron, 77*(1), 10–18. doi:10.1016/j.neuron.2012.12.023 PMID:23312512

Kettenmann, H., & Verkhratsky, A. (2008). Neuroglia: The 150 years after. *Trends in Neurosciences, 31*(12), 653–659. doi:10.1016/j.tins.2008.09.003 PMID:18945498

Kierdorf, K., & Prinz, M. (2013). Factors regulating microglia activation. *Frontiers in Cellular Neuroscience, 7*, 44. doi:10.3389/fncel.2013.00044 PMID:23630462

Kitamura, T., Tsuchihashi, Y., & Fujita, S. (1978). Initial response of silver-impregnated "resting microglia" to stab wounding in rabbit hippocampus. *Acta Neuropathologica*, *44*(1), 31–39. doi:10.1007/BF00691636 PMID:707043

Kopec, C. D., Li, B., Wei, W., Boehm, J., & Malinow, R. (2006). Glutamate receptor exocytosis and spine enlargement during chemically induced long-term potentiation. *The Journal of Neuroscience : The Official Journal of the Society for Neuroscience*, *26*(7), 2000–2009. doi:10.1523/JNEUROSCI.3918-05.2006 PMID:16481433

Kubota, Y., Takubo, K., Shimizu, T., Ohno, H., Kishi, K., Shibuya, M., Saya, H., & Suda, T. (2009). M-CSF inhibition selectively targets pathological angiogenesis and lymphangiogenesis. *The Journal of Experimental Medicine*, *206*(5), 1089–1102. doi:10.1084/jem.20081605 PMID:19398755

Kurz, H., & Christ, B. (1998). Embryonic CNS macrophages and microglia do not stem from circulating, but from extravascular precursors. *Glia*, *22*(1), 98–102. doi:10.1002/(SICI)1098-1136(199801)22:1<98::AID-GLIA10>3.0.CO;2-V PMID:9436792

Lang, C., Barco, A., Zablow, L., Kandel, E. R., Siegelbaum, S. A., & Zakharenko, S. S. (2004). Transient expansion of synaptically connected dendritic spines upon induction of hippocampal long-term potentiation. *Proceedings of the National Academy of Sciences of the United States of America*, *101*(47), 16665–16670. doi:10.1073/pnas.0407581101 PMID:15542587

Lee, M., Schwab, C., & Mcgeer, P. L. (2011). Astrocytes are GABAergic cells that modulate microglial activity. *Glia*, *59*(1), 152–165. doi:10.1002/glia.21087 PMID:21046567

Lee, S. H., Liu, L., Wang, Y. T., & Sheng, M. (2002). Clathrin adaptor AP2 and NSF interact with overlapping sites of GluR2 and play distinct roles in AMPA receptor trafficking and hippocampal LTD. *Neuron*, *36*(4), 661–674. doi:10.1016/S0896-6273(02)01024-3 PMID:12441055

Li, J., Ramenaden, E. R., Peng, J., Koito, H., Volpe, J. J., & Rosenberg, P. A. (2008). Tumor necrosis factor α mediates lipopolysaccharide-induced microglial toxicity to developing oligodendrocytes when astrocytes are present. *The Journal of Neuroscience : The Official Journal of the Society for Neuroscience*, *28*(20), 5321–5330. doi:10.1523/JNEUROSCI.3995-07.2008 PMID:18480288

Li, Q., & Barres, B. A. (2018). Microglia and macrophages in brain homeostasis and disease. *Nature Reviews. Immunology*, *18*(4), 225–242. doi:10.1038/nri.2017.125 PMID:29151590

Li, W.-W., Setzu, A., Zhao, C., & Franklin, R. J. M. (2005). Minocycline-mediated inhibition of microglia activation impairs oligodendrocyte progenitor cell responses and remyelination in a non-immune model of demyelination. *Journal of Neuroimmunology*, *158*(1–2), 58–66. doi:10.1016/j.jneuroim.2004.08.011 PMID:15589038

Liddelow, S. A., Guttenplan, K. A., Clarke, L. E., Bennett, F. C., Bohlen, C. J., Schirmer, L., Bennett, M. L., Münch, A. E., Chung, W.-S., Peterson, T. C., Wilton, D. K., Frouin, A., Napier, B. A., Panicker, N., Kumar, M., Buckwalter, M. S., Rowitch, D. H., Dawson, V. L., Dawson, T. M., & Barres, B. A. (2017). Neurotoxic reactive astrocytes are induced by activated microglia. *Nature*, *541*(7638), 481–487. doi:10.1038/nature21029 PMID:28099414

Lüscher, C., & Malenka, R. C. (2012). NMDA receptor-dependent long-term potentiation and long-term depression (LTP/LTD). *Cold Spring Harbor Perspectives in Biology*, *4*(6), a005710. doi:10.1101/cshperspect.a005710 PMID:22510460

Lynch, G. S., Dunwiddie, T., & Gribkoff, V. (1977). Heterosynaptic depression: A postsynaptic correlate of long-term potentiation. *Nature*, *266*(5604), 737–739. doi:10.1038/266737a0 PMID:195211

Majewska, A. K., Newton, J. R., & Sur, M. (2006). Remodeling of synaptic structure in sensory cortical areas in vivo. *The Journal of Neuroscience : The Official Journal of the Society for Neuroscience*, *26*(11), 3021–3029. doi:10.1523/JNEUROSCI.4454-05.2006 PMID:16540580

Maletic-Savatic, M., Malinow, R., & Svoboda, K. (1999). Rapid dendritic morphogenesis in CA1 hippocampal dendrites induced by synaptic activity. *Science*, *283*(5409), 1923–1927. doi:10.1126cience.283.5409.1923 PMID:10082466

Man, H.-Y., Lin, J. W., Ju, W. H., Ahmadian, G., Liu, L., Becker, L. E., Sheng, M., & Wang, Y. T. (2000). Regulation of AMPA receptor–mediated synaptic transmission by clathrin-dependent receptor internalization. *Neuron*, *25*(3), 649–662. doi:10.1016/S0896-6273(00)81067-3 PMID:10774732

Massey, P. V., & Bashir, Z. I. (2007). Long-term depression: Multiple forms and implications for brain function. *Trends in Neurosciences*, *30*(4), 176–184. doi:10.1016/j.tins.2007.02.005 PMID:17335914

Masuch, A., Shieh, C., van Rooijen, N., van Calker, D., & Biber, K. (2016). Mechanism of microglia neuroprotection: Involvement of P2X 7, TNF α, and valproic acid. *Glia*, *64*(1), 76–89. doi:10.1002/glia.22904 PMID:26295445

Matejuk, A., & Ransohoff, R. M. (2020). Crosstalk Between Astrocytes and Microglia : An Overview. *Frontiers in Immunology*, *11*(July), 1–11. doi:10.3389/fimmu.2020.01416 PMID:32765501

Matsuo, R., Yamagishi, M., Wakiya, K., Tanaka, Y., & Ito, E. (2013). Target innervation is necessary for neuronal polyploidization in the terrestrial slug Limax. *Developmental Neurobiology*. doi:10.1002/dneu PMID:23723165

Matsuzaki, M., Honkura, N., Ellis-Davies, G. C. R., & Kasai, H. (2004). Structural basis of long-term potentiation in single dendritic spines. *Nature*, *429*(6993), 761–766. doi:10.1038/nature02617 PMID:15190253

McKercher, S. R., Torbett, B. E., Anderson, K. L., Henkel, G. W., Vestal, D. J., Baribault, H., Klemsz, M., Feeney, A. J., Wu, G. E., Paige, C. J., & Maki, R. A. (1996). Targeted disruption of the PU. 1 gene results in multiple hematopoietic abnormalities. *The EMBO Journal*, *15*(20), 5647–5658. doi:10.1002/j.1460-2075.1996.tb00949.x PMID:8896458

Min, K.-J., Yang, M., Kim, S.-U., Jou, I., & Joe, E. (2006). Astrocytes induce hemeoxygenase-1 expression in microglia: A feasible mechanism for preventing excessive brain inflammation. *The Journal of Neuroscience : The Official Journal of the Society for Neuroscience*, *26*(6), 1880–1887. doi:10.1523/JNEUROSCI.3696-05.2006 PMID:16467537

Morris, G. P., Clark, I. A., Zinn, R., & Vissel, B. (2013). Neurobiology of Learning and Memory Microglia : A new frontier for synaptic plasticity, learning and memory, and neurodegenerative disease research. *Neurobiology of Learning and Memory, 105*, 40–53. doi:10.1016/j.nlm.2013.07.002 PMID:23850597

Mozzachiodi, R., & Byrne, J. H. (2010). More than synaptic plasticity: Role of nonsynaptic plasticity in learning and memory. *Trends in Neurosciences, 33*(1), 17–26. doi:10.1016/j.tins.2009.10.001 PMID:19889466

Nguyen, P. T., Dorman, L. C., Pan, S., Vainchtein, I. D., Han, R. T., Nakao-Inoue, H., Taloma, S. E., Barron, J. J., Molofsky, A. B., Kheirbek, M. A., & Molofsky, A. V. (2020). Microglial remodeling of the extracellular matrix promotes synapse plasticity. *Cell, 182*(2), 388–403. doi:10.1016/j.cell.2020.05.050 PMID:32615087

Nimmerjahn, A., Kirchhoff, F., & Helmchen, F. (2005). Resting microglial cells are highly dynamic surveillants of brain parenchyma in vivo. *Science, 308*(5726), 1314–1318. doi:10.1126cience.1110647 PMID:15831717

Nolte, C., Moller, T., Walter, T., & Kettenmann, H. (1996). Complement 5a controls motility of murine microglial cells in vitro via activation of an inhibitory G-protein and the rearrangement of the actin cytoskeleton. *Neuroscience, 73*(4), 1091–1107. doi:10.1016/0306-4522(96)00106-6 PMID:8809827

Pang, Y., Cai, Z., & Rhodes, P. G. (2003). Disturbance of oligodendrocyte development, hypomyelination and white matter injury in the neonatal rat brain after intracerebral injection of lipopolysaccharide. *Brain Research. Developmental Brain Research, 140*(2), 205–214. doi:10.1016/S0165-3806(02)00606-5 PMID:12586426

Parkhurst, C. N., Yang, G., Ninan, I., Savas, J. N., Yates, J. R. III, Lafaille, J. J., Hempstead, B. L., Littman, D. R., & Gan, W.-B. (2013). Microglia promote learning-dependent synapse formation through brain-derived neurotrophic factor. *Cell, 155*(7), 1596–1609. doi:10.1016/j.cell.2013.11.030 PMID:24360280

Pascual, O., Ben Achour, S., Rostaing, P., Triller, A., & Bessis, A. (2012). Microglia activation triggers astrocyte-mediated modulation of excitatory neurotransmission. *Proceedings of the National Academy of Sciences of the United States of America, 109*(4), E197–E205. doi:10.1073/pnas.1111098109 PMID:22167804

Peferoen, L., Kipp, M., van der Valk, P., van Noort, J. M., & Amor, S. (2014a). Oligodendrocyte-microglia cross-talk in the central nervous system. *Immunology, 141*(3), 302–313. doi:10.1111/imm.12163 PMID:23981039

Peferoen, L., Kipp, M., van der Valk, P., van Noort, J. M., & Amor, S. (2014b). Oligodendrocyte-microglia cross-talk in the central nervous system. *Immunology, 141*(3), 302–313. doi:10.1111/imm.12163 PMID:23981039

Petersen, M. A., & Dailey, M. E. (2004). Diverse microglial motility behaviors during clearance of dead cells in hippocampal slices. *Glia, 46*(2), 195–206. doi:10.1002/glia.10362 PMID:15042586

Ransohoff, R. M., & Brown, M. A. (2012). Innate immunity in the central nervous system. *The Journal of Clinical Investigation, 122*(4), 1164–1171. doi:10.1172/JCI58644 PMID:22466658

Ransohoff, R. M., & Cardona, A. E. (2010). The myeloid cells of the central nervous system parenchyma. *Nature*, *468*(7321), 253–262. doi:10.1038/nature09615 PMID:21068834

Redondo, R. L., & Morris, R. G. M. (2011). Making memories last: The synaptic tagging and capture hypothesis. *Nature Reviews. Neuroscience*, *12*(1), 17–30. doi:10.1038/nrn2963 PMID:21170072

Rothhammer, V., Borucki, D. M., Tjon, E. C., Takenaka, M. C., Chao, C.-C., Ardura-Fabregat, A., De Lima, K. A., Gutiérrez-Vázquez, C., Hewson, P., Staszewski, O., Blain, M., Healy, L., Neziraj, T., Borio, M., Wheeler, M., Dragin, L. L., Laplaud, D. A., Antel, J., Alvarez, J. I., & Quintana, F. J. (2018). Microglial control of astrocytes in response to microbial metabolites. *Nature*, *557*(7707), 724–728. doi:10.103841586-018-0119-x PMID:29769726

Roy, N. S., Wang, S., Jiang, L., Kang, J., Benraiss, A., Harrison-Restelli, C., Fraser, R. A. R., Couldwell, W. T., Kawaguchi, A., Okano, H., Nedergaard, M., & Goldman, S. A. (2000). In vitro neurogenesis by progenitor cells isolated from the adult human hippocampus. *Nature Medicine*, *6*(3), 271–277. doi:10.1038/73119 PMID:10700228

Saijo, K., & Glass, C. K. (2011). Microglial cell origin and phenotypes in health and disease. *Nature Reviews. Immunology*, *11*(11), 775–787. doi:10.1038/nri3086 PMID:22025055

Schafer, D. P., Lehrman, E. K., Kautzman, A. G., Koyama, R., Mardinly, A. R., Yamasaki, R., Ransohoff, R. M., Greenberg, M. E., Barres, B. A., & Stevens, B. (2012). Microglia sculpt postnatal neural circuits in an activity and complement-dependent manner. *Neuron*, *74*(4), 691–705. doi:10.1016/j.neuron.2012.03.026 PMID:22632727

Scott, E. W., Simon, M. C., Anastasi, J., & Singh, H. (1994). Requirement of transcription factor PU. 1 in the development of multiple hematopoietic lineages. *Science*, *265*(5178), 1573–1577. doi:10.1126cience.8079170 PMID:8079170

Sherwin, C., & Fern, R. (2005). Acute lipopolysaccharide-mediated injury in neonatal white matter glia: Role of TNF-α, IL-1β, and calcium. *The Journal of Immunology : Official Journal of the American Association of Immunologists*, *175*(1), 155–161. doi:10.4049/jimmunol.175.1.155 PMID:15972642

Sierra, A., Encinas, J. M., Deudero, J. J. P., Chancey, J. H., Enikolopov, G., Overstreet-Wadiche, L. S., Tsirka, S. E., & Maletic-Savatic, M. (2010). Microglia shape adult hippocampal neurogenesis through apoptosis-coupled phagocytosis. *Cell Stem Cell*, *7*(4), 483–495. doi:10.1016/j.stem.2010.08.014 PMID:20887954

Snyder, E. M., Nong, Y., Almeida, C. G., Paul, S., Moran, T., Choi, E. Y., Nairn, A. C., Salter, M. W., Lombroso, P. J., Gouras, G. K., & Greengard, P. (2005). Regulation of NMDA receptor trafficking by amyloid-β. *Nature Neuroscience*, *8*(8), 1051–1058. doi:10.1038/nn1503 PMID:16025111

Stence, N., Waite, M., & Dailey, M. E. (2001). Dynamics of microglial activation: A confocal time-lapse analysis in hippocampal slices. *Glia*, *33*(3), 256–266. doi:10.1002/1098-1136(200103)33:3<256::AID-GLIA1024>3.0.CO;2-J PMID:11241743

Stevens, B., Allen, N. J., Vazquez, L. E., Howell, G. R., Christopherson, K. S., Nouri, N., Micheva, K. D., Mehalow, A. K., Huberman, A. D., Stafford, B., Sher, A., Litke, A. M., Lambris, J. D., Smith, S. J., John, S. W. M., & Barres, B. A. (2007). The classical complement cascade mediates CNS synapse elimination. *Cell, 131*(6), 1164–1178. doi:10.1016/j.cell.2007.10.036 PMID:18083105

Streit, W. J. (2005). Microglia and neuroprotection: Implications for Alzheimer's disease. *Brain Research. Brain Research Reviews, 48*(2), 234–239. doi:10.1016/j.brainresrev.2004.12.013 PMID:15850662

Streit, W. J., Walter, S. A., & Pennell, N. A. (1999). Reactive microgliosis. *Progress in Neurobiology, 57*(6), 563–581. doi:10.1016/S0301-0082(98)00069-0 PMID:10221782

Szepesi, Z., Manouchehrian, O., Bachiller, S., & Deierborg, T. (2018). Bidirectional Microglia–Neuron Communication in Health and Disease. *Frontiers in Cellular Neuroscience, 12*(September), 1–26. doi:10.3389/fncel.2018.00323 PMID:30319362

Tanaka, T., Murakami, K., Bando, Y., & Yoshida, S. (2013). Minocycline reduces remyelination by suppressing ciliary neurotrophic factor expression after cuprizone-induced demyelination. *Journal of Neurochemistry, 127*(2), 259–270. doi:10.1111/jnc.12289 PMID:23647102

Thomas, W. E. (1992). Brain macrophages: Evaluation of microglia and their functions. *Brain Research. Brain Research Reviews, 17*(1), 61–74. doi:10.1016/0165-0173(92)90007-9 PMID:1638276

Tian, L., Ma, L., Kaarela, T., & Li, Z. (2012). Neuroimmune crosstalk in the central nervous system and its significance for neurological diseases. *Journal of Neuroinflammation, 9*(1), 1–10. doi:10.1186/1742-2094-9-155 PMID:22747919

Tremblay, M.-È., Lowery, R. L., & Majewska, A. K. (2010). Microglial interactions with synapses are modulated by visual experience. *PLoS Biology, 8*(11), e1000527. doi:10.1371/journal.pbio.1000527 PMID:21072242

Turrigiano, G. G. (2008). The self-tuning neuron: Synaptic scaling of excitatory synapses. *Cell, 135*(3), 422–435. doi:10.1016/j.cell.2008.10.008 PMID:18984155

Vainchtein, I. D., Chin, G., Cho, F. S., Kelley, K. W., Miller, J. G., Chien, E. C., Liddelow, S. A., Nguyen, P. T., Nakao-Inoue, H., Dorman, L. C., Akil, O., Joshita, S., Barres, B. A., Paz, J. T., Molofsky, A. B., & Molofsky, A. V. (2018). Astrocyte-derived interleukin-33 promotes microglial synapse engulfment and neural circuit development. *Science, 359*(6381), 1269–1273. doi:10.1126cience.aal3589 PMID:29420261

Wake, H., Moorhouse, A. J., Jinno, S., Kohsaka, S., & Nabekura, J. (2009). Resting microglia directly monitor the functional state of synapses in vivo and determine the fate of ischemic terminals. *The Journal of Neuroscience : The Official Journal of the Society for Neuroscience, 29*(13), 3974–3980. doi:10.1523/JNEUROSCI.4363-08.2009 PMID:19339593

Walton, N. M., Sutter, B. M., Laywell, E. D., Levkoff, L. H., Kearns, S. M., Marshall, G. P. II, Scheffler, B., & Steindler, D. A. (2006). Microglia instruct subventricular zone neurogenesis. *Glia, 54*(8), 815–825. doi:10.1002/glia.20419 PMID:16977605

Wang, C., Liu, F., Liu, Y.-Y., Zhao, C.-H., You, Y., Wang, L., Zhang, J., Wei, B., Ma, T., Zhang, Q., Zhang, Y., Chen, R., Song, H., & Yang, Z. (2011). Identification and characterization of neuroblasts in the subventricular zone and rostral migratory stream of the adult human brain. *Cell Research*, *21*(11), 1534–1550. doi:10.1038/cr.2011.83 PMID:21577236

Wang, C., Yue, H., Hu, Z., Shen, Y., Ma, J., Li, J., Wang, X.-D., Wang, L., Sun, B., Shi, P., Wang, L., & Gu, Y. (2020). Microglia mediate forgetting via complement-dependent synaptic elimination. *Science*, *367*(6478), 688–694. doi:10.1126cience.aaz2288 PMID:32029629

Wu, Y., Dissing-Olesen, L., MacVicar, B. A., & Stevens, B. (2015). Microglia: Dynamic mediators of synapse development and plasticity. *Trends in Immunology*, *36*(10), 605–613. doi:10.1016/j.it.2015.08.008 PMID:26431938

Xiao, M.-Y., Zhou, Q., & Nicoll, R. A. (2001). Metabotropic glutamate receptor activation causes a rapid redistribution of AMPA receptors. *Neuropharmacology*, *41*(6), 664–671. doi:10.1016/S0028-3908(01)00134-4 PMID:11640920

Zhou, Q., Homma, K. J., & Poo, M. M. (2004). Shrinkage of dendritic spines associated with long-term depression of hippocampal synapses. *Neuron*, *44*(5), 749–757. doi:10.1016/j.neuron.2004.11.011 PMID:15572107

Ziv, Y., Ron, N., Butovsky, O., Landa, G., Sudai, E., Greenberg, N., Cohen, H., Kipnis, J., & Schwartz, M. (2006). Immune cells contribute to the maintenance of neurogenesis and spatial learning abilities in adulthood. *Nature Neuroscience*, *9*(2), 268–275. doi:10.1038/nn1629 PMID:16415867

Chapter 5
Brain Aging and Glial Cells: What Is the Link?

Kamal Smimih
Faculty of Sciences and Techniques, Sultan Moulay Slimane University, Morocco

Chaima Azzouhri
Faculty of Sciences and Techniques, Sultan Moulay Slimane University, Morocco

Fatima Ez-Zahraa Saad
Independent Researcher, Morocco

Youssef Ait Hamdan
https://orcid.org/0009-0005-8616-8852
Higher Normal School, Cadi Ayyad University, Morocco

Jawad Laadraoui
https://orcid.org/0000-0002-9784-1265
Faculty of Sciences, Hassan II University of Casablanca, Casablanca, Morocco

Nadia Zouhairi
Faculty of Sciences and Techniques, Sultan Moulay Slimane University, Morocco

Mohamed Merzouki
https://orcid.org/0000-0002-9837-4312
Faculty of Sciences and Techniques, Sultan Moulay Slimane University, Morocco

ABSTRACT

Brain aging is a complicated and diversified process that includes an ongoing decline in cognitive performance as well as various cellular and molecular changes in the brain. During an individual's life, glial cells such as oligodendrocytes, microglia, and astrocytes are crucial for maintaining healthy brain function. Astrocytes support neurons but may become less efficient with age, contributing to cognitive decline. Microglia, the brain's immune cells, can turn overactive with age, causing inflammation which can lead to neuronal damage. Oligodendrocytes, responsible for myelin sheath maintenance, also undergo changes, affecting neural communication. With advancing age, changes in glial cell function may lead to neuroinflammation, oxidative damage, and abnormalities in neural circuitry, all of which contribute to age related cognitive decline and neurodegenerative disorders. This chapter delves into the intricate dynamics of brain aging, focusing on the profound influence of age-related changes in glial cells.

DOI: 10.4018/978-1-6684-9675-6.ch005

1. INTRODUCTION

Brain aging is characterized by a multitude of dynamic changes in glial cells, particularly astrocytes (Palmer and Ousman, 2018) and microglia (von Bernhardi, Eugenín-von Bernhardi, and Eugenín, 2015), which have far-reaching implications for neural function and age-related neurodegenerative diseases (Suksuphew and Noisa, 2015). Astrocytes, until thought to be inert supporters, reveal an unseen dynamism as they age (Matias, Morgado, and Gomes, 2019). Their cellular hypertrophy, the swelling of cell bodies and processes, and the reconfiguration of their structural components paint a vivid portrait of their evolution (Verkhratsky et al., 2019; Clarke et al., 2018; Stogsdill et al., 2017). In turn, microglia, the brain's vigilant sentinels, metamorphose in response to neuroinflammatory cues or cellular distress (Muzio, Viotti, and Martino, 2021). They transition from a state of constant surveillance, characterized by long, branched processes, to an activated, amoeboid form replete with pro-inflammatory attributes (Hickman et *al.*, 2013).

This change brings about long-lasting inflammation in the brain, which is a common sign of brain aging. It may significantly affect cognitive performance and potentially cause neurodegenerative diseases (Jin et *al.*, 2022). These changes in the shape of glial cells with the age course suggests that the aging brain is always changing, and this can affect the complex functioning of the brain.

This chapter explores the changing role of glial cells in the aging brain, exploring their morphological and functional changes. As well, it examines how these changes impact brain function, including synaptic activity, neurotransmitter regulation, and susceptibility to oxidative stress. Additionally, it highlights the glial cell contributions to age-related neurodegenerative diseases like Alzheimer's, Parkinson's, and multiple sclerosis, underscoring the significance of aging in these conditions.

Additionally, it investigates the delicate interplay of oxidative stress, inflammation, genetics and epigenetics, to understand the mechanisms causing these changes in order to figure out the mysteries of brain aging, enabling the development of new strategies for preserving the healthy aging and combatting age-related neurological disorders.

2. AGE-RELATED CHANGES IN GLIAL CELLS

2.1 Structural Alterations

2.1.1 Changes in glial cell morphology

The alteration in glial cell morphology, notably that seen in astrocytes and microglia, is one indicator of brain aging. Astrocytes show significant morphological changes with age. These changes include cellular hypertrophy, an increase in the size of the cell bodies and processes, and a reorganization of their structural components (Stogsdill et *al.*, 2017; Clarke et *al.*, 2018;Verkhratsky et *al.*, 2019). As a result, the intricate network of fine astrocytic processes that intimately associates with synapses and neurons becomes altered, impacting synaptic function and neurotransmitter regulation (Kim, Park, and Choi, 2019).

Microglia, the brain's resident immune cells, also undergo striking changes in morphology during aging. When they are young, microglia have a ramified shape, characterized by long, branched processes, facilitating their constant surveillance of the brain microenvironment (Vidal-Itriago et *al*,. 2022). However, in response to neuroinflammatory cues or cellular stress, microglia in aged brains transition into an activated, amoeboid form, marked by the retraction of their processes and the acquisition of pro-inflammatory features (Hickman et *al.*, 2013). This transformation can lead to a persistent state of chronic neuroinflammation, a prominent feature of brain aging that has implications for cognitive function and neurodegenerative diseases.

2.1.2 Reduction in glial cell density

One prominent aspect of brain aging is the reduction in glial cell density, notably affecting astrocytes and oligodendrocytes (J. J. Rodríguez et *al.*, 2009;Nave and Werner, 2014) Astrocytes, integral for maintaining brain homeostasis, neurotransmitter regulation, and structural support, exhibit a decline in their population density with advancing age (Clarke et *al.*, 2018). This reduction in astrocytic density may potentially compromise their pivotal roles in synaptic support, synaptic plasticity, and overall neural function (Popov et *al.*, 2021).

Similarly, oligodendrocytes, essential for myelinating axons and ensuring swift signal transmission, also experience age-related reductions in density (Nave and Werner 2014). This decrease in the numbers of oligodendrocytes may impair their ability to efficiently maintain and repair myelin sheaths, contributing to diminished signal conduction along axons in the aging brain. As well, decreased levels of cognitive performance are linked to the age-related degeneration of myelin, which is characterized by a reduction in thickness and density in certain brain areas such the prefrontal cortex and regions II and V of the corpus callosum in non-human primates (Sams, 2021).

2.2 Functional Changes

2.2.1 Impaired synaptic function and plasticity

Astrocytes are well-known for their crucial role in synaptic regulation and exhibit functional changes with advancing age. These changes encompass compromised glutamate clearance, diminished release of neurotrophic factors, and an altered capacity to modulate synaptic strength (Olabarria et al., 2010 ; Clarke and Barres, 2013). Such astrocytic impairments contribute to disruptions in synaptic function and plasticity, potentially underpinning age-related cognitive decline (Acosta, Anderson, and Anderson, 2017). Concurrently, microglia, the resident immune cells of the brain, undergo a phenotypic shift towards a pro-inflammatory state during aging, releasing inflammatory mediators that can perturb synaptic health and plasticity (Ransohoff and Perry, 2009 ; Hickman et *al.*, 2013). This chronic neuroinflammation is closely associated with synaptic dysfunction and may significantly contribute to cognitive impairments observed in aging and age-related neurodegenerative conditions (Filippo et *al.*, 2008;Singh, 2022). In general, these age-related glial cell modifications highlight their complex role in aging-related changes to synapse health and plasticity, which have significant effects on cognitive performance.

2.2.2 Altered neurotransmitter regulation

Neurotransmitter modulation in the aging brain has been linked to age-related changes in glial cells, especially astrocytes (VanGuilder et *al.*, 2010). Astrocytes play an important role in maintaining the balance of neurotransmitters like glutamate and gamma-aminobutyric acid (GABA). With advancing age, astrocytes exhibit diminished capacity for glutamate uptake, leading to increased synaptic glutamate levels and potential excitotoxicity (Olabarria et *al.*, 2010) (Henneberger et *al.*, 2020). Additionally, age-related astrocytic changes can affect the regulation of GABAergic transmission, contributing to an imbalance between excitation and inhibition (Kaiser et al. 2006) (Clarke and Barres 2013). These altered neurotransmitter dynamics underlie age-related cognitive impairments and increased susceptibility to neurodegenerative diseases (Scimemi 2014) (Buzsáki, Logothetis, and Singer 2013).

2.2.3 Enhanced susceptibility to oxidative stress

Glial cell changes associated with age are noted for their increased susceptibility to oxidative stress, a factor implicated in a variety of neurodegenerative diseases. Astrocytes, essential for maintaining redox homeostasis in the brain (Chen et *al.*, 2020), exhibit age-related changes that render them more vulnerable to oxidative insults. These alterations include decreased antioxidant enzyme activity, diminished glutathione levels, and impaired mitochondrial function (Bellaver et *al.*, 2017) (Robillard et *al.*, 2016). Additionally, microglia, the immune responders of the brain, can shift towards a pro-inflammatory state during aging, contributing to the release of reactive oxygen species (ROS) and perpetuating oxidative stress (Norden and Godbout, 2013;Salter and Stevens, 2017). Enhanced susceptibility to oxidative stress in glial cells may promote neuroinflammation and neuronal damage, potentially playing a crucial role in the progression of age-related neurodegenerative diseases such as Alzheimer's and Parkinson's (Sofroniew and Vinters, 2010;J. J. Rodríguez et *al.*, 2009;Hickman et *al.*, 2013).

3. GLIAL CELLS CONTRIBUTIONS TO AGE-RELATED NEURODEGENERATIVE DISEASES

3.1 Alzheimer's Disease

3.1.1 Amyloid-beta and tau pathology

Glial cells, such as astrocytes and microglia, have a big impact on Alzheimer's disease (AD). These glial cell contributions are significantly influenced by aging, which also intensifies the pathogenic mechanisms of AD. Astrocytes, typically involved in maintaining brain homeostasis, can become dysfunctional with age, impairing their ability to clear amyloid-beta (Aβ) peptides, thus promoting Aβ aggregation (Si et *al.*, 2023;Clarke and Barres, 2013). Aging also heightens microglial reactivity; while microglia are essential for Aβ clearance, their chronic activation in aging and AD leads to sustained neuroinflammation and neuronal damage (Hickman et *al.*, 2013 ; Hansen, Hanson, and Sheng, 2017). Furthermore, the interplay between Aβ and tau pathology in AD is exacerbated by aging, with glial cells modulating tau phosphorylation and propagation (Bolós et *al.*, 2017;Maphis et *al.*, 2015) Figure 1.

3.1.2 Astrocyte involvement in amyloid clearance

Astrocytes are integral to Aβ clearance, a central process in AD pathology. However, aging significantly impacts this vital function. In aging brains, astrocytes undergo functional and structural changes that impede their ability to efficiently clear Aβ. The age-related decline in the expression and function of Aβ-degrading enzymes, such as neprilysin and insulin-degrading enzyme, can hinder astrocytic Aβ degradation (Miners et *al.*, 2008; Kimura et *al.*, 2019). Moreover, aging-associated alterations in the expression of Aβ transporters like LRP1 and ABCA7 can affect Aβ efflux from the brain parenchyma, contributing to Aβ accumulation (Zlokovic, 2005; Aikawa, Holm, and Kanekiyo, 2018). Dysfunctional astrocytes in the aging brain also display reduced phagocytic activity and Aβ internalization, further hampering Aβ clearance (Koizumi et *al.*, 2007; Simpson et *al.*, 2010). These age-related impairments in astrocyte-mediated Aβ clearance mechanisms exacerbate Aβ aggregation and plaque formation in AD, emphasizing the critical role of astrocytes in the context of aging-related neurodegenerative diseases.

Figure 1. Schematic showing changes in glial cells between the adult and aging brain in AD

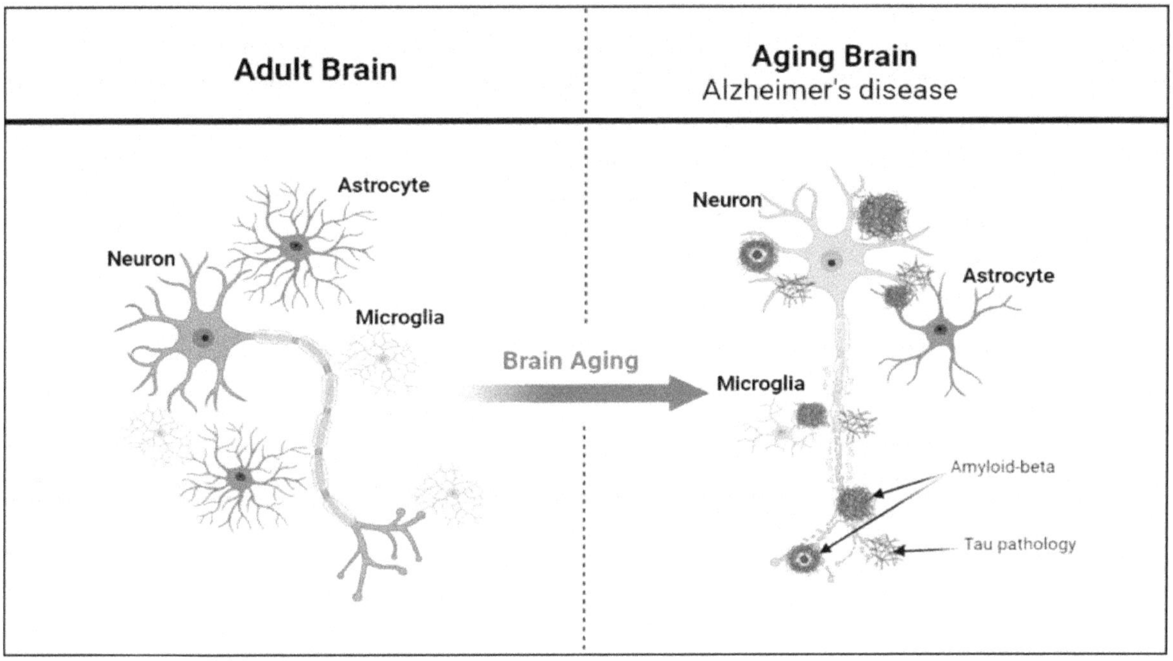

3.1.3 Microglial activation and neuroinflammation

AD is also characterized by microglial activation and chronic neuroinflammation. Microglia, the resident immune cells of the brain, undergo significant changes during aging, contributing to the pathogenesis of AD. Aging-associated microglial priming results in a heightened reactivity to pathological stimuli, leading to chronic release of pro-inflammatory cytokines, such as interleukin-1β and tumor necrosis factor-α (Norden and Godbout, 2013; Singh, 2022). Furthermore, microglia can become less efficient in phagocytosing and clearing Aβ plaques in the aging brain, exacerbating Aβ aggregation (Hickman et *al.*, 2013; Raj et *al.*, 2017). Dysregulated microglial responses can also contribute to tau hyperphosphorylation and propagation, further exacerbating neurodegeneration (Baik et *al.*, 2016; Maphis et *al.*, 2015). These age-related alterations in microglial function and their pro-inflammatory phenotype play a pivotal role in the progression of AD and other age-related neurodegenerative diseases, highlighting the critical impact of aging on glial cell contributions to neuroinflammation and pathology.

3.2 Parkinson's Disease

3.2.1 Alpha-synuclein aggregation

In Parkinson's disease (PD), glial cells, particularly astrocytes and microglia, play a significant role in the aggregation of alpha-synuclein (α-syn) protein, a hallmark of the disease (Gómez-Benito et *al.*, 2020). Aging profoundly influences these glial cell contributions, exacerbating PD pathology (Liu et *al.*, 2017). Astrocytes, which support neuronal function and maintain homeostasis, are essential in regulating extracellular α-syn levels (Sorrentino, Giasson, and Chakrabarty, 2019). However, with aging, astrocytes become less efficient in clearing α-syn aggregates and may even promote their accumulation through mechanisms such as impaired lysosomal function and altered protein degradation pathways(Booth, Hirst, and Wade-Martins, 2017). Microglia, the immune cells of the brain, exhibit age-related changes in phenotype and function, shifting towards a pro-inflammatory state. This chronic neuroinflammation in aging brains can perpetuate α-syn aggregation and contribute to the degeneration of dopaminergic neurons (C. Hansen et *al.*, 2011;Collier, Kanaan, and Kordower, 2011) Figure 2. The intricate interplay between glial cells and α-syn aggregation in the context of aging underscores their critical role in the pathogenesis of PD.

Figure 2. Schematic showing changes in glial cells between the adult and aging brain in PD

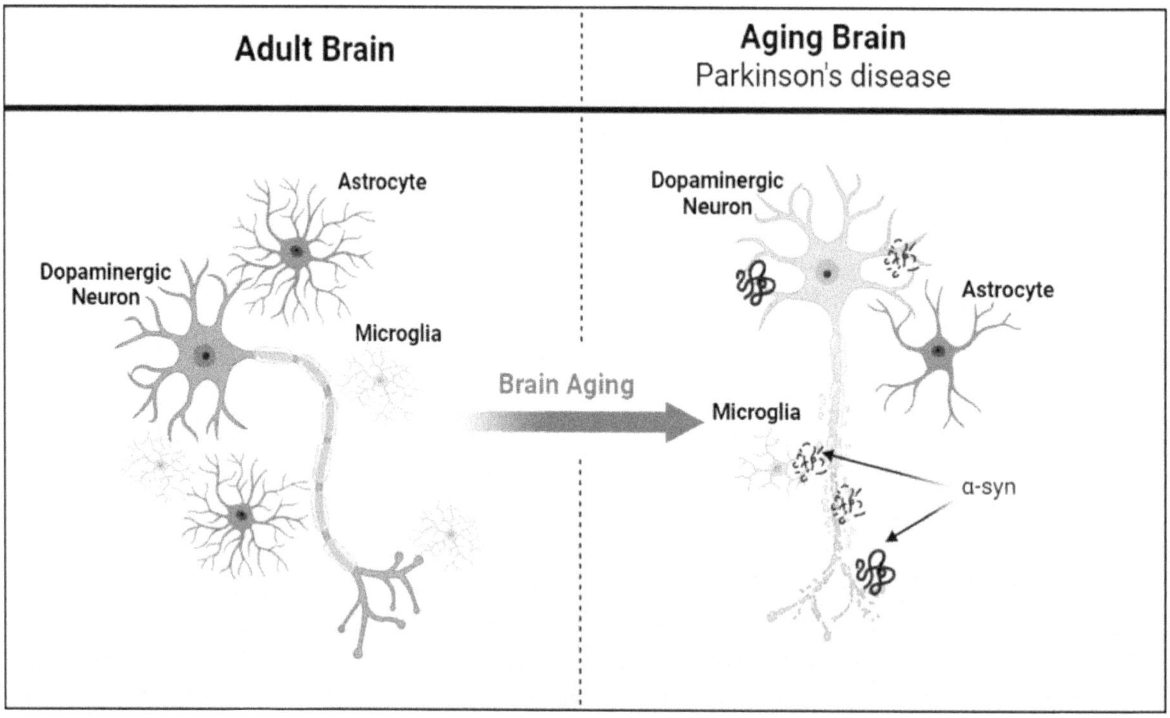

3.2.2 Microglial-mediated neuroinflammation

Microglia, the resident immune cells of the brain, play a crucial role in mediating neuroinflammation in PA, and the impact of aging significantly exacerbates this process. Age-related changes in microglia, termed "microglial priming," result in a heightened pro-inflammatory phenotype, characterized by increased production of cytokines, such as interleukin-1β and tumor necrosis factor-α (Norden and Godbout, 2013;Hirsch and Hunot, 2009). These age-related alterations in microglial function contribute to chronic neuroinflammation in the PD brain. Activated microglia can release ROS and pro-inflammatory mediators, further exacerbating dopaminergic neurodegeneration (Hickman et *al.*, 2013). Additionally, aging-related impairments in microglial phagocytic activity may hinder the clearance of α-syn aggregates, leading to their accumulation and propagation (Mao et *al.*, 2016;Rannikko, Weber, and Kahle, 2015). Understanding the intricate relationship between aging, microglial-mediated neuroinflammation, and α-syn pathology is crucial for unraveling the complexities of PD pathogenesis.

3.2.3 Astrocytic support and neuroprotection

Astrocytes, known for their essential role in supporting neuronal function and maintaining brain homeostasis, have been implicated in neuroprotection in the context of PD, and aging significantly influences these glial cell contributions. With advancing age, astrocytes can become less efficient in providing neuroprotection and supporting dopaminergic neurons, exacerbating PD pathology. Age-related astrocytic changes, including decreased secretion of neurotrophic factors like glial cell line-derived neurotrophic factor (GDNF) and brain-derived neurotrophic factor (BDNF), can contribute to the degeneration of dopaminergic neurons (Chung et *al.*, 2005;Rahmani et *al.* 2019). Additionally, impaired astrocytic glutamate clearance, which is crucial for preventing excitotoxicity, can lead to increased oxidative stress and neuronal damage in the aging brain (Booth, Hirst, and Wade-Martins, 2017).

3.3 Multiple Sclerosis

3.3.1 Oligodendrocyte damage and demyelination

In Multiple sclerosis (MS), a chronic autoimmune demyelinating disease of the central nervous system (CNS), the role of glial cells, particularly oligodendrocytes, is pivotal. The impact of aging on these glial cell contributions is substantial and significantly influences the disease course. Oligodendrocytes, responsible for myelinating axons and maintaining the integrity of the myelin sheath, are vulnerable to age-related changes. With advancing age, the regenerative capacity of oligodendrocyte precursor cells (OPCs) declines, hindering remyelination in response to demyelinating lesions (Shields et *al.*, 1999;Shen et *al.*, 2008). Additionally, age-related oligodendrocyte dysfunction may lead to decreased myelin maintenance and repair, exacerbating demyelination and axonal damage in MS (Kotter et *al.*, 2006;Franklin and ffrench-Constant, 2008). Furthermore, microglia, which actively participate in immune responses in the CNS, can adopt a pro-inflammatory phenotype with aging, potentially exacerbating neuroinflammation and demyelination (Fischer et *al.*, 2012). The intricate interplay between aging, glial cells, and demyelination underscores the critical impact of age-related factors on the pathogenesis of MS.

3.3.2 Microglial responses in autoimmune neuroinflammation

In MS, a complex autoimmune neuroinflammatory disease, microglia, the resident immune cells of the CNS, are central players. The impact of aging on microglial responses is particularly relevant in the context of MS, as aging significantly modulates the immune profile of these cells. With advancing age, microglia exhibit a state of chronic activation, termed microglial priming, characterized by an exaggerated pro-inflammatory phenotype and impaired anti-inflammatory responses (Norden and Godbout, 2013;Henry et *al.*, 2009). This age-related primed state can exacerbate neuroinflammation in MS, as microglia contribute to the production of pro-inflammatory cytokines and oxidative stress, perpetuating demyelination and axonal damage (Koning et *al.*, 2009). Furthermore, age-related impairments in microglial phagocytic activity can hinder the clearance of myelin debris, potentially contributing to sustained autoimmune responses (Miron et *al.*, 2013) Understanding the intricate dynamics between aging, microglial responses, and autoimmune neuroinflammation is crucial for unraveling the complexities of MS pathogenesis in the aging population.

4. MECHANISMS UNDERLYING AGE-RELATED CHANGES IN GLIAL CELLS

4.1 Oxidative Stress and Mitochondrial Dysfunction

Oxidative stress results from an imbalance between the production of ROS and the cell's ability to neutralize them with antioxidants (Pizzino et *al.*, 2017). In aging glial cells, this imbalance is exacerbated as ROS production increases due to factors such as inflammation and decreased efficiency of the antioxidant defense systems (Norden and Godbout, 2013;Angelova and Abramov, 2018). Glial cells, particularly microglia, generate ROS as part of their immune responses, which, when uncontrolled, can lead to chronic inflammation and neuronal damage (Hickman et *al.*, 2013).

Mitochondrial dysfunction, another hallmark of aging glial cells (Srivastava, 2017), involves a decline in mitochondrial membrane potential, impaired adenosine triphosphate (ATP) production, and an increase in ROS production within these organelles (Guo et *al.*, 2013). These age-related mitochondrial alterations have far-reaching consequences, as glial cells rely heavily on mitochondria for energy generation and cellular processes. In astrocytes, for example, dysfunctional mitochondria can impair their ability to regulate neurotransmitter levels and maintain ion homeostasis, impacting neuronal function (Booth, Hirst, and Wade-Martins, 2017).

Furthermore, age-related mitochondrial DNA (mtDNA) mutations can accumulate over time, leading to impaired mitochondrial function. This mtDNA damage not only contributes to oxidative stress but also affects the overall quality control of mitochondria in glial cells, further exacerbating mitochondrial dysfunction (Gustafsson, Falkenberg, and Larsson, 2016).

These age-related changes in oxidative stress and mitochondrial function have broad implications for neurodegenerative diseases, where glial cell dysfunction is often a contributing factor.

4.2 Inflammation And Cytokine Signaling

The mechanisms underpinning age-related changes in glial cells, particularly inflammation and cytokine signaling, are multifaceted and have significant implications for brain function during aging. Microglia, the brain's resident immune cells, undergo a distinct transformation with age (Norden and Godbout, 2013). In this primed state, microglia display an elevated pro-inflammatory phenotype characterized by increased expression of pro-inflammatory cytokines such as interleukin-1β (IL-1β), interleukin-6 (IL-6), and tumor necrosis factor-α (TNF-α)(Wang et *al.*, 2015). This heightened microglial reactivity can lead to chronic neuroinflammation, further exacerbating neuronal dysfunction in the aging brain (Wendimu and Hooks, 2022). Additionally, aging astrocytes contribute to neuroinflammation by releasing cytokines such as interleukin-6 (IL-6) and transforming growth factor-β (TGF-β), disrupting the delicate balance of neuroinflammatory responses and negatively impacting synaptic plasticity (Booth, Hirst, and Wade-Martins, 2017).

It is crucial to consider the systemic aspects of inflammation in the aging brain. The phenomenon of "inflamm-aging" represents a systemic low-grade inflammatory state that occurs with aging and significantly contributes to chronic neuroinflammation within the brain. Peripheral pro-inflammatory cytokines, including IL-6 and C-reactive protein (CRP), are often elevated in older individuals and can permeate the blood-brain barrier, further fueling neuroinflammatory processes (Franceschi and Campisi, 2014).

These intricate mechanisms of inflammation and cytokine signaling in aging glial cells have profound implications, as they are closely associated with the pathogenesis of various neurodegenerative diseases, cognitive decline, and overall brain health during aging.

4.3 Accumulation Of Cellular Damage and Proteostasis

Glial cells, including microglia and astrocytes, progressively accumulate cellular damage due to factors such as oxidative stress, DNA mutations, and impaired DNA repair mechanisms (Booth, Hirst, and Wade-Martins, 2017;Hickman et *al.*, 2013). This accumulation of damage results in functional impairments, including reduced phagocytic activity in microglia, compromised support functions in astrocytes, and impaired maintenance of ion homeostasis and neurotransmitter clearance (Hickman et *al.*, 2013;Booth, Hirst, and Wade-Martins, 2017;Norden and Godbout, 2013).

Age-related deficits in proteostasis mechanisms, such as autophagy and the ubiquitin-proteasome system, hamper the cell's ability to clear misfolded proteins and damaged organelles. In microglia, impaired proteostasis can hinder the clearance of neurotoxic protein aggregates, contributing to neuroinflammation in age-related neurodegenerative diseases(Mawuenyega et *al.*, 2010;José Julio Rodríguez, Noristani, and Verkhratsky, 2012;Norden and Godbout, 2013). Dysregulated proteostasis also affects astrocyte function, leading to reduced support for neurons and disrupted neurotransmitter regulation (Hickman et *al.* 2013;Booth, Hirst, and Wade-Martins, 2017).

These intricate mechanisms of accumulating cellular damage and proteostasis disruption underscore the critical role of these processes in glial cell dysfunction during aging and their profound implications for brain health.

4.4 Epigenetic and Genetic Factors

Age-related changes in glial cells are intimately tied to epigenetic and genetic factors, which wield significant influence over the cellular landscape of the aging brain (Delgado-Morales et *al.*, 2017). Epigenetic modifications, including DNA methylation and histone modifications, exert dynamic control over gene expression in glial cells across the lifespan (Brunet and Berger, 2014). However, during aging, these epigenetic marks can become dysregulated, leading to alterations in gene expression patterns that contribute to glial dysfunction (Spiegel, Sewal, and Rapp, 2014;Hickman et *al.*, 2013).

Epigenetic changes in glial cells can affect various processes, such as the regulation of inflammatory responses, the maintenance of synaptic plasticity, and the clearance of toxic protein aggregates (Rasheed et *al.*, 2021). For instance, alterations in DNA methylation patterns can lead to the upregulation of pro-inflammatory genes in microglia, contributing to chronic neuroinflammation (Hickman et *al*,. 2013). Additionally, epigenetic modifications can influence the expression of genes involved in synaptic function, potentially contributing to cognitive decline in aging (Spiegel, Sewal, and Rapp, 2014).

In parallel, genetic factors also play a substantial role in shaping age-related changes in glial cells. Genetic variations in genes associated with glial function, immune responses, and neurodegenerative disease susceptibility can modulate the vulnerability of glial cells to age-related alterations (Booth, Hirst, and Wade-Martins, 2017;Norden and Godbout, 2013).

The intricate interplay between genetics, epigenetics, and aging in glial cell biology results in a complex landscape of molecular and cellular changes. These factors converge to impact neuroinflammation, synaptic function, and the overall susceptibility to age-related neurodegenerative diseases.

5. CONCLUSION

In summary, the link between glial cells and brain aging is intricate and multidimensional, influenced by a variety of processes and mechanisms. Glial cells, including microglia, astrocytes, and oligodendrocytes, play essential roles in maintaining brain homeostasis and function throughout life. However, during the aging process, these cells undergo significant structural and functional alterations, ultimately contributing to age-related changes in the brain.

Age-related changes in glial cells morphology, density, and function have significant consequences for synaptic health, neurotransmitter regulation, and susceptibility to oxidative stress. These alterations contribute to impaired synaptic function, disrupted neurotransmitter balance, and enhanced susceptibility to neuroinflammatory processes. Glial cells also play significant roles in age-related neurodegenerative diseases such as AD, PD, and MS, where aging amplifies their contributions to pathology. The mechanisms underlying these age-related changes involve oxidative stress, inflammation, proteostasis disruption, and the interplay of epigenetic and genetic factors. Understanding these mechanisms is crucial for developing interventions to promote healthy brain aging and counteract age-related neurological disorders.

REFERENCES

Acosta, C., Anderson, H. D., & Anderson, C. M. (2017). 'Astrocyte Dysfunction in Alzheimer Disease'. *Journal of Neuroscience Research*, *95*(12), 2430–2447. doi:10.1002/jnr.24075 PMID:28467650

Aikawa, T., Holm, M.-L., & Kanekiyo, T. (2018). 'ABCA7 and Pathogenic Pathways of Alzheimer's Disease'. *Brain Sciences*, *8*(2), 27. doi:10.3390/brainsci8020027 PMID:29401741

Angelova, P. R., & Abramov, A. Y. (2018). 'Role of Mitochondrial ROS in the Brain: From Physiology to Neurodegeneration'. *FEBS Letters*, *592*(5), 692–702. doi:10.1002/1873-3468.12964 PMID:29292494

Baik, S. H., Kang, S., Son, S. M., & Mook-Jung, I. (2016). 'Microglia Contributes to Plaque Growth by Cell Death Due to Uptake of Amyloid β in the Brain of Alzheimer's Disease Mouse Model'. *Glia*, *64*(12), 2274–2290. doi:10.1002/glia.23074 PMID:27658617

Bellaver, B., Souza, D. G., Souza, D. O., & Quincozes-Santos, A. (2017). 'Hippocampal Astrocyte Cultures from Adult and Aged Rats Reproduce Changes in Glial Functionality Observed in the Aging Brain'. *Molecular Neurobiology*, *54*(4), 2969–2985. doi:10.100712035-016-9880-8 PMID:27026184

Bolós, M., Llorens-Martín, M., Perea, J. R., Jurado-Arjona, J., Rábano, A., Hernández, F., & Avila, J. (2017). 'Absence of CX3CR1 Impairs the Internalization of Tau by Microglia'. *Molecular Neurodegeneration*, *12*(1), 59. doi:10.118613024-017-0200-1 PMID:28810892

Booth, H. D. E., Hirst, W. D., & Wade-Martins, R. (2017). 'The Role of Astrocyte Dysfunction in Parkinson's Disease Pathogenesis'. *Trends in Neurosciences*, *40*(6), 358–370. doi:10.1016/j.tins.2017.04.001 PMID:28527591

Brunet, A., & Berger, S. L. (2014). 'Epigenetics of Aging and Aging-Related Disease'. *The Journals of Gerontology. Series A, Biological Sciences and Medical Sciences*, *69*(Suppl 1), S17–S20. doi:10.1093/gerona/glu042 PMID:24833581

Buzsáki, G., Logothetis, N., & Singer, W. (2013). 'Scaling Brain Size, Keeping Timing: Evolutionary Preservation of Brain Rhythms'. *Neuron*, *80*(3), 751–764. doi:10.1016/j.neuron.2013.10.002 PMID:24183025

Chen, Y., Qin, C., Huang, J., Tang, X., Liu, C., Huang, K., Xu, J., Guo, G., Tong, A., & Zhou, L. (2020). 'The Role of Astrocytes in Oxidative Stress of Central Nervous System: A Mixed Blessing'. *Cell Proliferation*, *53*(3), e12781. doi:10.1111/cpr.12781 PMID:32035016

Chung, C. Y., Seo, H., Sonntag, K. C., Brooks, A., Lin, L., & Isacson, O. (2005). 'Cell Type-Specific Gene Expression of Midbrain Dopaminergic Neurons Reveals Molecules Involved in Their Vulnerability and Protection'. *Human Molecular Genetics*, *14*(13), 1709–1725. doi:10.1093/hmg/ddi178 PMID:15888489

Clarke, L. E., & Barres, B. A. (2013). 'Emerging Roles of Astrocytes in Neural Circuit Development'. *Nature Reviews. Neuroscience*, *14*(5), 311–321. doi:10.1038/nrn3484 PMID:23595014

Clarke, L. E., Liddelow, S. A., Chakraborty, C., Münch, A. E., Heiman, M., & Barres, B. A. (2018). 'Normal Aging Induces A1-like Astrocyte Reactivity'. *Proceedings of the National Academy of Sciences of the United States of America*, *115*(8), E1896–E1905. doi:10.1073/pnas.1800165115 PMID:29437957

Collier, T. J., Kanaan, N. M., & Kordower, J. H. (2011). 'Ageing as a Primary Risk Factor for Parkinson's Disease: Evidence from Studies of Non-Human Primates'. *Nature Reviews. Neuroscience*, *12*(6), 359–366. doi:10.1038/nrn3039 PMID:21587290

Delgado-Morales, R., Agís-Balboa, R. C., Esteller, M., & Berdasco, M. (2017). 'Epigenetic Mechanisms during Ageing and Neurogenesis as Novel Therapeutic Avenues in Human Brain Disorders'. *Clinical Epigenetics*, *9*(1), 67. doi:10.118613148-017-0365-z PMID:28670349

Filippo, M. D., Sarchielli, P., Picconi, B., & Calabresi, P. (2008). 'Neuroinflammation and Synaptic Plasticity: Theoretical Basis for a Novel, Immune-Centred, Therapeutic Approach to Neurological Disorders'. *Trends in Pharmacological Sciences*, *29*(8), 402–412. doi:10.1016/j.tips.2008.06.005 PMID:18617277

Fischer, M. T., Sharma, R., Lim, J. L., Haider, L., Frischer, J. M., Drexhage, J., Don Mahad, M. B., van Horssen, J., & Lassmann, H. (2012). 'NADPH Oxidase Expression in Active Multiple Sclerosis Lesions in Relation to Oxidative Tissue Damage and Mitochondrial Injury'. *Brain, 135*(3), 886–899. doi:10.1093/brain/aws012 PMID:22366799

Franceschi, C., & Campisi, J. (2014). 'Chronic Inflammation (Inflammaging) and Its Potential Contribution to Age-Associated Diseases'. *The Journals of Gerontology. Series A, Biological Sciences and Medical Sciences, 69*(Suppl_1), S4–S9. doi:10.1093/gerona/glu057 PMID:24833586

Franklin, R. J. M., & ffrench-Constant, C. (2008). 'Remyelination in the CNS: From Biology to Therapy'. *Nature Reviews. Neuroscience, 9*(11), 839–855. doi:10.1038/nrn2480 PMID:18931697

Gómez-Benito, M., Granado, N., García-Sanz, P., Michel, A., Dumoulin, M., & Moratalla, R. (2020). 'Modeling Parkinson's Disease With the Alpha-Synuclein Protein'. *Frontiers in Pharmacology, 11*, 356. https://www.frontiersin.org/articles/10.3389/fphar.2020.00356. doi:10.3389/fphar.2020.00356 PMID:32390826

Guo, C., Sun, L., Chen, X., & Zhang, D. (2013). 'Oxidative Stress, Mitochondrial Damage and Neurodegenerative Diseases'. *Neural Regeneration Research, 8*(21), 2003–2014. doi:10.3969/j.issn.1673-5374.2013.21.009 PMID:25206509

Gustafsson, C. M., Falkenberg, M., & Larsson, N.-G. (2016). 'Maintenance and Expression of Mammalian Mitochondrial DNA'. *Annual Review of Biochemistry, 85*(1), 133–160. doi:10.1146/annurev-biochem-060815-014402 PMID:27023847

Hansen, C., Angot, E., Bergström, A.-L., Steiner, J. A., Pieri, L., Paul, G., Outeiro, T. F., Melki, R., Kallunki, P., Fog, K., Li, J.-Y., & Brundin, P. (2011). 'α-Synuclein Propagates from Mouse Brain to Grafted Dopaminergic Neurons and Seeds Aggregation in Cultured Human Cells'. *The Journal of Clinical Investigation, 121*(2), 715–725. doi:10.1172/JCI43366 PMID:21245577

Hansen, D. V., Hanson, J. E., & Sheng, M. (2017). 'Microglia in Alzheimer's Disease'. *The Journal of Cell Biology, 217*(2), 459–472. doi:10.1083/jcb.201709069 PMID:29196460

Henneberger, C., Bard, L., Panatier, A., Reynolds, J. P., Kopach, O., Medvedev, N. I., Minge, D., Herde, M. K., Anders, S., Kraev, I., Heller, J. P., Rama, S., Zheng, K., Jensen, T. P., Sanchez-Romero, I., Jackson, C. J., Janovjak, H., Ottersen, O. P., Nagelhus, E. A., & Rusakov, D. A. (2020). 'LTP Induction Boosts Glutamate Spillover by Driving Withdrawal of Perisynaptic Astroglia'. *Neuron, 108*(5), 919–936.e11. doi:10.1016/j.neuron.2020.08.030 PMID:32976770

Henry, C. J., Huang, Y., Wynne, A. M., & Godbout, J. P. (2009). 'Peripheral Lipopolysaccharide (LPS) Challenge Promotes Microglial Hyperactivity in Aged Mice That Is Associated with Exaggerated Induction of Both pro-Inflammatory IL-1β and Anti-Inflammatory IL-10 Cytokines'. *Brain, Behavior, and Immunity, 23*(3), 309–317. doi:10.1016/j.bbi.2008.09.002 PMID:18814846

Hickman, S. E., Kingery, N. D., Ohsumi, T. K., Borowsky, M. L., Wang, L., Means, T. K., & El Khoury, J. (2013). 'The Microglial Sensome Revealed by Direct RNA Sequencing'. *Nature Neuroscience, 16*(12), 1896–1905. doi:10.1038/nn.3554 PMID:24162652

Hirsch, E. C., & Hunot, S. (2009). Neuroinflammation in Parkinson's Disease: A Target for Neuroprotection? *Lancet Neurology*, *8*(4), 382–397. doi:10.1016/S1474-4422(09)70062-6 PMID:19296921

Jin, R., Aidan, K. Y. C., Wu, J., & Tatia, M. C. L. (2022). 'Relationships between Inflammation and Age-Related Neurocognitive Changes'. *International Journal of Molecular Sciences*, *23*(20), 12573. doi:10.3390/ijms232012573 PMID:36293430

Kaiser, M., Maletzki, I., Hülsmann, S., Holtmann, B., Schulz-Schaeffer, W., Kirchhoff, F., Bähr, M., & Neusch, C. (2006). 'Progressive Loss of a Glial Potassium Channel (KCNJ10) in the Spinal Cord of the SOD1 (G93A) Transgenic Mouse Model of Amyotrophic Lateral Sclerosis'. *Journal of Neurochemistry*, *99*(3), 900–912. doi:10.1111/j.1471-4159.2006.04131.x PMID:16925593

Kim, Y., Park, J., & Choi, Y. K. (2019). 'The Role of Astrocytes in the Central Nervous System Focused on BK Channel and Heme Oxygenase Metabolites: A Review'. *Antioxidants*, *8*(5), 121. doi:10.3390/antiox8050121 PMID:31060341

Kimura, T., Kimura, A. K., Ren, M., Monteiro, V., Xu, Y., Berno, B., Schlame, M., & Epand, R. M. (2019). 'Plasmalogen Loss Caused by Remodeling Deficiency in Mitochondria'. *Life Science Alliance*, *2*(4), e201900348. doi:10.26508/lsa.201900348 PMID:31434794

Koizumi, S., Shigemoto-Mogami, Y., Nasu-Tada, K., Shinozaki, Y., Ohsawa, K., Tsuda, M., Joshi, B. V., Jacobson, K. A., Kohsaka, S., & Inoue, K. (2007). 'UDP Acting at P2Y6 Receptors Is a Mediator of Microglial Phagocytosis'. *Nature*, *446*(7139), 1091–1095. doi:10.1038/nature05704 PMID:17410128

Koning, N., Swaab, D. F., Hoek, R. M., & Huitinga, I. (2009). 'Distribution of the Immune Inhibitory Molecules CD200 and CD200R in the Normal Central Nervous System and Multiple Sclerosis Lesions Suggests Neuron-Glia and Glia-Glia Interactions'. *Journal of Neuropathology and Experimental Neurology*, *68*(2), 159–167. doi:10.1097/NEN.0b013e3181964113 PMID:19151626

Kotter, M. R., Li, W.-W., Zhao, C., & Franklin, R. J. M. (2006). 'Myelin Impairs CNS Remyelination by Inhibiting Oligodendrocyte Precursor Cell Differentiation'. *The Journal of Neuroscience : The Official Journal of the Society for Neuroscience*, *26*(1), 328–332. doi:10.1523/JNEUROSCI.2615-05.2006 PMID:16399703

Liu, H., Yang, Y., Xia, Y., Zhu, W., Leak, R., Wei, Z., Wang, J., & Xiaoming, H. (2017). 'Aging of Cerebral White Matter'. *Ageing Research Reviews*, *34*, 64–76. . doi:10.1016/j.arr.2016.11.006

Mao, X., Ou, M. T., Karuppagounder, S. S., Kam, T.-I., Yin, X., Xiong, Y., Ge, P., Umanah, G. E., Brahmachari, S., Shin, J.-H., Kang, H. C., Zhang, J., Xu, J., Chen, R., Park, H., Andrabi, S. A., Kang, S. U., Gonçalves, R. A., Liang, Y., & Dawson, T. M. (2016). 'Pathological α-Synuclein Transmission Initiated by Binding Lymphocyte-Activation Gene 3'. *Science*, *353*(6307), aah3374. doi:10.1126cience.aah3374 PMID:27708076

Maphis, N., Xu, G., Kokiko-Cochran, O. N., Jiang, S., Cardona, A., Ransohoff, R. M., Lamb, B. T., & Bhaskar, K. (2015). 'Reactive Microglia Drive Tau Pathology and Contribute to the Spreading of Pathological Tau in the Brain'. *Brain*, *138*(6), 1738–1755. doi:10.1093/brain/awv081 PMID:25833819

Matias, I., Morgado, J., & Flávia, C. A. G. (2019). 'Astrocyte Heterogeneity: Impact to Brain Aging and Disease'. *Frontiers in Aging Neuroscience*, *11*, 59. https://www.frontiersin.org/articles/10.3389/fnagi.2019.00059. doi:10.3389/fnagi.2019.00059 PMID:30941031

Mawuenyega, K. G., Sigurdson, W., Ovod, V., Munsell, L., Kasten, T., Morris, J. C., Yarasheski, K. E., & Bateman, R. J. (2010). 'Decreased Clearance of CNS β-Amyloid in Alzheimer's Disease'. *Science*, *330*(6012), 1774–1774. doi:10.1126cience.1197623 PMID:21148344

Miners, J. S., Baig, S., Palmer, J., Palmer, L. E., Kehoe, P. G., & Love, S. (2008). 'SYMPOSIUM: Clearance of Aβ from the Brain in Alzheimer's Disease: Aβ-Degrading Enzymes in Alzheimer's Disease'. *Brain Pathology (Zurich, Switzerland)*, *18*(2), 240–252. doi:10.1111/j.1750-3639.2008.00132.x PMID:18363935

Miron, V. E., Boyd, A., Zhao, J.-W., Yuen, T. J., Ruckh, J. M., Shadrach, J. L., van Wijngaarden, P., Wagers, A. J., Williams, A., Franklin, R. J. M., & ffrench-Constant, C. (2013). 'M2 Microglia and Macrophages Drive Oligodendrocyte Differentiation during CNS Remyelination'. *Nature Neuroscience*, *16*(9), 1211–1218. doi:10.1038/nn.3469 PMID:23872599

Muzio, L., Viotti, A., & Martino, G. (2021). 'Microglia in Neuroinflammation and Neurodegeneration: From Understanding to Therapy'. *Frontiers in Neuroscience*, *15*(September), 742065. doi:10.3389/fnins.2021.742065 PMID:34630027

Nave, K.-A., & Werner, H. B. (2014). 'Myelination of the Nervous System: Mechanisms and Functions'. *Annual Review of Cell and Developmental Biology*, *30*(1), 503–533. doi:10.1146/annurev-cellbio-100913-013101 PMID:25288117

Norden, D. M., & Godbout, J. P. (2013). 'Review: Microglia of the Aged Brain: Primed to Be Activated and Resistant to Regulation: Increased Microglial Reactivity with Age'. *Neuropathology and Applied Neurobiology*, *39*(1), 19–34. doi:10.1111/j.1365-2990.2012.01306.x PMID:23039106

Olabarria, M., Noristani, H. N., Verkhratsky, A., & Rodríguez, J. J. (2010). 'Concomitant Astroglial Atrophy and Astrogliosis in a Triple Transgenic Animal Model of Alzheimer's Disease'. *Glia*, *58*(7), 831–838. doi:10.1002/glia.20967 PMID:20140958

Palmer, A. L., & Ousman, S. S. (2018). 'Astrocytes and Aging'. *Frontiers in Aging Neuroscience*, *10*, 337. https://www.frontiersin.org/articles/10.3389/fnagi.2018.00337. doi:10.3389/fnagi.2018.00337 PMID:30416441

Pizzino, G., Irrera, N., Cucinotta, M., Pallio, G., Mannino, F., Arcoraci, V., Squadrito, F., Altavilla, D., & Bitto, A. (2017). 'Oxidative Stress: Harms and Benefits for Human Health'. *Oxidative Medicine and Cellular Longevity*, *8416763*, 1–13. doi:10.1155/2017/8416763 PMID:28819546

Popov, A., Brazhe, A., Denisov, P., Sutyagina, O., Li, L., Lazareva, N., Verkhratsky, A., & Semyanov, A. (2021). 'Astrocyte Dystrophy in Ageing Brain Parallels Impaired Synaptic Plasticity'. *Aging Cell*, *20*(3), e13334. doi:10.1111/acel.13334 PMID:33675569

Rahmani, F., Saghazadeh, A., Rahmani, M., Teixeira, A. L., Rezaei, N., Aghamollaii, V., & Ardebili, H. E. (2019). 'Plasma Levels of Brain-Derived Neurotrophic Factor in Patients with Parkinson Disease: A Systematic Review and Meta-Analysis'. *Brain Research, 1704*(February), 127–136. doi:10.1016/j.brainres.2018.10.006 PMID:30296429

Raj, D., Yin, Z., Breur, M., Doorduin, J., Holtman, I. R., Olah, M., Mantingh-Otter, I. J., Van Dam, D., De Deyn, P. P., den Dunnen, W., Eggen, B. J. L., Amor, S., & Boddeke, E. (2017). 'Increased White Matter Inflammation in Aging- and Alzheimer's Disease Brain'. *Frontiers in Molecular Neuroscience, 10*, 206. https://www.frontiersin.org/articles/10.3389/fnmol.2017.00206. doi:10.3389/fnmol.2017.00206 PMID:28713239

Rannikko, E. H., Weber, S. S., & Kahle, P. J. (2015). Exogenous α-Synuclein Induces Toll-like Receptor 4 Dependent Inflammatory Responses in Astrocytes. *BMC Neuroscience, 16*(1), 57. doi:10.118612868-015-0192-0 PMID:26346361

Ransohoff, R. M., & Hugh Perry, V. (2009). Microglial Physiology: Unique Stimuli, Specialized Responses. *Annual Review of Immunology, 27*(1), 119–145. doi:10.1146/annurev.immunol.021908.132528 PMID:19302036

Rasheed, M., Liang, J., Wang, C., Deng, Y., & Chen, Z. (2021). 'Epigenetic Regulation of Neuroinflammation in Parkinson's Disease'. *International Journal of Molecular Sciences, 22*(9), 4956. doi:10.3390/ijms22094956 PMID:34066949

Robillard, K. N., Lee, K. M., Chiu, K. B., & MacLean, A. G. (2016). 'Glial Cell Morphological and Density Changes through the Lifespan of Rhesus Macaques'. *Brain, Behavior, and Immunity*, Microglia. *Physiology & Behavior, 55*(July), 60–69. doi:10.1016/j.bbi.2016.01.006 PMID:26851132

Rodríguez, J. J., Noristani, H. N., & Verkhratsky, A. (2012). 'The Serotonergic System in Ageing and Alzheimer's Disease'. *Progress in Neurobiology, 99*(1), 15–41. doi:10.1016/j.pneurobio.2012.06.010 PMID:22766041

Rodríguez, J. J., Olabarria, M., Chvatal, A., & Verkhratsky, A. (2009). 'Astroglia in Dementia and Alzheimer's Disease'. *Cell Death and Differentiation, 16*(3), 378–385. doi:10.1038/cdd.2008.172 PMID:19057621

Salter, M. W., & Stevens, B. (2017). 'Microglia Emerge as Central Players in Brain Disease'. *Nature Medicine, 23*(9), 1018–1027. doi:10.1038/nm.4397 PMID:28886007

Sams, E. C. (2021). 'Oligodendrocytes in the Aging Brain'. *Neuronal Signaling, 5*(3), NS20210008. doi:10.1042/NS20210008 PMID:34290887

Scimemi, A. (2014). 'Structure, Function, and Plasticity of GABA Transporters'. *Frontiers in Cellular Neuroscience, 8*. https://www.frontiersin.org/articles/10.3389/fncel.2014.00161. doi:10.3389/fncel.2014.00161 PMID:24987330

Shen, S., Sandoval, J., Swiss, V. A., Li, J., Dupree, J., Franklin, R. J. M., & Casaccia-Bonnefil, P. (2008). 'Age-Dependent Epigenetic Control of Differentiation Inhibitors Is Critical for Remyelination Efficiency'. *Nature Neuroscience, 11*(9), 1024–1034. doi:10.1038/nn.2172 PMID:19160500

Shields, S. A., Gilson, J. M., Blakemore, W. F., & Franklin, R. J. M. (1999). 'Remyelination Occurs as Extensively but More Slowly in Old Rats Compared to Young Rats Following Gliotoxin-Induced CNS Demyelination'. *Glia, 28*(1), 77–83. doi:10.1002/(SICI)1098-1136(199910)28:1<77::AID-GLIA9>3.0.CO;2-F PMID:10498825

Si, Z.-Z., Zou, C.-J., Mei, X., Li, X.-F., Luo, H., Shen, Y., Hu, J., Li, X.-X., Wu, L., & Liu, Y. (2023). 'Targeting Neuroinflammation in Alzheimer's Disease: From Mechanisms to Clinical Applications'. *Neural Regeneration Research, 18*(4), 708. doi:10.4103/1673-5374.353484 PMID:36204826

Simpson, J. E., Ince, P. G., Haynes, L. J., Theaker, R., Gelsthorpe, C., Baxter, L., Forster, G., Lace, G. L., Shaw, P. J., Matthews, F. E., Savva, G. M., Brayne, C., & Wharton, S. B. (2010). 'Population Variation in Oxidative Stress and Astrocyte DNA Damage in Relation to Alzheimer-Type Pathology in the Ageing Brain'. *Neuropathology and Applied Neurobiology, 36*(1), 25–40. doi:10.1111/j.1365-2990.2009.01030.x PMID:19422529

Singh, D. (2022). 'Astrocytic and Microglial Cells as the Modulators of Neuroinflammation in Alzheimer's Disease'. *Journal of Neuroinflammation, 19*(1), 206. doi:10.118612974-022-02565-0 PMID:35978311

Sofroniew, M. V., & Vinters, H. V. (2010). 'Astrocytes: Biology and Pathology'. *Acta Neuropathologica, 119*(1), 7–35. doi:10.100700401-009-0619-8 PMID:20012068

Sorrentino, Z. (2019). α-Synuclein and Astrocytes: Tracing the Pathways from Homeostasis to Neurodegeneration in Lewy Body Disease. *Acta Neuropathologica, 138*(1), 1–21. . doi:10.1007/s00401-019-01977-2

Spiegel, A. M., Sewal, A. S., & Rapp, P. R. (2014). 'Epigenetic Contributions to Cognitive Aging: Disentangling Mindspan and Lifespan'. *Learning & Memory (Cold Spring Harbor, N.Y.), 21*(10), 569–574. doi:10.1101/lm.033506.113 PMID:25227252

Srivastava, S. (2017). 'The Mitochondrial Basis of Aging and Age-Related Disorders'. *Genes, 8*(12), 398. doi:10.3390/genes8120398 PMID:29257072

Stogsdill, J. A., Ramirez, J., Di Liu, Y. H. K., Baldwin, K. T., Enustun, E., Ejikeme, T., Ji, R.-R., & Eroglu, C. (2017). 'Astrocytic Neuroligins Control Astrocyte Morphogenesis and Synaptogenesis'. *Nature, 551*(7679), 192–197. doi:10.1038/nature24638 PMID:29120426

Suksuphew, S., & Noisa, P. (2015). 'Neural Stem Cells Could Serve as a Therapeutic Material for Age-Related Neurodegenerative Diseases'. *World Journal of Stem Cells, 7*(2), 502–511. doi:10.4252/wjsc.v7.i2.502 PMID:25815135

VanGuilder, H. D., Yan, H., Farley, J. A., Sonntag, W. E., & Freeman, W. M. (2010). 'Aging Alters the Expression of Neurotransmission-Regulating Proteins in the Hippocampal Synaptoproteome'. *Journal of Neurochemistry, 113*(6), 1577–1588. doi:10.1111/j.1471-4159.2010.06719.x PMID:20374424

Verkhratsky, A., Parpura, V., Rodriguez-Arellano, J. J., & Zorec, R. (2019). Astroglia in Alzheimer's Disease. In Neuroglia in Neurodegenerative Diseases. Singapore: Springer. doi:10.1007/978-981-13-9913-8_11

Vidal-Itriago, A., Radford, R. A. W., Aramideh, J. A., Maurel, C., Scherer, N. M., Don, E. K., Lee, A., Chung, R. S., Graeber, M. B., & Morsch, M. (2022). Microglia Morphophysiological Diversity and Its Implications for the CNS. *Frontiers in Immunology*, *13*(October), 997786. doi:10.3389/fimmu.2022.997786 PMID:36341385

von Bernhardi, R., Eugenín-von Bernhardi, L., & Eugenín, J. (2015). Microglial Cell Dysregulation in Brain Aging and Neurodegeneration. *Frontiers in Aging Neuroscience*, *7*(July), 124. doi:10.3389/fnagi.2015.00124 PMID:26257642

Wang, W.-Y., Tan, M.-S., Yu, J.-T., & Tan, L. (2015). 'Role of Pro-Inflammatory Cytokines Released from Microglia in Alzheimer's Disease'. *Annals of Translational Medicine*, *3*(10), 136. doi:10.3978/j.issn.2305-5839.2015.03.49 PMID:26207229

Wendimu, M. Y., & Hooks, S. B. (2022). 'Microglia Phenotypes in Aging and Neurodegenerative Diseases'. *Cells*, *11*(13), 2091. doi:10.3390/cells11132091 PMID:35805174

Zlokovic, B. V. (2005). 'Neurovascular Mechanisms of Alzheimer's Neurodegeneration'. *Trends in Neurosciences*, *28*(4), 202–208. doi:10.1016/j.tins.2005.02.001 PMID:15808355

Chapter 6
Microglia and Neuroinflammation

Himel Mondal
https://orcid.org/0000-0001-6950-5857
All India Institute of Medical Sciences, Deoghar, India

Shaikat Mondal
https://orcid.org/0000-0001-5027-4657
Raiganj Governmnet Medical College, West Bengal, India

ABSTRACT

Microglia, specialized immune cells within the central nervous system (CNS), are central to maintaining CNS health and protecting against damage and infections. However, their improper activation can contribute to the development and progression of various neurological diseases, including Alzheimer's, Parkinson's, multiple sclerosis, autism, and traumatic brain injury. Neuroinflammation, a complex process involving microglia and other glial cells, is considered a key factor in the pathogenesis of many neurological disorders. To comprehensively address this, this chapter introduces microglia's essential role in CNS homeostasis and sets the stage for understanding the concept of neuroinflammation, range of neurological and psychiatric disorders, cutting-edge imaging and genetic tools, the potential therapeutic strategies for modulating microglial function and targeting neuroinflammation.

1. INTRODUCTION

Microglia, often referred to as the resident macrophages of the CNS, serve as the first line of defense against damage and infection. Their multifaceted role encompasses not only surveillance and immune response but also the preservation of homeostasis within the intricate neural circuits. However, the pendulum of microglial function swings delicately, and dysregulation in their activation can tip the balance, contributing to the pathogenesis of various neurological diseases.

DOI: 10.4018/978-1-6684-9675-6.ch006

Neuroinflammation—a dynamic process involving the activation of glial cells, prominently microglia, and recognized as a significant player in the genesis of numerous neurological disorders. The relationship of microglial activation and neuroinflammation is of utmost importance in understanding the pathology of many diseases like Alzheimer's disease to Parkinson's disease, multiple sclerosis, and depression.

Advancements in technology have bestowed upon researchers a powerful arsenal of tools for studying microglial function in vivo. The latest imaging and genetic techniques afford us a glimpse into the dynamic world of microglia, shedding light on their interactions with other cell types within the CNS. By understanding the intricacies of microglial function and neuroinflammation, we may uncover new avenues for therapeutic intervention.

This introductory chapter sets the stage for a comprehensive exploration of the symbiotic relationship between microglia and neuroinflammation—a relationship that, when harmonious, preserves neurological health, but when disrupted, can contribute to the intricate tapestry of neurological diseases.

2. INTRODUCTION TO MICROGLIA AND NEUROINFLAMMATION

2.1. Microglia

The central nervous system's (CNS) resident immune cells, or microglia, are crucial for preserving brain homeostasis and controlling immunological responses in the brain. Here, we go into some basic facts concerning microglia, such as their anatomy, origin, and purposes (Nayak et al., 2014).

Myeloid progenitor cells, which start off in the yolk sac during embryonic development, give rise to microglia cells. Early in embryogenesis, these progenitor cells go to the CNS where they give birth to progenitors of the microglia. After then, the precursors settle into the brain tissue and develop into adult microglia (Yin et al., 2017). A microglia's structure in the CNS is depicted in Figure 1.

Microglia have a characteristic shape with a tiny cell body and many, ramified processes that are both extremely active and numerous. Microglia has a highly branching phenotype while at rest, which aids in their ability to keep an eye on their environment. Continuously scanning the brain microenvironment for indications of damage, infection, or aberrant cellular activity, the processes are active (Morris et al., 2013).

The functions of microglia is summarized in Figure 2 and described briefly below.

As immunological sentinels, microglia keeps an eye out for any changes in the CNS. They are naturally able to identify and react to a variety of signals, including aberrant protein aggregates, damage-associated molecular patterns, and pathogen-associated molecular patterns (PAMPs). When microglia receives these signals, immunological responses are triggered by pattern recognition receptors (PRRs) produced on their surface, such as toll-like receptors (TLRs) (Kigerl et al., 2014).

Highly phagocytic microglia effectively engulfs and removes infections, dead cells, and other detritus from the brain. They are essential for preserving tissue integrity and eliminating noxious materials to stop tissue damage.

By generating and releasing several immune mediators, including cytokines, chemokines, and growth factors, microglia control immunological responses in the CNS. The balance of these chemicals, which can have both pro- and anti-inflammatory actions, is essential for preserving brain homeostasis (Guo et al., 2022).

Figure 1. Microglia in central nervous system

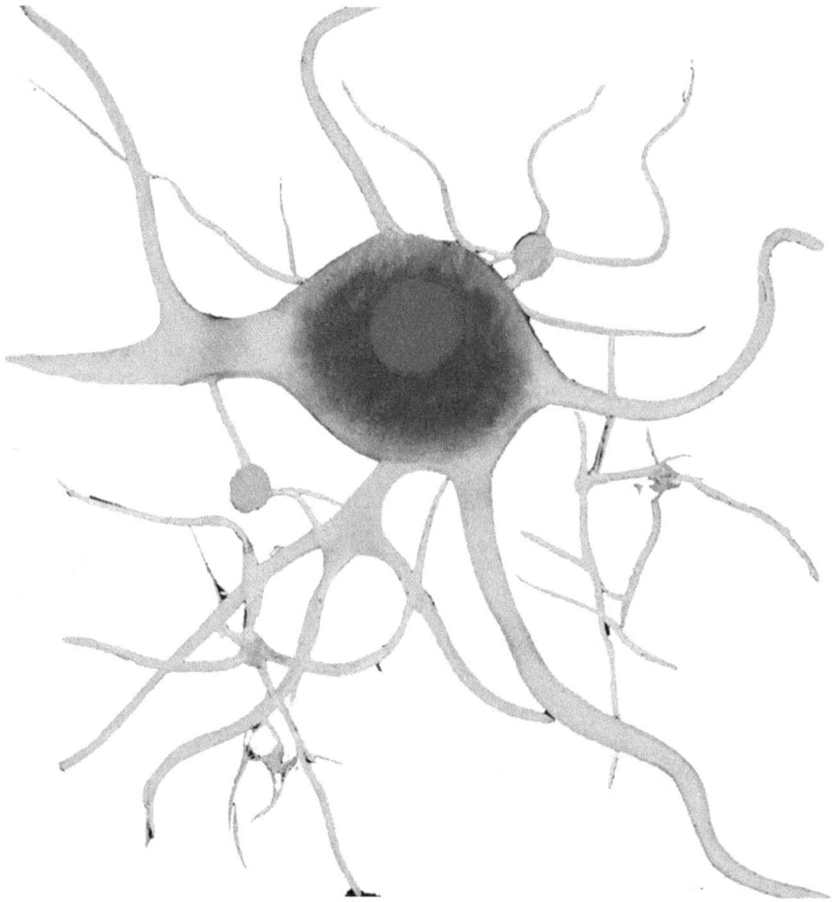

During development, microglia support synaptic plasticity and refinement by actively interacting with neurons. They take part in the removal of extra synapses, which enables the construction of precise neuronal circuits. Additionally, neurotropic substances that sustain neuronal survival and enhance synaptic function can be released by microglia (Konishi et al., 2019).

Neurotransmitter levels in the brain are regulated by microglia. They have the capacity to absorb and digest neurotransmitters, which has an impact on synaptic transmission and neuronal signalling.

2.2 Neuroinflammation

Inflammation that develops in the central nervous system (CNS) as a result of numerous traumas, such as infections, injuries, poisons, or chronic illnesses, is referred to as neuroinflammation. The process of neuroinflammation involves the migration of peripheral immune cells as well as the activation of resident immune cells in the CNS, predominantly microglia. The overall process of neuroinflammation is shown below (Michels et al., 2014).

Figure 2. Major functions of microglia

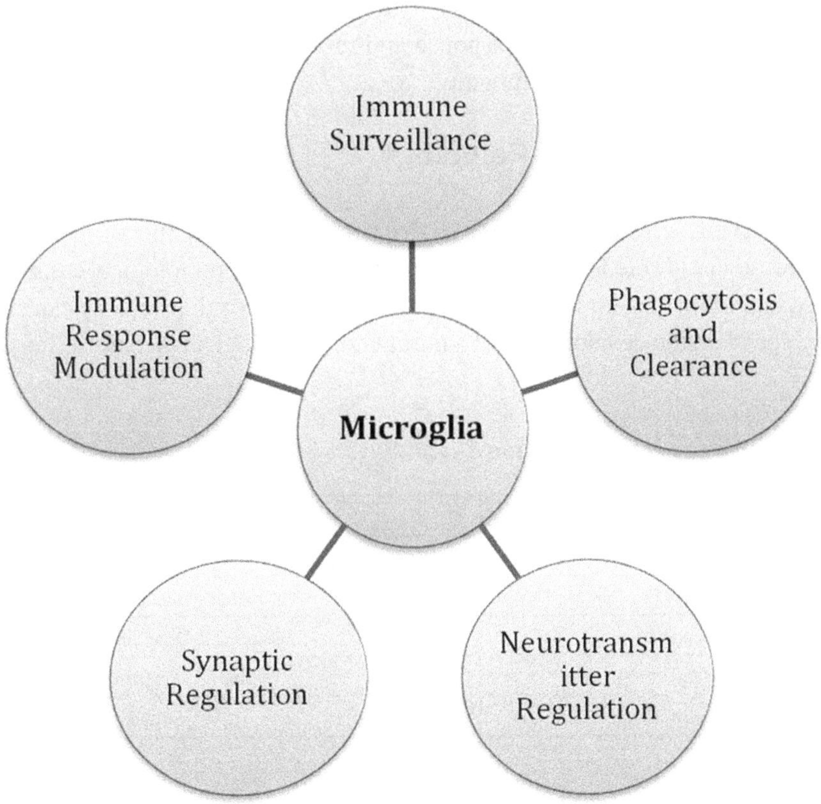

Infection by viruses, bacteria, or parasites, traumatic brain damage, stroke, neurodegenerative illnesses, autoimmune disorders, or exposure to toxins are just a few of the situations that can cause neuroinflammation. The CNS may become inflamed as a result of these occurrences.

The CNS's resident immune cells, called microglia, are the first to react to assaults to the brain. Microglia alters in morphology and function after activation. They adopt a more amoeboid form and produce cytokines, which are pro-inflammatory chemicals, including as tumor necrosis factor-alpha (TNF-alpha), interleukin-1 beta (IL-1 beta), and interleukin-6 (IL-6) (Laffer et al., 2019, Baskar et al., 2014).

Peripheral immune cells from the circulation, such as monocytes and T cells, can invade the CNS in more severe or persistent episodes of neuroinflammation. These immune cells add to the inflammatory response by being drawn to the site of inflammation by chemotactic signals (Dionisio-Santos et al., 2019).

The blood-brain barrier, which typically serves as a safeguard between the blood circulation and the CNS, can be weakened in neuroinflammatory diseases. This promotes neuroinflammation by allowing the passage of immune cells and inflammatory chemicals into the brain.

In a healthy environment, inflammation is a self-resolving process that is controlled by regulatory systems and anti-inflammatory chemicals. The clearance of neuroinflammation, however, may be compromised in several neurological conditions, resulting in persistent or chronic inflammation. It's crucial to remember that neuroinflammation is a complicated, dynamic process with both advantageous and disadvantageous impacts (Yong et al., 2019).

Chronic or severe neuroinflammation can cause tissue damage and the development of illness. Acute inflammation, on the other hand, is a protective reaction meant to get rid of toxic stimuli and promote tissue repair. To control the inflammatory response and restore CNS homeostasis, it is essential to understand the processes driving neuroinflammation.

2.3 Microglia in Physiology and Pathology

Within the central nervous system (CNS), microglia performs critical functions in both healthy and pathological circumstances [Table 1]. They have a dynamic role in immunological control, tissue healing, synaptic plasticity, and brain development. But deregulation of microglial activity can also cause neuroinflammation and speed up the development of a number of neurological diseases (Hanslik et al., 2021).

Table 1. Role of microglia in physiological and pathological conditions

Physiological	Pathological
Brain Development	Neuroinflammation
Synaptic Plasticity	Neurodegenerative Disorders
Neurotrophic Support	Traumatic Brain Injury (TBI)
Neurotrophic Support	Multiple Sclerosis (MS)
Immune Homeostasis	Stroke

2.4. Physiology

Microglia actively engages in shaping the growing brain throughout embryonic development. By consuming extra synapses through a process known as synaptic pruning, they aid in the improvement of brain circuits. This promotes proper brain growth by ensuring the creation of exact neural connections.

Synaptic plasticity control, which is crucial for memory, cognition, and learning, is aided by microglia. In response to environmental cues or neuronal activity, they can interact with synapses, modify synaptic transmission, and encourage synaptic remodelling.

Numerous neurotrophic and growth factors are released by microglia to support and promote the development and maintenance of neurons. These elements support the preservation of a wholesome neuronal environment and guarantee the correct operation of neural circuits.

By monitoring the brain microenvironment and reacting to any changes, microglia in a healthy CNS maintain immunological homeostasis. In order to maintain a delicate balance between pro-inflammatory and anti-inflammatory responses, they function as immune modulators. By doing so, the integrity of the tissue is maintained and severe inflammation is avoided.

2.5. Pathology

Neuroinflammation, a defining feature of several neurological illnesses, is mostly controlled by microglia. Chronic or severe neuroinflammation can cause neuronal injury and neurodegeneration, but acute inflammation is essential for tissue healing and pathogen removal (Muzio et al., 2021).

Neurodegenerative illnesses including Alzheimer's disease, Parkinson's disease, Huntington's disease, and amyotrophic lateral sclerosis (ALS) frequently exhibit microglial activation and neuroinflammation. In these conditions, activated microglia can create cytotoxic chemicals, accelerate the death of neurons, and contribute to the build-up of protein aggregates.

Microglia is activated after traumatic brain injury (TBI) and contributes to the neuroinflammatory response. Pro-inflammatory mediators are released when this happens, and these mediators may amplify secondary damage pathways, resulting in the loss of neuronal cells and functional deficits.

Microglia, together with invading immune cells, has a role in demyelination and neuroinflammation in Multiple Sclerosis (MS), an autoimmune disease of the central nervous system. Inflammatory chemicals that target myelin are released by activated microglia, which causes the myelin sheath that surrounds neurons to be destroyed.

Microglia gets activated and releases pro-inflammatory cytokines and reactive oxygen species in response to an ischemic insult during a stroke. Chronic or excessive microglial activation can worsen brain injury and impede functional recovery, but acute activation can aid in tissue regeneration and recovery.

The start and resolution of the inflammatory response in the CNS are both facilitated by microglia, which play a crucial role in neuroinflammation. They play a significant role in the pathophysiology of neurological illnesses due to their capacity to recognize and react to numerous inputs (Abe et al., 2020, Li et al., 2021). Potential approaches for creating therapeutic treatments include focusing on microglial activation and controlling their pro- and anti-inflammatory activities.

3. MICROGLIAL ACTIVATION AND NEUROINFLAMMATION

Numerous signalling channels and molecular interactions are involved in the intricate molecular processes that underlie microglial activation. There are a few typical essential actors and pathways involved in microglial activation, despite the fact that the precise processes might vary based on the particular stimuli and circumstances. Here are a few chemical processes that are known to activate microglia:

1) **Pattern Recognition Receptors (PRRs):** A variety of PRRs, including toll-like receptors (TLRs), NOD-like receptors (NLRs), and scavenger receptors, is expressed by microglia. The unique molecular patterns connected to infections (PAMPs) or cellular damage (DAMPs) is recognized by these receptors. When PAMPs or DAMPs bind to PRRs, intracellular signalling cascades are triggered, which activate the microglia (Pascual et al., 2021, Saijo et al., 2013).
2) **Nuclear Factor-kappa B (NF-κB) Pathway:** The transcription factor NF- κB is first transported into the nucleus upon activation of PRRs, notably TLRs. The expression of several pro-inflammatory genes, such as cytokines, chemokines, and adhesion molecules, is regulated by the NF- κB. A pro-inflammatory response in microglia must be induced by activating the NF- κB pathway (Frakes et al., 2014).

3) **Mitogen-Activated Protein Kinase (MAPK) Pathway:** Signals are sent from cell surface receptors to the nucleus through the MAPK pathway. When microglia are stimulated, MAPKs such p38 MAPK, c-Jun N-terminal kinase (JNK), and extracellular signal-regulated kinase (ERK) are activated. Pro-inflammatory chemicals are produced when the MAPK pathway is activated, which causes the phosphorylation and activation of transcription factors (Kim et al., 2004, Moens et al., 2013).
4) **Inflammasome Activation:** Multiprotein structures called inflammasomes detect danger signals within cells and start the production of pro-inflammatory cytokines including interleukin-1 (IL-1) and interleukin-18 (IL-18) (Hoffmann et al., 2020). The NLRP3 inflammasome is specifically linked to mediating neuroinflammatory responses in microglia. Caspase-1 is activated by the NLRP3 inflammasome, which then causes the maturation and release of pro-inflammatory cytokines (Freeman et al., 2016).
5) **Cytokine and Chemokine Signalling:** Different cytokines and chemokines that microglia can create and react to help in their activation and recruitment. Tumor necrosis factor-alpha (TNF-alpha), interleukin-1 (IL-1), and interferon-gamma (IFN-gamma) are a few examples of powerful microglia activators. Through these substances, autocrine and paracrine signalling can intensify the inflammatory response and maintain microglial activation (Chio et al., 2015).
6) **Purinergic signalling:** Purinergic receptors, such P2X and P2Y receptors are expressed by microglia and react to extracellular ATP produced during cellular injury or inflammation. Purinergic receptor activation results in intracellular calcium signaling, downstream signalling pathways, and cytokine release from the microglia (Calovi et al., 2019).
7) **Crosstalk with Other Cell Types:** Neurons, astrocytes, and endothelial cells are among the CNS cell types with which microglia interact. These interactions lead to the activation of the microglia by the release of soluble substances and direct cell-cell contact. An example is the production of neuronal-derived substances that activate microglia in response to neuronal damage or malfunction (Veremeyko et al., 2019).

4. THE ROLE OF MICROGLIA IN NEUROLOGICAL AND PSYCHIATRIC DISORDERS

Microglia, as the resident immune cells of the central nervous system (CNS), plays a critical role in both neurological and psychiatric disorders. Emerging research suggests that dysregulation of microglial function can contribute to the pathogenesis and progression of these disorders. Here, we will explore the role of microglia in various neurological and psychiatric conditions:

4. 1. Neurodegenerative Disorders

Neurodegenerative diseases including Alzheimer's disease (AD), Parkinson's disease (PD), Huntington's disease (HD), and amyotrophic lateral sclerosis (ALS) are characterized by microglial activation and neuroinflammation. In these circumstances, microglia take on an active and pro-inflammatory character in response to aberrant protein aggregates, neuronal injury, and cellular stress. When activated microglia continuously emits pro-inflammatory cytokines, reactive oxygen species, and other harmful chemicals, this can worsen neuroinflammation and cause neurodegeneration. The buildup of aberrant proteins, neuronal injury, and neuroinflammation are the hallmarks of these illnesses. Here, we give a quick overview of the illnesses and the distinctive functions of microglia.

1) Alzheimer's disease (AD): Microglia are directly linked to the development of -amyloid plaques and neurofibrillary tangles in the brain, which are the characteristic pathologies of AD. These protein clumps cause activation of microglia, which then trigger an immunological response. Microglia frequently has an altered phenotype in AD, which is defined by a persistent pro-inflammatory state. Tumor necrosis factor-alpha (TNF-alpha), interleukin-1 (IL-1), and other pro-inflammatory cytokines may be persistently released by activated microglia, which can result in neuroinflammation, synaptic dysfunction, and neuronal death. Additionally, microglia contributes to the phagocytosis necessary for amyloid clearance. However, in AD, the microglia's capacity to remove -amyloid is reduced, potentially accelerating the course of the illness (Hansen et al., 2018, Sarlus et al., 2017).
2) Parkinson's disease (PD): Motor symptoms in PD are caused by the specific loss of dopaminergic neurons in the substantia nigra. In response to neuronal damage and the buildup of -synuclein aggregates, which are indicative of PD, microglia is activated. Reactive oxygen species (ROS) and pro-inflammatory cytokines including interleukin-6 (IL-6) and interleukin-1 (IL-1) are released by activated microglia, which promote neuroinflammation and oxidative stress. Chronic microglial activation can hasten the death of dopaminergic neurons and accelerate the development of illness. Additionally, by releasing pro-inflammatory mediators, microglia can aid in the progression of -synuclein disease (Ho, 2019, Qian et al., 2008).
3) Huntington's disease (HD): In the striatum, medium spiny neurons exhibit progressive degeneration, which is a hallmark of HD. The mutant huntingtin protein clumps in the damaged neurons, activating microglia in response. Pro-inflammatory cytokines including TNF- and IL-1, as well as chemokines, are released by activated microglia, which promote neuroinflammation and neuronal dysfunction. Additionally, microglia are involved in the phagocytosis and removal of cellular waste, such as mutant huntingtin aggregates. The defective clearance systems in HD, however, can cause toxic proteins to build up and accelerate neurodegeneration (Palpagama et al., 2019, O 'Regan et al., 2021).
4) Amyotrophic Lateral Sclerosis (ALS): The gradual loss of motor neurons in the brainstem and spinal cord is a hallmark of ALS. In reaction to damage to the motor neuron and the buildup of protein aggregates like TDP-43 or SOD1, the microglia becomes activated. Neuroinflammation and neuronal death are facilitated by the production of pro-inflammatory cytokines, chemokines, and ROS by activated microglia. Additionally, microglia aid in tissue restoration by removing cellular waste. However, persistent microglial activation and the production of neurotoxic substances can aggravate neuroinflammation and advance the illness (Clarke et al., 2023).

4.2. Stroke

Microglia quickly gets activated after a stroke. Initially, the removal of cellular waste and promotion of tissue healing are favorable functions of microglia. However, too active microglia can cause secondary injury and neuroinflammation, which exacerbates neuronal death and functional impairment. They can also produce pro-inflammatory mediators at high levels (Qin et al., 2019).

Microglia responds to the damage quickly once ischemia begins. They alter in morphology, become active, and move towards the direction of the damage. Reactive oxygen species (ROS), cytokines, and chemokines are only a few of the signaling molecules released by activated microglia that support the inflammatory response.

In the ischemic region, microglia is essential for clearing away cellular debris and dead cells. They serve to clear away cell waste and promote tissue healing by phagocytosing injured neurons, astrocytes, and endothelial cells.

While microglia is helpful in clearing away debris, their activation after a stroke can also cause the release of cytokines and chemokines that promote inflammation. This inflammatory reaction can aggravate tissue injury and cause secondary harm. The recruitment and activation of immune cells can be further promoted by the production of pro-inflammatory chemicals such tumor necrosis factor-alpha (TNF-) and interleukin-1 beta (IL-1), which can exacerbate neuroinflammation (Taylor et al., 2013).

The blood-brain barrier (BBB), which controls the exchange of chemicals between the blood and the brain, is subject to manipulation by microglia. As a result of substances released by microglia during a stroke that might impair BBB integrity, immune cells and inflammatory chemicals are able to enter the brain parenchyma and cause neuroinflammation.

Microglia has a role in tissue repair and resolution after the initial inflammatory reaction. They exude trophic and anti-inflammatory substances that assist tissue remodeling and equilibrium restoration. In the ischemic region, microglia can encourage angiogenesis, neurogenesis, and synaptogenesis, speeding up the healing and repair processes (Dong et al., 2021).

The function of microglia goes beyond the acute stage of stroke. Microglia may continue to display an active phenotype during the chronic period and aid in the development of post-stroke neuroinflammation. This ongoing inflammation may hinder neuroplasticity and be a factor in stroke survivors' long-term cognitive and functional problems.

4.3 Multiple Sclerosis

The central nervous system (CNS) is affected by persistent inflammation, demyelination (loss of myelin sheaths), and neurodegeneration in multiple sclerosis (MS), an inflammatory condition. Microglia and invading immune cells aid in the degeneration of myelin sheaths in MS, an autoimmune illness marked by demyelination and neuroinflammation. Initiating and maintaining the immune response in the CNS, microglia function as antigen-presenting cells.

In the central nervous system (CNS), microglia serves as antigen-presenting cells (APCs), exposing autoreactive T lymphocytes to self-antigens. Microglia and T-cells are both activated in MS as a result of activated T-cells that have crossed the blood-brain barrier and come into contact with them. T-cells are exposed to myelin-derived antigens by microglia, which activates the immune system's defense against myelin (Zia et al., 2020).

Pro-inflammatory cytokines such interleukin-1 (IL-1), interleukin-6 (IL-6), tumor necrosis factor-alpha (TNF-), and interferon-gamma (IFN-) are released by activated microglia in MS. These cytokines aid in the attraction and activation of immune cells, such as macrophages and T-cells, which promotes demyelination and inflammation even further (Airas et al., 2022).

In MS, myelin removal involves microglia in two different ways. One way that microglia can assist in the removal of myelin breakdown products is by phagocytosing myelin fragments and injured cells. Scavenger receptors and complement receptors, for example, mediate this phagocytic action. On the other hand, persistent microglial activation might impede the removal procedure, aid in the build-up of myelin debris, and worsen neuroinflammation (Guerrero et al., 2020).

Multiple neurotoxic chemicals, including reactive oxygen species (ROS), nitric oxide (NO), and glutamate, are released by activated microglia in MS. These chemicals have the ability to cause neurodegeneration by causing oxidative stress, harming axons, and impairing neuronal function.

It may have an impact on MS's BBB's credibility. Microglia can contribute to the breakdown of tight connections between endothelial cells and increase immune cell infiltration into the CNS by releasing cytokines and other mediators (Luo et al., 2017).

Remyelination, the process of repairing damaged myelin sheaths, involves microglia. They expel substances that might either encourage or prevent remyelination. In order to facilitate remyelination, activated microglia emits substances that can draw in oligodendrocyte precursor cells (OPCs) and encourage their development into mature oligodendrocytes. However, chronic activation of microglia and the release of inhibitory factors can impede the remyelination process.

4.4. Neuropsychiatric Disorders

Microglia may have a role in the pathogenesis of neuropsychiatric illnesses such major depressive disorder (MDD), schizophrenia, bipolar disorder, and autism spectrum disorders, according to mounting evidence. Neuroinflammation, synaptic dysfunction, and altered brain connections, which are thought to play a role in the pathophysiology of various illnesses, can all be attributed to dysregulated microglial activity. Here are the illnesses.

1) Major Depressive Disorder (MDD): In postmortem brain analyses of people with MDD, aberrant microglia has been seen. Alterations to the microglia's shape, gene expression patterns, and immunological signaling are among these abnormalities. To promote neuroinflammation in MDD, activated microglia produce pro-inflammatory cytokines such interleukin-1 beta (IL-1 beta), interleukin-6 (IL-6), and tumor necrosis factor-alpha (TNF-alpha). Pro-inflammatory cytokines can interfere with synaptic function, alter neuroplasticity, and contribute to the pathogenesis of MDD (Wang et al., 2022, Jia et al., 2021).
2) Schizophrenia: Neuroinflammation and microglial activation have both been linked to the pathophysiology of schizophrenia. In the brains of people with schizophrenia, postmortem investigations have found increased microglial density and raised levels of pro-inflammatory cytokines. Synaptic dysfunction, disrupted neuronal connections, and decreased neurotransmission might result from dysregulated microglial activity and excessive release of pro-inflammatory chemicals, which can contribute to the cognitive and behavioral symptoms of schizophrenia (Gober et al., 2022, Snijders et al., 2021).
3) Bipolar Disorder: Bipolar illness has also been linked to immunological dysregulation and microglial activation. Bipolar illness patients' brains have altered immune-related gene expression and increased microglial density, according to postmortem research. Bipolar disorder's mood dysregulation and cognitive deficits may be influenced by abnormal microglial activity, which can also affect neural plasticity and hamper neurogenesis (Jakobsson et al., 2015, Sneeboer et al., 2019).
4) Autism Spectrum Disorders (ASD): ASD is a class of neurodevelopmental disorders distinguished by difficulties with social interaction and repetitive actions. In postmortem brain investigations of people with ASD, aberrant microglia, including altered shape and gene expression, has been

found. Neuroinflammation, synaptic dysfunction, and altered brain connections are hypothesized to be involved in the pathophysiology of ASD and can all be attributed to dysregulated microglial activity (Koyama et al., 2015, Rodriguez et al., 2011).

4.5. Traumatic Brain Injury (TBI)

Microglia undergoes activation and release pro-inflammatory cytokines and chemokines in response to TBI. The secondary damage cascade that results in neuroinflammation, neuronal loss, and long-term cognitive deficits can be aided by microglial activation.

Microglia responds quickly to an injury after a TBI by activating. They undergo morphological alterations and move in the direction of the wound. Numerous cues, including as the discharge of damage-associated molecular patterns (DAMPs) and inflammatory cytokines, cause this activation. Microglias that have been activated start an immunological response that can be both helpful and harmful to the healing process (Donat et al., 2017).

TBI triggers the production of pro-inflammatory cytokines from activated microglia, including interleukin-1 beta (IL-1), interleukin-6 (IL-6), and tumor necrosis factor-alpha (TNF-alpha). These cytokines help to attract and activate immune cells like neutrophils and macrophages, which exacerbates the inflammatory response. Although inflammation is an essential component of the body's early response, prolonged or severe inflammation can exacerbate neurological results and cause subsequent harm.

At the site of damage, microglia is essential for phagocytosing cellular debris and dead cells. They aid in clearing out injured neurons, axonal fragments, and byproducts of myelin breakdown. Clearing the damaged tissue and establishing an environment that supports healing and regeneration are both made possible by this phagocytic action (Todd et al., 2021).

The blood-brain barrier (BBB), which controls the exchange of molecules between the blood and the brain, may become compromised as a result of TBI. Following TBI, microglia actively contributes to regulating BBB integrity. They expel substances that might either increase or decrease BBB permeability. Modulation of the BBB can affect immune cell and molecule infiltration, which can further affect the inflammatory response and tissue healing procedures.

After a TBI, microglia plays a part in controlling neuroplasticity and synaptic remodeling. These processes are either promoted or inhibited by the factors they release. On the one hand, neuroplasticity and improved synaptic connection may be supported by substances released by microglia, such as brain-derived neurotrophic factor (BDNF) and insulin-like growth factor-1 (IGF-1). On the other hand, sustained microglial activation and the release of neurotoxic molecules can impede neuroplasticity and disrupt synaptic function (Liu et al., 2023, Saba et al., 2021).

5. IMAGING AND GENETIC TOOLS FOR STUDYING MICROGLIA

Advancements in imaging and genetic tools have significantly contributed to our understanding of microglial function in the central nervous system (CNS). These tools allow researchers to study microglia in vivo and investigate their dynamic interactions with other cell types. Here we discuss some of the latest imaging and genetic tools used to study microglia.

1) Two-Photon Microscopy: Two-photon microscopy is a powerful imaging technique that enables high-resolution imaging of microglia in live animals. It utilizes long-wavelength light to penetrate deep into the brain tissue, allowing researchers to visualize microglial morphology, dynamics, and interactions with other cells in real time. Two-photon microscopy has provided valuable insights into microglial motility, surveillance, and response to neuronal activity (Kondo et al., 2013).
2) Calcium Imaging: Calcium imaging is a technique used to monitor changes in intracellular calcium levels, which serve as an indicator of neuronal activity and microglial responses. Genetically encoded calcium indicators, such as GCaMP, can be selectively expressed in microglia using transgenic or viral approaches. This allows researchers to visualize and quantify microglial calcium dynamics in response to various stimuli or pathological conditions, providing insights into microglial activation and functional states (Tvrdik et al., 2017).
3) PET Imaging: Positron emission tomography (PET) imaging can provide information about microglial activation and inflammation in the CNS. Radioligands that bind to specific microglial markers, such as translocator protein (TSPO), can be used to track microglial activation in vivo. PET imaging allows non-invasive monitoring of microglial activity and has been applied in various neurological disorders, including neurodegenerative diseases and traumatic brain injury (Tronel et al., 2017).
4) Genetic Manipulation Techniques: Genetic tools are invaluable for understanding microglial function and the interactions between microglia and other cell types. Conditional knockout or transgenic approaches can be used to selectively modulate gene expression in microglia, allowing researchers to investigate the roles of specific genes or signaling pathways in microglial function. Moreover, cell-type-specific promoters, such as CX3CR1, can be utilized to drive expression of fluorescent proteins or genetically encoded sensors in microglia, enabling their visualization and tracking in vivo (Wieghofer et al., 2016).
5) Single-Cell RNA Sequencing: Single-cell RNA sequencing (scRNA-seq) has revolutionized our understanding of cellular heterogeneity and gene expression profiles in the CNS, including microglia. By analyzing gene expression at the single-cell level, scRNA-seq enables researchers to identify distinct microglial subsets and understand their functional states in different physiological and pathological conditions. This technique provides insights into the transcriptional changes occurring in microglia during development, aging, and disease (Pettas et al., 2022, Swanson et al., 2020).
6) CRISPR-Cas9 Genome Editing: The advent of CRISPR-Cas9 technology has facilitated precise genome editing in microglia. Researchers can use this technique to generate specific genetic modifications in microglia to study the impact on microglial function and their interactions with other cells. CRISPR-Cas9 allows for the introduction of genetic mutations, knockouts, or knock-ins, enabling the investigation of gene functions in microglia and their role in CNS disorders (Meneghini et al., 2021, Selvakumar et al., 2019).

6. THERAPEUTIC STRATEGIES FOR MODULATING MICROGLIAL FUNCTION

It has become clear that altering neuroinflammation and microglial activity is a possible treatment approach for a variety of neurological and psychiatric illnesses. The course of the illness and neurodegeneration are facilitated by abnormal microglial activation and neuroinflammation, which are frequent symptoms in many of these disorders. We now go through several therapy methods that concentrate on neuroinflammation and microglial activity.

Inflammatory inhibitors: Non-steroidal anti-inflammatory medicines (NSAIDs), corticosteroids, and other anti-inflammatory medications have all been studied for their ability to modify microglial activation and lessen neuroinflammation. These medications prevent the synthesis or activation of pro-inflammatory cytokines, which reduces inflammation. It is yet unclear, though, how well these drugs work to modify microglial activity and enhance therapeutic outcomes in neurological illnesses (Ozben et al., 2019).

Modulation of Microglial Activation States: Microglia can exhibit pro-inflammatory (M1-like) or anti-inflammatory (M2-like) phenotypes depending on their level of activation. To lessen neuroinflammation, microglia is thought to change their phenotype to one that is anti-inflammatory. In preclinical models of neurological diseases, a number of substances, including minocycline and PPAR-gamma agonists, have been found to support an anti-inflammatory microglial phenotype and have neuroprotective benefits (Orihuela et al., 2016, Guo et al., 2022).

Targeting Microglial Signaling Pathways: A number of signalling pathways, including the nuclear factor kappa B (NF-B) pathway, the mitogen-activated protein kinase (MAPK) pathway, and the Janus kinase/signal transducer and activator of transcription (JAK/STAT) pathway, regulate microglial activation and neuroinflammation. Using certain inhibitors or activators to modify these signalling pathways offers promise for controlling microglial activity and reducing excessive neuroinflammation (Popiolek-Barczyk et al., 2016, Guo et al., 2016).

Immunomodulatory Therapies: Immunomodulatory treatments have the ability to control neuroinflammation and microglial activity. To reduce neuroinflammation, monoclonal antibodies that target certain cytokines, such TNF- or IL-6, have been studied in neurodegenerative diseases. In order to modify immune responses and maybe affect microglial function, several strategies are being investigated, such as immune checkpoint inhibitors and adoptive cell therapy (Rashid et al., 2018, Muzio et al., 2021).

Neurotrophic elements Molecules known as neurotrophic factors help neurons develop, survive, and operate. They have demonstrated potential for modifying microglial activity and fostering neuroprotection. For example, BDNF (brain-derived neurotrophic factor) and GDNF (glial cell-derived neurotrophic factor) can affect microglial activation and support neuronal survival. Exogenous neurotrophic factor administration or methods to boost endogenous synthesis may have therapeutic advantages (Kimuraet al., 2016).

Targeting the Gut-Brain Axis: New research indicates that microglial activity and neuroinflammation may be influenced by the microbiota in the gut. Probiotics, prebiotics, and fecal microbiota transplantation are among methods that have showed promise in reducing neuroinflammation and enhancing neurological outcomes in a variety of illnesses. An innovative strategy to affect microglial activity indirectly is to modify the gut-brain axis (Liang et al., 2018).

The development of tailored therapies is difficult due to the intricacy of microglial biology and the variety of functions they play in various disease scenarios. These approaches show potential for the creation of novel treatments for neurological and psychiatric illnesses with more study and improvements in our knowledge of microglial activity.

7. CONCLUSION

In the intricate exploration of microglia and neuroinflammation, the pivotal role of microglia in maintaining CNS homeostasis and their intricate involvement in neurological diseases has been unveiled. The delicate equilibrium maintained by microglia and neuroinflammation, when disturbed, becomes a contributing factor to the pathogenesis of disorders such as Alzheimer's, Parkinson's, and multiple sclerosis.

Commencing with an introduction to the immune function of microglia in the CNS and the concept of neuroinflammation, subsequent sections delved into the molecular mechanisms governing microglial activation and the intricate connections to neurological and psychiatric disorders. Advancements in imaging and genetic tools disclosed dynamic interactions within the CNS, providing valuable insights.

The therapeutic potential of modulating microglial function to address neuroinflammation emerged as a promising avenue for neurological and psychiatric disorders. As we conclude, this chapter not only serves as an exploration of microglia and neuroinflammation but also sets the stage for future research, highlighting the ongoing quest to unravel the complexities within the microcosm of the central nervous system.

8. FUTURE DIRECTIONS

Microglia and neuroinflammation research is a fast developing topic with numerous open issues and promising directions for future investigation.

A great amount of variety is present in microglia, and different subpopulations may play various functional roles in both health and illness. These subpopulations' precise roles and possible treatment targets will become clearer with more characterization and knowledge of them (De Schepper et al., 2021).

Microglial activation is a dynamic process that might change as the illness progresses through various phases. Our understanding of disease processes will improve as we better comprehend the temporal dynamics of microglial activation and the elements that trigger their transitions between pro-inflammatory and anti-inflammatory states. This will also help us create targeted therapies (Zia et al., 2020).

Neurons, astrocytes, and oligodendrocytes are just a few of the CNS cell types that microglia interact with on a regular basis. Understanding how microglia contributes to both healthy and unhealthy neuronal function will be more fully understood by looking at the intricate crosstalk and signaling pathways involved in these interactions.

The clinical translation and optimization of treatment approaches that target neuroinflammation and microglial activity are still difficult tasks. These tactics need to be improved, therapeutic targets need to be found, and deliveries mechanisms need to be developed that selectively target microglia while limiting off-target effects. Translating the knowledge on microglia and neuroinflammation into viable therapeutics requires bridging the gap between preclinical investigations and clinical trials (McKee et al., 2023).

We can obtain a deeper comprehension of microglia's function in neuroinflammation and their potential as therapeutic targets by addressing these open issues and researching these areas. The creation of innovative therapies that use microglial activity to support brain health and treat a variety of CNS illnesses will be made possible by this information.

REFERENCES

Abe, N., Nishihara, T., Yorozuya, T., & Tanaka, J. (2020). Microglia and Macrophages in the Pathological Central and Peripheral Nervous Systems. *Cells*, *9*(9), 2132. doi:10.3390/cells9092132 PMID:32967118

Airas, L., & Yong, V. W. (2022). Microglia in multiple sclerosis - pathogenesis and imaging. *Current Opinion in Neurology*, *35*(3), 299–306. doi:10.1097/WCO.0000000000001045 PMID:35674072

Baskar Jesudasan, S. J., Todd, K. G., & Winship, I. R. (2014). Reduced inflammatory phenotype in microglia derived from neonatal rat spinal cord versus brain. *PLoS One*, *9*(6), e99443. doi:10.1371/journal.pone.0099443 PMID:24914808

Calovi, S., Mut-Arbona, P., & Sperlágh, B. (2019). Microglia and the Purinergic Signaling System. *Neuroscience*, *405*, 137–147. doi:10.1016/j.neuroscience.2018.12.021 PMID:30582977

Chio, C. C., Lin, M. T., & Chang, C. P. (2015). Microglial activation as a compelling target for treating acute traumatic brain injury. *Current Medicinal Chemistry*, *22*(6), 759–770. doi:10.2174/0929867321666141106124657 PMID:25386818

Clarke, B. E., & Patani, R. (2020). The microglial component of amyotrophic lateral sclerosis. *Brain*, *143*(12), 3526–3539. doi:10.1093/brain/awaa309 PMID:33427296

De Schepper, S., Crowley, G., & Hong, S. (2021). Understanding microglial diversity and implications for neuronal function in health and disease. *Developmental Neurobiology*, *81*(5), 507–523. doi:10.1002/dneu.22777 PMID:32757416

Dionisio-Santos, D. A., Olschowka, J. A., & O'Banion, M. K. (2019). Exploiting microglial and peripheral immune cell crosstalk to treat Alzheimer's disease. *Journal of Neuroinflammation*, *16*(1), 74. doi:10.118612974-019-1453-0 PMID:30953557

Donat, C. K., Scott, G., Gentleman, S. M., & Sastre, M. (2017). Microglial Activation in Traumatic Brain Injury. *Frontiers in Aging Neuroscience*, *9*, 208. doi:10.3389/fnagi.2017.00208 PMID:28701948

Dong, R., Huang, R., Wang, J., Liu, H., & Xu, Z. (2021). Effects of Microglial Activation and Polarization on Brain Injury After Stroke. *Frontiers in Neurology*, *12*, 620948. doi:10.3389/fneur.2021.620948 PMID:34276530

Frakes, A. E., Ferraiuolo, L., Haidet-Phillips, A. M., Schmelzer, L., Braun, L., Miranda, C. J., Ladner, K. J., Bevan, A. K., Foust, K. D., Godbout, J. P., Popovich, P. G., Guttridge, D. C., & Kaspar, B. K. (2014). Microglia induce motor neuron death via the classical NF-κB pathway in amyotrophic lateral sclerosis. *Neuron*, *81*(5), 1009–1023. doi:10.1016/j.neuron.2014.01.013 PMID:24607225

Freeman, L. C., & Ting, J. P. (2016). The pathogenic role of the inflammasome in neurodegenerative diseases. *Journal of Neurochemistry*, *136*(S1, Suppl 1), 29–38. doi:10.1111/jnc.13217 PMID:26119245

Gober, R., Ardalan, M., Shiadeh, S. M. J., Duque, L., Garamszegi, S. P., Ascona, M., Barreda, A., Sun, X., Mallard, C., & Vontell, R. T. (2022). Microglia activation in postmortem brains with schizophrenia demonstrates distinct morphological changes between brain regions. *Brain Pathology (Zurich, Switzerland)*, *32*(1), e13003. doi:10.1111/bpa.13003 PMID:34297453

Guerrero, B. L., & Sicotte, N. L. (2020). Microglia in Multiple Sclerosis: Friend or Foe? *Frontiers in Immunology*, *11*, 374. doi:10.3389/fimmu.2020.00374 PMID:32265902

Guo, C., Yang, L., Wan, C. X., Xia, Y. Z., Zhang, C., Chen, M. H., Wang, Z. D., Li, Z. R., Li, X. M., Geng, Y. D., & Kong, L. Y. (2016). Anti-neuroinflammatory effect of Sophoraflavanone G from Sophora alopecuroides in LPS-activated BV2 microglia by MAPK, JAK/STAT and Nrf2/HO-1 signaling pathways. *Phytomedicine*, *23*(13), 1629–1637. doi:10.1016/j.phymed.2016.10.007 PMID:27823627

Guo, L., Choi, S., Bikkannavar, P., & Cordeiro, M. F. (2022, March 17). Microglia: Key Players in Retinal Ageing and Neurodegeneration. *Frontiers in Cellular Neuroscience*, *16*, 804782. doi:10.3389/fncel.2022.804782 PMID:35370560

Guo, S., Wang, H., & Yin, Y. (2022). Microglia Polarization From M1 to M2 in Neurodegenerative Diseases. *Frontiers in Aging Neuroscience*, *14*, 815347. doi:10.3389/fnagi.2022.815347 PMID:35250543

Hansen, D. V., Hanson, J. E., & Sheng, M. (2018). Microglia in Alzheimer's disease. *The Journal of Cell Biology*, *217*(2), 459–472. doi:10.1083/jcb.201709069 PMID:29196460

Hanslik, K. L., Marino, K. M., & Ulland, T. K. (2021). Modulation of Glial Function in Health, Aging, and Neurodegenerative Disease. *Frontiers in Cellular Neuroscience*, *15*, 718324. doi:10.3389/fncel.2021.718324 PMID:34531726

Ho, M. S. (2019). Microglia in Parkinson's Disease. *Advances in Experimental Medicine and Biology*, *1175*, 335–353. doi:10.1007/978-981-13-9913-8_13 PMID:31583594

Hoffmann, S., & Beyer, C. (2020). A Fatal Alliance between Microglia, Inflammasomes, and Central Pain. *International Journal of Molecular Sciences*, *21*(11), 3764. doi:10.3390/ijms21113764 PMID:32466593

Jakobsson, J., Bjerke, M., Sahebi, S., Isgren, A., Ekman, C. J., Sellgren, C., Olsson, B., Zetterberg, H., Blennow, K., Pålsson, E., & Landén, M. (2015). Monocyte and microglial activation in patients with mood-stabilized bipolar disorder. Journal of psychiatry & neuroscience. *Journal of Psychiatry & Neuroscience*, *40*(4), 250–258. doi:10.1503/jpn.140183 PMID:25768030

Jia, X., Gao, Z., & Hu, H. (2021). Microglia in depression: Current perspectives. *Science China. Life Sciences*, *64*(6), 911–925. doi:10.100711427-020-1815-6 PMID:33068286

Kigerl, K. A., de Rivero Vaccari, J. P., Dietrich, W. D., Popovich, P. G., & Keane, R. W. (2014). Pattern recognition receptors and central nervous system repair. *Experimental Neurology*, *258*, 5–16. doi:10.1016/j.expneurol.2014.01.001 PMID:25017883

Kim, S. H., Smith, C. J., & Van Eldik, L. J. (2004). Importance of MAPK pathways for microglial pro-inflammatory cytokine IL-1 beta production. *Neurobiology of Aging*, *25*(4), 431–439. doi:10.1016/S0197-4580(03)00126-X PMID:15013563

Kimura, A., Namekata, K., Guo, X., Harada, C., & Harada, T. (2016). Neuroprotection, Growth Factors and BDNF-TrkB Signalling in Retinal Degeneration. *International Journal of Molecular Sciences*, *17*(9), 1584. doi:10.3390/ijms17091584 PMID:27657046

Kondo, S., & Okabe, S. (2013). In vivo two-photon microscopy of microglia. *Methods in Molecular Biology (Clifton, N.J.)*, *1041*, 319–335. doi:10.1007/978-1-62703-520-0_28 PMID:23813390

Konishi, H., Kiyama, H., & Ueno, M. (2019). Dual functions of microglia in the formation and refinement of neural circuits during development. *International Journal of Developmental Neuroscience, 77*(1), 18–25. doi:10.1016/j.ijdevneu.2018.09.009 PMID:30292872

Koyama, R., & Ikegaya, Y. (2015). Microglia in the pathogenesis of autism spectrum disorders. *Neuroscience Research, 100*, 1–5. doi:10.1016/j.neures.2015.06.005 PMID:26116891

Laffer, B., Bauer, D., Wasmuth, S., Busch, M., Jalilvand, T. V., Thanos, S., Meyer Zu Hörste, G., Loser, K., Langmann, T., Heiligenhaus, A., & Kasper, M. (2019). Loss of IL-10 Promotes Differentiation of Microglia to a M1 Phenotype. *Frontiers in Cellular Neuroscience, 13*, 430. doi:10.3389/fncel.2019.00430 PMID:31649508

Li, Y., Xia, Y., Yin, S., Wan, F., Hu, J., Kou, L., Sun, Y., Wu, J., Zhou, Q., Huang, J., Xiong, N., & Wang, T. (2021). Targeting Microglial α-Synuclein/TLRs/NF-kappaB/NLRP3 Inflammasome Axis in Parkinson's Disease. *Frontiers in Immunology, 12*, 719807. doi:10.3389/fimmu.2021.719807 PMID:34691027

Liang, S., Wu, X., & Jin, F. (2018). Gut-Brain Psychology: Rethinking Psychology From the Microbiota-Gut-Brain Axis. *Frontiers in Integrative Neuroscience, 12*, 33. doi:10.3389/fnint.2018.00033 PMID:30271330

Liu, N., Li, Y., Jiang, Y., Shi, S., Niamnud, A., Vodovoz, S. J., Katakam, P. V. G., Vidoudez, C., Dumont, A. S., & Wang, X. (2023). Establishment and Application of a Novel In Vitro Model of Microglial Activation in Traumatic Brain Injury. *The Journal of Neuroscience : The Official Journal of the Society for Neuroscience, 43*(2), 319–332. doi:10.1523/JNEUROSCI.1539-22.2022 PMID:36446585

Luo, C., Jian, C., Liao, Y., Huang, Q., Wu, Y., Liu, X., Zou, D., & Wu, Y. (2017). The role of microglia in multiple sclerosis. *Neuropsychiatric Disease and Treatment, 13*, 1661–1667. doi:10.2147/NDT.S140634 PMID:28721047

McKee, C. G., Hoffos, M., Vecchiarelli, H. A., & Tremblay, M. È. (2023). Microglia: A pharmacological target for the treatment of age-related cognitive decline and Alzheimer's disease. *Frontiers in Pharmacology, 14*, 1125982. doi:10.3389/fphar.2023.1125982 PMID:36969855

Meneghini, V., Peviani, M., Luciani, M., Zambonini, G., & Gritti, A. (2021). Delivery Platforms for CRISPR/Cas9 Genome Editing of Glial Cells in the Central Nervous System. *Frontiers in Genome Editing, 3*, 644319. doi:10.3389/fgeed.2021.644319 PMID:34713256

Michels, M., Danielski, L. G., Dal-Pizzol, F., & Petronilho, F. (2014). Neuroinflammation: Microglial activation during sepsis. *Current Neurovascular Research, 11*(3), 262–270. doi:10.2174/1567202611666140520122744 PMID:24845857

Moens, U., Kostenko, S., & Sveinbjørnsson, B. (2013). The Role of Mitogen-Activated Protein Kinase-Activated Protein Kinases (MAPKAPKs) in Inflammation. *Genes, 4*(2), 101–133. doi:10.3390/genes4020101 PMID:24705157

Morris, G. P., Clark, I. A., Zinn, R., & Vissel, B. (2013). Microglia: A new frontier for synaptic plasticity, learning and memory, and neurodegenerative disease research. *Neurobiology of Learning and Memory, 105*, 40–53. doi:10.1016/j.nlm.2013.07.002 PMID:23850597

Muzio, L., Viotti, A., & Martino, G. (2021). Microglia in Neuroinflammation and Neurodegeneration: From Understanding to Therapy. *Frontiers in Neuroscience*, *15*, 742065. doi:10.3389/fnins.2021.742065 PMID:34630027

Muzio, L., Viotti, A., & Martino, G. (2021). Microglia in Neuroinflammation and Neurodegeneration: From Understanding to Therapy. *Frontiers in Neuroscience*, *15*, 742065. doi:10.3389/fnins.2021.742065 PMID:34630027

Nayak, D., Roth, T. L., & McGavern, D. B. (2014). Microglia development and function. *Annual Review of Immunology*, *32*(1), 367–402. doi:10.1146/annurev-immunol-032713-120240 PMID:24471431

O'Regan, G. C., Farag, S. H., Casey, C. S., Wood-Kaczmar, A., Pocock, J. M., Tabrizi, S. J., & Andre, R. (2021). Human Huntington's disease pluripotent stem cell-derived microglia develop normally but are abnormally hyper-reactive and release elevated levels of reactive oxygen species. *Journal of Neuroinflammation*, *18*(1), 94. doi:10.118612974-021-02147-6 PMID:33874957

Orihuela, R., McPherson, C. A., & Harry, G. J. (2016). Microglial M1/M2 polarization and metabolic states. *British Journal of Pharmacology*, *173*(4), 649–665. doi:10.1111/bph.13139 PMID:25800044

Ozben, T., & Ozben, S. (2019). Neuro-inflammation and anti-inflammatory treatment options for Alzheimer's disease. *Clinical Biochemistry*, *72*, 87–89. doi:10.1016/j.clinbiochem.2019.04.001 PMID:30954437

Palpagama, T. H., Waldvogel, H. J., Faull, R. L. M., & Kwakowsky, A. (2019). The Role of Microglia and Astrocytes in Huntington's Disease. *Frontiers in Molecular Neuroscience*, *12*, 258. doi:10.3389/fnmol.2019.00258 PMID:31708741

Pascual, M., Calvo-Rodriguez, M., Núñez, L., Villalobos, C., Ureña, J., & Guerri, C. (2021). Toll-like receptors in neuroinflammation, neurodegeneration, and alcohol-induced brain damage. *IUBMB Life*, *73*(7), 900–915. doi:10.1002/iub.2510 PMID:34033211

Pettas, S., Karagianni, K., Kanata, E., Chatziefstathiou, A., Christoudia, N., Xanthopoulos, K., Sklaviadis, T., & Dafou, D. (2022). Profiling Microglia through Single-Cell RNA Sequencing over the Course of Development, Aging, and Disease. *Cells*, *11*(15), 2383. doi:10.3390/cells11152383 PMID:35954228

Popiolek-Barczyk, K., & Mika, J. (2016). Targeting the Microglial Signaling Pathways: New Insights in the Modulation of Neuropathic Pain. *Current Medicinal Chemistry*, *23*(26), 2908–2928. doi:10.2174/0929867323666160607120124 PMID:27281131

Qian, L., & Flood, P. M. (2008). Microglial cells and Parkinson's disease. *Immunologic Research*, *41*(3), 155–164. doi:10.100712026-008-8018-0 PMID:18512160

Qin, C., Zhou, L. Q., Ma, X. T., Hu, Z. W., Yang, S., Chen, M., Bosco, D. B., Wu, L. J., & Tian, D. S. (2019). Dual Functions of Microglia in Ischemic Stroke. *Neuroscience Bulletin*, *35*(5), 921–933. doi:10.100712264-019-00388-3 PMID:31062335

Rashid, K., Wolf, A., & Langmann, T. (2018). Microglia Activation and Immunomodulatory Therapies for Retinal Degenerations. *Frontiers in Cellular Neuroscience*, *12*, 176. doi:10.3389/fncel.2018.00176 PMID:29977192

Rodriguez, J. I., & Kern, J. K. (2011). Evidence of microglial activation in autism and its possible role in brain underconnectivity. *Neuron Glia Biology*, *7*(2-4), 205–213. doi:10.1017/S1740925X12000142 PMID:22874006

Saba, E. S., Karout, M., Nasrallah, L., Kobeissy, F., Darwish, H., & Khoury, S. J. (2021). Long-term cognitive deficits after traumatic brain injury associated with microglia activation. *Clinical Immunology (Orlando, Fla.)*, *230*, 108815. doi:10.1016/j.clim.2021.108815 PMID:34339843

Saijo, K., Crotti, A., & Glass, C. K. (2013). Regulation of microglia activation and deactivation by nuclear receptors. *Glia*, *61*(1), 104–111. doi:10.1002/glia.22423 PMID:22987512

Sarlus, H., & Heneka, M. T. (2017). Microglia in Alzheimer's disease. *The Journal of Clinical Investigation*, *127*(9), 3240–3249. doi:10.1172/JCI90606 PMID:28862638

Selvakumar, G. P., Ahmed, M. E., Raikwar, S. P., Thangavel, R., Kempuraj, D., Dubova, I., Saeed, D., Zahoor, H., Premkumar, K., Zaheer, S., Iyer, S., & Zaheer, A. (2019). CRISPR/Cas9 Editing of Glia Maturation Factor Regulates Mitochondrial Dynamics by Attenuation of the NRF2/HO-1 Dependent Ferritin Activation in Glial Cells. *Journal of Neuroimmune Pharmacology*, *14*(4), 537–550. doi:10.100711481-019-09833-6 PMID:30810907

Sneeboer, M. A. M., Snijders, G. J. L. J., Berdowski, W. M., Fernández-Andreu, A., van Mierlo, H. C., Berdenis van Berlekom, A., Litjens, M., Kahn, R. S., Hol, E. M., & de Witte, L. D.Psychiatric Donor Program of the Netherlands Brain Bank. (2019). Microglia in post-mortem brain tissue of patients with bipolar disorder are not immune activated. *Translational Psychiatry*, *9*(1), 153. doi:10.103841398-019-0490-x PMID:31127084

Snijders, G. J. L. J., van Zuiden, W., Sneeboer, M. A. M., Berdenis van Berlekom, A., van der Geest, A. T., Schnieder, T., MacIntyre, D. J., Hol, E. M., Kahn, R. S., & de Witte, L. D. (2021). A loss of mature microglial markers without immune activation in schizophrenia. *Glia*, *69*(5), 1251–1267. doi:10.1002/glia.23962 PMID:33410555

Swanson, M. E. V., Scotter, E. L., Smyth, L. C. D., Murray, H. C., Ryan, B., Turner, C., Faull, R. L. M., Dragunow, M., & Curtis, M. A. (2020). Identification of a dysfunctional microglial population in human Alzheimer's disease cortex using novel single-cell histology image analysis. *Acta Neuropathologica Communications*, *8*(1), 170. doi:10.118640478-020-01047-9 PMID:33081847

Taylor, R. A., & Sansing, L. H. (2013). Microglial responses after ischemic stroke and intracerebral hemorrhage. *Clinical & Developmental Immunology*, *746068*, 1–10. doi:10.1155/2013/746068 PMID:24223607

Todd, B. P., Chimenti, M. S., Luo, Z., Ferguson, P. J., Bassuk, A. G., & Newell, E. A. (2021). Traumatic brain injury results in unique microglial and astrocyte transcriptomes enriched for type I interferon response. *Journal of Neuroinflammation*, *18*(1), 151. doi:10.118612974-021-02197-w PMID:34225752

Tronel, C., Largeau, B., Santiago Ribeiro, M. J., Guilloteau, D., Dupont, A. C., & Arlicot, N. (2017). Molecular Targets for PET Imaging of Activated Microglia: The Current Situation and Future Expectations. *International Journal of Molecular Sciences*, *18*(4), 802. doi:10.3390/ijms18040802 PMID:28398245

Tvrdik, P., & Kalani, M. Y. S. (2017). In Vivo Imaging of Microglial Calcium Signaling in Brain Inflammation and Injury. *International Journal of Molecular Sciences*, *18*(11), 2366. doi:10.3390/ijms18112366 PMID:29117112

Veremeyko, T., Yung, A. W. Y., Dukhinova, M., Strekalova, T., & Ponomarev, E. D. (2019). The Role of Neuronal Factors in the Epigenetic Reprogramming of Microglia in the Normal and Diseased Central Nervous System. *Frontiers in Cellular Neuroscience*, *13*, 453. doi:10.3389/fncel.2019.00453 PMID:31680868

Wang, H., He, Y., Sun, Z., Ren, S., Liu, M., Wang, G., & Yang, J. (2022). Microglia in depression: An overview of microglia in the pathogenesis and treatment of depression. *Journal of Neuroinflammation*, *19*(1), 132. doi:10.118612974-022-02492-0 PMID:35668399

Wieghofer, P., & Prinz, M. (2016). Genetic manipulation of microglia during brain development and disease. *Biochimica et Biophysica Acta*, *1862*(3), 299–309. doi:10.1016/j.bbadis.2015.09.019 PMID:26432479

Yin, J., Valin, K. L., Dixon, M. L., & Leavenworth, J. W. (2017). The Role of Microglia and Macrophages in CNS Homeostasis, Autoimmunity, and Cancer. *Journal of Immunology Research*, *5150678*, 1–12. doi:10.1155/2017/5150678 PMID:29410971

Yong, H. Y. F., Rawji, K. S., Ghorbani, S., Xue, M., & Yong, V. W. (2019). The benefits of neuroinflammation for the repair of the injured central nervous system. *Cellular & Molecular Immunology*, *16*(6), 540–546. doi:10.103841423-019-0223-3 PMID:30874626

Zia, S., Rawji, K. S., Michaels, N. J., Burr, M., Kerr, B. J., Healy, L. M., & Plemel, J. R. (2020). Microglia Diversity in Health and Multiple Sclerosis. *Frontiers in Immunology*, *11*, 588021. doi:10.3389/fimmu.2020.588021 PMID:33240276

Zia, S., Rawji, K. S., Michaels, N. J., Burr, M., Kerr, B. J., Healy, L. M., & Plemel, J. R. (2020). Microglia Diversity in Health and Multiple Sclerosis. *Frontiers in Immunology*, *11*, 588021. doi:10.3389/fimmu.2020.588021 PMID:33240276

Chapter 7
Glial Cells Dysfunction and Chronic Pain

Hafida El Ghachi
https://orcid.org/0000-0002-2043-6202
Faculty of Sciences Semlalia, Cadi Ayyad University, Morocco

Asmaa Haj-khlifa
Faculty of Sciences Semlalia, Cadi Ayyad University, Morocco

Mjid Oukhrib
https://orcid.org/0000-0003-0199-4054
Faculty of Sciences Semlalia, Cadi Ayyad University, Morocco

Lahcen Tamegart
Faculty of Sciences, Abdelmalek Essaadi University, Morocco

Youssef Ait Hamdan
https://orcid.org/0009-0005-8616-8852
Higher Normal School, Cadi Ayyad University, Morocco

Abdelali Ben Maloui
Faculty of Sciences Semlalia, Cadi Ayyad University, Morocco

Najib Kissani
Department of Neurology, Medical Research Center, University Hospital Mohamed VI, Morocco

Halima Gamrani
Faculty of Sciences Semlalia, Cadi Ayyad University, Morocco

ABSTRACT

Chronic pain is a pathological health condition caused by the imbalance between inhibitory and excitatory signaling pain pathways. This clinical problem affects about 30% of adults worldwide, and current treatments are insufficient with side effects. The mechanisms underlying this disease were thought to be a result of neuronal dysfunction, however, several animal studies clearly showed that non-neuronal cells (glial cells), primarily microglia and astrocytes, and their interaction with neurons, had a key role in pain persistence. A long-lasting state of activated glia turns to a pathological element causing a painful sensation. This chapter aims to elucidate glial dysfunction by shedding light on peripheral and central glial cells involved in the pathogenesis of chronic pain.

DOI: 10.4018/978-1-6684-9675-6.ch007

INTRODUCTION

Chronic pain is a complex and debilitating condition that affects millions of people worldwide. It is characterized by persistent pain that lasts for weeks, months, or even years. It poses a substantial clinical challenge, impacting as much as 30% of adults globally. In the United States, this condition incurs an economic burden of over $600 billion annually in healthcare expenses, disability payments, and lost productivity (Huh et al., 2017).

Chronic pain is characterized by persistent or recurrent pain that lasts for at least three to six months, extending beyond the normal healing time of an injury or illness (Milligan & Watkins, 2009). It can be classified into different categories based on its underlying etiology, such as nociceptive, neuropathic, inflammatory, or functional pain. Nociceptive pain results from the activation of specialized pain-sensing nerve fibers in response to tissue damage or inflammation. Neuropathic pain, on the other hand, arises from abnormal neural processing due to damage or dysfunction of the nervous system. Inflammatory pain is associated with immune responses and tissue inflammation. Functional pain refers to pain without identifiable tissue damage or inflammation and is often linked to conditions like irritable bowel syndrome or fibromyalgia (Clauw et al., 2019).

Pain transmission involves a complex interplay of neural circuits and signaling molecules. Nociceptive signals are generated by specialized nerve endings called nociceptors, which respond to noxious stimuli such as heat, pressure, or chemical irritants. These signals are then relayed through the spinal cord to higher brain centers, including the thalamus and somatosensory cortex (Basbaum et al., 2009). The spinothalamic tract is crucial for transmitting pain information from the periphery to the brain. Within the central nervous system (CNS), glial cells are strategically positioned to modulate pain signals, contributing to the processing and modulation of pain perception (Basbaum et al., 2009).

While the exact mechanisms underlying chronic pain are not fully understood, research in recent years has highlighted the significant role of glial cells in the development and maintenance of this condition (Ji et al., 2013). Emerging evidence suggests that glial cells actively participate in the modulation of pain processing. For instance, microglia, when activated, release cytokines such as interleukin-1 beta (IL-1β) and tumor necrosis factor-alpha (TNF-α), which contribute to the sensitization of neurons and the enhancement of pain signals. Astrocytes, in turn, release gliotransmitters like glutamate and ATP, modulating synaptic transmission and contributing to synaptic plasticity. This glial-neuronal interaction potentiates pain pathways and can lead to the maintenance of chronic pain (Guo et al., 2007).

Glial cells, once considered merely supportive cells of the nervous system, are now recognized as key players in chronic pain pathways. This chapter will explore the dysfunctional involvement of glial cells in chronic pain, highlighting peripheral and central nervous system cells.

Neurobiological Basis of Pain Transmission

Under normal circumstances, pain functions as a protective and adaptive sensory system. It acts as a warning signal for the body, indicating tissue inflammation and damage, while also prompting behavioral changes that aid in wound healing and recovery (Basbaum et al., 2009).

In the peripheral nervous system, acute tissue insults like heat, cold, chemical, or mechanical injuries disturb the body's balance and stimulate pain receptors called nociceptors. A variety of inflammatory mediators, derived from damaged tissue or activated mast cells and neutrophils, are released locally in response to injury (Basbaum et al., 2009). These mediators include adenosine 5-triphosphate (ATP), bradykinin, prostaglandin E2, sodium, hydrogen, potassium, histamine and serotonin. These released substances interact with specific receptors on peripheral nociceptors, which leads to depolarizing the cell membrane and generating impulses within the Aδ and C fibers (Feizerfan & Sheh, 2015). Following tissue injury, activation of cyclooxygenase-2 and macrophage release of inflammatory mediators including interleukin-1B, interleukin-6, nerve growth factor and tumour necrosis factor alpha (TNF-alpha) contribute to the activation of C-fibers. These fibers then, in a retrograde process known as 'neurogenic inflammation', release substance P, calcitonin gene-related peptide (CGRP), neurokinin A, and nitric oxide (NO) leading to an 'inflammatory soup' that amplifies the process. This ultimately further activates 'silent' C fibers, reducing pain threshold and increasing excitability (Chapman et al., 2008; Feizerfan & Sheh, 2015). Additionally, tissue injury also triggers systemic inflammation, activating the sympathetic system and causing the release of noradrenaline, which further stimulates peripheral nociceptors. However, these changes are usually reversible once the healing of the tissues has been completed (Feizerfan & Sheh, 2015).

Within the spinal cord, primary afferent sensory fibers, whose cell bodies are located in the dorsal root ganglion (DRG), transmit the generated action potential from the injury site to the spinal cord. The sensory fibers, with the exception of those originating from the face, form synapses in the dorsal horn of the spinal cord, while the facial fibers form synapses in the caudate nucleus (Figure 1). C- and Aδ -fibers enter the spinal cord through the dorsal horn and form synapses in the reticular lamina I and II. However, Aδ fibers, which are involved in touch and pressure, synapse on lamina III-VI (Basbaum et al., 2009; Feizerfan & Sheh, 2015).

There are three groups of second-order neurons in the laminae I-V. The first group consists of proprioceptive cells, the second group of nociceptive specific cells that synapse on laminae I and II, and the third group of wide dynamic range (WDR) neurons that synapse on laminae V being quiescent neurons until sensitization (D'Mello & Dickenson, 2008). The release of excitatory substances, such as glutamate, substance P, CGRP and neurotrophic factors, facilitates the transmission of a nociceptive impulse from the primary afferent to the secondary afferent. The balance between excitatory and inhibitory functions determines the overall transmission of noxious stimuli to the brain (Feizerfan & Sheh, 2015).

The thalamus and cerebral cortex are the destinations of ascending pathways or secondary afferent neurons from the dorsal horn. The spinothalamic tract, which originates from lamina I, II and V, is involved in sensory discrimination and terminates in the somatosensory cortex. Other important ascending pathways are the spinoparabrachial and spinomesencephalic pathways, synapsing on lamina I and projecting to the brainstem and medulla to mediate nociception (Feizerfan & Sheh, 2015).

Nociception activates pain matrix in the brain that responds to nociceptive processing and includes multiple cortical areas. It is divided into two parts: the nociceptive cortical matrix and the second-order perceptual matrix (Garcia-Larrea & Peyron, 2013). The nociceptive cortical matrix, which is responsible for the earliest response to nociceptive input from the spinothalamic tract, consists of the posterior insula, medial parietal operculum and mid-cingulate cortex. The second-order perceptual matrix is believed to be responsible for determining and reflecting different individual responses to similar nociception and it includes the mid/anterior insula, anterior cingulate cortex, prefrontal cortex, the periaqueductal grey matter (PAG), rostroventromedial medulla (RVM), and reticular formation. Descending pathways arise from the PAG, RVM and adjacent reticular formation and secrete either norepinephrine or serotonin. Norepinephrine acts by stimulating a2 receptors, which inhibit neurotransmitter release from primary afferent neurons to minimize nociceptive transmission (Heinricher et al., 2009).

By eliminating acute nociceptive stimuli and completely healing tissue, normal homeostasis is restored and pain ceases. However, a series of complex pathophysiological changes in pain processing are initiated when nociceptive stimulation persists or is continuously repeated. These changes occur at multiple levels, from periphery to brain, ultimately leading to persistent pain (Feizerfan & Sheh, 2015). Top of Form

Figure 1. Schematic illustration of pain pathway
(Basbaum et al., 2009 ; Milligan & Watkins, 2009)

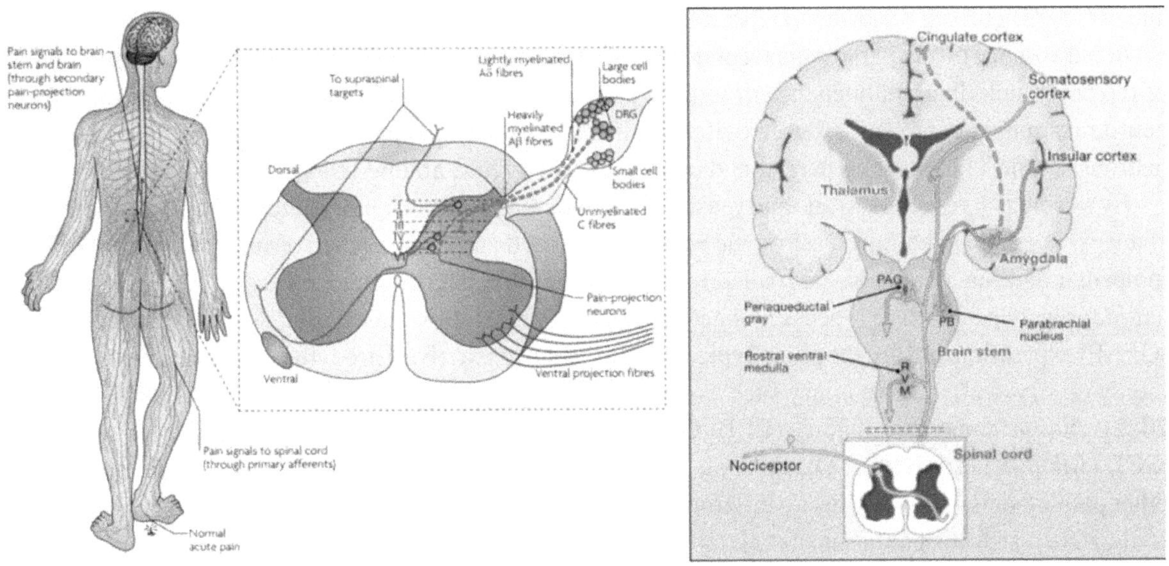

Role of Glial Cells Dysfunction in Pain Processing

Understanding the mechanisms behind chronic pain has advanced significantly over the last thirty years. It is now understood that chronic pain can result from localized neural plasticity either in the peripheral nervous system (PNS), known as peripheral sensitization, or in the central nervous system, known as central sensitization (Latremoliere & Woolf, 2009; Woolf, 1983). Initially, only peripheral and central nociceptive neurons were thought to be affected by synaptic changes. However, a growing body of evidence suggests that pain injury can induce activation of several non-neuronal cell types within the pain circuit, including immune cells and glial cells. This activation leads to local inflammation in both the PNS and the CNS, which is referred to as neuroinflammation. These activated cells are involved in bi-directional interactions with nociceptors and play an active role in the initiation and maintenance of chronic pain (Ji et al., 2016).

Astrocytes

Astrocytes are star-shaped glial cells that constitute a crucial component of the CNS with a percentage of 20-40% (Herculano-Houzel, 2014). These specialized cells play diverse and essential roles in maintaining the normal functioning of the brain and spinal cord. Astrocytes have a distinct appearance, featuring multiple branching processes that extend in various directions. Once thought to have mainly a structural support and a nutritional role, are now believed to be a key player in neural processes (Ben Haim & Rowitch, 2017). One of the primary characteristics of astrocytes is their ability to form a dense network of interconnected processes that wrap around synapses. These structures are known as tripartite synapses, comprising the pre-synaptic neuron, post-synaptic neuron, and the astrocytic processes in between. This close association allows astrocytes to modulate synaptic transmission by regulating neurotransmitter levels, uptake, and release (Araque et al., 1999). As mentioned by some studies, a single cortical astrocyte in the rodent brain can enwrap up to 140,000 synapses and contact 4-6 neuronal soma and 300-600 neuronal dendrites (Halassa et al., 2007; Ji et al., 2019)

In the absence of any pathological condition, astrocytes ensure several roles to the normal functioning of the brain including: maintenance of ion and extracellular pH balance, clearance of waste and excess neurotransmitters to prevent excitotoxicity and neuronal damage, supply glucose and provide energy to neurons, support blood-brain barrier and control of blood flow, among others (Ji et al., 2019).

However, as a response to an injury, astrocytes undergo numerous morphological and physiological changes. In many pathologies, astrocytic cells are activated and enter a reactive state with morphologic, molecular and functional changes called reactive astrogliosis. During this process, naive astrocytes undergo typical changes like process elongation, hypertrophy and increased glial fibrillary acidic protein (GFAP) expression that are the phenotypic characteristics of reactive astrocytes. These changes are commonly observed after nerve injury such as spinal nerve ligation (SNL) (Colburn et al., 1999; Guo et al., 2022), chronic constriction injury (CCI) (Garrison et al., 1991; Xue et al., 2017) and spinal cord injury (SCI) (Allahyari et al., 2022; Okada et al., 2018). Furthermore, this process has also been reported for other pain models, including tissue inflammation (Ducza et al., 2021), chemotherapy-induced pain (Luo et al., 2019), arthritis pain (Sagar et al., 2011) and chronic pain after castration (Ohmichi et al., 2014).

Reactive astrocytes can present different states. Some reactive states occur within minutes, such as the changes in the phosphorylation of signaling molecules or an increase in intracellular Ca2+. Others occur after hours or days (for example, astrocytic hypertrophy or translation regulation) (Lu & Gao, 2023). When a peripheral nerve is injured, astrocytic hypertrophy occurs 3 days later and lasts for several months (Li et al., 2020). Unlike astrocytes, microglia respond immediately to a stimulus and proliferate (Inoue & Tsuda, 2018), and this process reaches peak levels in the first week after nerve injury.

Naive astrocytes differentiate into different subsets, including reactive astrocytes and scar-forming astrocytes. Recent evidence suggests that reactive astrocytes are divided into two distinct phenotypes: A1 and A2 astrocytes. A1 astrocytes are toxic and cause rapid neurons death, while A2 astrocytes are neuroprotective by promoting tissue repair and survival of neurons (Liddelow et al., 2017; Miller, 2018). Activated astrocytes can promote the development and maintenance of chronic pain by releasing signaling molecules (Gao & Ji, 2010; Ji et al., 2016). Reactive and scarring astrocytes strongly express astrocyte marker proteins (GFAP, β-catenin, nestin and N-cadherin) but have their own marker genes. Reactive astrocyte marker genes include matrix metalloproteinase MMP-2, Plaur, MMP13, Axin2, Nes and Ctnnb1, and scar-forming astrocyte marker genes include Cdh2, Sox9 and chondroitin sulphate proteoglycan-related genes such as Csgalnact1, Chst11, Pcanwas, Acan and Slit2 (Li et al., 2019; Silver & Miller, 2004). These marker genes, combined with morphological features, can be used to determine the astrocytic phenotype. These findings may explain the dual effect of reactive astrocytes in central nervous system lesions and diseases.

The transformation of astrocytes from a normal to a reactive phenotype involves a variety of intercellular and intracellular signaling pathways (Karimi-Abdolrezaee & Billakanti, 2012). Four pathways have been proposed to contribute to astrogliosis: Gp130-JAK-STAT3 pathway, Notch-OLIG2 signaling pathway, TGFβ-RGMa-SMAD signaling pathway and Rac-GSPT1 signaling pathway (Li et al., 2019). Signaling molecules involved in astrocyte phenotypic transformation include pro-inflammatory cytokines (IL-1β, TNF-α and IL-6), gene transcription factors (signal transducer and activator of transcription 3 (STAT3)), extracellular signal-regulated kinase 1/2 (ERK1/2), oligodendrocyte transcription factor 2 (OLIG2), Sma- and Mad-related protein (SMAD) and G1 to S phase transition 1 (GSPT1)) and proteins (GFAP, connexins and aquaporin 4 (AQP4)) (Koyama, 2014). Interestingly, these signaling molecules are involved not only in astrocyte activation but also in the development of chronic pain, although it is not known whether their role in the development of chronic pain is mediated through the regulation of astrocyte activation (Li et al., 2019).

Many cell types, including neurons and glial cells such as microglia, oligodendrocytes, and inflammatory cells, can release signaling molecules that activate naive astrocytes. Cytokines and chemokines are secreted proteins that regulate immune responses and control immune cell trafficking. It is well known that cytokines and chemokines present in peripheral tissues, dorsal root ganglia (DRG), spinal cord and even the brain play a role in the pathogenesis of chronic pain. These molecules (e.g., TNF-α; IL-1β; CCL2) play an important role in the initiation and development of persistent pain and peripheral/central sensitization (for review see (Lu & Gao, 2023)).

At the supraspinal level, chronic pain results from peripheral and central sensitization (Lu & Gao, 2023). Central sensitization occurs not only in the spinal cord but also in supraspinal areas. Noxious stimuli are detected and converted into electrical signals, which are then transmitted from the DRG to the dorsal horn. The signal is then transmitted through the spinothalamic tract to the thalamus and then to the primary somatosensory cortex (S1) (Kuner & Kuner, 2021). Several areas such as the PAG, parabrachial nucleus, nucleus accumbens and anterior cingulate cortex (ACC) are involved in pain regulation (Chen et al., 2012). After nerve injury or inflammation, astrocytes undergo various changes in the brainstem, thalamus, ACC and S1 (Kim et al., 2016), and these changes contribute to central sensitization and chronic pain.

The spinal trigeminal nucleus is similar to the spinal dorsal horn and plays an important role in the transmission of trigeminal pain. Reactive astrocytes in the spinal trigeminal nucleus contribute to the pathogenesis of chronic orofacial pain (Okada-Ogawa et al., 2009). In reactive astrocytes, the TNF pathway is activated and increases the expression of GFAP, CX43 and IL-1β, leading to neuropathic pain in the trigeminal system (Guo et al., 2007). This process of astrocyte activation is initiated by the upregulation of several genes markers, including complement 3, complement factor B, MX dynamin-like GTPase 1 and S100a (Liddelow et al., 2017). Neurons and glia regulate this astrogliosis process through several signaling pathways, including JAK-STAT3, Notch-OLIG2 and TGFβ-SMAD (Chen et al., 2008; Koyama, 2014; Lu & Gao, 2023; Tsuda et al., 2011). In addition, after infraorbital nerve injury, reactive astrocytes are detected in the rostral ventromedial medulla (RVM), an important brainstem component for the descending pain modulation circuitry (Wei et al., 2008). Expression levels of the pro-inflammatory factors such as TNF-α and IL-1β are increased in RVM astrocytes. Intra-RVM administration of astrocytic inhibitors attenuates mechanical hyperalgesia and allodynia behaviors. In addition, TNFR1 and IL-1R are expressed in RVM neurons that express the NR1 subunit of the NMDA receptor. Injection of recombinant TNF-α or IL-1β increases NR1 phosphorylation and induces NMDAR-dependent allodynia (Wei et al., 2008). Supraspinal astrocytes are also involved in descending nociceptive modulation in a cancer-induced bone pain model. Ni et al. showed that activation of astrocytes in the ventrolateral PAG via the JNK pathway is involved in the alleviation of bone cancer pain in rats (Ni et al., 2019). Intrathecal administration of astrocytic cytotoxin or a JNK inhibitor reduces GFAP expression and mechanical allodynia. All of the above suggests a contribution of supraspinal astrocytes and central glial-neuronal interactions to the descending facilitation of chronic pain.

Astrocytes from other brain regions have also been reported to be involved in the pain matrix, such as the ACC and S1. ACC astrocytes play a role in the affective component of pain, including unpleasantness or aversion (Chen et al., 2012). SNL activates S1 astrocytes and increases mGluR5 expression. This leads to an increase in astrocytic Ca2+ transients in S1 and consequently the release of thrombospondin-1 from astrocytes, which promotes synaptogenesis and mechanical allodynia (Kim et al., 2016). In addition, Wahis et al. reported a novel function of astrocytes in the central nucleus of the amygdala (CeA) under conditions of neuropathic pain. They found that activation of oxytocin receptors (OTRs) in the CeA reduced anxiety-like behavior induced by neuropathic pain. Furthermore, suppression of OTRs in astrocytes in the lateral part of the CeA abolished the anxiolytic effects of OTR agonists (Wahis et al., 2021). These data highlight the central role of astrocyte-mediated oxytocin signaling in the regulation of emotional states in chronic pain conditions. It is important to note that the above research is mainly based on animal studies. Human astrocytes have several characteristics that differ from those of rodents. The human brain contains GFAP-positive astrocyte subtypes that are not expressed in rodents (Jurga et al., 2021). In addition, human cortical astrocytes are almost twice the size of rodent astrocytes (Oberheim et al., 2009). Real-time imaging studies of the brains of patients with chronic low back pain have shown glial activation in several regions, including the thalamus and S1 (Loggia et al., 2015). This research suggests the possibility of astrocytic activation in higher brain regions in humans. Therefore, determining whether there is neuronal activation associated with astrocyte activation in chronic pain may be important for future research.

During peripheral nerve injury or tissue damage, astrocytes enter a reactive state in response to various neurotransmitters or neuromodulators in the spinal cord or brain. Reactive astrocytes are generally accompanied by the activation of various intracellular signaling pathways. Therefore, the molecular mechanisms of astrocyte-microglia-neuron crosstalk in the spinal cord and brain under chronic pain conditions worth further investigation.

Microglia

Microglia, which mimic peripheral macrophages in their phagocytic capacity, are the archetypal resident immune cells of the CNS. They are heterogeneously distributed and comprise approximately 5-12% of CNS cells (Lawson et al., 1990). They are derived from bone marrow-derived monocytes that migrate during the perinatal period (Nakajima & Kohsaka, 2001). Resting microglia are not passive cells; they actively and continuously monitor their environment through their extended processes and help to promote the removal of debris, damaged cells or infectious agents (Hanisch & Kettenmann, 2007). However, microglia are also valued for their canonical 'glia-like' functions that help maintain the homeostasis and proper function of neurons in the CNS. To this end, microglia play specialized roles in synaptic pruning during development, maintenance of neuronal circuits and synaptic plasticity in the adult, and regulation of adult neurogenesis as part of normal physiological functions (Kierdorf & Prinz, 2017). Beyond homeostasis, microglia have been implicated in disease control and/or pathogenesis in a variety of neurodegenerative (e.g., Alzheimer's disease, Parkinson's disease, stroke) and neuropsychiatric (e.g., depression, anxiety) disorders (Crotti & Ransohoff, 2016; Salter & Stevens, 2017). When activated in disease conditions or following injury (e.g. peripheral nerve injury), microglia undergo 'microgliosis', which is characterized by profound morphological changes (hypertrophy and amoeboid shape), proliferation and functional changes correlating with changes in gene expression and function(Chen et al., 2018; Gilmore, 1975). These changes include the release of pro-inflammatory factors and increased expression of specific microglial proteins such as complement molecules CCR3/CD11b, major histocompatibility complex II [MHC II] and ionized calcium-binding adaptor molecule-1 [IBA1] (Ji et al., 2013). These slow changes often take hours to days to manifest; indeed, proliferation is already very prominent 2 days after spared nerve injury (Decosterd & Woolf, 2000).

Notably, although the astrocytic response appears to begin several days after injury and to be long-lasting, microglial activation often appears to be transient in the early phase and may even precede astrogliosis (Hald et al., 2009; Tanga et al., 2004). It is therefore plausible that a microglial response triggers astrocytic activation in the context of chronic pain, a hypothesis supported by the fact that reactive microglia-derived factors can activate astrocytes in vitro (Röhl et al., 2007). It is important to note, however, that microglial activation can also be prolonged in some models, such as neuropathy or focal burn injury (Chang et al., 2010). Notably, minocycline (a microglial inhibitor) prevents but does not reverse long-lasting pain, in contrast to fluorocitrate (an astrocyte-directed inhibitor) (Ledeboer et al., 2005; Raghavendra et al., 2003). This suggests that microglia may be transiently involved in the initiation of central changes leading to chronic pain, whereas astrocytic activation may be responsible for its long-term maintenance (Figure 2). The idea of a very upstream influence of microglia in the development of chronic pain is further supported by evidence showing that intrathecal injection of in vitro activated microglia induces thermal hyperalgesia and tactile allodynia, whereas astrocytes do not (Narita et al., 2006; Tsuda et al., 2003).

Interestingly, spinal microglial activation after nerve injury requires peripheral sensory afferent activity (Xie et al., 2009), with unmyelinated C-fibers and large A-fibers playing distinct roles in initiating and maintaining microglial activation (Hathway et al., 2009; Suter et al., 2009). Mechanistically, spinal microglial activation is induced by the release of several sensory neuron-derived inflammatory mediators, including (but not limited to) colon stimulatory factor 1 (CSF1), caspase-6, cytokines such as interleukin-1β and extracellular proteases (e.g. MMP9) (Abbadie et al., 2009; Berta et al., 2014; Donnelly et al., 2020; Guan et al., 2016). However, as discussed above, nerve injury-induced microgliosis is somewhat transient and self-limiting, decreasing within a few weeks of nerve injury and remaining at reduced levels at later stages (Kohno et al., 2018). Once microglial activation is reduced at later stages after nerve injury, inhibitors of microglia and inhibitors of activated microglial-derived cytokines are no longer effective in pain reduction (Raghavendra et al., 2003). It is important to note that although the activation of ipsilateral microglia is reduced from its peak after sciatic nerve injury, it remains above baseline or contralateral levels in rats, even at late stages (90 days after injury) (Echeverry et al., 2017).

Remarkably, a study using more aggressive chemo-ablative strategies (rather than inhibition with pharmacological microglial inhibitors) found that ablation of microglia 90 days after nerve injury attenuated pain hypersensitivity in male rats, despite a substantial reduction in microglia-derived pro-inflammatory cytokines (IL-1β, TNF, IL-6), which are typically associated with early-stage pain hypersensitivity. In contrast, although brain-derived neurotrophic factor BDNF expression in late stage microglia remains to be demonstrated, microglia may maintain pain hypersensitivity through the production of BDNF (Echeverry et al., 2017). This evidence suggests that microglia, and more broadly activated cytokines and microglia-derived pain mediators, have an active role in both initiation and maintenance of chronic pain after neural injury through these and other mechanisms.

There is a lack of understanding of the intracellular events that occur in glial cells prior to these changes. However, some important breakthroughs are now available. In particular, the mitogen-activated protein kinases (MAPKs) have emerged as important intracellular mediators in glial cells (Ji et al., 2009; Ji & Suter, 2007). MAPKs are divided into three major kinase families that are involved in many different pathways, namely extracellular signal-regulated kinases (ERKs), p38 kinases and c-jun N-terminal kinases (JNKs). The activation of individual MAPKs in the spinal cord following a pain-inducing peripheral injury is cell- and model-specific. For example, in neuropathic pain and focal burns, the p38 pathway is activated in microglia (Chang et al., 2010; Jin et al., 2003; Tsuda et al., 2004) and the JNK cascade is activated in astrocytes (Ma & Quirion, 2002; Zhuang et al., 2006). Conversely, it has been reported that morphine tolerance correlates with the activation of p38 in microglia, but with the activation of ERK in astrocytes (Wang et al., 2009). Importantly, intrathecal blockade of these pathways has been successfully used to prevent and reverse behavioral manifestations of chronic pain in several models (Jin et al., 2003; Tsuda et al., 2004; Wang et al., 2009; Zhuang et al., 2006). In particular, inhibition of astrocyte JNK activation in the spinal cord of neuropathic rats with a JNK inhibiting peptide (D-JNKI-1) prevented and reversed the onset of allodynia, indicating a key role for astrocyte JNK in the development of neuropathic pain (Zhuang et al., 2006). Taken together, these findings highlight the need to better understand the cell specificity and patterns of MAPK activation in pain, and suggest that spinal glial MAPKs may represent future pharmacological targets for pain (Gosselin et al., 2010).

Several molecules are upregulated after a peripheral nerve injury and positively contribute to promoting chronic pain. Purinergic receptors (e.g., P2X4, P2X7), Toll-like receptors (e.g., TLR2 and TLR4), complement components (e.g., C1q, C3 and C4), Chemokine receptors (CXCL1/ CX3CR1 and CCL2/ CCR2) are all contributors to microglia activation and therefore central sensitization in spinal cord pain circuits (Chen et al., 2018).

Oligodendrocytes

Oligodendrocytes constitute 45-75% of all glial cells and are the most prevalent type of glial cells in the central nervous system (CNS). They produce myelin sheaths that support and insulate axons in the CNS. Despite their wide distribution and importance, Oligodendrocytes have only recently begun to be implicated in pain (Donnelly et al., 2020). In an experimental model of neuropathic pain induced by chronic constriction nerve damage, Zarpelon et al. observed an upregulation of IL-33 mainly by oligodendrocytes in the dorsal spinal cord of adult mice, while mice lacking the IL-33 ST2 receptor experienced reduced pain. Furthermore, intrathecal administration of IL-33 induced hypersensitivity in naive mice and potentiated mechanical allodynia following nerve injury, dependent on TNF-α and IL-1β (Zarpelon et al., 2016). Postmortem spinal cord samples from patients with HIV-associated peripheral neuropathy show increased expression of oligodendrocyte markers-including NG2, Oligo2, and platelet-derived growth factor receptor a (PDGFRa and MBP)-reflecting persistent oligodendrocyte activation in chronic pain (Shi et al., 2016).

Additionally, a subgroup of patients with neuromyelitis optica, a demyelinating disease that causes pain in the optic nerve and spinal cord, possess autoantibodies that specifically target myelin oligodendrocyte glycoprotein (MOG) (Asseyer et al., 2018). It is worth noting that chronic pain affects a significant number of people with multiple sclerosis, which is also a demyelinating disease that results

Figure 2. Central glia activation following peripheral nerve injury
(Donnelly et al., 2020)

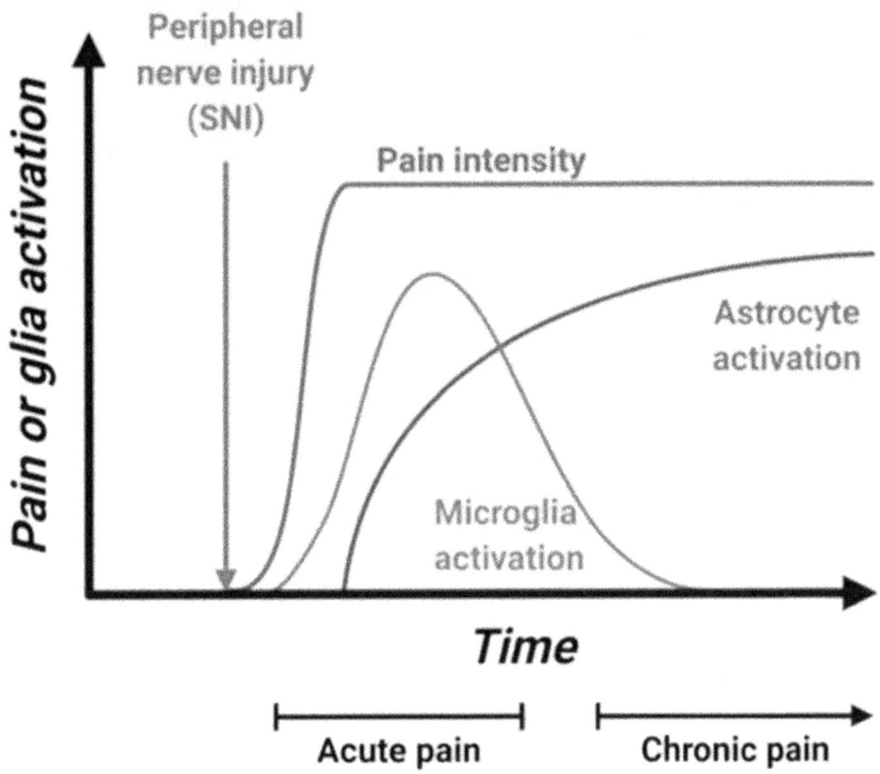

in autoimmune-mediated loss of oligodendrocytes (Urits et al., 2019). This suggests a plausible correlation between the destruction of oligodendrocytes and pain in humans. However, the removal of oligodendrocytes through experimental diphtheria toxin ablation in adult mice induced neuropathic pain for several weeks, as evidenced in 2014 research paper. The development of this pain was found to be independent of adaptive immune cells or reactive microglia and astrocytes (Gritsch et al., 2014). These contrasting outcomes suggest that oligodendrocytes, like other non-neuronal cells, have an active and contextual role in pain.

Schwann Cells

Besides the involvement of central microglia, astrocytes and oligodendrocytes in chronic pain, increasing evidence suggests that peripheral glial cells (Schwann and satellite glial cells, SGC) contribute to neuropathic pain (Campana, 2007). In response to painful stimuli, these peripheral glial cells are activated before the central glial cells activation and release a variety of inflammatory mediators that sensitize the nociceptors at the level of the axons (Schwann cells) and the cell bodies (SGCs). First, Schwann cells that form both myelin sheaths (around non-nociceptive Aβ and nociceptive Aδ fibers) and non-myelinating Remak bundles (around nociceptive C fibers) have been shown to respond to peripheral insult. A first consequence of their activation could be a change in the properties of the myelin, leading to an alteration in the conduction properties of the nociceptive fibers (Kobayashi et al., 2008). There is also increasing evidence for a direct role of Schwann cells in the generation and maintenance of neuropathic pain (Campana, 2007), directly through the release of inflammatory cytokines, and indirectly through the recruitment of immune cells like macrophages to the site of injury through the release of chemo-attractants. Following nerve injury, activated Schwann cells mediate the breakdown of the blood-nerve barrier through the secretion of matrix metalloproteinase 9 (MMP-9), which promotes the recruitment of immune cells from the vasculature and their subsequent release of additional pro-nociceptive mediators (Calvo et al., 2012). In the PNS, the Schwann cell is a dynamic player that isolates nerve fibers into myelinating and non-myelinating forms and dedifferentiates after injury to participate in repair processes, including phagocytosis. Schwann cells are analogous to CNS oligodendrocytes in function, but are very different in morphology and myelination properties (Ohara et al., 2009). Like SGCs, Schwann cells are neural crest derived. Schwann cells are also known to contribute to the inflammatory response that can alter nociceptive thresholds following nerve injury (Martini et al., 2008). However, there is good evidence that Schwann cells may not only produce inflammatory mediators but also release factors that facilitate healing in chronic pain conditions (Campana, 2007).

Satellite Glial Cells

Satellite glia (SGC) are another group of cells that are strategically positioned to modulate nociception, as they completely surround sensory neurons in spinal or trigeminal ganglia (Ohara et al., 2009). Although the functions of GSCs are poorly understood, they share many features with astrocytes in that they sense the activity of adjacent neurons and are thought to have regulatory functions (Hanani, 2005). They are found not only in sensory glia (DRG and TG) but also in sympathetic and parasympathetic ganglia. Like Schwann cells, SGCs are derived from neural crest cells. They are characterized by thin cell sheaths surrounding individual neurons. They share many similarities with astrocytes, although they do not have the same embryonic origin (Jessen & Mirsky, 2005): (1) they both express the glial markers GFAP, S100 and glutamine synthetase; and (2) they both form gap junctions, insulation, an ion sink and recycle neurotransmitters, but the morphology of the two cell types is very different. GSCs have a circular, flattened 'tortilla' morphology, whereas astrocytes are generally multipolar with very attenuated cytoplasmic processes. GSCs and astrocytes express many other common components, including ion channels such as the inwardly rectifying K+ channel, Kir4.1, and glutamate recycling components such as the glutamate-aspartate transporter (GLAST) (Hanani, 2005; Ohara et al., 2009). The number of GSCs in DRGs and TGs is much lower than that of astrocytes in the spinal cord. Unlike astrocytes, each GSC contacts only one neuron (Ji et al., 2013). Strikingly, the extracellular space between the SGC sheath and the associated neuronal plasma membrane is only 20 nm, allowing close interactions and efficient signaling between neurons and SGCs (Hanani, 2005). Emerging evidence suggests that SGCs are activated after painful injury and play an active role in the development of persistent pain (Dublin & Hanani, 2007; Hanani et al., 2002; Jasmin et al., 2010). SGCs also show increased coupling in persistent inflammatory and neuropathic pain (Dublin & Hanani, 2007; Zhang et al., 2009).

The mechanisms by which satellite cells sensitize sensory neurons are incompletely understood. However, an increased level of gap junction-based communication has been described following peripheral injection of complete Freund's adjuvant (CFA), painful colonic obstruction or nerve injury, suggesting a possible role for direct cell-cell coupling (Dublin & Hanani, 2007; Huang & Hanani, 2005). In addition, following nerve injury, SGC cells upregulate neurotrophin synthesis and act both as promoters of sympathetic budding within the ganglion (a mechanism thought to promote sensory neuron activity) (Pertin et al., 2007) and as direct sensitizers of nociceptive neurons (Zhou et al., 1999, 2000). Finally, after nerve or tissue injury, an increase in cytokine expression is observed in the DRG, supporting a sensitizing influence of inflammatory molecules released by the SGC on sensory neurons. In particular, it has been consistently demonstrated that tumor necrosis factor (TNF) alpha is upregulated in DRG satellite cells in parallel with the TNF receptor in DRG sensory neurons, suggesting an important role for TNF in peripheral sensitization (Miyagi et al., 2006). Nociceptive activity also induces the release of ATP from the neuronal soma to activate P2X7 in GSCs, leading to the release of TNF from GSCs and the subsequent increase in neuronal excitability (Zhang et al., 2007). Activation of P2X7 receptors in GSCs also reduces pain by downregulating P2X3 receptors in nociceptive neurons (Chen et al., 2008). In addition, sustained upregulation of matrix metalloproteinase-2 (MMP-2) in satellite cells has been shown to contribute to the maintenance of neuropathic pain through active cleavage of IL-1b and late activation of astrocytes (Kawasaki et al., 2008).

The Cross-Talk Between Microglia, Astrocytes, And Neurons In Chronic Pain Conditions

Astrocytes and microglia play a very important role in the regulation of neuronal function (Allen & Lyons, 2018). Communication between astrocytes, microglia and neurons in the spinal cord and brain facilitates central sensitization, manifested by increased neuronal responsiveness to normal or sub-threshold afferent inputs (Lu & Gao, 2023).

Microglia are activated by tissue or nerve damage (Inoue & Tsuda, 2018) and their morphology changes to an amoeboid form, accompanied by increased secretion of numerous inflammatory factors and microglial phagocytosis (Donnelly et al., 2020). When exposed to different extracellular/intracellular stimuli, microglia exhibit two phenotypes, M1 and M2 (Liu et al., 2020). The M1 phenotype is induced by TNF-α, IL-1β or other inflammatory mediators and microglia then secrete pro-inflammatory cytokines such as IL-6, IL-23 or chemokines (CCL2 and CCL5) to induce neuroinflammation leading to the maintenance of chronic pain (Inoue & Tsuda, 2018). The M2 phenotype is induced by IL-4 or IL-13 and is characterized by increased phagocytosis and production of growth factors such as insulin-like growth factor-1 and anti-inflammatory cytokines such as IL-10 (Cherry et al., 2014). The communication between astrocytes and microglia is partly mediated by secreted mediators such as growth factors, neurotransmitters, cytokines, chemokines, mediators of innate immunity, tissue damage molecules (e.g. ATP), mitogenic factors, NO, ROS and metabolic mediators such as amino acids that can be used for cellular metabolism and may also be responsible for tissue changes (Bezzi et al., 2001). The production and release of these mediators is normally under the control of key intracellular signaling pathways such as the MAPK pathway. Many chemokines, such as CXCL12, CXCL10, CXCL1 and CCL2, are mainly stored on astrocytes, and microglia express the appropriate chemokine receptors, such as CXCR4 or CCR2 (Hughes & Nibbs, 2018). This suggests a strong communication between microglia and astrocytes (Ji et al., 2019). Furthermore, Miyoshi and colleagues reported that IL-18 and their receptors are expressed in microglia and astrocytes, respectively, mediate microglia-astrocyte interaction in the spinal cord and promote neuropathic pain processing after nerve injury (Miyoshi et al., 2008). These findings suggest that activated microglia in the dorsal horn of the spinal cord are directly responsible for astrocytic activation after nerve injury.

Several studies have shown that neuronal signaling by microglia represents an important modulatory element in nociceptive circuits in the spinal cord (Inoue & Tsuda, 2018). In fact, it is the signaling between spinal microglia and second-order dorsal horn neurons that initiates the transmission and, ultimately, the perception of pain by the brain. Activation of microglial P2X4 receptors stimulates the synthesis and release of BDNF (Trang et al., 2009). PNI studies have shown that mice lacking P2X4 receptors show reduced microglial BDNF release and consequently suppressed mechanical allodynia due to altered BDNF signaling in the spinal cord. This suggests that the development of neuropathic pain requires BDNF release driven by P2X4 receptor activation.

Activation of astrocytes can lead to dysregulation of glutamate and GABA, resulting in an imbalance between excitatory and inhibitory neuronal inputs that amplifies pain signals. This occurs in multiple brain regions including S1, medial prefrontal, and ACC (Ishikawa et al., 2018; Kim et al., 2016). Similar to astrocyte-neuron crosstalk in the cortex, interactions between CXCL2 secreted by astrocytes and CXCR2 expressed by neurons in the spinal cord also contribute to neuropathic pain (Abbadie et al., 2009). Recent work has shown that astrocyte-neuron interactions mediated by CXCL1/CXCR2 in the supraspinal pain regulatory system facilitate chronic pain (Ni et al., 2019). Other studies have shown that CXCL12/CXCR4 and CXCL13/CXCR5-mediated astrocyte-neuron crosstalk maintains neuropathic pain through a central sensitization mechanism (Liu et al., 2019). These studies suggest a role for cytokines in glia-neuron cross-talk.

In addition to the glial changes that take place in the spinal cord and in the peripheral nerves, there are also significant changes that occur in several regions of the brain. The last step in nociceptive processing is the transmission of sensory information from the spinal cord to the brain. Lamina I neurons in the dorsal horn of the spinal cord receive noxious and thermal signals from the high-threshold periphery, with only less than 25% of these projecting neurons responding to low-threshold tactile stimuli (Keller et al., 2007). After PNI, the specificity of these neurons changes and the majority become sensitive to low-threshold stimuli, perhaps explained by an unmasking of polysynaptic connectivity in the dorsal horn, so that a low-threshold input can activate projection neurons in lamina I (Baba et al., 2003; Kohno et al., 2003), either by amplifying the excitatory input, by lowering the excitation threshold of the projection neurons, or by converting the inhibitory activity of the interneurons into excitatory activity (Coull et al., 2003). A similar effect is achieved by acute spinal application of ATP-stimulated microglia (Keller et al., 2007), suggesting that microglia may modulate spinal nociceptive output to the brain.

At the supraspinal level, glial activation has been reported in the thalamus after spinal cord nociception (Hains & Waxman, 2006; Zhao et al., 2007), in the cingulate cortex after sciatic nerve ligation (Kuzumaki et al., 2007; Narita et al., 2006), and in the nucleus accumbens after colonic inflammation (Sun et al., 2005). Importantly, the glial response observed in regions not directly involved in nociceptive processing, but rather in limbic circuits (e.g., prefrontal cortex), raises the possibility of glial involvement in the regulation of the emotional component of pain. In support of this hypothesis, there is growing evidence that disrupted astrocyte physiology in corticolimbic areas of the brain might play an important role in mood and affective disorders such as major depression (Gosselin et al., 2010; Rajkowska & Miguel-Hidalgo, 2007).

CONCLUSION

Chronic pain is a complex and multifaceted condition influenced by a combination of genetic, environmental, and psychological factors. The involvement of glial cells in chronic pain pathways adds another layer of complexity to its etiology. Dysfunctional glial cells can perpetuate chronic pain by sustaining a state of chronic inflammation, promoting neuronal sensitization, and inducing synaptic changes that enhance pain transmission. Understanding these intricate mechanisms may offer new opportunities for developing targeted therapies that address the role of glial cell dysfunction in chronic pain management.

REFERENCES

Abbadie, C., Bhangoo, S., De Koninck, Y., Malcangio, M., Melik-Parsadaniantz, S., & White, F. A. (2009). Chemokines and pain mechanisms. *Brain Research. Brain Research Reviews*, *60*(1), 125–134. doi:10.1016/j.brainresrev.2008.12.002 PMID:19146875

Allahyari, R. V., Heinsinger, N. M., Hwang, D., Jaffe, D. A., Rasouli, J., Shiers, S., Thomas, S. J., Price, T. J., Rostami, A., & Lepore, A. C. (2022). Response of Astrocyte Subpopulations Following Spinal Cord Injury. *Cells*, *11*(4), 721. doi:10.3390/cells11040721 PMID:35203371

Allen, N. J., & Lyons, D. A. (2018). Glia as architects of central nervous system formation and function. *Science*, *362*(6411), 181–185. doi:10.1126cience.aat0473 PMID:30309945

Araque, A., Parpura, V., Sanzgiri, R. P., & Haydon, P. G. (1999). Tripartite synapses: Glia, the unacknowledged partner. *Trends in Neurosciences*, *22*(5), 208–215. doi:10.1016/S0166-2236(98)01349-6 PMID:10322493

Asseyer, S., Schmidt, F., Chien, C., Scheel, M., Ruprecht, K., Bellmann-Strobl, J., Brandt, A. U., & Paul, F. (2018). Pain in AQP4-IgG-positive and MOG-IgG-positive neuromyelitis optica spectrum disorders. *Multiple Sclerosis Journal - Experimental, Translational and Clinical*, *4*(3), 2055217318796684. doi:10.1177/2055217318796684 PMID:30186620

Baba, H., Ji, R.-R., Kohno, T., Moore, K. A., Ataka, T., Wakai, A., Okamoto, M., & Woolf, C. J. (2003). Removal of GABAergic inhibition facilitates polysynaptic A fiber-mediated excitatory transmission to the superficial spinal dorsal horn. *Molecular and Cellular Neurosciences*, *24*(3), 818–830. doi:10.1016/S1044-7431(03)00236-7 PMID:14664828

Basbaum, A. I., Bautista, D. M., Scherrer, G., & Julius, D. (2009). Cellular and Molecular Mechanisms of Pain. *Cell*, *139*(2), 267–284. doi:10.1016/j.cell.2009.09.028 PMID:19837031

Ben Haim, L., & Rowitch, D. H. (2017). Functional diversity of astrocytes in neural circuit regulation. *Nature Reviews. Neuroscience*, *18*(1), 31–41. doi:10.1038/nrn.2016.159 PMID:27904142

Berta, T., Park, C.-K., Xu, Z.-Z., Xie, R.-G., Liu, T., Lü, N., Liu, Y.-C., & Ji, R.-R. (2014). Extracellular caspase-6 drives murine inflammatory pain via microglial TNF-α secretion. *The Journal of Clinical Investigation*, *124*(3), 1173–1186. doi:10.1172/JCI72230 PMID:24531553

Bezzi, P., Domercq, M., Vesce, S., & Volterra, A. (2001). Neuron-astrocyte cross-talk during synaptic transmission: Physiological and neuropathological implications. *Progress in Brain Research, 132*, 255–265. doi:10.1016/S0079-6123(01)32081-2 PMID:11544994

Calvo, M., Dawes, J. M., & Bennett, D. L. H. (2012). The role of the immune system in the generation of neuropathic pain. *Lancet Neurology, 11*(7), 629–642. doi:10.1016/S1474-4422(12)70134-5 PMID:22710756

Campana, W. M. (2007). Schwann cells: Activated peripheral glia and their role in neuropathic pain. *Brain, Behavior, and Immunity, 21*(5), 522–527. doi:10.1016/j.bbi.2006.12.008 PMID:17321718

Chang, Y.-W., Tan, A., Saab, C., & Waxman, S. (2010). Unilateral focal burn injury is followed by long-lasting bilateral allodynia and neuronal hyperexcitability in spinal cord dorsal horn. *The Journal of Pain, 11*(2), 119–130. doi:10.1016/j.jpain.2009.06.009 PMID:19744891

Chapman, C. R., Tuckett, R. P., & Song, C. W. (2008). Pain and stress in a systems perspective: Reciprocal neural, endocrine, and immune interactions. *The Journal of Pain, 9*(2), 122–145. doi:10.1016/j.jpain.2007.09.006 PMID:18088561

Chen, F.-L., Dong, Y.-L., Zhang, Z.-J., Cao, D.-L., Xu, J., Hui, J., Zhu, L., & Gao, Y.-J. (2012). Activation of astrocytes in the anterior cingulate cortex contributes to the affective component of pain in an inflammatory pain model. *Brain Research Bulletin, 87*(1), 60–66. doi:10.1016/j.brainresbull.2011.09.022 PMID:22004615

Chen, G., Zhang, Y.-Q., Qadri, Y. J., Serhan, C. N., & Ji, R.-R. (2018). Microglia in Pain: Detrimental and Protective Roles in Pathogenesis and Resolution of Pain. *Neuron, 100*(6), 1292–1311. doi:10.1016/j.neuron.2018.11.009 PMID:30571942

Chen, Y., Miles, D. K., Hoang, T., Shi, J., Hurlock, E., Kernie, S. G., & Lu, Q. R. (2008). The basic helix-loop-helix transcription factor olig2 is critical for reactive astrocyte proliferation after cortical injury. *The Journal of Neuroscience : The Official Journal of the Society for Neuroscience, 28*(43), 10983–10989. doi:10.1523/JNEUROSCI.3545-08.2008 PMID:18945906

Chen, Y., Zhang, X., Wang, C., Li, G., Gu, Y., & Huang, L.-Y. M. (2008). Activation of P2X7 receptors in glial satellite cells reduces pain through downregulation of P2X3 receptors in nociceptive neurons. *Proceedings of the National Academy of Sciences of the United States of America, 105*(43), 16773–16778. doi:10.1073/pnas.0801793105 PMID:18946042

Cherry, J. D., Olschowka, J. A., & O'Banion, M. K. (2014). Neuroinflammation and M2 microglia: The good, the bad, and the inflamed. *Journal of Neuroinflammation, 11*(1), 98. doi:10.1186/1742-2094-11-98 PMID:24889886

Clauw, D. J., Essex, M. N., Pitman, V., & Jones, K. D. (2019). Reframing chronic pain as a disease, not a symptom: Rationale and implications for pain management. *Postgraduate Medicine, 131*(3), 185–198. doi:10.1080/00325481.2019.1574403 PMID:30700198

Colburn, R. W., Rickman, A. J., & DeLeo, J. A. (1999). The effect of site and type of nerve injury on spinal glial activation and neuropathic pain behavior. *Experimental Neurology, 157*(2), 289–304. doi:10.1006/exnr.1999.7065 PMID:10364441

Coull, J. A. M., Boudreau, D., Bachand, K., Prescott, S. A., Nault, F., Sík, A., De Koninck, P., & De Koninck, Y. (2003). Trans-synaptic shift in anion gradient in spinal lamina I neurons as a mechanism of neuropathic pain. *Nature, 424*(6951), 938–942. doi:10.1038/nature01868 PMID:12931188

Crotti, A., & Ransohoff, R. M. (2016). Microglial Physiology and Pathophysiology: Insights from Genome-wide Transcriptional Profiling. *Immunity, 44*(3), 505–515. doi:10.1016/j.immuni.2016.02.013 PMID:26982357

D'Mello, R., & Dickenson, A. H. (2008). Spinal cord mechanisms of pain. *British Journal of Anaesthesia, 101*(1), 8–16. doi:10.1093/bja/aen088 PMID:18417503

Decosterd, I., & Woolf, C. J. (2000). Spared nerve injury: An animal model of persistent peripheral neuropathic pain. *Pain, 87*(2), 149–158. doi:10.1016/S0304-3959(00)00276-1 PMID:10924808

Donnelly, C. R., Andriessen, A. S., Chen, G., Wang, K., Jiang, C., Maixner, W., & Ji, R.-R. (2020). Central Nervous System Targets: Glial Cell Mechanisms in Chronic Pain. *Neurotherapeutics; the Journal of the American Society for Experimental NeuroTherapeutics, 17*(3), 846–860. doi:10.100713311-020-00905-7 PMID:32820378

Dublin, P., & Hanani, M. (2007). Satellite glial cells in sensory ganglia: Their possible contribution to inflammatory pain. *Brain, Behavior, and Immunity, 21*(5), 592–598. doi:10.1016/j.bbi.2006.11.011 PMID:17222529

Ducza, L., Szücs, P., Hegedűs, K., Bakk, E., Gajtkó, A., Wéber, I., & Holló, K. (2021). NLRP2 Is Overexpressed in Spinal Astrocytes at the Peak of Mechanical Pain Sensitivity during Complete Freund Adjuvant-Induced Persistent Pain. *International Journal of Molecular Sciences, 22*(21), 11408. doi:10.3390/ijms222111408 PMID:34768839

Echeverry, S., Shi, X. Q., Yang, M., Huang, H., Wu, Y., Lorenzo, L.-E., Perez-Sanchez, J., Bonin, R. P., De Koninck, Y., & Zhang, J. (2017). Spinal microglia are required for long-term maintenance of neuropathic pain. *Pain, 158*(9), 1792–1801. doi:10.1097/j.pain.0000000000000982 PMID:28746078

Feizerfan, A., & Sheh, G. (2015). Transition from acute to chronic pain. *Continuing Education in Anaesthesia, Critical Care & Pain, 15*(2), 98–102. doi:10.1093/bjaceaccp/mku044

Gao, Y.-J., & Ji, R.-R. (2010). Targeting astrocyte signaling for chronic pain. *Neurotherapeutics; the Journal of the American Society for Experimental NeuroTherapeutics, 7*(4), 482–493. doi:10.1016/j.nurt.2010.05.016 PMID:20880510

Garcia-Larrea, L., & Peyron, R. (2013). Pain matrices and neuropathic pain matrices. *Reviews in Pain, 154*(Suppl 1), S29–S43. doi:10.1016/j.pain.2013.09.001 PMID:24021862

Garrison, C. J., Dougherty, P. M., Kajander, K. C., & Carlton, S. M. (1991). Staining of glial fibrillary acidic protein (GFAP) in lumbar spinal cord increases following a sciatic nerve constriction injury. *Brain Research, 565*(1), 1–7. doi:10.1016/0006-8993(91)91729-K PMID:1723019

Gilmore, S. A. (1975). Proliferation of non-neuronal cells in spinal cords of irradiated, immature rats following transection of the sciatic nerve. *The Anatomical Record, 181*(4), 799–811. doi:10.1002/ar.1091810411 PMID:47232

Gosselin, R.-D., Suter, M. R., Ji, R.-R., & Decosterd, I. (2010). Glial cells and chronic pain. *The Neuroscientist, 16*(5), 519–531. doi:10.1177/1073858409360822 PMID:20581331

Gritsch, S., Lu, J., Thilemann, S., Wörtge, S., Möbius, W., Bruttger, J., Karram, K., Ruhwedel, T., Blanfeld, M., Vardeh, D., Waisman, A., Nave, K.-A., & Kuner, R. (2014). Oligodendrocyte ablation triggers central pain independently of innate or adaptive immune responses in mice. *Nature Communications, 5*(1), 5472. doi:10.1038/ncomms6472 PMID:25434649

Guan, Z., Kuhn, J. A., Wang, X., Colquitt, B., Solorzano, C., Vaman, S., Guan, A. K., Evans-Reinsch, Z., Braz, J., Devor, M., Abboud-Werner, S. L., Lanier, L. L., Lomvardas, S., & Basbaum, A. I. (2016). Injured sensory neuron-derived CSF1 induces microglial proliferation and DAP12-dependent pain. *Nature Neuroscience, 19*(1), 94–101. doi:10.1038/nn.4189 PMID:26642091

Guo, S., Song, Z., He, J., Yin, G., Zhu, J., Liu, H., Yang, L., Ji, X., Xu, X., Liu, Z., & Liu, J. (2022). Akt/Aquaporin-4 Signaling Aggravates Neuropathic Pain by Activating Astrocytes after Spinal Nerve Ligation in Rats. *Neuroscience, 482*, 116–131. doi:10.1016/j.neuroscience.2021.12.015 PMID:34942314

Guo, W., Wang, H., Watanabe, M., Shimizu, K., Zou, S., LaGraize, S. C., Wei, F., Dubner, R., & Ren, K. (2007). Glial-cytokine-neuronal interactions underlying the mechanisms of persistent pain. *The Journal of Neuroscience : The Official Journal of the Society for Neuroscience, 27*(22), 6006–6018. doi:10.1523/JNEUROSCI.0176-07.2007 PMID:17537972

Hains, B. C., & Waxman, S. G. (2006). Activated microglia contribute to the maintenance of chronic pain after spinal cord injury. *The Journal of Neuroscience : The Official Journal of the Society for Neuroscience, 26*(16), 4308–4317. doi:10.1523/JNEUROSCI.0003-06.2006 PMID:16624951

Halassa, M. M., Fellin, T., Takano, H., Dong, J.-H., & Haydon, P. G. (2007). Synaptic islands defined by the territory of a single astrocyte. *The Journal of Neuroscience : The Official Journal of the Society for Neuroscience, 27*(24), 6473–6477. doi:10.1523/JNEUROSCI.1419-07.2007 PMID:17567808

Hald, A., Nedergaard, S., Hansen, R. R., Ding, M., & Heegaard, A.-M. (2009). Differential activation of spinal cord glial cells in murine models of neuropathic and cancer pain. *European Journal of Pain (London, England), 13*(2), 138–145. doi:10.1016/j.ejpain.2008.03.014 PMID:18499488

Hanani, M. (2005). Satellite glial cells in sensory ganglia: From form to function. *Brain Research. Brain Research Reviews, 48*(3), 457–476. doi:10.1016/j.brainresrev.2004.09.001 PMID:15914252

Hanani, M., Huang, T. Y., Cherkas, P. S., Ledda, M., & Pannese, E. (2002). Glial cell plasticity in sensory ganglia induced by nerve damage. *Neuroscience, 114*(2), 279–283. doi:10.1016/S0306-4522(02)00279-8 PMID:12204197

Hanisch, U.-K., & Kettenmann, H. (2007). Microglia: Active sensor and versatile effector cells in the normal and pathologic brain. *Nature Neuroscience, 10*(11), 1387–1394. doi:10.1038/nn1997 PMID:17965659

Hathway, G. J., Vega-Avelaira, D., Moss, A., Ingram, R., & Fitzgerald, M. (2009). Brief, low frequency stimulation of rat peripheral C-fibres evokes prolonged microglial-induced central sensitization in adults but not in neonates. *Pain, 144*(1–2), 110–118. doi:10.1016/j.pain.2009.03.022 PMID:19410369

Heinricher, M. M., Tavares, I., Leith, J. L., & Lumb, B. M. (2009). Descending control of nociception: Specificity, recruitment and plasticity. *Brain Research. Brain Research Reviews*, *60*(1), 214–225. doi:10.1016/j.brainresrev.2008.12.009 PMID:19146877

Herculano-Houzel, S. (2014). The glia/neuron ratio: How it varies uniformly across brain structures and species and what that means for brain physiology and evolution. *Glia*, *62*(9), 1377–1391. doi:10.1002/glia.22683 PMID:24807023

Huang, T.-Y., & Hanani, M. (2005). Morphological and electrophysiological changes in mouse dorsal root ganglia after partial colonic obstruction. *American Journal of Physiology. Gastrointestinal and Liver Physiology*, *289*(4), G670–G678. doi:10.1152/ajpgi.00028.2005 PMID:15920014

Hughes, C. E., & Nibbs, R. J. B. (2018). A guide to chemokines and their receptors. *The FEBS Journal*, *285*(16), 2944–2971. doi:10.1111/febs.14466 PMID:29637711

Huh, Y., Ji, R.-R., & Chen, G. (2017). Neuroinflammation, Bone Marrow Stem Cells, and Chronic Pain. *Frontiers in Immunology*, *8*, 1014. https://www.frontiersin.org/articles/10.3389/fimmu.2017.01014. doi:10.3389/fimmu.2017.01014 PMID:28871264

Inoue, K., & Tsuda, M. (2018). Microglia in neuropathic pain: Cellular and molecular mechanisms and therapeutic potential. *Nature Reviews. Neuroscience*, *19*(3), 138–152. doi:10.1038/nrn.2018.2 PMID:29416128

Ishikawa, T., Eto, K., Kim, S. K., Wake, H., Takeda, I., Horiuchi, H., Moorhouse, A. J., Ishibashi, H., & Nabekura, J. (2018). Cortical astrocytes prime the induction of spine plasticity and mirror image pain. *Pain*, *159*(8), 1592–1606. doi:10.1097/j.pain.0000000000001248 PMID:29672449

Jasmin, L., Vit, J.-P., Bhargava, A., & Ohara, P. T. (2010). Can satellite glial cells be therapeutic targets for pain control? *Neuron Glia Biology*, *6*(1), 63–71. doi:10.1017/S1740925X10000098 PMID:20566001

Jessen, K. R., & Mirsky, R. (2005). The origin and development of glial cells in peripheral nerves. *Nature Reviews. Neuroscience*, *6*(9), 9. Advance online publication. doi:10.1038/nrn1746 PMID:16136171

Ji, R.-R., Berta, T., & Nedergaard, M. (2013). Glia and pain: Is chronic pain a gliopathy? *Pain*, *154*(Supplement 1), S10–S28. doi:10.1016/j.pain.2013.06.022 PMID:23792284

Ji, R.-R., Chamessian, A., & Zhang, Y.-Q. (2016). Pain regulation by non-neuronal cells and inflammation. *Science*, *354*(6312), 572–577. doi:10.1126cience.aaf8924 PMID:27811267

Ji, R.-R., Donnelly, C. R., & Nedergaard, M. (2019). Astrocytes in chronic pain and itch. *Nature Reviews. Neuroscience*, *20*(11), 667–685. doi:10.103841583-019-0218-1 PMID:31537912

Ji, R.-R., Gereau, R. W. IV, Malcangio, M., & Strichartz, G. R. (2009). MAP kinase and pain. *Brain Research. Brain Research Reviews*, *60*(1), 135–148. doi:10.1016/j.brainresrev.2008.12.011 PMID:19150373

Ji, R.-R., & Suter, M. R. (2007). P38 MAPK, microglial signaling, and neuropathic pain. *Molecular Pain*, *3*, 33. doi:10.1186/1744-8069-3-33 PMID:17974036

Jin, S.-X., Zhuang, Z.-Y., Woolf, C. J., & Ji, R.-R. (2003). P38 mitogen-activated protein kinase is activated after a spinal nerve ligation in spinal cord microglia and dorsal root ganglion neurons and contributes to the generation of neuropathic pain. *The Journal of Neuroscience : The Official Journal of the Society for Neuroscience, 23*(10), 4017–4022. doi:10.1523/JNEUROSCI.23-10-04017.2003 PMID:12764087

Jurga, A. M., Paleczna, M., Kadluczka, J., & Kuter, K. Z. (2021). Beyond the GFAP-Astrocyte Protein Markers in the Brain. *Biomolecules, 11*(9), 1361. doi:10.3390/biom11091361 PMID:34572572

Karimi-Abdolrezaee, S., & Billakanti, R. (2012). Reactive astrogliosis after spinal cord injury-beneficial and detrimental effects. *Molecular Neurobiology, 46*(2), 251–264. doi:10.100712035-012-8287-4 PMID:22684804

Kawasaki, Y., Xu, Z.-Z., Wang, X., Park, J. Y., Zhuang, Z.-Y., Tan, P.-H., Gao, Y.-J., Roy, K., Corfas, G., Lo, E. H., & Ji, R.-R. (2008). Distinct roles of matrix metalloproteases in the early- and late-phase development of neuropathic pain. *Nature Medicine, 14*(3), 331–336. doi:10.1038/nm1723 PMID:18264108

Keller, A. F., Beggs, S., Salter, M. W., & De Koninck, Y. (2007). Transformation of the output of spinal lamina I neurons after nerve injury and microglia stimulation underlying neuropathic pain. *Molecular Pain, 3*, 27. doi:10.1186/1744-8069-3-27 PMID:17900333

Kierdorf, K., & Prinz, M. (2017). Microglia in steady state. *The Journal of Clinical Investigation, 127*(9), 3201–3209. doi:10.1172/JCI90602 PMID:28714861

Kim, S. K., Hayashi, H., Ishikawa, T., Shibata, K., Shigetomi, E., Shinozaki, Y., Inada, H., Roh, S. E., Kim, S. J., Lee, G., Bae, H., Moorhouse, A. J., Mikoshiba, K., Fukazawa, Y., Koizumi, S., & Nabekura, J. (2016). Cortical astrocytes rewire somatosensory cortical circuits for peripheral neuropathic pain. *The Journal of Clinical Investigation, 126*(5), 1983–1997. doi:10.1172/JCI82859 PMID:27064281

Kobayashi, H., Chattopadhyay, S., Kato, K., Dolkas, J., Kikuchi, S.-I., Myers, R. R., & Shubayev, V. I. (2008). MMPs initiate Schwann cell-mediated MBP degradation and mechanical nociception after nerve damage. *Molecular and Cellular Neurosciences, 39*(4), 619–627. doi:10.1016/j.mcn.2008.08.008 PMID:18817874

Kohno, K., Kitano, J., Kohro, Y., Tozaki-Saitoh, H., Inoue, K., & Tsuda, M. (2018). Temporal Kinetics of Microgliosis in the Spinal Dorsal Horn after Peripheral Nerve Injury in Rodents. *Biological & Pharmaceutical Bulletin, 41*(7), 1096–1102. doi:10.1248/bpb.b18-00278 PMID:29962405

Kohno, T., Moore, K. A., Baba, H., & Woolf, C. J. (2003). Peripheral nerve injury alters excitatory synaptic transmission in lamina II of the rat dorsal horn. *The Journal of Physiology, 548*(Pt 1), 131–138. doi:10.1113/jphysiol.2002.036186 PMID:12576493

Koyama, Y. (2014). Signaling molecules regulating phenotypic conversions of astrocytes and glial scar formation in damaged nerve tissues. *Neurochemistry International, 78*, 35–42. doi:10.1016/j.neuint.2014.08.005 PMID:25180676

Kuner, R., & Kuner, T. (2021). Cellular Circuits in the Brain and Their Modulation in Acute and Chronic Pain. *Physiological Reviews, 101*(1), 213–258. doi:10.1152/physrev.00040.2019 PMID:32525759

Kuzumaki, N., Narita, M., Narita, M., Hareyama, N., Niikura, K., Nagumo, Y., Nozaki, H., Amano, T., & Suzuki, T. (2007). Chronic pain-induced astrocyte activation in the cingulate cortex with no change in neural or glial differentiation from neural stem cells in mice. *Neuroscience Letters*, *415*(1), 22–27. doi:10.1016/j.neulet.2006.12.057 PMID:17316991

Latremoliere, A., & Woolf, C. J. (2009). Central sensitization: A generator of pain hypersensitivity by central neural plasticity. *The Journal of Pain*, *10*(9), 895–926. doi:10.1016/j.jpain.2009.06.012 PMID:19712899

Lawson, L. J., Perry, V. H., Dri, P., & Gordon, S. (1990). Heterogeneity in the distribution and morphology of microglia in the normal adult mouse brain. *Neuroscience*, *39*(1), 151–170. doi:10.1016/0306-4522(90)90229-W PMID:2089275

Ledeboer, A., Sloane, E. M., Milligan, E. D., Frank, M. G., Mahony, J. H., Maier, S. F., & Watkins, L. R. (2005). Minocycline attenuates mechanical allodynia and proinflammatory cytokine expression in rat models of pain facilitation. *Pain*, *115*(1–2), 71–83. doi:10.1016/j.pain.2005.02.009 PMID:15836971

Li, T., Chen, X., Zhang, C., Zhang, Y., & Yao, W. (2019). An update on reactive astrocytes in chronic pain. *Journal of Neuroinflammation*, *16*(1), 140. doi:10.118612974-019-1524-2 PMID:31288837

Li, X., Li, M., Tian, L., Chen, J., Liu, R., & Ning, B. (2020). Reactive Astrogliosis: Implications in Spinal Cord Injury Progression and Therapy. *Oxidative Medicine and Cellular Longevity*, *9494352*, 1–14. doi:10.1155/2020/9494352 PMID:32884625

Liddelow, S. A., Guttenplan, K. A., Clarke, L. E., Bennett, F. C., Bohlen, C. J., Schirmer, L., Bennett, M. L., Münch, A. E., Chung, W.-S., Peterson, T. C., Wilton, D. K., Frouin, A., Napier, B. A., Panicker, N., Kumar, M., Buckwalter, M. S., Rowitch, D. H., Dawson, V. L., Dawson, T. M., & Barres, B. A. (2017). Neurotoxic reactive astrocytes are induced by activated microglia. *Nature*, *541*(7638), 481–487. doi:10.1038/nature21029 PMID:28099414

Liu, L.-R., Liu, J.-C., Bao, J.-S., Bai, Q.-Q., & Wang, G.-Q. (2020). Interaction of Microglia and Astrocytes in the Neurovascular Unit. *Frontiers in Immunology*, *11*, 1024. doi:10.3389/fimmu.2020.01024 PMID:32733433

Liu, Z.-Y., Song, Z.-W., Guo, S.-W., He, J.-S., Wang, S.-Y., Zhu, J.-G., Yang, H.-L., & Liu, J.-B. (2019). CXCL12/CXCR4 signaling contributes to neuropathic pain via central sensitization mechanisms in a rat spinal nerve ligation model. *CNS Neuroscience & Therapeutics*, *25*(9), 922–936. doi:10.1111/cns.13128 PMID:30955244

Loggia, M. L., Chonde, D. B., Akeju, O., Arabasz, G., Catana, C., Edwards, R. R., Hill, E., Hsu, S., Izquierdo-Garcia, D., Ji, R.-R., Riley, M., Wasan, A. D., Zürcher, N. R., Albrecht, D. S., Vangel, M. G., Rosen, B. R., Napadow, V., & Hooker, J. M. (2015). Evidence for brain glial activation in chronic pain patients. *Brain*, *138*(Pt 3), 604–615. doi:10.1093/brain/awu377 PMID:25582579

Lu, H.-J., & Gao, Y.-J. (2023). Astrocytes in Chronic Pain: Cellular and Molecular Mechanisms. *Neuroscience Bulletin*, *39*(3), 425–439. doi:10.100712264-022-00961-3 PMID:36376699

Luo, H., Liu, H.-Z., Zhang, W.-W., Matsuda, M., Lv, N., Chen, G., Xu, Z.-Z., & Zhang, Y.-Q. (2019). Interleukin-17 Regulates Neuron-Glial Communications, Synaptic Transmission, and Neuropathic Pain after Chemotherapy. *Cell Reports, 29*(8), 2384–2397.e5. doi:10.1016/j.celrep.2019.10.085 PMID:31747607

Ma, W., & Quirion, R. (2002). Partial sciatic nerve ligation induces increase in the phosphorylation of extracellular signal-regulated kinase (ERK) and c-Jun N-terminal kinase (JNK) in astrocytes in the lumbar spinal dorsal horn and the gracile nucleus. *Pain, 99*(1–2), 175–184. doi:10.1016/S0304-3959(02)00097-0 PMID:12237195

Martini, R., Fischer, S., López-Vales, R., & David, S. (2008). Interactions between Schwann cells and macrophages in injury and inherited demyelinating disease. *Glia, 56*(14), 1566–1577. doi:10.1002/glia.20766 PMID:18803324

Miller, S. J. (2018). Astrocyte Heterogeneity in the Adult Central Nervous System. *Frontiers in Cellular Neuroscience, 12*, 401. doi:10.3389/fncel.2018.00401 PMID:30524236

Milligan, E. D., & Watkins, L. R. (2009). Pathological and protective roles of glia in chronic pain. *Nature Reviews. Neuroscience, 10*(1), 23–36. doi:10.1038/nrn2533 PMID:19096368

Miyagi, M., Ohtori, S., Ishikawa, T., Aoki, Y., Ozawa, T., Doya, H., Saito, T., Moriya, H., & Takahashi, K. (2006). Up-regulation of TNFalpha in DRG satellite cells following lumbar facet joint injury in rats. *European Spine Journal: Official Publication of the European Spine Society, the European Spinal Deformity Society, and the European Section of the Cervical Spine Research Society, 15*(6), 953–958. doi:10.1007/s00586-005-1031-2

Miyoshi, K., Obata, K., Kondo, T., Okamura, H., & Noguchi, K. (2008). Interleukin-18-mediated microglia/astrocyte interaction in the spinal cord enhances neuropathic pain processing after nerve injury. *The Journal of Neuroscience : The Official Journal of the Society for Neuroscience, 28*(48), 12775–12787. doi:10.1523/JNEUROSCI.3512-08.2008 PMID:19036970

Nakajima, K., & Kohsaka, S. (2001). Microglia: Activation and their significance in the central nervous system. *Journal of Biochemistry, 130*(2), 169–175. doi:10.1093/oxfordjournals.jbchem.a002969 PMID:11481032

Narita, M., Narita, M., Kaneko, C., Hareyama, N., Miyatake, M., Shindo, K., Miyoshi, K., Nakajima, M., Nagumo, Y., Sato, F., Wachi, H., Seyama, Y., & Suzuki, T. (2006). Chronic pain-induced emotional dysfunction is associated with astrogliosis due to cortical delta-opioid receptor dysfunction. *Journal of Neurochemistry, 97*(5), 1369–1378. doi:10.1111/j.1471-4159.2006.03824.x PMID:16696849

Narita, M., Yoshida, T., Nakajima, M., Narita, M., Miyatake, M., Takagi, T., Yajima, Y., & Suzuki, T. (2006). Direct evidence for spinal cord microglia in the development of a neuropathic pain-like state in mice. *Journal of Neurochemistry, 97*(5), 1337–1348. doi:10.1111/j.1471-4159.2006.03808.x PMID:16606373

Ni, H., Wang, Y., An, K., Liu, Q., Xu, L., Zhu, C., Deng, H., He, Q., Wang, T., Xu, M., Zheng, Y., Huang, B., Fang, J., & Yao, M. (2019). Crosstalk between NFκB-dependent astrocytic CXCL1 and neuron CXCR2 plays a role in descending pain facilitation. *Journal of Neuroinflammation, 16*(1), 1. doi:10.118612974-018-1391-2 PMID:30606213

Ni, H.-D., Xu, L. S., Wang, Y., Li, H., An, K., Liu, M., Liu, Q., Deng, H., He, Q., Huang, B., Fang, J., & Yao, M. (2019). Astrocyte activation in the periaqueductal gray promotes descending facilitation to cancer-induced bone pain through the JNK MAPK signaling pathway. *Molecular Pain, 15*, 1744806919831909. doi:10.1177/1744806919831909 PMID:30700204

Oberheim, N. A., Takano, T., Han, X., He, W., Lin, J. H. C., Wang, F., Xu, Q., Wyatt, J. D., Pilcher, W., Ojemann, J. G., Ransom, B. R., Goldman, S. A., & Nedergaard, M. (2009). Uniquely hominid features of adult human astrocytes. *The Journal of Neuroscience : The Official Journal of the Society for Neuroscience, 29*(10), 3276–3287. doi:10.1523/JNEUROSCI.4707-08.2009 PMID:19279265

Ohara, P. T., Vit, J.-P., Bhargava, A., Romero, M., Sundberg, C., Charles, A. C., & Jasmin, L. (2009). Gliopathic Pain: When Satellite Glial Cells Go Bad. *The Neuroscientist, 15*(5), 450–463. doi:10.1177/1073858409336094 PMID:19826169

Ohmichi, M., Ohmichi, Y., Ohishi, H., Yoshimoto, T., Morimoto, A., Li, Y., Sakurai, H., Nakano, T., & Sato, J. (2014). Activated spinal astrocytes are involved in the maintenance of chronic widespread mechanical hyperalgesia after cast immobilization. *Molecular Pain, 10*, 6. doi:10.1186/1744-8069-10-6 PMID:24456903

Okada, S., Hara, M., Kobayakawa, K., Matsumoto, Y., & Nakashima, Y. (2018). Astrocyte reactivity and astrogliosis after spinal cord injury. *Neuroscience Research, 126*, 39–43. doi:10.1016/j.neures.2017.10.004 PMID:29054466

Okada-Ogawa, A., Suzuki, I., Sessle, B. J., Chiang, C.-Y., Salter, M. W., Dostrovsky, J. O., Tsuboi, Y., Kondo, M., Kitagawa, J., Kobayashi, A., Noma, N., Imamura, Y., & Iwata, K. (2009). Astroglia in medullary dorsal horn (trigeminal spinal subnucleus caudalis) are involved in trigeminal neuropathic pain mechanisms. *The Journal of Neuroscience : The Official Journal of the Society for Neuroscience, 29*(36), 11161–11171. doi:10.1523/JNEUROSCI.3365-09.2009 PMID:19741123

Pertin, M., Allchorne, A. J., Beggah, A. T., Woolf, C. J., & Decosterd, I. (2007). Delayed sympathetic dependence in the spared nerve injury (SNI) model of neuropathic pain. *Molecular Pain, 3*, 21. doi:10.1186/1744-8069-3-21 PMID:17672895

Raghavendra, V., Tanga, F., & DeLeo, J. A. (2003). Inhibition of microglial activation attenuates the development but not existing hypersensitivity in a rat model of neuropathy. *The Journal of Pharmacology and Experimental Therapeutics, 306*(2), 624–630. doi:10.1124/jpet.103.052407 PMID:12734393

Rajkowska, G., & Miguel-Hidalgo, J. J. (2007). Gliogenesis and glial pathology in depression. *CNS & Neurological Disorders - Drug Targets, 6*(3), 219–233. doi:10.2174/187152707780619326 PMID:17511618

Röhl, C., Lucius, R., & Sievers, J. (2007). The effect of activated microglia on astrogliosis parameters in astrocyte cultures. *Brain Research, 1129*(1), 43–52. doi:10.1016/j.brainres.2006.10.057 PMID:17169340

Sagar, D. R., Burston, J. J., Hathway, G. J., Woodhams, S. G., Pearson, R. G., Bennett, A. J., Kendall, D. A., Scammell, B. E., & Chapman, V. (2011). The contribution of spinal glial cells to chronic pain behaviour in the monosodium iodoacetate model of osteoarthritic pain. *Molecular Pain, 7*, 88. doi:10.1186/1744-8069-7-88 PMID:22093915

Salter, M. W., & Stevens, B. (2017). Microglia emerge as central players in brain disease. *Nature Medicine, 23*(9), 1018–1027. doi:10.1038/nm.4397 PMID:28886007

Shi, Y., Shu, J., Liang, Z., Yuan, S., & Tang, S.-J. (2016). EXPRESS: Oligodendrocytes in HIV-associated pain pathogenesis. *Molecular Pain, 12*, 1744806916656845. doi:10.1177/1744806916656845 PMID:27306410

Silver, J., & Miller, J. H. (2004). Regeneration beyond the glial scar. *Nature Reviews. Neuroscience, 5*(2), 146–156. doi:10.1038/nrn1326 PMID:14735117

Sun, Y.-N., Luo, J.-Y., Rao, Z.-R., Lan, L., & Duan, L. (2005). GFAP and Fos immunoreactivity in lumbo-sacral spinal cord and medulla oblongata after chronic colonic inflammation in rats. *World Journal of Gastroenterology, 11*(31), 4827–4832. doi:10.3748/wjg.v11.i31.4827 PMID:16097052

Suter, M. R., Berta, T., Gao, Y.-J., Decosterd, I., & Ji, R.-R. (2009). Large A-fiber activity is required for microglial proliferation and p38 MAPK activation in the spinal cord: Different effects of resiniferatoxin and bupivacaine on spinal microglial changes after spared nerve injury. *Molecular Pain, 5*, 53. doi:10.1186/1744-8069-5-53 PMID:19772627

Tanga, F. Y., Raghavendra, V., & DeLeo, J. A. (2004). Quantitative real-time RT-PCR assessment of spinal microglial and astrocytic activation markers in a rat model of neuropathic pain. *Neurochemistry International, 45*(2–3), 397–407. doi:10.1016/j.neuint.2003.06.002 PMID:15145554

Trang, T., Beggs, S., Wan, X., & Salter, M. W. (2009). P2X4-receptor-mediated synthesis and release of brain-derived neurotrophic factor in microglia is dependent on calcium and p38-mitogen-activated protein kinase activation. *The Journal of Neuroscience : The Official Journal of the Society for Neuroscience, 29*(11), 3518–3528. doi:10.1523/JNEUROSCI.5714-08.2009 PMID:19295157

Tsuda, M., Kohro, Y., Yano, T., Tsujikawa, T., Kitano, J., Tozaki-Saitoh, H., Koyanagi, S., Ohdo, S., Ji, R.-R., Salter, M. W., & Inoue, K. (2011). JAK-STAT3 pathway regulates spinal astrocyte proliferation and neuropathic pain maintenance in rats. *Brain, 134*(Pt 4), 1127–1139. doi:10.1093/brain/awr025 PMID:21371995

Tsuda, M., Mizokoshi, A., Shigemoto-Mogami, Y., Koizumi, S., & Inoue, K. (2004). Activation of p38 mitogen-activated protein kinase in spinal hyperactive microglia contributes to pain hypersensitivity following peripheral nerve injury. *Glia, 45*(1), 89–95. doi:10.1002/glia.10308 PMID:14648549

Tsuda, M., Shigemoto-Mogami, Y., Koizumi, S., Mizokoshi, A., Kohsaka, S., Salter, M. W., & Inoue, K. (2003). P2X4 receptors induced in spinal microglia gate tactile allodynia after nerve injury. *Nature, 424*(6950), 778–783. doi:10.1038/nature01786 PMID:12917686

Urits, I., Adamian, L., Fiocchi, J., Hoyt, D., Ernst, C., Kaye, A. D., & Viswanath, O. (2019). Advances in the Understanding and Management of Chronic Pain in Multiple Sclerosis: A Comprehensive Review. *Current Pain and Headache Reports, 23*(8), 59. doi:10.100711916-019-0800-2 PMID:31342191

Wahis, J., Baudon, A., Althammer, F., Kerspern, D., Goyon, S., Hagiwara, D., Lefevre, A., Barteczko, L., Boury-Jamot, B., Bellanger, B., Abatis, M., Da Silva Gouveia, M., Benusiglio, D., Eliava, M., Rozov, A., Weinsanto, I., Knobloch-Bollmann, H. S., Kirchner, M. K., Roy, R. K., ... Charlet, A. (2021). Astrocytes mediate the effect of oxytocin in the central amygdala on neuronal activity and affective states in rodents. *Nature Neuroscience*, *24*(4), 529–541. doi:10.103841593-021-00800-0 PMID:33589833

Wang, Z., Ma, W., Chabot, J.-G., & Quirion, R. (2009). Cell-type specific activation of p38 and ERK mediates calcitonin gene-related peptide involvement in tolerance to morphine-induced analgesia. *The FASEB Journal*, *23*(8), 2576–2586. doi:10.1096/fj.08-128348 PMID:19299480

Wei, F., Guo, W., Zou, S., Ren, K., & Dubner, R. (2008). Supraspinal glial-neuronal interactions contribute to descending pain facilitation. *The Journal of Neuroscience : The Official Journal of the Society for Neuroscience*, *28*(42), 10482–10495. doi:10.1523/JNEUROSCI.3593-08.2008 PMID:18923025

Woolf, C. J. (1983). Evidence for a central component of post-injury pain hypersensitivity. *Nature*, *306*(5944), 686–688. doi:10.1038/306686a0 PMID:6656869

Xie, W., Strong, J. A., & Zhang, J.-M. (2009). Early blockade of injured primary sensory afferents reduces glial cell activation in two rat neuropathic pain models. *Neuroscience*, *160*(4), 847–857. doi:10.1016/j.neuroscience.2009.03.016 PMID:19303429

Xue, P., Chen, L., Lu, X., Zhang, J., Bao, G., Xu, G., Sun, Y., Guo, X., Jiang, J., Gu, H., & Cui, Z. (2017). Vimentin Promotes Astrocyte Activation After Chronic Constriction Injury. *Journal of Molecular Neuroscience*, *63*(1), 91–99. doi:10.100712031-017-0961-6 PMID:28791619

Zarpelon, A. C., Rodrigues, F. C., Lopes, A. H., Souza, G. R., Carvalho, T. T., Pinto, L. G., Xu, D., Ferreira, S. H., Alves-Filho, J. C., McInnes, I. B., Ryffel, B., Quesniaux, V. F. J., Reverchon, F., Mortaud, S., Menuet, A., Liew, F. Y., Cunha, F. Q., Cunha, T. M., & Verri, W. A. Jr. (2016). Spinal cord oligodendrocyte-derived alarmin IL-33 mediates neuropathic pain. *The FASEB Journal*, *30*(1), 54–65. doi:10.1096/fj.14-267146 PMID:26310268

Zhang, H., Mei, X., Zhang, P., Ma, C., White, F. A., Donnelly, D. F., & Lamotte, R. H. (2009). Altered functional properties of satellite glial cells in compressed spinal ganglia. *Glia*, *57*(15), 1588–1599. doi:10.1002/glia.20872 PMID:19330845

Zhang, X., Chen, Y., Wang, C., & Huang, L.-Y. M. (2007). Neuronal somatic ATP release triggers neuron-satellite glial cell communication in dorsal root ganglia. *Proceedings of the National Academy of Sciences of the United States of America*, *104*(23), 9864–9869. doi:10.1073/pnas.0611048104 PMID:17525149

Zhao, P., Waxman, S. G., & Hains, B. C. (2007). Extracellular signal-regulated kinase-regulated microglia-neuron signaling by prostaglandin E2 contributes to pain after spinal cord injury. *The Journal of Neuroscience : The Official Journal of the Society for Neuroscience*, *27*(9), 2357–2368. doi:10.1523/JNEUROSCI.0138-07.2007 PMID:17329433

Zhou, X. F., Deng, Y. S., Chie, E., Xue, Q., Zhong, J. H., McLachlan, E. M., Rush, R. A., & Xian, C. J. (1999). Satellite-cell-derived nerve growth factor and neurotrophin-3 are involved in noradrenergic sprouting in the dorsal root ganglia following peripheral nerve injury in the rat. *The European Journal of Neuroscience*, *11*(5), 1711–1722. doi:10.1046/j.1460-9568.1999.00589.x PMID:10215925

Zhou, X. F., Deng, Y. S., Xian, C. J., & Zhong, J. H. (2000). Neurotrophins from dorsal root ganglia trigger allodynia after spinal nerve injury in rats. *The European Journal of Neuroscience*, *12*(1), 100–105. doi:10.1046/j.1460-9568.2000.00884.x PMID:10651864

Zhuang, Z.-Y., Wen, Y.-R., Zhang, D.-R., Borsello, T., Bonny, C., Strichartz, G. R., Decosterd, I., & Ji, R.-R. (2006). A peptide c-Jun N-terminal kinase (JNK) inhibitor blocks mechanical allodynia after spinal nerve ligation: Respective roles of JNK activation in primary sensory neurons and spinal astrocytes for neuropathic pain development and maintenance. *The Journal of Neuroscience : The Official Journal of the Society for Neuroscience*, *26*(13), 3551–3560. doi:10.1523/JNEUROSCI.5290-05.2006 PMID:16571763

Chapter 8
Astrocytes Functions and Their Involvement in Brain Injury:
A Focus on the Biomaterials' Role in Mitigating Traumatic Brain Injury

Youssef Ait Hamdan
https://orcid.org/0009-0005-8616-8852
University of Rennes, France & Higher Normal School, Cadi Ayyad University, Morocco

Mjid Oukhrib
https://orcid.org/0000-0003-0199-4054
Faculty of Sciences, Cadi Ayyad University, Morocco

Mohammed Rhazi
Higher Normal School, Cadi Ayyad University, Morocco

Abdelfettah Ait Baba
Faculty of Sciences, Cadi Ayyad University, Morocco

Bilal El-Mansoury
https://orcid.org/0000-0002-2583-7515
Faculty of Sciences, Chouaib Doukkali University, Morocco

Adil Elhousseini
Higher Normal school, Cadi Ayyad University, Morocco

Hamid Kabdy
Faculty of Sciences, Cadi Ayyad University, Morocco

Zainab Haïdar Ahmad
University of Rennes, France

Hassane Oudadesse
University of Rennes, France

ABSTRACT

Brain injury is damage to the brain caused by external forces. They are the leading cause of death and ill health in young men, with a high incidence worldwide. Glial cells make up the majority of central nervous system cells and play a key role in the brain's response to brain injury. Astrocytes play a crucial role in maintaining the brain's ionic and water balance, energy metabolism, regulation of the blood-brain barrier and immune response. In response to brain injury, astrocytes undergo morphological and behavioral changes that modify their protein expression. To sustain astrocytic responses in certain severe

DOI: 10.4018/978-1-6684-9675-6.ch008

situations, therapeutic strategies that make use of biomaterials represent a significant advance in the field of regenerative medicine, as they create a microenvironment more conducive to nerve tissue regeneration. This chapter examines the behavior and responses of astrocytes to brain injury in the context of traumatic brain injury (TBI), as well as looking at the implication of biomaterials in the treatment of TBI.

INTRODUCTION

Restoring neural tissue after damage caused by injury or disease remains a complex challenge that has yet to be resolved. Traumatic brain injury (TBI) is responsible for a significant number of deaths and disabilities worldwide, usually occurring as a result of an external mechanical force such as a road traffic accident, armed conflict, object penetration, explosion, among others (Blennow et al., 2016). TBI can be categorised as mild, moderate or severe based on clinical indicators such as state of consciousness, presence of amnesia, and other neurological symptoms (Blennow et al., 2016). Currently, it is generally recognised that repeated head trauma can lead to long-term effects, including the risk of accumulating neurodegenerative diseases such as Parkinson's disease, Alzheimer's disease (AD), chronic traumatic encephalopathy (CTE), among others (Kamins and Giza, 2016).

In the central nervous system (CNS), glial cells occupy more than half of the brain volume compared with neurons (Jessen, 2004). Several types of glial cells can be identified, including astrocytes, microglia, oligodendrocytes and oligodendrocyte progenitor cells (Peters, 2004). Astrocytes play a crucial role as response cells to disease and injury in the CNS. Astrocytes can detect lesions caused by head trauma and react by becoming reactive astrocytes, thus participating in damage and repair processes through their interactions with other cell types (Michinaga and Koyama, 2021; Vainchtein and Molofsky, 2020). TBI not only affects neuronal function but also disrupts brain function. The significant presence of neuroinflammation observed in TBI, as well as the demyelination of white matter pathways, highlights the essential role of glial cells in these traumatic brain injuries. Sometimes, astrocytes require assistance and support to cope with brain trauma. Scaffolds made from biomaterials can also be supported in the damaged region of the brain to facilitate and support the action of glial cells in creating an environment conducive to nerve regeneration (Gopalakrishnan et al., 2019). In this context, this chapter will look at astrocyte behavior and function in the context of brain injury as well as in the subsequent repair phase. We also examine the use of biomaterials to provide support to astrocytes in the treatment of these injuries.

ROLE OF ASTROCYTES IN BRAIN HEALTH

Astrocytes are a type of glial cell in the CNS that play a crucial role in maintaining brain health and supporting the function of neurons. They are highly versatile and perform a wide range of functions to ensure the proper functioning of the brain (Jessen, 2004). Here are some of the key functions of astrocytes:

1. Neurotransmitter Regulation

One of the important functions of astrocytes is the regulation of neurotransmitter levels in the brain (Kim et al., 2019). Neurotransmitters are chemicals that transmit signals between neurons, and astrocytes help to maintain the proper balance of these chemicals in the brain. Astrocytes also regulate synaptic transmission and play a role in the formation of new memories, long-term potentiation (LTP), and functional synaptic plasticity (Ota et al., 2013). In addition, astrocytes are involved in the regulation of brain metabolism, controlling blood-brain barrier permeability, and the neuro-repair process (Chiareli et al., 2021). Astrocytes cooperate with neurons on several levels, including neurotransmitter trafficking and recycling, ion homeostasis, and energy metabolism (Michinaga & Koyama, 2019) (Bélanger & Magistretti, 2022). In essence, astrocytes are pivotal in overseeing brain function and are indispensable for the maintenance of appropriate neurotransmitter levels in the brain.

2. Supporting the blood-brain barrier

Astrocytes play a crucial role in supporting the blood-brain barrier (BBB) (Abbott et al., 2006; Cabezas et al., 2014; Michinaga & Koyama, 2019). The BBB is formed by the endothelial cells that line cerebral microvessels and has an important role in maintaining a precisely regulated microenvironment for reliable neuronal signaling. Astrocytes are essential for the formation and maintenance of the BBB by providing secreted factors that lead to the adequate association between the cells of the BBB (Cabezas et al., 2014). Astrocytic endfeet express the potassium channel, Kir4.1, and aquaporin-4, which support the BBB function by controlling the ion and water balance (Kadry et al., 2020). Astrocytes also play a primary role in the expression of the tight junction proteins, including occludin, claudin-5, and zonula occludens-1 (ZO-1) in the BBB (Liu et al., 2018). In addition, astrocytes induce blood-brain barrier properties in endothelial cells (Kadry et al., 2020). In general, astrocytes are essential for the effective operation of the blood-brain barrier (BBB) and contribute significantly to preserving the integrity of the brain's microvasculature.

3. Controlling ion balance

Astrocytes play a crucial role in maintaining the proper balance of ions, including sodium, potassium, and calcium, in the extracellular space surrounding neurons. Astrocytes express various ion channels, including potassium channels, which are involved in spatial buffering of potassium ions (Bellot-Saez et al., 2017). Astrocytes also regulate the extracellular concentration of ions by taking up excess potassium ions and releasing them back into the extracellular space when needed (McNeill et al., 2021; Verkhratsky et al., 2020). In addition, astrocytes play a role in the regulation of intracellular sodium homeostasis (Mahmoud et al., 2019). Astrocytes also control the balance between glutamate uptake and release, which is important for maintaining glutamate homeostasis in the central nervous system (Mahmoud et al., 2019; McNeill et al., 2021).

4. Providing metabolic support to neurons

Astrocytes have a vital function in supplying metabolic assistance to neurons within the brain. They are responsible for maintaining the energy balance in the brain and providing nutrients to neurons. Astrocytes are involved in the regulation of glucose metabolism in the brain and play a critical role in the uptake and utilization of glucose by neurons (Beard et al., 2022). Astrocytes preferentially undergo glycolysis to produce pyruvate and lactate, which are then transported to neurons to be used as an energy source (MacVicar & Choi, 2017). Astrocytes also provide neurons with essential nutrients, such as amino acids and lipids, which are required for proper neuronal function (Chen et al., 2023; Turner & Adamson, 2011). In addition, astrocytes play a role in the regulation of neurotransmitter levels in the brain, which is important for maintaining proper neuronal function (MacVicar & Choi, 2017; Turner & Adamson, 2011).

5. Neuroprotection

Astrocytes play a crucial role in neuroprotection by promoting neuron survival, scavenging toxic molecules, and preventing excitotoxicity. Here are some ways in which astrocytes contribute to neuroprotection:

i. Promoting neuron survival: Astrocytes secrete various neurotrophic factors, such as brain-derived neurotrophic factor (BDNF), glial cell line-derived neurotrophic factor (GDNF), and nerve growth factor (NGF), which promote neuron survival and growth (Bélanger & Magistretti, 2009). These factors help to support the growth and survival of neurons, which is essential for maintaining proper brain function.
ii. Scavenging toxic molecules: Astrocytes are involved in the clearance of toxic molecules, such as reactive oxygen species (ROS), which can be harmful to neurons (Jiwaji et al., 2022). Astrocytes express various enzymes, such as superoxide dismutase (SOD) and glutathione peroxidase (GPx), which help to scavenge ROS and prevent oxidative damage to neurons. Astrocytes also play a role in the uptake and metabolism of glutamate, which is important for preventing excitotoxicity (Bélanger & Magistretti, 2009).
iii. Preventing excitotoxicity: Astrocytes play a crucial role in preventing excitotoxicity, which is a pathological process that can lead to neuronal death. Astrocytes regulate the extracellular concentration of glutamate by taking up excess glutamate and releasing it back into the extracellular space when needed. Astrocytes also express glutamate transporters, such as GLT-1, which are responsible for the majority of glutamate uptake in the brain. By regulating the extracellular concentration of glutamate, astrocytes help to prevent excitotoxicity and protect neurons from damage (Bélanger & Magistretti, 2009; Jiwaji et al., 2022).

In summary, **Figure 1** illustrates the roles that astrocytes play in neuroprotection, promoting the survival of neurons, eliminating harmful molecules and avoiding excitotoxicity. They are essential for maintaining optimal brain function and participate in various processes designed to protect neurons.

ASTROCYTE RESPONSES TO BRAIN INJURY

1. Astrocytic Reactivity

Astrocytes are vital components of a healthy CNS, serving various critical functions (Bélanger & Magistretti, 2009). In response to CNS damage or pathology, astrocytes undergo changes in gene expression, cellular structure, and function, collectively known as astrogliosis (Sofroniew, 2015). Despite the concept of astrogliosis being present since the 19th century, research on its mechanisms and functions

Figure 1. Different functions and roles of astrocytes in brain health

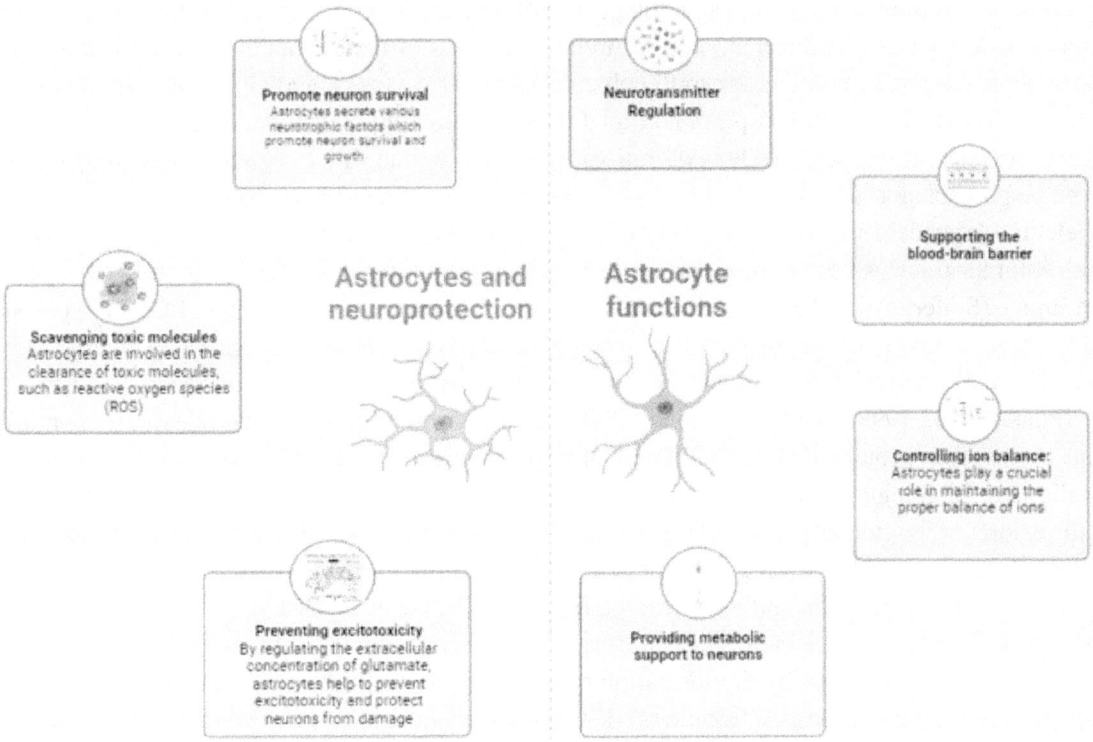

gained significant attention only in the last two decades, coinciding with an increased understanding of astrocytes' crucial roles in a healthy CNS (Sofroniew, 2015).

Astrogliosis encompasses a spectrum of potential modifications, including molecular, cellular, and functional changes in astrocytes in response to various CNS injuries and diseases (**Figure 2**). These changes vary based on the severity of the insult and are regulated by various intercellular and intracellular signaling molecules (Sharma et al., 2015). Astrogliosis can lead to both the acquisition of new functions and the loss of pre-existing ones, making it a complex and dynamic process that significantly impacts tissue repair and the outcomes of CNS injuries and pathologies. Different terms are sometimes used to describe astrocyte responses to CNS injuries or diseases, with variations among researchers. In this context, we will use "astrogliosis" and "reactive astrocytes" as general and inclusive terms, encompassing all manifestations of astrocyte responses to any type of CNS damage or disease.

Several molecular and morphological characteristics distinguish reactive astrocytes and reactive astrogliosis from non-reactive astrocytes. For example, the overexpression of glial fibrillary acidic protein (GFAP), the primary component of astrocyte intermediate filaments, is commonly used as a marker for reactive astrocytes (Eng et al., 2000). However, the expression of glutamine synthetase (GS), an enzyme crucial for glutamate-to-glutamine conversion in astrocytes, is altered in various neurological conditions, such as cerebral hypoxia, ischemia, epilepsy, and chronic neurodegenerative processes. GS changes are region-specific and not positively correlated with GFAP expression (Rose et al., 2013), making it an unreliable marker for reactive astrogliosis. GS plays a vital role in normal brain development and function, and its reduced expression and function may contribute to certain CNS disorders (Häberle et al., 2006).

The influence on astrocyte activation can be transmitted through the gp-130/signal transducer and activator of transcription 3 (STAT3) signaling pathway, leading to STAT3 phosphorylation and nuclear migration in astrocytes (Sriram et al., 2004). Additionally, this influence can indirectly affect other cell types such as microglia, neurons, or endothelial cells. Integrin-1-mediated signaling has an opposing effect on astrocyte activation and plays a crucial role in promoting their transformation into a mature, non-reactive state (Robel et al., 2009). Alterations in STAT3- or integrin-1-regulated signaling pathways may be responsible for inducing specific reactive astrogliosis characteristics following injury.

Calcium-dependent signaling via inositol 1,4,5-trisphosphate (IP3) and N-cadherin-associated processes within astrocytes are essential for normal reactive astrogliosis, preserving neuroprotection after brain injury (Seifert et al., 2006). Additionally, epidermal growth factor receptor (EGFR) expression, usually absent in non-reactive astrocytes, is observed in reactive astrocytes and activated microglia/macrophages. EGFR activation is a key player in transitioning astrocytes from a non-reactive to a reactive state (Planas et al., 1998; Erschbamer et al., 2007). EGFR ligands stimulate the release of chondroitin sulfate proteoglycans, and inhibiting EGFR signaling promotes axonal regeneration and functional recovery following CNS injury (Smith & Strunz, 2005; Koprivica et al., 2005). Moreover, astrocyte polarity and directional migration appear crucial for their ability to respond to injury, as astrocytes lacking the small RhoGTPase Cdc42, a key regulator of cell polarity, exhibit impaired recruitment to the injury site despite GFAP overexpression and hypertrophic responses (Robel et al., 2011).

Morphologically, reactive astrogliosis can vary significantly, ranging from subtle to pronounced reactions, often accompanied by the formation of glial scars, a well-known phenomenon among neuropathologists for generations (Sofroniew, 2015). Over the years, glial scars were generally attributed to astrocytes. However, more recently, it has been proposed that pericytes could play a crucial role in scar formation (Göritz et al., 2011). In response to CNS injuries or chronic neurodegenerative diseases, various pro-inflammatory cytokines.

Figure 2. Behavior and response of astrocytes in brain injury

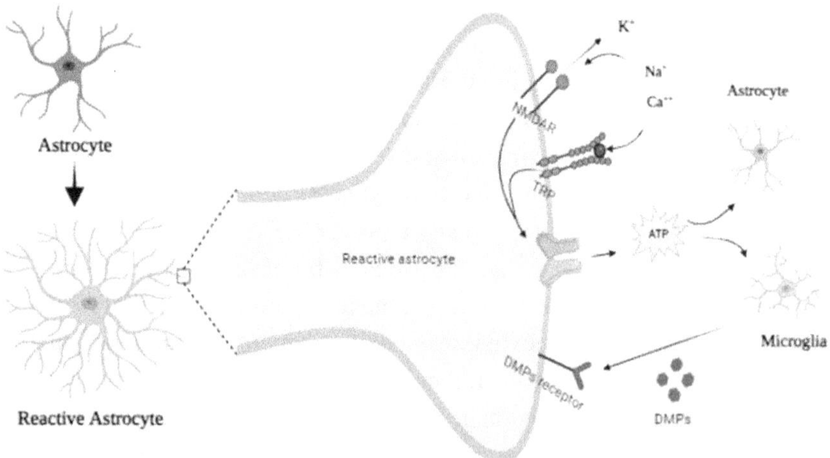

2. Inflammatory Response

The inflammatory response in the CNS involves the activation of various cell types, including astrocytes, in response to injury. Astrocytes are glial cells found in the CNS that typically support and protect neurons(sidoryk-wegrzynowicz et al., 2011). However, in response to injury, they can also release pro-inflammatory cytokines and other molecules. This release can serve both as a defense mechanism and as a potential contributor to secondary damage.(Pekny & Pekna, 2014)

As a defense mechanism, astrocytes detect signs of injury or inflammation within the CNS. When they identify damage or the presence of pathogens, they activate as a protective response to assist in eliminating the threat. The release of pro-inflammatory cytokines, such as interleukin-1β (IL-1β) and tumor necrosis factor-alpha (TNF-α), by astrocytes alerts the immune system to the problem, prompting other immune cells to migrate to the site of injury to help clear pathogens, damaged cells, and debris (Yu et al., 2021).

However, it's essential to note that the inflammatory response of astrocytes can sometimes become dysregulated, potentially leading to secondary damage. Excessive release of pro-inflammatory cytokines by astrocytes can result in chronic inflammation in the CNS, damaging healthy neurons and other cells. Prolonged inflammation can disrupt the blood-brain barrier, allowing immune cells and molecules to enter brain tissue, potentially worsening tissue damage and causing neuroinflammation (Burda et al., 2016).

In conditions such as neurodegenerative diseases (e.g., Alzheimer's, Parkinson's) or traumatic brain injuries, chronically activated astrocytes and sustained pro-inflammatory signaling can contribute to disease progression and worsen tissue damage over time. Understanding these mechanisms is crucial for developing therapeutic strategies to regulate the inflammatory response in the CNS (Cekanaviciute & Buckwalter, 2016; Boghdadi et al., 2020).

3. Scar Formation and Glial Scar

The formation of glial scars is a frequently observed response to injury and is primarily composed of astrocytes, as well as microglia, endothelial cells, fibroblasts, and extracellular matrix components (**Figure 3**) (Silver and Miller, 2004). Several mediators have been identified as being involved in triggering the formation of these glial scars, including TGF-β1, TGF-2, IFN-γ, FGF, and fibrinogen (DiProspero et al., 1997; Moon et Fawcett, 2001; Schachtrup et al., 2010). It is believed that these scars act as a physical barrier aimed at encapsulating damaged tissues, thus preventing toxic molecules and DAMPs (danger-associated molecular patterns) from entering healthy tissues, while also limiting access to invading cell types following an injury (**Figure 2**). However, it has also been demonstrated that they exert an inhibitory effect on axonal regrowth and regeneration (Ribotta et al., 2004; Silver and Miller, 2004).

Among the potential benefits of glial scar formation in brain repair is the isolation of damage. Glial scars help to isolate the damaged area, limiting the spread of injury-related factors to healthy neural tissues. This isolation prevents additional damage and protects adjacent neurons from toxic molecules. Furthermore, they form a barrier to the immune response by restricting the infiltration of immune cells. This can be both beneficial and detrimental, as it reduces inflammation and prevents further tissue damage, but it can also hinder certain aspects of the immune response necessary for tissue repair (Nicaise et al., 2022).

However, glial scar formation also comes with drawbacks. It can inhibit the regeneration of the central nervous system by creating physical and chemical barriers that impede axonal regeneration and synaptic plasticity, thus limiting functional recovery after an injury. Moreover, it can have long-term consequences by contributing to chronic neural dysfunction and inhibiting neural plasticity, thereby altering the brain's capacity to adapt and recover. Ultimately, the impact of glial scars on brain repair varies depending on the context and the extent of the injury. While they can offer short-term protection, their long-term effects on functional recovery are often mixed, and active research aims to develop strategies to modulate their formation (Fawcett & Asher, 1999) .

Figure 3. Reactive astrocytes in the formation of glial scar and where they form

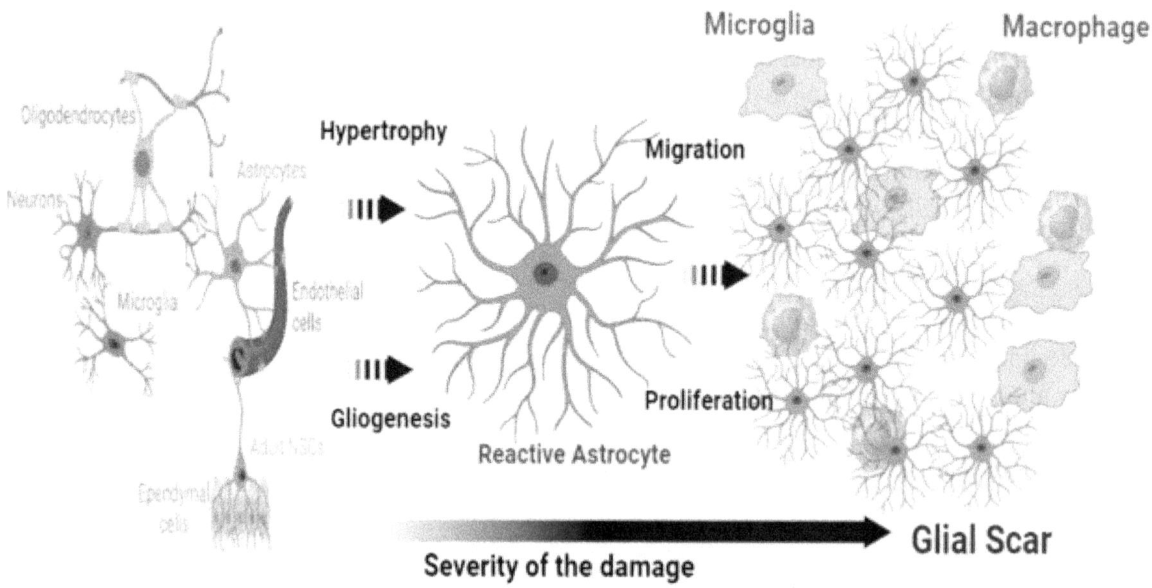

ASTROCYTES AND BRAIN INJURY RECOVERY

In the CNS, astrocytes are essential partners of neurons, in many brain diseases, astrocytes undergo morphological, molecular and functional changes to become reactive, and depending on the pathological context, these changes will have a significant impact on neurons (Poulot-becq-Giraudon et al., 2022). Animal studies now show the capacity of astrocytes in significant functional recovery after brain injury (Silver et al., 2015). Although astroglial cells are the most numerous cells in the SNC, their intervention in diseases that attack this nervous system remains limited. A study has highlighted the role of intermediate filaments in astroglial cells when the latter are subjected to severe mechanical or osmotic stress in cases of hypoxia and brain and spinal cord injury, when genetic ablation of these intermediate filaments is carried out. In mouse astrocytes, reactive gliosis is attenuated, the course of several CNS pathologies is modified, even signs of SNC regeneration become more prominent (Pekny & Pekna., 2004). Other studies reveal that astrocytes in spinal cord injuries express several molecules that promote axon growth, such that astrocyte scarring promotes SNC axon regeneration (Anderson et al., 2016). This stimulation of neurogenesis leads to the formation of scar tissue (Liberto et al., 2004).

In case of brain lesions, astrocytes play a key role in restoration of brain tissue by forming the glial scar; after severe SNC lesion, a significant deposition of extracellular matrix produced by activated astrocytes limits the extension of the lesion (Vedrenne, 2013). Additionally, research has shown that astrocytes play a vital role in the restoration of brain tissue after injury. This can promote new treatments allowing regeneration in the context of limiting damage to neurons during the reduction of blood flow following trauma or ischemia. Astrocytes are glial cells of the nervous system. These cells represent the majority of glial cells in the human cerebral cortex, in addition to their roles; mechanical support, nutrition and the regulation of nervous activity especially in the GABAergic system and the glutamate system (Endo et al., 2017). Studies show that astrocytes perform a variety of functions, including restoring damaged nerve tissue by minimizing inflammation at the injury site by surrounding nerve cells. On the surface of neural stem cells, the tyrosine kinase receptor ROR2 is highly expressed; however, the ROR2 gene is deactivated in the adult brain; results showed that ROR2 was significantly expressed in astrocytes at injured areas. This surface protein regulates the multiplication of neural stem cells, following this resonance, researchers propose that ROR2 controls the proliferation of astrocytes in injured areas. In this mouse study, inactivation of the ROR2 gene specifically in astrocytes resulted in a significant reduction in the number and density of proliferating astrocytes. Furthermore, activation of the ROR2 gene in cultured astrocytes was respected in response to basic fibroblast growth factor (BFGF) (Endo et al., 2017). In addition, astrocytes have a neuroprotective role following ischemic brain lesions, these glial cells have always been considered as defenders of neuronal function, it is a new pathophysiology pathway, during this operation, and after brain lesion focal, neurons and glial cells including astrocytes interact to determine the extent of the lesion in question. This protective pathway remains surprising as these cells protect or even help to regenerate nervous tissue after an acute attack (Trendelenburg et al., 2005). In the same context, astrocytic responses are considered early and important signs of the CNS following ischemic lesions (Panikar et al., 2005). In the case of a brain stab wound in adult rats, results showed that astrocyte reactivity was identified in the cerebral cortex, corpus callosum and deeper structures of the hemisphere, initially the area of reactivity. been limited to the vicinity of the wound, but subsequently this reactivity extends towards other areas. In light of these latter results, researchers proposed that astrocytes respond to mechanical disruption to injury, and that this response promotes the structural integrity of the injured tissue (Mathewson et al., 1985).

Post-embryonic neurogenesis is a hallmark feature of the vertebrate brain (Kaslin et al., 2008). Astrocytes, which are the main constituents of adult neurogenic niches, evolve as important regulators of neurogenesis in two ways: as neural stem cells (NSCs) and as niche cells (Schneider and al., 2019). During this neurogenesis, microglia and astrocytes play a leading role. Microglia, through their phagocytic capacity and their interactions with neurons, maintain the environment of the neurogenic niche. While astrocytes manipulate neurogenesis by releasing different soluble factors. such as adenosine triphosphate and lactate (Araki et al., 2021). In humans, a study which shed light on neurogenesis in the hippocampus of adults, however astrocytes regulate this neurogenesis by their juxtacrine signaling (cell-cell contact) to activate NSC this leads to the formation of neurons that affect learning and memory (Ashton et al., 2012). These researchers found that ephrin-B2 from rodent hippocampal astrocytes regulates neurogenesis in vivo via EphB4 receptors on NSCs to activate B-catenin (Ashton et al., 2012). However hippocampal neurogenesis is controlled by environmental conditions in the neurogenic niche, this process continues throughout life and plays a fundamental role in the maintenance of spatial cognitive function (Araki et al., 2021).

Astrocytes can be epigenetically reprogrammed to become neurons either naturally in vivo, or artificially via exogenous treatments to restore neurons lost due to disease or injury (Griffiths et al., 2020). Neurogenesis and regeneration are dependent on various factors, such as environmental stimuli, regional specificity, homology between species, etc. Although the precise mechanisms remain an area of intense research, astrocytes play a crucial role in the restoration of damaged nerve tissue, particularly inflammation at the site of injury, through their interaction with surrounding nerve cells. Astrocytes also have an impact on neurogenesis through their juxtracrine signaling. Furthermore, it is possible to epigenetically reprogram astrocytes to transform them into neurons, either naturally in vivo or artificially, opening the way to potential therapeutic applications.

ROLE OF BIOMATERIALS IN TBI TREATMENT

Biomaterials have long been of great interest in medical treatments. Initially, the aim was to provide mechanical and structural support for the treatment of age-related conditions, such as hip and dental prostheses. However, it is essential to note that these materials were practical solutions, not intended to replace more complex living tissues (Hench, 1998). Advances in biology have paved the way for 2nd-generation biomaterials, enabling them to exert some control over the environment (Yamamuro et al, 1990 ; Nasiri et al., 2016). 3rd-generation biomaterials have been developed to emit instructive molecular signals, respond to physiological conditions and finally adopt the property of biodegradability at the level of regenerated host tissues. These cutting-edge biomaterials are particularly well suited to applications in neural tissue engineering and TBI treatment. They aim to achieve goals such as tissue repair and regeneration, including the possibility of specifically targeting the inflammatory response (Maclean et al., 2018). Polymers of natural origin have interesting physicochemical and biological properties. They have the ability to maintain their functionality for a certain period of time and then degrade, making them effectively biodegradable in an in vivo environment. These biopolymers generally fall into two broad categories: polysaccharides, including chitosan and hyaluronic acid, and proteins, of which collagen is the typical representative (Kornev et al., 2018).

Chitosan (β-(1→4)-2-amino-D-glucose) is a linear polysaccharide derived from partial deacetylation of chitin. Chitosan has many benefits as a brain-targeted drug vehicle (drug carrier), especially nasal drug delivery. Along with its capacity to penetrate the BBB, it can also regulate release, cling to mucus, and open tight junctions in the nasal membrane. These qualities promote its use in drug delivery from the nose to the brain (Mohammed et al., 2017; Yao et al., 2018). The versatility of chitosan, which can be used for drug delivery in the shape of microspheres, capsules, hydrogels, conjugates, nanoparticles, films, beads, or tablets, is another benefit (Hamdan *et al.*, 2023a). Currently, chitosan and its derivatives as plausible molecules with powerful biological and therapeutic features to target neurodegenerative disorders are well documented in several studies (Hao *et al.*, 2017; Hamdan et al., 2023b). Chitosan is suitable for a wide range of biomedical uses, including drug transport, gene delivery, tissue engineering, and regenerative therapies, among others, thanks to all these characteristics. Within our team, trials for tissue engineering by formation of hydrogels for therapeutic use have been studied with a terrestrial source of beetle for the first time (Hamdan et al., 2023a). However, the low solubility of this polysaccharide limits its use in food and biomedical applications. Chitosan-based derivatives (oligosacharides, or chemically-modified derivatives) are water-soluble and could be useful in the future.

Hyaluronic acid (HA) is a natural polysaccharide composed of two alternating units, β-1,4-D-glucuronic acid and β-1,3-N-acetyl-D-glucosamine. This is one of the most fundamental and physiologically relevant constituents of the brain's extracellular matrix (ECM). As an essential component of MECC, HA-based hydrogels exhibit mechanical characteristics and rheological behavior similar to those of cerebral tissue (Khaing et al., 2015). Implantation of HA into cortical defects controls neuronal cell adhesion and promotes the creation of new neuronal networks. After 6 weeks, there was a reduction in foreign body response and inhibition of glial scar formation at the implantation site, suggesting excellent compatibility of this hydrogel with cerebral tissue (Yao et al., 2018).

Collagen is an essential component of the ECM, accounting for some 25% of the proteins found in the human body (Friess, 1998). Sprout has developed great interest in the field of neurorepair due to its high biodegradability and remarkable biocompatibility, as demonstrated by various in vivo tests (Ucar et al., 2018). Implantation of a glycosaminoglycan (collagen-GAG) scaffold after TBI significantly reduces lesion size compared with the control group. Cellular markers such as doublecortin and neuronal nuclear protein indicating the significant presence of migrating neuroblasts and mature neurons in regenerated tissue after collagen-GAG implantation, leading to functional recovery from TBI (Huang et al., 2012). As far as synthetic polymers are concerned, materials such as polyesters like poly(lactic acid) (PLA), poly(glycolic acid) (PGA) and polyanhydrides also offer extremely attractive properties. The PLGA scaffold promoted the survival of neural stem cells (NSPCs) for over 2 weeks, helping them to develop into neurons, establish synaptic connections and display synaptic activity. In addition, PLGA-based nanoparticles encapsulated over 80% of stromal cell-derived factor 1 (SDF-1) and released it over several days. Injection of these SDF-1-loaded nanoparticles at the site of injury attracted large numbers of NSPCs to the damaged area (Xiong et al., 2009; Zamproni et al., 2017). PCL is a synthetic polymer frequently employed in tissue engineering due to its availability, biological compatibility and ease of processing (Wu et al., 2016). PCL nanofibers support, maintain and regulate NSPC differentiation. These PCL nanofibers enable NSPCs to differentiate into neuronal cells and guide axonal growth by emitting guidance signals on contact (Nisbet et al.,2008). Following implantation of a PCL scaffold at brain injury sites, the scaffold did not elicit a glial healing response or visible signs of foreign body injury (Nisbet et al., 2008).

In addition, other work has shown that the use of a materials-based hydrogel (epidermal growth factor and astrocytes) enhanced astrocyte proliferation and activation, suggesting that this hydrogel can transform astrocytes into neuro-supportive phenotypes, promoting neuroprotection and neuroplasticity (Chan et al., 2019). Figure 4 will schematically summarize the combination of 3D printing technology with biomaterial-based tissue engineering and its role in supporting and promoting astrocytes in TBI.

Figure 4. Implication of biomaterials and 3D printing technology in the management of TBI

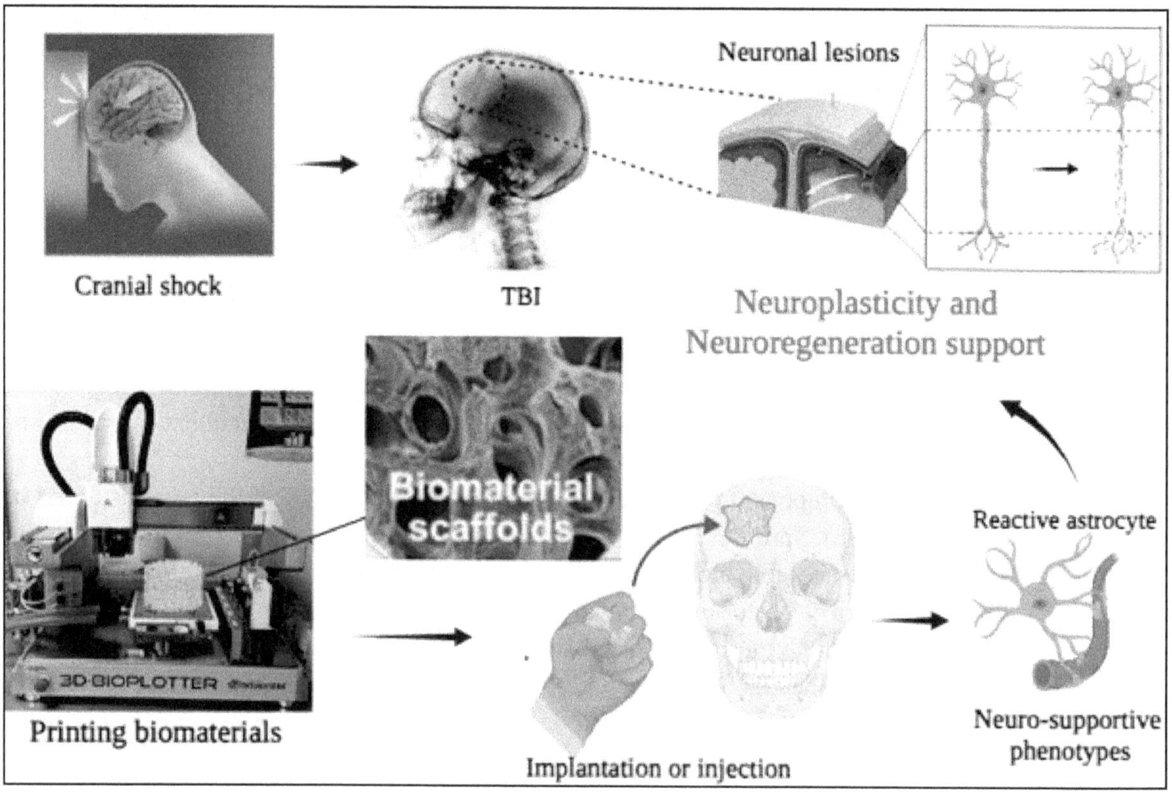

CONCLUSION

Accidents across the globe represent a major injury as far as TBI is concerned. The pathological process that is triggered after a TBI is extremely complex, involving the interrelation of various cell types and being influenced by environmental modifications. The adverse consequences associated with TBI are highly variable in humans. For this reason, detailed analysis of the specific functions of astrocytes and their behavior, as well as the mechanisms underlying the pathological features of TBI at different times, in different locations and in different types of lesions could facilitate the development of appropriate interventions for the treatment of TBI. Biomaterial-based therapeutic approaches offer effective and circumventable solutions by using scaffolds to support astrocytes in order to limit and counteract brain damage. It would be desirable and advisable to consider potential avenues for future studies as well as the implications arising from a better understanding of the contribution of astrocytes to brain health and recovery.

REFERENCES

Abbott, N. J., Rönnbäck, L., & Hansson, E. (2006). Astrocyte–endothelial interactions at the blood–brain barrier. *Nature Reviews. Neuroscience*, *7*(1), 1. doi:10.1038/nrn1824 PMID:16371949

Araki, T., Ikegaya, Y., & Koyama, R. (2021). The effects of microglia-and astrocyte-derived factors on neurogenesis in health and disease. *The European Journal of Neuroscience*, *54*(5), 5880–5901. doi:10.1111/ejn.14969 PMID:32920880

Ashton, R. S., Conway, A., Pangarkar, C., Bergen, J., Lim, K. I., Shah, P., Bissell, M., & Schaffer, D. V. (2012). Astrocytes regulate adult hippocampal neurogenesis through ephrin-B signaling. *Nature Neuroscience*, *15*(10), 1399–1406. doi:10.1038/nn.3212 PMID:22983209

Balasingam, V., Tejada-Berges, T., Wright, E., Bouckova, R., & Yong, V. W. (1994). Reactive astrogliosis in the neonatal mouse brain and its modulation by cytokines. *The Journal of Neuroscience : The Official Journal of the Society for Neuroscience*, *14*(2), 846–856. doi:10.1523/JNEUROSCI.14-02-00846.1994 PMID:8301364

Beard, E., Lengacher, S., Dias, S., Magistretti, P. J., & Finsterwald, C. (2022). Astrocytes as Key Regulators of Brain Energy Metabolism : New Therapeutic Perspectives. *Frontiers in Physiology*, *12*, 825816. doi:10.3389/fphys.2021.825816 PMID:35087428

Bélanger, M., & Magistretti, P. J. (2009). The role of astroglia in neuroprotection. *Dialogues in Clinical Neuroscience*, *11*(3), 281–295. doi:10.31887/DCNS.2009.11.3/mbelanger PMID:19877496

Bélanger, M., & Magistretti, P. J. (2009). The role of astroglia in neuroprotection. *Dialogues in Clinical Neuroscience*, *11*(3), 281–295. doi:10.31887/DCNS.2009.11.3/mbelanger PMID:19877496

Bélanger, M., & Magistretti, P. J. (2022). The role of astroglia in neuroprotection. *Dialogues in Clinical Neuroscience*. PMID:19877496

Bellot-Saez, A., Kékesi, O., Morley, J. W., & Buskila, Y. (2017). Astrocytic modulation of neuronal excitability through K+ spatial buffering. *Neuroscience and Biobehavioral Reviews*, *77*, 87–97. doi:10.1016/j.neubiorev.2017.03.002 PMID:28279812

Blennow, K., Brody, D. L., Kochanek, P. M., Levin, H., McKee, A., Ribbers, G. M., Yaffe, K., & Zetterberg, H. (2016). Traumatic brain injuries. *Nature Reviews. Disease Primers*, *2*(1), 1–19. doi:10.1038/nrdp.2016.84 PMID:27853132

Boghdadi, A. G., Teo, L., & Bourne, J. A. (2020). The Neuroprotective Role of Reactive Astrocytes after Central Nervous System Injury. *Journal of Neurotrauma*, *37*(5), 681–691. doi:10.1089/neu.2019.6938 PMID:32031052

Buffo, A., Rolando, C., & Ceruti, S. (2010). Astrocytes in the damaged brain : Molecular and cellular insights into their reactive response and healing potential. *Biochemical Pharmacology*, *79*(2), 77–89. doi:10.1016/j.bcp.2009.09.014 PMID:19765548

Burda, J. E., Bernstein, A. M., & Sofroniew, M. V. (2016). Astrocyte roles in traumatic brain injury. *Experimental neurology, 275*(03), 305-315. doi:10.1016/j.expneurol.2015.03.020

Cabezas, R., Ávila, M., Gonzalez, J., El-Bachá, R. S., Báez, E., García-Segura, L. M., Jurado Coronel, J. C., Capani, F., Cardona-Gomez, G. P., & Barreto, G. E. (2014). Astrocytic modulation of blood brain barrier : Perspectives on Parkinson's disease. *Frontiers in Cellular Neuroscience*, *8*, 211. doi:10.3389/fncel.2014.00211 PMID:25136294

Chan, S. J., Niu, W., Hayakawa, K., Hamanaka, G., Wang, X., Cheah, P. S., Guo, S., Yu, Z., Arai, K., Selim, M. H., Kurisawa, M., Spector, M., & Lo, E. H. (2019). Promoting neuro-supportive properties of astrocytes with epidermal growth factor hydrogels. *Stem Cells Translational Medicine*, *8*(12), 1242–1248. doi:10.1002ctm.19-0159 PMID:31483567

Chen, Z., Yuan, Z., Yang, S., Zhu, Y., Xue, M., Zhang, J., & Leng, L. (2023). Brain Energy Metabolism : Astrocytes in Neurodegenerative Diseases. *CNS Neuroscience & Therapeutics*, *29*(1), 24–36. doi:10.1111/cns.13982 PMID:36193573

Chiareli, R. A., Carvalho, G. A., Marques, B. L., Mota, L. S., Oliveira-Lima, O. C., Gomes, R. M., Birbrair, A., Gomez, R. S., Simão, F., Klempin, F., Leist, M., & Pinto, M. C. X. (2021). The Role of Astrocytes in the Neurorepair Process. *Frontiers in Cell and Developmental Biology*, *9*, 665795. https://www.frontiersin.org/articles/10.3389/fcell.2021.665795. doi:10.3389/fcell.2021.665795 PMID:34113618

Endo, M., Ubulkasim, G., Kobayashi, C., Onishi, R., Aiba, A., & Minami, Y. (2017). Critical role of Ror2 receptor tyrosine kinase in regulating cell cycle progression of reactive astrocytes following brain injury. *Glia*, *65*(1), 182–197. doi:10.1002/glia.23086 PMID:27726178

Eng, L. F., Ghirnikar, R. S., & Lee, Y. L. (2000). Glial fibrillary acidic protein : GFAP-thirty-one years (1969–2000). *Neurochemical Research*, *25*(9/10), 1439–1451. doi:10.1023/A:1007677003387 PMID:11059815

Fawcett, J. W., & Asher, R. A. (1999). The glial scar and central nervous system repair. *Brain Research Bulletin*, *49*(6), 377–391. doi:10.1016/S0361-9230(99)00072-6 PMID:10483914

Friess, W. (1998). Collagen–biomaterial for drug delivery. *European Journal of Pharmaceutics and Biopharmaceutics*, *45*(2), 113–136. doi:10.1016/S0939-6411(98)00017-4 PMID:9704909

Gopalakrishnan, A., Shankarappa, S. A., & Rajanikant, G. K. (2019). Hydrogel scaffolds: Towards restitution of ischemic stroke-injured brain. *Translational Stroke Research*, *10*(1), 1–18. doi:10.100712975-018-0655-6 PMID:30151667

Göritz, C., Dias, D. O., Tomilin, N., Barbacid, M., Shupliakov, O., & Frisén, J. (2011). A pericyte origin of spinal cord scar tissue. *Science*, *333*(6039), 238–242. doi:10.1126cience.1203165 PMID:21737741

Griffiths, B. B., Bhutani, A., & Stary, C. M. (2020). Adult neurogenesis from reprogrammed astrocytes. *Neural Regeneration Research*, *15*(6), 973. doi:10.4103/1673-5374.270292 PMID:31823866

Häberle, J., Görg, B., Toutain, A., Rutsch, F., Benoist, J.-F., Gelot, A., Suc, A.-L., Koch, H. G., Schliess, F., & Häussinger, D. (2006). Inborn error of amino acid synthesis : Human glutamine synthetase deficiency. *Journal of Inherited Metabolic Disease*, *29*(2-3), 352–358. doi:10.100710545-006-0256-5 PMID:16763901

Hamdan, Y. A., Elouali, S., Eladlani, N., Lefeuvre, B., Oudadesse, H., & Rhazi, M. (2023a). Investigation on Akis granulifera (Coleoptera, Sahlberg, 1823) as a potential source of chitin and chitosan: Extraction, characterization and hydrogel formation. *International Journal of Biological Macromolecules*, *252*, 126292. doi:10.1016/j.ijbiomac.2023.126292 PMID:37573901

Hamdan, Y. A., Elouali, S., Maloui, A. B., El-Mansoury, B., Oudadesse, H., & Rhazi, M. (2023b). Chitosan and Its Derivatives as Potential Neuro-Protective Agents for the Treatment of Parkinson's Disease. In Experimental and Clinical Evidence of the Neuropathology of Parkinson's Disease (pp. 253-274). IGI Global. doi:10.4018/978-1-6684-5156-4.ch015

Hao, C., Wang, W., Wang, S., Zhang, L., & Guo, Y. (2017). An overview of the protective effects of chitosan and acetylated chitosan oligosaccharides against neuronal disorders. *Marine Drugs*, *15*(4), 89. doi:10.3390/md15040089 PMID:28333077

Hao, C., Wang, W., Wang, S., Zhang, L., & Guo, Y. (2017). An Overview of the Protective Effects of Chitosan and Acetylated Chitosan Oligosaccharides against Neuronal Disorders. MDPI. . doi:10.3390/md15040089

He, Y., Hakvoort, T. B., Vermeulen, J. L., Lamers, W. H., & Van Roon, M. A. (2007). Glutamine synthetase is essential in early mouse embryogenesis. *Developmental Dynamics*, *236*(7), 1865–1875. doi:10.1002/dvdy.21185 PMID:17557305

Hench, L. L. (1998). Biomaterials: A forecast for the future. *Biomaterials*, *19*(16), 1419–1423. doi:10.1016/S0142-9612(98)00133-1 PMID:9794512

Huang, K. F., Hsu, W. C., Chiu, W. T., & Wang, J. Y. (2012). Functional improvement and neurogenesis after collagen-GAG matrix implantation into surgical brain trauma. *Biomaterials*, *33*(7), 2067–2075. doi:10.1016/j.biomaterials.2011.11.040 PMID:22169139

Jessen, K. R. (2004). Glial cells. *The International Journal of Biochemistry & Cell Biology*, *36*(10), 1861–1867. doi:10.1016/j.biocel.2004.02.023 PMID:15203098

Jessen, K. R. (2004). Glial cells. *The International Journal of Biochemistry & Cell Biology*, *36*(10), 1861–1867. doi:10.1016/j.biocel.2004.02.023 PMID:15203098

Jiwaji, Z., Tiwari, S. S., Avilés-Reyes, R. X., Hooley, M., Hampton, D., Torvell, M., Johnson, D. A., McQueen, J., Baxter, P., Sabari-Sankar, K., Qiu, J., He, X., Fowler, J., Febery, J., Gregory, J., Rose, J., Tulloch, J., Loan, J., Story, D., & Hardingham, G. E. (2022). Reactive astrocytes acquire neuroprotective as well as deleterious signatures in response to Tau and Aß pathology. *Nature Communications*, *13*(1), 1. doi:10.103841467-021-27702-w PMID:35013236

Kadry, H., Noorani, B., & Cucullo, L. (2020). A blood–brain barrier overview on structure, function, impairment, and biomarkers of integrity. *Fluids and Barriers of the CNS*, *17*(1), 69. doi:10.118612987-020-00230-3 PMID:33208141

Kamins, J., & Giza, C. C. (2016). Concussion—mild traumatic brain injury: Recoverable injury with potential for serious sequelae. *Neurosurgery Clinics of North America*, *27*(4), 441–452. doi:10.1016/j.nec.2016.05.005 PMID:27637394

Kaslin, J., Ganz, J., & Brand, M. (2008). Proliferation, neurogenesis and regeneration in the non-mammalian vertebrate brain. *Philosophical Transactions of the Royal Society of London. Series B, Biological Sciences*, *363*(1489), 101–122. doi:10.1098/rstb.2006.2015 PMID:17282988

Khaing, Z. Z., & Seidlits, S. K. (2015). Hyaluronic acid and neural stem cells: Implications for biomaterial design. *Journal of Materials Chemistry. B*, *3*(40), 7850–7866. doi:10.1039/C5TB00974J PMID:32262899

Kim, Y., Park, J., & Choi, Y. K. (2019). The Role of Astrocytes in the Central Nervous System Focused on BK Channel and Heme Oxygenase Metabolites : A Review. *Antioxidants*, *8*(5), 121. doi:10.3390/antiox8050121 PMID:31060341

Kornev, V. A., Grebenik, E. A., Solovieva, A. B., Dmitriev, R. I., & Timashev, P. S. (2018). Hydrogel-assisted neuroregeneration approaches towards brain injury therapy: A state-of-the-art review. *Computational and Structural Biotechnology Journal*, *16*, 488–502. doi:10.1016/j.csbj.2018.10.011 PMID:30455858

Liu, C.-Y., Yang, Y., Ju, W.-N., Wang, X., & Zhang, H.-L. (2018). Emerging Roles of Astrocytes in Neuro-Vascular Unit and the Tripartite Synapse With Emphasis on Reactive Gliosis in the Context of Alzheimer's Disease. *Frontiers in Cellular Neuroscience*, *12*, 193. https://www.frontiersin.org/articles/10.3389/fncel.2018.00193. doi:10.3389/fncel.2018.00193 PMID:30042661

MacVicar, B. A., & Choi, H. B. (2017). Astrocytes Provide Metabolic Support for Neuronal Synaptic Function in Response to Extracellular K+. *Neurochemical Research*, *42*(9), 2588–2594. doi:10.100711064-017-2315-8 PMID:28664400

Mahmoud, S., Gharagozloo, M., Simard, C., & Gris, D. (2019). Astrocytes Maintain Glutamate Homeostasis in the CNS by Controlling the Balance between Glutamate Uptake and Release. *Cells*, *8*(2), 2. doi:10.3390/cells8020184 PMID:30791579

Mathewson, A. J., & Berry, M. (1985). Observations on the astrocyte response to a cerebral stab wound in adult rats. *Brain Research*, *327*(1-2), 61–69. doi:10.1016/0006-8993(85)91499-4 PMID:3986520

McNeill, J., Rudyk, C., Hildebrand, M. E., & Salmaso, N. (2021). Ion Channels and Electrophysiological Properties of Astrocytes : Implications for Emergent Stimulation Technologies. *Frontiers in Cellular Neuroscience*, *15*, 644126. doi:10.3389/fncel.2021.644126 PMID:34093129

Michinaga, S., & Koyama, Y. (2019). Dual roles of astrocyte-derived factors in regulation of blood-brain barrier function after brain damage. *International Journal of Molecular Sciences*, *20*(3), 571. doi:10.3390/ijms20030571 PMID:30699952

Michinaga, S., & Koyama, Y. (2019). Dual Roles of Astrocyte-Derived Factors in Regulation of Blood-Brain Barrier Function after Brain Damage. *International Journal of Molecular Sciences*, *20*(3), 571. doi:10.3390/ijms20030571 PMID:30699952

Mohammed, M. A., Syeda, J. T., Wasan, K. M., & Wasan, E. K. (2017). An overview of chitosan nanoparticles and its application in non-parenteral drug delivery. *Pharmaceutics*, *9*(4), 53. doi:10.3390/pharmaceutics9040053 PMID:29156634

Nasiri, N., Ceramidas, A., Mukherjee, S., Panneerselvan, A., Nisbet, D. R., & Tricoli, A. (2016). Ultra-porous nanoparticle networks: A biomimetic coating morphology for enhanced cellular response and infiltration. *Scientific Reports*, *6*(1), 24305. doi:10.1038rep24305 PMID:27076035

Nicaise, A. M., D'Angelo, A., Ionescu, R.-B., Krzak, G., Willis, C. M., & Pluchino, S. (2022). The role of neural stem cells in regulating glial scar formation and repair. *Cell and Tissue Research*, *387*(3), 399–414. doi:10.100700441-021-03554-0 PMID:34820704

Nisbet, D. R., Rodda, A. E., Horne, M. K., Forsythe, J. S., & Finkelstein, D. I. (2009). Neurite infiltration and cellular response to electrospun polycaprolactone scaffolds implanted into the brain. *Biomaterials*, *30*(27), 4573–4580. doi:10.1016/j.biomaterials.2009.05.011 PMID:19500836

Nisbet, D. R., Yu, L. M. Y., Zahir, T., Forsythe, J. S., & Shoichet, M. S. (2008). Characterization of neural stem cells on electrospun poly (ε-caprolactone) submicron scaffolds: Evaluating their potential in neural tissue engineering. *Journal of Biomaterials Science. Polymer Edition*, *19*(5), 623–634. doi:10.1163/156856208784089652 PMID:18419941

Ota, Y., Zanetti, A. T., & Hallock, R. M. (2013). The Role of Astrocytes in the Regulation of Synaptic Plasticity and Memory Formation. *Neural Plasticity*, *185463*, 1–11. doi:10.1155/2013/185463 PMID:24369508

Panickar, K. S., & Norenberg, M. D. (2005). Astrocytes in cerebral ischemic injury: Morphological and general considerations. *Glia*, *50*(4), 287–298. doi:10.1002/glia.20181 PMID:15846806

Pekny, M., & Pekna, M. (2014). Astrocyte Reactivity and Reactive Astrogliosis : Costs and Benefits. *Physiological Reviews*, *94*(4), 1077–1098. doi:10.1152/physrev.00041.2013 PMID:25287860

Peters, A. (2004). A fourth type of neuroglial cell in the adult central nervous system. *Journal of Neurocytology*, *33*(3), 345–357. doi:10.1023/B:NEUR.0000044195.64009.27 PMID:15475689

Planas, A. M., Justicia, C., Soriano, M. A., & Ferrer, I. (1998). Epidermal growth factor receptor in proliferating reactive glia following transient focal ischemia in the rat brain. *Glia*, *23*(2), 120–129. doi:10.1002/(SICI)1098-1136(199806)23:2<120::AID-GLIA3>3.0.CO;2-A PMID:9600380

Robel, S., Bardehle, S., Lepier, A., Brakebusch, C., & Götz, M. (2011). Genetic deletion of cdc42 reveals a crucial role for astrocyte recruitment to the injury site in vitro and in vivo. *The Journal of Neuroscience : The Official Journal of the Society for Neuroscience*, *31*(35), 12471–12482. doi:10.1523/JNEUROSCI.2696-11.2011 PMID:21880909

Robel, S., Mori, T., Zoubaa, S., Schlegel, J., Sirko, S., Faissner, A., Goebbels, S., Dimou, L., & Götz, M. (2009). Conditional deletion of β1-integrin in astroglia causes partial reactive gliosis. *Glia*, *57*(15), 1630–1647. doi:10.1002/glia.20876 PMID:19373938

Rose, C. F., Verkhratsky, A., & Parpura, V. (2013). Astrocyte glutamine synthetase : Pivotal in health and disease. *Biochemical Society Transactions*, *41*(6), 1518–1524. doi:10.1042/BST20130237 PMID:24256247

Schneider, J., Karpf, J., & Beckervordersandforth, R. (2019). Role of astrocytes in the neurogenic niches. *Astrocytes: Methods and Protocols*, 19-33.

Seifert, G., Schilling, K., & Steinhäuser, C. (2006). Astrocyte dysfunction in neurological disorders : A molecular perspective. *Nature Reviews. Neuroscience*, *7*(3), 194–206. doi:10.1038/nrn1870 PMID:16495941

Sharma, K., Zhang, G., & Li, S. (2015). Astrogliosis and Axonal Regeneration. In K.-F. So & X.-M. Xu (Eds.), *Neural Regeneration* (pp. 181–196). Academic Press. doi:10.1016/B978-0-12-801732-6.00011-2

Sidoryk-Wegrzynowicz, M., Wegrzynowicz, M., Lee, E., Bowman, A. B., & Aschner, M. (2011). Role of Astrocytes in Brain Function and Disease. *Toxicologic pathology, 39*(1), 115-123. doi:10.1177/0192623310385254

Smith, G. M., & Strunz, C. (2005). Growth factor and cytokine regulation of chondroitin sulfate proteoglycans by astrocytes. *Glia, 52*(3), 209–218. doi:10.1002/glia.20236 PMID:15968632

Sofroniew, M. V. (2015). Astrogliosis. *Cold Spring Harbor Perspectives in Biology, 7*(2), a020420. doi:10.1101/cshperspect.a020420 PMID:25380660

Sriram, K., Benkovic, S. A., Hebert, M. A., Miller, D. B., & O'Callaghan, J. P. (2004). Induction of gp130-related cytokines and activation of JAK2/STAT3 pathway in astrocytes precedes up-regulation of glial fibrillary acidic protein in the 1-methyl-4-phenyl-1, 2, 3, 6-tetrahydropyridine model of neurodegeneration : Key signaling pathway for astrogliosis in vivo? *The Journal of Biological Chemistry, 279*(19), 19936–19947. doi:10.1074/jbc.M309304200 PMID:14996842

Trendelenburg, G., & Dirnagl, U. (2005). Neuroprotective role of astrocytes in cerebral ischemia: Focus on ischemic preconditioning. *Glia, 50*(4), 307–320. doi:10.1002/glia.20204 PMID:15846804

Turner, D. A., & Adamson, D. C. (2011). Neuronal-Astrocyte Metabolic Interactions : Understanding the Transition into Abnormal Astrocytoma Metabolism. *Journal of Neuropathology and Experimental Neurology, 70*(3), 167–176. doi:10.1097/NEN.0b013e31820e1152 PMID:21293295

Ucar, B., & Humpel, C. (2018). Collagen for brain repair: Therapeutic perspectives. *Neural Regeneration Research, 13*(4), 595. doi:10.4103/1673-5374.230273 PMID:29722301

Vainchtein, I. D., & Molofsky, A. V. (2020). Astrocytes and microglia: In sickness and in health. *Trends in Neurosciences, 43*(3), 144–154. doi:10.1016/j.tins.2020.01.003 PMID:32044129

Vedrenne, N. (2013). *Roles of different astrocyte populations in repair mechanisms after central nervous system injury*[Dissertation, Limoges].

Verkhratsky, A., Untiet, V., & Rose, C. R. (2020). Ionic signalling in astroglia beyond calcium. *The Journal of Physiology, 598*(9), 1655–1670. doi:10.1113/JP277478 PMID:30734296

Wu, Y., Sriram, G., Fawzy, A. S., Fuh, J. Y., Rosa, V., Cao, T., & Wong, Y. S. (2016). Fabrication and evaluation of electrohydrodynamic jet 3D printed polycaprolactone/chitosan cell carriers using human embryonic stem cell-derived fibroblasts. *Journal of Biomaterials Applications, 31*(2), 181–192. doi:10.1177/0885328216652537 PMID:27252227

Xiong, Y., Zeng, Y. S., Zeng, C. G., Du, B. L., He, L. M., Quan, D. P., Zhang, W., Wang, J.-M., Wu, J.-L., Li, Y., & Li, J. (2009). Synaptic transmission of neural stem cells seeded in 3-dimensional PLGA scaffolds. *Biomaterials, 30*(22), 3711–3722. doi:10.1016/j.biomaterials.2009.03.046 PMID:19375792

Yamamuro, T., Hench, L. L., & Wilson, J. (1990). CRC handbook of bioactive ceramics. (No Title). Maclean, F. L., Horne, M. K., Williams, R. J., & Nisbet, D. R. (2018). Biomaterial systems to resolve brain inflammation after traumatic injury. *APL Bioengineering, 2*(2).

Yao, Z. A., Chen, F. J., Cui, H. L., Lin, T., Guo, N., & Wu, H. G. (2018). Efficacy of chitosan and sodium alginate scaffolds for repair of spinal cord injury in rats. *Neural Regeneration Research*, *13*(3), 502. doi:10.4103/1673-5374.228756 PMID:29623937

Yao, Z. A., Chen, F. J., Cui, H. L., Lin, T., Guo, N., & Wu, H. G. (2018). Efficacy of chitosan and sodium alginate scaffolds for repair of spinal cord injury in rats. *Neural Regeneration Research*, *13*(3), 502. doi:10.4103/1673-5374.228756 PMID:29623937

Yu, G., Zhang, Y., & Ning, B. (2021). Reactive Astrocytes in Central Nervous System Injury : Subgroup and Potential Therapy. *Frontiers in Cellular Neuroscience*, *15*, 792764. doi:10.3389/fncel.2021.792764 PMID:35002629

Zamproni, L. N., Mundim, M. V., Porcionatto, M. A., & des Rieux, A. (2017). Injection of SDF-1 loaded nanoparticles following traumatic brain injury stimulates neural stem cell recruitment. *International Journal of Pharmaceutics*, *519*(1-2), 323–331. doi:10.1016/j.ijpharm.2017.01.036 PMID:28115261

Chapter 9
Parkinson's Disease Involving Glial Cell Dysfunction

Mjid Oukhrib
https://orcid.org/0000-0003-0199-4054
Faculty of Sciences Semlalia, Cadi Ayyad University, Morocco

Lahcen Tamegart
Faculty of Science, Abdelmalek Essaâdi University, Morocco

Hafida El Ghachi
https://orcid.org/0000-0002-2043-6202
Faculty of Sciences Semlalia, Cadi Ayyad University, Morocco

Asmaa Haj-Khlifa
Faculty of Sciences Semlalia, Cadi Ayyad University, Morocco

Youssef Ait Hamdan
https://orcid.org/0009-0005-8616-8852
University of Rennes, France & Higher Normal School, Morocco

Abdelali Ben Maloui
Faculty of Sciences Semlalia, Cadi Ayyad University, Morocco

Halima Gamrani
Faculty of Sciences Semlalia, Cadi Ayyad University, Morocco

Mohamed Chraa
Faculty of Medicine and Pharmacy, Cadi Ayyad University, Morocco

ABSTRACT

It is well known that the decline in the viability of dopamine neurons slowly leads to the appearance of various symptoms characteristic of Parkinson's disease (PD). These symptoms depend mainly on dysfunction of nigrostriatal dopaminergic denervation. The cause of neuronal death has not yet been elucidated, although there are several hypotheses suggesting different factors that may trigger it. One possible mechanism of neurodegeneration is the establishment of chronic inflammation in the central nervous system, where glial cells are key regulators of inflammatory responses. They also play a phagocytic role, engulfing synapses, apoptotic cells, cellular debris and released toxic proteins. An imbalance in the activation of these cells can lead to an overproduction of cytotoxic factors, which contribute to the death of dopamine neurons. As PD involves not only the loss of dopamine neurons, but also the dysfunction of glial cells, whose loss or excessive activation can contribute to neuronal death, there is a need to better understand the role of these cells in PD in order to develop effective therapies.

DOI: 10.4018/978-1-6684-9675-6.ch009

INTRODUCTION

Parkinson's disease (PD) is the second most common neurodegenerative disorder, affecting around 10 million people worldwide (Pringsheim et al., 2014). Clinically, PD is diagnosed by the recognition of cardinal symptoms of the motor system, such as resting tremor, bradykinesia, muscle rigidity and postural instability (Kim et al., 2011). Motor disorders are due to the progressive degradation of dopaminergic neurons in the substantia nigra pars compacta (SNpc) and striatum along the nigrostriatal pathway (Langston, 2006; Lees et al., 2009). Although less studied than motor symptoms, a number of studies show that non-motor symptoms of PD play an important role in predicting the emergence of clinical motor symptoms in PD (Dujardin et al., 1999; Weintraub & Burn, 2011).

There have been no substantial changes in the treatment pathway for PD in recent years, and the gold standard treatment has been levodopa since its inception in the early 1960s (Witt & Fahn, 2016). However, it is important to note that levodopa is only beneficial in the early stages of the disease, as continued use can lead to undesirable side effects such as dyskinesia, dependence and depression (Jankovic & Aguilar, 2008; Jimenez-Shahed et al., 2016). Dopamine agonists and inhibitors of dopamine-degrading enzymes have been used as alternative means of controlling or delaying the progression of PD, but with little efficacy in the advanced stages of the disease (Dextera & Jenner, 2013; Rascol et al., 2003). In addition, for people with significant motor difficulties, surgical procedures such as deep brain stimulation of the medial globus pallidus, subthalamic nucleus and pedunculopontine nucleus have been used (Okun, 2012). Stem cell-based methods of treating PD have been presented as a new treatment pathway. These procedures are being further investigated to establish their efficacy in the treatment of PD patients. Several types of stem cells have been studied and have shown promise for the treatment of PD (Goodarzi et al., 2015.; Zhang et al., 2017), including embryonic stem cells (ESCs) (Brederlau et al., 2006), neural stem cells (NSCs) (Daadi et al., 2012), mesenchymal stem cells (MSCs) (Z. Zhang et al., 2008) and induced pluripotent stem cells (iPSCs) (Savchenko et al., 2018).

Glial cells have recently emerged as a promising source for the development of therapeutic approaches for neurodegenerative diseases such as PD (Bradl & Lassmann, 2010; J. Wang et al., 2013). It remains unclear whether glial cells play a role in the development of PD or whether they represent a viable therapeutic option. Indeed, previous research has emphasised a dual function, indicating that glial cells may switch from a neuroprotective to a neurodegenerative profile throughout the development and progression of PD, although this remains to be proven (De Miranda et al., 2018; Yue et al., 2018).

This chapter will attempt to review what is currently known about several glial cell types. It will also identify their important tasks in healthy brains and in brains affected by PD, and determine whether these glial cells play a role in the onset and progression of PD, and whether they are possible therapeutic targets and agents for the treatment of the disease.

THE GLIAL SYSTEM: ROLES AND FUNCTIONS

In terms of cellular diversity, the CNS is made up of several cell types in addition to neurons, namely glial cells, whose function has not been well understood. For many years, these glial cells were considered only as support cells maintaining the viability of neurons. From the point of view of cellular and functional diversity, glial cells are superior to neurons (Fields et al., 2014), and are divided mainly into three essential groups: astrocytes, microglia and oligodendrocytes, whose cells can regulate neuronal activity and development (Clarke & Barres, 2013; Jäkel & Dimou, 2017; Postuma et al., 2015; Subhramanyam et al., 2019). Studies have shown that astrocytes and microglia play important roles as promoters of axonal growth, modulators of morphological plasticity of neuronal receptive terminals, and in dendritic extension (Bitzer-Quintero et al., 2012; Vinet et al., 2012). Glial cells generally communicate with each other and respond to various stimuli via intercellular diffusion of specific gliotransmitters, such as glutamate and ATP, and via intracellular waves of calcium (Ca2+) (Araque et al., 2000; Fields & Stevens-Graham, 2002; Hirase et al., 2004; Bazargani & Attwell, 2016).

Astrocytes

In mammals, astrocytes represent the most common class of glial cells with a percentage of 20 to 40% and show functional heterogeneity (Colombo et al., 2016). The functions of astrocytes go beyond providing structural support for neurons (Banker, 1980; Schreiner et al., 2015). They are also responsible for metabolic support and energy regulation via the secretion of neurotrophic factors (Perea et al., 2009; Bosson et al., 2015), modulation of synaptic transmission and neuronal excitability (Henneberger et al., 2010; van Deijk et al., 2017), maintenance of the integrity of the blood-brain barrier (Yamamizu et al., 2017), elimination of dead cells, regulation of the extracellular balance of ions, fluids and transmitters, and scar formation (Perea et al., 2009; Khakh & Sofroniew, 2015; Colombo et al., 2016;).

Astrocytes secrete molecules important for synapse formation and maturation during embryonic development, such as d-serine, ATP, BDNF and glypicans (Cao et al., 2013; Z. Tan et al., 2017), while one study showed that resident and fully differentiated astrocytes in the hippocampus also participate in synaptogenesis processes (Catalani et al., 2002).

Astrocytes are key players in hippocampal neurogenesis (Sultan et al., 2015; Terrillion et al., 2017), and actively participate in the formation and integration of adult granule neurons by releasing specific growth and neurotrophic factors (Moss et al., 2016; Seri et al., 2001; Sultan et al., 2015).

Furthermore, astrocytes have been shown to participate in the differentiation and maturation of neurons, particularly dopaminergic (DA) neurons, through their ability to produce and discharge growth factors, such as basic fibroblast growth factor (bFGF), bFGF is an essential neurotrophic factor that plays a crucial role in the specialisation and embryonic development of the neuronal lineage, as demonstrated in ESCs (Domingues et al., 2020).

As already mentioned, astrocytes are not excitable and, unlike neurons, do not propagate action potentials. However, they can detect neuronal inputs via ion channels, neurotransmitter receptors and transporters (Bosson et al., 2015; Perea et al., 2009). This results in modulation of their response through a Ca2+ signaling mechanism (Khakh & Sofroniew, 2015; Srinivasan et al., 2015), which regulates the metabolic and trophic support of neurons, as well as neuroinflammatory factors (Blanco-Suárez et al., 2017).

Several functional neurotransmitter receptors, such as gamma-aminobutyric acid (GABA) or glutamate, are expressed, leading to the propagation of a Ca2+ flux within astrocytic populations (Bosson et al., 2015; Perea et al., 2009). The flow of Ca2+ thus generated regulates the secretion of specific gliotransmitters (Di Castro et al., 2011; Volterra et al., 2014). Axonal receptor activation results from the release by astrocytes of various gliotransmitters, such as GABA, glutamate, ATP and D-serine (Perea et al., 2016).

Astrocytes express many of the same receptors as neurons and release gliotransmitters, neurotrophic factors, and neuroinflammatory factors, enabling them to regulate synaptic transmission and neuronal excitability. These properties give astrocytes the ability to influence the activity of neuronal networks (Blanco-Suárez et al., 2017).

Microglial Cells

Microglial cells are considered to be the innate immune cells that reside in the healthy brain and are the first to respond to pathological insults (Baufeld et al., 2018; Heneka et al., 2015). Microglial cells are present throughout the brain and their proportion varies from 5 to 12% of the total cell population in the mouse brain, depending on location. They have a variety of morphological aspects, including branched longitudinal shapes, branched radial shapes and compact round shapes (Lawson et al., 1990). As immune cells, microglia are capable of chemotaxis, cytokine production and phagocytosis (Michell-Robinson et al., 2015; Shigemoto-Mogami et al., 2014).

In pathological conditions, microglial cells change their shape into a soma with long, thin branched processes and adopt a less branched morphology with fewer, thicker processes, enabling them to adapt and react easily to a new condition (Sierra et al., 2010; Mosser et al., 2017). The involvement of these microglial cells in host homeostasis and defense mechanisms by participating in three essential functions (S. Hickman et al., 2018). Firstly, they detect changes in their environment using sensors encoded by different genes (S. E. Hickman et al., 2013). The second is to maintain myelin homeostasis and remodel synapses after cell migration to damaged sites (Zhan et al., 2014; S. Hickman et al., 2018). The third function is to protect against noxious stimuli, such as damage-associated molecular patterns (DAMPs) and pathogen-associated molecular patterns (PAMPs). Using cellular receptors, such as Toll-like receptors (TLRs), viral receptors and nuclear oligomerisation domain receptors, microglia have the ability to detect DAMPs and PAMPs (Glass et al., 2010; Stephenson et al., 2018).

While responding to these stimuli, microglial cells produce pro-inflammatory cytokines, namely the interleukins IL-1β and IL-16, the tumour necrosis factor (TNF-α), and chemokines including C-motif chemokine ligand 2 (CCL2) and IL-18 in order to reinforce cellular intervention and eliminate pathological agents (Glass et al., 2010; S. Hickman et al., 2018). Although neuroinflammation is considered a neuroprotective mechanism, it can induce neurotoxicity when sustained and is associated with neurodegeneration (S. Hickman et al., 2018). In addition, microglia priming due to chronic stress and ageing leads to an exaggerated inflammatory response and dystrophic morphology (Niraula et al., 2017).

With regard to heterogeneity, microglial cells are the subject of intense debate; studies have indicated that these cells can be classified phenotypically as M1 or M2 microglial cells (Antonietta Ajmone-Cat et al., 2013; Tang et al., 2016). M1 cells appear to react to infections and lesions and constitute the initial defense against tissue damage, whereas M2 cells are considered to be the main functional cells capable of reducing the pro-inflammatory response and encouraging attenuation of inflammation and tissue repair by producing anti-inflammatory molecules (Ponomarev et al., 2007; Tang et al., 2016). On the other hand, studies have shown that microglial cells display great diversity and do not follow a precise M1 or M2 phenotype, in both homeostatic and stimulated states (Ransohoff, 2016). Furthermore, M1 and M2 markers are apparent in microglia during neurodevelopment and persist into adulthood, making it impossible to assert that M1 or M2 phenotypes are exclusive in resting or activated states (Hammond et al., 2019; Ransohoff, 2016).

The capacity for phagocytosis is one of the best understood functions of microglia. Microglia are ready to trigger programmed cell death and eliminate cellular waste when an unfavourable stimulus enters the CNS (Nayak et al., 2014; He et al., 2018). A specific receptor expressed on the cell surface and in downstream signalling pathways is required for the phagocytic activity of microglia. However, new insights into the functions of microglial cells have emerged in recent years, demonstrating that they are not just immune cells of the brain. In fact, microglial cells are also involved in the establishment of synaptic networks and the survival of neurons, thereby supporting the homeostatic function of the CNS. In addition, research has shown that microglial cells have a paracrine action capable of promoting neuronal differentiation and maturation during CNS development. For example, insulin-like growth factor-1 (IGF-1) secreted by surrounding microglia was essential for the survival of layer V cortical neurons throughout postnatal development, according to Ueno et al (Ueno et a2013) and Nayak et al (2014). These findings were supported by evidence that suppression or exacerbation of microglial activity induced cell death in layer V cortical neurons, indicating a supportive role for microglia in neuronal survival and differentiation (Ueno et al., 2013; Schafer & Stevens, 2015).

As with astrocytes, microglia secrete molecules important for synapse formation and maturation during embryonic development, such as D-serine, ATP, BDNF and glypicans (Cao et al., 2013; Parkhurst et al., 2013; Z. Tan et al., 2017), as well as other trophic factors such as nerve growth factor (NGF) and fibroblast growth factor (FGF). Similarly, microglia have been shown to play a role in synapse maturation and pruning, both of which are essential for the proper development of excitatory synapse transmission. By regulating the expansion of dopaminergic axons in the forebrain, microglia in the prenatal mouse brain regulate the wiring of forebrain circuits (Squarzoni et al., 2014; Tay et al., 2017). It is well known that microglia move towards the lesion and participate in the phagocytic elimination of cellular debris (Weinhard et al., 2018). The ability of microglial cells to continuously extend and retract functional processes in the extracellular space of a damaged brain has also been described recently. It can therefore be hypothesised that microglial cells may be involved in the control of synapse maturation and function (Davalos et al., 2005; Nimmerjahn et al., 2005; Weinhard et al., 2018).

Oligodendrocytes

Oligodendrocytes are the third category of glial cells present in the CNS. As highly specialised cells, oligodendrocytes perform a variety of vital tasks, including myelination of CNS axons, support of axon metabolism and promotion of neuroplasticity (Bradl & Lassmann, 2010; McKenzie et al., 2014; Nave & Werner, 2014). This myelination speeds up and improves the transmission of electrical signals along axons, which is essential for rapid and efficient nerve communication (Simons & Nave, 2016). However, due to the complex design of these cells and their high metabolic requirements, their functionality is easily compromised.

Oligodendrocytes are considered to be the most vulnerable cells in the central nervous system due to their high metabolic requirements, which make them particularly vulnerable to oxidative stress (Peferoen et al., 2014; Giacci et al., 2018). In addition, it was previously thought that oligodendrocytes were not involved in inflammatory situations, but recent data have shown the opposite (Peferoen et al., 2014; Zeis et al., 2016). This indicates that immune mediators such as interleukin (IL-8), a crucial cytokine involved in the recruitment of microglia, may be produced by oligodendrocytes (Balabanov et al., 2007). In addition, BDNF mRNA is expressed by oligodendrocytes, and studies have examined how BDNF expression varies in response to disease or injury. BDNF expression was found to be elevated in oligodendrocytes in a model of spinal cord injury (Smith et al., 2013).

In addition, oligodendrocytes are essential for cognitive function (Nickel & Gu, 2018). It has recently been shown that the development of new oligodendrocytes is essential for learning motor skills (Xiao et al., 2016). In addition, motor learning promotes the flexibility of human myelin, which is an example of a learning-related increase in myelination (Hofstetter et al., 2013; Lakhani et al., 2016).

GLIAL CELL DYSFUNCTION IN PD

The Role of Glial Cells in The Pathogenesis of PD

Glial cells have been recognised as important players in disease progression in the context of neurodegenerative disorders of the central nervous system, particularly PD (Gu et al., 2010; Chitnis & Weiner, 2017). Indeed, they are linked to the onset of neuroinflammation and degenerative processes in neuronal environments. However, it has been proposed that glial cell activation does not contribute directly to the pathophysiological mechanisms underlying the disease, but rather is a secondary consequence of neuronal loss. A number of proteins, including α-synuclein (PARK1 and PARK4), parkin (PARK2), DJ-1 (PARK7) and ATPase 13A2 (ATP13A2 gene), play a role in controlling the activation of microglia and astrocytes. These proteins are encoded by genes associated with familial types of PD (Glass et al., 2010; Oksanen et al., 2019).

The SNpc of PD patients have been shown to contain reactive astrocytes (Miklossy et al., 2006). Dopaminergic neurodegeneration is accompanied by abnormal astrocyte function. The development of PD and the biology of astrocytes involve several genes encoding various proteins, such as; α-synuclein (encoded by the SNCA gene), Ca2+-independent phospholipase A2 (encoded by the PLA2G6 gene), Parkin (encoded by the PARK2 gene), DJ-1 (encoded by the PARK7 gene), β-glucocerebrosidase (encoded by the GBA gene) and lysosomal ATPase type 5 (encoded by the ATP13A2 gene) (Booth et al., 2017). Interferon (IFN) and TLR4 signals are used by the DJ-1 protein to control astrocyte activation (J. Kim et al., 2013; Oksanen et al., 2019). The NLPR3 inflammasome may not activate if ATP13A2 levels are maintained (Booth et al., 2017; Oksanen et al., 2019).

In the SNpc of PD patients, activation of microglial cells is increasingly pronounced (Rocha et al., 2015). Moreover, in the early stages of PD, the degree of microglial activation corresponds to the final dopaminergic loss (Ouchi et al., 2005). Dopaminergic neurons are often surrounded by activated microglia, which promote inflammation (Tang & Le, 2015). Dying dopaminergic neurons generate α-synuclein aggregates, which activate microglia and cause them to take on a pro-inflammatory character (W. Zhang et al., 2005). The pro-inflammatory cytokines TNF-α, NO and IL-1 generated by microglia can modify the neuroinflammatory process in PD due to the overexpression of α-synuclein, which causes microglia to adopt a reactive pro-inflammatory phenotype (Hirsch & Hunot, 2009; Rojanathammanee et al., 2011). A TLR ligand called lipopolysaccharide can kill dopaminergic neurons by inducing microglia to display a pro-inflammatory state (Tang et al., 2016). Also known as MPTP, this neurotoxin damages dopaminergic neurons by inducing mitochondrial dysfunction and indirectly activating microglia (Dauer et al., 2003; Tang et al., 2016).

NEUROINFLAMMATION INDUCED BY GLIAL CELLS

Astrocyte-Mediated Neuroinflammation in PD

Enormous studies show that astrocytes also play a crucial role in the neuroinflammatory pathways of PD. In response to inflammatory stimuli such as LPS, IL-1 and TNF-α, astrocytes in vivo and in vitro produce pro-inflammatory cytokines (Saijo et al., 2009; Tanaka et al., 2013). Reactive astrogliosis has been observed in various animal models of PD and is characterised by cell body hypertrophy with cell extension and increased levels of glial fibrillary acidic protein (GFAP) expression. Astrogliosis is also observed in the affected brain areas of PD patients, indicating that astrocytes may be involved in the immunological processes of PD (Yamada et al., 1992).

Astrocytic responses after stimulation are slower than microglial activation. Following immunological triggers such as systemic lipopolysaccharide (LPS) therapy and α-synuclein aggregation, microglia are able to trigger an inflammatory response. Subsequently, various chemical substances, including pro-inflammatory mediators generated by activated microglia, can activate astrocytes. The astrocytes then reinforce these immunological signals (Saijo et al., 2009). During neurodegeneration, the increased loss of DA neurons in SNpc is largely due to uncontrolled neuroinflammation caused by the simultaneous activation of microglia and astrocytes (Glass et al., 2010; Saijo et al., 2009). In vitro administration of α-synuclein significantly increases TNF-α and IL-6 expression in astrocytes cultured outside the body (Fellner et al., 2013). In mice, excessive production of mutant -synuclein in astrocytes causes microglial activation, extensive astrogliosis and degeneration of DA neurons and motor neurons (Gu et al., 2010).

The dopaminergic neuronal system appears to deteriorate progressively with age, according to the results of post-mortem research in humans and animals (ANTONINI & LEENDERS, 1993). It is unclear whether the reduction in the number of dopamine D2 receptors (Drd2) in the ageing brain has a substantial impact on cognitive ability and contributes to the development of early to mid-stage PD. Changes in neuronal cells are the main cause of this decline. In contrast, glial Drd2 is only marginally involved in total levels. It is likely that glial Drd2, although extremely low in the brain, is more sensitive to changes in the microenvironment, as astroglial and microglial cells are the 'sensors' of the brain and constantly monitor brain activity. However, the absence of glial Drd2 can have significant physiological effects. The anti-inflammatory protein alpha-B crystallin is significantly reduced in the CNS when astrocytic Drd2 is suppressed (Liu et al., 2015; Shao et al., 2017) (**Fig 1**). It is possible that reduced levels of Drd2, perhaps in glial cells, in the ageing brain may have a negative impact on immunological homeostasis and exacerbate the progression of PD.

Microglia-Mediated Inflammation in PD

The main cell types involved in inflammatory responses in the CNS are microglial cells, as previously mentioned (Saijo et al., 2009). The discovery by McGeer's team in 1988 of activated microglia in human post-mortem brain tissue SNpc provided the first indication that neuroinflammation played a role in the pathophysiology of PD (McGeer et al., 1988). Other studies show that microglia are highly activated in several regions of the PD brain (Bartels et al., 2010; Gerhard et al., 2006). In addition, several types of animal models of PD show significant activation of microglia in the SNpc and striatum (Tansey et al., 2007; Leal et al., 2013). In the midbrain of PD patients, further biochemical studies indicate higher amounts of pro-inflammatory mediators such as IL-1β, TNF-α and IFN-γ. These results clearly suggest that immunological elements are involved in the aetiology of PD.

Due to a number of immunomodulators, including CD200, CD47, CX3CL1, CD22 and CD95, microglia remain in a dormant state under normal conditions. Neuronal cells are the main producers of these compounds (Chang et al., 2000; Sheridan & Murphy, 2013). The fact that these chemicals are only expressed by microglia in the central nervous system is remarkable and underlines the critical role that interactions between neurons and microglia play in the control of neuroinflammation (Chang et al., 2000; Sheridan & Murphy, 2013). In animal models of PD, the absence of CX3CL1 or CX3CR1 increases the neurotoxicity induced by systemic lipopolysaccharide treatment and amplifies the death of DA neurons in the SNpc (Cardona et al., 2006; Pabon et al., 2011). Altered CD200-CD200R communication increases microglia activation and exacerbates degeneration of DA neurons in animal models of PD in rats (X. J. Wang et al., 2011; S. Zhang et al., 2011).

In the early stages of neurodegeneration, it has been proposed that activated microglia may be beneficial to the host (Cagnin et al., 2006; Hu et al., 2012). One of these hypotheses is the suppression in vivo of Jmjd3, which is necessary for the polarisation of M2 microglia. In an animal model of PD, this study showed a significant effect of microglial hyperactivation and an increase in the death of dopamine neurons (Tang et al., 2013). This suggests that M2 microglia have a protective role in this process. But persistent microglial hyperactivation in the PD brain dramatically increases the production of a wide range of pro-inflammatory cytokines, including TNF-α, IL-1β, interleukin-6 and IFN- γ, accelerating the degradation of nigral DA neurons (Ferrari et al., 2006; Mccoy et al., 2006).

MMP-3, a protein produced by degenerating neurons, plays a crucial role, at least in vitro, in controlling the activation states of microglia. In microglia-neuron cocultures, overexpression of MMP-3 significantly activates microglia and enhances the response to oxidative stress (Y. S. Kim et al., 2007). This demonstrates that MMP-3 is a key player in the degeneration of dopaminergic neurons and that microglia contribute substantially to neuroinflammation in the pathogenesis of PD.

ATP, a purinergic neurotransmitter, can effectively influence microglial activities (Koizumi et al., 2013; George et al., 2015). ATP, which is produced by injured neurons and neighbouring astrocytes, controls the migration of microglia to inflamed and damaged areas (George et al., 2015). In addition, ATP binds to the P2Y receptor, which is predominantly expressed by microglia in the brain, and causes increased production of TNF-α, IL-1β and nitric oxide (Davalos et al., 2005).

Activation Phenotypes of Microglia in PD

There is increasing evidence that microglia have two unique activation phenotypes, known as the alternative M1 and M2 phenotypes. The production of different ranges of cytokines characterises these different states of microglia activation (Pisanu et al., n.d.). M1 activation of microglia is induced by LPS/IFN-γ treatment, whereas M2 activation is triggered by IL-4/IL-13 treatment. M1 activation of microglia is characterised by the production of pro-inflammatory cytokines such as IL-1β, IL-6, IL-12, TNF-α, as well as other cytotoxic chemicals such as reactive oxygen species, superoxide and NO. These chemicals help to amplify pro-inflammatory responses after injury or infection. M2 microglia act as an immunosuppressant, opposing conventional M1 microglia and promoting tissue repair. M2 microglia produce anti-inflammatory cytokines such as TGF-β, IL-4, IL-10 and IL-13. The different gene expression profile of each type of microglia activation can be used to identify it. CD206, Arg1, YM1 and RELM-α, for example, have been identified as distinctive genes expressed during the M2 phase of mouse microglia (Tang et al., 2016). Cytokines can influence Arg1, YM1 and RELM-α levels. Following stimulation by IL-4, their concentrations increase significantly in microglia in primary culture or in frontal and striatal cortical areas of the mouse brain (Pepe et al., 2014).

Environmental contaminants and misfolded proteins activate microglia towards the M1 phenotype in animal models of PD (Ha et al., 2012; H. G. Kim et al., 2012). A progressive decrease in CD206 expression caused by chronic MPTP administration indicates that microglial M2 phase activation is downregulated in the development of PD (Pisanu et al., 2014). IL-4 treatment, on the other hand, enhances microglial expression of histone H3K27me3 demethylase (Jmjd3), which is involved in the control of epigenetic chromosomal alterations and leads to a variety of human disorders. Knockdown of Jmjd3 significantly reduces the expression of M2 marker genes, including CD206 and Arg1, in the N9 microglial cell line, demonstrating that Jmjd3 is required for M2 microglia polarisation. Knockdown of Jmjd3 in vivo exacerbates the loss of DA neurons in the SNpc of the MPTP-induced animal model of PD by reversing M2 activation of microglia (Tang et al., 2013).

Research suggests that the majority of activated microglia express M2-associated genes in the early stages of lesions of various types. However, as the disease progresses, M1 signature genes become more common (Hu et al., 2012) (**Fig 1**). These findings highlight the importance of balancing the different phenotypes of microglia activation in PD. Manipulation of the microglia activation state transition appears to be a viable technique to intervene in the development of PD. Although current data suggests that M1 and M2 play various roles in the aetiology of PD in animal models, comparable results have not yet been obtained in humans. Further studies on this subject are therefore needed in the future.

Figure 1. The inflammatory pathways involved in the pathogenesis of PD

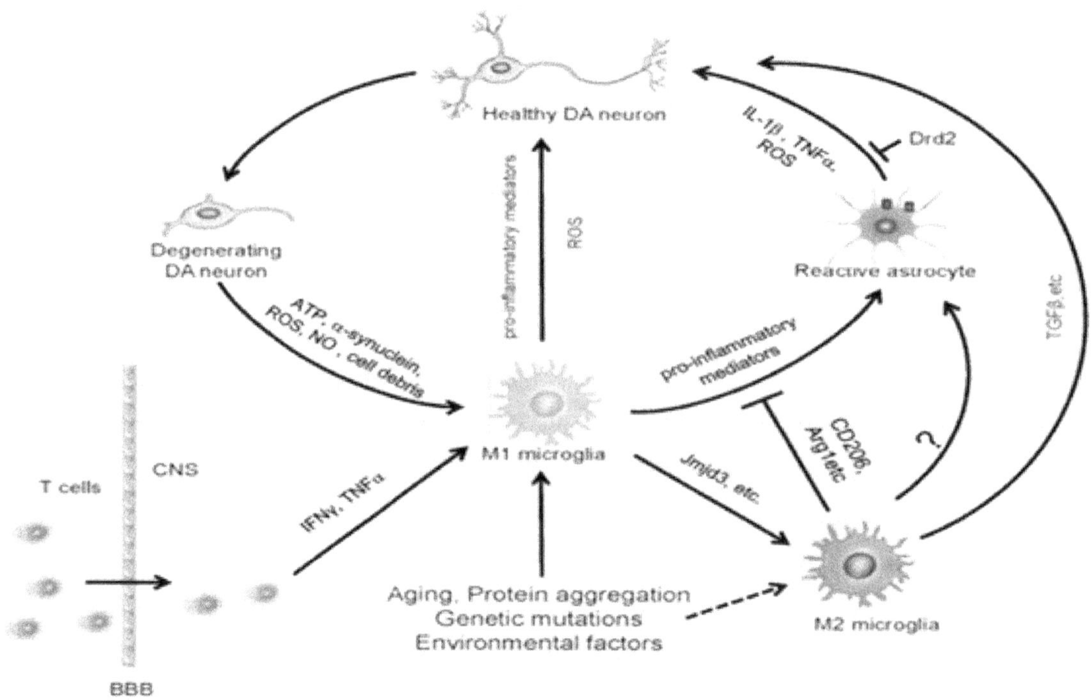

In pathological situations, microglia become active and acquire the M1 phenotype. Pro-inflammatory mediators from M1 microglia activate astrocytes, leading to increased production of pro-inflammatory factors, superoxide radicals and nitric oxide, all of which contribute to the degeneration of DA neurons. DA neurodegeneration can lead to the production of chemicals that stimulate glial cells and cause inflammation. A subset of microglia may become activated and acquire the M2 phenotype in the later stages of PD. This results in the release of anti-inflammatory substances such as TGF-β, which have a neuroprotective effect in PD (modify according to Wang et al., 2015).

THERAPEUTIC APPROACHES TO PD TARGETING GLIAL CELLS

Current drugs for PD only improve motor symptoms and do not prevent or treat the disease. Therefore, the main challenge is to design a strategy that provides neuroprotection or modifies the disease itself. In this context, targeting neurons and/or glial cells could be an interesting avenue for developing new therapies for PD.

Anti-Inflammatory Therapies

Various anti-inflammatory therapies, including ibuprofen, dexamethasone, minocycline, amantadine, vasoactive intestinal peptide, pituitary adenylate cyclase-activating peptide, TGF-β and IL-10, have been shown in animal models to inhibit dopaminergic cell death (Zhu et al., 2002; Manthripragada et al., 2011). Nevertheless, the consequences of the use of anti-inflammatory drugs in patients with PDare inconclusive. With the exception of aspirin, one meta-analysis found that non-steroidal anti-inflammatory drugs could reduce the incidence of PD (Gagne & Power, 2010). However, another meta-analysis showed that non-steroidal anti-inflammatory drugs have no significant effect on the risk of PD, with only ibuprofen having a modest protective effect (Samii et al., 2009). NLY01, a glucagon-like peptide-1 receptor agonist, protected against dopaminergic cell loss and behavioural impairment in a mouse model of sporadic PD (Yun et al., 2018). This neuroprotective effect was found to be due to the prevention of conversion of astrocytes to the neurotoxic phenotype by microglia (Yun et al., 2018). Minocycline, whose neuroprotective properties have been demonstrated in numerous in vivo and in vitro studies, did not slow the progression of early PD in a randomised clinical trial lasting 12 to 18 months (NINDS NET-PD Investigators, 2006; Cankaya et al., 2019).

It has been proposed that reducing the M1 phenotype in microglial cells may reduce the release of pro-inflammatory cytokines such as IFN-γ, TNF-α, and IL-1β (Price et al., 2009). TNF has already been targeted in animal models of PD to minimise the toxicity associated with the M1 phenotype. A possible technique would be to target molecules involved in activating the anti-inflammatory M2 microglial phenotype or encouraging the transition from the pro-inflammatory M1 phenotype to the M2 phenotype. The applications of IL-10 and beta interferon are promising in this field (Price et al., 2009).

Cannabinoid type 2 (CB2) receptors are a therapeutic alternative in PD because they are often detected in the microglia activated in the disease (Cassano et al., 2017). Pharmacological stimulation of CB2 receptors in microglia resulted in a reduction in microglial activation, which attenuated functional motor deficits in an MPTP animal model of PD (Price et al., 2009; Cassano et al., 2017). Similarly, histone deacetylase (HDAC) is a possible therapeutic target for PD due to its function in glial cell regulation and its effects on synuclein aggregation (Tan et al., 2018; Garbes et al., 2013). HDAC inhibitors have been proposed as a potential therapeutic option for PD because studies have shown that they limit the production of pro-inflammatory cytokines by microglia, thereby reducing inflammation (Faraco et al., 2009; Tan et al., 2018).

Gene Therapy to Regulate Glial Cell Behaviour

Several astrocyte-based treatments are currently under development, including astrocyte transplantation and pharmacological correction of astrocyte dysfunction. With regard to the first technique, human astrocytes corresponding to the disease can be obtained from primary sources or by differentiation of

ESCs (embryonic stem cells) or iPSCs (induced pluripotent stem cells) (Santos et al., 2017; Li et al., 2018). Various procedures for the production of astrocytes from ESCs and iPSCs have been developed in recent years(Krencik & Zhang, 2011; Jones et al., 2017). Once the stem cells have developed into astrocytes, they can be transplanted to replace defective cells or to stimulate the survival of existing neurons. Although research into these procedures is limited, current evidence suggests that it is possible to generate astrocytes from stem cells in PD(di Domenico et al., 2019).

Furthermore, astrocytes produced from stem cells can be used for disease modelling and drug development. They can be used to discover new chemical substances that protect the CNS against aggressions such as ROS (Gupta et al., 2011; Thorne et al., 2016). One example is the use of human-ESC-derived astrocytes in a phenotypic experiment to discover compounds that protect against oxidative stress (Thorne et al., 2016). Given that oxidative stress plays a well-known role in the pathogenesis of many neurological disorders, the identification of protective chemicals against stress could significantly contribute to slowing or stopping the development of disease.

Recently, glial cell transplantation has emerged as a potential technique for CNS regeneration methods (White & Barry, 2015). However, cell transplantation techniques are not universally accepted and other methods, such as the use of the glial cell secretome (Lev et al., 2013), have been proposed and studied. The glial cell secretome has already been analysed and identified as a potential therapeutic strategy for CNS diseases such as PD(Jha et al., 2013; White & Barry, 2015). Indeed, from a clinical point of view, the glial secretome could be an effective tool not only as a therapeutic method, but also as a reliable source of diagnostic and prognostic biomarkers, as well as for the discovery of therapeutic targets (M. Chang et al., 2003; Suk, 2014).

CONCLUSION

Although the involvement of glial cells in PD has been the subject of less research than that of neurons, their role in the disease is now a burning issue. However, the precise importance of these cells in the pathogenesis of PD is still unknown. Previous research suggests that glial cells can protect or damage dopaminergic neurons depending on their state of activity. It is therefore plausible that the loss of dopaminergic neurons leads to glial cell dysfunction. Although neither hypothesis has been proven, it is likely that they are both viable and act in concert, revealing new approaches and perspectives for understanding the pathophysiology of PD and potential therapeutic alternatives. A better understanding of the influence of glial cells on the progression of PD could lay the foundations for future therapeutic discoveries targeting the multiple curative mechanisms of the disease.

REFERENCES

Antonietta Ajmone-Cat, M., Mancini, M., De Simone, R., Cilli, P., & Minghetti, L. (2013). Microglial polarization and plasticity: Evidence from organotypic hippocampal slice cultures. *Glia*, *61*(10), 1698–1711. doi:10.1002/glia.22550 PMID:23918452

Antonini, A., & Leenders, K. L. (1993). Dopamine D2 Receptors in Normal Human Brain: Effect of Age Measured by Positron Emission Tomography (PET) and [11C]-Raclopride. *Annals of the New York Academy of Sciences*, *695*(1), 81–85. doi:10.1111/j.1749-6632.1993.tb23033.x PMID:8239318

Araque, A., Li, N., Doyle, R. T., & Haydon, P. G. (2000). SNARE Protein-Dependent Glutamate Release from Astrocytes. *The Journal of Neuroscience : The Official Journal of the Society for Neuroscience*, *20*(2), 666–673. doi:10.1523/JNEUROSCI.20-02-00666.2000 PMID:10632596

Balabanov, R., Strand, K., Goswami, R., McMahon, E., Begolka, W., Miller, S. D., & Popko, B. (2007). Interferon-γ-Oligodendrocyte Interactions in the Regulation of Experimental Autoimmune Encephalomyelitis. *The Journal of Neuroscience : The Official Journal of the Society for Neuroscience*, *27*(8), 2013–2024. doi:10.1523/JNEUROSCI.4689-06.2007 PMID:17314297

Banker, G. A. (1980). Trophic Interactions Between Astroglial Cells and Hippocampal Neurons in Culture. *Science*, *209*(4458), 809–810. doi:10.1126cience.7403847 PMID:7403847

Bartels, A. L., Willemsen, A. T. M., Doorduin, J., de Vries, E. F. J., Dierckx, R. A., & Leenders, K. L. (2010). [11C]-PK11195 PET: Quantification of neuroinflammation and a monitor of anti-inflammatory treatment in Parkinson's disease? *Parkinsonism & Related Disorders*, *16*(1), 57–59. doi:10.1016/j.parkreldis.2009.05.005 PMID:19487152

Baufeld, C., O'Loughlin, E., Calcagno, N., Madore, C., & Butovsky, O. (2018). Differential contribution of microglia and monocytes in neurodegenerative diseases. *Journal of Neural Transmission (Vienna, Austria)*, *125*(5), 809–826. doi:10.100700702-017-1795-7 PMID:29063348

Bazargani, N., & Attwell, D. (2016). Astrocyte calcium signaling: the third wave. *Nature Neuroscience*, *19*(2), 182–189. doi:10.1038/nn.4201

Bitzer-Quintero, O. (2012). Immune system in the brain: a modulatory role on dendritic spine morphophysiology? *Hindawi.ComOK Bitzer-Quintero, I González-BurgosNeural Plasticity, 2012.* Hindawi. https://www.hindawi.com/journals/np/2012/348642/abs/

Blanco-Suárez, E. (2017). Role of astrocyte–synapse interactions in CNS disorders. *Wiley Online LibraryE Blanco-Suárez, ALM Caldwell, NJ AllenThe Journal of Physiology.* Wiley. https://physoc.onlinelibrary.wiley.com/doi/abs/10.1113/JP270988

Booth, H. D. E., Hirst, W. D., & Wade-Martins, R. (2017). The Role of Astrocyte Dysfunction in Parkinson's Disease Pathogenesis. *Trends in Neurosciences*, *40*(6), 358–370. doi:10.1016/j.tins.2017.04.001 PMID:28527591

Bosson, A., Boisseau, S., Buisson, A., Savasta, M., & Albrieux, M. (2015). Disruption of dopaminergic transmission remodels tripartite synapse morphology and astrocytic calcium activity within substantia nigra pars reticulata. *Glia*, *63*(4), 673–683. doi:10.1002/glia.22777 PMID:25511180

Bradl, M., & Lassmann, H. (2010). Oligodendrocytes: Biology and pathology. *Acta Neuropathologica*, *119*(1), 37–53. doi:10.100700401-009-0601-5 PMID:19847447

Brederlau, A., Correia, A. S., Anisimov, S. V., Elmi, M., Paul, G., Roybon, L., Morizane, A., Bergquist, F., Riebe, I., Nannmark, U., Carta, M., Hanse, E., Takahashi, J., Sasai, Y., Funa, K., Brundin, P., Eriksson, P. S., & Li, J.-Y. (2006). Transplantation of Human Embryonic Stem Cell-Derived Cells to a Rat Model of Parkinson's Disease: Effect of In Vitro Differentiation on Graft Survival and Teratoma Formation. *Stem Cells (Dayton, Ohio)*, *24*(6), 1433–1440. doi:10.1634temcells.2005-0393 PMID:16556709

Cagnin, A., Kassiou, M., Meikle, S. R., & Banati, R. B. (2006). In vivo evidence for microglial activation in neurodegenerative dementia. *Acta Neurologica Scandinavica*, *114*(s185, SUPPL. 185), 107–114. doi:10.1111/j.1600-0404.2006.00694.x PMID:16866919

Cankaya, S., Cankaya, B., Kilic, U., Kilic, E., & Yulug, B. (2019). The therapeutic role of minocycline in Parkinson's disease. *Drugs in Context*, *8*, 1–14. doi:10.7573/dic.212553 PMID:30873213

Cao, X., Li, L. P., Wang, Q., Wu, Q., Hu, H. H., Zhang, M., Fang, Y. Y., Zhang, J., Li, S. J., Xiong, W. C., Yan, H. C., Gao, Y. B., Liu, J. H., Li, X. W., Sun, L. R., Zeng, Y. N., Zhu, X. H., & Gao, T. M. (2013). Astrocyte-derived ATP modulates depressive-like behaviors. *Nature Medicine*, *19*(6), 773–777. doi:10.1038/nm.3162

Cardona, A. E., Pioro, E. P., Sasse, M. E., Kostenko, V., Cardona, S. M., Dijkstra, I. M., Huang, D. R., Kidd, G., Dombrowski, S., Dutta, R., Lee, J. C., Cook, D. N., Jung, S., Lira, S. A., Littman, D. R., & Ransohoff, R. M. (2006). Control of microglial neurotoxicity by the fractalkine receptor. *Nature Neuroscience*, *9*(7), 917–924. doi:10.1038/nn1715

Cassano, T., Calcagnini, S., Pace, L., De Marco, F., Romano, A., & Gaetani, S. (2017). Cannabinoid receptor 2 signaling in neurodegenerative disorders: From pathogenesis to a promising therapeutic target. *Frontiers in Neuroscience*, *11*(FEB), 242186. doi:10.3389/fnins.2017.00030 PMID:28210207

Catalani, A., Sabbatini, M., & Consoli, C. (n.d.). *Glial fibrillary acidic protein immunoreactive astrocytes in developing rat hippocampus.* Elsevier. https://www.sciencedirect.com/science/article/pii/S0047637401003566

Chang, R., Hudson, P., Wilson, B., & Haddon, L. (n.d.). *Influence of neurons on lipopolysaccharide-stimulated production of nitric oxide and tumor necrosis factor-α by cultured glia.* Elsevier. https://www.sciencedirect.com/science/article/pii/S0006899399022556

Chang, M., Son, H., & Lee, Y. (n.d.). *Neurons and astrocytes secrete factors that cause stem cells to differentiate into neurons and astrocytes, respectively.* Elsevier. https://www.sciencedirect.com/science/article/pii/S104474310300068X

Chitnis, T., & Weiner, H. L. (2017). CNS inflammation and neurodegeneration. *The Journal of Clinical Investigation*, *127*(10), 3577–3587. doi:10.1172/JCI90609 PMID:28872464

Clarke, L. E., & Barres, B. A. (2013). Erratum: Emerging roles of astrocytes in neural circuit development (Nature Reviews Neuroscience (2013) 14 (311-321)). *Nature Reviews. Neuroscience*, *14*(6), 442. doi:10.1038/nrn3506

Colombo, E. (2016). *Astrocytes: key regulators of neuroinflammation.* Cell.com. https://www.cell.com/trends/immunology/fulltext/S1471-4906(16)30072-2

Daadi, M. M., Grueter, B. A., Malenka, R. C., Redmond, D. E., & Steinberg, G. K. (2012). Dopaminergic Neurons from Midbrain-Specified Human Embryonic Stem Cell-Derived Neural Stem Cells Engrafted in a Monkey Model of Parkinson's Disease. *PLoS One*, *7*(7), e41120. doi:10.1371/journal.pone.0041120 PMID:22815935

Dauer, W., & Neuron, S. (2003). *Parkinson's disease: mechanisms and models*. Cell.com. https://www.cell.com/fulltext/S0896-6273(03)00568-3

Davalos, D., Grutzendler, J., Yang, G., Kim, J. V, Zuo, Y., Jung, S., Littman, D. R., Dustin, M. L., & Gan, W.-B. (2005). ATP mediates rapid microglial response to local brain injury in vivo. *Nature Neuroscience*. doi:10.1038/nn1472

De Miranda, B. R., Rocha, E. M., Bai, Q., El Ayadi, A., Hinkle, D., Burton, E. A., & Timothy Greenamyre, J. (2018). Astrocyte-specific DJ-1 overexpression protects against rotenone-induced neurotoxicity in a rat model of Parkinson's disease. *Neurobiology of Disease*, *115*, 101–114. doi:10.1016/j.nbd.2018.04.008 PMID:29649621

Dextera, D. T., & Jenner, P. (2013). Parkinson disease: From pathology to molecular disease mechanisms. *Free Radical Biology & Medicine*, *62*, 132–144. doi:10.1016/j.freeradbiomed.2013.01.018 PMID:23380027

Di Castro, M. A., Chuquet, J., Liaudet, N., Bhaukaurally, K., Santello, M., Bouvier, D., Tiret, P., & Volterra, A. (2011). Local Ca2+ detection and modulation of synaptic release by astrocytes. *Nature Neuroscience, 14*(10), 1276–1284. doi:10.1038/nn.2929

di Domenico, A., Carola, G., Calatayud, C., Pons-Espinal, M., Muñoz, J. P., Richaud-Patin, Y., Fernandez-Carasa, I., Gut, M., Faella, A., Parameswaran, J., Soriano, J., Ferrer, I., Tolosa, E., Zorzano, A., Cuervo, A. M., Raya, A., & Consiglio, A. (2019). Patient-Specific iPSC-Derived Astrocytes Contribute to Non-Cell-Autonomous Neurodegeneration in Parkinson's Disease. *Stem Cell Reports*, *12*(2), 213–229. doi:10.1016/j.stemcr.2018.12.011 PMID:30639209

Domingues, A. V., Pereira, I. M., Vilaça-Faria, H., Salgado, A. J., Rodrigues, A. J., & Teixeira, F. G. (2020). Glial cells in Parkinson's disease: protective or deleterious? In Cellular and Molecular Life Sciences, 77(24). doi:10.100700018-020-03584-x

Dujardin, K., Degreef, J. F., Rogelet, P., Defebvre, L., & Destee, A. (1999). Impairment of the supervisory attentional system in early untreated patients with Parkinson's disease. *Journal of Neurology*, *246*(9), 783–788. doi:10.1007004150050455 PMID:10525975

Faraco, G., Pittelli, M., Cavone, L., Fossati, S., Porcu, M., Mascagni, P., Fossati, G., Moroni, F., & Chiarugi, A. (2009). Histone deacetylase (HDAC) inhibitors reduce the glial inflammatory response in vitro and in vivo. *Neurobiology of Disease*, *36*(2), 269–279. doi:10.1016/j.nbd.2009.07.019 PMID:19635561

Fellner, L., Irschick, R., Schanda, K., Reindl, M., Klimaschewski, L., Poewe, W., Wenning, G. K., & Stefanova, N. (2013). *Toll-like receptor 4 is required for α-synuclein dependent activation of microglia and astroglia.* Wiley. doi:10.1002/glia.22437

Ferrari, C. C., Pott Godoy, M. C., Tarelli, R., Chertoff, M., Depino, A. M., & Pitossi, F. J. (2006). Progressive neurodegeneration and motor disabilities induced by chronic expression of IL-1β in the substantia nigra. *Neurobiology of Disease, 24*(1), 183–193. doi:10.1016/j.nbd.2006.06.013 PMID:16901708

Fields, R. D., Araque, A., Johansen-Berg, H., Lim, S. S., Lynch, G., Nave, K. A., Nedergaard, M., Perez, R., Sejnowski, T., & Wake, H. (2014). Glial biology in learning and cognition. *The Neuroscientist, 20*(5), 426–431. doi:10.1177/1073858413504465 PMID:24122821

Fields, R. D., & Stevens-Graham, B. (2002). New Insights into Neuron-Glia Communication. *Science, 298*(5593), 556–562. doi:10.1126cience.298.5593.556 PMID:12386325

Gagne, J. J., & Power, M. C. (2010). *Anti-inflammatory drugs and risk of Parkinson disease A meta-analysis*. Neurology. www.neurology.org

Garbes, L., Riessland, M., & Wirth, B. (n.d.). *Histone Acetylation as a Potential Therapeutic Target in Motor Neuron Degenerative Diseases*.

George, J., Gonçalves, F. Q., Cristóvão, G., Rodrigues, L., Meyer Fernandes, J. R., Gonçalves, T., Cunha, R. A., & Gomes, C. A. (2015). Different danger signals differently impact on microglial proliferation through alterations of ATP release and extracellular metabolism. *Glia, 63*(9), 1636–1645. doi:10.1002/glia.22833 PMID:25847308

Gerhard, A., Pavese, N., Hotton, G., Turkheimer, F., Es, M., Hammers, A., Eggert, K., Oertel, W., Banati, R. B., & Brooks, D. J. (2006). In vivo imaging of microglial activation with PET in idiopathic Parkinson's disease. *Neurobiology of Disease, 21*(2), 404–412. doi:10.1016/j.nbd.2005.08.002 PMID:16182554

Giacci, M. K., Bartlett, C. A., Smith, N. M., Iyer, K. S., Toomey, L. M., Jiang, H., Guagliardo, P., Kilburn, M. R., & Fitzgerald, M. (2018). Oligodendroglia Are Particularly Vulnerable to Oxidative Damage after Neurotrauma In Vivo. *The Journal of Neuroscience : The Official Journal of the Society for Neuroscience, 38*(29), 6491–6504. doi:10.1523/JNEUROSCI.1898-17.2018 PMID:29915135

Glass, C., Saijo, K., Winner, B., & Marchetto. (2010). *Mechanisms underlying inflammation in neurodegeneration.* Cell.com. https://www.cell.com/fulltext/S0092-8674(10)00168-6

Goodarzi, P., & Aghayan, H. (2015). Stem cell-based approach for the treatment of Parkinson's disease. *Journal of the Islamic Republic of Iran.* https://www.ncbi.nlm.nih.gov/pmc/articles/PMC4431356/

Gu, X. L., Long, C. X., Sun, L., Xie, C., Lin, X., & Cai, H. (2010). Astrocytic expression of Parkinson's disease-related A53T -synuclein causes neurodegeneration in mice. *Molecular Brain, 3*(1), 12. doi:10.1186/1756-6606-3-12 PMID:20409326

Gupta, K., Patani, R., Baxter, P., Serio, A., Story, D., Tsujita, T., Hayes, J. D., Pedersen, R. A., Hardingham, G. E., & Chandran, S. (2011). Human embryonic stem cell derived astrocytes mediate non-cell-autonomous neuroprotection through endogenous and drug-induced mechanisms. *Cell Death & Differentiation, 5, 19*(5), 779–787. doi:10.1038/cdd.2011.154

Ha, S. K., Moon, E., Lee, P., Ryu, J. H., Oh, M. S., & Kim, S. Y. (2012). Acacetin attenuates neuroinflammation via regulation the response to LPS stimuli in Vitro and in Vivo. *Neurochemical Research, 37*(7), 1560–1567. doi:10.100711064-012-0751-z PMID:22447574

Hammond, T. R., Dufort, C., Dissing-Olesen, L., Giera, S., Young, A., Wysoker, A., Walker, A. J., Gergits, F., Segel, M., Nemesh, J., Marsh, S. E., Saunders, A., Macosko, E., Ginhoux, F., Chen, J., Franklin, R. J. M., Piao, X., McCarroll, S. A., & Stevens, B. (2019). Single-Cell RNA Sequencing of Microglia throughout the Mouse Lifespan and in the Injured Brain Reveals Complex Cell-State Changes. *Immunity, 50*(1), 253–271.e6. doi:10.1016/j.immuni.2018.11.004 PMID:30471926

He, H., Zhou, Y., Zhou, Y., Zhuang, J., He, X., Wang, S., & Lin, W. (2018). *Dexmedetomidine Mitigates Microglia-Mediated Neuroinflammation through Upregulation of Programmed Cell Death Protein 1 in a Rat Spinal Cord Injury Model.* Mary Ann Libert Publications. doi:10.1089/neu.2017.5625

Heneka, M. T., Carson, M. J., El Khoury, J., Landreth, G. E., Brosseron, F., Feinstein, D. L., Jacobs, A. H., Wyss-Coray, T., Vitorica, J., Ransohoff, R. M., Herrup, K., Frautschy, S. A., Finsen, B., Brown, G. C., Verkhratsky, A., Yamanaka, K., Koistinaho, J., Latz, E., Halle, A., & Kummer, M. P. (2015). Neuroinflammation in Alzheimer's disease. *Lancet Neurology, 14*(4), 388–405. doi:10.1016/S1474-4422(15)70016-5 PMID:25792098

Henneberger, C., Papouin, T., Oliet, S. H. R., & Rusakov, D. A. (2010). Long-term potentiation depends on release of d-serine from astrocytes. *Nature, 463*(7278), 232–236. doi:10.1038/nature08673

Hickman, S., Izzy, S., Sen, P., Morsett, L., & El Khoury, J. (2018). Microglia in neurodegeneration. In Nature Neuroscience (Vol. 21, Issue 10, pp. 1359–1369). Nature Publishing Group. doi:10.103841593-018-0242-x

Hickman, S. E., Kingery, N. D., Ohsumi, T. K., Borowsky, M. L., Wang, L. C., Means, T. K., & El Khoury, J. (2013). The microglial sensome revealed by direct RNA sequencing. *Nature Neuroscience, 16*(12), 1896–1905. doi:10.1038/nn.3554

Hirase, H., Qian, L., Barthó, P., & Buzsáki, G. (2004). Calcium Dynamics of Cortical Astrocytic Networks In Vivo. *PLoS Biology, 2*(4), e96. doi:10.1371/journal.pbio.0020096 PMID:15094801

Hirsch, E. C., & Hunot, S. (2009). Neuroinflammation in Parkinson's disease: A target for neuroprotection? *Lancet Neurology, 8*(4), 382–397. doi:10.1016/S1474-4422(09)70062-6 PMID:19296921

Hofstetter, S., Tavor, I., Moryosef, S. T., & Assaf, Y. (2013). Short-term learning induces white matter plasticity in the fornix. *The Journal of Neuroscience : The Official Journal of the Society for Neuroscience, 33*(31), 12844–12850. doi:10.1523/JNEUROSCI.4520-12.2013 PMID:23904619

Hu, X., Li, P., Guo, Y., Wang, H., Leak, R. K., Chen, S., Gao, Y., & Chen, J. (2012). Microglia/macrophage polarization dynamics reveal novel mechanism of injury expansion after focal cerebral ischemia. *Stroke, 43*(11), 3063–3070. doi:10.1161/STROKEAHA.112.659656 PMID:22933588

Jäkel, S., & Dimou, L. (2017). Glial cells and their function in the adult brain: A journey through the history of their ablation. *Frontiers in Cellular Neuroscience, 11*, 235525. doi:10.3389/fncel.2017.00024 PMID:28243193

Jankovic, J., & Aguilar, L. G. (2008). Current approaches to the treatment of Parkinson’s disease. *Neuropsychiatric Disease and Treatment, 4*(4), 743–757. doi:10.2147/NDT.S2006 PMID:19043519

Jha, M. K., Seo, M., Kim, J. H., Kim, B. G., Cho, J. Y., & Suk, K. (2013). The secretome signature of reactive glial cells and its pathological implications. *Biochimica et Biophysica Acta (BBA) -. Biochimica et Biophysica Acta. Proteins and Proteomics, 1834*(11), 2418–2428. doi:10.1016/j.bbapap.2012.12.006 PMID:23269363

Jimenez-Shahed, J., Telkes, I., Viswanathan, A., & Ince, N. F. (2016). GPi oscillatory activity differentiates tics from the resting state, voluntary movements, and the unmedicated Parkinsonian state. *Frontiers in Neuroscience, 10*(SEP), 204266. doi:10.3389/fnins.2016.00436 PMID:27733815

Jones, V. C., Atkinson-Dell, R., Verkhratsky, A., & Mohamet, L. (2017). Aberrant iPSC-derived human astrocytes in Alzheimer's disease. *Cell Death & Disease, 8*(3), e2696–e2696. doi:10.1038/cddis.2017.89

Khakh, B. S., & Sofroniew, M. V. (2015). Diversity of astrocyte functions and phenotypes in neural circuits. *Nature Neuroscience, 18*(7), 942–952. doi:10.1038/nn.4043

Kim, H. G., Ju, M. S., Ha, S. K., Lee, H., Lee, H., Kim, S. Y., & Oh, M. S. (2012). Acacetin Protects Dopaminergic Cells against 1-Methyl-4-phenyl-1,2,3,6-tetrahydropyridine-Induced Neuroinflammation in Vitro and in Vivo. *Biological & Pharmaceutical Bulletin, 35*(8), 1287–1294. doi:10.1248/bpb.b12-00127 PMID:22863927

Kim, H. J., Kim, H. J., Lee, J. Y., Yun, J. Y., Kim, S. Y., Park, S. S., & Jeon, B. S. (2011). Phenotype analysis in patients with early onset Parkinson's disease with and without parkin mutations. *Journal of Neurology, 258*(12), 2260–2267. doi:10.100700415-011-6110-1 PMID:21625934

Kim, J., Choi, D., Jeong, H., Kim, J., & Kim, D. (n.d.). *DJ-1 facilitates the interaction between STAT1 and its phosphatase, SHP-1, in brain microglia and astrocytes: a novel anti-inflammatory function of DJ-1.* Elsevier. https://www.sciencedirect.com/science/article/pii/S0969996113002283

Kim, Y. S., Choi, D. H., Block, M. L., Lorenzl, S., Yang, L., Kim, Y. J., Sugama, S., Cho, B. P., Hwang, O., Browne, S. E., Kim, S. Y., Hong, J., Flint Beal, M., & Joh, T. H. (2007). A pivotal role of matrix metalloproteinase-3 activity in dopaminergic neuronal degeneration via microglial activation. *The FASEB Journal, 21*(1), 179–187. doi:10.1096/fj.06-5865com PMID:17116747

Koizumi, S., Ohsawa, K., Inoue, K., & Glia, S. K. (2013). *Purinergic receptors in microglia: functional modal shifts of microglia mediated by P2 and P1 receptors.* Wiley doi:10.1002/glia.22358

Krencik, R., & Zhang, S. C. (2011). Directed differentiation of functional astroglial subtypes from human pluripotent stem cells. *Nature Protocols, 6*(11), 1710–1717. doi:10.1038/nprot.2011.405

Lakhani, B., Borich, M. R., Jackson, J. N., Wadden, K. P., Peters, S., Villamayor, A., MacKay, A. L., Vavasour, I. M., Rauscher, A., & Boyd, L. A. (2016). Motor Skill Acquisition Promotes Human Brain Myelin Plasticity. *Neural Plasticity, 2016*, 1–7. doi:10.1155/2016/7526135 PMID:27293906

Langston, J. W. (2006). The parkinson's complex: Parkinsonism is just the tip of the iceberg. *Annals of Neurology, 59*(4), 591–596. doi:10.1002/ana.20834 PMID:16566021

Lawson, L. J., & Perry, S. (1990). *Heterogeneity In The Distribution And Morphology Of M~Croglia In The Normal Adult Mouse Ibrain.* Wiley.

Leal, M. C., Casabona, J. C., Puntel, M., & Pitossi, F. (2013). Interleukin-1beta and TNF-alpha: Reliable targets for protective therapies in Parkinson's disease? *Frontiers in Cellular Neuroscience*, *7*(APR), 45874. doi:10.3389/FNCEL.2013.00053/BIBTEX

Lees, A. J., Hardy, J., & Revesz, T. (2009). Parkinson's disease. *Lancet*, *373*(9680), 2055–2066. doi:10.1016/S0140-6736(09)60492-X PMID:19524782

Lev, N., Barhum, Y., Ben-Zur, T., Melamed, E., Steiner, I., & Offen, D. (2013). Knocking out DJ-1 attenuates astrocytes neuroprotection against 6-hydroxydopamine toxicity. *Journal of Molecular Neuroscience*, *50*(3), 542–550. doi:10.100712031-013-9984-9 PMID:23536331

Li, X., Tao, Y., Bradley, R., Du, Z., Tao, Y., Kong, L., Dong, Y., Jones, J., Yan, Y., Harder, C. R. K., Friedman, L. M., Bilal, M., Hoffmann, B., & Zhang, S. C. (2018). Fast Generation of Functional Subtype Astrocytes from Human Pluripotent Stem Cells. *Stem Cell Reports*, *11*(4), 998–1008. doi:10.1016/j.stemcr.2018.08.019 PMID:30269954

Liu, Y., Zhou, Q., Tang, M., Fu, N., & Shao, W. (2015). *Upregulation of alphaB-crystallin expression in the substantia nigra of patients with Parkinson's disease*. Elsevier. https://www.sciencedirect.com/science/article/pii/S0197458015000482

Manthripragada, A. D., Schernhammer, E. S., Qiu, J., Friis, S., Wermuth, L., Olsen, J. H., & Ritz, B. (2011). *Non-steroidal anti-inflammatory drug use and the risk of Parkinson's disease*. Karger. doi:10.1159/000325653

Mccoy, M. K., Martinez, T. N., Ruhn, K. A., Szymkowski, D. E., Smith, C. G., Botterman, B. R., Tansey, K. E., & Tansey, M. G. (2006). Blocking soluble tumor necrosis factor signaling with dominant-negative tumor necrosis factor inhibitor attenuates loss of dopaminergic neurons. *Social Neuroscience*. doi:10.1523/JNEUROSCI.1504-06.2006 PMID:16971520

McGeer, P., Itagaki, S., & Boyes, B. (1988). *Reactive microglia are positive for HLA-DR in the substantia nigra of Parkinson's and Alzheimer's disease brains*. AAN Enterprises. https://n.neurology.org/content/38/8/1285.short

McKenzie, I. A., Ohayon, D., Li, H., De Faria, J. P., Emery, B., Tohyama, K., & Richardson, W. D. (2014). Motor skill learning requires active central myelination. *Science*, *346*(6207), 318–322. doi:10.1126cience.1254960 PMID:25324381

Michell-Robinson, M. A., Touil, H., Healy, L. M., Owen, D. R., Durafourt, B. A., Bar-Or, A., Antel, J. P., & Moore, C. S. (2015). Roles of microglia in brain development, tissue maintenance and repair. *Brain*, *138*(5), 1138–1159. doi:10.1093/brain/awv066 PMID:25823474

Miklossy, J., Doudet, D., & Schwab, C. (2006). *Role of ICAM-1 in persisting inflammation in Parkinson disease and MPTP monkeys*. Elsevier. https://www.sciencedirect.com/science/article/pii/S0014488605003730

Moss, J., Gebara, E., Bushong, E. A., Sánchez-Pascual, I., O'Laoi, R., El M'Gharia, I., Kocher-Braissant, J., Ellisman, M. H., & Toni, N. (2016). Fine processes of Nestin-GFP-positive radial glia-like stem cells in the adult dentate gyrus ensheathe local synapses and vasculature. *Proceedings of the National Academy of Sciences of the United States of America*, *113*(18), E2536–E2545. doi:10.1073/pnas.1514652113 PMID:27091993

Mosser, C. A., Baptista, S., Arnoux, I., & Audinat, E. (2017). Microglia in CNS development: Shaping the brain for the future. *Progress in Neurobiology, 149–150*, 1–20. doi:10.1016/j.pneurobio.2017.01.002 PMID:28143732

Nave, K. A., & Werner, H. B. (2014). Myelination of the Nervous System: Mechanisms and Functions. *Annurey Cellbio, 30*, 503–533. doi:10.1146/annurev-cellbio-100913-013101

Nayak, D., Roth, T. L., & McGavern, D. B. (2014). Microglia Development and Function*. *Annurey Cellbio, 32*, 367–402. doi:10.1146/annurev-immunol-032713-120240

Nickel, M., & Gu, C. (2018). Regulation of central nervous system myelination in higher brain functions. *Neural Plasticity, 2018*, 1–12. doi:10.1155/2018/6436453 PMID:29692804

Nimmerjahn, A., Kirchhoff, F., & Helmchen, F. (2005). Neuroscience: Resting microglial cells are highly dynamic surveillants of brain parenchyma in vivo. *Science, 308*(5726), 1314–1318. doi:10.1126cience.1110647 PMID:15831717

Niraula, A. & Sheridan, J. (n.d.). Microglia priming with aging and stress. *Nature.* https://www.nature.com/articles/npp2016185

Oksanen, M., Lehtonen, S., Jaronen, M., Goldsteins, G., Hämäläinen, R. H., & Koistinaho, J. (2019). Astrocyte alterations in neurodegenerative pathologies and their modeling in human induced pluripotent stem cell platforms. *Cellular and Molecular Life Sciences, 76*(14), 2739–2760. doi:10.100700018-019-03111-7 PMID:31016348

Okun, M. S. (2012). Deep-Brain Stimulation for Parkinson's Disease. *NEJM, 367*(16), 1529–1538. doi:10.1056/NEJMct1208070

Ouchi, Y., Yoshikawa, E., Sekine, Y., Futatsubashi, M., Kanno, T., Ogusu, T., & Torizuka, T. (2005). Microglial activation and dopamine terminal loss in early Parkinson's disease. *Annals of Neurology, 57*(2), 168–175. doi:10.1002/ana.20338 PMID:15668962

Pabon, M. M., Bachstetter, A. D., Hudson, C. E., Gemma, C., & Bickford, P. C. (2011). CX3CL1 reduces neurotoxicity and microglial activation in a rat model of Parkinson's disease. *Journal of Neuroinflammation, 8*(1), 9. doi:10.1186/1742-2094-8-9 PMID:21266082

Parkhurst, C. N., Yang, G., Ninan, I., Savas, J. N., Yates, J. R. III, Lafaille, J. J., Hempstead, B. L., Littman, D. R., & Gan, W. B. (2013). Microglia Promote Learning-Dependent Synapse Formation through Brain-Derived Neurotrophic Factor. *Cell, 155*(7), 1596–1609. doi:10.1016/j.cell.2013.11.030 PMID:24360280

Peferoen, L., Kipp, M., van der Valk, P., van Noort, J. M., & Amor, S. (2014). Oligodendrocyte-microglia cross-talk in the central nervous system. *Immunology, 141*(3), 302–313. doi:10.1111/imm.12163 PMID:23981039

Pepe, G., Calderazzi, G., De Maglie, M., Villa, A. M., & Vegeto, E. (2014). Heterogeneous induction of microglia M2a phenotype by central administration of interleukin-4. *Journal of Neuroinflammation, 11*(1), 211. doi:10.118612974-014-0211-6 PMID:25551794

Perea, G., Gómez, R., Mederos, S., Covelo, A., Ballesteros, J. J., Schlosser, L., Hernández-Vivanco, A., Martín-Fernández, M., Quintana, R., Rayan, A., Díez, A., Fuenzalida, M., Agarwal, A., Bergles, D. E., Bettler, B., Manahan-Vaughan, D., Martín, E. D., Kirchhoff, F., & Araque, A. (2016). Activity-dependent switch of gabaergic inhibition into glutamatergic excitation in astrocyte-neuron networks. *ELife, 5*(DECEMBER2016). doi:10.7554/eLife.20362

Perea, G., Navarrete, M., & Araque, A. (2009). Tripartite synapses: Astrocytes process and control synaptic information. *Trends in Neurosciences, 32*(8), 421–431. doi:10.1016/j.tins.2009.05.001 PMID:19615761

Pisanu, A., Lecca, D., Mulas, G., & Wardas, J. (2014). *Dynamic changes in pro-and anti-inflammatory cytokines in microglia after PPAR-γ agonist neuroprotective treatment in the MPTPp mouse model of.* Elsevier. https://www.sciencedirect.com/science/article/pii/S0969996114002423

Ponomarev, E. D., Maresz, K., Tan, Y., & Dittel, B. N. (2007). CNS-Derived Interleukin-4 Is Essential for the Regulation of Autoimmune Inflammation and Induces a State of Alternative Activation in Microglial Cells. *The Journal of Neuroscience : The Official Journal of the Society for Neuroscience, 27*(40), 10714–10721. doi:10.1523/JNEUROSCI.1922-07.2007 PMID:17913905

Postuma, R. B., Iranzo, A., Hogl, B., Arnulf, I., Ferini-Strambi, L., Manni, R., Miyamoto, T., Oertel, W., Dauvilliers, Y., El Ju, Y., Puligheddu, M., Sonka, K., Pelletier, A., Santamaria, J., Frauscher, B., Leu-Semenescu, S., Zucconi, M., Terzaghi, M., Miyamoto, M., & Montplaisir, J. Y. (2015). Risk factors for neurodegeneration in idiopathic rapid eye movement sleep behavior disorder: A multicenter study. *Annals of Neurology, 77*(5), 830–839. doi:10.1002/ana.24385 PMID:25767079

Price, D. A., Martinez, A. A., Seillier, A., Koek, W., Acosta, Y., Fernandez, E., Strong, R., Lutz, B., Marsicano, G., Roberts, J. L., & Giuffrida, A. (2009). WIN55,212-2, a cannabinoid receptor agonist, protects against nigrostriatal cell loss in the 1-methyl-4-phenyl-1,2,3,6-tetrahydropyridine mouse model of Parkinson's disease. *The European Journal of Neuroscience, 29*(11), 2177–2186. doi:10.1111/j.1460-9568.2009.06764.x PMID:19490092

Pringsheim, T., Jette, N., Frolkis, A., & Steeves, T. D. L. (2014). The prevalence of Parkinson's disease: A systematic review and meta-analysis. *Movement Disorders, 29*(13), 1583–1590. doi:10.1002/mds.25945 PMID:24976103

Ransohoff, R. M. (2016). A polarizing question: do M1 and M2 microglia exist? *Nature Neuroscience 2, 19*(8), 987–991. doi:10.1038/nn.4338

Rascol, O., Payoux, P., Ory, F., Ferreira, J. J., Brefel-Courbon, C., & Montastruc, J. L. (2003). Limitations of current Parkinson's disease therapy. *Annals of Neurology, 53*(S3), S3–S15. doi:10.1002/ana.10513 PMID:12666094

Rocha, N. P., De Miranda, A. S., & Teixeira, A. L. (2015). Insights into neuroinflammation in Parkinson's disease: From biomarkers to anti-inflammatory based therapies. *BioMed Research International, 2015*, 1–12. doi:10.1155/2015/628192 PMID:26295044

Rojanathammanee, L., Murphy, E. J., & Combs, C. K. (2011). Expression of mutant alpha-synuclein modulates microglial phenotype in vitro. *Journal of Neuroinflammation, 8*(1), 44. doi:10.1186/1742-2094-8-44 PMID:21554732

Saijo, K., Winner, B., Carson, C. T., Collier, J. G., Boyer, L., Rosenfeld, M. G., Gage, F. H., & Glass, C. K. (2009). A Nurr1/CoREST Pathway in Microglia and Astrocytes Protects Dopaminergic Neurons from Inflammation-Induced Death. *Cell, 137*(1), 47–59. doi:10.1016/j.cell.2009.01.038 PMID:19345186

Samii, A., Etminan, M., Wiens, M. O., & Jafari, S. (2009). NSAID use and the risk of parkinsons disease: Systematic review and meta-analysis of observational studies. *Drugs & Aging*, *26*(9), 769–779. doi:10.2165/11316780-000000000-00000 PMID:19728750

Santos, R., Vadodaria, K. C., Jaeger, B. N., Mei, A., Lefcochilos-Fogelquist, S., Mendes, A. P. D., Erikson, G., Shokhirev, M., Randolph-Moore, L., Fredlender, C., Dave, S., Oefner, R., Fitzpatrick, C., Pena, M., Barron, J. J., Ku, M., Denli, A. M., Kerman, B. E., Charnay, P., & Gage, F. H. (2017). Differentiation of Inflammation-Responsive Astrocytes from Glial Progenitors Generated from Human Induced Pluripotent Stem Cells. *Stem Cell Reports*, *8*(6), 1757–1769. doi:10.1016/j.stemcr.2017.05.011 PMID:28591655

Savchenko, E., Marote, A., Russ, K., Collin, A., Goldwurm, S., Roybon, L., & Pomeshchik, Y. (2018). Generation of a human induced pluripotent stem cell line (CSC-42) from a patient with sporadic form of Parkinson's disease. *Stem Cell Research*, *27*, 78–81. doi:10.1016/j.scr.2018.01.002 PMID:29334629

Schafer, D. P., & Stevens, B. (2015). Microglia Function in Central Nervous System Development and Plasticity. *Cold Spring Harbor Perspectives in Biology*, *7*(10), a020545. doi:10.1101/cshperspect.a020545 PMID:26187728

Schreiner, B., Romanelli, E., Liberski, P., Ingold-Heppner, B., Sobottka-Brillout, B., Hartwig, T., Chandrasekar, V., Johannssen, H., Zeilhofer, H. U., Aguzzi, A., Heppner, F., Kerschensteiner, M., & Becher, B. (2015). Astrocyte Depletion Impairs Redox Homeostasis and Triggers Neuronal Loss in the Adult CNS. *Cell Reports*, *12*(9), 1377–1384. doi:10.1016/j.celrep.2015.07.051 PMID:26299968

Seri, B., Manuel García-Verdugo, J., Mcewen, B. S., & Alvarez-Buylla, A. (2001). Astrocytes give rise to new neurons in the adult mammalian hippocampus. *Journal of Neuroscience*. https://www.jneurosci.org/content/21/18/7153.short

Shao, W., Zhang, S., Tang, M., Zhang, X., & Zhou, Z. (2013). Suppression of neuroinflammation by astrocytic dopamine D2 receptors via αB-crystallin. *Nature*. doi:10.1038/nature11748

Sheridan, G. K., & Murphy, K. J. (2013). Neuron–glia crosstalk in health and disease: fractalkine and CX3CR1 take centre stage. *Open Biology, 3*(1 DEC). doi:10.1098/rsob.130181

Shigemoto-Mogami, Y., Hoshikawa, K., Goldman, J. E., Sekino, Y., & Sato, K. (2014). Microglia Enhance Neurogenesis and Oligodendrogenesis in the Early Postnatal Subventricular Zone. *The Journal of Neuroscience : The Official Journal of the Society for Neuroscience*, *34*(6), 2231–2243. doi:10.1523/JNEUROSCI.1619-13.2014 PMID:24501362

Sierra, A., Encinas, J. M., Deudero, J. J. P., Chancey, J. H., Enikolopov, G., Overstreet-Wadiche, L. S., Tsirka, S. E., & Maletic-Savatic, M. (2010). Microglia Shape Adult Hippocampal Neurogenesis through Apoptosis-Coupled Phagocytosis. *Cell Stem Cell*, *7*(4), 483–495. doi:10.1016/j.stem.2010.08.014 PMID:20887954

Simons, M., & Nave, K. A. (2016). Oligodendrocytes: Myelination and axonal support. *Cold Spring Harbor Perspectives in Biology*, *8*(1), a020479. doi:10.1101/cshperspect.a020479 PMID:26101081

Smith, C. M., Cooksey, E., & Duncan, I. D. (2013). Myelin loss does not lead to axonal degeneration in a long-lived model of chronic demyelination. *The Journal of Neuroscience : The Official Journal of the Society for Neuroscience*, *33*(6), 2718–2727. doi:10.1523/JNEUROSCI.4627-12.2013 PMID:23392698

Squarzoni, P., Oller, G., Hoeffel, G., Pont-Lezica, L., Rostaing, P., Low, D., Bessis, A., Ginhoux, F., & Garel, S. (2014). Microglia Modulate Wiring of the Embryonic Forebrain. *Cell Reports*, *8*(5), 1271–1279. doi:10.1016/j.celrep.2014.07.042 PMID:25159150

Srinivasan, R., Huang, B. S., Venugopal, S., Johnston, A. D., Chai, H., Zeng, H., Golshani, P., & Khakh, B. S. (2015). Ca2+ signaling in astrocytes from Ip3r2−/− mice in brain slices and during startle responses in vivo. *Nature Neuroscience, 18*(5), 708–717. doi:10.1038/nn.4001

Stephenson, J., Nutma, E., van der Valk, P., & Amor, S. (2018). Inflammation in CNS neurodegenerative diseases. *Immunology*, *154*(2), 204–219. doi:10.1111/imm.12922 PMID:29513402

Subhramanyam, C. S., Wang, C., Hu, Q., & Dheen, S. T. (2019). Microglia-mediated neuroinflammation in neurodegenerative diseases. In *Seminars in Cell and Developmental Biology* (Vol. 94, pp. 112–120). Elsevier Ltd., doi:10.1016/j.semcdb.2019.05.004

Suk, K. (2014). Combined analysis of the glia secretome and the CSF proteome: neuroinflammation and novel biomarkers. *EPR, 7*(2), 263–274. doi:10.1586/epr.10.6

Sultan, S., Li, L., Moss, J., Petrelli, F., Cassé, F., Gebara, E., Lopatar, J., Pfrieger, F. W., Bezzi, P., Bischofberger, J., & Toni, N. (2015). Synaptic Integration of Adult-Born Hippocampal Neurons Is Locally Controlled by Astrocytes. *Neuron*, *88*(5), 957–972. doi:10.1016/j.neuron.2015.10.037 PMID:26606999

Tan, Y., Delvaux, E., Nolz, J., Coleman, P. D., Chen, S., & Mastroeni, D. (2018). Upregulation of histone deacetylase 2 in laser capture nigral microglia in Parkinson's disease. *Neurobiology of Aging*, *68*, 134–141. doi:10.1016/j.neurobiolaging.2018.02.018 PMID:29803514

Tan, Z., Liu, Y., Xi, W., Lou, H. F., Zhu, L., Guo, Z., Mei, L., & Duan, S. (2017). Glia-derived ATP inversely regulates excitability of pyramidal and CCK-positive neurons. *Nature Communications, 8*(1), 1–14. doi:10.1038/ncomms13772

Tanaka, T., Kai, S., Matsuyama, T., Adachi, T., Fukuda, K., & Hirota, K. (2013). General anesthetics inhibit LPS-induced IL-1β expression in glial cells. *PLoS One*, *8*(12), e82930. doi:10.1371/journal.pone.0082930 PMID:24349401

Tang, Y. (2016). *Differential roles of M1 and M2 microglia in neurodegenerative diseases*. Springer. doi:10.1007/s12035-014-9070-5

Tang, Y., & Le, W. (2015). Differential Roles of M1 and M2 Microglia in Neurodegenerative Diseases. *Molecular Neurobiology, 53*(2), 1181–1194. doi:10.1007/s12035-014-9070-5

Tang, Y., Li, T., Li, J., Yang, J., Liu, H., Zhang, X. J., & Le, W. (2013). Jmjd3 is essential for the epigenetic modulation of microglia phenotypes in the immune pathogenesis of Parkinson's disease. *Cell Death & Differentiation, 21*(3), 369–380. doi:10.1038/cdd.2013.159

Tansey, M., & McCoy, M. (n.d.). *Neuroinflammatory mechanisms in Parkinson's disease: potential environmental triggers, pathways, and targets for early therapeutic intervention*. Elsevier. https://www.sciencedirect.com/science/article/pii/S0014488607002701

Tay, T. L., Savage, J. C., Hui, C. W., Bisht, K., & Tremblay, M. È. (2017). Microglia across the lifespan: From origin to function in brain development, plasticity and cognition. *The Journal of Physiology, 595*(6), 1929–1945. doi:10.1113/JP272134 PMID:27104646

Terrillion, C. E., Abazyan, B., Yang, Z., Crawford, J., Shevelkin, A. V., Jouroukhin, Y., Yoo, K. H., Cho, C. H., Roychaudhuri, R., Snyder, S. H., Jang, M. H., & Pletnikov, M. V. (2017). DISC1 in Astrocytes Influences Adult Neurogenesis and Hippocampus-Dependent Behaviors in Mice. *Neuropsychopharmacology, 42*(11), 2242–2251. doi:10.1038/npp.2017.129

The NINDS NET-PD Investigators. (2006). A randomized, double-blind, futility clinical trial of creatine and minocycline in early Parkinson disease. *Neurology*. PMID:16481597

Thorne, N., Malik, N., Shah, S., Zhao, J., Class, B., Aguisanda, F., Southall, N., Xia, M., McKew, J. C., Rao, M., & Zheng, W. (2016). High-Throughput Phenotypic Screening of Human Astrocytes to Identify Compounds That Protect Against Oxidative Stress. *Stem Cells Translational Medicine, 5*(5), 613–627. doi:10.5966ctm.2015-0170 PMID:27034412

Ueno, M., Fujita, Y., Tanaka, T., Nakamura, Y., Kikuta, J., Ishii, M., & Yamashita, T. (2013). Layer V cortical neurons require microglial support for survival during postnatal development. *Nature Neuroscience, 16*(5), 543–551. doi:10.1038/nn.3358

van Deijk, A. L. F., Camargo, N., Timmerman, J., Heistek, T., Brouwers, J. F., Mogavero, F., Mansvelder, H. D., Smit, A. B., & Verheijen, M. H. G. (2017). Astrocyte lipid metabolism is critical for synapse development and function in vivo. *Glia, 65*(4), 670–682. doi:10.1002/glia.23120 PMID:28168742

Vinet, J., van Weering, H. R., Heinrich, A., Kälin, R. E., Wegner, A., Brouwer, N., Heppner, F. L., van Rooijen, N., & Boddeke, W. G. M. H., & Biber, K. (2012). Neuroprotective function for ramified microglia in hippocampal excitotoxicity. *Journal of Neuroinflammation*. http://www.jneuroinflammation.com/content/9/1/27

Volterra, A., Liaudet, N., & Savtchouk, I. (2014). Astrocyte Ca2+ signalling: an unexpected complexity. *Nature Reviews Neuroscience, 15*(5), 327–335. doi:10.1038/nrn3725

Wang, J., Song, N., Jiang, H., Wang, J., & Xie, J. (2013). Pro-inflammatory cytokines modulate iron regulatory protein 1 expression and iron transportation through reactive oxygen/nitrogen species production in ventral mesencephalic neurons. *Biochimica et Biophysica Acta. Molecular Basis of Disease, 1832*(5), 618–625. doi:10.1016/j.bbadis.2013.01.021 PMID:23376588

Wang, Q., Liu, Y., & Zhou, J. (2015). Neuroinflammation in Parkinson's disease and its potential as therapeutic target. In Translational Neurodegeneration, 4(1). BioMed Central Ltd. doi:10.118640035-015-0042-0

Wang, X. J., Zhang, S., Yan, Z. Q., Zhao, Y. X., Zhou, H. Y., Wang, Y., Lu, G. Q., & Zhang, J. D. (2011). Impaired CD200–CD200R-mediated microglia silencing enhances midbrain dopaminergic neurodegeneration: Roles of aging, superoxide, NADPH oxidase, and p38 MAPK. *Free Radical Biology & Medicine, 50*(9), 1094–1106. doi:10.1016/j.freeradbiomed.2011.01.032 PMID:21295135

Weinhard, L., Di Bartolomei, G., Bolasco, G., Machado, P., Schieber, N. L., Neniskyte, U., Exiga, M., Vadisiute, A., Raggioli, A., Schertel, A., Schwab, Y., & Gross, C. T. (2018). Microglia remodel synapses by presynaptic trogocytosis and spine head filopodia induction. *Nature Communications, 9*(1), 1–14. doi:10.1038/s41467-018-03566-5

Weintraub, D., & Burn, D. J. (2011). Parkinson's disease: The quintessential neuropsychiatric disorder. *Movement Disorders, 26*(6), 1022–1031. doi:10.1002/mds.23664 PMID:21626547

White, R. E., & Barry, D. S. (2015). The emerging roles of transplanted radial glial cells in regenerating the central nervous system. *Neural Regeneration Research, 10*(10), 1548–1551. doi:10.4103/1673-5374.165317 PMID:26692835

Witt, P. A. L., & Fahn, S. (2016). Levodopa therapy for Parkinson disease. *Neurology, 86*(14, Supplement 1), S3–S12. doi:10.1212/WNL.0000000000002509 PMID:27044648

Xiao, L., Ohayon, D., Mckenzie, I. A., Sinclair-Wilson, A., Wright, J. L., Fudge, A. D., Emery, B., Li, H., & Richardson, W. D. (2016). Rapid production of new oligodendrocytes is required in the earliest stages of motor-skill learning. *Nature Neuroscience, 19*(9), 1210–1217. doi:10.1038/nn.4351 PMID:27455109

Yamada, T., Kawamata, T., Walker, D. G., & McGeer, P. L. (1992). Vimentin immunoreactivity in normal and pathological human brain tissue. *Acta Neuropathologica, 84*(2), 157–162. doi:10.1007/BF00311389 PMID:1523971

Yamamizu, K., Iwasaki, M., Takakubo, H., Sakamoto, T., Ikuno, T., Miyoshi, M., Kondo, T., Nakao, Y., Nakagawa, M., Inoue, H., & Yamashita, J. K. (2017). RETRACTED: In Vitro Modeling of Blood-Brain Barrier with Human iPSC-Derived Endothelial Cells, Pericytes, Neurons, and Astrocytes via Notch Signaling. *Stem Cell Reports, 8*(3), 634–647. doi:10.1016/j.stemcr.2017.01.023 PMID:28238797

Yue, P., Gao, L., Wang, X., Ding, X., & Teng, J. (2018). Pretreatment of glial cell-derived neurotrophic factor and geranylgeranylacetone ameliorates brain injury in Parkinson's disease by its anti-apoptotic and anti-oxidative property. *Journal of Cellular Biochemistry, 119*(7), 5491–5502. doi:10.1002/jcb.26712 PMID:29377238

Yun, S., Kam, T., Panicker, N., Kim, S., & Oh, Y. (2018). *Block of A1 astrocyte conversion by microglia is neuroprotective in models of Parkinson's disease.* Nature. https://www.nature.com/articles/s41591-018-0051-5

Zeis, T., Enz, L., & Schaeren-Wiemers, N. (2016). The immunomodulatory oligodendrocyte. *Brain Research, 1641*, 139–148. doi:10.1016/j.brainres.2015.09.021 PMID:26423932

Zhan, Y., Paolicelli, R., & Sforazzini, F. (2014). *Deficient neuron-microglia signaling results in impaired functional brain connectivity and social behavior*. Nature. https://www.nature.com/articles/nn.3641

Zhang, Q., Chen, W., Tan, S., & Lin, T. (2017). *Stem Cells for Modeling and Therapy of Parkinson's Disease.* Mary Ann Liebert Publishers. doi:10.1089/hum.2016.116

Zhang, S., Wang, X. J., Tian, L. P., Pan, J., Lu, G. Q., Zhang, Y. J., Ding, J. Q., & Di Chen, S. (2011). CD200-CD200R dysfunction exacerbates microglial activation and dopaminergic neurodegeneration in a rat model of Parkinson's disease. *Journal of Neuroinflammation, 8*(1), 154. doi:10.1186/1742-2094-8-154 PMID:22053982

Zhang, W., Wang, T., Pei, Z., Miller, D. S., Wu, X., Block, M. L., Wilson, B., Zhang, W., Zhou, Y., Hong, J.-S., & Zhang, J. (2005). Aggregated α-synuclein activates microglia: A process leading to disease progression in Parkinson's disease. *The FASEB Journal, 19*(6), 533–542. doi:10.1096/fj.04-2751com PMID:15791003

Zhang, Z., Wang, X., & Wang, S. (2008). Isolation and characterization of mesenchymal stem cells derived from bone marrow of patients with Parkinson's disease. *In Vitro Cellular & Developmental Biology. Animal, 44*(5–6), 169–177. doi:10.100711626-008-9093-1 PMID:18401666

Zhu, Y., Yang, G. Y., Ahlemeyer, B., Pang, L., Che, X. M., Culmsee, C., Klumpp, S., & Krieglstein, J. (2002). Transforming Growth Factor-β1 Increases Bad Phosphorylation and Protects Neurons Against Damage. *The Journal of Neuroscience : The Official Journal of the Society for Neuroscience, 22*(10), 3898–3909. doi:10.1523/JNEUROSCI.22-10-03898.2002 PMID:12019309

Chapter 10
Amyotrophic Lateral Sclerosis Involving Gliopathy:
Insights Into the Underlying Mechanisms

Asmaa Haj-Khlifa
Faculty of Science Semlalia, Cadi Ayyad University, Morocco

Mjid Oukhrib
Faculty of Science Semlalia, Cadi Ayyad University, Morocco

Hafida El Ghachi
https://orcid.org/0000-0002-2043-6202
Faculty of Science Semlalia, Cadi Ayyad University, Morocco

Halima Gamrani
Laboratory of Neuroscience, Pharmacology and Environment, Faculty of Science Semlalia, Cadi Ayyad University, Morocco

Moulay Mustapha Bouyatas
Polydisciplinary Faculty of Safi, Cadi Ayyad University, Morocco

ABSTRACT

Amyotrophic Lateral Sclerosis (ALS) is a devastating neurodegenerative disease, causing death within two to five years following diagnosis. ALS is a disease characterized by motor neuron degeneration and destruction of the neuromuscular junction (NMJ). To date, numerous genetic mutations have been associated with ALS, but the mutation in the gene coding for superoxide dismutase (SOD1) has been the most extensively studied. Nevertheless, studies carried out on rodent models and ALS patients have shown that damage to glial cells contributes directly to the development and progression of the disease. However, few studies have focused on the properties of glial cells in ALS. The accumulation of knowledge on the active role and pathological mechanisms of each glial type in the disease must be carefully applied to better understand the ALS pathophysiology, but also for the development of targeted therapy for glial cells. Therefore, in this chapter the authors aimed to compile the most recent information on how each type of glial cell contributes to the development of ALS.

DOI: 10.4018/978-1-6684-9675-6.ch010

1. INTRODUCTION

Amyotrophic lateral sclerosis is a fatal neurodegenerative disease for which, unfortunately, there is currently no effective treatment. The disease is characterized by progressive motor neuron death, primarily affecting motor neurons in the cortex, brain stem, and spinal cord (Viader, 2023). The result is a loss of voluntary motor function, although sphincter control and oculomotricity remain unaffected. In the typical form, the motor deficit is associated with cramps, fasciculations, amyotrophy, and pyramidal syndrome. Affected patients do not all present the same symptoms, the same temporality of onset or the same order, nor the same speed of aggravation (Fonollosa and Tourlet, 2023). 10% of ALS cases are hereditary, while the remainders are considered sporadic and the cause has yet to be discovered. 20% of hereditary cases of ALS are caused by mutations in the gene coding for superoxide dismutase 1 (SOD1) (Rosen et al; 1993; Pasinelli and Brown, 2006). In its unmutated form, the enzyme encoded by this gene neutralizes cell-damaging oxygenated free radicals. Mutations in SOD1 found in ALS patients lead to a change in the structure of the enzyme, which then becomes toxic (Inserm, 2017).

Mitochondria also appear to play a role in the disease. These small cell-specific energy centers malfunction in the motor neurons of SOD1-deficient mice, leading to a reduction in the amount of energy produced. Motor neurons (MNs) then develop alternative survival strategies, such as the production of energy from fatty acids, which leads to the synthesis of ketone bodies toxic in large quantities, or cholesterol. Moreover, some patients exhibit hypermetabolism, i.e. increased use of glucose or fatty acids by the body, which can generate significant weight loss and worsen their prognosis (Inserm, 2017).

Despite the specific loss of MNs in ALS, the results of several studies suggest that non-neuronal cells contribute significantly to the disease, particularly glial cells. Expression of the mutated SOD1 gene in MNs had an impact on the development and early progression of the disease, while expression of the SOD1 gene in central nervous system (CNS) glial cells affected the development and progression of the disease. Indeed, studies conducted in animal models of ALS have highlighted a chronic inflammatory state in which astrocytes, microglial cells, and macrophages present in the motor neuron environment play a deleterious role and participate in disease progression (Clement et al; 2003; Arbour, 2016). These mechanisms could therefore constitute a therapeutic target. In this chapter, we will discuss more data implicating glial cells in ALS pathology.

2. THE INVOLVEMENT OF ASTROCYTES IN THE PATHOGENESIS OF ALS

Astrocytes form the main glial cell population of the nervous system (Chneiweiss, 2002). They play a critical role in CNS development and physiology, being involved in many relevant aspects of neuronal function, such as neuronal trophic support, differentiation, and neuronal survival (Araque et al; 1999). A strong glial reaction has been observed which generally surrounds the upper and lower motor neurons in ALS patients (Kushner et al; 1991; Schipper, 1998). The presence of activated astrocytes has been confirmed in tissue from patients with ALS brain and spinal cord (Philips and Rothstein, 2014). Astrocyte activation becomes increasingly prominent in advanced stages of the disease (Keller et al; 2009). All these observations have led to the study of the potential role of astrocytes in the pathogenesis and progression of ALS, which involves different mechanisms that can lead to the loss of homeostatic functions or the acquisition of toxic functions.

Figure 1. Contribution of glial cells in ALS pathology

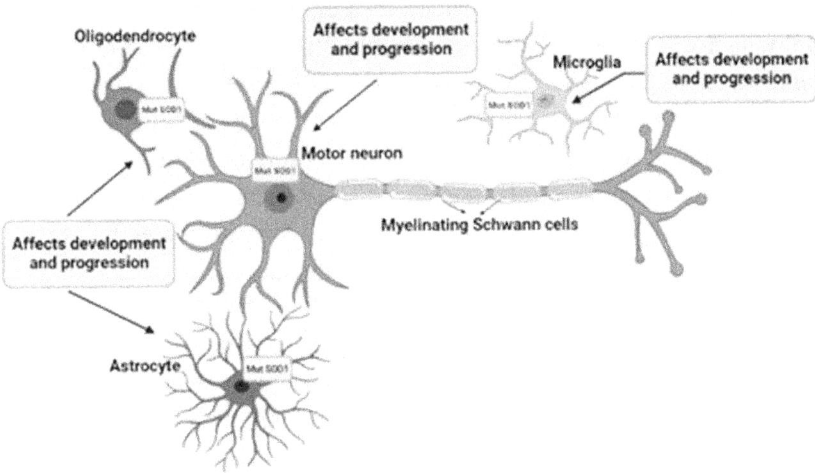

One of the major functions of astrocytes is to maintain a low extracellular concentration of glutamate. Glutamate mediates most excitatory neurotransmissions in the mammalian CNS. Astrocytes remove excess glutamate from synaptic clefts mainly by using the glutamate transporter EAAT2/GLT-1 (Rothstein et al; 1995). Uptake of glutamate from the synaptic cleft between the presynaptic and postsynaptic neuron by the glutamate transporters EAAT2 and EAAT1 expressed by astrocytes will block excessive postsynaptic stimulation of glutamate receptors and thus prevent motor neuron cell death, a process called glutamate-mediated excitotoxicity (Fray et al; 1998; Howland et al; 2002). Dysfunction of the glutamate transport system can lead to excessive glutamate accumulation in the synapse and neurodegeneration. Consequently, a significant increase in glutamate levels in the cerebrospinal fluid of ALS patients has been reported (Rothstein et al; 1990; Rothstein et al; 1992). Furthermore, astrocytic overexpression of the EAAT2 transporter in G93A transgenic mice rescued motor deficits but did not influence lifespan (Guo et al; 2003). Conversely, reduction of EAAT2 in astrocytes potentiated motor neuron loss and slightly reduced survival (Pardo et al; 2006). In a rat model, there is a localized loss of glutamate transporters surrounding motor neurons in presymptomatic animals, and an almost total loss in the end stage of the disease. In contrast, in symptomatic and presymptomatic mice or mice carrying the G93A SOD1 mutation, a less pronounced and delayed astrocytic reactivity is observed, and neither the EAAT1 transporters nor the EAAT2 transporters seem to be altered (Liu et al; 2004).

Recent investigations have demonstrated that mutant astrocytes expressing SODl release toxic factors that can especially kill wild-type and mutant motor neurons in culture. Reactive astrocytes have been implicated in the generation of free radicals and neuronal death in culture. Thus, it appears that astrocytes must be induced to produce inducible nitric oxide synthase (iNOS) by LPS or cytokines in order to promote neuronal death. The damage to the mitochondrial complexes of co-cultured cortical neurons is caused by astrocyte production of nitric oxide or peroxynitrite, and this also increases the excitotoxicity caused by NMDA (Bolan˜os et al; 1995; Stewart et al; 2000). According to (Liu et al; 2004), the ability of reactive astrocytes to produce nitric oxide is crucial for the induction of motor neuron apoptosis in a co-culture model. Additionally, nitrotyrosine labeling has been observed in apoptosizing motor neurons in vitro in SOD-1 mutant mice and in patients with sporadic and familial ALS (Este´vez et al; 1999;

Sasaki et al; 2001). Moreover, a second aspect of the deleterious behavior of astrocytes in ALS is the insufficient release of neurotrophic factors considered essential for maintaining neuronal health. Glia-derived neurotrophic factor (GDNF), ciliary neurotrophic factor (CNTF), brain-derived neurotrophic factor (BDNF), and vascular endothelial growth factor (VEGF) are all released by astrocytes and can protect motor neurons (Ekestern, 2004; Dewil et al; 2007). This loss of neurotrophins could be a direct or indirect cause of neuronal death. In addition to this, it has been found that elevated levels of D-serine are particularly detected in the spinal cords of both ALS patients, and transgenic mice (Sasabe et al; 2007). Since D-serine is a co-agonist of NMDA receptors and can be released by astrocytes, this release could contribute to exacerbating excitotoxicity. However, the abnormal release of astrocytic D-serine remains still to be validated (Henneberger et al; 2010).

Several studies have attempted to identify the mechanisms of toxicity exerted by astrocytes. Indeed, these mechanisms fall into two broad categories: a first according to which astrocytes release various molecules leading to a neurotoxic cascade affecting MNs that demonstrate a decrease in the expression of major histocompatibility complex type I (MHC-I) molecules due to factors secreted by astrocytes (Song et al; 2016). MHC-I molecules, a self-recognition system, act as inhibitory ligands for certain cells (Natural Killer cells). In this way, reduced expression of MHC-I molecules enables cytotoxic mechanisms, normally inhibited, to unfold. In a second category, astrocytes themselves show functional abnormalities, manifested in poor communication between them. Intercellular communication between astrocytes is essential for the proper functioning of the nervous system, and is mediated by the presence of gap junctions. These gap junctions allow astrocytes to be coupled and form a glial syncytium that has been shown to have notably physiological implications in ion homeostasis and synaptic transmission. However, under pathological conditions, the expression of certain protein components essential to the channels forming these gap junctions, such as connexins, as well as the function of the latter, are altered (Arbour; 2016).

In contrast, healthy astrocytes play a protective role for MNs via secreted products. A reduced concentration of neurtrophins has been found in the cerebrospinal fluid (CSF) of ALS patients and its supplementation has been shown to protect MNs (Krakora et al; 2013; Mishra et al; 2016; Shruthi et al; 2017; Izrael et al; 2020). Astrocytes release extracellular vesicles as well (Verkhratsky et al; 2016) that could target nearby or long-distance targets with selective potential for neurons (Venturini et al; 2019). For instance, astrocyte-derived extracellular vesicles have been shown to be positive for neuroglobin, a protein functioning as a neuroprotectant against cell injury (Venturini et al; 2019). In addition, other neuroprotective groups of proteins secreted by astrocytes include metalloproteases and their inhibitors) or immunomodulatory factors (Gardner and Ghorpade; 2003; Jha et al; 2019).

New intervention trials are being conducted, and astrocytes are now an interesting target for therapy. The transcription factor, Nrf2, regulates the expression of genes containing the antioxidant response element (ARE), which are preferentially activated in astrocytes. In vitro protection for neighboring neurons was achieved by activating ARE/Nrf2 in astrocytes, which also increased the lifespan of ALS mice (Vargas et al; 2008). However, neither the beginning nor the development of the disease was impacted in SOD1 mutant animals in a study in which proliferating astrocytes were the object of selective ablation (Lepore et al; 2008a). Furthermore, it has been demonstrated that astrocyte ablation in injury models does not help outcome, as astrocytes may play a protective role (Faulkner et al; 2004). Nevertheless, transplantation of healthy glial precursor cells, which subsequently differentiated into astrocytes, proved neuroprotective and prolonged survival time in the SOD1 mutant model (Lepore el al; 2008b).

In summary, the SOD1 mutation could cause both a loss of function (release of neurotrophic factors, abnormal regulation of GluR2 subunits and the glutamate transporter) and a gain of function (release of toxic factors and excessive release of D-serine) in astrocytes. Moreover, it's necessary to mention that an important functional dialogue takes place between astrocytes and microglial cells, since selective down regulation of mutated SOD1 gene expression in astrocytes delayed microglial activation and strongly slowed disease development in SOD1G37R mice (Yamanaka et al; 2008).

3. MICROGLIA AND ALS

Microglial cells are the main immune cells of the CNS. Microglia have a crucial function in CNS homeostasis, since they constitute the first level of defense against pathogens, contributing to immune responses, but are also involved in tissue repair (Ginhoux and Merad, 2011). When activated by a disruptive element, "activated" microglial cells alter their morphology, migrate towards distress signals or the damaged site, proliferate, clear debris and release a variety of pro-inflammatory molecules (Beers et al; 2006; Hensley et al; 2002). Microgliosis is a pathological characteristic observed in the tissues of patients suffering from diseases of the nervous system, including ALS. A study of ALS patients who are still alive using positron emission tomography (PET) imaging not only confirmed this microglial activation in patients' brains, but also correlated it with the severity of damage to the upper MNs (Turner et al; 2004). It is important to note that microglial activation could be triggered, among other things, by mSOD1 secretion (Urushitani et al; 2006; Zhao et al; 2010). However, analysis of microgliosis in the CNS of ALS patients depends largely on post-mortem tissue studies, which demonstrate increased microgliosis in the motor cortex, brainstem motor nuclei, corticospinal tract and ventral horn of the spinal cord (Kawamata et al; 1992; Corcia et al; 2012). Furthermore, studies in animal models have reported that in vivo resident microglia rises as the disease progresses, and their activation states show a continuum between the two classic microglial phenotypes – neuroprotective M2 and toxic M1 (Filipi et al; 2020).

Microgliosis and the CNS's inflammatory response may have either positive or negative consequences. In addition, it has been shown that ALS mice models exhibit a high level of several pro-inflammatory molecules (Alexianu et al; 2001; Hensley et al; 2003; Henkel et al; 2005). M2 microglia secrete high levels of IL-4, IL-10 and IGF-1, block the release of pro-inflammatory cytokines and support Th2 functions. In addition, M2 participates in the resolution of inflammation through endocytic clearance and promotes tissue repair, which can improve neuronal survival. In contrast, activated M1 microglia are generated in response to IFN-γ or lipopolysaccharide (LPS), eliciting Th1 responses and subsequently inducing neuronal damage by secreting pro-inflammatory cytokines (TNF-α and IL-1β), nitric oxide (NO) and reducing the release of trophic factors. Studies by (Almer et al; 1999; Hensley et al; 2002) confirmed the release of pro-inflammatory cytokines, chemokines, anti-inflammatory cytokines, neurotrophic factors and reactive oxygen species by activated microglia. Oxidative stress-mediated damage has been suggested as a potential contributor to motor neuron degeneration in ALS. Interestingly, activated microglia is also a powrful source of extracellular and intracellular ROS that could lead to neuronal damage (Block et al; 2007).

The identity of the factors that activate microglia was examined. The innate immune system is the first line of defense against invading pathogens, which are primarily recognized by toll-like receptors (TLRs). SOD1 mutant mice were found to have elevated levels of innate immune receptors such as TLR2 and CD14. Moreover, bone marrow deficient in MyD88 (myeloid differentiation factor 88), an adaptor protein essential for transmitting most TLR signaling accelerated disease progression in SOD1G37R mice (Kang and Rivest, 2007).

However, the impact of acquired immunity in ALS has recently been the subject of extensive studies. T cells have been detected in ALS mice models and as well as in ALS patients (Engelhardt et al; 1993) which has suggested that acquired immunity may be involved in this disease. Besides, in ALS animals, genetic deletion of CD4+ T cells or functional T cells increased disease development (Beers et al; 2008; Chiu et al; 2008). In addition, these studies revealed that the presence of CD4+ T cells stabilized the state of microglial activation, with a decrease in the level of pro-inflammatory cytokines and an increase in the neurotrophic factor, IGF-1. Another recent study reported that the transfer of activated CD4+CD25+ cells prolonged the lifespan of SOD1 mutant mice (Banerjee et al; 2008). These findings support a protective role of a particular T lymphocyte population through modulated microglial activation.

In vivo studies have consolidated the role of microglial cells in disease progression. Reducing SOD1 expression in microglia and macrophages slowed disease progression in SOD1G37R (Boillee et al; 2006) and SOD1G85R mice (Wang et al; 2009). A complementary approach, using microglia/macrophage replacement by bone marrow transplantation, demonstrated that wild-type microglia/macrophages could delay disease progression in SOD1G93A mice (Beers et al; 2006). Yet, since the reduction of proliferating microglia by around half did not affect the rate of progression in the same strain of SOD1 mice, it is inferred that disease progression does not impact the proliferation of mutant microglia in initial response to injury (Gowing et al; 2008).

At the molecular level, nuclear factor-kappa B (NF-kB), the master regulator of inflammation, is selectively activated by mutant proteins, such as TDP-43 and FUS. Accordingly, having the ability to effectively alter microglial phenotypes, enhancing the anti-inflammatory properties and inhibiting or reducing M1 toxicity, could be a promising therapeutic approach for ALS (Geloso et al; 2017).

After all, the cellular mechanisms underlying this microglial toxicity are still under investigation. Prevention of microglial activation has been achieved through the use of minocycline, a tetracycline antibiotic that inhibits microglial activation, or by inhibiting certain signalling pathways or the release of neurotoxic molecules. Unfortunately, these interventions did not have a significant impact on the disease (Philips and Robberecht, 2011). In fact, most of these therapeutic strategies lack spatial and temporal specificity, as they need to target, globally, both the beneficial and detrimental processes of the inflammatory response (Philips and Rothstein, 2014). Moreover, it seems that the other components of the inflammatory response must also be taken into consideration, since the dialogue between glial cells and T lymphocytes is proving to be a key element in the control of microglial and astrocytic toxicity (Beers et al; 2008; Chiu et al; 2008). In light of this, it may be suggested that an effective therapeutic strategy for ALS could be achieved by interfering with different pathways in different cell types.

4. OLIGODENDROCYTES

Oligodendrocytes are the myelinating cells of the CNS. They play a crucial role in both generating and maintaining the myelin sheath around the axon allowing rapid salutary conduction of action potentials. They also provide metabolic support for motor neurons (Bradl and Lassmann, 2010). Some motor neurons can be considerably long (> 3 feet) and are particularly dependent on metabolic substrates provided by oligodendrocytes. Oligodendrocytes have been found to express the monocarboxylate transporter 1 (MCT1) through which metabolic substrates such as ketone bodies, lactate and hydrogen ions are exported

Figure 2. Release of molecules by M1 and M2 microglia that can affect motor neuron survival

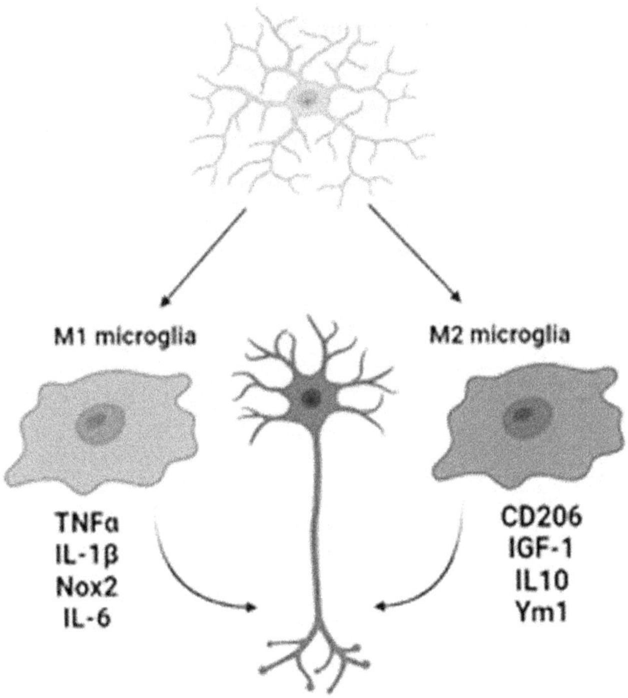

and then supplied to neurons, particularly in regions of close neuronal contact such as paranodal junctions (Lee et al; 2012). MCT1 first releases lactate, while MCT2 then allows lactate to be taken up by neurons (Rafiki et al; 2003). Mice with decreased MCT1-mediated transport exhibit widespread axonal degeneration, suggesting inadequate metabolic support of oligodendrocytes to neurons. This suggests that neurons themselves are largely dependent on this oligodendrocyte-derived lactate (Lee et al; 2012). Furthermore, mice lacking the expression of specific oligodendrocyte markers such as Proteolipid protein (PLP) and CNPase develop axonal degeneration independent of generalized demyelination, again suggesting that oligodendrocytes apart from their myelin sheath-generating function, play a decisive role in providing generalized support for axons (Griffiths et al; 1998; Lappe-Siefke et al; 2003). Recent investigations have shown the involvement of oligodendrocytes in ALS, they carry the same ALS-inducing mutations, the same protein aggregates were found in neurons and in spinal cord oligodendrocytes of ALS patients. It was revealed that oligodendrocytes are severely damaged during the disease and that their degeneration occurs before MN death (Ferraiuolo et al; 2016). In an attempt to compensate for the loss of oligodendrocytes, it was reported that progenitor cells were very proliferative, but also failed to reach maturation. As a result, motor fibers in both mouse models and in the spinal cords of ALS patients show clear signs of demyelination (Wong et al; 1995). It remains unclear why newly generated oligodendrocytes fail to function properly, but their disappearance contributes to motor neuron loss, as evidenced by the fact that partial prevention of oligodendroglial damage significantly delays disease in the ALS rodent model (Kang et al; 2013). Despite compensation for the loss of oligodendrocytes, clear oligodendrocyte dysfunction was observed, manifested by reduced expression of many markers associated with oligodendrocyte biology such as Myelin basic protein (MBP), a component of the myelin sheath, as well as MCT1 expression (Lee et al; 2012; Kang et al; 2013; Philips et al; 2013).

In some preliminary studies, it has been demonstrated that oxidative stress can lead to the death of oligodendrocytes, malfunction of oligodendrocyte progenitor cells, and demyelination of gray matter in the spinal cord and motor cortex in ALS. Furthermore, results from immunocytochemistry of malondialdehyde (MDA) and oxidized phospholipids prove the pathogenic relationship between oxidative stress and oligodendrocyte death (Gong et al; 2022). Thus, further studies should focus on the relationship between oxidative stress damage and oligodendrocytes in ALS.

Importantly, oligodendroglial damage has not only been observed in an ALS mouse model, but also in human ALS tissue (Kang et al; 2013). In addition to providing evidence of oligodendrocyte damage in ALS disease, the results of these studies showed that oligodendrocytes have an important role in the pathogenesis of ALS (Kang et al; 2013; Philips et al; 2013).

5. SCHWANN CELLS

Schwann cells are responsible for myelination of the peripheral nervous system. They are closely associated with motor neuron axons and are important for axonal development, regeneration, and homeostasis (Fink et al; 1999). To date, little is known about the involvement of Schwann cells in ALS disease. Studies have shown changes in peripheral myelin along motor neuron axons which are most likely due to axonal degeneration (Perrie et al; 1993). Indeed, studies investigating the in vivo impact of Schwann cells in the pathogenesis of ALS, either through the specific expression or ablation of mutant SOD1 in these cells, have provided divergent results. For example, high specific expression of SOD1G93A in Schwann cells led to no motor phenotype, loss of MNs, or axonal degeneration (Turner et al; 2010). While a second approach: ablation of SOD1, specifically in these cells, led to two different scenarios depending on the type of SOD1 enzymatic activity: 1: ablation of SOD1G37R proved to be neurotoxic and caused substantial acceleration of the disease (Lobsiger et al; 2009), while 2: ablation of SOD1G85R was neuroprotective, delaying loss of MNs, activation of astrocytes and microglial cells, and delaying disease development and progression (Wang et al; 2012). It should be noted that, due to methodological constraints (lack of specific markers), these studies have also targeted a small proportion of non-myelinating Schwann cells (Lobsiger et al; 2009). A close relationship between Schwann cells and motor neuron axons warrants further study to help us better understand the pathology of ALS. Considering the anatomical proximity of myelinating Schwann cells and MNs axons, these cells clearly deserve more attention in ALS research, and elucidation of their contribution to the disease would help us better understand its pathophysiology.

6. CONCLUSION

In summary, it seems that the development and progression of ALS depend on the complex interactions between neurons and their surrounding cells. Indeed, the combination of dysfunctions of these different cell types, both neuronal and non-neuronal, as well as the synergy that develops between them, both appear to be crucial for the full expression of the disease. It's not hard to imagine that cellular interactions could begin focally in the MNs and subsequently trigger a cascade of changes spreading to other cell types. In fact, a recent study carried out a spatiotemporal analysis of actively translated mRNAs in the ribosomes of different cells. Their results suggest that ALS pathogenesis would involve a temporal cascade of specific damage initiating in MNs and then spreading to neighboring glial cells (Sun et al; 2015). Thus, each cell would contribute to the initiation and progression of the disease in a spatially and temporally coordinated manner. In the light of the data presented in this chapter, the degeneration of MNs therefore appears to be the result of toxicity emanating as much from the expression of mutated proteins within them as in the cells surrounding them, the glial cells (Lobsiger and Cleveland, 2007). A better understanding of the molecular pathology in glial cells will contribute to the development of therapies that will slow the progression of ALS for the benefit of both sporadic and familiar ALS patients.

REFERENCES

Alexianu, M. E., Kozovska, M., & Appel, S. H. (2001). Immune reactivity in a mouse model of familial ALS correlates with disease progression. *Neurology*, *57*(7), 1282–1289. doi:10.1212/WNL.57.7.1282 PMID:11591849

Almer, G., Vukosavic, S., Romero, N., & Przedborski, S. (1999). Inducible nitric oxide synthase upregulation in a transgenic mouse model of familial amyotrophic lateral sclerosis. *Journal of Neurochemistry*, *72*(6), 2415–2425. doi:10.1046/j.1471-4159.1999.0722415.x PMID:10349851

Araque, A., Sanzgiri, R. P., Parpura, V., & Haydon, P. G. (1999). Astrocyte-induced modulation of synaptic transmission. *Canadian Journal of Physiology and Pharmacology*, *77*(9), 699–706. doi:10.1139/y99-076 PMID:10566947

Banerjee, R., Mosley, R. L., Reynolds, A. D., Dhar, A., Jackson-Lewis, V., Gordon, P. H., Przedborski, S., & Gendelman, H. E. (2008). Adaptive immune neuroprotection in G93A-SOD1 amyotrophic lateral sclerosis mice. *PLoS One*, *3*(7), e2740. doi:10.1371/journal.pone.0002740 PMID:18648532

Beers, D. R., Henkel, J. S., Xiao, Q., Zhao, W., Wang, J., Yen, A. A., Siklos, L., McKercher, S. R., & Appel, S. H. (2006). Wild-type microglia extend survival in PU.1 knockout mice with familial amyotrophic lateral sclerosis. *Proceedings of the National Academy of Sciences of the United States of America*, *103*(43), 16021–16026. doi:10.1073/pnas.0607423103 PMID:17043238

Beers, D. R., Henkel, J. S., Zhao, W., Wang, J., & Appel, S. H. (2008). CD4+ T cells support glial neuroprotection, slow disease progression, and modify glial morphology in an animal model of inherited ALS. *Proceedings of the National Academy of Sciences of the United States of America*, *105*(40), 15558–15563. doi:10.1073/pnas.0807419105 PMID:18809917

Block, M. L., Zecca, L., & Hong, J. S. (2007). Microglia-mediated neurotoxicity: Uncovering the molecular mechanisms. *Nature Reviews. Neuroscience*, *8*(1), 57–69. doi:10.1038/nrn2038 PMID:17180163

Boillée, S., Yamanaka, K., Lobsiger, C. S., Copeland, N. G., Jenkins, N. A., Kassiotis, G., Kollias, G., & Cleveland, D. W. (2006). Onset and progression in inherited ALS determined by motor neurons and microglia. *Science*, *312*(5778), 1389–1392. doi:10.1126cience.1123511 PMID:16741123

Bolaños, J. P., Heales, S. J., Land, J. M., & Clark, J. B. (1995). Effect of peroxynitrite on the mitochondrial respiratory chain: Differential susceptibility of neurones and astrocytes in primary culture. *Journal of Neurochemistry*, *64*(5), 1965–1972. doi:10.1046/j.1471-4159.1995.64051965.x PMID:7722484

Bradl, M., & Lassmann, H. (2010). Oligodendrocytes: Biology and pathology. *Acta Neuropathologica, 119*(1), 37–53. doi:10.100700401-009-0601-5 PMID:19847447

Chiu, I. M., Chen, A., Zheng, Y., Kosaras, B., Tsiftsoglou, S. A., Vartanian, T. K., Brown, R. H. Jr, & Carroll, M. C. (2008). T lymphocytes potentiate endogenous neuroprotective inflammation in a mouse model of ALS. *Proceedings of the National Academy of Sciences of the United States of America, 105*(46), 17913–17918. doi:10.1073/pnas.0804610105 PMID:18997009

Chneiweiss, H. (2002). Les astrocytes contrôlent la neurogenèse dans le système nerveux central adulte. *M/S: médecine sciences, 18*(11), 1065–1066. https://doi.org/ doi:10.1051/medsci/200218111065

Clement, A. M., Nguyen, M. D., Roberts, E. A., Garcia, M. L., Boillée, S., Rule, M., McMahon, A. P., Doucette, W., Siwek, D., Ferrante, R. J., Brown, R. H. Jr, Julien, J. P., Goldstein, L. S., & Cleveland, D. W. (2003). Wild-type nonneuronal cells extend survival of SOD1 mutant motor neurons in ALS mice. *Science, 302*(5642), 113–117. doi:10.1126cience.1086071 PMID:14526083

Corcia, P., Tauber, C., Vercoullie, J., Arlicot, N., Prunier, C., Praline, J., Nicolas, G., Venel, Y., Hommet, C., Baulieu, J. L., Cottier, J. P., Roussel, C., Kassiou, M., Guilloteau, D., & Ribeiro, M. J. (2012). Molecular imaging of microglial activation in amyotrophic lateral sclerosis. *PLoS One*, *7*(12), e52941. doi:10.1371/journal.pone.0052941 PMID:23300829

Danielle, A. (2016). *Altérations des cellules de Schwann périsynaptiques à la jonction neuromusculaire: implications pour la sclérose latérale amyotrophique*. Département de Neurosciences Faculté de Médecine, Université de Montréal.

Dewil, M., Lambrechts, D., Sciot, R., Shaw, P. J., Ince, P. G., Robberecht, W., & Van den Bosch, L. (2007). Vascular endothelial growth factor counteracts the loss of phospho-Akt preceding motor neurone degeneration in amyotrophic lateral sclerosis. *Neuropathology and Applied Neurobiology*, *33*(5), 499–509. doi:10.1111/j.1365-2990.2007.00850.x PMID:17854437

Ekestern, E. (2004). Neurotrophic factors and amyotrophic lateral sclerosis. *Neurodegenerative Diseases, 1*(2-3), 88–100. doi:10.1159/000080049 PMID:16908980

Engelhardt, J. I., Tajti, J., & Appel, S. H. (1993). Lymphocytic Infiltrates in the Spinal Cord in Amyotrophic Lateral Sclerosis. *Archives of Neurology*, *50*(1), 30–36. doi:10.1001/archneur.1993.00540010026013 PMID:8093428

Estévez, A. G., Crow, J. P., Sampson, J. B., Reiter, C., Zhuang, Y., Richardson, G. J., Tarpey, M. M., Barbeito, L., & Beckman, J. S. (1999). Induction of nitric oxide-dependent apoptosis in motor neurons by zinc-deficient superoxide dismutase. *Science, 286*(5449), 2498–2500. doi:10.1126cience.286.5449.2498 PMID:10617463

Faulkner, J. R., Herrmann, J. E., Woo, M. J., Tansey, K. E., Doan, N. B., & Sofroniew, M. V. (2004). Reactive astrocytes protect tissue and preserve function after spinal cord injury. *The Journal of Neuroscience : The Official Journal of the Society for Neuroscience*, *24*(9), 2143–2155. doi:10.1523/JNEUROSCI.3547-03.2004 PMID:14999065

Ferraiuolo, L., Meyer, K., Sherwood, T. W., Vick, J., Likhite, S., Frakes, A., Miranda, C. J., Braun, L., Heath, P. R., Pineda, R., Beattie, C. E., Shaw, P. J., Askwith, C. C., McTigue, D., & Kaspar, B. K. (2016). Oligodendrocytes contribute to motor neuron death in ALS via SOD1-dependent mechanism. *Proceedings of the National Academy of Sciences of the United States of America, 113*(42), E6496–E6505. doi:10.1073/pnas.1607496113 PMID:27688759

Filipi, T., Hermanova, Z., Tureckova, J., Vanatko, O., & Anderova, A. M. (2020). Glial Cells-The Strategic Targets in Amyotrophic Lateral Sclerosis Treatment. *Journal of Clinical Medicine, 9*(1), 261. doi:10.3390/jcm9010261 PMID:31963681

Fink, T., Davey, D. F., & Ansselin, A. D. (1999). Glutaminergic and adrenergic receptors expressed on adult guinea pig Schwann cells in vitro. *Canadian Journal of Physiology and Pharmacology, 77*(3), 204–210. doi:10.1139/y99-008 PMID:10535694

Fonollosa, A., & Tourlet, C. (2023).Pathogenèse et évolution de la sclérose latérale amyotrophique, *Kinésithérapie, la Revue, 23*. doi:10.1016/j.kine.2023.01.008

Fray, A. E., Ince, P. G., Banner, S. J., Milton, I. D., Usher, P. A., Cookson, M. R., & Shaw, P. J. (1998). The expression of the glial glutamate transporter protein EAAT2 in motor neuron disease: An immunohistochemical study. *The European Journal of Neuroscience, 10*(8), 2481–2489. doi:10.1046/j.1460-9568.1998.00273.x PMID:9767379

Gardner, J., & Ghorpade, A. (2003). Tissue inhibitor of metalloproteinase (TIMP)-1: The TIMPed balance of matrix metalloproteinases in the central nervous system. *Journal of Neuroscience Research, 74*(6), 801–806. doi:10.1002/jnr.10835 PMID:14648584

Geloso, M. C., Corvino, V., Marchese, E., Serrano, A., Michetti, F., & D'Ambrosi, N. (2017). The Dual Role of Microglia in ALS: Mechanisms and Therapeutic Approaches. *Frontiers in Aging Neuroscience, 9*, 242. doi:10.3389/fnagi.2017.00242 PMID:28790913

Ginhoux, F., & Merad, M. (2011). Les cellules de la microglie - Leurs origines extra-embryonnaires enfin révélées [Microglia arise from extra-embryonic yolk sac primitive progenitors]. *Medecine sciences. Ms, 27*(8-9), 719–724. doi:10.1051/medsci/2011278013 PMID:21880259

Gong, Z., Ba, L., & Zhang, M. (2022). Dysfunction of the oligodendrocytes in amyotrophic lateral sclerosis. *Journal of Biomedical Research, 36*(5), 336–342. doi:10.7555/JBR.36.20220009 PMID:36106501

Gowing, G., Philips, T., Van Wijmeersch, B., Audet, J. N., Dewil, M., Van Den Bosch, L., Billiau, A. D., Robberecht, W., & Julien, J. P. (2008). Ablation of proliferating microglia does not affect motor neuron degeneration in amyotrophic lateral sclerosis caused by mutant superoxide dismutase. *The Journal of Neuroscience : The Official Journal of the Society for Neuroscience, 28*(41), 10234–10244. doi:10.1523/JNEUROSCI.3494-08.2008 PMID:18842883

Griffiths, I., Klugmann, M., Anderson, T., Yool, D., Thomson, C., Schwab, M. H., Schneider, A., Zimmermann, F., McCulloch, M., Nadon, N., & Nave, K. A. (1998). Axonal swellings and degeneration in mice lacking the major proteolipid of myelin. *Science, 280*(5369), 1610–1613. doi:10.1126cience.280.5369.1610 PMID:9616125

Guo, H., Lai, L., Butchbach, M. E., Stockinger, M. P., Shan, X., Bishop, G. A., & Lin, C. L. (2003). Increased expression of the glial glutamate transporter EAAT2 modulates excitotoxicity and delays the onset but not the outcome of ALS in mice. *Human Molecular Genetics*, *12*(19), 2519–2532. doi:10.1093/hmg/ddg267 PMID:12915461

Henkel, J. S., Beers, D. R., Siklós, L., & Appel, S. H. (2006). The chemokine MCP-1 and the dendritic and myeloid cells it attracts are increased in the mSOD1 mouse model of ALS. *Molecular and Cellular Neurosciences*, *31*(3), 427–437. doi:10.1016/j.mcn.2005.10.016 PMID:16337133

Henneberger, C., Papouin, T., Oliet, S. H., & Rusakov, D. A. (2010). Long-term potentiation depends on release of D-serine from astrocytes. *Nature*, *463*(7278), 232–236. doi:10.1038/nature08673 PMID:20075918

Hensley, K., Fedynyshyn, J., Ferrell, S., Floyd, R. A., Gordon, B., Grammas, P., Hamdheydari, L., Mhatre, M., Mou, S., Pye, Q. N., Stewart, C., West, M., West, S., & Williamson, K. S. (2003). Message and protein-level elevation of tumor necrosis factor alpha (TNF alpha) and TNF alpha-modulating cytokines in spinal cords of the G93A-SOD1 mouse model for amyotrophic lateral sclerosis. *Neurobiology of Disease*, *14*(1), 74–80. doi:10.1016/S0969-9961(03)00087-1 PMID:13678668

Hensley, K., Floyd, R. A., Gordon, B., Mou, S., Pye, Q. N., Stewart, C., West, M., & Williamson, K. (2002). Temporal patterns of cytokine and apoptosis-related gene expression in spinal cords of the G93A-SOD1 mouse model of amyotrophic lateral sclerosis. *Journal of Neurochemistry*, *82*(2), 365–374. doi:10.1046/j.1471-4159.2002.00968.x PMID:12124437

Howland, D. S., Liu, J., She, Y., Goad, B., Maragakis, N. J., Kim, B., Erickson, J., Kulik, J., DeVito, L., Psaltis, G., DeGennaro, L. J., Cleveland, D. W., & Rothstein, J. D. (2002). Focal loss of the glutamate transporter EAAT2 in a transgenic rat model of SOD1 mutant-mediated amyotrophic lateral sclerosis (ALS). *Proceedings of the National Academy of Sciences of the United States of America*, *99*(3), 1604–1609. doi:10.1073/pnas.032539299 PMID:11818550

Inserm. (2017). Sclérose latérale amyotrophique (SLA) / maladie de Charcot. *Une dégénérescence des motoneurones encore incurable.* https://www.inserm.fr/dossier/sclerose-laterale-amyotrophique-sla-maladie-charcot/.

Izrael, M., Slutsky, S. G., & Revel, M. (2020). Rising Stars: Astrocytes as a Therapeutic Target for ALS Disease. *Frontiers in Neuroscience*, *14*, 824. doi:10.3389/fnins.2020.00824 PMID:32848579

Jha, M. K., Jo, M., Kim, J. H., & Suk, K. (2019). Microglia-Astrocyte Crosstalk: An Intimate Molecular Conversation. *The Neuroscientist*, *25*(3), 227–240. doi:10.1177/1073858418783959 PMID:29931997

Kang, J., & Rivest, S. (2007). MyD88-deficient bone marrow cells accelerate onset and reduce survival in a mouse model of amyotrophic lateral sclerosis. *The Journal of Cell Biology*, *179*(6), 1219–1230. doi:10.1083/jcb.200705046 PMID:18086918

Kang, S. H., Li, Y., Fukaya, M., Lorenzini, I., Cleveland, D. W., Ostrow, L. W., Rothstein, J. D., & Bergles, D. E. (2013). Degeneration and impaired regeneration of gray matter oligodendrocytes in amyotrophic lateral sclerosis. *Nature Neuroscience*, *16*(5), 571–579. doi:10.1038/nn.3357 PMID:23542689

Kawamata, T., Akiyama, H., Yamada, T., & McGeer, P. L. (1992). Immunologic reactions in amyotrophic lateral sclerosis brain and spinal cord tissue. *American Journal of Pathology*, *140*(3), 691–707. PMID:1347673

Keller, A. F., Gravel, M., & Kriz, J. (2009). Live imaging of amyotrophic lateral sclerosis pathogenesis: Disease onset is characterized by marked induction of GFAP in Schwann cells. *Glia*, *57*(10), 1130–1142. doi:10.1002/glia.20836 PMID:19115383

Krakora, D., Mulcrone, P., Meyer, M., Lewis, C., Bernau, K., Gowing, G., Zimprich, C., Aebischer, P., Svendsen, C. N., & Suzuki, M. (2013). Synergistic effects of GDNF and VEGF on lifespan and disease progression in a familial ALS rat model. *Molecular therapy: the journal of the American Society of Gene Therapy, 21*(8), 1602–1610. doi:10.1038/mt.2013.108

Kushner, P. D., Stephenson, D. T., & Wright, S. (1991). Reactive astrogliosis is widespread in the subcortical white matter of amyotrophic lateral sclerosis brain. *Journal of Neuropathology and Experimental Neurology*, *50*(3), 263–277. doi:10.1097/00005072-199105000-00008 PMID:2022968

Lappe-Siefke, C., Goebbels, S., Gravel, M., Nicksch, E., Lee, J., Braun, P. E., Griffiths, I. R., & Nave, K. A. (2003). Disruption of Cnp1 uncouples oligodendroglial functions in axonal support and myelination. *Nature Genetics*, *33*(3), 366–374. doi:10.1038/ng1095 PMID:12590258

Lee, Y., Morrison, B. M., Li, Y., Lengacher, S., Farah, M. H., Hoffman, P. N., Liu, Y., Tsingalia, A., Jin, L., Zhang, P. W., Pellerin, L., Magistretti, P. J., & Rothstein, J. D. (2012). Oligodendroglia metabolically support axons and contribute to neurodegeneration. *Nature*, *487*(7408), 443–448. doi:10.1038/nature11314 PMID:22801498

Lepore, A. C., Dejea, C., Carmen, J., Rauck, B., Kerr, D. A., Sofroniew, M. V., & Maragakis, N. J. (2008a). Selective ablation of proliferating astrocytes does not affect disease outcome in either acute or chronic models of motor neuron degeneration. *Experimental Neurology*, *211*(2), 423–432. doi:10.1016/j.expneurol.2008.02.020 PMID:18410928

Lepore, A. C., Rauck, B., Dejea, C., Pardo, A. C., Rao, M. S., Rothstein, J. D., & Maragakis, N. J. (2008b). Focal transplantation-based astrocyte replacement is neuroprotective in a model of motor neuron disease. *Nature Neuroscience*, *11*(11), 1294–1301. doi:10.1038/nn.2210 PMID:18931666

Liu, Q. S., Xu, Q., Kang, J., & Nedergaard, M. (2004). Astrocyte activation of presynaptic metabotropic glutamate receptors modulates hippocampal inhibitory synaptic transmission. *Neuron Glia Biology*, *1*(4), 307–316. doi:10.1017/S1740925X05000190 PMID:16755304

Lobsiger, C. S., Boillee, S., McAlonis-Downes, M., Khan, A. M., Feltri, M. L., Yamanaka, K., & Cleveland, D. W. (2009). Schwann cells expressing dismutase active mutant SOD1 unexpectedly slow disease progression in ALS mice. *Proceedings of the National Academy of Sciences of the United States of America*, *106*(11), 4465–4470. doi:10.1073/pnas.0813339106 PMID:19251638

Lobsiger, C. S., & Cleveland, D. W. (2007). Glial cells as intrinsic components of non-cell-autonomous neurodegenerative disease. *Nature Neuroscience*, *10*(11), 1355–1360. doi:10.1038/nn1988 PMID:17965655

Pardo, A. C., Wong, V., Benson, L. M., Dykes, M., Tanaka, K., Rothstein, J. D., & Maragakis, N. J. (2006). Loss of the astrocyte glutamate transporter GLT1 modifies disease in SOD1(G93A) mice. *Experimental Neurology, 201*(1), 120–130. doi:10.1016/j.expneurol.2006.03.028 PMID:16753145

Pasinelli, P., & Brown, R. H. (2006). Molecular biology of amyotrophic lateral sclerosis: Insights from genetics. *Nature Reviews. Neuroscience, 7*(9), 710–723. doi:10.1038/nrn1971 PMID:16924260

Perrie, W. T., Lee, G. T., Curtis, E. M., Sparke, J., Buller, J. R., & Rossi, M. L. (1993). Changes in the myelinated axons of femoral nerve in amyotrophic lateral sclerosis. *Journal of Neural Transmission. Supplementum, 39*, 223–233. PMID:8360662

Philips, T., Bento-Abreu, A., Nonneman, A., Haeck, W., Staats, K., Geelen, V., Hersmus, N., Küsters, B., Van Den Bosch, L., Van Damme, P., Richardson, W. D., & Robberecht, W. (2013). Oligodendrocyte dysfunction in the pathogenesis of amyotrophic lateral sclerosis. *Brain, 136*(Pt 2), 471–482. doi:10.1093/brain/aws339 PMID:23378219

Philips, T., & Robberecht, W. (2011). Neuroinflammation in amyotrophic lateral sclerosis: Role of glial activation in motor neuron disease. *Lancet Neurology, 10*(3), 253–263. doi:10.1016/S1474-4422(11)70015-1 PMID:21349440

Philips, T., & Rothstein, J. D. (2014). Glial cells in amyotrophic lateral sclerosis. *Experimental neurology, 262m* 111–120. doi:10.1016/j.expneurol.2014.05.015

Rafiki, A., Boulland, J. L., Halestrap, A. P., Ottersen, O. P., & Bergersen, L. (2003). Highly differential expression of the monocarboxylate transporters MCT2 and MCT4 in the developing rat brain. *Neuroscience, 122*(3), 677–688. doi:10.1016/j.neuroscience.2003.08.040 PMID:14622911

Rosen, D. R., Siddique, T., Patterson, D., Figlewicz, D. A., Sapp, P., Hentati, A., Donaldson, D., Goto, J., O'Regan, J. P., Deng, H. X., Rahmani, Z., Krizus, A., McKenna-Yasek, D., Cayabyab, A., Gaston, S. M., Berger, R., Tanzi, R. E., Halperin, J. J., Herzfeldt, B., & Brown, R. H. Jr. (1993). Mutations in Cu/Zn superoxide dismutase gene are associated with familial amyotrophic lateral sclerosis. *Nature, 362*(6415), 59–62. doi:10.1038/362059a0 PMID:8446170

Rothstein, J. D., Martin, L. J., & Kuncl, R. W. (1992). Decreased glutamate transport by the brain and spinal cord in amyotrophic lateral sclerosis. *The New England Journal of Medicine, 326*(22), 1464–1468. doi:10.1056/NEJM199205283262204 PMID:1349424

Rothstein, J. D., Tsai, G., Kuncl, R. W., Clawson, L., Cornblath, D. R., Drachman, D. B., Pestronk, A., Stauch, B. L., & Coyle, J. T. (1990). Abnormal excitatory amino acid metabolism in amyotrophic lateral sclerosis. *Annals of Neurology, 28*(1), 18–25. doi:10.1002/ana.410280106 PMID:2375630

Rothstein, J. D., Van Kammen, M., Levey, A. I., Martin, L. J., & Kuncl, R. W. (1995). Selective loss of glial glutamate transporter GLT-1 in amyotrophic lateral sclerosis. *Annals of Neurology, 38*(1), 73–84. doi:10.1002/ana.410380114 PMID:7611729

Sasabe, J., Chiba, T., Yamada, M., Okamoto, K., Nishimoto, I., Matsuoka, M., & Aiso, S. (2007). D-serine is a key determinant of glutamate toxicity in amyotrophic lateral sclerosis. *The EMBO Journal, 26*(18), 4149–4159. doi:10.1038j.emboj.7601840 PMID:17762863

Sasaki, S., Warita, H., Abe, K., & Iwata, M. (2001). Inducible nitric oxide synthase (iNOS) and nitrotyrosine immunoreactivity in the spinal cords of transgenic mice with a G93A mutant SOD1 gene. *Journal of Neuropathology and Experimental Neurology*, *60*(9), 839–846. doi:10.1093/jnen/60.9.839 PMID:11556540

Schipper, H. M. (1996). Astrocytes, brain aging, and neurodegeneration. *Neurobiology of Aging*, *17*(3), 467–480. doi:10.1016/0197-4580(96)00014-0 PMID:8725909

Shruthi, S., Sumitha, R., Varghese, A. M., Ashok, S., Chandrasekhar Sagar, B. K., Sathyaprabha, T. N., Nalini, A., Kramer, B. W., Raju, T. R., Vijayalakshmi, K., & Alladi, P. A. (2017). Brain-Derived Neurotrophic Factor Facilitates Functional Recovery from ALS-Cerebral Spinal Fluid-Induced Neurodegenerative Changes in the NSC-34 Motor Neuron Cell Line. *Neurodegenerative Diseases*, *17*(1), 44–58. doi:10.1159/000447559 PMID:27617773

Song, S., Miranda, C. J., Braun, L., Meyer, K., Frakes, A. E., Ferraiuolo, L., Likhite, S., Bevan, A. K., Foust, K. D., McConnell, M. J., Walker, C. M., & Kaspar, B. K. (2016). Major histocompatibility complex class I molecules protect motor neurons from astrocyte-induced toxicity in amyotrophic lateral sclerosis. *Nature Medicine*, *22*(4), 397–403. doi:10.1038/nm.4052 PMID:26928464

Stewart, V. C., Sharpe, M. A., Clark, J. B., & Heales, S. J. (2000). Astrocyte-derived nitric oxide causes both reversible and irreversible damage to the neuronal mitochondrial respiratory chain. *Journal of Neurochemistry*, *75*(2), 694–700. doi:10.1046/j.1471-4159.2000.0750694.x PMID:10899944

Sun, S., Sun, Y., Ling, S. C., Ferraiuolo, L., McAlonis-Downes, M., Zou, Y., Drenner, K., Wang, Y., Ditsworth, D., Tokunaga, S., Kopelevich, A., Kaspar, B. K., Lagier-Tourenne, C., & Cleveland, D. W. (2015). Translational profiling identifies a cascade of damage initiated in motor neurons and spreading to glia in mutant SOD1-mediated ALS. *Proceedings of the National Academy of Sciences of the United States of America*, *112*(50), E6993–E7002. doi:10.1073/pnas.1520639112 PMID:26621731

Turner, B. J., Ackerley, S., Davies, K. E., & Talbot, K. (2010). Dismutase-competent SOD1 mutant accumulation in myelinating Schwann cells is not detrimental to normal or transgenic ALS model mice. *Human Molecular Genetics*, *19*(5), 815–824. doi:10.1093/hmg/ddp550 PMID:20008901

Turner, M. R., Cagnin, A., Turkheimer, F. E., Miller, C. C., Shaw, C. E., Brooks, D. J., Leigh, P. N., & Banati, R. B. (2004). Evidence of widespread cerebral microglial activation in amyotrophic lateral sclerosis: An [11C](R)-PK11195 positron emission tomography study. *Neurobiology of Disease*, *15*(3), 601–609. doi:10.1016/j.nbd.2003.12.012 PMID:15056468

Urushitani, M., Sik, A., Sakurai, T., Nukina, N., Takahashi, R., & Julien, J. P. (2006). Chromogranin-mediated secretion of mutant superoxide dismutase proteins linked to amyotrophic lateral sclerosis. *Nature Neuroscience*, *9*(1), 108–118. doi:10.1038/nn1603 PMID:16369483

Vargas, M. R., Johnson, D. A., Sirkis, D. W., Messing, A., & Johnson, J. A. (2008). Nrf2 activation in astrocytes protects against neurodegeneration in mouse models of familial amyotrophic lateral sclerosis. *The Journal of Neuroscience : The Official Journal of the Society for Neuroscience*, *28*(50), 13574–13581. doi:10.1523/JNEUROSCI.4099-08.2008 PMID:19074031

Venturini, A., Passalacqua, M., Pelassa, S., Pastorino, F., Tedesco, M., Cortese, K., Gagliani, M. C., Leo, G., Maura, G., Guidolin, D., Agnati, L. F., Marcoli, M., & Cervetto, C. (2019). Exosomes From Astrocyte Processes: Signaling to Neurons. *Frontiers in Pharmacology*, *10*, 1452. doi:10.3389/fphar.2019.01452 PMID:31849688

Verkhratsky, A., Matteoli, M., Parpura, V., Mothet, J. P., & Zorec, R. (2016). Astrocytes as secretory cells of the central nervous system: Idiosyncrasies of vesicular secretion. *The EMBO Journal*, *35*(3), 239–257. doi:10.15252/embj.201592705 PMID:26758544

Viader, F. (2023, March). La sclérose latérale amyotrophique: une maladie neurodégénérative emblématiqueAmyotrophic lateral sclerosis: An emblematic neurodegenerative disease. *Bulletin de l'Académie Nationale de Médecine*, *207*(3), 272–286. doi:10.1016/j.banm.2023.01.002

Wang, L., Pytel, P., Feltri, M. L., Wrabetz, L., & Roos, R. P. (2012). Selective knockdown of mutant SOD1 in Schwann cells ameliorates disease in G85R mutant SOD1 transgenic mice. *Neurobiology of Disease*, *48*(1), 52–57. doi:10.1016/j.nbd.2012.05.014 PMID:22668777

Wang, L., Sharma, K., Grisotti, G., & Roos, R. P. (2009). The effect of mutant SOD1 dismutase activity on non-cell autonomous degeneration in familial amyotrophic lateral sclerosis. *Neurobiology of Disease*, *35*(2), 234–240. doi:10.1016/j.nbd.2009.05.002 PMID:19442735

Wong, P. C., Pardo, C. A., Borchelt, D. R., Lee, M. K., Copeland, N. G., Jenkins, N. A., Sisodia, S. S., Cleveland, D. W., & Price, D. L. (1995). An adverse property of a familial ALS-linked SOD1 mutation causes motor neuron disease characterized by vacuolar degeneration of mitochondria. *Neuron*, *14*(6), 1105–1116. doi:10.1016/0896-6273(95)90259-7 PMID:7605627

Yamanaka, K., Chun, S. J., Boillee, S., Fujimori-Tonou, N., Yamashita, H., Gutmann, D. H., Takahashi, R., Misawa, H., & Cleveland, D. W. (2008). Astrocytes as determinants of disease progression in inherited amyotrophic lateral sclerosis. *Nature Neuroscience*, *11*(3), 251–253. doi:10.1038/nn2047 PMID:18246065

Zhao, W., Beers, D. R., Henkel, J. S., Zhang, W., Urushitani, M., Julien, J. P., & Appel, S. H. (2010). Extracellular mutant SOD1 induces microglial-mediated motoneuron injury. *Glia*, *58*(2), 231–243. doi:10.1002/glia.20919 PMID:19672969

Chapter 11
Oligodendrogliopathy in Multiple Sclerosis

Abdelali Ben Maloui
Faculty of Sciences Semlalia, Cadi Ayyad University, Morocco

Lahcen Tamegart
Abdelmalek Essaadi University, Morocco

Mjid Oukhrib
https://orcid.org/0000-0003-0199-4054
Cadi Ayyad University, Morocco

Hafida El Ghachi
https://orcid.org/0000-0002-2043-6202
Faculty of Sciences Semlalia, Cadi Ayyad University, Morocco

Youssef Ait Hamdan
Cadi Ayyad University, Morocco

Rachid Soulimani
Lorraine University, France

Halima Gamrani
Cadi Ayyad University, Morocco

ABSTRACT

Multiple sclerosis (MS) is a complex autoimmune disease of the central nervous system (CNS), which is characterized by demyelination and neurodegeneration. Oligodendrocytes have an essential role in maintaining the integrity of myelin, the protective sheath that surrounds nerve fibers and is indispensable for efficient signal transmission. Nevertheless, in multiple sclerosis, oligodendrocytes become dysfunctional, leading to myelin deterioration and axonal degeneration. Recent research suggests that metabolic changes, including mitochondrial dysfunction and alterations in glucose and lipid metabolism, contribute significantly to the pathogenesis of MS. Mitochondrial dysfunction is observed in both immune cells and CNS oligodendrocytes of MS patients. Impaired mitochondrial function leads to energy deficits, affecting crucial processes such as impulse transmission and axonal transport, ultimately contributing to neurodegeneration. Understanding the complex relationship between these mechanisms is crucial to the development of an effective treatment for MS.

DOI: 10.4018/978-1-6684-9675-6.ch011

INTRODUCTION

Multiple sclerosis (MS) is a chronic autoimmune disease wich is characterized by demyelination of nerve fibers in the central nervous system. Oligodendrocytes, specialized glial cells are responsible for the production and maintenance of myelin, have an essential role in the pathogenesis of multiple sclerosis. Often referred to as oligodendrogliopathy, this phenomenon involves the destruction of normal oligodendrocyte function following an inappropriate attack by the immune system. The ensuing inflammatory response leads to myelin degradation, resulting in altered nerve signal transmission and a wide range of neurological symptoms.

The complex interaction between immune cells, oligodendrocytes and myelin in MS is the subject of ongoing research. Studies have elucidated the mechanisms by which immune cells infiltrate central nervous system and trigger an inflammatory cascade, ultimately targeting oligodendrocytes and their myelin sheaths. As oligodendrogliopathy progresses, areas of demyelination give rise to characteristic lesions and scar tissue formation, contributing to the clinical heterogeneity observed in multiple sclerosis patients.

MULTIPLE SCLEROSIS

Multiple sclerosis (MS) is a chronic and autoimmune disease of the central nervous system (CNS), manifested by inflammation, demyelinization and extensive focal lesions in the white and gray matter, and axonal loss (Lassmann et al., 2018). The prevalence of the disease in developed and developing countries is increasing exponentially, and generally occurs in young adults (Dobson R., et al., 2019). Moreover, women appear to be two to three times more likely to be affected by the disease, due to an earlier but less progressive phenotype than men (Harbo et al., 2013).

The current classification of MS includes four major phases: clinically isolated syndrome, relapsing-remitting phase (RRMS), secondary progressive phase (SPMS) and primary progressive phase (PPMS). Nevertheless, there is considerable overlap between these phases, and the transitions between them remain unclear (Dimitriou et al., 2023). RRMS, which affects around 80% of patients, is characterised by inflammation, demyelination and intermittent episodes of neurological dysfunction. Over time, untreated patients may progress to the MS phase due to continued disease deterioration and fewer clinical relapses. Conversely, around 10% of patients present with a PPMS phenotype from the outset, without prior clinical relapse. This progression may be reflected in increasing physical disability, cognitive impairment or other neurological symptoms. A recent finding, called "progression independent of relapse activity" (PIRA), has been observed in some people with multiple sclerosis, where the disease progresses without the acute relapses typical of the relapsing-remitting phase. PIRA appears to be related to chronic inflammatory tissue damage involving substantial early white matter atrophy (Cagol A., et al., 2022).

Histologically, focal lesions were found in the white and gray matter, comprising the cortex, basal ganglia, brainstem and spinal cord. Blood-brain barrier (BBB) disruption, axonal death and demyelinating lesions result in significant progressive neurodegeneration. Oligodendrocyte precursor cells and oligodendrocytes are largely affected by this disease (Stys P. K., et al., 2019). As demyelination spreads through tissues, microglia and inflammatory cells are recruited to remove myelin debris and combat the hostile environment. Two types of microglia are present in the brains of people with MS: pro-inflammatory microglia and anti-inflammatory microglia (Garg N., et al., 2015).

The most effective way of recreating the inflammatory appearance of MS is the classic autoimmune mouse model, also known as experimental autoimmune encephalitis (EAE). In that model, specific antigens or peptides are introduced to activate the innate immune system and autoreactive T cells, leading to various disease phenotypes and MS-like pathology (Bjelobaba I., et al., 2018).

Currently, the treatment strategy for multiple sclerosis is based on the utilization of disease-modifying therapies (DMT) and symptomatic treatments. The neurological dysfunction noted in these patients is a fluctuating and disabling experience that can impact quality of life and represent a heavy burden for patients (De Angelis F., et al., 2018). DMT treatment involves modulating the inflammatory component of MS over the longer term, thereby reducing relapse rates and disease progress. Many disease-modifying drugs are approved by the Food and Drug Administration (FDA), and more are still in development (De Angelis F., et al., 2018).

Although progress has been made with these drugs, they have often been associated with mild to serious adverse effects and have failed to prevent the axonal damage observed in the progressive form of the disease. DMTs can be classified as being recombinant cytokines, complex peptide mixtures, monoclonal antibodies and small molecules (Winkelmann A., et al., 2016). More than a dozen DMTs are approved for the treatment of MS, each with a distinct mechanism of action. Approved drugs include interferon beta, glatiramer acetate, teriflunomide, dimethyl fumarate, fingolimod, natalizumab, ocrelizumab, siponimod, cladribine and alemtuzumab (Piehl F., 2021; Travers B. S., et al., 2022).

Despite the availability of many effective and well-known immunotherapies, a considerable proportion of people with multiple sclerosis (MS) still experience severe and disabling symptoms. Although symptomatic, evidence-based treatments are rare, they are a good strategy for relieving the symptoms associated with multiple sclerosis and preventing the secondary problems that can result.

The remyelination process is a key component of both neuroprotective and a regenerative strategy designed to treat MS (Stangel M., et al., 2017). Myelin is a crucial structure that surrounds and protects axons. It is also implicated in the exchange and maintenance of metabolites. Remyelination can succeed in the early stages of the disease, thanks to the survival of oligodendrocytes and their progenitors (OPCs). However, neurons, oligodendrocytes and OPCs continue to decline as the disease progresses. Loss of these cell types and the inflammatory environment lead to failure of remyelination (Nave K. A., 2010). There are currently several strategies to treat the pathophysiology of multiple sclerosis, including various approaches like remyelination, cell therapy and metabolic regulation.

Remyelination techniques in the context of MS imply the application of various drugs and therapeutic strategies to stimulate the production of new myelin or support the repair of existing myelin. Many studies have identified various compounds likely to enhance oligodendrocyte recruitment, survival and differentiation, thereby promoting remyelination. A number of these drugs target specific brain receptors, notably muscarinic acetylcholine receptors (such as M1) and histamine receptors (such as H1 and H3). Examples include benztropine, clemastine, quetiapine, ivermectin and GSK239512 (Plemel J. R., et al., 2017; Amin M., and Hersh C. M., 2023).

Furthermore, opicinumab is a very interesting target for remyelination, as it is a humanized monoclonal antibody specifically designed to antagonize the effects of leucine-rich neuronal repeat protein 1 (LINGO-1). LINGO-1 is a glycoprotein found on the surface of neurons and oligodendrocytes within the central nervous system (CNS). This inhibits essential processes such as oligodendrocyte differentiation, myelination, neuronal survival and axonal regeneration. Many in vitro and in vivo studies have indicated that blocking LINGO-1 can facilitate axon remyelination. However, clinical trials have shown mixed results (Kremer D., et al., 2019).

As well as using drugs as a therapeutic strategy, cell therapy also offers several advantages in remyelination therapies. By inserting specific cells or stem cells that have the potential to differentiate into oligodendrocytes, cell therapy targets direct repair of damaged myelin in the CNS. In addition, cell therapies may exert immunomodulatory effects, reduce inflammation and create an environment conducive to remyelination (Plemel J. R., et al., 2017). Currently, several clinical trials are underway to explore the potential effects of these cells. Some trials aim to modulate inflammation, thereby creating a microenvironment conducive to remyelination, while others focus on using OPCs to reconstitute endogenous niches or replace cells affected by genetic or age-related factors (López-Muguruza E., et al., 2022). Promising cell types under investigation include mesenchymal stem cells and neural stem cells (Alanazi, A., et al., 2022).

OLIGODENDROGLIA

The CNS consists of a tightly woven network of specialised cells like neurons and glial cells (Yellen G., 2018). Until the beginning of the XXth century, glial cells used to be defined as "Nervenkitt" because, in that era, they were considered only as the glue that holds neuronal tissue together. However, an exhaustive structural analysis of the brain by Pio del Rio-Hortega led to the description and differentiation of a cell subtype known as neuroglia (Mergenthaler P., et al., 2013). They are widely distributed in white and gray matter. Unlike neurons, these cells are unable to extend their axons, but are composed of processes that enable them to reach multiple axons (Bonora M., et al., 2012). Advances in microscopy and staining techniques have made it possible to classify this neuroglia into three cell types: astrocytes, microglia and oligodendrocytes (OL) (Beirowski B., 2022). The cells involved in the early stages of oligodendrocyte development had not yet been described, for lack of suitable markers. Despite this, Raff, Miller and Noble made an important discovery in 1983 when they identified oligodendrocyte precursor cells (OPCs), establishing them as the fourth most important glial subtype in the brain (Kuhn S., et al., 2019).

Oligodendrocyte Precursor Cells

Throughout development, OPCs are the main source of mature, myelin-generating OLs (Pérez-Cerdá F., et al., 2015). Although they are highly proliferative, OPCs account for only around 5-10% of the total cell population in the brain, with lower numbers in the gray matter (Bonora M., et al., 2012). OPCs are also known for their high expression of the platelet-derived growth factor receptor alpha subunit (PDGFR-α) and NG2 markers, the latest explaining why OPCs are also known as NG2-positive cells (Figure 1) (Bonora M., et al., 2012).

OPCs emerge during development from the brain and spinal cord ventricular zone, where they proliferate and migrate to different regions in order to differentiate into oligodendrocytes. While OPC proliferation peaks during early postnatal stages, this rate declines over time (Skaper S. D., 2019). Nevertheless, OPCs continue to be one of the most active dividing cell types in adult central nervous systems (Dimou L., et al., 2008). Adult OPCs were described as having the capacity to generate myelinating OLs in the CNS, particularly in demyelinating lesions (Wang J. Q., et al., 2023). The live imaging reveals that the lost of one OPC results in the rapid division of a neighboring OPC. This implies that all OPCs can enter the cell cycle and that lost cells are replaced not by the mobilization of SVZ progenitors or a subset of highly proliferative OPCs, but rather by the collective homeostatic behavior of the entire OPC population

Figure 1. Visual representation of the different stages of oligodendrocyte differentiation and maturation, and their distinctive markers

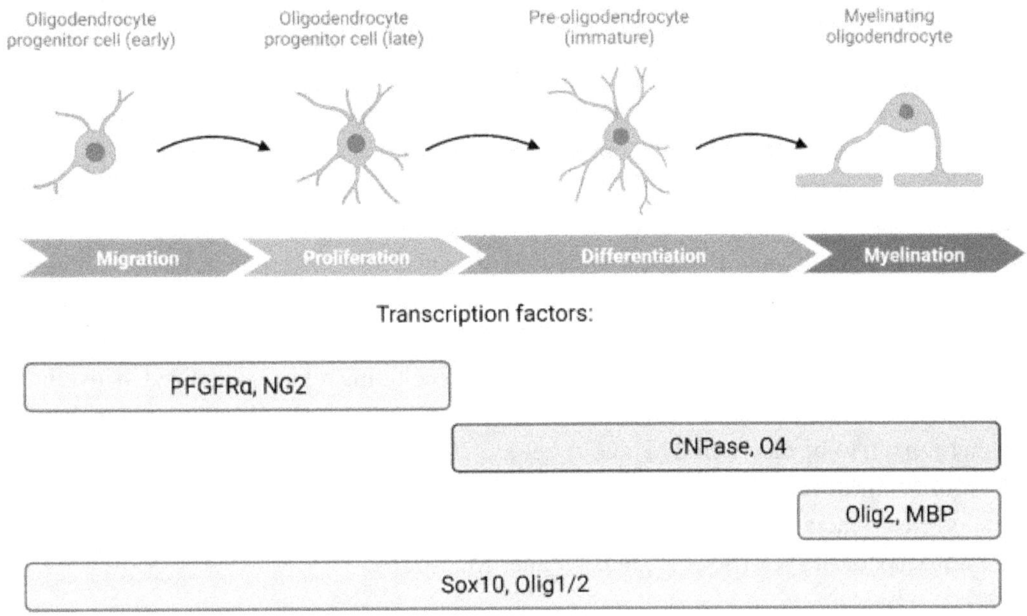

(Nishiyama A., et al., 2021). The differentiation of OPCs into immature OLs is controlled by a series of extrinsic and intrinsic signals. Once OPCs have adopted a more mature phenotype, they lose their high proliferative capacity, making this step critical (Akay L. A., et al., 2021).

Although OPCs are mostly known for their role in oligodendrocyte formation, they also perform many other functions in the brain (Kuhn S., et al., 2019). Under homeostatic conditions, OPCs have acquired a more active contributory role. OPCs express multiple ion channels such as voltage-dependent Na+ channels and ionotropic glutamate and GABA receptors (Chamberlain K. A., et al., 2016). OPCs thus have the ability to receive synaptic inputs from neurons and retain unique electrical properties, but whether they are capable of generating action potentials remains unknown. Other studies have highlighted the role that OPCs can play in maintaining BBB integrity and mediating neuroinflammation, as well as other functions. Genetic ablation of NG2-positive OPCs not only led to a reduction in the density and branching of the vascular network in the brain, but also to microglial overactivation followed by neuronal death due to severe neuroinflammation (Dimou L., et al., 2008).

Oligodendrocytes

Oligodendrocytes (OLs), as a type of glia involved in a wide range of neurological functions, are responsible for the production of myelin sheaths around neuronal axons (Pérez-Cerdá F., et al., 2015). The differentiation of OPCs into mature, myelin-producing OLs is divided into four stages: proliferative OPCs, immature oligodendrocyte precursor cells (pre-OLCs), differentiated OLCs and myelinating OLCs (Marisca R., et al., 2020). During their development, pre-OLCs initiate the construction of myelin outgrowths on specific axons, leading to the loss of their bipolar morphology and the transition to a more advanced, mature phenotype. At this stage, oligodendrocytes show strong expression of CNPase, O4 and O1 (Figure 1). Later, mature oligodendrocytes will be able to produce multiple concentric membrane layers around axons, as well as myelin proteins. The Olig2 marker is a cell lineage marker specific to mature oligodendrocytes (34). And, combined with the transcription factor Sox10, it unambiguously identifies the oligodendroglial lineage (Falcão A. M., et al., 2018).

As previously mentioned, oligodendrocytes are mainly renowned for their ability to produce myelin in the white matter of the CNS. The myelination process varies according to brain region, with oligodendrocytes often isolating a large number of axons. In some cases, however, they may encompass a single axon, as is the case in the ventral spinal cord (Akay L. A., et al., 2021). The complex interaction between axons and myelinating oligodendrocytes is supported by compelling evidence suggesting that ion balance is controlled by channels located at the interface between the myelin sheath and the axon (Yi C., et al., 2023). As a widespread signaling molecule and second messenger, the regulation of intracellular calcium concentration ([Ca2+]i) in oligodendrocytes, including regions localized in myelin sheaths, has a direct impact on the process of myelin formation and remodeling, and may also influence other functions that remain to be discovered (Tiane A., et al., 2019). In addition, oligodendrocytes perform a variety of non-canonical functions, including metabolic and trophic support of axons. One strategy for doing so is to release lactate via the monocarboxylate transporter 1, which is then harnessed by axons to generate mitochondrial ATP (Sock E., and Wegner M., 2021).

MYELIN IN THE CNS

The human CNS is essentially devoid of myelin at birth, but after birth the oligodendrocyte population increases dramatically, resulting in extensive myelination during the first years of childhood. The myelination process persists throughout adolescence and into adulthood, following a distinct pattern in space and time, corresponding to the development and preservation of functional neural circuits (Suminaite D., et al., 2019).

The existence of myelin sheets facilitates the rapid diffusion of action potentials through a mechanism known as saltatory conduction. To achieve this, voltage-gated Na+ channels are concentrated in small gaps between neighboring myelin sheets, known as nodes of Ranvier, and act as electrical insulators. Fully myelinated axons along their entire length conduct impulses much faster than unmyelinated axons (Yi C., et al., 2023). Furthermore, adequate insulation depends on the specific types and proportions of lipids and proteins that make up myelin.

Myelin membrane retains the essential properties of lipid bilayers found in cell membranes, but its structure and composition are unique. Like other cell membranes, the myelin sheath consists mainly of water, lipids and protein molecules (Figure 2). However, the proportions of these components in myelin differ from those generally observed in a normal cell membrane (Fünfschilling U., et al., 2012). While most biological membranes have a balanced ratio of proteins to lipids, the myelin sheath has a higher proportion of lipids, constituting around 80% of its dry weight. This lipid/protein ratio in myelin plays a role in the compact arrangement of the myelin sheath through non-covalent interactions between lipids and myelin proteins (Williamson J. M., et al., 2018).

Figure 2. The structure and molecular makeup of the myelin membrane in the CNS

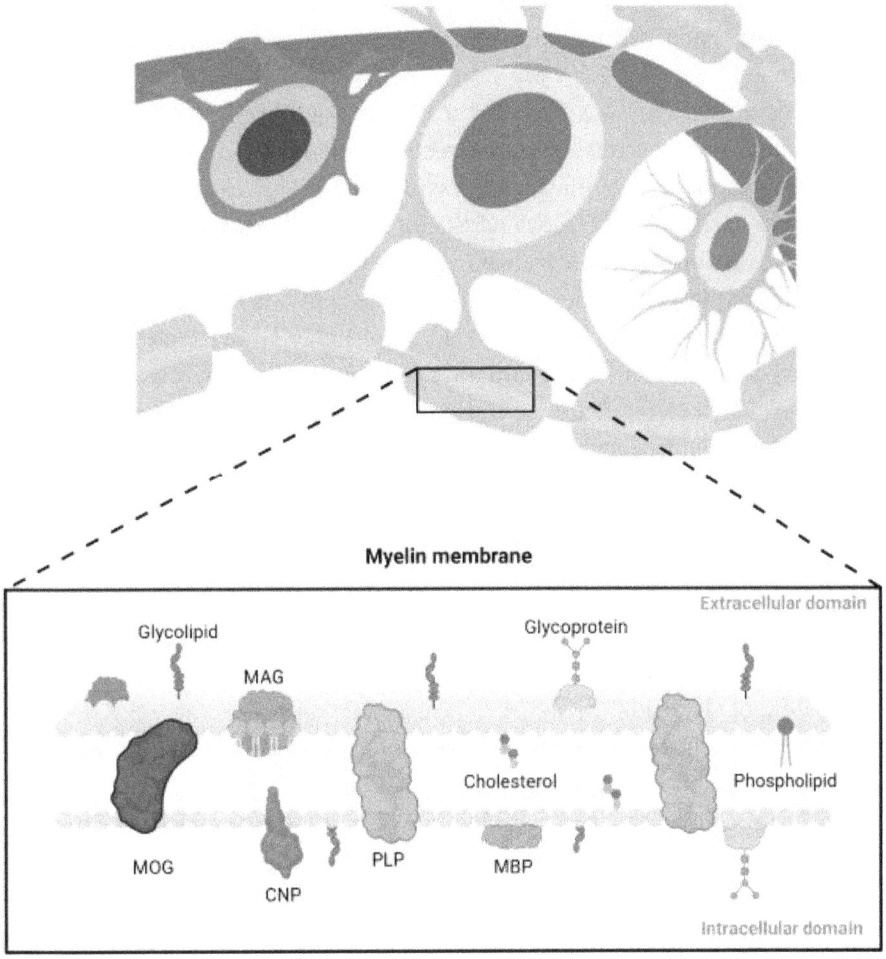

METABOLIC PROFILE OF OLIGODENDROGLIA AND MYELIN

The CNS is particularly vulnerable to metabolic changes as it is thought to have no energy reserves (Yellen G., 2018). Furthermore, although the brain represents a small proportion of body weight, its daily energy consumption hovers around 20%. Neurons absorb most of this energy expenditure, as they require large amounts of energy to ensure the rapid conduction of axonal impulses (Ruskamo S., et al., 2022).

Glucose, the main energy substrate used by brain cells, is involved in many essential functions such as ATP production, neurotransmitter synthesis and oxidative stress management (Boggs J. M. et al., 2006). Brain cells generate ATP from glucose mainly via two metabolic pathways: glycolysis in the cytosol and oxidative phosphorylation (OX-PHOS) in the mitochondria. The ATP obtained by OX-PHOS is associated with NADH production from the citric acid cycle (TCA) (Jahn O., et al., 2020).

OLs prefer to use glycolysis rather than OX-PHOS to supply their high ATP demand. Under favorable, nutrient-rich conditions, OPCs and OLs of human origin rely much more heavily on glycolysis to generate ATP. However, when confronted with stress or pathological conditions, oligodendrocytes have the ability to withdraw their processes from the myelin sheath, adopting a reduced metabolic state as a survival mechanism. Nevertheless, this retraction leads to destabilization of the compact axonal myelin sheath. OPCs, on the other hand, are unable to retract their processes and therefore rapidly undergo cell death under stress (Rone M. B., et al., 2016).

The high glycolytic activity of OLs can lead to the production of pyruvate in the cytosol, which provides metabolic support for neuronal oxidative metabolism. However, mitochondria need more time to utilize the pyruvate generated, resulting in excess pyruvate. This excess pyruvate is converted into lactate, the end product of glycolysis (Li S. and Sheng Z. H., 2023). Lactate is then released into the periaxonal space via the monocarboxylate transporter 1 (MCT1), and taken up by axons via the neuronal transporter MCT2. This pathway by which oligodendrocytes provide trophic support to axons is known as the oligodendrocyte-axon lactate shuttle (Mot A. I., et al., 2022). High lactate production and transport can have a positive impact on axons. The myelin that envelops these axons creates specialized channels called "myelin channels", which facilitate the transport of lactate and other beneficial factors to axons, thereby promoting the maintenance and health of axonal structures (Simons M., and Nave K. A., 2016).

In addition to providing metabolic fuel for oligodendrocytes, lactate can also be used by them for lipid production. In vitro experiments have shown that lactate can enhance the differentiation process of mouse OPCs. In the event of low glucose availability or hypoglycemic conditions, the proliferation and differentiation capacity of oligodendrocytes is diminished, and their ability to produce myelin is inhibited. It can therefore be concluded that the proper functioning of lactate exchange between oligodendrocyte and axon is essential for the differentiation of cells derived from the oligodendrocyte stem cell line (Rinholm J. E., et al., 2011).

Nevertheless, when oxygen is present, OLs are particularly resistant to glucose withdrawal, revealing the existence of other mitochondria-dependent energy sources. Drosophila have the ability to switch to GA degradation in the mitochondria when carbohydrates are in short supply. Experiments in Drosophila subjected to long-term nutritional stress have demonstrated that glial cells switch from glycolysis to β-oxidation to supply nutrients to axons. Furthermore, glucose deprivation and inhibition of β-oxidation in oligodendrocytes induce rapid neurodegeneration. The results emphasize that glial cells can rely on the degradation of glial fatty acids or peripheral lipid stores, which can be converted to ketone bodies, to maintain brain function and ensure survival under adverse conditions. Furthermore, they also suggest that ketone bodies could serve as an alternative metabolic fuel for neurons (McMullen E., et al., 2023).

Since the 1960s, it was known that the two primary ketone bodies, acetoacetate and β-hydroxybutyrate, serve as metabolic fuels for the brain (Owen O. E., et al., 1967). But their capacity to replace glucose in the oxidative metabolism of neurons has only recently been discovered (McMullen E., et al., 2023). The liver plays a central role in the production and storage of ketone bodies, using acetyl-CoA derived from imported fatty acids for ketogenesis. Nevertheless, it is thought that local lipid reserves in the Drosophila cortex and glia have the capacity to use their own reserves to generate ketone bodies and transport them to areas subject to nutritional deprivation. Consequently, in times of starvation, these newly synthesized ketone bodies can be taken up by neurons and used as energy substrates to support memory formation (Silva B., et al., 2022).

Constant myelination requires constant synthesis of lipids, which are supplied by oligodendrocytes. Lipid metabolites can also serve as energy reserves in the event of hypoglycemia (Asadollahi E., et al., 2022). The autophagic-lysosomal pathway is also responsible for recycling myelin lipids, resulting in the release of fatty acids that can be used to produce new myelin lipids. But when glucose availability is insufficient, LOs can redirect fatty acids released during myelin degradation to β-oxidation, resulting in the production of acetyl-CoA, which serves as a substrate for OX-PHOS. Experiments in a mouse optic nerve model have demonstrated that ATP production from lipids enables greater allocation of glucose-derived metabolites to axons, thereby promoting their preservation (Li S., & Sheng Z. H., 2023).

Peroxisomes, small organelles found in the cytosol of most eukaryotic cells, are also involved in a range of functions involving the metabolism and detoxification of reactive oxygen species. Like mitochondria, they are essential for the β-oxidation of fatty acids, particularly very long-chain fatty acids (VLCFAs) (Kassmann C. M., et al., 2007). Peroxisomes are highly abundant and can be identified in many CNS cells, particularly glial cells. A number of recent studies have demonstrated the presence of peroxisomes in the innermost tongue of the myelin sheath. Inhibition of peroxins, proteins involved in the biosynthesis of these organelles, leads to their dysregulation and, consequently, to white matter abnormalities. These findings have led to the hypothesis that peroxisomes play an important role in the direct support of axons and the maintenance of myelin. Furthermore, peroxisome inhibition studies have concluded that myelin-forming glial cell peroxisomes undergo β-oxidation to shorten the VLCFA chain for subsequent transfer to axonal mitochondria (Kassmann C. M., 2014).

BIOENERGY ALTERATIONS IN MULTIPLE SCLEROSIS

Numerous studies using a variety of biological samples have already provided compelling evidence of metabolome abnormalities in MS patients. Cerebrospinal fluid (CSF) samples from MS patients have shown increased levels of lactate, while also demonstrating perturbations in glucose and energy metabolism. Studies using mass spectrometry-based metabolomics showed alterations in lipid and fatty acid metabolism, with high concentrations of circulating free fatty acids and fatty acid oxidation products (Oliveira E. M. L. D., et al., 2019).

Mitochondrial Dysfunction and Oxidative Stress

Two hypotheses are suggested to explain both the origin and pathological features of MS. The first is that the cascade of events begins in the periphery, where dysregulated autoreactive T cells enter the CNS along with macrophages and B cells. This triggers a targeted attack on myelin sheaths, leading to the progression of RRMS and CNS damage. This triggers a targeted attack on myelin sheaths, leading to the progression of RRMS and CNS lesions. This triggers a targeted attack on myelin sheaths, leading to RRMS and CNS lesions. Conversely, the second hypothesis asserts that MS is a neurodegenerative disease, with the initial dysfunction or autoinflammatory behavior occurring in the CNS itself. According to this hypothesis, there is a chronic elimination of antigenic cellular components, leading to an inflammatory response and subsequent degeneration of the myelin sheath (DeMarshall C., et al., 2017).

Secondly, the hypothesis that MS is a neurodegenerative disease is supported by the possibility of mitochondrial dysfunction. This dysfunction could lead to energy deficiency and interfere with processes such as impulse transmission, axonal transport and ion trafficking (Stys P. K., et al., 2005). Furthermore, impaired mitochondrial function not only leads to energy deficits, but also triggers the activation of inflammatory cells in the CNS, further accelerating the progression of neurodegeneration (Butler R., et al., 2022). MS patients are characterized by reduced ATP production, potentially resulting from decreased activity of mitochondrial electron transport chain complexes (Barcelos I. P. D., et al., 2019). Nevertheless, it should be noted that in some cases, the activity of complex V is sometimes enhanced as a compensatory mechanism for the reduced activity of complex I. Despite this compensatory effort, disruption of mitochondrial function ultimately leads to reduced energy production (Butler R., et al., 2022).

In addition to its involvement in altered bioenergetics, mitochondrial dysfunction is strongly associated with oxygen metabolism and the production of reactive oxygen species (ROS). As the brain uses a substantial proportion of inhaled air (20%) and consumes a significant amount of tissue oxygen (90%) for energy production, this metabolic process leads to the production of harmful ROS (Adiele R. C. and Adiele C. A., 2019). ROS and their reactive products attack all classes of biomolecules, including lipids. In particular, the CNS, whose membranes have high levels of polyunsaturated fatty acids, makes neuronal cells highly sensitive and vulnerable to the damaging effects of these reactive species. The interaction between ROS and nitric oxide results in the generation of highly reactive peroxynitrites (Butler R., et al., 2022). This by-product specifically endangers OPCs due to their limited antioxidant defense mechanisms. As a result, OPCs are unable to transform into myelinating oligodendrocytes, leading to impaired myelin production and maintenance in the context of MS (Spaas J., et al., 2021).

Inflammation And Bioenergetic Interaction

Mononuclear phagocytes (MPs) constitute the CNS surveillance system and are primarily responsible for local immune surveillance. Resident MPs include microglia, which account for around 5-10% of all brain cells, and macrophages, which reside in perivascular spaces and choroid plexuses. Neurodegeneration leads to a shift in the balance of microglial activation towards a pro-inflammatory state. This leads to the production of chemokines/cytokines, such as tumor necrosis factor alpha (TNF-α), interleukin (IL)-6, IL-1β and IL-12, which have pro-inflammatory properties (Butler R., et al., 2022).

At the same time, secretion of these pro-inflammatory cytokines has adverse effects on crucial mitochondrial components, leading to a decline in mitochondrial respiratory chain function and exacerbating neurodegeneration. In addition, the enzymes responsible for the tricarboxylic acid cycle and oxidative phosphorylation in mitochondria are affected by the presence of these pro-inflammatory mediators in the inflamed CNS. Although evidence suggests that inflammatory factors impact mitochondrial dynamics, the specific molecular mechanisms by which these mediators induce damage are still poorly understood (van Horssen J., et al., 2019).

In addition, macrophages and microglia are recognized for their role in myelin clearance. However, under inflammatory conditions and when they adopt their pro-inflammatory state, they lose this ability, resulting in increased oxidative stress due to disruption of myelin clearance. Myelin is also an essential barrier against oxidative stress. But in multiple sclerosis (MS), disruption of myelin clearance leads to the accumulation of damaged myelin, exposing nerve fibers to increased oxidative stress, which disrupts nerve signal transmission and results in the neurological symptoms characteristic of MS (Castellanos D. B., et al., 2021).

Metabolism of Glucose in MS

There is evidence of altered glucose metabolism in MS, with reduced glucose uptake and utilization in certain brain regions. This metabolic dysfunction can impact on the energy supply of neurons and other cells, potentially contributing to neurodegeneration. Glucose-deprived conditions are thought to have detrimental effects on OPCs, resulting in fewer and finer processes, while oligodendrocytes tend to shift to a more glycolytic state to survive (Rosko L., et al., 2019).

Furthermore, lactate levels in the serum and CSF of MS patients vary according to clinical stage. During the early stages of pathogenesis, lactate levels tend to be lower, but as the disease progresses, an increase in lactate expression can be observed. This suggests that there is an elevation in extra-mitochondrial glucose metabolism in MS patients, which may also be associated with impaired mitochondrial function (Regenold W. T., et al., 2008). In view of the above data, fluctuations in lactate levels can potentially serve as diagnostic criteria, as they could indicate disease progression (Fonalledas-Perelló M. A., et al., 2008).

Lipid Metabolism Dysregulation in MS

In the brain, lipids play an essential role in a variety of processes, including neurogenesis, signal transduction, neuronal communication, membrane compartmentalization and modulation of gene expression. Because of their essential structural and physiological functions, any changes in lipid metabolism may indicate abnormal brain function (Grassi S., et al., 2020). Scientific evidence indicates that lipid metabolism of the arachidonic acid pathway is altered and undergoes changes in the context of multiple sclerosis pathology. In addition, specific lipid abnormalities have been reported, such as deficiencies in 18:2 and 20:4 FAs, as well as in total PUFAs (polyunsaturated fatty acids), accompanied by compensatory increases in saturated fatty acids with shorter carbon chains (Ferreira H. B., et al., 2020).

Using the non-targeted lipidomic approach, studies have revealed distinct lipid signatures in multiple sclerosis (MS) patients compared to healthy controls. MS patients have a unique phospholipidomic profile, characterized by significant reductions in key phospholipids, notably phosphatidylethanolamine (PE) and phosphatidylcholine (PC) species, which play a crucial role in antioxidant functions. Consequently, certain phospholipids show promise as potential biomarkers for clinical applications in multiple sclerosis (MS) (Ferreira H. B., et al., 2021).

Both macroautophagy and lysosome-mediated degradation are essential processes that contribute significantly to myelin renewal and regeneration. However, pathological conditions and aging often lead to a decrease in the efficiency of these mechanisms. As a result, this disruption has a negative impact on the overall well-being of oligodendrocytes, reducing their ability to synthesize fatty acids essential for myelin maintenance and energy production (Aber E. R., et al., 2022).

Furthermore, sphingolipids, important components of lipid bilayers with functional and structural roles, have been implicated in the pathological processes of MS (Giussani P., et al., 2021). Recent research highlights alterations in sphingolipid pathways that may contribute to oligodendrocyte damage, suggesting that dysregulated anti-inflammatory and pro-inflammatory lipids are potential contributors to multiple sclerosis pathology. Ceramide lipids, such as sphingosine, are associated with oligodendrocyte damage and acute demyelination (Dasgupta S., and Ray S. K., 2017). Ceramide release from stressed oligodendrocytes could trigger autoimmune responses following active demyelination, which could serve as a diagnostic and prognostic marker (Podbielska M., et al., 2021).

There are also studies linking cholesterol derivatives, known as oxysterols, to inflammatory demyelination in multiple sclerosis. As the brain is made up of around 25% cholesterol, which is higher than in other organs, oxysterols, particularly 24(S)-hydroxycholesterol, may have an impact on CNS cells through disruption of the BBB (Mukhopadhyay S., et al., 2017). Oxysterols influence lipid synthesis by acting on transcription factors, such as sterol regulatory element binding proteins (SREBPs), which regulate genes involved in lipid homeostasis (Podbielska M., et al., 2021).

Overall, considering the complex interplay between bioenergetics and lipid metabolism, the exact mechanisms underlying these interactions remain complex and require further investigation. Further research is needed to determine whether the bioenergetic irregularities observed in MS specifically affect certain lipid groups, rather than influencing the whole range. This would open up possibilities for targeting specific lipid types while sparing others. Such discoveries could pave the way for interventions targeting specific lipid categories while sparing others.

HOW ALTERED BIOENERGETICS CAN INFLUENCE OLIGODENDROGLIOPATHY AND MS PROGRESSION

The bioenergetics can have a significant impact on the progression of MS by altering the function of the axon-myelin unit. Indeed, disruption of the axon-myelin unit can occur acutely, for example during transient ischemia, or chronically in MS and other neurodegenerative diseases, and involves myelin NMDA receptors, a key element that regulates the supply of energy substrates between oligodendrocytes/myelin and axons (Saab A. S., et al., 2016). In this way, the glutamate released by axons during action potential propagation activates NMDA receptors in oligodendrocytes, which translocate the glucose transporter GLUT1 into the plasma membrane, promoting glycolysis and the shuttling of lactate to axons. This mechanism fine-tunes axonal energy requirements during neuron-to-neuron communication, and is impaired during dysmyelination and demyelination, leading to mitochondrial dysfunction and, ultimately, axonal damage. Detailed mechanisms would include: first, energy failure due to the inability of oligodendrocytes to generate lactate for export to the axon, with reduced ATP synthesis in axonal mitochondria ; second, a consequent failure of ion transporters to maintain Na+ and K+ and, consequently, impaired action potential propagation; and thirdly, axonal Ca2+ dyshomeostasis leading to increased activation of calpains, phospholipases and other enzymes, ultimately resulting in structural axonal damage.

Consequently, demyelinated axons in focal plaques of relapsing-remitting MS may undergo transient energy deficits compromising their function (signal transmission) and structure. Partial remyelination in shadow plaques can restore energy supply and prevent the fatal fate of compromised axons, but not in acute and chronic plaques without overt remyelination. Consequently, long-term bioenergetic deficits can lead to axon loss, anterograde and retrograde neuronal degeneration and MS progression.

CONCLUSION

In this chapter, we present the metabolic profile of oligodendrocytes and myelin, as well as myelin lipid metabolism, and summarize how and where metabolic alterations occur in MS. These include energy deficits in axons, leading to slower signal propagation, and oligodendrocyte dysfunction in rebuilding myelin after demyelination, all of this leading inexorably to the progression of multiple sclerosis.

REFERENCES

Aber, E. R., Griffey, C. J., Davies, T., Li, A. M., Yang, Y. J., Croce, K. R., Goldman, J. E., Grutzendler, J., Canman, J. C., & Yamamoto, A. (2022). Oligodendroglial macroautophagy is essential for myelin sheath turnover to prevent neurodegeneration and death. *Cell Reports*, *41*(3), 111480. doi:10.1016/j.celrep.2022.111480

Adiele, R. C., & Adiele, C. A. (2019). Metabolic defects in multiple sclerosis. *Mitochondrion*, *44*, 7–14. doi:10.1016/j.mito.2017.12.005

Akay, L. A., Effenberger, A. H., & Tsai, L. H. (2021). Cell of all trades: Oligodendrocyte precursor cells in synaptic, vascular, and immune function. *Genes & Development*, *35*(3-4), 180–198. doi:10.1101/gad.344218.120

Alanazi, A., Alassiri, M., Jawdat, D., & Almalik, Y. (2022). Mesenchymal stem cell therapy: A review of clinical trials for multiple sclerosis. *Regenerative Therapy*, *21*, 201–209. doi:10.1016/j.reth.2022.07.003

Amin, M., & Hersh, C. M. (2023). Updates and advances in multiple sclerosis neurotherapeutics. *Neurodegenerative Disease Management*, *14*(1), 47–70. doi:10.2217/nmt-2021-0058

AsadollahiE.TrevisiolA.SaabA. S.LooserZ. J.DibajP.KuschK.NaveK. A. (2022). Myelin lipids as nervous system energy reserves. bioRxiv. doi:10.1101/2022.02.24.481621

Barcelos, I. P. D., Troxell, R. M., & Graves, J. S. (2019). Mitochondrial dysfunction and multiple sclerosis. *Biology (Basel)*, *8*(2), 37. doi:10.3390/biology8020037

Beirowski, B. (2022). Emerging evidence for compromised axonal bioenergetics and axoglial metabolic coupling as drivers of neurodegeneration. *Neurobiology of Disease*, *170*, 105751. doi:10.1016/j.nbd.2022.105751

Bjelobaba, I., Begovic-Kupresanin, V., Pekovic, S., & Lavrnja, I. (2018). Animal models of multiple sclerosis: Focus on experimental autoimmune encephalomyelitis. *Journal of Neuroscience Research*, *96*(6), 1021–1042. doi:10.1002/jnr.24224

Boggs, J. M. (2006). Myelin basic protein: A multifunctional protein. *Cellular and Molecular Life Sciences*, *63*(17), 1945–1961. doi:10.100700018-006-6094-7

Bonora, M., Patergnani, S., Rimessi, A., De Marchi, E., Suski, J. M., Bononi, A., Giorgi, C., Marchi, S., Missiroli, S., Poletti, F., Wieckowski, M. R., & Pinton, P. (2012). ATP synthesis and storage. *Purinergic Signalling*, *8*(3), 343–357. doi:10.100711302-012-9305-8

Butler, R., Bradford, D., & Rodgers, K. E. (2022). Analysis of shared underlying mechanism in neurodegenerative disease. *Frontiers in Aging Neuroscience*, *14*, 1006089. doi:10.3389/fnagi.2022.1006089

Cagol, A., Schaedelin, S., Barakovic, M., Benkert, P., Todea, R. A., Rahmanzadeh, R., Galbusera, R., Lu, P.-J., Weigel, M., Melie-Garcia, L., Ruberte, E., Siebenborn, N., Battaglini, M., Radue, E.-W., Yaldizli, Ö., Oechtering, J., Sinnecker, T., Lorscheider, J., Fischer-Barnicol, B., & Granziera, C. (2022). Association of brain atrophy with disease progression independent of relapse activity in patients with relapsing multiple sclerosis. *JAMA Neurology*, *79*(7), 682–692. doi:10.1001/jamaneurol.2022.1025

Castellanos, D. B., Martín-Jiménez, C. A., Rojas-Rodríguez, F., Barreto, G. E., & González, J. (2021). Brain lipidomics as a rising field in neurodegenerative contexts: Perspectives with Machine Learning approaches. *Frontiers in Neuroendocrinology*, *61*, 100899. doi:10.1016/j.yfrne.2021.100899

Chamberlain, K. A., Nanescu, S. E., Psachoulia, K., & Huang, J. K. (2016). Oligodendrocyte regeneration: Its significance in myelin replacement and neuroprotection in multiple sclerosis. *Neuropharmacology*, *110*, 633–643. doi:10.1016/j.neuropharm.2015.10.010

Dasgupta, S., & Ray, S. K. (2017). Diverse biological functions of sphingolipids in the CNS: Ceramide and sphingosine regulate myelination in developing brain but stimulate demyelination during pathogenesis of multiple sclerosis. *Journal of Neurology and Psychology*, *5*(1).

De Angelis, F., John, N. A., & Brownlee, W. J. (2018). Disease-modifying therapies for multiple sclerosis. *bmj, 363*.

DeMarshall, C., Goldwaser, E. L., Sarkar, A., Godsey, G. A., Acharya, N. K., Thayasivam, U., Belinka, B. A., & Nagele, R. G. (2017). Autoantibodies as diagnostic biomarkers for the detection and subtyping of multiple sclerosis. *Journal of Neuroimmunology*, *309*, 51–57. doi:10.1016/j.jneuroim.2017.05.010

Dimitriou, N. G., Meuth, S. G., Martinez-Lapiscina, E. H., Albrecht, P., & Menge, T. (2023). Treatment of patients with multiple sclerosis transitioning between relapsing and progressive disease. *CNS Drugs*, *37*(1), 69–92. doi:10.100740263-022-00977-3

Dimou, L., Simon, C., Kirchhoff, F., Takebayashi, H., & Götz, M. (2008). Progeny of Olig2-expressing progenitors in the gray and white matter of the adult mouse cerebral cortex. *The Journal of Neuroscience : The Official Journal of the Society for Neuroscience*, *28*(41), 10434–10442. doi:10.1523/JNEUROSCI.2831-08.2008

Dobson, R., & Giovannoni, G. (2019). Road, W.; Unit, PN; Square, C. *Multiple Sclerosis—A review. European Journal of Neurology*, *26*, 27–40. doi:10.1111/ene.13819

Falcão, A. M., van Bruggen, D., Marques, S., Meijer, M., Jäkel, S., Agirre, E., Samudyata, Floriddia, E. M., Vanichkina, D. P., ffrench-Constant, C., Williams, A., Guerreiro-Cacais, A. O., & Castelo-Branco, G. (2018). Disease-specific oligodendrocyte lineage cells arise in multiple sclerosis. *Nature Medicine*, *24*(12), 1837–1844. doi:10.103841591-018-0236-y

Ferreira, H. B., Melo, T., Monteiro, A., Paiva, A., Domingues, P., & Domingues, M. R. (2021). Serum phospholipidomics reveals altered lipid profile and promising biomarkers in multiple sclerosis. *Archives of Biochemistry and Biophysics*, *697*, 108672. doi:10.1016/j.abb.2020.108672

Ferreira, H. B., Neves, B., Guerra, I. M., Moreira, A., Melo, T., Paiva, A., & Domingues, M. R. (2020). An overview of lipidomic analysis in different human matrices of multiple sclerosis. *Multiple Sclerosis and Related Disorders*, *44*, 102189. doi:10.1016/j.msard.2020.102189

Fonalledas-Perelló, M. A., Valero-Politi, J., Lizarraga-Dallo, M. A., & Segura-Cardona, R. (2008). The cerebrospinal fluid lactate is decreased in early stages of multiple sclerosis. *Puerto Rico Health Sciences Journal*, *27*(2).

Fünfschilling, U., Supplie, L. M., Mahad, D., Boretius, S., Saab, A. S., Edgar, J., Brinkmann, B. G., Kassmann, C. M., Tzvetanova, I. D., Möbius, W., Diaz, F., Meijer, D., Suter, U., Hamprecht, B., Sereda, M. W., Moraes, C. T., Frahm, J., Goebbels, S., & Nave, K. A. (2012). Glycolytic oligodendrocytes maintain myelin and long-term axonal integrity. *Nature*, *485*(7399), 517–521. doi:10.1038/nature11007

Garg, N., & Smith, T. W. (2015). An update on immunopathogenesis, diagnosis, and treatment of multiple sclerosis. *Brain and Behavior*, *5*(9), e00362. doi:10.1002/brb3.362

Giussani, P., Prinetti, A., & Tringali, C. (2021). The role of Sphingolipids in myelination and myelin stability and their involvement in childhood and adult demyelinating disorders. *Journal of Neurochemistry*, *156*(4), 403–414. doi:10.1111/jnc.15133

Grassi, S., Giussani, P., Mauri, L., Prioni, S., Sonnino, S., & Prinetti, A. (2020). Lipid rafts and neurodegeneration: Structural and functional roles in physiologic aging and neurodegenerative diseases: Thematic Review Series: Biology of Lipid Rafts. *Journal of Lipid Research*, *61*(5), 636–654. doi:10.1194/jlr.TR119000427

Harbo, H. F., Gold, R., & Tintoré, M. (2013). Sex and gender issues in multiple sclerosis. *Therapeutic Advances in Neurological Disorders*, *6*(4), 237–248. doi:10.1177/1756285613488434

Jahn, O., Siems, S. B., Kusch, K., Hesse, D., Jung, R. B., Liepold, T., Uecker, M., Sun, T., & Werner, H. B. (2020). The CNS myelin proteome: Deep profile and persistence after post-mortem delay. *Frontiers in Cellular Neuroscience*, *14*, 239. doi:10.3389/fncel.2020.00239

Kassmann, C. M. (2014). Myelin peroxisomes–essential organelles for the maintenance of white matter in the nervous system. *Biochimie*, *98*, 111–118. doi:10.1016/j.biochi.2013.09.020

Kassmann, C. M., Lappe-Siefke, C., Baes, M., Brügger, B., Mildner, A., Werner, H. B., Natt, O., Michaelis, T., Prinz, M., Frahm, J., & Nave, K. A. (2007). Axonal loss and neuroinflammation caused by peroxisome-deficient oligodendrocytes. *Nature Genetics*, *39*(8), 969–976. doi:10.1038/ng2070

Kremer, D., Göttle, P., Flores-Rivera, J., Hartung, H. P., & Küry, P. (2019). Remyelination in multiple sclerosis: From concept to clinical trials. *Current Opinion in Neurology*, *32*(3), 378–384. doi:10.1097/WCO.0000000000000692

Kuhn, S., Gritti, L., Crooks, D., & Dombrowski, Y. (2019). Oligodendrocytes in development, myelin generation and beyond. *Cells*, *8*(11), 1424. doi:10.3390/cells8111424

Lassmann, H. (2018). Multiple sclerosis pathology. *Cold Spring Harbor Perspectives in Medicine*, *8*(3), a028936. doi:10.1101/cshperspect.a028936

Li, S., & Sheng, Z. H. (2023). Oligodendrocyte-derived transcellular signaling regulates axonal energy metabolism. *Current Opinion in Neurobiology*, *80*, 102722. doi:10.1016/j.conb.2023.102722

López-Muguruza, E., Villar-Gómez, N., Matias-Guiu, J. A., Selma-Calvo, B., Moreno-Jiménez, L., Sancho-Bielsa, F., Lopez-Carbonero, J., Benito-Martín, M. S., García-Flores, S., Bonel-García, N., Kamal, O. M.-F., Ojeda-Hernández, D., Matías-Guiu, J., & Gómez-Pinedo, U. (2022). The integration of cell therapy and biomaterials as treatment strategies for remyelination. *Life (Basel, Switzerland)*, *12*(4), 474. doi:10.3390/life12040474

Marisca, R., Hoche, T., Agirre, E., Hoodless, L. J., Barkey, W., Auer, F., Castelo-Branco, G., & Czopka, T. (2020). Functionally distinct subgroups of oligodendrocyte precursor cells integrate neural activity and execute myelin formation. *Nature Neuroscience*, *23*(3), 363–374. doi:10.103841593-019-0581-2

McMullen, E., Hertenstein, H., Strassburger, K., Deharde, L., Brankatschk, M., & Schirmeier, S. (2023). Glycolytically impaired Drosophila glial cells fuel neural metabolism via β-oxidation. *Nature Communications*, *14*(1), 2996. doi:10.103841467-023-38813-x

Mergenthaler, P., Lindauer, U., Dienel, G. A., & Meisel, A. (2013). Sugar for the brain: The role of glucose in physiological and pathological brain function. *Trends in Neurosciences*, *36*(10), 587–597. doi:10.1016/j.tins.2013.07.001

Mot, A. I., Depp, C., & Nave, K. A. (2022). An emerging role of dysfunctional axon-oligodendrocyte coupling in neurodegenerative diseases. *Dialogues in Clinical Neuroscience*.

Mukhopadhyay, S., Fellows, K., Browne, R. W., Khare, P., Krishnan Radhakrishnan, S., Hagemeier, J., Weinstock-Guttman, B., Zivadinov, R., & Ramanathan, M. (2017). Interdependence of oxysterols with cholesterol profiles in multiple sclerosis. *Multiple Sclerosis*, *23*(6), 792–801. doi:10.1177/1352458516666187

Nave, K. A. (2010). Myelination and support of axonal integrity by glia. *Nature*, *468*(7321), 244–252. doi:10.1038/nature09614

Nishiyama, A., Shimizu, T., Sherafat, A., & Richardson, W. D. (2021, August). Life-long oligodendrocyte development and plasticity. [). Academic Press.]. *Seminars in Cell & Developmental Biology*, *116*, 25–37. doi:10.1016/j.semcdb.2021.02.004

Oliveira, E. M. L. D., Montani, D. A., Oliveira-Silva, D., Rodrigues-Oliveira, A. F., Matas, S. L. D. A., Fernandes, G. B. P., Silva, I. D. C. G., & Lo Turco, E. G. (2019). Multiple sclerosis has a distinct lipid signature in plasma and cerebrospinal fluid. *Arquivos de Neuro-Psiquiatria*, *77*(10), 696–704. doi:10.1590/0004-282x20190122

Owen, O. E., Morgan, A. P., Kemp, H. G., Sullivan, J. M., Herrera, M. G., & Cahill, G. J. Jr. (1967). Brain metabolism during fasting. *The Journal of Clinical Investigation*, *46*(10), 1589–1595. doi:10.1172/JCI105650

Pérez-Cerdá, F., Sánchez-Gómez, M. V., & Matute, C. (2015). Pío del Río Hortega and the discovery of the oligodendrocytes. *Frontiers in Neuroanatomy*, *9*, 92. doi:10.3389/fnana.2015.00092

Piehl, F. (2021). Current and emerging disease-modulatory therapies and treatment targets for multiple sclerosis. *Journal of Internal Medicine*, *289*(6), 771–791. doi:10.1111/joim.13215

Plemel, J. R., Liu, W. Q., & Yong, V. W. (2017). Remyelination therapies: A new direction and challenge in multiple sclerosis. *Nature Reviews. Drug Discovery*, *16*(9), 617–634. doi:10.1038/nrd.2017.115

Podbielska, M., O'Keeffe, J., & Pokryszko-Dragan, A. (2021). New insights into multiple sclerosis mechanisms: Lipids on the track to control inflammation and neurodegeneration. *International Journal of Molecular Sciences*, *22*(14), 7319. doi:10.3390/ijms22147319

Regenold, W. T., Phatak, P., Makley, M. J., Stone, R. D., & Kling, M. A. (2008). Cerebrospinal fluid evidence of increased extra-mitochondrial glucose metabolism implicates mitochondrial dysfunction in multiple sclerosis disease progression. *Journal of the Neurological Sciences*, *275*(1-2), 106–112. doi:10.1016/j.jns.2008.07.032

Rinholm, J. E., Hamilton, N. B., Kessaris, N., Richardson, W. D., Bergersen, L. H., & Attwell, D. (2011). Regulation of oligodendrocyte development and myelination by glucose and lactate. *The Journal of Neuroscience : The Official Journal of the Society for Neuroscience*, *31*(2), 538–548. doi:10.1523/JNEUROSCI.3516-10.2011

Rone, M. B., Cui, Q. L., Fang, J., Wang, L. C., Zhang, J., Khan, D., Bedard, M., Almazan, G., Ludwin, S. K., Jones, R., Kennedy, T. E., & Antel, J. P. (2016). Oligodendrogliopathy in multiple sclerosis: Low glycolytic metabolic rate promotes oligodendrocyte survival. *The Journal of Neuroscience : The Official Journal of the Society for Neuroscience*, *36*(17), 4698–4707. doi:10.1523/JNEUROSCI.4077-15.2016

Rosko, L., Smith, V. N., Yamazaki, R., & Huang, J. K. (2019). Oligodendrocyte bioenergetics in health and disease. *The Neuroscientist*, *25*(4), 334–343. doi:10.1177/1073858418793077

Ruskamo, S., Raasakka, A., Pedersen, J. S., Martel, A., Škubník, K., Darwish, T., Porcar, L., & Kursula, P. (2022). Human myelin proteolipid protein structure and lipid bilayer stacking. *Cellular and Molecular Life Sciences*, *79*(8), 419. doi:10.100700018-022-04428-6

Saab, A. S., Tzvetavona, I. D., Trevisiol, A., Baltan, S., Dibaj, P., Kusch, K., Möbius, W., Goetze, B., Jahn, H. M., Huang, W., Steffens, H., Schomburg, E. D., Pérez-Samartín, A., Pérez-Cerdá, F., Bakhtiari, D., Matute, C., Löwel, S., Griesinger, C., Hirrlinger, J., & Nave, K. A. (2016). Oligodendroglial NMDA receptors regulate glucose import and axonal energy metabolism. *Neuron*, *91*(1), 119–132. doi:10.1016/j.neuron.2016.05.016

Silva, B., Mantha, O. L., Schor, J., Pascual, A., Plaçais, P. Y., Pavlowsky, A., & Preat, T. (2022). Glia fuel neurons with locally synthesized ketone bodies to sustain memory under starvation. *Nature Metabolism*, *4*(2), 213–224. doi:10.103842255-022-00528-6

Simons, M., & Nave, K. A. (2016). Oligodendrocytes: Myelination and axonal support. *Cold Spring Harbor Perspectives in Biology*, *8*(1), a020479. doi:10.1101/cshperspect.a020479

Skaper, S. D. (2019). Oligodendrocyte precursor cells as a therapeutic target for demyelinating diseases. *Progress in Brain Research*, *245*, 119–144. doi:10.1016/bs.pbr.2019.03.013

Sock, E., & Wegner, M. (2021). Using the lineage determinants Olig2 and Sox10 to explore transcriptional regulation of oligodendrocyte development. *Developmental Neurobiology*, *81*(7), 892–901. doi:10.1002/dneu.22849

Spaas, J., Van Veggel, L., Schepers, M., Tiane, A., Van Horssen, J., Wilson, D. M. III, Moya, P. R., Piccart, E., Hellings, N., Eijnde, B. O., Derave, W., Schreiber, R., & Vanmierlo, T. (2021). Oxidative stress and impaired oligodendrocyte precursor cell differentiation in neurological disorders. *Cellular and Molecular Life Sciences*, *78*(10), 4615–4637. doi:10.100700018-021-03802-0

Stangel, M., Kuhlmann, T., Matthews, P. M., & Kilpatrick, T. J. (2017). Achievements and obstacles of remyelinating therapies in multiple sclerosis. *Nature Reviews. Neurology*, *13*(12), 742–754. doi:10.1038/nrneurol.2017.139

Stys, P. K. (2005). General mechanisms of axonal damage and its prevention. *Journal of the Neurological Sciences*, *233*(1-2), 3–13. doi:10.1016/j.jns.2005.03.031

Stys, P. K., & Tsutsui, S. (2019). Recent advances in understanding multiple sclerosis. *F1000 Research*, 8.

Suminaite, D., Lyons, D. A., & Livesey, M. R. (2019). Myelinated axon physiology and regulation of neural circuit function. *Glia*, *67*(11), 2050–2062. doi:10.1002/glia.23665

Tiane, A., Schepers, M., Rombaut, B., Hupperts, R., Prickaerts, J., Hellings, N., van den Hove, D., & Vanmierlo, T. (2019). From OPC to oligodendrocyte: An epigenetic journey. *Cells*, *8*(10), 1236. doi:10.3390/cells8101236

Travers, B. S., Tsang, B. K., & Barton, J. L. (2022). Multiple sclerosis: Diagnosis, disease-modifying therapy and prognosis. *Australian Journal of General Practice*, *51*(4), 199–206. doi:10.31128/AJGP-07-21-6103

van Horssen, J., van Schaik, P., & Witte, M. (2019). Inflammation and mitochondrial dysfunction: A vicious circle in neurodegenerative disorders? *Neuroscience Letters*, *710*, 132931. doi:10.1016/j.neulet.2017.06.050

Wang, J. Q., Gao, M. Y., Gao, R., Zhao, K. H., Zhang, Y., & Li, X. (2023). Oligodendrocyte lineage cells: Advances in development, disease, and heterogeneity. *Journal of Neurochemistry*, *164*(4), 468–480. doi:10.1111/jnc.15728

Williamson, J. M., & Lyons, D. A. (2018). Myelin dynamics throughout life: An ever-changing landscape? *Frontiers in Cellular Neuroscience*, *12*, 424. doi:10.3389/fncel.2018.00424

Winkelmann, A., Loebermann, M., Reisinger, E. C., Hartung, H. P., & Zettl, U. K. (2016). Disease-modifying therapies and infectious risks in multiple sclerosis. *Nature Reviews. Neurology*, *12*(4), 217–233. doi:10.1038/nrneurol.2016.21

Yellen, G. (2018). Fueling thought: Management of glycolysis and oxidative phosphorylation in neuronal metabolism. *The Journal of Cell Biology*, *217*(7), 2235–2246. doi:10.1083/jcb.201803152

Yi, C., Verkhratsky, A., & Niu, J. (2023). Pathological potential of oligodendrocyte precursor cells: Terra incognita. *Trends in Neurosciences*, *46*(7), 581–596. doi:10.1016/j.tins.2023.04.003

Chapter 12
Hepatic Encephalopathy as a Gliopathy:
The Mechanisms of Alzheimer Type II Astrocytosis

Bilal El-Mansoury
https://orcid.org/0000-0002-2583-7515
Faculty of Sciences, Chouaib Doukkali University, Morocco

Abdelaati El Khiat
Higher Institute of Nursing Professions and Health Techniques, Ouarzazate, Morocco & Cadi Ayyad University, Morocco

Omar El Hiba
Faculty of Sciences, Chouaib Doukkali University, Morocco

El Baz Soraia
Faculty of Science Semlalia, Cadi Ayyad University, Morocco

Arumugam Radhakrishnan Jayakumar
Miller School of Medicine, University of Miami, USA

ABSTRACT

Hepatic encephalopathy (HE) is defined as a wide spectrum of neuropsychiatric abnormalities caused by liver dysfunction (acute or chronic) and/or portal-systemic shunting. The histopathologic hallmark of HE is astrocyte swelling following acute liver failure (ALF) or the presence of the so-called Alzheimer type 2 astrocytosis under chronic liver disease. HE can be classified according to the undelaying cause into three types: type A as an essential component of acute liver failure, type B as a consequence of porto-systemic shunts in the absence of liver dysfunction, and type C in patients with liver cirrhosis and porto-systemic bypass. While ammonia, manganese, proinflammatory cytokines, and other precipitating factors play a role in the pathogenesis of HE, the exact mechanisms leading to the development of HE are not fully elucidated. This chapter provides a brief overview of HE with a focus on the mechanisms of AT2A.

DOI: 10.4018/978-1-6684-9675-6.ch012

INTRODUCTION

Hepatic encephalopathy (HE) is broadly defined as brain disorder brought on by impaired liver function and/or portal-systemic shunting (El-Mansoury et al., 2023), which manifests as a wide spectrum of neurocognitive and neuropsychiatric abnormalities (El Khiat et al., 2023), ranging from subclinical alterations to coma (El-Mansoury et al., 2023). Four axes of guidelines were established in 2014 by the European Association for the Study of the Liver (EASL) and the American Association for the Study of Liver Diseases (AASLD) to improve the categorization and management of HE (Dharel & Bajaj, 2015). These guidelines were focusing on etiology, severity, time course, and precipitating factors.

Although the exact cause of HE is unknown, several underlying mechanisms are thought to play a role in its pathogenesis. The development of HE is significantly influenced by the gut microbiota (GM). The human GM is made up of thousands of microorganisms that play a tremendous role in maintaining the hosts' health by performing a variety of physiological tasks and shielding them from diseases and aggressors (El-Mansoury et al., 2023). The GM generate several toxic substances including ammonia. It is highly established that ammonia plays a central role in the development of HE.

It is highly established that astrocytes play a variety of roles in the brain, such as supplying the surrounding neurons by nutrients and mechanical support, controlling ion transport and neurotransmitter uptake, and playing a crucial role in the blood-brain barrier (BBB) integrity (Aldridge et al., 2015). In the context of liver disease, ammonia is insufficiently eliminated, leading to its build up in the systemic circulation (hyperammonemia), consequently crossing the BBB, and ultimately reaching the brain tissue. Astrocytes play a neuroprotective role in ammonia detoxification through the action of the glutamine synthetase–mediated reaction (ammonia and glutamate are converted to glutamine), and thus, they are the primary target cells in HE. It is noteworthy that following acute liver failure (ALF), astrocyte typically take a swollen shape (astrocyte swelling), and in chronic liver disease (CLD) circumstances the so-called Alzheimer type II astrocytes (AT2A) is found in the brain of patients as well as animal models of HE (chronic HE). Indeed, AT2A was demonstrated following CLD in several brain regions including the cerebral cortex, basal ganglia, and pontine nuclei. Therefore, it is considered as a gliopathy (Gelpi et al., 2020). This chapter provides an overview on hepatic anatomy and function as well as the common complication of CLD; hepatic encephalopathy. Additionally, this chapter focuses mainly on the mechanisms of AT2A following chronic HE.

ANATOMY AND PHYSIOLOGY OF THE LIVER

Anatomy of Liver

External Anatomy of Liver

The external anatomy of liver is characterized by the gross landmarks including the vena cava the gallbladder, and the hepatic ligaments (Fig.1). The liver is the largest gland in the body, making about 2.5% of the total weight of the body and weighing about 1500 g in adult stage. It performs a variety of functions that are essential for preserving appropriate physiological homeostasis (Juza & Pauli, 2014). The two lobes of the liver are commonly described using two different descriptions: morphologic anatomy and functional anatomy (Couinaud's segmental classification). This later is more detailed and is based on the portal vein segmentation.

Regarding its anatomical position, the liver is located in the upper right part of the abdominal cavity (the right hypochondriac region), beneath the diaphragm, and on top of the stomach. It is protected by the rib cage and maintained in place by ligamentous attachments (avascular attachments) that are peritoneal reflections (Fig.1). First, the falciform ligament which originates at or close to the umbilicus and extends along the umbilical fissure to the anterior portion of the liver. It then extends along the umbilical fissure to the anterior portion of the liver (Abdel-Misih & Bloomston, 2010). Moreover, the ligamentum teres, also known as the round ligament, is located within the lower border of the falciform ligament. It is an element of the obliterated umbilical vein (ductus venosus), which continues from the umbilicus into the umbilical fissure, where it connects to the left branch of the portal vein in continuity with the ligamentum venosum (Abdel-Misih & Bloomston, 2010). This later is located within a fissure on the liver's inferior surface between the left lobe anteriorly and the caudate lobe posteriorly, where it is also invested by the gastrohepatic ligament. Along the diaphragmatic surface, which is the uncovered portion of the liver made up of peritoneal reflections of the diaphragm, the coronary ligament is located anteriorly and posteriorly to a convex area (the cranial aspect of the liver). Additionally, the left and right triangular ligaments are formed by the convergence of these regions to the left and right of the liver, respectively. Furthermore, the liver is attached to the retroperitoneum by the right coronary and right triangular ligaments, which extend posteriorly and caudally toward the right kidney (Abdel-Misih & Bloomston, 2010) (Fig.1).

Figure 1. Macroscopic anatomy of the liver: (a) Anterior and (b) posterior views
(Ayyildiz, 2015)

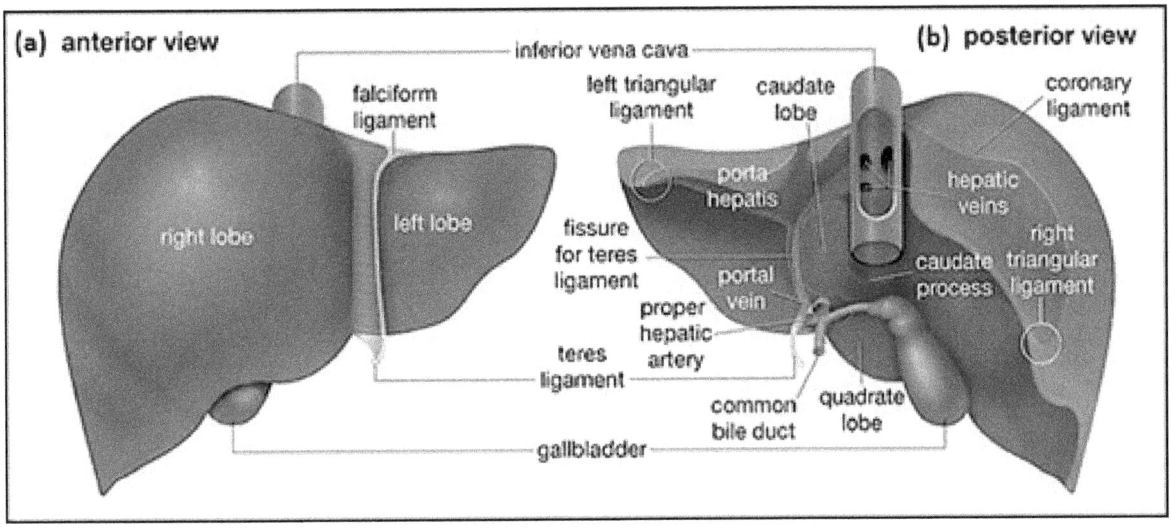

Internal Anatomy of Liver

Hepatic Vasculature

The liver is hypervascularized and receives up to 25% of total cardiac output at rest. It receives blood supplies from both; the hepatic artery which constitute around 25% to 30% of the blood supply (Mulaikal & Emond, 2012), and the portal vein (fig.2a), responsible for the remaining 70% to 75%, which provides the bulk of the nutritive blood supply to the liver, carrying partially oxygenated blood from the gut, which is rich in chemicals absorbed after digestion (Wallace et al., 2008). The arterial and portal blood eventually mixes within the hepatic sinusoids, and then drained into the systemic circulation through the hepatic venous system (Blumgart & Belghiti, 2007).

Generally, the common hepatic artery takes its origin from the celiac axis, the splenic artery, and the left gastric artery (Abdel-Misih & Bloomston, 2010). In the liver, the hepatic artery is divided to supply the right and left lobes (Mulaikal & Emond, 2012). While, the portal vein is consisted of the anastomosis of the splenic vein and the superior mesenteric vein. The portal vein receives drainage from other venous branches, such as the tributaries of the right gastric and pancreaticoduodenal veins, the coronary (left gastric) vein, and the cystic vein (Abdel-Misih & Bloomston, 2010). It is noteworthy that the superior mesenteric vein carries nutrients and toxins from gastrointestinal tract to the portal circulation, whereas the splenic vein provides pancreatic hormones (Mulaikal & Emond, 2012).

Figure 2. Hepatic vasculature: (a) Common hepatic arterial configuration, (b) portal vein and the hepatic venous vasculature inflow. HA, hepatic artery
(Abdel-Misih & Bloomston, 2010).

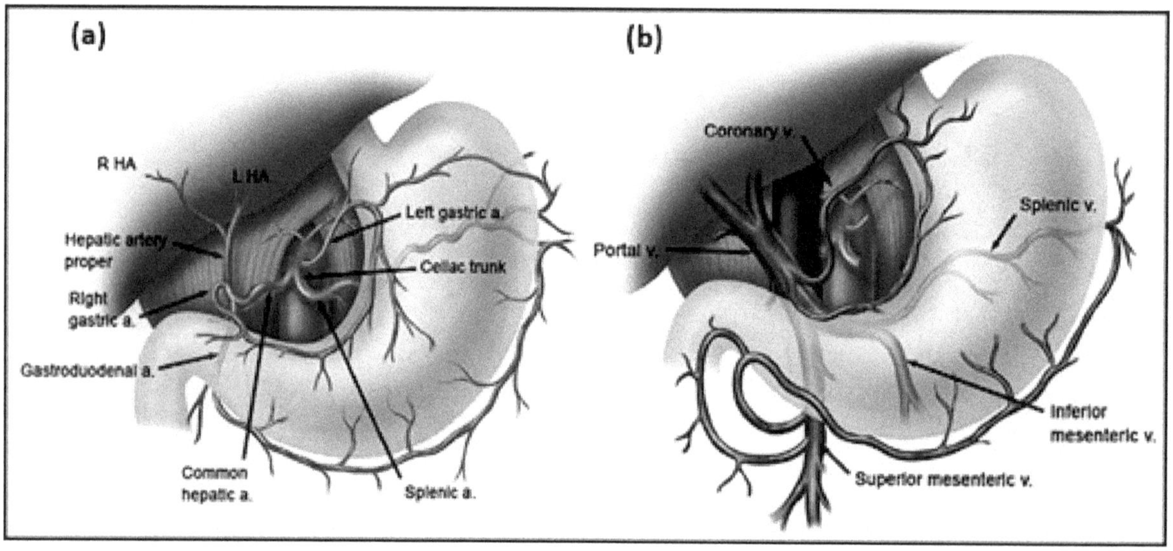

Hepatic Lobule

The basic architectural unit of the liver is known as the hepatic lobule (fig.3), which is made up of hepatocyte plates surrounded by sinusoidal capillaries that point in the direction of the central efferent vein. The liver lobule (liver parenchyma), is approximately hexagon-shaped, with the hepatic artery, bile duct, and portal vein marking each of the six peripheral corners (Si-Tayeb et al., 2010). The lobule has thin-walled sinusoidal capillaries draining the portal triad toward a central vein (Figure 3).

Figure 3. Microscopic structure of the liver; the hepatic lobule
(Ayyildiz, 2015)

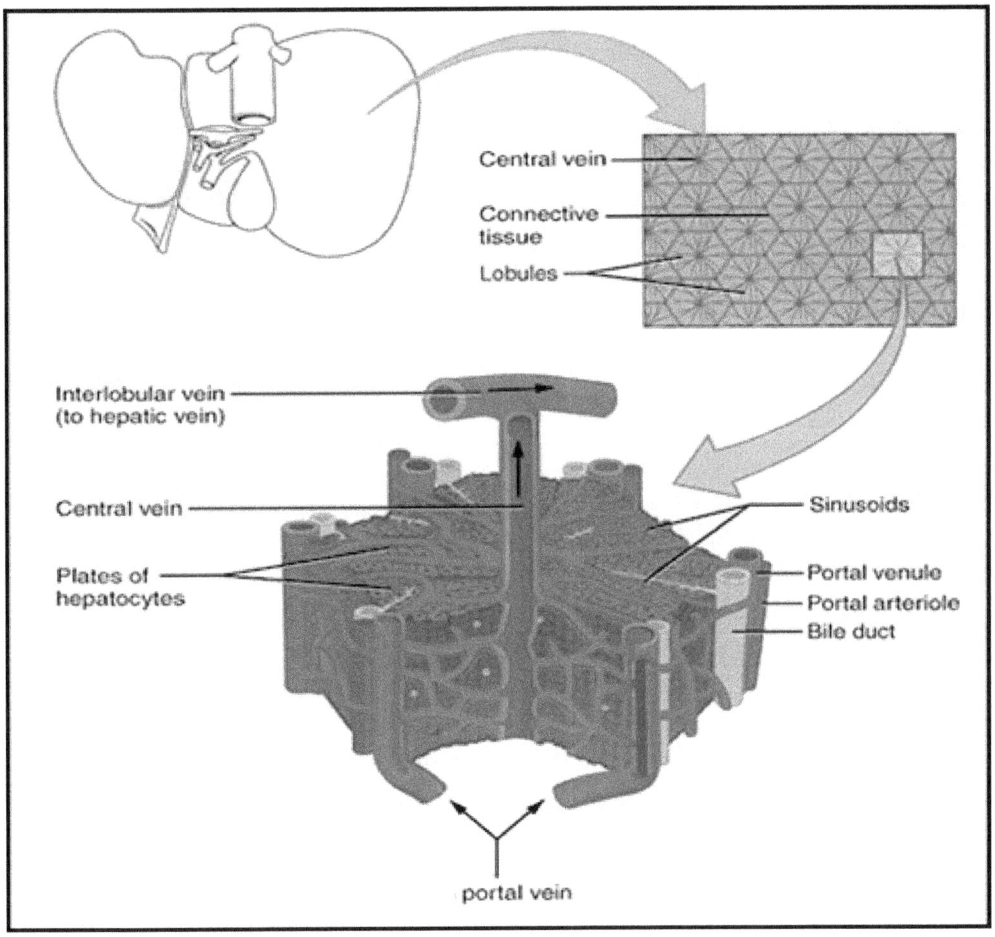

Functions of the Liver

The liver makes up 2%–5% of the total weight of the body. Since hepatocytes make up the majority of the liver cells, the liver is easily accessible for protein purification by classical biochemical processes. These include the isolation of hormones, growth factors, signaling molecules, transcription factors, and growth factors, making the liver an essential organ that plays a large variety of roles in the body (Si-Tayeb et al., 2010). Furthermore, the liver demonstrates exocrine and endocrine functions. A number of hormones, including thrombopoietin, angiotensinogen, and insulin-like growth factors, are secreted during endocrine processes. On the other hand, bile is primarily in the liver, and released in the intestines ensuring an important role in digestion. The liver also maintains the storage of glycogen, detoxifies drugs, regulates metabolism, synthesis and transport of cholesterol, catabolizes amino acids and produces urea, and secretes a wide range of major serum proteins, such as albumin, apolipoproteins, complement, and clotting factors (Si-Tayeb et al., 2010).

CHRONIC LIVER DISEASE

Chronic liver disease (CLD) is a significant cause of morbidity and mortality worldwide. The prevalence of CLD is increased in recent years. There is a wide range of etiologies of CLD including metabolic disorders, toxins, medications (acetaminophen toxicity), autoimmune diseases (I.e., primary biliary cirrhosis, primary sclerosing cholangitis, autoimmune hepatitis), alcoholic liver disease, chronic viral hepatitis (I.e., including hepatitis B and C), and genetic causes (I.e., hereditary hemochromatosis and Wilson disease) (Sharma & Nagalli, 2022). CLD occurs as result of progressive deterioration of the liver for more than six months (Sharma & Nagalli, 2022). The hallmark of CLD is the gradual and diffuse accumulation of scar tissue, also known as fibrous tissue, in response to persistent liver inflammation or injury. In the early stages of development, fibrosis can be reversible even though it is typically irreversible (Sharma & Nagalli, 2022).

Liver Fibrosis

The onset of liver fibrosis is usually insidious, if left untreated, fibrosis can progress to cirrhosis, ultimately leading to liver failure, related morbidity and mortality and possible death (Bataller et al., 2005; Tsukada et al., 2006; Sharma & Nagalli, 2022). Fibrosis refers to an excessive accumulation of extracellular matrix (ECM) proteins as a result of chronic liver injury (Sharma & Nagalli, 2022), which is the most common characteristic of almost all CLDs regardless of the cause.

Hepatic fibrosis is caused by three major protein families: glycoproteins, collagens, and proteoglycans (Tsukada et al., 2006). Indeed, the fibrotic liver has been shown to express higher levels of a number of glycoproteins (including hyaluronic acid, fibronectin, laminin, merosin, tenascin, and nidogen), increased deposition of many proteoglycans (including heparin, dermatan, chondroitin sulfates, perlecan, syndecan, biglycan, and decorin) and increased deposition of collagen, specifically the fibrillar collagens. The two most frequently related and most greatly increased forms of collagens in the liver are types I and III, despite the fact that the number of distinct collagens are identified in the liver (Tsukada et al., 2006).

The accumulation of ECM proteins (mostly collagens Type I and Type III) (Berumen et al., 2022) is caused by an imbalance between the synthesis of ECM proteins (fibrogenesis) and their degradation (fibrolysis) leading to the formation of fibrous scar, which can ultimately compromise normal liver function. Liver fibrosis is primarily caused by the chronically activated wound-healing response. This wound-healing process is triggered by an orderly series of biological processes involving cells and soluble substances that are targeted at repairing the tissue lesions (Pinzani, 2015). Regardless the etiology of liver injury, the common molecular pathways of liver fibrosis include activation of hepatic stellate cells

(HSCs), persistent inflammation with cytokine production, and ultimately hepatocyte death (Berumen et al., 2022). Importantly, it appears that the major cell type involved in the pathophysiology of fibrosis is the HSC (Berumen et al., 2022). In short, the hepatic cells' injury-induced response triggers the production of substances like TGF-ß1 and monocyte chemoattractant protein 1 (MCP1), which promote inflammation and activate HSCs into myofibroblasts that produce collagen. Importantly, myofibroblasts are the main target of antifibrotic therapy since they have been demonstrated to be a crucial mechanism in the development of liver fibrosis (Berumen et al., 2022). In the other hand, the excess production of ECM proteins could be attributed to decreased activity of ECM-removal which can be resulted from an overexpression of their specific inhibitors (TIMPs), consequently participating to this ECM accumulation (Bataller et al., 2005). As a result, the proteins build up in the liver, thus producing a scar tissue, which deforms the hepatic architecture (Bataller et al., 2005). This eventually modifies the liver's typical function and results in liver damage.

Liver Cirrhosis

Cirrhosis is defined as an advanced stage of fibrosis (Pinzani, 2015), or a final stage of CLD that results in disruption of liver architecture, the formation of widespread regenerative nodules of liver parenchyma, vascular reorganization, neo-angiogenesis, and deposition of ECM-proteins (Sharma & Nagalli, 2022), and consequently distrupted hepatic function. Clinically, cirrhosis has been classified into two main stages: compensated and decompensated. Even though the liver may appear normal in the early stages of cirrhosis (compensated cirrhosis), the disease progresses and presents surface nodularity (spectrum of hepatocellular nodules), the majority of which are regenerative, due to the liver's special ability to regenerate in cirrhosis. Moreover, the most prominent morphologic characteristic of liver cirrhosis is caudate lobe enlargement (Reiner, 2021). However, it should be noted that in 25% of cirrhosis, the liver shape and contour appear normal on computed tomography (CT) scans or Magnetic Resonance Imaging (MRI) (Reiner, 2021).

In most patients, the progression from fibrosis to cirrhosis occurs generally within an interval of 15 to 20 years. Liver cirrhosis can present different manifestations, from asymptomatic to total liver failure. About 40% of patients with cirrhosis have no symptoms at all and they may not experience any significant consequences. However, as the liver gradually deteriorates and complications begin to appear which can include ascites, esophageal varices, jaundice, hepatic encephalopathy, and renal failure which are among the common clinical consequences of cirrhosis. Additionally, the end stage of cirrhosis (decompensated cirrhosis) is associated with short term survival and a high risk factor for developing hepatocellular carcinoma (Bataller et al., 2005; Tsukada et al., 2006).

It should be noted that the pathophysiology of liver cirrhosis varies depending on the underlying etiology (Böttcher & Pinzani, 2017). Despite the fact that cirrhosis is often the outcome of progressive fibrogenesis, there are several patterns of fibrotic growth that vary depending on the origin of fibrogenic cells and the underlying conditions that cause the fibrosis (Böttcher & Pinzani, 2017). Therefore, this however, changes the perception of liver cirrhosis and creates new challenges in treating this disease.

HEPATIC ENCEPHALOPATHY: DEFINITION AND EPIDEMIOLOGY

HE is a wide range of neurocognitive and neuropsychiatric abnormalities arising from impaired liver function and/or portal-systemic blood shunting. It typically affects 30–45% of patients and is the second most prevalent consequence of decompensated liver cirrhosis after ascites (Bohra et al., 2020). HE is characterized by a wide spectrum of neuropsychiatric abnormalities ranging from hardly noticeable symptoms to a severe ones (hepatic coma) (T. V Shulyatnikova & Shavrin, 2017). Owing to varying degrees of disease manifestation severity, several underlying causes and the distinction between covert and overt forms of HE, determining the incidence and prevalence of HE can be challenging (Elsaid & Rustgi, 2020). It has been observed that 20–80% of cirrhosis patients experience HE. About 10–14% of patients with cirrhosis had overt HE at the time of diagnosis. This increases to roughly 10–50% in individuals who have had a transjugular intrahepatic portosystemic shunt (TIPS) and 16–21% in patients with decompensated cirrhosis. Ultimately, overt HE is expected to occur in 30–40% of cirrhosis patients over the clinical course of the disease (Elsaid & Rustgi, 2020). It has been reported that severe forms of HE are associated with reduced survival rate to two years in patients with liver cirrhosis (Bohra et al., 2020). Even though HE has a significant predictive value for determining the risk of a fatal result in liver cirrhosis, only a small percentage of cirrhotic patients have a clinical diagnosis of it since the majority of cases are covert (Deutsch-Link et al., 2022).

CLINICAL MANIFESTATIONS AND CLASSIFICATION OF HEPATIC ENCEPHALOPATHY

Clinical Manifestations

HE can be associated with a wide range of neurologic, psychological, and musculoskeletal symptoms. Patients may only have mild symptoms in the beginning, such as disturbances in sleep-wake cycles, but as the symptoms worsen, the patients may experience disinhibition, impatience, and apathy. If treatment is not received, patients may experience worsening symptoms, including disorientation, memory loss, slurred speech, confusion, and eventually coma (Karin Weissenborn, 1998; Wiltfang et al., 1998). A common symptom of HE is asterixis, which is not unique to HE as it can also occur in other causes of encephalopathy (K Weissenborn et al., 2005). Patients with minimal HE (covert HE) have abnormal results in psychometric testing, subtle personality changes (Jaeger et al., 2019), may also show minor issues with coordination (i.e., as changes in their handwriting) (Vilstrup et al., 2014), reduced ability to drive, and inability or difficulty performing other complex tasks (Jaeger et al., 2019), without the overtly visible signs of HE (María Ortiz et al., 2005). As a result, mild HE is thought to be a significant predictor of the development of overt HE and negatively affects cognitive abilities and health-related quality of life in patients with cirrhosis (Bajaj et al., 2007; Prasad et al., 2007). Alternatively, overt HE is diagnosed only after all other potential causes of changed mental status have been ruled out. Overt HE is more noticeable during clinical presentation, with higher severity of symptoms (Jaeger et al., 2019).

In patients with TIPS, HE usually becomes clinically apparent 2–3 weeks after TIPS insertion (Aj, 1994). Furthermore, under cirrhosis, HE patients exhibit Parkinsonism symptoms (El Hiba et al., 2023), including slowed speech, bradykinesia, muscle rigidity, and tremors (Kabaria et al., 2021). Since these symptoms are observed in about 4% of patients, it appears that they are more common than previously thought (Tryc et al., 2013).

Classification

HE has been classified into four primary axes according to national guidelines: the underlying cause, the time-course of the disease, the degree of disease manifestation, and the presence of precipitating factors (Vilstrup et al., 2014). Firstly, HE classified into three types: A, B and C according to the underlying cause. Type A HE is observed following acute liver failure, Type B HE is associated with portal-systemic bypass with liver parenchyma damage, and Type C HE is caused by liver cirrhosis with portal hypertension or systemic shunting (Kabaria et al., 2021). Type C HE is the most frequent type, and patients with this type often exhibit symptoms of liver failure, including ascites, jaundice, spider telangiectasias, and palmar erythema (Mumtaz et al., 2010). Moreover, based on the severity of the clinical manifestations, type C HE is subdivided into covert HE (including minimal HE and West-Haven grade I HE) and overt HE (West-Haven grades II-IV HE) (Vilstrup et al., 2014). The severity of HE is generally classified using the West Haven Criteria (WHC) and the International Society for Hepatic Encephalopathy and Nitrogen Metabolism (ISHEN) criteria (Table 1) (Kabaria et al., 2021). Patients with higher WHC grades being associated with more severe HE. The mortality rate is high for overt HE patients regardless of grade. Furthermore, patients can be further categorized based on the precipitating factors and time course of the disease (the third and fourth axes) (Vilstrup et al., 2014). Three further subcategories of timing exist: episodic, recurring, and persistent. Recurrent HE is defined as having episodes that happen less frequently than six months apart, while persistent HE is characterized by altered behavior that always occurs in the context of recurrent HE. Infections, gastrointestinal bleeding, overdiuresis, irregular electrolytes, and constipation are some examples of precipitating causes. It is worth mentioning that 90% of individuals with HE can be resolved by treating triggering causes (Vilstrup et al., 2014).

By integrating both subjective and objective data, alternative grading methods, including the HE Scoring Algorithm (HESA), were created to offer a more accurate grading of HE severity (Hassanein et al., 2008) (table 1). Another scoring system is the Clinical HE Staging Scale (CHESS), which comprises of nine clinical items and linearly determines severity using simple terms and questions (M Ortiz et al., 2007) (Table 1).

Table 1. Various criteria to define severity of hepatic encephalopathy

<table>
<tr><th>WHC Grade</th><th>ISHEN</th><th>HESA</th><th>CHESS</th><th>Clinical manifestations</th></tr>
<tr><td colspan="4">Unimpaired</td><td>No encephalopathy at all.
No history of HE.</td></tr>
<tr><td>Minimal</td><td rowspan="2">Covert HE</td><td>Grade 0</td><td>0-1</td><td>Alteration in psychomotor speed/executive functions.
No indication of a change in clinical status.</td></tr>
<tr><td>Grade I</td><td>Grade I</td><td>0-3</td><td>Anxiety, attention span deficit, struggling with simple mathematic skills, changed sleep pattern, unaware of disabilities.</td></tr>
<tr><td>Grade II</td><td rowspan="3">Overt HE</td><td>Grade II</td><td>1-6</td><td>Lethargy, not oriented to time, personality changes, bizarre behvior, dyspraxia, asterixis.</td></tr>
<tr><td>Grade III</td><td>Grade III</td><td>3-6</td><td>Somnolence, semistupor, responsive to stimuli, increasing confusion and disorientation.</td></tr>
<tr><td>Grade IV</td><td>Grade IV</td><td>9</td><td>Coma</td></tr>
</table>

The data are reported as mean±SD or median (and interquartile range) as appropriate. HRS: hepatorenal syndrome; MAP: mean arterial pressure: NA, not available.

PATHOGENESIS AND PATHOPHYSIOLOGY OF HEPATIC ENCEPHALOPATHY

Regarding the pathophysiology of HE, the ammoniacal theory is the most supported. It is well known that hepatic dysfunction causes elevated ammonia concentrations and reduced urea metabolism, which permits ammonia to build up in the bloodstream and brain. The findings of research on neuropathology, spectroscopy, and neurochemistry continue to support the hypothesis that ammonia plays a significant role in the pathophysiology HE under both acute and chronic liver failure (Butterworth, 2002). Astrocytes are the most abundant cells of the CNS and are the cell type which is most commonly affected in patients with HE owing to the exclusive localization of glutamine synthase (GS) within the CNS to astrocytes (Martinez-Hernandez et al., 1977; Michael D Norenberg & Martinez-Hernandez, 1979). Damage to astrocytes in HE is characterized by either cell swelling (following ALF) or Alzheimer Type II astrocytosis (AT2A) (following CLD). According to reports, patients with ALF have brain-blood ammonia concentration ratios (usually on the order of 2) that can increase by as much as fourfold. Additionally, arterial blood ammonia concentrations are a reliable indicators of cerebral herniation in these patients (Butterworth, 2002).

Although neurotoxicity secondary to hyperammonemia remains central in the development of HE, several novel pathogenetic mechanisms have been discovered in the past ten years. For example, worse outcomes have been linked to systemic inflammation, increased levels of pro-inflammatory cytokines, and activated microglia in both patient and animal models of HE (fig.4) (Jaeger et al., 2019) as indicative of neuroinflammation (Butterworth, 2019). Indeed, microglial cell activation has been reported in neuropathology investigations. This leads to a proinflammatory state that attracts monocytes and elevates the production of cytokine in the brain (Butterworth, 2013). Moreover, other gut-derived toxins have been proposed. These include benzodiazepine like substances, neurotoxic short- and medium-chain fatty acids, phenols, and mercaptans and manganese, among others (fig. 4). These toxins are well known for their negative effects on the brain and these might work synergistically leading to the development of HE (Butterworth, 2008).

Following liver disease, there is an increase of circulating bile acids, ammonia, manganese, and pro-inflammatory markers in the bloodstream, among others. These variables are elevated systemically, which raises their concentrations in the brain and causes oxidative and metabolic stress in addition to neuroinflammation, all of which promote the development of HE.

HEPATIC ENCEPHALOPATHY AS ASTROGLIOPATHY: ASTROCYTE SWELLING AND ALZHEIMER TYPE 2 ASTROCYTES

Three classic astrocyte responses to CNS damage include cell swelling, reactive astrocytosis (astrogliosis), and AT2A. Astrocytes swelling is an acute change resulting from impaired cell volume regulation. While, reactive astrocytosis is the response of astrocytes to destructive injury which may benefit the injured nervous system by participating in diverse biological processes (Eddleston & Mucke, 1993), whereas the AT2A is typically observed following metabolic/physiologic disorders and early phase of ischemia, trauma, and other acute injuries. However, AT2A is primarily detected in chronic HE.

Figure 4. Pathogenesis of HE

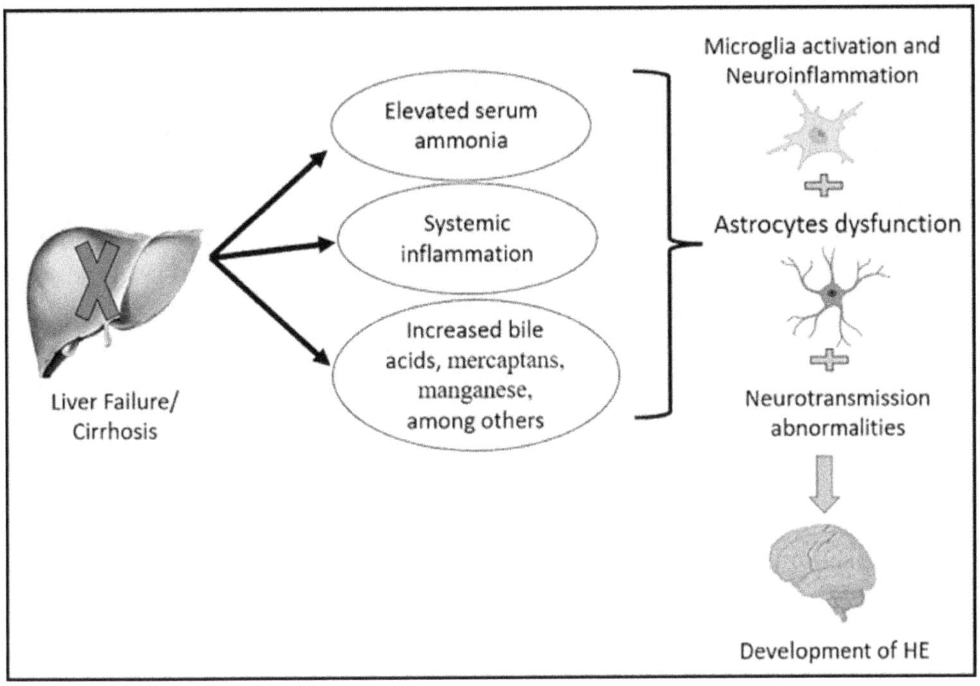

Astrocytes Swelling and Brain Edema

In ALF, brain edema is the principal pathologic feature. In states of hyperammonemia, ammonia detoxification within astrocytes leads to an intracellular accumulation of glutamine (Aldridge et al., 2015). When too much glutamine is produced by astrocytes, it is taken up by mitochondria and broken down into ammonia and glutamate by phosphate-activated glutaminase (PAG). Ammonia accumulation in the mitochondria can lead to oxidative stress and consequently, astrocyte swelling (Aldridge et al., 2015), which ultimately, leads to cerebral edema (Blei et al., 1994; Michael D Norenberg & Martinez-Hernandez, 1979; Swaminathan et al., 2018). Cerebral edema leads to the development of intracranial hypertension (ICH), and consequently brain herniation which is a leading cause of death in ALF.

A variety of studies have demonstrated low-grade cerebral edema in cirrhotic patients via indirect measurements of water content using MRI techniques (Aggarwal et al., 1994; Poveda et al., 2010). Indeed, low-grade astrocyte swelling was reported following CLD (Häussinger et al., 2000), which, even in the absence of clinically obvious ICH, may have important functional repercussions. It has also been proposed that cerebral function may be altered by impairment of the cross-talk between enlarged astrocytes and neurons (Kimelberg, 2005). Another evidence in support of the existence of both vasogenic and cytotoxic edema following CLD, was reported by Chavarria et al. (2011) in cirrhotic patients waiting for liver transplantation (Chavarria et al., 2011).

Alzheimer Type 2 Astrocytes

The pathomorphology of the brain damage in chronic HE indicated that the main histopathologic hallmark is the detection of AT2A in some brain regions (T. V. A. Shulyatnikova & Tumanskyi, 2023). AT2A are characterized by a large, pale nuclei, margination of chromatin patterns, and often, prominent nucleoli than in resting or reactive fibrillary or gemistocytic astrocytes (Fig. 5A), and scarce cellular processes. Oftentimes, AT2A form pairs—doublets or even triplets (Fig. 5B) (Gelpi et al., 2020). AT2A also manifested alterations in the expression of key astrocytic proteins such glial fibrillary acidic protein (GFAP), glutamate transporters, and "peripheral-type"(mitochondrial) benzodiazepine receptors (PTBR) (Butterworth, 2002). AT2A was first noted by von Hosslin and Alzheimer in 1912. They reported finding two aberrant astrocyte types in a patient with Westphal-Strumpell pseudosclerosis (also known as hepatolenticular degeneration or Wilson disease), one of which the authors of subsequent studies referred to as "Alzheimer type II astrocytes" (Agarwal & Mais, 2020). It was first noted in 1914 and subsequently supported by other publications that AT2A is present in a range of acquired non-Wilsonian forms of liver disease (Agarwal & Mais, 2020). The significance of astroglial changes in the pathogenesis of HE was then ultimately highlighted by Adams and Foley's famous clinicopathologic investigation in 1953 (Michael D Norenberg, 1987b). In 1987, Dr. Norenberg suggested that the degree of astroglia dysfunction is typically correlates well with the severity of the clinical symptoms of HE (Norenberg, 1987).

Figure 5. (a) Micrographs showing AT2A have the characteristic enlarged cell nuclei and clear chromatin; (b) AT2A forming a triplet of nuclei (black arrow). Scale bar = 20 μm
(Gelpi et al., 2020)

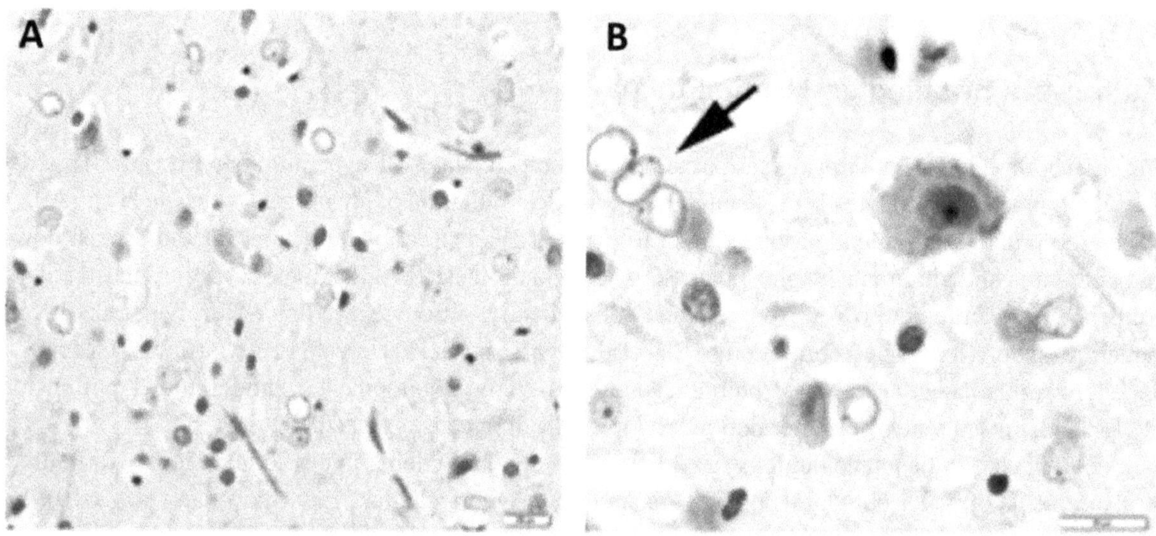

Ammonia and Alzheimer Type 2 Astrocytes

Following CLD, astrocytes exhibit the characteristic morphological features of AT2A (Coltart et al., 2013), and this is thought to be the cause of neurologic dysfunction in HE (Agarwal & Mais, 2020).

Several molecules including high ammonia level (hyperammonemia) are believed to contribute to development of AT2A. As far as is currently known, the process starts with intestinal and bacterial enzymes breaking down intestinal glutamine and urea into ammonia. This enters the portal circulation and, in the normal processes, the liver converts it to urea. In the context of liver disease, blood either bypasses the liver or passes through a dysfunctional liver; ammonia may be partially or incompletely converted to urea (or not converted at all in the case of hepatocytes bypassed by portosystemic shunts), which results in hyperammonemia (Blei et al., 1994; Michael D Norenberg & Martinez-Hernandez, 1979; K V Rama Rao et al., 2005; Swaminathan et al., 2018). At higher levels (hyperammonemia), ammonia crosses the BBB reaching the brain tissue. In the brain, there is an ineffective urea cycle that relies upon astrocyte glutamine synthetase for ammonia detoxification, thus astrocytes play a neuroprotective role in hyperammonemia.

Experimental evidence demonstrates that AT2A can be induced by ammonia in vitro, and observational studies confirm that AT2A are observed in animals and human patients with HE (Butterworth, 2010; Frederick, 2011; Michael D Norenberg, 1987a). Indeed, AT2A can be induced in laboratory animals when hepatic failure or bypass is induced, and astrocyte cultures, when exposed to excess ammonia. Otherwise, it should be noted that another important factor that supports the vital role of ammonia is that any condition that causes elevated brain or blood ammonia level results in the formation of AT2A. Indeed, in addition to different experimental animal models of hyperammonemia, AT2A is observed in the brains of individuals with congenital abnormalities of the urea cycle enzymes that result in persistent hyperammonemia (Gibson et al., 1974; M D Norenberg, 1977), and in astrocyte cultures chronically exposed to hyperammonemia (Aldridge et al., 2015). Additionally, bile duct ligated rats (BDL; model of type C HE) fed with an ammoniagenic diet, cerebral ammonia levels increase with evidence of the development of AT2A (Aldridge et al., 2015). The appearance of AT2A, therefore, is thought to be primarily the result of induction stimulated by the presence of ammonia (Butterworth, 2010; Frederick, 2011). Additionally, it has been shown that with liver cirrhosis progression, AT2A exhibited in the thalamus, the cerebellum, the striatum and the cerebral cortex, which was positively correlated with histochemical ammonia expression in these brain regions (Shulyatnikova & Tumanskyi, 2023).

Manganese and Alzheimer Type 2 Astrocytes

Despite the fact that ammonia plays an important role in the development of AT2A, manganese may also play an important role. According to studies, astrocytes have a high affinity, high capacity, and unique transport system for accumulating manganese. One important effect of manganese exposure is the development of AT2A in both humans and primates (Chandra, 1972; Pentschew et al., 1963). Hazell et al. (2006) have performed an interesting study in which they reported AT2A in both cortical and sub-cortical structures in rat treated with manganese chloride during four days (Hazell et al., 2006). Histological assessment of brains from rats treated with manganese for 4 days revealed the presence of cells exhibiting nuclear pallor, chromatin margination, and swollen (white) processes (Fig. 6) (Hazell et al., 2006). However, co-treatment with the manganese chelator 1,2-cyclohexylenedinitrilotetraacetic acid or the antioxidant N-acetylcysteine totally stopped this pathology, suggesting that the oxidative stress brought on by this metal may be the mechanism underlying the observed cellular change (AT2A) (Hazell et al., 2006). More importantly, Hazell et al (1999) have reported an upregulation of the PTBR following exposure of cultured astrocytes to manganese (Hazell et al., 1999), and at both the mRNA and protein level were reported in a sub-acute rat model (Hazell et al., 2003). It has also been demonstrated that mice treated with the neurosteroid tetrahydroprogesterone leads to development of AT2A (Michael D Norenberg et al., 1997), suggesting that the observed AT2A may be largely explained by enhanced neurosteroid production brought on by manganese-induced PTBR overexpression. Another potential contributing component to the AT2A could be the mitochondrial permeability transition (MPT), as evidenced by a recent study conducted on cultured astrocytes after manganese treatment. This process can potentially be inhibited through antioxidant treatment (Kakulavarapu V Rama Rao & Norenberg, 2004). This would imply that manganese's effects on mitochondria function through the generation of reactive oxygen species, which is in line with data showing the metal causes oxidative stress in astrocytes (Chen & Liao, 2002; Erikson et al., 2004), consequently, leading to the development of AT2A.

Figure 6. Photomicrographs of cerebral cortex in manganese treated rats and control rats. (A) Representative control rat showing a normal astrocyte (black arrow). (B) Representative 4-day manganese-treated rat showing two astrocytes exhibiting nuclear pallor, chromatin margination, and swollen processes characteristic of AT2A (black arrow). Scale bar = 20 μm
(Hazell et al., 2006)

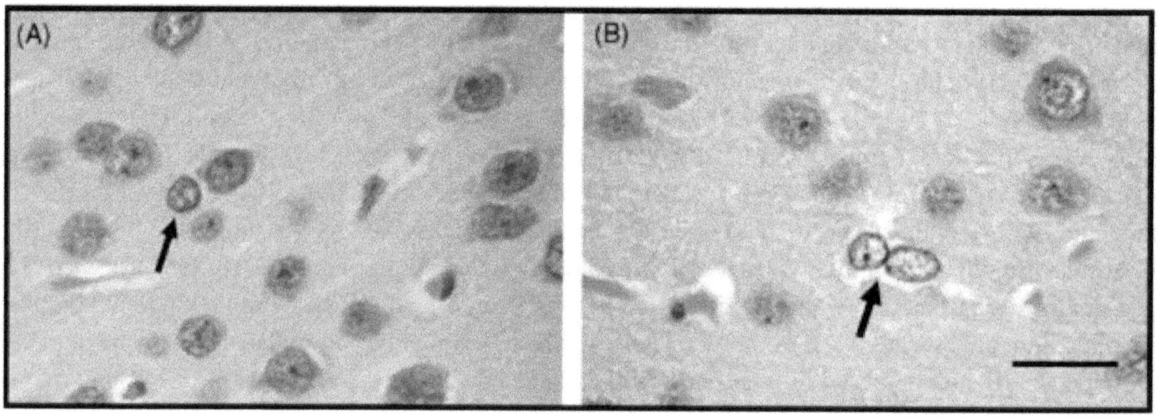

CONCLUSION

HE is wide range of brain abnormalities rising from impaired liver function. This disorder targets especially astrocytes cells leading to astrocytes swelling (following ALF) or AT2A (following CLD), albeit recent studies have shown the involvement of microglial cells and neuronal degeneration. The mechanisms of AT2A are not fully understood, however ammonia and manganese could play a crucial role in the development of this pathology.

REFERENCES

Abdel-Misih, S. R. Z., & Bloomston, M. (2010). Liver anatomy. *Surgical Clinics*, *90*(4), 643–653. PMID:20637938

Agarwal, A. N., & Mais, D. D. (2020). *Sensitivity and Specificity of Alzheimer Type II Astrocytes in Hepatic Encephalopathy*. ARPA. doi:10.5858/arpa.2018-0455-OA

Aggarwal, S., Kramer, D., Yonas, H., Obrist, W., Kang, Y., Martin, M., & Policare, R. (1994). Cerebral hemodynamic and metabolic changes in fulminant hepatic failure: A retrospective study. *Hepatology (Baltimore, Md.)*, *19*(1), 80–87. doi:10.1002/hep.1840190114 PMID:8276371

Aj, S. (1994). Portosystemic encephalopathy after transjugular intrahepatic portosystemic shunt: Results of a prospective controlled study. *Hepatology (Baltimore, Md.)*, *20*, 46–55. PMID:8020904

Aldridge, D. R., Tranah, E. J., & Shawcross, D. L. (2015). Pathogenesis of hepatic encephalopathy: Role of ammonia and systemic inflammation. *Journal of Clinical and Experimental Hepatology*, *5*(S1), S7–S20. doi:10.1016/j.jceh.2014.06.004 PMID:26041962

Ayyildiz, M. (2015). *Effect of Preservation Solution and Period on the Mechanical and Histological Properties of Liver*.

Bajaj, J. S., Saeian, K., Verber, M. D., Hischke, D., Hoffmann, R. G., Franco, J., Varma, R. R., & Rao, S. M. (2007). Inhibitory control test is a simple method to diagnose minimal hepatic encephalopathy and predict development of overt hepatic encephalopathy. *Official Journal of the American College of Gastroenterology| ACG*, *102*(4), 754–760.

Bataller, R., Brenner, D. A., Bataller, R., & Brenner, D. A. (2005). Liver fibrosis Science in medicine Liver fibrosis. *The Journal of Clinical Investigation*, *115*(April), 209–218. doi:10.1172/JCI24282

Berumen, J., Baglieri, J., Kisseleva, T., & Mekeel, K. (2022). LIVER FIBROSIS. *Pathophysiology and Clinical Implications.*, *13*(1), 1–25. doi:10.1002/wsbm.1499.LIVER PMID:32713091

Blei, A. T., Olafsson, S., Therrien, G., & Butterworth, R. F. (1994). Ammonia-induced brain edema and intracranial hypertension in rats after portacaval anastomosis. *Hepatology (Baltimore, Md.)*, *19*(6), 1437–1444. doi:10.1002/hep.1840190619 PMID:8188174

Blumgart, L. H., & Belghiti, J. (2007). *Surgery of the liver, biliary tract, and pancreas* (Vol. 1). Saunders Elsevier Philadelphia.

Bohra, A., Worland, T., Hui, S., Terbah, R., Farrell, A., & Robertson, M. (2020). Prognostic significance of hepatic encephalopathy in patients with cirrhosis treated with current standards of care. *World Journal of Gastroenterology*, *26*(18), 2221–2231. doi:10.3748/wjg.v26.i18.2221 PMID:32476788

Böttcher, K., & Pinzani, M. (2017). Pathophysiology of liver fi brosis and the methodological barriers to the development of anti- fi brogenic agents ☆. *Advanced Drug Delivery Reviews*, *121*, 3–8. doi:10.1016/j.addr.2017.05.016 PMID:28600202

Butterworth, R. F. (2002). Pathophysiology of Hepatic Encephalopathy : A New Look at Ammonia. *Metabolic Brain Disease*, *17*(4), 221–227. doi:10.1023/A:1021989230535 PMID:12602499

Butterworth, R. F. (2008). *The concept of synergism*. *38*(Table 1), 116–121. doi:10.1111/j.1872-034X.2008.00436.x

Butterworth, R. F. (2010). Altered glial–neuronal crosstalk: Cornerstone in the pathogenesis of hepatic encephalopathy. *Neurochemistry International*, *57*(4), 383–388. doi:10.1016/j.neuint.2010.03.012 PMID:20350577

Butterworth, R. F. (2013). The liver–brain axis in liver failure: Neuroinflammation and encephalopathy. *Nature Reviews. Gastroenterology & Hepatology*, *10*(9), 522–528. doi:10.1038/nrgastro.2013.99 PMID:23817325

Butterworth, R. F. (2019). Hepatic Encephalopathy in Cirrhosis: Pathology and Pathophysiology. *Drugs*, *79*(S1, s1), 17–21. doi:10.100740265-018-1017-0 PMID:30706423

Chandra, S. V. (1972). Histological and histochemical changes in experimental manganese encephalopathy in rabbits. *Archiv fur Toxikologie*, *29*(1), 29–38. doi:10.1007/BF00316512 PMID:4261594

Chavarria, L., Alonso, J., Garcia-Martinez, R., Aymerich, F. X., Huerga, E., Jacas, C., Vargas, V., Cordoba, J., & Rovira, A. (2011). Biexponential analysis of diffusion-tensor imaging of the brain in patients with cirrhosis before and after liver transplantation. *AJNR. American Journal of Neuroradiology*, *32*(8), 1510–1517. doi:10.3174/ajnr.A2533 PMID:21700786

Chen, C.-J., & Liao, S.-L. (2002). Oxidative stress involves in astrocytic alterations induced by manganese. *Experimental Neurology*, *175*(1), 216–225. doi:10.1006/exnr.2002.7894 PMID:12009774

Coltart, I., Tranah, T. H., & Shawcross, D. L. (2013). Inflammation and hepatic encephalopathy. *Archives of Biochemistry and Biophysics*, *536*(2), 189–196. doi:10.1016/j.abb.2013.03.016 PMID:23583306

Deutsch-Link, S., Moon, A. M., Jiang, Y., Barritt, A. S. IV, & Tapper, E. B. (2022). Serum ammonia in cirrhosis: Clinical impact of hyperammonemia, utility of testing, and national testing trends. *Clinical Therapeutics*, *44*(3), e45–e57. doi:10.1016/j.clinthera.2022.01.008 PMID:35125217

Dharel, N., & Bajaj, J. S. (2015). Definition and nomenclature of hepatic encephalopathy. *Journal of Clinical and Experimental Hepatology*, *5*, S37–S41. doi:10.1016/j.jceh.2014.10.001 PMID:26041955

Eddleston, M., & Mucke, L. (1993). Molecular profile of reactive astrocytes—Implications for their role in neurologic disease. *Neuroscience*, *54*(1), 15–36. doi:10.1016/0306-4522(93)90380-X PMID:8515840

El Hiba, O., El-Mansoury, B., Chatoui, H., Benammi, H., Balzano, T., Merzouki, M., El Hidan, M. A., Radhakrishnan, J., Bahi, L., & Idrissi, H. R. (2023). Clinical and Experimental Evidence of Cirrhosis-Related Parkinsonism. In *Experimental and Clinical Evidence of the Neuropathology of Parkinson's Disease* (pp. 102–119). IGI Global. doi:10.4018/978-1-6684-5156-4.ch007

El Khiat, A., El-Mansoury, B., El Hiba, O., Tamegart, L., Rais, H., Fdil, N., Sellami, S., El Mokhtar, M. A., Kissani, N., & Gamrani, H. (2023). P14 Altered Locomotor Behavior Associated With Purkinje Cells and Astrocytes Dysfunctions in Rat With Acute Hepatic Encephalopathy Involving the Oxidative Stress and Neuroinflammation. *Official Journal of the American College of Gastroenterology| ACG*, *118*(9S), S11.

El-Mansoury, Bilal, Smimih, K., Hamdan, Y. A., El Khiat, A., Elouali, S., El Fatimi, H., Draoui, A., El Hiba, O., & Jayakumar, A. R. (2023). The Interplay Between Environmental Pollutants, Gut Microbiota, and Infections: Current Concepts and Therapies. *Ecological and Evolutionary Perspectives on Infections and Morbidity*, 173–207.

El-Mansoury, B, Smimih, K., Aimrane, A., Draoui, A., Ferssiwi, A., Bitar, A., Jayakumar, A. R., & El Hiba, O. (2023). P8 Cognitive Impairment Associated With Cortical and Hippocampal Microglia Activation and Altered Synaptic Function in a Rat Model of Chronic Hepatic Encephalopathy. *Official Journal of the American College of Gastroenterology| ACG, 118*(9S), S7.

Elsaid, M. I., & Rustgi, V. K. (2020). Epidemiology of hepatic encephalopathy. *Clinics in Liver Disease, 24*(2), 157–174. doi:10.1016/j.cld.2020.01.001 PMID:32245524

Erikson, K. M., Dobson, A. W., Dorman, D. C., & Aschner, M. (2004). Manganese exposure and induced oxidative stress in the rat brain. *The Science of the Total Environment, 334*, 409–416. doi:10.1016/j.scitotenv.2004.04.044 PMID:15504526

Frederick, R. T. (2011). Current concepts in the pathophysiology and management of hepatic encephalopathy. *Gastroenterology & Hepatology, 7*(4), 222. PMID:21857820

Gelpi, E., Rahimi, J., Klotz, S., Schmid, S., Ricken, G., Forcen-vega, S., Budka, H., & Kovacs, G. G. (2020). The autophagic marker p62 highlights Alzheimer type II astrocytes in metabolic / hepatic encephalopathy. *Neuropathology, 40*(4), 358–366. doi:10.1111/neup.12660 PMID:32483828

Gibson, G. E., Zimber, A., Krook, L., Richardson, E. P. Jr, & Visek, W. J. (1974). Brain histology and behavior of mice injected with urease. *Journal of Neuropathology and Experimental Neurology, 33*(2), 201–211. doi:10.1097/00005072-197404000-00001 PMID:4822569

Hassanein, T. I., Hilsabeck, R. C., & Perry, W. (2008). Introduction to the hepatic encephalopathy scoring algorithm (HESA). *Digestive Diseases and Sciences, 53*(2), 529–538. doi:10.100710620-007-9895-0 PMID:17710551

Häussinger, D., Kircheis, G., Fischer, R., Schliess, F., & vom Dahl, S. (2000). Hepatic encephalopathy in chronic liver disease: A clinical manifestation of astrocyte swelling and low-grade cerebral edema? *Journal of Hepatology, 32*(6), 1035–1038. doi:10.1016/S0168-8278(00)80110-5 PMID:10898326

Hazell, A. S., Desjardins, P., & Butterworth, R. F. (1999). Chronic exposure of rat primary astrocyte cultures to manganese results in increased binding sites for the 'peripheral-type'benzodiazepine receptor ligand 3H-PK 11195. *Neuroscience Letters, 271*(1), 5–8. doi:10.1016/S0304-3940(99)00489-9 PMID:10471200

Hazell, A. S., Normandin, L., Nguyen, B., & Kennedy, G. (2003). Upregulation of 'peripheral-type'benzodiazepine receptors in the globus pallidus in a sub-acute rat model of manganese neurotoxicity. *Neuroscience Letters, 349*(1), 13–16. doi:10.1016/S0304-3940(03)00649-9 PMID:12946575

Hazell, A. S., Normandin, L., Norenberg, M. D., Kennedy, G., & Yi, J.-H. (2006). Alzheimer type II astrocytic changes following sub-acute exposure to manganese and its prevention by antioxidant treatment. *Neuroscience Letters, 396*(3), 167–171. doi:10.1016/j.neulet.2005.11.064 PMID:16384640

Jaeger, V., Demorrow, S., & Mcmillin, M. (2019). The Direct Contribution of Astrocytes and Microglia to the Pathogenesis of Hepatic Encephalopathy. *Journal of Clinical and Translational Hepatology*, *7*(X), 352–361. doi:10.14218/JCTH.2019.00025 PMID:31915605

Juza, R. M., & Pauli, E. M. (2014). Clinical and surgical anatomy of the liver: A review for clinicians. *Clinical Anatomy (New York, N.Y.)*, *27*(5), 764–769. doi:10.1002/ca.22350 PMID:24453062

Kabaria, S., Dalal, I., Gupta, K., & Bhurwal, A. (2003, July). Carlos, Minacapelli, D., Catalano, C., & Rustgi, V. (2021). Hepatic encephalopathy: A review. *Annals of Hepatology*, *2*(3), 122–130. Advance online publication. doi:10.1016/S1665-2681(19)32137-4

Kimelberg, H. K. (2005). Astrocytic swelling in cerebral ischemia as a possible cause of injury and target for therapy. *Glia*, *50*(4), 389–397. doi:10.1002/glia.20174 PMID:15846797

Martinez-Hernandez, A., Bell, K. P., & Norenberg, M. D. (1977). Glutamine synthetase: Glial localization in brain. *Science*, *195*(4284), 1356–1358. doi:10.1126cience.14400 PMID:14400

Mulaikal, T. A., & Emond, J. C. (2012). *Liver Anesthesiology and Critical Care Medicine*. Liver Anesthesiology and Critical Care Medicine., doi:10.1007/978-1-4614-5167-9

Mumtaz, K., Ahmed, U. S., Abid, S., Baig, N., Hamid, S., & Jafri, W. (2010). Precipitating factors and the outcome of hepatic encephalopathy in liver cirrhosis. *Journal of the College of Physicians and Surgeons—Pakistan*, *20*(8), 514. PMID:20688015

Norenberg, M. D. (1977). A light and electron microscopic study of experimental portal-systemic (ammonia) encephalopathy. Progression and reversal of the disorder. *Laboratory Investigation*, *36*(6), 618–627. https://europepmc.org/abstract/MED/559221 PMID:559221

Norenberg, M. D. (1987a). The role of astrocytes in hepatic encephalopathy. *Neurochemical Pathology*, *6*(1-2), 13–33. doi:10.1007/BF02833599 PMID:3306480

Norenberg, Michael D. (1987b). The Role of Astrocytes in Hepatic Encephalopathy. *Neurochernical Pathology, 6*.

Norenberg, Michael D, Itzhak, Y., & Bender, A. S. (1997). The peripheral benzodiazepine receptor and neurosteroids in hepatic encephalopathy. *Advances in Cirrhosis, Hyperammonemia, and Hepatic Encephalopathy*, 95–111.

Norenberg, M. D., & Martinez-Hernandez, A. (1979). Fine structural localization of glutamine synthetase in astrocytes of rat brain. *Brain Research*, *161*(2), 303–310. doi:10.1016/0006-8993(79)90071-4 PMID:31966

Ortiz, M., Cordoba, J., Doval, E., Jacas, C., Pujadas, F., Esteban, R., & Guardia, J. (2007). Development of a clinical hepatic encephalopathy staging scale. *Alimentary Pharmacology & Therapeutics*, *26*(6), 859–867. doi:10.1111/j.1365-2036.2007.03394.x PMID:17767470

Ortiz, M., Jacas, C., & Córdoba, J. (2005). Minimal hepatic encephalopathy: Diagnosis, clinical significance and recommendations. *Journal of Hepatology*, *42*(1), S45–S53. doi:10.1016/j.jhep.2004.11.028 PMID:15777572

Pentschew, A., Ebner, F. F., & Kovatch, R. M. (1963). Experimental manganese encephalopathy in monkeys: A preliminary report. *Journal of Neuropathology and Experimental Neurology*, *22*(3), 488–499. doi:10.1097/00005072-196307000-00010 PMID:14045007

Pinzani, M. (2015). *Pathophysiology of Liver Fibrosis*. 492–497. doi:10.1159/000374096

Poveda, M.-J., Bernabeu, Á., Concepción, L., Roa, E., de Madaria, E., Zapater, P., Pérez-Mateo, M., & Jover, R. (2010). Brain edema dynamics in patients with overt hepatic encephalopathy: A magnetic resonance imaging study. *NeuroImage*, *52*(2), 481–487. doi:10.1016/j.neuroimage.2010.04.260 PMID:20451628

Prasad, S., Dhiman, R. K., Duseja, A., Chawla, Y. K., Sharma, A., & Agarwal, R. (2007). Lactulose improves cognitive functions and health-related quality of life in patients with cirrhosis who have minimal hepatic encephalopathy. *Hepatology (Baltimore, Md.)*, *45*(3), 549–559. doi:10.1002/hep.21533 PMID:17326150

Rao, K. V., Panickar, K. S., Jayakumar, A. R., & Norenberg, M. D. (2005). Astrocytes protect neurons from ammonia toxicity. *Neurochemical Research*, *30*(10), 1311–1318. doi:10.100711064-005-8803-2 PMID:16341593

Rao, Kakulavarapu V Rama, & Norenberg, M. D. (2004). Manganese induces the mitochondrial permeability transition in cultured astrocytes. *Journal of Biological Chemistry, 279*(31), 32333–32338.

Reiner, K. S. & C. S. (2021). *Diffuse Liver Disease: Cirrhosis, Focal Lesions in Cirrhosis, and Vascular Liver Disease*. Springer. doi:10.1007/978-3-319-75019-4

Sharma, A., & Nagalli, S. (2022). Chronic liver disease. In *StatPearls* [Internet]. StatPearls Publishing.

Shulyatnikova, T. V., & Shavrin, V. A. (2017). Modern view on hepatic encephalopathy: Basic terms and concepts of pathogenesis. *Патологія*, *14*(3), 371–380. doi:10.14739/2310-1237.2017.3.118773

Shulyatnikova, T. V. A., & Tumanskyi, V. O. (2023). Ammonia level and Alzheimer type 2 astrocytes in the brain of deceased patients with liver cirrhosis of the varying degree. *Pathologia*, *20*(1), 36–44. doi:10.14739/2310-1237.2023.1.276453

Si-Tayeb, K., Lemaigre, F. P., & Duncan, S. A. (2010). Organogenesis and development of the liver. *Developmental Cell*, *18*(2), 175–189. doi:10.1016/j.devcel.2010.01.011 PMID:20159590

Swaminathan, M., Ellul, M. A., & Cross, T. J. S. (2018). Hepatic encephalopathy: Current challenges and future prospects. *Hepatic Medicine : Evidence and Research*, *10*, 1–11. doi:10.2147/HMER.S118964 PMID:29606895

Tryc, A. B., Goldbecker, A., Berding, G., Rümke, S., Afshar, K., Shahrezaei, G. H., Pflugrad, H., Barg-Hock, H., Strassburg, C. P., Hecker, H., & Weissenborn, K. (2013). Cirrhosis-related Parkinsonism: Prevalence, mechanisms and response to treatments. *Journal of Hepatology*, *58*(4), 698–705. doi:10.1016/j.jhep.2012.11.043 PMID:23220368

Tsukada, S., Parsons, C. J., & Rippe, R. A. (2006). *Mechanisms of liver fibrosis B*. Elsevier. doi:10.1016/j.cca.2005.06.014

Vilstrup, H., Amodio, P., Bajaj, J., Cordoba, J., Ferenci, P., Mullen, K. D., Weissenborn, K., & Wong, P. (2014). Hepatic encephalopathy in chronic liver disease: 2014 Practice Guideline by the American Association for the Study of Liver Diseases and the European Association for the Study of the Liver. *Hepatology (Baltimore, Md.)*, *60*(2), 715–735. doi:10.1002/hep.27210 PMID:25042402

Wallace, K., Burt, A. D., & Wright, M. C. (2008). Liver fibrosis. *Biochemical Journal*. doi:10.1042/BJ20071570

Weissenborn, K. (1998). Diagnosis of encephalopathy. *Digestion*, *59*(Suppl. 2), 22–24. doi:10.1159/000051415 PMID:9718414

Weissenborn, K., Bokemeyer, M., Krause, J., Ennen, J., & Ahl, B. (2005). Neurological and neuropsychiatric syndromes associated with liver disease. *AIDS (London, England)*, *19*(Suppl 3), S93–S98. doi:10.1097/01.aids.0000192076.03443.6d PMID:16251835

Wiltfang, J., Nolte, W., Weissenborn, K., Kornhuber, J., & Rüther, E. (1998). Psychiatric aspects of portal-systemic encephalopathy. *Metabolic Brain Disease*, *13*(4), 379–389. doi:10.1023/A:1020697127718 PMID:10206828

Chapter 13
Astroglia and Acute Hepatic Encephalopathy:
The Mechanisms of Cytotoxic Brain Edema

Omar El Hiba
Faculty of sciences, Chouaib Doukkali University, Morocco

Bilal El-Mansoury
https://orcid.org/0000-0002-2583-7515
Faculty of Sciences, Chouaib Doukkali University, Morocco

Michael Norenberg
Miller School of Medicine, University of Miami, USA

Arumugam Jayakumar
Miller School of Medicine, University of Miami, USA

ABSTRACT

Acute hepatic encephalopathy (AHE) is a fatal neuro-psychological complication associated with acute liver failure (ALF). While different etiologies were established, over dosage of acetaminophen consumption is still the most frequent. The neuropathological basis of AHE induced death is still not fully understood; however, a body line of evidence sustains the role of rapid onset and progress of brain edema leading to intracranial hypertension, and finally brain herniation and death. The role of astrocytes in the pathogenesis of HE, in general, is well sustained through their role as a detoxifying component of the central nervous system, especially from peripheral generated ammonia, leading to astrocyte swelling, generally associated to brain edema onset. The current chapter will describe the pathomechanisms underlying astrocyte swelling in AHE, with the eventual evolution into macroscopic brain edema and the deathly herniation in humans as well as animal models of AHE.

DOI: 10.4018/978-1-6684-9675-6.ch013

INTRODUCTION

Hepatic encephalopathy (HE) is a wide range of neuropsychiatric abnormalities resulted from impaired liver function (El-Mansoury et al. 2023; El Hiba et al. 2023; El Khiat et al. 2023). Acute liver failure (ALF) is defined as the pathological condition that refers to severe liver dysfunction which evolves rapidly leading to the onset of neurological complications commonly called acute hepatic encephalopathy (AHE), with very poor prognostic and which generally associated to high risk of coma and mortality, reaching up to 80%, while it represents approximately 0.1% of total death cases and 6% of liver related deaths. Otherwise, clinical observations in patients with ALF, support the presence of extended organs dysfunctions including kidneys, lungs, hearth and the circulatory system (Lee 2022).

AHE is well known as the common neuropsychiatric syndrome occurring in patients with ALF, which includes disturbances in consciousness, personality and intellectual capacity. Such neuro-clinical features are generally associated with severe biochemical abnormalities such as high blood ammonia levels, altered renal and hepatic markers, and peripheral inflammation.

The high mortality risk in AHE has been associated particularly to increased intracranial pressure (ICP) resulting in brain herniation. The pathophysiology of ICP may involve the development of brain edema which refers to excessive accumulation of fluid into the intracellular (cytotoxic edema) or extracellular (vasogenic edema) spaces of the brain parenchyma.

Particularly, cytotoxic edema involves astrocytic swelling, which is the most important histological hallmark observed in the brain tissue of patients with ALF (Kato et al. 1992) as well as in animal models of AHE (El Khiat et al. 2022a). While astrocyte swelling has been associated to several factors mainly ammonia, neuroinflammation and oxidative stress.

The clinical observation in patients with ALF sustains the evidence of a correlation between the severity of brain edema and blood and brain ammonia levels. While the establishment of such correlation has been confirmed in the in vivo and in vitro animal models of hyperammonemia (Norenberg et al. 2007).

The present chapter provides an updated overview of the pathophysiological mechanisms of astrocytic swelling underlying the development of cytotoxic brain edema in AHE, with special focus on the clinical observations and the experimental data.

BRAIN EDEMA: CLINICAL FEATURES

By definition, cerebral edema refers to increased amount of water within the brain tissue. Knowing that the brain bone is not extensible, water accumulation leads to progressive rise in the intracranial pressure (ICP), and therefore, the onset of intracranial hypertension (ICH). Such abnormality is generally associated to severe neurological complications due to decreased cerebral blood flow, tissue perfusion and ischemia (Larsen 1996).

ICH has been associated to acute forms of hepatic failures since the first description by Ware et al., in 1971(Ware, D'Agostino, and Combes 1971). Such neurological complication was found in up to 80% patients with ALF at the coma stage (fig. 1, fig.2), while studies on animal models of ALF appeared to well mimic such clinical feature.

Figure 1. (A) Rapid brain edema onset in a patient with acute liver failure; (B) The same patient with normal TC scanner at the admission eight hours earlier
(Vaquero, Chung, and Blei 2003)

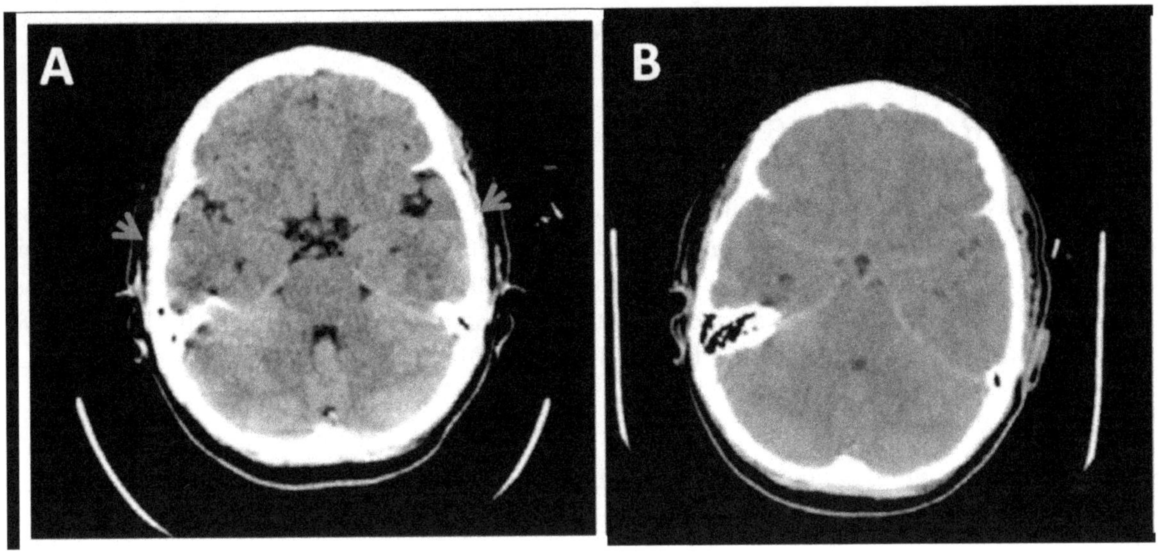

Figure 2. Association between brain volume and intracranial pressure (ICP). Increased brain volume due to edema leads to increased ICP. Even at the same range of volume increase (A and B), the lower brain compliance results in higher increases of pressure in B. In the acute case, the curve is steeper due to a lesser capacity of compensation.
(Vaquero et al. 2003)

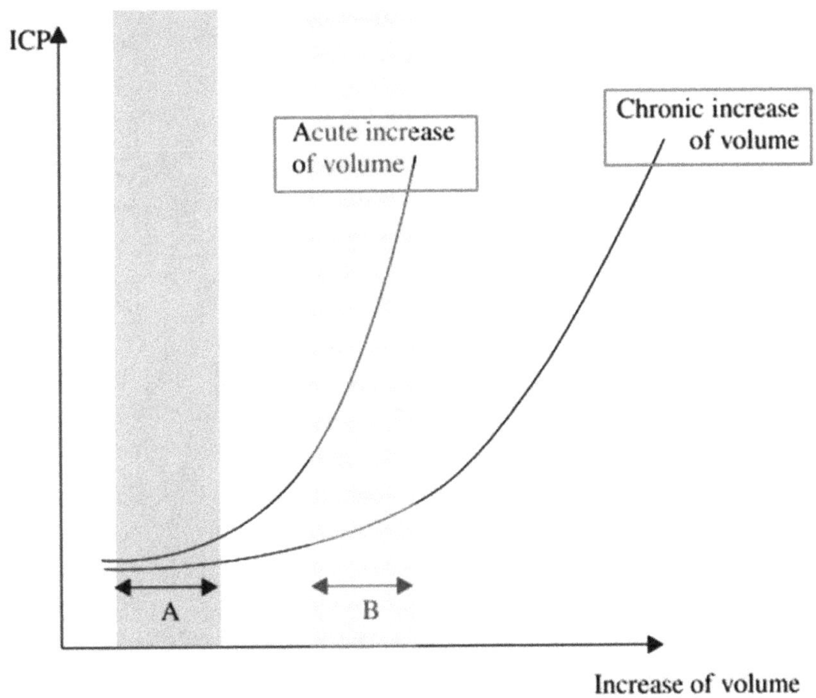

Otherwise, impaired autoregulation of cerebral blood flow which is known to be well documented in HE patients with the acute form may result from a possible vasodilatation of cerebral arteries due to increased intracranial blood volume also known as cerebral hyperemia or the so-called luxury perfusion (Larsen et al. 1996, 2000; Strauss et al. 1998).

TYPES OF BRAIN EDEMAS

Traditionally, two types of brain edema have been proposed with two different mechanisms of onset including the vasogenic and cytotoxic types. The first refers to abnormal accumulation of fluid within the extracellular space of the cerebral parenchyma as a result of altered blood brain barrier (BBB) permeability. Whereas, the cytotoxic type is generally featured by intracellular accumulation of fluid inducing cell swelling (fig. 3)(Simard et al. 2007; Thrane, Thrane, and Nedergaard 2014).

Figure 3. Schematic representation of the mechanisms of brain edemas development

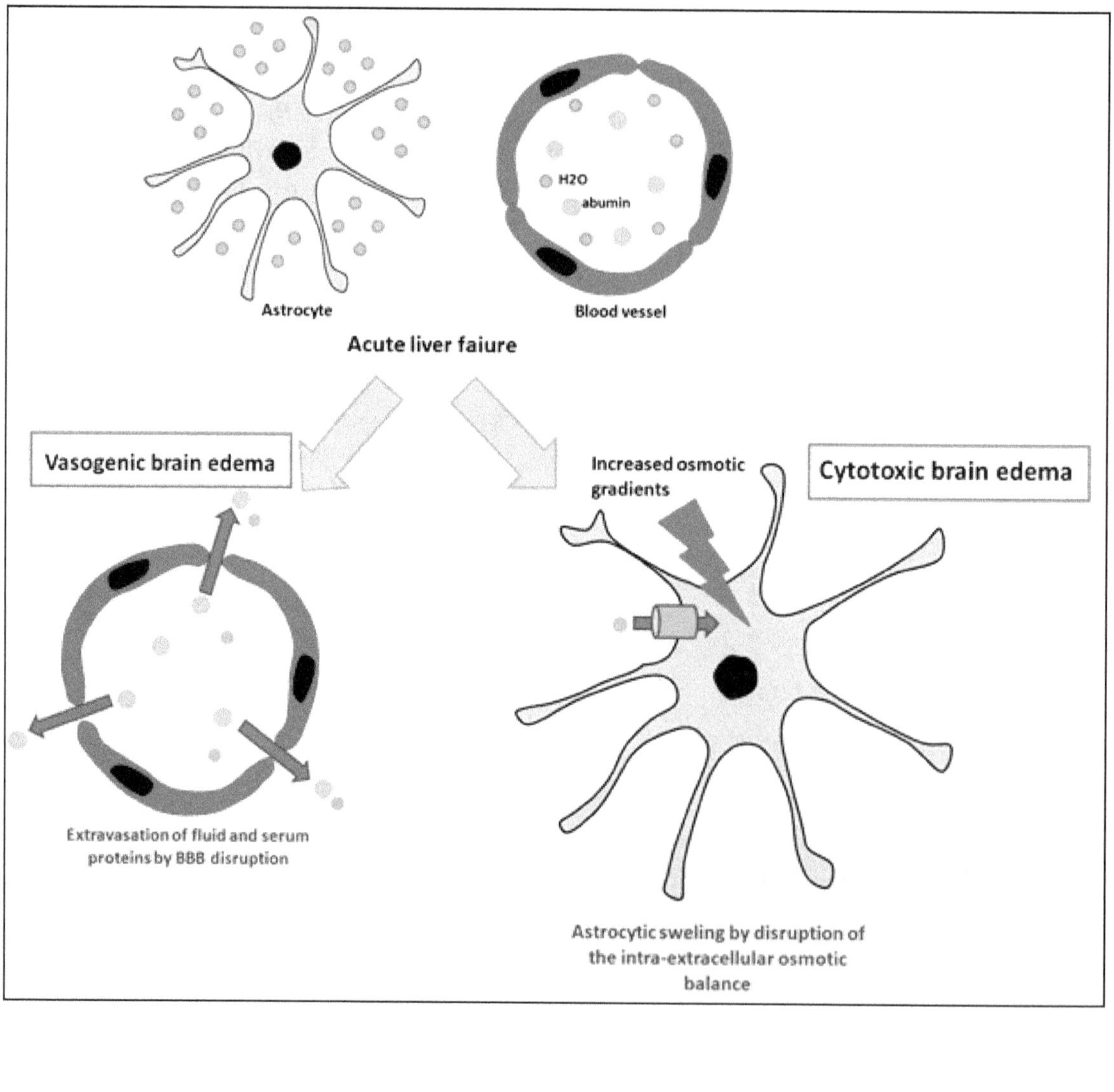

In the case of vasogenic brain edema, following acute liver injury induced AHE, tight junctions of the endothelial cells are disrupted as a result of several factors of central and peripheral origins (inflammatory cytokines, oxidative stress…) as well, activated glial cells release vascular permeability and inflammatory factors leading to increased BBB permeability which in result causes extravasation of fluids and albumin and there accumulation into the extracellular space of the brain parenchyma. On the other side, cytotoxic brain edema changes the cytosolic osmolarity disrupt the osmotic gradients leading to water entry through specific water channels (Aquaporins) and therefore, increased cell volume and astrocytic swelling.

MECHANISMS INVOLVED IN THE CYTOTOXIC BRAIN EDEMA IN ACUTE HEPATIC ENCEPHALOPATHY

Cytotoxic edema refers to a net build-up of brain water that occurs in the cellular compartments of the brain. Cytotoxic brain edema represents the major neuropathological finding following ALF (Rama Rao et al. 2014). Astrocyte swelling and modification of astrocytic ramifications were reported in several experimental studies (fig.4) (El Khiat et al. 2022b), as well as in patients with AHE and cultured astrocytes. It is believed that ammonia, inflammation/infection, and hyponatremia are the three primary circulating components involved in the development of cytotoxic brain edema (Blei 2008).Because only astrocytes in the brain detoxify ammonia, it has been hypothesized that glutamine builds up in astrocytes (the glutamine hypothesis), microglia activation and the subsequent neuroinflammation as well as oxidative/nitrosative stress and mitochondria abnormalities in astrocytes, all play a significant role in the development of astrocytes swelling and the subsequent brain edema following acute HE. More recently, ion channels and transporters dysfunctions are reported to have a critical role in the development of astrocyte swelling/brain edema.

Ion Channels and Transporters Deregulations

Additionally, cell volume is known to be regulated by ion channels and transporters; a disruption in these systems may cause cell swelling. Indeed, recently, Jayakumar and colleagues reported that ammonia, a major etiological factor in ALF, activated the Na–K–Cl cotransporter-1 (NKCC1) in cultured astrocytes, and that blocking this activation reduced astrocyte swelling. These findings imply that NKCC1 activation plays a significant role in the mechanism underlying brain edema in ALF (Jayakumar et al., 2011). Moreover, other investigations reported upregulation of ATP-dependent, nonselective cation (NCCa-ATP) channel in cultured astrocytes treated with ammonia (5 mM) using the sulfonylurea receptor type 1 (SUR1) protein expression (a marker of NCCa-ATP channel activity) (Jayakumar et al., 2014). While, blocking SUR1 activity reduced ammonia-induced cell swelling (Jayakumar et al., 2014). More importantly, the authors showed that the observed increase of SUR1 expression in cultured astrocytes with ammonia was decreased when co-treating theses astrocytes cultures with aNF-κB inhibitor; BAY 11–7082(Jayakumar et al., 2014). Suggesting that ammonia induced overexpression of SUR1 leading to cell swelling might be mediated through the NF-κB pathway.

Figure 4. Photomicrographs of typical astrocytes in the cerebellar granular layer immunolabelled with GFAP showing modifications of the astrocytic form in rats treated with TAA in comparison with controls.A: CTR, B: TAA12h (12 hours following TAA administration); C: 24 h TAA (24 hours following TAA administration), D; TAA36h (36 hours following TAA administration)
(El Khiat et al. 2022b)

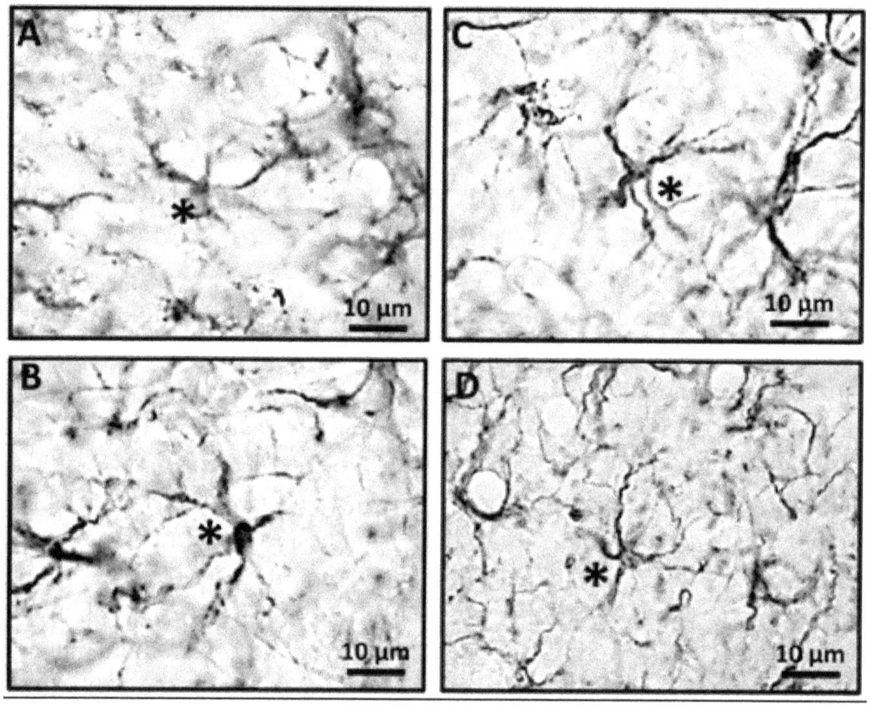

Furthermore, in rat model of ALF induced by thioacetamide (TAA), the expression of aquaporin-4 (AQP4) in perivascular astrocytic end-feet was increased (Rama Rao et al., 2010), indicating that such increase could be critical for the development of brain edema in ALF. Additionally, Rao et al. (2014) reported increased brain water content in different models (wild type mice) of ALF induced by TAA or acetaminophen (APAP), while in AQP4-null mice the brain water content did not undergo any changes (Rao et al. 2014). Moreover, it has been demonstrated that AQP4 gene suppression in cultured astrocytes inhibits ammonia-induced cell swelling (Rama Rao et al., 2010). Additionally, Eefsen et al. (2010) reported an upregulation of AQP4 in mice with ALF (Eefsen et al. 2010). More recently, rat model of ALF induced by APAP administration, the immunohistochemical study of AQP4 expression showed a significant increase of this protein in the cerebral cortex, hippocampus and striatum (Shulyatnikova and Tumanskiy 2022), supporting the role of AQP4 in the brain edema associated with ALF.

Ammonia and Glutamine Hypothesis

The glutamine hypothesis is predicated on the observation that the brain converts ammonia through detoxification into glutamine, whose osmotic actions in astrocytes may be responsible for the onset of brain edema (Blei and Larsen 2001). In the brain, ammonia detoxification takes place in astrocytes. After all, the brain does not have a full urea cycle, thus ammonia is removed by the amidation of glutamate to generate glutamine, which is facilitated by glutamine synthetase (GS), an enzyme that is only present in astrocytes (Martinez-Hernandez et al., 1977; Norenberg, 1987). Thus, the glutamine hypothesis sustains that the accumulation of glutamine during the generation of acute hyperammonemia causes astrocytes swelling (Blei and Larsen 2001). Indeed, both experimental models and humans with ALF are associated with a significant increase in brain glutamine concentrations, which are produced in astrocytes through the amidation of glutamate by ammonia (McConnell et al. 1995; Swain, Butterworth, and Blei 1992). Glutamine has the ability to act as an organic osmolyte; its intracellular concentration can vary greatly in order to preserve cell volume without negatively impacting its structure or function (Strange 2004).Nevertheless, following ALF, the measurement of glutamine levels directly in the brain, reveals significantly rise which often up to sixfold(Blei 2008).It has been observed that rats six weeks post-PCA maintain a stable level of brain organic osmolytes, with increased glutamine while myo-inositol and taurine were decreased (Cordoba, Gottstein, and Blei 1996). Interestingly, in an experimental setting, normal rats' cerebral edema caused by ammonia does not occur when glutamine production is inhibited with methionine-sulfoximine (MSO) (Takahashi et al. 1991). Indeed, further studies confirmed that MSO completely prevents brain edema and ammonia-induced astrocyte swelling(Willard-Mack et al. 1996).

Oxidative/Nitrosative Stress and Mitochondria Abnormalities in Astrocytes

Oxidative stress has been proposed in the reactive nitrogen and oxygen species (RNOS) may be produced as a consequence of astrocyte swelling, which may then exacerbate swelling on its own. The induction of nitric oxide synthase via N-Mythyl-D-asparate glutamate receptors (NMDA), an increase in intracellular calcium, and an increase in the expression of peripheral benozodiazepine receptors are among the mechanisms for the synthesis of RNOS that have been examined in isolated glial cells (Butterworth 2000). While astrocytes have been recognized as a major source of RNOS, additional research has revealed that endothelial cells, fibroblasts, neurons, and microglia are also involved in the synthesis of RNOS in HE (Jayakumar et al. 2012; Rao, Brahmbhatt, and Norenberg 2013; Yus et al. 1974; Zemtsova et al. 2011). Noteworthy, glutamine itself has been linked in experiments on isolated glial cells to the production of RNOS by its accumulation in mitochondria and consequent activation of the mitochondrial permeability transition (Norenberg 2003).Briefly, glutaminase (GLS) I converts glutamine that has accumulated in the cytoplasm into glutamate when it enters the mitochondria. Glutamate dehydrogenase (GDH) then oxidatively deaminates glutamate to α-ketoglutarate (αKG), which supplies energy and fuel for the tricarboxylic acid cycle (TCA). Free ammonia is released in both phases. Ammonia in the mitochondria changes the pH of the mitochondria and prevents the respiratory chain and other mitochondrial enzymes and biochemical activities, which increases the production of RNOS (Murthy et al. 2001). Thus, in oxidative/nitrosative stress changes cause alteration in mitochondrial permeability(Bai et al. 2001), which is accompanied by mitochondrial fracture and swelling (Drews et al. 2020; Görg et al. 2015). Changes in mitochondrial activity impact astrocyte function, leading to cytotoxic brain edema (Oksanen et al. 2019).

Furthermore, variations in the expression of osmotic transporter proteins also indicate the relationship between astrocyte swelling and oxidative/nitrosative stress (Oenarto et al. 2014). Studies undertaken in vitro on rat astrocytes have demonstrated that downregulations in the mRNA levels of taurine transporter protein (TAUT), and sodium-dependent inositol transporter protein (SMIT) are linked to swelling caused by HE-triggered stimuli (Fu et al., 2012; Oenarto et al., 2014). Additionally, the cerebral cortex of rats undergoing acute or chronic hyperammonemia treatment has been shown to have downregulated SMIT mRNA (Oenarto et al. 2014).Moreover, Rao et al. reported that ammonia increases cellular water intake in astrocytes by upregulating the water channel protein AQ4 in the plasma membrane, which is reliant on oxidative/nitrosative stress (Elsherbini et al. 2022; Rao et al. 2003).

In the other hand, brain dysfunction may already be exacerbated by reactive oxygen species (ROS) produced extraneously in HE. In spite of the fact that the cellular source of ROS in peripheral blood is still unknown, Bosoi et al. demonstrated that bile duct ligated (BDL) rats had elevated arterial plasma levels of hydrogen peroxide. Removing ROS from peripheral blood prevented brain edema, whereas raising ROS levels induced brain edema(Bosoi et al. 2012; Bosoi, Tremblay, and Rose 2014). These results imply that hyperammonemia brought on by liver disease and ROS that originate outside of the brain may work in combination to generate cerebral osmotic stress (Lu 2023).

Hyponatremia and Brain Edema

It has been reported that hyponatremia is associated with increased HE occurrence and influences mortality after ALF(Cárdenas et al. 2014; Guevara et al. 2009, 2010; Liotta et al. 2017). Hyponatremia refers to a low serum sodium concentration to a level below 136 mmol/l(Córdoba and García-martinez 2010). In clinical practice, it is the most prevalent electrolyte imbalance that is observed. Serum sodium is the primary factor influencing plasma osmolality, which in turn impacts cell volume(Kengne and Decaux 2017).According to reports, a reduction in sodium concentration causes a drop in serum osmotic pressure. As a compensatory mechanism, factors that control osmotic pressure, like the previously mentioned glutamine and myo-inositol, move to the extracellular space, inducing cellular edema (Córdoba and García-martinez 2010; Heins and Zwingmann 2010). Clinically, acute hyponatremia can induce brain edema by itself (Fraser and Arieff 1997). About 25% of subjects in ALF patients had hyponatremia, which is particularly common in those with severe encephalopathy (Ong et al. 2003). Indeed, hyponatremia has also been shown to operate as a HE-aggravating factor, cause astrocyte dysfunction, and the subsequent brain edema (Córdoba and García-martinez 2010; Heins and Zwingmann 2010; Iwasa and Takei 2015). More importantly, the use of hypertonic saline in ALF patients who had an ICP bolt inserted and were thought to be at risk of cerebral edema. With serum sodium levels between 145 and 150 mmol/l, the active treatment group experienced fewer ICH episodes than the control group (Wendon and Lee 2008).Furthermore, in a rat model of ammonia-induced brain edema, hyponatremia exacerbates ammonia-induced brain edema (Córdoba, Gottstein, and Blei 1998; Wendon and Lee 2008).Additionally, hyponatremia was linked to cytotoxic brain edema in the acute hyponatremic encephalopathy model (Sevick et al. 1992). All this data suggests that hyponatremia might play an important role in the development of cytotoxic brain edema following ALF.

Neuroinflammation and Astrocyte Swelling

Neuroinflammation has been shown recently as an important contributor to the development of both acute and chronic HE (Jayakumar, Rao, and Norenberg 2015). Fist evidence on the involvement of inflammation in the pathogenesis of HE comes from the study by Gans et al.(1971) in which they reported that FHF was associated with peripheral infections and inflammation (Gans et al. 1971). Notably, it is well known that, under normal circumstances, blood-derived cytokines—likely originating from necrotic liver and/or sepsis—freely cross the BBB(Banks, Kastin, and Broadwell 1995).It has been reported that interleukin-6 (IL-6) plasma levels were increased in both AHF and in chronic HE(Izumi et al. 1995). Furthermore, elevated serum concentrations of TNF-a, IL-6, and IL-1B, together with their variations between arterio-venous regions, were in line with a brain cytokine efflux in individuals suffering from AHE (Wigmore et al. 1998; Wright et al. 2007). All this compelling evidence supports the role of neuroinflammation in HE.

Role of Neuroinflammation in Cytotoxic Brain Edema

The first evidence supporting the role of neuroinflammation in AHE was established by Chung et al.(2001) (Chung, Gottstein, and Blei 2001). The authors showed the anti-inflammatory agent, indomethacin (a potent inhibitor of cyclooxygenase-2), has a protective effect against the development of brain edema in rats with a PCA-infused with ammonia(Chung et al. 2001). Additionally, LPS-induced endotoxemia in rats was shown to aggravate the brain edema and encephalopathy associated with ALF (Wright et al. 2007). While, indomethacin was also shown the have a positive effect in normalizing the ICP in patients AHE(Clemmesen, Hansen, and Larsen 1997).In addition, the antibiotic minocycline reduced the degree of encephalopathy and stopped cerebral edema from developing in experimental ALF (Jiang, Desjardins, and Butterworth 2009). Of note, minocycline can induceliver damage on its own (Böcker, Estler, and Ludewig-Sandig 1991; Elkayam, Yaron, and Caspi 1999), ruling out any therapeutic utility in the management of acute HE.

Furthermore, transgenic mice lacking in TNF-a, IL-1b, and IL-6 receptors provide more evidence that neuroinflammation plays a role in brain edema. These mice also exhibit less cerebral edema than wild type mice (Bémeur et al. 2010). More importantly, cultured astrocytes treated with inflammatory cytokines(TNF-a, IL-1b, IL-6 and IFN-g)were induced astrocyte swelling(Chung et al. 2001). This astrocyte swilling was aggravated when astrocytes were pre-treated with ammonia (Jayakumar et al. 2015). Further research supporting an astrocytic inflammatory response in vitro comes from recent reports showing that ammonia activated inflammatory factors in cultured astrocytes, including constitutive nitric oxide synthase (cNOS) and NADPH oxidase (NOX), and that this led to the production of reactive oxygen and nitrogen species (RONS), phospholipase-2 (PLA2), and mitogen-activated protein kinases (MAPKs) (Jayakumar et al. 2006, 2009). While,blocking these factors was associated with reduced astrocytes swelling brought on by ammonia in cultured astrocytes (Jayakumar et al. 2006, 2009). These findings clearly point to a synergistic relationship between ammonia, reactive oxygen/nitrogen species, and inflammatory cytokines in t development of the development of cytotoxic brain edemain the context of AHE (Rama Rao, Jayakumar, Tong, Alvarez, et al. 2010).

The Contribution of Microglia Activation in the Development of Cytotoxic Brain Edema

It has been suggested recently that the development of astrocyte swelling/brain edema in AHE is related to microglial activation and the subsequent production and release of pro-inflammatory cytokines (Jiang et al. 2009; Rodrigo et al. 2010; Wilkinson et al. 1974; Wright et al. 2007)including IL-1a, IL-1b and TNF-a (Garden and Möller 2006). Microglia activation can be triggered by various stimuli, including as glutamate, extracellular potassium, LPS, and pro-inflammatory cytokines that result from liver necrosis and/or sepsis (Garden and Möller 2006).In addition, experimental models of hyperammonemia were shown to be associated with microglia activation(Görg, Bidmon, and Häussinger 2013),as well as in autopsied brain tissue of patients with ALF(Butterworth 2011).Interestingly, following ALF, the increased levels of IL-1b, TNF-a, and IL-6 mRNAs have been identified in the rats' brains. The degree of these pro-inflammatory cytokines' expression was positively associated with the beginning of brain edema and the advancement of encephalopathy(Jiang et al. 2009). While, treatment with minocycline inhibited the elevation of TNF-a, IL-6, and IL-1b, as well as microglial activation, which was linked to a concurrent reduction in encephalopathy and brain edema. These results support the hypothesis that microglial activation plays a role in the development of cerebral edema in AHE(Jiang et al. 2009). In addition, it was found that primary cultures of microglia exposed to ammonia (1 mM for 72 h or 5 mM for 24 h) produced and released cytokines such as IL-6 and TNF-a (Rao et al. 2013).This suggests that ammonia could trigger microglia activation and the subsequent release of proinflammatory cytokines which could play a role in the development of cytotoxic brain edema in AHE.

The Involvement of Cerebral Endothelial Cells in Astrocytes Swelling/Brain Edema in AHE

The cerebral endothelial cells (ECs) are shown recently to play a role in the development of astrocyte swelling/brain edema following AHE. Considering that ECs can also activate NF-KB, cyclooxygenase-2 (COX2), and inducible nitric oxide synthase (iNOS), among other pro-inflammatory molecules (Feuerstein, Wang, and Barone 1998), it is possible that these cells might contribute to the development of astrocyte swelling/brain edema in AHE. It has been recently reported that ECs cultures treated with ammonia were associated with the release of cytokines and ROS(Arumugam R Jayakumar et al. 2014; Jayakumar and Norenberg 2013a; Skowrońska et al. 2012), and the activation ofNF-kB(Arumugam R Jayakumar et al. 2014; Jayakumar et al. 2012; Jayakumar and Norenberg 2013b). The cytokines (i.e., IL-1b and TNF-a) receptorshave been identified the brain's ECs(Hashimoto et al. 1991), and that activation of these receptors subsequently stimulated the synthesis and the release of additional cytokines,exacerbating brain edema inAHE(Matsumura and Kobayashi 2004). Additionally, ammonia, cytokines, and LPS were added to primary cultures of rat brain endothelial cells to produce conditioned medium (CM), which was then added to cultured astrocytes and the cell volume was measured. The researchers showed that astrocytes exposed to CM from ECs treated with ammonia resulted in a notable increase in cell swelling. This effect was further amplified when astrocytes were exposed to CM from ECs treated with a combination of ammonia, LPS, and cytokines (Arumugam R Jayakumar et al. 2014; Jayakumar et al. 2012).Taken together these results, ECs could activate a multitude of inflammatory factors responsible for the development of cytotoxic brain edema linked to AHE.

CONCLUSION

AHE is a complex neuropsychiatric condition of great concerns. Cytotoxic brain edema or astrocyte swelling is the most common neuropathological finding in both patients and animal models of ALF as well as in cultured astrocytes. It is linked to a high mortality rate because of brain herniation resulted from ICP. The development of cytotoxic brain edema is attributed to oxidative stress, mitochondria abnormalities and neuroinflammation in addition to ion channels and transporters dysfunction along with hyponatremia.

REFERENCES

Bai, G., Rama Rao, K. V., Murthy, C. R. K., Panickar, K. S., Jayakumar, A. R., & Norenberg, M. D. (2001). Ammonia Induces the Mitochondrial Permeability Transition in Primary Cultures of Rat Astrocytes. *Journal of Neuroscience Research*, *66*(5), 981–991. doi:10.1002/jnr.10056 PMID:11746427

Banks, W. A., Kastin, A. J., & Broadwell, R. D. (1995). Passage of Cytokines across the Blood-Brain Barrier. *Neuroimmunomodulation*, *2*(4), 241–248. doi:10.1159/000097202 PMID:8963753

Bémeur, C., Qu, H., Desjardins, P., & Butterworth, R. F. (2010). IL-1 or TNF Receptor Gene Deletion Delays Onset of Encephalopathy and Attenuates Brain Edema in Experimental Acute Liver Failure. *Neurochemistry International*, *56*(2), 213–215. doi:10.1016/j.neuint.2009.11.010 PMID:19931338

Blei, A. T. (2001). Pathophysiology of Brain Edema in Fulminant Hepatic Failure, Revisited. *Metabolic Brain Disease*, *16*(1–2), 85–94. doi:10.1023/A:1011670713730 PMID:11726092

Blei, A. T. (2008). Brain Edema in Acute Liver Failure. *Critical Care Clinics*, *24*(1), 99–114. doi:10.1016/j.ccc.2007.11.001 PMID:18241781

Böcker, R., Estler, C. J., & Ludewig-Sandig, D. (1991). Evaluation of the Hepatotoxic Potential of Minocycline. *Antimicrobial Agents and Chemotherapy*, *35*(7), 1434–1436. doi:10.1128/AAC.35.7.1434 PMID:1929305

Bosoi, C. R., Tremblay, M., & Rose, C. F. (2014). Induction of Systemic Oxidative Stress Leads to Brain Oedema in Portacaval Shunted Rats. *Liver International*, *34*(9), 1322–1329. doi:10.1111/liv.12414 PMID:25354203

Bosoi, C. R., Yang, X., Huynh, J., Parent-Robitaille, C., Jiang, W., Tremblay, M., & Rose, C. F. (2012). Systemic Oxidative Stress Is Implicated in the Pathogenesis of Brain Edema in Rats with Chronic Liver Failure. *Free Radical Biology & Medicine*, *52*(7), 1228–1235. doi:10.1016/j.freeradbiomed.2012.01.006 PMID:22300646

Butterworth, R. F. (2000). The Astrocytic ('Peripheral-Type') Benzodiazepine Receptor: Role in the Pathogenesis of Portal-Systemic Encephalopathy. *Neurochemistry International*, *36*(4–5), 411–416. doi:10.1016/S0197-0186(99)00132-1 PMID:10733008

Butterworth, R. F. (2011). Neuroinflammation in Acute Liver Failure: Mechanisms and Novel Therapeutic Targets. *Neurochemistry International*, *59*(6), 830–836. doi:10.1016/j.neuint.2011.07.014 PMID:21864609

Cárdenas, A., Solà, E., Rodríguez, E., Barreto, R., Graupera, I., Pavesi, M., Saliba, F., Welzel, T. M., Martinez-Gonzalez, J., Gustot, T., Bernardi, M., Arroyo, V., & Ginès, P. (2014). Hyponatremia Influences the Outcome of Patients with Acute-on-Chronic Liver Failure: An Analysis of the CANONIC Study. *Critical Care*, *18*(6), 1–8. doi:10.118613054-014-0700-0 PMID:25643318

Chung, C., Gottstein, J., & Blei, A. T. (2001). Indomethacin Prevents the Development of Experimental Ammonia-Induced Brain Edema in Rats after Portacaval Anastomosis. *Hepatology (Baltimore, Md.)*, *34*(2), 249–254. doi:10.1053/jhep.2001.26383 PMID:11481608

Clemmesen, J. O., Hansen, B. A., & Larsen, F. S. (1997). Indomethacin Normalizes Intracranial Pressure in Acute Liver Failure: A Twenty-three-year-old Woman Treated with Indomethacin. *Hepatology (Baltimore, Md.)*, *26*(6), 1423–1425. doi:10.1002/hep.510260608 PMID:9397981

Córdoba, J. & García-martinez, R. (2010). *Hyponatremic and Hepatic Encephalopathies : Similarities, Differences and Coexistence*. 73–80.

Cordoba, J., Gottstein, J., & Blei, A. T. (1996). Glutamine, Myo-inositol, and Organic Brain Osmolytes after Portocaval Anastomosis in the Rat: Implications for Ammonia-induced Brain Edema. *Hepatology (Baltimore, Md.)*, *24*(4), 919–923. doi:10.1053/jhep.1996.v24.pm0008855198 PMID:8855198

Córdoba, J., Gottstein, J., & Blei, A. T. (1998). Chronic Hyponatremia Exacerbates Ammonia-Induced Brain Edema in Rats after Portacaval Anastomosis. *Journal of Hepatology*, *29*(4), 589–594. doi:10.1016/S0168-8278(98)80154-2 PMID:9824268

Drews, L., Zimmermann, M., Westhoff, P., Brilhaus, D., Poss, R. E., Bergmann, L., Wiek, C., Brenneisen, P., Piekorz, R. P., Mettler-Altmann, T., Weber, A. P. M., & Reichert, A. S. (2020). Ammonia Inhibits Energy Metabolism in Astrocytes in a Rapid and Glutamate Dehydrogenase 2-Dependent Manner. *Disease Models & Mechanisms*, *13*(10), dmm047134. doi:10.1242/dmm.047134 PMID:32917661

Eefsen, M., Jelnes, P., Schmidt, L. E., Ben Vainer, H. C. B., & Larsen, F. S. (2010). Brain Expression of the Water Channels Aquaporin-1 and-4 in Mice with Acute Liver Injury, Hyperammonemia and Brain Edema. *Metabolic Brain Disease*, *25*(3), 315–323. doi:10.100711011-010-9213-y PMID:20938728

El Hiba, O., El-Mansoury, B., Chatoui, H., Benammi, H., Balzano, T., Merzouki, M., Abdelmonaim El Hidan, M., Radhakrishnan, J., Bahi, L., & Idrissi, H. R. (2023). Clinical and Experimental Evidence of Cirrhosis-Related Parkinsonism. Experimental and Clinical Evidence of the Neuropathology of Parkinson's Disease. IGI Global. doi:10.4018/978-1-6684-5156-4.ch007

El Khiat, A., El Hiba, O., Tamegart, L., Rais, H., Fdil, N., Sellami, S., Ait El Mokhtar, M., & Gamrani, H. (2022a). Time Dependent Alteration of Locomotor Behavior in Rat with Acute Liver Failure Induced Cerebellar Neuroinflammation and Neuro-Astroglial Damage. *Journal of Chemical Neuroanatomy*, *119*, 102055. doi:10.1016/j.jchemneu.2021.102055 PMID:34863855

El Khiat, A., El-Mansoury, B., El Hiba, O., Tamegart, L., Rais, H., Fdil, N., Sellami, S., Ait El Mokhtar, M., Kissani, N., & Gamrani, H. (2023). P14 Altered Locomotor Behavior Associated With Purkinje Cells and Astrocytes Dysfunctions in Rat With Acute Hepatic Encephalopathy Involving the Oxidative Stress and Neuroinflammation. *Official Journal of the American College of Gastroenterology| ACG*, *118*(9S), S11.

El-Mansoury, B., Smimih, K., Aimrane, A., Draoui, A., Ferssiwi, A., Bitar, A., Jayakumar, A., & El Hiba, O. (2023). P8 Cognitive Impairment Associated With Cortical and Hippocampal Microglia Activation and Altered Synaptic Function in a Rat Model of Chronic Hepatic Encephalopathy. *Official Journal of the American College of Gastroenterology*, *118*(9S), S7.

Elkayam, O., Yaron, M., & Caspi, D. (1999). Minocycline-Induced Autoimmune Syndromes: An Overview. Seminars in arthritis and rheumatism, 28. Elsevier.

Elsherbini, D. M. A., Ghoneim, F. M., El-Mancy, E. M., Ebrahim, H. A., El-Sherbiny, M., El-Shafey, M., Al-Serwi, R. H., & Elsherbiny, N. M. (2022). Astrocytes Profiling in Acute Hepatic Encephalopathy: Possible Enrolling of Glial Fibrillary Acidic Protein, Tumor Necrosis Factor-Alpha, Inwardly Rectifying Potassium Channel (Kir 4.1) and Aquaporin-4 in Rat Cerebral Cortex. *Frontiers in Cellular Neuroscience, 16*, 896172. doi:10.3389/fncel.2022.896172 PMID:36060277

Feuerstein, G. Z., Wang, X., & Barone, F. C. (1998). The Role of Cytokines in the Neuropathology of Stroke and Neurotrauma. *Neuroimmunomodulation, 5*(3–4), 143–159. doi:10.1159/000026331 PMID:9730680

Fraser, C. L., & Arieff, A. I. (1997). Epidemiology, Pathophysiology, and Management of Hyponatremic Encephalopathy. *The American Journal of Medicine, 102*(1), 67–77. doi:10.1016/S0002-9343(96)00274-4 PMID:9209203

Fu, H., Li, B., Hertz, L., & Peng, L. (2012). Contributions in Astrocytes of SMIT1/2 and HMIT to Myo-Inositol Uptake at Different Concentrations and PH. *Neurochemistry International, 61*(2), 187–194. doi:10.1016/j.neuint.2012.04.010 PMID:22564531

Gans, H., Mori, K., Lindsey, E., Kaster, B., Richter, D., Quinlan, R., Dineen, P. A., & Tan, B. H. (1971). Septicemia as a Manifestation of Acute Liver Failure. *Surgery, Gynecology & Obstetrics, 132*(5), 783–790. PMID:5555405

Garden, G. A., & Möller, T. (2006). Microglia Biology in Health and Disease. *Journal of Neuroimmune Pharmacology, 1*(2), 127–137. doi:10.100711481-006-9015-5 PMID:18040779

Görg, B., Bidmon, H.-J., & Häussinger, D. (2013). Gene Expression Profiling in the Cerebral Cortex of Patients with Cirrhosis with and without Hepatic Encephalopathy. *Hepatology (Baltimore, Md.), 57*(6), 2436–2447. doi:10.1002/hep.26265 PMID:23325665

Görg, B., Karababa, A., Shafigullina, A., Bidmon, H. J., & Häussinger, D. (2015). Ammonia-induced Senescence in Cultured Rat Astrocytes and in Human Cerebral Cortex in Hepatic Encephalopathy. *Glia, 63*(1), 37–50. doi:10.1002/glia.22731 PMID:25092802

Guevara, M., Baccaro, M. E., Ríos, J., Martín-Llahí, M., Uriz, J., Ruiz del Arbol, L., Planas, R., Monescillo, A., Guarner, C., Crespo, J., Bañares, R., Arroyo, V., & Ginès, P. (2010). Risk Factors for Hepatic Encephalopathy in Patients with Cirrhosis and Refractory Ascites: Relevance of Serum Sodium Concentration. *Liver International, 30*(8), 1137–1142. doi:10.1111/j.1478-3231.2010.02293.x PMID:20602681

Guevara, M., Baccaro, M., Torre, A., Gómez-Ansón, B., Ríos, J., Torres, F., Rami, L., Monté-Rubio, G., Martín-Llahí, M., Arroyo, V. (2009). Hyponatremia Is a Risk Factor of Hepatic Encephalopathy in Patients with Cirrhosis: A Prospective Study with Time-Dependent Analysis. *Official Journal of the American College of Gastroenterology, 104*(6), 1382–89.

Hashimoto, M. Y. S. F. T. Y., Ishikawa, Y., Yokota, S., Goto, F., Bando, T., Sakakibara, Y., & Iriki, M. (1991). Action Site of Circulating Interleukin-1 on the Rabbit Brain. *Brain Research, 540*(1–2), 217–223. doi:10.1016/0006-8993(91)90510-3 PMID:2054614

Heins, J., & Zwingmann, C. (2010). Organic Osmolytes in Hyponatremia and Ammonia Toxicity. *Metabolic Brain Disease, 25*(1), 81–89. doi:10.100711011-010-9170-5 PMID:20195726

Iwasa, M., & Takei, Y. (2015). Pathophysiology and Management of Hepatic Encephalopathy 2014 Update: Ammonia Toxicity and Hyponatremia. *Hepatology Research*, *45*(12), 1155–1162. doi:10.1111/hepr.12495 PMID:25604571

Izumi, S., Hughes, R. D., Langley, P. G., Pernambuco, J. R. B., & Williams, R. (1995). Acute Phase Response after Liver Transplantation for Fulminant Hepatic Failure and Cirrhosis. *Transplant International*, *8*(5), 340–345. doi:10.1111/j.1432-2277.1995.tb01533.x PMID:7576014

Jayakumar, A. R., & Norenberg, M. D. (2013a). Endothelial-Astrocytic Interactions in Acute Liver Failure. *Metabolic Brain Disease*, *28*(2), 183–186. doi:10.100711011-012-9344-4 PMID:23099995

Jayakumar, A. R. & Norenberg, M. (2013b). *Endothelial-Astrocytic Interactions in Acute Liver Failure*. NIH.

Jayakumar, A. R., Panickar, K. S., & Ch, R. K. (2006). Oxidative Stress and Mitogen-Activated Protein Kinase Phosphorylation Mediate Ammonia-Induced Cell Swelling and Glutamate Uptake Inhibition in Cultured Astrocytes. *The Journal of Neuroscience : The Official Journal of the Society for Neuroscience*, *26*(18), 4774–4784. doi:10.1523/JNEUROSCI.0120-06.2006 PMID:16672650

Jayakumar, A. R., Rama Rao, K. V., & Norenberg, M. D. (2015). Neuroinflammation in Hepatic Encephalopathy: Mechanistic Aspects. *Journal of Clinical and Experimental Hepatology*, *5*(March), S21–S28. doi:10.1016/j.jceh.2014.07.006 PMID:26041953

Jayakumar, A. R., Rama Rao, K. V., Tong, X. Y., & Norenberg, M. D. (2009). Calcium in the Mechanism of Ammonia-induced Astrocyte Swelling. *Journal of Neurochemistry*, *109*(s1), 252–257. doi:10.1111/j.1471-4159.2009.05842.x PMID:19393035

Jayakumar, A. R., Tong, X. Y., Curtis, K. M., Ruiz-Cordero, R., Abreu, M. T., & Norenberg, M. D. (2014). Increased Toll-like Receptor 4 in Cerebral Endothelial Cells Contributes to the Astrocyte Swelling and Brain Edema in Acute Hepatic Encephalopathy. *Journal of Neurochemistry*, *128*(6), 890–903. doi:10.1111/jnc.12516 PMID:24261962

Jayakumar, A. R., Tong, X. Y., Ospel, J., & Norenberg, M. D. (2012). Role of Cerebral Endothelial Cells in the Astrocyte Swelling and Brain Edema Associated with Acute Hepatic Encephalopathy. *Neuroscience*, *218*, 305–316. doi:10.1016/j.neuroscience.2012.05.006 PMID:22609932

Jayakumar, A. R., Valdes, V., & Norenberg, M. D. (2011). The Na–K–Cl Cotransporter in the Brain Edema of Acute Liver Failure. *Journal of Hepatology*, *54*(2), 272–278. doi:10.1016/j.jhep.2010.06.041 PMID:21056502

Jayakumar, A. R., Valdes, V., Tong, X. Y., Shamaladevi, N., Gonzalez, W., & Norenberg, M. D. (2014). Sulfonylurea Receptor 1 Contributes to the Astrocyte Swelling and Brain Edema in Acute Liver Failure. *Translational Stroke Research*, *5*(1), 28–37. doi:10.100712975-014-0328-z PMID:24443056

Jiang, W., Desjardins, P., & Butterworth, R. F. (2009). Direct Evidence for Central Proinflammatory Mechanisms in Rats with Experimental Acute Liver Failure: Protective Effect of Hypothermia. *Journal of Cerebral Blood Flow and Metabolism*, *29*(5), 944–952. doi:10.1038/jcbfm.2009.18 PMID:19259110

Kato, M., Hughes, R. D., Keays, R. T., & Williams, R. (1992). Electron Microscopic Study of Brain Capillaries in Cerebral Edema from Fulminant Hepatic Failure. *Hepatology (Baltimore, Md.)*, *15*(6), 1060–1066. doi:10.1002/hep.1840150615 PMID:1592344

Kengne, F. G., & Decaux, G. (2017). Hyponatremia and the Brain. *Kidney International Reports*, *3*(1), 24–35. doi:10.1016/j.ekir.2017.08.015 PMID:29340311

Larsen, F. S. (1996). Cerebral Circulation in Liver Failure: Ohm's Law in Force. Seminars in liver disease, 16. Thieme Medical Publishers, Inc.

Larsen, F. S., Hansen, B. A., Pott, F., Ejlersen, E., Secher, N. H., Paulson, O. B., & Knudsen, G. M. (1996). Dissociated Cerebral Vasoparalysis in Acute Liver Failure: A Hypothesis of Gradual Cerebral Hyperaemia. *Journal of Hepatology*, *25*(2), 145–151. doi:10.1016/S0168-8278(96)80066-3 PMID:8878774

Larsen, F. S., Strauss, G., Møller, K., & Hansen, B. A. (2000). Regional Cerebral Blood Flow Autoregulation in Patients with Fulminant Hepatic Failure. *Liver Transplantation*, *6*(6), 795–800. doi:10.1053/jlts.2000.18705 PMID:11084071

Lee, W. (2022). Acute Liver Failure. *Yamada's Textbook of Gastroenterology,* 1889–1905.

Liotta, E. M., Romanova, A. L., Lizza, B. D., Rasmussen-torvik, L. J., Kim, M., Francis, B., Sangha, R. S., Carroll, T. J., Ganger, D., Ladner, D. P., Naidech, A. M., Paparello, J. J., Prabhakaran, S., Sorond, F. A., & Maas, M. B. (2017). Osmotic Shifts, Cerebral Edema, and Neurologic Deterioration in Severe Hepatic Encephalopathy. *Critical Care Medicine*, 1–10. PMID:29341965

Lu, K. (2023). Cellular Pathogenesis of Hepatic Encephalopathy: An Update. *Biomolecules*, *13*(2), 396. doi:10.3390/biom13020396 PMID:36830765

Martinez-Hernandez, A., Bell, K. P., & Norenberg, M. D. (1977). Glutamine Synthetase: Glial Localization in Brain. *Science*, *195*(4284), 1356–1358. doi:10.1126cience.14400 PMID:14400

Matsumura, K., & Kobayashi, S. (2004). Signaling the Brain in Inflammation: The Role of Endothelial Cells. *Frontiers in Bioscience*, *9*(5), 2819–2826. doi:10.2741/1439 PMID:15353317

McConnell, J. R., Antonson, D. L., Ong, C. S., Chu, W.-K., Fox, I. J., Heffron, T. G., Langnas, A. N., & Shaw, B. W. Jr. (1995). Proton Spectroscopy of Brain Glutamine in Acute Liver Failure. *Hepatology (Baltimore, Md.)*, *22*(1), 69–74. doi:10.1002/hep.1840220111 PMID:7601435

Murthy, Ch. R. K., Rama Rao, K. V., Bai, G., & Norenberg, M. D. (2001). Ammonia-induced Production of Free Radicals in Primary Cultures of Rat Astrocytes. *Journal of Neuroscience Research*, *66*(2), 282–288. doi:10.1002/jnr.1222 PMID:11592125

Norenberg, M. D. (1987). The Role of Astrocytes in Hepatic Encephalopathy. *Neurochemical Pathology*, *6*(1-2), 13–33. doi:10.1007/BF02833599 PMID:3306480

Norenberg, M. D. (2003). Oxidative and Nitrosative Stress in Ammonia Neurotoxicity. *Hepatology (Baltimore, Md.)*, *37*(2), 245–248. doi:10.1053/jhep.2003.50087 PMID:12540772

Norenberg, M. D., Jayakumar, A. R., Rama Rao, K. V., & Panickar, K. S. (2007). New Concepts in the Mechanism of Ammonia-Induced Astrocyte Swelling. *Metabolic Brain Disease*, *22*(3-4), 219–234. doi:10.100711011-007-9062-5 PMID:17823859

Oenarto, J., Görg, B., Moos, M., Bidmon, H.-J., & Häussinger, D. (2014). Expression of Organic Osmolyte Transporters in Cultured Rat Astrocytes and Rat and Human Cerebral Cortex. *Archives of Biochemistry and Biophysics*, *560*, 59–72. doi:10.1016/j.abb.2014.06.024 PMID:25004465

Oksanen, M., Lehtonen, S., Jaronen, M., Goldsteins, G., Hämäläinen, R. H., & Koistinaho, J. (2019). Astrocyte Alterations in Neurodegenerative Pathologies and Their Modeling in Human Induced Pluripotent Stem Cell Platforms. *Cellular and Molecular Life Sciences*, *76*(14), 2739–2760. doi:10.100700018-019-03111-7 PMID:31016348

Ong, J. P., Aggarwal, A., Krieger, D., Easley, K. A., Karafa, M. T., Van Lente, F., Arroliga, A. C., & Mullen, K. D. (2003). Correlation between Ammonia Levels and the Severity of Hepatic Encephalopathy. *The American Journal of Medicine*, *114*(3), 188–193. doi:10.1016/S0002-9343(02)01477-8 PMID:12637132

Rao, K. V. (2003). Increased Aquaporin-4 Expression in Ammonia-Treated Cultured Astrocytes. *Neuroreport*, *14*(18), 2379–2382. doi:10.1097/00001756-200312190-00018 PMID:14663195

Rao, K. V. (2014). Aquaporin-4 Deletion in Mice Reduces Encephalopathy and Brain Edema in Experimental Acute Liver Failure. *Neurobiology of Disease*, *63*, 222–228. doi:10.1016/j.nbd.2013.11.018 PMID:24321433

Rao, K. V., Brahmbhatt, M., & Norenberg, M. D. (2013). Microglia Contribute to Ammonia-Induced Astrocyte Swelling in Culture. *Metabolic Brain Disease*, *28*(2), 139–143. doi:10.100711011-012-9339-1 PMID:23065046

Rao, Rama, R., Kakulavarapu, V., Arumugam, R., Jayakumar, X., Tong, V., Alvarez, M., & Michael D. (2010). Marked Potentiation of Cell Swelling by Cytokines in Ammonia-Sensitized Cultured Astrocytes. *Journal of Neuroinflammation*, *7*(1), 1–8. PMID:20942959

Rao, R. (2010). Brain Aquaporin-4 in Experimental Acute Liver Failure. *Journal of Neuropathology and Experimental Neurology*, *69*(9), 869–879. doi:10.1097/NEN.0b013e3181ebe581 PMID:20720509

Rao, R. (2014). Aquaporin-4 Deletion in Mice Reduces Encephalopathy and Brain Edema in Experimental Acute Liver Failure. *Neurobiology of Disease*, *63*, 222–228. doi:10.1016/j.nbd.2013.11.018 PMID:24321433

Rodrigo, R., Cauli, O., Gomez–Pinedo, U., Agusti, A., Hernandez–Rabaza, V., Garcia–Verdugo, J.-M., & Felipo, V. (2010). Hyperammonemia Induces Neuroinflammation That Contributes to Cognitive Impairment in Rats with Hepatic Encephalopathy. *Gastroenterology*, *139*(2), 675–684. doi:10.1053/j.gastro.2010.03.040 PMID:20303348

Sevick, J., Kanda, F., Mintorovitch, J., Arieff, A. I., Kucharczyk, J., Tsuruda, J. S., Norman, D., & Moseley, M. E. (1992). Cytotoxic Brain Edema : Assessment with Diffusion-Weighted MR Imaging. *Radiology*, *185*(3), 687–690. doi:10.1148/radiology.185.3.1438745 PMID:1438745

Shulyatnikova, T. V., & Tumanskiy, V. O. (2022). Immunohistochemical Study of the Brain Aquaporin-4 in the Rat Acute Liver Failure Model. *The Art of Medication*, *1455*(iii), 103–108. doi:10.21802/artm.2022.1.21.103

Simard, J., Kent, T. A., Chen, M., Tarasov, K. V., & Gerzanich, V. (2007). Brain Oedema in Focal Ischaemia: Molecular Pathophysiology and Theoretical Implications. *Lancet Neurology*, *6*(3), 258–268. doi:10.1016/S1474-4422(07)70055-8 PMID:17303532

Skowrońska, M., Zielińska, M., Wójcik-Stanaszek, L., Ruszkiewicz, J., Milatovic, D., Aschner, M., & Albrecht, J. (2012). Ammonia Increases Paracellular Permeability of Rat Brain Endothelial Cells by a Mechanism Encompassing Oxidative/Nitrosative Stress and Activation of Matrix Metalloproteinases. *Journal of Neurochemistry*, *121*(1), 125–134. doi:10.1111/j.1471-4159.2012.07669.x PMID:22260250

Strange, K. (2004). Cellular Volume Homeostasis. *Advances in Physiology Education*, *28*(4), 155–159. doi:10.1152/advan.00034.2004 PMID:15545344

Strauss, G., Hansen, B. A., Knudsen, G. M., & Larsen, F. S. (1998). Hyperventilation Restores Cerebral Blood Flow Autoregulation in Patients with Acute Liver Failure. *Journal of Hepatology*, *28*(2), 199–203. doi:10.1016/0168-8278(88)80006-0 PMID:9514532

Swain, M., Butterworth, R. F., & Blei, A. T. (1992). Ammonia and Related Amino Acids in the Pathogenesis of Brain Edema in Acute Ischemic Liver Failure in Rats. *Hepatology (Baltimore, Md.)*, *15*(3), 449–453. doi:10.1002/hep.1840150316 PMID:1544626

Takahashi, H., Koehler, R. C., Brusilow, S. W., & Traystman, R. J. (1991). Inhibition of Brain Glutamine Accumulation Prevents Cerebral Edema in Hyperammonemic Rats. *American Journal of Physiology. Heart and Circulatory Physiology*, *261*(3), H825–H829. doi:10.1152/ajpheart.1991.261.3.H825 PMID:1679605

Thrane, A. S., Thrane, V. R., & Nedergaard, M. (2014). Drowning Stars: Reassessing the Role of Astrocytes in Brain Edema. *Trends in Neurosciences*, *37*(11), 620–628. doi:10.1016/j.tins.2014.08.010 PMID:25236348

Vaquero, J., Chung, C., & Blei, A. T. (2003). Brain Edema in Acute Liver Failure. A Window to the Pathogenesis of Hepatic Encephalopathy. *Annals of Hepatology*, *2*(1), 12–22. doi:10.1016/S1665-2681(19)32153-2 PMID:15094701

Ware, A. J., D'Agostino, A. N., & Combes, B. (1971). Cerebral Edema: A Major Complication of Massive Hepatic Necrosis. *Gastroenterology*, *61*(6), 877–884. doi:10.1016/S0016-5085(19)33402-X PMID:5125688

Wendon, J., & William Lee, Æ. (2008). Encephalopathy and Cerebral Edema in the Setting of Acute Liver Failure : Pathogenesis and Management. *Neurocritical Care*, *9*(1), 97–102. doi:10.100712028-008-9123-6 PMID:18688582

Wigmore, S. J., Walsh, T. S., Lee, A., & Ross, J. A. (1998). Pro-Inflammatory Cytokine Release and Mediation of the Acute Phase Protein Response in Fulminant Hepatic Failure. *Intensive Care Medicine*, *24*(3), 224–229. doi:10.1007001340050554 PMID:9565803

Wilkinson, S. P., Gazzard, B. G., Arroyo, V., Moodie, H., & Williams, R. (1974). Relation of Renal Impairment and Haemorrhagic Diathesis to Endotoxaemia in Fulminant Hepatic Failure. *Lancet, 303*(7857), 521–524. doi:10.1016/S0140-6736(74)92711-1 PMID:4131970

Willard-Mack, C. L., Koehler, R. C., Hirata, T., Cork, L. C., Takahashi, H., Traystman, R. J., & Brusilow, S. W. (1996). Inhibition of Glutamine Synthetase Reduces Ammonia-Induced Astrocyte Swelling in Rat. *Neuroscience, 71*(2), 589–599. doi:10.1016/0306-4522(95)00462-9 PMID:9053810

Wright, G., Shawcross, D., Olde Damink, S. W. M., & Jalan, R. (2007). Brain Cytokine Flux in Acute Liver Failure and Its Relationship with Intracranial Hypertension. *Metabolic Brain Disease, 22*(3-4), 375–388. doi:10.100711011-007-9071-4 PMID:17899343

Zemtsova, I., Görg, B., Keitel, V., Bidmon, H. J., Schrör, K., & Häussinger, D. (2011). Microglia Activation in Hepatic Encephalopathy in Rats and Humans. *Hepatology (Baltimore, Md.), 54*(1), 204–215. doi:10.1002/hep.24326 PMID:21452284

Chapter 14
Glioblastoma:
Physiopathology and Complications

Dipsanu Paul
The Assam Royal Global University, India

Dharmeswar Barhoi
https://orcid.org/0000-0002-3184-990X
The Assam Royal Global University, India

ABSTRACT

Glioblastoma is the most aggressive primary malignant brain tumor in adults. Complex genetic and molecular changes that cause unchecked cell proliferation, invasion of the surrounding brain tissue, and angiogenesis are the hallmarks of the physiopathology of glioblastoma. Although there are treatment options for this deadly tumor that include surgery, radiation, and chemotherapy, the blood-brain barrier and the tumor's infiltrative nature restrict their effectiveness, frequently leading to tumor recurrence and illness progression. To create new therapeutic approaches and enhance patient outcomes, it is crucial to comprehend the physiopathology of glioblastoma and its associated consequences. To improve treatment and quality of life for patients with glioblastoma, further research is required to clarify molecular causes, discover therapeutic targets, and address the difficulties provided by comorbidities.

INTRODUCTION

Glial cells are the supporting cells that surround and feed neurons in the brain, giving birth to the extremely aggressive kind of brain cancer known as glioblastoma (GBM); making up 50% of the glioma subtypes and more than 60% of all brain tumors (Aly et al., 2020).GBM generally only affects the Central Nervous System because of its fast development and the CNS's unique design, which is bordered by the blood-brain barrier and dura mater and lacks lymphatic channels (De Barros et al., 2019). GBMs can appear in the brain stem, cerebellum, and spinal cord, even though they mostly only happen in the brain. The four lobes of the brain—frontal (25%), temporal (20%), parietal (13%), and occipital (3%), account for sixty-one percent of all primary gliomas [American Association of Neuroscience Nurses (AANN),2014]

DOI: 10.4018/978-1-6684-9675-6.ch014

(Ahmad et al., 2012). GBMs were once assumed to only come from glial cells, but data now reveals that they may come from a variety of cell types with characteristics resembling those of neural stem cells.

These cells are in various phases of development from stem cells to neurons to glia, with molecular changes in signaling pathways acting largely as a determinant of phenotypic variance rather than different cell types at the beginning (Ali et al., 2015). GBMs can be classed as primary, or de novo, emerging without a known predecessor, or secondary, where a low-grade tumor changes over time into a GBM. Most GBMs are primary, and patients with primary GBMs have worse prognoses than patients with secondary GBMs. Primary GBM patients more frequently have older ages (Ambartsumian et al., 2019). The age-adjusted incidence rate is 3.2 per 100,000 people on average (Fabi et al., 2004). A 5-year survival rate of 5% was recorded for GBM in the United States alone in 2016, where an estimated 12,120 patients received the diagnosis (Omuro et al., 2007). The incidence rises sharply after the age of 54 and peaks at 15.24 per 100,000 people between the ages of 75 and 84 years (Omuro et al., 2007). The median age of persons diagnosed with GBM has increased to 64 years over the past few decades as a result of the disease being more common in older people and people living longer in industrialized countries.

GENETIC ANOMALIES AND MOLECULAR PATHWAYS DISCOVERED IN GBM

Tumour growth in GBM cells is aided by increased expression of cell surface membrane receptors that govern intracellular signal transduction pathways controlling proliferation and cell cycle abnormalities, including an increase in DNA repair proteins and aberrant cell death processes (Barzegar Behrooz et al., 2019). The Cancer Genome Atlas (TCGA) research network's comprehensive study of the genomic alterations revealed that the most frequently disrupted signaling cascades in GBM include modifications to pathways that are connected to receptor tyrosine kinase (RTK) signaling, TP53 pathway, CDK4/RB1 pathway.

Receptor Tyrosine Kinase (RTK) Signaling in Glioblastoma

More than 80% of primary GBMs have mutations or amplifications of RTK, including the insulin-like growth factor receptor (IGFR-1), the basic fibroblast growth factor receptor 1 (FGFR-1), and the epidermal growth factor receptor (EGFR)(Bezecny, 2014). These structurally related proteins work together to control a convoluted signaling network that powers and controls numerous cellular processes. The RAS/MAPK pathway, which promotes cellular proliferation, differentiation, and migration in gliomas, and the PI3K/AKT/mTOR pathway, which primarily works to promote cell proliferation and survival by advancing the cell cycle and inhibiting apoptosis, are the two main signaling pathways used by RTK (Bezecny, 2014). PTEN, a tumor suppressor gene and an antagonist of this pathway, controls the activity of PI3K (Brown et al., 2017). A major contributor to resistance to EGFR therapies is PTEN loss, which is present in 36% of gliomas. Loss of PTEN may cause a dramatic upregulation of this pathway (Del Vecchio et al., 2013).

The majority of genetic alterations, which affect nearly 57% of GBM tumors, include EGFR mutations, rearrangements, alternative splicing, and focal amplifications (Gomez-Manzano et al., 1996). A classical subtype of GBM patient will almost always have EGFR amplification, whereas secondary GBMs are extremely uncommon to have this mutation (Sherr& Roberts, 1999). EGFR inhibitors, such as gefitinib and erlotinib, have sadly not been successfully introduced into clinical trials for patients with GBMs despite the high frequency of EGFR gene amplification (Chow et al., 2017). Additionally, EGFRvIII mutation, which results from an in-frame deletion of exons 2–7 and promotes constitutive and ligand-independent receptor activity, is present in 50% of patients with EGFR amplification (Clarke & Kumar, 2005). The prognostic value of EGFRvIII, however, is still debatable. For instance, Shinojima et al. have demonstrated that EGFRvIII expression in the presence of EGFR amplification is a potent predictor of poor survival and prognosis (Coffey et al., 2001). However, in patients with newly diagnosed GBM treated with gross total resection and conventional radio chemotherapy (TMZ), Montano et al. prospectively examined the association between EGFRvIII expression and overall survival. Notably, EGFRvIII found that patients had noticeably longer overall survival in this case.

In terms of additional RTK alterations, high-level PDGFRA amplification has been found in 13% of secondary GBMs, and amplification and/or mutation of EGFR were found in nearly 50% of these tumors (Haas-Kogan et al., 2005). Even though PDGFRA is closely linked to GBM, glivec anti-PDGFRA therapy only had a modest clinical impact (Reardon et al., 2009). Despite being much less common in GBMs, alterations like c-Met amplification and FGFR mutations have been detected in 2% of the GBMs that have been studied (Kuhnt et al., 2011).

The simultaneous expression of mutations in the downstream elements of growth factor receptor pathways is possible with all these activating genetic abnormalities in numerous RTKs inside a single GBM. Mutations in either the catalytic (PIK3CA) or regulatory (PIK3R1) domains of PI3K may activate this oncogenic pathway, which is the most potent in GBM (Parsons et al., 2008). Nearly 10% of the GBMs in the TCGA analysis exhibited PIK3R1 mutations, which have not been reported to be widely expressed in any other cancers (Johnson & O'Neill, 2012). The tumor suppressor gene NF1, which produces the RAS-negative regulator neurofibromin, is mutated or deleted in 15% to 18% of primary GBMs (mesenchymal subtype), according to more recent research (Di Carlo et al., 2019).

Alteration of Tumor Protein 53 (TP53) Pathway in Glioblastoma

The target genes for cell cycle arrest in the G1 and/or G2 phase, cell death and differentiation, DNA repair, and neovascularization are all controlled by the p53 protein encoded by the TP53 tumor suppressor gene on chromosome 17q13.1 (Dmytriyeva et al., 2012). TP53 may become inactive indirectly through mutation or deletion, or directly through harm to associated genes (Dong et al., 2017). TP53 mutations were initially more commonly found in secondary GBMs (65% versus 28%) (Franceschi et al., 2007) than primary GBMs, but more recent TCGA data have revealed that TP53 mutations are present in a significant number of primary GBMs. The overall frequency of genetic alterations in the TP53 pathway was found in 87% of GBMs, with 35% of them being caused by TP53 mutations or homozygous deletion(Johnson & O'Neill, 2012).

Retinoblastoma Pathway (RB 1 Pathway)

To promote the change from G1 to S-phase, the RB1 protein is crucial for controlling the cell cycle. The CDKN2A (p16INK4a) protein binds to CDK4 cyclin and inhibits the CDK4/cyclin D1 complex, which prevents this transition from being blocked (Padfield et al., 2015). Loss of RB1's usual function may result from altered expression of the p16INK4a, CDK4, or RB1 genes. Both primary and secondary GBMs are frequently found to have this pathway inactivated (Eyupoglu et al., 2006).

PUTATIVE BIOMARKERS FOR EARLY DETECTION OF GLIOBLASTOMA

The potential markers CD133, CD70 (CD27 L), S100A4, SOX-2, and Nestin are frequently used to identify or separate glioblastoma stem cells (GSCs). The roles of these markers and their participation in various important cellular signaling pathways may be better understood with further research, which could lead to the development of novel GBM treatments.

Figure 1. Biomarkers network axes in glioblastoma signaling pathways

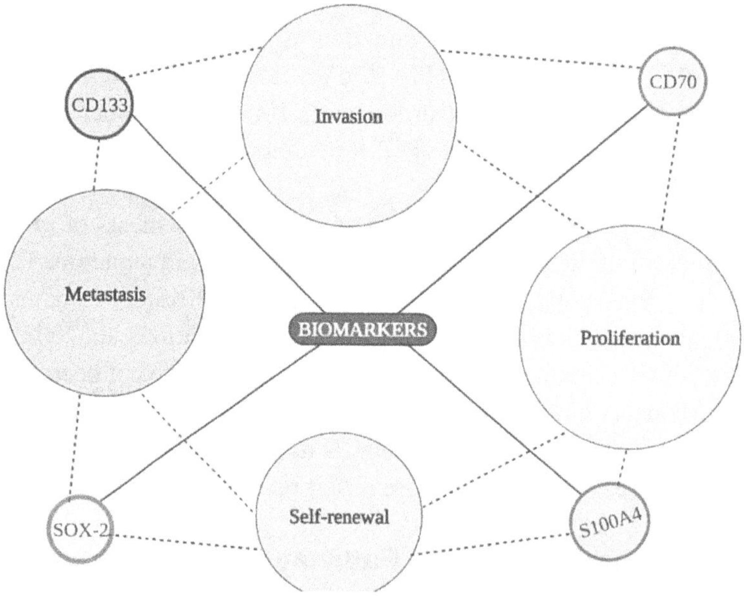

Prominin-1

CD133, also called Prominin-1, is a glycoprotein encoded by the PROM1 gene (Furnari et al., 2015) and found in the membrane protrusions of both healthy and malignant cells (Furnari et al., 2007). The PROM1 gene, which is found on chromosome 4p15, contains 37 exons and codes for a 120 kDa protein with five transmembrane domains (Keles et al., 1999). Numerous studies have shown that CD133 is a significant cancer stem cell marker involved in tumorigenesis and cancer progression in a variety of tumors such as gliomas, colorectal cancer, pancreatic cancer, ovarian cancer, prostate carcinoma, and hepatocellular carcinoma (Guzauskas et al., 2018). The upregulation of CD133 in glioblastoma has been extensively associated with the self-renewal of CSCs and their resistance to temozolomide (TMZ) through the activation of the c-Jun N-terminal kinase (JNK) signaling pathway (Garros-Regulez et al., 2016). There is still debate over CD133's suitability as a GSC marker despite the fact that it has been identified as one of the markers strongly associated with GSCs because of differential glycosylation, which can lead to variable epitope masking (Ge et al., 2017).

Cluster of Differentiation 70 (CD70)

CD70 belongs to the TNF receptor family as a type II transmembrane protein and is also referred to as CD27L (Gensert et al., 2007) which is another glycosylated transmembrane protein of TNF. CD27 is the sole identified ligand that binds to CD70, and it has been linked with poor prognosis in certain solid tumors and lymphomas (Ohgaki & Kleihues, 2009). Lymphomas and some solid tumors express CD70, and its interaction with CD27 controls the cytotoxic T cells' function. Recent studies have found that CD70 expression is not detectable in normal tissues from various organs but is strongly expressed in glioma tissues, indicating that CD70 expression is typically limited to tumors (Gopal et al., 2016).

S100A4

The calcium-binding protein S100A4 also referred to as FSP1/mts-1/metastasin(Ohgaki&Kleihues, 2007) is a member of the S100 protein family and is essential for epithelial-mesenchymal transition (EMT) (Ohgaki et al., 2004). In the central nervous system, this protein exhibits a wide range of trophic factor activity (Hooker et al., 2010) S100A4 is noticeably overexpressed in damaged human and rodent brains, according to studies (Mellinghoff et al., 2005). In many cancers, including glioblastoma, S100A4 has been identified as a prognostic indicator, metastasis promoter, and regulator (Ohgaki & Kleihues, 2007).

SOX-2

SRY-box 2 (SOX-2) is a transcription factor that belongs to the SOX family of the sry-related high-mobility group (HMG) boxes (Dimou et al., 2020). It was first discovered in 1994 and is a gene that codes for a 317 amino acid protein and is found on chromosome 3q26.3-q27(Huse & Aldape, 2014). Different solid tumors, such as glioblastomas, melanomas, lung cancer, prostate cancer, and breast cancer, have been found to express SOX-2 abnormally (Wang et al., 2015). SOX-2 expression is frequently increased in gliomas, where it is known to be essential for growth and survival and to be strongly correlated with relapse following chemotherapy or radiotherapy (Jin et al., 2018)also inhibition of SOX-2 causes differentiated glioblastoma cells to undergo cellular senescence (Johanns et al., 2016).

PRE-CLINICAL AND CLINICAL STUDIES OF GLIOBLASTOMA

Clinical trials have shown low efficacy due to the difficulty in developing drugs that specifically target important oncogenes, epigenetic enzymes, and pathways found in glioblastoma cells (Anton et al., 2012); it has also been challenging to find drugs that target specific targets because of the poor stability and unfavorable pharmacokinetic characteristics of many drugs or molecules. A Lys 27-to-methionine (K27M) mutation in one allele of the H3F3A gene, which codes for the histone H3 variant H3.3, has been found in 60% of high-grade pediatric glioma cases recently (Vousden & Lane et al., 2007). Therefore, investigating the precise histone and DNA alterations connected to glioblastoma may aid in broadening the scope of the current investigation into the epigenetic causes of gliomagenesis.

HDAC (Histone deacetylase) inhibitors are a class of medications that focus on the epigenetic changes in tumor cells (Kazama et al., 2018). Epigenetic modifications are those that modify the cellular phenotype and gene expression without altering the DNA sequence. Histone modification is a crucial mechanism in the development, growth, and resistance to therapy of tumors (Kelly & Cowley et al., 2013) by regulating the structure and function of chromatin, histone proteins play a critical part in regulating gene expression. The histone code, which refers to how these histone proteins regulate chromatin remodeling, is based on post-translational modifications of histone tails, such as acetylation, methylation, ubiquitination, and phosphorylation (Kazama et al., 2018). When used alone or in conjunction with radiation therapy, HDAC inhibitors work to balance out the activity of the enzyme histone acetyltransferase (HAT) and HDAC. This improves the body's inherent capacity to fight cancer. Since HDAC inhibitors can cross the blood-brain barrier (BBB) (Anton et al., 2012) they are particularly useful in the treatment of neurologic cancers.

By causing cell-cycle arrest, intrinsic and extrinsic apoptotic mechanisms, mitotic cell death, autophagic cell death, producing reactive oxygen species (ROS), obstructing angiogenesis, and enhancing natural killer (NK) cell-mediated tumor immunity, HDAC inhibitors have been shown to be effective in preclinical and clinical settings. It is crucial to go over the basic workings of HDAC inhibitor therapy as well as the preclinical and clinical applications of HDAC inhibitors in the treatment of glioblastoma.

Figure 2. The role of HDAC and HAT on chromatin structure

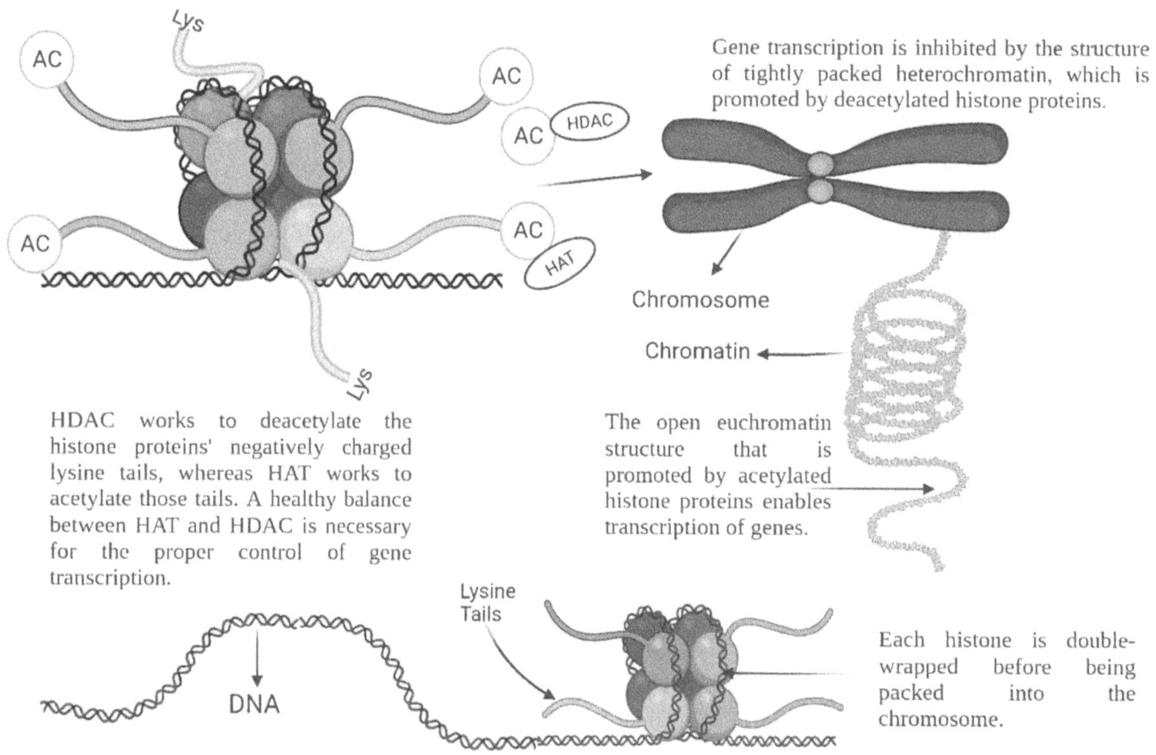

HAT and HDAC Regulation to Therapy

Histone acetyltransferases (HATs) and histone deacetylases (HDACs), which add and remove acetyl groups from lysine residues, respectively, act in opposition to each other to control histone acetylation, a crucial aspect of epigenetic regulation. Histone lysine residues are acetylated by HATs while acetyl groups are removed from lysine tails by HDACs (Li et al., 2017)while increased HDAC activity results in decreased gene transcription, increased HAT activity enhances gene transcription (Lin et al., 2021). HDACs can control a large number of genes' expression by interacting directly with transcription factors like E2f, Stat3, p53, retinoblastoma protein, NFkB, and TFIIE. HDACs also aim for nonhistone proteins, which are crucial for maintaining cellular homeostasis and are involved in processes like cell cycle progression, differentiation, and apoptosis. Tumor cells frequently have these pathways damaged, making them vulnerable to HDAC inhibitor therapy. The roles of HDAC and HAT on chromatin structure are shown in Figure 2 and it is done using BioRender.com.

Short-Chain Fatty Acids

In preclinical studies for the treatment of glioblastoma, sodium butyrate, a class I and IIa HDAC inhibitor, is being investigated. It has also advanced to phase II clinical trials as an endogenous antibiotic. Sodium butyrate causes cell cycle arrest by activating p16 and p21, stabilizing p21 mRNA, and stabilizing p21 mRNA (Vousden & Lane, 2007)by up-regulating CD81, it also causes astrocyte growth arrest and differentiation (Miller et al., 2019) It causes apoptosis in glioblastoma cells by up-regulating the Bad protein. Additionally, it has been discovered that sodium butyrate inhibits the growth of tumor vasculature by lowering the expression of vascular endothelial growth factor (VEGF) (Lacroix et al., 2001).

Benzamides

Entinostat (MS-275), a type I HDAC inhibitor, is undergoing phase III clinical trials for the treatment of breast cancer and is presently being tested in pre-clinical studies for the treatment of glioblastoma. According to studies, the combination of entinostat and temozolomide can stop the growth of glioma cells by increasing p21 and causing G_0/G_1 cell cycle arrest (Nakada et al., 2011). Apoptosis may also result from it. Entinostat has been demonstrated to cause mitochondrial and ROS-induced cell death in other cancer cell lines, and it may also do so in glioblastoma. Less than 0.001% of an injected dose of entinostat per cubic cm is distributed in the brain tissue, despite the fact that it can cross the blood-brain barrier (Davis, 2016). Therefore, to increase the efficacy of entinostat therapy, drug-targeting techniques may be required.

Hydroxamate Derivatives

CBHA (m-carboxy cinnamic acid bis-hydroxide). Pre-clinical trials for the treatment of glioblastoma are being conducted with the HDAC 3 inhibitor CBHA. In vitro, CBHA combined with retinoids induced apoptosis in neuroblastoma cells and combined with retinoic acid inhibited neuroblastoma in mice with severe combined immunodeficiency (SCID) (Oppenlander et al., 2014).

HDAC-42, also known as AR-42. Currently in phase I clinical trials for acute myeloid leukemia, AR-42 inhibits a number of classical HDACs. Inhibiting tumorigenesis in an in vivo model, according to research on glioblastoma cell lines, is possible because AR-42 inhibits the activity of the enzyme telomerase (Cohen et al., 2005).

Subric bis-hydroxamate, or SBHA is an HDAC 1 and 3 inhibitor called SBHA is presently undergoing pre-clinical testing. In a study on human glioma cells, it was discovered that SBHA increased p21 and decreased Cdc-2, which resulted in the downregulation of survivin and the x-linked inhibitor of apoptosis protein (XIAP), which in turn caused TRAIL to induce apoptosis (Navari et al., 2016). HDAC inhibitors with the molecular targets HDAC 1, 3, and 8 include Scriptaid. It is currently being tested in preclinical studies for the treatment of glioblastoma. In one study, it was discovered that it activates Jun N-terminal kinase (JNK), which reduces cell proliferation and induces apoptosis in glioblastoma cells (Shinojima et al., 2003).

TREATMENT STRATEGIES FOR GLIOBLASTOMA

Managing GBM is a multifaceted process that is tailored to the individual patient and can be quite expensive. It has been approximated that the total cost of treating GBM for each patient is currently around USD 95,000 on average (Ostrom et al., 2014). Imaging techniques such as positron emission tomography (PET) or magnetic resonance imaging (MRI) are used to map the location and size of GBM in the brain (Ostrom et al., 2015).

Surgical Excision

Depending on the size and location of the tumor post-diagnosis patients typically undergo surgical resection of the tumor mass followed by chemo and radiotherapy (XRT) (Wen & Kesari, 2008). According to clinical reports, patients benefit from aggressive surgical tumor resection (Pankratova et al., 2018). Numerous studies have revealed that patients who have extensive surgical tumor debulking—that is, removal of the tumor more than 1-2 cm from its boundary—have longer overall survival (OS) times than those who only have less extensive removal of the tumor (Perry et al., 2017).

Anti-Glioblastoma Drugs

Patients with GBM are typically treated with XRT and temozolomide (TMZ) daily in conjunction with surgical intervention as the first line of treatment (Phillips et al., 2006)with a dose ranging from 54 to 60 Gy, XRT is typically administered over a period of six weeks (Pratt et al., 2017). Due to the poor efficacy of many agents, the use of chemotherapy in the treatment of GBM has historically been a contentious topic, but the release of TMZ in 2005 has since made it a viable option. An increase in survival time from an average of 9 to 10 months to 14 months has been observed when TMZ is added to the XRT regimen (Abounader & Laterra, 2005). Furthermore, randomized trials have shown that XRT plus TMZ results in significantly longer overall survival rates over 5 years compared to XRT alone (10.9% vs. 1.9%, respectively) (Faisal et al., 2023).

Tumour Treating Fields

A novel non-invasive treatment method known as "tumor treating fields" (TTF) involves exposing the tumor to low-intensity (1-2 V/cm) and intermediate-frequency (100 kHz to 1 MHz) electric fields. In GBM cells that are dividing quickly, this method messes up mitotic processes, which leads to cell death. Mild dermatitis and localized skin irritation are frequent side effects of TTF treatment, which involves placing electrodes on the scalp that has been shaved (Rezaianzadeh et al., 2012). For both newly diagnosed and recurrent GBM, TTF has been approved for use (Mittal et al., 2017). TTF is a promising treatment option because studies have shown that it can significantly increase overall survival when compared to radiochemotherapy alone (Parisi et al., 2015).

Radiation and Chemotherapy

After surgical resection, fractionated radiation and concurrent chemotherapy with TMZ are the standard treatments for GBM patients. Hypofractionated radiation plus chemotherapy has been shown to prolong median overall survival (median OS) in patients over the age of 65 (9.3 months vs. 7.6 months) and was well tolerated (Sawa et al., 2002). Patients who are elderly or have poor performance status should be given this treatment protocol, according to recommendations. The gold standard for GBM chemotherapy is temozolomide, an oral cytotoxic DNA-alkylating medication.

Table 1. Treatment modality of Glioblastoma and its side effects

Treatment Modality	Side Effects	References
Surgery	Surgical site infection	Ali et al., 2015
	Wound healing complications	Szopa et al., 2017
Radiation	Therapy Fatigue	Sharma et al., 2013
	Hair loss	Parisi et al., 2015
	Skin changes (redness, dryness, or itching)	Hirsch et al., 2017
	Cognitive changes (memory loss, attention difficulties)	Sturm et al., 2014
Chemotherapy	Nausea and vomiting	Cohen et al., 2005
	Fatigue	Stupp, 2005
	Low blood cell counts (anemia, leukopenia, thrombocytopenia)	Shi et al., 2022
	Increased risk of infection	Sharma et al 2013
Targeted Therapies	Skin rash	Miller et al., 2019
	Gastrointestinal disturbances (diarrhea, nausea)	Miller et al., 2019
	Fatigue	Miller et al., 2019
	Liver toxicity	Miller et al., 2019

Note: The frequent adverse reactions to treatment for glioblastoma are summarized in the table above. It is crucial to remember that each person will experience side effects differently, and some side effects may not be listed here.

SIDE EFFECTS OF GLIOBLASTOMA TREATMENT

With both disease and treatment, there are frequent side effects. Radiation dose and treatment fields as well as extensive surgery can have neurological effects (Table 1). Fatigue, skin changes, and hair loss are some of the acute radiation side effects, whereas leukoencephalopathy and cognitive changes are rare long-term side effects. It is crucial to continuously check for temozolomide-related side effects and toxicity, such as nausea or vomiting, fatigue, and hematologic effects (such as thrombocytopenia, lymphopenia, and neutropenia). In addition to routine laboratory tests like complete blood counts and liver function to check for severe liver toxicity, which can be fatal, prophylactic antibiotics for Pneumocystis pneumonia should be taken into consideration. Sometimes it may be necessary to delay or reduce the dose.

PHYSIOLOGICAL CHANGES IN GLIOBLASTOMA PATIENT

The majority of people with GBM (glioblastoma) typically have a short clinical history, lasting between three and six months. However, the clinical history may span a number of years when the tumor is derived from a low-grade astrocytoma (Shabason et al., 2011). Patients with GBM may experience a range of signs and symptoms that are mainly caused by two different types of mechanisms which includes primary and secondary type of effects.

In primary condition brain tissue destruction occurs when necrosis leads to the destruction of brain cells, resulting in various symptoms like focal neural deficits (40-60%) and cognitive impairments. The specific signs and symptoms caused by the tumor depend on the areas of the brain affected by the malignancy. In the case of some GBM patients, a serious condition may occur known as brain tissue destruction this happens when necrosis leads to the destruction of brain cells, resulting in various symptoms like focal neural deficits (40-60%) and cognitive impairments. The specific signs and symptoms caused by the tumor depend on the areas of the brain affected by the malignancy (Shabason et al., 2011).

In the case of secondary mechanism increased intracranial pressure occurs in between 30 and 50 percent of GBM (glioblastoma) patients, the onset of headaches is directly correlated with the tumor's slow growth and the edema that has accumulated around it. Headaches, which are regarded as a characteristic feature, are brought on by a shift in the contents of the skull as a result of this (Shabason et al., 2011). GBM patients have these types of typical headaches with no significant pattern and regularity, these can lead to vomiting and papilledema in some cases.

Numerous studies have been conducted on how cancer cells and the vasculature interact. Although vessel co-option has received less attention, angiogenesis is still widely acknowledged as the key process in this interaction. The term "vessel co-option" describes a non-angiogenic mechanism by which tumors take advantage of pre-existing blood vessels. Blood and/or lymph vessel invasion (LBVI), also known as a vascular invasion (VI), is the process by which tumor cells infiltrate the lumen of blood and/or lymph vessels, resulting in the presence of circulating tumor cells. In patients without lymph node involvement who have breast carcinoma, bladder cancer, non-small cell lung cancer, or colorectal cancer, the presence of vascular invasion has been identified as a significant prognostic factor for poor clinical outcomes (Schwartz et al., 2020). Lymph node-negative patients who are at higher risk can be identified by including vascular invasion in the conventional prognostic factors and may be candidates for adjuvant therapy. So as a whole, the main key differences between a healthy brain and an affected brain are the cell growth and division that are carefully controlled in the case of a healthy brain by a sophisticated signaling system. This guarantees that old cells are eliminated through programmed cell death and that new cells are produced as needed. In contrast, glioblastoma cells multiply quickly and uncontrollably, which results in the growth of a tumor.

Also, the blood-brain barrier is a defence mechanism that controls how things get from the bloodstream to brain tissue. It contributes to preserving a stable environment for ideal brain performance. The blood-brain barrier may be damaged in the case of glioblastoma-affected patients, allowing cancer cells to travel from the bloodstream into the brain and vice versa. The safety of the environment around the brain is compromised by this breach.

The immune system responds to inflammation when glioblastoma is present in the brain. Even though it is a defence mechanism, sustained and chronic inflammation can harm healthy brain tissue. The symptoms and development of glioblastoma may be influenced by this immune response, which may also have an adverse effect on brain function.

Figure 3. Difference between a healthy brain and a glioblastoma-affected brain

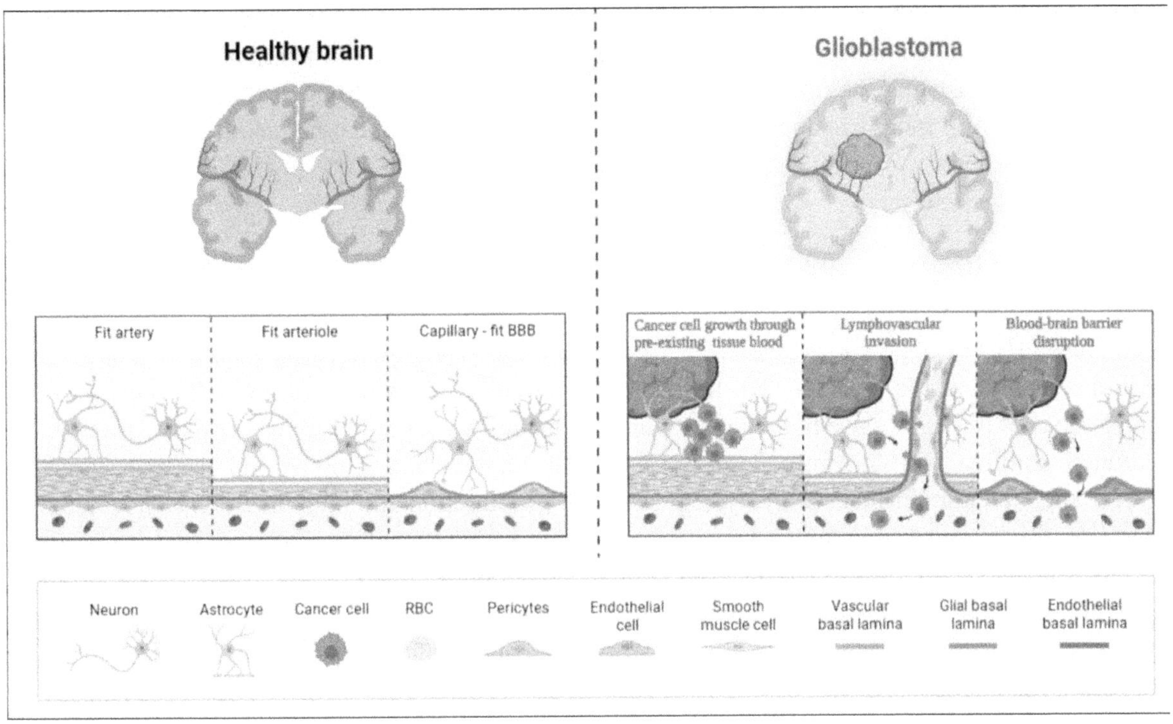

So, understanding the physiological differences between a normal brain and a glioblastoma-affected brain is very much crucial for early detection, accurate diagnosis, and effective treatment of individuals.

A FUTURE PERSPECTIVE ON GLIOBLASTOMA

Glioblastoma is a highly aggressive brain tumor associated with poor prognosis and limited treatment options. However, ongoing research and technological advancements have opened new possibilities for the future of glioblastoma treatment. This article explores the advancements and novel strategies that hold promise in overcoming the challenges associated with glioblastoma.

Targeted therapies are one area of development. Specific molecular targets and signaling pathways that are essential for the development and growth of tumors have been identified through the molecular characterization of glioblastoma. Tyrosine kinase inhibitors and monoclonal antibodies, two types of targeted agents, have been created to interfere with these targets and pathways, improving treatment outcomes (Sawa et al., 2002). Targeted therapies for the treatment of glioblastoma are made even more effective by personalized medicine methods based on unique genomic profiles (Phillips et al., 2006).

Another potential treatment option for glioblastoma is gene therapy. Viral vectors make it easier to deliver therapeutic genes into tumor cells, while CRISPR-Cas9 gene editing technology allows the precise modification of genes linked to cancer (Li et al., 2017). Gene therapies are intended to stop the growth of tumor cells, cause apoptosis, or improve the efficiency of other therapeutic approaches (Wen & Kesari, 2008). Ongoing research aims to enhance the safety and effectiveness of these methods.

In conclusion, the future of glioblastoma treatment is promising with advancements and novel strategies such as targeted therapies, immunotherapies, gene therapies, and combination treatments. Ongoing research and collaborative efforts among researchers, clinicians, and pharmaceutical companies are crucial for translating these advancements into clinical practice. Although challenges remain, the pursuit of innovative treatment approaches offers hope for improved outcomes and a brighter future for glioblastoma patients.

CONCLUSION

In conclusion, a thorough knowledge of the physiopathology and complications associated with glioblastoma, as well as the physiological function of glial cells, offers important new insights into the complexities of this aggressive brain tumor. Glial cell mutations in glioblastoma's origin cause them to malfunction normally, which promotes the growth of the tumor. The aggressive and invasive characteristics of the tumor are influenced by dysregulated cell proliferation and immune evasion. The tumor's effects on the surrounding brain tissue cause complications that result in cognitive and neurological deficits. Understanding the critical function of glial cells in preserving brain homeostasis and promoting neuronal function is essential. In order to create therapies that effectively target signaling pathways and the tumor microenvironment and ultimately improve patient outcomes, more research is required to fully understand these connections.

REFERENCES

Abounader, R., & Laterra, J. (2005). Scatter factor/hepatocyte growth factor in brain tumor growth and angiogenesis. *Neuro-Oncology*, *7*(4), 436–451. doi:10.1215/S1152851705000050 PMID:16212809

Ahmad, M., Hamid, A., Hussain, A., Majeed, R., Qurishi, Y., Bhat, J. A., Najar, R. A., Qazi, A. K., Zargar, M. A., Singh, S. K., & Saxena, A. K. (2012). Understanding histone deacetylases in the cancer development and treatment: An epigenetic perspective of cancer chemotherapy. *DNA and Cell Biology*, *31*(S1), S-62–S-71. doi:10.1089/dna.2011.1575 PMID:22462686

Ali, M., Bennardo, M., Almenawer, S. A., Zagzoog, N., Smith, A. A., Dao, D., BHSc, Ajani, O., Farrokhyar, F., & Singh, S. K. (2015). Exploring predictors of surgery and comparing operative treatment approaches for pediatric intracranial arachnoid cysts: A case series of 83 patients. *Journal of Neurosurgery. Pediatrics*, *16*(3), 275–282. doi:10.3171/2015.2.PEDS14612 PMID:26067335

Aly, A., Singh, P., Korytowsky, B., Ling, Y. L., Kale, H. P., Dastani, H. B., & Norden, A. D. (2020). Survival, costs, and health care resource use by line of therapy in US Medicare patients with newly diagnosed glioblastoma: A retrospective observational study. *Neuro-Oncology Practice*, *7*(2), 164–175. PMID:32626585

Ambartsumian, N., Klingelhöfer, J., & Grigorian, M. (2019). The multifaceted S100A4 protein in cancer and inflammation. *Calcium-binding proteins of the EF-hand superfamily: from basics to medical applications*, 339-365.

Anton, K., Baehring, J. M., & Mayer, T. (2012). Glioblastoma multiforme: Overview of current treatment and future perspectives. *Hematology/Oncology Clinics of North America*, *26*(4), 825–853. doi:10.1016/j.hoc.2012.04.006 PMID:22794286

Barzegar Behrooz, A., Syahir, A., & Ahmad, S. (2019). CD133: Beyond a cancer stem cell biomarker. *Journal of Drug Targeting*, *27*(3), 257–269. doi:10.1080/1061186X.2018.1479756 PMID:29911902

Bezecny, P. (2014). Histone deacetylase inhibitors in glioblastoma: Pre-clinical and clinical experience. *Medical Oncology (Northwood, London, England)*, *31*(6), 985. doi:10.100712032-014-0985-5 PMID:24838514

Brown, D. V., Filiz, G., Daniel, P. M., Hollande, F., Dworkin, S., Amiridis, S., Kountouri, N., Ng, W., Morokoff, A. P., & Mantamadiotis, T. (2017). Expression of CD133 and CD44 in glioblastoma stem cells correlates with cell proliferation, phenotype stability and intra-tumor heterogeneity. *PLoS One*, *12*(2), e0172791. doi:10.1371/journal.pone.0172791 PMID:28241049

Chow, K. H., Park, H. J., George, J., Yamamoto, K., Gallup, A. D., Graber, J. H., Chen, Y., Jiang, W., Steindler, D. A., Neilson, E. G., Kim, B. Y. S., & Yun, K. (2017). S100A4 is a biomarker and regulator of glioma stem cells that is critical for mesenchymal transition in glioblastoma. *Cancer Research*, *77*(19), 5360–5373. doi:10.1158/0008-5472.CAN-17-1294 PMID:28807938

Clarke, C. R. A. (2005). *Neurological diseases in Kumar & Clark Clinical Medicine*. Kumar P and Clark M.

Coffey, D. C., Kutko, M. C., Glick, R. D., Butler, L. M., Heller, G., Rifkind, R. A, & La Quaglia, M. P. (2001). The histone deacetylase inhibitor, CBHA, inhibits growth of human neuroblastoma xenografts in vivo, alone and synergistically with all-trans retinoic acid. *Cancer Research*, *61*(9), 3591–3594. PMID:11325825

Cohen, M. H., Johnson, J. R., & Pazdur, R. (2005). Food and Drug Administration Drug approval summary: Temozolomide plus radiation therapy for the treatment of newly diagnosed glioblastoma multiforme. *Clinical Cancer Research*, *11*(19), 6767–6771. doi:10.1158/1078-0432.CCR-05-0722 PMID:16203762

Davis, M. E. (2016). Glioblastoma: Overview of disease and treatment. *Clinical Journal of Oncology Nursing*, *20*(5), S2–S8. doi:10.1188/16.CJON.S1.2-8 PMID:27668386

De Barros, A., Attal, J., Roques, M., Nicolau, J., Sol, J. C., Cohen-Jonathan-Moyal, E., & Roux, F. E. (2019). Impact on survival of early tumor growth between surgery and radiotherapy in patients with de novo glioblastoma. *Journal of Neuro-Oncology*, *142*(3), 489–497. doi:10.100711060-019-03120-3 PMID:30783874

Del Vecchio, C. A., Giacomini, C. P., Vogel, H., Jensen, K. C., Florio, T., Merlo, A., Pollack, J. R., & Wong, A. J. (2013). EGFRvIII gene rearrangement is an early event in glioblastoma tumorigenesis and expression defines a hierarchy modulated by epigenetic mechanisms. *Oncogene*, *32*(21), 2670–2681. doi:10.1038/onc.2012.280 PMID:22797070

Di Carlo, D. T., Cagnazzo, F., Benedetto, N., Morganti, R., & Perrini, P. (2019). Multiple high-grade gliomas: Epidemiology, management, and outcome. A systematic review and meta-analysis. *Neurosurgical Review*, *42*(2), 263–275. doi:10.100710143-017-0928-7 PMID:29138949

Dimou, J., Beland, B., & Kelly, J. (2020). Supramaximal resection: A systematic review of its safety, efficacy and feasibility in glioblastoma. *Journal of Clinical Neuroscience*, *72*, 328–334. doi:10.1016/j.jocn.2019.12.021 PMID:31864830

Dmytriyeva, O., Pankratova, S., Owczarek, S., Sonn, K., Soroka, V., Ridley, C. M., Marsolais, A., Lopez-Hoyos, M., Ambartsumian, N., Lukanidin, E., Bock, E., Berezin, V., & Kiryushko, D. (2012). The metastasis-promoting S100A4 protein confers neuroprotection in brain injury. *Nature Communications*, *3*(1), 1197. doi:10.1038/ncomms2202 PMID:23149742

Dong, H., Hao, X., Cui, B., & Guo, M. (2017). MiR-429 suppresses glioblastoma multiforme by targeting SOX2. *Cell Biochemistry and Function*, *35*(5), 260–268. doi:10.1002/cbf.3271 PMID:28749077

Eyupoglu, I. Y., Hahnen, E., Trankle, C., Savaskan, N. E., Siebzehnrubl, F. A., Buslei, R., Lemke, D., Wick, W., Fahlbusch, R., & Blumcke, I. (2006). Experimental therapy of malignant gliomas using the inhibitor of histone deacetylase MS-275. *Molecular Cancer Therapeutics*, *5*(5), 1248–1255. doi:10.1158/1535-7163.MCT-05-0533 PMID:16731757

Fabi, A., Vidiri, A., Carapella, C., Pace, A., Occhipinti, E., Caroli, F., & Cognetti, F. (2004). Bone metastasis from glioblastoma multiforme without central nervous system relapse: A case report. *Anticancer Research*, *24*(4), 2563–2566. PMID:15330215

Faisal, S. M., Castro, M. G., & Lowenstein, P. R. (2023). Adenovirus-mediated combined cytotoxic and immune-stimulatory gene therapy using dual-vector Ad-hCMV-TK and Ad-hCMV-Flt3L: translational explorations from rodents to glioma patients. *Molecular Therapy*.

Franceschi, E., Cavallo, G., Lonardi, S., Magrini, E., Tosoni, A., Grosso, D., Scopece, L., Blatt, V., Urbini, B., Pession, A., Tallini, G., Crinò, L., & Brandes, A. A. (2007). Gefitinib in patients with progressive high-grade gliomas: A multicentre phase II study by Gruppo Italiano Cooperativo di Neuro-Oncologia (GICNO). *British Journal of Cancer*, *96*(7), 1047–1051. doi:10.1038j.bjc.6603669 PMID:17353924

Furnari, F. B., Cloughesy, T. F., Cavenee, W. K., & Mischel, P. S. (2015). Heterogeneity of epidermal growth factor receptor signaling networks in glioblastoma. *Nature Reviews. Cancer*, *15*(5), 302–310. doi:10.1038/nrc3918 PMID:25855404

Furnari, F. B., Fenton, T., Bachoo, R. M., Mukasa, A., Stommel, J. M., Stegh, A., Hahn, W. C., Ligon, K. L., Louis, D. N., Brennan, C., Chin, L., DePinho, R. A., & Cavenee, W. K. (2007). Malignant astrocytic glioma: Genetics, biology, and paths to treatment. *Genes & Development*, *21*(21), 2683–2710. doi:10.1101/gad.1596707 PMID:17974913

Garros-Regulez, L., Aldaz, P., Arrizabalaga, O., Moncho-Amor, V., Carrasco-Garcia, E., Manterola, L., Moreno-Cugnon, L., Barrena, C., Villanua, J., Ruiz, I., Pollard, S., Lovell-Badge, R., Sampron, N., Garcia, I., & Matheu, A. (2016). mTOR inhibition decreases SOX2-SOX9 mediated glioma stem cell activity and temozolomide resistance. *Expert Opinion on Therapeutic Targets*, *20*(4), 393–405. doi:10.1517/14728222.2016.1151002 PMID:26878385

Ge, H., Mu, L., Jin, L., Yang, C., Chang, Y., Long, Y., DeLeon, G., Deleyrolle, L., Mitchell, D. A., Kubilis, P. S., Lu, D., Qi, J., Gu, Y., Lin, Z., & Huang, J. (2017). Tumor associated CD70 expression is involved in promoting tumor migration and macrophage infiltration in GBM. *International Journal of Cancer, 141*(7), 1434–1444. doi:10.1002/ijc.30830 PMID:28612394

Gensert, J. M., Baranova, O. V., Weinstein, D. E., & Ratan, R. R. (2007). CD81, a cell cycle regulator, is a novel target for histone deacetylase inhibition in glioma cells. *Neurobiology of Disease, 26*(3), 671–680. doi:10.1016/j.nbd.2007.03.008 PMID:17481908

Gomez-Manzano, C., Fueyo, J., Kyritsis, A. P., Steck, P. A., Roth, J. A., McDonnell, T. J, & Alfred Yung, W. K. (1996). Adenovirus-mediated transfer of the p53 gene produces rapid and generalized death of human glioma cells via apoptosis. *Cancer Research, 56*(4), 694–699. PMID:8630997

Gopal, K., Gupta, N., Zhang, H., Alshareef, A., Alqahtani, H., Bigras, G., Lewis, J., Douglas, D., Kneteman, N., Lavasanifar, A., & Lai, R. (2016). Oxidative stress induces the acquisition of cancer stem-like phenotype in breast cancer detectable by using a Sox2 regulatory region-2 (SRR2) reporter. *Oncotarget, 7*(3), 3111–3127. doi:10.18632/oncotarget.6630 PMID:26683522

Guzauskas, G. F., Salzberg, M., & Wang, B. C. (2018). Estimated lifetime survival benefit of tumor treating fields and temozolomide for newly diagnosed glioblastoma patients. *CNS Oncology, 7*(3), CNS23. doi:10.2217/cns-2018-0010 PMID:30124334

Haas-Kogan, D. A., Prados, M. D., Tihan, T., Eberhard, D. A., Jelluma, N., Arvold, N. D., Baumber, R., Lamborn, K. R., Kapadia, A., Malec, M., Berger, M. S., & Stokoe, D. (2005). Epidermal growth factor receptor, protein kinase B/Akt, and glioma response to erlotinib. *Journal of the National Cancer Institute, 97*(12), 880–887. doi:10.1093/jnci/dji161 PMID:15956649

Hirsch, T., Rothoeft, T., Teig, N., Bauer, J. W., Pellegrini, G., De Rosa, L., Scaglione, D., Reichelt, J., Klausegger, A., Kneisz, D., Romano, O., Secone Seconetti, A., Contin, R., Enzo, E., Jurman, I., Carulli, S., Jacobsen, F., Luecke, T., Lehnhardt, M., & De Luca, M. (2017). Regeneration of the entire human epidermis using transgenic stem cells. *Nature, 551*(7680), 327–332. doi:10.1038/nature24487 PMID:29144448

Huse, J. T., & Aldape, K. D. (2014). The evolving role of molecular markers in the diagnosis and management of diffuse glioma. *Clinical Cancer Research, 20*(22), 5601–5611. doi:10.1158/1078-0432.CCR-14-0831 PMID:25398843

Jin, L., Ge, H., Long, Y., Yang, C., Chang, Y., Mu, L., Sayour, E. J., De Leon, G., Wang, Q. J., Yang, J. C., Kubilis, P. S., Bao, H., Xia, S., Lu, D., Kong, Y., Hu, L., Shang, Y., Jiang, C., Nie, J., & Huang, J. (2018). CD70, a novel target of CAR T-cell therapy for gliomas. *Neuro-Oncology, 20*(1), 55–65. doi:10.1093/neuonc/nox116 PMID:28651374

Johanns, T. M., Miller, C. A., Dorward, I. G., Tsien, C., Chang, E., Perry, A., Uppaluri, R., Ferguson, C., Schmidt, R. E., Dahiya, S., Ansstas, G., Mardis, E. R., & Dunn, G. P. (2016). Immunogenomics of hypermutated glioblastoma: A patient with germline POLE deficiency treated with checkpoint blockade immunotherapy. *Cancer Discovery, 6*(11), 1230–1236. doi:10.1158/2159-8290.CD-16-0575 PMID:27683556

Johnson, D. R., & O'Neill, B. P. (2012). Glioblastoma survival in the United States before and during the temozolomide era. *Journal of Neuro-Oncology*, *107*(2), 359–364. doi:10.100711060-011-0749-4 PMID:22045118

Kazama, S., Kishikawa, J., Kiyomatsu, T., Kawai, K., Nozawa, H., Ishihara, S., & Watanabe, T. (2018). Expression of the stem cell marker CD133 is related to tumor development in colorectal carcinogenesis. *Asian Journal of Surgery*, *41*(3), 274–278. doi:10.1016/j.asjsur.2016.12.002 PMID:28190751

Keles, G. E., Anderson, B., & Berger, M. S. (1999). The effect of extent of resection on time to tumor progression and survival in patients with glioblastoma multiforme of the cerebral hemisphere. *Surgical Neurology*, *52*(4), 371–379. doi:10.1016/S0090-3019(99)00103-2 PMID:10555843

Kelly, R. D., & Cowley, S. M. (2013). The physiological roles of histone deacetylase (HDAC) 1 and 2: Complex co-stars with multiple leading parts. *Biochemical Society Transactions*, *41*(3), 741–749. doi:10.1042/BST20130010 PMID:23697933

Kuhnt, D., Becker, A., Ganslandt, O., Bauer, M., Buchfelder, M., & Nimsky, C. (2011). Correlation of the extent of tumor volume resection and patient survival in surgery of glioblastoma multiforme with high-field intraoperative MRI guidance. *Neuro-Oncology*, *13*(12), 1339–1348. doi:10.1093/neuonc/nor133 PMID:21914639

Lacroix, M., Abi-Said, D., Fourney, D. R., Gokaslan, Z. L., Shi, W., DeMonte, F., Lang, F. F., McCutcheon, I. E., Hassenbusch, S. J., Holland, E., Hess, K., Michael, C., Miller, D., & Sawaya, R. (2001). A multivariate analysis of 416 patients with glioblastoma multiforme: Prognosis, extent of resection, and survival. *Journal of Neurosurgery*, *95*(2), 190–198. doi:10.3171/jns.2001.95.2.0190 PMID:11780887

Li, B., McCrudden, C. M., Yuen, H. F., Xi, X., Lyu, P., Chan, K. W., Zhang, S. D., & Kwok, H. F. (2017). CD133 in brain tumor: The prognostic factor. *Oncotarget*, *8*(7), 11144–11159. doi:10.18632/oncotarget.14406 PMID:28055976

Lin, H., Li, G., Peng, X., Deng, A., Ye, L., Shi, L., Wang, T., & He, J. (2021). The use of CRISPR/Cas9 as a tool to study human infectious viruses. *Frontiers in Cellular and Infection Microbiology*, *11*, 590989. doi:10.3389/fcimb.2021.590989 PMID:34513721

Miller, A. M., Shah, R. H., Pentsova, E. I., Pourmaleki, M., Briggs, S., Distefano, N., Zheng, Y., Skakodub, A., Mehta, S. A., Campos, C., Hsieh, W.-Y., Selcuklu, S. D., Ling, L., Meng, F., Jing, X., Samoila, A., Bale, T. A., Tsui, D. W. Y., Grommes, C., & Mellinghoff, I. K. (2019). Tracking tumour evolution in glioma through liquid biopsies of cerebrospinal fluid. *Nature*, *565*(7741), 654–658. doi:10.103841586-019-0882-3 PMID:30675060

Mittal, S., Klinger, N. V., Michelhaugh, S. K., Barger, G. R., Pannullo, S. C., & Juhász, C. (2017). Alternating electric tumor treating fields for treatment of glioblastoma: Rationale, preclinical, and clinical studies. *Journal of Neurosurgery*, *128*(2), 414–421. doi:10.3171/2016.9.JNS16452 PMID:28298023

Nakada, M., Kita, D., Watanabe, T., Hayashi, Y., Teng, L., Pyko, I. V., & Hamada, J. I. (2011). Aberrant signaling pathways in glioma. *Cancers (Basel)*, *3*(3), 3242–3278. doi:10.3390/cancers3033242 PMID:24212955

Navari, R. M. (2016). Olanzapine for the prevention of chemotherapy-induced nausea and vomiting. *Management of Chemotherapy-Induced Nausea and Vomiting: New Agents and New Uses of Current Agents*, 107-120. Springer.

Ohgaki, H., Dessen, P., Jourde, B., Horstmann, S., Nishikawa, T., Di Patre, P. L., Burkhard, C., Schüler, D., Probst-Hensch, N. M., Maiorka, P. C., Baeza, N., Pisani, P., Yonekawa, Y., Yasargil, M. G., Lütolf, U. M., & Kleihues, P. (2004). Genetic pathways to glioblastoma: A population-based study. *Cancer Research*, *64*(19), 6892–6899. doi:10.1158/0008-5472.CAN-04-1337 PMID:15466178

Ohgaki, H., & Kleihues, P. (2007). Genetic pathways to primary and secondary glioblastoma. *American Journal of Pathology*, *170*(5), 1445–1453. doi:10.2353/ajpath.2007.070011 PMID:17456751

Ohgaki, H., & Kleihues, P. (2009). Genetic alterations and signaling pathways in the evolution of gliomas. *Cancer Science*, *100*(12), 2235–2241. doi:10.1111/j.1349-7006.2009.01308.x PMID:19737147

Omuro, A. M., Faivre, S., & Raymond, E. (2007). Lessons learned in the development of targeted therapy for malignant gliomas. *Molecular Cancer Therapeutics*, *6*(7), 1909–1919. doi:10.1158/1535-7163.MCT-07-0047 PMID:17620423

Oppenlander, M. E., Wolf, A. B., Snyder, L. A., Bina, R., Wilson, J. R., Coons, S. W., Ashby, L. S., Brachman, D., Nakaji, P., Porter, R. W., Smith, K. A., Spetzler, R. F., & Sanai, N. (2014). An extent of resection threshold for recurrent glioblastoma and its risk for neurological morbidity. *Journal of Neurosurgery*, *120*(4), 846–853. doi:10.3171/2013.12.JNS13184 PMID:24484232

Ostrom, Q. T., Bauchet, L., Davis, F. G., Deltour, I., Fisher, J. L., Langer, C. E., Pekmezci, M., Schwartzbaum, J. A., Turner, M. C., Walsh, K. M., Wrensch, M. R., & Barnholtz-Sloan, J. S. (2014). The epidemiology of glioma in adults: A "state of the science" review. *Neuro-Oncology*, *16*(7), 896–913. doi:10.1093/neuonc/nou087 PMID:24842956

Ostrom, Q. T., Gittleman, H., Fulop, J., Liu, M., Blanda, R., Kromer, C., Wolinsky, Y., Kruchko, C., & Barnholtz-Sloan, J. S. (2015). CBTRUS statistical report: Primary brain and central nervous system tumors diagnosed in the United States in 2008-2012. *Neuro-Oncology*, *17*(suppl_4), iv1–iv62. doi:10.1093/neuonc/nov189 PMID:26511214

Padfield, E., Ellis, H. P., & Kurian, K. M. (2015). Current therapeutic advances targeting EGFR and EGFRvIII in glioblastoma. *Frontiers in Oncology*, *5*, 5. doi:10.3389/fonc.2015.00005 PMID:25688333

Pankratova, S., Klingelhofer, J., Dmytriyeva, O., Owczarek, S., Renziehausen, A., Syed, N., Porter, A. E., Dexter, D. T., & Kiryushko, D. (2018). The S100A4 protein signals through the ErbB4 receptor to promote neuronal survival. *Theranostics*, *8*(14), 3977–3990. doi:10.7150/thno.22274 PMID:30083275

Parisi, S., Corsa, P., Raguso, A., Perrone, A., Cossa, S., Munafò, T., Sanpaolo, G., Donno, E., Clemente, M. A., Piombino, M., Parisi, F., & Valle, G. (2015). Temozolomide and radiotherapy versus radiotherapy alone in high grade gliomas: A very long-term comparative study and literature review. *BioMed Research International*, *2015*, 2015. doi:10.1155/2015/620643 PMID:25815327

Parsons, D. W., Jones, S., Zhang, X., Lin, J. C. H., Leary, R. J., Angenendt, P., & Kinzler, K. W. (2008). An integrated genomic analysis of human glioblastoma multiforme. *science, 321*(5897), 1807-1812.

Perry, J. R., Laperriere, N., O'Callaghan, C. J., Brandes, A. A., Menten, J., Phillips, C., Fay, M., Nishikawa, R., Cairncross, J. G., Roa, W., Osoba, D., Rossiter, J. P., Sahgal, A., Hirte, H., Laigle-Donadey, F., Franceschi, E., Chinot, O., Golfinopoulos, V., Fariselli, L., & Mason, W. P. (2017). Short-course radiation plus temozolomide in elderly patients with glioblastoma. *The New England Journal of Medicine, 376*(11), 1027–1037. doi:10.1056/NEJMoa1611977 PMID:28296618

Phillips, H. S., Kharbanda, S., Chen, R., Forrest, W. F., Soriano, R. H., Wu, T. D., Misra, A., Nigro, J. M., Colman, H., Soroceanu, L., Williams, P. M., Modrusan, Z., Feuerstein, B. G., & Aldape, K. (2006). Molecular subclasses of high-grade glioma predict prognosis, delineate a pattern of disease progression, and resemble stages in neurogenesis. *Cancer Cell, 9*(3), 157–173. doi:10.1016/j.ccr.2006.02.019 PMID:16530701

Pratt, D., Pittaluga, S., Palisoc, M., Fetsch, P., Xi, L., Raffeld, M., Gilbert, M. R., & Quezado, M. (2017). Expression of CD70 (CD27L) is associated with epithelioid and sarcomatous features in IDH-Wild-Type glioblastoma. *Journal of Neuropathology and Experimental Neurology, 76*(8), 697–708. doi:10.1093/jnen/nlx051 PMID:28789475

Reardon, D. A., Dresemann, G., Taillibert, S., Campone, M., van den Bent, M., Clement, P., Blomquist, E., Gordower, L., Schultz, H., Raizer, J., Hau, P., Easaw, J., Gil, M., Tonn, J., Gijtenbeek, A., Schlegel, U., Bergstrom, P., Green, S., Weir, A., & Nikolova, Z. (2009). Multicentre phase II studies evaluating imatinib plus hydroxyurea in patients with progressive glioblastoma. *British Journal of Cancer, 101*(12), 1995–2004. doi:10.1038j.bjc.6605411 PMID:19904263

Rezaianzadeh, A., Talei, A., Rajaeefard, A., Hasanzadeh, J., Tabatabai, H., Tahmasebi, S., & Mousavizadeh, A. (2012). Vascular invasion as an independent prognostic factor in lymph node negative invasive breast cancer. *Asian Pacific Journal of Cancer Prevention, 13*(11), 5767–5772. doi:10.7314/APJCP.2012.13.11.5767 PMID:23317254

Sawa, H., Murakami, H., Ohshima, Y., Murakami, M., Yamazaki, I., Tamura, Y., Mima, T., Satone, A., Ide, W., Hashimoto, I., & Kamada, H. (2002). Histone deacetylase inhibitors such as sodium butyrate and trichostatin A inhibit vascular endothelial growth factor (VEGF) secretion from human glioblastoma cells. *Brain Tumor Pathology, 19*(2), 77–81. doi:10.1007/BF02478931 PMID:12622137

Shabason, J. E., Tofilon, P. J., & Camphausen, K. (2011). Grand rounds at the National Institutes of Health: HDAC inhibitors as radiation modifiers, from bench to clinic. *Journal of Cellular and Molecular Medicine, 15*(12), 2735–2744. doi:10.1111/j.1582-4934.2011.01296.x PMID:21362133

Sharma, N. L., Groselj, B., Hamdy, F. C., & Kiltie, A. E. (2013). The emerging role of histone deacetylase (HDAC) inhibitors in urological cancers. *BJU International, 111*(4), 537–542. doi:10.1111/j.1464-410X.2012.11647.x PMID:23551441

Sherr, C. J., & Roberts, J. M. (1999). CDK inhibitors: Positive and negative regulators of G1-phase progression. *Genes & Development, 13*(12), 1501–1512. doi:10.1101/gad.13.12.1501 PMID:10385618

Shi, D. D., Savani, M. R., Levitt, M. M., Wang, A. C., Endress, J. E., Bird, C. E., Buehler, J., Stopka, S. A., Regan, M. S., Lin, Y.-F., Puliyappadamba, V. T., Gao, W., Khanal, J., Evans, L., Lee, J. H., Guo, L., Xiao, Y., Xu, M., Huang, B., & McBrayer, S. K. (2022). De novo pyrimidine synthesis is a targetable vulnerability in IDH mutant glioma. *Cancer Cell, 40*(9), 939–956. doi:10.1016/j.ccell.2022.07.011 PMID:35985343

Shinojima, N., Tada, K., Shiraishi, S., Kamiryo, T., Kochi, M., Nakamura, H., & Ushio, Y. (2003). Prognostic value of epidermal growth factor receptor in patients with glioblastoma multiforme. *Cancer Research, 63*(20), 6962–6970. PMID:14583498

Shwartz, Y., Gonzalez-Celeiro, M., Chen, C. L., Pasolli, H. A., Sheu, S. H., Fan, S. M. Y., Shamsi, F., Assaad, S., Lin, E. T.-Y., Zhang, B., Tsai, P.-C., He, M., Tseng, Y.-H., Lin, S.-J., & Hsu, Y. C. (2020). Cell types promoting goosebumps form a niche to regulate hair follicle stem cells. *Cell, 182*(3), 578–593. doi:10.1016/j.cell.2020.06.031 PMID:32679029

Stupp, R., Mason, W. P., van den Bent, M. J., Weller, M., Fisher, B., Taphoorn, M. J. B., Belanger, K., Brandes, A. A., Marosi, C., Bogdahn, U., Curschmann, J., Janzer, R. C., Ludwin, S. K., Gorlia, T., Allgeier, A., Lacombe, D., Cairncross, J. G., Eisenhauer, E., & Mirimanoff, R. O. (2005). Radiotherapy plus concomitant and aduuvant temozolomide for globlastoma. *The New England Journal of Medicine, 352*(10), 987–996. doi:10.1056/NEJMoa043330 PMID:15758009

Sturm, D., Bender, S., Jones, D. T., Lichter, P., Grill, J., Becher, O., Hawkins, C., Majewski, J., Jones, C., Costello, J. F., Iavarone, A., Aldape, K., Brennan, C. W., Jabado, N., & Pfister, S. M. (2014). Paediatric and adult glioblastoma: Multiform (epi) genomic culprits emerge. *Nature Reviews. Cancer, 14*(2), 92–107. doi:10.1038/nrc3655 PMID:24457416

Szopa, W., Burley, T. A., Kramer-Marek, G., & Kaspera, W. (2017). Diagnostic and therapeutic biomarkers in glioblastoma: Current status and future perspectives. *BioMed Research International, 2017*, 2017. doi:10.1155/2017/8013575 PMID:28316990

Vousden, K. H., & Lane, D. P. (2007). p53 in health and disease. *Nature Reviews. Molecular Cell Biology, 8*(4), 275–283. doi:10.1038/nrm2147 PMID:17380161

Wang, J., Su, H. K., Zhao, H. F., Chen, Z. P., & To, S. S. T. (2015). Progress in the application of molecular biomarkers in gliomas. *Biochemical and Biophysical Research Communications, 465*(1), 1–4. doi:10.1016/j.bbrc.2015.07.148 PMID:26253473

Wen, P. Y., & Kesari, S. (2008). Malignant gliomas in adults. *The New England Journal of Medicine, 359*(5), 492–507. doi:10.1056/NEJMra0708126 PMID:18669428

Chapter 15
Glial Cell Biology and Their Multifaceted Functions in Alzheimer's Disease:
New Therapeutic Prospects

Laadraoui Jawad
https://orcid.org/0000-0002-9784-1265
Aïn Chock Faculty of Sciences, Hassan II University of Casablanca, Casablanca, Morocco

Hamid Kabdy
Cadi Ayyad University, Morocco

Hajar Azraida
https://orcid.org/0000-0002-2208-8264
Independent Researcher, Morocco

Aimrane Abdelmohcine
Independent Researcher, Morocco

Kamal Smimih
Faculty of Sciences and Techniques, Sultan Moulay Slimane University, Morocco

ABSTRACT

Although the pathophysiology of Alzheimer's disease (AD) is exceedingly complex and poorly understood, the illness is nonetheless of great interest to the scientific community. Recent advances in AD research have allowed for the possibility that further treatment advantages might be found, which would assist patients all around the world. However, recent studies suggest that glial cells, such as microglia, astrocytes, oligodendrocytes, and oligodendrocyte progenitor cells (NG2 glia), are linked to the pathogenesis of AD and may offer several potential therapeutic targets against AD. Previous research on AD has focused primarily on neurons. Glial cells are essential to the structural integrity of neurons and are necessary for regulating homeostasis (concentration of ions, neurotransmitters) within the central nervous system. This chapter investigates the following topics: (i) the function of glial cells in the pathogenesis of AD; (ii) the intricate functions of the constituent parts; and (iii) prospective therapeutic targets that may one day improve the quality of life for AD patients.

DOI: 10.4018/978-1-6684-9675-6.ch015

INTRODUCTION

Alzheimer's disease (AD), a crippling neurological condition, is characterized by hyperphosphorylated tau protein aggregation in the brain and β-amyloid (Aβ)-induced senile plaques, which eventually cause patients to lose their cognitive function and develop dementia (Cummings et al., 2014). Age is one of the biggest risk factors for AD, which is why it is becoming common to see older people worldwide. In 2010, there were 35 million AD patients worldwide; by 2030, that number is expected to increase to 65 million. When compared to females, human males have about twice as many cases of AD, and around the age of 85, over 50% of males have the disease (Cummings et al., 2014). Considering the age at which the disease manifests itself, there are two subcategories of Alzheimer's disease. Sporadic AD (SAD) develops at a much later age than familial AD (FAD), which manifests itself earlier in life.

The amyloid cascade theory, initially put forward in 1992, holds that aberrant amyloid beta (Aβ) deposition in numerous brain regions is an essential stage leading to neuronal death and AD (Hardy & Higgins, 1992). According to O'brien & Wong, (2011), β-secretase (BACE1) and γ-secretase are involved in the production of A from Aβ from amyloid precursor protein (APP). Aside from having a negative impact on neighboring neurons, Aβ accumulation in the brain can also result in a number of processes, including tau hyperphosphorylation and the formation of neurofibrillary tangles (NFTs), which are important players in the neurodegenerative processes associated with AD (Mohandas et al., 2009; Niedowicz et al., 2011). There are presently no therapies or medications that may reverse or stop the progression of AD, despite countless research illuminating the cellular and molecular processes underlying its pathogenesis (Poon et al., 2020).

Multiple studies currently suggest that inflammation may be the primary neuropathological hallmark causing neurodegeneration in AD, despite the fact that the amyloid cascade theory is still the most popular explanation for how AD pathogenesis occurs. Increased cytokine levels have also been seen in the brains of AD patients (Ojala et al., 2009) and animal models of the disease (Benzing et al., 1999; Apelt et al., 2001; Patel et al., 2005). Numerous studies have also shown that the mechanism of neurodegeneration-associated inflammatory signaling heavily depends on the activation of glial cells (Wang et al., 2015). The creation and maintenance of synapses, the creation of the axonal myelin sheath, the maintenance of normal levels of neurotransmitters, and mitochondrial activities are just a few of the critical jobs that glial cells play in the central nervous system (CNS) (Wong et al., 2020). According to Overmyer et al (1999), microglia, the main immune cell in the CNS, are intimately linked to aging. Microglia's contributions in the development of AD have been the subject of several research (Swanson et al., 2020; Schwabe et al.,2020; Streit et al., 2004). Alterations in microglial gene expression and the emergence of dystrophic microglia, a sign of cellular senescence, have been related to aging (Streit et al., 2004). Glial cells are thought to be directly related to cognitive impairments and the pathogenesis of AD, according to an increasing body of research (Dossi et al., 2018; Uddin & Lim, 2022). In the brains of people with AD, reactive astrocytes and microglia have been observed in higher concentrations close to senile plaques (McGeer et al., 1987; Zotova et al., 2011), suggesting these immune cells may be involved in the etiology of AD.

There is established correlation between the degree of cognitive impairment and brain shrinkage and the level of glial cell activation and their interaction (Cagnin et al., 2001). Therefore, it seems sense to hypothesize that glial neuroinflammatory responses, namely those of microglia and astrocyte, worsen the neurodegeneration linked to AD.

In this chapter, we discuss the interrelationship between glial cells and AD, it also focuses on potential therapeutic interventions that would target these cells.

THE FUNCTIONS OF GLIAL CELLS IN ALZHEIMER'S DISEASE

Neurons and glial cells are the two types of cells that make up the nervous system. Since many years, it has been unclear how many glial cells there are in total compared to neurons, although modern cell counting techniques, such the isotropic fractionator, indicate that there may be less glial cells overall (Von Bartheld et al., 2016). Among the glial cells that are active in the nervous system are oligodendrocytes, polydendrocytes (NG2 glia), astrocytes, Schwann cells, satellite cells, microglia, ependymal cells, and enteric glial cells. These glial cells help produce myelin sheaths around the nervous system's axons, which are important for cell-to-cell transmission, maintain synapses, and maintain homeostasis in the neighborhood of neurons (Jessen, 2004). Neurodegenerative illnesses like AD may proceed more quickly due to homeostatic disturbance (Jessen, 2004). Through in vivo study using a variety of animal models, recent studies have highlighted the significance of glial cells in AD research.

Astrocytes in Alzheimer's Disease

Throughout AD, astrocytes experience complicated and contradictory alterations that are region-specific. While astrocyte populations in close proximity to amyloid beta plaques are known to become reactive as the illness progresses, huge populations of astrocytes are expected to atrophy across the brain (Rodríguez et al., 2014). The early astroglial phenotype in the 3XTg AD model is widespread dystrophy (Rodríguez et al., 2014).

Astrocyte degeneration, which is characterized by a decrease in astrocyte volume and surface area as well as a reduction in protoplasmic processes, has been shown in animal models of AD. The hippocampal astrodegeneration, entorhinal cortex (EC) and medial prefrontal cortex (mPFC) in 3XTg AD animals first appear early in disease, before amyloid beta presence (Olabarria et al., 2010; Yeh et al., 2011; Kulijewicz-Nawrot et al., 2012). Similar to this, in the hippocampus of the PDAPP animal model, astroglial atrophy is shown before the appearance of amyloid beta (Beauquis et al., 2014). In the latter stages of the illness, the presence of A activates astrocytes, which causes reactive astrogliosis in the regions around the plaques (Hou et al., 2011; Olabarria et al., 2010).

A wide range of immunological mediators may be released by and responded to by astrocytes in both the resting and reactive states, which have a crucial role in controlling the inflammatory response in the brain (Farina et al., 2007). In particular, astrocytes release a variety of cytokines that can cause inflammation, including IFN, IL-1, TNF, IL-6, and TGF (Constam et al., 1992). In transgenic mouse models of AD and human AD brain tissues, several of these proinflammatory cytokines are increased (Abbas et al., 2002; Tan et al., 2002).

IFN is a powerful regulatory cytokine that increases in the AD brain and activates microglia while promoting inflammation (Huberman et al., 1994). Although microglia and astrocytes can also release it, T cells and natural killer cells are the main producers (Fultz et al., 1993; Frost & Li, 2017). The levels of TNFα, a cytokine that supports inflammation during the acute period, are higher in brain, CSF, and AD serum (Tarkowski et al., 1999). When compared to normal Tg2576 mice, CD40-deficient Tg2576 animals exhibit lower levels of BACE activity, Aβ burden, and gliosis, underlining the role of TNF in the pathogenesis of AD (Tan et al., 2002). The plasma, cerebrospinal fluid, and brain of AD patients have been found to have high levels of IL-6, which might have anti- and pro-inflammatory properties (Shibata et al., 2002; Baranowska-Bik et al., 2008). One of the first cytokines to be released in the aftermath of injury, IL-1β plays a key role in mediating inflammation in addition to cell proliferation, differentiation, and death. Near the locations of amyloid plaques, it is present in significant concentrations (Griffin et al., 1989).

Microglia in Alzheimer's Disease

Microglia can be stimulated by Aβ and Tau. Microglia congregate and develop into diverse activated phenotypes in the region of Aβ deposition, inflammatory molecule production is elevated, and TREM2 and CD11b are strongly expressed close to the synapses. Microglial synapses, as well as phagocytosis and secretion, are two additional mechanisms by which Tau-like proteins can be transferred. Microglia activation is likewise strongly reliant on toll-like receptors (TLR4, TLR1, etc.). The co-receptors of TLR include the CD14 and NOD-like receptors, which include the NLRP3 inflammasome and are PRRs with detectable characteristics in AD models (Sarris et al., 2021). These results have also been supported by various additional investigations. Aggregated -ap 1-42 activates microglia, alters their branching patterns, raises the indicator of proliferation, and triggers the production of TNF, whereas -ap 25-35 is amorphous and does not activate microglia. Nitric oxide (NO) release was not induced by any peptide, although their toxicity was examined. Certain cytokines have been linked to the activation of microglia, according to several research. For instance, the previously described TNF also has an opposite effect on microglia and encourages their activation (Casal et al., 2002). Additionally, ifn- can be overexpressed to improve the standard activation of microglia (Minogue et al., 2014).

Malmsten, Park, and Wu et al. state that increased production of proteins such as Aβ25–35, chromogranin A (CGA), and Aβf promotes microglia activation. The increase in Aβf also leads to a significant increase in AChE activity, which in turn affects the levels of AChE, BuChE, and ChAT (Figure 1) (Wu et al., 2013; Malmsten et al., 2014). Interestingly, it was discovered that cognitive impairment and more advanced disease stages in AD are linked to greater CSF PGRN. There is a correlation between the increase in CSF PGRN and the rise in TREM2 (triggering receptor 2) soluble on myeloid cells. Despite the fact that CSF PGRN is not an AD diagnostic biomarker, it may, together with sTREM2, represent the disease's activation of microglia (Suárez-Calvet et al., 2018). Additionally, the activation of microglia can be mediated by the presence of several relevant targets in cells, and the particular overexpression of CD200 suppresses microglial activation and secretion. Finally, Yasuno et al (2012) observed that the large area of DAA1106 binding to PBR was considerably enhanced in MCI patients, suggesting that the activation of microglia may begin prior to the onset of dementia. PET scans were done on a small number of AD patients, normal persons and MCI patients. Additionally, research has shown that the magnitude of the impact as assessed by PAM is enormous, and also the mediation model supports the activation of microglia via Tau protein buildup. Additionally, a common variation (rs2997325) was discovered, which impacts PAM as well as the body's activation of microglia (Felsky et al., 2019). Activation of microglia in AD is significant because it can promote phagocytosis and Aβ clearance. However, if microglia are overactivated, they can release an excessive quantity of inflammatory mediators via the NF-kB pathway, which can damage synapses and nerves and cause neuroinflammatory reactions. In order to limit overactivation, the homologous receptors expressed on microglia collaborate with several neuronal immunomodulators, such as CX3CL1/CX3CLR1, CD200/CD200R, CD172A/CD47, CD22/

CD45 and others (Minogue et al., 2014). It is important to note that telomere shortening and excessive cholesterol both contribute to the activation of microglia (Park et al., 2013; Rolyan et al., 2011). However, in pharmacological research, Pls prevents TLR4 from being endocytosed and prevents downstream caspases from activating, which prevents the activation of microglia (Ali et al., 2019). It was shown that minocycline also suppressed microglia activation in a mouse expressing mutant human progerin-1 (PS1) and mutant human APP (Biscaro et al., 2012).

Figure 1. After being deposited, abeta (Aβ) attaches to TLP and enters to the cell by endocytosis. Consequently, the activation of inflammatory bodies (NLRP3) leads to the release of several inflammatory factors through the NF-KB pathway. The protein known as tumor necrosis factor (TNF), which engages and stimulates microglia, is released in high quantities when NLRP3 and CAPase-1, which functions as TNF's downstream route, are both active. Numerous inflammatory cytokines are released, which damages neurons and encourages the formation of Aβ
(Guan et al., 2022)

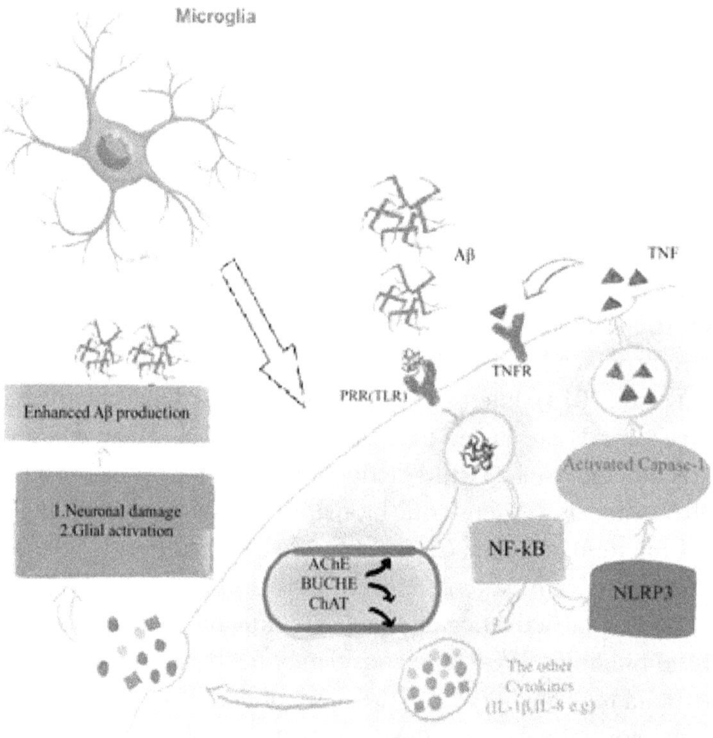

Oligodendrocytes in AD

Some forms of AD showed myelin degradation before amyloid and tau pathology appeared, which can be attributed to the oligodendrocytes' vulnerability to this neurodegenerative illness. In its early stages, the loss of myelin sheaths frequently appears to be the first indication of the sickness. A large body of research has indicated that myelin is associated with AD because oligodendrocyte sensitivity during Alzheimer's disease pathology easily causes myelin to break down and the myelin sheath to disappear.

There are several cellular partners engaged in the procedure of myelin modification, and aging itself is a significant contributing element. Several hyper signal breakouts in T2-weighted images (T2WI) with chronic cerebral hypoperfusion, frequently linked to carotid stenosis, are indicators that may be seen on brain MRI frequently (Lorenzini et al., 2020). Animal models and human pathology both show that these changes are more pronounced at the stage of MCI than they are at the established AD stage. Post-mortem individuals with myelin abnormalities frequently have a relationship between oligodendrocyte losses and axonal changes (Nasrabady et al., 2018a). Even before aggregates in the form of amyloid plaques emerge, the buildup of Aβ peptides is regarded as a princeps factor in the neurodegenerative process (Horiuchi et al., 2020; Cai & Xiao, 2016). Numerous research has looked at how oligodendrocytes react to the presence of amyloid peptides or plaques. This effect is amplified by the existence of the APOE ε4 allele in the individual's genomic baggage. These cells undergo spontaneous involution with aging, and their loss is close to 25% from the age of 50 (Mueller & Weiner, 2009; Liu et al., 2013). Additionally, the severity of the disease is correlated with the significance of an increase in the expression of myelinating genes in oligodendrocytes from Alzheimer's patients (Gouw et al., 2008; Iordanishvili et al., 2019). The myelin sheath is thicker and has different architecture in transgenic animals that overexpress the APP (Wu et al., 2017; Gu et al., 2018; Ferreira et al., 2020).

OPCs are found in the brain, even in the adult stage, and are mostly derived from the ventricular and subventricular areas of the brain (De la Fuente et al., 2020; Sulehria et al., 2020; Sams, 2021). According to spatial-temporal factors, OPCs regulate the angiogenesis, vascularization, and myelination of the white matter, all of which support the stability, functionality, and integrity of the axons (Papuć et al., 2020); Kuhn et al., 2019). Real synaptic connections are formed, ionic homeostasis is regulated, and OPC activity is regulated by a variety of neurotransmitters ((Hamilton et al., 2017); Fontenas et al., 2019). The release of neurotrophic factors, such as brain-derived neurotrophic factors (BDNF) and neuroregulin 1, whose action is regulated by neuronal activity, target these cells as well (Hu et al., 2012). The relationship between early oligodendrocyte changes in AD, changes in myelination, and early disease symptoms has been the subject of several investigations. Most frequently, defects in oligodendrocyte development are linked to disruption of excitotoxic excitotoxic oxidative stress events, which is mediated by glutamatergic metabotropic receptors, which are present in high levels in oligodendrocyte precursors (Butt et al., 2019). Free radicals would be accentuated by additional variables such excessive iron ion levels, glutathione cycle abnormalities, and mitochondrial chain disorders brought on by the toxicity of Aβ peptides (Butt et al., 2019). Axonal survival for functional and metabolic exchanges with the myelin envelope is at the forefront of mitochondria diseases (Xu et al., 2001; Griggs et al., 2017).

The level of brain activity, which influences the electrical characteristics of axons, is directly correlated with myelination. The whole myelin-axon, which forms a pair with several functional and metabolic linkages, is rapidly affected by the proteinopathy caused by Aβ-peptide poisoning (Stadelmann et al., 2019). Oligomeric Aβ peptides and hyperphosphorylated tau proteins are the principal targets of proteinopathies' toxicity, which results in degenerations in AD (Allegri, 2020; Jellinger, 2020). Senile plaques are infrequently seen in hydrophobic white matter, which does not encourage the accumulation of harmful oligomers. As a result, it is mostly these latter that cause myelin to become toxic. The present treatment approaches for removing plaque do not appear to have a significant impact on clinical symptomatology (Vaz & Silvestre, 2020). In reality, the reason why such therapies are ineffective is still up for debate, as is the extent to which the therapeutic techniques for plaque removal have clinical impact.

With the drugs investigated, it has been feasible to reduce the buildup of abnormal amyloid peptides without significantly improving patients' cognitive impairment, which has been the focus of therapeutic research against AD for more than 20 years. In many instances, the amyloid hypothesis updated is contested, occasionally in favor of basic changes to the myelin sheath (Frisoni et al., 2022). From a metabolic perspective, oligodendrocytes are quite active, especially when myelination or remyelination is taking place. A myelination issue might be caused by a cellular respiratory anomaly that is connected to hypovascularization or ischemia (Nasrabady et al., 2018b). Elderly people or those with early-onset AD symptoms frequently have vascular diseases that impact the white matter (Nasrabady et al., 2018b). The fact that the human brain is primarily myelinated may help to explain why it is susceptible to neurodegeneration (Timmler & Simons, 2019).

The human CNS is heavily composed of oligodendrocytes, notably in the neocortex, where they make up roughly 75% of the glial cells (Edwards et al., 2018). They are regarded as being extremely fragile, and their density rapidly declines in the brains of the elderly beginning at the age of 50 (You et al., 2019). In numerous hippocampal areas, a substantial loss of oligodendrocytes has been seen using a variety of labeling techniques, which is unrelated to the amount of amyloid deposits (DeFlitch et al., 2022). These unique depopulations are likely to occur before the brain connectome's disarray, which precedes the onset of AD, and a subsequent demyelination surrounding amyloid peptide deposit breakouts (Kövari et al., 2007). Given that the amyloid oligodendrocytes actively contribute in the creation of cholesterol forming synaptic connections, these dysfunctions are significantly related with anomalies in lipoprotein metabolism (Marangon et al., 2020). They release apolipoproteins E and J, which, depending on the alleles implicated in the beginning of AD, are serious risk factors (Johnson, 2020). It's interesting to note that mice's motor learning appears to depend heavily on the generation of new oligodendrocytes (Long & Corfas, 2014). Numerous research has been carried out to define the changes detectable by MRI methods in the structure and architecture of myelin during the development of AD, particularly in the early stages characterized by moderate, deteriorating memory problems (MCI). Schematically, the data demonstrated that cerebral myelin levels had fallen very early on, along with oligodendrocyte and axon losses, microglial activations, and dilated perivascular areas in the white matter. The main focus of these research, which compare MCI patients and controls, is on the differences between the aqueous contents of the intra- and extra-cellular spaces at the periventricular level. Myelin structure appears to be altered very early and oligodendrocytic re-myelination is inhibited by increasing ischemia with vascular and energy losses linked to the toxicity of several proteinopathies, including amyloidosis Aβ (Kavroulakis et al., 2018; Nasrabady et al., 2018b; Ota et al., 2019).

THE AMYLOID HYPOTHESIS AND THE INVOLVEMENT OF NEURONS GLIAL CELLS

Neuroinflammation

Astrocytes and microglia have a critical part in the pathogenesis of AD, which is one of the primary causes driving neuroinflammation. The effector component of the tumor necrosis factor receptor-associated factor 6 (TRAF6) is induced when A binds to the toll-like receptor 4 (TLR4) on microglia and astrocytic cells (Okun et al., 2009; Shi et al., 2016; He et al., 2020). Additionally, it has been noted that LRP1 knockdown or silencing increased nuclear NF-κB p65 subunit expression as well as TLR4,

MyD88, TRAF6, neuroinflammation, microgliosis, and astrogliosis in the hippocampus and cerebral cortex in WT and APP/PS1 double-transgenic mice (He et al., 2020). The expression of CD33, another microglial membrane receptor, rises in the AD brain, where it interacts with the sialic acid residues of Aβ to reduce Aβ clearance. Therefore, CD33 deletion in the APP/PS1 mice reduced the development of A plaque and the deterioration in cognitive function, respectively (Griciuc et al., 2013; Perea et al., 2020). Additionally, TREM, a type I transmembrane glycoprotein produced in microglial cells, directly interacts with various forms of Aβ, with soluble Aβ 42 oligomers showing the highest affinity (Lessard et al., 2018). In the 5xFAD mouse model, the increased TREM2 expression led to a reduction in the amyloid load and neurobehavioral impairments (Lee & Chung, 2019). While Tau competed with the neuronal ligand CX3CL1, it was discovered that microglia-mediated phagocytosis of Tau evolved through its interaction with the microglial receptor, CX3CR1. This was supported by the findings of CX3CR1-deficient animals and microglial cells, which showed abnormal Tau absorption and breakdown (Bolós et al., 2017). Studies on CC-chemokine receptor 5 (CCR5−/−)-deficient mice, macrophage inflammatory protein-1 (MIP-1α−/−), and Tg2576 transgenic mice models of AD showed that the NF-B and mitogen-activated protein kinase (MAPK) pathways later cause astrocytes and microglia to switch from an anti-inflammatory to a pro-inflammatory phenotype. This encouraged the pathogenesis of AD by promoting the production of cytokines such as interleukin (IL6, IL1α, and IL1β), tumor necrosis factor (TNF), granulocyte-macrophage colony-stimulating factor, and macrophage inflammatory proteins. As a result, there is an increase in NO synthesis, lipid peroxidation, and ultimately neuronal injury (Asiimwe et al., 2016) (figure 2). These cytokines also cause Nitric Oxide Synthase activity in the glial cells. One of the most extensively researched cell types for neuroinflammation and AD is the astrocyte, and the activation of its reactive phenotype entails an elevated production of the protein glial fibrillary acidic protein (GFAP). Additionally, it has been observed that GFAP and GFAP-immunoreactive astrocytes are often prevalent close to A plaques among the various GFAP isoforms (Kamphuis et al., 2014).

In response to A, the astroglial and microglial TLR4 signaling pathways MyD88→TRAF6→MAPK, to NF-B, AP1, and inflammatory signals are activated. Additionally, elevated amounts of reactive oxygen species (ROS) and NOS cause neurodegeneration and oxidative stress.

Glutamate Signaling

As seen in the APPsw/PSEN1DE9 mice, amyloid fibrils enhance the activity of metabotropic glutamate receptor 1 (mGluRl), which is crucial for the control of synaptic plasticity and cognition (Bie et al., 2019; Li et al., 2020). In the CNS's synaptic and extrasynaptic regions, astrocytes that express glutamate transporters actively take up glutamate (Rose et al., 2018). By increasing glutamine synthase activity, the astrocytic glutamate-aspartate transporter (GLAST) and glutamate transporter-1 (GLT-1) lower the glutamate load in the synaptic cleft and encourage glutamine synthesis. A D-[3H]-aspartic acid uptake experiment in cultured hippocampal astrocytes and neurons revealed that a decrease in these astrocytic transporter levels hastens the development of cognitive impairments in AD. According to reports, Aβ1-42 causes aberrant GLT-1 localization and internalization in astrocytes, preventing the clearance of amounts of extracellular glutamate generated during synaptic transmission. Oxidative stress and inflammatory pathways are involved in this. As seen in astrocytes from acute mouse hippocampus slices, Aβ1-42 also produces inflammation-related transcriptional change in GLT-1 and GLAST levels, mostly in astrocytes (Scimemi et al., 2013). An increase in extrasynaptic NMDAR-mediated excitotoxicity caused by a subsequent spread of glutamate from one synaptic region to another limit's normal neuronal survival,

Figure 2. AD, oxidative stress, neuroinflammation, and glial Toll-like receptor 4 (TLR4)

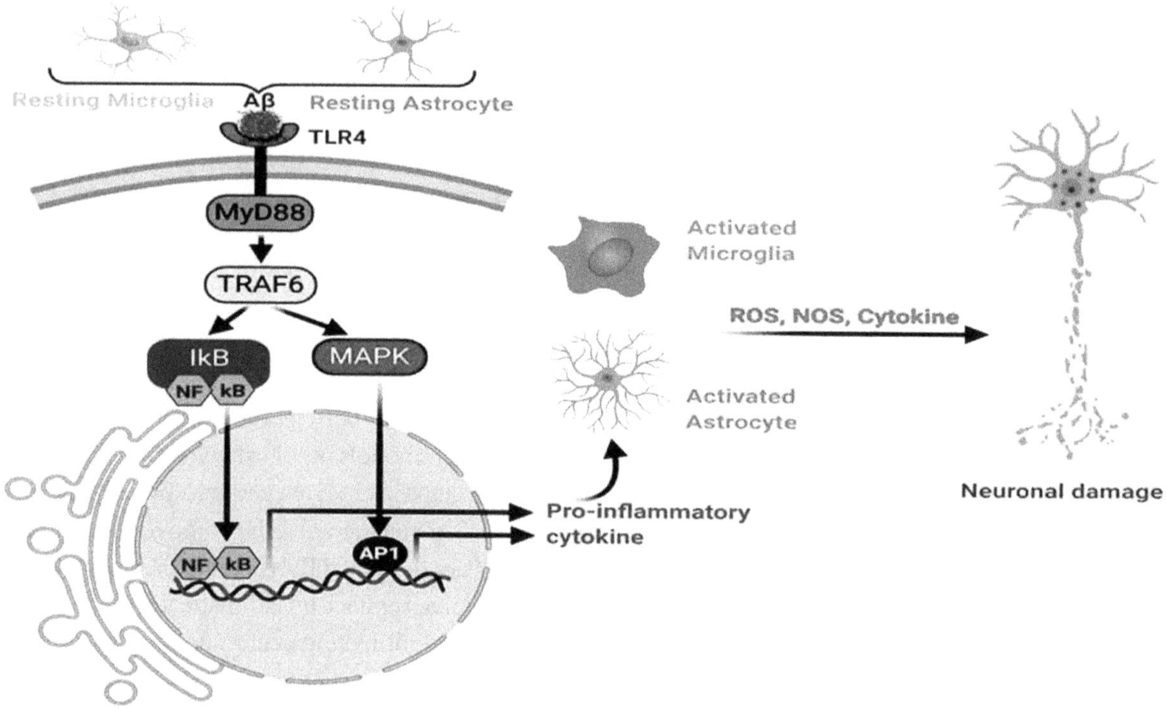

activity, and networking, and subsequently synaptic transmission (Findley et al., 2019). Hippocampal long-term potentiation is inhibited by an associated immediate neuronal Ca2+ influx that involves the activation of NMDARs that contain GluN2B (Liu et al., 2019). In addition to extracellular glutamate dyshomeostasis, a clear connection has been shown between altered astrocytic GLT1 and complement proteins involved in microglial synaptic remodeling in AD. The C1q protein, which was seen in the aged C1q-deficient mice by Stephan et al. (2013), induces an increased synaptic classical complement cascade in the CNS, and microglia facilitates synaptic clearance by phagocytosis (Luchena et al., 2018; Stephan et al., 2013). Furthermore, GLT1 decreases the constitutive glutamatergic intensity, amplitude, and frequency of spontaneous micro excitatory synaptic current in hippocampal CA1 neurons as well as presynaptic metabotropic glutamate receptor activity. As documented in primary glial cells and in the hippocampus and cortical areas of APP transgenic mice, the connection between astrocytes and microglia through complement activation here plays a critical role in amyloid pathogenesis (Lian et al., 2016). Additionally, considerable activation of the complement proteins C3 and C1q, which control innate immunity, microglia-mediated synaptic pruning, and neuronal survival, occurs in the brain and aids in the development of AD (Hansen et al., 2018).

Lipoprotein Dysfunction

According to Dai et al. (2018) and Huentelman et al. (2019), APP, PSEN1 or PSEN2 with polymorphic gene expression, altered peripheral lipid metabolism, and dysfunctional cholesterol and lipoprotein metabolism all contribute to familial AD. According to Ramanathan et al. (2015), LRP1 significantly contributes to vascular dysfunctions. However, Ashok et al. (2016) and He et al. (2020), have demonstrate that LRP1 controls Aβ generation via altering the regulation of proteases, growth factors, and eventual inflammation. LRP1 is widely distributed in brain tissue, and its neuronal deletion activates the NF-κB and MAPK signaling pathways as well as astrogliosis and microglial reactivity. This has been supported by the finding that transgenic AD mice have lower hippocampus LRP1 expression (He et al., 2020).

Late-onset AD (LOAD), which accounts for 95% of AD cases, exhibits a different link with the isoforms of the APOE gene in contrast to early-onset AD, which includes genetic abnormalities in the APP, PSEN1 and PSEN2 genes (Corder et al., 1993). The main genetic risk factor for LOAD is APOE-4 carrier in the brain, and each APO-E-ε4 allele lowers the age of AD onset (Liu et al., 2013). The ApoE isoform contributes significantly to the pathophysiology of AD, which is marked by a rise in gliosis, an accumulation of A, and hyperphosphorylation of tau. According to studies done on human individuals, the APOE variation of ε4 is mostly released by the brain's astroglial cells and is linked to innate immunological responses (Gale et al., 2014).

The APOE4 produced by astrocytes and microglia promotes the synthesis and accumulation of A in neurons. Additionally, as seen in the post-mortem paraffin-embedded human frontal cortex tissue samples collected from the University of Southern California AD Research Center (Halliday et al., 2016), APOE4 derived from the pericytes and other cells of the NVU induces BBB degradation and malfunction. The pathogenicity of AD is determined by the APOE isoforms in the following order: apoE4 > apoE3 > apoE2. On the other hand, apoE4 appears to have a lower ability to remove and degrade A by enzymatic means when compared to the other isoforms (Figure 3).

There has also been evidence of a direct relationship between LRP1 and Tau, whereby tau's potential contact with microtubules at its microtubule binding repeat region-controlled tau endocytosis in neurons. Additionally, LRP1 loss in mice and neuroblastoma cell line resulted in decreased LRP1 interaction, which in turn inhibited tau neuronal propagation (Rauch et al., 2020). LRP1 thus serves as a common component that controls the pathophysiology of both A and tau in AD.

Astrocytic and microglial APOE, with APOE4 > ApoE3 being more involved, contribute to increased vascular injury and act as risk factors for Aβ production. Conversely, APOE3 > ApoE4 facilitates the breakdown and removal of Aβ. Thus, AD pathogenesis is promoted by dysregulated APOE3 and APOE4 levels.

CONTRIBUTION OF ASTROGLIOSIS MICROGLIOSIS AXIS (AMA) IN THE PROGRESSION OF AD

In a variety of experimental settings, functional investigations of microgliosis and astrogliosis and their connection to AD have been conducted. Although understanding the cellular and molecular processes underpinning AD pathogenesis depends on resolving the timing of AD pathology, the link between gliosis and AD pathogenesis is complicated and remains unresolved. Astrocyte activation and a rise in Aβ levels, for instance, are shown to peak when clinical symptoms start to develop, even before the moderate cognitive impairment phase. In contrast, neuronal loss, tau pathology, and microglial activation appear later in the course of the illness and are associated with more severe clinical symptoms (Jack et al., 2010). The evidence supporting astrocyte and microglia communication during the development of AD is summarized here (Carter et al., 2012).

Figure 3. AD, Glial cells, and APOE

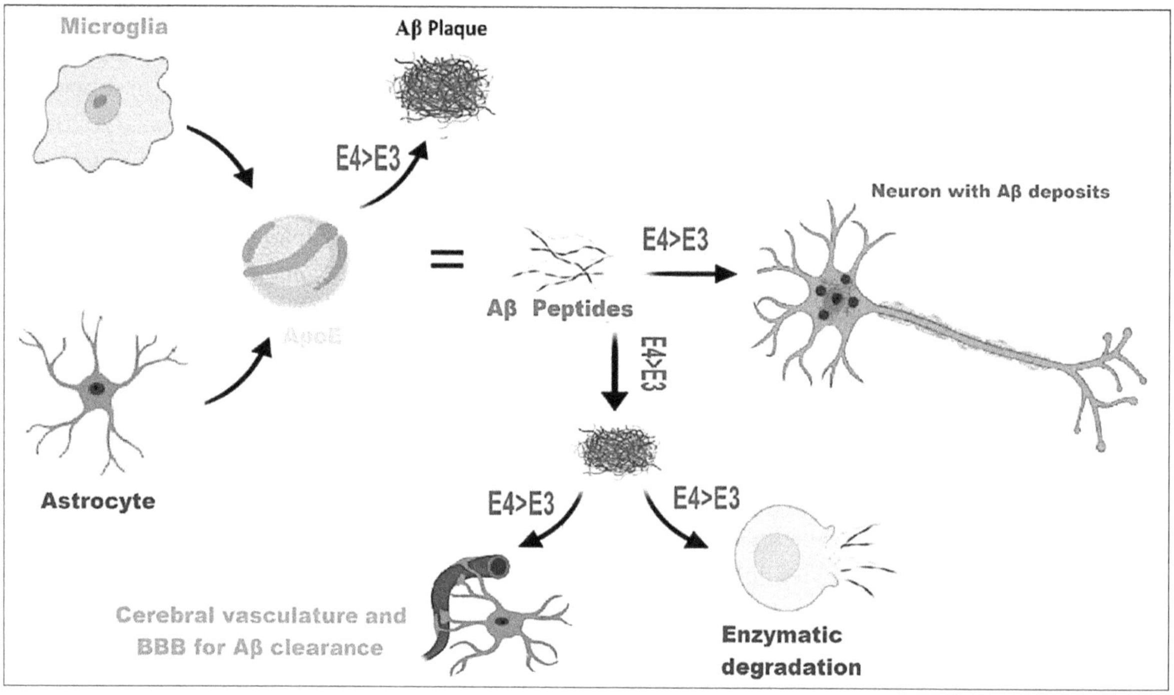

CONVENTIONAL AND NOVEL EXPERIMENTAL MODELS UTILIZED FOR INVESTIGATING THE ROLE OF GLIOSIS IN AD

Proper and controlled models are necessary to evaluate the relationship between activated microglia and reactive astrocytes in neurodegenerative disorders such as AD. Here, we describe the methodology utilized to evaluate and control astrocyte and microglial activity using recently developed models that might be useful in determining the function of gliosis in AD.

Models of Reactive Astrocytes That Modulate MAO-B

Due to the elevated expression of this molecule in astrocytes in the brains of AD patients (Emilsson et al., 2002) and close to A plaques in animal models of AD monoamine oxidase B (MAO-B) has been linked to the etiology of AD. Reactive astrocytes' abnormal synthesis and release of GABA as a result of MAO-B activation diminish the likelihood that granule cells would spike through presynaptic GABA receptors. In addition, MAO-B generates hydrogen peroxide, a kind of ROS, which adds to the pathophysiology of AD. As a result, it would be predicted that reducing ROS-induced oxidative stress through the regulation of MAO-B activity would slow the development of the illness. These factors have led to the use of MAO-B inhibitors to pharmacologically prevent astrogliosis in experimental AD models. In APP/PS1 mice, the MAO-B inhibitor selegiline has been shown to decrease astrocyte reactivity, and this effect is linked to a decrease in memory impairment. The MAO-B inhibitor sembragiline can counteract the heightened astrocyte reactivity seen in transgenic rats that conditionally overexpress MAO-B in GFAP-positive astrocytes, which is linked with an increase in ROS production and neuronal death

(Borroni et al., 2017). It is obvious that more research is needed to determine the therapeutic benefits of MAO-B inhibitors on astrogliosis in AD patients.

Models Of HSV, AAV, and Adenovirus-Mediated Reactive Astrocytes

Through the production of acute and chronic CNS disorders, viruses have been linked to the genesis of AD (Deatly et al., 1991; Itabashi et al., 1997). According to recent research by Eimer et al., herpes simplex virus-1 (HSV-1) infection accelerates the development of AD and A deposition through protective amyloidosis. It has been proposed that one of the major cellular mediators of virus-induced CNS disease is reactive astrocytes. HSV-1 has been shown to activate astrocytes in the CNS (Ma et al., 1994) and it has been demonstrated that the astrocyte protein AXL allows Zika virus to infect these cells (Chen et al., 2018). It was recently shown that HSV-1 activates GSK-3, which then phosphorylates the amyloid precursor protein (APP) (Piacentini et al., 2015). Although the mechanisms behind virus-induced reactive astrogliosis are yet unknown, the strong correlation between such infections and AD suggests that further research into this connection will need to use virus-mediated experimental models.

The AAV2/5-gfp104-eGFP virus has been demonstrated to increase the population of reactive astrocytes in a titer-dependent manner, as has the AAV virus previously utilized to produce reactive astrocytes in vivo (Ortinski et al., 2010). In this model, researchers discovered that reactive astrocytes alter network hyperexcitability, block inhibitory neurotransmission, and downregulate glutamate synthetase expression levels. Similar to this, Woo et al. demonstrated that the adenovirus Adeno-GFAP-GFP causes astrocyte reactivity in the dentate gyrus area of the mouse hippocampus (Woo et al., 2017), with astrocytes in this region being hypertrophied following viral infection in comparison to astrocytes in uninfected animals. Therefore, these models might be easily used to cause focused reactive astrogliosis in particular areas of interest in the brain. The use of such models must be done very carefully in order to distinguish between the impacts of reactive astrogliosis and virally-induced inflammatory responses.

Reactive Astrocyte AD Models That Are Vimentin- and GFAP-Modulated

Since hypertrophied processes and enhanced expression of intermediate filaments such as GFAP and vimentin define astrocyte reactivity, the involvement of astrocyte intermediate filaments in AD has been investigated by breeding genetically impaired mice with transgenic mice carrying the APP gene. Although research employing these models have demonstrated that GFAP and vimentin KO in APP/PS1 mice is linked with an approximately two-fold increase in A plaque development at 8 and 12-months of age, the validity of these models is still up for debate. The fact that APP processing and the levels of soluble and interstitial fluid A were unaltered in these triple transgenic animals implies that the deletion of astrocyte intermediate filaments (IF) affects A degradation rather than A production. The morphology of astrocytes around A plaques was significantly altered in GFAP and vimentin-deficient rats, with limited process hypertrophy and no interaction with neighboring A plaques. Additionally, the neurite dystrophy in these mice increased noticeably. These results suggest that A plaque-related dystrophic neurites are attenuated by changes in astrocyte morphology associated with activation (Kraft et al., 2013; Kamphuis et al., 2015). However, care must be taken when interpreting these observations since intermediate filament KO seems to have other consequences, such as preventing GFAP/vimentin KO animals from exhibiting elevations in endothelin B receptor protein expression that are typically seen in reactive astrocytes (Wilhelmsson et al., 2004).

Modulation of Inflammatory Signaling Pathways in Responsive Astrocytes in Animal Models

By genetically altering astrocytes using AAV-GFAP-VIVIT, which is an artificial peptide that prevents the physical connection between calcineurin and NFAT, astrocyte reactivity can be prevented. It prevents astrocytes from activating NFAT, which lowers cytokine release and neuroinflammation. By using this method in vivo on APP/PS1 mice, AD pathology is considerably reduced, amyloid levels are decreased, and cognitive abilities are improved, showing that reactive astrocytes have a harmful effect (Zhang et al., 2014). Studies conducted in vitro confirm this finding by demonstrating that astrocyte VIVIT expression reduces the neurotoxic effects of activated astrocytes on nearby neurons (Sama et al., 2008; Abdul et al., 2009). Another method to prevent astrocyte reactivity is to overexpress IKK's dominant negative form or to knockout or decrease IKK expression in these cells (Hsiao et al., 2013). Such strategies block the NF-B pathway in inflammatory astrocytes. Last but not least, deletion of aquaporin 4 (AQP4) has also been used to inhibit astrocytic activities, and loss of this channel has been demonstrated to worsen brain A buildup and memory problems in APP/PS1 mice (Xu et al., 2015). However, given the small number of research using these models, it is unclear if and how the reactivity of reactive astrocytes affects AD development.

Animal Experimentation to Control AD Microglia Activity

Blocking microglial activation is expected to slow the course of the illness since microglial hyper-activation is thought to be neurotoxic. To ascertain the role of microglia activation in the pathogenesis of AD, pharmacological and genetic modification of microglial activation in animal models of AD has been carried out. According to Asraf et al., inhibiting angiotensin-converting enzyme (ACE) and preventing the production of angiotensin II (Ang II) by captopril reduced LPS-induced NO release and controlled iNOS, TNF-α, and IL-10 in BV2 microglia cells. Additionally, the 5X FAD in vivo AD mouse model was used using this technology. Microglial activation was reduced and Aβ load was reduced after a two-month intranasal captopril therapy, showing that microglial activation worsens AD pathogenesis (Asraf et al., 2018). The most highly expressed gene in microglial culture, NFAT 2c isoform, was found to be inhibited or deleted by Manocha et al., 2017 which controlled microglial activation and prevented the release of cytokines from microglia. Cytokines and microgliosis were decreased when NFATc2 KO mice were mated with APP/PS1 mice. Plaque burden, however, was unaffected (Manocha et al., 2017). On the other hand, mice and AD have both been linked to the phagocytic functions of microglial triggering receptor expressed on myeloid cells 2 (TREM2). A decreased microglial inflammatory response is brought on by TREM2 deficiency (Sieber et al., 2013). Researchers have demonstrated that TREM2-deficient microglia are unsuccessful at either clustering around or eliminating fibrillary Aβ by crossing TREM2 deficient animals with AD mouse models (Schmid et al., 2002). These results imply that AD pathogenesis is aggravated by microglial activation.

POTENTIAL TARGETS OF INTEREST FOCUSED ON GLIAL CELLS

Potential Therapeutic Targets With Glial Orientation

Approaches based on recent discoveries may help glial-focused AD treatment develop. For instance, evidence shows that an immunization may be able to reverse the morphological modifications of microglia, which may lower total A deposit levels (Rusznák et al., 2018). Genetic studies have demonstrated a significant relationship between AD etiology and polymorphisms in the genes that encode microglial proteins, in addition to morphology. According to recent research employing APPSwe/PS1dE9 mice, TREM2 (Triggering Receptor Expressed on Myeloid Cells) is raised in microglia and becomes more prominent as AD proceeds (Jiang et al., 2020). Although TREM2 was first identified as an AD risk gene, this study shown that TREM2 may actually promote A phagocytosis, which would enhance patient cognitive skills by halting the progression of AD by increasing neuroprotection. Additionally, a number of recent studies using TREM2-deficient animals demonstrate that this defect may cause a notable rise in A-induced disease (Kawabori et al., 2015; Wang et al., 2016). Ulrich et al. 2017 just released a very thorough analysis of the role of TREM2 in AD that examines the expanding research on TREM2 in AD. This review offers some new information that is relevant to the overall development of AD (Ulrich et al., 2017) and emphasizes the innate immune system's larger role in neurodegenerative illness. As a result, TREM2 may one day be used as a viable therapeutic target in AD treatments that focus on glial cells. Additionally, research using APP/PS1 mice has demonstrated that blocking the CaN/NFAT pathway in astrocytes reduces the rate of astrogliosis, which in turn enhances cognitive abilities (Furman et al., 2012). In 2012, Furman et al. 2012 demonstrated that they could target the hippocampus astrocytes in APP/PS1 mice by employing adeno-associated virus (AAV) vectors that included the astrocyte-specific Gfa2 promoter. The vectors promoted the production of a peptide called VIVIT, which blocks the calcineurin/NFAT pathway and decreases astrocytic activation. The mice had decreased glial activation, reduced Aβ levels, and improvements in synaptic and cognitive function after receiving this therapy for a number of months. These positive findings call for more research into cutting-edge astrocyte-based treatments for AD.

Potential Treatments for Alzheimer's Disease Using Antioxidants

Oligodendrocytes and neurons are more vulnerable to oxidative stress than microglia or astrocytes, and both lipid peroxidation and protein oxidation play a significant role in AD. A-mediated Ca2+ disruption in astrocytes has been demonstrated to cause the death of these neurons, which are dependent on astrocytes for the provision of glutathione (GSH) precursors. Along with maintaining low GSH levels, oligodendrocytes have increased iron contents, which may reduce their capacity to scavenge reactive oxygen species (ROS). As a result, a plausible connection is made between the CNS's antioxidant capacity and A-mediated effects. While Trolox, a vitamin E derivative, has been demonstrated to lessen cell death in astrocytes and neurons (Bai & Lipski, 2014), N-acetylcystein has been found to decrease A-induced cellular death in oligodendrocytes (Xu et al., 2001). Additionally, curcumin has been demonstrated to raise GSH levels by boosting the activity of glutamate-cystein ligase, an enzyme that produces GSH, potentially providing greater neuroprotection against AD (Lavoie et al., 2009). It's important to emphasize, nevertheless, that there is no proof that giving individuals with moderate cognitive impairment vitamin E in the form of alpha-tocopherol can stop the onset of dementia or even enhance their cognitive abilities (Farina et al., 2007; Fritsche et al., 2017).

WNT Pathway Stimulation for The Treatment of Alzheimer's Disease

Since Aβ action increases GSK-3β and phosphorylated β-catenin expression levels, stimulating the Wnt/-catenin pathway may be a useful tactic to reduce AD pathology (Xu et al., 2011). Studies conducted in vitro and in vivo, notably those involving 3xTg-AD mice, have demonstrated that the GSK-3β inhibitor TWS119 has reduced the impairment of neuronal myelination (Malm et al., 2007; Desai et al., 2011). Similar activation of this signaling system may help astrocytes lessen their toxicity from glutamate. It was shown that pyrrolidine dithiocarbamate (PDTC) stimulation enhanced GLT-1 levels and decreased tau protein phosphorylation, which is another significant recognized component along with Aβ accumulation, in APP/PS1dE9 mice (Malm et al., 2007). Additionally, recent research on APPswePS1dE9 mice revealed that rosiglitazone and lithium stimulated the brain similarly and decreased astrogliosis, the number of activated microglia, and the amount of Aβ plaque (Toledo et el., 2010). Nonetheless, research on hAD samples showed that activating this pathway comes at a price since it raises AD neuroinflammation, which negates the improved Aβ clearance by having the same microglia phagocytose the inflammation (Halleskog et al., 2011). This is because stimulation of this pathway increases levels of β-catenin, which in turn increases levels of activated microglia.

Polyamines' Function in Alzheimer's Disease and Potential Therapy

Polyamines (PAs) are a family of macromolecules that are common in all living cells and are relevant to the development of AD. DNA, RNA, proteins, and other negatively charged molecules interact with polyamines, which are polycations. Putrescine, spermidine, and spermine are examples of polyamines that play crucial roles in numerous biochemical mechanisms play a role in the division and multiplication of cells, and have been linked to serious human disorders including AD (Minois, 2014; Handa et al., 2018). Glial cells and neurons communicate with one another via polyamines. Stressful conditions like brain injury and ischemia make this communication more noticeable. Further research is necessary to further understand the storage processes of these PAs. The control of neuronal and glial receptors and channels in many neurodegenerative diseases is known to be a function of Pas. However, the advancement of AD has also been linked to polyamine precursors such ornithine, arginine, and proline as well as PA products like agmatine, acetyl-spermine, etc. For instance, research using autopsied human brain tissue showed that ornithine decarboxylase, a crucial regulator of PA biosynthesis, was significantly upregulated in the temporal cortex, significantly downregulated in the occipital cortex, and unaffected in the hippocampus in AD, which was also supported by animal studies. It is now possible to extrapolate that aberrant PA system activity is implicated in AD because it has been shown that ornithine decarboxylase is important in mature brain function (Morrison et al., 1998). L-arginine is an amino acid that is semi-essential to life and possesses a number of bioactive metabolites, making it another precursor to PA. It has also been demonstrated that L-arginine is considerably changed in AD brain tissue. Further evidence of age-related variations in tissue concentrations of L-arginine and its metabolites, including putrescine, L-citrulline, spermidine, L-ornithine, spermine, agmatine, glutamate, glutamine and gamma-aminobutyric acid was provided in this work by Liu et al. These modifications happened in a metabolite- and location-specific way. Together, these results demonstrate that PA metabolism is vulnerable to change in AD brains compared to normal brains and may be interesting for further research (Yi et al., 2009; Liu et al., 2013). Another recent investigation on the PA precursor L-ornithine and the PA system enzyme L-ornithine decarboxylase (ODC) revealed that ODC was highly overexpressed in select neurons and in alterations in the central nervous system that are unique to AD (Bernstein & Müller, 1999). One of the specialized enzymes involved in the synthesis of polyamines, spermidine synthase, is in charge of catalyzing the creation of the precursor to spermine, spermidine, from the PA putrescine. Polyclonal antibodies were used in cell localization tests on murine brain tissue, and the results showed that PAs have relatively little localisation in neurons, whereas astrocytes showed the most prominent localization. Additionally, oligodendrocytes showed some localisation that may be a secondary source of extracellular Pas (Laube et al., 1997). Additionally, there is proof that retinal glial cells have PAs that are K+ channel-active. These PAs' production may be enhanced in pathological circumstances, and they may influence K+ signaling, which in turn affects glial cell function and proliferation (Biedermann et al., 1998). The absence of PA synthesis in neurons and synapses has been demonstrated in studies on adult brain tissue, but this is not the case in glial cells, where there is adequate expression of the required enzymes (Krauss et al., 2007).

Both inwardly rectifying K+ channels and glutamate receptors are known to be modulated by the PAs spermidine and spermine. By buffering extracellular K+ and glutamate, glial cells often act as a support system for neurons (Orkand et al., 1966; Kucheryavykh et al., 2007; Olsen et al., 2015). The malfunction of a key astrocyte K+ channel, Kir4.1, is known to emerge in the early stages of pathogenesis in a number of neurodegenerative illnesses (Olsen et al., 2015). PAs are in charge of modulating glial Kir channels. Studies using the Novikoff cell line and the A172 human glioblastoma cell line have shown that PAs suppress the downregulation of Cx43-mediated gap junction communication. This is mainly caused by increased [Ca2+] and [H+], which frequently go hand in hand with ischemia and other disorders. These gap junctional communications that suggest that Cx43 gap junctions may be controlled by spermine were considerably decreased by the siRNA Cx43 deletion (Kucheryavykh et al., 2017). Excitability and Ca2+ levels in glial cells and neurons may be impacted by PAs that regulate a number of cationic channels and significant receptors (Skatchkov et al., 2014). Multiple neurodegenerative disorders, including AD, have been demonstrated to target these receptors and channels (Skatchkov et al., 2015; Kucheryavykh et al., 2017). Another element of PAs was identified by metabolic profiling of AD brains, which showed a considerable drop in PA concentrations (Liu et al., 2015).

Since it has been found by in vitro spectroscopy and AFM imaging that PAs enhance Aβ-peptide fibrillation and control Aβ aggregation formation, hence relieving Aβ-induced toxicity (Luo et al., 2016), polyamine targeted treatment may prove to be beneficial. In a recent research, Aβ-peptide aggregates were the target of spermidine conjugates bearing variations of variably ordered 3,5-dibenzylidenepiperidin-4-one as bioactive motifs. It was shown that these substances provided neuroprotection through their anti-aggregation effect. When all of these data are considered, an intriguing line of reasoning about the functions of PAs in AD treatment approaches emerges (Simoni et al., 2016). The appropriate characterization of possible neuroprotective pathways may be crucial in identifying novel treatment targets because progressive neurodegeneration is a major issue when it comes to stress, ischemia, or any CNS-related disease like AD (Skatchkov et al., 2014).

Alternative Glial-Oriented Approaches

Engrafting neural stem cells (NSCs) in AD has great potential since NSCs are capable of releasing a variety of advantageous substances, including neurotrophic substances that encourage CNS regeneration as well as moving to damage sites and developing into neural cells. Studies conducted on 3xTg-AD mice reveal that engraftment increased the quantity of oligodendrocytes and astrocytes, as well as brain-derived neurotrophic factor (BDNF), but had no effect on tau protein phosphorylation or Aβ deposition (Blurton-Jones et al., 2009). Production of Cortical GABAergic interneurons from embryonic medial ganglionic eminence (MGE) cells is another intriguing area of study that has been explored (Tricoire et al., 2011). The hippocampus and cortex of the brain create, move, and disseminate immature progenitor cortical interneurons. According to studies, it is possible to remove this specific component from rodent embryos and transplant it into adult animals. The transplant recipients showed promise in treating a number of neurodegenerative diseases, and it's possible that these cells could soon be a popular treatment option for AD (Valente et al., 2013; Gilani et al., 2014). Additionally, another team of researchers demonstrated that sodium phenylbutyrate can stimulate enhanced BDNF expression in 5xFAD mice and repair synapses, enhancing cognitive performance (Corbett et al., 2013). In addition to reducing oxidative stress and cell death in vitro, deliberate overexpression of glial cell-derived neurotrophic factor in 3xTg-AD mice showed preservation of learning capacity and memory in vivo (Revilla et al., 2014). Table 1 offers an overview of the prospective therapeutics that target glia.

Table 1. A list of the substances that might be used as AD treatments

Agent	Mechanism of action
Trolox	Decreased astrocyte and neuronal death (Bai & Lipski, 2014)
N-acetylcystein	Decreased oligodendrocyte mortality (Xu et al., 2001)
Curcumin	Increased levels of glutathione (GSH) in astrocytes and neurons (Lavoie et al., 2009)
TWS119	Augmented myelination (Desai et al., 2011)
Pyrrolidine dithiocarbamate (PDTC)	Increased glutamate transporter 1 levels (Malm et al., 2007)
Lithium/rosiglitayone	Decreased AB load (Toledo & Inestrosa, 2010)
Neural stem cells	Enhanced cognitive abilities (Blurton-Jones et al., 2009)
Sodium phenylbutyrate	Enhanced cognitive abilities (Corbett et al., 2013)

CONCLUSION

Although the cause of neurodegeneration is still unknown, current research has shown that the onset and course of the illness depend heavily on the dysfunction of neuron-glia connections and the absence of glial support. The production of amyloid plaques, cerebrovascular injury, and AD pathogenesis are all significantly influenced by glial cells through a variety of pathogenic routes and processes. Dendritic damage, synaptic nerve injury, altered neurotransmission, and general cortical and hippocampal atrophy are the outcomes of the process. More research is needed to answer important concerns about neuronal-glia interactions, such as why do astrocytes touch some synapses but not others and how does this communication alter over the course of development or as a disease progresses.

REFERENCES

Abbas, N., Bednar, I., Mix, E., Marie, S., Paterson, D., Ljungberg, A., Morris, C., Winblad, B., Nordberg, A., & Zhu, J. (2002). Up-regulation of the inflammatory cytokines IFN-γ and IL-12 and down-regulation of IL-4 in cerebral cortex regions of APPSWE transgenic mice. *Journal of Neuroimmunology*, *126*(1-2), 50–57. doi:10.1016/S0165-5728(02)00050-4 PMID:12020956

Abdul, H. M., Sama, M. A., Furman, J. L., Mathis, D. M., Beckett, T. L., Weidner, A. M., Patel, E. S., Baig, I., Murphy, M. P., Levine, H. III, Kraner, S. D., & Norris, C. M. (2009). Cognitive decline in Alzheimer's disease is associated with selective changes in calcineurin/NFAT signaling. *BaigJournal of Neuroscience, 2009. Social Neuroscience*, *29*(41), 12957–12969. Advance online publication. doi:10.1523/JNEUROSCI.1064-09.2009 PMID:19828810

Ali, F., Hossain, M. S., Sejimo, S., & Akashi, K. (2019). Plasmalogens inhibit endocytosis of toll-like receptor 4 to attenuate the inflammatory signal in microglial cells. *Molecular Neurobiology*, *56*(5), 3404–3419. doi:10.100712035-018-1307-2 PMID:30128650

Allegri, R. F. (2020). Moving from neurodegenerative dementias, to cognitive proteinopathies, replacing "where" by "what".... *Dementia & Neuropsychologia*, *14*(3), 237–242. doi:10.1590/1980-57642020dn14-030005 PMID:32973977

Apelt, J. research, R. S.-B., (2001). *β-amyloid-induced glial expression of both pro-and anti-inflammatory cytokines in cerebral cortex of aged transgenic Tg2576 mice with Alzheimer plaque pathology*. Elsevier. https://www.sciencedirect.com/science/article/pii/S0006899300031760

Ashok, A., Rai, N. K., Raza, W., Pandey, R., & Bandyopadhyay, S. (2016). Chronic cerebral hypoperfusion-induced impairment of Abeta clearance requires HB-EGF-dependent sequential activation of HIF1alpha and MMP9. *Neurobiology of Disease*, *95*, 179–193. doi:10.1016/j.nbd.2016.07.013 PMID:27431094

Asiimwe, N., Yeo, S. G., Kim, M.-S., Jung, J., & Jeong, N. Y. (2016). Nitric oxide : Exploring the contextual link with Alzheimer's disease. *Oxidative Medicine and Cellular Longevity*, *2016*, 2016. doi:10.1155/2016/7205747 PMID:28096943

Asraf, K., Torika, N., Apte, R. N., & Fleisher-Berkovich, S. (2018). Microglial Activation Is Modulated by Captopril : In Vitro and in Vivo Studies. *Frontiers in Cellular Neuroscience*, *12*, 116. doi:10.3389/fncel.2018.00116 PMID:29765306

Bai, J.-Z., & Lipski, J. (2014). Involvement of TRPV4 channels in Aβ40-induced hippocampal cell death and astrocytic Ca2+ signalling. *Neurotoxicology*, *41*, 64–72. doi:10.1016/j.neuro.2014.01.001 PMID:24457011

Baranowska-Bik, A., Bik, W., Wolinska-Witort, E., Martynska, L., Chmielowska, M., Barcikowska, M., & Baranowska, B. (2008). Plasma beta amyloid and cytokine profile in women with Alzheimer's disease. *Neuroendocrinology Letters, 2008•nel. Altamira.Sk*, *29*(1), 18283248–18290108.

Beauquis, J., Vinuesa, A., Pomilio, C., Pavía, P., Galván, V., & Saravia, F. (2014). Neuronal and glial alterations, increased anxiety, and cognitive impairment before hippocampal amyloid deposition in PDAPP mice, model of Alzheimer's disease. *Hippocampus*, *24*(3), 257–269. doi:10.1002/hipo.22219 PMID:24132937

Benzing, W. C., Wujek, J. R., Ward, E. K., Shaffer, D., Ashe, K. H., Younkin, S. G., & Brunden, K. R. (1999). Evidence for glial-mediated inflammation in aged APPSW transgenic mice. *Neurobiology of Aging*, *20*(6), 581–589. doi:10.1016/S0197-4580(99)00065-2 PMID:10674423

Bernstein, H.-G., & Müller, M. (1999). The cellular localization of the L-ornithine decarboxylase/polyamine system in normal and diseased central nervous systems. *Progress in Neurobiology*, *57*(5), 485–505. doi:10.1016/S0301-0082(98)00065-3 PMID:10215098

Bie, B., Wu, J., Foss, J. F., & Naguib, M. (2019). Activation of mGluR1 Mediates C1q-Dependent Microglial Phagocytosis of Glutamatergic Synapses in Alzheimer's Rodent Models. *Molecular Neurobiology*, *56*(8), 5568–5585. doi:10.100712035-019-1467-8 PMID:30652266

Biedermann, B., Skatchkov, S. N., Brunk, I., Bringmann, A., Pannicke, T., Bernstein, H.-G., Faude, F., Germer, A., Veh, R., & Reichenbach, A. (1998). Spermine/spermidine is expressed by retinal glial (Müller) cells and controls distinct K+ channels of their membrane. *Glia*, *23*(3), 209–220. doi:10.1002/(SICI)1098-1136(199807)23:3<209::AID-GLIA4>3.0.CO;2-# PMID:9633806

Biscaro, B., Lindvall, O., Tesco, G., Ekdahl, C. T., & Nitsch, R. M. (2012). Inhibition of microglial activation protects hippocampal neurogenesis and improves cognitive deficits in a transgenic mouse model for Alzheimer's disease. *Neurodegenerative Diseases*, *9*(4), 187–198. doi:10.1159/000330363 PMID:22584394

Blurton-Jones, M., Kitazawa, M., Martinez-Coria, H., Castello, N. A., Müller, F.-J., Loring, J. F., Yamasaki, T. R., Poon, W. W., Green, K. N., & LaFerla, F. M. (2009). Neural stem cells improve cognition via BDNF in a transgenic model of Alzheimer disease. *Proceedings of the National Academy of Sciences of the United States of America*, *106*(32), 13594–13599. doi:10.1073/pnas.0901402106 PMID:19633196

Bolós, M., Llorens-Martín, M., Perea, J. R., Jurado-Arjona, J., Rábano, A., Hernández, F., & Avila, J. (2017). Absence of CX3CR1 impairs the internalization of Tau by microglia. *Molecular Neurodegeneration*, *12*(1), 1–14. doi:10.118613024-017-0200-1 PMID:28810892

Borroni, E., Bohrmann, B., Grueninger, F., Prinssen, E., Nave, S., Loetscher, H., Chinta, S. J., Rajagopalan, S., Rane, A., Siddiqui, A., Ellenbroek, B., Messer, J., Pähler, A., Andersen, J. K., Wyler, R., & Cesura, A. M. (2017). Sembragiline: A novel, selective monoamine oxidase type b inhibitor for the treatment of Alzheimer's disease. *The Journal of Pharmacology and Experimental Therapeutics*, *362*(3), 413–423. doi:10.1124/jpet.117.241653 PMID:28642233

Butt, A. M., De La Rocha, I. C., & Rivera, A. (2019). Oligodendroglial cells in Alzheimer's disease. *Neuroglia in neurodegenerative diseases*, 325-333.

Cagnin, A., Brooks, D. J., Kennedy, A. M., Gunn, R. N., Myers, R., Turkheimer, F. E., Jones, T., & Banati, R. B. (2001). In-vivo measurement of activated microglia in dementia. *Lancet*, *358*(9280), 461–467. doi:10.1016/S0140-6736(01)05625-2 PMID:11513911

Cai, Z., & Xiao, M. (2016). Oligodendrocytes and Alzheimer's disease. *The International Journal of Neuroscience*, *126*(2), 97–104. doi:10.3109/00207454.2015.1025778 PMID:26000818

Carter, S. F., Schöll, M., Almkvist, O., Wall, A., Engler, H., Långström, B., & Nordberg, A. (2012). Evidence for astrocytosis in prodromal Alzheimer disease provided by 11C-deuterium-L-deprenyl : A multitracer PET paradigm combining 11C-Pittsburgh compound B and 18F-FDG. *Journal of Nuclear Medicine*, *53*(1), 37–46. doi:10.2967/jnumed.110.087031 PMID:22213821

Casal, C., Serratosa, J., & Tusell, J. M. (2002). Relationship between β-AP peptide aggregation and microglial activation. *Brain Research*, *928*(1-2), 76–84. doi:10.1016/S0006-8993(01)03362-5 PMID:11844474

Chen, J., Yang, Y., Yang, Y., Zou, P., Chen, J., He, Y., Shui, S., Cui, Y., Bai, R., Liang, Y., Hu, Y., Jiang, B., Lu, L., Zhang, X., Liu, J., & Xu, J. (2018). AXL promotes Zika virus infection in astrocytes by antagonizing type I interferon signalling. *Nature Microbiology*, *3*(3), 302–309. doi:10.103841564-017-0092-4 PMID:29379210

Constam, D. B., Philipp, J., Malipiero, U. V., Ten Dijke, P., Schachner, M., & Fontana, A. (1992). Differential expression of transforming growth factor-beta 1,-beta 2, and-beta 3 by glioblastoma cells, astrocytes, and microglia. *Journal of Immunology (Baltimore, Md.: 1950)*, *148*(5), 1404-1410.

Corbett, G. T., Roy, A., & Pahan, K. (2013). Sodium phenylbutyrate enhances astrocytic neurotrophin synthesis via protein kinase C (PKC)-mediated activation of cAMP-response element-binding protein (CREB) : Implications for Alzheimer disease therapy. *The Journal of Biological Chemistry*, *288*(12), 8299–8312. doi:10.1074/jbc.M112.426536 PMID:23404502

Corder, E. H., Saunders, A. M., Strittmatter, W. J., Schmechel, D. E., Gaskell, P. C., Small, G. W., Roses, A. D., Haines, J. L., & Pericak-Vance, M. A. (1993). Gene dose of apolipoprotein E type 4 allele and the risk of Alzheimer's disease in late onset families. *Science*, *261*(5123), 921–923. doi:10.1126cience.8346443 PMID:8346443

Cummings, J. L., Morstorf, T., & Zhong, K. (2014). Alzheimer's disease drug-development pipeline : Few candidates, frequent failures. *Alzheimer's Research & Therapy*, *6*(4), 1–7. doi:10.1186/alzrt269 PMID:25024750

Dai, J., Johnson, E. C. B., Dammer, E. B., Duong, D. M., Gearing, M., Lah, J. J., Levey, A. I., Wingo, T. S., & Seyfried, N. T. (2018). Effects of APOE genotype on brain proteomic network and cell type changes in Alzheimer's disease. *Frontiers in Molecular Neuroscience*, *11*, 454. doi:10.3389/fnmol.2018.00454 PMID:30618606

De la Fuente, A. G., Queiroz, R. M., Ghosh, T., McMurran, C. E., Cubillos, J. F., Bergles, D. E., Fitzgerald, D. C., Jones, C. A., Lilley, K. S., Glover, C. P., & Franklin, R. J. M. (2020). Changes in the oligodendrocyte progenitor cell proteome with ageing. *Molecular & Cellular Proteomics*, *19*(8), 1281–1302. doi:10.1074/mcp.RA120.002102 PMID:32434922

Deatly, A. M., Haase, A. T., & Ball, M. J. (1991). Herpes Simplex Virus Type 1 Transcription during Latent Infections of Mouse and Man : Implications for Dementia. In E. Kurstak (Ed.), *Psychiatry and Biological Factors* (pp. 277–286). Springer. doi:10.1007/978-1-4684-5811-4_25

DeFlitch, L., Gonzalez-Fernandez, E., Crawley, I., & Kang, S. H. (2022). Age and Alzheimer's Disease-Related oligodendrocyte changes in hippocampal subregions. *Frontiers in Cellular Neuroscience*, *16*, 847097. doi:10.3389/fncel.2022.847097 PMID:35465615

Desai, M. K., Guercio, B. J., Narrow, W. C., & Bowers, W. J. (2011). An Alzheimer's disease-relevant presenilin-1 mutation augments amyloid-beta-induced oligodendrocyte dysfunction. *Glia, 59*(4), 627–640. doi:10.1002/glia.21131 PMID:21294162

Dossi, E., Vasile, F., & Rouach, N. (2018). Human astrocytes in the diseased brain. *Brain Research Bulletin, 136*, 139–156. doi:10.1016/j.brainresbull.2017.02.001 PMID:28212850

Edwards, L. J., Kirilina, E., Mohammadi, S., & Weiskopf, N. (2018). Microstructural imaging of human neocortex in vivo. *NeuroImage, 182*, 184–206. doi:10.1016/j.neuroimage.2018.02.055 PMID:29588229

Emilsson, L., Saetre, P., Balciuniene, J., Castensson, A., Cairns, N., & Jazin, E. E. (2002). Increased monoamine oxidase messenger RNA expression levels in frontal cortex of Alzheimer's disease patients. *Neuroscience Letters, 326*(1), 56–60. doi:10.1016/S0304-3940(02)00307-5 PMID:12052537

Farina, C., Aloisi, F., & Meinl, E. (2007). Astrocytes are active players in cerebral innate immunity. *Trends in Immunology, 28*(3), 138–145. doi:10.1016/j.it.2007.01.005 PMID:17276138

Felsky, D., Roostaei, T., Nho, K., Risacher, S. L., Bradshaw, E. M., Petyuk, V., Schneider, J. A., Saykin, A., Bennett, D. A., & De Jager, P. L. (2019). Neuropathological correlates and genetic architecture of microglial activation in elderly human brain. *Nature Communications, 10*(1), 409. doi:10.103841467-018-08279-3 PMID:30679421

Ferreira, S., Pitman, K. A., Wang, S., Summers, B. S., Bye, N., Young, K. M., & Cullen, C. L. (2020). Amyloidosis is associated with thicker myelin and increased oligodendrogenesis in the adult mouse brain. *CullenJournal of Neuroscience Research. Journal of Neuroscience Research, 98*(10), 1905–1932. doi:10.1002/jnr.24672 PMID:32557778

Findley, C. A., Bartke, A., Hascup, K. N., & Hascup, E. R. (2019). Amyloid Beta-Related Alterations to Glutamate Signaling Dynamics During Alzheimer's Disease Progression. *ASN Neuro, 11*, 175909141985554. doi:10.1177/1759091419855541 PMID:31213067

Fontenas, L., Welsh, T. G., Piller, M., Coughenour, P., Gandhi, A. V., Prober, D. A., & Kucenas, S. (2019). The neuromodulator adenosine regulates oligodendrocyte migration at motor exit point transition zones. *Cell Reports, 27*(1), 115–128. doi:10.1016/j.celrep.2019.03.013 PMID:30943395

Frisoni, G. B., Altomare, D., Thal, D. R., Ribaldi, F., van der Kant, R., Ossenkoppele, R., Blennow, K., Cummings, J., van Duijn, C., Nilsson, P. M., Dietrich, P.-Y., Scheltens, P., & Dubois, B. (2022). The probabilistic model of Alzheimer disease : The amyloid hypothesis revised. *Nature Reviews. Neuroscience, 23*(1), 53–66. doi:10.103841583-021-00533-w PMID:34815562

Fritsche, S., Wang, X., Antioxidants, C. J. (2017). Recent advances in our understanding of tocopherol biosynthesis in plants: an overview of key genes, functions, and breeding of vitamin E improved crops. *Antioxidants, mdpi.Com*. doi:10.3390/antiox6040099

Frost, G. R., & Li, Y. M. (2017). The role of astrocytes in amyloid production and Alzheimer's disease. *Open Biology, 7*(12), 170228. doi:10.1098/rsob.170228 PMID:29237809

Fultz, M. J., Barber, S. A., Dieffenbach, C. W., & Vogel, S. N. (1993). Induction of IFN-γ in macrophages by lipopolysaccharide. *International Immunology*, *5*(11), 1383–1392. doi:10.1093/intimm/5.11.1383 PMID:8260452

Furman, J. L., Sama, D. M., Gant, J. C., Beckett, T. L., Murphy, M. P., Bachstetter, A. D., Van Eldik, L. J., & Norris, C. M. (2012). Targeting Astrocytes Ameliorates Neurologic Changes in a Mouse Model of Alzheimer's Disease. *The Journal of Neuroscience : The Official Journal of the Society for Neuroscience*, *32*(46), 16129–16140. doi:10.1523/JNEUROSCI.2323-12.2012 PMID:23152597

Gale, S. C., Gao, L., Mikacenic, C., Coyle, S. M., Rafaels, N., Murray Dudenkov, T., Madenspacher, J. H., Draper, D. W., Ge, W., Aloor, J. J., Azzam, K. M., Lai, L., Blackshear, P. J., Calvano, S. E., Barnes, K. C., Lowry, S. F., Corbett, S., Wurfel, M. M., & Fessler, M. B. (2014). APOepsilon4 is associated with enhanced *in vivo* innate immune responses in human subjects. *The Journal of Allergy and Clinical Immunology*, *134*(1), 127–134. doi:10.1016/j.jaci.2014.01.032 PMID:24655576

Gilani, A. I., Chohan, M. O., Inan, M., Schobel, S. A., Chaudhury, N. H., Samuel, P., Chuhma, N., Glickstein, S., Merker, R. J., Xu, Q., Small, S. A., Anderson, S. A., Ross, M. E., & Moore, H. (2014). Interneuron precursor transplants in adult hippocampus reverse psychosis-relevant features in a mouse model of hippocampal disinhibition. *Proceedings of the National Academy of Sciences of the United States of America*, *111*(20), 7450–7455. doi:10.1073/pnas.1316488111 PMID:24794528

Gouw, A. A., Seewann, A., Vrenken, H., Van Der Flier, W. M., Rozemuller, J. M., Barkhof, F., Scheltens, P., & Geurts, J. J. G. (2008). Heterogeneity of white matter hyperintensities in Alzheimer's disease : Post-mortem quantitative MRI and neuropathology. *Brain*, *131*(12), 3286–3298. doi:10.1093/brain/awn265 PMID:18927145

Griciuc, A., Serrano-Pozo, A., Parrado, A. R., Lesinski, A. N., Asselin, C. N., Mullin, K., Hooli, B., Choi, S. H., Hyman, B. T., & Tanzi, R. E. (2013). Alzheimer's disease risk gene CD33 inhibits microglial uptake of amyloid beta. *Neuron*, *78*(4), 631–643. doi:10.1016/j.neuron.2013.04.014 PMID:23623698

Griffin, W. S., Stanley, L. C., Ling, C., White, L., MacLeod, V., Perrot, L. J., White, C. L. III, & Araoz, C. (1989). Brain interleukin 1 and S-100 immunoreactivity are elevated in Down syndrome and Alzheimer disease. *Proceedings of the National Academy of Sciences of the United States of America*, *86*(19), 7611–7615. doi:10.1073/pnas.86.19.7611 PMID:2529544

Griggs, R. & Yermakov, L. (2017). *Formation and disruption of functional domains in myelinated CNS axons*. Elsevier.

Gu, L., Wu, D., Tang, X., Qi, X., Li, X., Bai, F., Chen, X., Ren, Q., & Zhang, Z. (2018). Myelin changes at the early stage of 5XFAD mice. *Brain Research Bulletin*, *137*, 285–293. doi:10.1016/j.brainresbull.2017.12.013 PMID:29288735

Guan, Y.-H., Zhang, L.-J., Wang, S.-Y., Deng, Y.-D., Zhou, H.-S., Chen, D.-Q., & Zhang, L.-C. (2022). The role of microglia in Alzheimer's disease and progress of treatment. *Ibrain*, *8*(1), 37–47. doi:10.1002/ibra.12023 PMID:37786418

Halleskog, C., Mulder, J., Dahlström, J., Mackie, K., Hortobágyi, T., Tanila, H., Kumar Puli, L., Färber, K., Harkany, T., & Schulte, G. (2011). WNT signaling in activated microglia is proinflammatory. *Glia*, *59*(1), 119–131. doi:10.1002/glia.21081 PMID:20967887

Halliday, M. R., Rege, S. V., Ma, Q., Zhao, Z., Miller, C. A., Winkler, E. A., & Zlokovic, B. V. (2016). Accelerated pericyte degeneration and blood-brain barrier breakdown in apolipoprotein E4 carriers with Alzheimer's disease. *Journal of Cerebral Blood Flow and Metabolism*, *36*(1), 216–227. doi:10.1038/jcbfm.2015.44 PMID:25757756

Hamilton, N. B., Clarke, L. E., Lorena Arancibia-Carcamo, I., Kougioumtzidou, E., Matthey, M., Arad, R. K., Whiteley, L., Bergersen, L. H., Richardson, W. D., & Attwell, D. (2017). Endogenous GABA controls oligodendrocyte lineage cell number, myelination, and CNS internode length. *Glia*, *65*(2), 309–321. doi:10.1002/glia.23093 PMID:27796063

Handa, A. K., Fatima, T., & Mattoo, A. K. (2018). Polyamines: Bio-Molecules with Diverse Functions in Plant and Human Health and Disease. *Frontiers in Chemistry*, *6*, 10. doi:10.3389/fchem.2018.00010 PMID:29468148

Hansen, D. V., Hanson, J. E., & Sheng, M. (2018). Microglia in Alzheimer's disease. *The Journal of Cell Biology*, *217*(2), 459–472. doi:10.1083/jcb.201709069 PMID:29196460

Hardy, J. A., & Higgins, G. A. (1992). Alzheimer's disease : The amyloid cascade hypothesis. *Science*, *256*(5054), 184–185. doi:10.1126cience.1566067 PMID:1566067

He, Y., Ruganzu, J. B., Jin, H., Peng, X., Ji, S., Ma, Y., Zheng, L., & Yang, W. (2020). LRP1 knockdown aggravates Aβ1–42-stimulated microglial and astrocytic neuroinflammatory responses by modulating TLR4/NF-κB/MAPKs signaling pathways. *Experimental Cell Research*, *394*(2), 112166. doi:10.1016/j.yexcr.2020.112166 PMID:32645395

Horiuchi, M., Maezawa, I., Itoh, A., Wakayama, K., & L. J.-N. (2012). *Amyloid β1–42 oligomer inhibits myelin sheet formation in vitro*. Elsevier. https://www.sciencedirect.com/science/article/pii/S0197458010002216?casa_token=ej2OJHtzdogAAAAA:IgCWwtmXoIHp6Z-

Hou, L., Liu, Y., Wang, X., Ma, H., He, J., Zhang, Y., Yu, C., Guan, W., & Ma, Y. (2011). The effects of amyloid-β42 oligomer on the proliferation and activation of astrocytes in vitro. *In Vitro Cellular & Developmental Biology. Animal*, *47*(8), 573–580. doi:10.100711626-011-9439-y PMID:21858610

Hsiao, H.-Y., Chen, Y.-C., Chen, H.-M., Tu, P.-H., & Chern, Y. (2013). A critical role of astrocyte-mediated nuclear factor-κB-dependent inflammation in Huntington's disease. *Human Molecular Genetics*, *22*(9), 1826–1842. doi:10.1093/hmg/ddt036 PMID:23372043

Hu, J.-G., Wu, X.-J., Feng, Y.-F., Xi, G.-M., Wang, Z.-H., Zhou, J.-S., & Lü, H.-Z. (2012). PDGF-AA and bFGF mediate B104CM-induced proliferation of oligodendrocyte precursor cells. *International Journal of Molecular Medicine*, *30*(5), 1113–1118. doi:10.3892/ijmm.2012.1110 PMID:22922759

Huberman, M., Shalit, F., Roth-Deri, I., Gutman, B., Brodie, C., Kott, E., & Sredni, B. (1994). Correlation of cytokine secretion by mononuclear cells of Alzheimer patients and their disease stage. *Journal of Neuroimmunology*, *52*(2), 147–152. doi:10.1016/0165-5728(94)90108-2 PMID:8034754

Huentelman, M., De Both, M., Jepsen, W., Piras, I. S., Talboom, J. S., Willeman, M., Reiman, E. M., Hardy, J., & Myers, A. J. (2019). Common BACE2 polymorphisms are associated with altered risk for Alzheimer's disease and CSF amyloid biomarkers in APOE epsilon4 non-carriers. *Scientific Reports*, *9*(1), 9640. doi:10.103841598-019-45896-4 PMID:31270419

Iordanishvili, E., Schall, M., Loução, R., & Neuroimage, M. Z. (2019). *Quantitative MRI of cerebral white matter hyperintensities: a new approach towards understanding the underlying pathology*. Elsevier.

Itabashi, S., Arai, H., Matsui, T., Higuchi, S., & Sasaki, H. (1997). Herpes simplex virus and risk of Alzheimer's disease. *Lancet*, *349*(9058), 1102. doi:10.1016/S0140-6736(05)62325-2 PMID:9107270

Jack, C. R. Jr, Knopman, D. S., Jagust, W. J., Shaw, L. M., Aisen, P. S., Weiner, M. W., Petersen, R. C., & Trojanowski, J. Q. (2010). Hypothetical model of dynamic biomarkers of the Alzheimer's pathological cascade. *Lancet Neurology*, *9*(1), 119–128. doi:10.1016/S1474-4422(09)70299-6 PMID:20083042

Jellinger, K. A. (2020). Neuropathological assessment of the Alzheimer spectrum. *Journal of Neural Transmission (Vienna, Austria)*, *127*(9), 1229–1256. doi:10.100700702-020-02232-9 PMID:32740684

Jessen, K. R. (2004). Glial cells. *The International Journal of Biochemistry & Cell Biology*, *36*(10), 1861–1867. doi:10.1016/j.biocel.2004.02.023 PMID:15203098

Jiang, K., Wang, J., Wu, F., Xue, Q., Yao, Q., Zhang, J., Chen, Y., Zhang, G., Zhu, Z., Yan, H., Zhu, L., & Yip, H. (2020). Dopant-Free Organic Hole-Transporting Material for Efficient and Stable Inverted All-Inorganic and Hybrid Perovskite Solar Cells. *Advanced Materials*, *32*(16), 1908011. doi:10.1002/adma.201908011 PMID:32115824

Johnson, L. A. (2020). APOE and metabolic dysfunction in Alzheimer's disease. *International Review of Neurobiology*, *154*, 131–151. doi:10.1016/bs.irn.2020.02.002 PMID:32739002

Kamphuis, W., Kooijman, L., Orre, M., Stassen, O., Pekny, M., & Hol, E. M. (2015). GFAP and vimentin deficiency alters gene expression in astrocytes and microglia in wild-type mice and changes the transcriptional response of reactive glia in mouse model for Alzheimer's disease. *Glia*, *63*(6), 1036–1056. doi:10.1002/glia.22800 PMID:25731615

Kamphuis, W., Middeldorp, J., Kooijman, L., Sluijs, J. A., Kooi, E. J., Moeton, M., Freriks, M., Mizee, M. R., & Hol, E. M. (2014). Glial fibrillary acidic protein isoform expression in plaque related astrogliosis in Alzheimer's disease. *Neurobiology of Aging*, *35*(3), 492–510. doi:10.1016/j.neurobiolaging.2013.09.035 PMID:24269023

Kavroulakis, E., Simos, P. G., Kalaitzakis, G., Maris, T. G., Karageorgou, D., Zaganas, I., Panagiotakis, S., Basta, M., Vgontzas, A., & Papadaki, E. (2018). Myelin content changes in probable Alzheimer's disease and mild cognitive impairment : Associations with age and severity of neuropsychiatric impairment. *Journal of Magnetic Resonance Imaging*, *47*(5), 1359–1372. doi:10.1002/jmri.25849 PMID:28861929

Kawabori, M., Kacimi, R., Kauppinen, T., Calosing, C., Kim, J. Y., Hsieh, C. L., Nakamura, M. C., & Yenari, M. A. (2015). Triggering Receptor Expressed on Myeloid Cells 2 (TREM2) Deficiency Attenuates Phagocytic Activities of Microglia and Exacerbates Ischemic Damage in Experimental Stroke. *The Journal of Neuroscience : The Official Journal of the Society for Neuroscience*, *35*(8), 3384–3396. doi:10.1523/JNEUROSCI.2620-14.2015 PMID:25716838

Kövari, E., Gold, G., Herrmann, F. R., Canuto, A., Hof, P. R., Bouras, C., & Giannakopoulos, P. (2007). Cortical microinfarcts and demyelination affect cognition in cases at high risk for dementia. *Neurology*, *68*(12), 927–931. doi:10.1212/01.wnl.0000257094.10655.9a PMID:17372128

Kraft, A. W., Hu, X., Yoon, H., Yan, P., Xiao, Q., Wang, Y., Gil, S. C., Brown, J., Wilhelmsson, U., Restivo, J. L., Cirrito, J. R., Holtzman, D. M., Kim, J., Pekny, M., & Lee, J. (2013). Attenuating astrocyte activation accelerates plaque pathogenesis in APP/PS1 mice. *The FASEB Journal*, *27*(1), 187–198. doi:10.1096/fj.12-208660 PMID:23038755

Krauss, M., Weiss, T., Langnaese, K., Richter, K., Kowski, A., Veh, R. W., & Laube, G. (2007). Cellular and subcellular rat brain spermidine synthase expression patterns suggest region-specific roles for polyamines, including cerebellar pre-synaptic function. *Journal of Neurochemistry*, *103*(2), 679–693. doi:10.1111/j.1471-4159.2007.04770.x PMID:17635671

Kucheryavykh, L. Y., Benedikt, J., Cubano, L. A., Skatchkov, S. N., Bukauskas, F. F., & Kucheryavykh, Y. V. (2017). Polyamines preserve connexin43-mediated gap junctional communication during intracellular hypercalcemia and acidosis. *Neuroreport*, *28*(4), 208–213. doi:10.1097/WNR.0000000000000746 PMID:28134630

Kuhn, S., Gritti, L., Crooks, D., & Dombrowski, Y. (2019). Oligodendrocytes in development, myelin generation and beyond. *Cells*, *8*(11), 1424. doi:10.3390/cells8111424 PMID:31726662

Kulijewicz-Nawrot, M., Verkhratsky, A., Chvá Tal, A., Syková, E., & Rodríguez, J. J. (2012). Astrocytic cytoskeletal atrophy in the medial prefrontal cortex of a triple transgenic mouse model of Alzheimer's disease. *Journal of Anatomy*, *221*(3), 252–262. doi:10.1111/j.1469-7580.2012.01536.x

Laube, G., Ru¨, R., & Veh, R. W. (1997). Astrocytes, not neurons, show most prominent staining for spermidine/spermine-like immunoreactivity in adult rat brain. *Glia*, *19*(2), 171–179. doi:10.1002/(SICI)1098-1136(199702)19:2<171::AID-GLIA8>3.0.CO;2-3 PMID:9034833

Lavoie, S., Chen, Y., Dalton, T. P., Gysin, R., Cuénod, M., Steullet, P., & Do, K. Q. (2009). Curcumin, quercetin, and tBHQ modulate glutathione levels in astrocytes and neurons : Importance of the glutamate cysteine ligase modifier subunit. *Journal of Neurochemistry*, *108*(6), 1410–1422. doi:10.1111/j.1471-4159.2009.05908.x PMID:19183254

Lee, E., & Chung, W.-S. (2019). Glial control of synapse number in healthy and diseased brain. *Frontiers in Cellular Neuroscience*, *13*, 42. doi:10.3389/fncel.2019.00042 PMID:30814931

Lessard, C. B., Malnik, S. L., Zhou, Y., Ladd, T. B., Cruz, P. E., Ran, Y., Mahan, T. E., Chakrabaty, P., Holtzman, D. M., Ulrich, J. D., Colonna, M., & Golde, T. E. (2018). High-affinity interactions and signal transduction between Aβ oligomers and TREM 2. *EMBO Molecular Medicine*, *10*(11), e9027. doi:10.15252/emmm.201809027 PMID:30341064

Li, S., & Selkoe, D. J. (2020). A mechanistic hypothesis for the impairment of synaptic plasticity by soluble Aβ oligomers from Alzheimer's brain. *Journal of Neurochemistry*. *Journal of Neurochemistry*, *154*(6), 583–597. doi:10.1111/jnc.15007 PMID:32180217

Lian, H., Litvinchuk, A., Chiang, A. C.-A., Aithmitti, N., Jankowsky, J. L., & Zheng, H. (2016). Astrocyte-Microglia Cross Talk through Complement Activation Modulates Amyloid Pathology in Mouse Models of Alzheimer's Disease. *The Journal of Neuroscience : The Official Journal of the Society for Neuroscience, 36*(2), 577–589. doi:10.1523/JNEUROSCI.2117-15.2016 PMID:26758846

Liu, C. C., Kanekiyo, T., Xu, H., & Bu, G. (2013). Apolipoprotein E and Alzheimer disease: Risk, mechanisms and therapy. *Nature Reviews. Neurology, 9*(2), 106–118. doi:10.1038/nrneurol.2012.263 PMID:23296339

Liu, J., Chang, L., Song, Y., Li, H., & Wu, Y. (2019). The Role of NMDA Receptors in Alzheimer's Disease. *Frontiers in Neuroscience, 13*, 43. doi:10.3389/fnins.2019.00043 PMID:30800052

Liu, J.-H., Wang, W., Wu, H., Gong, X., & Moriguchi, T. (2015). Polyamines function in stress tolerance : From synthesis to regulation. *Frontiers in Plant Science, 6*, 827. doi:10.3389/fpls.2015.00827 PMID:26528300

Long, P., & Corfas, G. (2014). To learn is to myelinate. *Science, 346*(6207), 298–299. doi:10.1126cience.1261127 PMID:25324370

Lorenzini, L., Fernandez, M., Baldassarro, V. A., Bighinati, A., Giuliani, A., Calzà, L., & Giardino, L. (2020). White matter and neuroprotection in Alzheimer's dementia. *Molecules (Basel, Switzerland), 25*(3), 503. doi:10.3390/molecules25030503 PMID:31979414

Luchena, C., Zuazo-Ibarra, J., Alberdi, E., Matute, C., & Capetillo-Zarate, E. (2018). Contribution of Neurons and Glial Cells to Complement-Mediated Synapse Removal during Development, Aging and in Alzheimer's Disease. *Mediators of Inflammation, 2018*, 1–12. doi:10.1155/2018/2530414 PMID:30533998

Luo, J., Wärmländer, S., & A. G.-J. (2016). Cross-interactions between the Alzheimer disease amyloid-β peptide and other amyloid proteins: a further aspect of the amyloid cascade hypothesis. *ASBMB.*

Ma, K.-C., Nie, X. J., Höög, A., Olsson, Y., & Zhang, W. W. (1994). Reactive astrocytes in viral infections of the human brain express endothelin-like immunoreactivity. *Journal of the Neurological Sciences, 126*(2), 184–192. doi:10.1016/0022-510X(94)90271-2 PMID:7531760

Malm, T. M., Iivonen, H., Goldsteins, G., Keksa-Goldsteine, V., Ahtoniemi, T., Kanninen, K., Salminen, A., Auriola, S., Van Groen, T., Tanila, H., & Koistinaho, J. (2007). Pyrrolidine Dithiocarbamate Activates Akt and Improves Spatial Learning in APP/PS1 Mice without Affecting β-Amyloid Burden. *The Journal of Neuroscience : The Official Journal of the Society for Neuroscience, 27*(14), 3712–3721. doi:10.1523/JNEUROSCI.0059-07.2007 PMID:17409235

Malmsten, L., Vijayaraghavan, S., Hovatta, O., Marutle, A., & Darreh-Shori, T. (2014). Fibrillar β-amyloid 1-42 alters cytokine secretion, cholinergic signalling and neuronal differentiation. *Journal of Cellular and Molecular Medicine, 18*(9), 1874–1888. doi:10.1111/jcmm.12343 PMID:25109373

Manocha, G. D., Ghatak, A., Puig, K. L., Kraner, S. D., Norris, C. M., & Combs, C. K. (2017). NFATc2 Modulates Microglial Activation in the AβPP/PS1 Mouse Model of Alzheimer's Disease. *Journal of Alzheimer's Disease, 58*(3), 775–787. doi:10.3233/JAD-151203 PMID:28505967

Marangon, D., Boccazzi, M., Lecca, D., & Fumagalli, M. (2020). Regulation of oligodendrocyte functions: Targeting lipid metabolism and extracellular matrix for myelin repair. *Journal of Clinical Medicine*, *9*(2), 470. doi:10.3390/jcm9020470 PMID:32046349

McGeer, P. L., Itagaki, S., Tago, H., & McGeer, E. G. (1987). Reactive microglia in patients with senile dementia of the Alzheimer type are positive for the histocompatibility glycoprotein HLA-DR. *Neuroscience Letters*, *79*(1-2), 195–200. doi:10.1016/0304-3940(87)90696-3 PMID:3670729

Minogue, A. M., Jones, R. S., Kelly, R. J., McDonald, C. L., Connor, T. J., & Lynch, M. A. (2014). Age-associated dysregulation of microglial activation is coupled with enhanced blood-brain barrier permeability and pathology in APP/PS1 mice. *Neurobiology of Aging*, *35*(6), 1442–1452. doi:10.1016/j.neurobiolaging.2013.12.026 PMID:24439957

Minois, N. (2014). Molecular basis of the 'anti-aging' effect of spermidine and other natural polyamines-a mini-review. *Gerontology*, *60*(4), 319–326. doi:10.1159/000356748 PMID:24481223

Mohandas, E., Rajmohan, V., & Raghunath, B. (2009). Neurobiology of Alzheimer's disease. *Indian Journal of Psychiatry*, *51*(1), 55. doi:10.4103/0019-5545.44908 PMID:19742193

Morrison, L. D., Cao, X.-C., & Kish, S. J. (1998). Ornithine decarboxylase in human brain : Influence of aging, regional distribution, and Alzheimer's disease. *Journal of Neurochemistry*, *71*(1), 288–294. doi:10.1046/j.1471-4159.1998.71010288.x PMID:9648877

Mueller, S. G., & Weiner, M. W. (2009). Selective effect of age, Apo e4, and Alzheimer's disease on hippocampal subfields. *Hippocampus*, *19*(6), 558–564. doi:10.1002/hipo.20614 PMID:19405132

Nasrabady, S. E., Rizvi, B., Goldman, J. E., & Brickman, A. M. (2018a). White matter changes in Alzheimer's disease : A focus on myelin and oligodendrocytes. *Acta Neuropathologica Communications*, *6*(1), 1–10. doi:10.118640478-018-0515-3 PMID:29499767

Nasrabady, S. E., Rizvi, B., Goldman, J. E., & Brickman, A. M. (2018b). White matter changes in Alzheimer's disease : A focus on myelin and oligodendrocytes. *Acta Neuropathologica Communications*, *6*(1), 1–10. doi:10.118640478-018-0515-3 PMID:29499767

Ojala, J., Alafuzoff, I., Herukka, S.-K., van Groen, T., Tanila, H., & Pirttilä, T. (2009). Expression of interleukin-18 is increased in the brains of Alzheimer's disease patients. *Neurobiology of Aging*, *30*(2), 198–209. doi:10.1016/j.neurobiolaging.2007.06.006 PMID:17658666

Okun, E., Griffioen, K. J., Lathia, J. D., Tang, S.-C., Mattson, M. P., & Arumugam, T. V. (2009). Toll-like receptors in neurodegeneration. *Brain Research. Brain Research Reviews*, *59*(2), 278–292. doi:10.1016/j.brainresrev.2008.09.001 PMID:18822314

Olabarria, M., Noristani, H. N., Verkhratsky, A., & Rodríguez, J. J. (2010). Concomitant astroglial atrophy and astrogliosis in a triple transgenic animal model of Alzheimer's disease. *Glia*, *58*(7), 831–838. doi:10.1002/glia.20967 PMID:20140958

Olsen, M. L., Khakh, B. S., Skatchkov, S. N., Zhou, M., Lee, C. J., & Rouach, N. (2015). New insights on astrocyte ion channels : Critical for homeostasis and neuron-glia signaling. *The Journal of Neuroscience : The Official Journal of the Society for Neuroscience, 35*(41), 13827–13835. doi:10.1523/JNEUROSCI.2603-15.2015 PMID:26468182

Orkand, R. K., Nicholls, J. G., & Kuffler, S. W. (1966). Effect of nerve impulses on the membrane potential of glial cells in the central nervous system of amphibia. *Journal of Neurophysiology, 29*(4), 788–806. doi:10.1152/jn.1966.29.4.788 PMID:5966435

Ortinski, P. I., Dong, J., Mungenast, A., Yue, C., Takano, H., Watson, D. J., Haydon, P. G., & Coulter, D. A. (2010). Selective induction of astrocytic gliosis generates deficits in neuronal inhibition. *Nature Neuroscience, 13*(5), 584–591. doi:10.1038/nn.2535 PMID:20418874

Ota, M., Sato, N., & Kimura, Y. (2019). Changes of myelin organization in patients with Alzheimer's disease shown by q-space myelin map imaging. *Matsuda Dementia and Geriatric Cognitive Disorders Extra.* Karger. https://karger.com/dee/article/9/1/24/102937

Overmyer, M., Helisalmi, S., Soininen, H., Laakso, M., Riekkinen, P. Sr, & Alafuzoff, I. (1999). Reactive microglia in aging and dementia : An immunohistochemical study of postmortem human brain tissue. *Acta Neuropathologica, 97*(4), 383–392. doi:10.1007004010051002 PMID:10208278

Papuć, E., Science, K. R.-A. (2020). The role of myelin damage in Alzheimer's disease pathology. *Termedia. Archives of Medical Science, 16*(2), 345–351. doi:10.5114/aoms.2018.76863

Park, S. H., Kim, J. H., Choi, K. H., Jang, Y. J., Bae, S. S., Choi, B. T., & Shin, H. K. (2013). Hypercholesterolemia accelerates amyloid β-induced cognitive deficits. *International Journal of Molecular Medicine, 31*(3), 577–582. doi:10.3892/ijmm.2013.1233 PMID:23314909

Patel, N., Paris, D., Mathura, V., & A. Q.-J. (2005). Inflammatory cytokine levels correlate with amyloid load in transgenic mouse models of Alzheimer's disease. *Journal of Neuroinflammation.* https://link.springer.com/article/10.1186/1742-2094-2-9

Perea, J., Bolós, M., & Biomolecules, J. A. (2020). Microglia in Alzheimer's disease in the context of tau pathology. *Mdpi.ComJR Perea, M Bolós, J AvilaBiomolecules •mdpi. Biomolecules, 10*(10), 1–26. doi:10.3390/biom10101439 PMID:33066368

Piacentini, R., Li Puma, D. D., Ripoli, C., Elena Marcocci, M., De Chiara, G., Garaci, E., Teresa Palamara, A., & Grassi, C. (2015). Herpes Simplex Virus type-1 infection induces synaptic dysfunction in cultured cortical neurons via GSK-3 activation and intraneuronal amyloid-β protein accumulation. *Scientific Reports, 5*(1), 15444. doi:10.1038rep15444 PMID:26487282

Poon, C. H., Wang, Y., Fung, M.-L., Zhang, C., & Lim, L. W. (2020). Rodent models of amyloid-beta feature of Alzheimer's disease : Development and potential treatment implications. *Aging and Disease, 11*(5), 1235. doi:10.14336/AD.2019.1026 PMID:33014535

Ramanathan, A., Nelson, A. R., Sagare, A. P., & Zlokovic, B. V. (2015). Impaired vascular-mediated clearance of brain amyloid beta in Alzheimer's disease: The role, regulation and restoration of LRP1. *Frontiers in Aging Neuroscience, 7*, 136. doi:10.3389/fnagi.2015.00136 PMID:26236233

Rauch, J. N., Luna, G., Guzman, E., Audouard, M., Challis, C., Sibih, Y. E., Leshuk, C., Hernandez, I., Wegmann, S., Hyman, B. T., Gradinaru, V., Kampmann, M., & Kosik, K. S. (2020). LRP1 is a master regulator of tau uptake and spread. *Nature, 580*(7803), 381–385. doi:10.103841586-020-2156-5 PMID:32296178

Revilla, S., Ursulet, S., Álvarez-López, M. J., Castro-Freire, M., Perpiñá, U., García-Mesa, Y., Bortolozzi, A., Giménez-Llort, L., Kaliman, P., Cristòfol, R., Sarkis, C., & Sanfeliu, C. (2014). Lenti-GDNF gene therapy protects against Alzheimer's disease-like neuropathology in 3xTg-AD mice and MC65 cells. *CNS Neuroscience & Therapeutics, 20*(11), 961–972. doi:10.1111/cns.12312 PMID:25119316

Rodríguez, J. J., Yeh, C.-Y., Terzieva, S., Olabarria, M., Kulijewicz-Nawrot, M., & Verkhratsky, A. (2014). Complex and region-specific changes in astroglial markers in the aging brain. *Neurobiology of Aging, 35*(1), 15–23. doi:10.1016/j.neurobiolaging.2013.07.002 PMID:23969179

Rolyan, H., Scheffold, A., Heinrich, A., & Brain, Y. B.-N. (2011). Telomere shortening reduces Alzheimer's disease amyloid pathology in mice. *Academic.* https://academic.oup.com/brain/article-abstract/134/7/2044/293996

Rose, C. R., Felix, L., Zeug, A., Dietrich, D., Reiner, A., & Henneberger, C. (2018). Astroglial Glutamate Signaling and Uptake in the Hippocampus. *Frontiers in Molecular Neuroscience, 10*, 451. doi:10.3389/fnmol.2017.00451 PMID:29386994

Rusznák, Z., Sengul, G., Paxinos, G., Kim, W. S., & Fu, Y. (2018). Odor Enrichment Increases Hippocampal Neuron Numbers in Mouse. *Experimental Neurobiology, 27*(2), 94–102. doi:10.5607/en.2018.27.2.94 PMID:29731675

Sama, M. A., Mathis, D. M., Furman, J. L., Abdul, H. M., Artiushin, I. A., Kraner, S. D., & Norris, C. M. (2008). Interleukin-1β-dependent Signaling between Astrocytes and Neurons Depends Critically on Astrocytic Calcineurin/NFAT Activity. *The Journal of Biological Chemistry, 283*(32), 21953–21964. doi:10.1074/jbc.M800148200 PMID:18541537

Sams, E. C. (2021). Oligodendrocytes in the aging brain. *Neuronal Signaling, 5*(3), NS20210008. doi:10.1042/NS20210008 PMID:34290887

Sarris, J., Marx, W., Ashton, M. M., Ng, C. H., Galvao-Coelho, N., Ayati, Z., Zhang, Z.-J., Kasper, S., Ravindran, A., Harvey, B. H., Lopresti, A., Mischoulon, D., Amsterdam, J., Yatham, L. N., & Berk, M. (2021). Plant-based Medicines (Phytoceuticals) in the Treatment of Psychiatric Disorders : A Meta-review of Meta-analyses of Randomized Controlled Trials: Les médicaments à base de plantes (phytoceutiques) dans le traitement des troubles psychiatriques: une méta-revue des méta-analyses d'essais randomisés contrôlés. *Canadian Journal of Psychiatry, 66*(10), 849–862. doi:10.1177/0706743720979917 PMID:33596697

Schmid, C. D., Sautkulis, L. N., Danielson, P. E., Cooper, J., Hasel, K. W., Hilbush, B. S., Sutcliffe, J. G., & Carson, M. J. (2002). Heterogeneous expression of the triggering receptor expressed on myeloid cells-2 on adult murine microglia : Microglial expression of TREM-2. *Journal of Neurochemistry, 83*(6), 1309–1320. doi:10.1046/j.1471-4159.2002.01243.x PMID:12472885

Schwabe, T., & Srinivasan, K. (2020). *Shifting paradigms: The central role of microglia in Alzheimer's disease.* Elsevier. https://www.sciencedirect.com/science/article/pii/S0969996120302370

Scimemi, A., Meabon, J. S., Woltjer, R. L., Sullivan, J. M., Diamond, J. S., & Cook, D. G. (2013). Amyloid-β_{1-42} Slows Clearance of Synaptically Released Glutamate by Mislocalizing Astrocytic GLT-1. *The Journal of Neuroscience : The Official Journal of the Society for Neuroscience, 33*(12), 5312–5318. doi:10.1523/JNEUROSCI.5274-12.2013 PMID:23516295

Shi, S., Liang, D., Chen, Y., Xie, Y., Wang, Y., Wang, L., Wang, Z., & Qiao, Z. (2016). Gx-50 reduces β-amyloid-induced TNF-α, IL-1β, NO, and PGE2 expression and inhibits NF-κB signaling in a mouse model of Alzheimer's disease. *European Journal of Immunology, 2016•Wiley Online Library, 46*(3), 665–676. doi:10.1002/eji.201545855

Shibata, N., Ohnuma, T., Takahashi, T., Baba, H., Ishizuka, T., Ohtsuka, M., Ueki, A., Nagao, M., & Arai, H. (2002). Effect of IL-6 polymorphism on risk of Alzheimer disease : Genotype-phenotype association study in Japanese cases. *American Journal of Medical Genetics, 114*(4), 436–439. doi:10.1002/ajmg.10417 PMID:11992567

Sieber, M. W., Jaenisch, N., Brehm, M., Guenther, M., Linnartz-Gerlach, B., Neumann, H., Witte, O. W., & Frahm, C. (2013). Attenuated Inflammatory Response in Triggering Receptor Expressed on Myeloid Cells 2 (TREM2) Knock-Out Mice following Stroke. *PLoS One, 8*(1), e52982. doi:10.1371/journal.pone.0052982 PMID:23301011

Simoni, E., Caporaso, R., Bergamini, C., Fiori, J., Fato, R., Miszta, P., Filipek, S., Caraci, F., Giuffrida, M. L., Andrisano, V., Minarini, A., Bartolini, M., & Rosini, M. (2016). Polyamine conjugation as a promising strategy to target amyloid aggregation in the framework of Alzheimer's disease. *ACS Medicinal Chemistry Letters, 7*(12), 1145–1150. doi:10.1021/acsmedchemlett.6b00339 PMID:27994754

Skatchkov, S. N., Bukauskas, F. F., Benedikt, J., Inyushin, M., & Kucheryavykh, Y. V. (2015). Intracellular spermine prevents acid-induced uncoupling of Cx43 gap junction channels. *Neuroreport, 26*(9), 528–532. doi:10.1097/WNR.0000000000000385 PMID:26011388

Skatchkov, S. N., Woodbury-Fariña, M. A., & Eaton, M. (2014). The role of glia in stress : Polyamines and brain disorders. *Psychiatria Clinica, 37*(4), 653–678. PMID:25455070

Stadelmann, C., Timmler, S., Barrantes-Freer, A., & Simons, M. (2019). Myelin in the central nervous system : Structure, function, and pathology. *Physiological Reviews, 99*(3), 1381–1431. doi:10.1152/physrev.00031.2018 PMID:31066630

Stephan, A. H., Madison, D. V., Mateos, J. M., Fraser, D. A., Lovelett, E. A., Coutellier, L., Kim, L., Tsai, H.-H., Huang, E. J., Rowitch, D. H., Berns, D. S., Tenner, A. J., Shamloo, M., & Barres, B. A. (2013). A dramatic increase of C1q protein in the CNS during normal aging. *Journal of Neuroscience. Social Neuroscience, 33*(33), 13460–13474. doi:10.1523/JNEUROSCI.1333-13.2013 PMID:23946404

Streit, W. J., Sammons, N. W., Kuhns, A. J., & Sparks, D. L. (2004). Dystrophic microglia in the aging human brain. *Glia, 45*(2), 208–212. doi:10.1002/glia.10319 PMID:14730714

Suárez-Calvet, M., Capell, A., Araque Caballero, M. Á., Morenas-Rodríguez, E., Fellerer, K., Franzmeier, N., Kleinberger, G., Eren, E., Deming, Y., Piccio, L., Karch, C. M., Cruchaga, C., Paumier, K., Bateman, R. J., Fagan, A. M., Morris, J. C., Levin, J., Danek, A., Jucker, M., ... Haass, C. (2018). CSF progranulin increases in the course of Alzheimer's disease and is associated with sTREM 2, neurodegeneration and cognitive decline. *EMBO Molecular Medicine*, *10*(12), e9712. doi:10.15252/emmm.201809712 PMID:30482868

Sulehria, T., Corbett, A. M., Sharma, N., Nagarajan, D., Abushamma, A., Gagle, S., & Johnson, A. (n.d.). Increasing Progenitor Cell Proliferation in the Sub-Ventricular Zone: a therapeutic treatment for progressive multiple sclerosis? *Ingentaconnect.* doi:10.2174/1872211314999201117130123

Swanson, M. E., Scotter, E. L., Smyth, L. C., Murray, H. C., Ryan, B., Turner, C., Faull, R. L., Dragunow, M., & Curtis, M. A. (2020). Identification of a dysfunctional microglial population in human Alzheimer's disease cortex using novel single-cell histology image analysis. *Acta Neuropathologica Communications*, *8*(1), 1–16. doi:10.118640478-020-01047-9 PMID:33081847

Tan, J., Town, T., Crawford, F., Mori, T., DelleDonne, A., Crescentini, R., Obregon, D., Flavell, R. A., & Mullan, M. J. (2002). Role of CD40 ligand in amyloidosis in transgenic Alzheimer's mice. *Nature Neuroscience*, *5*(12), 1288–1293. doi:10.1038/nn968 PMID:12402041

Tarkowski, E., Blennow, K., Wallin, A., & Tarkowski, A. (1999). Intracerebral production of tumor necrosis factor-α, a local neuroprotective agent, in Alzheimer disease and vascular dementia. *Journal of Clinical Immunology*, *19*(4), 223–230. doi:10.1023/A:1020568013953 PMID:10471976

Timmler, S., & Simons, M. (2019). Grey matter myelination. *Glia*, *67*(11), 2063–2070. doi:10.1002/glia.23614 PMID:30860619

Toledo, E. M., & Inestrosa, N. C. (2010). Activation of Wnt signaling by lithium and rosiglitazone reduced spatial memory impairment and neurodegeneration in brains of an APPswe/PSEN1ΔE9 mouse model of Alzheimer's disease. *Molecular Psychiatry*, *15*(3), 272–285. doi:10.1038/mp.2009.72 PMID:19621015

Tong, H., Zhang, X., Meng, X., Xu, P., Zou, X., & Qu, S. (2017). Amyloid-beta peptide decreases expression and function of glutamate transporters in nervous system cells. *The International Journal of Biochemistry & Cell Biology*, *85*, 75–84. doi:10.1016/j.biocel.2017.01.017 PMID:28189841

Tricoire, L., Pelkey, K. A., Erkkila, B. E., Jeffries, B. W., Yuan, X., & McBain, C. J. (2011). A blueprint for the spatiotemporal origins of mouse hippocampal interneuron diversity. *The Journal of Neuroscience : The Official Journal of the Society for Neuroscience*, *31*(30), 10948–10970. doi:10.1523/JNEUROSCI.0323-11.2011 PMID:21795545

Uddin, M. S., & Lim, L. W. (2022). Glial cells in Alzheimer's disease: From neuropathological changes to therapeutic implications. *Ageing Research Reviews*, *78*, 101622. doi:10.1016/j.arr.2022.101622 PMID:35427810

Ulrich, J., Ulland, T., Colonna, M., & Neuron, D. H. (2017). Elucidating the role of TREM2 in Alzheimer's disease. *Cell.* doi:10.1016/j.neuron.2017.02.042

Valente, M. F., Romariz, S., Calcagnotto, M. E., Ruiz, L., Mello, L. E., Frussa-Filho, R., & Longo, B. M. (2013). Postnatal transplantation of interneuronal precursor cells decreases anxiety-like behavior in adult mice. *Cell Transplantation, 22*(7), 1237–1247. doi:10.3727/096368912X657422 PMID:23031356

Vaz, M., & Silvestre, S. (2020). Alzheimer's disease : Recent treatment strategies. *European Journal of Pharmacology, 887*, 173554. doi:10.1016/j.ejphar.2020.173554 PMID:32941929

Von Bartheld, C. S., Bahney, J., & Herculano-Houzel, S. (2016). The search for true numbers of neurons and glial cells in the human brain : A review of 150 years of cell counting. *The Journal of Comparative Neurology, 524*(18), 3865–3895. doi:10.1002/cne.24040 PMID:27187682

Wang, W.-Y., Tan, M.-S., Yu, J.-T., & Tan, L. (2015). Role of pro-inflammatory cytokines released from microglia in Alzheimer's disease. *Annals of Translational Medicine, 3*(10). PMID:26207229

Wang, Y., Ulland, T., & Ulrich, J. (2016). TREM2-mediated early microglial response limits diffusion and toxicity of amyloid plaques. *Journal of Experimental Medicine.* https://rupress.org/jem/article-abstract/213/5/667/42126

Wilhelmsson, U., Li, L., Pekna, M., Berthold, C.-H., Blom, S., Eliasson, C., Renner, O., Bushong, E., Ellisman, M., Morgan, T. E., & Pekny, M. (2004). Absence of Glial Fibrillary Acidic Protein and Vimentin Prevents Hypertrophy of Astrocytic Processes and Improves Post-Traumatic Regeneration. *The Journal of Neuroscience : The Official Journal of the Society for Neuroscience, 24*(21), 5016–5021. doi:10.1523/JNEUROSCI.0820-04.2004 PMID:15163694

Wong, K. Y., Roy, J., Fung, M. L., Heng, B. C., Zhang, C., & Lim, L. W. (2020). Relationships between mitochondrial dysfunction and neurotransmission failure in Alzheimer's disease. *Aging and Disease, 11*(5), 1291. doi:10.14336/AD.2019.1125 PMID:33014538

Woo, J., Im, S.-K., Chun, H., Jung, S.-Y., Oh, S.-J., Choi, N., Lee, C. J., & Hur, E.-M. (2017). Functional Characterization of Resting and Adenovirus-Induced Reactive Astrocytes in Three-Dimensional Culture. *Experimental Neurobiology, 26*(3), 158–167. doi:10.5607/en.2017.26.3.158 PMID:28680301

Wu, Y., Ma, Y., Liu, Z., Geng, Q., & Chen, Z. (2017). *Alterations of myelin morphology and oligodendrocyte development in early stage of Alzheimer's disease mouse model*. Elsevier.

Wu, Z., Sun, L., Hashioka, S., Yu, S., & C. S.-N. (2013). *Differential pathways for interleukin-1β production activated by chromogranin A and amyloid β in microglia*. Elsevier.

Xu, J., Chen, S., Ahmed, S. H., Chen, H., Ku, G., Goldberg, M. P., & Hsu, C. Y. (2001). Amyloid-β peptides are cytotoxic to oligodendrocytes. *The Journal of Neuroscience : The Official Journal of the Society for Neuroscience, 21*(1), RC118. doi:10.1523/JNEUROSCI.21-01-j0001.2001 PMID:11150354

Xu, J.-P., Zhao, J., & Li, S. (2011). Roles of NG2 glial cells in diseases of the central nervous system. *Neuroscience Bulletin, 27*(6), 413–421. doi:10.100712264-011-1838-2 PMID:22108818

Xu, Z., Xiao, N., Chen, Y., Huang, H., Marshall, C., Gao, J., Cai, Z., Wu, T., Hu, G., & Xiao, M. (2015). Deletion of aquaporin-4 in APP/PS1 mice exacerbates brain Aβ accumulation and memory deficits. *Molecular Neurodegeneration, 10*(1), 58. doi:10.118613024-015-0056-1 PMID:26526066

Yasuno, F., Kosaka, J., Ota, M., Higuchi, M., Ito, H., Fujimura, Y., Nozaki, S., Takahashi, S., Mizukami, K., Asada, T., & Suhara, T. (2012). Increased binding of peripheral benzodiazepine receptor in mild cognitive impairment–dementia converters measured by positron emission tomography with [11C] DAA1106. *Psychiatry Research: Neuroimaging*, *203*(1), 67–74. doi:10.1016/j.pscychresns.2011.08.013 PMID:22892349

Yeh, C.-Y., Vadhwana, B., Verkhratsky, A., & Rodríguez, J. J. (2011). Early astrocytic atrophy in the entorhinal cortex of a triple transgenic animal model of Alzheimer's disease. *ASN Neuro*, *3*(5), AN20110025. doi:10.1042/AN20110025 PMID:22103264

Yi, J., Horky, L. L., Friedlich, A. L., Shi, Y., Rogers, J. T., & Huang, X. (2009). L-arginine and Alzheimer's disease. *International Journal of Clinical and Experimental Pathology*, *2*(3), 211. PMID:19079617

You, Y., Joseph, C., Wang, C., Gupta, V., Liu, S., Yiannikas, C., Chua, B. E., Chitranshi, N., Shen, T., Dheer, Y., Invernizzi, A., Borotkanics, R., Barnett, M., Graham, S. L., & Klistorner, A. (2019). Demyelination precedes axonal loss in the transneuronal spread of human neurodegenerative disease. *Brain*, *142*(2), 426–442. doi:10.1093/brain/awy338 PMID:30668642

Zhang, Y., Chen, K., Sloan, S. A., Bennett, M. L., Scholze, A. R., O'Keeffe, S., Phatnani, H. P., Guarnieri, P., Caneda, C., Ruderisch, N., Deng, S., Liddelow, S. A., Zhang, C., Daneman, R., Maniatis, T., Barres, B. A., & Wu, J. Q. (2014). An RNA-Sequencing Transcriptome and Splicing Database of Glia, Neurons, and Vascular Cells of the Cerebral Cortex. *The Journal of Neuroscience : The Official Journal of the Society for Neuroscience*, *34*(36), 11929–11947. doi:10.1523/JNEUROSCI.1860-14.2014 PMID:25186741

Zotova, E., Holmes, C., Johnston, D., Neal, J. W., Nicoll, J. A. R., & Boche, D. (2011). Microglial alterations in human Alzheimer's disease following Aβ42 immunization. *Neuropathology and Applied Neurobiology*, *37*(5), 513–524. doi:10.1111/j.1365-2990.2010.01156.x PMID:21166690

Chapter 16
Astrocytes Reprogramming for Neurodegenerative Disease Management

Lahcen Tamegart
Abdelmalek Essaâdi University, Morocco

Rabii Hilali
Abdelmalek Essaâdi University, Morocco

Hajar Ouaki
Biology Department, Higher School of Teachers (ENS), Mohammed V University in Rabat, Morocco

Fatima Rachidi
Mohammed V University in Rabat, Morocco

Ibtissam Chakir
Sultan Moulay Slimane University, Morocco

Rabia Ajana
Abdelmalek Essaadi University, Morocco

Dia Eddine Oudghiri
https://orcid.org/0000-0001-7849-132X
Abdelmalek Essaadi University, Morocco

Ahmed Draoui
https://orcid.org/0000-0002-7700-059X
Cadi Ayyad University, Morocco

Mjid Oukhrib
https://orcid.org/0000-0003-0199-4054
Faculty of Sciences Semlalia, Cadi Ayyad University, Morocco

Hafida El Ghachi
https://orcid.org/0000-0002-2043-6202
Cadi Ayyad University, Morocco

Abdelali Ben Maloui
Cadi Ayyad University, Morocco

Abdelaati El Khiat
Independent Researcher, Morocco

Halima Gamrani
Cadi Ayyad University, Morocco

ABSTRACT

Neurodegenerative disorders (NDD) are chronic conditions that lead to nerve cells degeneration in the central nervous system. The most common NDD are Alzheimer's disease (AD), Parkinson's disease (PD), amyotrophic lateral sclerosis (ALS), multiple sclerosis (MS), and Huntington's disease (HD). Parkinson's disease (PD) is recognized as the second most common neurodegenerative disease after Alzheimer's

DOI: 10.4018/978-1-6684-9675-6.ch016

disease (AD) and Huntington disease (HD). Astrocytes reprogramming has been recently proposed as one of the most important treatments to stop the progression of these neurological diseases. By targeting the microRNA, astrocytes in the animal brain could be reprogrammed into various functional neural systems. Activation of endogenous genes could be a revolutionary therapeutic approach for treating neurological diseases. Reprogrammed astrocytes could also be used to generate various types of neurons lost in case of Huntington's disease, Parkinson's disease, and Alzheimer's disease. The ability to reprogram astrocytes into functional neurons was classified as a new advancement for PD, AD, and HD treatments.

INTRODUCTION

Neurodegenerative diseases (NDD) are serious, chronic conditions that damage and destroy the nerve cells of the central nervous system, affecting millions of people worldwide. The most common neurodegenerative diseases are Parkinson's disease (PD), Alzheimer disease (AD) and Huntington's disease (HD). Parkinson's disease is a progressive neurodegenerative disorder characterized mainly by the motor features (Jankovic, 2008). Impairment of voluntary motor control leads to signs of bradykinesia, rigidity, postural instability and tremor. Additionally, the disease is characterised by the presence of other non-motor symptoms (El-Mansoury et al., 2023). Alzheimer's disease is a neurodegenerative disorder with an obvious degeneration of the neuronal cells. In order to clarify the mechanisms controlling AD progression, two hypotheses in general were proposed; the deposition of beta-amyloid (Aβ) protein between neurons (Hamd-Ghadareh et al., 2022; Hu et al., 2022), and the presence of neurofibrillary tangles (NFT) accumulation within the neurons (Kim and Park, 2020; Schneider et al., 2022). In addition, recent studies suggest that environmental factors including metal ions play an important role in disease pathogenesis (Lee et al., 2018; Tian et al., 2022; Z. Zhou et al., 2022). Huntington's disease (HD) was defined for the first time by Georges Huntingtonin 1872. The neuropathology of HD characterised by a dysfunction and death of specific neurons within the brain. The prevalence of Huntington's disease is about 5 cases per 100,000 individuals. The frequency of occurrence is the same for both men and women. A study revealed that motor, cognitive, and behavioral dysfunction disorders are all impacted by HD (Zuccato and Cattaneo, 2014).

At the moment, symptomatic treatment is in general; the only therapeutic approach for AD. This kind of treatment was based on using acetylcholinesterase (AChE) inhibitors to enhance ACh bioavailability at the synaptic area (Pasandideh and Arasteh, 2021; Richter et al., 2018). Food and Drug Administration (FAD) approved recently five molecules (donepezil, galantamine, rivastigmine, memantine and memantine combined with donepezil). A six drug aducanumab, under FDA review waiting for potential approval (Wiley, 2021). In addition various treatment were used for treatment of PD and HD, but all these current treatment approaches of NDD still limited then a new therapeutic approach based on reprogrammed astrocytes could be a potential treatment for these disorders.

Various studies revealed the application of reprogrammed astrocytes for treatment of NNDs such as PD (Dopamine), AD (Glutamate and Acetylcholine) and HD (GABA) (Guo et al., 2014) **(Figure 1)**. Astrocytes are mainly reprogrammed into glutamatergic neurons, whereas NG2 cells can be reprogrammed into both glutamatergic and GABAergic neurons. Such different cell fates after reprogramming by the same transcription factor may provide important clues regarding the lineage relationship between neurons and glial cells astrocytes in culture can be reprogrammed into functional neurons after expressing NeuroD1. The in vivo reprogramming of reactive glial cells into functional neurons after brain injury or in diseased mouse brain could potentially provide a therapeutic approach (Guo et al., 2014) **(Figure 1)**.

Figure 1. Astrocytes reprogramming for treatment Parkinson, Alzheimer, and Huntington diseases

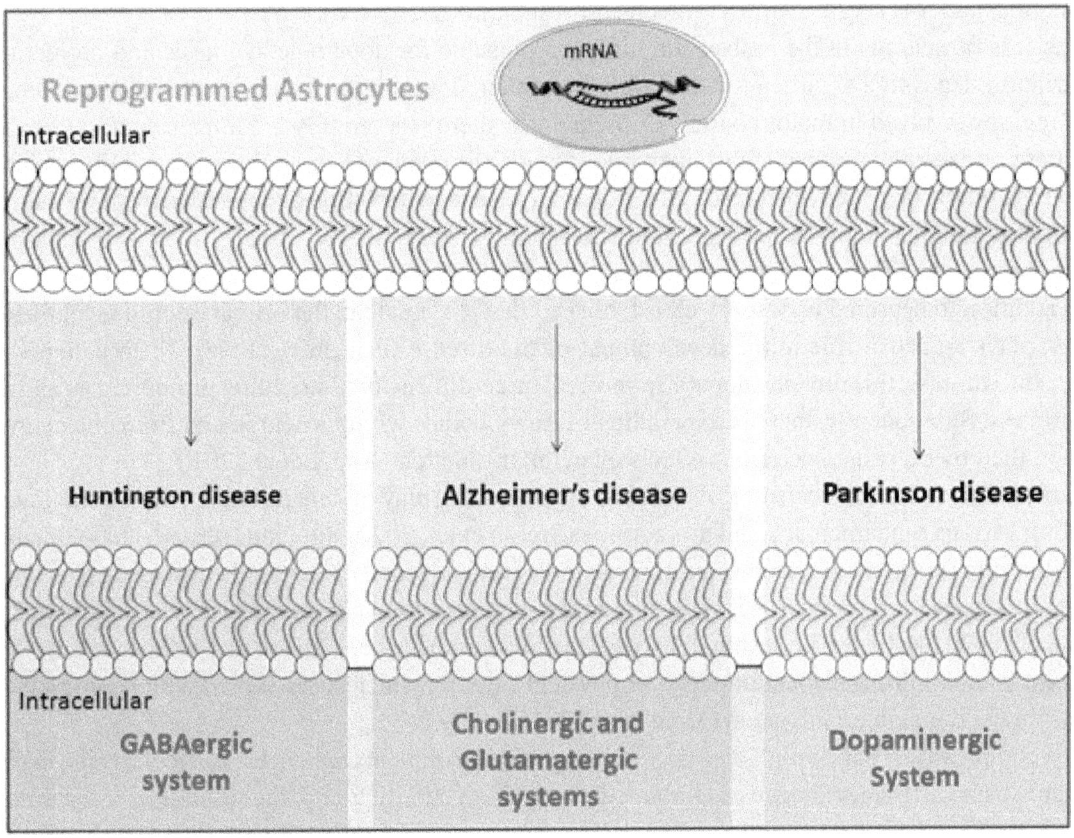

PARKINSON DISEASE AND TREATMENT WITH ASTROCYTE REPROGRAMMING

Parkinson Disease

Parkinson's disease is the second most common neurodegenerative disorder after Alzheimer's disease, and its prevalence increases with age (Defebvre and Vérin, 2020; Frisardi et al., 2016). According to the World Health Organization, global estimates indicate that in 2019, over 8,5 million people were affected by Parkinson's disease. Previous studies have shown that both incidence and prevalence of this condition are approximately two times higher in men than women (Haaxma et al., 2007).

Pathophysiology of PD

Neuropathology of Parkinson's disease is characterized by the loss of dopaminergic neurons in the substantia nigra, along with the presence of intracellular inclusions in surviving neurons known as Lewy bodies (Prigent et al., 2017). However, it's important to note that Parkinson's disease affects other regions of brain and involves various types of neurons, including GABAergic, noradrenergic, glutamatergic, cholinergic, and 5-HTergic neurons (Hirsch and Hunot, 2009).

The loss of neurons in the susbstantia nigra, responsible for dopamine synthesis and projecting to the striatum, leads to a significant reduction in dopamine levels in the caudate nucleus, putamen, and brain regions involved in motor control. Consequently, there is a progressive impairment in the ability to initiate or execute voluntary movement, along with an imbalance in cholinergic and dopaminergic neurotransmitters (Wei and Shetty, 2021). It is, therefore, due to depletion of dopamine by more than 80% in the basal ganglia and sustantia nigra, the characteristic symptoms of disease, such as bradykinesia, muscle rigidity, and resting tremors, manifest were observed (Capriotti and Terzakis, 2016).

In addition to neuronal loss, it is well established that α-synuclein, the primary component of Lewy bodies, plays an active role in the development of the disease (Prigent et al., 2017). In patients with Parkinson's disease, this protein adopts an abnormal three-dimensional structure, promoting an aggregation process. Subsequently, there is a spreading of Lewy bodies within a neuron and from one neuron to another, thereby exerting deleterious effects on cellular function (Angot et al., 2010).

Parkinson's disease appears to result from a complex interplay of genetic and environmental factors affecting various fundamental cellular processes (Simon et al., 2020). Previous research has shown that exposure to toxic chemicals such as pesticides, industrial solvents, or certain heavy metals, as well as mild to moderate head traumas, has been associated with increased risk of the disease (Goldman, 2014; Tanner et al., 2014). However, several lifestyle-related factor have been linked to a reduced risk of developing Parkinson's disease, including regular physical activity, which is associated with neuroprotective effects on the dopaminergic system (Yang et al., 2015).

The complexity of Parkinson's disease is accompanied by clinical challenges, particularly the inability to establish an early-stage diagnosis (Kalia and Lang, 2015). Particularly, the instability to establish an early-stage diagnosis (Kalia and Lang, 2015). The treatment of Parkinson's disease is symptomatic and primarily relies on the use of dopaminergic medications. This includes dopatherapy through the administration of L-DOPA, which serves as a precursor to dopamine. This molecule has the ability to cross the blood-barrier and, under the action of the enzyme aromatic L-amino acid decarboxylase (ADDC), it is converted into dopamine, making it available to deficient neurons (LeWitt, 2008). Other treatment options involve dopaminergic agonists such as ropinirole and pramipexole, as well as enzymatic inhibitors like catechol-O-methyltransferase (COMT), monoamine oxidase (MOA), and aldehyde dehydrogenase (ALDH). These inhibitors reduce the degradation of L-DOPA, thereby increasing its bioavailability and duration of action (Hayes, 2019).

In addition to medication-based treatment, surgery has become an essential option of care available to patients with Parkinson's disease. Deep brain stimulation (DBS) by precisely implanting electrodes at various levels:

- In the thalamus, which facilitates the management of resting tremors (Schuurman et al., 2008).
- In the subthalamic nucleus, where motor outcomes are highly favorable, with the potential benefits exceeding 60% (Benabid et al., 2009).

- In the Globus pallidus internus, leading to a reduction in dyskinesia by a range of 40 to 80% (Moro et al., 2010).

In Vivo Astrocytes Reprogramming for Treatment Of PD

Numerous alternative therapies for Parkinson's disease are currently in the research, development, and pre-clinical validation phases. One such approach is the in vivo reprogramming of astrocytes. In fact, astrocytes offer several advantages for in vivo reprogramming within the brain. These non-neuronal cells play a crucial role in supporting and regulating the neuronal environment and exhibit high plasticity in altering cell fate (Yu et al., 2020).

Astrocyte reprogramming involves genetic environment manipulation to transform astrocytes in vivo into dopaminergic neurons with the aim of restoring lost neurons and preventing disease progression. This transformation entails the overexpression of transcription factors mediated by a viral vector, along with microRNAs (Talifu et al., 2023).

Recent reprogramming studies have provided substantial evidence that transformed central nervous system astrocytes mature progressively and generate action potentials. Based on in vivo studies, Rivetti Di Val Cervo et al., 2017 demonstrated that astrocytes can be transformed into dopaminergic neurons using vectors that overexpress the transcription factors NeuroD1, Ascl1, and Lmx1a, along with microRNA 2018. This approach leads to improved motor function in murine models of Parkinson's disease.

Another study conducted by Zhou et al., 2020 reported an improvement in the motor function of mouse models of Parkinson's disease after injecting an AVV (adeno-associated virus) vector into the striatum of this mice. This vector carried RNA targeting the polypyrimidine tract-binding protein 1 (PTBP1) through CRISPR/CasRX, dopaminergic neuronal markers.

The research findings by (Giehrl-Schwab et al., 2022) suggest that reprogramming astrocytes into GABAergic neurons using AVV vectors carrying intein-split-dCas9 in combination with activators (AVV-dCas) mitigate aspects of voluntary motor behavior in a toxin-induced murine model of Parkinson's disease. This paves the way for a novel therapeutic approach beyond dopamine restoration.

ALZHEIMER'S DISEASE AND TREATMENT WITH ASTROCYTE REPROGRAMMING

Alzheimer's Disease

Alzheimer's disease is a progressive neurodegenerative disorder (Masters et al., 2015). It is characterized by a progressive loss of neurons, which results in a loss of memory, language, reasoning, and cognitive functions (Ballard et al., 2011; Scheltens et al., 2021). Alzheimer's disease is the leading cause of dementia in the elderly (Chen et al., 2021).

The exact cause of Alzheimer's disease is not yet fully understood, but it is likely to be due to a combination of genetic and environmental factors(Armstrong, 2013).There aremany factors that may contribute to Alzheimer's disease like age(Frigerio et al., 2019; Neu et al., 2017), gender (women are more likely than men to develop the disease.) (Frigerio et al., 2019), level of education (it has been demonstrated that people with a higher level of education have a lower risk of developing the disease)

(Meng et al., 2020), cardiovascular risk factors (high blood pressure, diabetes, obesity, and smoking), head trauma and exposure to toxins and inflammation (Tini et al., 2020).

Alzheimer's disease (AD) is a neurodegenerative disorder that affect millions of people worldwide. Clinically defined by gradual cognitive decline with impairments in executive function, language, praxis, and visual processing. (AD) is the main cause of dementia and is one of the most lethal diseases of this century (Cai et al., 2012).

Pathophysiology of AD

Alzheimer's disease is also caused by the accumulation of two abnormal proteins in the brain: amyloid beta and tau(Muralidar et al., 2020; Walsh and Selkoe, 2020). Amyloid beta is a protein normally produced by the brain(Selkoe, 2015). However, in Alzheimer's disease, this protein accumulates in the form of plaques between neurons,these plaques disrupt communication between neurons, causing their death (Muralidar et al., 2020; Selkoe, 2015; Walsh and Selkoe, 2020; Yang et al., 2017). On the other hand, tau is a protein that helps maintain the integrity of neurons. So, this protein accumulates as tangles inside neurons. These tangles disrupt the functioning of neurons and lead to their death (Muralidar et al., 2020; Selkoe, 2015).

Symptoms of Alzheimer's disease appear gradually and worsen over time (Cummings et al., 2015). Primary symptoms frequently include are memory loss, word-finding difficulties, and behavioral changes (Barnes et al., 2015). As the disease progresses, people with the disease may have difficulty remembering recent events, following a conversation, or performing simple tasks (Joe and Ringman, 2019). In the advanced stages of the disease, people suffering from the disease may require help with all aspects of their quotidian lives (Lanctôt et al., 2017).The symptoms of Alzheimer's disease differ from one person to another. Some people canpresent only mild symptoms in the early stages, while others have more severe symptoms (Dubois et al., 2016).

There is no single test for diagnosing Alzheimer's disease (Rasmussen and Langerman, 2019). The diagnosis is generally based on a clinical examination to assess the patient's cognitive and behavioral symptoms, an evaluation of cognitive abilities through neuropsychological assessment (Warren et al., 2023), a brain imaging examination to identify brain lesions, and biological tests, like the determination of amyloid beta protein in cerebrospinal fluid, which is used to confirm the diagnosis of Alzheimer's disease (Villemagne et al., 2017; Villemagne and Rowe, 2011). However, these tests are not always available or reliable (Schilling et al., 2022).

Pathologically the symptoms of Alzheimer's disease (AD) can be correlated to the loss of synapses, neurons, neuropil and an overall reduction of grey matter (Tanzi and Bertram, 2005). Pathophysiologically, AD is defined by the presence in the brain of patients ofextracellular amyloid-β (aβ) plaques and intracellular neurofibrillary tangles (NFts), which are composed of hyperphosphorylated tau proteins (Selkoe and Hardy, 2016). Aβ is a normal by-product of cellular metabolism generated by the cleavage of the transmembrane amyloid precursor protein (APP) by the aspartyl proteases β- and γ-secretase (Plant et al., 2003). The accumulation of Aβ in the brain is considered an early event in AD pathogenesis that lead to neuroinflammation and neurodegeneration. However,Numerous clinical studies have by now documented that the presynaptic protein α-synuclein (αSyn), mainly associated with synucleinopathies like Parkinson's disease (PD), dementia with Lewy bodies (DLB) and multiple system atrophy (MSA), is involved in the pathophysiology of AD. Lewy-related pathology (LRP), primarily comprised of αSyn, is present in a majority of autopsied AD brains, and higher levels of αSyn in the cerebrospinal fluid (CSF) of patients with AD have been linked to cognitive decline(Twohig and Nielsen, 2019).

In Vivo Astrocytes Reprogramming for Treatment Of AD

To treat Alzheimer's disease in a curative way is still unavailable (Abeysinghe et al., 2020). On the other hand, certain drugs can help slow the development of the disease and related symptoms (Loera-Valencia et al., 2019).

A new therapeutic approach for Alzheimer's disease involves reprogramming astrocytes in vivo to become neuroprotective factor-secreting cells (C. Zhou et al., 2022). This approach is based on the idea that astrocytes can be reprogrammed to produce factors that can help protect neurons against the deleterious effects of abnormal proteins (Kehtari, 2022).Several techniques can be used to reprogram astrocytes in vivo. One technique is to use viruses to deliver transcription factors that can induce the expression of specific genes in astrocytes (Yu et al., 2020). Another is to use chemical agents to induce epigenetic changes in astrocytes (Talifu et al., 2023).

In vivo reprogramming of astrocytes is a promising therapeutic approach for Alzheimer's disease. Clinical studies are currently underway to evaluate the efficacy of this approach in Alzheimer's patients (Pardo-Moreno et al., 2022; Yu et al., 2021).

There is no certain way to prevent Alzheimer's disease. But there are things you can do to reduce the risk, as maintaining a healthy lifestyle by eating a balanced diet (Cremonini et al., 2019), exercising physical activity regularly (Panza et al., 2018), avoiding alcohol (Reale et al., 2020), tobacco abuse (Rhodes et al., 2022), stimulating the brain by learning new things (Jakobs et al., 2020), practicing mental activities, and ensuring close social relationships (Zhang et al., 2021).

Alzheimer's disease is a critical illness that can have a significant impact on the lives of sufferers and their loved ones. It is important to understand the causes, symptoms, and treatments of the disease so that you can take the necessary steps to prevent or manage it.

Reactive astrocytes exhibit functional diversity in AD pathology, promoting neuroprotection through Aβ degradation and clearance by secreting proteolytic enzymes such as metalloendopeptidases (NEP, ECE and IDE) and matrix metalloproteinases (MMP2 and MMP9) that cleave Aβ (**Figure 1**) and also contributing to pathological processes triggered by Aβ (Nalivaeva et al., 2012). Reactive glial cells generated after brain injury or in an AD model can be directly reprogrammed into functional neurons by a single transcription factor, NeuroD1, in mouse brain in vivo (Lee et al., 2022).

HUNTINGTON DISEASE AND TREATMENT WITH ASTROCYTE REPROGRAMMING

Huntington Disease

Cell-autonomous and non-cell-autonomous pathways could lead to HD (Creus-Muncunill and Ehrlich, 2019). Due to the disease's heterogeneous appearance and lack of a family history, the diagnosis of HD might occasionally be delayed (Keum et al., 2016). A CAG repeat expansion in the huntingtin gene (Htt) results in an enlarged polyglutamine region in exon 1 of the protein which is the hereditary disease. Any person bearing the mutation is certain to develop the disease if the triplet has more than 40 copies.

Figure 2. Astrocytes reprogramming for treatment Alzheimer's disease

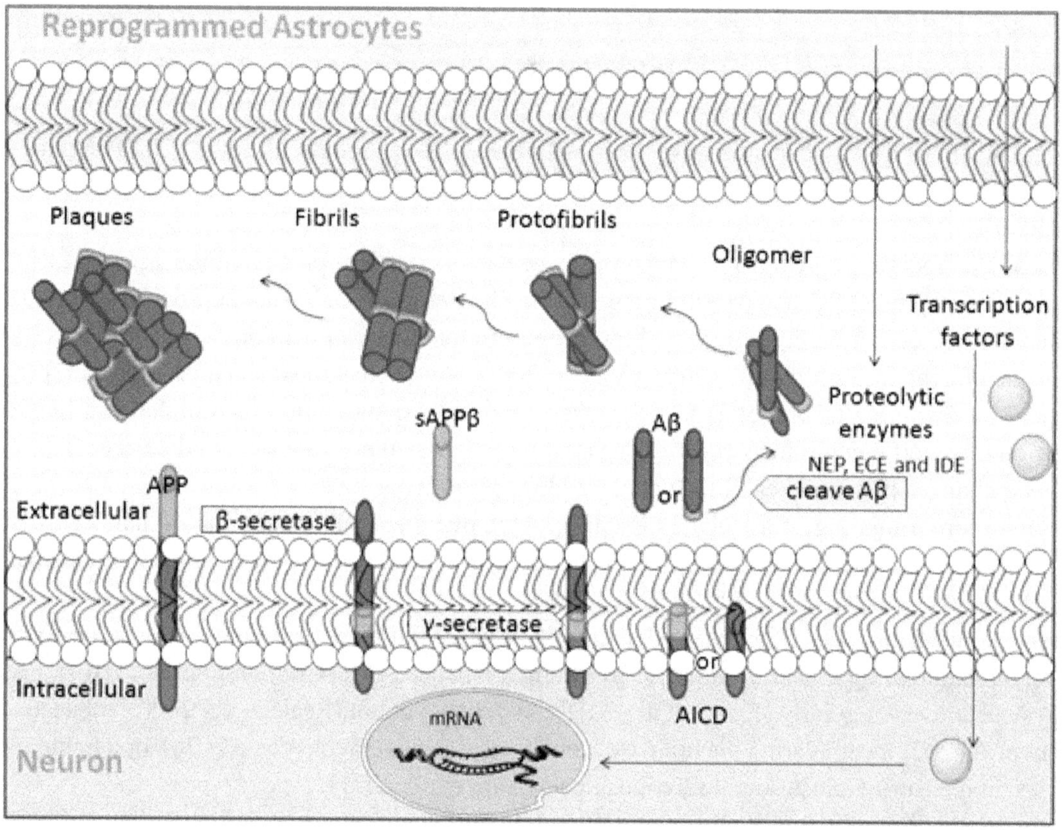

The usual functions of this protein, which include producing and transporting BDNF (for brain-derived neurotrophic factor), a substance involved in the survival of neurons, are altered as a result of this mutation, which also results in the acquisition of additional harmful properties. Its production in the brain and movement to the striatum are insufficient in HD. They decrease striatal neuron survival and prevent the development of functional connections between these two areas (Gauthier et al., 2004; Zuccato et al., 2008). The striatum is a part of the brain that controls behaviour, cognition, and motor skills. Htt is a protein that can be found in both human and rodent tissues, but it is more prevalent in the central nervous system (CNS), where it is mostly present in the cytoplasm of brain cells. It affects apoptosis, protein trafficking, axonal transport, postsynaptic signaling, and transcriptional control during development (Saudou and Humbert, 2016). Thus, the dysfunction of numerous intracellular pathways is caused by a loss of function of the normal protein and a harmful gain of function of the mutant huntingtin (Sharp et al., 1995). All of these disorders have some characteristics in common, such as delayed onset, selective neuronal susceptibility, lifetime expression of disease-related proteins, aberrant protein processing and aggregation, and cellular toxic effects on both cell types. While the cerebellum is typically spared from the neurodegenerative process, mutant Htt (mHtt) creates soluble fragments and aggregates that appear to be toxic to neurons, especially to medium spiny neurons in the striatum. These aggregates then spread to other locations, such as the cerebral cortex (Ross and Tabrizi, 2011).

Pathophysiology of HD

HD is caused by genetic mutations that lead to loss of gamma-aminobutyric acid (GABAergic) releasing neurons in the striatum, leading to motor, cognitive, and psychiatric dysfunction. The prevalence of HD in the population ranges from 0.02 to 1 per 10,000 people (Rawlins et al., 2016; Rikani et al., 2014). The course of the disease is defined by a clinically silent or asymptomatic period, followed by a symptomatic period in which the patient's condition gradually worsens. Movement, cognitive and psychiatric disorders vary in number, duration and intensity depending on the patient (Jimenez-Sanchez et al., 2017). Chorea is the most common. It is manifested by sudden, unpredictable and involuntary movements that gradually spread to all muscles. About 90% of patients experience this symptom. Chorea is mainly caused by fatigue, but can also be increased by stress, emotions or concentration (Rosenblatt, 2022). Other motor symptoms include abnormal posture (dystonia), muscle stiffness, and slow movements (bradykinesia). This can greatly affect daily activities: problems with chewing, swallowing, balance, joint deformities, etc. Patients' motor function gradually declines over time, to the point where they can no longer perform certain movements, move around alone, feed themselves, or even communicate. The patient's cognitive function also gradually declines, reaching the dementia stage. Often precedes motor symptoms of depression and the risk of suicide is high in these patients. Anxiety or lethargy mental states accompanied by delusions (in about 10% of cases), hallucinations (rare) or obsessions (in more than 15% of patients) may be observed. Irritability and even threats of aggression can harm social relationships. Psychiatric symptoms, such as depression, psychosis, and obsessive-compulsive disorder, are also common in HD and are particularly distressing for patients. Other common and disabling signs of HD are less well known: weight loss, sleep disturbances, skeletal muscle atrophy, heart failure, and loss of circadian rhythm (Arenas et al., 1998). Patients' neurological impairment was assessed using the Unified Huntington's Disease Rating Scale (UHDRS). The most vulnerable area is the striatum, where glial cells impair their ability to support neurons and keep them healthy. Changes in calcium and potassium levels produced by astrocytes are used in neural communication (Reiner et al., 1988).

Most factors associated with HD (Rosas et al., 2008):

- Aggregation of mutant huntingtin or disruption of transcription
- Huntingtin is cleaved into toxic fragments
- Mitochondrial and cytosolic dysfunction astrocytes and microglia in HD
- Alterations in HD neuronal homeostasis and synaptic plasticity in HD
- Ubiquitin–Proteasome system and autophagy
- Alterations in gene expression beyond transcription
- Impaired protein degradation system.

In Vivo Astrocytes Reprogramming for Treatment Of HD

Neuronal identity reprogramming strategies aim to treat age-related neurodegenerative disorders with newly generated neurons that regenerate neuronal structure and functional circuits in the body. Analysis of the specific transcriptional and epigenetic characteristics of different regions and neurons may improve the efficiency of reprogramming (Talifu et al., 2023). The goal is to transplant functional cells to replace destroyed nerve cells. Cellular reprogramming technology offers a unique opportunity to generate induced pluripotent stem cells (iPSCs) from patient-specific cells (Cho et al., 2019). Reprogramming astrocytes into neurons using transcription factors (TFs) and microRNAs (miRNAs). Converting astrocytes into GABAergic neurons to replenish lost neurons is a promising example of reprogramming for the treatment of HD. When NeuroD1 and Dlx2 were overexpressed in a group of HD model mice, astrocytes could be reprogrammed into GABAergic neurons. These modified neurons showed similar electrophysiological characteristics to the original neurons, reduced dysfunction, and extended the lifespan of the mouse model (Wu et al., 2020). In vivo cell conversion technology to reprogram striatal astrocytes into GABAergic neurons in R6/2 and YAC128 HD mouse models through AAV-mediated ectopic expression of factors NeuroD1 and Dlx2 transcription. Early stage neurons converted into astrocytes show action potentials and synaptic events, and project their axons to the target globuspallidus and substantia nigra in a manner depends on time. NeuroD1 and Dlx2 reprogram striatal astrocytes into GABAergic neurons: together have a synergistic effect in converting striatal astrocytes into neurons. HD-iPSCs are capable of generating phenotypically normal functional neurons in vitro and can survive and differentiate into neurons in the adult mouse brain in vivo after implantation graft. Surprisingly, when HD-iPSCs were asked to differentiate into an astrocyte lineage, the presence of electron-clear cytoplasmic vacuoles in F- astrocytes was observed. HD-iPSC and D-HD-iPSC, this is talked about more in D-HD. -astrocytes. HD-iPSCs can generate GABAergic striatal neurons, the type of neurons most susceptible to degeneration in HD. HD-iPSC-derived neural progenitors were transplanted into an HD mouse model of unilateral excitotoxic early-stage injury and significant behavioral recovery was observed in the mice implanted. It is essential to carefully evaluate the long-term effects of transformed neurons before applying this technology to treat human diseases (Carter et al., 2014; Juopperi et al., 2012; Khakh et al., 2017).

CONCLUSION

In conclusion, no therapies could regenerate neurons in the patient's brains. In addition, the ability to reprogram astrocytes into functional neurons was classified as a new advancement for PD, AD and HD treatments. Reprogrammed astrocytes could also be used to generate various types of neurons lost in case of Huntington's disease, Parkinson's disease, and Alzheimer's disease. Imbalances between various neurotransmitters (DA, serotonin and Ach) could lead to motor and non-motor symptoms, Astrocyte reprogramming has the ability to increase some neurotransmitters levels and the restore the proper balance needed for neuronal function.

REFERENCES

Abeysinghe, A., Deshapriya, R., & Udawatte, C. (2020). Alzheimer's disease; a review of the pathophysiological basis and therapeutic interventions. *Life Sciences*, *256*, 117996. doi:10.1016/j.lfs.2020.117996 PMID:32585249

Angot, E., Steiner, J. A., Hansen, C., Li, J.-Y., & Brundin, P. (2010). Are synucleinopathies prion-like disorders? *Lancet Neurology*, *9*(11), 1128–1138. doi:10.1016/S1474-4422(10)70213-1 PMID:20846907

Arenas, J., Campos, Y., Ribacoba, R., Martín, M. A., Rubio, J. C., Ablanedo, P., & Cabello, A. (1998). Complex I defect in muscle from patients with Huntington's disease. *Annals of Neurology*, *43*(3), 397–400. doi:10.1002/ana.410430321 PMID:9506560

Armstrong, R. A. (2013). What causes Alzheimer's disease? *Folia Neuropathologica*, *51*, 169–188. doi:10.5114/fn.2013.37702 PMID:24114635

Ballard, C., Gauthier, S., Corbett, A., Brayne, C., Aarsland, D., & Jones, E. (2011). Alzheimer's disease. *Lancet*, *377*(9770), 1019–1031. doi:10.1016/S0140-6736(10)61349-9 PMID:21371747

Barnes, J., Dickerson, B. C., Frost, C., Jiskoot, L. C., Wolk, D., & van der Flier, W. M. (2015). Alzheimer's disease first symptoms are age dependent: Evidence from the NACC dataset. *Alzheimer's & Dementia*, *11*(11), 1349–1357. doi:10.1016/j.jalz.2014.12.007 PMID:25916562

Benabid, A. L., Chabardes, S., Mitrofanis, J., & Pollak, P. (2009). Deep brain stimulation of the subthalamic nucleus for the treatment of Parkinson's disease. *Lancet Neurology*, *8*(1), 67–81. doi:10.1016/S1474-4422(08)70291-6 PMID:19081516

Cai, H., Cong, W., Ji, S., Rothman, S., Maudsley, S., & Martin, B. (2012). Metabolic dysfunction in Alzheimer's disease and related neurodegenerative disorders. *Current Alzheimer Research*, *9*(1), 5–17. doi:10.2174/156720512799015064 PMID:22329649

Capriotti, T., & Terzakis, K. (2016). Parkinson Disease. *Home Healthcare Now*, *34*(6), 300–307. doi:10.1097/NHH.0000000000000398 PMID:27243427

Carter, R. L., Chen, Y., Kunkanjanawan, T., Xu, Y., Moran, S. P., Putkhao, K., Yang, J., Huang, A. H. C., Parnpai, R., & Chan, A. W. S. (2014). Reversal of cellular phenotypes in neural cells derived from Huntington's disease monkey-induced pluripotent stem cells. *Stem Cell Reports*, *3*(4), 585–593. doi:10.1016/j.stemcr.2014.07.011 PMID:25358787

Chen, W.-Q., Wu, F.-F., Lv, H.-B., Xing, W.-T., Liu, Q., Liu, J.-P., Ge, Y.-G., & Lu, Y.-Q. (2021). Whether cognitive behavioral therapy is effective for Alzheimer's disease: A protocol for systematic review and network meta-analysis. *Medicine*, *100*(1), 100. doi:10.1097/MD.0000000000023945 PMID:33429752

Cho, I. K., Yang, B., Forest, C., Qian, L., & Chan, A. W. S. (2019). Amelioration of Huntington's disease phenotype in astrocytes derived from iPSC-derived neural progenitor cells of Huntington's disease monkeys. *PLoS One*, *14*(3), e0214156. doi:10.1371/journal.pone.0214156 PMID:30897183

Cremonini, A. L., Caffa, I., Cea, M., Nencioni, A., Odetti, P., & Monacelli, F. (2019). Nutrients in the prevention of Alzheimer's disease. *Oxidative Medicine and Cellular Longevity*, *2019*, 2019. doi:10.1155/2019/9874159 PMID:31565158

Creus-Muncunill, J., & Ehrlich, M. E. (2019). Cell-autonomous and non-cell-autonomous pathogenic mechanisms in Huntington's disease: Insights from in vitro and in vivo models. *Neurotherapeutics; the Journal of the American Society for Experimental NeuroTherapeutics*, *16*(4), 957–978. doi:10.100713311-019-00782-9 PMID:31529216

Cummings, J. L., Isaacson, R. S., Schmitt, F. A., & Velting, D. M. (2015). A practical algorithm for managing Alzheimer's disease: What, when, and why? *Annals of Clinical and Translational Neurology*, *2*(3), 307–323. doi:10.1002/acn3.166 PMID:25815358

Defebvre, L., & Vérin, M. (2020). *La maladie de Parkinson*. Elsevier Health Sciences.

Dubois, B., Padovani, A., Scheltens, P., Rossi, A., & Dell'Agnello, G. (2016). Timely diagnosis for Alzheimer's disease: A literature review on benefits and challenges. *Journal of Alzheimer's Disease*, *49*(3), 617–631. doi:10.3233/JAD-150692 PMID:26484931

El-Mansoury, B., Ben Maloui, A., Draoui, A., Hamdan, Y. A., Oukhrib, M., Smimih, K., Elmourid, A., El Khiat, A., Boulbaroud, S., & Meftah, I. (2023). Non-Motor Symptoms in Parkinson's Disease: The Other Side of the Disease. In *Experimental and Clinical Evidence of the Neuropathology of Parkinson's Disease* (pp. 25–45). IGI Global. doi:10.4018/978-1-6684-5156-4.ch003

Frigerio, C. S., Wolfs, L., Fattorelli, N., Thrupp, N., Voytyuk, I., Schmidt, I., Mancuso, R., Chen, W.-T., Woodbury, M. E., & Srivastava, G. (2019). The major risk factors for Alzheimer's disease: Age, sex, and genes modulate the microglia response to Aβ plaques. *Cell Reports*, *27*(4), 1293–1306. doi:10.1016/j.celrep.2019.03.099 PMID:31018141

Frisardi, V., Santamato, A., & Cheeran, B. (2016). Parkinson's disease: new insights into pathophysiology and rehabilitative approaches. Park. Dis.

Gauthier, L. R., Charrin, B. C., Borrell-Pagès, M., Dompierre, J. P., Rangone, H., Cordelières, F. P., De Mey, J., MacDonald, M. E., Leßmann, V., Humbert, S., & Saudou, F. (2004). Huntingtin controls neurotrophic support and survival of neurons by enhancing BDNF vesicular transport along microtubules. *Cell*, *118*(1), 127–138. doi:10.1016/j.cell.2004.06.018 PMID:15242649

Giehrl-Schwab, J., Giesert, F., Rauser, B., Lao, C. L., Hembach, S., Lefort, S., Ibarra, I. L., Koupourtidou, C., Luecken, M. D., Truong, D. J., Fischer-Sternjak, J., Masserdotti, G., Prakash, N., Ninkovic, J., Hölter, S. M., Vogt Weisenhorn, D. M., Theis, F. J., Götz, M., & Wurst, W. (2022). Parkinson's disease motor symptoms rescue by CRISPRa-reprogramming astrocytes into GABAergic neurons. *EMBO Molecular Medicine*, *14*(5), e14797. doi:10.15252/emmm.202114797 PMID:35373464

Goldman, S. M. (2014). Environmental toxins and Parkinson's disease. *Annual Review of Pharmacology and Toxicology*, *54*(1), 141–164. doi:10.1146/annurev-pharmtox-011613-135937 PMID:24050700

Guo, Z., Zhang, L., Wu, Z., Chen, Y., Wang, F., & Chen, G. (2014). In vivo direct reprogramming of reactive glial cells into functional neurons after brain injury and in an Alzheimer's disease model. *Cell Stem Cell*, *14*(2), 188–202. doi:10.1016/j.stem.2013.12.001 PMID:24360883

Haaxma, C. A., Bloem, B. R., Borm, G. F., Oyen, W. J. G., Leenders, K. L., Eshuis, S., Booij, J., Dluzen, D. E., & Horstink, M. W. I. M. (2007). Gender differences in Parkinson's disease. *Journal of Neurology, Neurosurgery, and Psychiatry*, *78*(8), 819–824. doi:10.1136/jnnp.2006.103788 PMID:17098842

Hamd-Ghadareh, S., Salimi, A., Parsa, S., & Mowla, S. J. (2022). Development of three-dimensional semi-solid hydrogel matrices for ratiometric fluorescence sensing of Amyloid β peptide and imaging in SH-SY5 cells: Improvement of point of care diagnosis of Alzheimer's disease biomarker. *Biosensors & Bioelectronics*, *199*, 113895. doi:10.1016/j.bios.2021.113895 PMID:34968953

Hayes, M. T. (2019). Parkinson's disease and parkinsonism. *The American Journal of Medicine*, *132*(7), 802–807. doi:10.1016/j.amjmed.2019.03.001 PMID:30890425

Hirsch, E. C., & Hunot, S. (2009). Neuroinflammation in Parkinson's disease: A target for neuroprotection? *Lancet Neurology*, *8*(4), 382–397. doi:10.1016/S1474-4422(09)70062-6 PMID:19296921

Hu, S., Yang, C., Li, Y., Luo, Q., & Luo, H. (2022). Nanozyme sensor array based on manganese dioxide for the distinction between multiple amyloid β peptides and their dynamic aggregation process. *Biosensors & Bioelectronics, 199*, 113881. doi:10.1016/j.bios.2021.113881 PMID:34915216

Jakobs, M., Lee, D. J., & Lozano, A. M. (2020). Modifying the progression of Alzheimer's and Parkinson's disease with deep brain stimulation. *Neuropharmacology, 171*, 107860. doi:10.1016/j.neuropharm.2019.107860 PMID:31765650

Jankovic, J. (2008). Parkinson's disease: Clinical features and diagnosis. *Journal of Neurology, Neurosurgery, and Psychiatry, 79*(4), 368–376. doi:10.1136/jnnp.2007.131045 PMID:18344392

Jimenez-Sanchez, M., Licitra, F., Underwood, B. R., & Rubinsztein, D. C. (2017). Huntington's disease: Mechanisms of pathogenesis and therapeutic strategies. *Cold Spring Harbor Perspectives in Medicine, 7*(7), 7. doi:10.1101/cshperspect.a024240 PMID:27940602

Joe, E., Ringman, J.M., 2019. Cognitive symptoms of Alzheimer's disease: clinical management and prevention. bmj 367.

Juopperi, T. A., Kim, W. R., Chiang, C.-H., Yu, H., Margolis, R. L., Ross, C. A., Ming, G., & Song, H. (2012). Astrocytes generated from patient induced pluripotent stem cells recapitulate features of Huntington's disease patient cells. *Molecular Brain, 5*(1), 1–14. doi:10.1186/1756-6606-5-17 PMID:22613578

Kalia, L. V., & Lang, A. E. (2015). Parkinson's disease. *Lancet, 386*(9996), 896–912. doi:10.1016/S0140-6736(14)61393-3 PMID:25904081

Kehtari, T. (2022). *In Vivo Direct Reprogramming of Reactive Astrocytes into Neurons in a Rat Model of Alzheimer's Disease.*

Keum, J. W., Shin, A., Gillis, T., Mysore, J. S., Elneel, K. A., Lucente, D., Hadzi, T., Holmans, P., Jones, L., & Orth, M. (2016). The HTT CAG-expansion mutation determines age at death but not disease duration in Huntington disease. *American Journal of Human Genetics, 98*(2), 287–298. doi:10.1016/j.ajhg.2015.12.018 PMID:26849111

Khakh, B. S., Beaumont, V., Cachope, R., Munoz-Sanjuan, I., Goldman, S. A., & Grantyn, R. (2017). Unravelling and exploiting astrocyte dysfunction in Huntington's disease. *Trends in Neurosciences, 40*(7), 422–437. doi:10.1016/j.tins.2017.05.002 PMID:28578789

Kim, K., & Park, C. B. (2020). Femtomolar sensing of Alzheimer's tau proteins by water oxidation-coupled photoelectrochemical platform. *Biosensors & Bioelectronics, 154*, 112075. doi:10.1016/j.bios.2020.112075 PMID:32056970

Lanctôt, K. L., Amatniek, J., Ancoli-Israel, S., Arnold, S. E., Ballard, C., Cohen-Mansfield, J., Ismail, Z., Lyketsos, C., Miller, D. S., Musiek, E., Osorio, R. S., Rosenberg, P. B., Satlin, A., Steffens, D., Tariot, P., Bain, L. J., Carrillo, M. C., Hendrix, J. A., Jurgens, H., & Boot, B. (2017). Neuropsychiatric signs and symptoms of Alzheimer's disease: New treatment paradigms. *Alzheimer's & Dementia: Translational Research & Clinical Interventions, 3*(3), 440–449. doi:10.1016/j.trci.2017.07.001 PMID:29067350

Lee, H.-G., Wheeler, M. A., & Quintana, F. J. (2022). Function and therapeutic value of astrocytes in neurological diseases. *Nature Reviews. Drug Discovery*, *21*(5), 339–358. doi:10.103841573-022-00390-x PMID:35173313

Lee, M.-C., Yu, W.-C., Shih, Y.-H., Chen, C.-Y., Guo, Z.-H., Huang, S.-J., Chan, J. C. C., & Chen, Y.-R. (2018). Zinc ion rapidly induces toxic, off-pathway amyloid-β oligomers distinct from amyloid-β derived diffusible ligands in Alzheimer's disease. *Scientific Reports*, *8*(1), 1–16. doi:10.103841598-018-23122-x PMID:29555950

LeWitt, P. A. (2008). Levodopa for the treatment of Parkinson's disease. *The New England Journal of Medicine*, *359*(23), 2468–2476. doi:10.1056/NEJMct0800326 PMID:19052127

Loera-Valencia, R., Cedazo-Minguez, A., Kenigsberg, P. A., Page, G., Duarte, A. I., Giusti, P., Zusso, M., Robert, P., Frisoni, G. B., Cattaneo, A., Zille, M., Boltze, J., Cartier, N., Buee, L., Johansson, G., & Winblad, B. (2019). Current and emerging avenues for Alzheimer's disease drug targets. *Journal of Internal Medicine*, *286*(4), 398–437. doi:10.1111/joim.12959 PMID:31286586

Masters, C. L., Bateman, R., Blennow, K., Rowe, C. C., Sperling, R. A., & Cummings, J. L. (2015). Alzheimer's disease. *Nature Reviews. Disease Primers*, *1*(1), 15056. doi:10.1038/nrdp.2015.56 PMID:27188934

Meng, Q., Lin, M.-S., & Tzeng, I.-S. (2020). Relationship between exercise and Alzheimer's disease: A narrative literature review. *Frontiers in Neuroscience*, *14*, 131. doi:10.3389/fnins.2020.00131 PMID:32273835

Moro, E., Lozano, A. M., Pollak, P., Agid, Y., Rehncrona, S., Volkmann, J., Kulisevsky, J., Obeso, J. A., Albanese, A., Hariz, M. I., Quinn, N. P., Speelman, J. D., Benabid, A. L., Fraix, V., Mendes, A., Welter, M.-L., Houeto, J.-L., Cornu, P., Dormont, D., & Lang, A. E. (2010). Long-term results of a multicenter study on subthalamic and pallidal stimulation in Parkinson's disease. *Movement Disorders*, *25*(5), 578–586. doi:10.1002/mds.22735 PMID:20213817

Muralidar, S., Ambi, S. V., Sekaran, S., Thirumalai, D., & Palaniappan, B. (2020). Role of tau protein in Alzheimer's disease: The prime pathological player. *International Journal of Biological Macromolecules*, *163*, 1599–1617. doi:10.1016/j.ijbiomac.2020.07.327 PMID:32784025

Nalivaeva, N. N., Beckett, C., Belyaev, N. D., & Turner, A. J. (2012). Are amyloid-degrading enzymes viable therapeutic targets in Alzheimer's disease? *Journal of Neurochemistry*, *120*(s1), 167–185. doi:10.1111/j.1471-4159.2011.07510.x PMID:22122230

Neu, S. C., Pa, J., Kukull, W., Beekly, D., Kuzma, A., Gangadharan, P., Wang, L.-S., Romero, K., Arneric, S. P., Redolfi, A., Orlandi, D., Frisoni, G. B., Au, R., Devine, S., Auerbach, S., Espinosa, A., Boada, M., Ruiz, A., Johnson, S. C., & Toga, A. W. (2017). Apolipoprotein E genotype and sex risk factors for Alzheimer disease: A meta-analysis. *JAMA Neurology*, *74*(10), 1178–1189. doi:10.1001/jamaneurol.2017.2188 PMID:28846757

Panza, G. A., Taylor, B. A., MacDonald, H. V., Johnson, B. T., Zaleski, A. L., Livingston, J., Thompson, P. D., & Pescatello, L. S. (2018). Can exercise improve cognitive symptoms of Alzheimer's disease? *Journal of the American Geriatrics Society*, *66*(3), 487–495. doi:10.1111/jgs.15241 PMID:29363108

Pardo-Moreno, T., González-Acedo, A., Rivas-Domínguez, A., García-Morales, V., García-Cozar, F. J., Ramos-Rodríguez, J. J., & Melguizo-Rodríguez, L. (2022). Therapeutic approach to Alzheimer's disease: Current treatments and new perspectives. *Pharmaceutics*, *14*(6), 1117. doi:10.3390/pharmaceutics14061117 PMID:35745693

Parkinson, J. (2002). An essay on the shaking palsy. *The Journal of Neuropsychiatry and Clinical Neurosciences*, *14*(2), 223–236. doi:10.1176/jnp.14.2.223 PMID:11983801

Pasandideh, S., & Arasteh, A. (2021). Evaluation of antioxidant and inhibitory properties of Citrus aurantium L. on the acetylcholinesterase activity and the production of amyloid nano–bio fibrils. *International Journal of Biological Macromolecules*, *182*, 366–372. doi:10.1016/j.ijbiomac.2021.04.043 PMID:33848544

Plant, L. D., Boyle, J. P., Smith, I. F., Peers, C., & Pearson, H. A. (2003). The production of amyloid β peptide is a critical requirement for the viability of central neurons. *The Journal of Neuroscience : The Official Journal of the Society for Neuroscience*, *23*(13), 5531–5535. doi:10.1523/JNEUROSCI.23-13-05531.2003 PMID:12843253

Prigent, A., Lionnet, A., Corbillé, A.-G., & Derkinderen, P. (2017). Neuropathologie et physiopathologie de la maladie de Parkinson: Focus sur l'α-synucléine. *Presse Medicale (Paris, France)*, *46*(2), 182–186. doi:10.1016/j.lpm.2016.11.010

Rasmussen, J., & Langerman, H. (2019). Alzheimer's disease–why we need early diagnosis. *Degenerative Neurological and Neuromuscular Disease*, *9*, 123–130. doi:10.2147/DNND.S228939 PMID:31920420

Rawlins, M. D., Wexler, N. S., Wexler, A. R., Tabrizi, S. J., Douglas, I., Evans, S. J. W., & Smeeth, L. (2016). The prevalence of Huntington's disease. *Neuroepidemiology*, *46*(2), 144–153. doi:10.1159/000443738 PMID:26824438

Reale, M., Costantini, E., Jagarlapoodi, S., Khan, H., Belwal, T., & Cichelli, A. (2020). Relationship of wine consumption with Alzheimer's disease. *Nutrients*, *12*(1), 206. doi:10.3390/nu12010206 PMID:31941117

Reiner, A., Albin, R. L., Anderson, K. D., D'Amato, C. J., Penney, J. B., & Young, A. B. (1988). Differential loss of striatal projection neurons in Huntington disease. *Proceedings of the National Academy of Sciences of the United States of America*, *85*(15), 5733–5737. doi:10.1073/pnas.85.15.5733 PMID:2456581

Rhodes, A., Inker, J., Richardson, J., & Zanjani, F. (2022). Alzheimer's Disease Prevention Health Coaching. J. *Prev. Alzheimer's Dis.*, *9*, 277–285. PMID:35543001

Richter, N., Beckers, N., Onur, O. A., Dietlein, M., Tittgemeyer, M., Kracht, L., Neumaier, B., Fink, G. R., & Kukolja, J. (2018). Effect of cholinergic treatment depends on cholinergic integrity in early Alzheimer's disease. *Brain*, *141*(3), 903–915. doi:10.1093/brain/awx356 PMID:29309600

Rikani, A. A., Choudhry, Z., Choudhry, A. M., Rizvi, N., Ikram, H., Mobassarah, N. J., & Tuli, S. (2014). The mechanism of degeneration of striatal neuronal subtypes in Huntington disease. *Annals of Neurosciences*, *21*, 112. doi:10.5214/ans.0972.7531.210308 PMID:25206077

Rivetti Di Val Cervo, P., Romanov, R. A., Spigolon, G., Masini, D., Martín-Montañez, E., Toledo, E. M., La Manno, G., Feyder, M., Pifl, C., Ng, Y.-H., Sánchez, S. P., Linnarsson, S., Wernig, M., Harkany, T., Fisone, G., & Arenas, E. (2017). Induction of functional dopamine neurons from human astrocytes in vitro and mouse astrocytes in a Parkinson's disease model. *Nature Biotechnology*, *35*(5), 444–452. doi:10.1038/nbt.3835 PMID:28398344

Rosas, H. D., Salat, D. H., Lee, S. Y., Zaleta, A. K., Pappu, V., Fischl, B., Greve, D., Hevelone, N., & Hersch, S. M. (2008). Cerebral cortex and the clinical expression of Huntington's disease: Complexity and heterogeneity. *Brain*, *131*(4), 1057–1068. doi:10.1093/brain/awn025 PMID:18337273

Rosenblatt, A. (2022). Neuropsychiatry of Huntington's disease. *Dialogues in Clinical Neuroscience*. PMID:17726917

Ross, C. A., & Tabrizi, S. J. (2011). Huntington ' s disease : From molecular pathogenesis to clinical treatment. *Lancet Neurology*, *10*(1), 83–98. doi:10.1016/S1474-4422(10)70245-3 PMID:21163446

Saudou, F., & Humbert, S. (2016). The biology of huntingtin. *Neuron*, *89*(5), 910–926. doi:10.1016/j.neuron.2016.02.003 PMID:26938440

Scheltens, P., De Strooper, B., Kivipelto, M., Holstege, H., Chételat, G., Teunissen, C. E., Cummings, J., & van der Flier, W. M. (2021). Alzheimer's disease. *Lancet*, *397*(10284), 1577–1590. doi:10.1016/S0140-6736(20)32205-4 PMID:33667416

Schilling, L. P., Balthazar, M. L. F., Radanovic, M., Forlenza, O. V., Silagi, M. L., Smid, J., Barbosa, B. J. A. P., Frota, N. A. F., de Souza, L. C., & Vale, F. A. C. (2022). Diagnosis of Alzheimer's disease: Recommendations of the Scientific Department of Cognitive Neurology and Aging of the Brazilian Academy of Neurology. *Dementia & Neuropsychologia*, *16*, 25–39. doi:10.1590/1980-5764-dn-2022-s102pt PMID:36533157

Schneider, M. E., Guillade, L., Correa-Duarte, M. A., & Moreira, F. T. C. (2022). Tau proteinsTau 181 Protein detection in Early-Stage Alzheimer's Disease. *Bioelectrochemistry (Amsterdam, Netherlands)*, 108057. doi:10.1016/j.bioelechem.2022.108057

Schuurman, P. R., Bosch, D. A., Merkus, M. P., & Speelman, J. D. (2008). Long-term follow-up of thalamic stimulation versus thalamotomy for tremor suppression. *Movement Disorders*, *23*(8), 1146–1153. doi:10.1002/mds.22059 PMID:18442104

Selkoe, D. J. (2015). Alzheimer disease. In *Rosenberg's Molecular and Genetic Basis of Neurological and Psychiatric Disease* (pp. 753–768). Elsevier. doi:10.1016/B978-0-12-410529-4.00067-X

Selkoe, D. J., & Hardy, J. (2016). The amyloid hypothesis of Alzheimer's disease at 25 years. *EMBO Molecular Medicine*, *8*(6), 595–608. doi:10.15252/emmm.201606210 PMID:27025652

Sharp, A. H., Loev, S. J., Schilling, G., Li, S., Li, X., Bao, J., Wagster, M. V., Kotzuk, J. A., Steiner, J. P., Lo, A., Hedreen, J., Sisodia, S., Snyder, S. H., Dawson, T. M., Ryugo, D. K., & Ross, C. A. (1995). Widespread Expression of Huntington ' s Disease Gene (IT15). *Protein Product*, *14*, 1065–1074. PMID:7748554

Simon, D. K., Tanner, C. M., & Brundin, P. (2020). Parkinson Disease Epidemiology, Pathology, Genetics, and Pathophysiology. *Clinics in Geriatric Medicine*, *36*(1), 1–12. doi:10.1016/j.cger.2019.08.002 PMID:31733690

Talifu, Z., Liu, J.-Y., Pan, Y.-Z., Ke, H., Zhang, C.-J., Xu, X., Gao, F., Yu, Y., Du, L.-J., & Li, J.-J. (2023). In vivo astrocyte-to-neuron reprogramming for central nervous system regeneration: A narrative review. *Neural Regeneration Research*, *18*(4), 750. doi:10.4103/1673-5374.353482 PMID:36204831

Tanner, C. M., Goldman, S. M., Ross, G. W., & Grate, S. J. (2014). The disease intersection of susceptibility and exposure: Chemical exposures and neurodegenerative disease risk. *Alzheimer's & Dementia*, *10*(3S), S213–S225. doi:10.1016/j.jalz.2014.04.014 PMID:24924672

Tanzi, R. E., & Bertram, L. (2005). Twenty years of the Alzheimer's disease amyloid hypothesis: A genetic perspective. *Cell*, *120*(4), 545–555. doi:10.1016/j.cell.2005.02.008 PMID:15734686

Tian, Y., Tian, Y., Yuan, Z., Zeng, Y., Wang, S., Fan, X., Yang, D., & Yang, M. (2022). Iron Metabolism in Aging and Age-Related Diseases. *International Journal of Molecular Sciences*, *23*(7), 3612. doi:10.3390/ijms23073612 PMID:35408967

Tini, G., Scagliola, R., Monacelli, F., La Malfa, G., Porto, I., Brunelli, C., & Rosa, G. M. (2020). Alzheimer's disease and cardiovascular disease: A particular association. *Cardiology Research and Practice*, *2020*, 2020. doi:10.1155/2020/2617970 PMID:32454996

Twohig, D., & Nielsen, H. M. (2019). α-synuclein in the pathophysiology of Alzheimer's disease. *Molecular Neurodegeneration*, *14*(1), 1–19. doi:10.118613024-019-0320-x PMID:31186026

Villemagne, V. L., Dore, V., Bourgeat, P., Burnham, S. C., Laws, S., Salvado, O., Masters, C. L., & Rowe, C. C. (2017). Aβ-amyloid and tau imaging in dementia. In *Seminars in Nuclear Medicine* (pp. 75–88). Elsevier. doi:10.1053/j.semnuclmed.2016.09.006

Villemagne, V. L., & Rowe, C. C. (2011). Amyloid imaging. *International Psychogeriatrics*, *23*(S2), S41–S49. doi:10.1017/S1041610211000895 PMID:21729418

Walsh, D. M., & Selkoe, D. J. (2020). Amyloid β-protein and beyond: The path forward in Alzheimer's disease. *Current Opinion in Neurobiology*, *61*, 116–124. doi:10.1016/j.conb.2020.02.003 PMID:32197217

Warren, S. L., Reid, E., Whitfield, P., Helal, A. M., Abo Hamza, E. G., Tindle, R., Moustafa, A. A., & Hamid, M. S. (2023). Cognitive and behavioral abnormalities in individuals with Alzheimer's disease, mild cognitive impairment, and subjective memory complaints. *Current Psychology (New Brunswick, N.J.)*, 1–11. doi:10.100712144-023-04281-1

Wei, Z.-Y. D., & Shetty, A. K. (2021). Treating Parkinson's disease by astrocyte reprogramming: Progress and challenges. *Science Advances*, *7*(26), eabg3198. doi:10.1126ciadv.abg3198 PMID:34162545

Wiley, J. (2021). Alzheimer's disease facts and figures. *Alzheimer's & Dementia*, *17*(3), 327–406. doi:10.1002/alz.12328 PMID:33756057

Wu, Z., Parry, M., Hou, X.-Y., Liu, M.-H., Wang, H., Cain, R., Pei, Z.-F., Chen, Y.-C., Guo, Z.-Y., Abhijeet, S., & Chen, G. (2020). Gene therapy conversion of striatal astrocytes into GABAergic neurons in mouse models of Huntington's disease. *Nature Communications*, *11*(1), 1105. doi:10.103841467-020-14855-3 PMID:32107381

Yang, F., Trolle Lagerros, Y., Bellocco, R., Adami, H.-O., Fang, F., Pedersen, N. L., & Wirdefeldt, K. (2015). Physical activity and risk of Parkinson's disease in the Swedish National March Cohort. *Brain*, *138*(2), 269–275. doi:10.1093/brain/awu323 PMID:25410713

Yang, T., Li, S., Xu, H., Walsh, D. M., & Selkoe, D. J. (2017). Large soluble oligomers of amyloid β-protein from Alzheimer brain are far less neuroactive than the smaller oligomers to which they dissociate. *The Journal of Neuroscience : The Official Journal of the Society for Neuroscience*, *37*(1), 152–163. doi:10.1523/JNEUROSCI.1698-16.2016 PMID:28053038

Yu, T.-W., Lane, H.-Y., & Lin, C.-H. (2021). Novel therapeutic approaches for Alzheimer's disease: An updated review. *International Journal of Molecular Sciences*, *22*(15), 8208. doi:10.3390/ijms22158208 PMID:34360973

Yu, X., Nagai, J., & Khakh, B. S. (2020). Improved tools to study astrocytes. *Nature Reviews. Neuroscience*, *21*(3), 121–138. doi:10.103841583-020-0264-8 PMID:32042146

Zhang, X.-X., Tian, Y., Wang, Z.-T., Ma, Y.-H., Tan, L., & Yu, J.-T. (2021). The epidemiology of Alzheimer's disease modifiable risk factors and prevention. *The Journal of Prevention of Alzheimer's Disease*, *8*, 313–321. doi:10.14283/jpad.2021.15 PMID:34101789

Zhou, C., Ni, W., Zhu, T., Dong, S., Sun, P., & Hua, F. (2022). Cellular reprogramming and its potential application in Alzheimer's disease. *Frontiers in Neuroscience*, *16*, 884667. doi:10.3389/fnins.2022.884667 PMID:35464309

Zhou, H., Su, J., Hu, X., Zhou, C., Li, H., Chen, Z., Xiao, Q., Wang, B. O., Wu, W., Sun, Y., Zhou, Y., Tang, C., Liu, F., Wang, L., Feng, C., Liu, M., Li, S., Zhang, Y., Xu, H., ... Yang, H. (2020). Glia-to-neuron conversion by CRISPR-CasRx alleviates symptoms of neurological disease in mice. *Cell*, *181*(3), 590–603. doi:10.1016/j.cell.2020.03.024 PMID:32272060

Zhou, Z., Chen, S., Huang, Y., Gu, B., Li, J., Wu, C., Yin, P., Zhang, Y., & Li, H. (2022). Simultaneous visualization and quantification of copper (II) ions in Alzheimer's disease by a near-infrared fluorescence probe. *Biosensors & Bioelectronics*, *198*, 113858. doi:10.1016/j.bios.2021.113858 PMID:34871835

Zuccato, C. & Cattaneo, E. (2014). Huntington's disease. *Neurotrophic factors,* 357–409. Research Gate.

Zuccato, C., Marullo, M., Conforti, P., MacDonald, M. E., Tartari, M., & Cattaneo, E. (2008). Systematic assessment of BDNF and its receptor levels in human cortices affected by Huntington's disease. *Brain Pathology (Zurich, Switzerland)*, *18*(2), 225–238. doi:10.1111/j.1750-3639.2007.00111.x PMID:18093249

Chapter 17
Involvement of Glial Cells in the Pathophysiology and Treatment of Depression

Hanane El Fatimi
Mohammed VI University of Health Sciences, Morocco

Loubna Khalki
Mohammed VI University of Health Sciences, Morocco

ABSTRACT

Depression is a complex neuropsychiatric illness affecting millions worldwide. Furthermore, its exact cause remains uncertain. Theories include the monoamine hypothesis, changes in hormonal systems, inflammation, immune alterations, neurogenesis issues, and environmental factors. Recent research has highlighted the role of glial cells, particularly microglia and astrocytes, in depression. This chapter explores the implications of glial cells in depression, shedding light on their involvement in neuroinflammation, synaptic plasticity, and the regulation of mood-related circuits. An intricate interplay between glial cells, proinflammatory cytokines, and neuronal processes appears to underlie the etiology and progression of depression. Understanding the dynamic interactions between glial cells and neurons in the context of depression offers promising avenues for novel therapeutic interventions targeting this debilitating disorder's neuroinflammatory components.

INTRODUCTION

Depression is a potentially life-threatening disorder that affects hundreds of millions of people all over the world. It can occur at any age from childhood to late life and is a tremendous cost to society as this disorder causes severe distress and disruption of life and, if left untreated, can be fatal (Brigitta, 2002). The exact mechanisms of depression are not fully understood. The first major depression hypothesis was formulated about 30 years ago, and the main symptoms of the disease are due to monoaminergic systems dysfunction; norepinephrine (NE), serotonin (5-hydroxytryptamine; 5-HT), and/or dopamine (DA). It

DOI: 10.4018/978-1-6684-9675-6.ch017

is highly established that the monoaminergic system regulates mood, attention, motivation, and fatigue. Hence, it is clearly that its dysfunction can lead to many behavioral symptoms such as psychomotor agitation or retardation (Brigitta, 2002). Indeed, the abnormal functional and behavioral effects of depression or mania can result from alterations in neurotransmitter synthesis, storage, or release, as well as from disturbed sensitivity of their receptors or intracellular messenger functions (Brigitta, 2002). Along with this, there is evidence that other systems are involved in triggering depressive episodes, such as the neuroendocrine axis (corticotropic axis), peptidergic (substance P), and hormonal (melatonin) systems. All these systems share the common property of affecting cell proliferation in specific cerebral regions such as the hippocampus which is one of the regions of the brain where neural progenitor cells continue to divide and give rise to new neurons in adult animals and it is involved in memory and learning function.

Meanwhile, the focus of research has shifted from monoamines toward other molecular mechanisms, and some factors other than monoamine deficiency systems must be taken into account when describing the neurobiological basis of major depression (Czéh & Lucassen, 2007). Recently, it has been hypothesized that neurotrophic dysregulation (Duman & Li, 2012) and glial cells degeneration or dysfunction, particularly astrocytes contributes to the pathogenesis of depression (Śmiałowska et al., 2013). This chapter will describe the involvement of glial cells in the pathogenesis and treatment of depression.

GLIAL CELLS DYSFUNCTION IN DEPRESSION

Glial cells are composed of distinct populations of oligodendrocytes (OLs), microglia, and astrocytes. It is known that glia (predominantly astrocytes) are thought to support neurons trophically, regulate neuronal metabolism, form synapses, and participate in neurotransmission in addition to their traditional functions in myelin formation, inflammatory processes, and neuronal migration (radial glia).

Recently, depression (Ménard et al., 2016) and neurodegenerative disorders (Capani et al., 2016; Maragakis & Rothstein, 2006), have been linked to the degradation or malfunction of glial cells, particularly astrocytes. Indeed, postmortem studies of depressed suicides show hypertrophy of astrocytes in the white matter (WM) of the anterior cingulate cortex (Torres-Platas et al., 2011), and decreased packing density or number of glial cell populations in frontolimbic brain regions (Rajkowska & A Stockmeier, 2013). It is important to note that a recent immunohistochemistry study of glial fibrillary acidic protein (GFAP) suggested that astrocytes are responsible for the overall glial pathology in Major depressive disorder (MDD) (Miguel-Hidalgo et al., 2000). There is a statically significant correlation between age and GFAP immunoreactivity among subjects with MDD. Hence, a significant reduction in the population of reactive astroglia is found in a small subgroup of young subjects with MDD, as compared to young control subjects and older subjects with MDD. Indeed, most of the participants in this subgroup of younger adults with MDD committed suicide. According to recent research, the levels of the protein GFAP are also lower in these young adults with MDD when compared to age-matched control subjects, and GFAP levels are positively correlated with both the age at death and the age of depression onset (Miguel-Hidalgo et al., 2000).

On the other hand, OLs may contribute to the cellular pathology of depression. In fact, OLs exhibit ultrastructural abnormalities and a decrease in their density and immunoreactivity in both the dorsolateral prefrontal and anterior frontal cortex in patients with MDD (Orlovskaya et al., 2000; Uranova et al., 2001). Additionally, changes in glial size and shape as well as decrease in glial density have been linked to depression-related monoamine and glutamate system dysfunction. For instance, almost all of the receptor systems, ion channels, and transporters present in neurons are expressed by astrocytes. Therefore, 5-HT, NE, or DA neurotransmission may be influenced by the postsynaptic monoaminergic receptors found on the cell bodies and processes of glial cells. Moreover, astroglia are crucial in controlling NMDA receptor activity as they are the main sites where glial transporters take up glutamate. In order to prevent cell death in vitro and to provide neurons with energy, astrocytes control the amounts of extracellular glutamate in the body. In addition, functional neuroimaging may be able to detect the effects of astrocyte disease in MDD, which may serve to indirectly enhance glutamate-mediated neuronal excitotoxicity (Stockmeier & Rajkowska, 2004).

Alteration In Neurons and Glia in The Cerebral Cortex

Glial cell density and quantity have unexpectedly been shown to be significantly reduced in the most common cell defect observed in mood disorders including depressive disorder. In fact, the anterior cingulate cortex, dorsolateral prefrontex (DLPFC), and orbitofrontal cortex (OFC), of MDD patients have all consistently shown decrease in glial cells according to different laboratories. In the supracallosal region of the anterior cingulate cortex, analysis of glial cells revealed a 22% reduction in glial cell density in MDD (D. Cotter et al., 2001). Specially patients who had a family history of mood disorders, there was a significant loss of 20% to 40% in the number and density of glial cells in the subgenual area of the anterior cingulate cortex (Öngür et al., 1998; Rajkowska et al., 1999). In addition, the DLPFC and OFC of patients with MDD also showed a noticeable decrease in cell-packing densities ranging from 11% to 30% in both regions, in comparison to an age- and gender-matched control group. Moreover, other morphometric studies examining glial cell counts and size measurements in the two regions also reported the same results (D. Cotter et al., 2002; Rajkowska et al., 1999). Glial cells reductions were most pronounced in layers III and VI of the DLPFC and caudal OFC, while the effect on the rostral section of the OFC was rather marginal (Fig.1) (Rajkowska et al., 1999).

Alterations in Neurons and Glial Cells in the Hippocampus

It is worth mentioning that, the hippocampus also appears to exhibit glial dysfunction in depressive disorders, beyond the frontal cortex. Indeed, some preclinical and neuroimaging studies have linked MDD pathogenesis to the hippocampus formation. Moreover, the hippocampal cellular integrity and apoptosis of depressed, steroid-treated, and healthy control participants have been studied (Choudary et al., 2005; Lucassen, Müller, et al., 2001). Evidence of a mild increase in DNA fragmentation linked to apoptosis and necrotic cell death in the majority of the depressives, which was found in the dentate gyrus, CA1, and CA4 was reported (Lucassen, Müller, et al., 2001). In addition, CA1 and CA2 showed decreased astrocyte immunoreactivity for cellular GFAP and the neuron-specific phosphoprotein B50 (or GAP-45) (M. B. Müller et al., 2001). There is a strong evidence that patients with MDD have noticeable abnormalities in their dentate gyrus and CA regions.

Importantly, current research on the hippocampus in a sizable number of subjects with MDD and age-matched healthy control subjects reveals a notable rise in the density of glial cells in all hippocampal CA subfields and the granule cell layer of the dentate gyrus (Stockmeier et al., 2003). However, glial cell processes, dendrites, and proximal axons form a lattice around the cell bodies of neurons to form the neuropil. Other postmortem findings showed reduced dendritic spine density on neurons and reduced apical dendritic arborization in the subiculum in a small sample of mixed patients with bipolar disorder

Figure 1. Changes in neuronal size and size-dependent density in layer II of rostral OFC in a 73-year-old female with MDD compared to a 71-year-old psychiatrically normal female control subject
(Rajkowska et al., 1999)

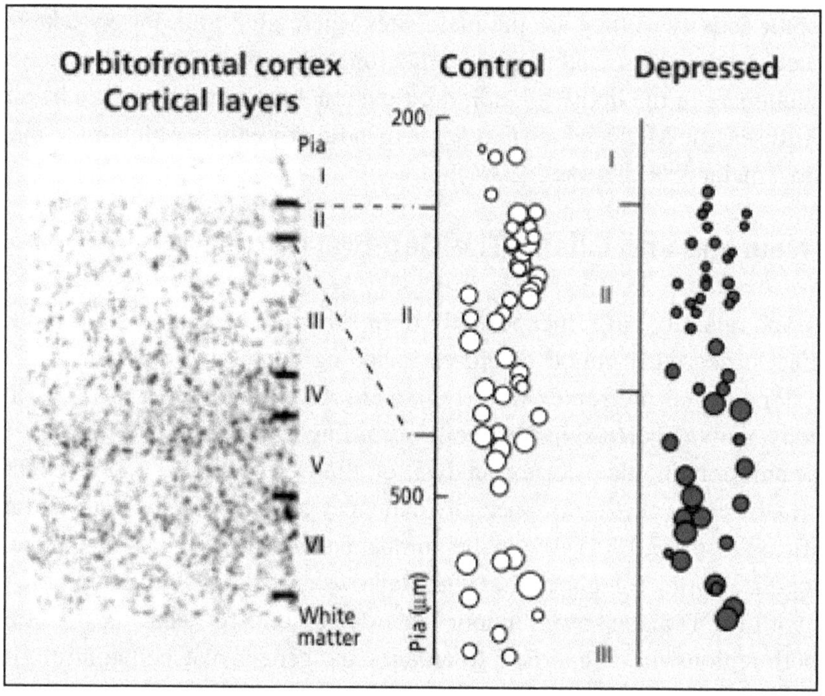

or depression lend support to the idea that neuropil loss in the hippocampus occurring in MDD (Rosoklija et al., 2000). Therefore, a loss in neuropil including dendritic branching, dendritic spine complexity, and glial processes may be a key factor in the decreased volume of the hippocampus observed in some cases of patients with depression.

COMPROMISED GLIAL CELLS IN DEPRESSION

Glial cells have predominantly been regarded as the "glue" of the brain since their discovery more than 150 years ago (Virchow, 1858), and they are usually considered to be of relatively little importance in brain physiology. However, glial cell pathology is implicated as a potential contributor to a wide range of neuropsychiatric disorders in light of relatively recent discoveries clarifying the crucial function of glial cells in a plethora of physiological processes. Astrocytes, OLs, and microglia are the three primary subgroups into which glial cells have traditionally been divided. In contrast to microglia, which come from the immune system, astrocyte, and oligodendrocyte lineage cells are formed from brain stem cells. According to current knowledge, each of these kinds of glia performs a wide range of physiological functions, from regulating the metabolism of amino acid neurotransmitters and brain energy supply to initiating synaptic remodeling (Eroglu & Barres, 2010).

It is worth mentioning that, the neurotrophic hypothesis of stress-related mood disorders first emerged over 20 years ago in addition to the "classical" monoamine theory of depression (Duman & Monteggia,

2006). Moreover, numerous studies have shown that stress and antidepressant therapy have the opposite effect on the expression of particular neurotrophic factors in the limbic brain regions involved in the regulation of mood and cognitive processes (Autry & Monteggia, 2012). In addition, patients with depression have been found to have altered expression of several different neurotrophic factors, such as brain-derived neurotrophic factor (BDNF) (Bocchio-Chiavetto et al., 2010; Molendijk et al., 2014), nerve growth factor (de Azevedo Cardoso et al., 2014; Pallavi et al., 2013), neurotrophin-3 (Otsuki et al., 2008; Pallavi et al., 2013), vascular endothelial growth factor (Berent et al., 2014; Fornaro et al., 2013), and insulin-like growth factor-1 (Sievers et al., 2014; Weber-Hamann et al., 2009). However, the neurotrophic theory implicated the decreased neurotrophic factors in stress-induced depression and proposed that antidepressants could restore such abnormality (Duman & Monteggia, 2006). In addition, the neurotrophic factor glial cell line-derived neurotrophic factor (GDNF) is intimately linked to depression. According to Boku et al. (2013), GDNF has a significant role in the proliferation, differentiation, maturation, and survival of astrocytes (Boku et al., 2013), as well as many neurons (A. Pascual et al., 2008) including the hippocampus (Irala et al., 2016) and other neurons, which will undoubtedly have an impact on MDD. Moreover, to activate the intracellular tyrosine kinase domain and the subsequent signaling protein cascade, GDNF first binds to the GDNF-family receptor a1 (GFRa1) before forming a complex with the receptor tyrosine kinase receptor (Airaksinen & Saarma, 2002), GDNF promotes the differentiation of dopaminergic neurons (Christophersen et al., 2007; Lin et al., 1993) and serotonergic neurons (Ducray et al., 2006), and can increase neurite growth in a variety of neuronal types (Ducray et al., 2006; Takaku et al., 2013; Wakeman et al., 2014). In addition, GDNF has also been demonstrated to protect mesencephalic neuron-derived cells (Ortiz-Ortiz et al., 2011), serotonergic neurons from dorsal raphe nucleus (Hochstrasser et al., 2011), mesencephalic dopaminergic neurons (Jaumotte & Zigmond, 2014), and glial cells (Uzdensky et al., 2013) from oxidative or neuro-inflammatory injury. In the recent past, researchers have looked at GDNF activity alterations in depressed patients (Takebayashi et al., 2006), it has been reported that the whole blood GDNF levels were found to be significantly lower in patients with serious MDD who had undergone remission (de Azevedo Cardoso et al., 2014; Diniz et al., 2012; Pallavi et al., 2015; Tseng et al., 2013; X. Zhang et al., 2008, 2014).

Furthermore, BDNF expression in the hippocampus of patients with depression has been studied, and alterations in these factors might be related to changes in cell density and volume in these patients. Preliminary research suggests that long-term antidepressant use may control the levels of BDNF in the human brain. Additionally, in an immunohistochemical study of subjects with MDD, Only those individuals taking antidepressant drugs exhibit increased immunoreactivity of BDNF in the dentate gyrus and hilus in an immunohistochemistry examination of participants with MDD (Chen et al., 2001). Indeed, it was found that suicide victims who had either MDD or other mental conditions had significantly lower levels of BDNF in their hippocampus and DLPFC (Dwivedi et al., 2003).

Microglial Cells and Depression

Microglial cells are a crucial component of innate immune responses and associated brain functions (Ginhoux et al., 2010). Moreover, overstimulating microglia causes negative effects by causing them to release too many inflammatory mediators, such as the pro-inflammatory cytokines including tumor necrosis factor (TNF-) and interleukin-1 (IL-1), which are harmful. Recently, it has become clear that the microglia play other roles in brain plasticity, such as removing apoptotic cells and unneeded synaptic connections as well as shaping and altering the developing and mature central nervous system (CNS) (Schafer et al., 2012).

It is highly established that, depression is also tightly associated with alterations of microglia (Yirmiya et al., 2015). One of the primary pathophysiologies of depression has been thought to be the excessive and persistent activation of microglia and pro-inflammatory cytokines (Fenn et al., 2014). Moreover, several studies suggested that frontolimbic areas of depressive patients or suicide attempters have altered morphology of the microglia (Schnieder et al., 2014; Setiawan et al., 2015; Steiner et al., 2008; Torres-Platas et al., 2014). Additionally, anterior cingulate cortex and thalamus of suicide victims have shown signs of microgliosis (Steiner et al., 2008), demonstrating that MDD may be connected with changes in microglial function. Likewise, repeated stress in rodents also triggers microglia dystrophy (Kreisel et al., 2014; Milior et al., 2016; Ślusarczyk et al., 2015). Furthermore, microglia function is affected by all the signaling systems traditionally associated with depression, namely glucocorticoids (Ros-Bernal et al., 2011), BDNF (Gomes et al., 2013), 5-HT (Krabbe et al., 2012; N. Müller & Schwarz, 2007), and NE or DA (Färber et al., 2005). Importantly, several antidepressants including ketamine (Walker et al., 2013), fluoxetine (Y. C. Chung et al., 2011), or citalopram (Su et al., 2015) have highlighted the role of microglia in depression. Thus, the available evidence indicates that microglia dysfunction is a core event in depression, affecting synaptic plasticity either directly (Paolicelli et al., 2011; Zhan et al., 2014) or indirectly through its ability to control astrocytic function (O. Pascual et al., 2012).

Oligodendrocytes and Depression

Oligodendrocytes (OLs) are well known as the myelinating cells of the CNS, a significant fraction of these cells are "satellite cells" in the cerebral cortex, where they coexist closely with the cell bodies of pyramidal neurons (Peters & Feldman, 1976). They are specialized glial cells in the CNS whose main functions include producing the myelin sheath that wraps and electrically insulates neuronal axons, supporting axons metabolic needs, and taking part in neuroplasticity processes. Their lineage in the CNS includes immature precursors (OPCs) and mature OLs (Baumann & Pham-Dinh, 2001; X. E. Zhou et al., 2021). They are also extremely susceptible to glutamate excitotoxicity, which is mediated via AMPA/kainate receptors (Mcdonald et al., 1998).

In the CNS, oligodendrocyte progenitor cells (OPCs) primarily differentiate into OLs, producing myelin sheaths that enwrap segments of axons known as internodal regions of nerve fibers, they create a unique structure known as WM, which is a significant part of the CNS. Moreover, grey matter (GM) in the brain is mostly made up of neuronal cell bodies, dendrites, and typical synaptic structures for processing local information. In contrast, WM pathways are a portion of physically distinct neural structures that mediate special talents and/or behaviors (B. Zhou et al., 2021). According to a recent study by Edgar and Sibille (Edgar & Sibille, 2012), the possible effect altered OLs shape and function may have on brain circuitry and how this may affect mood regulation in subjects with psychiatric disorders is especially discussed (Sanacora & Banasr, 2013). It is important to note that changed neuronal traffic in specific WM tracts has the potential to lead to the onset of complex behavioral dysfunctions like depression. Indeed, researchers have discovered through the use of magnetic resonance imaging that decreased WM hyperintensities (WMHs) and abnormalities in myelin integrity begin to affect the prefrontal regions of the brain in the early stages of depression and continue to get worse as the disease progresses without obvious changes in OPCs in the WM (Cole et al., 2012). Moreover, the PFC, a crucial brain region engaged in complex emotional and cognitive behaviors, is now known to have very sensitive oligodendrogenesis and myelination (J. Liu, Wang, et al., 2012; Makinodan et al., 2012). It has been demonstrated that myelin-related genes such as myelin oligodendrocyte glycoprotein (Mog) and ermin (Ermn) were dramatically downregulated in depressed states brought on by stress in mice (J. Liu, Dietz, et al., 2012).

Additionally, several lines of evidence from animal research have suggested that WM anomalies may be crucial in the pathophysiological pathways underlying depression, which is relevant to structural changes in the depressed human brain. Gao et al. discovered that in a chronic unpredictable stress (CUS) model of depression in rats, the WM volume, the total length of myelinated fibers, and the mean width of myelin sheaths in the WM of the CUS rats were dramatically reduced in comparison to the control group (Gao et al., 2017). In fact, wang et al. detected a significant decrease in proteins related to myelin and OLs and discovered that stress caused depressive-like behaviors and WM impairments in UCMS animals. In other neurological and psychiatric illnesses, impaired WM integrity is more frequently associated with depressed symptoms. The depressive phenotype may therefore be influenced by the interruption of OL lineage cell proliferation and differentiation in WM (B. Zhou et al., 2021). Indeed, MDD is a disabling disorder and suicide is its most serious consequence (Blair-West et al., 1997). Consequently, there has been an increase in interest in recent years in studying the neurological mechanisms behind serious depression and how they relate to suicide (Monkul et al., 2007). Hence, the results of the qPCR were validated as decreased expression of QKI protein, which likewise functions as a myelination regulator in the postnatal CNS and is extremely important (Ebersole et al., 1996; Zhao et al., 2010), was also noticed in the OFC, indicating a unique role for QKI in myelination-related impairments in the etiology of mental disorders. Moreover, QKI is also crucial for determining myelin OLs in MDD suicide victims (Klempan et al., 2009).Indeed, recent microarray analysis of postmortem tissue from depressed people showed that transcriptional genes related to myelination or the OL lineage were present (Pantazatos et al., 2017), especially genes that are important for myelin structure (CNP, MAG, MAL, MOG, MOBP, PMP22, PLLP, PLP1), were significantly downregulated when the symptoms worsened (Aston et al., 2005; Pantazatos et al., 2017).

Astrocytes Dysfunction in Depression

The most prevalent type of glial cells are called astrocytes, and depending on whether they are found in the grey or white brain matter, they are further classified into protoplasmic and fibrous subtypes (Oberheim et al., 2012). It is known that astrocytes support the structure of the brain, regulate pH and ion homeostasis in extracellular space, and play metabolic roles in the uptake of glucose and glutamate, storage of glycogen, clearance of K, supply of energy and nutrients to the neurons, clearance of neuronal waste, and monitoring of peripheral changes in blood composition through their end feet on blood vessels. In addition, a new research has shown that astrocytes are dynamic regulators of synaptic strength, synaptogenesis, and neuronal formation in the adult dentate gyrus (Haydon, 2001; Nedergaard et al., 2003). Therefore, changes in astrocyte structure are anticipated to have a significant functional impact on communication between neurons and glia, as well as on changes in hippocampal volume.

It is worth mentioning that the primary protein of astrocyte intermediate filaments and a particular marker for astrocytes is GFAP (Montgomery, 1994; Patanow et al., 1996). In both health and disease, astrocytes with GFAP positivity perform a variety of tasks, such as directing neuronal migration and producing extracellular matrix proteins and adhesion molecules (Coyle & Schwarcz, 2000; Fatemi et al., 2002a,b). In fact, increases in GFAP production could indicate neurodegeneration, reactive damage, or even astrogliosis (Coyle & Schwarcz, 2000; Fatemi et al., 2002a,b;). Indeed, many studies have found diminished GFAP expression in postmortem brain tissue from the hippocampus (M. B. Müller et al., 2001), PFC (Si et al., 2004; Webster et al., 2001), anterior cingulate, amygdala (Altshuler et al., 2010), and even cerebellum (Fatemi et al., 2004), indicating a unique role for astrocytes in the disease.

Moreover, recent studies showed that there has been mounting proof that altering astroglial function or development may be a factor in several neuropsychiatric illnesses (Molofsky et al., 2012). In fact, the expanding body of postmortem research points to an association between mood disorders and glial cell pathology, particularly astrocyte pathology (Sanacora et al., 2003). Thus, the astrocytic disease may be an early contributor to the underlying pathophysiology of mood disorders, as shown by the fact that the lowered DLPFC levels of GFAP were most pronounced in MDD participants 60 years of age and older (Miguel-Hidalgo et al., 2000). Research of depressed patients and completed suicides have revealed decreased expression of numerous elements crucial to the control and function of astrocytes. Indeed, SLC1A2, SLC1A3, and GLUL, the genes encoding for the two excitatory amino acid transporters (EAATs) that are primarily expressed in astrocytes and the important enzyme responsible for converting glutamate to glutamine in the astrocyte, were found to be significantly downregulated in the anterior cingulate cortex (area 24) and left DLPFC (areas 9 and 46) from subjects with MDD (Choudary et al., 2005). In a subsequent study, showed that the messenger RNA expression of several astrocyte-specific transcripts, including those encoding for GFAP, S100 calcium-binding protein, fibroblast growth factor receptor 3, gap junction proteins (connexins) Cx30 and Cx43, and aquaporin-4, was significantly reduced in the locus coeruleus of MDD patients (Bernard et al., 2011). In the OFC and the DLPFC of suicide completers, similar findings of downregulated Cx30 and Cx43 expression as well as lower expression of the astrocyte-specific tropomyosin-related kinase-B receptor (TrkB.1) isoform were identified (Ernst et al., 2009, 2011).

Additionally, the majority of research examining post-mortem brain tissues from adults with MDD finds that frontolimbic structures contain fewer astrocytic-like components (D. Cotter et al., 2002; Medina et al., 2016; Nagy et al., 2015; Öngür et al., 1998; Rajkowska et al., 1999; Torres-Platas et al., 2016). The claim that astrocyte alteration in frontolimbic areas is related to depression is supported by a variety of lines of evidence (Altshuler et al., 2010; L. Peng et al., 2015; Rajkowska & A Stockmeier, 2013). Alongside this, astrocytic shape changes, which are characterized by hypertrophy cell bodies (D. Cotter et al., 2002; Rajkowska et al., 1999; Torres-Platas et al., 2011, 2016), and a change in the density of certain astrocytic markers, like GFAP, in frontolimbic areas (Gittins & Harrison, 2011; Miguel-Hidalgo et al., 2000; Schlicht et al., 2007; Si et al., 2004), connexins (Ernst et al., 2011; Miguel-Hidalgo et al., 2014; Sun et al., 2012), aquaporin-4 (Rajkowska et al., 2013), GLT-1 and glutamine synthase (Choudary et al., 2005) and an increased release of S100β (Grabe et al., 2001; Schroeter et al., 2008). However, animal studies demonstrating that the selective death of frontocortical astrocytes with the gliotoxin L-aminoadipic acid is sufficient to cause a depressive-like phenotype establish a causal relationship between astrocytic dysfunction and depression (Banasr & Duman, 2008). Moreover, a depressive-like phenotype also develops when astrocytes are functionally inhibited, such as when synaptobrevin-2 is downregulated, which reduces astrocytic vesicular release (Cao et al., 2013), upon altering connexin-mediated gap-junctions (Sun et al., 2012), upon knocking out IP3-receptor type-2 (Cao et al., 2013) or aquaporin-4 (Kong et al., 2014) or inhibiting astrocytic glutamate transporters (Bechtholt-Gompf et al., 2010; John et al., 2012). Notably, astrocytic function is affected by the signaling systems assumed as traditional culprits of depression, namely glucocorticoids (Yin et al., 2013), BDNF (W.-X. Liu et al.,

2016; Ye et al., 2011), 5-HT (Hertz et al., 2015), NE (Madrigal et al., 2009; Pankratov & Lalo, 2015) or DA (Shao et al., 2013).

It is highly established that the most important excitatory neurotransmitter in the mammalian brain is glutamate. Glutamic acid-induced stimulation is in balance with the inhibitory effects of -aminobutyric acid (GABA) which is the brain primary inhibitory amino acid neurotransmitter. Recent studies found an imbalance between excitation and inhibition in the brains of patients suffering from mental illnesses such as anxiety and depression (Śmiałowska et al., 2013). It has been hypothesized that depression and anxiety are linked to rising glutamatergic excitement and falling GABAergic inhibition. Furthermore, the brains and cerebrospinal fluid (CSF) of patients with depression were shown to have higher glutamate levels (Hashimoto et al., 2007). As glial cells, especially astrocytes, play a very important role in the maintenance of the glutamate/GABA balance. Moreover, astrocytes have a key role in controlling the local concentration of ions and neuroactive chemicals, which is essential for maintaining CNS homeostasis (Largo et al., 1996). Indeed, numerous research teams discovered bidirectional communication between neurons and astrocytes in the 1990s, and the term "tripartite synapse" was proposed because astrocytes represent the third component of the synapse (Araque et al., 1999; Rodnight & Gottfried, 2013). Indeed, in the glutamatergic synapses, the most common type of synapses in the CNS, astrocytes regulate the levels of extracellular glutamate (Wierońska & Pilc, 2009), to maintain the balance between the excitatory and inhibitory neurotransmission there. In other words, synaptic NMDAR-mediated glutamatergic neurotransmission has positive consequences like synaptic plasticity, learning, and memory, but extrasynaptic NMDAR-mediated glutamatergic neurotransmission has negative effects like excitotoxicity (Hardingham & Bading, 2010). It is worth mentioning that, the majority of antidepressant medications therapeutic effects and the pathogenesis of depression both depend heavily on glutamatergic neurotransmission (Feyissa et al., 2009). Furthermore, treatments alleviating depressive symptoms can recover astrocytic function, such as selective serotonin reuptake inhibitors (SSRIs) (Czéh et al., 2006; Schipke et al., 2011) or electroconvulsive shock (Iwata et al., 2011), and some are even critically dependent on astrocytic function, such as fluoxetine (Kong et al., 2009) or deep brain stimulation (Etiévant et al., 2015). Thus, it has been proposed that antidepressants may act by reducing glutamate release and/or postsynaptic excitability (Paul & Skolnick, 2003). In fact, studies using NMDA receptor antagonists lend pharmacological support to the glutamatergic theory of depression. For instance, functional NMDA receptor antagonists have been shown to have an antidepressant-like effect for the first time by Trullas and Skolnick that functional antagonists of NMDA receptor exhibit antidepressant like activity (Trullas & Skolnick, 1990).

GLIAL CELLS ASSOCIATED INFLAMMATION AS A CONTRIBUTOR TO THE DEVELOPMENT OF DEPRESSION

It is worth mentioning that depression is also tightly associated with inflammation (Yirmiya et al., 2015). About 30 years ago, the first indication that inflammation and depression are related was discovered. Indeed, recent research has shown a connection between inflammation and depression. Accordingly, patients with depression have elevated levels of peripheral inflammatory markers (Howren et al., 2009; Raison et al., 2006). On the other hand, promoting inflammation results in a depressive-like illness behavior (Dantzer et al., 2008). Moreover, depression is frequently experienced by patients with autoimmune and inflammatory illnesses (Kiecolt-Glaser et al., 2015). Thus, when someone is under social stress, their natural immune system can become active (Bullmore, 2018). In fact, recent research indicated that stress primarily activates the immune system in two ways, including the hypothalamic-pituitary-adrenal (HPA) axis and the sympathetic nervous system, both of which have immunomodulatory properties (Miller & Raison, 2016). Moreover, the expression of inflammatory cytokines, such as IFN-, IL-1, IL-2, IL-3, IL-5, IL-8, IL-9, IL-12A, IL-13, IL-15, and TNF, is also significantly altered in depressed patients (Shelton et al., 2011)Bullmore, 2018; Dantzer et al., 2008; Miller & Raison, 2016). In addition, in depression activated microglia are a primary source of the release of proinflammatory cytokines such IL-1, IL-6, and IL-12 (Enache et al., 2019; Miller & Raison, 2016).

GLIAL CELLS, SYNAPTIC PLASTICITY AND DEPRESSION

Recently, abnormalities of glial cells function have been shown to contribute to the impairments of structural plasticity and overall pathophysiology of mood disorders (D. R. Cotter et al., 2001; Coyle & Schwarcz, 2000). Numerous theories have been investigated to explain depressive-related behaviours, including chronic stress, failure of adult neurogenesis, altered neuroplasticity, dysfunction of monoaminergic systems, and hereditary variables (HK, 2001; Nestler et al., 2002). Indeed, recent studies suggest that a major underlying mechanism in the development of the illness may be defective neurons and glial plasticity. It is worth mentioning that glial cells have been shown to play a role in at least certain types of depression in the modulation of brain neuron-glia networks. In fact, Rial et al. investigate the many synaptic connections between glial and neuronal cells, which can be compromised in depressive-like behaviors (Rial et al., 2016). Additionally, Jo et al. discussed the importance of glial-mediated immune modulation through cytokine signaling that may trigger depressive episodes (Jo et al., 2015).

Since astrocyte networks contain gap junction communication systems, it is reasonable to assume that these systems will exhibit some plasticity. As a result, these systems will be susceptible to alterations brought on by pathological conditions in both glial cell populations and other populations of brain cells (Medina et al., 2016). However, the majority of these structural plasticity alterations are temporary and typically reversible after enough recovery intervals, after ceasing exposure to stress or corticosteroids, or as a result of antidepressant therapy (Heine et al., 2004; Lucassen et al., 2004; Lucassen, Vollmann-Honsdorf, et al., 2001). In light of the research findings, it may be possible to modulate brain networks using these novel types of neuro-glia communication, and manipulation of these networks may lead to the development of autologous therapies for depression (Fig. 2) (Oliveira et al., 2016).

MODELS OF GLIAL / NEURONAL CELL PATHOLOGY IN DEPRESSION

Based on all the results on glial and neuronal pathology in numerous research, a hypothetical system for the advancement of cell disease in depression has been developed (Fig. 1). The purpose of this model is to explain how cellular modifications, glutamate, glucocorticoids, and neurotrophic factors interact to cause depression. Hence, it is hypothesised that in early stages of depressive disorder, a confluence of hereditary and environmental (such as stressor) variables may result in the pathology of glial cells, which might lead later to neurons dysfunction. Moreover, early glial disease may be caused by a variety of reasons, including a lack of neurotrophic and angiogenic factors as well as stress-related increases in glucocorticoids. In animal studies, glucocorticoids in excess and psychosocial stress decrease the growth

Figure 2. Schematic representation of cellular pathways by which glial cells may influence neuronal plasticity

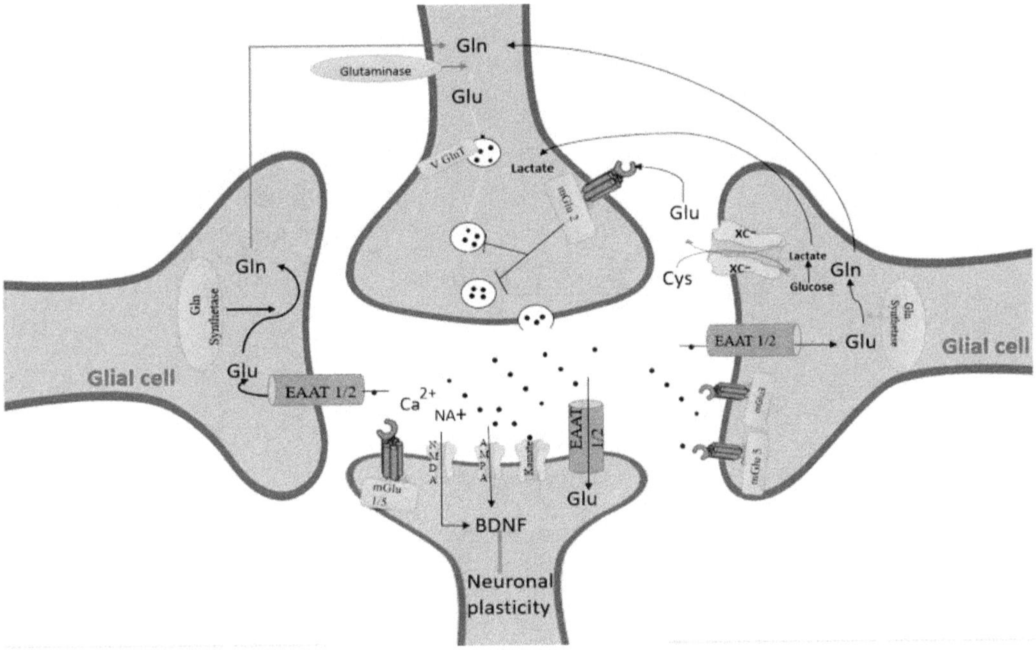

of astrocytes and OLs, and in depressed patients, glucocorticoids change the expression of GFAP (M. B. Müller et al., 2001). Additionally, in animal stress models linked to depression, decreases in neurotrophic and angiogenic factors were also seen. Indeed, by boosting the expression of neurotrophic and glia-derived substances as well as other glial markers, persistent antidepressant treatment or electroconvulsive convulsions can correct this stress-induced deficiency (Angelucci et al., 2002; Chen et al., 2001; Duman & Monteggia, 2006; Malberg et al., 2000; Newton et al., 2003, 2006) and hypothesizes that glia could be used as new therapeutic targets for antidepressant treatment.

Furthermore, it is hypothesized that the loss of glial cells in the early stages of depression may result in glial glutamate transporter expression being lessened, which would then result in less glutamate being taken up from synaptic clefts (Kugaya & Sanacora, 2005). It is known in a depressed individual's; an excess of extracellular glutamate may contribute to neuronal damage or neuronal death. Thus, it is conceivable that depression in older people may be caused by neuronal damage that leads to glial proliferation. Indeed, glia proliferation and reductions in neuronal density have been seen in the OFC of older depressed individuals as compared to younger depressed individuals (Rajkowska & Miguel-Hidalgo, 2007). Neuroimaging and postmortem studies lend credence to the idea that younger depressive individuals have fewer glial cells, which decrease the number of glial glutamate transporters while increasing the amount of extracellular glutamate. Additionally, according to magnetic resonance spectroscopy results, there are reports of elevated glutamate levels in depressed patients (Kugaya & Sanacora, 2005; Sanacora et al., 2003), as well as decreased gene expression for glial markers of glutamate neurotransmission in homogenates from the postmortem frontal brain of depressed individuals (Fig. 3) (Choudary et al., 2005).

Figure 3. Hypothetical scheme of the progression of cell pathology in depression from young adulthood to old age
(Rajkowska & Miguel-Hidalgo, 2008)

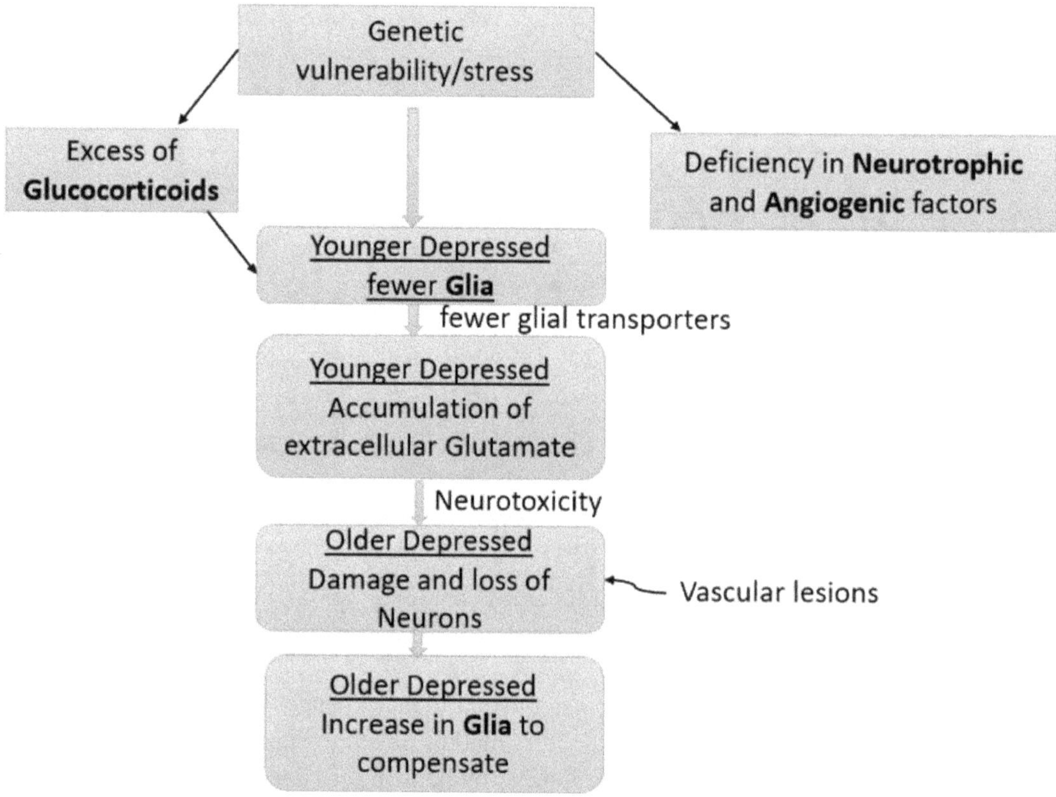

TARGETING GLIAL CELLS IN THE TREATMENT OF DEPRESSION

Glial cells can be targeted in order to treat depression. Broadly speaking, changes in astrocytes, which regulate a variety of brain circuits and behavioral features, have also been connected to depression treatment (Nagai et al., 2021). Moreover, research conducted in vitro and in vivo on astrocytes revealed the presence of the serotonin transporter SERT, the noradrenaline transporter NET, and the uptake 2 transporter (Inazu et al., 2003; Schildkraut & Mooney, 2004). The capacity of astrocytes to detect monoamine neurotransmission and reuptake DA, NA, or 5-HT is essential because the majority of current pharmaceutical antidepressant treatments target these neurotransmitters. Astrocytic receptor expression and function are altered when antidepressants cause a surge in NA in the synaptic cleft, according to multiple studies. In the rat forebrain, for instance, long-term administration of the tricyclic antidepressant imipramine reduces the density of 1 adrenergic receptors at the membrane of astrocytes a finding that was once thought to be a characteristic of the antidepressant response (Sapena et al., 1996) . Additionally, the extracellular monoamine transporter uptake 2 can retake the NA that is present in the synaptic cleft into neighbouring glia. Once in astrocytes, catechol O-methyltransferase (COMT), a potent inhibitor of this transporter, converts NA into normetanephrine (Männistö & Kaakkola, 1999). Similarly, SSRIs and tricyclic antidepressants downregulate the expression of astrocytic serotonin sodium-dependent transporter

SERT, which increases the concentration of 5-HT in the synaptic cleft. Consequently, astrocytes actively contribute to the rise in extracellular NA and 5-HT concentrations brought on by antidepressants of the NA and/or 5-HT reuptake inhibitor class (Schildkraut & Mooney, 2004). In addition, it has been reported that fluoxetine reversed psychosocial stress-induced decreases in the number of astrocytes (Czéh et al., 2006). Moreover, another found that riluzole can prevent the decrease in GFAP mRNA expression in the rat prefrontal cortex following exposure to CUS (Naskar et al., 2022). In fact, citalopram, fluoxetine, and SSRI antidepressants, are believed to have antidepressant effects because they can activate calcium signaling in astrocytes in the PFC (Schipke et al., 2011).

Indeed, TCAs and SSRIs have also been shown to have an impact on neuroinflammation and microglial activation. Minocycline, melation, FCPR16, pioglitazone, iptakalim, and caffeine are among the other substances that affect microglial polarisation and depressive-like behaviour (Nieto-Quero et al., 2021). Moreover, tricyclics, such as imipramine and clomipramine, decreased the quantity of activated hippocampus microglia, thereby reversing the depressed behaviours brought on by LPS (Lipopolysaccharide), LH (Learned helplessness), and CSDS (chronic social defeat stress). The activation of anti-inflammatory microglia by these conventional antidepressants results in the release of anti-inflammatory cytokines and neurotrophic factors, indicating that clinical antidepressants may have an anti-inflammatory effect (Gong et al., 2019; Iwata et al., 2016; Jiang et al., 2020) . The degree of cytokines is highly connected with how well patients respond to antidepressant therapy. For instance, patients who are resistant to treatment typically have higher levels of IL-6 and TNF-α than those who respond (Lanquillon et al., 2000). However, some findings supported contradictory theories that antidepressants occasionally raise the amount of inflammation in the brain. For example, citalopram treatment increased TNF-α and IFN-γ levels in the PFC, whereas ibuprofen, an anti-inflammatory drug, counteracted these effects(Warner-Schmidt et al., 2011). The monoamine oxidase inhibitor phenelzine has been demonstrated to upregulate the production of iNOS, TNF-α, and IL-6 in LPS-treated BV-2 cells, hence augmenting the microglia-mediated immune responses.

It has been confirmed that in LPS-activated primary microglia cells, phenelzine also elevated the levels of NO, TNF-α, and IL-6 through the NF-κB signalling pathway (H.-S. Chung et al., 2012). Ketamine has demonstrated potential antidepressant effects through its ability to reduce neuroinflammation mediated by microglia both in vivo and in vitro (Tan et al., 2017). The fast and long-lasting antidepressant effects of (R)-ketamine, an isomer of ketamine, were prevented by partial depletion of microglia with PLX3397, indicating that microglial activation may play a role in the antidepressant effects of ketamine (Tan et al., 2017). Minocycline is a potential agent for the treatment of depression (Pae et al., 2008) . Chronic minocycline treatment inhibited microglia and HPA axis hyperactivity, hence clearly alleviating depressive-like symptoms. In stress models, minocycline's potent neuroprotective effects were primarily achieved via upregulating BDNF-mediated synaptic plasticity and regulating the pro- and anti-inflammatory balance in the PFC and HIP (C. Zhang et al., 2019). Meanwhile, long-term minocycline therapy reversed CMS-induced alterations in behaviour, hippocampus LTP, CD11b expression, NLRP3 inflammasome, BDNF, and p-GluR1 levels (Z. Peng et al., 2020). All this compelling evidence suggest the involvement of glial cell in depression. Thus, a deeper comprehension of the pathophysiological mechanisms underlying the interactions between these and neurons in depression may hold the key to its prevention, and the development of new medications.

CONCLUSION

Overall, research on the role of glial cells in depression has been increased recently. Importantly, new research indicates that glial cells, notably astrocytes and microglia, are the most important players in the pathophysiology of depression. It is worth mentioning that the precise function of glial cells in depression is still not fully elucidated, and more research is required to comprehend the intricate relationships between these cells and neurons, as well as other contributors in the pathogenesis of depressive disorders. Hence, there is growing interest in the idea of using glial cell targeting as a treatment for depression. Glial cell research will continue to provide new and improved treatment options for depression and associated symptoms.

REFERENCES

Airaksinen, M. S., & Saarma, M. (2002). The GDNF family: Signalling, biological functions and therapeutic value. *Nature Reviews. Neuroscience*, *3*(5), 383–394. doi:10.1038/nrn812 PMID:11988777

Altshuler, L. L., Abulseoud, O. A., Foland-Ross, L., Bartzokis, G., Chang, S., Mintz, J., Hellemann, G., & Vinters, H. V. (2010). Amygdala astrocyte reduction in subjects with major depressive disorder but not bipolar disorder. *Bipolar Disorders*, *12*(5), 541–549. doi:10.1111/j.1399-5618.2010.00838.x PMID:20712756

Angelucci, F., Aloe, L., Jiménez-Vasquez, P., & Mathé, A. A. (2002). Electroconvulsive stimuli alter the regional concentrations of nerve growth factor, brain-derived neurotrophic factor, and glial cell line-derived neurotrophic factor in adult rat brain. *The Journal of ECT*, *18*(3), 138–143. doi:10.1097/00124509-200209000-00005 PMID:12394532

Araque, A., Parpura, V., Sanzgiri, R. P., & Haydon, P. G. (1999). Tripartite synapses: Glia, the unacknowledged partner. *Trends in Neurosciences*, *22*(5), 208–215. doi:10.1016/S0166-2236(98)01349-6 PMID:10322493

Aston, C., Jiang, L., & Sokolov, B. P. (2005). Transcriptional profiling reveals evidence for signaling and oligodendroglial abnormalities in the temporal cortex from patients with major depressive disorder. *Molecular Psychiatry*, *10*(3), 309–322. doi:10.1038j.mp.4001565 PMID:15303102

Autry, A. E., & Monteggia, L. M. (2012). Brain-derived neurotrophic factor and neuropsychiatric disorders. *Pharmacological Reviews*, *64*(2), 238–258. doi:10.1124/pr.111.005108 PMID:22407616

Banasr, M., & Duman, R. S. (2008). Glial loss in the prefrontal cortex is sufficient to induce depressive-like behaviors. *Biological Psychiatry*, *64*(10), 863–870. doi:10.1016/j.biopsych.2008.06.008 PMID:18639237

Baumann, N., & Pham-Dinh, D. (2001). Biology of oligodendrocyte and myelin in the mammalian central nervous system. *Physiological Reviews*, *81*(2), 871–927. doi:10.1152/physrev.2001.81.2.871 PMID:11274346

Bechtholt-Gompf, A. J., Walther, H. V., Adams, M. A., Carlezon, W. A. Jr, Öngür, D., & Cohen, B. M. (2010). Blockade of astrocytic glutamate uptake in rats induces signs of anhedonia and impaired spatial memory. *Neuropsychopharmacology*, *35*(10), 2049–2059. doi:10.1038/npp.2010.74 PMID:20531459

Berent, D., Macander, M., Szemraj, J., Orzechowska, A., & Galecki, P. (2014). Vascular endothelial growth factor A gene expression level is higher in patients with major depressive disorder and not affected by cigarette smoking, hyperlipidemia or treatment with statins. *Acta Neurobiologiae Experimentalis*, *74*(1), 82–90. PMID:24718046

Bernard, R., Kerman, I. A., Thompson, R. C., Jones, E. G., Bunney, W. E., Barchas, J. D., Schatzberg, A. F., Myers, R. M., Akil, H., & Watson, S. J. (2011). Altered expression of glutamate signaling, growth factor, and glia genes in the locus coeruleus of patients with major depression. *Molecular Psychiatry*, *16*(6), 634–646. doi:10.1038/mp.2010.44 PMID:20386568

Blair-West, G. W., Mellsop, G. W., & Eyeson-Annan, M. L. (1997). Down-rating lifetime suicide risk in major depression. *Acta Psychiatrica Scandinavica*, *95*(3), 259–263. doi:10.1111/j.1600-0447.1997.tb09629.x PMID:9111861

Bocchio-Chiavetto, L., Bagnardi, V., Zanardini, R., Molteni, R., Gabriela Nielsen, M., Placentino, A., Giovannini, C., Rillosi, L., Ventriglia, M., Riva, M. A., & Gennarelli, M. (2010). Serum and plasma BDNF levels in major depression: A replication study and meta-analyses. *The World Journal of Biological Psychiatry*, *11*(6), 763–773. doi:10.3109/15622971003611319 PMID:20334574

Boku, S., Nakagawa, S., Takamura, N., Kato, A., Takebayashi, M., Hisaoka-Nakashima, K., Omiya, Y., Inoue, T., & Kusumi, I. (2013). GDNF facilitates differentiation of the adult dentate gyrus-derived neural precursor cells into astrocytes via STAT3. *Biochemical and Biophysical Research Communications*, *434*(4), 779–784. doi:10.1016/j.bbrc.2013.04.011 PMID:23603362

Brigitta, B. (2002). Pathophysiology of depression and mechanisms of treatment. *Dialogues in Clinical Neuroscience*, *4*(1), 7–20. doi:10.31887/DCNS.2002.4.1/bbondy PMID:22033824

Bullmore, E. (2018). Inflamed depression. *Lancet*, *392*(10154), 1189–1190. doi:10.1016/S0140-6736(18)32356-0 PMID:30712704

Cao, X., Li, L.-P., Wang, Q., Wu, Q., Hu, H.-H., Zhang, M., Fang, Y.-Y., Zhang, J., Li, S.-J., Xiong, W.-C., Yan, H.-C., Gao, Y.-B., Liu, J.-H., Li, X.-W., Sun, L.-R., Zeng, Y.-N., Zhu, X.-H., & Gao, T.-M. (2013). Astrocyte-derived ATP modulates depressive-like behaviors. *Nature Medicine*, *19*(6), 773–777. doi:10.1038/nm.3162 PMID:23644515

Capani, F., Quarracino, C., Caccuri, R., & Sica, R. E. P. (2016). Astrocytes as the main players in primary degenerative disorders of the human central nervous system. *Frontiers in Aging Neuroscience*, *8*, 45. doi:10.3389/fnagi.2016.00045 PMID:26973519

Chen, B., Dowlatshahi, D., MacQueen, G. M., Wang, J.-F., & Young, L. T. (2001). Increased hippocampal BDNF immunoreactivity in subjects treated with antidepressant medication. *Biological Psychiatry*, *50*(4), 260–265. doi:10.1016/S0006-3223(01)01083-6 PMID:11522260

Choudary, P. V., Molnar, M., Evans, S. J., Tomita, H., Li, J. Z., Vawter, M. P., Myers, R. M., Bunney, W. E. Jr, Akil, H., Watson, S. J., & Jones, E. G. (2005). Altered cortical glutamatergic and GABAergic signal transmission with glial involvement in depression. *Proceedings of the National Academy of Sciences of the United States of America*, *102*(43), 15653–15658. doi:10.1073/pnas.0507901102 PMID:16230605

Christophersen, N. S., Grønborg, M., Petersen, T. N., Fjord-Larsen, L., Jørgensen, J. R., Juliusson, B., Blom, N., Rosenblad, C., & Brundin, P. (2007). Midbrain expression of Delta-like 1 homologue is regulated by GDNF and is associated with dopaminergic differentiation. *Experimental Neurology, 204*(2), 791–801. doi:10.1016/j.expneurol.2007.01.014 PMID:17320866

Chung, H.-S., Kim, H., & Bae, H. (2012). Phenelzine (monoamine oxidase inhibitor) increases production of nitric oxide and proinflammatory cytokines via the NF-κB pathway in lipopolysaccharide-activated microglia cells. *Neurochemical Research, 37*(10), 2117–2124. doi:10.100711064-012-0833-y PMID:22763802

Chung, Y. C., Kim, S. R., Park, J.-Y., Chung, E. S., Park, K. W., Won, S. Y., Bok, E., Jin, M., Park, E. S., & Yoon, S.-H. (2011). Fluoxetine prevents MPTP-induced loss of dopaminergic neurons by inhibiting microglial activation. *Neuropharmacology, 60*(6), 963–974. doi:10.1016/j.neuropharm.2011.01.043 PMID:21288472

Cole, J., Chaddock, C. A., Farmer, A. E., Aitchison, K. J., Simmons, A., McGuffin, P., & Fu, C. H. Y. (2012). White matter abnormalities and illness severity in major depressive disorder. *The British Journal of Psychiatry, 201*(1), 33–39. doi:10.1192/bjp.bp.111.100594 PMID:22576724

Cotter, D., Mackay, D., Chana, G., Beasley, C., Landau, S., & Everall, I. P. (2002). Reduced neuronal size and glial cell density in area 9 of the dorsolateral prefrontal cortex in subjects with major depressive disorder. *Cerebral Cortex (New York, N.Y.), 12*(4), 386–394. doi:10.1093/cercor/12.4.386 PMID:11884354

Cotter, D., Mackay, D., Landau, S., Kerwin, R., & Everall, I. (2001). Reduced glial cell density and neuronal size in the anterior cingulate cortex in major depressive disorder. *Archives of General Psychiatry, 58*(6), 545–553. doi:10.1001/archpsyc.58.6.545 PMID:11386983

Cotter, D. R., Pariante, C. M., & Everall, I. P. (2001). Glial cell abnormalities in major psychiatric disorders: The evidence and implications. *Brain Research Bulletin, 55*(5), 585–595. doi:10.1016/S0361-9230(01)00527-5 PMID:11576755

Coyle, J. T., & Schwarcz, R. (2000). Mind glue: Implications of glial cell biology for psychiatry. *Archives of General Psychiatry, 57*(1), 90–93. doi:10.1001/archpsyc.57.1.90 PMID:10632240

Czéh, B., & Lucassen, P. J. (2007). What causes the hippocampal volume decrease in depression? Are neurogenesis, glial changes and apoptosis implicated? *European Archives of Psychiatry and Clinical Neuroscience, 257*(5), 250–260. doi:10.100700406-007-0728-0 PMID:17401728

Czéh, B., Simon, M., Schmelting, B., Hiemke, C., & Fuchs, E. (2006). Astroglial plasticity in the hippocampus is affected by chronic psychosocial stress and concomitant fluoxetine treatment. *Neuropsychopharmacology, 31*(8), 1616–1626. doi:10.1038j.npp.1300982 PMID:16395301

Dantzer, R., O'connor, J. C., Freund, G. G., Johnson, R. W., & Kelley, K. W. (2008). From inflammation to sickness and depression: When the immune system subjugates the brain. *Nature Reviews. Neuroscience, 9*(1), 46–56. doi:10.1038/nrn2297 PMID:18073775 de Azevedo Cardoso, T., Mondin, T. C., Wiener, C. D., Marques, M. B., Fucolo, B., Pinheiro, R. T., de Souza, L. D. M., da Silva, R. A., Jansen, K., & Oses, J. P. (2014). Neurotrophic factors, clinical features and gender differences in depression. *Neurochemical Research*, *39*(8), 1571–1578. doi:10.100711064-014-1349-4 PMID:24899094

Diniz, B. S., Teixeira, A. L., Miranda, A. S., Talib, L. L., Gattaz, W. F., & Forlenza, O. V. (2012). Circulating Glial-derived neurotrophic factor is reduced in late-life depression. *Journal of Psychiatric Research*, *46*(1), 135–139. doi:10.1016/j.jpsychires.2011.09.007 PMID:21963280

Ducray, A., Krebs, S. H., Schaller, B., Seiler, R. W., Meyer, M., & Widmer, H. R. (2006). GDNF family ligands display distinct action profiles on cultured GABAergic and serotonergic neurons of rat ventral mesencephalon. *Brain Research*, *1069*(1), 104–112. doi:10.1016/j.brainres.2005.11.056 PMID:16380100

Duman, R. S., & Li, N. (2012). A neurotrophic hypothesis of depression: Role of synaptogenesis in the actions of NMDA receptor antagonists. *Philosophical Transactions of the Royal Society of London. Series B, Biological Sciences*, *367*(1601), 2475–2484. doi:10.1098/rstb.2011.0357 PMID:22826346

Duman, R. S., & Monteggia, L. M. (2006). A neurotrophic model for stress-related mood disorders. *Biological Psychiatry*, *59*(12), 1116–1127. doi:10.1016/j.biopsych.2006.02.013 PMID:16631126

Dwivedi, Y., Rizavi, H. S., Conley, R. R., Roberts, R. C., Tamminga, C. A., & Pandey, G. N. (2003). Altered gene expression of brain-derived neurotrophic factor and receptor tyrosine kinase B in postmortem brain of suicide subjects. *Archives of General Psychiatry*, *60*(8), 804–815. doi:10.1001/archpsyc.60.8.804 PMID:12912764

Ebersole, T. A., Chen, Q. I., Justice, M. J., & Artzt, K. (1996). The quaking gene product necessary in embryogenesis and myelination combines features of RNA binding and signal transduction proteins. *Nature Genetics*, *12*(3), 260–265. doi:10.1038/ng0396-260 PMID:8589716

Edgar, N., & Sibille, E. (2012). A putative functional role for oligodendrocytes in mood regulation. *Translational Psychiatry*, *2*(5), e109–e109. doi:10.1038/tp.2012.34 PMID:22832953

Enache, D., Pariante, C. M., & Mondelli, V. (2019). Markers of central inflammation in major depressive disorder: A systematic review and meta-analysis of studies examining cerebrospinal fluid, positron emission tomography and post-mortem brain tissue. *Brain, Behavior, and Immunity*, *81*, 24–40. doi:10.1016/j.bbi.2019.06.015 PMID:31195092

Ernst, C., Chen, E. S., & Turecki, G. (2009). Histone methylation and decreased expression of TrkB.T1 in orbital frontal cortex of suicide completers. *Molecular Psychiatry*, *14*(9), 830–832. doi:10.1038/mp.2009.35 PMID:19696771

Ernst, C., Nagy, C., Kim, S., Yang, J. P., Deng, X., Hellstrom, I. C., Choi, K. H., Gershenfeld, H., Meaney, M. J., & Turecki, G. (2011). Dysfunction of astrocyte connexins 30 and 43 in dorsal lateral prefrontal cortex of suicide completers. *Biological Psychiatry*, *70*(4), 312–319. doi:10.1016/j.biopsych.2011.03.038 PMID:21571253

Eroglu, C., & Barres, B. A. (2010). Regulation of synaptic connectivity by glia. *Nature*, *468*(7321), 223–231. doi:10.1038/nature09612 PMID:21068831

Etiévant, A., Oosterhof, C., Bétry, C., Abrial, E., Novo-Perez, M., Rovera, R., Scarna, H., Devader, C., Mazella, J., Wegener, G., Sánchez, C., Dkhissi-Benyahya, O., Gronfier, C., Coizet, V., Beaulieu, J. M., Blier, P., Lucas, G., & Haddjeri, N. (2015). Astroglial control of the antidepressant-like effects of prefrontal cortex deep brain stimulation. *EBioMedicine*, *2*(8), 898–908. doi:10.1016/j.ebiom.2015.06.023 PMID:26425697

Färber, K., Pannasch, U., & Kettenmann, H. (2005). Dopamine and noradrenaline control distinct functions in rodent microglial cells. *Molecular and Cellular Neurosciences*, *29*(1), 128–138. doi:10.1016/j.mcn.2005.01.003 PMID:15866053

Fatemi, S. H., Emamian, E. S., Sidwell, R. W., Kist, D. A., Stary, J. M., Earle, J. A., & Thuras, P. (2002). Human influenza viral infection in utero alters glial fibrillary acidic protein immunoreactivity in the developing brains of neonatal mice. *Molecular Psychiatry*, *7*(6), 633–640. doi:10.1038j.mp.4001046 PMID:12140787

Fatemi, S. H., Laurence, J. A., Araghi-Niknam, M., Stary, J. M., Schulz, S. C., Lee, S., & Gottesman, I. I. (2004). Glial fibrillary acidic protein is reduced in cerebellum of subjects with major depression, but not schizophrenia. *Schizophrenia Research*, *69*(2–3), 317–323. doi:10.1016/j.schres.2003.08.014 PMID:15469203

Fenn, K. M., Evans, S. B., McCorkle, R., DiGiovanna, M. P., Pusztai, L., Sanft, T., Hofstatter, E. W., Killelea, B. K., Knobf, M. T., Lannin, D. R., Abu-Khalaf, M., Horowitz, N. R., & Chagpar, A. B. (2014). Impact of financial burden of cancer on survivors' quality of life. *Journal of Oncology Practice / American Society of Clinical Oncology*, *10*(5), 332–338. doi:10.1200/JOP.2013.001322 PMID:24865220

Feyissa, A. M., Chandran, A., Stockmeier, C. A., & Karolewicz, B. (2009). Reduced levels of NR2A and NR2B subunits of NMDA receptor and PSD-95 in the prefrontal cortex in major depression. *Progress in Neuro-Psychopharmacology & Biological Psychiatry*, *33*(1), 70–75. doi:10.1016/j.pnpbp.2008.10.005 PMID:18992785

Fornaro, M., Rocchi, G., Escelsior, A., Contini, P., Ghio, M., Colicchio, S., De Berardis, D., Amore, M., Fornaro, P., & Martino, M. (2013). VEGF plasma level variations in duloxetine-treated patients with major depression. *Journal of Affective Disorders*, *151*(2), 590–595. doi:10.1016/j.jad.2013.06.055 PMID:23871390

Gao, Y., Ma, J., Tang, J., Liang, X., Huang, C., Wang, S., Chen, L., Wang, F., Tan, C., Chao, F., Zhang, L., Qiu, X., Luo, Y.-M., Xiao, Q., Du, L., Xiao, Q., & Tang, Y. (2017). White matter atrophy and myelinated fiber disruption in a rat model of depression. *The Journal of Comparative Neurology*, *525*(8), 1922–1933. doi:10.1002/cne.24178 PMID:28118485

Ginhoux, F., Greter, M., Leboeuf, M., Nandi, S., See, P., Gokhan, S., Mehler, M. F., Conway, S. J., Ng, L. G., Stanley, E. R., Samokhvalov, I. M., & Merad, M. (2010). Fate mapping analysis reveals that adult microglia derive from primitive macrophages. *Science*, *330*(6005), 841–845. doi:10.1126cience.1194637 PMID:20966214

Gittins, R. A., & Harrison, P. J. (2011). A morphometric study of glia and neurons in the anterior cingulate cortex in mood disorder. *Journal of Affective Disorders, 133*(1–2), 328–332. doi:10.1016/j.jad.2011.03.042 PMID:21497910

Gomes, C., Ferreira, R., George, J., Sanches, R., Rodrigues, D. I., Gonçalves, N., & Cunha, R. A. (2013). Activation of microglial cells triggers a release of brain-derived neurotrophic factor (BDNF) inducing their proliferation in an adenosine A2A receptor-dependent manner: A2A receptor blockade prevents BDNF release and proliferation of microglia. *Journal of Neuroinflammation, 10*(1), 1–13. doi:10.1186/1742-2094-10-16 PMID:23363775

Gong, W., Zhang, S., Zong, Y., Halim, M., Ren, Z., Wang, Y., Ma, Y., Li, B., Ma, L., Zhou, G., Yu, J., Zhang, J., & Liu, Q. (2019). Involvement of the microglial NLRP3 inflammasome in the anti-inflammatory effect of the antidepressant clomipramine. *Journal of Affective Disorders, 254*, 15–25. doi:10.1016/j.jad.2019.05.009 PMID:31082627

Grabe, H. J., Ahrens, N., Rose, H.-J., Kessler, C., & Freyberger, H. J. (2001). Neurotrophic factor S100beta in major depression. *Neuropsychobiology, 44*(2), 88–90. doi:10.1159/000054922 PMID:11490177

Hardingham, G. E., & Bading, H. (2010). Synaptic versus extrasynaptic NMDA receptor signalling: Implications for neurodegenerative disorders. *Nature Reviews. Neuroscience, 11*(10), 682–696. doi:10.1038/nrn2911 PMID:20842175

Hashimoto, K., Sawa, A., & Iyo, M. (2007). Increased levels of glutamate in brains from patients with mood disorders. *Biological Psychiatry, 62*(11), 1310–1316. doi:10.1016/j.biopsych.2007.03.017 PMID:17574216

Haydon, P. G. (2001). GLIA: Listening and talking to the synapse. *Nature Reviews. Neuroscience, 2*(3), 185–193. doi:10.1038/35058528 PMID:11256079

Heine, V. M., Maslam, S., Zareno, J., Joëls, M., & Lucassen, P. J. (2004). Suppressed proliferation and apoptotic changes in the rat dentate gyrus after acute and chronic stress are reversible. *The European Journal of Neuroscience, 19*(1), 131–144. doi:10.1046/j.1460-9568.2003.03100.x PMID:14750971

Hertz, L., Rothman, D. L., Li, B., & Peng, L. (2015). Chronic SSRI stimulation of astrocytic 5-HT2B receptors change multiple gene expressions/editings and metabolism of glutamate, glucose and glycogen: A potential paradigm shift. *Frontiers in Behavioral Neuroscience, 9*, 25. PMID:25750618

Hochstrasser, T., Ullrich, C., Sperner-Unterweger, B., & Humpel, C. (2011). Inflammatory stimuli reduce survival of serotonergic neurons and induce neuronal expression of indoleamine 2, 3-dioxygenase in rat dorsal raphe nucleus organotypic brain slices. *Neuroscience, 184*, 128–138. doi:10.1016/j.neuroscience.2011.03.070 PMID:21501664

Howren, M. B., Lamkin, D. M., & Suls, J. (2009). Associations of depression with C-reactive protein, IL-1, and IL-6: A meta-analysis. *Psychosomatic Medicine, 71*(2), 171–186. doi:10.1097/PSY.0b013e3181907c1b PMID:19188531

Inazu, M., Takeda, H., & Matsumiya, T. (2003). Functional expression of the norepinephrine transporter in cultured rat astrocytes. *Journal of Neurochemistry, 84*(1), 136–144. doi:10.1046/j.1471-4159.2003.01514.x PMID:12485410

Irala, D., Bonafina, A., Fontanet, P. A., Alsina, F. C., Paratcha, G., & Ledda, F. (2016). The GDNF-GFRα1 complex promotes the development of hippocampal dendritic arbors and spines via NCAM. *Development (Cambridge, England)*, *143*(22), 4224–4235. doi:10.1242/dev.140350 PMID:27707798

Iwata, M., Ishida, H., Kaneko, K., & Shirayama, Y. (2016). Learned helplessness activates hippocampal microglia in rats: A potential target for the antidepressant imipramine. *Pharmacology, Biochemistry, and Behavior*, *150*, 138–146. doi:10.1016/j.pbb.2016.10.005 PMID:27769904

Iwata, M., Shirayama, Y., Ishida, H., Hazama, G., & Nakagome, K. (2011). Hippocampal astrocytes are necessary for antidepressant treatment of learned helplessness rats. *Hippocampus*, *21*(8), 877–884. doi:10.1002/hipo.20803 PMID:20572198

Jaumotte, J. D., & Zigmond, M. J. (2014). Comparison of GDF5 and GDNF as neuroprotective factors for postnatal dopamine neurons in ventral mesencephalic cultures. *Journal of Neuroscience Research*, *92*(11), 1425–1433. doi:10.1002/jnr.23425 PMID:24916473

Jiang, N., Lv, J., Wang, H., Huang, H., Wang, Q., Lu, C., Zeng, G., & Liu, X. (2020). Ginsenoside Rg1 ameliorates chronic social defeat stress-induced depressive-like behaviors and hippocampal neuroinflammation. *Life Sciences*, *252*, 117669. doi:10.1016/j.lfs.2020.117669 PMID:32298740

Jo, W. K., Zhang, Y., Emrich, H. M., & Dietrich, D. E. (2015). Glia in the cytokine-mediated onset of depression: Fine tuning the immune response. *Frontiers in Cellular Neuroscience*, *9*(JULY), 1–13. doi:10.3389/fncel.2015.00268 PMID:26217190

John, C. S., Smith, K. L., Van't Veer, A., Gompf, H. S., Carlezon, W. A. Jr, Cohen, B. M., Öngür, D., & Bechtholt-Gompf, A. J. (2012). Blockade of astrocytic glutamate uptake in the prefrontal cortex induces anhedonia. *Neuropsychopharmacology*, *37*(11), 2467–2475. doi:10.1038/npp.2012.105 PMID:22739467

Kiecolt-Glaser, J. K., Derry, H. M., & Fagundes, C. P. (2015). Inflammation: Depression fans the flames and feasts on the heat. *The American Journal of Psychiatry*, *172*(11), 1075–1091. doi:10.1176/appi.ajp.2015.15020152 PMID:26357876

Klempan, T. A., Ernst, C., Deleva, V., Labonte, B., & Turecki, G. (2009). Characterization of QKI gene expression, genetics, and epigenetics in suicide victims with major depressive disorder. *Biological Psychiatry*, *66*(9), 824–831. doi:10.1016/j.biopsych.2009.05.010 PMID:19545858

Kong, H., Sha, L., Fan, Y., Xiao, M., Ding, J., Wu, J., & Hu, G. (2009). Requirement of AQP4 for antidepressive efficiency of fluoxetine: Implication in adult hippocampal neurogenesis. *Neuropsychopharmacology*, *34*(5), 1263–1276. doi:10.1038/npp.2008.185 PMID:18923397

Krabbe, G., Matyash, V., Pannasch, U., Mamer, L., Boddeke, H. W. G. M., & Kettenmann, H. (2012). Activation of serotonin receptors promotes microglial injury-induced motility but attenuates phagocytic activity. *Brain, Behavior, and Immunity*, *26*(3), 419–428. doi:10.1016/j.bbi.2011.12.002 PMID:22198120

Kreisel, T., Frank, M. G., Licht, T., Reshef, R., Ben-Menachem-Zidon, O., Baratta, M. V., Maier, S. F., & Yirmiya, R. (2014). Dynamic microglial alterations underlie stress-induced depressive-like behavior and suppressed neurogenesis. *Molecular Psychiatry*, *19*(6), 699–709. doi:10.1038/mp.2013.155 PMID:24342992

Kugaya, A., & Sanacora, G. (2005). Beyond monoamines: Glutamatergic function in mood disorders. *CNS Spectrums*, *10*(10), 808–819. doi:10.1017/S1092852900010403 PMID:16400244

Lanquillon, S., Krieg, J.-C., Bening-Abu-Shach, U., & Vedder, H. (2000). Cytokine production and treatment response in major depressive disorder. *Neuropsychopharmacology*, *22*(4), 370–379. doi:10.1016/S0893-133X(99)00134-7 PMID:10700656

Largo, C., Cuevas, P., Somjen, G. G., Del Rio, R. M., & Herreras, O. (1996). The effect of depressing glial function in rat brain in situ on ion homeostasis, synaptic transmission, and neuron survival. *The Journal of Neuroscience : The Official Journal of the Society for Neuroscience*, *16*(3), 1219–1229. doi:10.1523/JNEUROSCI.16-03-01219.1996 PMID:8558250

Laurence, J. A., & Fatemi, S. H. (2005). Glial fibrillary acidic protein is elevated in superior frontal, parietal and cerebellar cortices of autistic subjects. *Cerebellum (London, England)*, *4*(3), 206–210. doi:10.1080/14734220500208846 PMID:16147953

Lin, L.-F. H., Doherty, D. H., Lile, J. D., Bektesh, S., & Collins, F. (1993). GDNF: A glial cell line-derived neurotrophic factor for midbrain dopaminergic neurons. *Science*, *260*(5111), 1130–1132. doi:10.1126cience.8493557 PMID:8493557

Liu, J., Dietz, K., DeLoyht, J. M., Pedre, X., Kelkar, D., Kaur, J., Vialou, V., Lobo, M. K., Dietz, D. M., Nestler, E. J., Dupree, J., & Casaccia, P. (2012). Impaired adult myelination in the prefrontal cortex of socially isolated mice. *Nature Neuroscience*, *15*(12), 1621–1623. doi:10.1038/nn.3263 PMID:23143512

Liu, J., Wang, L., Tan, J., Ji, P., Gauthier, S., Zhang, Y., Ma, T., & Liu, S. (2012). Burden, anxiety and depression in caregivers of veterans with dementia in Beijing. *Archives of Gerontology and Geriatrics*, *55*(3), 560–563. doi:10.1016/j.archger.2012.05.014 PMID:22749403

Liu, W.-X., Wang, J., Xie, Z.-M., Xu, N., Zhang, G.-F., Jia, M., Zhou, Z.-Q., Hashimoto, K., & Yang, J.-J. (2016). Regulation of glutamate transporter 1 via BDNF-TrkB signaling plays a role in the anti-apoptotic and antidepressant effects of ketamine in chronic unpredictable stress model of depression. *Psychopharmacology*, *233*(3), 405–415. doi:10.100700213-015-4128-2 PMID:26514555

Lucassen, P. J., Fuchs, E., & Czéh, B. (2004). Antidepressant treatment with tianeptine reduces apoptosis in the hippocampal dentate gyrus and temporal cortex. *Biological Psychiatry*, *55*(8), 789–796. doi:10.1016/j.biopsych.2003.12.014 PMID:15050859

Lucassen, P. J., Müller, M. B., Holsboer, F., Bauer, J., Holtrop, A., Wouda, J., Hoogendijk, W. J. G., De Kloet, E. R., & Swaab, D. F. (2001). Hippocampal apoptosis in major depression is a minor event and absent from subareas at risk for glucocorticoid overexposure. *American Journal of Pathology*, *158*(2), 453–468. doi:10.1016/S0002-9440(10)63988-0 PMID:11159183

Lucassen, P. J., Vollmann-Honsdorf, G. K., Gleisberg, M., Czéh, B., De Kloet, E. R., & Fuchs, E. (2001). Chronic psychosocial stress differentially affects apoptosis in hippocampal subregions and cortex of the adult tree shrew. *The European Journal of Neuroscience*, *14*(1), 161–166. doi:10.1046/j.0953-816x.2001.01629.x PMID:11488960

Madrigal, J. L. M., Leza, J. C., Polak, P., Kalinin, S., & Feinstein, D. L. (2009). Astrocyte-derived MCP-1 mediates neuroprotective effects of noradrenaline. *The Journal of Neuroscience : The Official Journal of the Society for Neuroscience*, *29*(1), 263–267. doi:10.1523/JNEUROSCI.4926-08.2009 PMID:19129402

Makinodan, M., Rosen, K. M., Ito, S., & Corfas, G. (2012). A critical period for social experience–dependent oligodendrocyte maturation and myelination. *Science*, *337*(6100), 1357–1360. doi:10.1126cience.1220845 PMID:22984073

Malberg, J. E., Eisch, A. J., Nestler, E. J., & Duman, R. S. (2000). Chronic antidepressant treatment increases neurogenesis in adult rat hippocampus. *The Journal of Neuroscience : The Official Journal of the Society for Neuroscience*, *20*(24), 9104–9110. doi:10.1523/JNEUROSCI.20-24-09104.2000 PMID:11124987

Manji, H. K., Drevets, W. C., & Charney, D. S.HK. (2001). Drevets WC., Charney DS. The cellular neurobiology of depression. *Nature Medicine*, *7*(5), 541–547. doi:10.1038/87865 PMID:11329053

Männistö, P. T., & Kaakkola, S. (1999). Catechol-O-methyltransferase (COMT): Biochemistry, molecular biology, pharmacology, and clinical efficacy of the new selective COMT inhibitors. *Pharmacological Reviews*, *51*(4), 593–628. PMID:10581325

Maragakis, N. J., & Rothstein, J. D. (2006). Mechanisms of disease: Astrocytes in neurodegenerative disease. *Nature Clinical Practice. Neurology*, *2*(12), 679–689. doi:10.1038/ncpneuro0355 PMID:17117171

Mcdonald, J. W., Althomsons, S. P., Hyrc, K. L., Choi, D. W., & Goldberg, M. P. (1998). Oligodendrocytes from forebrain are highly vulnerable to AMPA/kainate receptor-mediated excitotoxicity. *Nature Medicine*, *4*(3), 291–297. doi:10.1038/nm0398-291 PMID:9500601

Medina, A., Watson, S. J., Bunney, W. Jr, Myers, R. M., Schatzberg, A., Barchas, J., Akil, H., & Thompson, R. C. (2016). Evidence for alterations of the glial syncytial function in major depressive disorder. *Journal of Psychiatric Research*, *72*, 15–21. doi:10.1016/j.jpsychires.2015.10.010 PMID:26519765

Ménard, C., Hodes, G. E., & Russo, S. J. (2016). Pathogenesis of depression: Insights from human and rodent studies. *Neuroscience*, *321*, 138–162. doi:10.1016/j.neuroscience.2015.05.053 PMID:26037806

Miguel-Hidalgo, J. J., Baucom, C., Dilley, G., Overholser, J. C., Meltzer, H. Y., Stockmeier, C. A., & Rajkowska, G. (2000). Glial fibrillary acidic protein immunoreactivity in the prefrontal cortex distinguishes younger from older adults in major depressive disorder. *Biological Psychiatry*, *48*(8), 861–873. doi:10.1016/S0006-3223(00)00999-9 PMID:11063981

Miguel-Hidalgo, J. J., Wilson, B. A., Hussain, S., Meshram, A., Rajkowska, G., & Stockmeier, C. A. (2014). Reduced connexin 43 immunolabeling in the orbitofrontal cortex in alcohol dependence and depression. *Journal of Psychiatric Research*, *55*, 101–109. doi:10.1016/j.jpsychires.2014.04.007 PMID:24774648

Milior, G., Lecours, C., Samson, L., Bisht, K., Poggini, S., Pagani, F., Deflorio, C., Lauro, C., Alboni, S., Limatola, C., Branchi, I., Tremblay, M.-E., & Maggi, L. (2016). Fractalkine receptor deficiency impairs microglial and neuronal responsiveness to chronic stress. *Brain, Behavior, and Immunity*, *55*, 114–125. doi:10.1016/j.bbi.2015.07.024 PMID:26231972

Miller, A. H., & Raison, C. L. (2016). The role of inflammation in depression: From evolutionary imperative to modern treatment target. *Nature Reviews. Immunology*, *16*(1), 22–34. doi:10.1038/nri.2015.5 PMID:26711676

Molendijk, M. L., Spinhoven, P., Polak, M., Bus, B. A. A., Penninx, B., & Elzinga, B. M. (2014). Serum BDNF concentrations as peripheral manifestations of depression: Evidence from a systematic review and meta-analyses on 179 associations (N= 9484). *Molecular Psychiatry*, *19*(7), 791–800. doi:10.1038/mp.2013.105 PMID:23958957

Molofsky, A. V., Krenick, R., Ullian, E., Tsai, H., Deneen, B., Richardson, W. D., Barres, B. A., & Rowitch, D. H. (2012). Astrocytes and disease: A neurodevelopmental perspective. *Genes & Development*, *26*(9), 891–907. doi:10.1101/gad.188326.112 PMID:22549954

Monkul, E. S., Hatch, J. P., Nicoletti, M. A., Spence, S., Brambilla, P., de Lacerda, A. L. T., Sassi, R. B., Mallinger, A. G., Keshavan, M. S., & Soares, J. C. (2007). Fronto-limbic brain structures in suicidal and non-suicidal female patients with major depressive disorder. *Molecular Psychiatry*, *12*(4), 360–366. doi:10.1038j.mp.4001919 PMID:17389903

Montgomery, D. L. (1994). Astrocytes: Form, functions, and roles in disease. *Veterinary Pathology*, *31*(2), 145–167. doi:10.1177/030098589403100201 PMID:8203078

Müller, M. B., Lucassen, P. J., Yassouridis, A., Hoogendijk, W. J. G., Holsboer, F., & Swaab, D. F. (2001). Neither major depression nor glucocorticoid treatment affects the cellular integrity of the human hippocampus. *The European Journal of Neuroscience*, *14*(10), 1603–1612. doi:10.1046/j.0953-816x.2001.01784.x PMID:11860455

Müller, N., & Schwarz, M. J. (2007). The immune-mediated alteration of serotonin and glutamate: Towards an integrated view of depression. *Molecular Psychiatry*, *12*(11), 988–1000. doi:10.1038j.mp.4002006 PMID:17457312

Nagai, J., Yu, X., Papouin, T., Cheong, E., Freeman, M. R., Monk, K. R., Hastings, M. H., Haydon, P. G., Rowitch, D., Shaham, S., & Khakh, B. S. (2021). Behaviorally consequential astrocytic regulation of neural circuits. *Neuron*, *109*(4), 576–596. doi:10.1016/j.neuron.2020.12.008 PMID:33385325

Nagy, C., Suderman, M., Yang, J., Szyf, M., Mechawar, N., Ernst, C., & Turecki, G. (2015). Astrocytic abnormalities and global DNA methylation patterns in depression and suicide. *Molecular Psychiatry*, *20*(3), 320–328. doi:10.1038/mp.2014.21 PMID:24662927

Naskar, S., Datta, S., & Chattarji, S. (2022). Riluzole prevents stress-induced spine plasticity in the hippocampus but mimics it in the amygdala. *Neurobiology of Stress*, *18*, 100442. doi:10.1016/j.ynstr.2022.100442 PMID:35330860

Nedergaard, M., Ransom, B., & Goldman, S. A. (2003). New roles for astrocytes: Redefining the functional architecture of the brain. *Trends in Neurosciences*, *26*(10), 523–530. doi:10.1016/j.tins.2003.08.008 PMID:14522144

Nestler, E. J., Gould, E., & Manji, H. (2002). Preclinical models: Status of basic research in depression. *Biological Psychiatry*, *52*(6), 503–528. doi:10.1016/S0006-3223(02)01405-1 PMID:12361666

Newton, S. S., Collier, E. F., Hunsberger, J., Adams, D., Terwilliger, R., Selvanayagam, E., & Duman, R. S. (2003). Gene profile of electroconvulsive seizures: Induction of neurotrophic and angiogenic factors. *The Journal of Neuroscience : The Official Journal of the Society for Neuroscience*, *23*(34), 10841–10851. doi:10.1523/JNEUROSCI.23-34-10841.2003 PMID:14645477

Newton, S. S., Girgenti, M. J., Collier, E. F., & Duman, R. S. (2006). Electroconvulsive seizure increases adult hippocampal angiogenesis in rats. *The European Journal of Neuroscience*, *24*(3), 819–828. doi:10.1111/j.1460-9568.2006.04958.x PMID:16930411

Nieto-Quero, A., Chaves-Pena, P., Santín, L. J., Pérez-Martín, M., & Pedraza, C. (2021). Do changes in microglial status underlie neurogenesis impairments and depressive-like behaviours induced by psychological stress? A systematic review in animal models. *Neurobiology of Stress*, *15*, 100356. doi:10.1016/j.ynstr.2021.100356 PMID:34355047

Oberheim, N. A., Goldman, S. A., & Nedergaard, M. (2012). Heterogeneity of astrocytic form and function. *Astrocytes: Methods and Protocols*, 23–45.

Oliveira, J. F., Gomes, C. A., Vaz, S. H., Sousa, N., & Pinto, L. (2016). Editorial: Glial plasticity in depression. *Frontiers in Cellular Neuroscience*, *10*(JUN), 1–2. doi:10.3389/fncel.2016.00163 PMID:27378858

Öngür, D., Drevets, W. C., & Price, J. L. (1998). Glial reduction in the subgenual prefrontal cortex in mood disorders. *Proceedings of the National Academy of Sciences of the United States of America*, *95*(22), 13290–13295. doi:10.1073/pnas.95.22.13290 PMID:9789081

Orlovskaya, D. D., Vostrikov, V. M., Rachmanova, V. I., & Uranova, N. A. (2000). Decreased numerical density of oligodendroglial cells in postmortem prefrontal cortex in schizophrenia, bipolar affective disorder and major depression. *Schizophrenia Research*, *1*(41), 105–106. doi:10.1016/S0920-9964(00)90551-6

Ortiz-Ortiz, M. A., Morán, J. M., Ruiz-Mesa, L. M., Bonmatty, R. G., & Fuentes, J. M. (2011). Protective effect of the glial cell line-derived neurotrophic factor (GDNF) on human mesencephalic neuron-derived cells against neurotoxicity induced by paraquat. *Environmental Toxicology and Pharmacology*, *31*(1), 129–136. doi:10.1016/j.etap.2010.09.013 PMID:21787677

Otsuki, K., Uchida, S., Watanuki, T., Wakabayashi, Y., Fujimoto, M., Matsubara, T., Funato, H., & Watanabe, Y. (2008). Altered expression of neurotrophic factors in patients with major depression. *Journal of Psychiatric Research*, *42*(14), 1145–1153. doi:10.1016/j.jpsychires.2008.01.010 PMID:18313696

Pae, C.-U., Marks, D. M., Han, C., & Patkar, A. A. (2008). Does minocycline have antidepressant effect? *Biomedicine and Pharmacotherapy*, *62*(5), 308–311. doi:10.1016/j.biopha.2007.12.005 PMID:18267354

Pallavi, P., Sagar, R., Mehta, M., Sharma, S., Subramanium, A., Shamshi, F., Sengupta, U., Pandey, R. M., & Mukhopadhyay, A. K. (2015). Serum cytokines and anxiety in adolescent depression patients: Gender effect. *Psychiatry Research*, *229*(1–2), 374–380. doi:10.1016/j.psychres.2015.06.036 PMID:26163725

Pallavi, P., Sagar, R., Mehta, M., Sharma, S., Subramanium, A., Shamshi, F., Sengupta, U., Qadri, R., Pandey, R. M., & Mukhopadhyay, A. K. (2013). Serum neurotrophic factors in adolescent depression: Gender difference and correlation with clinical severity. *Journal of Affective Disorders*, *150*(2), 415–423. doi:10.1016/j.jad.2013.04.033 PMID:23769609

Pankratov, Y., & Lalo, U. (2015). Role for astroglial α1-adrenoreceptors in gliotransmission and control of synaptic plasticity in the neocortex. *Frontiers in Cellular Neuroscience*, *9*, 230. doi:10.3389/fncel.2015.00230 PMID:26136663

Pantazatos, S. P., Huang, Y. Y., Rosoklija, G. B., Dwork, A. J., Arango, V., & Mann, J. J. (2017). Whole-transcriptome brain expression and exon-usage profiling in major depression and suicide: Evidence for altered glial, endothelial and ATPase activity. *Molecular Psychiatry*, *22*(5), 760–773. doi:10.1038/mp.2016.130 PMID:27528462

Paolicelli, R. C., Bolasco, G., Pagani, F., Maggi, L., Scianni, M., Panzanelli, P., Giustetto, M., Ferreira, T. A., Guiducci, E., Dumas, L., Ragozzino, D., & Gross, C. T. (2011). Synaptic pruning by microglia is necessary for normal brain development. *Science*, *333*(6048), 1456–1458. doi:10.1126cience.1202529 PMID:21778362

Pascual, A., Hidalgo-Figueroa, M., Piruat, J. I., Pintado, C. O., Gomez-Diaz, R., & López-Barneo, J. (2008). Absolute requirement of GDNF for adult catecholaminergic neuron survival. *Nature Neuroscience*, *11*(7), 755–761. doi:10.1038/nn.2136 PMID:18536709

Pascual, O., Ben Achour, S., Rostaing, P., Triller, A., & Bessis, A. (2012). Microglia activation triggers astrocyte-mediated modulation of excitatory neurotransmission. *Proceedings of the National Academy of Sciences of the United States of America*, *109*(4), E197–E205. doi:10.1073/pnas.1111098109 PMID:22167804

Patanow, C. M., Day, J. R., & Billingsley, M. L. (1996). Alterations in hippocampal expression of SNAP-25, GAP-43, stannin and glial fibrillary acidic protein following mechanical and trimethyltin-induced injury in the rat. *Neuroscience*, *76*(1), 187–202. doi:10.1016/S0306-4522(96)00335-1 PMID:8971771

Paul, I. A., & Skolnick, P. (2003). Glutamate and depression: Clinical and preclinical studies. *Annals of the New York Academy of Sciences*, *1003*(1), 250–272. doi:10.1196/annals.1300.016 PMID:14684451

Peng, L., Verkhratsky, A., Gu, L., & Li, B. (2015). Targeting astrocytes in major depression. *Expert Review of Neurotherapeutics*, *15*(11), 1299–1306. doi:10.1586/14737175.2015.1095094 PMID:26471936

Peng, Z., Zhang, C., Yan, L., Zhang, Y., Yang, Z., Wang, J., & Song, C. (2020). EPA is more effective than DHA to improve depression-like behavior, glia cell dysfunction and hippcampal apoptosis signaling in a chronic stress-induced rat model of depression. *International Journal of Molecular Sciences*, *21*(5), 1769. doi:10.3390/ijms21051769 PMID:32150824

Peters, A., & Feldman, M. L. (1976). The projection of the lateral geniculate nucleus to area 17 of the rat cerebral cortex. I. General description. *Journal of Neurocytology*, *5*(1), 63–84. doi:10.1007/BF01176183 PMID:1249593

Raison, C. L., Capuron, L., & Miller, A. H. (2006). Cytokines sing the blues: Inflammation and the pathogenesis of depression. *Trends in Immunology*, *27*(1), 24–31. doi:10.1016/j.it.2005.11.006 PMID:16316783

Rajkowska, G., Hughes, J., Stockmeier, C. A., Miguel-Hidalgo, J. J., & Maciag, D. (2013). Coverage of blood vessels by astrocytic endfeet is reduced in major depressive disorder. *Biological Psychiatry*, *73*(7), 613–621. doi:10.1016/j.biopsych.2012.09.024 PMID:23146357

Rajkowska, G., & Miguel-Hidalgo, J. (2007). Gliogenesis and glial pathology in depression. *CNS & Neurological Disorders-Drug Targets (Formerly Current Drug Targets-CNS & Neurological Disorders), 6*(3), 219–233.

Rajkowska, G., & Miguel-Hidalgo, J. (2008). Gliogenesis and Glial Pathology in Depression. *CNS & Neurological Disorders - Drug Targets, 6*(3), 219–233. doi:10.2174/187152707780619326 PMID:17511618

Rajkowska, G., Miguel-Hidalgo, J. J., Wei, J., Dilley, G., Pittman, S. D., Meltzer, H. Y., Overholser, J. C., Roth, B. L., & Stockmeier, C. A. (1999). Morphometric evidence for neuronal and glial prefrontal cell pathology in major depression. *Biological Psychiatry, 45*(9), 1085–1098. doi:10.1016/S0006-3223(99)00041-4 PMID:10331101

Rajkowska, G., & Stockmeier, A. (2013). Astrocyte pathology in major depressive disorder: Insights from human postmortem brain tissue. *Current Drug Targets, 14*(11), 1225–1236. doi:10.2174/13894501113149990156 PMID:23469922

Rial, D., Lemos, C., Pinheiro, H., Duarte, J. M., Gonçalves, F. Q., Real, J. I., Prediger, R. D., Gonçalves, N., Gomes, C. A., Canas, P. M., Agostinho, P., & Cunha, R. A. (2016). Depression as a glial-based synaptic dysfunction. *Frontiers in Cellular Neuroscience, 9*(JAN2016), 1–11. doi:10.3389/fncel.2015.00521

Rodnight, R. B., & Gottfried, C. (2013). Morphological plasticity of rodent astroglia. *Journal of Neurochemistry, 124*(3), 263–275. doi:10.1111/jnc.12087 PMID:23278277

Ros-Bernal, F., Hunot, S., Herrero, M. T., Parnadeau, S., Corvol, J.-C., Lu, L., Alvarez-Fischer, D., Carrillo-de Sauvage, M. A., Saurini, F., Coussieu, C., Kinugawa, K., Prigent, A., Höglinger, G., Hamon, M., Tronche, F., Hirsch, E. C., & Vyas, S. (2011). Microglial glucocorticoid receptors play a pivotal role in regulating dopaminergic neurodegeneration in parkinsonism. *Proceedings of the National Academy of Sciences of the United States of America, 108*(16), 6632–6637. doi:10.1073/pnas.1017820108 PMID:21467220

Rosoklija, G., Toomayan, G., Ellis, S. P., Keilp, J., Mann, J. J., Latov, N., Hays, A. P., & Dwork, A. J. (2000). Structural abnormalities of subicular dendrites in subjects with schizophrenia and mood disorders: Preliminary findings. *Archives of General Psychiatry, 57*(4), 349–356. doi:10.1001/archpsyc.57.4.349 PMID:10768696

Sanacora, G., & Banasr, M. (2013). From pathophysiology to novel antidepressant drugs: Glial contributions to the pathology and treatment of mood disorders. *Biological Psychiatry, 73*(12), 1172–1179. doi:10.1016/j.biopsych.2013.03.032 PMID:23726152

Sanacora, G., Rothman, D. L., Mason, G., & Krystal, J. H. (2003). Clinical studies implementing glutamate neurotransmission in mood disorders. *Annals of the New York Academy of Sciences, 1003*(1), 292–308. doi:10.1196/annals.1300.018 PMID:14684453

Sapena, R., Morin, D., Zini, R., Morin, C., & Tillement, J.-P. (1996). Desipramine treatment differently down-regulates β-adrenoceptors of freshly isolated neurons and astrocytes. *European Journal of Pharmacology, 300*(1–2), 159–162. doi:10.1016/0014-2999(96)00060-X PMID:8741184

Schafer, D. P., Lehrman, E. K., Kautzman, A. G., Koyama, R., Mardinly, A. R., Yamasaki, R., Ransohoff, R. M., Greenberg, M. E., Barres, B. A., & Stevens, B. (2012). Microglia sculpt postnatal neural circuits in an activity and complement-dependent manner. *Neuron, 74*(4), 691–705. doi:10.1016/j.neuron.2012.03.026 PMID:22632727

Schildkraut, J. J., & Mooney, J. J. (2004). Toward a rapidly acting antidepressant: The normetanephrine and extraneuronal monoamine transporter (uptake 2) hypothesis. *The American Journal of Psychiatry, 161*(5), 909–911. doi:10.1176/appi.ajp.161.5.909 PMID:15121658

Schipke, C. G., Heuser, I., & Peters, O. (2011). Antidepressants act on glial cells: SSRIs and serotonin elicit astrocyte calcium signaling in the mouse prefrontal cortex. *Journal of Psychiatric Research, 45*(2), 242–248. doi:10.1016/j.jpsychires.2010.06.005 PMID:20619420

Schlicht, K., Büttner, A., Siedler, F., Scheffer, B., Zill, P., Eisenmenger, W., Ackenheil, M., & Bondy, B. (2007). Comparative proteomic analysis with postmortem prefrontal cortex tissues of suicide victims versus controls. *Journal of Psychiatric Research, 41*(6), 493–501. doi:10.1016/j.jpsychires.2006.04.006 PMID:16750834

Schnieder, T. P., Trencevska, I., Rosoklija, G., Stankov, A., Mann, J. J., Smiley, J., & Dwork, A. J. (2014). Microglia of prefrontal white matter in suicide. *Journal of Neuropathology and Experimental Neurology, 73*(9), 880–890. doi:10.1097/NEN.0000000000000107 PMID:25101704

Schroeter, M. L., Abdul-Khaliq, H., Krebs, M., Diefenbacher, A., & Blasig, I. E. (2008). Serum markers support disease-specific glial pathology in major depression. *Journal of Affective Disorders, 111*(2–3), 271–280. doi:10.1016/j.jad.2008.03.005 PMID:18430474

Setiawan, E., Wilson, A. A., Mizrahi, R., Rusjan, P. M., Miler, L., Rajkowska, G., Suridjan, I., Kennedy, J. L., Rekkas, P. V., Houle, S., & Meyer, J. H. (2015). Role of translocator protein density, a marker of neuroinflammation, in the brain during major depressive episodes. *JAMA Psychiatry, 72*(3), 268–275. doi:10.1001/jamapsychiatry.2014.2427 PMID:25629589

Shao, W., Zhang, S., Tang, M., Zhang, X., Zhou, Z., Yin, Y., Zhou, Q., Huang, Y., Liu, Y., Wawrousek, E., Chen, T., Li, S., Xu, M., Zhou, J., Hu, G., & Zhou, J. (2013). Suppression of neuroinflammation by astrocytic dopamine D2 receptors via αB-crystallin. *Nature, 494*(7435), 90–94. doi:10.1038/nature11748 PMID:23242137

Shelton, R. C., Claiborne, J., Sidoryk-Wegrzynowicz, M., Reddy, R., Aschner, M., Lewis, D. A., & Mirnics, K. (2011). Altered expression of genes involved in inflammation and apoptosis in frontal cortex in major depression. *Molecular Psychiatry, 16*(7), 751–762. doi:10.1038/mp.2010.52 PMID:20479761

Si, X., Miguel-Hidalgo, J. J., O'Dwyer, G., Stockmeier, C. A., & Rajkowska, G. (2004). Age-dependent reductions in the level of glial fibrillary acidic protein in the prefrontal cortex in major depression. *Neuropsychopharmacology, 29*(11), 2088–2096. doi:10.1038j.npp.1300525 PMID:15238995

Sievers, C., Auer, M. K., Klotsche, J., Athanasoulia, A. P., Schneider, H. J., Nauck, M., Völzke, H., John, U., Schulz, A., Freyberger, H. J., Friedrich, N., Biffar, R., Stalla, G. K., Wallaschofski, H., & Grabe, H. J. (2014). IGF-I levels and depressive disorders: Results from the Study of Health in Pomerania (SHIP). *European Neuropsychopharmacology, 24*(6), 890–896. doi:10.1016/j.euroneuro.2014.01.008 PMID:24507017

Ślusarczyk, J., Trojan, E., Głombik, K., Budziszewska, B., Kubera, M., Lasoń, W., Popiołek-Barczyk, K., Mika, J., Wędzony, K., & Basta-Kaim, A. (2015). Prenatal stress is a vulnerability factor for altered morphology and biological activity of microglia cells. *Frontiers in Cellular Neuroscience, 9*, 82. doi:10.3389/fncel.2015.00082 PMID:25814933

Śmiałowska, M., Szewczyk, B., Woźniak, M., Wawrzak-Wleciał, A., & Domin, H. (2013). Glial degeneration as a model of depression. *Pharmacological Reports, 65*(6), 1572–1579. doi:10.1016/S1734-1140(13)71518-4 PMID:24553005

Steiner, J., Bielau, H., Brisch, R., Danos, P., Ullrich, O., Mawrin, C., Bernstein, H.-G., & Bogerts, B. (2008). Immunological aspects in the neurobiology of suicide: Elevated microglial density in schizophrenia and depression is associated with suicide. *Journal of Psychiatric Research, 42*(2), 151–157. doi:10.1016/j.jpsychires.2006.10.013 PMID:17174336

Stockmeier, C. A., Mahajan, G., Konick, L. C., Overholser, J. C., Jurjus, G. J., Meltzer, H. Y., Friedman, L., & Rajkowska, G. (2003). Neuronal and glial density is increased and neuronal soma size is decreased in hippocampus in major depressive disorder (MDD). *Biological Psychiatry, 53*(8), 70S–70S.

Stockmeier, C. A., & Rajkowska, G. (2004). Cellular abnormalities in depression: Evidence from postmortem brain tissue. *Dialogues in Clinical Neuroscience, 6*(2), 185–197. doi:10.31887/DCNS.2004.6.2/cstockmeier PMID:22033633

Su, F., Yi, H., Xu, L., & Zhang, Z. (2015). Fluoxetine and S-citalopram inhibit M1 activation and promote M2 activation of microglia in vitro. *Neuroscience, 294*, 60–68. doi:10.1016/j.neuroscience.2015.02.028 PMID:25711936

Sun, J.-D., Liu, Y., Yuan, Y.-H., Li, J., & Chen, N.-H. (2012). Gap junction dysfunction in the prefrontal cortex induces depressive-like behaviors in rats. *Neuropsychopharmacology, 37*(5), 1305–1320. doi:10.1038/npp.2011.319 PMID:22189291

Takaku, S., Yanagisawa, H., Watabe, K., Horie, H., Kadoya, T., Sakumi, K., Nakabeppu, Y., Poirier, F., & Sango, K. (2013). GDNF promotes neurite outgrowth and upregulates galectin-1 through the RET/PI3K signaling in cultured adult rat dorsal root ganglion neurons. *Neurochemistry International, 62*(3), 330–339. doi:10.1016/j.neuint.2013.01.008 PMID:23340048

Takebayashi, M., Hisaoka, K., Nishida, A., Tsuchioka, M., Miyoshi, I., Kozuru, T., Hikasa, S., Okamoto, Y., Shinno, H., Morinobu, S., & Yamawaki, S. (2006). Decreased levels of whole blood glial cell line-derived neurotrophic factor (GDNF) in remitted patients with mood disorders. *The International Journal of Neuropsychopharmacology, 9*(5), 607–612. doi:10.1017/S1461145705006085 PMID:16191208

Tan, S., Wang, Y., Chen, K., Long, Z., & Zou, J. (2017). Ketamine alleviates depressive-like behaviors via down-regulating inflammatory cytokines induced by chronic restraint stress in mice. *Biological & Pharmaceutical Bulletin*, *40*(8), 1260–1267. doi:10.1248/bpb.b17-00131 PMID:28769008

Torres-Platas, S. G., Cruceanu, C., Chen, G. G., Turecki, G., & Mechawar, N. (2014). Evidence for increased microglial priming and macrophage recruitment in the dorsal anterior cingulate white matter of depressed suicides. *Brain, Behavior, and Immunity*, *42*, 50–59. doi:10.1016/j.bbi.2014.05.007 PMID:24858659

Torres-Platas, S. G., Hercher, C., Davoli, M. A., Maussion, G., Labonté, B., Turecki, G., & Mechawar, N. (2011). Astrocytic hypertrophy in anterior cingulate white matter of depressed suicides. *Neuropsychopharmacology*, *36*(13), 2650–2658. doi:10.1038/npp.2011.154 PMID:21814185

Torres-Platas, S. G., Nagy, C., Wakid, M., Turecki, G., & Mechawar, N. (2016). Glial fibrillary acidic protein is differentially expressed across cortical and subcortical regions in healthy brains and downregulated in the thalamus and caudate nucleus of depressed suicides. *Molecular Psychiatry*, *21*(4), 509–515. doi:10.1038/mp.2015.65 PMID:26033239

Trullas, R., & Skolnick, P. (1990). Functional antagonists at the NMDA receptor complex exhibit antidepressant actions. *European Journal of Pharmacology*, *185*(1), 1–10. doi:10.1016/0014-2999(90)90204-J PMID:2171955

Tseng, P.-T., Lee, Y., & Lin, P.-Y. (2013). Age-associated decrease in serum glial cell line-derived neurotrophic factor levels in patients with major depressive disorder. *Progress in Neuro-Psychopharmacology & Biological Psychiatry*, *40*, 334–339. doi:10.1016/j.pnpbp.2012.09.009 PMID:23022671

Uranova, N., Orlovskaya, D., Vikhreva, O., Zimina, I., Kolomeets, N., Vostrikov, V., & Rachmanova, V. (2001). Electron microscopy of oligodendroglia in severe mental illness. *Brain Research Bulletin*, *55*(5), 597–610. doi:10.1016/S0361-9230(01)00528-7 PMID:11576756

Uzdensky, A., Komandirov, M., Fedorenko, G., & Lobanov, A. (2013). Protection effect of GDNF and neurturin on photosensitized crayfish neurons and glial cells. *Journal of Molecular Neuroscience*, *49*(3), 480–490. doi:10.100712031-012-9858-6 PMID:22847529

Virchow, R. L. K. (1858). *Die CELLULARPATHOLOGIE in ihrer Begründung auf physiologische und pathologische Gewebelehre* [Cellular Pathology As Based Upon Physiological and Pathological Histology]. Hirschwald.

Wakeman, D. R., Redmond, D. E. Jr, Dodiya, H. B., Sladek, J. R. Jr, Leranth, C., Teng, Y. D., Samulski, R. J., & Snyder, E. Y. (2014). Human neural stem cells survive long term in the midbrain of dopamine-depleted monkeys after GDNF overexpression and project neurites toward an appropriate target. *Stem Cells Translational Medicine*, *3*(6), 692–701. doi:10.5966ctm.2013-0208 PMID:24744393

Walker, A. K., Budac, D. P., Bisulco, S., Lee, A. W., Smith, R. A., Beenders, B., Kelley, K. W., & Dantzer, R. (2013). NMDA receptor blockade by ketamine abrogates lipopolysaccharide-induced depressive-like behavior in C57BL/6J mice. *Neuropsychopharmacology*, *38*(9), 1609–1616. doi:10.1038/npp.2013.71 PMID:23511700

Warner-Schmidt, J. L., Vanover, K. E., Chen, E. Y., Marshall, J. J., & Greengard, P. (2011). Antidepressant effects of selective serotonin reuptake inhibitors (SSRIs) are attenuated by antiinflammatory drugs in mice and humans. *Proceedings of the National Academy of Sciences of the United States of America, 108*(22), 9262–9267. doi:10.1073/pnas.1104836108 PMID:21518864

Weber-Hamann, B., Blum, W. F., Kratzsch, J., Gilles, M., Heuser, I., & Deuschle, M. (2009). Insulin-like growth factor-I (IGF-I) serum concentrations in depressed patients: Relationship to saliva cortisol and changes during antidepressant treatment. *Pharmacopsychiatry, 42*(01), 23–28. doi:10.1055-0028-1085442 PMID:19153943

Webster, M. J., Knable, M. B., Johnston-Wilson, N., Nagata, K., Inagaki, M., & Yolken, R. H. (2001). Immunohistochemical localization of phosphorylated glial fibrillary acidic protein in the prefrontal cortex and hippocampus from patients with schizophrenia, bipolar disorder, and depression. *Brain, Behavior, and Immunity, 15*(4), 388–400. doi:10.1006/brbi.2001.0646 PMID:11782105

Wierońska, J. M., & Pilc, A. (2009). Metabotropic glutamate receptors in the tripartite synapse as a target for new psychotropic drugs. *Neurochemistry International, 55*(1–3), 85–97. doi:10.1016/j.neuint.2009.02.019 PMID:19428811

Ye, Y., Wang, G., Wang, H., & Wang, X. (2011). Brain-derived neurotrophic factor (BDNF) infusion restored astrocytic plasticity in the hippocampus of a rat model of depression. *Neuroscience Letters, 503*(1), 15–19. doi:10.1016/j.neulet.2011.07.055 PMID:21843601

Yin, Y., Zhang, X., Li, Z., Deng, L., Jiao, G., Zhang, B., Xie, P., Mu, H., Qiao, W., & Zou, J. (2013). Glucocorticoid receptor β regulates injury-mediated astrocyte activation and contributes to glioma pathogenesis via modulation of β-catenin/TCF transcriptional activity. *Neurobiology of Disease, 59*, 165–176. doi:10.1016/j.nbd.2013.07.013 PMID:23906498

Yirmiya, R., Rimmerman, N., & Reshef, R. (2015). Depression as a microglial disease. *Trends in Neurosciences, 38*(10), 637–658. doi:10.1016/j.tins.2015.08.001 PMID:26442697

Zhan, Y., Paolicelli, R. C., Sforazzini, F., Weinhard, L., Bolasco, G., Pagani, F., Vyssotski, A. L., Bifone, A., Gozzi, A., Ragozzino, D., & Gross, C. T. (2014). Deficient neuron-microglia signaling results in impaired functional brain connectivity and social behavior. *Nature Neuroscience, 17*(3), 400–406. doi:10.1038/nn.3641 PMID:24487234

Zhang, C., Zhang, Y.-P., Li, Y.-Y., Liu, B.-P., Wang, H.-Y., Li, K.-W., Zhao, S., & Song, C. (2019). Minocycline ameliorates depressive behaviors and neuro-immune dysfunction induced by chronic unpredictable mild stress in the rat. *Behavioural Brain Research, 356*, 348–357. doi:10.1016/j.bbr.2018.07.001 PMID:30003978

Zhang, X., Ru, B., Sha, W., Xin, W., Zhou, H., & Zhang, Y. (2014). Performance on the W isconsin card-sorting test and serum levels of glial cell line-derived neurotrophic factor in patients with major depressive disorder. *Asia-Pacific Psychiatry, 6*(3), 302–307. doi:10.1111/appy.12120 PMID:24436216

Zhang, X., Zhang, Z., Xie, C., Xi, G., Zhou, H., Zhang, Y., & Sha, W. (2008). Effect of treatment on serum glial cell line-derived neurotrophic factor in depressed patients. *Progress in Neuro-Psychopharmacology & Biological Psychiatry, 32*(3), 886–890. doi:10.1016/j.pnpbp.2008.01.004 PMID:18262703

Zhao, L., Mandler, M. D., Yi, H., & Feng, Y. (2010). Quaking I controls a unique cytoplasmic pathway that regulates alternative splicing of myelin-associated glycoprotein. *Proceedings of the National Academy of Sciences of the United States of America, 107*(44), 19061–19066. doi:10.1073/pnas.1007487107 PMID:20956316

Zhou, B., Zhu, Z., Ransom, B. R., & Tong, X. (2021). Oligodendrocyte lineage cells and depression. *Molecular Psychiatry, 26*(1), 103–117. doi:10.103841380-020-00930-0 PMID:33144710

Zhou, X. E., Suino-Powell, K., Schultz, C. R., Aleiwi, B., Brunzelle, J. S., Lamp, J., Vega, I. E., Ellsworth, E., Bachmann, A. S., & Melcher, K. (2021). Structural basis of binding and inhibition of ornithine decarboxylase by 1-amino-oxy-3-aminopropane. *The Biochemical Journal, 478*(23), 4137–4149. doi:10.1042/BCJ20210647 PMID:34796899

Chapter 18
Therapeutic Strategies for Lysosomal Storage Diseases by Targeting Glial Cells

Sabir Es-Said
https://orcid.org/0000-0003-2246-5645
Faculty of Medicine, Cadi Ayyad University, Morocco

Fdil Naima
Faculty of Medicine, Cadi Ayad University, Morocco

ABSTRACT

Lysosomal storage disorders (LSDs) are a group of about 70 life-threatening conditions caused by genetic defects affecting lysosomal components. The underscoring molecular deficiency leads to widespread cellular dysfunction through most tissues in the body, including peripheral organs and the central nervous system (CNS). Clinical and experimental strategies encompassing enzyme replacement, gene and cell therapies, substrate reduction, and chemical chaperones are showing considerable potential in attenuating the peripheral pathology. Microglial and astrocyte activation is a hallmark of many LSDs that affect the CNS, which often precedes and predicts regions where eventual neuron loss will occur. However, the timing, intensity, and duration of neuroinflammation may ultimately dictate the impact on CNS homeostasis. For example, a transient inflammatory response following CNS insult/injury can be neuroprotective, as glial cells attempt to remove the insult and provide trophic support to neurons.

INTRODUCTION

Lysosomal storage diseases (LSDs) are monogenic hereditary metabolic diseases affecting the functioning of a lysosomal enzyme (or its activator) involved in the degradation of complex molecules, or more rarely of a transport or membrane protein. This results in an accumulation of metabolites in lysosomes leading to cell damage: disruption of autophagy, oxidative stress, inflammation and increased apoptosis, disruption of cell signaling pathways. LSDs are rare or very rare diseases individually, since their

DOI: 10.4018/978-1-6684-9675-6.ch018

incidence is between 1 in 40,000 and 1 in 1,000,000. However, their incidence as a group is greater, on the order of 1 in 5000 live births (Al-Jasmi et al., 2013; Platt et al., 2018; Poupětová et al., 2010). They include about 70 different pathologies that can be classified according to the nature of the accumulated compound or the mechanism involved (Kingma et al., 2015). The mode of transmission is most often autosomal recessive except for Fabry, Hunter and Danon diseases transmitted in an X-linked mode. Clinical orientation is essential but is not sufficient to establish the diagnosis. Specialized biochemistry and molecular biology laboratories, attached to rare disease reference centers and forming an integral part of the health sector, carry out biological examinations contributing to the diagnosis. This is classically based on the first-line biochemical study: study of the metabolite accumulated in biological fluids (plasma, urine, etc.) often by tandem mass spectrometry (MSMS) and/or measurements of enzymatic activities in the blood mostly. The molecular study confirms the diagnosis in a second time.

Attenuated forms can present a significant diagnostic problem, although severe phenotypes are easily detected. It is well known that musculoskeletal issues often prompt a patient to seek medical attention for the first time. Timely diagnosis and therapy initiation are crucial to prevent irreversible organ damage and poor quality of life, as several of these conditions can be efficiently treated with enzyme replacement. As a result, every physician needs to be knowledgeable with the hallmark signs and symptoms of lysosomal storage disorders.

Many LSDs have modest to severe involvement of the central nervous system (CNS). Post-mortem tissues and animal illness models, respectively, depict the last stage of CNS pathology and partially replicate the disease. Over the past few decades, induced pluripotent stem cell (hiPSC)-based human models have been widely used to study LSD disease in various tissues and organs, including the central nervous system. A potentially useful method for defining the impact of pathological storage on neurodevelopment, survival, and function of neurons and glial cells in neurodegenerative LSDs is neural stem/progenitor cells (NSCs) generated from patient-specific hiPSCs (hiPS-NSCs). The study of neuron-glia functional and dysfunctional interactions is also being aided by the development of novel 2D co-culture systems and 3D hiPSC-based models. These developments are helping to define the role of neurodevelopment and neuroinflammation in the onset and progression of the disease, which has significant implications for treatment timing and efficacy (Luciani et al., 2020).

THE LYSOSOMAL SYSTEM: FUNCTIONS AND USES

The lysosome is an organelle which can be spherical, ovoid, or sometimes tubular, and having a variable size between 0.1 and 2 µm, present in most plant and animal cells (Appelqvist et al., 2013). In humans, lysosomes are present in around a hundred copies in all cells of the body, except in red blood cells. Two distinct classes of proteins are present at the lysosome: a) acid hydrolases, approximately 50 in number, activated by the acidic pH between 4.5 and 5 at the lumen of the lysosome, b) membrane proteins, which are 25 in number (Mindell, 2012). These enzymes and proteins are all essential for the lysosome to perform its functions. These consist in particular of degrading macromolecules, polysaccharides, glycosaminoglycans and complex lipids into their elementary molecules respectively: amino acids, monosaccharides and free fatty acids. Lysosomal degradation products are transported outside the lysosome via specific transporters located in the lysosomal membrane (Saftig & Klumperman, 2009), or via vesicular membrane trafficking for reuse in biosynthetic pathways (Ruivo et al., 2009). (Figure 1).

Figure 1. Lysosomal membrane trafficking pathways
(Samie & Xu, 2014)

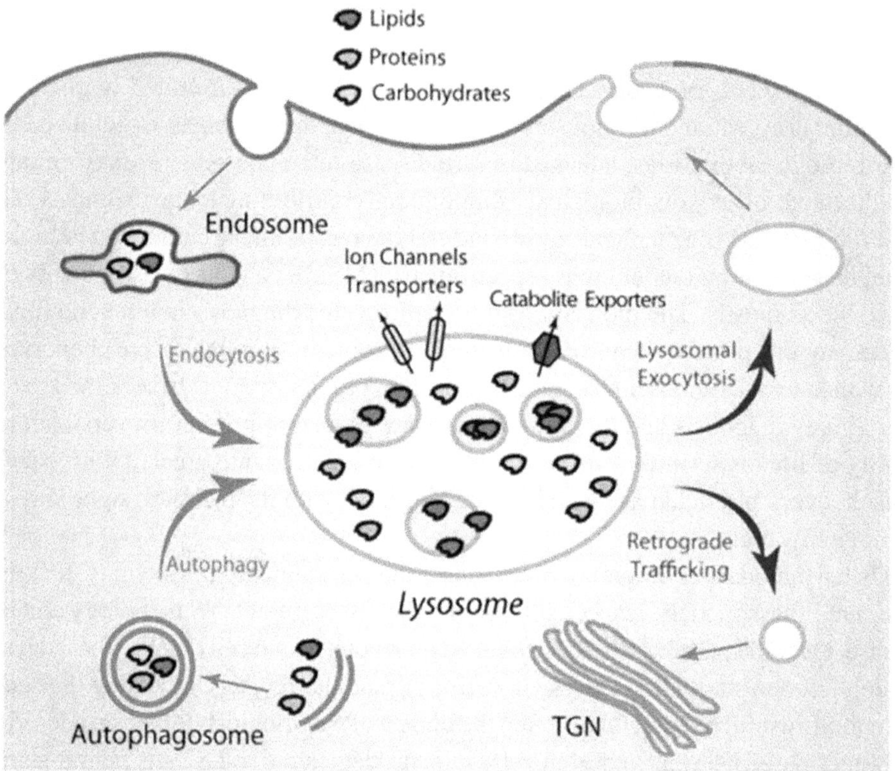

THERAPEUTIC STRATEGIES OF METACHROMATIC LEUKODYSTROPHY TARGETING GLIAL CELLS

Leukodystrophies (LDs) are a class of hereditary disorders in which problems in the synthesis and/or maintenance of myelin within the central and occasionally peripheral nervous systems are exclusively or predominantly caused by molecular abnormalities of glial cells. The term leukodystrophies (LDs) (leuko-white, dystrophy defective nutrition) refers to a group on inherited diseases in which molecular abnormalities of glial cells are responsible for exclusive or predominant defects in myelin sheath formation and/or maintenance within the brain, spinal cord and often also the peripheral nerves. The white matter of the brain and spinal cord is a complex structure, which is composed of a vast number of axons, sheathed with a compact, lipid-rich membrane, the myelin. Besides myelinated axons, white matter contains glial cells (astrocytes, microglial cells and oligodendrocytes) that play a major role in structural and trophic support of myelin and axons

A significant group of leukodystrophies (LDs) known as de/dysmyelinating LDs, such as X-linked adrenoleukodystrophy (X-ALD) (Moser et al., 2007; Mosser et al., 1993), metachromatic LD (MLD) (Polten et al., 1991; von Figura & Kresse, 1972), and globoid cell LD (GLD), also referred to as Krabbe disease (Wenger et al., 2000), all have been characterized through the identification of enzyme or transporter defects in lysosomes and peroxisomes. Additionally, these LDs are primarily characterized by the destruction of myelin. LDs were originally categorized in neuropathological investigation based on the staining properties of myelin breakdown products.

Three main categories can be distinguished from the childhood white matter disorders based on their underlying pathophysiology: 1) the hypomyelinating disorders, which are characterized by reduced or abnormal myelin formation; 2) the metabolic demyelinations, which are characterized by loss of myelin following initial normal formation and are frequently linked to the accumulation of metabolic substrates and the appearance of toxic metabolites; and 3) the non-metabolic disorders of myelin, which include ischemic injury, which frequently causes extensive tissue loss and destruction. All of these cell-based treatment approaches seek to replace the aberrant glial cells with oligodendrocytes that can produce functional myelin. They also hope to act as delivery systems for wild-type enzymes, whose replacement may be enough to allow the deficient host cells to produce or maintain normal myelin (Osorio & Goldman, 2016)

The understanding of the physiopathology of LD has significantly expanded in light of recent findings, especially with regard to the function of different glial cells in the demyelinating process. While it was long believed that the only cause of primary demyelination of the central nervous system (CNS) was an injury directly to the oligodendrocyte, which is the cell responsible for creating and maintaining the myelin sheath, genes associated with Alexander disease have since been found (Brenner et al., 2001), childhood ataxia with central hypomyelination/vanishing white matter syndrome (Schiffmann & Elroy-Stein, 2006) and Canavan disease (Kaul et al., 1994), have reported the involvement of astrocytes in myelin formation and maintenance (De Keyser et al., 2008).

The gene therapy approaches that have been developed thus far to address three specific LDs focus on intracerebral delivery of adeno-associated virus (AAV) vectors that primarily target neurons, or on hematopoietic stem cell (HSC) and lentiviral vectors to target microglia precursors. The gene therapy strategy in two of the LDs (MLD and GLD) also capitalizes on the possibility of oligodendrocyte cross-correction by engineered/genetically corrected cells (microglia or neurons). None of the newly developed gene therapy vectors have been able to effectively and selectively target oligodendrocytes following intracerebral or intravenous injection, despite notable advancements in this area. The discovery or development of novel AAV vectors that transduce far better astrocytes in vivo has been made possible by recent screening of novel AAV serotypes or their engineering using highthroughput random diversification and selection methodologies (Koerber et al., 2009). Nevertheless, the clinics have not yet used those novel AAV vectors.

The potential to alter a subgroup of glial cells called microglia in the brain underpins HSC gene therapy for CNS diseases. These cells are widely dispersed throughout the central nervous system and can make up as much as 20% of all glial cells. Myeloid precursor cells, which enter the brain and spinal cord during development, give rise to resident microglial cells (Perry et al., 2010). Despite their similarities to monocytes and macrophages, microglia may be an ontogenically separate population within the mononuclear phagocyte system, according to recent research (Ginhoux et al., 2010). Sensing changes in brain tissue, brain microglia carry out both innate and adaptive immune responses. However, in a healthy brain, microglia are more than just macrophages. We're still learning about other roles, like as the ability to maintain synaptic integrity, respond to external stressors (pain), and influence neurogenesis (Graeber, 2010).

THERAPEUTIC STRATEGIES OF NIEMANN–PICK TYPE C DISEASE TARGETING GLIAL CELLS

Niemann-Pick disease is a type of metabolic disease that is inherited and falls under the broad category of lysosomal storage disorders (LSD), which are characterized by deficiencies in one or more lysosomal activities (Toledano-Zaragoza & Ledesma, 2020). NPC affects an estimated 1–9:100000 people (https://www.orpha.net/ accessed on 16 August 2021). A genetically determined abnormal function of lysosomal proteins is the common denominator among the group of disorders known as Niemann-Pick disease (types A, B, and C). Type A and Type B Niemann-Pick diseases are caused by mutations in the SMPD1 gene. The lysosomal enzyme acid sphingomyelinase activity, which breaks down the lipid sphingomyelin in ceramide and phosphorylcholine, is deficient as a result of these. Sphingomyelin will build up if this enzyme malfunctions or fails to act adequately (Schuchman & Wasserstein, 2015). On the other hand, a mutation in the NPC1 and NPC2 genes, which are found on chromosomes 18 (locus 18q11-q12) and 14 (locus 14q24.3), respectively, results in Niemann-Pick type C (NPC). The NPC2 gene encodes a protein that binds and transports cholesterol and has been demonstrated to interact closely with NPC1. The NPC1 gene encodes a protein found in the membranes of the late endosome/lysosome compartment (LE/L).

About 95% of NPC cases are caused by genetic mutations in the NPC1 gene, while 5% are caused by mutations in the NPC2 gene. A deficiency in the intracellular transport of endocytosed cholesterol, caused by mutations in both genes, leads to the sequestration of unesterified cholesterol, sphingosine, glycosphingolipids, and sphingomyelin in the LE/L compartment of several human tissues, including the brain (Vanier, 2010). The clinical manifestations of this lipidosis are mostly related to the progressive, broad neurological degeneration since the phenotypic displays a variety of symptoms, including hepatosplenomegaly, motor dysfunctions, cerebellar ataxia, seizures, and dementia (Imrie et al., 2007). Before they pass away, patients have symptoms for several years, and the cost of treating neurological conditions rises with their severity. The identification of extremely disease-specific biomarkers and biochemical assays like cholesterol esterification and Filipin staining of patient fibroblasts can be used to make the diagnosis; nevertheless, genetic testing is necessary for ultimate validation (Wassif et al., 2016).

Myelin Defects in NPC1

Numerous clinical data points to myelin abnormalities in NPC patients. When compared to age-matched controls, an analysis of six adult patients showed a broad loss in white matter, which is suggestive of poor myelination (Walterfang et al., 2010). In another study, the myelin water fraction (MWF) in the white matter of two adult patients with NPC was shown to be lower than that of 15 healthy control participants using diffusion tensor imaging (DTI), a magnetic resonance imaging technique used to detect the diffusion of water in white matter (Davies-Thompson et al., 2016), indicative of reduced myelination; in addition, the less clinically affected patient showed focal reductions in MWF in the corpus callosum in contrast to the more extensive and widespread reductions across entire fiber tracts observed in the more affected subject. Another 9-year-old patient with a fast progressing variant of NPC displayed extremely significant myelin loss, which greatly outweighs the degree of neuronal destruction (Kodachi et al., 2017). These data support the hypothesis that there may be a relationship between the severity of myelination abnormalities and the disease's advancement, and that treating this pathological event could help treat NPC symptoms.

Despite being the primary players in myelination, it is important to remember that oligodendrocytes interact with both astrocytes and microglia, two types of glial cells, to maintain homeostasis in the myelin sheath (Ishibashi et al., 2006).

Because of the difficulty in translating the numerous molecular abnormalities seen in NPC experimental models to the clinical situation, preclinical research on NPC is less advanced than that of other neurodegenerative illnesses (Fog & Kirkegaard, 2019). Positively, a great deal of information has been gleaned from preclinical research done on NPC models, both in vitro and in vivo, regarding the molecular pathways that contribute to the pathogenesis of the disease. Clinical research into nonprimary childhood cancer (NPC) is difficult because of its rarity, which makes large-scale clinical trials impractical. This is true of all orphan conditions. The challenges also stem from the fact that many young patients do not receive an early diagnosis of NPC and that treatment needs to be effective in both the brain and the visceral organs. With the exception of miglustat, there is presently no effective treatment for NPC disease due to these and other factors. Dysmyelination is one of the newest and most intriguing phenotypes taken into consideration among the several outcome measures used in NPC preclinical and clinical investigations, as it is a characteristic of NPC in both patients and animal models (Zhou et al., 2020).

THERAPEUTIC STRATEGIES OF MUCOPOLYSACCHARIDOSIS TARGETING GLIAL CELLS

The mucopolysaccharidoses (MPS) are a class of lysosomal storage disorders, defined by the lack or impairment of an enzyme necessary for the intra-lysosomal breakdown of glycosaminoglycans (GAGs) (McKusick & Neufeld, 1983). Due to these enzyme deficits, GAGs build up in the lysosomes and are excreted in significant levels in the urine, which causes a multisystem clinical disease that affects the organs, tissues, and cells. The seven major types of MPS are caused by 11 different enzymatic defects, which vary significantly in terms of age at onset, severity, and loss of development skills such as speech and learning. The majority of these types are characterized by multisystem involvement, which includes skeletal abnormalities, macrocephaly, loss of speech and learning, unusual hair growth, corneal clouding, hepatosplenomegaly, pneumonias, and mental retardation (Muenzer, 2004; Neufeld & Muenzer, 2001).

Treatment for mucopolysaccharidosis must be started as early as possible to be effective because these conditions are progressing. Thus, it is essential to identify patients as soon as possible. For certain MPS forms, hematopoietic stem cell transplants and enzyme replacement therapy (ERT) are currently available (Parenti et al., 2015), and substrate reduction and gene therapies are in development (Gilkes & Heldermon, 2014). Since they are uncommon disorders, early identification based on early symptom detection is difficult and has yielded only patchy results; as such, screening strategies, such as newborn screening (NBS), may be more successful.

Mucopolysaccharidosis Type III

Mucopolysaccharidosis type III (MPS III, Sanfilippo syndrome) is a progressive neurlogical lysosomal storage disorder, resulted from biallelic mutations in one of four genes encoding enzymes involved in the breakdown of heparan sulfate (HS). According to the defective enzyme, there are four subtypes of Sanfilippo syndrome: MPSIIIA (OMIM#252900) is caused by a lack of N-sulfoglucosamine sulfohydrolase (Kresse & Neufeld, 1972); deficit in α-N-acetylglucosaminidase causes MPSIIIB (OMIM#252920) (von

Figura & Kresse, 1972); heparan-α-glucosaminide N-acetyltransferase (HGSNAT) deficiency causes MPSIIIC (OMIM#252930) (Klein et al., 1978); and MPSIIID (OMIM#252940) is caused by N-acetylglucosamine-6-sulfatase deficiency (Kresse et al., 1980). Progressive multisystem somatic disease is also observed, albeit less frequently than in MPS I (Hurler disease) or MPS II (Hunter disease), even though the central nervous system degradation is the main characteristic. While MPSIIIC and D appear to be significantly rarer, MPSIIIA is the most prevalent kind of Sanfilippo syndrome in north-west Europe and Australia, and type B is the most common type in south-east Europe and Brazil (Coelho et al., 1997; Meikle et al., 1999; Valstar et al., 2008).

While the clinical symptoms of all Sanfilippo syndrome subtypes are identical and largely manifest as progressive neurodegenerative diseases, MPSIIIC appears to have a gentler clinical trajectory, despite the disease's very varied severity (Delgadillo et al., 2013). The disease slowly progresses to neurological alteration characterized by speech impairment that precedes loss of motor function (Héron et al., 2011). Affected people exhibit serious behavioral issues during this time, including hyperactivity that gets worse due to sleep abnormalities. One of the primary characteristics of MPSIIIC illness is neurological abnormalities (Héron et al., 2011). A recent study examined whether mice with HGSNAT deficiency have lysosomal disease in their brains. The age of start and progression of the lysosomal pathology affecting the central nervous system were assessed in animals that were 2, 12, and 22 months old. GAG accumulation in all examined brain areas increased gradually in HGSNAT-deficient male and female mice, ranging from 4% to 80% over values seen in healthy littermates. The most frontal areas of the brain showed the greatest accumulation in both sexes, where the corpulent GAG was easily observed as early as two months of life (Marcó et al., 2016). A few years ago, it was proposed to explain the presence of large storage vesicles in glial cells compared to neurons with minimal evidence of storage disease, or the presence of foamy macrophages in nearly all tissues of MPS animal models, by suggesting that macrophages might buffer storage overload by absorbing GAGs from the interstitial space or blood (Jolly et al., 2007). On the other hand, it is possible that the liver and spleen turnover GAGs more quickly than other tissues; in this case, the absence of enzymatic activity has a considerably greater effect on substrate accumulation.

Mucopolysaccharidosis Type II

Hunter syndrome (MPS II) is the only MPS inherited in an X-linked manner, and lack of iduronate sulfatase in this disease leads to dermatan sulfate and heparan sulfate build up (Matheus et al., 2004). Hunter syndrome, commonly known as mucopolysaccharidosis type II (MPS II, MIM # 309900), is a rare genetic illness that is inherited as an X-linked trait. Its incidence rate varies between 0.38 and 1.09 per 100,000 live neonates in Brazil and Portugal, respectively. In general, the incidence of MPS II is lower in European countries than in East Asian countries, where MPS II incidence makes up around 50% of all mucopolysaccharidoses (MPSs) (Khan et al., 2017).

Clinically, MPS II is divided into two subtypes: Hunter A (severe) and Hunter B (moderate). Dysostosis multiplex, developmental delay, childhood-onset progressive neurological symptoms, and somatic involvement are the characteristics of the former. The latter has a later onset and a less severe course of the illness. Patients with various forms of MPS have recently undergone bone marrow transplantation; nevertheless, while this treatment improves the physical aspects, the improvement of neurological symptoms is still insufficient (Seto et al., 2001; Takahashi et al., 2001). Neuropathologically, patients with MPS II and IIIB exhibit hydrocephalus, moderate gliosis in the white matter, perivascular space dilatation, and/or neuronal edema (Tamagawa et al., 1985). Though uncommon transmission into the brain has contributed to the failure of contemporary enzyme replacement therapy to cure neurological diseases, the pathophysiology of intelligence impairment in MPS patients remains obscure (Muenzer et al., 2007).

CONCLUSION

LSDs frequently cause neurodegeneration, which can result in permanent impairment and even death. The first step in developing and testing treatment solutions for these deadly diseases is understanding their causes in animal models, and there have been significant advancements in this area in recent years. Alterations in glial cells and distinct neuronal cell compartments have been linked to common pathogenic molecular pathways that extend beyond lysosomal dysfunction. Animal models have demonstrated the efficacy of gene therapy in addressing the underlying abnormality in many onogenic illnesses. Whether this tactic applies to the human brain as a whole has not yet been established.

The enormous advancements in molecular neuroscience and the current state of CNS disease treatment clearly differ from one another. For the most part, lysosomal storage diseases (LSDs) remain incurable despite the fact that the underlying gene is known in majority of these cases. For this class of illnesses, gene therapy is being investigated as a potential therapeutic strategy. In order to restore therapeutic levels of the protein in affected tissues, gene therapy for liposomal diseases (LSDs) aims to introduce the DNA corresponding to the lysosomal enzyme or protein deficiency into a cell. Approaches to ex vivo and in vivo gene therapy have been investigated in experiments. Cross-correction is the foundation of the ex vivo gene treatment of LSDs approach. Transferring the protein from the genetically altered donor cells to the recipient cells that lack protein is essential for a successful course of treatment.

REFERENCES

Al-Jasmi, F. A., Tawfig, N., Berniah, A., Ali, B. R., Taleb, M., Hertecant, J. L., Bastaki, F., & Souid, A. K. (2013). *Prevalence and novel mutations of lysosomal storage disorders in United Arab Emirates: LSD in UAE*. Epub 2013/02/23.

Appelqvist, H., Wäster, P., Kågedal, K., & Öllinger, K. (2013). The lysosome: From waste bag to potential therapeutic target. *Journal of Molecular Cell Biology*, *5*(4), 214–226. doi:10.1093/jmcb/mjt022 PMID:23918283

Brenner, M., Johnson, A. B., Boespflug-Tanguy, O., Rodriguez, D., Goldman, J. E., & Messing, A. (2001). Mutations in GFAP, encoding glial fibrillary acidic protein, are associated with Alexander disease. *Nature Genetics*, *27*(1), 117–120. doi:10.1038/83679 PMID:11138011

Coelho, J. C., Wajner, M., Burin, M. G., Vargas, C. R., & Giugliani, R. (1997). Selective screening of 10,000 high-risk Brazilian patients for the detection of inborn errors of metabolism. *European Journal of Pediatrics*, *156*(8), 650–654. doi:10.1007004310050685 PMID:9266201

Davies-Thompson, J., Vavasour, I., Scheel, M., Rauscher, A., & Barton, J. J. S. (2016). Reduced myelin water in the white matter tracts of patients with Niemann-pick disease type C. *AJNR. American Journal of Neuroradiology*, *37*(8), 1487–1489. doi:10.3174/ajnr.A4719 PMID:26939636

De Keyser, J., Mostert, J. P., & Koch, M. W. (2008). Dysfunctional astrocytes as key players in the pathogenesis of central nervous system disorders. *Journal of the Neurological Sciences*, *267*(1–2), 3–16. doi:10.1016/j.jns.2007.08.044 PMID:17935736

Delgadillo, V., O'Callaghan, M. del M., Gort, L., Coll, M. J., & Pineda, M. (2013). Natural history of Sanfilippo syndrome in Spain. *Orphanet Journal of Rare Diseases*, *8*(1), 1–11. doi:10.1186/1750-1172-8-189 PMID:24314109

Fog, C. K., & Kirkegaard, T. (2019). Animal models for Niemann-Pick type C: Implications for drug discovery & development. *Expert Opinion on Drug Discovery*, *14*(5), 499–509. doi:10.1080/17460441.2019.1588882 PMID:30887840

Gilkes, J. A., & Heldermon, C. D. (2014). Mucopolysaccharidosis III (Sanfilippo Syndrome)-disease presentation and experimental therapies. *Pediatric Endocrinology Reviews; PER*, *12*, 133–140. PMID:25345095

Ginhoux, F., Greter, M., Leboeuf, M., Nandi, S., See, P., Gokhan, S., Mehler, M. F., Conway, S. J., Ng, L. G., Stanley, E. R., Samokhvalov, I. M., & Merad, M. (2010). Fate mapping analysis reveals that adult microglia derive from primitive macrophages. *Science*, *330*(6005), 841–845. doi:10.1126cience.1194637 PMID:20966214

Graeber, M. B. (2010). Changing face of microglia. *Science*, *330*(6005), 783–788. doi:10.1126cience.1190929 PMID:21051630

Héron, B., Mikaeloff, Y., Froissart, R., Caridade, G., Maire, I., Caillaud, C., Levade, T., Chabrol, B., Feillet, F., Ogier, H., Valayannopoulos, V., Michelakakis, H., Zafeiriou, D., Lavery, L., Wraith, E., Danos, O., Heard, J.-M., & Tardieu, M. (2011). Incidence and natural history of mucopolysaccharidosis type III in France and comparison with United Kingdom and Greece. *American Journal of Medical Genetics. Part A*, *155*(1), 58–68. doi:10.1002/ajmg.a.33779 PMID:21204211

Imrie, J., Dasgupta, S., Besley, G. T. N., Harris, C., Heptinstall, L., Knight, S., Vanier, M. T., Fensom, A. H., Ward, C., Jacklin, E., Whitehouse, C., & Wraith, J. E. (2007). The natural history of Niemann–Pick disease type C in the UK. *Journal of Inherited Metabolic Disease*, *30*(1), 51–59. doi:10.100710545-006-0384-7 PMID:17160617

Ishibashi, T., Dakin, K. A., Stevens, B., Lee, P. R., Kozlov, S. V., Stewart, C. L., & Fields, R. D. (2006). Astrocytes promote myelination in response to electrical impulses. *Neuron*, *49*(6), 823–832. doi:10.1016/j.neuron.2006.02.006 PMID:16543131

Jolly, R. D., Johnstone, A. C., Norman, E. J., Hopwood, J. J., & Walkley, S. U. (2007). Pathology of mucopolysaccharidosis IIIA in Huntaway dogs. *Veterinary Pathology*, *44*(5), 569–578. doi:10.1354/vp.44-5-569 PMID:17846229

Kaul, R., Gao, G. P., Aloya, M., Balamurugan, K., Petrosky, A., Michals, K., & Matalon, R. (1994). Canavan disease: Mutations among Jewish and non-Jewish patients. *American Journal of Human Genetics*, *55*(1), 34. PMID:8023850

Khan, S. A., Peracha, H., Ballhausen, D., Wiesbauer, A., Rohrbach, M., Gautschi, M., Mason, R. W., Giugliani, R., Suzuki, Y., Orii, K. E., Orii, T., & Tomatsu, S. (2017). Epidemiology of mucopolysaccharidoses. *Molecular Genetics and Metabolism*, *121*(3), 227–240. doi:10.1016/j.ymgme.2017.05.016 PMID:28595941

Kingma, S. D. K., Bodamer, O. A., & Wijburg, F. A. (2015). Epidemiology and diagnosis of lysosomal storage disorders; challenges of screening. *Best Practice & Research. Clinical Endocrinology & Metabolism*, *29*(2), 145–157. doi:10.1016/j.beem.2014.08.004 PMID:25987169

Klein, U., Kresse, H., & Von Figura, K. (1978). Sanfilippo syndrome type C: deficiency of acetyl-CoA: alpha-glucosaminide N-acetyltransferase in skin fibroblasts. *Proceedings of the National Academy of Sciences of the United States of America*, *75*(10), 5185–5189. doi:10.1073/pnas.75.10.5185 PMID:33384

Kodachi, T., Matsumoto, S., Mizuguchi, M., Osaka, H., Kanai, N., Nanba, E., Ohno, K., & Yamagata, T. (2017). Severe demyelination in a patient with a late infantile form of Niemann-Pick disease type C. *Neuropathology*, *37*(5), 426–430. doi:10.1111/neup.12380 PMID:28387450

Koerber, J. T., Klimczak, R., Jang, J.-H., Dalkara, D., Flannery, J. G., & Schaffer, D. V. (2009). Molecular evolution of adeno-associated virus for enhanced glial gene delivery. *Molecular Therapy*, *17*(12), 2088–2095. doi:10.1038/mt.2009.184 PMID:19672246

Kresse, H., & Neufeld, E. F. (1972). The Sanfilippo A corrective factor: Purification and mode of action. *The Journal of Biological Chemistry*, *247*(7), 2164–2170. doi:10.1016/S0021-9258(19)45506-3 PMID:4259567

Kresse, H., Paschke, E., Von Figura, K., Gilberg, W., & Fuchs, W. (1980). Sanfilippo disease type D: Deficiency of N-acetylglucosamine-6-sulfate sulfatase required for heparan sulfate degradation. *Proceedings of the National Academy of Sciences of the United States of America*, *77*(11), 6822–6826. doi:10.1073/pnas.77.11.6822 PMID:6450420

Luciani, M., Gritti, A., & Meneghini, V. (2020). Human iPSC-based models for the development of therapeutics targeting neurodegenerative lysosomal storage diseases. *Frontiers in Molecular Biosciences*, *7*, 224. doi:10.3389/fmolb.2020.00224 PMID:33062642

Marcó, S., Pujol, A., Roca, C., Motas, S., Ribera, A., Garcia, M., Molas, M., Villacampa, P., Melia, C. S., & Sanchez, V. (2016). Progressive neurologic and somatic disease in a novel mouse model of human mucopolysaccharidosis type IIIC. *Disease Models & Mechanisms*, *9*(9), 999–1013. PMID:27491071

Matheus, M. G., Castillo, M., Smith, J. K., Armao, D., Towle, D., & Muenzer, J. (2004). Brain MRI findings in patients with mucopolysaccharidosis types I and II and mild clinical presentation. *Neuroradiology*, *46*(8), 666–672. doi:10.100700234-004-1215-1 PMID:15205860

McKusick, V. A., & Neufeld, E. F. (1983). The mucopolysaccharide storage diseases. In J. B. Stanbury, J. B. Wyngaarden, D. S. Fredrickson, J. L. Goldstein, & M. S. Brown (Eds.), *The Metabolic Basis of Inherited Disease*. McGraw-Hill.

Meikle, P. J., Hopwood, J. J., Clague, A. E., & Carey, W. F. (1999). Prevalence of lysosomal storage disorders. *Journal of the American Medical Association, 281*(3), 249–254. doi:10.1001/jama.281.3.249 PMID:9918480

Mindell, J. A. (2012). Lysosomal acidification mechanisms. *Annual Review of Physiology, 74*(1), 69–86. doi:10.1146/annurev-physiol-012110-142317 PMID:22335796

Moser, H. W., Mahmood, A., & Raymond, G. V. (2007). X-linked adrenoleukodystrophy. *Nature Clinical Practice. Neurology, 3*(3), 140–151. doi:10.1038/ncpneuro0421 PMID:17342190

Mosser, J., Douar, A.-M., Sarde, C.-O., Kioschis, P., Feil, R., Moser, H., Poustka, A.-M., Mandel, J.-L., & Aubourg, P. (1993). Putative X-linked adrenoleukodystrophy gene shares unexpected homology with ABC transporters. *Nature, 361*(6414), 726–730. doi:10.1038/361726a0 PMID:8441467

Muenzer, J. (2004). The mucopolysaccharidoses: A heterogeneous group of disorders with variable pediatric presentations. *The Journal of Pediatrics, 144*(5), S27–S34. doi:10.1016/j.jpeds.2004.01.052 PMID:15126981

Muenzer, J., Gucsavas-Calikoglu, M., McCandless, S. E., Schuetz, T. J., & Kimura, A. (2007). A phase I/II clinical trial of enzyme replacement therapy in mucopolysaccharidosis II (Hunter syndrome). *Molecular Genetics and Metabolism, 90*(3), 329–337. doi:10.1016/j.ymgme.2006.09.001 PMID:17185020

Neufeld, E. F., & Muenzer, J. (2001). The mucopolysaccharidoses. The Metabolic and Molecular Bases of Inherited Disease. New York: McGraw-Hill.

Osorio, M. J., & Goldman, S. A. (2016). Glial progenitor cell-based treatment of the childhood leukodystrophies. *Experimental Neurology, 283*, 476–488. doi:10.1016/j.expneurol.2016.05.010 PMID:27170209

Parenti, G., Andria, G., & Ballabio, A. (2015). Lysosomal storage diseases: From pathophysiology to therapy. *Annual Review of Medicine, 66*(1), 471–486. doi:10.1146/annurev-med-122313-085916 PMID:25587658

Perry, V. H., Nicoll, J. A. R., & Holmes, C. (2010). Microglia in neurodegenerative disease. *Nature Reviews. Neurology, 6*(4), 193–201. doi:10.1038/nrneurol.2010.17 PMID:20234358

Platt, F. M., d'Azzo, A., Davidson, B. L., Neufeld, E. F., & Tifft, C. J. (2018). Lysosomal storage diseases. *Nature Reviews. Disease Primers, 4*(1), 27. doi:10.103841572-018-0025-4 PMID:30275469

Polten, A., Fluharty, A. L., Fluharty, C. B., Kappler, J., von Figura, K., & Gieselmann, V. (1991). Molecular basis of different forms of metachromatic leukodystrophy. *The New England Journal of Medicine, 324*(1), 18–22. doi:10.1056/NEJM199101033240104 PMID:1670590

Poupětová, H., Ledvinová, J., Berná, L., Dvořáková, L., Kožich, V., & Elleder, M. (2010). The birth prevalence of lysosomal storage disorders in the Czech Republic: Comparison with data in different populations. *Journal of Inherited Metabolic Disease, 33*(4), 387–396. doi:10.100710545-010-9093-7 PMID:20490927

Ruivo, R., Anne, C., Sagné, C., & Gasnier, B. (2009). Molecular and cellular basis of lysosomal transmembrane protein dysfunction. *Biochimica et Biophysica Acta (BBA)-. Molecular Cell Research, 1793*(4), 636–649. PMID:19146888

Saftig, P., & Klumperman, J. (2009). Lysosome biogenesis and lysosomal membrane proteins: Trafficking meets function. *Nature Reviews. Molecular Cell Biology*, *10*(9), 623–635. doi:10.1038/nrm2745 PMID:19672277

Samie, M. A., & Xu, H. (2014). Lysosomal exocytosis and lipid storage disorders. *Journal of Lipid Research*, *55*(6), 995–1009. doi:10.1194/jlr.R046896 PMID:24668941

Schiffmann, R., & Elroy-Stein, O. (2006). Childhood ataxia with CNS hypomyelination/vanishing white matter disease—A common leukodystrophy caused by abnormal control of protein synthesis. *Molecular Genetics and Metabolism*, *88*(1), 7–15. doi:10.1016/j.ymgme.2005.10.019 PMID:16378743

Schuchman, E. H., & Wasserstein, M. P. (2015). Types A and B niemann-pick disease. *Best Practice & Research. Clinical Endocrinology & Metabolism*, *29*(2), 237–247. doi:10.1016/j.beem.2014.10.002 PMID:25987176

Seto, T., Kono, K., Morimoto, K., Inoue, Y., Shintaku, H., Hattori, H., Matsuoka, O., Yamano, T., & Tanaka, A. (2001). Brain magnetic resonance imaging in 23 patients with mucopolysaccharidoses and the effect of bone marrow transplantation. *Annals of Neurology*, *50*(1), 79–92. doi:10.1002/ana.1098 PMID:11456314

Takahashi, Y., Sukegawa, K., Aoki, M., Ito, A., Suzuki, K., Sakaguchi, H., Watanabe, M., Isogai, K., Mizuno, S., Hoshi, H., Kuwata, K., Tomatsu, S., Kato, S., Ito, T., Kondo, N., & Orii, T. (2001). Evaluation of accumulated mucopolysaccharides in the brain of patients with mucopolysaccharidoses by 1H-magnetic resonance spectroscopy before and after bone marrow transplantation. *Pediatric Research*, *49*(3), 349–355. doi:10.1203/00006450-200103000-00008 PMID:11228260

Tamagawa, K., Morimatsu, Y., Fujisawa, K., Hara, A., & Taketomi, T. (1985). Neuropathological study and chemico-pathoiogical correlation in sibling cases of Sanfilippo syndrome type B. *Brain & Development*, *7*(6), 599–609. doi:10.1016/S0387-7604(85)80008-5 PMID:3938624

Toledano-Zaragoza, A., & Ledesma, M. D. (2020). Addressing neurodegeneration in lysosomal storage disorders: Advances in Niemann Pick diseases. *Neuropharmacology*, *171*, 107851. doi:10.1016/j.neuropharm.2019.107851 PMID:31734384

Valstar, M. J., Ruijter, G. J. G., Van Diggelen, O. P., Poorthuis, B. J., & Wijburg, F. A. (2008). Sanfilippo syndrome: A mini-review. *Journal of Inherited Metabolic Disease*, *31*(2), 240–252. doi:10.100710545-008-0838-5 PMID:18392742

Vanier, M. T. (2010). Niemann-Pick disease type C. *Orphanet Journal of Rare Diseases*, *5*(1), 1–18. doi:10.1186/1750-1172-5-16 PMID:20525256

von Figura, K., & Kresse, H. (1972). The Sanfilippo B corrective factor: A N-acetyl-α-D-glucosaminidase. *Biochemical and Biophysical Research Communications*, *48*(2), 262–269. doi:10.1016/S0006-291X(72)80044-5 PMID:4261365

Walterfang, M., Fahey, M., Desmond, P., Wood, A., Seal, M. L., Steward, C., Adamson, C., Kokkinos, C., Fietz, M., & Velakoulis, D. (2010). White and gray matter alterations in adults with Niemann-Pick disease type C: A cross-sectional study. *Neurology*, *75*(1), 49–56. doi:10.1212/WNL.0b013e3181e6210e PMID:20484681

Wassif, C. A., Cross, J. L., Iben, J., Sanchez-Pulido, L., Cougnoux, A., Platt, F. M., Ory, D. S., Ponting, C. P., Bailey-Wilson, J. E., Biesecker, L. G., & Porter, F. D. (2016). High incidence of unrecognized visceral/neurological late-onset Niemann-Pick disease, type C1, predicted by analysis of massively parallel sequencing data sets. *Genetics in Medicine*, *18*(1), 41–48. doi:10.1038/gim.2015.25 PMID:25764212

Wenger, D. A., Rafi, M. A., Luzi, P., Datto, J., & Costantino-Ceccarini, E. (2000). Krabbe disease: Genetic aspects and progress toward therapy. *Molecular Genetics and Metabolism*, *70*(1), 1–9. doi:10.1006/mgme.2000.2990 PMID:10833326

Zhou, Y., Borchelt, D., Bauson, J. C., Fazio, S., Miles, J. R., Tavori, H., & Notterpek, L. (2020). Subcellular diversion of cholesterol by gain-and loss-of-function mutations in PMP22. *Glia*, *68*(11), 2300–2315. doi:10.1002/glia.23840 PMID:32511821

Chapter 19
Investigation of Potential Neuroprotective Role of Chitosan-Based Biomaterials and Their Derivatives by Targeting Glial Cells

Youssef Ait Hamdan
https://orcid.org/0009-0005-8616-8852
University of Rennes, France & Higher Normal School, Cadi Ayyad University, Morocco

Mohammed Rhazi
Higher Normal School, Cadi Ayyad University, Morocco

Hamid Kabdy
Faculty of Sciences, Cadi Ayyad University, Morocco

Abdelali Benmaloui
Faculty of Sciences Semlalia, Morocco

Samia Elouali
Higher Normal School, Cadi Ayyad University, Morocco

Jawad Laadraoui
https://orcid.org/0000-0002-9784-1265
Faculty of Sciences, Hassan II University, Morocco

Zainab Haidar Ahmad
University of Rennes, France

Nadia Eladlani
Higher Normal school, Cadi Ayyad University, Morocco

Hassane Oudadesse
University of Rennes, France

ABSTRACT

Glial cells, although often overlooked in comparison with neurons, play a crucial role in the structure and function of the nervous system. Their loss can lead to impairment of functions such as learning and memory, contributing to neurodegenerative diseases. To combat these diseases effectively, therapeutic strategies must restore the function of neuronal and glial cells. Natural resources offer promising thera-

DOI: 10.4018/978-1-6684-9675-6.ch019

peutic molecules for the treatment of neurodegenerative diseases. Chitosan (CTS), chitooligosaccharide (COS) and their derivatives from arthropod exoskeletons have attracted the attention of neuroscientists because of their neuroprotective effects and their ability to transport therapeutic substances across the blood-brain barrier (BBB). This chapter explores the role of natural resources in the fight against neuronal diseases.

INTRODUCTION

In the nervous system, there are two main cell types: neurons and glial cells. Surprisingly, glial cells account for up to 90% of all cells in the human brain. Although the term "glia" derives from the Greek word for "glue", now the understanding these cells do more than simply hold nerve cells together, and should not simply be regarded as passive support for the neurons of the nervous system (Parpura *et al.*, 2012). In fact, recent convincing studies suggest that glial cells play an active and crucial role in brain development and function, far beyond what we previously thought (Allen & Barres, 2005; Seifert *et al.*, 2006; Parpura *et al.*, 2012). Given their crucial role in the functioning of the nervous system, any pathological alteration of glial cells would have significant and damaging consequences for the brain. Many neurodegenerative disorders are at the root of glial cell dysfunction. As a result, one of the major challenges in the treatment of these neurological disorders and neurodegenerative diseases lies in the fact that some drugs do not cross the Blood Brain Barrier (BBB) to reach brain tissue (Gao, 2016). Consequently, it is very important to develop new drugs and approaches to target these disorders.

Based on neuroprotective mechanisms, several neuroprotective agents can be used to treat neurological disorders (Pellicciari et al., 1998; Chandrasekaran et at., 2003) such as antioxidants (Pellicciari et al., 1998; Behl & Moosmann, 2002) and agents with anti-inflammatory effects (Agnello et al, 2002; Gao et al., 2003). Natural substances, as a source of potential therapeutic molecules, have not received much attention in the treatment of neurodegenerative diseases, although they can play an important role. For example, the marine environment is known for its nooks and crannies in the structures of bioactive compounds with promising neuroprotective biological activities (Alonso et al., 2005). Recent reports on chitosan (CTS) and its derivatives as plausible molecules to target neurodegenerative disorders are well documented in several studies (Hao et al., 2017; Hamdan et al., 2023). This chapter discusses the use of CTS derivatives to treat glial cell-related neurological disorders. It focuses on neuroprotective mechanisms and controlled drug release, with potential therapeutic implications.

GLIAL CELLS DISEASE AND THEIR PATHOGENESIS

The glial cells of the central nervous system play a fundamental role in maintaining the balance of nerve tissues and contribute to the protection and support of neurons. They fulfill several essential functions, including providing nutrients and oxygen to neurons, removing dead cells, fighting against pathogens, and producing myelin. There are two categories of glial cells: non-myelinating cells, such as ependymocytes that line the ventricular cavities, microglia that act as sentinels in the inflammatory response of the central nervous system, and astrocytes that contribute to cerebral homeostasis. Additionally, there are myelinating cells, the oligodendrocytes, which are responsible for forming the myelin sheath around axons.

ASTROCYTES AND THEIR PATHOGENESIS

Astrocytes are glial cells that play important roles in the nutrition, homeostasis, and protection of the nervous system. They have a star-like appearance (Araque et al., 1999) and occupy a prominent and versatile position in the brain (Perez-Nievas & Serrano-Pozo, 2018). They were first described by Rudolph Virchow in 1846 (Molofsky & Deneen, 2015). Structurally, they provide vital support to neurons and other brain cells by forming a complex network of processes that envelop synapses, blood vessels, and neuronal cell bodies (Perez-Nievas & Serrano-Pozo, 2018). On an ionic level, astrocytes are essential for maintaining the balance of ion exchanges, particularly potassium, in the central nervous system (sidoryk-wegrzynowicz et al., 2011; Martinez-Banaclocha, 2020). They act as a "spatial buffer" by redistributing potassium released into the extracellular environment. Additionally, astrocytes actively contribute to the formation of the blood-brain barrier, protecting the brain by maintaining a stable environment against potentially harmful substances circulating in the blood(Beenhakker & Huguenard, 2010). Their role does not stop there; astrocytes play a crucial role in the uptake and recycling of neurotransmitters, eliminating excess neurotransmitters released during synaptic transmissions for smooth neuronal communication. For example, the rapid absorption of gamma-aminobutyric acid (GABA) prevents the inhibitory signal from spreading to other regions (Kinney & Spain, 2002). Regarding energy metabolism, astrocytes capture and metabolize glucose, which is the main source of energy for neurons. The glycogen they store can be released as glucose to meet the energy needs of active neurons (Pellerin, 2008).

Their contributions to the proper functioning of the nervous system are diverse, and ongoing research continues to reveal other astrocytic functions that significantly contribute to their overall role in the nervous system. According to current scientific evidence, astrocytes play a crucial role in the process of neurorepair and neurogenesis, and they are essential for restoring local balance after brain injury (Acosta-Martínez, 2020).

Gliomas are primary brain tumors that originate from glial cells, mainly astrocytes. They represent the most common form of central nervous system tumors in adults, resulting from uncontrolled proliferation of astrocytes. Gliomas exhibit significant tumor heterogeneity, making their management complex, but advancements in research offer promising prospects for new targeted therapies (Baumann & Pham-dinh, 2002).

Astrocytes play a crucial role in neurodegenerative diseases (Phatnani & Maniatis, 2015). For example, in Alzheimer's disease, they contribute to inflammation and neuronal damage due to the accumulation of amyloid proteins (Perez-Nievas & Serrano-Pozo, 2018). In Parkinson's disease, they are involved in inflammation and neurodegeneration. Astrocytopathies are a group of rare and hereditary disorders that affect astrocytes, leading to alterations in cellular metabolism and other functions. Dysfunctions of astrocytes are also associated with other neurological and psychiatric disorders, opening up possibilities for new therapeutic targets.

In addition to the specific diseases mentioned earlier, astrocyte dysfunctions are associated with various other neurological and psychiatric disorders (Stobart & Anderson, 2013). Neurological conditions like epilepsy often involve altered astrocytic activity, which can contribute to excessive neuronal excitability and epileptic seizures (SIRI et al., 2013). Similarly, psychiatric disorders such as depression and schizophrenia are associated with abnormalities in astrocyte function, including altered neurotransmission, metabolic deficits, and faulty regulation of glutamate release, a key neurotransmitter in the brain (SIRI et al., 2013). Astrocytes also play a significant role in regulating cerebral blood flow, and dysfunctions in this function can contribute to strokes and cerebral ischemia. Moreover, deficiencies in the detoxification and waste clearance mechanisms by astrocytes may be linked to the accumulation of abnormal proteins in the brain, contributing to various neurodegenerative diseases (Phatnani & Maniatis, 2015).

OLIGODENDROCYTES AND THEIR PATHOGENESIS

Oligodendrocytes are cells in the central nervous system (CNS) responsible for myelination. They represent the final stage of a cellular lineage that undergoes a complex and precisely coordinated program of proliferation, migration, differentiation, and myelination to ultimately produce the insulating sheath surrounding axons (Brinkmann et al., 2008). However, oligodendrocytes play a crucial role as producers of essential myelin in the central nervous system. Myelin is an insulating sheath that surrounds the axons of neurons, allowing for rapid and efficient transmission of electrical signals (Barres, 2008). Oligodendrocytes are responsible for the formation and maintenance of this myelin sheath, thus playing a key role in neuronal function. During brain development, oligodendrocytes generate multiple cytoplasmic extensions, which they wrap around the axons of neurons, forming segments of myelin. These segments, called internodes, are interrupted by nodes of Ranvier, where myelin is absent, allowing for rapid propagation of electrical signals along the axon (Del Bigio, 2010). Dysfunctions of oligodendrocytes and degradation of myelin are associated with several demyelinating diseases, such as multiple sclerosis. In multiple sclerosis, the immune system attacks the myelin, leading to the degeneration of oligodendrocytes and impairment of nerve transmission.

Oligodendrocytopathies encompass a group of demyelinating disorders resulting from alterations or dysfunction of oligodendrocytes, the cells responsible for producing myelin in the central nervous system. These conditions can have severe consequences on brain and nervous system function due to the loss or degradation of myelin, which insulates neurons' axons (Bouchat et al., 2018). Consequently, demyelination leads to slow and inadequate transmission of electrical signals along the axons. This can manifest as various symptoms, including motor impairments, coordination problems, cognitive deficits, sensory alterations, and disturbances in sphincter control (Nakahara et al., 2012). A well-known oligodendrocytopathy is multiple sclerosis, an autoimmune disease where the immune system attacks myelin, resulting in disseminated demyelinating lesions in the brain and spinal cord. MS causes a variety of neurological symptoms and can significantly impact patients' quality of life (Viader, 2023). Other rare oligodendrocytopathies include inherited genetic disorders such as leukodystrophy, characterized by progressive demyelination due to specific metabolic defects in oligodendrocytes (Barboura et al., 2010). Early and accurate diagnosis of oligodendrocytopathies is crucial for appropriate management and targeted therapeutic interventions (Pellerin et al., 2008).

MICROGLIA AND THEIR PATHOGENESIS

Depending on the anatomical region, microglia constitute 0.5% to 16.6% of the total cell population in the human brain and 5% to 12% in the mouse brain (Gomes-Leal, 2012). Under physiological conditions, the number and functions of microglia are closely regulated by the local microenvironment and interactions with neighboring cells. In response to an insult, microglia can adopt different functional states, modifying their proliferation (Gómez-Nicola et al., 2013) morphology (Cuadros & Navascués, 1998), phagocytic activity (Sierra et al., 2013), antigen presentation -(Jimenez-Ferrer et al., 2017), and release of inflammatory factors such as cytokines and chemokines (Kettenmann et al., 2011). In addition to their role in immune defense, microglia also participate in brain remodeling and repair processes. They play a role in the formation and elimination of synapses, contributing to brain plasticity and learning. Moreover, they may be involved in regulating neurogenesis, the process of creating new neurons (Kwon & Koh, 2020).

However, prolonged and excessive activation of microglia can be damaging to the brain, leading to chronic inflammation and neuronal damage. Dysfunctions of microglia have been associated with various neurological and neurodegenerative diseases, such as Alzheimer's disease, Parkinson's disease, and multiple sclerosis. Microglia and brain inflammation play a central role in neurodegenerative diseases, particularly in Alzheimer's disease and Parkinson's disease (Bachiller et al., 2018).

In Alzheimer's disease, microglial inflammation is strongly implicated in disease progression (Henneman et al., 2009). Amyloid plaques, characteristic of this disease, trigger excessive and prolonged microglial activation (Venegas & Heneka, 2019). This leads to the release of pro-inflammatory cytokines and free radicals, which damage healthy neurons and exacerbate brain lesions. The chronic inflammation induced by microglia contributes to neurodegeneration, synaptic loss, and cognitive impairment in Alzheimer's patients (Ulland et al., 2017).

In Parkinson's disease, the interaction between microglia and dopaminergic neurons is a crucial aspect (Bachiller et al., 2018). The progressive death of dopaminergic neurons in the substantia nigra of the brain is a hallmark of Parkinson's disease (Brettschneider et al., 2015). Activated microglia play a paradoxical role; they may have beneficial effects in attempting to protect dopaminergic neurons in response to injury (Perry, 2012). The activation of microglia in Parkinson's disease has been extensively studied and documented. In 1988, Mcgeer et al. (1988a) demonstrated reactivity of the human leukocyte antigen (HLA) receptor in the substantia nigra pars compacta (SNpc) of patients with Parkinson's disease. In addition to microglial activation, various neuroinflammatory factors have been identified in the brains of Parkinson's patients (Qian et al., 2011). Elevated levels of TNF-α, IL-1β, IL-6, and interferon-γ (IFN-γ) have been detected in the striatum and substantia nigra of Parkinson's patients compared to healthy controls (Parent & Parent, 2010). Another pro-inflammatory molecule, NOS, has been measured in glial cells near dopaminergic neurons in the brains of Parkinson's patients (Hunot et al., 1996). The production of nitric oxide by glial cells can be toxic to dopaminergic neurons (Boje & Arora, 1992). In fact, in vitro inhibition of NOS protects neurons in co-culture with lipopolysaccharide (LPS)-activated microglial cells (Boje and Arora, 1992). Additionally, the production of different pro-inflammatory cytokines may have a toxic effect on neurons expressing their receptors. For example, TNF-α is highly expressed in dopaminergic neurons of Parkinson's patients, which could exert deleterious effects in an inflammatory context (Mogi et al., 2000).

Strong evidence has highlighted the involvement of the innate immune system and microglial response in the progression of neurodegenerative diseases, such as Parkinson's disease and Alzheimer's disease. Microglial cells can assume various functional activation states, playing an active role in brain homeostasis. However, uncontrolled inflammatory responses by microglia can have detrimental effects. Signaling pathways, such as JAK/STAT and NFκB, govern the pro-inflammatory microglial response, and molecules such as TNF-α, IL-1β, and iNOS are primarily induced by these signaling pathways. This activation is associated with deleterious effects in neurodegenerative diseases, amplifying the inflammatory response and increasing damage to neurons (Bachiller et al., 2018).

EPENDYMOCYTES AND THEIR PATHOGENESIS

The ependymal cells are multiciliated cells of neuroepithelial origin that line the spinal cord and cerebral ventricles (Del Bigio, 2010). They are derived from radial glial cells in the embryo between embryonic day 14 and 16 (Spassky et al., 2005). Initially, they form as immobile monopolar epithelial cells, and then during the first two postnatal weeks, they develop into mobile multiciliated cells. The complete differentiation and maturation of ependymal cells are still not fully understood (Deng et al., 2023). Ependymal cells play a crucial role as regulators of cerebrospinal fluid (CSF), an essential fluid that bathes the brain and spinal cord (Lechtreck et al., 2008). These epithelial cells line the cerebral ventricles and the central canal of the spinal cord, forming the walls of these CSF-filled cavities (Deng et al., 2023). Their main function is to actively produce and secrete CSF, which serves several vital roles in the central nervous system. CSF acts as a cushion, protecting the brain and spinal cord from shocks and trauma. It also transports nutrients, hormones, and metabolic waste, facilitating exchanges between the brain and the rest of the body. Additionally, ependymal cells are involved in regulating the chemical composition of CSF. They actively control the levels of ions, glucose, and other molecules in the fluid, thus maintaining a stable and supportive environment for optimal neurological function (Nascimento et al., 2018).

Ependymocytes play a crucial role in the central nervous system by fulfilling several essential functions. One of their main tasks is to create a beating movement to maintain a normal flow of cerebrospinal fluid, which is vital for protecting and ensuring the proper functioning of the brain and spinal cord (Nakayama & Kohno, 1974). Additionally, they are involved in the transport of various substances, expressing genes associated with transport function and contributing to the transport of glucose and fructose (Chiba & Sakata-Yanagimoto, 2020). CD9-positive ependymocytes have been identified as capable of generating neurospheres, highlighting their role in regulating neural stem cells (Horiguchi et al., 2020). Overall, these cells exhibit a complex array of essential functions for the functioning and balance of the central nervous system. The Stroke, one of the most prevalent cerebrovascular diseases, stands as the leading cause of morbidity and mortality worldwide (Kelly & Rothwell, 2020). Following a stroke, there is a decrease in the number of ependymal cells in the neurogenic niche at 6 and 16 weeks, and ectopic ependymal cells migrate to the brain's striatum, contributing to scar formation (Danilov et al., 2012). Furthermore, subarachnoid hemorrhage leads to ependymal cell desquamation due to vasospasms in the choroid artery and rupture of the subependymal basal membrane (Yolas et al., 2016).

In general, hydrocephalus encompasses any increase in cerebrospinal fluid (CSF), including cerebral edema. More specifically, it is defined as ventricular hypertrophy that results in accelerated head growth or necessitates surgical intervention (Schrander-Stumpel & Fryns, 1998). This condition is characterized by excessive accumulation of CSF in the cerebral ventricles, leading to increased intracranial pressure. Several molecules present in ependymal cells are associated with hydrocephalus. For example, the elimination of the Daple protein, associated with a high leucine content (Ccdc88c), induces hydrocephalus and ependymal cilia misorientation (Oshita et al., 2003). In nematodes, flies, and mice, cilium assembly and function are regulated by the Regulator of Cilium Formation (RFX) family (Bonnafe et al., 2004). RFX3 inactivation in mice has been shown to cause hydrocephalus, although without Sylvian aqueduct stenosis, characterized by hypoplasia of the secretory ependymal cells of the subcommissural organ and alterations in the number rather than the ultrastructure of cilia (Baas et al., 2006). Another crucial molecule is Cdc42, a small GTPase that regulates numerous cellular functions, including cell polarity (Etienne-Manneville, 2004). Cdc42 inactivation leads to hydrocephalus, postnatal death, and disrupts ependymal cell differentiation, resulting in aqueductal stenosis (Peng et al., 2013).

Connections have been established between ependymal cells and other brain diseases. For instance, after brain injury, inflammatory cytokines upregulate parvalbumin (PV) expression, enhancing ependymal cell movement and adhesion, thus promoting ependymal layer repair and closure of leaks (Szabolcsi & Celio, 2015). In sections of the cervical spinal cord from patients who died from severe central nervous system trauma, a significant increase in nestin-positive ependymal cells was observed, suggesting that some ependymal cells have the potential for neurogenesis in response to brain injury (Cawsey et al., 2015). Ubiquitin (UbI) positive inclusions have been detected in ependymal cells, particularly in the central canal, and their presence is more common in patients with neurological disorders, although not specific to any particular disease. Furthermore, in a rat model of acute aseptic neuroinflammation, microglial activation affected the viability of ependymal cells, leading to the death of some of these cells (Fernández-Arjona et al., 2021).

In the context of Alzheimer's disease, irregular and thickened basal membranes are also observed, similar to aging (Marques et al., 2013). Accumulations of deposits such as Biondi ring tangles, lipofuscin deposits, and lipid droplets rich in oleic acid side chains selectively form in ependymal cells, contributing to the deterioration of the neurogenic niche in the SVZ, resulting in increased suppression of neurogenesis and affecting striatal and hippocampal functions. Additionally, the expression of aquaporins such as AQP1 and cerebrospinal fluid secretion is also reduced (Muthusamy et al., 2015).

CHITIN, CHITOSAN, AND CHITOOLIGOSACCHARIDES

Chitin (β-1,4-linked 2-acetamido-2-deoxy-D-glucose) is an extremely prevalent polymer which found throughout the biosphere. Animals both on land and at sea, insects, and microbes are sources of chitin. For instance, arthropod exoskeletons are primarily made of chitin-based composites in more than 90% of animal and insect species (Rinaudo, 2006). Recently, researchers have looked into a number of marine sources of chitin (Rhazi et al., 2000) and about 21,120,000 tons of chitin-based refuse are thrown out annually in northern Morocco (Arrouze et al., 2019). It is widely recognized that environment is a rich source of chemical compounds that are biologically active and may have beneficial effects, such as neuroprotective ones (Nweet al., 2009). Accordingly, numerous studies, CTS is one of the biologically active substances with strong neuroprotective qualities, and its effects can be attributed to a variety of its traits.

Depending on the application objective, it is essential to take into account the source and method of obtaining CTS. The traits of the finished product are determined by these elements. Its purity, molecular weight (Mw), crystallinity, and level of deacetylation (DD) are crucial in biomedical uses (Younes & Rinaudo, 2015). The biological characteristics of CTS, such as its biocompatibility, non-toxicity, non-allergenicity, and biodegradability, as well as its antifungal, antibacterial, antioxidant, anticancer, and anti-inflammatory activities, are overshadowed by these factors. It is interesting to note that these properties have also been acknowledged: immunoadjuvant, anti-thrombogenic, and anti-cholesterolemic (Kim &Rajapakse, 2005), and applications in the treatment of covid-19 (Ait Hamdan et al, 2022).

CTS (β-(1→4)-2-amino-D-glucose) is a linear polysaccharide derived from partial deacetylation of chitin (Figure 1) (Kim &Rajapakse, 2005; Je & Kim, 2005). The low solubility of this polysaccharide, however, restricts its use in food and biomedical uses (Kumar, MRN, 2000). Some scholars are interested in the transformation of CTS into oligosaccharides because of this restriction (Jeon, 2001). Due to their shorter chain length and free amino groups in the D-glucosamine units, chitooligosaccharides or chitosan oligomer (COS), the oligosaccharide version of CTS, are easily soluble in water (Figure 1) (Jeon, 2000). These characteristics enable COS to engage in an aqueous environment with negatively charged polymers, macromolecules, and polyanions (Turan &Nagata, 2006; Prabaharan, 2008).

Figure 1. Chemical structure of chitin, chitosan and chitosan oligomer (COS)

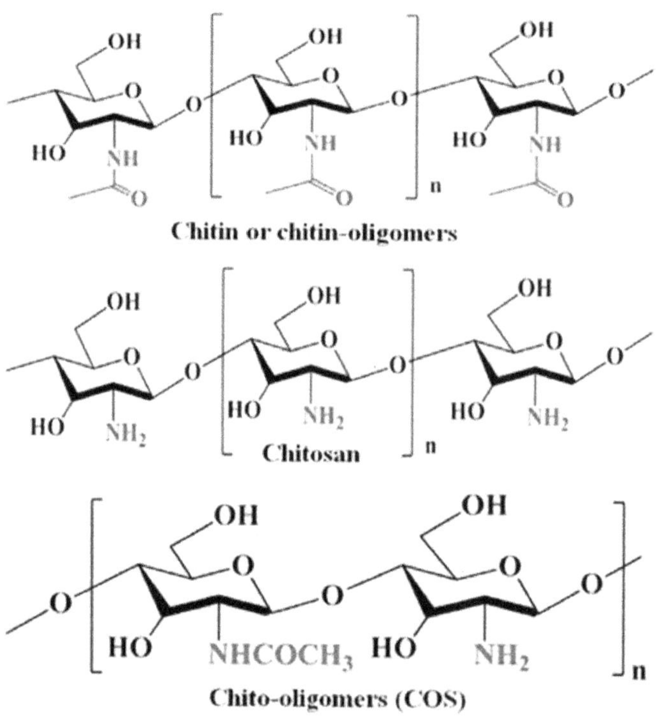

Furthermore, because of its active functional groups, CTS has undergone extensive modification to suit a variety of biological and medical requirements. CTS units have a main hydroxyl group at C-6, a secondary hydroxyl group at C-3, and an amino/acetamido group at C-2 in their molecular structure. Through the use of these groups, CTS can be subjected to a number of chemical transformations, such as acylation, N-phthaloylation, tosylation, reductive alkylation, O-carboxymethylation, N-carboxyalkylation, silylation, and graft copolymerization (Ashrafi and Azadi, 2007; Saboktakin, 2011; Lanjhiyana et al., 2013). These modifications impart the properties of the final products. For instance, CTS used peptides derived from laminin in its grafts to promote neurite development and neuron attachment (Kuo & Chiu, 2011). CTS has many benefits as a brain-targeted drug vehicle (drug carrier), especially nasal drug delivery. Along with its capacity to penetrate the BBB, it can also regulate release, cling to mucus, and open tight junctions in the nasal membrane. These qualities promote its use in drug delivery from the nose to the brain (Mohammed et al., 2017; Ali and Ahmed, 2018). The versatility of CTS, which can be used for drug delivery in the shape of microspheres, capsules, hydrogels, conjugates, nanoparticles, films, beads, or tablets, is another benefit (Yao et al., 2018). CTS is suitable for a wide range of biomedical uses, including drug transport, gene delivery, tissue engineering, and regenerative therapies, among others, thanks to all these characteristics. Moreover, Polymer nanoparticles have also received a lot of interest as a promising CNS drug delivery system due to their capacity to target the brain and their controlled release characteristic (Fazil et al., 2012). CTS nanoparticles are characterized by their positively charged surfaces and good mucosal adhesion and these features may enhance the effectiveness of intravenous or intranasal drug delivery to the brain (Babu & Ramesh, 2017).

CHITOSAN, COS, AND NEUROPROTECTIVE PROPERTIES

Neuroprotection involves various mechanisms like antioxidants, anti-inflammatory agents, anti-excitotoxicity, apoptosis inhibition, β-amyloid suppression, and acetylcholinesterase inhibition.

ANTI-OXIDATIVE STRESS ACTION

Numerous studies have demonstrated the critical role that ROS-induced oxidative stress plays in the pathogenesis of neurological and neurodegenerative disorders. ROS are by-products of aerobic respiration, and the amount they generate is tightly controlled by cellular antioxidant substances and enzymes, but the excessive production of ROS causes cell death (Dro, 2002). Thus, a potential therapeutic goal in NDD is the suppression of free radicals. Hao and colleagues showed that pretreatment with COS significantly reduced the release of lactate dehydrogenase and the production of reactive oxygen species in PC12 cells (Hao et al., 2015). Additionally, xu et al. showed that COS can avoid apoptosis by reducing the rise in the Bax/Bcl-2 ratio and activation of caspase-3, which both contribute to glutamate-induced cell death. Similar to this, COS was shown to have anti-neurotoxic properties against Cu (II)-induced neurotoxicity in cultured cortex rat neurons (Xu et al.,2010). The pretreated COS demonstrated obvious concentration-dependent attenuation of Cu (II) toxicity to cortical neurons. After the 2',7'-dichlorofluorescin assay, COS were found to depreciate the Cu (II)-induced elevation. Hence, COS protect against Cu(II)-induced neurotoxicity in primary cortical neurons by interfering with the increase of intracellular reactive oxygen species (Xu et al.,2010).

SUPPRESSION OF Β-AMYLOID FORMATION

The primary element of senile plaques in AD patients is the -amyloid (A) peptide. Amyloid precursor proteins (APP) can be broken down into A peptides by proteolysis enzymes like - and -secretase (Evin et al., 2016; Dai et al.,2013). Dai et al. showed that Pretreatment with COS greatly reduced the amount of cell death brought on by exposure to A, and COS also reduced ROS production (Okamura et al.,2004; Dai et al., 2015). This suggests that COS may have anti-Aβ fibrillogenesis properties. Furthermore, CTS derivatives have been shown to successfully suppress-secretase activity, with the aminoethyl derivative (AE-CTS) demonstrating the strongest inhibitory activity in comparison to the other derivatives (Je & Kim, 2005). Similarly, byun et al. reported that deacetylated CTS could clearly inhibit β-amyloid formation by blocking β- secretase activity (Byun et al., 2005).

ACETYLCHOLINESTERASE INHIBITORY ACTIVITY

Acetylcholine (ACh) deficiency has been linked the pathogenesis of AD (Khodagholi et al., 2010). Therefore, ACh inhibition by acetylcholinesterase (AChE), enzyme that catalyzes its degradation, is used as a target among the symptomatic treatments of AD (Khodagholi et al., 2010; Dai et al., 2013; Okamura et al., 2004). Due to their soluble nature, A class of naturally occurring, non-toxic carbohydrates known as COSs are rapidly absorbed by the body and have significant biological activity. Acetylcholinesterase (AChE) has been demonstrated to be inhibited by COS and its derivatives, including dimethyl-COS and diethyl-COS, with diethyl-COS having the strongest AChE inhibitory action as a result of its hydrophobicity (Jiang et al., 2009). The effectiveness of COS as AChEIs will also be improved by chemical modification, and their inhibitory activity relies on the hydrophobicity of the group added to them as well as the level of deacetylation. These findings imply that COS and its compounds may be useful as non-toxic AChE inhibitors for the treatment of AD (Jiang et al., 2009).

ANTI-NEUROINFLAMMATORY

The immune system of the body usually responds to irritation or infections by causing inflammation (Kim & Joh, 2006; Aktas et al., 2007). Neuroinflammation is the medical term for this inflammatory process in the brain. The beginning and pathogenesis of neurodegeneration are significantly influenced by neuroinflammation. Interesting connections demonstrate a link between NDD and chronic inflammation (Kim & Joh, 2006). In AD pathology, chronic inflammation is associated with Aβ plaques formation and interleukin-1β (IL-1β) (Kim et al., 2002). Kim et al. showed that CTS inhibited pro-inflammatory cytokine production in human astrocytoma cells activated by Aβ peptide 25-35 (Aβ25-35) and interleukin-1β (IL-1β). It is interesting to note that pretreatment with CTS greatly decreased the expression of iNOS and the secretion of the pro-inflammatory cytokines TNF-alpha and IL-6 in astrocytoma (Kim et al., 2002). Fang et al. investigated the protective effect and the mechanism of COS on retinal ischemia and reperfusion injury, and they found that a high dose of COS is effective in reversing the reduction of the b-wave ratio in the electroretinogram and retinal thickness brought on by retinal ischemia and reperfusion injury (Fang et al., 2015). Similar to this, COS decreased the expression of inflammatory mediators p53 and Bax while increasing the expression of Bcl-2, reducing retinal oxidative damage and the number of apoptotic cells (Khodagholi et al., 2010). Additionally, Kalmar et al. (2009) investigated the anti-neuroinflammatory effects of CTS and its derivatives on NT2 neuronal cells and discovered that CTS exerts its anti-neuroinflammatory action by upregulating heat shock protein 70 (Hsp-70) and inhibiting nuclear factor kappa-light-chain-enhancer of activated B cells (NF-Kb) activation(Kalmar & Greensmith, 2009). By attaching to NF-B and then inhibiting it, Hsp-70's anti-inflammatory site mechanism is mediated (Kim & Joh, 2006). It is therefore necessary to conduct additional research on the anti-neuroinflammatory effect of CTS and its derivatives in other glial cells, especially microglial cells.

ANTI-APOPTOSIS ACTION

NDD can be as result of cell apoptosis (process of programmed cell death) either directly or indirectly. Although is a fundamental event in early development, apoptosis (Twomey & McCarthy, 2005; Fadeel & Orrenius, 2005) is associated with various conditions. In NDD, apoptosis could have an important role in their development, and hence, targeting it could alleviate or stop neurodegeneration (Vila & Przedborski, 2003). Several studies have suggested that COS and its derivatives could inhibit neuronal cell apoptosis. According to Wang et al. (2016), pretreatment with CTS significantly increased cell viability and decreased the dose-dependent lactate dehydrogenase release caused by dibutyltin (DBT) (Wang et al., 2016). Additionally, CTS reduced intracellular ROS production, mitochondrial membrane potential disturbance, and cell apoptosis that were brought on by DBT. Moreover, CTS, also, inhibited DBT-induced caspase-9 and -3 activation at the mRNA and protein expression level. Therefore, by blocking the mitochondria-dependent mechanism, CTS might prevent DBT from causing apoptosis in PC12 cells (Wang et al., 2016). Furthermore, Koo *et al.*, showed that water-soluble high molecular weight CTS could inhibit serum deprivation-induced cell apoptosis in human astrocytes. Hence, CTS derivatives might enhance or stop neurological disorders involving glutamate-induced neuronal cell death (Kooet al., 2002).

Table 1. Mode of neuroprotection by CTS and its derivatives against neurological diseases

Neuronal diseases	Chitosan & COS	Mode of action	References
PD;	*Chitosans*	*Anti-apoptosis action;*	(Khodagholi et al., 2010 ; Wang et al., 2016 ; Cho et al., 2010)
Spinal cord injury		*Anti-oxidative stress*	
		Anti-BACE-1	(Dai et al., 2013 ; Dai et al., 2015 ; Byun et al., 2005 ; Zhou et al., 2008 ; Jia et al., 2016)
AD		*Anti-neuroinflammation;*	
		Anti-apoptosis action	
PD	*COS*	*Anti-exitotoxicaction;*	(Zhou et al., 2008)
		Anti-apoptosis action	
HD		*Anti-exitotoxic action;*	
		Anti-oxidative stres	(Xu et al., 2010)
Nerve crush injury		*Promoting nerve regeneration*	(Jiang et al., 2009 ; Fang et al., 2015 ; Gong et al., 2009)
		Anti-neuroinflammation	
AD	*Peracetylated chitosan*	*Anti-oxidative stress;*	(Hao et al., 2015)
PD	*COS*	*Anti-apoptosis action*	
AD	*Water-soluble chitosans*	*Anti-BACE-1*	(Kim et al., 2002 ; Koo et al., 2002)
		Anti-apoptosis	
AD	*COS*	*Anti-AChE and BACE-1*	(Je &Kim, 2005 ; Byun et al., 2005 ; Yoon et al., 2009)

ANTI-EXCITOTOXIC ACTION

Glutamate is the most abundant excitatory neurotransmitter in the CNS (Choi, 1988) [23]. It ensures several physiologic functions; however, in several conditions, its elevated concentrations within the synaptic cleft can induce neuronal damage. Zhou *et al.*, have shown that COS (M.W. 800) has protective effects against neurotoxicity induced by glutamate neurons' cultures of the hippocampus (Zhou et al., 2008). Moreover, COS has been shown to prevent intracellular calcium Ca2+ concentration elevation induced by glutamate and may stop the activation of caspase-3 induced by glutamate (Zhou et al., 2008). In addition, Dai and colleagues found that COS can inhibit amyloid beta (Aβ) aggregation in a dose-dependent manner (Dai et al., 2015). The inclusion of COS may reduce the neurotoxicity caused by A1-42 in rat cortical neurons. As a result, COS may be able to stop glutamate-induced neurotoxicity both in vitro and in vivo, which may help to delay the onset of neurodegenerative disorders. Table 1 summarizes the possible neuroprotective benefits of CTS and its derivatives against neurological illnesses, and Figure 2 shows their anti-neuronal dysfunction pathways.

Figure 2. The schematic drawing of the anti-neuronal disorder mechanisms of CTS and COS

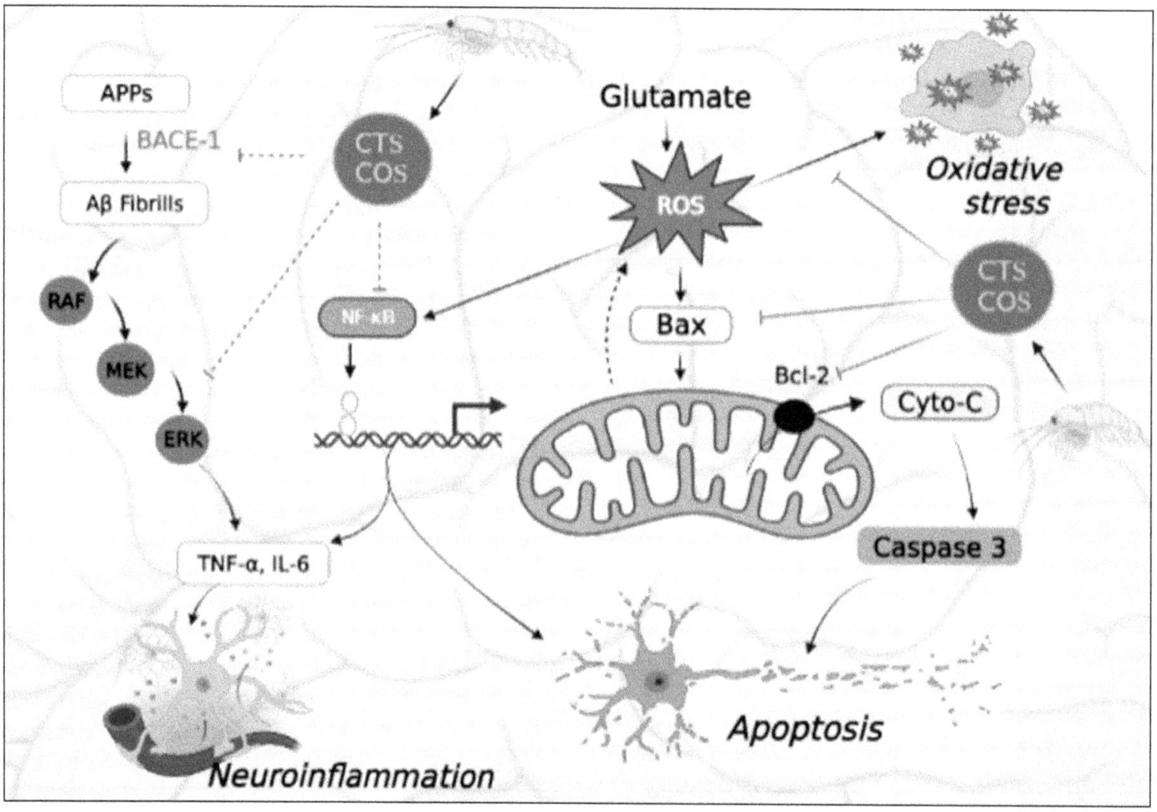

CHITOSAN DRUG DELIVERY SYSTEMS FOR NEURONAL DISORDERS IN GLIAL CELLS

CTS has drawn the attention of researchers for its non-invasive and non-toxic quality for drug delivery. It is considered non-toxic and often used in drug administration. Recently, many studies have been designed to develop drug delivery systems using CTS to treat various neurological diseases via glial cells; several forms and derivatives of chitosan have been used for various uses, it has been extensively studied in brain drug delivery (Shuangwen Yu et al, 2019). In general, polymer nanoparticles such as chitosan, have attracted much attention as a promising central nervous system drug delivery system for its brain targeting ability and controlled release property (Fazil et al., 2012), they have been used extensively in treatment of brain disease including gliomas, PD, AD, etc (Sridhar et al., 2018, Xu et al., 2017). Also, CTS nanoparticles are characterized by their positively charged surfaces and their good adhesion to mucous membranes which gives it the possibility of improving the cerebral administration of drugs, whether intranasally or intravenously. However, the low water solubility of CTS is a major limitation to its application (Babu and Ramesh, 2017). Moreover, the encapsulation efficiency of hydrophilic and hydrophobic drugs is low, which could be overcome by modifying the CTS or optimizing the preparation method (Babu and Ramesh, 2017).

Intranasal administration based on CTS bypasses the BBB and allows drugs to be delivered directly to the brain; this administration is considered relevant given its effectiveness in targeting glial cells and neurons in several neuronal diseases. In the treatment of Alzheimer's disease, several difficulties hinder the penetration of drugs through the BBB, insufficient clearance of the Aβ peptide and the massive release of inflammatory factors, must be urgently overcome. To address these issues, Yang et al. have developed special and novel nanoparticles composed of CTS and bovine serum albumin (BSA) to improve the penetration of drugs through the BBB. (Yang et al., 2018)

During drug delivery, CTS-SLNs (chitosan-Solid lipid nanoparticles) have several advantages regarding the NTB (nose to brain) pathway, protection against degradation in the nasal cavity, high entrapment, high stability, effective mucoadhesive properties, controlled release, high safety profile and increased penetration through epithelial mucus (Rassu et al., 2017). In a study by Cometa et al., they investigated the impact of CTS-GSLs on intranasal dopamine delivery to the brain in the management of Parkinson's disease. They showed that dopamine-loaded CTS-SLNs were smaller than their non-CTS counterparts, had a moderately positive zeta potential value, and had a high scavenging efficiency (81%) (Cometa et al., 2020).

As Elkomy et al. mentioned in a research paper, CTS-NLCs (nanostructured lipid transporters) are also considered potential tools to protect drugs from degradation in the nasal cavity and to stimulate the transfer of drugs from the nose to the brain (Elkomy et al., 2020). In the NTB pathway, CTS-NLCs have many advantages, such as high trapping, high stability, good safety and low toxicity profiles, incredible mucoadhesive characteristics, which enhance penetration through epithelial mucus and prolong the retention time in the nasal cavity (Abo El-Enin et al., 2022). Gartziandia et al. investigated the role of CTS-NLCs in brain protein delivery by intranasal administration by in vitro assays; they demonstrated the biocompatibility of nanocarriers and their cellular uptake by 16HBE14o cells. Moreover, when the particles were incubated with erythrocytes, no hemagglutination or hemolysis process was detected and no toxicity signals emerged in the nasal mucosa of mice after CTS-NLC administration. Finally, the study of biodistribution of CTS-NLCs loaded with the near-infrared dye showed efficient cerebral transfer of the particles after intranasal administration (Gartziandia et al., 2016). Singh et al. also examined the effects of the Glycol CTS-NLC coating on cerebral administration of asenapine (an antipsychotic drug) via the intranasal route of administration in Charles-Foster rats, they showed a systemic bioavailability to the brain between 2.3 and 4 times higher NLC coated with chitosan glycol than asenapine solution after intranasal administration (Singh et al., 2018). As previously reported in the literature, several authors have used CTS-NLCs to deliver drugs to the brain intranasally, such as glial cell-derived neurotrophic factor (GDNF) (Gartziandia et al., 2016, Hernando et al. 2018)

By local injection, the hydrogel of chitosan makes it possible to carry out a prolonged and also well controlled release of drugs, this serves to target specific areas in the nervous system, in particular the glial cells by a targeted distribution. Advanced therapies that combine cells with biomaterial-based carriers are recognized as an emerging and powerful method to treat challenging diseases, such as spinal cord injury (SCI). By improving survival and engraftment of transplanted cells, biomimetic hydrogels can be properly engineered to encapsulate cells and localize them to the injured site in a minimally invasive manner. In this work, Boido et al. developped a CTS-based hydrogels to harbor mesenchymal stem cells (MS-CTS), since their paracrine action can therapeutically enhance SCI regeneration, limiting glial scar formation and reducing death. cell at the injured site. Then they studied a case of an injectable and highly permeable CTS-based hydrogel. The MS-CTS were mixed with the hydrogel solution before gelation. MS-CTS viability was not affected by the CS hydrogel and the encapsulated MS-CTS were able to release MS-CTS vesicles as well as maintain their antioxidant characteristics (Boido et al., 2019). Other uses also for active targeting which involves specific ligands targeting receptors which are glial cells (Béduneau et al., 2007), which makes it possible to increase the efficiency of the delivery and the specificity of the targeting based on delivery systems with base of chitosan and Modulation of glial activation in which chitosan can intervene in the regulation of the expression of specific genes in glial cells, this by using it in the delivery of nucleic acids such as small interfering RNAs (Lin et al., 2020).

Stability and formulation continue to be of great interest, in terms of preserving the integrity of drugs while promoting controlled release through BBB to glial cells and neurons (Figure 3).

CONCLUSION

Bioactive compounds from natural resources display neuroprotective properties. CTS and its derivatives display a range of potential neuroprotective activities, which can help slow the development of certain pathologies. In addition, chitosan-based drug delivery systems for neuronal disorders in glial cells have attracted much interest in the CNS, as they cross the BBB to target neurological disorders and neurodegenerative diseases associated with glial cell dysfunction. Interest in CTS and its derivatives in the treatment of glial pathologies therefore appears to be an emerging field, and CTS-based neurodegenerative therapies hold great promise. However, further research is needed to better understand the potential involvement of CTS and its derivatives in the treatment of neurological disorders, as well as their implication in various applications aimed at protecting the CNS.

Figure 3. Chitosan drug delivery systems for neuronal disorders in glial cells

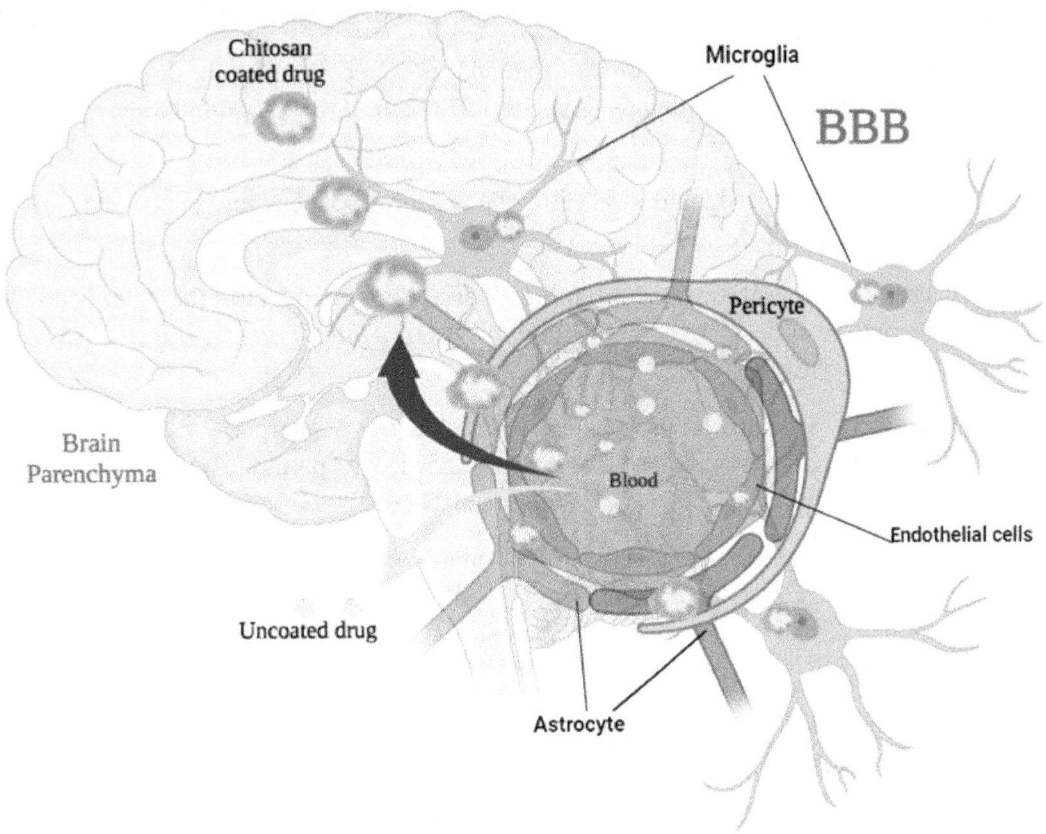

REFERENCES

Abo El-Enin, H. A., Elkomy, M. H., Naguib, I. A., Ahmed, M. F., Alsaidan, O. A., Alsalahat, I., Ghoneim, M. M., & Eid, H. M. (2022). Lipid nanocarriers overlaid with chitosan for brain delivery of berberine via the nasal route. *Pharmaceuticals (Basel, Switzerland), 15*(3), 281. doi:10.3390/ph15030281 PMID:35337079

Acosta-Martínez, M. (2020). Shaping Microglial Phenotypes Through Estrogen Receptors : Relevance to Sex-Specific Neuroinflammatory Responses to Brain Injury and Disease. *The Journal of Pharmacology and Experimental Therapeutics, 375*(1), 223–236. doi:10.1124/jpet.119.264598 PMID:32513838

Agnello, D., Bigini, P., Villa, P., Mennini, T., Cerami, A., Brines, ML et Ghezzi, P. (2002). L'érythropoïétine exerce un effet anti-inflammatoire sur le SNC dans un modèle d'encéphalomyélite auto-immune expérimentale. *Recherche sur le cerveau, 952*(1), 128-134.

Ait Hamdan, Y., El Amerany, F., Desbrières, J., Aghrinane, A., Oudadesse, H., & Rhazi, M. (2022). The evolution of the global COVID-19 epidemic in Morocco and understanding the different therapeutic approaches of chitosan in the control of the pandemic. *Polymer Bulletin*, 1–27. PMID:36466080

Aktas, O., Ullrich, O., Infante-Duarte, C., Nitsch, R., & Zipp, F. (2007). Neuronal damage in brain inflammation. *Archives of Neurology, 64*(2), 185–189. doi:10.1001/archneur.64.2.185 PMID:17296833

Ali, A., & Ahmed, S. (2018). A review on chitosan and its nanocomposites in drug delivery. *International Journal of Biological Macromolecules, 109*, 273–286. doi:10.1016/j.ijbiomac.2017.12.078 PMID:29248555

Allen, NJ, et Barres, BA (2005). Signalisation entre glie et neurones: focus sur la plasticité synaptique. *Opinion actuelle en neurobiologie, 15*(5), 542-548.

Alonso, D., Castro, A., & Martinez, A. (2005). Marine compounds for the therapeutic treatment of neurological disorders. *Expert Opinion on Therapeutic Patents, 15*(10), 1377–1386. doi:10.1517/13543776.15.10.1377

Araque, A., Parpura, V., Sanzgiri, R. P., Haydon, P. G., Araque, A., Parpura, V., Sanzgiri, R. P., & Haydon, P. G. (1999). Tripartite synapses : Glia, the unacknowledged partner. *Trends in Neurosciences, 22*(5), 208–215. doi:10.1016/S0166-2236(98)01349-6 PMID:10322493

Arrouze, F., Desbrieres, J., Lidrissi Hassani, S., & Tolaimate, A. (2021). Investigation of β-chitin extracted from cuttlefish: Comparison with squid β-chitin. *Polymer Bulletin, 78*(12), 7219–7239. doi:10.100700289-020-03466-z

Ashrafi, H., & Azadi, A. (2016). Chitosan-based hydrogel nanoparticle amazing behaviors during transmission electron microscopy. *International Journal of Biological Macromolecules, 84*, 31–34. doi:10.1016/j.ijbiomac.2015.11.089 PMID:26658231

Baas, D., Meiniel, A., Benadiba, C., Bonnafe, E., Meiniel, O., Reith, W., & Durand, B. (2006). A deficiency in RFX3 causes hydrocephalus associated with abnormal differentiation of ependymal cells. *The European Journal of Neuroscience, 24*(4), 1020–1030. doi:10.1111/j.1460-9568.2006.05002.x PMID:16930429

Babu, A., & Ramesh, R. (2017). Multifaceted applications of chitosan in cancer drug delivery and therapy. *Marine Drugs, 15*(4), 96. doi:10.3390/md15040096 PMID:28346381

Babu, A., & Ramesh, R. (2017). Multifaceted applications of chitosan in cancer drug delivery and therapy. *Marine Drugs, 15*(4), 96. doi:10.3390/md15040096 PMID:28346381

Bachiller, S., Jiménez-Ferrer, I., Paulus, A., Yang, Y., Swanberg, M., Deierborg, T., & Boza-Serrano, A. (2018). Microglia in Neurological Diseases : A Road Map to Brain-Disease Dependent-Inflammatory Response. *Frontiers in Cellular Neuroscience, 12*, 488. https://www.frontiersin.org/articles/10.3389/fncel.2018.00488. doi:10.3389/fncel.2018.00488 PMID:30618635

Barboura, I., Ferchichi, S., Dandana, A., Jaidane, Z., Khelifa, S. B., Chahed, H., Mansour, R. B., Chebel, S., Maire, I., & Miled, A. (2010). Metachromatic leucodystrophy Clinical, biological, and therapeutic aspects. *Annales de Biologie Clinique, 68*(4), 385–391. doi:10.1684/abc.2010.0448 PMID:20650733

Barres, B. A. (2008). The Mystery and Magic of Glia : A Perspective on Their Roles in Health and Disease. *Neuron, 60*(3), 430–440. doi:10.1016/j.neuron.2008.10.013 PMID:18995817

Baumann, N., & Pham-dinh, D. (2002). Astrocytes. In V. S. Ramachandran (Ed.), *Encyclopedia of the Human Brain* (pp. 251–268). Academic Press., doi:10.1016/B0-12-227210-2/00042-X

Béduneau, A., Saulnier, P., & Benoit, J. P. (2007). Active targeting of brain tumors using nanocarriers. *Biomaterials*, *28*(33), 4947–4967. doi:10.1016/j.biomaterials.2007.06.011 PMID:17716726

Beenhakker, M. P., & Huguenard, J. R. (2010). Astrocytes as Gatekeepers of GABAB Receptor Function. *The Journal of Neuroscience : The Official Journal of the Society for Neuroscience*, *30*(45), 15262–15276. doi:10.1523/JNEUROSCI.3243-10.2010 PMID:21068331

Behl, C. et Moosmann, B. (2002). Neuroprotection antioxydante dans la maladie d'Alzheimer comme approche préventive et thérapeutique. *Biologie et médecine des radicaux libres, 33*(2), 182-191.

Beilina, A., & Cookson, M. R. (2016). Genes associated with Parkinson's disease: Regulation of autophagy and beyond. *Journal of Neurochemistry, 139*(S1), 91–107. doi:10.1111/jnc.13266 PMID:26223426

Boido, M., Ghibaudi, M., Gentile, P., Favaro, E., Fusaro, R., & Tonda-Turo, C. (2019). Chitosan-based hydrogel to support the paracrine activity of mesenchymal stem cells in spinal cord injury treatment. *Scientific Reports*, *9*(1), 6402. doi:10.103841598-019-42848-w PMID:31024032

Boje, K. M., & Arora, P. K. (1992). Microglial-produced nitric oxide and reactive nitrogen oxides mediate neuronal cell death. *Brain Research*, *587*(2), 250–256. doi:10.1016/0006-8993(92)91004-X PMID:1381982

Bonnafe, E., Touka, M., AitLounis, A., Baas, D., Barras, E., Ucla, C., Moreau, A., Flamant, F., Dubruille, R., Couble, P., Collignon, J., Durand, B., & Reith, W. (2004). The Transcription Factor RFX3 Directs Nodal Cilium Development and Left-Right Asymmetry Specification. *Molecular and Cellular Biology*, *24*(10), 4417–4427. doi:10.1128/MCB.24.10.4417-4427.2004 PMID:15121860

Bouchat, J., Couturier, B., Marneffe, C., Gankam-Kengne, F., Balau, B., De Swert, K., Brion, J.-P., Poncelet, L., Gilloteaux, J., & Nicaise, C. (2018). Regional oligodendrocytopathy and astrocytopathy precede myelin loss and blood–brain barrier disruption in a murine model of osmotic demyelination syndrome. *Glia*, *66*(3), 606–622. doi:10.1002/glia.23268 PMID:29168586

Brettschneider, J., Tredici, K. D., Lee, V. M.-Y., & Trojanowski, J. Q. (2015). Spreading of pathology in neurodegenerative diseases : A focus on human studies. *Nature Reviews. Neuroscience*, *16*(2), 2. doi:10.1038/nrn3887 PMID:25588378

Brinkmann, B. G., Agarwal, A., Sereda, M. W., Garratt, A. N., Müller, T., Wende, H., Stassart, R. M., Nawaz, S., Humml, C., Velanac, V., Radyushkin, K., Goebbels, S., Fischer, T. M., Franklin, R. J., Lai, C., Ehrenreich, H., Birchmeier, C., Schwab, M. H., & Nave, K. A. (2008). Neuregulin-1/ErbB Signaling Serves Distinct Functions in Myelination of the Peripheral and Central Nervous System. *Neuron*, *59*(4), 581–595. doi:10.1016/j.neuron.2008.06.028 PMID:18760695

Byun, H. G., Kim, Y. T., Park, P. J., Lin, X., & Kim, S. K. (2005). Chitooligosaccharides as a novel β-secretase inhibitor. *Carbohydrate Polymers*, *61*(2), 198–202. doi:10.1016/j.carbpol.2005.05.003

Cawsey, T., Duflou, J., Weickert, C. S., & Gorrie, C. A. (2015). Nestin-Positive Ependymal Cells Are Increased in the Human Spinal Cord after Traumatic Central Nervous System Injury. *Journal of Neurotrauma, 32*(18), 1393–1402. doi:10.1089/neu.2014.3575 PMID:25599268

Chandrasekaran, K., Mehrabian, Z., Spinnewyn, B., Chinopoulos, C., Drieu, K., & Fiskum, G. (2003). Neuroprotective effects of bilobalide, a component of Ginkgo biloba extract (EGb 761®) in global brain ischemia and in excitotoxicity-induced neuronal death. *Pharmacopsychiatry, 36*(S 1), 89-94.

Chiba, S., & Sakata-Yanagimoto, M. (2020). Advances in understanding of angioimmunoblastic T-cell lymphoma. *Leukemia, 34*(10), 10. doi:10.103841375-020-0990-y PMID:32704161

Choi, D. W. (1988). Glutamate neurotoxicity and diseases of the nervous system. *Neuron, 1*(8), 623–634. doi:10.1016/0896-6273(88)90162-6 PMID:2908446

Cuadros, M. A., & Navascués, J. (1998). The origin and differentiation of microglial cells during development. *Progress in Neurobiology, 56*(2), 173–189. doi:10.1016/S0301-0082(98)00035-5 PMID:9760700

Dai, X., Chang, P., Zhu, Q., Liu, W., Sun, Y., Zhu, S., & Jiang, Z. (2013). Chitosan oligosaccharides protect rat primary hippocampal neurons from oligomeric β-amyloid 1-42-induced neurotoxicity. *Neuroscience Letters, 554*, 64–69. doi:10.1016/j.neulet.2013.08.046 PMID:23999027

Dai, X., Hou, W., Sun, Y., Gao, Z., Zhu, S., & Jiang, Z. (2015). Chitosan oligosaccharides inhibit/disaggregate fibrils and attenuate amyloid β-mediated neurotoxicity. *International Journal of Molecular Sciences, 16*(5), 10526–10536. doi:10.3390/ijms160510526 PMID:26006224

Danilov, A. I., Kokaia, Z., & Lindvall, O. (2012). Ectopic ependymal cells in striatum accompany neurogenesis in a rat model of stroke. *Neuroscience, 214*, 159–170. doi:10.1016/j.neuroscience.2012.03.062 PMID:22531371

Del Bigio, M. R. (2010). Ependymal cells: Biology and pathology. *Acta Neuropathologica, 119*(1), 55–73. doi:10.100700401-009-0624-y PMID:20024659

Dröge, W. (2002). Free radicals in the physiological control of cell function. *Physiological Reviews, 82*(1), 47–95. doi:10.1152/physrev.00018.2001 PMID:11773609

Elkomy, M. H., Ali, A. A., & Eid, H. M. (2022). Chitosan on the surface of nanoparticles for enhanced drug delivery: A comprehensive review. *Journal of Controlled Release, 351*, 923–940. doi:10.1016/j.jconrel.2022.10.005 PMID:36216174

Elkomy, M. H., Ali, A. A., & Eid, H. M. (2022). Chitosan on the surface of nanoparticles for enhanced drug delivery: A comprehensive review. *Journal of Controlled Release, 351*, 923–940. doi:10.1016/j.jconrel.2022.10.005 PMID:36216174

Etienne-Manneville, S. (2004). Cdc42—The centre of polarity. *Journal of Cell Science, 117*(8), 1291–1300. doi:10.1242/jcs.01115 PMID:15020669

Evin, G. (2016). Future therapeutics in Alzheimer's disease: Development status of BACE inhibitors. *BioDrugs, 30*(3), 173–194. doi:10.100740259-016-0168-3 PMID:27023706

Fadeel, B., & Orrenius, S. (2005). Apoptosis: A basic biological phenomenon with wide-ranging implications in human disease. *Journal of Internal Medicine*, *258*(6), 479–517. doi:10.1111/j.1365-2796.2005.01570.x PMID:16313474

Fang, I. M., Yang, C. M., & Yang, C. H. (2015). Chitosan oligosaccharides prevented retinal ischemia and reperfusion injury via reduced oxidative stress and inflammation in rats. *Experimental Eye Research*, *130*, 38–50. doi:10.1016/j.exer.2014.12.001 PMID:25479043

Fazil, M., Md, S., Haque, S., Kumar, M., Baboota, S., Sahni, J., & Ali, J. (2012). Development and evaluation of rivastigmine loaded chitosan nanoparticles for brain targeting. *European Journal of Pharmaceutical Sciences*, *47*(1), 6–15. doi:10.1016/j.ejps.2012.04.013 PMID:22561106

Fazil, M., Md, S., Haque, S., Kumar, M., Baboota, S., Sahni, J., & Ali, J. (2012). Development and evaluation of rivastigmine loaded chitosan nanoparticles for brain targeting. *European Journal of Pharmaceutical Sciences*, *47*(1), 6–15. doi:10.1016/j.ejps.2012.04.013 PMID:22561106

Fernández-Arjona, M. del M., León-Rodríguez, A., López-Ávalos, M. D., & Grondona, J. M. (2021). Microglia activated by microbial neuraminidase contributes to ependymal cell death. *Fluids and Barriers of the CNS*, *18*(1), 15. doi:10.118612987-021-00249-0 PMID:33757539

Gao, H. (2017). Perspectives on dual targeting delivery systems for brain tumors. *Journal of Neuroimmune Pharmacology*, *12*(1), 6–16. doi:10.100711481-016-9687-4 PMID:27270720

Gao, HM, Liu, B., Zhang, W. et Hong, JS (2003). Nouvelle thérapie anti-inflammatoire pour la maladie de Parkinson. *Tendances des sciences pharmacologiques, 24*(8), 395-401.

Gartziandia, O., Herrán, E., Ruiz-Ortega, J. A., Miguelez, C., Igartua, M., Lafuente, J. V., Pedraz, J. L., Ugedo, L., & Hernández, R. M. (2016). Intranasal administration of chitosan-coated nanostructured lipid carriers loaded with GDNF improves behavioral and histological recovery in a partial lesion model of Parkinson's disease. *Journal of Biomedical Nanotechnology*, *12*(12), 2220–2280. doi:10.1166/jbn.2016.2313 PMID:29372975

Gomes-Leal, W. (2012). Microglial physiopathology: How to explain the dual role of microglia after acute neural disorders? *Brain and Behavior*, *2*(3), 345–356. doi:10.1002/brb3.51 PMID:22741103

Gómez-Nicola, D., Fransen, N. L., Suzzi, S., & Perry, V. H. (2013). Regulation of Microglial Proliferation during Chronic Neurodegeneration. *The Journal of Neuroscience : The Official Journal of the Society for Neuroscience*, *33*(6), 2481–2493. doi:10.1523/JNEUROSCI.4440-12.2013 PMID:23392676

Hamdan, Y. A., Elouali, S., Ben Maloui, A., El-Mansoury, B., Oudadesse, H., & Rhazi, M. (2023). Chitosan and Its Derivatives as Potential Neuro-Protective Agents for the Treatment of Parkinson's Disease. In A. Draoui, O. El Hiba, & A. Jayakumar (Eds.), *Experimental and Clinical Evidence of the Neuropathology of Parkinson's Disease* (pp. 253–274). IGI Global. doi:10.4018/978-1-6684-5156-4.ch015

Hao, C., Gao, L., Zhang, Y., Wang, W., Yu, G., Guan, H., Zhang, L., & Li, C. (2015). Acetylated chitosan oligosaccharides act as antagonists against glutamate-induced PC12 cell death via Bcl-2/Bax signal pathway. *Marine Drugs*, *13*(3), 1267–1289. doi:10.3390/md13031267 PMID:25775423

Hao, C., Wang, W., Wang, S., Zhang, L., & Guo, Y. (2017). An overview of the protective effects of chitosan and acetylated chitosan oligosaccharides against neuronal disorders. *Marine Drugs, 15*(4), 89. doi:10.3390/md15040089 PMID:28333077

Henneman, W. J. P., Sluimer, J. D., Barnes, J., van der Flier, W. M., Sluimer, I. C., Fox, N. C., Scheltens, P., Vrenken, H., & Barkhof, F. (2009). Hippocampal atrophy rates in Alzheimer disease : Added value over whole brain volume measures. *Neurology, 72*(11), 999–1007. doi:10.1212/01.wnl.0000344568.09360.31 PMID:19289740

Hernando, S., Herran, E., Figueiro-Silva, J., Pedraz, J. L., Igartua, M., Carro, E., & Hernandez, R. M. (2020). In vitro investigations on dopamine loaded solid lipid nanoparticles. *Journal of Pharmaceutical and Biomedical Analysis, 185*, 113257. doi:10.1016/j.jpba.2020.113257 PMID:32199326

Horiguchi, S., Fujita, Y., Watanabe, S., Xue, Y., & Nagamatsu, K. (2020). End-to-End Speaker Diarization for an Unknown Number of Speakers with Encoder-Decoder Based Attractors (arXiv:2005.09921). arXiv. /arXiv.2005.09921 doi:10.21437/Interspeech.2020-1022

Hunot, S., Boissière, F., Faucheux, B., Brugg, B., Mouatt-Prigent, A., Agid, Y., & Hirsch, E. C. (1996). Nitric oxide synthase and neuronal vulnerability in parkinson's disease. *Neuroscience, 72*(2), 355–363. doi:10.1016/0306-4522(95)00578-1 PMID:8737406

Je, J. Y., & Kim, S. K. (2005). Water-soluble chitosan derivatives as a BACE1 inhibitor. *Bioorganic & Medicinal Chemistry, 13*(23), 6551–6555. doi:10.1016/j.bmc.2005.07.004 PMID:16084727

Je, J. Y., & Kim, S. K. (2005). Water-soluble chitosan derivatives as a BACE1 inhibitor. *Bioorganic & Medicinal Chemistry, 13*(23), 6551–6555. doi:10.1016/j.bmc.2005.07.004 PMID:16084727

Jeon, Y. J., Park, P. J., & Kim, S. K. (2001). Antimicrobial effect of chitooligosaccharides produced by bioreactor. *Carbohydrate Polymers, 44*(1), 71–76. doi:10.1016/S0144-8617(00)00200-9

Jeon, Y. J., Shahidi, F., & Kim, S. K. (2000). Preparation of chitin and chitosan oligomers and their applications in physiological functional foods. *Food Reviews International, 16*(2), 159–176. doi:10.1081/FRI-100100286

Jiang, M., Zhuge, X., Yang, Y., Gu, X., & Ding, F. (2009). The promotion of peripheral nerve regeneration by chitooligosaccharides in the rat nerve crush injury model. *Neuroscience Letters, 454*(3), 239–243. doi:10.1016/j.neulet.2009.03.042 PMID:19429091

Jimenez-Ferrer, I., Jewett, M., Tontanahal, A., Romero-Ramos, M., & Swanberg, M. (2017). Allelic difference in Mhc2ta confers altered microglial activation and susceptibility to α-synuclein-induced dopaminergic neurodegeneration. *Neurobiology of Disease, 106*, 279–290. doi:10.1016/j.nbd.2017.07.016 PMID:28736195

Kalmar, B., & Greensmith, L. (2009). Induction of heat shock proteins for protection against oxidative stress. *Advanced Drug Delivery Reviews, 61*(4), 310–318. doi:10.1016/j.addr.2009.02.003 PMID:19248813

Kelly, D., & Rothwell, P. M. (2020). Disentangling the multiple links between renal dysfunction and cerebrovascular disease. *Journal of Neurology, Neurosurgery, and Psychiatry, 91*(1), 88–97. doi:10.1136/jnnp-2019-320526 PMID:31511306

Kettenmann, H., Hanisch, U.-K., Noda, M., & Verkhratsky, A. (2011). Physiology of Microglia. *Physiological Reviews, 91*(2), 461–553. doi:10.1152/physrev.00011.2010 PMID:21527731

Khodagholi, F., Eftekharzadeh, B., Maghsoudi, N., & Rezaei, P. F. (2010). Chitosan prevents oxidative stress-induced amyloid β formation and cytotoxicity in NT2 neurons: Involvement of transcription factors Nrf2 and NF-κB. *Molecular and Cellular Biochemistry, 337*(1-2), 39–51. doi:10.100711010-009-0284-1 PMID:19844776

Kim, M. S., Sung, M. J., Seo, S. B., Yoo, S. J., Lim, W. K., & Kim, H. M. (2002). Water-soluble chitosan inhibits the production of pro-inflammatory cytokine in human astrocytoma cells activated by amyloid β peptide and interleukin-1β. *Neuroscience Letters, 321*(1-2), 105–109. doi:10.1016/S0304-3940(02)00066-6 PMID:11872267

Kim, S. K., & Rajapakse, N. (2005). Enzymatic production and biological activities of chitosan oligosaccharides (COS): A review. *Carbohydrate Polymers, 62*(4), 357–368. doi:10.1016/j.carbpol.2005.08.012

Kim, Y. S., & Joh, T. H. (2006). Microglia, major player in the brain inflammation: Their roles in the pathogenesis of Parkinson's disease. *Experimental & Molecular Medicine, 38*(4), 333–347. doi:10.1038/emm.2006.40 PMID:16953112

Kinney, G. A., & Spain, W. J. (2002). Synaptically Evoked GABA Transporter Currents in Neocortical Glia. *Journal of Neurophysiology, 88*(6), 2899–2908. doi:10.1152/jn.00037.2002 PMID:12466416

Koo, H. N., Jeong, H. J., Hong, S. H., Choi, J. H., An, N. H., & Kim, H. M. (2002). High molecular weight water-soluble chitosan protects against apoptosis induced by serum starvation in human astrocytes. *The Journal of Nutritional Biochemistry, 13*(4), 245–249. doi:10.1016/S0955-2863(01)00218-2 PMID:11988407

Kumar, M. N. R. (2000). A review of chitin and chitosan applications. *Reactive & Functional Polymers, 46*(1), 1–27. doi:10.1016/S1381-5148(00)00038-9

Kumirska, J., Weinhold, M. X., Czerwicka, M., & Stepnowski, P. (2011). Influence of the chemical structure and physicochemical properties of chitin-and chitosan-based materials on their biomedical activity. *Biomedical Engineering, Trends in Materials Science*, 25-64.

Kuo, Y. C., & Chiu, K. H. (2011). Inverted colloidal crystal scaffolds with laminin-derived peptides for neuronal differentiation of bone marrow stromal cells. *Biomaterials, 32*(3), 819–831. doi:10.1016/j.biomaterials.2010.09.057 PMID:20974492

Kwon, H. S., & Koh, S.-H. (2020). Neuroinflammation in neurodegenerative disorders : The roles of microglia and astrocytes. *Translational Neurodegeneration, 9*(1), 42. doi:10.118640035-020-00221-2 PMID:33239064

Lanjhiyana, S. K., Bajpayee, P., Kesavan, K., Lanjhiyana, S., & Muthu, M. S. (2013). Chitosan–sodium alginate blended polyelectrolyte complexes as potential multiparticulate carrier system: Colon-targeted delivery and gamma scintigraphic imaging. *Expert Opinion on Drug Delivery, 10*(1), 5–15. doi:10.1517/17425247.2013.734805 PMID:23106236

Lechtreck, K.-F., Delmotte, P., Robinson, M. L., Sanderson, M. J., & Witman, G. B. (2008). Mutations in Hydin impair ciliary motility in mice. *The Journal of Cell Biology*, *180*(3), 633–643. doi:10.1083/jcb.200710162 PMID:18250199

Lin, J., Jo, S. B., Kim, T. H., Kim, H. W., & Chew, S. Y. (2020). RNA interference in glial cells for nerve injury treatment. *Journal of Tissue Engineering*, *11*, 2041731420939224. doi:10.1177/2041731420939224 PMID:32670539

Marques, F., Sousa, J. C., Sousa, N., & Palha, J. A. (2013). Blood-brain-barriers in aging and in Alzheimer's disease. *Molecular Neurodegeneration*, *8*(1), 38. doi:10.1186/1750-1326-8-38 PMID:24148264

Martinez-Banaclocha, M. (2020). Astroglial Isopotentiality and Calcium-Associated Biomagnetic Field Effects on Cortical Neuronal Coupling. *Cells*, *9*(2), 2. doi:10.3390/cells9020439 PMID:32069981

Mogi, M., Kondo, A., Kinpara, K., & Togari, A. (2000). Anti-apoptotic action of nerve growth factor in mouse osteoblastic cell line. *Life Sciences*, *67*(10), 1197–1206. doi:10.1016/S0024-3205(00)00705-0 PMID:10954053

Mohammed, M. A., Syeda, J. T., Wasan, K. M., & Wasan, E. K. (2017). An overview of chitosan nanoparticles and its application in non-parenteral drug delivery. *Pharmaceutics*, *9*(4), 53. doi:10.3390/pharmaceutics9040053 PMID:29156634

Molofsky, A. V., & Deneen, B. (2015). Astrocyte development : A Guide for the Perplexed. *Glia*, *63*(8), 1320–1329. doi:10.1002/glia.22836 PMID:25963996

Muthusamy, N., Sommerville, L. J., Moeser, A. J., Stumpo, D. J., Sannes, P., Adler, K., Blackshear, P. J., Weimer, J. M., & Ghashghaei, H. T. (2015). MARCKS-dependent mucin clearance and lipid metabolism in ependymal cells are required for maintenance of forebrain homeostasis during aging. *Aging Cell*, *14*(5), 764–773. doi:10.1111/acel.12354 PMID:26010231

Nakahara, J., Maeda, M., Aiso, S., & Suzuki, N. (2012). Current Concepts in Multiple Sclerosis : Autoimmunity Versus Oligodendrogliopathy. *Clinical Reviews in Allergy & Immunology*, *42*(1), 26–34. doi:10.100712016-011-8287-6 PMID:22189514

Nakayama, Y., & Kohno, K. (1974). Number and polarity of the ependymal cilia in the central canal of some vertebrates. *Journal of Neurocytology*, *3*(4), 449–458. doi:10.1007/BF01098732 PMID:4436691

Nascimento, D. M., Nunes, Y. L., Figueirêdo, M. C. B., de Azeredo, H. M. C., Aouada, F. A., Feitosa, J. P. A., Rosa, M. F., & Dufresne, A. (2018). Nanocellulose nanocomposite hydrogels : Technological and environmental issues. *Green Chemistry*, *20*(11), 2428–2448. doi:10.1039/C8GC00205C

Nwe, N., Furuike, T., & Tamura, H. (2009). The mechanical and biological properties of chitosan scaffolds for tissue regeneration templates are significantly enhanced by chitosan from Gongronella butleri. *Materials (Basel)*, *2*(2), 374–398. doi:10.3390/ma2020374

Okamura, N., Suemoto, T., Shiomitsu, T., Suzuki, M., Shimadzu, H., Akatsu, H., Yamamoto, T., Arai, H., Sasaki, H., Yanai, K., Staufenbiel, M., Kudo, Y., & Sawada, T. (2004). A novel imaging probe for in vivo detection of neuritic and diffuse amyloid plaques in the brain. *Journal of Molecular Neuroscience*, *24*(2), 247–255. doi:10.1385/JMN:24:2:247 PMID:15456938

Oshita, A., Kishida, S., Kobayashi, H., Michiue, T., Asahara, T., Asashima, M., & Kikuchi, A. (2003). Identification and characterization of a novel Dvl-binding protein that suppresses Wnt signalling pathway. *Genes to Cells*, *8*(12), 1005–1017. doi:10.1111/j.1365-2443.2003.00692.x PMID:14750955

Parent, M., & Parent, A. (2010). Substantia nigra and Parkinson's disease : A brief history of their long and intimate relationship. *The Canadian Journal of Neurological Sciences*, *37*(3), 313–319. doi:10.1017/S0317167100010209 PMID:20481265

Parpura, V., Heneka, MT, Montana, V., Oliet, SH, Schousboe, A., Haydon, PG, & Verkhratsky, A. (2012). Cellules gliales en (patho)physiologie. *Tourillon de neurochimie, 121*(1), 4-27.

Pellerin, B. A., Wollheim, W. M., Feng, X., & Vörösmarty, C. J. (2008). The application of electrical conductivity as a tracer for hydrograph separation in urban catchments. *Hydrological Processes*, *22*(12), 1810–1818. doi:10.1002/hyp.6786

Pellicciari, R., Costantino, G., Marinozzi, M., & Natalini, B. (1998). Modulation of glutamate receptor pathways in the search for new neuroprotective agents. *Il Farmaco*, *53*(4), 255–261. doi:10.1016/S0014-827X(98)00018-4 PMID:9658582

Peng, X., Lin, Q., Liu, Y., Jin, Y., Druso, J. E., Antonyak, M. A., Guan, J.-L., & Cerione, R. A. (2013). Inactivation of Cdc42 in embryonic brain results in hydrocephalus with ependymal cell defects in mice. *Protein & Cell*, *4*(3), 231–242. doi:10.100713238-012-2098-2 PMID:23150167

Perez-Nievas, B. G., & Serrano-Pozo, A. (2018). Deciphering the astrocyte reaction in Alzheimer's disease. *Frontiers in Aging Neuroscience*, *10*, 114. doi:10.3389/fnagi.2018.00114 PMID:29922147

Phatnani, H., & Maniatis, T. (2015). Astrocytes in Neurodegenerative Disease. *Cold Spring Harbor Perspectives in Biology*, *7*(6), a020628. doi:10.1101/cshperspect.a020628 PMID:25877220

Prabaharan, M. (2008). Review Paper: Chitosan derivatives as promising materials for controlled. *Drug Deliv J Biomat Applicat*, *23*(1), 5–36. PMID:18593819

Qian, L., Winfree, E., & Bruck, J. (2011). Neural network computation with DNA strand displacement cascades. *Nature*, *475*(7356), 7356. doi:10.1038/nature10262 PMID:21776082

Rassu, G., Soddu, E., Posadino, A. M., Pintus, G., Sarmento, B., Giunchedi, P., & Gavini, E. (2017). Nose-to-brain delivery of BACE1 siRNA loaded in solid lipid nanoparticles for Alzheimer's therapy. *Colloids and Surfaces. B, Biointerfaces*, *152*, 296–301. doi:10.1016/j.colsurfb.2017.01.031 PMID:28126681

Rinaudo, M. (2006). Chitine et chitosane: Propriétés et applications. *Progrès de la science des polymères, 31*(7), 603-632.

Saboktakin, M. R., Tabatabaie, R. M., Maharramov, A., & Ramazanov, M. A. (2011). Synthesis and characterization of pH-dependent glycol chitosan and dextran sulfate nanoparticles for effective brain cancer treatment. *International Journal of Biological Macromolecules*, *49*(4), 747–751. doi:10.1016/j.ijbiomac.2011.07.006 PMID:21782844

Schrander-Stumpel, C., & Fryns, J.-P. (1998). Congenital hydrocephalus : Nosology and guidelines for clinical approach and genetic counselling. *European Journal of Pediatrics*, *157*(5), 355–362. doi:10.1007004310050830 PMID:9625330

Seifert, G., Schilling, K., & Steinhäuser, C. (2006). Astrocyte dysfunction in neurological disorders: A molecular perspective. *Nature Reviews. Neuroscience*, *7*(3), 194–206. doi:10.1038/nrn1870 PMID:16495941

Sierra, A., Abiega, O., Shahraz, A., & Neumann, H. (2013). Janus-faced microglia : Beneficial and detrimental consequences of microglial phagocytosis. *Frontiers in Cellular Neuroscience*, *7*. https://www.frontiersin.org/articles/10.3389/fncel.2013.00006. doi:10.3389/fncel.2013.00006 PMID:23386811

Singh, S. K., Hidau, M. K., Gautam, S., Gupta, K., Singh, K. P., Singh, S. K., & Singh, S. (2018). Glycol chitosan functionalized asenapine nanostructured lipid carriers for targeted brain delivery: Pharmacokinetic and teratogenic assessment. *International Journal of Biological Macromolecules*, *108*, 1092–1100. doi:10.1016/j.ijbiomac.2017.11.031 PMID:29126941

Spassky, N., Merkle, F. T., Flames, N., Tramontin, A. D., García-Verdugo, J. M., & Alvarez-Buylla, A. (2005). Adult Ependymal Cells Are Postmitotic and Are Derived from Radial Glial Cells during Embryogenesis. *The Journal of Neuroscience : The Official Journal of the Society for Neuroscience*, *25*(1), 10–18. doi:10.1523/JNEUROSCI.1108-04.2005 PMID:15634762

Sridhar, V., Gaud, R., Bajaj, A., & Wairkar, S. (2018). Pharmacocinétique et pharmacodynamique des nanoparticules de sélégiline administrées par voie intranasale avec une administration cérébrale améliorée dans la maladie de Parkinson. Nanomédecine: Nanotechnologie. *Biology and Medicine (Aligarh)*, *14*(8), 2609–2618.

Stobart, J. L., & Anderson, C. M. (2013). Multifunctional role of astrocytes as gatekeepers of neuronal energy supply. *Frontiers in Cellular Neuroscience*, *7*, 38. doi:10.3389/fncel.2013.00038 PMID:23596393

Szabolcsi, V., & Celio, M. R. (2015). De novo expression of parvalbumin in ependymal cells in response to brain injury promotes ependymal remodeling and wound repair. *Glia*, *63*(4), 567–594. doi:10.1002/glia.22768 PMID:25421913

Turan, K., & Nagata, K. (2006). Chitosan-DNA nanoparticles: The effect of cell type and hydrolysis of chitosan on in vitro DNA transfection. *Pharmaceutical Development and Technology*, *11*(4), 503–512. doi:10.1080/10837450600940873 PMID:17101521

Twomey, C., & McCarthy, J. V. (2005). Pathways of apoptosis and importance in developement. *Journal of Cellular and Molecular Medicine*, *9*(2), 345–359. doi:10.1111/j.1582-4934.2005.tb00360.x PMID:15963254

Ulland, T. K., Song, W. M., Huang, S. C.-C., Ulrich, J. D., Sergushichev, A., Beatty, W. L., Loboda, A. A., Zhou, Y., Cairns, N. J., Kambal, A., Loginicheva, E., Gilfillan, S., Cella, M., Virgin, H. W., Unanue, E. R., Wang, Y., Artyomov, M. N., Holtzman, D. M., & Colonna, M. (2017). TREM2 Maintains Microglial Metabolic Fitness in Alzheimer's Disease. *Cell*, *170*(4), 649–663.e13. doi:10.1016/j.cell.2017.07.023 PMID:28802038

Venegas, C., & Heneka, M. T. (2019). Inflammasome-mediated innate immunity in Alzheimer's disease. *The FASEB Journal*, *33*(12), 13075–13084. doi:10.1096/fj.201900439 PMID:31702392

Viader, F. (2023). Amyotrophic lateral sclerosis: An emblematic neurodegenerative disease. *Bulletin de l'Académie Nationale de Médecine*, *207*(3), 272–286. doi:10.1016/j.banm.2023.01.002

Vila, M., & Przedborski, S. (2003). Targeting programmed cell death in neurodegenerative diseases. *Nature Reviews. Neuroscience*, *4*(5), 365–375. doi:10.1038/nrn1100 PMID:12728264

Wang, X., Miao, J., Yan, C., Ge, R., Liang, T., Liu, E., & Li, Q. (2016). Chitosan attenuates dibutyltin-induced apoptosis in PC12 cells through inhibition of the mitochondria-dependent pathway. *Carbohydrate Polymers*, *151*, 996–1005. doi:10.1016/j.carbpol.2016.06.053 PMID:27474647

Wolf, N. I., Ffrench-Constant, C., & van der Knaap, M. S. (2021). Leucodystrophies hypomyélinisantes – Démêler la biologie de la myéline. *Nature Reviews. Neurology*, *17*(2), 88–103. doi:10.103841582-020-00432-1 PMID:33324001

Xu, W., Huang, H. C., Lin, C. J., & Jiang, Z. F. (2010). Chitooligosaccharides protect rat cortical neurons against copper induced damage by attenuating intracellular level of reactive oxygen species. *Bioorganic & Medicinal Chemistry Letters*, *20*(10), 3084–3088. doi:10.1016/j.bmcl.2010.03.105 PMID:20399653

Xu, Y., Asghar, S., Yang, L., Li, H., Wang, Z., Ping, Q., & Xiao, Y. (2017). Lactoferrin-coated polysaccharide nanoparticles based on chitosan hydrochloride/hyaluronic acid/PEG for treating brain glioma. *Carbohydrate Polymers*, *157*, 419–428. doi:10.1016/j.carbpol.2016.09.085 PMID:27987946

Yang, R., Zheng, Y., Wang, Q., & Zhao, L. (2018). Curcumin-loaded chitosan–bovine serum albumin nanoparticles potentially enhanced Aβ 42 phagocytosis and modulated macrophage polarization in Alzheimer's disease. *Nanoscale Research Letters*, *13*(1), 1–9. doi:10.118611671-018-2759-z PMID:30350003

Yao, Z. A., Chen, F. J., Cui, H. L., Lin, T., Guo, N., & Wu, H. G. (2018). Efficacy of chitosan and sodium alginate scaffolds for repair of spinal cord injury in rats. *Neural Regeneration Research*, *13*(3), 502. doi:10.4103/1673-5374.228756 PMID:29623937

Yolas, C., Ozdemir, N. G., Kanat, A., Aydin, M. D., Keles, P., Kepoglu, U., Aydin, N., & Gundogdu, C. (2016). Uncovering a New Cause of Obstructive Hydrocephalus Following Subarachnoid Hemorrhage: Choroidal Artery Vasospasm–Related Ependymal Cell Degeneration and Aqueductal Stenosis—First Experimental Study. *World Neurosurgery*, *90*, 484–491. doi:10.1016/j.wneu.2016.03.049 PMID:27020981

Younes, I., & Rinaudo, M. (2015). Chitin and chitosan preparation from marine sources. Structure, properties and applications. *Marine Drugs*, *13*(3), 1133–1174. doi:10.3390/md13031133 PMID:25738328

Yu, S., Xu, X., Feng, J., Liu, M., & Hu, K. (2019). Chitosan and chitosan coating nanoparticles for the treatment of brain disease. *International Journal of Pharmaceutics*, *560*, 282–293. doi:10.1016/j.ijpharm.2019.02.012 PMID:30772458

Zhou, S., Yang, Y., Gu, X., & Ding, F. (2008). Chitooligosaccharides protect cultured hippocampal neurons against glutamate-induced neurotoxicity. *Neuroscience Letters*, *444*(3), 270–274. doi:10.1016/j.neulet.2008.08.040 PMID:18755243

Chapter 20
Chitosan Polysaccharides: Modulation of Neuroinflammation

Youssef Ait Hamdan
https://orcid.org/0009-0005-8616-8852
University of Rennes, France & Higher Normal School, Cadi Ayyad University, Morocco

Hassane Oudadesse
University of Rennes, France

Samia Elouali
Higher Normal school, Cadi Ayyad University, Morocco

Kamal Smimih
Faculty of Sciences and Techniques, Sultan Moulay Slimane University, Mococco

Hafida El Ghachi
https://orcid.org/0000-0002-2043-6202
Faculty of Sciences, Cadi Ayyad University, Morocco

Ayieton Jeraude Stive David Gnimassoun
Higher Normal School, Cadi Ayyad University, Morocco

Ramadane Abakar Ramdane
Nice University Hospital, France

Nadia Eladlani
Higher Normal school, Cadi Ayyad University, Morocco

Mohammed Rhazi
Higher Normal school, Cadi Ayyad University, Morocco

ABSTRACT

Neuroinflammation is a complex, well-orchestrated process involving various groups of CNS glial cells and immune cells. Glial cells play a remarkable structural and functional role in the nervous system compared to neurons. However, neuroinflammation at the central level is a key player in various neurological disorders, including neurodegenerative diseases and CNS lesions. Therapeutic approaches to combat human neurodegenerative diseases must therefore restore neuronal and glial cell function. Natural resources are a source of potential therapeutic molecules for the treatment of neurodegenerative diseases. Currently, chitosan and its derivatives from arthropod exoskeletons are endowed with powerful anti-neuroinflammatory properties, as well as the ability to transport therapeutic substances across the BBB. This chapter discusses possible therapeutic options, the mechanism and role of chitosan in alleviating neuroinflammation at the central level, and the resulting diseases, in particular glial cell disorders.

DOI: 10.4018/978-1-6684-9675-6.ch020

INTRODUCTION

Neuroinflammation is a feature of the central nervous system (CNS), shared by a range of disorders, spanning from acute injury to neurodegenerative diseases and neuropsychiatric conditions. These neuro-inflammatory disorders and their consequences contribute substantially to the global burden of morbidity and mortality, with remarkable economic repercussions on healthcare expenditure (Aarli et al., 2014).

While classical inflammation is a beneficial tissue repair response, prolonged or inappropriate inflammation can aggravate or even cause the death of neighboring neuronal and glial cells (Neher et al., 2012), (Brown & Vilalta, 2015). In the CNS, glial cells provide essential support and regulate the activity of neurons, which form the main signal transmission pathways in the nervous system. Ultimately, pathological neuroinflammation leads to the deterioration of axons and their supporting cells. In the presence of a localized lesion, microglial cells signal astrocytes to create a glial scar, causing them to differentiate, migrate, multiply and secrete carbohydrate-based fibrous matrices (Sabelström et al., 2013) (Meletis et al., 2008). This creates a physical and chemical barrier that isolates the lesion, hindering axon regeneration (Fehlings & Tighe, 2008). Microglia and astrocytes often dominate the environment within inflammation, creating cytotoxic conditions conducive to neuronal death. This dynamic leads to the loss of adjacent biological circuits, resulting in neuronal and cognitive disorders (Tsui et al., 2019).

Despite current treatments aimed at alleviating the specific symptoms of certain disorders, no cure is currently available for neuro-inflammatory conditions. Although the cause, pathology and symptoms of these disorders differ considerably, they all share a central inflammatory component. On the basis of neuroprotective mechanisms, several neuroprotective agents can be used to tackle neuroinflammation and the neurological disorders that result (Pellicciari et al.,1998), (K. Chandrasekaran, 2003) such as antioxidants (Pellicciari et al., 1998), (Behl & Moosmann, 2002) and agents with anti-inflammatory effects (Agnello et al., 2002), (Gao et al., 2003). However, one of the major challenges in the treatment of neurological disorders and neurodegenerative diseases lies in the fact that some drugs do not cross the blood-brain barrier (BBB) to reach brain tissue (Gao, 2017).

Consequently, it is very important to develop new drugs and approaches that have both anti-inflammatory properties and can easily cross the BBB to target these disorders. As a source of potential therapeutic molecules, bioresources have not received much attention in the treatment of neurodegenerative diseases, although they can play an important role. For example, the marine environment is known for its nooks and crannies in the structures of bioactive compounds with promising neuroprotective biological activities (Alonso et al., 2005). Currently, chitosan (CTS) and its derivatives as plausible molecules with powerful biological and therapeutic features to target neurodegenerative disorders are well documented in several studies (Hao et al., 2017; Hamdan et al., 2023). This chapter reviews recent advances in the use and applications of CTS and its derivatives for the modulation of central neuroinflammation and the prevention of neurological and neurodegenerative disorders. It will focus on the mechanisms and role of these marine derivatives in combination with other molecules for therapeutic use, forming effective preventive and curative approaches.

NEUROINFLAMMATION AND GLIAL CELLS

Neuroinflammation is an inflammatory response that occurs in the nervous system in response to various stimuli such as infections, tissue injuries, toxins, or neurodegenerative disorders (Estera et al., 2021), (Mishra et al., 2021). Its role is to promote tissue repair and activate the immune system (Shi et al., 2021). However, it can also have detrimental effects, leading to neurotoxicity and neuronal degeneration (Asslih et al., 2021).

The inflammatory process in the CNS involves different categories of cells, including microglial cells and astrocytes (Figure 1), both of which belong to the glial cell category (Lassmann, 2020), (Shabab et al., 2017). These cells interact in a complex and dynamic manner, playing a critical role in immune responses. However, this cellular interaction can also be a source of pathology, contributing to various neurological disorders.

Figure 1. Representation of pathology of astrocytes and microglia in inflammation

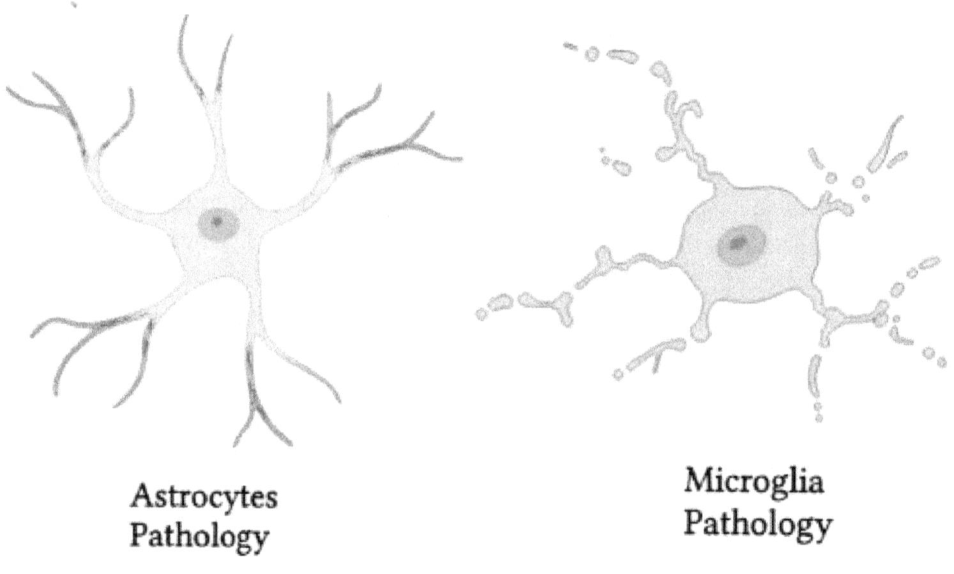

This intricate network of communication and cooperation primarily involves microglial cells and astrocytes, with a bidirectional interaction (Nosi et al., 2021). Microglia and astrocytes can adopt pro-inflammatory or anti-inflammatory phenotypes, and their interaction within the neurovascular unit (NVU) contributes to neuroinflammation (Liu et al., 2020). The astrocyte-secreted and frizzled-related protein 1 (SFRP1) promotes and supports microglial activation, resulting in chronic inflammation (Rueda-Carrasco et al., 2020). The activation of microglia and astrocytes is heterogeneous, and their phenotypic distribution can change as neurodegenerative diseases progress (Kwon & Koh, 2020). For example, in Alzheimer's disease (AD), activated microglia play a beneficial role in clearing beta-amyloid, but as the disease advances, they begin to secrete pro-inflammatory cytokines such as IL-1 and TNF-α, thereby contributing to neurodegeneration (Kaur et al., 2019). Prolonged activation of microglial cells under the pro-inflammatory phenotype leads to the accumulation of pro-inflammatory cytokines in the central nervous system, resulting in excessive inflammation (Mishra et al., 2021), (Kwon & Koh, 2020). This neuroinflammation disrupts homeostasis and is associated with various neurological disorders. In the context of neurodegenerative diseases, it is hypothesized that neuroinflammation contributes to the development and aggregation of altered proteins, such as tau in AD and α-synuclein in Parkinson's disease (PD) (Mohan et al., 2021; Saitgareeva et al., 2020).

Neuroinflammation can also alter the function of the BBB, leading to increased permeability and allowing molecules to enter the central nervous system (Takata et al., 2021; Yamanaka et al., 2021). Additionally, the accumulation of inflammatory cytokines can contribute to BBB dysfunction (van Vliet et al., 2020). This disruption of the BBB can result in the entry of oxidants into the central nervous system, leading to oxidative stress and damage (Kölliker-Frers et al., 2021). Studies have shown that overactivation of microglia by lipopolysaccharide (LPS) can lead to BBB collapse and alterations in tight junction proteins (Edison, 2021), (Shigemoto-Mogami et al., 2018). Similarly, sustained inflammation can lead to microglia phagocytosing astrocytic end feet, thereby affecting BBB function (Haruwaka et al., 2019). Under conditions such as sporadic small vessel disease (SVD), increased microglial activation and BBB permeability have been observed, even in regions of seemingly normal white matter (da Fonseca et al., 2014). In summary, pro-inflammatory cytokines stemming from microglial overactivation can disrupt BBB integrity, potentially contributing to neuroinflammation and neurodegenerative disorders.

CHITOSAN AND ITS DERIVATIVES

It is widely recognized that environment is a rich source of chemical compounds that are biologically active and may have beneficial effects, such as neuroprotective ones (Nwe et al., 2009). Accordingly, numerous studies, CTS is one of the biologically active substances with strong neuroprotective qualities, and its effects can be attributed to a variety of its traits. CTS (β-(1→4)-2-amino-D-glucose) is a linear polysaccharide derived from partial deacetylation of chitin (S. K. Kim & Rajapakse, 2005; Je & Kim, 2005). Chitin (β-1,4-linked 2-acetamido-2-deoxy-D-glucose) is an extremely prevalent polymer which found throughout the biosphere. Animals both on land and at sea, insects, and microbes are sources of chitin. For instance, arthropod exoskeletons are primarily made of chitin-based composites in more than 90% of animal and insect species (Rinaudo, 2019). Recently, researchers have looked into a number of marine sources of chitin (Rhazi et al., 2000) and about 21,120,000 tons of chitin-based refuse are thrown out annually in northern Morocco (Arrouze et al., 2019). The source and method of obtaining CTS are very considerable. Its purity, molecular weight (Mw), crystallinity and level of deacetylation (DD) are crucial for many biomedical applications (Younes & Rinaudo, 2015). The biological characteristics of CTS, such as its biocompatibility, non-toxicity, non-allergenicity, and biodegradability, as well as its antifungal, antibacterial, antioxidant, anticancer, and anti-inflammatory activities, are overshadowed by these factors. It is interesting to note that these properties have also been acknowledged: immunoadjuvant, anti-thrombogenic, and anti-cholesterolemic (Ravi Kumar, 2000), and applications in the treatment of covid-19 (Ait Hamdan et al., 2022).

The low solubility of this polysaccharide, however, restricts its use in food and biomedical use (Kumar, 2000). Some scholars are interested in the transformation of CTS into oligosaccharides because of this restriction (Jeon, Y.-J.; Park, P.-J.; Kim, 2001). Due to their shorter chain length and free amino groups in the D-glucosamine units, chitooligosaccharides (COS), the oligosaccharide version of CTS (Figure 2), are easily soluble in water (Jeon et al., 2000). These characteristics enable COS to engage in an aqueous environment with negatively charged polymers, macromolecules, and polyanions (Turan & Nagata, 2006; Prabaharan, 2008).

Furthermore, because of its active functional groups, CTS has undergone extensive modification to suit a variety of biological and medical requirements. CTS units have a main hydroxyl group at C-6, a secondary hydroxyl group at C-3, and an amino/acetamido group at C-2 in their molecular structure. Through the use of these groups, CTS can be subjected to a number of chemical transformations, such as acylation, N-phthaloylation, tosylation, reductive alkylation, O-carboxymethylation, N-carboxyalkylation, silylation, and graft copolymerization (Ashrafi & Azadi, 2016; Lanjhiyana et al., 2013; Saboktakin et al., 2011).

Figure 2. Chemical structure of chitosan and chitosan oligomers (COS)

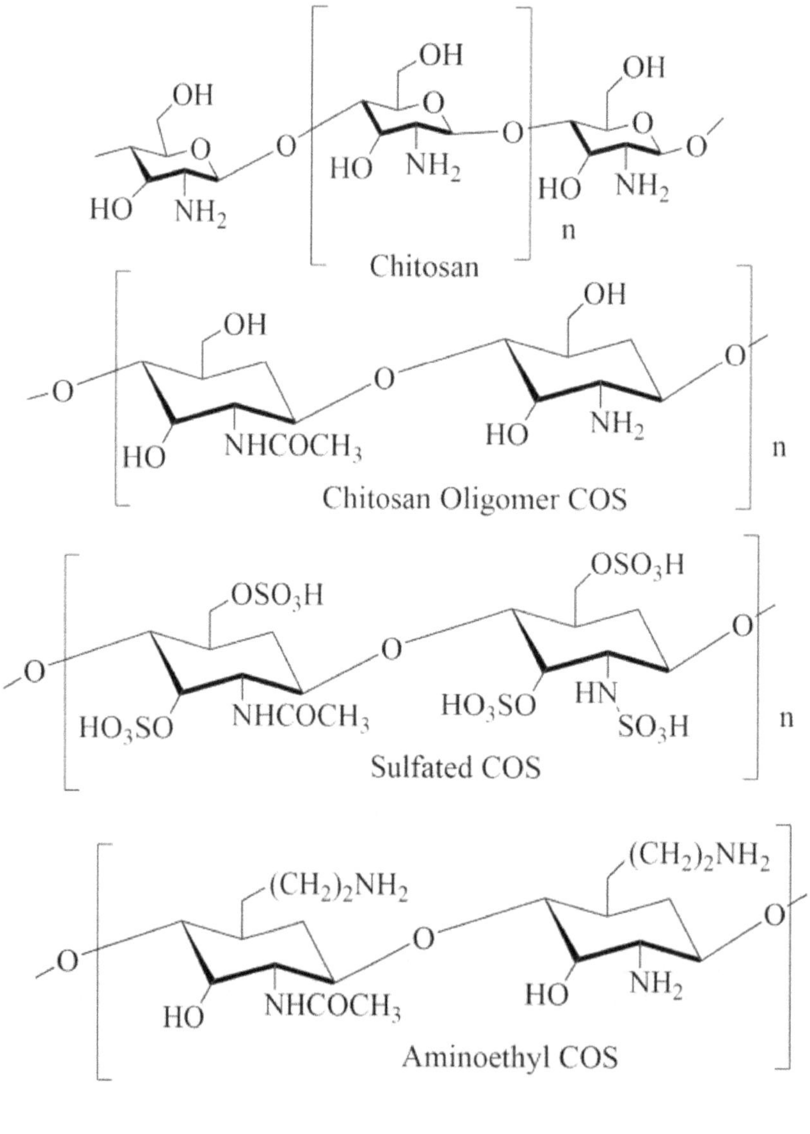

These modifications impart the properties of the final products. For instance, CTS used peptides derived from laminin in its grafts to promote neurite development and neuron attachment (Kuo & Chiu, 2011). CTS has many benefits as a brain-targeted drug vehicle (drug carrier), especially nasal drug delivery. Along with its capacity to penetrate the BBB, it can also regulate release, cling to mucus, and open tight junctions in the nasal membrane. These qualities promote its use in drug delivery from the nose to the brain (Ali & Ahmed, 2018), (Mohammed et al., 2017). The versatility of CTS, which can be used for drug delivery in the shape of microspheres, capsules, hydrogels, conjugates, nanoparticles, films, beads, or tablets, is another benefit (Yao et al., 2018). CTS is suitable for a wide range of biomedical uses, including drug transport, gene delivery, tissue engineering, and regenerative therapies, among others, thanks to all these characteristics. Moreover, Polymer nanoparticles have also received a lot of interest as a promising CNS drug delivery system due to their capacity to target the brain and their controlled release characteristic (Fazil et al., 2012). CTS nanoparticles are characterized by their positively charged surfaces and good mucosal adhesion and these features may enhance the effectiveness of intravenous or intranasal drug delivery to the brain (Babu & Ramesh, 2017).

ANTI-INFLAMMATORY ACTIVITY OF CHITOSAN AND ITS DERIVATIVES

Inflammation is the first defense reaction to infection or injury. It occurs in a specific tissue in response to a particular set of immune and inflammatory cells. Its aim is to restore the structural and functional integrity of the tissue after exposure to an undesirable disturbing factor (Gallo et al., 2017; Kim, 2018). Countless studies have examined the anti-inflammatory and pro-inflammatory properties of CTS and its derivatives. Davydova and colleagues examined the anti-inflammatory effect of high molecular weight (115 kDa) and low molecular weight (5,2 kDa) CTS. Both CTS samples showed a clear increase in the production of IL-10, an anti-inflammatory cytokine, in the blood of the animals tested. In addition, these samples also showed an ability to reduce the progression of colitis (Davydova et al., 2016).

The researchers concluded that the main contribution to CTS anti-inflammatory effect comes from its molecular structural elements, independently of its molecular weight. In addition, Friedman and colleagues observed that CTS-alginate nanoparticles possessed the ability to inhibit inflammatory cytokines and chemokines induced by some bacteria. The results showed that these CTS-alginate nanoparticles effectively inhibited bacteria-induced cytokine production in human monocytes and keratinocytes in a dose-dependent manner (Friedman et al., 2013).

In addition to their ability to inhibit, they have also demonstrated remarkable specificity in terms of their potential for controlled drug release for topical treatments. Research by Oliveira and his team examined the inhibition of pro-inflammatory cytokines and the anti-inflammatory properties of CTS film (Oliveira et al., 2012). Findings from the results indicate a reduction in TNF-α (pro-inflammatory cytokines) after 3 to 10 days of cell culture on the CTS film, in parallel with a significant increase in the anti-inflammatory cytokines IL-10 and TGF-β1. Many researchers have highlighted the anti-inflammatory properties of COS, although the precise mechanism has yet to be fully elucidated. Chung and colleagues examined two varieties of COS, one with a high molecular weight (70 kDa) and the other with a low molecular weight (<1 kDa), and compared their anti-inflammatory efficacy (Chung et al., 2012).

Low-molecular-weight COS exhibited a notable inhibitory effect on the cytokines IL-4, IL-13 and TNF-α, suggesting a potential for reducing allergic inflammation in vivo. Li and colleagues proposed a mechanism to explain lipopolysaccharide-induced NF-κB-dependent inflammatory gene expression mediated by COS. This response was linked to a reduction in core translocation of NF-κB (Li et al., 2014). The study focuses on NF-κB, a crucial transcription factor that regulates pro-inflammatory responses. In a similar study by (Ma et al., 2011), the beneficial impact of COS pretreatment on reducing LPS-induced NF-κB and AP-1 activation in macrophages was demonstrated. This underlines the potential of COS as an inhibitor of NF-κB- and AP-1-mediated inflammatory responses in macrophages. This potential is further corroborated by its ability to suppress LPS-triggered expression of c-fos, a proto-oncogene, in a concentration-dependent manner. Research by Yang and his team has described different varieties of COS with varying molecular weights: COS-A (10 kDa < PM < 20 kDa) and COS-C (1 kDa < PM < 3 kDa) (Yang et al., 2010). Both forms of COS strongly inhibited LPS-induced nitric oxide production in RAW 264.7 cells, with reductions of 50.2% and 44.1% respectively, without toxicity. Compared with COS-A, the lower-molecular-weight COS-C showed greater inhibitory activity, even at lower concentrations. Furthermore, Li and colleagues studied the pro- and anti-inflammatory effects of COS obtained by enzymatic hydrolysis with chitosanase on cytokines (Li et al., 2018).

MODULATION OF NEUROINFLAMMATION BY CHITOSAN AND ITS DERIVATIVES

The immune system of the body usually responds to irritation or infections by causing inflammation (Kim & Joh, 2006 ; Aktas et al., 2007). Neuroinflammation is the medical term for this inflammatory process in the brain. The beginning and pathogenesis of neurodegeneration are significantly influenced by neuroinflammation. Interesting connections demonstrate a link between and chronic inflammation (Kim & Joh, 2006). In AD pathology, chronic inflammation is associated with Aβ plaques formation and interleukin-1β (IL-1β) (Kim et al., 2002). Kim et al. showed that CTS inhibited pro-inflammatory cytokine production in human astrocytoma cells activated by Aβ peptide 25-35 (Aβ25-35) and interleukin-1β (IL-1β) (Kim et al., 2002) . It is interesting to note that pretreatment with CTS greatly decreased the expression of iNOS and the secretion of the pro-inflammatory cytokines TNF-alpha and IL-6 in astrocytoma (Kim et al., 2002). Fang et al. investigated the protective effect and the mechanism of COS on retinal ischemia and reperfusion injury, and they found that a high dose of COS is effective in reversing the reduction of the b-wave ratio in the electroretinogram and retinal thickness brought on by retinal ischemia and reperfusion injury (Fang et al., 2015). Similar to this, COS decreased the expression of inflammatory mediators p53 and Bax while increasing the expression of Bcl-2, reducing retinal oxidative damage and the number of apoptotic cells (Khodagholi et al., 2010) . Additionally, Kalmar et al. (2009) investigated the anti-neuroinflammatory effects of CTS and its derivatives on NT2 neuronal cells and discovered that CTS exerts its anti-neuroinflammatory action by upregulating heat shock protein 70 (Hsp-70) and inhibiting nuclear factor kappa-light-chain-enhancer of activated B cells (NF-Kb) activation (Kalmar & Greensmith, 2009). By attaching to NF-Kb and then inhibiting it, Hsp-70's anti-inflammatory site mechanism is mediated (Kim & Joh, 2006). It is therefore necessary to conduct additional research on the anti-neuroinflammatory effect of CTS and its derivatives in other glial cells, especially microglial cells.

Figure 3. Schematic representation of the modulation of neuroinflammation by CTS and COS

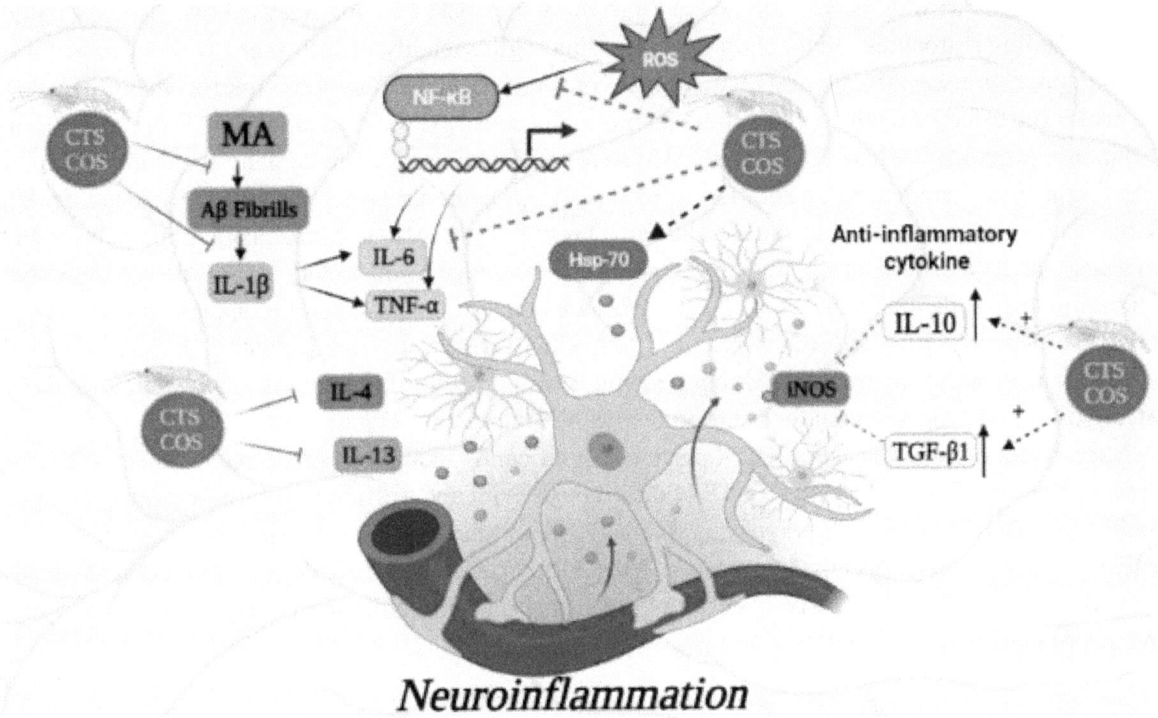

Research conducted by Louw and his team revealed that CTS can form complexes with microRNAs, and penetrate microglia and macrophages, and thus reduce the production of inflammatory cytokines (Louw et al., 2016). Investigations carried out by Mattotti and his team have led to a better understanding of the reactions of glial cells to substrates made from CTS in culture medium. Their observations revealed an inverse relationship between the ability of cells to adhere to the substrate and the amount of serum present in the medium (Mattotti et al., 2017).

The study by Kim and his team evaluated the use of CTS implants in the spinal cord (Kim et al., 2011). They compared different degrees of CTS deacetylation (78% and 85%) with other biomaterials such as polyglycolide and expanded poly(tetrafluoroethylene). Implanted as sheets in the spinal cord of uninjured rats, these biomaterials were monitored for 6 months to assess tissue responses and degradation. Microglia activation was noted after one month for Gore-Tex and CTS, but diminished after 6 months. Vicryl showed degradation, while Gore-Tex showed no degradation. CTS implants underwent chain degradation, without significantly altering structural integrity (Kim et al., 2011). There are indications that CTS can activate macrophages by binding to n-acetylglucosamine receptors. This could intensify inflammatory responses and restrict its potential as a scaffold in the central nervous system (Nisbet et al., 2008 ; Crompton et al., 2006).

Cerqueira and his team investigated the potential use of nanoparticles consisting of carboxymethylchitosan polyamidoamine (CMCht/PAMAM) dendrimers loaded with methylprednisolone (MP), a synthetic corticosteroid with anti-inflammatory and antioxidant properties. The aim was to alleviate neurological deficits resulting from spinal cord injury (Cerqueira et al., 2013) . The study showed that

CMCht/PAMAM nanoparticles efficiently transport biological agents into cells by endocytosis in rats. Methylprednisolone was successfully loaded into these nanoparticles and released over 14 days. Glial cells, including astrocytes, oligodendrocytes and microglia, readily absorb them without toxicity at certain concentrations. Tests on rats with spinal cord injuries show functional improvement thanks to the prolonged effect of methylprednisolone on microglia density (Pereira et al., 2011 ; Tsui et al., 2019).

Natural polymers such as CTS and its derivatives are often used to create nano structures aimed at targeting neuroinflammation. In addition, CTS has shown neuronal protection in AD by reducing Aβ, AchE, oxidative stress and neuroinflammation (Ouyang et al., 2017). A promising approach for PD involves a chitosan-coated nanoemulsion of ropinirole and nigella oil to reduce NF-κB signaling and inhibit lipid peroxidation (Nehal et al., 2021). COS are being studied for their therapeutic effects in neuronal disorders. Specific COSs (COS_2, COS_3, COS_4, COS_5) were tested in mice with PD. COS improved behavior and protected neurons by reducing α-synuclein and neuroinflammation. COS_3 was particularly effective, showing the potential for neuroprotective health products (Wang et al., 2022).

Microinjections of chitosan/miR-124 particles significantly reduced positive macrophages in the injured spinal cord. Positively charged CTS of different sizes interacts efficiently with negatively charged double-stranded RNA to form polyplexes. CTS/siRNA particles are widely used as non-toxic siRNA delivery vectors for various cells (Louw et al., 2016). Furthermore, understanding the intracellular signaling pathways of CTS and its derivatives are of great importance in order to investigate their underlying anti-neuroinflammatory effect.

CONCLUSION

Natural resources offer extremely rich bioactive compounds that can have neuroprotective effects, including CTS and its derivatives. In this context, CTS and its derivatives have shown anti-neuroinflammatory activity, which is mediated by modulation of inflammation status at the central level, and consequently prevents glial cell damage. In addition to their central modulation, these polysaccharides are documented to cross the BBB to provide a controlled drug delivery system to treat inflammation at the glial level. Interest in CTS and its derivatives in modulating inflammation and protecting glial cells by preventing the development of neurological disorders therefore appears to be an emerging field, and CTS-based therapies hold great promise. However, further research is needed to better understand the potential involvement of CTS and its derivatives in the treatment of neurological disorders caused by glial cell inflammation, as well as their implication in various applications aimed at protecting and regenerating CNS tissues.

REFERENCES

Aarli, J. A., Abramsky, O., Chin, J. H., & Vora, N. (2014). Section Editors Global Perspectives *THE GLOBAL BURDEN OF NEUROLOGIC DISEASES*. WHO. https://www.who.int/nmh/publications/best_buys_summary/

Agnello, D., Bigini, P., Villa, P., Mennini, T., Cerami, A., Brines, M. L., & Ghezzi, P. (2002). E rythropoietin exerts an anti-inflammatory effect on the CNS in a model of experimental autoimmune encephalomyelitis.[). *Brain Research, 952*(1), 128–134. www.elsevier.com/locate/bres. doi:10.1016/S0006-8993(02)03239-0 PMID:12363412

Ait Hamdan, Y., El Amerany, F., Desbrières, J., Aghrinane, A., Oudadesse, H., & Rhazi, M. (2022). The evolution of the global COVID-19 epidemic in Morocco and understanding the different therapeutic approaches of chitosan in the control of the pandemic. In *Polymer Bulletin*. Springer Science and Business Media Deutschland GmbH., doi:10.100700289-022-04579-3

Aktas, O., Ullrich, O., Infante-Duarte, C., Nitsch, R., & Zipp, F. (2007). Neuronal Damage in Brain Inflammation. *Archives of Neurology, 64*(2), 185–189. doi:10.1001/archneur.64.2.185 PMID:17296833

Ali, A., and Ahmed, S. (2018). A review on chitosan and its nanocomposites in drug delivery. *Int. J. Biol. Macromol., 109,* 273–286. https://doi.org/. 12.078 doi:10.1016/j.ijbiomac.2017

Alonso, D., Castro, A., & Martinez, A. (2005). Marine compounds for the therapeutic treatment of neurological disorders. In Expert Opinion on Therapeutic Patents (Vol. 15, Issue 10, pp. 1377–1386). doi:10.1517/13543776.15.10.1377

Arrouze, F., Desbrieres, J., Rhazi, M., Essahli, M., & Tolaimate, A. (2019). Valorization of chitins extracted from North Morocco shrimps: Comparison of chitin reactivity and characteristics. *Journal of Applied Polymer Science, 136*(30), 47804. doi:10.1002/app.47804

Ashrafi, H., & Azadi, A. (2016). Chitosan-based hydrogel nanoparticle amazing behaviors during transmission electron microscopy. *International Journal of Biological Macromolecules, 84,* 31–34. doi:10.1016/j.ijbiomac.2015.11.089 PMID:26658231

Asslih, S., Damri, O., & Agam, G. (2021). Neuroinflammation as a common denominator of complex diseases (Cancer, diabetes type 2, and neuropsychiatric disorders). In International Journal of Molecular Sciences, 22(11). MDPI. doi:10.3390/ijms22116138

Babu, A., & Ramesh, R. (2017). Multifaceted applications of chitosan in cancer drug delivery and therapy. In Marine Drugs. MDPI AG. doi:10.3390/md15040096

Behl, C., & Moosmann, B. (2002). Antioxidant neuroprotection in Alzheimer's disease as preventive and therapeutic approach. *Free Radical Biology & Medicine, 33*(2), 182–191. doi:10.1016/S0891-5849(02)00883-3 PMID:12106814

Brown, G. C., & Vilalta, A. (2015). How microglia kill neurons. *Brain Research, 1628,* 288–297. doi:10.1016/j.brainres.2015.08.031 PMID:26341532

Cerqueira, S. R., Oliveira, J. M., Silva, N. A., Leite-Almeida, H., Ribeiro-Samy, S., Almeida, A., Mano, J. F., Sousa, N., Salgado, A. J., & Reis, R. L. (2013). Microglia response and in vivo therapeutic potential of methylprednisolone-loaded dendrimer nanoparticles in spinal cord injury. *Small, 9*(5), 738–749. doi:10.1002mll.201201888 PMID:23161735

Chandrasekaran, K. (2003). neuroprotective-effects-of-bilobalide-a-component-of-ginkgo-bilo-2003. Pharmacopsychiatry, S89–S94.

Chung, M. J., Park, J. K., & Park, Y. (2012). Anti-inflammatory effects of low-molecular weight chitosan oligosaccharides in IgE–antigen complex-stimulated RBL-2H3 cells and asthma model mice. *International Immunopharmacology*, *12*(2), 453–459. doi:10.1016/j.intimp.2011.12.027 PMID:22266066

Crompton, K. E., Tomas, D., Finkelstein, D. I., Marr, M., Forsythe, J. S., & Horne, M. K. (2006). Inflammatory response on injection of chitosan/GP to the brain. *Journal of Materials Science. Materials in Medicine*, *17*(7), 633–639. doi:10.100710856-006-9226-6 PMID:16770548

da Fonseca, A. C. C., Matias, D., Garcia, C., Amaral, R., Geraldo, L. H., Freitas, C., & Lima, F. R. S. (2014). The impact of microglial activation on blood-brain barrier in brain diseases. In Frontiers in Cellular Neuroscience. Frontiers Media S.A. doi:10.3389/fncel.2014.00362

Davydova, V. N., Kalitnik, A. A., Markov, P. A., Volod'ko, A. V., Popov, S. V., & Ermak, I. M. (2016). Cytokine-inducing and anti-inflammatory activity of chitosan and its low-molecular derivative. *Applied Biochemistry and Microbiology*, *52*(5), 476–482. doi:10.1134/S0003683816050070 PMID:29513410

Edison, P. (2021). Microglial activation and blood-brain barrier leakage: Chicken and egg? In Brain, 144(5). doi:10.1093/brain/awab149

Estera, L. A., Walsh, S. P., Headen, J. A., Williamson, R. E., & Kalinski, A. L. (2021). Neuroinflammation: Breaking barriers and bridging gaps. *Neuroscience Research*. doi:10.1016/j.neures.2021.11.001 PMID:34748905

Fang, I. M., Yang, C. M., & Yang, C. H. (2015). Chitosan oligosaccharides prevented retinal ischemia and reperfusion injury via reduced oxidative stress and inflammation in rats. *Experimental Eye Research*, *130*, 38–50. doi:10.1016/j.exer.2014.12.001 PMID:25479043

Fazil, M., Md, S., Haque, S., Kumar, M., Baboota, S., Sahni, J. K., & Ali, J. (2012). Development and evaluation of rivastigmine loaded chitosan nanoparticles for brain targeting. *European Journal of Pharmaceutical Sciences*, *47*(1), 6–15. doi:10.1016/j.ejps.2012.04.013 PMID:22561106

Fehlings, M. G., & Tighe, A. (2008). Spinal cord injury: The promise of translational research. *Neurosurgical Focus*, *25*(5), E1. doi:10.3171/FOC.2008.25.11.E1 PMID:18980469

Friedman, A. J., Phan, J., Schairer, D. O., Champer, J., Qin, M., Pirouz, A., Blecher-Paz, K., Oren, A., Liu, P. T., Modlin, R. L., & Kim, J. (2013). Antimicrobial and Anti-Inflammatory Activity of Chitosan–Alginate Nanoparticles: A Targeted Therapy for Cutaneous Pathogens. *The Journal of Investigative Dermatology*, *133*(5), 1231–1239. doi:10.1038/jid.2012.399 PMID:23190896

Gallo, J., Raska, M., Kriegova, E., & Goodman, S. B. (2017). Inflammation and its resolution and the musculoskeletal system. *Journal of Orthopaedic Translation*, *10*, 52–67. doi:10.1016/j.jot.2017.05.007 PMID:28781962

Gao, H. (2017). Perspectives on Dual Targeting Delivery Systems for Brain Tumors. *Journal of Neuroimmune Pharmacology*, *12*(1), 6–16. doi:10.100711481-016-9687-4 PMID:27270720

Gao, H. M., Liu, B., Zhang, W., & Hong, J. S. (2003). Novel anti-inflammatory therapy for Parkinson's disease. *Trends in Pharmacological Sciences*, *24*(8), 395–401. doi:10.1016/S0165-6147(03)00176-7 PMID:12915048

Hamdan, Y. A., Elouali, S., Ben Maloui, A., El-Mansoury, B., Oudadesse, H., & Rhazi, M. (2023). *Chitosan and Its Derivatives as Potential Neuro-Protective Agents for the Treatment of Parkinson's Disease.*, doi:10.4018/978-1-6684-5156-4.ch015

Hao, C., Wang, W., Wang, S., Zhang, L., & Guo, Y. (2017). An overview of the protective effects of chitosan and acetylated chitosan oligosaccharides against neuronal disorders. In Marine Drugs. MDPI AG. doi:10.3390/md15040089

Haruwaka, K., Ikegami, A., Tachibana, Y., Ohno, N., Konishi, H., Hashimoto, A., Matsumoto, M., Kato, D., Ono, R., Kiyama, H., Moorhouse, A. J., Nabekura, J., & Wake, H. (2019). Dual microglia effects on blood brain barrier permeability induced by systemic inflammation. *Nature Communications, 10*(1), 5816. doi:10.103841467-019-13812-z PMID:31862977

Je, J. Y., & Kim, S. K. (2005). Water-soluble chitosan derivatives as a BACE1 inhibitor. *Bioorganic & Medicinal Chemistry, 13*(23), 6551–6555. doi:10.1016/j.bmc.2005.07.004 PMID:16084727

Jeon, Y.-J., Park, P.-J., & Kim, S.-K. (2001). Antimicrobial effect of chitooligosaccharides produced by bioreactor. *Carbohydrate Polymers, 44*(1), 71–76. doi:10.1016/S0144-8617(00)00200-9

Jeon, Y. J., Shahidi, F., & Kim, S. K. (2000). Preparation of chitin and chitosan oligomers and their applications in physiological functional foods. *Food Reviews International, 16*(2), 159–176. doi:10.1081/FRI-100100286

Kalmar, B., & Greensmith, L. (2009). Induction of heat shock proteins for protection against oxidative stress. In Advanced Drug Delivery Reviews, 61(4), 310–318. doi:10.1016/j.addr.2009.02.003

Kaur, D., Sharma, V., & Deshmukh, R. (2019). Activation of microglia and astrocytes: a roadway to neuroinflammation and Alzheimer's disease. In Inflammopharmacology, 27(4), 663–677. Birkhauser Verlag AG. doi:10.100710787-019-00580-x

Khodagholi, F., Eftekharzadeh, B., Maghsoudi, N., & Rezaei, P. F. (2010). Chitosan prevents oxidative stress-induced amyloid β formation and cytotoxicity in NT2 neurons: Involvement of transcription factors Nrf2 and NF-κB. In Molecular and Cellular Biochemistry (Vol. 337, Issues 1–2, pp. 39–51). Springer Netherlands. doi:10.100711010-009-0284-1

Kim, H., Tator, C. H., & Shoichet, M. S. (2011). Chitosan implants in the rat spinal cord: Biocompatibility and biodegradation. *Journal of Biomedical Materials Research - Part A, 97* A(4), 395–404. doi:10.1002/jbm.a.33070

Kim, M. S., Sung, M. J., Seo, S. B., Yoo, S. J., Lim, W. K., & Kim, H. M. (2002). Water-soluble chitosan inhibits the production of pro-inflammatory cytokine in human astrocytoma cells activated by amyloid β peptide and interleukin-1β. *Neuroscience Letters, 321*(1–2), 105–109. doi:10.1016/S0304-3940(02)00066-6 PMID:11872267

Kim, S. (2018). Competitive Biological Activities of Chitosan and Its Derivatives: Antimicrobial, Antioxidant, Anticancer, and Anti-Inflammatory Activities. *International Journal of Polymer Science, 2018*, 1–13. doi:10.1155/2018/1708172

Kim, S. K., & Rajapakse, N. (2005). Enzymatic production and biological activities of chitosan oligosaccharides (COS): A review. *Carbohydrate Polymers*, *62*(4), 357–368. doi:10.1016/j.carbpol.2005.08.012

Kim, Y. S., & Joh, T. H. (2006). Microglia, major player in the brain inflammation: Their roles in the pathogenesis of Parkinson's disease. *Experimental & Molecular Medicine*, *38*(4), 333–347. doi:10.1038/emm.2006.40 PMID:16953112

Kölliker-Frers, R., Udovin, L., Otero-Losada, M., Kobiec, T., Herrera, M. I., Palacios, J., Razzitte, G., & Capani, F. (2021). Neuroinflammation: An Integrating Overview of Reactive-Neuroimmune Cell Interactions in Health and Disease. In Mediators of Inflammation. Hindawi Limited. doi:10.1155/2021/9999146

Kuo, Y. C., & Chiu, K. H. (2011). Inverted colloidal crystal scaffolds with lamininderived peptides for neuronal differentiation of bone marrow stromal cells. *Biomaterials*, *32*(3), 819–831. doi:10.1016/j.biomaterials.2010.09.057 PMID:20974492

Kwon, H. S., & Koh, S. H. (2020). Neuroinflammation in neurodegenerative disorders: the roles of microglia and astrocytes. In Translational Neurodegeneration, 9(1). BioMed Central Ltd. doi:10.118640035-020-00221-2

Lanjhiyana, S. K., Bajpayee, P., Kesavan, K., Lanjhiyana, S., & Muthu, M. S. (2013). Chitosan-sodium alginate blended polyelectrolyte complexes as potential multiparticulate carrier system: Colon-targeted delivery and gamma scintigraphic imaging. *Expert Opinion on Drug Delivery*, *10*(1), 5–15. doi:10.1517/17425247.2013.734805 PMID:23106236

Lassmann, H. (2020). Pathology of inflammatory diseases of the nervous system: Human disease versus animal models. *Glia*, *68*(4), 830–844. doi:10.1002/glia.23726 PMID:31605512

Li, Y., Chen, L., Liu, Y., Zhang, Y., Liang, Y., & Mei, Y. (2018). Anti-inflammatory effects in a mouse osteoarthritis model of a mixture of glucosamine and chitooligosaccharides produced by bi-enzyme single-step hydrolysis. *Scientific Reports*, *8*(1), 5624. doi:10.103841598-018-24050-6 PMID:29618773

Li, Y., Liu, H., Xu, Q. S., Du, Y. G., & Xu, J. (2014). Chitosan oligosaccharides block LPS-induced O-GlcNAcylation of NF-κB and endothelial inflammatory response. *Carbohydrate Polymers*, *99*, 568–578. doi:10.1016/j.carbpol.2013.08.082 PMID:24274545

Liu, L. R., Liu, J. C., Bao, J. S., Bai, Q. Q., & Wang, G. Q. (2020). Interaction of Microglia and Astrocytes in the Neurovascular Unit. In *Frontiers in Immunology* (Vol. 11). Frontiers Media S.A. doi:10.3389/fimmu.2020.01024

Louw, A. M., Kolar, M. K., Novikova, L. N., Kingham, P. J., Wiberg, M., Kjems, J., & Novikov, L. N. (2016). Chitosan polyplex mediated delivery of miRNA-124 reduces activation of microglial cells in vitro and in rat models of spinal cord injury. *Nanomedicine; Nanotechnology, Biology, and Medicine*, *12*(3), 643–653. doi:10.1016/j.nano.2015.10.011 PMID:26582736

Ma, P., Liu, H. T., Wei, P., Xu, Q. S., Bai, X. F., Du, Y. G., & Yu, C. (2011). Chitosan oligosaccharides inhibit LPS-induced over-expression of IL-6 and TNF-α in RAW264.7 macrophage cells through blockade of mitogen-activated protein kinase (MAPK) and PI3K/Akt signaling pathways. *Carbohydrate Polymers*, *84*(4), 1391–1398. doi:10.1016/j.carbpol.2011.01.045

Mattotti, M., Alvarez, Z., Delgado, L., Mateos-Timoneda, M. A., Aparicio, C., Planell, J. A., Alcántara, S., & Engel, E. (2017). Differential neuronal and glial behavior on flat and micro patterned chitosan films. *Colloids and Surfaces. B, Biointerfaces*, *158*, 569–577. doi:10.1016/j.colsurfb.2017.07.045 PMID:28750339

Meletis, K., Barnabé-Heider, F., Carlén, M., Evergren, E., Tomilin, N., Shupliakov, O., & Frisén, J. (2008). Spinal cord injury reveals multilineage differentiation of ependymal cells. *PLoS Biology*, *6*(7), 1494–1507. doi:10.1371/journal.pbio.0060182 PMID:18651793

Mishra, A., Bandopadhyay, R., Singh, P. K., Mishra, P. S., Sharma, N., & Khurana, N. (2021a). Neuroinflammation in neurological disorders: pharmacotherapeutic targets from bench to bedside. In Metabolic Brain Disease. Springer. doi:10.100711011-021-00806-4

Mishra, A., Bandopadhyay, R., Singh, P. K., Mishra, P. S., Sharma, N., & Khurana, N. (2021b). Neuroinflammation in neurological disorders: pharmacotherapeutic targets from bench to bedside. In Metabolic Brain Disease. Springer. doi:10.100711011-021-00806-4

Mohammed, M. A., Syeda, J., Wasan, K. M., & Wasan, E. K. (2017). An overview of chitosan nanoparticles and its application in non-parenteral drug delivery. *Pharmaceutics*, *9*(4), 53–79. doi:10.3390/pharmaceutics9040053 PMID:29156634

Mohan, V., Edamakanti, C. R., & Sharma, V. (2021). Role of Neuroinflammation in Neurodegenerative Disorders. *The Molecular Immunology of Neurological Diseases*, 41–49. doi:10.1016/B978-0-12-821974-4.00001-7

Nehal, N., Nabi, B., Rehman, S., Pathak, A., Iqubal, A., Khan, S. A., Yar, M. S., Parvez, S., Baboota, S., & Ali, J. (2021). Chitosan coated synergistically engineered nanoemulsion of Ropinirole and nigella oil in the management of Parkinson's disease: Formulation perspective and In vitro and In vivo assessment. *International Journal of Biological Macromolecules*, *167*, 605–619. doi:10.1016/j.ijbiomac.2020.11.207 PMID:33278450

Neher, J. J., Neniskyte, U., & Brown, G. C. (2012). Primary phagocytosis of neurons by inflamed microglia: Potential roles in neurodegeneration. *Frontiers in Pharmacology*, *3*(FEB). doi:10.3389/fphar.2012.00027 PMID:22403545

Nisbet, D. R., Crompton, K. E., Horne, M. K., Finkelstein, D. I., & Forsythe, J. S. (2008). Neural tissue engineering of the CNS using hydrogels. In Journal of Biomedical Materials Research - Part B Applied Biomaterials, 87(1). doi:10.1002/jbm.b.31000

Nosi, D., Lana, D., Giovannini, M. G., Delfino, G., & Zecchi-Orlandini, S. (2021). Neuroinflammation: Integrated nervous tissue response through intercellular interactions at the "whole system" scale. In Cells, 10(5). MDPI. doi:10.3390/cells10051195

Nwe, N., Furuike, T., & Tamura, H. (2009). The mechanical and biological properties of chitosan scaffolds for tissue regeneration templates are significantly enhanced by chitosan from Gongronella butleri. *Materials (Basel)*, *2*(2), 374–398. doi:10.3390/ma2020374

Oliveira, M. I., Santos, S. G., Oliveira, M. J., Torres, A. L., & Barbosa, M. A. (2012). Chitosan drives anti-inflammatory macrophage polarisation and pro-inflammatory dendritic cell stimulation. *European Cells & Materials*, *24*, 136–153. doi:10.22203/eCM.v024a10 PMID:22828991

Ouyang, Q. Q., Zhao, S., Li, S. D., & Song, C. (2017). Application of chitosan, chitooligosaccharide, and their derivatives in the treatment of Alzheimer's disease. In Marine Drugs. MDPI AG. doi:10.3390/md15110322

Pellicciari, R., Costantino, G., Marinozzi, M., & Natalini, B. (1998). Modulation of glutamate receptor pathways in the search for new neuroprotective agents. *Il Farmaco*, *53*(53), 255–261. doi:10.1016/S0014-827X(98)00018-4 PMID:9658582

Pereira, V. H., Salgado, A. J., Oliveira, J. M., Cerqueira, S. R., Frias, A. M., Fraga, J. S., Roque, S., Falcão, A. M., Marques, F., Neves, N. M., Mano, J. F., Reis, R. L., & Sousa, N. (2011). In vivo biodistribution of carboxymethylchitosan/poly(amidoamine) dendrimer nanoparticles in rats. *Journal of Bioactive and Compatible Polymers*, *26*(6), 619–627. doi:10.1177/0883911511425567

Prabaharan, M. (2008). Review paper: Chitosan derivatives as promising materials for controlled drug delivery. In Journal of Biomaterials Applications, 23(1). doi:10.1177/0885328208091562

Ravi Kumar, M. N. V. (2000). A review of chitin and chitosan applications. *Reactive & Functional Polymers*, *46*(1), 1–27. doi:10.1016/S1381-5148(00)00038-9

Rhazi, M., Desbriè, J., Tolaimate, A., Alagui, A., & Vottero, P. (2000). *Investigation of different natural sources of chitin: influence of the source and deacetylation process on the physicochemical characteristics of chitosan.*

Rinaudo, M. P. S. (2019). *From Chitin to Chitosan.* Glycopedia.Eu. http://www.glycopedia.eu/IMG/pdf/from_chitin_to_chitosan.pdf.

Rueda-CarrascoJ.Inés MateoM.BorrotoA.Jesús Martin-BermejoM.KummerM. P.SchwartzS.López-AtalayaJ. P.AlarconB.HenekaM. T.EsteveP.BovolentaP. (2020). SFRP1 shapes astrocyte to microglia cross-talk in acute and chronic neuroinflammation. BiorXiv. doi:10.1101/2020.03.10.982579

Sabelström, H., Stenudd, M., Réu, P., Dias, D. O., Elfineh, M., Zdunek, S., Damberg, P., Göritz, C., & Frisén, J. (2013). Resident neural stem cells restrict tissue damage and neuronal loss after spinal cord injury in mice. *Science*, *342*(6158), 637–640. doi:10.1126cience.1242576 PMID:24179227

Saboktakin, M. R., Tabatabaie, R. M., Maharramov, A., & Ramazanov, M. A. (2011). Synthesis and characterization of pH-dependent glycol chitosan and dextran sulfate nanoparticles for effective brain cancer treatment. *International Journal of Biological Macromolecules*, *49*(4), 747–751. doi:10.1016/j.ijbiomac.2011.07.006 PMID:21782844

Saitgareeva, A. R., Bulygin, K. V., Gareev, I. F., Beylerli, O. A., & Akhmadeeva, L. R. (2020). The role of microglia in the development of neurodegeneration. *Neurological Sciences*, *41*(12), 3609–3615. doi:10.100710072-020-04468-5 PMID:32458252

Shabab, T., Khanabdali, R., Moghadamtousi, S. Z., Kadir, H. A., & Mohan, G. (2017). Neuroinflammation pathways: a general review. In International Journal of Neuroscience (Vol. 127, Issue 7, pp. 624–633). Taylor and Francis Ltd. doi:10.1080/00207454.2016.1212854

Shi, Y., Zhang, X., Zhang, H., Ge, L., & Zhao, L. (2021). Research progress in pathogenesis and treatment of neuroinflammation. *E3S Web of Conferences, 292*. doi:10.1051/e3sconf/202129203100

Shigemoto-Mogami, Y., Hoshikawa, K., & Sato, K. (2018). Activated microglia disrupt the blood-brain barrier and induce chemokines and cytokines in a rat in vitro model. *Frontiers in Cellular Neuroscience, 12*, 494. doi:10.3389/fncel.2018.00494 PMID:30618641

Takata, F., Nakagawa, S., Matsumoto, J., & Dohgu, S. (2021). Blood-Brain Barrier Dysfunction Amplifies the Development of Neuroinflammation: Understanding of Cellular Events in Brain Microvascular Endothelial Cells for Prevention and Treatment of BBB Dysfunction. In Frontiers in Cellular Neuroscience (Vol. 15). Frontiers Media S.A. doi:10.3389/fncel.2021.661838

Tsui, C., Koss, K., Churchward, M. A., & Todd, K. G. (2019). Biomaterials and glia: Progress on designs to modulate neuroinflammation. *Acta Biomaterialia, 83*, 13–28. doi:10.1016/j.actbio.2018.11.008 PMID:30414483

Turan, K., & Nagata, K. (2006). Chitosan-DNA nanoparticles: The effect of cell type and hydrolysis of chitosan on in vitro DNA transfection. *Pharmaceutical Development and Technology, 11*(4), 503–512. doi:10.1080/10837450600940873 PMID:17101521

van Vliet, E. A., Ndode-Ekane, X. E., Lehto, L. J., Gorter, J. A., Andrade, P., Aronica, E., Gröhn, O., & Pitkänen, A. (2020). Long-lasting blood-brain barrier dysfunction and neuroinflammation after traumatic brain injury. *Neurobiology of Disease, 145*, 105080. doi:10.1016/j.nbd.2020.105080 PMID:32919030

Wang, B., Wang, L., Qu, Y., Lu, J., & Xia, W. (2022). Chitosan oligosaccharides exert neuroprotective effects via modulating the PI3K/Akt/Bcl-2 pathway in a Parkinsonian model. *Food & Function, 13*(10), 5838–5853. doi:10.1039/D1FO04374A PMID:35545086

Yamanaka, G., Suzuki, S., Morishita, N., Takeshita, M., Kanou, K., Takamatsu, T., Suzuki, S., Morichi, S., Watanabe, Y., Ishida, Y., Go, S., Oana, S., Kashiwagi, Y., & Kawashima, H. (2021). Role of neuroinflammation and blood-brain barrier permutability on migraine. In International Journal of Molecular Sciences, 22(16). MDPI AG. doi:10.3390/ijms22168929

Yang, E. J., Kim, J. G., Kim, J. Y., Kim, S. C., Lee, N. H., & Hyun, C. G. (2010). Anti-inflammatory effect of chitosan oligosaccharides in RAW 264.7 cells. *Open Life Sciences, 5*(1), 95–102. doi:10.247811535-009-0066-5

Yao, Z. A., Chen, F. J., Cui, H. L., Lin, T., Guo, N., & Wu, H. G. (2018). Efficacy of chitosan and sodium alginate scaffolds for repair of spinal cord injury in rats. *Neural Regeneration Research, 13*(3), 502–509. doi:10.4103/1673-5374.228756 PMID:29623937

Younes, I., & Rinaudo, M. (2015). Chitin and chitosan preparation from marine sources. Structure, properties and applications. *Marine Drugs, 13*(3), 1133–1174. doi:10.3390/md13031133 PMID:25738328

Chapter 21
Chitosan-Based Biomaterials for Tissue Engineering of Glial Cells

Youssef Ait Hamdan
https://orcid.org/0009-0005-8616-8852
University of Rennes, France & Higher Normal School, Cadi Ayyad University, Morocco

Hassane Oudadesse
University of Rennes, France

Laadraoui Jawad
https://orcid.org/0000-0002-9784-1265
Faculty of Sciences, Hassan II University, Morocco

Kamal Smimih
Faculty of Sciences and Techniques, Sultan Moulay Slimane University, Morocco

Hamid Kabdy
Faculty of Sciences and Techniques, Sultan Moulay Slimane University, Morocco

Samia Elouali
Higher Normal school, Cadi Ayyad University, Morocco

Nadia Eladlani
Higher Normal school, Cadi Ayyad University, Morocco

Mohammed Rhazi
Higher Normal school, Cadi Ayyad University, Morocco

ABSTRACT

Glial cells play a remarkable structural and functional role in the nervous system compared to neurons. However, damage to glial cells leads to degeneration and the onset of numerous neurodegenerative diseases. It is well known that the central nervous system (CNS) has a limited capacity for tissue regeneration. The use of chitosan-based biomaterials has emerged as a potential alternative solution to overcome these limitations. Interest in these biomaterials for CNS biomedical implementation has increased due to their ability to cross the BBB, their mucoadherence, and especially their hydrogel-forming capacity. In addition, their ability to form porous scaffolds and to carry cells and biomolecules has offered a means of achieving glial cell regeneration. Thus, this chapter aims to bring together recent work that highlights the potential of chitosan-based biomaterials suitable for glial cell tissue engineering.

DOI: 10.4018/978-1-6684-9675-6.ch021

INTRODUCTION

Neurological and neurodegenerative diseases (NDD) are a major public health problem. Neurodegeneration refers to the progressive loss of structure and function of the central nervous system (CNS), leading in particular to cognitive dysfunction and motor impairment (Amor et al., 2014). Neurodegeneration, or the death of neuronal cells, is the histopathological hallmark of neurodegenerative diseases such as Alzheimer's disease (AD), multiple sclerosis (MS), amyotrophic lateral sclerosis (ALS) and Parkinson's disease (PD). Neurodegeneration has also been reported in stroke, neurotrophic infection, prion disease, traumatic brain and spinal cord injury, as well as in neuropsychiatric and genetic disorders (Choi, 1988; Amor et al., 2014).

During both normal and pathological development of the nervous system, glial cells play an essential role in the remodeling of neural tissue. They are also involved in tissue regeneration after injury (Kim et al., 2016; Song et al., 2018). However, in certain pathological conditions, their contribution to neuronal regeneration may be limited. Consequently, various strategies have been developed to safeguard neurons from damage, apoptosis, and neurodegeneration, aiming to achieve neuroprotection (Pangestuti & kim, 2010; Kostrzewa &Segura-Aguilar, 2003). Neuroprotection seeks to minimize neurodegeneration or impairment of neurons after CNS injury, aiming to maintain neuronal integrity and functionality as much as possible (Tucci & Bagetta, 2008). Based on these mechanisms, several neuroprotective agents have been explored for the management of neurological conditions (Pellicciari et al., 1998; Chandrasekaran et al., 2003).

Potential therapeutic biomolecules have not received much attention in the treatment of neurodegenerative disorders, although they could play an important role. For example, the marine environment is known for its nooks and crannies in the structures of bioactive compounds with promising applications (Alonso et al., 2005). Recent reports on chitosan (CTS) and its derivatives as plausible molecules to target neurodegenerative disorders are well documented in several studies (Hao et al., 2017; Hamdan et al. 2023a). This chapter, reviews recent advances in the use and applications of CTS in neuronal tissue engineering involving glial cells to target neurological and neurodegenerative disorders.

GLIAL CELL DEGENERATION AND PATHOLOGY

Neurodegenerative diseases are conventionally characterized by progressive neurological disorders marked by the selective loss of distinct populations of neurons. Parkinson's disease, Alzheimer's disease (AD), epilepsy, and amyotrophic lateral sclerosis (ALS) are among the conditions historically associated with the degeneration of dopaminergic, cholinergic, GABAergic, and motor neurons, respectively (Davies and Maloney, 1976; Leenders et al., 1990; Xu-Friedman et al., 2001; Buskila et al., 2019b). However, a growing body of evidence suggests that dysfunction in glial cells may also play a significant role in the pathology of these diseases. In this section, we will provide an overview of how glial dysfunction contributes specifically to various neurodegenerative diseases (Figure 1).

Figure 1. The physiological roles of glial cell

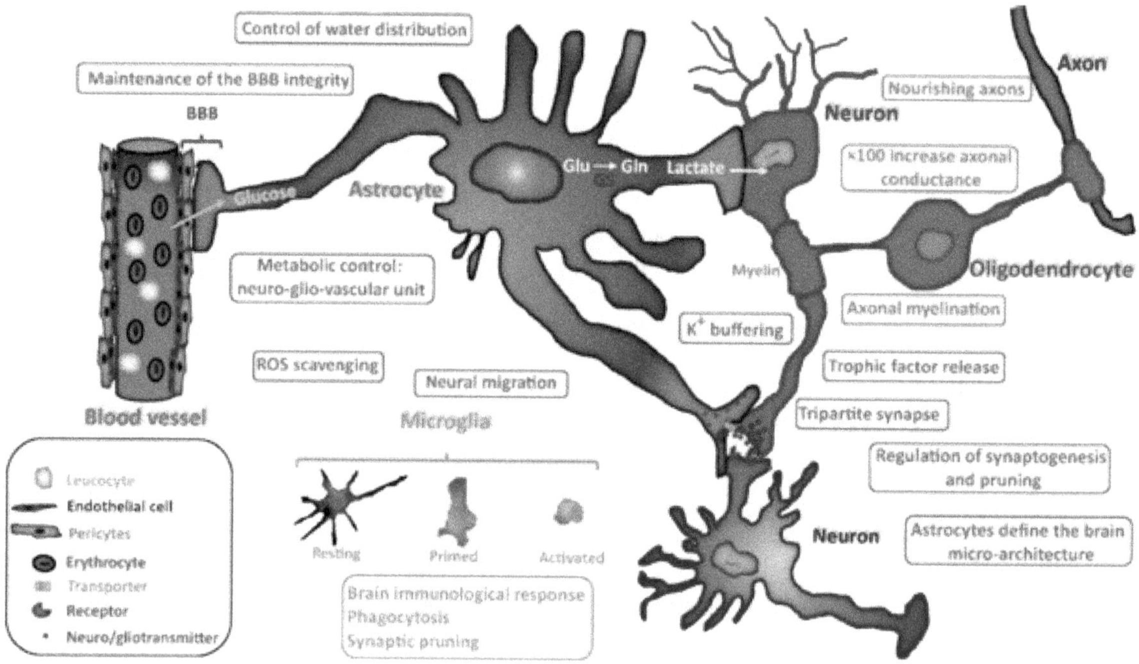

Epilepsy

Epilepsy is an intricate degenerative neurological condition characterized by recurrent seizures, encompassing both focal and generalized seizures (Devinsky et al., 2013). Previous investigations into epilepsy predominantly focused on neurons; however, recent findings indicate the involvement of glial cells in the disease's pathology.

A notable feature of acquired epilepsy is the formation of glial scars through hypertrophic astrocytes (Das et al., 2012; Do Nascimento et al., 2012), resulting in a reduction of gap-junction connectivity between these astrocytes (Wallraff et al., 2006). Indeed, Bedner et al. (2015) recently demonstrated that disruptions in astrocytic gap-junction coupling occur within 4 hours following status epilepticus. In alignment with this study, acute brain slices from transgenic mice with deficient astrocytic coupling displayed spontaneous epileptiform activity and heightened synaptic transmission (Wallraff et al., 2006; Pannasch et al., 2011).

Dysfunctional astrocytes could play a pivotal role in generating and propagating epileptic seizures (Seifert et al., 2006) due to their inability to effectively clear K+ ions. Elevated extracellular K+ levels have been linked to increased synchronous oscillatory activity (Bellot-Saez et al., 2018) and the induction of epileptiform activity in brain slices (Gabriel et al., 2004). The regulation of K+ buffering in astrocytes is also influenced by Kir4.1 channels located in astrocytic end feet (Zurolo et al., 2012). Notably, genetic variations in the KCNJ10 gene encoding Kir4.1 channels have been associated with susceptibility to seizures (Ferraro et al., 2004; Wallraff et al., 2006). Furthermore, dysfunctional astrocytes with reduced Kir4.1 channel expression exhibited impaired extracellular K+ and glutamate clearance, contributing to neuronal hyperexcitability (Patel et al., 2019).

While an ongoing debate exists regarding whether gliosis is a consequence or a causative factor of seizures (Patel et al., 2019), evidence supporting the latter suggests that astrogliosis alone can trigger spontaneous seizures even in the absence of other pathologies (Robel et al., 2015).

It has been proposed for a considerable duration that astrocytic processes extend toward synapses that release high levels of glutamate, functioning as a mechanism to counteract spill-over and hinder hyper-excitability (Ventura and Harris, 1999). Indeed, prior research indicated that individuals with epilepsy displayed reduced levels of glutamine synthetase (GS) in the hippocampus (Petroff et al., 2002), leading to glutamate accumulation within astrocytes and the extracellular space. Likewise, epilepsy animal models exhibited a notable decrease in GS activity during the chronic epileptogenic phase (Kang et al., 2006; Bidmon et al., 2008). Consequently, it has been postulated that the decline in GS activity primarily contributes to heightened extracellular glutamate levels and the onset of seizures.

Research has confirmed that astrocytes undergo alterations in morphology and domain organization in cases of neurodegenerative disease, with a tenfold increase in overlapping astrocytic processes observed in mouse epilepsy models (Golarai et al., 2008). A recent study indicated that as epilepsy progresses, the tripartite synapse experiences molecular, structural, and functional modifications in the hippocampus (Clarkson et al., 2020). This study highlighted not only a decrease in GLT-1 expression density but also an increased wrapping of astrocyte processes around synapses, aiming to reduce excitability by enhancing glutamate uptake (Clarkson et al., 2020).

Huntington's Disease (HD)

HD is an inherited brain disorder characterized by the progressive degeneration of specific neuronal populations within the corticostriatal circuit and malfunctioning glial cells (Tong et al., 2014; Jiang et al., 2017). One of the initial astrocytic dysfunctions observed in HD is a reduction in glutamate transporter expression (Faideau et al., 2010), leading to diminished uptake of extracellular glutamate in both the striatum and cerebral cortex (Przedborski et al., 2000; Behrens et al., 2002; Shin et al., 2005; Hassel et al., 2008). Additionally, the mutated huntingtin protein has been linked to decreased Kir4.1 channel expression and impaired astrocytic K+ clearance capacity (Tong et al., 2014), resulting in hyperexcitability among local striatal neurons (Tong et al., 2014; Benraiss et al., 2016). Notably, the restoration of astrocytic Kir4.1 channel expression in an HD mouse model rescued glutamate signaling, underscoring the significance of maintaining K+ balance in HD development (Jiang et al., 2016).

Recent work by the Khakh group highlighted a significant reduction in astrocyte morphology and spatial territory within the striatum, preceding synaptic dysfunction in an HD mouse model (Octeau et al., 2018). Dysfunctional glial cells have been identified as pivotal in HD progression, as engrafting mutant huntingtin-expressing human glial progenitor cells into healthy mice led to the manifestation of disease phenotypes (Benraiss et al., 2016). Furthermore, the introduction of healthy human glial cells into transgenic HD mice markedly increased survival rates, reduced motor deterioration, and reinstated K+ balance (Benraiss et al., 2016).

Amyotrophic Lateral Sclerosis (ALS)

ALS stands as the most prevalent form of motor neuron disease, impacting the brain and spinal cord and leading to gradual degradation of both upper and lower motor neurons responsible for muscle movement, respiration, and speech control (Martin et al., 2017). Astrocytes are believed to contribute to ALS progression by exhibiting aberrant functionality in terms of glutamate transporters and extracellular K+ concentration, culminating in heightened extracellular glutamate levels and excitotoxicity (Rothstein et al., 1995; Do-Ha et al., 2018). Notably, SOD1 mutations, known to potentially trigger oxidative stress (Barber et al., 2006) and frequently detected in individuals affected by familial ALS (Tang et al., 2019), have been associated with diminished expression of EAAT2 transporters within astrocytes (Pedersen et al., 1998).

Additionally, Gong et al. (2000) demonstrated that the presence of mutated SOD1 is required in both neurons and astrocytes for the progression of ALS, as the expression of mutated SOD1 solely in neurons did not suffice to initiate ALS (Pramatarova et al., 2001). However, alternative in vivo investigations have indicated that transplanted glial-restricted precursor cells into wild-type rats alone can induce motor neuron demise (Papadeas et al., 2011). These observations underscore that ALS doesn't exclusively stem from either astrocytes or neurons, but rather, both cell types play integral roles in the pathogenesis of ALS (Rothstein et al., 1995; Do-Ha et al., 2018).

Frontotemporal Dementia

Frontotemporal dementia (FTD) stands as the predominant dementia type affecting individuals aged 65 and below (Hogan et al., 2016), characterized by degeneration of the frontal and anterior temporal lobes (Broe et al., 2004). While the precise role of astrocytes in FTD remains incompletely elucidated, a meta-analysis of FTD cases unveiled an escalation in astrocytic cell death with disease progression, juxtaposed with consistent neuronal apoptosis (Broe et al., 2004). These findings suggest that the observed neurodegeneration in FTD might be facilitated by the decline in astrocytic support. A recent study by Hallmann et al. (2017) demonstrated altered whole-genome expression in FTD astrocytes, including the upregulation of genes linked to synapse organization and transmission (e.g., CHRNA1). Furthermore, reduced expression of EAAT2 has been documented in the frontal cortex of FTD patients compared to controls (Umoh et al., 2018), implying that deficiencies in astrocytic glutamate uptake could contribute to excitotoxicity and neuronal degeneration (see Figure 2).

FTD progression is accompanied by changes in proteins, neuroinflammation, and vascular dysfunction (Martinac et al., 2001; Garbuzova-Davis et al., 2012), with prominent involvement of microglia and astrocytes (Philips and Robberecht, 2011). Notably, microglial activation has been linked to neuronal loss and TAU deposition, suggesting its occurrence in the early stages of the disease (Schofield et al., 2003). Moreover, autopsy samples from FTD patients exhibited elevated levels of reactive astrocytes displaying signs of deterioration, likely contributing to neuronal cell demise (Su et al., 2000).

Parkinson's Disease

Parkinson's disease (PD) ranks as the second most prevalent neurodegenerative disorder (de Lau and Breteler, 2006), presenting symptoms such as resting tremors and bradykinesia (Salarian et al., 2007). A hallmark of PD is the loss of dopaminergic neurons in the substantia nigra pars compacta (SNpc) and the emergence of α-synuclein deposits, known as Lewy bodies (Kordower et al., 2013). Recent work by Bosson et al. (2015) indicated an elevation in the count of glutamatergic synapses, concomitant with changes in astrocytic morphology within the SNr of a PD rat model. Moreover, the intracerebroventricular injection of α-synuclein led to an increase in astrocyte numbers, heightened density of excitatory synapses, and elevated levels of TGF-b1 in the striatum of the injected animals (Diniz et al., 2019). Correspondingly, larger astrocytic spatial territories were observed in the striatum of parkinsonian monkeys (Villalba and Smith, 2011), implying that astrocytic remodeling acts as a compensatory mechanism to counter the heightened extracellular glutamate levels and overall neuronal hyperactivity seen in PD.

Neurodegeneration of dopaminergic neurons has also been linked to alterations in the glial antioxidative system. Solano et al. (2008) demonstrated that dysfunction of the PD-associated protein Parkin could lead to a decrease in the synthesis of the antioxidant glutathione (GSH) by astrocytes, underpinning the reduced GSH levels observed in PD patients (Baillet et al., 2010). Previous studies have indicated that preventing oxidative stress can safeguard dopaminergic neurons in an animal model of PD (Chen et al., 2009; Kitamura et al., 2011), further underscoring the role of glial malfunction in dopaminergic neuron degeneration.

It is well established that glial cells in post-mortem PD patients exhibit elevated levels of α-synuclein accumulation (Wakabayashi et al., 2000). Direct neuron-glia communication plays a pivotal role in PD progression, involving the release of α-synuclein at axon terminal sites where it is taken up by astrocytes, inducing an inflammatory response (Lee et al., 2010). Although it is generally believed that astrocytes clear α-synuclein from synapses as a protective mechanism (Booth et al., 2017), recent research postulates that malfunctioning astrocytes exacerbate PD pathology by accumulating and transmitting α-synuclein to neighboring dopamine-producing neurons, thereby intensifying neurodegeneration (di Domenico et al., 2019). Indeed, cultured neurons derived from healthy individuals exhibited heightened susceptibility to neuronal death upon contact with iPSC-derived astrocytes from PD patients. However, when healthy astrocytes were co-cultured with PD patient neurons, α-synuclein did not accumulate, and neuronal function was restored (di Domenico et al., 2019). These investigations underscore the role of dysfunctional astrocytes in the propagation of α-synuclein, accumulating abnormal levels due to impaired protein degradation, thereby potentially amplifying the spread of PD in the brain. Given that the SNpc region possesses the lowest astrocyte-to-neuron ratio in the central nervous system (Damier et al., 1993), this discrepancy might explain the susceptibility of dopaminergic neurons to PD due to their comparatively limited astrocytic support (de Majo et al., 2020).

Alzheimer's Disease

Alzheimer's disease (AD), recognized as the most widespread form of dementia globally (Ferri et al., 2005), represents a progressive neurodegenerative condition characterized by the accumulation of β-amyloid (Aβ) plaques, intracellular neurofibrillary tangles (NFTs) featuring hyper-phosphorylated tau, and activated glial cells encircling senile plaques (Morley et al., 2018). While Aβ deposits and tau primarily emanate from neurons, genes associated with the onset of late-onset Alzheimer's, including apolipoprotein (APOE), apolipoprotein J (APOJ), and sortilin-related-receptor-1 (SORL), are mainly expressed by glial cells (Arranz and De Strooper, 2019), indicating the involvement of both glia and neurons in the disease's pathogenesis. Notably, a recent study by Habib et al. (2020) identified a subset of disease-associated astrocytes in an AD mouse model, which emerged in early disease stages and increased with disease progression.

The presence of Aβ deposits in the brain leads to an early loss of synapses and neurons (Buskila et al., 2013), abnormal activation of microglia, and oxidative stress (Leyns et al., 2019). In recent findings, the group led by Beth Stevens reported that phagocytic microglia play a role in mediating synaptic loss during the early phases of AD through the complement system, highlighting an anomalous activation of the synaptic pruning pathway (Hong et al., 2016b). Furthermore, mounting evidence suggests that glutamatergic communication between neurons and astrocytes at the tripartite synapse is disrupted as AD progresses, leading to increased glutamate release from both neurons and astrocytes (Pirttimaki et al., 2013; Talantova et al., 2013). Studies conducted in an AD animal model have demonstrated that astrocytes in regions such as the entorhinal cortex, prefrontal cortex, and hippocampus undergo morphological and functional changes. These alterations encompass a reduction in the number of astrocytes expressing the enzyme glutamine synthase, atrophy, and a decrease in the surface area, volume, and complexity of astrocytic processes (Beauquis et al., 2013, 2014; Kulijewicz-Nawrot et al., 2013). Furthermore, astrocytes during the pre-symptomatic AD phase exhibit diminished vesicle trafficking (Stenovec et al., 2016) and reduced expression of AQP4 and GLT-1, leading to disturbances in water and glutamate equilibrium (Hoshino et al., 2017) and elevated glutamate release from astrocytes, culminating in hyperexcitability and cellular demise (Pirttimaki et al., 2013; Talantova et al., 2013).

Multiple Sclerosis

Multiple sclerosis (MS) is a disease in which the myelin sheath of axons degenerates, resulting in a wide range of symptoms and functional impairments including pain, muscle weakness, and tremors (Trapp et al., 1998). The canonical view of this disease progression is that CD4+ T-lymphocytes cross the BBB into the parenchyma which causes a cascade of several events leading to inflammatory lesion and demyelination (Laplaud et al., 2004; Bahbouhi et al., 2009). However, a secondary hypothesis states that the disease is neurally initiated and begins in the CNS, caused by metabolic changes affecting neurons and glia which leads to BBB breakdown (Chaudhuri and Behan, 2004). Although somewhat controversial, evidence supporting this theory suggests that neuronal loss and inflammation are present at the very early stages of MS (Filippi et al., 2003). If it is the case that MS begins in the CNS, neuroprotective interventions may be effective as future MS treatments. However, the autoimmune vs. CNS-derived theories of MS are still the source of an ongoing debate in the literature.

The classification of different neurodegenerative diseases is based on certain neurological symptoms expressed in each disorder, however, there is considerable clinical and pathological comorbidity across neurodegenerative disorders. Patients with FTD may present with ALS or parkinsonian syndromes and vice versa (Lomen-Hoerth et al., 2002; Padovani et al., 2007). Also, major risk factors that lead to glial malfunction, such as cardiovascular dysfunction, TBI, and stroke are common in different neurodegenerative diseases, suggesting that similarities between different disorders stem from common glial dysfunction.

CHITOSAN AND TISSUE ENGINEERING IN CNS

Chitosan (CTS) is the derivative of chitin by the deacetylation reaction with a degree of more than 50% (Tolaimate et al., 2023). As a result of this reaction, CTS will have free, functional amine groups that promote complete solubility in acidic organic solvents. CTS is used in many applications such as food, medicine, and textile due to its solubility, biocompatibility, and biodegradability properties (Sahariah & Másson, 2017; Yi et al., 2005; Van Vlierberghe et al., 2011). In addition, it is considered one of the most widely used biomaterials for the manufacture of hydrogels for biomedical applications such as tissue engineering, injectable materials and implants, and wound regeneration (Badhe et al., 2017; Kong et al., 2019; Zou et al., 2019; Hamdan et al., 2023b). CTS (β-(1→4)-2-amino-D-glucose) is a linear polysaccharide derived from partial deacetylation of chitin (Figure 2) (Kim & Rajapakse, 2005; Je & Kim 2005). The low solubility of this polysaccharide, however, restricts its use in food and biomedical uses (Kumar, 2000).

Figure 2. Chemical structure of chitosan

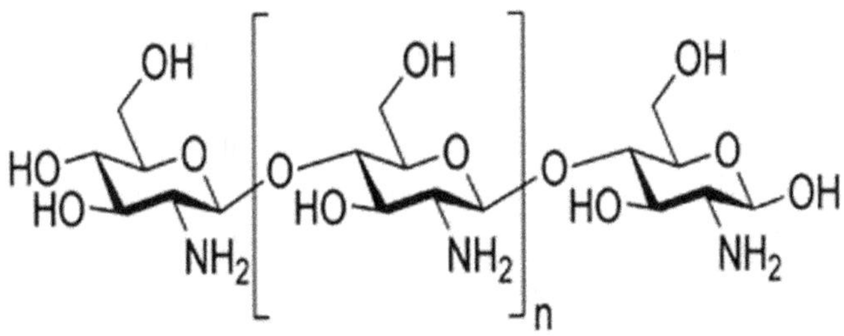

It is widely recognized that environment is a rich source of chemical compounds that are biologically active and may have beneficial effects, such as neuroprotective ones (New et al., 2009). Accordingly, numerous studies, CTS is one of the biologically active substances with strong neuroprotective qualities, and its effects can be attributed to a variety of its traits (Table 1).

Table 1. therapeutic properties of chitosan against neurological and neurodegenerative diseases

Properties	Description
Biocompatibility	No toxicity, no rejection, no inflammatory reaction.
Structure	Presence of functional groups (-OH, -COOH and -NH2). Presence of flexibility thanks to its macromolecular chains. Its molecular weight is modifiable.
Mucoadhesive property	Adhesion properties to mucus surfaces
Stability	Stability assured by the presence of charges on its surface.
Loading capacity	Encapsulation of drugs and molecules for therapeutic use.
Release kinetics	Good drugs release within a suitable period.
Nature	It is biologically active and can carry a large number of pharmacological activities.
Mechanical property	It is characterized by high absorbing property, flexibility and elasticity.
Porosity	Highly permeable. It can interact with mucosal cells of the respiratory tract, opening the tight junction and increasing the drug's permeation.
Biodegradability	Biodegradable in nature. Its degradation products did not show any kind of toxicity and can easily be eliminated from the body.
Drug delivery systems	Possible by using chitosan nanoparticles and microspheres to deliver therapeutic agents to the brain, to improve drug targeting and efficacy.
Nerve regeneration	Chitosan scaffolds promote nerve regeneration and tissue repair after nerve damage.
Wound healing	Chitosan's ability to improve wound healing associated with ulcers or skin wounds, as in the case of diabetic neuropathy.
Antioxidant and anti-inflammatory	These effects may indirectly contribute to neurological health by reducing oxidative stress and inflammation, which are implicated in various neurological disorders

Depending on the application objective, it is essential to take into account the source and method of obtaining CTS. The traits of the finished product are determined by these elements. Its purity, molecular weight (Mw), crystallinity, and level of deacetylation (DD) are crucial in biomedical uses (Younes & Rinaudo 2015). For instance, CTS used peptides derived from laminin in its grafts to promote neurite development and neuron attachment (Kuo & Chiu, 2011). Also, CTS has many benefits as a brain-targeted drug vehicle (drug carrier), especially nasal drug delivery. Along with its capacity to penetrate the BBB, it can also regulate release, cling to mucus, and open tight junctions in the nasal membrane. These qualities promote its use in drug delivery from the nose to the brain (Ali & Ahmed, 2018; Mohammed et al., 2017).

The versatility of CTS, which can be used for drug delivery in the shape of microspheres, capsules, hydrogels, conjugates, nanoparticles, films, beads, or tablets, is another benefit (Yao et al.,2018). CTS is suitable for a wide range of biomedical uses, including drug transport, gene delivery, tissue engineering, and regenerative therapies, among others, thanks to all these characteristics. Different CTS-based polymers are being developed with the goal of promoting neural cell adhesion, proliferation, and differentiation in tissue engineering and regenerative medicine in the CNS. For more effective methods of tissue substitution, these CTS-based biomaterials can be used as scaffolds to replicate the extracellular matrix and microenvironment of the body. Thus, for the success of biomaterials, attributes like biocompatibility, biodegradability, mechanical strength, architecture, and cell adhesion capability become vital (Gnavi et al., 2013). Ricks and coworkers have outpointed the characteristics of CTS-based scaffolds for the nervous system regeneration (Ricks et al., 2014). They claimed that it is simple to alter the physicochemical characteristics of CTS and modified CTS to create particular structural features for scaffolds. Therefore, by rebuilding missing tissues and reconnecting disconnected neuronal processes, the implantation of CTS-based biomaterials into the CNS provides a great way to address the strength of its regenerative capacity. Another method for enhancing regenerative treatments is the incorporation of stem cells and biomolecules into these biomaterials in these scaffolds. Biomaterials therefore support cell therapy by acting as transport systems that encourage cell survival and engraftment as well as by giving the cells a self-sufficient microenvironment. The transplanted cells can also be removed from the harmed host tissue (Ricks et al., 2014). As delivery systems that encourage cell survival and engraftment, biomaterials therefore support cell therapy. The ability to segregate the implanted cells from the injured host tissue and the provision of a distinct milieu by biomaterials are both advantages.

Several strategies have been developed in recent years in order to increase cell adhesion, differentiation, and viability on CTS-based scaffolds. Different mixtures of biopolymers and CTS have been used to modulate the microstructure of the scaffolds and their properties. In the recent development of polyaniline-chloride/CTS-based bio-nanocomposites, Abasi et al. found that physical characteristics of the scaffolds, such as electrical conductivity and shape, have a stronger impact on cell-substrate interactions than molecular affinity (Abasi et al., 2019). Similarly, Sung et al. have studied neuronal cells on flat, micro- and nano-textured CTS substrates and found that cell adhesion increases on flat CTS surfaces (Sung et al., 2015). It is well known that internal structure and surface design of the scaffolds is critical for cell adhesion and proliferation. Using 3D printing technology for nerve tissue repair has become increasingly of great importance. In rats with spinal cord injury (SCI), Sun and coworkers have printed a 3D collagen and CTS-based scaffold with a specific structure and they observed nerve fiber regeneration and functional recovery after its implantation (Sun et al., 2019). Rao *et al.* have developed neurotrophin-3-CTS tubes that promoted neuroprotection, neurogenesis, revascularization, and anti-inflammation on SCI conditions and they reported that CTS-based hydrogels have shown remarkable potential of the CNS repair (Rao et al., 2018). Indeed, they have been demonstrated in many animal models to boost the survival rate of injured neurons and offer a favorable microenvironment for axon regrowth (Nawrotek et al., 2017). In another study, the development of CTS-based hydrogel to target rat spinal cord injury, have shown axonal regrowth and sustained recovery of locomotor function (Chedly et al., 2017). Further studies of advantages of CTS hydrogels are needed to define their therapeutic potential in different models of NDD.

CHITOSAN-BASED SCAFFOLDING AND CELL THERAPY FOR GLIAL CELLS

The CNS comprises neurons as its functional units, distinguished by their unique capacity to transmit rapid electrical signals called action. However, neurons are susceptible to damage. Most fully developed neurons are post-mitotic and can be identified through markers such as β-III-tubulin (Tuj-1) and microtubule-associated protein 2 (MAP-2). Conversely, the primary cells driving regeneration are astrocytes. These cells play vital roles within the CNS during normal functioning, development, and post-injury recovery (White et al, 2008).

Astrocytes offer physical and trophic support to neurons, displaying either permissive behavior, aiding growth and guidance, as observed during cerebral cortex development, or acting as potent inhibitors, such as in the formation of the glial scar following injury (White et al, 2008). Consequently, when designing biomaterials for CNS tissue engineering, it becomes crucial to examine not only neuronal responses, but also those of glial cells, as well as the interactions between these two cell types.

In a study by Marta Mattotti et al., the reactions of glial and neuronal cells to CTS were investigated, both in flat and micrometrically patterned forms. The findings indicated differing responses of neuronal and glial cells when exposed to flat and micro-patterned CTS films. CTS, even when uncoated, proves to be a favorable substrate for the attachment and growth of both glial cells and neurons, triggering distinct reactions in each cell type. Notably, CTS facilitates the maturation of glial cells, implying astroglial activation even in the presence of linear patterns (Marta Mattotti et al., 2017)

Interestingly, these mature or reactive glial cells exhibited permissiveness towards axonal growth. Axons organized and aligned along the patterned grooves, and the dimensions of these linear topographic patterns influenced both neurite and cell responses. Specifically, patterns with a width of 10 micrometers induced the fasciculation of axons, a targeted response for certain applications. This suggests that line-patterned CTS holds promise as a substrate for CNS tissue engineering, particularly when precise axonal orientation is required (Watanabe, 2018).

Nonetheless, the progression of these damaging processes involves the loss of extracellular matrix as well as neuronal and glial cells. This loss of tissue creates an inhospitable environment for transplanted cells, leading to poor engraftment and limited cell viability (Boisserand et al., 2016). In recent years, the integration of diverse biomaterials into CNS cell therapy has shown promise in enhancing cell survival, integration, and differentiation (Führmann and Shoichet, 2018). Within this context, CTS-based biomaterials have been combined with stem/precursor cells to establish a path toward neuroregeneration. These biopolymeric structures not only serve as carriers for delivering cells and offering physical support but also play a role in modulating the biological microenvironment to steer axonal growth and facilitate the integration of healthy tissue into the injured area (Boni et al., 2018).

Using 3D printing technology for nerve tissue repair has become increasingly of great importance (Figure 3). The stiffness of the scaffolds should fall within the range of 0.1–1 kPa to accurately mimic the pliability of soft tissues like the brain. It is widely recognized that stiffness exerts a pronounced influence on the response and function of stem cells (Liang et al., 2019). Furthermore, it has been established that cell size impacts cellular reactions to matrix stiffness within 3D cultures, particularly in the case of larger cells, such as numerous human stem cells (Bao et al., 2019).

Figure 3. Schematic representation of 3D printing technology for glial cell repair

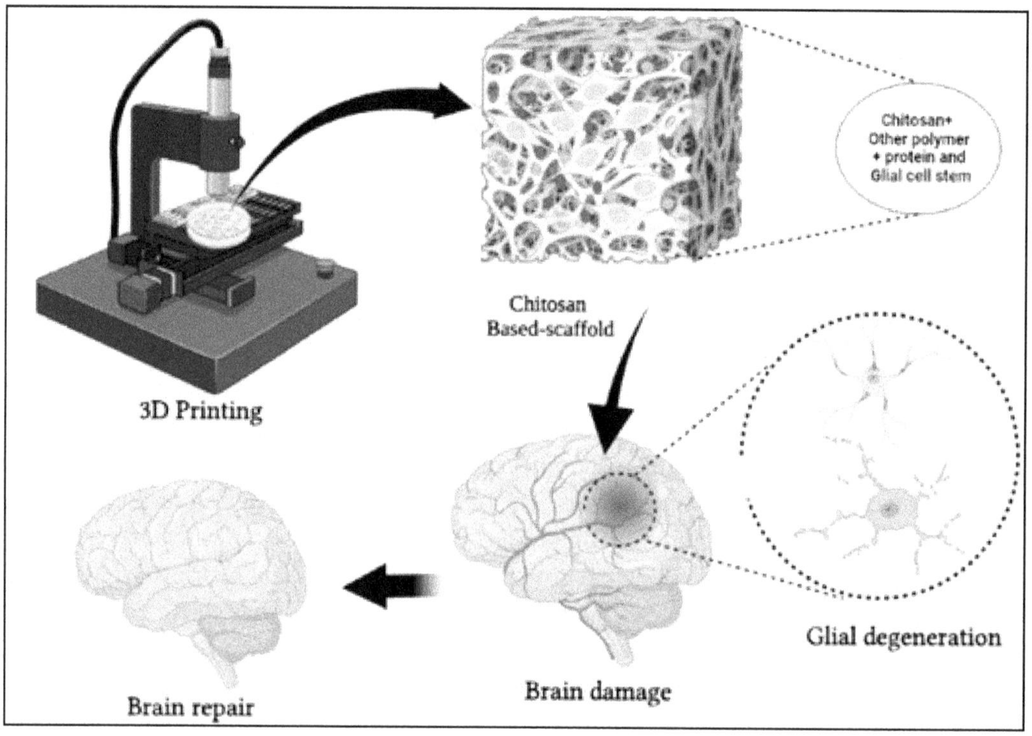

In addition to its versatility, CTS and its derivatives have demonstrated superiority as substrates for cell therapy when compared to other polymers. For instance, Scanga et al. (2010) highlighted the capacity of adult murine neural progenitor cells (NPCs) to proliferate and differentiate into the three major neural cell types when cultured on CTS hydrogel films. In contrast, NPC differentiation did not occur on poly (oligoethylene oxide dimethacrylate-co-2-amino ethyl methacrylate) or its blend with poly(vinyl alcohol), nor on poly(glycerol dimethacrylate-co-2-amino ethyl methacrylate) (Scanga et al., 2010). Similarly, Kim et al. (2016) observed more favorable functional improvements in rats with spinal cord injuries (SCI) following mesenchymal stem cell (MSC) transplantation on CTS scaffolds compared to PLGA scaffolds. Furthermore, they investigated intralesional injection of the same cells and contrasted it with scaffold-based transplantation in rats. The use of scaffolds resulted in higher MSC engraftment (Kim et al., 2016). Moreover, the culture of rat PC12 cells and human neural stem cells on CTS has exhibited superior outcomes compared to electrospun nanofibers derived from cellulose acetate or polyethersulfone, as demonstrated by Du et al. (2014).

The inclusion of trophic factors such as FGF-2, NGF, PDGF, and bFGF has also exhibited the ability to enhance the stemness of neural stem cells and promote their differentiation and proliferation when combined with CTS scaffolds or microspheres (Hoveizi et al., 2015; Skop et al., 2016; Moore et al., 2018). More recently, Skop and colleagues devised a cell-scaffold approach utilizing a radial glial neural precursor cell line engineered to conditionally secrete insulin-like growth factor I. This modified cell line was affixed to a CTS-based microsphere scaffold and introduced into the lesion site of adult rats with traumatic brain injury (TBI). The researchers observed encouraging differentiation across all three major neural cell types neurons, astrocytes, and oligodendrocytes along with an enhanced propensity for neuronal differentiation. These observed effects translated into a restoration of somatosensory function. However, it's important to note that the presence of insulin-like growth factor I did not correlate with a higher rate of cell retention or an improved capacity for cell replacement. As a result, the precise mechanism through which functional recovery is enhanced requires further elucidation in forthcoming investigations (Skop et al., 2019). In the neural network, glial cells are vital but often overlooked. While neurons shine, glial cells support neuron function, maintain nervous system structure, aid signal transmission, and regulate ion balance (Charcot, 1886). Yet, these glial cells – astrocytes, oligodendrocytes, microglia – also relate to neurological disorders. From multiple sclerosis to Parkinson's, glial cells affect inflammation, demyelination, and key neurological mechanisms. In this context, the emergence of new therapeutic strategies offers glimmers of hope for treating these complex conditions. Among these promising approaches, the use of CTS-based materials in combination with cell therapy is garnering increasing interest. By capitalizing on the unique properties of CTS and harnessing the regenerative potential of glial cells, this approach opens new perspectives to alleviate the devastating effects of many neurological diseases (Eslahi et al., 2021). At the heart of the complex interactions between CTS and cells, this biomaterial acts as a biochemical and mechanical mediator, guiding cell differentiation and influencing signalling pathways (Islam et al., 2020).

In the context of nerve regeneration, CTS appears particularly promising. Its adhesive capacity facilitates interaction with glial and neuronal cells, thereby promoting the cellular connections essential for nerve tissue regeneration (Y. Kim et al., 2023). In addition, the controlled release properties of CTS allow the progressive diffusion of growth factors and bioactive molecules, stimulating axonal growth and nerve repair (Y. Kim et al., 2023). In this respect, glioma, the most common primary brain tumour in adolescents, is a fatal disease worldwide. Despite advances in radiotherapy and chemotherapy, improvement rates in glioma patients remain insufficient. This highlights the need to rethink and improve existing therapeutic approaches, and even to discover new treatment platforms. CTS and its derivatives appear to be promising targets in the fight against cancer, as they favour prolonged release and better internalisation of chemotherapeutic agents. These properties translate into enhanced cytotoxicity for cancer cells (Adhikari & Yadav, 2018), making CTS a potentially suitable candidate for the treatment of various cancers, including glioma. In the specific context of glioma, CTS and its derivatives can inhibit cancer by targeting associated signalling pathways, such as MMP-9, VEGF, TRAIL and nuclear factor-κB (Eslahi et al., 2021). In addition, these compounds can be used as vectors to transport different therapeutic agents to the tumour region (Adhikari & Yadav, 2018).

Beyond their anti-glioma effects, CTS and its derivatives can be employed to culture glioma cancer cells, thereby providing a deeper understanding of the pathogenesis of this disease. Furthermore, three-dimensional CTS scaffolds emerge as potential targets, offering highly advantageous drug screening platforms. In this context, our synthesis underscores the anti-glioma effects of CTS, as well as its application as an innovative therapeutic delivery system in glioma management (Eslahi et al., 2021). Advanced therapies combining cells with biomaterial-based supports are emerging as a potent method for treating spinal cord injuries. Research has demonstrated that hydrogels, particularly those derived from CTS, can enhance the survival and engraftment of transplanted cells. These biomimetic hydrogels can be designed to encapsulate cells and target them minimally invasively towards the injured area. Furthermore, they can be developed to accommodate mesenchymal stem cells, whose paracrine action can therapeutically stimulate stem cell regeneration, curtail glial scar formation, and reduce cellular mortality at the injury site (Boido et al., 2019).

These materials exhibit anti-inflammatory (Chang et al., 2019; Mohyuddin et al., 2021), and antioxidant properties (Ngo & Kim, 2014) that, in synergy with mesenchymal stem cells, orchestrate the shift from inflammation towards the establishment of a less hostile environment post-trauma. This innovative approach holds the promise of laying the groundwork for a more effective and targeted treatment of spinal cord injuries, thereby offering hope for enhancing the quality of life for those affected (Eslahi et al., 2021). An intriguing aspect of this interaction lies in the CTS role in promoting cell survival and alignment. Research has highlighted its positive impact on Schwann cells, a crucial subtype of glial cells. This suggests that CTS could potentially enhance glial cell viability and proper alignment, thereby reinforcing their essential functions (Boecker et al., 2019; Hao et al., 2017).

Furthermore, CTS exhibits the capacity to modulate the activation of microglial cells, a key player in the immune response of the central nervous system. Furthermore, CTS exhibits a captivating capacity to induce differentiation. Particularly, interactions with Schwann cells have yielded compelling outcomes(Yuan et al., 2004). It has been demonstrated that CTS induces the differentiation of mesenchymal stromal cells into phenotypes resembling Schwann cells (Gnavi et al., 2013). Alternatively, it creates an environment conducive to the flourishing of Schwann cells, potentially enhancing glial cell differentiation and, consequently, fostering an environment conducive to neuroregeneration (Boecker et al., 2019).-Understanding the delicate interplay between CTS and glial cells is complex and context-dependent. This interaction varies with applications and conditions. As our understanding deepens, CTS therapeutic potential in neurological disorders requires ongoing research to fully harness its transformative capabilities.

CONCLUSION

Several approaches have been devised to enhance cell adhesion, differentiation, and viability on scaffolds based on chitosan (CTS). The biological activities of CTS and its derivatives have garnered attention in the field of neuronal regenerative medicine. Additionally, CTS plays a role in neuro-regeneration through stem cell transplantation using CTS-based carriers, offering promising prospects. As a result, the interest in employing CTS and its derivatives for addressing neurological disorders is a burgeoning area, with CTS-based therapies for neurodegenerative conditions being actively advocated. Nevertheless, further research is required to comprehensively understand the potential role of CTS and its derivatives in treating neurological disorders and their implications in various applications focused on safeguarding and rejuvenating CNS tissue.

REFERENCES

Abasi, S., Aggas, J. R., & Guiseppi-Elie, A. (2019). Physiochemical and morphological dependent growth of NIH/3T3 and PC-12 on polyaniline-chloride/chitosan bionanocomposites. *Materials Science and Engineering C, 99*, 1304–1312. doi:10.1016/j.msec.2019.02.018 PMID:30889665

Adhikari, H. S., & Yadav, P. N. (2018). Anticancer Activity of Chitosan, Chitosan Derivatives, and Their Mechanism of Action. *International Journal of Biomaterials, 2952085*, 1–29. doi:10.1155/2018/2952085 PMID:30693034

Ali, A., & Ahmed, S. (2018). A review on chitosan and its nanocomposites in drug delivery. *International Journal of Biological Macromolecules, 109*, 273–286. doi:10.1016/j.ijbiomac.2017.12.078 PMID:29248555

Alonso, D., Castro, A., & Martinez, A. (2005). Marine compounds for the therapeutic treatment of neurological disorders. *Expert Opinion on Therapeutic Patents, 15*(10), 1377–1386. doi:10.1517/13543776.15.10.1377

Amor, S., Peferoen, L. A., Vogel, D. Y., Breur, M., van der Valk, P., Baker, D., & van Noort, J. M. (2014). Inflammation in neurodegenerative diseases–an update. *Immunology, 142*(2), 151–166. doi:10.1111/imm.12233 PMID:24329535

Arranz, A. M., & De Strooper, B. (2019). The role of astroglia in Alzheimer's disease: Pathophysiology and clinical implications. *Lancet Neurology, 18*(4), 406–414. doi:10.1016/S1474-4422(18)30490-3 PMID:30795987

Badhe, RV, Bijukumar, D., Chejara, DR, Mabrouk, M., Choonara, YE, Kumar, P., ... & Pillay, V. (2017). Un échafaudage macroporeux biomimétique bicouche composite chitosane-gélatine pour l'ingénierie tissulaire des vaisseaux sanguins. *Polymères glucidiques, 157*, 1215-1225.

Bahbouhi, B., Berthelot, L., Pettré, S., Michel, L., Wiertlewski, S., Weksler, B., Romero, I.-A., Miller, F., Couraud, P.-O., Brouard, S., Laplaud, D.-A., & Soulillou, J.-P. (2009). Peripheral blood CD4+ T lymphocytes from multiple sclerosis patients are characterized by higher PSGL-1 expression and transmigration capacity across a human blood-brain barrier-derived endothelial cell line. *Journal of Leukocyte Biology, 86*(5), 1049–1063. doi:10.1189/jlb.1008666 PMID:19696154

Baillet, A., Chanteperdrix, V., Trocmé, C., Casez, P., Garrel, C., & Besson, G. (2010). The role of oxidative stress in amyotrophic lateral sclerosis and Parkinson's disease. *Neurochemical Research, 35*(10), 1530–1537. doi:10.100711064-010-0212-5 PMID:20535556

Bao, M., Xie, J., Katoele, N., Hu, X., Wang, B., Piruska, A., & Huck, W. T. S. (2019). Cellular volume and matrix stiffness direct stem cell behavior in a 3D microniche. *ACS Applied Materials & Interfaces, 11*(2), 1754–1759. doi:10.1021/acsami.8b19396 PMID:30584755

Barber, S. C., Mead, R. J., & Shaw, P. J. (2006). Oxidative stress in ALS: A mechanism of neurodegeneration and a therapeutic target. *Biochimica et Biophysica Acta, 1762*(11-12), 1051–1067. doi:10.1016/j.bbadis.2006.03.008 PMID:16713195

Beauquis, J., Pavía, P., Pomilio, C., Vinuesa, A., Podlutskaya, N., Galvan, V., & Saravia, F. (2013). Environmental enrichment prevents astroglial pathological changes in the hippocampus of APP transgenic mice, model of Alzheimer's disease. *Experimental Neurology*, *239*, 28–37. doi:10.1016/j.expneurol.2012.09.009 PMID:23022919

Behrens, P. F., Franz, P., Woodman, B., Lindenberg, K. S., & Landwehrmeyer, G. B. (2002). Impaired glutamate transport and glutamate-glutamine cycling: Downstream effects of the Huntington mutation. *Brain*, *125*(8), 1908–1922. doi:10.1093/brain/awf180 PMID:12135980

Benraiss, A., Wang, S., Herrlinger, S., Li, X., Chandler-Militello, D., Mauceri, J., Burm, H. B., Toner, M., Osipovitch, M., Jim Xu, Q., Ding, F., Wang, F., Kang, N., Kang, J., Curtin, P. C., Brunner, D., Windrem, M. S., Munoz-Sanjuan, I., Nedergaard, M., & Goldman, S. A. (2016). Human glia can both induce and rescue aspects of disease phenotype in Huntington disease. *Nature Communications*, *7*(1), 11758. doi:10.1038/ncomms11758 PMID:27273432

Bidmon, H. J., Görg, B., Palomero-Gallagher, N., Schleicher, A., Häussinger, D., Speckmann, E. J., & Zilles, K. (2008). Glutamine synthetase becomes nitrated and its activity is reduced during repetitive seizure activity in the pentylentetrazole model of epilepsy. *Epilepsia*, *49*(10), 1733–1748. doi:10.1111/j.1528-1167.2008.01642.x PMID:18479397

Boecker, A., Daeschler, S. C., Kneser, U., & Harhaus, L. (2019). Relevance and Recent Developments of Chitosan in Peripheral Nerve Surgery. *Frontiers in Cellular Neuroscience*, *13*, 104. doi:10.3389/fncel.2019.00104 PMID:31019452

Boisserand, L. S., Kodama, T., Papassin, J., Auzely, R., Moisan, A., Rome, C., & Detante, O. (2016). Biomaterial applications in cell-based therapy in experimental stroke. *Stem Cells International*, *6810562*, 1–14. doi:10.1155/2016/6810562 PMID:27274738

Boni, R., Ali, A., Shavandi, A., & Clarkson, A. N. (2018). Current and novel polymeric biomaterials for neural tissue engineering. *Journal of Biomedical Science*, *25*(1), 90. doi:10.118612929-018-0491-8 PMID:30572957

Booth, H. D. E., Hirst, W. D., & Wade-Martins, R. (2017). The role of astrocyte dysfunction in Parkinson's disease pathogenesis. *Trends in Neurosciences*, *40*(6), 358–370. doi:10.1016/j.tins.2017.04.001 PMID:28527591

Bosson, A., Boisseau, S., Buisson, A., Savasta, M., & Albrieux, M. (2015). Disruption of dopaminergic transmission remodels tripartite synapse morphology and astrocytic calcium activity within substantia nigra pars reticulata. *Glia*, *63*(4), 673–683. doi:10.1002/glia.22777 PMID:25511180

Broe, M., Kril, J., & Halliday, G. M. (2004). Astrocytic degeneration relates to the severity of disease in frontotemporal dementia. *Brain*, *127*(10), 2214–2220. doi:10.1093/brain/awh250 PMID:15282215

Buskila, Y., Crowe, S. E., & Ellis-Davies, G. C. R. (2013). Synaptic deficits in layer 5 neurons precede overt structural decay in 5xFAD mice. *Neuroscience*, *254*, 152–159. doi:10.1016/j.neuroscience.2013.09.016 PMID:24055684

Buskila, Y., Kékesi, O., Bellot-Saez, A., Seah, W., Berg, T., Trpceski, M., Yerbury, J. J., & Ooi, L. (2019b). Dynamic interplay between H-current and M-current controls motoneuron hyperexcitability in amyotrophic lateral sclerosis. *Cell Death & Disease*, *10*(4), 310. doi:10.103841419-019-1538-9 PMID:30952836

Chandrasekaran, K., Mehrabian, Z., Spinnewyn, B., Chinopoulos, C., Drieu, K., & Fiskum, G. (2003). Neuroprotective effects of bilobalide, a component of Ginkgo biloba extract (EGb 761®) in global brain ischemia and in excitotoxicity-induced neuronal death. *Pharmacopsychiatry, 36*(S 1), 89-94.

Chang, S.-H., Lin, Y.-Y., Wu, G.-J., Huang, C.-H., & Tsai, G. J. (2019). Effect of chitosan molecular weight on anti-inflammatory activity in the RAW 264.7 macrophage model. *International Journal of Biological Macromolecules*, *131*, 167–175. doi:10.1016/j.ijbiomac.2019.02.066 PMID:30771390

Chaudhuri, A., & Behan, P. O. (2004). Multiple sclerosis is not an autoimmune disease. *Archives of Neurology*, *61*(10), 1610–1612. doi:10.1001/archneur.61.10.1610 PMID:15477520

Chedly, J., Soares, S., Montembault, A., von Boxberg, Y., Veron-Ravaille, M., Mouffle, C., Benassy, M.-N., Taxi, J., David, L., & Nothias, F. (2017). Physical chitosan microhydrogels as scaffolds for spinal cord injury restoration and axon regeneration. *Biomaterials*, *138*, 91–107. doi:10.1016/j.biomaterials.2017.05.024 PMID:28554011

Chen, P. C., Vargas, M. R., Pani, A. K., Smeyne, R. J., Johnson, D. A., Kan, Y. W., & Johnson, J. A. (2009). Nrf2-mediated neuroprotection in the MPTP mouse model of Parkinson's disease: Critical role for the astrocyte. *Proceedings of the National Academy of Sciences of the United States of America*, *106*(8), 2933–2938. doi:10.1073/pnas.0813361106 PMID:19196989

Choi, D. W. (1988). Neurotoxicité du glutamate et maladies du système nerveux. *Neurone*, *1*(8), 623–634. doi:10.1016/0896-6273(88)90162-6

Clarkson, C., Smeal, R. M., Hasenoehrl, M. G., White, J. A., Rubio, M. E., & Wilcox, K. S. (2020). Ultrastructural and functional changes at the tripartite synapse during epileptogenesis in a model of temporal lobe epilepsy. *Experimental Neurology*, *326*, 113196. doi:10.1016/j.expneurol.2020.113196 PMID:31935368

Damier, P., Hirsch, E. C., Zhang, P., Agid, Y., & Javoy-Agid, F. (1993). Glutathione peroxidase, glial cells and Parkinson's disease. *Neuroscience*, *52*(1), 1–6. doi:10.1016/0306-4522(93)90175-F PMID:8433802

Das, A., Wallace, G. C. IV, Holmes, C., McDowell, M. L., Smith, J. A., Marshall, J. D., Bonilha, L., Edwards, J. C., Glazier, S. S., Ray, S. K., & Banik, N. L. (2012). Hippocampal tissue of patients with refractory temporal lobe epilepsy is associated with astrocyte activation, inflammation, and altered expression of channels and receptors. *Neuroscience*, *220*, 237–246. doi:10.1016/j.neuroscience.2012.06.002 PMID:22698689

Davies, P., & Maloney, A. J. F. (1976). Selective loss of central cholinergic neurons in Alzheimer's disease. *Lancet*, *308*(8000), 1403. doi:10.1016/S0140-6736(76)91936-X PMID:63862

de Lau, L. M. L., & Breteler, M. M. B. (2006). Epidemiology of Parkinson's disease. *Lancet Neurology*, *5*(6), 525–535. doi:10.1016/S1474-4422(06)70471-9 PMID:16713924 de Majo, M., Koontz, M., Rowitch, D., & Ullian, E. M. (2020). An update on human astrocytes and their role in development and disease. *Glia*, *68*(4), 685–704. doi:10.1002/glia.23771 PMID:31926040

Devinsky, O., Vezzani, A., Najjar, S., De Lanerolle, N. C., & Rogawski, M. A. (2013). Glia and epilepsy: Excitability and inflammation. *Trends in Neurosciences*, *36*(3), 174–184. doi:10.1016/j.tins.2012.11.008 PMID:23298414

di Domenico, A., Carola, G., Calatayud, C., Pons-Espinal, M., Muñoz, J. P., Richaud-Patin, Y., Fernandez-Carasa, I., Gut, M., Faella, A., Parameswaran, J., Soriano, J., Ferrer, I., Tolosa, E., Zorzano, A., Cuervo, A. M., Raya, A., & Consiglio, A. (2019). Patient-specific iPSC-derived astrocytes contribute to non-cell-autonomous neurodegeneration in Parkinson's disease. *Stem Cell Reports*, *12*(2), 213–229. doi:10.1016/j.stemcr.2018.12.011 PMID:30639209

Diniz, L. P., Matias, I., Araujo, A. P. B., Garcia, M. N., Barros-Aragão, F. G. Q., Alves-Leon, S. V., de Souza, J. M., Foguel, D., Figueiredo, C. P., Braga, C., Romão, L., & Gomes, F. C. A. (2019). α-synuclein oligomers enhance astrocyte-induced synapse formation through TGF-β1 signaling in a Parkinson's disease model. *Journal of Neurochemistry*, *150*(2), 138–157. doi:10.1111/jnc.14710 PMID:31009074

Do-Ha, D., Buskila, Y., & Ooi, L. (2018). Impairments in motor neurons, interneurons and astrocytes contribute to hyperexcitability in ALS: Underlying mechanisms and paths to therapy. *Molecular Neurobiology*, *55*(2), 1410–1418. doi:10.100712035-017-0392-y PMID:28160214

Do Nascimento, A. L., Dos Santos, N. F., Campos Pelágio, F., Aparecida Teixeira, S., De Moraes Ferrari, E. A., & Langone, F. (2012). Neuronal degeneration and gliosis time-course in the mouse hippocampal formation after pilocarpine-induced status epilepticus. *Brain Research*, *1470*, 98–110. doi:10.1016/j.brainres.2012.06.008 PMID:22781142

Du, J., Tan, E., Kim, H. J., Zhang, A., Bhattacharya, R., & Yarema, K. J. (2014). Comparative evaluation of chitosan, cellulose acetate, and polyethersulfone nanofiber scaffolds for neural differentiation. *Carbohydrate Polymers*, *99*, 483–490. doi:10.1016/j.carbpol.2013.08.050 PMID:24274534

Eslahi, M., Dana, P. M., Asemi, Z., Hallajzadeh, J., Mansournia, M. A., & Yousefi, B. (2021). The effects of chitosan-based materials on glioma : Recent advances in its applications for diagnosis and treatment. *International Journal of Biological Macromolecules*, *168*, 124–129. doi:10.1016/j.ijbiomac.2020.11.180 PMID:33275978

Faideau, M., Kim, J., Cormier, K., Gilmore, R., Welch, M., Auregan, G., Dufour, N., Guillermier, M., Brouillet, E., Hantraye, P., Deglon, N., Ferrante, R. J., & Bonvento, G. (2010). in vivo expression of polyglutamine-expanded huntingtin by mouse striatal astrocytes impairs glutamate transport: A correlation with Huntington's disease subjects. *Human Molecular Genetics*, *19*(15), 3053–3067. doi:10.1093/hmg/ddq212 PMID:20494921

Ferri, C. P., Prince, M., Brayne, C., Brodaty, H., Fratiglioni, L., Ganguli, M., Hall, K., Hasegawa, K., Hendrie, H., Huang, Y., Jorm, A., Mathers, C., Menezes, P. R., Rimmer, E., & Scazufca, M. (2005). Global prevalence of dementia: A Delphi consensus study. *Lancet*, *366*(9503), 2112–2117. doi:10.1016/S0140-6736(05)67889-0 PMID:16360788

Filippi, M., Bozzali, M., Rovaris, M., Gonen, O., Kesavadas, C., & Ghezzi, A. (2003). Evidence for widespread axonal damage at the earliest clinical stage of multiple sclerosis. *Brain*, *126*(2), 433–437. doi:10.1093/brain/awg038 PMID:12538409

Führmann, T., & Shoichet, M. S. (2018). The role of biomaterials in overcoming barriers to regeneration in the central nervous system. *Biomedical Materials (Bristol, England)*, *13*(5), 050201. doi:10.1088/1748-605X/aac2f6 PMID:29864020

Gabriel, S., Njunting, M., Pomper, J. K., Merschhemke, M., Sanabria, E. R. G., Eilers, A., Kivi, A., Zeller, M., Meencke, H.-J., Cavalheiro, E. A., Heinemann, U., & Lehmann, T.-N. (2004). Stimulus and potassium-induced epileptiform activity in the human dentate gyrus from patients with and without hippocampal sclerosis. *The Journal of Neuroscience : The Official Journal of the Society for Neuroscience*, *24*(46), 10416–10430. doi:10.1523/JNEUROSCI.2074-04.2004 PMID:15548657

Garbuzova-Davis, S., Hernandez-Ontiveros, D. G., Rodrigues, M. C. O., Haller, E., Frisina-Deyo, A., Mirtyl, S., Sallot, S., Saporta, S., Borlongan, C. V., & Sanberg, P. R. (2012). Impaired blood-brain/spinal cord barrier in ALS patients. *Brain Research*, *1469*, 114–128. doi:10.1016/j.brainres.2012.05.056 PMID:22750125

Gnavi, S., Barwig, C., Freier, T., Haastert-Talini, K., Grothe, C., & Geuna, S. (2013). The use of chitosan-based scaffolds to enhance regeneration in the nervous system. *International Review of Neurobiology*, *109*, 1–62. doi:10.1016/B978-0-12-420045-6.00001-8 PMID:24093605

Golarai, G., Greenwood, A., Feeney, D., & Connor, J. (2008). Physiological and structural evidence for hippocampal involvement in persistent seizure susceptibility after traumatic brain injury. *The Journal of Neuroscience : The Official Journal of the Society for Neuroscience*, *21*(21), 8523–8537. doi:10.1523/JNEUROSCI.21-21-08523.2001 PMID:11606641

Habib, N., McCabe, C., Medina, S., Varshavsky, M., Kitsberg, D., Dvir-Szternfeld, R., Green, G., Dionne, D., Nguyen, L., Marshall, J. L., Chen, F., Zhang, F., Kaplan, T., Regev, A., & Schwartz, M. (2020). Disease-associated astrocytes in Alzheimer's disease and aging. *Nature Neuroscience*, *23*(6), 701–706. doi:10.103841593-020-0624-8 PMID:32341542

Hallmann, A. L., Araúzo-Bravo, M. J., Mavrommatis, L., Ehrlich, M., Röpke, A., Brockhaus, J., Missler, M., Sterneckert, J., Schöler, H. R., Kuhlmann, T., Zaehres, H., & Hargus, G. (2017). Astrocyte pathology in a human neural stem cell model of frontotemporal dementia caused by mutant TAU protein. *Scientific Reports*, *7*(1), 42991. doi:10.1038rep42991 PMID:28256506

Hamdan, Y. A., Elouali, S., Eladlani, N., Lefeuvre, B., Oudadesse, H., & Rhazi, M. (2023b). Investigation on Akis granulifera (Coleoptera, Sahlberg, 1823) as a potential source of chitin and chitosan: Extraction, characterization and hydrogel formation. *International Journal of Biological Macromolecules*, *252*, 126292. doi:10.1016/j.ijbiomac.2023.126292 PMID:37573901

Hamdan, Y. A., Elouali, S., Maloui, A. B., El-Mansoury, B., Oudadesse, H., & Rhazi, M. (2023a). Chitosan and Its Derivatives as Potential Neuro-Protective Agents for the Treatment of Parkinson's Disease. In Experimental and Clinical Evidence of the Neuropathology of Parkinson's Disease (pp. 253-274). IGI Global. doi:10.4018/978-1-6684-5156-4.ch015

Hao, C., Wang, W., Wang, S., Zhang, L., & Guo, Y. (2017). An overview of the protective effects of chitosan and acetylated chitosan oligosaccharides against neuronal disorders. *Marine Drugs, 15*(4), 89. doi:10.3390/md15040089 PMID:28333077

Hassel, B., Tessler, S., Faull, R. L. M., & Emson, P. C. (2008). Glutamate uptake is reduced in prefrontal cortex in Huntington's disease. *Neurochemical Research, 33*(2), 232–237. doi:10.100711064-007-9463-1 PMID:17726644

Hogan, D. B., Jetté, N., Fiest, K. M., Roberts, J. I., Pearson, D., Smith, E. E., Roach, P., Kirk, A., Pringsheim, T., & Maxwell, C. J. (2016). The prevalence and incidence of frontotemporal dementia: A systematic review. *The Canadian Journal of Neurological Sciences, 43*(S1), S96–S109. doi:10.1017/cjn.2016.25 PMID:27307130

Hong, S., Dissing-Olesen, L., & Stevens, B. (2016b). New insights on the role of microglia in synaptic pruning in health and disease. *Current Opinion in Neurobiology, 36*, 128–134. doi:10.1016/j.conb.2015.12.004 PMID:26745839

Hoshino, K., Hayakawa, M., & Morimoto, Y. (2017). Minocycline prevents the impairment of hippocampal long-term potentiation in the septic mouse. *Shock (Augusta, Ga.), 48*(2), 209–214. doi:10.1097/SHK.0000000000000847 PMID:28187038

Hoveizi, E., Tavakol, S., & Ebrahimi-Barough, S. (2015). Neuroprotective effect of transplanted neural precursors embedded on PLA/CS scaffold in an animal model of multiple sclerosis. *Molecular Neurobiology, 51*(3), 1334–1342. doi:10.100712035-014-8812-8 PMID:25081586

Je, J. Y., & Kim, S. K. (2005). Water-soluble chitosan derivatives as a BACE1 inhibitor. *Bioorganic & Medicinal Chemistry, 13*(23), 6551–6555. doi:10.1016/j.bmc.2005.07.004 PMID:16084727

Jiang, X., Zhang, H., Duan, F., & Quan, X. (2017). Identify Huntington's disease associated genes based on restricted Boltzmann machine with RNA-seq data. *BMC Bioinformatics, 18*(1), 447. doi:10.118612859-017-1859-6 PMID:29020921

Kaminsky, N., Bihari, O., Kanner, S., & Barzilai, A. (2016). Connecting malfunctioning glial cells and brain degenerative disorders. *Genomics, Proteomics & Bioinformatics, 14*(3), 155–165. doi:10.1016/j.gpb.2016.04.003 PMID:27245308

Kang, T. C., Kim, D. S., Kwak, S. E., Kim, J. E., Won, M. H., Kim, D. W., Choi, S.-Y., & Kwon, O.-S. (2006). Epileptogenic roles of astroglial death and regeneration in the dentate gyrus of experimental temporal lobe epilepsy. *Glia, 54*(4), 258–271. doi:10.1002/glia.20380 PMID:16845674

Kim, S. K., & Rajapakse, N. (2005). Enzymatic production and biological activities of chitosan oligosaccharides (COS): A review. *Carbohydrate Polymers, 62*(4), 357–368. doi:10.1016/j.carbpol.2005.08.012

Kim, S. Y., Porter, B. E., Friedman, A., & Kaufer, D. (2016). A potential role for glia-derived extracellular matrix remodeling in postinjury epilepsy. *Journal of Neuroscience Research, 94*(9), 794–803. doi:10.1002/jnr.23758 PMID:27265805

Kim, Y., Zharkinbekov, Z., Raziyeva, K., Tabyldiyeva, L., Berikova, K., Zhumagul, D., Temirkhanova, K., & Saparov, A. (2023). Chitosan-Based Biomaterials for Tissue Regeneration. *Pharmaceutics*, *15*(3), 3. doi:10.3390/pharmaceutics15030807 PMID:36986668

Kim, Y. C., Kim, Y. H., Kim, J. W., & Ha, K. Y. (2016). Transplantation of mesenchymal stem cells for acute spinal cord injury in rats: Comparative study between Intralesional injection and scaffold based transplantation. *Journal of Korean Medical Science*, *31*(9), 1373–1382. doi:10.3346/jkms.2016.31.9.1373 PMID:27510379

Kitamura, Y., Watanabe, S., Taguchi, M., Takagi, K., Kawata, T., Takahashi-Niki, K., Yasui, H., Maita, H., Iguchi-Ariga, S. M. M., & Ariga, H. (2011). Neuroprotective effect of a new DJ-1-binding compound against neurodegeneration in Parkinson's disease and stroke model rats. *Molecular Neurodegeneration*, *6*(1), 48. doi:10.1186/1750-1326-6-48 PMID:21740546

Kong, X., Fu, J., Shao, K., Wang, L., Lan, X., & Shi, J. (2019). Biomimetic hydrogel for rapid and scar-free healing of skin wounds inspired by the healing process of oral mucosa. *Acta Biomaterialia*, *100*, 255–269. doi:10.1016/j.actbio.2019.10.011 PMID:31606531

Kordower, J. H., Olanow, C. W., Dodiya, H. B., Chu, Y., Beach, T. G., Adler, C. H., Halliday, G. M., & Bartus, R. T. (2013). Disease duration and the integrity of the nigrostriatal system in Parkinson's disease. *Brain*, *136*(8), 2419–2431. doi:10.1093/brain/awt192 PMID:23884810

Kostrzewa, R. M., & Segura-Aguilar, J. (2003). Novel mechanisms and approaches in the study of neurodegeneration and neuroprotection. A review. *Neurotoxicity Research*, *5*(6), 375–383. doi:10.1007/BF03033166 PMID:14715440

Kulijewicz-Nawrot, M., Syková, E., Chvátal, A., Verkhratsky, A., & Rodríguez, J. J. (2013). Astrocytes and glutamate homoeostasis in Alzheimer's disease: A decrease in glutamine synthetase, but not in glutamate transporter-1, in the prefrontal cortex. *ASN Neuro*, *5*(4), 273–282. doi:10.1042/AN20130017 PMID:24059854

Kumar, M. N. R. (2000). A review of chitin and chitosan applications. *Reactive & Functional Polymers*, *46*(1), 1–27. doi:10.1016/S1381-5148(00)00038-9

Kuo, Y. C., & Chiu, K. H. (2011). Inverted colloidal crystal scaffolds with laminin-derived peptides for neuronal differentiation of bone marrow stromal cells. *Biomaterials*, *32*(3), 819–831. doi:10.1016/j.biomaterials.2010.09.057 PMID:20974492

Laplaud, D. A., Ruiz, C., Wiertlewski, S., Brouard, S., Berthelot, L., & Guillet, M. (2004). Blood T-cell receptor β chain transcriptome in multiple sclerosis. Characterization of the T cells with altered CDR3 length distribution. *Brain*, *127*(5), 981–995. doi:10.1093/brain/awh119 PMID:14985265

Lee, H. J., Suk, J. E., Patrick, C., Bae, E. J., Cho, J. H., Rho, S., Hwang, D., Masliah, E., & Lee, S.-J. (2010). Direct transfer of α-synuclein from neuron to astroglia causes inflammatory responses in synucleinopathies. *The Journal of Biological Chemistry*, *285*(12), 9262–9272. doi:10.1074/jbc.M109.081125 PMID:20071342

Leenders, K. L., Salmon, E. P., Tyrrell, P., Perani, D., Brooks, D. J., Sager, H., Jones, T., Marsden, C. D., & Frackowiak, R. S. J. (1990). The nigrostriatal dopaminergic system assessed in vivo by positron emission tomography in healthy volunteer subjects and patients with Parkinson's disease. *Archives of Neurology*, *47*(12), 1290–1298. doi:10.1001/archneur.1990.00530120034007 PMID:2123623

Leyns, C. E. G., Gratuze, M., Narasimhan, S., Jain, N., Koscal, L. J., Jiang, H., Manis, M., Colonna, M., Lee, V. M. Y., Ulrich, J. D., & Holtzman, D. M. (2019). TREM2 function impedes tau seeding in neuritic plaques. *Nature Neuroscience*, *22*(8), 1217–1222. doi:10.103841593-019-0433-0 PMID:31235932

Liang, K., Bae, K. H., & Kurisawa, M. (2019). Recent advances in the design of injectable hydrogels for stem cell-based therapy. *Journal of Materials Chemistry. B*, *7*(24), 3775–3791. doi:10.1039/C9TB00485H

Lomen-Hoerth, C., Anderson, T., & Miller, B. (2002). The overlap of amyotrophic lateral sclerosis and frontotemporal dementia. *Neurology*, *59*(7), 1077–1079. doi:10.1212/WNL.59.7.1077 PMID:12370467

Martin, S., Al Khleifat, A., & Al-Chalabi, A. (2017). What causes amyotrophic lateral sclerosis? *F1000 Research*, *6*, 371. doi:10.12688/f1000research.10476.1 PMID:28408982

Martinac, J. A., Craft, D. K., Su, J. H., Kim, R. C., & Cotman, C. W. (2001). Astrocytes degenerate in frontotemporal dementia: Possible relation to hypoperfusion. *Neurobiology of Aging*, *22*(2), 195–207. doi:10.1016/S0197-4580(00)00231-1 PMID:11182469

Mattotti, M., Alvarez, Z., Delgado, L., Mateos-Timoneda, M. A., Aparicio, C., Planell, J. A., Alcántara, S., & Engel, E. (2017). Differential neuronal and glial behavior on flat and micro patterned chitosan films. *Colloids and Surfaces. B, Biointerfaces*, *158*, 569–577. doi:10.1016/j.colsurfb.2017.07.045 PMID:28750339

Mohammed, M. A., Syeda, J. T., Wasan, K. M., & Wasan, E. K. (2017). An overview of chitosan nanoparticles and its application in non-parenteral drug delivery. *Pharmaceutics*, *9*(4), 53. doi:10.3390/pharmaceutics9040053 PMID:29156634

Moore, L., Skop, N. B., Rothbard, D. E., Corrubia, L. R., & Levison, S. W. (2018). Tethered growth factors on biocompatible scaffolds improve stemness of cultured rat and human neural stem cells and growth of oligodendrocyte progenitors. *Methods (San Diego, Calif.)*, *133*, 54–64. doi:10.1016/j.ymeth.2017.08.015 PMID:28887084

Morley, J. E., Farr, S. A., & Nguyen, A. D. (2018). Alzheimer disease. *Clinics in Geriatric Medicine*, *34*(4), 591–601. doi:10.1016/j.cger.2018.06.006 PMID:30336989

Nawrotek, K., Marqueste, T., Modrzejewska, Z., Zarzycki, R., Rusak, A., & Decherchi, P. (2017). T hermogelling chitosan lactate hydrogel improves functional recovery after a C2 spinal cord hemisection in rat. *Journal of Biomedical Materials Research. Part A*, *105*(7), 2004–2019. doi:10.1002/jbm.a.36067 PMID:28324618

Ngo, D.-H., & Kim, S.-K. (2014). Chapter Two—Antioxidant Effects of Chitin, Chitosan, and Their Derivatives. In S.-K. Kim (Ed.), *Advances in Food and Nutrition Research* (Vol. 73, pp. 15–31). Academic Press. doi:10.1016/B978-0-12-800268-1.00002-0

Nwe, N., Furuike, T., & Tamura, H. (2009). The mechanical and biological properties of chitosan scaffolds for tissue regeneration templates are significantly enhanced by chitosan from Gongronella butleri. *Materials (Basel)*, *2*(2), 374–398. doi:10.3390/ma2020374

Octeau, J. C., Chai, H., Jiang, R., Bonanno, S. L., Martin, K. C., & Khakh, B. S. (2018). An optical neuron-astrocyte proximity assay at synaptic distance scales. *Neuron*, *98*(1), 49.e9–66.e9. doi:10.1016/j.neuron.2018.03.003 PMID:29621490

Ojeda-Hernández, D. D., Canales-Aguirre, A. A., Matias-Guiu, J., Gomez-Pinedo, U., & Mateos-Díaz, J. C. (2020). Potential of chitosan and its derivatives for biomedical applications in the central nervous system. *Frontiers in Bioengineering and Biotechnology*, *8*, 389. doi:10.3389/fbioe.2020.00389 PMID:32432095

Padovani, A., Agosti, C., Premi, E., Bellelli, G., & Borroni, B. (2007). Extrapyramidal symptoms in frontotemporal dementia: Prevalence and clinical correlations. *Neuroscience Letters*, *422*(1), 39–42. doi:10.1016/j.neulet.2007.05.049 PMID:17574750

Pangestuti, R., & Kim, SK (2010). Neuroprotective properties of chitosan and its derivatives. *marine drugs*, *8*(7), 2117-2128.

Pannasch, U., Vargová, L., Reingruber, J., Ezan, P., Holcman, D., Giaume, C., Syková, E., & Rouach, N. (2011). Astroglial networks scale synaptic activity and plasticity. *Proceedings of the National Academy of Sciences of the United States of America*, *108*(20), 8467–8472. doi:10.1073/pnas.1016650108 PMID:21536893

Papadeas, S. T., Kraig, S. E., O'Banion, C., Lepore, A. C., & Maragakis, N. J. (2011). Astrocytes carrying the superoxide dismutase 1 (SOD1 G93A) mutation induce wild-type motor neuron degeneration in vivo. *Proceedings of the National Academy of Sciences of the United States of America*, *108*(43), 17803–17808. doi:10.1073/pnas.1103141108 PMID:21969586

Patel, D. C., Tewari, B. P., Chaunsali, L., & Sontheimer, H. (2019). Neuron-glia interactions in the pathophysiology of epilepsy. *Nature Reviews. Neuroscience*, *20*(5), 282–297. doi:10.103841583-019-0126-4 PMID:30792501

Pedersen, W. A., Fu, W., Keller, J. N., Markesbery, W. R., Appel, S., Smith, R. G., Kasarskis, E., & Mattson, M. P. (1998). Protein modification by the lipid peroxidation product 4-hydroxynonenal in the spinal cords of amyotrophic lateral sclerosis patients. *Annals of Neurology*, *44*(5), 819–824. doi:10.1002/ana.410440518 PMID:9818940

Pellicciari, R., Costantino, G., Marinozzi, M., & Natalini, B. (1998). Modulation of glutamate receptor pathways in the search for new neuroprotective agents. *Il Farmaco*, *53*(4), 255–261. doi:10.1016/S0014-827X(98)00018-4 PMID:9658582

Petroff, O. A. C., Errante, L. D., Rothman, D. L., Kim, J. H., & Spencer, D. D. (2002). Glutamate-glutamine cycling in the epileptic human hippocampus. *Epilepsia*, *43*(7), 703–710. doi:10.1046/j.1528-1157.2002.38901.x PMID:12102672

Philips, T., & Robberecht, W. (2011). Neuroinflammation in amyotrophic lateral sclerosis: Role of glial activation in motor neuron disease. *Lancet Neurology*, *10*(3), 253–263. doi:10.1016/S1474-4422(11)70015-1 PMID:21349440

Pirttimaki, T. M., Codadu, N. K., Awni, A., Pratik, P., Nagel, D. A., Hill, E. J., Dineley, K. T., & Parri, H. R. (2013). α7 nicotinic receptor-mediated astrocytic gliotransmitter release: Aβ effects in a preclinical Alzheimer's mouse model. *PLoS One*, *8*(11), e81828. doi:10.1371/journal.pone.0081828 PMID:24312364

Pramatarova, A., Laganière, J., Roussel, J., Brisebois, K., & Rouleau, G. A. (2001). Neuron-specific expression of mutant superoxide dismutase 1 in transgenic mice does not lead to motor impairment. *The Journal of Neuroscience : The Official Journal of the Society for Neuroscience*, *21*(10), 3369–3374. doi:10.1523/JNEUROSCI.21-10-03369.2001 PMID:11331366

Przedborski, S., Jackson-Lewis, V., Djaldetti, R., Liberatore, G., Vila, M., & Vukosavic, S. (2000). The parkinsonian toxin MPTP: Action and mechanism. *Restorative Neurology and Neuroscience*, *16*, 135–142. PMID:12671216

Rao, J. S., Zhao, C., Zhang, A., Duan, H., Hao, P., Wei, R. H., Shang, J., Zhao, W., Liu, Z., Yu, J., Fan, K. S., Tian, Z., He, Q., Song, W., Yang, Z., Sun, Y. E., & Li, X. (2018). NT3-chitosan enables de novo regeneration and functional recovery in monkeys after spinal cord injury. *Proceedings of the National Academy of Sciences of the United States of America*, *115*(24), E5595–E5604. doi:10.1073/pnas.1804735115 PMID:29844162

Ricks, C. B., Shin, S. S., Becker, C., & Grandhi, R. (2014). Extracellular matrices, artificial neural scaffolds and the promise of neural regeneration. *Neural Regeneration Research*, *9*(17), 1573. doi:10.4103/1673-5374.141778 PMID:25368641

Robel, S., Buckingham, S. C., Boni, J. L., Campbell, S. L., Danbolt, N. C., Riedemann, T., Sutor, B., & Sontheimer, H. (2015). Reactive astrogliosis causes the development of spontaneous seizures. *The Journal of Neuroscience : The Official Journal of the Society for Neuroscience*, *35*(8), 3330–3345. doi:10.1523/JNEUROSCI.1574-14.2015 PMID:25716834

Rothstein, J. D., Van Kammen, M., Levey, A. I., Martin, L. J., & Kuncl, R. W. (1995). Selective loss of glial glutamate transporter GLT-1 in amyotrophic lateral sclerosis. *Annals of Neurology*, *38*(1), 73–84. doi:10.1002/ana.410380114 PMID:7611729

Sahariah, P., & Másson, M. (2017). Antimicrobial chitosan and chitosan derivatives: A review of the structure–activity relationship. *Biomacromolecules*, *18*(11), 3846–3868. doi:10.1021/acs.biomac.7b01058 PMID:28933147

Salarian, A., Russmann, H., Wider, C., Burkhard, P. R., Vingerhoets, F. J. G., & Aminian, K. (2007). Quantification of tremor and bradykinesia in Parkinson's disease using a novel ambulatory monitoring system. *IEEE Transactions on Biomedical Engineering*, *54*(2), 313–322. doi:10.1109/TBME.2006.886670 PMID:17278588

Scanga, V. I., Goraltchouk, A., Nussaiba, N., Shoichet, M. S., & Morshead, C. M. (2010). Biomaterials for neural-tissue engineering — Chitosan supports the survival, migration, and differentiation of adult-derived neural stem and progenitor cells. *Canadian Journal of Chemistry*, *88*(3), 277–287. doi:10.1139/v09-171

Schofield, E., Kersaitis, C., Shepherd, C. E., Kril, J. J., & Halliday, G. M. (2003). Severity of gliosis in Pick's disease and frontotemporal lobar degeneration: Tau-positive glia differentiate these disorders. *Brain, 126*(4), 827–840. doi:10.1093/brain/awg085 PMID:12615642

Seifert, G., Schilling, K., & Steinhäuser, C. (2006). Astrocyte dysfunction in neurological disorders: A molecular perspective. *Nature Reviews. Neuroscience, 7*(3), 194–206. doi:10.1038/nrn1870 PMID:16495941

Shin, J. Y., Fang, Z. H., Yu, Z. X., Wang, C. E., Li, S. H., & Li, X. J. (2005). Expression of mutant huntingtin in glial cells contributes to neuronal excitotoxicity. *The Journal of Cell Biology, 171*(6), 1001–1012. doi:10.1083/jcb.200508072 PMID:16365166

Skop, N. B., Calderon, F., Cho, C. H., Gandhi, C. D., & Levison, S. W. (2016). Optimizing a multifunctional microsphere scaffold to improve neural precursor cell transplantation for traumatic brain injury repair. *Journal of Tissue Engineering and Regenerative Medicine, 10*(10), E419–E432. doi:10.1002/term.1832 PMID:27730762

Skop, N. B., Singh, S., Antikainen, H., Saqcena, C., Calderon, F., Rothbard, D. E., Cho, C. H., Gandhi, C. D., Levison, S. W., & Dobrowolski, R. (2019). Subacute transplantation of native and genetically engineered neural progenitors seeded on microsphere scaffolds promote repair and functional recovery after traumatic brain injury. *ASN Neuro, 11*, 1759091419830186. doi:10.1177/1759091419830186 PMID:30818968

Solano, R. M., Casarejos, M. J., Menéndez-Cuervo, J., Rodriguez-Navarro, J. A., De Yébenes, J. G., & Mena, M. A. (2008). Glial dysfunction in parkin null mice: Effects of aging. *The Journal of Neuroscience : The Official Journal of the Society for Neuroscience, 28*(3), 598–611. doi:10.1523/JNEUROSCI.4609-07.2008 PMID:18199761

Song, I., & Dityatev, A. (2018). Crosstalk between glia, extracellular matrix and neurons. *Brain Research Bulletin, 136*, 101–108. doi:10.1016/j.brainresbull.2017.03.003 PMID:28284900

Stenovec, M., Trkov, S., Lasič, E., Terzieva, S., Kreft, M., Rodríguez Arellano, J. J., Parpura, V., Verkhratsky, A., & Zorec, R. (2016). Expression of familial Alzheimer disease presenilin 1 gene attenuates vesicle traffic and reduces peptide secretion in cultured astrocytes devoid of pathologic tissue environment. *Glia, 64*(2), 317–329. doi:10.1002/glia.22931 PMID:26462451

Su, J. H., Nichol, K. E., Sitch, T., Sheu, P., Chubb, C., Miller, B. L., Tomaselli, K. J., Kim, R. C., & Cotman, C. W. (2000). DNA damage and activated caspase-3 expression in neurons and astrocytes: Evidence for apoptosis in frontotemporal dementia. *Experimental Neurology, 163*(1), 9–19. doi:10.1006/exnr.2000.7340 PMID:10785439

Sun, Y., Yang, C., Zhu, X., Wang, J. J., Liu, X. Y., Yang, X. P., An, X.-W., Liang, J., Dong, H.-J., Jiang, W., Chen, C., Wang, Z.-G., Sun, H.-T., Tu, Y., Zhang, S., Chen, F., & Li, X. H. (2019). 3D printing collagen/chitosan scaffold ameliorated axon regeneration and neurological recovery after spinal cord injury. *Journal of Biomedical Materials Research. Part A, 107*(9), 1898–1908. doi:10.1002/jbm.a.36675 PMID:30903675

Sung, C. Y., Yang, C. Y., Chen, W. S., Wang, Y. K., Yeh, J. A., & Cheng, C. M. (2015). Probing neural cell behaviors through micro-/nano-patterned chitosan substrates. *Biofabrication*, *7*(4), 045007. doi:10.1088/1758-5090/7/4/045007 PMID:26685015

Talantova, M., Sanz-Blasco, S., Zhang, X., Xia, P., Akhtar, M. W., Okamoto, S. I., Dziewczapolski, G., Nakamura, T., Cao, G., Pratt, A. E., Kang, Y.-J., Tu, S., Molokanova, E., McKercher, S. R., Hires, S. A., Sason, H., Stouffer, D. G., Buczynski, M. W., Solomon, J. P., & Lipton, S. A. (2013). Aβ induces astrocytic glutamate release, extrasynaptic NMDA receptor activation and synaptic loss. *Proceedings of the National Academy of Sciences of the United States of America*, *110*(27), E2518–E2527. doi:10.1073/pnas.1306832110 PMID:23776240

Tang, L., Ma, Y., Liu, X. L., Chen, L., & Fan, D. S. (2019). Better survival in female SOD1-mutant patients with ALS: A study of SOD1-related natural history. *Translational Neurodegeneration*, *8*(1), 2. doi:10.118640035-018-0142-8 PMID:30637102

Tolaimate, A., Desbrieres, J., Rhazi, M., & Alagui, A. (2003). Contribution to the preparation of chitins and chitosans with controlled physico-chemical properties. *Polymer*, *44*(26), 7939–7952. doi:10.1016/j.polymer.2003.10.025

Tong, X., Ao, Y., Faas, G. C., Nwaobi, S. E., Xu, J., Haustein, M. D., Anderson, M. A., Mody, I., Olsen, M. L., Sofroniew, M. V., & Khakh, B. S. (2014). Astrocyte Kir4.1 ion channel deficits contribute to neuronal dysfunction in Huntington's disease model mice. *Nature Neuroscience*, *17*(5), 694–703. doi:10.1038/nn.3691 PMID:24686787

Trapp, B. D., Peterson, J., Ransohoff, R. M., Rudick, R., Mörk, S., & Bö, L. (1998). Axonal transection in the lesions of multiple sclerosis. *The New England Journal of Medicine*, *338*(5), 278–285. doi:10.1056/NEJM199801293380502 PMID:9445407

Tucci, P., & Bagetta, G. (2008). How to study neuroprotection? *Cell Death and Differentiation*, *15*(6), 1084–1085. doi:10.1038/cdd.2008.32

Umoh, M. E., Dammer, E. B., Dai, J., Duong, D. M., Lah, J. J., Levey, A. I., Gearing, M., Glass, J. D., & Seyfried, N. T. (2018). A proteomic network approach across the ALS - FTD disease spectrum resolves clinical phenotypes and genetic vulnerability in human brain. *EMBO Molecular Medicine*, *10*(1), 48–62. doi:10.15252/emmm.201708202 PMID:29191947

Van Vlierberghe, S., Dubruel, P., & Schacht, E. (2011). Hydrogels à base de biopolymères comme échafaudages pour les applications d'ingénierie tissulaire: Une revue. *Biomacromolecules*, *12*(5), 1387–1408. doi:10.1021/bm200083n PMID:21388145

Ventura, R., & Harris, K. M. (1999). Three-dimensional relationships between hippocampal synapses and astrocytes. *The Journal of Neuroscience : The Official Journal of the Society for Neuroscience*, *19*(16), 6897–6906. doi:10.1523/JNEUROSCI.19-16-06897.1999 PMID:10436047

Villalba, R. M., & Smith, Y. (2011). Neuroglial plasticity at striatal glutamatergic synapses in Parkinson's disease. *Frontiers in Systems Neuroscience*, *5*, 68. doi:10.3389/fnsys.2011.00068 PMID:21897810

Wakabayashi, K., Hayashi, S., Yoshimoto, M., Kudo, H., & Takahash, H. (2000). NACP αsynuclein-positive filamentous inclusions in astrocytes and oligodendrocytes of Parkinson's disease brains. *Acta Neuropathologica*, *99*(1), 14–20. doi:10.1007/PL00007400 PMID:10651022

Wallraff, A., Köhling, R., Heinemann, U., Theis, M., Willecke, K., & Steinhäuser, C. (2006). The impact of astrocytic gap junctional coupling on potassium buffering in the hippocampus. *The Journal of Neuroscience : The Official Journal of the Society for Neuroscience*, *26*(20), 5438–5447. doi:10.1523/JNEUROSCI.0037-06.2006 PMID:16707796

White, R. E., & Jakeman, L. B. (2008). Don't fence me in: Harnessing the beneficial roles of astrocytes for spinal cord repair. *Restorative Neurology and Neuroscience*, *26*(2-3), 197–214. doi:10.1016/j.colsurfb.2017.07.045 PMID:18820411

Xu-Friedman, M. A., Harris, K. M., & Regehr, W. G. (2001). Three-dimensional comparison of ultrastructural characteristics at depressing and facilitating synapses onto cerebellar Purkinje cells. *The Journal of Neuroscience : The Official Journal of the Society for Neuroscience*, *21*(17), 6666–6672. doi:10.1523/JNEUROSCI.21-17-06666.2001 PMID:11517256

Yao, Z. A., Chen, F. J., Cui, H. L., Lin, T., Guo, N., & Wu, H. G. (2018). Efficacy of chitosan and sodium alginate scaffolds for repair of spinal cord injury in rats. *Neural Regeneration Research*, *13*(3), 502. doi:10.4103/1673-5374.228756 PMID:29623937

Yi, H., Wu, L. Q., Bentley, W. E., Ghodssi, R., Rubloff, G. W., Culver, J. N., & Payne, G. F. (2005). Biofabrication with chitosan. *Biomacromolecules*, *6*(6), 2881–2894. doi:10.1021/bm050410l PMID:16283704

Younes, I., & Rinaudo, M. (2015). Chitin and chitosan preparation from marine sources. Structure, properties and applications. *Marine Drugs*, *13*(3), 1133–1174. doi:10.3390/md13031133 PMID:25738328

Yuan, Y., Zhang, P., Yang, Y., Wang, X., & Gu, X. (2004). The interaction of Schwann cells with chitosan membranes and fibers in vitro. *Biomaterials*, *25*(18), 4273–4278. doi:10.1016/j.biomaterials.2003.11.029 PMID:15046917

Zou, L., Zhang, Y., Liu, X., Chen, J., & Zhang, Q. (2019). Biomimetic mineralization on natural and synthetic polymers to prepare hybrid scaffolds for bone tissue engineering. *Colloids and Surfaces. B, Biointerfaces*, *178*, 222–229. doi:10.1016/j.colsurfb.2019.03.004 PMID:30870789

Zurolo, E., de Groot, M., Iyer, A., Anink, J., van Vliet, E. A., Heimans, J. J., Reijneveld, J. C., Gorter, J. A., & Aronica, E. (2012). Regulation of Kir4.1 expression in astrocytes and astrocytic tumors: A role for interleukin-1 β. *Journal of Neuroinflammation*, *9*(1), 280. doi:10.1186/1742-2094-9-280 PMID:23270518

Compilation of References

Aarli, J. A., Abramsky, O., Chin, J. H., & Vora, N. (2014). Section Editors Global Perspectives *THE GLOBAL BURDEN OF NEUROLOGIC DISEASES.* WHO. https://www.who.int/nmh/publications/best_buys_summary/

Aarum, J., Sandberg, K., Haeberlein, S. L. B., & Persson, M. A. A. (2003). Migration and differentiation of neural precursor cells can be directed by microglia. *Proceedings of the National Academy of Sciences of the United States of America*, *100*(26), 15983–15988. doi:10.1073/pnas.2237050100 PMID:14668448

Abasi, S., Aggas, J. R., & Guiseppi-Elie, A. (2019). Physiochemical and morphological dependent growth of NIH/3T3 and PC-12 on polyaniline-chloride/chitosan bionanocomposites. *Materials Science and Engineering C*, *99*, 1304–1312. doi:10.1016/j.msec.2019.02.018 PMID:30889665

Abbadie, C., Bhangoo, S., De Koninck, Y., Malcangio, M., Melik-Parsadaniantz, S., & White, F. A. (2009). Chemokines and pain mechanisms. *Brain Research. Brain Research Reviews*, *60*(1), 125–134. doi:10.1016/j.brainresrev.2008.12.002 PMID:19146875

Abbas, N., Bednar, I., Mix, E., Marie, S., Paterson, D., Ljungberg, A., Morris, C., Winblad, B., Nordberg, A., & Zhu, J. (2002). Up-regulation of the inflammatory cytokines IFN-γ and IL-12 and down-regulation of IL-4 in cerebral cortex regions of APPSWE transgenic mice. *Journal of Neuroimmunology*, *126*(1-2), 50–57. doi:10.1016/S0165-5728(02)00050-4 PMID:12020956

Abbott, N. J. (2002). Astrocyte-endothelial interactions and blood-brain barrier permeability. *Journal of Anatomy*, *200*(6), 629–638. doi:10.1046/j.1469-7580.2002.00064.x PMID:12162730

Abbott, N. J., Rönnbäck, L., & Hansson, E. (2006). 'Astrocyte–Endothelial Interactions at the Blood–Brain Barrier'. *Nature Reviews. Neuroscience*, *7*(1), 41–53. doi:10.1038/nrn1824 PMID:16371949

Abdel-Misih, S. R. Z., & Bloomston, M. (2010). Liver anatomy. *Surgical Clinics*, *90*(4), 643–653. PMID:20637938

Abdul, H. M., Sama, M. A., Furman, J. L., Mathis, D. M., Beckett, T. L., Weidner, A. M., Patel, E. S., Baig, I., Murphy, M. P., Levine, H. III, Kraner, S. D., & Norris, C. M. (2009). Cognitive decline in Alzheimer's disease is associated with selective changes in calcineurin/NFAT signaling. *BaigJournal of Neuroscience, 2009. Social Neuroscience*, *29*(41), 12957–12969. Advance online publication. doi:10.1523/JNEUROSCI.1064-09.2009 PMID:19828810

Abe, N., Nishihara, T., Yorozuya, T., & Tanaka, J. (2020). Microglia and Macrophages in the Pathological Central and Peripheral Nervous Systems. *Cells*, *9*(9), 2132. doi:10.3390/cells9092132 PMID:32967118

Aber, E. R., Griffey, C. J., Davies, T., Li, A. M., Yang, Y. J., Croce, K. R., Goldman, J. E., Grutzendler, J., Canman, J. C., & Yamamoto, A. (2022). Oligodendroglial macroautophagy is essential for myelin sheath turnover to prevent neurodegeneration and death. *Cell Reports*, *41*(3), 111480. doi:10.1016/j.celrep.2022.111480

Abeysinghe, A., Deshapriya, R., & Udawatte, C. (2020). Alzheimer's disease; a review of the pathophysiological basis and therapeutic interventions. *Life Sciences*, *256*, 117996. doi:10.1016/j.lfs.2020.117996 PMID:32585249

Abo El-Enin, H. A., Elkomy, M. H., Naguib, I. A., Ahmed, M. F., Alsaidan, O. A., Alsalahat, I., Ghoneim, M. M., & Eid, H. M. (2022). Lipid nanocarriers overlaid with chitosan for brain delivery of berberine via the nasal route. *Pharmaceuticals (Basel, Switzerland)*, *15*(3), 281. doi:10.3390/ph15030281 PMID:35337079

Abounader, R., & Laterra, J. (2005). Scatter factor/hepatocyte growth factor in brain tumor growth and angiogenesis. *Neuro-Oncology*, *7*(4), 436–451. doi:10.1215/S1152851705000050 PMID:16212809

Abraham, W. C., & Bear, M. F. (1996). Metaplasticity : Plasticity of synaptic. *Trends in Neurosciences*, *19*(4), 126–130. doi:10.1016/S0166-2236(96)80018-X PMID:8658594

Acosta, C., Anderson, H. D., & Anderson, C. M. (2017). 'Astrocyte Dysfunction in Alzheimer Disease'. *Journal of Neuroscience Research*, *95*(12), 2430–2447. doi:10.1002/jnr.24075 PMID:28467650

Acosta-Martínez, M. (2020). Shaping Microglial Phenotypes Through Estrogen Receptors : Relevance to Sex-Specific Neuroinflammatory Responses to Brain Injury and Disease. *The Journal of Pharmacology and Experimental Therapeutics*, *375*(1), 223–236. doi:10.1124/jpet.119.264598 PMID:32513838

Adhikari, H. S., & Yadav, P. N. (2018). Anticancer Activity of Chitosan, Chitosan Derivatives, and Their Mechanism of Action. *International Journal of Biomaterials*, *2952085*, 1–29. doi:10.1155/2018/2952085 PMID:30693034

Adiele, R. C., & Adiele, C. A. (2019). Metabolic defects in multiple sclerosis. *Mitochondrion*, *44*, 7–14. doi:10.1016/j.mito.2017.12.005

Agarwal, A. N., & Mais, D. D. (2020). *Sensitivity and Specificity of Alzheimer Type II Astrocytes in Hepatic Encephalopathy.* ARPA. doi:10.5858/arpa.2018-0455-OA

Aggarwal, S., Kramer, D., Yonas, H., Obrist, W., Kang, Y., Martin, M., & Policare, R. (1994). Cerebral hemodynamic and metabolic changes in fulminant hepatic failure: A retrospective study. *Hepatology (Baltimore, Md.)*, *19*(1), 80–87. doi:10.1002/hep.1840190114 PMID:8276371

Agnello, D., Bigini, P., Villa, P., Mennini, T., Cerami, A., Brines, ML et Ghezzi, P. (2002). L'érythropoïétine exerce un effet anti-inflammatoire sur le SNC dans un modèle d'encéphalomyélite auto-immune expérimentale. *Recherche sur le cerveau*, *952*(1), 128-134.

Agnello, D., Bigini, P., Villa, P., Mennini, T., Cerami, A., Brines, M. L., & Ghezzi, P. (2002). E rythropoietin exerts an anti-inflammatory effect on the CNS in a model of experimental autoimmune encephalomyelitis.[). *Brain Research*, *952*(1), 128–134. www.elsevier.com/locate/bres. doi:10.1016/S0006-8993(02)03239-0 PMID:12363412

Ahmad, M., Hamid, A., Hussain, A., Majeed, R., Qurishi, Y., Bhat, J. A., Najar, R. A., Qazi, A. K., Zargar, M. A., Singh, S. K., & Saxena, A. K. (2012). Understanding histone deacetylases in the cancer development and treatment: An epigenetic perspective of cancer chemotherapy. *DNA and Cell Biology*, *31*(S1), S-62–S-71. doi:10.1089/dna.2011.1575 PMID:22462686

Aikawa, T., Holm, M.-L., & Kanekiyo, T. (2018). 'ABCA7 and Pathogenic Pathways of Alzheimer's Disease'. *Brain Sciences*, *8*(2), 27. doi:10.3390/brainsci8020027 PMID:29401741

Airaksinen, M. S., & Saarma, M. (2002). The GDNF family: Signalling, biological functions and therapeutic value. *Nature Reviews. Neuroscience*, *3*(5), 383–394. doi:10.1038/nrn812 PMID:11988777

Airas, L., & Yong, V. W. (2022). Microglia in multiple sclerosis - pathogenesis and imaging. *Current Opinion in Neurology*, *35*(3), 299–306. doi:10.1097/WCO.0000000000001045 PMID:35674072

Ait Hamdan, Y., El Amerany, F., Desbrières, J., Aghrinane, A., Oudadesse, H., & Rhazi, M. (2022). The evolution of the global COVID-19 epidemic in Morocco and understanding the different therapeutic approaches of chitosan in the control of the pandemic. In *Polymer Bulletin*. Springer Science and Business Media Deutschland GmbH., doi:10.100700289-022-04579-3

Ait Hamdan, Y., El Amerany, F., Desbrières, J., Aghrinane, A., Oudadesse, H., & Rhazi, M. (2022). The evolution of the global COVID-19 epidemic in Morocco and understanding the different therapeutic approaches of chitosan in the control of the pandemic. *Polymer Bulletin*, 1–27. PMID:36466080

Aj, S. (1994). Portosystemic encephalopathy after transjugular intrahepatic portosystemic shunt: Results of a prospective controlled study. *Hepatology (Baltimore, Md.)*, *20*, 46–55. PMID:8020904

Akay, L. A., Effenberger, A. H., & Tsai, L. H. (2021). Cell of all trades: Oligodendrocyte precursor cells in synaptic, vascular, and immune function. *Genes & Development*, *35*(3-4), 180–198. doi:10.1101/gad.344218.120

Aktas, O., Ullrich, O., Infante-Duarte, C., Nitsch, R., & Zipp, F. (2007). Neuronal damage in brain inflammation. *Archives of Neurology*, *64*(2), 185–189. doi:10.1001/archneur.64.2.185 PMID:17296833

Alanazi, A., Alassiri, M., Jawdat, D., & Almalik, Y. (2022). Mesenchymal stem cell therapy: A review of clinical trials for multiple sclerosis. *Regenerative Therapy*, *21*, 201–209. doi:10.1016/j.reth.2022.07.003

Aldridge, D. R., Tranah, E. J., & Shawcross, D. L. (2015). Pathogenesis of hepatic encephalopathy: Role of ammonia and systemic inflammation. *Journal of Clinical and Experimental Hepatology*, *5*(S1), S7–S20. doi:10.1016/j.jceh.2014.06.004 PMID:26041962

Alexianu, M. E., Kozovska, M., & Appel, S. H. (2001). Immune reactivity in a mouse model of familial ALS correlates with disease progression. *Neurology*, *57*(7), 1282–1289. doi:10.1212/WNL.57.7.1282 PMID:11591849

Ali, A., and Ahmed, S. (2018). A review on chitosan and its nanocomposites in drug delivery. *Int. J. Biol. Macromol., 109,* 273–286. https://doi.org/. 12.078 doi:10.1016/j.ijbiomac.2017

Ali, A., & Ahmed, S. (2018). A review on chitosan and its nanocomposites in drug delivery. *International Journal of Biological Macromolecules*, *109*, 273–286. doi:10.1016/j.ijbiomac.2017.12.078 PMID:29248555

Ali, F., Hossain, M. S., Sejimo, S., & Akashi, K. (2019). Plasmalogens inhibit endocytosis of toll-like receptor 4 to attenuate the inflammatory signal in microglial cells. *Molecular Neurobiology*, *56*(5), 3404–3419. doi:10.100712035-018-1307-2 PMID:30128650

Ali, M., Bennardo, M., Almenawer, S. A., Zagzoog, N., Smith, A. A., Dao, D., BHSc, Ajani, O., Farrokhyar, F., & Singh, S. K. (2015). Exploring predictors of surgery and comparing operative treatment approaches for pediatric intracranial arachnoid cysts: A case series of 83 patients. *Journal of Neurosurgery. Pediatrics*, *16*(3), 275–282. doi:10.3171/2015.2.PEDS14612 PMID:26067335

Al-Jasmi, F. A., Tawfig, N., Berniah, A., Ali, B. R., Taleb, M., Hertecant, J. L., Bastaki, F., & Souid, A. K. (2013). *Prevalence and novel mutations of lysosomal storage disorders in United Arab Emirates: LSD in UAE*. Epub 2013/02/23.

Allahyari, R. V., Heinsinger, N. M., Hwang, D., Jaffe, D. A., Rasouli, J., Shiers, S., Thomas, S. J., Price, T. J., Rostami, A., & Lepore, A. C. (2022). Response of Astrocyte Subpopulations Following Spinal Cord Injury. *Cells*, *11*(4), 721. doi:10.3390/cells11040721 PMID:35203371

Allaman, I., Bélanger, M., & Magistretti, P. J. (2011). Astrocyte-neuron metabolic relationships: For better and for worse. *Trends in Neurosciences*, *34*(2), 76–87. doi:10.1016/j.tins.2010.12.001 PMID:21236501

Allegri, R. F. (2020). Moving from neurodegenerative dementias, to cognitive proteinopathies, replacing "where" by "what"…. *Dementia & Neuropsychologia, 14*(3), 237–242. doi:10.1590/1980-57642020dn14-030005 PMID:32973977

Allen, N.J. (2014). Astrocyte regulation of synaptic behavior. *Annual Review of Cell and Developmental Biology, 30*(1), 439–63. https://doi.org/ bio-100913-013053 doi:10.1146/annual-cell

Allen, NJ, et Barres, BA (2005). Signalisation entre glie et neurones: focus sur la plasticité synaptique. *Opinion actuelle en neurobiologie, 15*(5), 542-548.

Allen, N. J., Bennett, M. L., Foo, L. C., Wang, G. X., Chakraborty, C., Smith, S. J., & Barres, B. A. (2012). Astrocyte glypicans 4 and 6 promote formation of excitatory synapses via GluA1 AMPA receptors. *Nature, 486*(7403), 410–414. doi:10.1038/nature11059 PMID:22722203

Allen, N. J., & Eroglu, C. (2017). 'Cell Biology of Astrocyte-Synapse Interactions'. *Neuron, 96*(3), 697–708. doi:10.1016/j.neuron.2017.09.056 PMID:29096081

Allen, N. J., & Lyons, D. A. (2018). Glia as architects of central nervous system formation and function. *Science, 362*(6411), 181–185. doi:10.1126cience.aat0473 PMID:30309945

Almer, G., Vukosavic, S., Romero, N., & Przedborski, S. (1999). Inducible nitric oxide synthase up-regulation in a transgenic mouse model of familial amyotrophic lateral sclerosis. *Journal of Neurochemistry, 72*(6), 2415–2425. doi:10.1046/j.1471-4159.1999.0722415.x PMID:10349851

Alonso, D., Castro, A., & Martinez, A. (2005). Marine compounds for the therapeutic treatment of neurological disorders. *Expert Opinion on Therapeutic Patents, 15*(10), 1377–1386. doi:10.1517/13543776.15.10.1377

Altshuler, L. L., Abulseoud, O. A., Foland-Ross, L., Bartzokis, G., Chang, S., Mintz, J., Hellemann, G., & Vinters, H. V. (2010). Amygdala astrocyte reduction in subjects with major depressive disorder but not bipolar disorder. *Bipolar Disorders, 12*(5), 541–549. doi:10.1111/j.1399-5618.2010.00838.x PMID:20712756

Alvarez, J. I., Dodelet-Devillers, A., Kebir, H., Ifergan, I., Fabre, P. J., Terouz, S., Sabbagh, M., Wosik, K., Bourbonnière, L., Bernard, M., van Horssen, J., de Vries, H. E., Charron, F., & Prat, A. (2011). 'The Hedgehog Pathway Promotes Blood-Brain Barrier Integrity and CNS Immune Quiescence'. *Science, 334*(6063), 1727–1731. doi:10.1126cience.1206936 PMID:22144466

Alvarez, J. I., Katayama, T., & Prat, A. (2013). Glial influence on the blood-brain barrier. *Glia, 61*(12), 1939–1958. doi:10.1002/glia.22575 PMID:24123158

Aly, A., Singh, P., Korytowsky, B., Ling, Y. L., Kale, H. P., Dastani, H. B., & Norden, A. D. (2020). Survival, costs, and health care resource use by line of therapy in US Medicare patients with newly diagnosed glioblastoma: A retrospective observational study. *Neuro-Oncology Practice, 7*(2), 164–175. PMID:32626585

Ambartsumian, N., Klingelhöfer, J., & Grigorian, M. (2019). The multifaceted S100A4 protein in cancer and inflammation. *Calcium-binding proteins of the EF-hand superfamily: from basics to medical applications*, 339-365.

Amin, M., & Hersh, C. M. (2023). Updates and advances in multiple sclerosis neurotherapeutics. *Neurodegenerative Disease Management, 14*(1), 47–70. doi:10.2217/nmt-2021-0058

Amor, S., Peferoen, L. A., Vogel, D. Y., Breur, M., van der Valk, P., Baker, D., & van Noort, J. M. (2014). Inflammation in neurodegenerative diseases–an update. *Immunology, 142*(2), 151–166. doi:10.1111/imm.12233 PMID:24329535

Amor, S., Puentes, F., Baker, D., & Van Der Valk, P. (2010). Inflammation in neurodegenerative diseases. *Immunology, 129*(2), 154–169. doi:10.1111/j.1365-2567.2009.03225.x PMID:20561356

Andersson, M., Blomstrand, F., & Hanse, E. (2007). Astrocytes play a critical role in transient heterosynaptic depression in the rat hippocampal CA1 region. *The Journal of Physiology, 585*(3), 843–852. doi:10.1113/jphysiol.2007.142737 PMID:17962333

Angelova, P. R., & Abramov, A. Y. (2018). 'Role of Mitochondrial ROS in the Brain: From Physiology to Neurodegeneration'. *FEBS Letters, 592*(5), 692–702. doi:10.1002/1873-3468.12964 PMID:29292494

Angelucci, F., Aloe, L., Jiménez-Vasquez, P., & Mathé, A. A. (2002). Electroconvulsive stimuli alter the regional concentrations of nerve growth factor, brain-derived neurotrophic factor, and glial cell line-derived neurotrophic factor in adult rat brain. *The Journal of ECT, 18*(3), 138–143. doi:10.1097/00124509-200209000-00005 PMID:12394532

Angot, E., Steiner, J. A., Hansen, C., Li, J.-Y., & Brundin, P. (2010). Are synucleinopathies prion-like disorders? *Lancet Neurology, 9*(11), 1128–1138. doi:10.1016/S1474-4422(10)70213-1 PMID:20846907

Antonietta Ajmone-Cat, M., Mancini, M., De Simone, R., Cilli, P., & Minghetti, L. (2013). Microglial polarization and plasticity: Evidence from organotypic hippocampal slice cultures. *Glia, 61*(10), 1698–1711. doi:10.1002/glia.22550 PMID:23918452

Antonini, A., & Leenders, K. L. (1993). Dopamine D2 Receptors in Normal Human Brain: Effect of Age Measured by Positron Emission Tomography (PET) and [11C]-Raclopride. *Annals of the New York Academy of Sciences, 695*(1), 81–85. doi:10.1111/j.1749-6632.1993.tb23033.x PMID:8239318

Anton, K., Baehring, J. M., & Mayer, T. (2012). Glioblastoma multiforme: Overview of current treatment and future perspectives. *Hematology/Oncology Clinics of North America, 26*(4), 825–853. doi:10.1016/j.hoc.2012.04.006 PMID:22794286

Apelt, J. research, R. S.-B., (2001). *β-amyloid-induced glial expression of both pro-and anti-inflammatory cytokines in cerebral cortex of aged transgenic Tg2576 mice with Alzheimer plaque pathology.* Elsevier. https://www.sciencedirect.com/science/article/pii/S0006899300031760

Appelqvist, H., Wäster, P., Kågedal, K., & Öllinger, K. (2013). The lysosome: From waste bag to potential therapeutic target. *Journal of Molecular Cell Biology, 5*(4), 214–226. doi:10.1093/jmcb/mjt022 PMID:23918283

Araki, T., Ikegaya, Y., & Koyama, R. (2021). The effects of microglia-and astrocyte-derived factors on neurogenesis in health and disease. *The European Journal of Neuroscience, 54*(5), 5880–5901. doi:10.1111/ejn.14969 PMID:32920880

Araque, A., Carmignoto, G., Haydon, P. G., Oliet, S. H., Robitaille, R., & Volterra, A. (2014). Gliotransmitters travel in time and space. *Neuron, 81*(4), 728–739. doi:10.1016/j.neuron.2014.02.007 PMID:24559669

Araque, A., Li, N., Doyle, R. T., & Haydon, P. G. (2000). SNARE Protein-Dependent Glutamate Release from Astrocytes. *The Journal of Neuroscience : The Official Journal of the Society for Neuroscience, 20*(2), 666–673. doi:10.1523/JNEUROSCI.20-02-00666.2000 PMID:10632596

Araque, A., Parpura, V., Sanzgiri, R. P., & Haydon, P. G. (1999). 'Tripartite Synapses: Glia, the Unacknowledged Partner'. *Trends in Neurosciences, 22*(5), 208–215. doi:10.1016/S0166-2236(98)01349-6 PMID:10322493

Araque, A., Sanzgiri, R. P., Parpura, V., & Haydon, P. G. (1999). Astrocyte-induced modulation of synaptic transmission. *Canadian Journal of Physiology and Pharmacology, 77*(9), 699–706. doi:10.1139/y99-076 PMID:10566947

Arenas, J., Campos, Y., Ribacoba, R., Martín, M. A., Rubio, J. C., Ablanedo, P., & Cabello, A. (1998). Complex I defect in muscle from patients with Huntington's disease. *Annals of Neurology, 43*(3), 397–400. doi:10.1002/ana.410430321 PMID:9506560

Armstrong, R. A. (2013). What causes Alzheimer's disease? *Folia Neuropathologica, 51*, 169–188. doi:10.5114/fn.2013.37702 PMID:24114635

Arnett, H. A., Mason, J., Marino, M., Suzuki, K., Matsushima, G. K., & Ting, J. P.-Y. (2001). TNFα promotes proliferation of oligodendrocyte progenitors and remyelination. *Nature Neuroscience*, *4*(11), 1116–1122. doi:10.1038/nn738 PMID:11600888

Arnold, T., & Betsholtz, C. (2013). Erratum to: The importance of microglia in the development of the vasculature in the central nervous system. *Vascular Cell*, *5*, 1–7.

Arranz, A. M., & De Strooper, B. (2019). The role of astroglia in Alzheimer's disease: Pathophysiology and clinical implications. *Lancet Neurology*, *18*(4), 406–414. doi:10.1016/S1474-4422(18)30490-3 PMID:30795987

Arrouze, F., Desbrieres, J., Lidrissi Hassani, S., & Tolaimate, A. (2021). Investigation of β-chitin extracted from cuttlefish: Comparison with squid β-chitin. *Polymer Bulletin*, *78*(12), 7219–7239. doi:10.100700289-020-03466-z

Arrouze, F., Desbrieres, J., Rhazi, M., Essahli, M., & Tolaimate, A. (2019). Valorization of chitins extracted from North Morocco shrimps: Comparison of chitin reactivity and characteristics. *Journal of Applied Polymer Science*, *136*(30), 47804. doi:10.1002/app.47804

AsadollahiE.TrevisiolA.SaabA. S.LooserZ. J.DibajP.KuschK.NaveK. A. (2022). Myelin lipids as nervous system energy reserves. bioRxiv. doi:10.1101/2022.02.24.481621

Asanuma, M., Miyazaki, I., Murakami, S., Diaz-Corrales, F. J., & Ogawa, N. (2014). Striatal Astrocytes Act as a Reservoir for L-DOPA. *PLoS One*, *9*(9), e106362. doi:10.1371/journal.pone.0106362 PMID:25188235

Ashok, A., Rai, N. K., Raza, W., Pandey, R., & Bandyopadhyay, S. (2016). Chronic cerebral hypoperfusion-induced impairment of Abeta clearance requires HB-EGF-dependent sequential activation of HIF1alpha and MMP9. *Neurobiology of Disease*, *95*, 179–193. doi:10.1016/j.nbd.2016.07.013 PMID:27431094

Ashrafi, H., & Azadi, A. (2016). Chitosan-based hydrogel nanoparticle amazing behaviors during transmission electron microscopy. *International Journal of Biological Macromolecules*, *84*, 31–34. doi:10.1016/j.ijbiomac.2015.11.089 PMID:26658231

Ashton, R. S., Conway, A., Pangarkar, C., Bergen, J., Lim, K. I., Shah, P., Bissell, M., & Schaffer, D. V. (2012). Astrocytes regulate adult hippocampal neurogenesis through ephrin-B signaling. *Nature Neuroscience*, *15*(10), 1399–1406. doi:10.1038/nn.3212 PMID:22983209

Ashwell, K. W. S., Holländer, H., Streit, W., & Stone, J. (1989). The appearance and distribution of microglia in the developing retina of the rat. *Visual Neuroscience*, *2*(5), 437–448. doi:10.1017/S0952523800012335 PMID:2487081

Asiimwe, N., Yeo, S. G., Kim, M.-S., Jung, J., & Jeong, N. Y. (2016). Nitric oxide : Exploring the contextual link with Alzheimer's disease. *Oxidative Medicine and Cellular Longevity*, *2016*, 2016. doi:10.1155/2016/7205747 PMID:28096943

Asraf, K., Torika, N., Apte, R. N., & Fleisher-Berkovich, S. (2018). Microglial Activation Is Modulated by Captopril : In Vitro and in Vivo Studies. *Frontiers in Cellular Neuroscience*, *12*, 116. doi:10.3389/fncel.2018.00116 PMID:29765306

Asseyer, S., Schmidt, F., Chien, C., Scheel, M., Ruprecht, K., Bellmann-Strobl, J., Brandt, A. U., & Paul, F. (2018). Pain in AQP4-IgG-positive and MOG-IgG-positive neuromyelitis optica spectrum disorders. *Multiple Sclerosis Journal - Experimental, Translational and Clinical*, *4*(3), 2055217318796684. doi:10.1177/2055217318796684 PMID:30186620

Asslih, S., Damri, O., & Agam, G. (2021). Neuroinflammation as a common denominator of complex diseases (Cancer, diabetes type 2, and neuropsychiatric disorders). In International Journal of Molecular Sciences, 22(11). MDPI. doi:10.3390/ijms22116138

Aston, C., Jiang, L., & Sokolov, B. P. (2005). Transcriptional profiling reveals evidence for signaling and oligodendroglial abnormalities in the temporal cortex from patients with major depressive disorder. *Molecular Psychiatry*, *10*(3), 309–322. doi:10.1038j.mp.4001565 PMID:15303102

Atwood, H. L., & Karunanithi, S. (2002). Diversification of synaptic strength: Presynaptic elements. *Nature Reviews. Neuroscience*, *3*(7), 497–516. doi:10.1038/nrn876 PMID:12094207

Autry, A. E., & Monteggia, L. M. (2012). Brain-derived neurotrophic factor and neuropsychiatric disorders. *Pharmacological Reviews*, *64*(2), 238–258. doi:10.1124/pr.111.005108 PMID:22407616

Ayyildiz, M. (2015). *Effect of Preservation Solution and Period on the Mechanical and Histological Properties of Liver*.

Baas, D., Meiniel, A., Benadiba, C., Bonnafe, E., Meiniel, O., Reith, W., & Durand, B. (2006). A deficiency in RFX3 causes hydrocephalus associated with abnormal differentiation of ependymal cells. *The European Journal of Neuroscience*, *24*(4), 1020–1030. doi:10.1111/j.1460-9568.2006.05002.x PMID:16930429

Baba, H., Ji, R.-R., Kohno, T., Moore, K. A., Ataka, T., Wakai, A., Okamoto, M., & Woolf, C. J. (2003). Removal of GABAergic inhibition facilitates polysynaptic A fiber-mediated excitatory transmission to the superficial spinal dorsal horn. *Molecular and Cellular Neurosciences*, *24*(3), 818–830. doi:10.1016/S1044-7431(03)00236-7 PMID:14664828

Babu, A., & Ramesh, R. (2017). Multifaceted applications of chitosan in cancer drug delivery and therapy. *Marine Drugs*, *15*(4), 96. doi:10.3390/md15040096 PMID:28346381

Bachiller, S., Jiménez-Ferrer, I., Paulus, A., Yang, Y., Swanberg, M., Deierborg, T., & Boza-Serrano, A. (2018). Microglia in Neurological Diseases : A Road Map to Brain-Disease Dependent-Inflammatory Response. *Frontiers in Cellular Neuroscience*, *12*, 488. https://www.frontiersin.org/articles/10.3389/fncel.2018.00488. doi:10.3389/fncel.2018.00488 PMID:30618635

Bachstetter, A. D., Morganti, J. M., Jernberg, J., Schlunk, A., Mitchell, S. H., Brewster, K. W., Hudson, C. E., Cole, M. J., Harrison, J. K., Bickford, P. C., & Gemma, C. (2011). Fractalkine and CX3CR1 regulate hippocampal neurogenesis in adult and aged rats. *Neurobiology of Aging*, *32*(11), 2030–2044. doi:10.1016/j.neurobiolaging.2009.11.022 PMID:20018408

Badhe, RV, Bijukumar, D., Chejara, DR, Mabrouk, M., Choonara, YE, Kumar, P., ... & Pillay, V. (2017). Un échafaudage macroporeux biomimétique bicouche composite chitosane-gélatine pour l'ingénierie tissulaire des vaisseaux sanguins. *Polymères glucidiques*, *157*, 1215-1225.

Bahbouhi, B., Berthelot, L., Pettré, S., Michel, L., Wiertlewski, S., Weksler, B., Romero, I.-A., Miller, F., Couraud, P.-O., Brouard, S., Laplaud, D.-A., & Soulillou, J.-P. (2009). Peripheral blood CD4+ T lymphocytes from multiple sclerosis patients are characterized by higher PSGL-1 expression and transmigration capacity across a human blood-brain barrier-derived endothelial cell line. *Journal of Leukocyte Biology*, *86*(5), 1049–1063. doi:10.1189/jlb.1008666 PMID:19696154

Bai, G., Rama Rao, K. V., Murthy, C. R. K., Panickar, K. S., Jayakumar, A. R., & Norenberg, M. D. (2001). Ammonia Induces the Mitochondrial Permeability Transition in Primary Cultures of Rat Astrocytes. *Journal of Neuroscience Research*, *66*(5), 981–991. doi:10.1002/jnr.10056 PMID:11746427

Bai, J.-Z., & Lipski, J. (2014). Involvement of TRPV4 channels in Aβ40-induced hippocampal cell death and astrocytic Ca2+ signalling. *Neurotoxicology*, *41*, 64–72. doi:10.1016/j.neuro.2014.01.001 PMID:24457011

Baik, S. H., Kang, S., Son, S. M., & Mook-Jung, I. (2016). 'Microglia Contributes to Plaque Growth by Cell Death Due to Uptake of Amyloid β in the Brain of Alzheimer's Disease Mouse Model'. *Glia*, *64*(12), 2274–2290. doi:10.1002/glia.23074 PMID:27658617

Baillet, A., Chanteperdrix, V., Trocmé, C., Casez, P., Garrel, C., & Besson, G. (2010). The role of oxidative stress in amyotrophic lateral sclerosis and Parkinson's disease. *Neurochemical Research, 35*(10), 1530–1537. doi:10.100711064-010-0212-5 PMID:20535556

Bajaj, J. S., Saeian, K., Verber, M. D., Hischke, D., Hoffmann, R. G., Franco, J., Varma, R. R., & Rao, S. M. (2007). Inhibitory control test is a simple method to diagnose minimal hepatic encephalopathy and predict development of overt hepatic encephalopathy. *Official Journal of the American College of Gastroenterology| ACG, 102*(4), 754–760.

Balabanov, R., Strand, K., Goswami, R., McMahon, E., Begolka, W., Miller, S. D., & Popko, B. (2007). Interferon-γ-Oligodendrocyte Interactions in the Regulation of Experimental Autoimmune Encephalomyelitis. *The Journal of Neuroscience : The Official Journal of the Society for Neuroscience, 27*(8), 2013–2024. doi:10.1523/JNEUROSCI.4689-06.2007 PMID:17314297

Balasingam, V., Tejada-Berges, T., Wright, E., Bouckova, R., & Yong, V. W. (1994). Reactive astrogliosis in the neonatal mouse brain and its modulation by cytokines. *The Journal of Neuroscience : The Official Journal of the Society for Neuroscience, 14*(2), 846–856. doi:10.1523/JNEUROSCI.14-02-00846.1994 PMID:8301364

Ballard, C., Gauthier, S., Corbett, A., Brayne, C., Aarsland, D., & Jones, E. (2011). Alzheimer's disease. *Lancet, 377*(9770), 1019–1031. doi:10.1016/S0140-6736(10)61349-9 PMID:21371747

Banasr, M., & Duman, R. S. (2008). Glial loss in the prefrontal cortex is sufficient to induce depressive-like behaviors. *Biological Psychiatry, 64*(10), 863–870. doi:10.1016/j.biopsych.2008.06.008 PMID:18639237

Banerjee, R., Mosley, R. L., Reynolds, A. D., Dhar, A., Jackson-Lewis, V., Gordon, P. H., Przedborski, S., & Gendelman, H. E. (2008). Adaptive immune neuroprotection in G93A-SOD1 amyotrophic lateral sclerosis mice. *PLoS One, 3*(7), e2740. doi:10.1371/journal.pone.0002740 PMID:18648532

Banerjee, R., Raj, A., Potdar, C., Pal, P. K., Yadav, R., Kamble, N., Holla, V., & Datta, I. (2023). Astrocytes Differentiated from LRRK2-I1371V Parkinson's-Disease-Induced Pluripotent Stem Cells Exhibit Similar Yield but Cell-Intrinsic Dysfunction in Glutamate Uptake and Metabolism, ATP Generation, and Nrf2-Mediated Glutathione Machinery. *Cells, 12*(12), 1592. doi:10.3390/cells12121592 PMID:37371062

Banker, G. A. (1980). Trophic Interactions Between Astroglial Cells and Hippocampal Neurons in Culture. *Science, 209*(4458), 809–810. doi:10.1126cience.7403847 PMID:7403847

Bankston, A. N., Mandler, M. D., & Feng, Y. (2013). Oligodendroglia and neurotrophic factors in neurodegeneration. *Neuroscience Bulletin, 29*(2), 216–228. doi:10.100712264-013-1321-3 PMID:23558590

Banks, W. A., Kastin, A. J., & Broadwell, R. D. (1995). Passage of Cytokines across the Blood-Brain Barrier. *Neuroimmunomodulation, 2*(4), 241–248. doi:10.1159/000097202 PMID:8963753

Bao, M., Xie, J., Katoele, N., Hu, X., Wang, B., Piruska, A., & Huck, W. T. S. (2019). Cellular volume and matrix stiffness direct stem cell behavior in a 3D microniche. *ACS Applied Materials & Interfaces, 11*(2), 1754–1759. doi:10.1021/acsami.8b19396 PMID:30584755

Baranowska-Bik, A., Bik, W., Wolinska-Witort, E., Martynska, L., Chmielowska, M., Barcikowska, M., & Baranowska, B. (2008). Plasma beta amyloid and cytokine profile in women with Alzheimer's disease. *Neuroendocrinology Letters, 2008•nel. Altamira.Sk, 29*(1), 18283248–18290108.

Barber, S. C., Mead, R. J., & Shaw, P. J. (2006). Oxidative stress in ALS: A mechanism of neurodegeneration and a therapeutic target. *Biochimica et Biophysica Acta, 1762*(11-12), 1051–1067. doi:10.1016/j.bbadis.2006.03.008 PMID:16713195

Barboura, I., Ferchichi, S., Dandana, A., Jaidane, Z., Khelifa, S. B., Chahed, H., Mansour, R. B., Chebel, S., Maire, I., & Miled, A. (2010). Metachromatic leucodystrophy Clinical, biological, and therapeutic aspects. *Annales de Biologie Clinique*, *68*(4), 385–391. doi:10.1684/abc.2010.0448 PMID:20650733

Barcelos, I. P. D., Troxell, R. M., & Graves, J. S. (2019). Mitochondrial dysfunction and multiple sclerosis. *Biology (Basel)*, *8*(2), 37. doi:10.3390/biology8020037

Barker, A. J., Koch, S. M., Reed, J., Barres, B. A., & Ullian, E. M. (2008). Developmental Control of Synaptic Receptivity. *The Journal of Neuroscience : The Official Journal of the Society for Neuroscience*, *28*(33), 8150–8160. doi:10.1523/JNEUROSCI.1744-08.2008 PMID:18701677

Barker, A. J., & Ullian, E. M. (2008). New roles for astrocytes in developing synaptic circuits. *Communicative & Integrative Biology*, *1*(2), 207–211. doi:10.4161/cib.1.2.7284 PMID:19513261

Barnes, J., Dickerson, B. C., Frost, C., Jiskoot, L. C., Wolk, D., & van der Flier, W. M. (2015). Alzheimer's disease first symptoms are age dependent: Evidence from the NACC dataset. *Alzheimer's & Dementia*, *11*(11), 1349–1357. doi:10.1016/j.jalz.2014.12.007 PMID:25916562

Barres, B. A. (2008). The mystery and magic of glia: A perspective on their roles in health and disease. *Neuron*, *60*(3), 430–440. doi:10.1016/j.neuron.2008.10.013 PMID:18995817

Bartels, A. L., Willemsen, A. T. M., Doorduin, J., de Vries, E. F. J., Dierckx, R. A., & Leenders, K. L. (2010). [11C]-PK11195 PET: Quantification of neuroinflammation and a monitor of anti-inflammatory treatment in Parkinson's disease? *Parkinsonism & Related Disorders*, *16*(1), 57–59. doi:10.1016/j.parkreldis.2009.05.005 PMID:19487152

Barzegar Behrooz, A., Syahir, A., & Ahmad, S. (2019). CD133: Beyond a cancer stem cell biomarker. *Journal of Drug Targeting*, *27*(3), 257–269. doi:10.1080/1061186X.2018.1479756 PMID:29911902

Basbaum, A. I., Bautista, D. M., Scherrer, G., & Julius, D. (2009). Cellular and Molecular Mechanisms of Pain. *Cell*, *139*(2), 267–284. doi:10.1016/j.cell.2009.09.028 PMID:19837031

Baskar Jesudasan, S. J., Todd, K. G., & Winship, I. R. (2014). Reduced inflammatory phenotype in microglia derived from neonatal rat spinal cord versus brain. *PLoS One*, *9*(6), e99443. doi:10.1371/journal.pone.0099443 PMID:24914808

Bataller, R., Brenner, D. A., Bataller, R., & Brenner, D. A. (2005). Liver fibrosis Science in medicine Liver fibrosis. *The Journal of Clinical Investigation*, *115*(April), 209–218. doi:10.1172/JCI24282

Bataveljić, D., Nikolić, L., Milosević, M., Todorović, N., & Andjus, P. R. (2012). Changes in the astrocytic aquaporin-4 and inwardly rectifying potassium channel expression in the brain of the amyotrophic lateral sclerosis SOD1(G93A) rat model. *Glia*, *60*(12), 1991–2003. doi:10.1002/glia.22414 PMID:22987392

Baufeld, C., O'Loughlin, E., Calcagno, N., Madore, C., & Butovsky, O. (2018). Differential contribution of microglia and monocytes in neurodegenerative diseases. *Journal of Neural Transmission (Vienna, Austria)*, *125*(5), 809–826. doi:10.100700702-017-1795-7 PMID:29063348

Baumann, N., & Pham-Dinh, D. (2001). Biology of oligodendrocyte and myelin in the mammalian central nervous system. *Physiological Reviews*, *81*(2), 871–927. doi:10.1152/physrev.2001.81.2.871 PMID:11274346

Baumann, N., & Pham-dinh, D. (2002). Astrocytes. In V. S. Ramachandran (Ed.), *Encyclopedia of the Human Brain* (pp. 251–268). Academic Press., doi:10.1016/B0-12-227210-2/00042-X

Bazargani, N., & Attwell, D. (2016). Astrocyte calcium signaling: the third wave. *Nature Neuroscience*, *19*(2), 182–189. doi:10.1038/nn.4201

Beard, E., Lengacher, S., Dias, S., Magistretti, P. J., & Finsterwald, C. (2022). Astrocytes as Key Regulators of Brain Energy Metabolism : New Therapeutic Perspectives. *Frontiers in Physiology, 12*, 825816. doi:10.3389/fphys.2021.825816 PMID:35087428

Bear, M. F. (1999). Homosynaptic long-term depression: A mechanism for memory? *Proceedings of the National Academy of Sciences of the United States of America, 96*(17), 9457–9458. doi:10.1073/pnas.96.17.9457 PMID:10449713

Beauquis, J., Vinuesa, A., Pomilio, C., Pavía, P., Saravia, F. (2014). Alteraciones hipocampales y cambios cognitivos preceden al depósito de placas amiloides en un modelo murino de la enfermedad de Alzheimer [Hippocampal and cognitive alterations precede amyloid deposition in a mouse model for Alzheimer's disease]. *Medicina (B Aires), 74*(4), 282-6.

Beauquis, J., Pavía, P., Pomilio, C., Vinuesa, A., Podlutskaya, N., Galvan, V., & Saravia, F. (2013). Environmental enrichment prevents astroglial pathological changes in the hippocampus of APP transgenic mice, model of Alzheimer's disease. *Experimental Neurology, 239*, 28–37. doi:10.1016/j.expneurol.2012.09.009 PMID:23022919

Beauquis, J., Vinuesa, A., Pomilio, C., Pavía, P., Galván, V., & Saravia, F. (2014). Neuronal and glial alterations, increased anxiety, and cognitive impairment before hippocampal amyloid deposition in PDAPP mice, model of Alzheimer's disease. *Hippocampus, 24*(3), 257–269. doi:10.1002/hipo.22219 PMID:24132937

Bechtholt-Gompf, A. J., Walther, H. V., Adams, M. A., Carlezon, W. A. Jr, Öngür, D., & Cohen, B. M. (2010). Blockade of astrocytic glutamate uptake in rats induces signs of anhedonia and impaired spatial memory. *Neuropsychopharmacology, 35*(10), 2049–2059. doi:10.1038/npp.2010.74 PMID:20531459

Béduneau, A., Saulnier, P., & Benoit, J. P. (2007). Active targeting of brain tumors using nanocarriers. *Biomaterials, 28*(33), 4947–4967. doi:10.1016/j.biomaterials.2007.06.011 PMID:17716726

Beenhakker, M. P., & Huguenard, J. R. (2010). Astrocytes as Gatekeepers of GABAB Receptor Function. *The Journal of Neuroscience : The Official Journal of the Society for Neuroscience, 30*(45), 15262–15276. doi:10.1523/JNEUROSCI.3243-10.2010 PMID:21068331

Beers, D. R., Henkel, J. S., Xiao, Q., Zhao, W., Wang, J., Yen, A. A., Siklos, L., McKercher, S. R., & Appel, S. H. (2006). Wild-type microglia extend survival in PU.1 knockout mice with familial amyotrophic lateral sclerosis. *Proceedings of the National Academy of Sciences of the United States of America, 103*(43), 16021–16026. doi:10.1073/pnas.0607423103 PMID:17043238

Beers, D. R., Henkel, J. S., Zhao, W., Wang, J., & Appel, S. H. (2008). CD4+ T cells support glial neuroprotection, slow disease progression, and modify glial morphology in an animal model of inherited ALS. *Proceedings of the National Academy of Sciences of the United States of America, 105*(40), 15558–15563. doi:10.1073/pnas.0807419105 PMID:18809917

Behl, C. et Moosmann, B. (2002). Neuroprotection antioxydante dans la maladie d'Alzheimer comme approche préventive et thérapeutique. *Biologie et médecine des radicaux libres, 33*(2), 182-191.

Behl, C., & Moosmann, B. (2002). Antioxidant neuroprotection in Alzheimer's disease as preventive and therapeutic approach. *Free Radical Biology & Medicine, 33*(2), 182–191. doi:10.1016/S0891-5849(02)00883-3 PMID:12106814

Behrens, P. F., Franz, P., Woodman, B., Lindenberg, K. S., & Landwehrmeyer, G. B. (2002). Impaired glutamate transport and glutamate-glutamine cycling: Downstream effects of the Huntington mutation. *Brain, 125*(8), 1908–1922. doi:10.1093/brain/awf180 PMID:12135980

Beilina, A., & Cookson, M. R. (2016). Genes associated with Parkinson's disease: Regulation of autophagy and beyond. *Journal of Neurochemistry, 139*(S1), 91–107. doi:10.1111/jnc.13266 PMID:26223426

Beirowski, B. (2022). Emerging evidence for compromised axonal bioenergetics and axoglial metabolic coupling as drivers of neurodegeneration. *Neurobiology of Disease*, *170*, 105751. doi:10.1016/j.nbd.2022.105751

Bélanger, M., & Magistretti, P. J. (2009). The role of astroglia in neuroprotection. *Dialogues in Clinical Neuroscience*, *11*(3), 281–295. doi:10.31887/DCNS.2009.11.3/mbelanger PMID:19877496

Bellaver, B., Souza, D. G., Souza, D. O., & Quincozes-Santos, A. (2017). 'Hippocampal Astrocyte Cultures from Adult and Aged Rats Reproduce Changes in Glial Functionality Observed in the Aging Brain'. *Molecular Neurobiology*, *54*(4), 2969–2985. doi:10.100712035-016-9880-8 PMID:27026184

Bellot-Saez, A., Kékesi, O., Morley, J. W., & Buskila, Y. (2017). Astrocytic modulation of neuronal excitability through K+ spatial buffering. *Neuroscience and Biobehavioral Reviews*, *77*, 87–97. doi:10.1016/j.neubiorev.2017.03.002 PMID:28279812

Bémeur, C., Qu, H., Desjardins, P., & Butterworth, R. F. (2010). IL-1 or TNF Receptor Gene Deletion Delays Onset of Encephalopathy and Attenuates Brain Edema in Experimental Acute Liver Failure. *Neurochemistry International*, *56*(2), 213–215. doi:10.1016/j.neuint.2009.11.010 PMID:19931338

Ben Haim, L., & Rowitch, D. H. (2017). Functional diversity of astrocytes in neural circuit regulation. *Nature Reviews. Neuroscience*, *18*(1), 31–41. doi:10.1038/nrn.2016.159 PMID:27904142

Benabid, A. L., Chabardes, S., Mitrofanis, J., & Pollak, P. (2009). Deep brain stimulation of the subthalamic nucleus for the treatment of Parkinson's disease. *Lancet Neurology*, *8*(1), 67–81. doi:10.1016/S1474-4422(08)70291-6 PMID:19081516

Bennett, F. C., & Molofsky, A. V. (2019). The immune system and psychiatric disease: A basic science perspective. *Clinical and Experimental Immunology*, *197*(3), 294–307. doi:10.1111/cei.13334 PMID:31125426

Benraiss, A., Wang, S., Herrlinger, S., Li, X., Chandler-Militello, D., Mauceri, J., Burm, H. B., Toner, M., Osipovitch, M., Jim Xu, Q., Ding, F., Wang, F., Kang, N., Kang, J., Curtin, P. C., Brunner, D., Windrem, M. S., Munoz-Sanjuan, I., Nedergaard, M., & Goldman, S. A. (2016). Human glia can both induce and rescue aspects of disease phenotype in Huntington disease. *Nature Communications*, *7*(1), 11758. doi:10.1038/ncomms11758 PMID:27273432

Benveniste, E. N. (1997). Role of macrophages/microglia in multiple sclerosis and experimental allergic encephalomyelitis. *Journal of Molecular Medicine*, *75*(3), 165–173. doi:10.1007001090050101 PMID:9106073

Benzing, W. C., Wujek, J. R., Ward, E. K., Shaffer, D., Ashe, K. H., Younkin, S. G., & Brunden, K. R. (1999). Evidence for glial-mediated inflammation in aged APPSW transgenic mice. *Neurobiology of Aging*, *20*(6), 581–589. doi:10.1016/S0197-4580(99)00065-2 PMID:10674423

Berent, D., Macander, M., Szemraj, J., Orzechowska, A., & Galecki, P. (2014). Vascular endothelial growth factor A gene expression level is higher in patients with major depressive disorder and not affected by cigarette smoking, hyperlipidemia or treatment with statins. *Acta Neurobiologiae Experimentalis*, *74*(1), 82–90. PMID:24718046

Bernardinelli, Y., Muller, D., & Nikonenko, I. (2014). 'Astrocyte-Synapse Structural Plasticity'. *Neural Plasticity*, *2014*, 1–13. doi:10.1155/2014/232105 PMID:24511394

Bernardinelli, Y., Randall, J., Janett, E., Nikonenko, I., König, S., Jones, E. V., Flores, C. E., Murai, K. K., Bochet, C. G., Holtmaat, A., & Muller, D. (2014). Activity-dependent structural plasticity of perisynaptic astrocytic domains promotes excitatory synapse stability. *Current Biology*, *24*(15), 79–88. doi:10.1016/j.cub.2014.06.025 PMID:25042585

Bernard, R., Kerman, I. A., Thompson, R. C., Jones, E. G., Bunney, W. E., Barchas, J. D., Schatzberg, A. F., Myers, R. M., Akil, H., & Watson, S. J. (2011). Altered expression of glutamate signaling, growth factor, and glia genes in the locus coeruleus of patients with major depression. *Molecular Psychiatry, 16*(6), 634–646. doi:10.1038/mp.2010.44 PMID:20386568

Bernstein, H.-G., & Müller, M. (1999). The cellular localization of the L-ornithine decarboxylase/polyamine system in normal and diseased central nervous systems. *Progress in Neurobiology, 57*(5), 485–505. doi:10.1016/S0301-0082(98)00065-3 PMID:10215098

Berta, T., Park, C.-K., Xu, Z.-Z., Xie, R.-G., Liu, T., Lü, N., Liu, Y.-C., & Ji, R.-R. (2014). Extracellular caspase-6 drives murine inflammatory pain via microglial TNF-α secretion. *The Journal of Clinical Investigation, 124*(3), 1173–1186. doi:10.1172/JCI72230 PMID:24531553

Berumen, J., Baglieri, J., Kisseleva, T., & Mekeel, K. (2022). LIVER FIBROSIS. *Pathophysiology and Clinical Implications., 13*(1), 1–25. doi:10.1002/wsbm.1499.LIVER PMID:32713091

Bessis, A., Béchade, C., Bernard, D., & Roumier, A. (2007). Microglial control of neuronal death and synaptic properties. *Glia, 55*(3), 233–238. doi:10.1002/glia.20459 PMID:17106878

Bezecny, P. (2014). Histone deacetylase inhibitors in glioblastoma: Pre-clinical and clinical experience. *Medical Oncology (Northwood, London, England), 31*(6), 985. doi:10.100712032-014-0985-5 PMID:24838514

Bezzi, P., Domercq, M., Vesce, S., & Volterra, A. (2001). Neuron-astrocyte cross-talk during synaptic transmission: Physiological and neuropathological implications. *Progress in Brain Research, 132*, 255–265. doi:10.1016/S0079-6123(01)32081-2 PMID:11544994

Biber, K., Neumann, H., Inoue, K., & Boddeke, H. W. G. M. (2007). Neuronal 'On'and 'Off'signals control microglia. *Trends in Neurosciences, 30*(11), 596–602. doi:10.1016/j.tins.2007.08.007 PMID:17950926

Bidmon, H. J., Görg, B., Palomero-Gallagher, N., Schleicher, A., Häussinger, D., Speckmann, E. J., & Zilles, K. (2008). Glutamine synthetase becomes nitrated and its activity is reduced during repetitive seizure activity in the pentylentetrazole model of epilepsy. *Epilepsia, 49*(10), 1733–1748. doi:10.1111/j.1528-1167.2008.01642.x PMID:18479397

Bie, B., Wu, J., Foss, J. F., & Naguib, M. (2019). Activation of mGluR1 Mediates C1q-Dependent Microglial Phagocytosis of Glutamatergic Synapses in Alzheimer's Rodent Models. *Molecular Neurobiology, 56*(8), 5568–5585. doi:10.100712035-019-1467-8 PMID:30652266

Biedermann, B., Skatchkov, S. N., Brunk, I., Bringmann, A., Pannicke, T., Bernstein, H.-G., Faude, F., Germer, A., Veh, R., & Reichenbach, A. (1998). Spermine/spermidine is expressed by retinal glial (Müller) cells and controls distinct K+ channels of their membrane. *Glia, 23*(3), 209–220. doi:10.1002/(SICI)1098-1136(199807)23:3<209::AID-GLIA4>3.0.CO;2-# PMID:9633806

Biscaro, B., Lindvall, O., Tesco, G., Ekdahl, C. T., & Nitsch, R. M. (2012). Inhibition of microglial activation protects hippocampal neurogenesis and improves cognitive deficits in a transgenic mouse model for Alzheimer's disease. *Neurodegenerative Diseases, 9*(4), 187–198. doi:10.1159/000330363 PMID:22584394

Bitzer-Quintero, O. (2012). Immune system in the brain: a modulatory role on dendritic spine morphophysiology? *Hindawi.ComOK Bitzer-Quintero, I González-BurgosNeural Plasticity, 2012.* Hindawi. https://www.hindawi.com/journals/np/2012/348642/abs/

Bjelobaba, I., Begovic-Kupresanin, V., Pekovic, S., & Lavrnja, I. (2018). Animal models of multiple sclerosis: Focus on experimental autoimmune encephalomyelitis. *Journal of Neuroscience Research, 96*(6), 1021–1042. doi:10.1002/jnr.24224

Blair-West, G. W., Mellsop, G. W., & Eyeson-Annan, M. L. (1997). Down-rating lifetime suicide risk in major depression. *Acta Psychiatrica Scandinavica, 95*(3), 259–263. doi:10.1111/j.1600-0447.1997.tb09629.x PMID:9111861

Blanco-Suárez, E. (2017). Role of astrocyte–synapse interactions in CNS disorders. *Wiley Online LibraryE Blanco-Suárez, ALM Caldwell, NJ AllenThe Journal of Physiology.* Wiley. https://physoc.onlinelibrary.wiley.com/doi/abs/10.1113/JP270988

Blanco-Suarez, E., Liu, T. F., Kopelevich, A., & Allen, N. J. (2018). Astrocyte-secreted chordin-like 1 drives synapse maturation and limits plasticity by increasing synaptic GluA2 AMPA receptors. *Neuron, 100*(5), 1116–1132. doi:10.1016/j.neuron.2018.09.043 PMID:30344043

Blank, T., & Prinz, M. (2012). *Microglia as Modulators of Cognition and Neuropsychiatric Disorders., 000*, 1–9. doi:10.1002/glia.22372 PMID:22740320

Blei, A. T. (2001). Pathophysiology of Brain Edema in Fulminant Hepatic Failure, Revisited. *Metabolic Brain Disease, 16*(1–2), 85–94. doi:10.1023/A:1011670713730 PMID:11726092

Blei, A. T. (2008). Brain Edema in Acute Liver Failure. *Critical Care Clinics, 24*(1), 99–114. doi:10.1016/j.ccc.2007.11.001 PMID:18241781

Blei, A. T., Olafsson, S., Therrien, G., & Butterworth, R. F. (1994). Ammonia-induced brain edema and intracranial hypertension in rats after portacaval anastomosis. *Hepatology (Baltimore, Md.), 19*(6), 1437–1444. doi:10.1002/hep.1840190619 PMID:8188174

Blennow, K., Brody, D. L., Kochanek, P. M., Levin, H., McKee, A., Ribbers, G. M., Yaffe, K., & Zetterberg, H. (2016). Traumatic brain injuries. *Nature Reviews. Disease Primers, 2*(1), 1–19. doi:10.1038/nrdp.2016.84 PMID:27853132

Blinzinger, K., & Kreutzberg, G. (1968). Displacement of synaptic terminals from regenerating motoneurons by microglial cells. *Zeitschrift für Zellforschung und Mikroskopische Anatomie, 85*(2), 145–157. doi:10.1007/BF00325030 PMID:5706753

Bliss, T. V. P., & Collingridge, G. L. (1993). A synaptic model of memory: Long-term potentiation in the hippocampus. *Nature, 361*(6407), 31–39. doi:10.1038/361031a0 PMID:8421494

Bliss, T. V. P., & Lømo, T. (1973). Long-lasting potentiation of synaptic transmission in the dentate area of the anaesthetized rabbit following stimulation of the perforant path. *The Journal of Physiology, 232*(2), 331–356. doi:10.1113/jphysiol.1973.sp010273 PMID:4727084

Block, M. L., Zecca, L., & Hong, J. S. (2007). Microglia-mediated neurotoxicity: Uncovering the molecular mechanisms. *Nature Reviews. Neuroscience, 8*(1), 57–69. doi:10.1038/nrn2038 PMID:17180163

Blumgart, L. H., & Belghiti, J. (2007). *Surgery of the liver, biliary tract, and pancreas* (Vol. 1). Saunders Elsevier Philadelphia.

Blurton-Jones, M., Kitazawa, M., Martinez-Coria, H., Castello, N. A., Müller, F.-J., Loring, J. F., Yamasaki, T. R., Poon, W. W., Green, K. N., & LaFerla, F. M. (2009). Neural stem cells improve cognition via BDNF in a transgenic model of Alzheimer disease. *Proceedings of the National Academy of Sciences of the United States of America, 106*(32), 13594–13599. doi:10.1073/pnas.0901402106 PMID:19633196

Bocchio-Chiavetto, L., Bagnardi, V., Zanardini, R., Molteni, R., Gabriela Nielsen, M., Placentino, A., Giovannini, C., Rillosi, L., Ventriglia, M., Riva, M. A., & Gennarelli, M. (2010). Serum and plasma BDNF levels in major depression: A replication study and meta-analyses. *The World Journal of Biological Psychiatry, 11*(6), 763–773. doi:10.3109/15622971003611319 PMID:20334574

Böcker, R., Estler, C. J., & Ludewig-Sandig, D. (1991). Evaluation of the Hepatotoxic Potential of Minocycline. *Antimicrobial Agents and Chemotherapy*, *35*(7), 1434–1436. doi:10.1128/AAC.35.7.1434 PMID:1929305

Boecker, A., Daeschler, S. C., Kneser, U., & Harhaus, L. (2019). Relevance and Recent Developments of Chitosan in Peripheral Nerve Surgery. *Frontiers in Cellular Neuroscience*, *13*, 104. doi:10.3389/fncel.2019.00104 PMID:31019452

Boggs, J. M. (2006). Myelin basic protein: A multifunctional protein. *Cellular and Molecular Life Sciences*, *63*(17), 1945–1961. doi:10.100700018-006-6094-7

Boghdadi, A. G., Teo, L., & Bourne, J. A. (2020). The Neuroprotective Role of Reactive Astrocytes after Central Nervous System Injury. *Journal of Neurotrauma*, *37*(5), 681–691. doi:10.1089/neu.2019.6938 PMID:32031052

Bohra, A., Worland, T., Hui, S., Terbah, R., Farrell, A., & Robertson, M. (2020). Prognostic significance of hepatic encephalopathy in patients with cirrhosis treated with current standards of care. *World Journal of Gastroenterology*, *26*(18), 2221–2231. doi:10.3748/wjg.v26.i18.2221 PMID:32476788

Boido, M., Ghibaudi, M., Gentile, P., Favaro, E., Fusaro, R., & Tonda-Turo, C. (2019). Chitosan-based hydrogel to support the paracrine activity of mesenchymal stem cells in spinal cord injury treatment. *Scientific Reports*, *9*(1), 6402. doi:10.103841598-019-42848-w PMID:31024032

Boillée, S., Yamanaka, K., Lobsiger, C. S., Copeland, N. G., Jenkins, N. A., Kassiotis, G., Kollias, G., & Cleveland, D. W. (2006). Onset and progression in inherited ALS determined by motor neurons and microglia. *Science*, *312*(5778), 1389–1392. doi:10.1126cience.1123511 PMID:16741123

Boisserand, L. S., Kodama, T., Papassin, J., Auzely, R., Moisan, A., Rome, C., & Detante, O. (2016). Biomaterial applications in cell-based therapy in experimental stroke. *Stem Cells International*, *6810562*, 1–14. doi:10.1155/2016/6810562 PMID:27274738

Boje, K. M., & Arora, P. K. (1992). Microglial-produced nitric oxide and reactive nitrogen oxides mediate neuronal cell death. *Brain Research*, *587*(2), 250–256. doi:10.1016/0006-8993(92)91004-X PMID:1381982

Boku, S., Nakagawa, S., Takamura, N., Kato, A., Takebayashi, M., Hisaoka-Nakashima, K., Omiya, Y., Inoue, T., & Kusumi, I. (2013). GDNF facilitates differentiation of the adult dentate gyrus-derived neural precursor cells into astrocytes via STAT3. *Biochemical and Biophysical Research Communications*, *434*(4), 779–784. doi:10.1016/j.bbrc.2013.04.011 PMID:23603362

Bolaños, J. P., Heales, S. J., Land, J. M., & Clark, J. B. (1995). Effect of peroxynitrite on the mitochondrial respiratory chain: Differential susceptibility of neurones and astrocytes in primary culture. *Journal of Neurochemistry*, *64*(5), 1965–1972. doi:10.1046/j.1471-4159.1995.64051965.x PMID:7722484

Bolós, M., Llorens-Martín, M., Perea, J. R., Jurado-Arjona, J., Rábano, A., Hernández, F., & Avila, J. (2017). 'Absence of CX3CR1 Impairs the Internalization of Tau by Microglia'. *Molecular Neurodegeneration*, *12*(1), 59. doi:10.118613024-017-0200-1 PMID:28810892

Bonifati, V., Rizzu, P., van Baren, M. J., Schaap, O., Breedveld, G. J., Krieger, E., Dekker, M. C., Squitieri, F., Ibanez, P., Joosse, M., van Dongen, J. W., Vanacore, N., van Swieten, J. C., Brice, A., Meco, G., van Duijn, C. M., Oostra, B. A., & Heutink, P. (2003). Mutations in the DJ-1 gene associated with autosomal recessive early-onset parkinsonism. *Science*, *299*(5604), 256–259. doi:10.1126cience.1077209 PMID:12446870

Boni, R., Ali, A., Shavandi, A., & Clarkson, A. N. (2018). Current and novel polymeric biomaterials for neural tissue engineering. *Journal of Biomedical Science*, *25*(1), 90. doi:10.118612929-018-0491-8 PMID:30572957

Bonnafe, E., Touka, M., AitLounis, A., Baas, D., Barras, E., Ucla, C., Moreau, A., Flamant, F., Dubruille, R., Couble, P., Collignon, J., Durand, B., & Reith, W. (2004). The Transcription Factor RFX3 Directs Nodal Cilium Development and Left-Right Asymmetry Specification. *Molecular and Cellular Biology*, *24*(10), 4417–4427. doi:10.1128/MCB.24.10.4417-4427.2004 PMID:15121860

Bonora, M., Patergnani, S., Rimessi, A., De Marchi, E., Suski, J. M., Bononi, A., Giorgi, C., Marchi, S., Missiroli, S., Poletti, F., Wieckowski, M. R., & Pinton, P. (2012). ATP synthesis and storage. *Purinergic Signalling*, *8*(3), 343–357. doi:10.100711302-012-9305-8

Booth, H. D. E., Hirst, W. D., & Wade-Martins, R. (2017). The Role of astrocyte dysfunction in Parkinson's disease pathogenesis. Trends in Neurosciences, 40, 358–370. https://doi.org/. tins.2017.04.001 doi:10.1016/j

Booth, H. D. E., Hirst, W. D., & Wade-Martins, R. (2017). 'The Role of Astrocyte Dysfunction in Parkinson's Disease Pathogenesis'. *Trends in Neurosciences*, *40*(6), 358–370. doi:10.1016/j.tins.2017.04.001 PMID:28527591

Borroni, E., Bohrmann, B., Grueninger, F., Prinssen, E., Nave, S., Loetscher, H., Chinta, S. J., Rajagopalan, S., Rane, A., Siddiqui, A., Ellenbroek, B., Messer, J., Pähler, A., Andersen, J. K., Wyler, R., & Cesura, A. M. (2017). Sembragiline: A novel, selective monoamine oxidase type b inhibitor for the treatment of Alzheimer's disease. *The Journal of Pharmacology and Experimental Therapeutics*, *362*(3), 413–423. doi:10.1124/jpet.117.241653 PMID:28642233

Bosoi, C. R., Tremblay, M., & Rose, C. F. (2014). Induction of Systemic Oxidative Stress Leads to Brain Oedema in Portacaval Shunted Rats. *Liver International*, *34*(9), 1322–1329. doi:10.1111/liv.12414 PMID:25354203

Bosoi, C. R., Yang, X., Huynh, J., Parent-Robitaille, C., Jiang, W., Tremblay, M., & Rose, C. F. (2012). Systemic Oxidative Stress Is Implicated in the Pathogenesis of Brain Edema in Rats with Chronic Liver Failure. *Free Radical Biology & Medicine*, *52*(7), 1228–1235. doi:10.1016/j.freeradbiomed.2012.01.006 PMID:22300646

Bosson, A., Boisseau, S., Buisson, A., Savasta, M., & Albrieux, M. (2015). Disruption of dopaminergic transmission remodels tripartite synapse morphology and astrocytic calcium activity within substantia nigra pars reticulata. *Glia*, *63*(4), 673–683. doi:10.1002/glia.22777 PMID:25511180

Bosworth, A. P., & Allen, N. J. (2017). The diverse actions of astrocytes during synaptic development. *Current Opinion in Neurobiology*, *47*, 38–43. doi:10.1016/j.conb.2017.08.017 PMID:28938161

Böttcher, K., & Pinzani, M. (2017). Pathophysiology of liver fi brosis and the methodological barriers to the development of anti- fi brogenic agents ☆. *Advanced Drug Delivery Reviews*, *121*, 3–8. doi:10.1016/j.addr.2017.05.016 PMID:28600202

Bouchat, J., Couturier, B., Marneffe, C., Gankam-Kengne, F., Balau, B., De Swert, K., Brion, J.-P., Poncelet, L., Gilloteaux, J., & Nicaise, C. (2018). Regional oligodendrocytopathy and astrocytopathy precede myelin loss and blood–brain barrier disruption in a murine model of osmotic demyelination syndrome. *Glia*, *66*(3), 606–622. doi:10.1002/glia.23268 PMID:29168586

Bradl, M., & Lassmann, H. (2010). Oligodendrocytes: Biology and pathology. *Acta Neuropathologica*, *119*(1), 37–53. doi:10.100700401-009-0601-5 PMID:19847447

Braun, P. E., & Brostoff, S. W. (1977). Proteins of Myelin. In P. Morell (Ed.), *Myelin* (pp. 201–231). Springer US. doi:10.1007/978-1-4615-7514-6_6

Brederlau, A., Correia, A. S., Anisimov, S. V., Elmi, M., Paul, G., Roybon, L., Morizane, A., Bergquist, F., Riebe, I., Nannmark, U., Carta, M., Hanse, E., Takahashi, J., Sasai, Y., Funa, K., Brundin, P., Eriksson, P. S., & Li, J.-Y. (2006). Transplantation of Human Embryonic Stem Cell-Derived Cells to a Rat Model of Parkinson's Disease: Effect of In Vitro Differentiation on Graft Survival and Teratoma Formation. *Stem Cells (Dayton, Ohio)*, *24*(6), 1433–1440. doi:10.1634temcells.2005-0393 PMID:16556709

Brenner, M., Johnson, A. B., Boespflug-Tanguy, O., Rodriguez, D., Goldman, J. E., & Messing, A. (2001). Mutations in GFAP, encoding glial fibrillary acidic protein, are associated with Alexander disease. *Nature Genetics*, *27*(1), 117–120. doi:10.1038/83679 PMID:11138011

Brettschneider, J., Tredici, K. D., Lee, V. M.-Y., & Trojanowski, J. Q. (2015). Spreading of pathology in neurodegenerative diseases : A focus on human studies. *Nature Reviews. Neuroscience*, *16*(2), 2. doi:10.1038/nrn3887 PMID:25588378

Brigitta, B. (2002). Pathophysiology of depression and mechanisms of treatment. *Dialogues in Clinical Neuroscience*, *4*(1), 7–20. doi:10.31887/DCNS.2002.4.1/bbondy PMID:22033824

Brinkmann, B. G., Agarwal, A., Sereda, M. W., Garratt, A. N., Müller, T., Wende, H., Stassart, R. M., Nawaz, S., Humml, C., Velanac, V., Radyushkin, K., Goebbels, S., Fischer, T. M., Franklin, R. J., Lai, C., Ehrenreich, H., Birchmeier, C., Schwab, M. H., & Nave, K. A. (2008). Neuregulin-1/ErbB Signaling Serves Distinct Functions in Myelination of the Peripheral and Central Nervous System. *Neuron*, *59*(4), 581–595. doi:10.1016/j.neuron.2008.06.028 PMID:18760695

Broe, M., Kril, J., & Halliday, G. M. (2004). Astrocytic degeneration relates to the severity of disease in frontotemporal dementia. *Brain*, *127*(10), 2214–2220. doi:10.1093/brain/awh250 PMID:15282215

Brosnan, C. F., & Raine, C. S. (2013). 'The Astrocyte in Multiple Sclerosis Revisited'. *Glia*, *61*(4), 453–465. doi:10.1002/glia.22443 PMID:23322421

Brown, D. V., Filiz, G., Daniel, P. M., Hollande, F., Dworkin, S., Amiridis, S., Kountouri, N., Ng, W., Morokoff, A. P., & Mantamadiotis, T. (2017). Expression of CD133 and CD44 in glioblastoma stem cells correlates with cell proliferation, phenotype stability and intra-tumor heterogeneity. *PLoS One*, *12*(2), e0172791. doi:10.1371/journal.pone.0172791 PMID:28241049

Brown, G. C., & Vilalta, A. (2015). How microglia kill neurons. *Brain Research*, *1628*, 288–297. doi:10.1016/j.brainres.2015.08.031 PMID:26341532

Brunet, A., & Berger, S. L. (2014). 'Epigenetics of Aging and Aging-Related Disease'. *The Journals of Gerontology. Series A, Biological Sciences and Medical Sciences*, *69*(Suppl 1), S17–S20. doi:10.1093/gerona/glu042 PMID:24833581

Buffo, A., Rolando, C., & Ceruti, S. (2010). Astrocytes in the damaged brain : Molecular and cellular insights into their reactive response and healing potential. *Biochemical Pharmacology*, *79*(2), 77–89. doi:10.1016/j.bcp.2009.09.014 PMID:19765548

Bullmore, E. (2018). Inflamed depression. *Lancet*, *392*(10154), 1189–1190. doi:10.1016/S0140-6736(18)32356-0 PMID:30712704

Burda, J. E., Bernstein, A. M., & Sofroniew, M. V. (2016). Astrocyte roles in traumatic brain injury. *Experimental neurology*, *275*(03), 305-315. doi:10.1016/j.expneurol.2015.03.020

Bushong, E. A., Martone, M. E., & Ellisman, M. H. (2004). 'Maturation of Astrocyte Morphology and the Establishment of Astrocyte Domains during Postnatal Hippocampal Development'. *International Journal of Developmental Neuroscience*, *22*(2), 73–86. doi:10.1016/j.ijdevneu.2003.12.008 PMID:15036382

Buskila, Y., Crowe, S. E., & Ellis-Davies, G. C. R. (2013). Synaptic deficits in layer 5 neurons precede overt structural decay in 5xFAD mice. *Neuroscience*, *254*, 152–159. doi:10.1016/j.neuroscience.2013.09.016 PMID:24055684

Buskila, Y., Kékesi, O., Bellot-Saez, A., Seah, W., Berg, T., Trpceski, M., Yerbury, J. J., & Ooi, L. (2019b). Dynamic interplay between H-current and M-current controls motoneuron hyperexcitability in amyotrophic lateral sclerosis. *Cell Death & Disease*, *10*(4), 310. doi:10.103841419-019-1538-9 PMID:30952836

Butler, R., Bradford, D., & Rodgers, K. E. (2022). Analysis of shared underlying mechanism in neurodegenerative disease. *Frontiers in Aging Neuroscience*, *14*, 1006089. doi:10.3389/fnagi.2022.1006089

Butovsky, O., Ziv, Y., Schwartz, A., Landa, G., Talpalar, A. E., Pluchino, S., Martino, G., & Schwartz, M. (2006). Microglia activated by IL-4 or IFN-γ differentially induce neurogenesis and oligodendrogenesis from adult stem/progenitor cells. *Molecular and Cellular Neurosciences*, *31*(1), 149–160. doi:10.1016/j.mcn.2005.10.006 PMID:16297637

Butt, A. M., De La Rocha, I. C., & Rivera, A. (2019). Oligodendroglial cells in Alzheimer's disease. *Neuroglia in neurodegenerative diseases*, 325-333.

Butt, A. M., Papanikolaou, M., & Rivera, A. 2019. 'Physiology of Oligodendroglia'. In Neuroglia in Neurodegenerative Diseases. Springer Singapore. doi:10.1007/978-981-13-9913-8_5

Butt, A. M. (2006). Neurotransmitter-mediated calcium signaling in oligodendrocyte physiology and pathology. *Glia*, *54*(7), 666–675. doi:10.1002/glia.20424 PMID:17006895

Butterworth, R. F. (2008). *The concept of synergism. 38*(Table 1), 116–121. doi:10.1111/j.1872-034X.2008.00436.x

Butterworth, R. F. (2000). The Astrocytic ('Peripheral-Type') Benzodiazepine Receptor: Role in the Pathogenesis of Portal-Systemic Encephalopathy. *Neurochemistry International*, *36*(4–5), 411–416. doi:10.1016/S0197-0186(99)00132-1 PMID:10733008

Butterworth, R. F. (2002). Pathophysiology of Hepatic Encephalopathy : A New Look at Ammonia. *Metabolic Brain Disease*, *17*(4), 221–227. doi:10.1023/A:1021989230535 PMID:12602499

Butterworth, R. F. (2010). Altered glial–neuronal crosstalk: Cornerstone in the pathogenesis of hepatic encephalopathy. *Neurochemistry International*, *57*(4), 383–388. doi:10.1016/j.neuint.2010.03.012 PMID:20350577

Butterworth, R. F. (2011). Neuroinflammation in Acute Liver Failure: Mechanisms and Novel Therapeutic Targets. *Neurochemistry International*, *59*(6), 830–836. doi:10.1016/j.neuint.2011.07.014 PMID:21864609

Butterworth, R. F. (2013). The liver–brain axis in liver failure: Neuroinflammation and encephalopathy. *Nature Reviews. Gastroenterology & Hepatology*, *10*(9), 522–528. doi:10.1038/nrgastro.2013.99 PMID:23817325

Butterworth, R. F. (2019). Hepatic Encephalopathy in Cirrhosis: Pathology and Pathophysiology. *Drugs*, *79*(S1, s1), 17–21. doi:10.100740265-018-1017-0 PMID:30706423

Buzsáki, G., Logothetis, N., & Singer, W. (2013). 'Scaling Brain Size, Keeping Timing: Evolutionary Preservation of Brain Rhythms'. *Neuron*, *80*(3), 751–764. doi:10.1016/j.neuron.2013.10.002 PMID:24183025

Byun, H. G., Kim, Y. T., Park, P. J., Lin, X., & Kim, S. K. (2005). Chitooligosaccharides as a novel β-secretase inhibitor. *Carbohydrate Polymers*, *61*(2), 198–202. doi:10.1016/j.carbpol.2005.05.003

Cabezas, R., Avila, M., Gonzalez, J., El-Bachá, R. S., Báez, E., García-Segura, L. M., Jurado Coronel, J. C., Capani, F., Cardona-Gomez, G. P., & Barreto, G. E. (2014). Astrocytic modulation of the blood-brain barrier: Perspectives on Parkinson's disease. *Frontiers in Cellular Neuroscience*, *8*, 211. doi:10.3389/fncel.2014.00211 PMID:25136294

Cagnin, A., Brooks, D. J., Kennedy, A. M., Gunn, R. N., Myers, R., Turkheimer, F. E., Jones, T., & Banati, R. B. (2001). In-vivo measurement of activated microglia in dementia. *Lancet*, *358*(9280), 461–467. doi:10.1016/S0140-6736(01)05625-2 PMID:11513911

Cagnin, A., Kassiou, M., Meikle, S. R., & Banati, R. B. (2006). In vivo evidence for microglial activation in neurodegenerative dementia. *Acta Neurologica Scandinavica*, *114*(s185, SUPPL. 185), 107–114. doi:10.1111/j.1600-0404.2006.00694.x PMID:16866919

Cagol, A., Schaedelin, S., Barakovic, M., Benkert, P., Todea, R. A., Rahmanzadeh, R., Galbusera, R., Lu, P.-J., Weigel, M., Melie-Garcia, L., Ruberte, E., Siebenborn, N., Battaglini, M., Radue, E.-W., Yaldizli, Ö., Oechtering, J., Sinnecker, T., Lorscheider, J., Fischer-Barnicol, B., & Granziera, C. (2022). Association of brain atrophy with disease progression independent of relapse activity in patients with relapsing multiple sclerosis. *JAMA Neurology*, *79*(7), 682–692. doi:10.1001/jamaneurol.2022.1025

Cai, H., Cong, W., Ji, S., Rothman, S., Maudsley, S., & Martin, B. (2012). Metabolic dysfunction in Alzheimer's disease and related neurodegenerative disorders. *Current Alzheimer Research*, *9*(1), 5–17. doi:10.2174/156720512799015064 PMID:22329649

Cai, Z., & Xiao, M. (2016). Oligodendrocytes and Alzheimer's disease. *The International Journal of Neuroscience*, *126*(2), 97–104. doi:10.3109/00207454.2015.1025778 PMID:26000818

Calovi, S., Mut-Arbona, P., & Sperlágh, B. (2019). Microglia and the Purinergic Signaling System. *Neuroscience*, *405*, 137–147. doi:10.1016/j.neuroscience.2018.12.021 PMID:30582977

Calvo, M., Dawes, J. M., & Bennett, D. L. H. (2012). The role of the immune system in the generation of neuropathic pain. *Lancet Neurology*, *11*(7), 629–642. doi:10.1016/S1474-4422(12)70134-5 PMID:22710756

Campana, W. M. (2007). Schwann cells: Activated peripheral glia and their role in neuropathic pain. *Brain, Behavior, and Immunity*, *21*(5), 522–527. doi:10.1016/j.bbi.2006.12.008 PMID:17321718

Cankaya, S., Cankaya, B., Kilic, U., Kilic, E., & Yulug, B. (2019). The therapeutic role of minocycline in Parkinson's disease. *Drugs in Context*, *8*, 1–14. doi:10.7573/dic.212553 PMID:30873213

Cao, X., Li, L. P., Wang, Q., Wu, Q., Hu, H. H., Zhang, M., Fang, Y. Y., Zhang, J., Li, S. J., Xiong, W. C., Yan, H. C., Gao, Y. B., Liu, J. H., Li, X. W., Sun, L. R., Zeng, Y. N., Zhu, X. H., & Gao, T. M. (2013). Astrocyte-derived ATP modulates depressive-like behaviors. *Nature Medicine*, *19*(6), 773–777. doi:10.1038/nm.3162

Capani, F., Quarracino, C., Caccuri, R., & Sica, R. E. P. (2016). Astrocytes as the main players in primary degenerative disorders of the human central nervous system. *Frontiers in Aging Neuroscience*, *8*, 45. doi:10.3389/fnagi.2016.00045 PMID:26973519

Capriotti, T., & Terzakis, K. (2016). Parkinson Disease. *Home Healthcare Now*, *34*(6), 300–307. doi:10.1097/NHH.0000000000000398 PMID:27243427

Cárdenas, A., Solà, E., Rodríguez, E., Barreto, R., Graupera, I., Pavesi, M., Saliba, F., Welzel, T. M., Martinez-Gonzalez, J., Gustot, T., Bernardi, M., Arroyo, V., & Ginès, P. (2014). Hyponatremia Influences the Outcome of Patients with Acute-on-Chronic Liver Failure: An Analysis of the CANONIC Study. *Critical Care*, *18*(6), 1–8. doi:10.118613054-014-0700-0 PMID:25643318

Cardona, A. E., Pioro, E. P., Sasse, M. E., Kostenko, V., Cardona, S. M., Dijkstra, I. M., Huang, D. R., Kidd, G., Dombrowski, S., Dutta, R., Lee, J. C., Cook, D. N., Jung, S., Lira, S. A., Littman, D. R., & Ransohoff, R. M. (2006). Control of microglial neurotoxicity by the fractalkine receptor. *Nature Neuroscience*, *9*(7), 917–924. doi:10.1038/nn1715

Carlson, B. M. (2019). Tissues. In *The Human Body* (pp. 27–63). Elsevier. doi:10.1016/B978-0-12-804254-0.00002-8

Carter, R. L., Chen, Y., Kunkanjanawan, T., Xu, Y., Moran, S. P., Putkhao, K., Yang, J., Huang, A. H. C., Parnpai, R., & Chan, A. W. S. (2014). Reversal of cellular phenotypes in neural cells derived from Huntington's disease monkey-induced pluripotent stem cells. *Stem Cell Reports*, *3*(4), 585–593. doi:10.1016/j.stemcr.2014.07.011 PMID:25358787

Carter, S. F., Schöll, M., Almkvist, O., Wall, A., Engler, H., Långström, B., & Nordberg, A. (2012). Evidence for astrocytosis in prodromal Alzheimer disease provided by 11C-deuterium-L-deprenyl : A multitracer PET paradigm combining 11C-Pittsburgh compound B and 18F-FDG. *Journal of Nuclear Medicine*, *53*(1), 37–46. doi:10.2967/jnumed.110.087031 PMID:22213821

Casal, C., Serratosa, J., & Tusell, J. M. (2002). Relationship between β-AP peptide aggregation and microglial activation. *Brain Research*, *928*(1-2), 76–84. doi:10.1016/S0006-8993(01)03362-5 PMID:11844474

Cassano, T., Calcagnini, S., Pace, L., De Marco, F., Romano, A., & Gaetani, S. (2017). Cannabinoid receptor 2 signaling in neurodegenerative disorders: From pathogenesis to a promising therapeutic target. *Frontiers in Neuroscience*, *11*(FEB), 242186. doi:10.3389/fnins.2017.00030 PMID:28210207

Castellanos, D. B., Martín-Jiménez, C. A., Rojas-Rodríguez, F., Barreto, G. E., & González, J. (2021). Brain lipidomics as a rising field in neurodegenerative contexts: Perspectives with Machine Learning approaches. *Frontiers in Neuroendocrinology*, *61*, 100899. doi:10.1016/j.yfrne.2021.100899

Catalani, A., Sabbatini, M., & Consoli, C. (n.d.). *Glial fibrillary acidic protein immunoreactive astrocytes in developing rat hippocampus*. Elsevier. https://www.sciencedirect.com/science/article/pii/S0047637401003566

Cawsey, T., Duflou, J., Weickert, C. S., & Gorrie, C. A. (2015). Nestin-Positive Ependymal Cells Are Increased in the Human Spinal Cord after Traumatic Central Nervous System Injury. *Journal of Neurotrauma*, *32*(18), 1393–1402. doi:10.1089/neu.2014.3575 PMID:25599268

Cecchini, M. G., Dominguez, M. G., Mocci, S., Wetterwald, A., Felix, R., Fleisch, H., Chisholm, O., Hofstetter, W., Pollard, J. W., & Stanley, E. R. (1994). Role of colony stimulating factor-1 in the establishment and regulation of tissue macrophages during postnatal development of the mouse. *Development (Cambridge, England)*, *120*(6), 1357–1372. doi:10.1242/dev.120.6.1357 PMID:8050349

Cerqueira, S. R., Oliveira, J. M., Silva, N. A., Leite-Almeida, H., Ribeiro-Samy, S., Almeida, A., Mano, J. F., Sousa, N., Salgado, A. J., & Reis, R. L. (2013). Microglia response and in vivo therapeutic potential of methylprednisolone-loaded dendrimer nanoparticles in spinal cord injury. *Small*, *9*(5), 738–749. doi:10.1002mll.201201888 PMID:23161735

Chai, H., Diaz-Castro, B., Shigetomi, E., Monte, E., Octeau, J. C., Yu, X., Cohn, W., Rajendran, P. S., Vondriska, T. M., Whitelegge, J. P., Coppola, G., & Khakh, B. S. (2017). Neural Circuit-Specialized Astrocytes: Transcriptomic, Proteomic, Morphological, and Functional Evidence. *Neuron*, *95*(3), 531–549.e9. doi:10.1016/j.neuron.2017.06.029 PMID:28712653

Chamberlain, K. A., Nanescu, S. E., Psachoulia, K., & Huang, J. K. (2016). Oligodendrocyte regeneration: Its significance in myelin replacement and neuroprotection in multiple sclerosis. *Neuropharmacology*, *110*, 633–643. doi:10.1016/j.neuropharm.2015.10.010

Chandra, S. V. (1972). Histological and histochemical changes in experimental manganese encephalopathy in rabbits. *Archiv fur Toxikologie*, *29*(1), 29–38. doi:10.1007/BF00316512 PMID:4261594

Chandrasekaran, K. (2003). neuroprotective-effects-of-bilobalide-a-component-of-ginkgo-bilo-2003. Pharmacopsychiatry, S89–S94.

Chandrasekaran, K., Mehrabian, Z., Spinnewyn, B., Chinopoulos, C., Drieu, K., & Fiskum, G. (2003). Neuroprotective effects of bilobalide, a component of Ginkgo biloba extract (EGb 761®) in global brain ischemia and in excitotoxicity-induced neuronal death. *Pharmacopsychiatry, 36*(S 1), 89-94.

Chang, M., Son, H., & Lee, Y. (n.d.). *Neurons and astrocytes secrete factors that cause stem cells to differentiate into neurons and astrocytes, respectively*. Elsevier. https://www.sciencedirect.com/science/article/pii/S104474310300068X

Chang, R., Hudson, P., Wilson, B., & Haddon, L. (n.d.). *Influence of neurons on lipopolysaccharide-stimulated production of nitric oxide and tumor necrosis factor-α by cultured glia*. Elsevier. https://www.sciencedirect.com/science/article/pii/S0006899399022556

Chang, S.-H., Lin, Y.-Y., Wu, G.-J., Huang, C.-H., & Tsai, G. J. (2019). Effect of chitosan molecular weight on anti-inflammatory activity in the RAW 264.7 macrophage model. *International Journal of Biological Macromolecules*, *131*, 167–175. doi:10.1016/j.ijbiomac.2019.02.066 PMID:30771390

Chang, Y.-W., Tan, A., Saab, C., & Waxman, S. (2010). Unilateral focal burn injury is followed by long-lasting bilateral allodynia and neuronal hyperexcitability in spinal cord dorsal horn. *The Journal of Pain*, *11*(2), 119–130. doi:10.1016/j.jpain.2009.06.009 PMID:19744891

Chan, S. J., Niu, W., Hayakawa, K., Hamanaka, G., Wang, X., Cheah, P. S., Guo, S., Yu, Z., Arai, K., Selim, M. H., Kurisawa, M., Spector, M., & Lo, E. H. (2019). Promoting neuro-supportive properties of astrocytes with epidermal growth factor hydrogels. *Stem Cells Translational Medicine*, *8*(12), 1242–1248. doi:10.1002ctm.19-0159 PMID:31483567

Chapman, C. R., Tuckett, R. P., & Song, C. W. (2008). Pain and stress in a systems perspective: Reciprocal neural, endocrine, and immune interactions. *The Journal of Pain*, *9*(2), 122–145. doi:10.1016/j.jpain.2007.09.006 PMID:18088561

Chaudhuri, A., & Behan, P. O. (2004). Multiple sclerosis is not an autoimmune disease. *Archives of Neurology*, *61*(10), 1610–1612. doi:10.1001/archneur.61.10.1610 PMID:15477520

Chavarria, L., Alonso, J., Garcia-Martinez, R., Aymerich, F. X., Huerga, E., Jacas, C., Vargas, V., Cordoba, J., & Rovira, A. (2011). Biexponential analysis of diffusion-tensor imaging of the brain in patients with cirrhosis before and after liver transplantation. *AJNR. American Journal of Neuroradiology*, *32*(8), 1510–1517. doi:10.3174/ajnr.A2533 PMID:21700786

Checchin, D., Sennlaub, F., Levavasseur, E., Leduc, M., & Chemtob, S. (2006). Potential role of microglia in retinal blood vessel formation. *Investigative Ophthalmology & Visual Science*, *47*(8), 3595–3602. doi:10.1167/iovs.05-1522 PMID:16877434

Chedly, J., Soares, S., Montembault, A., von Boxberg, Y., Veron-Ravaille, M., Mouffle, C., Benassy, M.-N., Taxi, J., David, L., & Nothias, F. (2017). Physical chitosan microhydrogels as scaffolds for spinal cord injury restoration and axon regeneration. *Biomaterials*, *138*, 91–107. doi:10.1016/j.biomaterials.2017.05.024 PMID:28554011

Chen, B., Dowlatshahi, D., MacQueen, G. M., Wang, J.-F., & Young, L. T. (2001). Increased hippocampal BDNF immunoreactivity in subjects treated with antidepressant medication. *Biological Psychiatry*, *50*(4), 260–265. doi:10.1016/S0006-3223(01)01083-6 PMID:11522260

Chen, C.-J., & Liao, S.-L. (2002). Oxidative stress involves in astrocytic alterations induced by manganese. *Experimental Neurology*, *175*(1), 216–225. doi:10.1006/exnr.2002.7894 PMID:12009774

Chen, F.-L., Dong, Y.-L., Zhang, Z.-J., Cao, D.-L., Xu, J., Hui, J., Zhu, L., & Gao, Y.-J. (2012). Activation of astrocytes in the anterior cingulate cortex contributes to the affective component of pain in an inflammatory pain model. *Brain Research Bulletin*, *87*(1), 60–66. doi:10.1016/j.brainresbull.2011.09.022 PMID:22004615

Chen, G., Zhang, Y.-Q., Qadri, Y. J., Serhan, C. N., & Ji, R.-R. (2018). Microglia in Pain: Detrimental and Protective Roles in Pathogenesis and Resolution of Pain. *Neuron*, *100*(6), 1292–1311. doi:10.1016/j.neuron.2018.11.009 PMID:30571942

Chen, J., Tan, Z., Zeng, L., Zhang, X., He, Y., Gao, W., Wu, X., Li, Y., Bu, B., Wang, W., & Duan, S. (2013). Heterosynaptic long-term depression mediated by ATP released from astrocytes. *Glia*, *61*(2), 178–191. doi:10.1002/glia.22425 PMID:23044720

Chen, J., Yang, Y., Yang, Y., Zou, P., Chen, J., He, Y., Shui, S., Cui, Y., Bai, R., Liang, Y., Hu, Y., Jiang, B., Lu, L., Zhang, X., Liu, J., & Xu, J. (2018). AXL promotes Zika virus infection in astrocytes by antagonizing type I interferon signalling. *Nature Microbiology*, *3*(3), 302–309. doi:10.103841564-017-0092-4 PMID:29379210

Chen, P. C., Vargas, M. R., Pani, A. K., Smeyne, R. J., Johnson, D. A., Kan, Y. W., & Johnson, J. A. (2009). Nrf2-mediated neuroprotection in the MPTP mouse model of Parkinson's disease: Critical role for the astrocyte. *Proceedings of the National Academy of Sciences of the United States of America*, *106*(8), 2933–2938. doi:10.1073/pnas.0813361106 PMID:19196989

Chen, W.-Q., Wu, F.-F., Lv, H.-B., Xing, W.-T., Liu, Q., Liu, J.-P., Ge, Y.-G., & Lu, Y.-Q. (2021). Whether cognitive behavioral therapy is effective for Alzheimer's disease: A protocol for systematic review and network meta-analysis. *Medicine*, *100*(1), 100. doi:10.1097/MD.0000000000023945 PMID:33429752

Chen, W., Zhang, X. I. A., & Huang, W. (2016). Role of neuroinflammation in neurodegenerative diseases. *Molecular Medicine Reports*, *13*(4), 3391–3396. doi:10.3892/mmr.2016.4948 PMID:26935478

Chen, Y., Bourne, J., Pieribone, V. A., & Fitzsimonds, R. M. (2004). The role of actin in the regulation of dendritic spine morphology and bidirectional synaptic plasticity. *Neuroreport*, *15*(5), 829–832. doi:10.1097/00001756-200404090-00018 PMID:15073524

Chen, Y., Miles, D. K., Hoang, T., Shi, J., Hurlock, E., Kernie, S. G., & Lu, Q. R. (2008). The basic helix-loop-helix transcription factor olig2 is critical for reactive astrocyte proliferation after cortical injury. *The Journal of Neuroscience : The Official Journal of the Society for Neuroscience*, *28*(43), 10983–10989. doi:10.1523/JNEUROSCI.3545-08.2008 PMID:18945906

Chen, Y., Qin, C., Huang, J., Tang, X., Liu, C., Huang, K., Xu, J., Guo, G., Tong, A., & Zhou, L. (2020). 'The Role of Astrocytes in Oxidative Stress of Central Nervous System: A Mixed Blessing'. *Cell Proliferation*, *53*(3), e12781. doi:10.1111/cpr.12781 PMID:32035016

Chen, Y., & Swanson, R. A. (2003). Astrocytes and brain injury. *Journal of Cerebral Blood Flow and Metabolism*, *23*(2), 137–149. doi:10.1097/01.WCB.0000044631.80210.3C PMID:12571445

Chen, Y., Zhang, X., Wang, C., Li, G., Gu, Y., & Huang, L.-Y. M. (2008). Activation of P2X7 receptors in glial satellite cells reduces pain through downregulation of P2X3 receptors in nociceptive neurons. *Proceedings of the National Academy of Sciences of the United States of America*, *105*(43), 16773–16778. doi:10.1073/pnas.0801793105 PMID:18946042

Chen, Z., Yuan, Z., Yang, S., Zhu, Y., Xue, M., Zhang, J., & Leng, L. (2023). Brain Energy Metabolism : Astrocytes in Neurodegenerative Diseases. *CNS Neuroscience & Therapeutics*, *29*(1), 24–36. doi:10.1111/cns.13982 PMID:36193573

Cherry, J. D., Olschowka, J. A., & O'Banion, M. K. (2014). Neuroinflammation and M2 microglia: The good, the bad, and the inflamed. *Journal of Neuroinflammation*, *11*(1), 98. doi:10.1186/1742-2094-11-98 PMID:24889886

Chever, O., Lee, C. Y., & Rouach, N. (2014). Astroglial connexin43 hemichannels tune basal excitatory synaptic transmission. *The Journal of Neuroscience : The Official Journal of the Society for Neuroscience*, *34*(34), 11228–11232. doi:10.1523/JNEUROSCI.0015-14.2014 PMID:25143604

Chew, L.-J., Fusar-Poli, P., & Schmitz, T. (2013). Oligodendroglial alterations and the role of microglia in white matter injury: Relevance to schizophrenia. *Developmental Neuroscience*, *35*(2–3), 102–129. doi:10.1159/000346157 PMID:23446060

Chiareli, R. A., Carvalho, G. A., Marques, B. L., Mota, L. S., Oliveira-Lima, O. C., Gomes, R. M., Birbrair, A., Gomez, R. S., Simão, F., Klempin, F., Leist, M., & Pinto, M. C. X. (2021). The Role of Astrocytes in the Neurorepair Process. *Frontiers in Cell and Developmental Biology*, *9*, 665795. https://www.frontiersin.org/articles/10.3389/fcell.2021.665795. doi:10.3389/fcell.2021.665795 PMID:34113618

Chiba, S., & Sakata-Yanagimoto, M. (2020). Advances in understanding of angioimmunoblastic T-cell lymphoma. *Leukemia, 34*(10), 10. doi:10.103841375-020-0990-y PMID:32704161

Chio, C. C., Lin, M. T., & Chang, C. P. (2015). Microglial activation as a compelling target for treating acute traumatic brain injury. *Current Medicinal Chemistry, 22*(6), 759–770. doi:10.2174/0929867321666141106124657 PMID:25386818

Chitnis, T., & Weiner, H. L. (2017). CNS inflammation and neurodegeneration. *The Journal of Clinical Investigation, 127*(10), 3577–3587. doi:10.1172/JCI90609 PMID:28872464

Chiu, I. M., Chen, A., Zheng, Y., Kosaras, B., Tsiftsoglou, S. A., Vartanian, T. K., Brown, R. H. Jr, & Carroll, M. C. (2008). T lymphocytes potentiate endogenous neuroprotective inflammation in a mouse model of ALS. *Proceedings of the National Academy of Sciences of the United States of America, 105*(46), 17913–17918. doi:10.1073/pnas.0804610105 PMID:18997009

Chneiweiss, H. (2002). Les astrocytes contrôlent la neurogenèse dans le système nerveux central adulte. *M/S: médecine sciences, 18*(11), 1065–1066. https://doi.org/ doi:10.1051/medsci/200218111065

Cho, I. K., Yang, B., Forest, C., Qian, L., & Chan, A. W. S. (2019). Amelioration of Huntington's disease phenotype in astrocytes derived from iPSC-derived neural progenitor cells of Huntington's disease monkeys. *PLoS One, 14*(3), e0214156. doi:10.1371/journal.pone.0214156 PMID:30897183

Choi, D. W. (1988). Glutamate neurotoxicity and diseases of the nervous system. *Neuron, 1*(8), 623–634. doi:10.1016/0896-6273(88)90162-6 PMID:2908446

Choudary, P. V., Molnar, M., Evans, S. J., Tomita, H., Li, J. Z., Vawter, M. P., Myers, R. M., Bunney, W. E. Jr, Akil, H., Watson, S. J., & Jones, E. G. (2005). Altered cortical glutamatergic and GABAergic signal transmission with glial involvement in depression. *Proceedings of the National Academy of Sciences of the United States of America, 102*(43), 15653–15658. doi:10.1073/pnas.0507901102 PMID:16230605

Chow, K. H., Park, H. J., George, J., Yamamoto, K., Gallup, A. D., Graber, J. H., Chen, Y., Jiang, W., Steindler, D. A., Neilson, E. G., Kim, B. Y. S., & Yun, K. (2017). S100A4 is a biomarker and regulator of glioma stem cells that is critical for mesenchymal transition in glioblastoma. *Cancer Research, 77*(19), 5360–5373. doi:10.1158/0008-5472.CAN-17-1294 PMID:28807938

Christophersen, N. S., Grønborg, M., Petersen, T. N., Fjord-Larsen, L., Jørgensen, J. R., Juliusson, B., Blom, N., Rosenblad, C., & Brundin, P. (2007). Midbrain expression of Delta-like 1 homologue is regulated by GDNF and is associated with dopaminergic differentiation. *Experimental Neurology, 204*(2), 791–801. doi:10.1016/j.expneurol.2007.01.014 PMID:17320866

Christopherson, K. S., Ullian, E. M., Stokes, C. C. A., Mullowney, C. E., Hell, J. W., Agah, A., Lawler, J., Mosher, D. F., Bornstein, P., & Barres, B. A. (2005). 'Thrombospondins Are Astrocyte-Secreted Proteins That Promote CNS Synaptogenesis'. *Cell, 120*(3), 421–433. doi:10.1016/j.cell.2004.12.020 PMID:15707899

Chung, C. Y., Seo, H., Sonntag, K. C., Brooks, A., Lin, L., & Isacson, O. (2005). 'Cell Type-Specific Gene Expression of Midbrain Dopaminergic Neurons Reveals Molecules Involved in Their Vulnerability and Protection'. *Human Molecular Genetics, 14*(13), 1709–1725. doi:10.1093/hmg/ddi178 PMID:15888489

Chung, C., Gottstein, J., & Blei, A. T. (2001). Indomethacin Prevents the Development of Experimental Ammonia-Induced Brain Edema in Rats after Portacaval Anastomosis. *Hepatology (Baltimore, Md.), 34*(2), 249–254. doi:10.1053/jhep.2001.26383 PMID:11481608

Chung, H.-S., Kim, H., & Bae, H. (2012). Phenelzine (monoamine oxidase inhibitor) increases production of nitric oxide and proinflammatory cytokines via the NF-κB pathway in lipopolysaccharide-activated microglia cells. *Neurochemical Research, 37*(10), 2117–2124. doi:10.100711064-012-0833-y PMID:22763802

Chung, M. J., Park, J. K., & Park, Y. (2012). Anti-inflammatory effects of low-molecular weight chitosan oligosaccharides in IgE–antigen complex-stimulated RBL-2H3 cells and asthma model mice. *International Immunopharmacology, 12*(2), 453–459. doi:10.1016/j.intimp.2011.12.027 PMID:22266066

Chung, Y. C., Kim, S. R., Park, J.-Y., Chung, E. S., Park, K. W., Won, S. Y., Bok, E., Jin, M., Park, E. S., & Yoon, S.-H. (2011). Fluoxetine prevents MPTP-induced loss of dopaminergic neurons by inhibiting microglial activation. *Neuropharmacology, 60*(6), 963–974. doi:10.1016/j.neuropharm.2011.01.043 PMID:21288472

Cirillo, C., Capoccia, E., Iuvone, T., Cuomo, R., Sarnelli, G., Steardo, L., & Esposito, G. (2015). S100B inhibitor pentamidine attenuates reactive gliosis and reduces neuronal loss in a mouse model of Alzheimer's disease. *BioMed Research International, 2015*, 508342. doi:10.1155/2015/508342 PMID:26295040

Clarke, B. E., & Patani, R. (2020). The microglial component of amyotrophic lateral sclerosis. *Brain, 143*(12), 3526–3539. doi:10.1093/brain/awaa309 PMID:33427296

Clarke, C. R. A. (2005). *Neurological diseases in Kumar & Clark Clinical Medicine*. Kumar P and Clark M.

Clarke, L. E., & Barres, B. A. (2013). 'Emerging Roles of Astrocytes in Neural Circuit Development'. *Nature Reviews. Neuroscience, 14*(5), 311–321. doi:10.1038/nrn3484 PMID:23595014

Clarke, L. E., & Barres, B. A. (2013). Erratum: Emerging roles of astrocytes in neural circuit development (Nature Reviews Neuroscience (2013) 14 (311-321)). *Nature Reviews. Neuroscience, 14*(6), 442. doi:10.1038/nrn3506

Clarke, L. E., Liddelow, S. A., Chakraborty, C., Münch, A. E., Heiman, M., & Barres, B. A. (2018). Normal aging induces A1-like astrocyte reactivity. *Proceedings of the National Academy of Sciences of the United States of America, 115*(8), E1896–E1905. doi:10.1073/pnas.1800165115 PMID:29437957

Clarkson, C., Smeal, R. M., Hasenoehrl, M. G., White, J. A., Rubio, M. E., & Wilcox, K. S. (2020). Ultrastructural and functional changes at the tripartite synapse during epileptogenesis in a model of temporal lobe epilepsy. *Experimental Neurology, 326*, 113196. doi:10.1016/j.expneurol.2020.113196 PMID:31935368

Clauw, D. J., Essex, M. N., Pitman, V., & Jones, K. D. (2019). Reframing chronic pain as a disease, not a symptom: Rationale and implications for pain management. *Postgraduate Medicine, 131*(3), 185–198. doi:10.1080/00325481.2019.1574403 PMID:30700198

Clement, A. M., Nguyen, M. D., Roberts, E. A., Garcia, M. L., Boillée, S., Rule, M., McMahon, A. P., Doucette, W., Siwek, D., Ferrante, R. J., Brown, R. H. Jr, Julien, J. P., Goldstein, L. S., & Cleveland, D. W. (2003). Wild-type nonneuronal cells extend survival of SOD1 mutant motor neurons in ALS mice. *Science, 302*(5642), 113–117. doi:10.1126cience.1086071 PMID:14526083

Clemmesen, J. O., Hansen, B. A., & Larsen, F. S. (1997). Indomethacin Normalizes Intracranial Pressure in Acute Liver Failure: A Twenty-three-year-old Woman Treated with Indomethacin. *Hepatology (Baltimore, Md.), 26*(6), 1423–1425. doi:10.1002/hep.510260608 PMID:9397981

Coelho, J. C., Wajner, M., Burin, M. G., Vargas, C. R., & Giugliani, R. (1997). Selective screening of 10,000 high-risk Brazilian patients for the detection of inborn errors of metabolism. *European Journal of Pediatrics, 156*(8), 650–654. doi:10.1007004310050685 PMID:9266201

Coffey, D. C., Kutko, M. C., Glick, R. D., Butler, L. M., Heller, G., Rifkind, R. A, & La Quaglia, M. P. (2001). The histone deacetylase inhibitor, CBHA, inhibits growth of human neuroblastoma xenografts in vivo, alone and synergistically with all-trans retinoic acid. *Cancer Research, 61*(9), 3591–3594. PMID:11325825

Cohen, M. H., Johnson, J. R., & Pazdur, R. (2005). Food and Drug Administration Drug approval summary: Temozolomide plus radiation therapy for the treatment of newly diagnosed glioblastoma multiforme. *Clinical Cancer Research, 11*(19), 6767–6771. doi:10.1158/1078-0432.CCR-05-0722 PMID:16203762

Colburn, R. W., Rickman, A. J., & DeLeo, J. A. (1999). The effect of site and type of nerve injury on spinal glial activation and neuropathic pain behavior. *Experimental Neurology, 157*(2), 289–304. doi:10.1006/exnr.1999.7065 PMID:10364441

Cole, J., Chaddock, C. A., Farmer, A. E., Aitchison, K. J., Simmons, A., McGuffin, P., & Fu, C. H. Y. (2012). White matter abnormalities and illness severity in major depressive disorder. *The British Journal of Psychiatry, 201*(1), 33–39. doi:10.1192/bjp.bp.111.100594 PMID:22576724

Collier, T. J., Kanaan, N. M., & Kordower, J. H. (2011). 'Ageing as a Primary Risk Factor for Parkinson's Disease: Evidence from Studies of Non-Human Primates'. *Nature Reviews. Neuroscience, 12*(6), 359–366. doi:10.1038/nrn3039 PMID:21587290

Colombo, E. (2016). *Astrocytes: key regulators of neuroinflammation.* Cell.com. https://www.cell.com/trends/immunology/fulltext/S1471-4906(16)30072-2

Colombo, J. A., & Reisin, H. D. (2004). Interlaminarastroglia of the cerebral cortex: A marker of the primate brain. *Brain Research, 1006*(1), 126–131. doi:10.1016/j.brainres.2004.02.003 PMID:15047031

Colonna, M., & Butovsky, O. (2017a). *Microglia Function in the Central Nervous System During Health and Neurodegeneration.*

Colonna, M., & Butovsky, O. (2017). 'Microglia Function in the Central Nervous System During Health and Neurodegeneration'. *Annual Review of Immunology, 35*(1), 441–468. doi:10.1146/annurev-immunol-051116-052358 PMID:28226226

Coltart, I., Tranah, T. H., & Shawcross, D. L. (2013). Inflammation and hepatic encephalopathy. *Archives of Biochemistry and Biophysics, 536*(2), 189–196. doi:10.1016/j.abb.2013.03.016 PMID:23583306

Constam, D. B., Philipp, J., Malipiero, U. V., Ten Dijke, P., Schachner, M., & Fontana, A. (1992). Differential expression of transforming growth factor-beta 1,-beta 2, and-beta 3 by glioblastoma cells, astrocytes, and microglia. *Journal of Immunology (Baltimore, Md.: 1950), 148*(5), 1404-1410.

Cooke, S. F., & Bliss, T. V. P. (2006). Plasticity in the human central nervous system. *Brain, 129*(7), 1659–1673. doi:10.1093/brain/awl082 PMID:16672292

Corbett, G. T., Roy, A., & Pahan, K. (2013). Sodium phenylbutyrate enhances astrocytic neurotrophin synthesis via protein kinase C (PKC)-mediated activation of cAMP-response element-binding protein (CREB) : Implications for Alzheimer disease therapy. *The Journal of Biological Chemistry, 288*(12), 8299–8312. doi:10.1074/jbc.M112.426536 PMID:23404502

Corcia, P., Tauber, C., Vercoullie, J., Arlicot, N., Prunier, C., Praline, J., Nicolas, G., Venel, Y., Hommet, C., Baulieu, J. L., Cottier, J. P., Roussel, C., Kassiou, M., Guilloteau, D., & Ribeiro, M. J. (2012). Molecular imaging of microglial activation in amyotrophic lateral sclerosis. *PLoS One, 7*(12), e52941. doi:10.1371/journal.pone.0052941 PMID:23300829

Corder, E. H., Saunders, A. M., Strittmatter, W. J., Schmechel, D. E., Gaskell, P. C., Small, G. W., Roses, A. D., Haines, J. L., & Pericak-Vance, M. A. (1993). Gene dose of apolipoprotein E type 4 allele and the risk of Alzheimer's disease in late onset families. *Science, 261*(5123), 921–923. doi:10.1126cience.8346443 PMID:8346443

Córdoba, J. & García-martinez, R. (2010). *Hyponatremic and Hepatic Encephalopathies : Similarities, Differences and Coexistence*. 73–80.

Cordoba, J., Gottstein, J., & Blei, A. T. (1996). Glutamine, Myo-inositol, and Organic Brain Osmolytes after Portocaval Anastomosis in the Rat: Implications for Ammonia-induced Brain Edema. *Hepatology (Baltimore, Md.)*, *24*(4), 919–923. doi:10.1053/jhep.1996.v24.pm0008855198 PMID:8855198

Córdoba, J., Gottstein, J., & Blei, A. T. (1998). Chronic Hyponatremia Exacerbates Ammonia-Induced Brain Edema in Rats after Portacaval Anastomosis. *Journal of Hepatology*, *29*(4), 589–594. doi:10.1016/S0168-8278(98)80154-2 PMID:9824268

Corkrum, M., Covelo, A., Lines, J., Bellocchio, L., Pisansky, M., Loke, K., Quintana, R., Rothwell, P. E., Lujan, R., Marsicano, G., Martin, E. D., Thomas, M. J., Kofuji, P., & Araque, A. (2020, March 18). Dopamine-Evoked Synaptic Regulation in the Nucleus Accumbens Requires Astrocyte Activity. *Neuron*, *105*(6), 1036–1047.e5. doi:10.1016/j.neuron.2019.12.026 PMID:31954621

Cornell, J., Salinas, S., Huang, H., & Zhou, M. (2022). *Microglia regulation of synaptic plasticity and learning and memory, 17*(4), 705–716.

Cotter, D. R., Pariante, C. M., & Everall, I. P. (2001). Glial cell abnormalities in major psychiatric disorders: The evidence and implications. *Brain Research Bulletin*, *55*(5), 585–595. doi:10.1016/S0361-9230(01)00527-5 PMID:11576755

Cotter, D., Mackay, D., Chana, G., Beasley, C., Landau, S., & Everall, I. P. (2002). Reduced neuronal size and glial cell density in area 9 of the dorsolateral prefrontal cortex in subjects with major depressive disorder. *Cerebral Cortex (New York, N.Y.)*, *12*(4), 386–394. doi:10.1093/cercor/12.4.386 PMID:11884354

Cotter, D., Mackay, D., Landau, S., Kerwin, R., & Everall, I. (2001). Reduced glial cell density and neuronal size in the anterior cingulate cortex in major depressive disorder. *Archives of General Psychiatry*, *58*(6), 545–553. doi:10.1001/archpsyc.58.6.545 PMID:11386983

Coull, J. A. M., Boudreau, D., Bachand, K., Prescott, S. A., Nault, F., Sík, A., De Koninck, P., & De Koninck, Y. (2003). Trans-synaptic shift in anion gradient in spinal lamina I neurons as a mechanism of neuropathic pain. *Nature*, *424*(6951), 938–942. doi:10.1038/nature01868 PMID:12931188

Covelo, A., & Araque, A. (2018). Neuronal activity determines distinct gliotransmitter release from a single astrocyte. *eLife*, *7*, e32237. doi:10.7554/eLife.32237 PMID:29380725

Coyle, J. T., & Schwarcz, R. (2000). Mind glue: Implications of glial cell biology for psychiatry. *Archives of General Psychiatry*, *57*(1), 90–93. doi:10.1001/archpsyc.57.1.90 PMID:10632240

Cremonini, A. L., Caffa, I., Cea, M., Nencioni, A., Odetti, P., & Monacelli, F. (2019). Nutrients in the prevention of Alzheimer's disease. *Oxidative Medicine and Cellular Longevity*, *2019*, 2019. doi:10.1155/2019/9874159 PMID:31565158

Creus-Muncunill, J., & Ehrlich, M. E. (2019). Cell-autonomous and non-cell-autonomous pathogenic mechanisms in Huntington's disease: Insights from in vitro and in vivo models. *Neurotherapeutics; the Journal of the American Society for Experimental NeuroTherapeutics*, *16*(4), 957–978. doi:10.100713311-019-00782-9 PMID:31529216

Crompton, K. E., Tomas, D., Finkelstein, D. I., Marr, M., Forsythe, J. S., & Horne, M. K. (2006). Inflammatory response on injection of chitosan/GP to the brain. *Journal of Materials Science. Materials in Medicine*, *17*(7), 633–639. doi:10.100710856-006-9226-6 PMID:16770548

Crotti, A., & Ransohoff, R. M. (2016). Microglial Physiology and Pathophysiology: Insights from Genome-wide Transcriptional Profiling. *Immunity*, *44*(3), 505–515. doi:10.1016/j.immuni.2016.02.013 PMID:26982357

Cserép, C., Pósfai, B., & Dénes, Á. (2021). Shaping Neuronal Fate: Functional Heterogeneity of Direct Microglia-Neuron Interactions. *Neuron*, *109*(2), 222–240. doi:10.1016/j.neuron.2020.11.007 PMID:33271068

Cuadros, M. A., Martin, C., Coltey, P., Almendros, A., & Navascués, J. (1993). First appearance, distribution, and origin of macrophages in the early development of the avian central nervous system. *The Journal of Comparative Neurology*, *330*(1), 113–129. doi:10.1002/cne.903300110 PMID:8468399

Cuadros, M. A., & Navascués, J. (1998). The origin and differentiation of microglial cells during development. *Progress in Neurobiology*, *56*(2), 173–189. doi:10.1016/S0301-0082(98)00035-5 PMID:9760700

Cuellar-Santoyo, A. O., Ruiz-Rodríguez, V. M., Mares-Barbosa, T. B., Patrón-Soberano, A., Howe, A. G., Portales-Pérez, D. P., Miquelajáuregui Graf, A., & Estrada-Sánchez, A. M. (2023). Revealing the contribution of astrocytes to glutamatergic neuronal transmission. *Frontiers in Cellular Neuroscience*, *16*, 1037641. doi:10.3389/fncel.2022.1037641 PMID:36744061

Cummings, J. L., Isaacson, R. S., Schmitt, F. A., & Velting, D. M. (2015). A practical algorithm for managing Alzheimer's disease: What, when, and why? *Annals of Clinical and Translational Neurology*, *2*(3), 307–323. doi:10.1002/acn3.166 PMID:25815358

Cummings, J. L., Morstorf, T., & Zhong, K. (2014). Alzheimer's disease drug-development pipeline : Few candidates, frequent failures. *Alzheimer's Research & Therapy*, *6*(4), 1–7. doi:10.1186/alzrt269 PMID:25024750

Czéh, B., & Lucassen, P. J. (2007). What causes the hippocampal volume decrease in depression? Are neurogenesis, glial changes and apoptosis implicated? *European Archives of Psychiatry and Clinical Neuroscience*, *257*(5), 250–260. doi:10.100700406-007-0728-0 PMID:17401728

Czéh, B., Simon, M., Schmelting, B., Hiemke, C., & Fuchs, E. (2006). Astroglial plasticity in the hippocampus is affected by chronic psychosocial stress and concomitant fluoxetine treatment. *Neuropsychopharmacology*, *31*(8), 1616–1626. doi:10.1038j.npp.1300982 PMID:16395301

D'Mello, R., & Dickenson, A. H. (2008). Spinal cord mechanisms of pain. *British Journal of Anaesthesia*, *101*(1), 8–16. doi:10.1093/bja/aen088 PMID:18417503

da Fonseca, A. C. C., Matias, D., Garcia, C., Amaral, R., Geraldo, L. H., Freitas, C., & Lima, F. R. S. (2014). The impact of microglial activation on blood-brain barrier in brain diseases. In Frontiers in Cellular Neuroscience. Frontiers Media S.A. doi:10.3389/fncel.2014.00362

Daadi, M. M., Grueter, B. A., Malenka, R. C., Redmond, D. E., & Steinberg, G. K. (2012). Dopaminergic Neurons from Midbrain-Specified Human Embryonic Stem Cell-Derived Neural Stem Cells Engrafted in a Monkey Model of Parkinson's Disease. *PLoS One*, *7*(7), e41120. doi:10.1371/journal.pone.0041120 PMID:22815935

Dai, J., Johnson, E. C. B., Dammer, E. B., Duong, D. M., Gearing, M., Lah, J. J., Levey, A. I., Wingo, T. S., & Seyfried, N. T. (2018). Effects of APOE genotype on brain proteomic network and cell type changes in Alzheimer's disease. *Frontiers in Molecular Neuroscience*, *11*, 454. doi:10.3389/fnmol.2018.00454 PMID:30618606

Dai, X., Chang, P., Zhu, Q., Liu, W., Sun, Y., Zhu, S., & Jiang, Z. (2013). Chitosan oligosaccharides protect rat primary hippocampal neurons from oligomeric β-amyloid 1-42-induced neurotoxicity. *Neuroscience Letters*, *554*, 64–69. doi:10.1016/j.neulet.2013.08.046 PMID:23999027

Dai, X., Hou, W., Sun, Y., Gao, Z., Zhu, S., & Jiang, Z. (2015). Chitosan oligosaccharides inhibit/disaggregate fibrils and attenuate amyloid β-mediated neurotoxicity. *International Journal of Molecular Sciences*, *16*(5), 10526–10536. doi:10.3390/ijms160510526 PMID:26006224

Dallérac, G., & Rouach, N. (2016). Astrocytes as new targets to improve cognitive functions. *Progress in Neurobiology, 144*, 48–67. doi:10.1016/j.pneurobio.2016.01.003 PMID:26969413

Damier, P., Hirsch, E. C., Zhang, P., Agid, Y., & Javoy-Agid, F. (1993). Glutathione peroxidase, glial cells and Parkinson's disease. *Neuroscience, 52*(1), 1–6. doi:10.1016/0306-4522(93)90175-F PMID:8433802

Danbolt, N. C. (2001). 'Glutamate Uptake'. *Progress in Neurobiology, 65*(1), 1–105. doi:10.1016/S0301-0082(00)00067-8 PMID:11369436

Daneman, R., Zhou, L., Kebede, A. A., & Barres, B. A. (2010). 'Pericytes Are Required for Blood–Brain Barrier Integrity during Embryogenesis'. *Nature, 468*(7323), 562–566. doi:10.1038/nature09513 PMID:20944625

Danielle, A. (2016). *Altérations des cellules de Schwann périsynaptiques à la jonction neuromusculaire: implications pour la sclérose latérale amyotrophique*. Département de Neurosciences Faculté de Médecine, Université de Montréal.

Danilov, A. I., Kokaia, Z., & Lindvall, O. (2012). Ectopic ependymal cells in striatum accompany neurogenesis in a rat model of stroke. *Neuroscience, 214*, 159–170. doi:10.1016/j.neuroscience.2012.03.062 PMID:22531371

Dantzer, R., O'connor, J. C., Freund, G. G., Johnson, R. W., & Kelley, K. W. (2008). From inflammation to sickness and depression: When the immune system subjugates the brain. *Nature Reviews. Neuroscience, 9*(1), 46–56. doi:10.1038/nrn2297 PMID:18073775

Das, A., Wallace, G. C. IV, Holmes, C., McDowell, M. L., Smith, J. A., Marshall, J. D., Bonilha, L., Edwards, J. C., Glazier, S. S., Ray, S. K., & Banik, N. L. (2012). Hippocampal tissue of patients with refractory temporal lobe epilepsy is associated with astrocyte activation, inflammation, and altered expression of channels and receptors. *Neuroscience, 220*, 237–246. doi:10.1016/j.neuroscience.2012.06.002 PMID:22698689

Dasgupta, S., & Ray, S. K. (2017). Diverse biological functions of sphingolipids in the CNS: Ceramide and sphingosine regulate myelination in developing brain but stimulate demyelination during pathogenesis of multiple sclerosis. *Journal of Neurology and Psychology, 5*(1).

Dauer, W., & Neuron, S. (2003). *Parkinson's disease: mechanisms and models*. Cell.com. https://www.cell.com/fulltext/S0896-6273(03)00568-3

Davalos, D., Grutzendler, J., Yang, G., Kim, J. V., Zuo, Y., Jung, S., Littman, D. R., Dustin, M. L., & Gan, W.-B. (2005). ATP mediates rapid microglial response to local brain injury in vivo. *Nature Neuroscience, 8*(6), 752–758. doi:10.1038/nn1472 PMID:15895084

Davies, P., & Maloney, A. J. F. (1976). Selective loss of central cholinergic neurons in Alzheimer's disease. *Lancet, 308*(8000), 1403. doi:10.1016/S0140-6736(76)91936-X PMID:63862

Davies-Thompson, J., Vavasour, I., Scheel, M., Rauscher, A., & Barton, J. J. S. (2016). Reduced myelin water in the white matter tracts of patients with Niemann-pick disease type C. *AJNR. American Journal of Neuroradiology, 37*(8), 1487–1489. doi:10.3174/ajnr.A4719 PMID:26939636

Davis, M. E. (2016). Glioblastoma: Overview of disease and treatment. *Clinical Journal of Oncology Nursing, 20*(5), S2–S8. doi:10.1188/16.CJON.S1.2-8 PMID:27668386

Davydova, V. N., Kalitnik, A. A., Markov, P. A., Volod'ko, A. V., Popov, S. V., & Ermak, I. M. (2016). Cytokine-inducing and anti-inflammatory activity of chitosan and its low-molecular derivative. *Applied Biochemistry and Microbiology, 52*(5), 476–482. doi:10.1134/S0003683816050070 PMID:29513410

De Angelis, F., John, N. A., & Brownlee, W. J. (2018). Disease-modifying therapies for multiple sclerosis. *bmj, 363*.

de Azevedo Cardoso, T., Mondin, T. C., Wiener, C. D., Marques, M. B., Fucolo, B., Pinheiro, R. T., de Souza, L. D. M., da Silva, R. A., Jansen, K., & Oses, J. P. (2014). Neurotrophic factors, clinical features and gender differences in depression. *Neurochemical Research*, *39*(8), 1571–1578. doi:10.100711064-014-1349-4 PMID:24899094

De Barros, A., Attal, J., Roques, M., Nicolau, J., Sol, J. C., Cohen-Jonathan-Moyal, E., & Roux, F. E. (2019). Impact on survival of early tumor growth between surgery and radiotherapy in patients with de novo glioblastoma. *Journal of Neuro-Oncology*, *142*(3), 489–497. doi:10.100711060-019-03120-3 PMID:30783874

De Keyser, J., Mostert, J. P., & Koch, M. W. (2008). Dysfunctional astrocytes as key players in the pathogenesis of central nervous system disorders. *Journal of the Neurological Sciences*, *267*(1–2), 3–16. doi:10.1016/j.jns.2007.08.044 PMID:17935736

De la Fuente, A. G., Queiroz, R. M., Ghosh, T., McMurran, C. E., Cubillos, J. F., Bergles, D. E., Fitzgerald, D. C., Jones, C. A., Lilley, K. S., Glover, C. P., & Franklin, R. J. M. (2020). Changes in the oligodendrocyte progenitor cell proteome with ageing. *Molecular & Cellular Proteomics*, *19*(8), 1281–1302. doi:10.1074/mcp.RA120.002102 PMID:32434922

de Lau, L. M. L., & Breteler, M. M. B. (2006). Epidemiology of Parkinson's disease. *Lancet Neurology*, *5*(6), 525–535. doi:10.1016/S1474-4422(06)70471-9 PMID:16713924

De Luca, S. N., Soch, A., Sominsky, L., Nguyen, T.-X., Bosakhar, A., & Spencer, S. J. (2020). Glial remodeling enhances short-term memory performance in Wistar rats. *Journal of Neuroinflammation*, *17*(1), 1–18. doi:10.118612974-020-1729-4 PMID:32028971

de Majo, M., Koontz, M., Rowitch, D., & Ullian, E. M. (2020). An update on human astrocytes and their role in development and disease. *Glia*, *68*(4), 685–704. doi:10.1002/glia.23771 PMID:31926040

De Miranda, B. R., Rocha, E. M., Bai, Q., El Ayadi, A., Hinkle, D., Burton, E. A., & Timothy Greenamyre, J. (2018). Astrocyte-specific DJ-1 overexpression protects against rotenone-induced neurotoxicity in a rat model of Parkinson's disease. *Neurobiology of Disease*, *115*, 101–114. doi:10.1016/j.nbd.2018.04.008 PMID:29649621

De Schepper, S., Crowley, G., & Hong, S. (2021). Understanding microglial diversity and implications for neuronal function in health and disease. *Developmental Neurobiology*, *81*(5), 507–523. doi:10.1002/dneu.22777 PMID:32757416

De Zeeuw, C. I., & Hoogland, T. M. (2015). Reappraisal of Bergmann glial cells as modulators of cerebellar circuit function. *Frontiers in Cellular Neuroscience*, *9*, 246. doi:10.3389/fncel.2015.00246 PMID:26190972

Deatly, A. M., Haase, A. T., & Ball, M. J. (1991). Herpes Simplex Virus Type 1 Transcription during Latent Infections of Mouse and Man : Implications for Dementia. In E. Kurstak (Ed.), *Psychiatry and Biological Factors* (pp. 277–286). Springer. doi:10.1007/978-1-4684-5811-4_25

Decosterd, I., & Woolf, C. J. (2000). Spared nerve injury: An animal model of persistent peripheral neuropathic pain. *Pain*, *87*(2), 149–158. doi:10.1016/S0304-3959(00)00276-1 PMID:10924808

Defebvre, L., & Vérin, M. (2020). *La maladie de Parkinson*. Elsevier Health Sciences.

DeFlitch, L., Gonzalez-Fernandez, E., Crawley, I., & Kang, S. H. (2022). Age and Alzheimer's Disease-Related oligodendrocyte changes in hippocampal subregions. *Frontiers in Cellular Neuroscience*, *16*, 847097. doi:10.3389/fncel.2022.847097 PMID:35465615

Del Bigio, M. R. (2010). 'Ependymal Cells: Biology and Pathology'. *Acta Neuropathologica*, *119*(1), 55–73. doi:10.100700401-009-0624-y PMID:20024659

Del Vecchio, C. A., Giacomini, C. P., Vogel, H., Jensen, K. C., Florio, T., Merlo, A., Pollack, J. R., & Wong, A. J. (2013). EGFRvIII gene rearrangement is an early event in glioblastoma tumorigenesis and expression defines a hierarchy modulated by epigenetic mechanisms. *Oncogene*, *32*(21), 2670–2681. doi:10.1038/onc.2012.280 PMID:22797070

Delgadillo, V., O'Callaghan, M. del M., Gort, L., Coll, M. J., & Pineda, M. (2013). Natural history of Sanfilippo syndrome in Spain. *Orphanet Journal of Rare Diseases*, *8*(1), 1–11. doi:10.1186/1750-1172-8-189 PMID:24314109

Delgado-Morales, R., Agís-Balboa, R. C., Esteller, M., & Berdasco, M. (2017). 'Epigenetic Mechanisms during Ageing and Neurogenesis as Novel Therapeutic Avenues in Human Brain Disorders'. *Clinical Epigenetics*, *9*(1), 67. doi:10.118613148-017-0365-z PMID:28670349

DeMarshall, C., Goldwaser, E. L., Sarkar, A., Godsey, G. A., Acharya, N. K., Thayasivam, U., Belinka, B. A., & Nagele, R. G. (2017). Autoantibodies as diagnostic biomarkers for the detection and subtyping of multiple sclerosis. *Journal of Neuroimmunology*, *309*, 51–57. doi:10.1016/j.jneuroim.2017.05.010

Desai, M. K., Guercio, B. J., Narrow, W. C., & Bowers, W. J. (2011). An Alzheimer's disease-relevant presenilin-1 mutation augments amyloid-beta-induced oligodendrocyte dysfunction. *Glia*, *59*(4), 627–640. doi:10.1002/glia.21131 PMID:21294162

Deutsch-Link, S., Moon, A. M., Jiang, Y., Barritt, A. S. IV, & Tapper, E. B. (2022). Serum ammonia in cirrhosis: Clinical impact of hyperammonemia, utility of testing, and national testing trends. *Clinical Therapeutics*, *44*(3), e45–e57. doi:10.1016/j.clinthera.2022.01.008 PMID:35125217

Devinsky, O., Vezzani, A., Najjar, S., De Lanerolle, N. C., & Rogawski, M. A. (2013). Glia and epilepsy: Excitability and inflammation. *Trends in Neurosciences*, *36*(3), 174–184. doi:10.1016/j.tins.2012.11.008 PMID:23298414

Dewil, M., Lambrechts, D., Sciot, R., Shaw, P. J., Ince, P. G., Robberecht, W., & Van den Bosch, L. (2007). Vascular endothelial growth factor counteracts the loss of phospho-Akt preceding motor neurone degeneration in amyotrophic lateral sclerosis. *Neuropathology and Applied Neurobiology*, *33*(5), 499–509. doi:10.1111/j.1365-2990.2007.00850.x PMID:17854437

Dextera, D. T., & Jenner, P. (2013). Parkinson disease: From pathology to molecular disease mechanisms. *Free Radical Biology & Medicine*, *62*, 132–144. doi:10.1016/j.freeradbiomed.2013.01.018 PMID:23380027

Dharel, N., & Bajaj, J. S. (2015). Definition and nomenclature of hepatic encephalopathy. *Journal of Clinical and Experimental Hepatology*, *5*, S37–S41. doi:10.1016/j.jceh.2014.10.001 PMID:26041955

Di Carlo, D. T., Cagnazzo, F., Benedetto, N., Morganti, R., & Perrini, P. (2019). Multiple high-grade gliomas: Epidemiology, management, and outcome. A systematic review and meta-analysis. *Neurosurgical Review*, *42*(2), 263–275. doi:10.100710143-017-0928-7 PMID:29138949

Di Castro, M. A., Chuquet, J., Liaudet, N., Bhaukaurally, K., Santello, M., Bouvier, D., Tiret, P., & Volterra, A. (2011). Local Ca2+ detection and modulation of synaptic release by astrocytes. *Nature Neuroscience, 14*(10), 1276–1284. doi:10.1038/nn.2929

di Domenico, A., Carola, G., Calatayud, C., Pons-Espinal, M., Muñoz, J. P., Richaud-Patin, Y., Fernandez-Carasa, I., Gut, M., Faella, A., Parameswaran, J., Soriano, J., Ferrer, I., Tolosa, E., Zorzano, A., Cuervo, A. M., Raya, A., & Consiglio, A. (2019). Patient-Specific iPSC-Derived Astrocytes Contribute to Non-Cell-Autonomous Neurodegeneration in Parkinson's Disease. *Stem Cell Reports*, *12*(2), 213–229. doi:10.1016/j.stemcr.2018.12.011 PMID:30639209

Diaz-Aparicio, I., Paris, I., Sierra-Torre, V., Plaza-Zabala, A., Rodríguez-Iglesias, N., Márquez-Ropero, M., Beccari, S., Huguet, P., Abiega, O., Alberdi, E., Matute, C., Bernales, I., Schulz, A., Otrokocsi, L., Sperlagh, B., Happonen, K. E., Lemke, G., Maletic-Savatic, M., Valero, J., & Sierra, A. (2020). Microglia actively remodel adult hippocampal neurogenesis through the phagocytosis secretome. *The Journal of Neuroscience : The Official Journal of the Society for Neuroscience*, *40*(7), 1453–1482. doi:10.1523/JNEUROSCI.0993-19.2019 PMID:31896673

Dienel, G. A. (2013). Astrocytic Energetics during Excitatory Neurotransmission: What Are Contributions of Glutamate Oxidation and Glycolysis? *Neurochemistry International*, *63*(4), 244–258. doi:10.1016/j.neuint.2013.06.015 PMID:23838211

Dimitriou, N. G., Meuth, S. G., Martinez-Lapiscina, E. H., Albrecht, P., & Menge, T. (2023). Treatment of patients with multiple sclerosis transitioning between relapsing and progressive disease. *CNS Drugs*, *37*(1), 69–92. doi:10.100740263-022-00977-3

Dimou, J., Beland, B., & Kelly, J. (2020). Supramaximal resection: A systematic review of its safety, efficacy and feasibility in glioblastoma. *Journal of Clinical Neuroscience*, *72*, 328–334. doi:10.1016/j.jocn.2019.12.021 PMID:31864830

Dimou, L., Simon, C., Kirchhoff, F., Takebayashi, H., & Götz, M. (2008). Progeny of Olig2-expressing progenitors in the gray and white matter of the adult mouse cerebral cortex. *The Journal of Neuroscience : The Official Journal of the Society for Neuroscience*, *28*(41), 10434–10442. doi:10.1523/JNEUROSCI.2831-08.2008

Diniz, B. S., Teixeira, A. L., Miranda, A. S., Talib, L. L., Gattaz, W. F., & Forlenza, O. V. (2012). Circulating Glial-derived neurotrophic factor is reduced in late-life depression. *Journal of Psychiatric Research*, *46*(1), 135–139. doi:10.1016/j.jpsychires.2011.09.007 PMID:21963280

Diniz, L. P., Matias, I., Araujo, A. P. B., Garcia, M. N., Barros-Aragão, F. G. Q., Alves-Leon, S. V., de Souza, J. M., Foguel, D., Figueiredo, C. P., Braga, C., Romão, L., & Gomes, F. C. A. (2019). α-synuclein oligomers enhance astrocyte-induced synapse formation through TGF-β1 signaling in a Parkinson's disease model. *Journal of Neurochemistry*, *150*(2), 138–157. doi:10.1111/jnc.14710 PMID:31009074

Dionisio-Santos, D. A., Olschowka, J. A., & O'Banion, M. K. (2019). Exploiting microglial and peripheral immune cell crosstalk to treat Alzheimer's disease. *Journal of Neuroinflammation*, *16*(1), 74. doi:10.118612974-019-1453-0 PMID:30953557

Dmytriyeva, O., Pankratova, S., Owczarek, S., Sonn, K., Soroka, V., Ridley, C. M., Marsolais, A., Lopez-Hoyos, M., Ambartsumian, N., Lukanidin, E., Bock, E., Berezin, V., & Kiryushko, D. (2012). The metastasis-promoting S100A4 protein confers neuroprotection in brain injury. *Nature Communications*, *3*(1), 1197. doi:10.1038/ncomms2202 PMID:23149742

Do Nascimento, A. L., Dos Santos, N. F., Campos Pelágio, F., Aparecida Teixeira, S., De Moraes Ferrari, E. A., & Langone, F. (2012). Neuronal degeneration and gliosis time-course in the mouse hippocampal formation after pilocarpine-induced status epilepticus. *Brain Research*, *1470*, 98–110. doi:10.1016/j.brainres.2012.06.008 PMID:22781142

Dobson, R., & Giovannoni, G. (2019). Road, W.; Unit, PN; Square, C. *Multiple Sclerosis—A review. European Journal of Neurology*, *26*, 27–40. doi:10.1111/ene.13819

Do, H. (1949). *The organization of behavior.*

Do-Ha, D., Buskila, Y., & Ooi, L. (2018). Impairments in motor neurons, interneurons and astrocytes contribute to hyperexcitability in ALS: Underlying mechanisms and paths to therapy. *Molecular Neurobiology*, *55*(2), 1410–1418. doi:10.100712035-017-0392-y PMID:28160214

Domingues, A. V., Pereira, I. M., Vilaça-Faria, H., Salgado, A. J., Rodrigues, A. J., & Teixeira, F. G. (2020). Glial cells in Parkinson's disease: protective or deleterious? In Cellular and Molecular Life Sciences, 77(24). doi:10.100700018-020-03584-x

Donat, C. K., Scott, G., Gentleman, S. M., & Sastre, M. (2017). Microglial Activation in Traumatic Brain Injury. *Frontiers in Aging Neuroscience*, *9*, 208. doi:10.3389/fnagi.2017.00208 PMID:28701948

Dong, H., Hao, X., Cui, B., & Guo, M. (2017). MiR-429 suppresses glioblastoma multiforme by targeting SOX2. *Cell Biochemistry and Function*, *35*(5), 260–268. doi:10.1002/cbf.3271 PMID:28749077

Dong, R., Huang, R., Wang, J., Liu, H., & Xu, Z. (2021). Effects of Microglial Activation and Polarization on Brain Injury After Stroke. *Frontiers in Neurology*, *12*, 620948. doi:10.3389/fneur.2021.620948 PMID:34276530

Donnelly, C. R., Andriessen, A. S., Chen, G., Wang, K., Jiang, C., Maixner, W., & Ji, R.-R. (2020). Central Nervous System Targets: Glial Cell Mechanisms in Chronic Pain. *Neurotherapeutics; the Journal of the American Society for Experimental NeuroTherapeutics*, *17*(3), 846–860. doi:10.100713311-020-00905-7 PMID:32820378

Dossi, E., Vasile, F., & Rouach, N. (2018). Human astrocytes in the diseased brain. *Brain Research Bulletin*, *136*, 139–156. doi:10.1016/j.brainresbull.2017.02.001 PMID:28212850

Dowling, C., & Allen, N. J. (2018). Mice lacking glypican 4 display juvenile hyperactivity and adult social interaction deficits. *Brain Plasticity*, *4*(2), 197–209. doi:10.3233/BPL-180079 PMID:30598870

Drews, L., Zimmermann, M., Westhoff, P., Brilhaus, D., Poss, R. E., Bergmann, L., Wiek, C., Brenneisen, P., Piekorz, R. P., Mettler-Altmann, T., Weber, A. P. M., & Reichert, A. S. (2020). Ammonia Inhibits Energy Metabolism in Astrocytes in a Rapid and Glutamate Dehydrogenase 2-Dependent Manner. *Disease Models & Mechanisms*, *13*(10), dmm047134. doi:10.1242/dmm.047134 PMID:32917661

Dröge, W. (2002). Free radicals in the physiological control of cell function. *Physiological Reviews*, *82*(1), 47–95. doi:10.1152/physrev.00018.2001 PMID:11773609

Dublin, P., & Hanani, M. (2007). Satellite glial cells in sensory ganglia: Their possible contribution to inflammatory pain. *Brain, Behavior, and Immunity*, *21*(5), 592–598. doi:10.1016/j.bbi.2006.11.011 PMID:17222529

Dubois, B., Padovani, A., Scheltens, P., Rossi, A., & Dell'Agnello, G. (2016). Timely diagnosis for Alzheimer's disease: A literature review on benefits and challenges. *Journal of Alzheimer's Disease*, *49*(3), 617–631. doi:10.3233/JAD-150692 PMID:26484931

Ducray, A., Krebs, S. H., Schaller, B., Seiler, R. W., Meyer, M., & Widmer, H. R. (2006). GDNF family ligands display distinct action profiles on cultured GABAergic and serotonergic neurons of rat ventral mesencephalon. *Brain Research*, *1069*(1), 104–112. doi:10.1016/j.brainres.2005.11.056 PMID:16380100

Ducza, L., Szücs, P., Hegedűs, K., Bakk, E., Gajtkó, A., Wéber, I., & Holló, K. (2021). NLRP2 Is Overexpressed in Spinal Astrocytes at the Peak of Mechanical Pain Sensitivity during Complete Freund Adjuvant-Induced Persistent Pain. *International Journal of Molecular Sciences*, *22*(21), 11408. doi:10.3390/ijms222111408 PMID:34768839

Dudvarski Stankovic, N., Teodorczyk, M., Ploen, R., Zipp, F., & Schmidt, M. H. H. (2016). Microglia–blood vessel interactions: A double-edged sword in brain pathologies. *Acta Neuropathologica*, *131*(3), 347–363. doi:10.100700401-015-1524-y PMID:26711460

Du, J., Tan, E., Kim, H. J., Zhang, A., Bhattacharya, R., & Yarema, K. J. (2014). Comparative evaluation of chitosan, cellulose acetate, and polyethersulfone nanofiber scaffolds for neural differentiation. *Carbohydrate Polymers*, *99*, 483–490. doi:10.1016/j.carbpol.2013.08.050 PMID:24274534

Dujardin, K., Degreef, J. F., Rogelet, P., Defebvre, L., & Destee, A. (1999). Impairment of the supervisory attentional system in early untreated patients with Parkinson's disease. *Journal of Neurology*, *246*(9), 783–788. doi:10.1007004150050455 PMID:10525975

Duman, R. S., & Li, N. (2012). A neurotrophic hypothesis of depression: Role of synaptogenesis in the actions of NMDA receptor antagonists. *Philosophical Transactions of the Royal Society of London. Series B, Biological Sciences*, *367*(1601), 2475–2484. doi:10.1098/rstb.2011.0357 PMID:22826346

Duman, R. S., & Monteggia, L. M. (2006). A neurotrophic model for stress-related mood disorders. *Biological Psychiatry*, *59*(12), 1116–1127. doi:10.1016/j.biopsych.2006.02.013 PMID:16631126

Dutrillaux, A. M., Xie, H., & Dutrillaux, B. (2007). 'High Chromosomal Polymorphism and Heterozygosity in *Cyclocephala Tridentata* from Guadeloupe: Chromosome Comparison with Some Other Species of Dynastinae (Coleoptera: Scarabaeidae)'. *Cytogenetic and Genome Research*, *119*(3–4), 248–254. doi:10.1159/000112070 PMID:18253038

Dwivedi, Y., Rizavi, H. S., Conley, R. R., Roberts, R. C., Tamminga, C. A., & Pandey, G. N. (2003). Altered gene expression of brain-derived neurotrophic factor and receptor tyrosine kinase B in postmortem brain of suicide subjects. *Archives of General Psychiatry*, *60*(8), 804–815. doi:10.1001/archpsyc.60.8.804 PMID:12912764

Ebersole, T. A., Chen, Q. I., Justice, M. J., & Artzt, K. (1996). The quaking gene product necessary in embryogenesis and myelination combines features of RNA binding and signal transduction proteins. *Nature Genetics*, *12*(3), 260–265. doi:10.1038/ng0396-260 PMID:8589716

Echeverry, S., Shi, X. Q., Yang, M., Huang, H., Wu, Y., Lorenzo, L.-E., Perez-Sanchez, J., Bonin, R. P., De Koninck, Y., & Zhang, J. (2017). Spinal microglia are required for long-term maintenance of neuropathic pain. *Pain*, *158*(9), 1792–1801. doi:10.1097/j.pain.0000000000000982 PMID:28746078

Eddleston, M., & Mucke, L. (1993). Molecular profile of reactive astrocytes—Implications for their role in neurologic disease. *Neuroscience*, *54*(1), 15–36. doi:10.1016/0306-4522(93)90380-X PMID:8515840

Edgar, N., & Sibille, E. (2012). A putative functional role for oligodendrocytes in mood regulation. *Translational Psychiatry*, *2*(5), e109–e109. doi:10.1038/tp.2012.34 PMID:22832953

Edison, P. (2021). Microglial activation and blood-brain barrier leakage: Chicken and egg? In Brain, 144(5). doi:10.1093/brain/awab149

Edwards, L. J., Kirilina, E., Mohammadi, S., & Weiskopf, N. (2018). Microstructural imaging of human neocortex in vivo. *NeuroImage*, *182*, 184–206. doi:10.1016/j.neuroimage.2018.02.055 PMID:29588229

Eefsen, M., Jelnes, P., Schmidt, L. E., Ben Vainer, H. C. B., & Larsen, F. S. (2010). Brain Expression of the Water Channels Aquaporin-1 and-4 in Mice with Acute Liver Injury, Hyperammonemia and Brain Edema. *Metabolic Brain Disease*, *25*(3), 315–323. doi:10.100711011-010-9213-y PMID:20938728

Ehninger, D., & Kempermann, G. (2008). Neurogenesis in the adult hippocampus. *Cell and Tissue Research*, *331*(1), 243–250. doi:10.100700441-007-0478-3 PMID:17938969

Ekdahl, C. T. (2012). Microglial activation-tuning and pruning adult neurogenesis. *Frontiers in Pharmacology, 3 MAR*(March), 1–9. doi:10.3389/fphar.2012.00041

Ekestern, E. (2004). Neurotrophic factors and amyotrophic lateral sclerosis. *Neurodegenerative Diseases*, *1*(2-3), 88–100. doi:10.1159/000080049 PMID:16908980

El Hiba, O., El-Mansoury, B., Chatoui, H., Benammi, H., Balzano, T., Merzouki, M., El Hidan, M. A., Radhakrishnan, J., Bahi, L., & Idrissi, H. R. (2023). Clinical and Experimental Evidence of Cirrhosis-Related Parkinsonism. In *Experimental and Clinical Evidence of the Neuropathology of Parkinson's Disease* (pp. 102–119). IGI Global. doi:10.4018/978-1-6684-5156-4.ch007

El Khiat, A., El-Mansoury, B., El Hiba, O., Tamegart, L., Rais, H., Fdil, N., Sellami, S., Ait El Mokhtar, M., Kissani, N., & Gamrani, H. (2023). P14 Altered Locomotor Behavior Associated With Purkinje Cells and Astrocytes Dysfunctions in Rat With Acute Hepatic Encephalopathy Involving the Oxidative Stress and Neuroinflammation. *Official Journal of the American College of Gastroenterology| ACG, 118*(9S), S11.

El Khiat, A., El-Mansoury, B., El Hiba, O., Tamegart, L., Rais, H., Fdil, N., Sellami, S., El Mokhtar, M. A., Kissani, N., & Gamrani, H. (2023). P14 Altered Locomotor Behavior Associated With Purkinje Cells and Astrocytes Dysfunctions in Rat With Acute Hepatic Encephalopathy Involving the Oxidative Stress and Neuroinflammation. *Official Journal of the American College of Gastroenterology| ACG, 118*(9S), S11.

El Khiat, A., El Hiba, O., Tamegart, L., Rais, H., Fdil, N., Sellami, S., Ait El Mokhtar, M., & Gamrani, H. (2022a). Time Dependent Alteration of Locomotor Behavior in Rat with Acute Liver Failure Induced Cerebellar Neuroinflammation and Neuro-Astroglial Damage. *Journal of Chemical Neuroanatomy, 119*, 102055. doi:10.1016/j.jchemneu.2021.102055 PMID:34863855

Elkayam, O., Yaron, M., & Caspi, D. (1999). Minocycline-Induced Autoimmune Syndromes: An Overview. Seminars in arthritis and rheumatism, 28. Elsevier.

Elkomy, M. H., Ali, A. A., & Eid, H. M. (2022). Chitosan on the surface of nanoparticles for enhanced drug delivery: A comprehensive review. *Journal of Controlled Release, 351*, 923–940. doi:10.1016/j.jconrel.2022.10.005 PMID:36216174

El-Mansoury, B, Smimih, K., Aimrane, A., Draoui, A., Ferssiwi, A., Bitar, A., Jayakumar, A. R., & El Hiba, O. (2023). P8 Cognitive Impairment Associated With Cortical and Hippocampal Microglia Activation and Altered Synaptic Function in a Rat Model of Chronic Hepatic Encephalopathy. *Official Journal of the American College of Gastroenterology| ACG, 118*(9S), S7.

El-Mansoury, B. El Hiba, O., Agnaou, M., Jayakumar, A. R., El Khiat, A., Kahim, K., Boulbaroud, S., Zouhairi, N., Tibssirte, O., & Bouighajd, H. (2023). Neuropathology of Parkinson's Disease: An Update of the Recent Findings. In Experimental and Clinical Evidence of the Neuropathology of Parkinson's Disease (pp. 82–101). IGI Global.

El-Mansoury, B., Laadraoui, J., El Khiat, A., El Hiba, O., Jayakumar, A. R., Ferssiwi, A., & Bitar, A. (2023). *Ataxia in Multiple Sclerosis: From Current Understanding to Therapy.*

El-Mansoury, B., Smimih, K., Aimrane, A., Draoui, A., Ferssiwi, A., Bitar, A., Jayakumar, A., & El Hiba, O. (2023). P8 Cognitive Impairment Associated With Cortical and Hippocampal Microglia Activation and Altered Synaptic Function in a Rat Model of Chronic Hepatic Encephalopathy. *Official Journal of the American College of Gastroenterology, 118*(9S), S7.

El-Mansoury, Bilal, Smimih, K., Hamdan, Y. A., El Khiat, A., Elouali, S., El Fatimi, H., Draoui, A., El Hiba, O., & Jayakumar, A. R. (2023). The Interplay Between Environmental Pollutants, Gut Microbiota, and Infections: Current Concepts and Therapies. *Ecological and Evolutionary Perspectives on Infections and Morbidity*, 173–207.

El-Mansoury, B., Ben Maloui, A., Draoui, A., Hamdan, Y. A., Oukhrib, M., Smimih, K., Elmourid, A., El Khiat, A., Boulbaroud, S., & Meftah, I. (2023). Non-Motor Symptoms in Parkinson's Disease: The Other Side of the Disease. In *Experimental and Clinical Evidence of the Neuropathology of Parkinson's Disease* (pp. 25–45). IGI Global. doi:10.4018/978-1-6684-5156-4.ch003

Elsaid, M. I., & Rustgi, V. K. (2020). Epidemiology of hepatic encephalopathy. *Clinics in Liver Disease*, *24*(2), 157–174. doi:10.1016/j.cld.2020.01.001 PMID:32245524

Elsherbini, D. M. A., Ghoneim, F. M., El-Mancy, E. M., Ebrahim, H. A., El-Sherbiny, M., El-Shafey, M., Al-Serwi, R. H., & Elsherbiny, N. M. (2022). Astrocytes Profiling in Acute Hepatic Encephalopathy: Possible Enrolling of Glial Fibrillary Acidic Protein, Tumor Necrosis Factor-Alpha, Inwardly Rectifying Potassium Channel (Kir 4.1) and Aquaporin-4 in Rat Cerebral Cortex. *Frontiers in Cellular Neuroscience*, *16*, 896172. doi:10.3389/fncel.2022.896172 PMID:36060277

Emilsson, L., Saetre, P., Balciuniene, J., Castensson, A., Cairns, N., & Jazin, E. E. (2002). Increased monoamine oxidase messenger RNA expression levels in frontal cortex of Alzheimer's disease patients. *Neuroscience Letters*, *326*(1), 56–60. doi:10.1016/S0304-3940(02)00307-5 PMID:12052537

Enache, D., Pariante, C. M., & Mondelli, V. (2019). Markers of central inflammation in major depressive disorder: A systematic review and meta-analysis of studies examining cerebrospinal fluid, positron emission tomography and post-mortem brain tissue. *Brain, Behavior, and Immunity*, *81*, 24–40. doi:10.1016/j.bbi.2019.06.015 PMID:31195092

Endo, M., Ubulkasim, G., Kobayashi, C., Onishi, R., Aiba, A., & Minami, Y. (2017). Critical role of Ror2 receptor tyrosine kinase in regulating cell cycle progression of reactive astrocytes following brain injury. *Glia*, *65*(1), 182–197. doi:10.1002/glia.23086 PMID:27726178

Engelhardt, J. I., Tajti, J., & Appel, S. H. (1993). Lymphocytic Infiltrates in the Spinal Cord in Amyotrophic Lateral Sclerosis. *Archives of Neurology*, *50*(1), 30–36. doi:10.1001/archneur.1993.00540010026013 PMID:8093428

Engert, F., & Bonhoeffer, T. (1999). Dendritic spine changes associated with hippocampal long-term synaptic plasticity. *Nature*, *399*(6731), 66–70. doi:10.1038/19978 PMID:10331391

Eng, L. F., Ghirnikar, R. S., & Lee, Y. L. (2000). Glial fibrillary acidic protein : GFAP-thirty-one years (1969–2000). *Neurochemical Research*, *25*(9/10), 1439–1451. doi:10.1023/A:1007677003387 PMID:11059815

Erikson, K. M., Dobson, A. W., Dorman, D. C., & Aschner, M. (2004). Manganese exposure and induced oxidative stress in the rat brain. *The Science of the Total Environment*, *334*, 409–416. doi:10.1016/j.scitotenv.2004.04.044 PMID:15504526

Eriksson, P. S., Perfilieva, E., Björk-Eriksson, T., Alborn, A.-M., Nordborg, C., Peterson, D. A., & Gage, F. H. (1998). Neurogenesis in the adult human hippocampus. *Nature Medicine*, *4*(11), 1313–1317. doi:10.1038/3305 PMID:9809557

Ernst, C., Chen, E. S., & Turecki, G. (2009). Histone methylation and decreased expression of TrkB. T1 in orbital frontal cortex of suicide completers. *Molecular Psychiatry*, *14*(9), 830–832. doi:10.1038/mp.2009.35 PMID:19696771

Ernst, C., Nagy, C., Kim, S., Yang, J. P., Deng, X., Hellstrom, I. C., Choi, K. H., Gershenfeld, H., Meaney, M. J., & Turecki, G. (2011). Dysfunction of astrocyte connexins 30 and 43 in dorsal lateral prefrontal cortex of suicide completers. *Biological Psychiatry*, *70*(4), 312–319. doi:10.1016/j.biopsych.2011.03.038 PMID:21571253

Eroglu, C., & Barres, B. A. (2010). Regulation of synaptic connectivity by glia. *Nature*, *468*(7321), 223–231. doi:10.1038/nature09612 PMID:21068831

Eslahi, M., Dana, P. M., Asemi, Z., Hallajzadeh, J., Mansournia, M. A., & Yousefi, B. (2021). The effects of chitosan-based materials on glioma : Recent advances in its applications for diagnosis and treatment. *International Journal of Biological Macromolecules*, *168*, 124–129. doi:10.1016/j.ijbiomac.2020.11.180 PMID:33275978

Estera, L. A., Walsh, S. P., Headen, J. A., Williamson, R. E., & Kalinski, A. L. (2021). Neuroinflammation: Breaking barriers and bridging gaps. *Neuroscience Research*. doi:10.1016/j.neures.2021.11.001 PMID:34748905

Estévez, A. G., Crow, J. P., Sampson, J. B., Reiter, C., Zhuang, Y., Richardson, G. J., Tarpey, M. M., Barbeito, L., & Beckman, J. S. (1999). Induction of nitric oxide-dependent apoptosis in motor neurons by zinc-deficient superoxide dismutase. *Science, 286*(5449), 2498–2500. doi:10.1126cience.286.5449.2498 PMID:10617463

Etienne-Manneville, S. (2004). Cdc42—The centre of polarity. *Journal of Cell Science, 117*(8), 1291–1300. doi:10.1242/jcs.01115 PMID:15020669

Etiévant, A., Oosterhof, C., Bétry, C., Abrial, E., Novo-Perez, M., Rovera, R., Scarna, H., Devader, C., Mazella, J., Wegener, G., Sánchez, C., Dkhissi-Benyahya, O., Gronfier, C., Coizet, V., Beaulieu, J. M., Blier, P., Lucas, G., & Haddjeri, N. (2015). Astroglial control of the antidepressant-like effects of prefrontal cortex deep brain stimulation. *EBioMedicine, 2*(8), 898–908. doi:10.1016/j.ebiom.2015.06.023 PMID:26425697

Evin, G. (2016). Future therapeutics in Alzheimer's disease: Development status of BACE inhibitors. *BioDrugs, 30*(3), 173–194. doi:10.100740259-016-0168-3 PMID:27023706

Eyupoglu, I. Y., Hahnen, E., Trankle, C., Savaskan, N. E., Siebzehnrubl, F. A., Buslei, R., Lemke, D., Wick, W., Fahlbusch, R., & Blumcke, I. (2006). Experimental therapy of malignant gliomas using the inhibitor of histone deacetylase MS-275. *Molecular Cancer Therapeutics, 5*(5), 1248–1255. doi:10.1158/1535-7163.MCT-05-0533 PMID:16731757

Fabi, A., Vidiri, A., Carapella, C., Pace, A., Occhipinti, E., Caroli, F., & Cognetti, F. (2004). Bone metastasis from glioblastoma multiforme without central nervous system relapse: A case report. *Anticancer Research, 24*(4), 2563–2566. PMID:15330215

Fadeel, B., & Orrenius, S. (2005). Apoptosis: A basic biological phenomenon with wide-ranging implications in human disease. *Journal of Internal Medicine, 258*(6), 479–517. doi:10.1111/j.1365-2796.2005.01570.x PMID:16313474

Faideau, M., Kim, J., Cormier, K., Gilmore, R., Welch, M., Auregan, G., Dufour, N., Guillermier, M., Brouillet, E., Hantraye, P., Deglon, N., Ferrante, R. J., & Bonvento, G. (2010). in vivo expression of polyglutamine-expanded huntingtin by mouse striatal astrocytes impairs glutamate transport: A correlation with Huntington's disease subjects. *Human Molecular Genetics, 19*(15), 3053–3067. doi:10.1093/hmg/ddq212 PMID:20494921

Faisal, S. M., Castro, M. G., & Lowenstein, P. R. (2023). Adenovirus-mediated combined cytotoxic and immune-stimulatory gene therapy using dual-vector Ad-hCMV-TK and Ad-hCMV-Flt3L: translational explorations from rodents to glioma patients. *Molecular Therapy.*

Falcão, A. M., van Bruggen, D., Marques, S., Meijer, M., Jäkel, S., Agirre, E., Samudyata, Floriddia, E. M., Vanichkina, D. P., ffrench-Constant, C., Williams, A., Guerreiro-Cacais, A. O., & Castelo-Branco, G. (2018). Disease-specific oligodendrocyte lineage cells arise in multiple sclerosis. *Nature Medicine, 24*(12), 1837–1844. doi:10.103841591-018-0236-y

Falcone, C., McBride, E. L., Hopkins, W. D., Hof, P. R., Manger, P. R., Sherwood, C. C., Noctor, S. C., & Martínez-Cerdeño, V. (2022). Redefining varicose projection astrocytes in primates. *Glia, 70*(1), 145–154. doi:10.1002/glia.24093 PMID:34533866

Falcone, C., Wolf-Ochoa, M., Amina, S., Hong, T., Vakilzadeh, G., Hopkins, W. D., Hof, P. R., Sherwood, C. C., Manger, P. R., Noctor, S. C., & Martínez-Cerdeño, V. (2019). Cortical interlaminar astrocytes across the therian mammal radiation. *The Journal of Comparative Neurology, 527*(10), 1654–1674. doi:10.1002/cne.24605 PMID:30552685

Fang, I. M., Yang, C. M., & Yang, C. H. (2015). Chitosan oligosaccharides prevented retinal ischemia and reperfusion injury via reduced oxidative stress and inflammation in rats. *Experimental Eye Research, 130*, 38–50. doi:10.1016/j.exer.2014.12.001 PMID:25479043

Fantin, A., Vieira, J. M., Gestri, G., Denti, L., Schwarz, Q., Prykhozhij, S., Peri, F., Wilson, S. W., & Ruhrberg, C. (2010). Tissue macrophages act as cellular chaperones for vascular anastomosis downstream of VEGF-mediated endothelial tip cell induction. *Blood*, *116*(5), 829–840. doi:10.1182/blood-2009-12-257832 PMID:20404134

Faraco, G., Pittelli, M., Cavone, L., Fossati, S., Porcu, M., Mascagni, P., Fossati, G., Moroni, F., & Chiarugi, A. (2009). Histone deacetylase (HDAC) inhibitors reduce the glial inflammatory response in vitro and in vivo. *Neurobiology of Disease*, *36*(2), 269–279. doi:10.1016/j.nbd.2009.07.019 PMID:19635561

Färber, K., Pannasch, U., & Kettenmann, H. (2005). Dopamine and noradrenaline control distinct functions in rodent microglial cells. *Molecular and Cellular Neurosciences*, *29*(1), 128–138. doi:10.1016/j.mcn.2005.01.003 PMID:15866053

Farhy-Tselnicker, I., & Allen, N. J. (2018). Astrocytes, neurons, synapses: A tripartite view on cortical circuit development. *Neural Development*, *13*(1), 1–12. doi:10.118613064-018-0104-y PMID:29712572

Farhy-Tselnicker, I., van Casteren, A. C., Lee, A., Chang, V. T., Aricescu, A. R., & Allen, N. J. (2017). Astrocyte-secreted glypican 4 regulates release of neuronal pentraxin 1 from axons to induce functional synapse formation. *Neuron*, *96*(2), 428–445. doi:10.1016/j.neuron.2017.09.053 PMID:29024665

Farina, C., Aloisi, F., & Meinl, E. (2007). Astrocytes are active players in cerebral innate immunity. *Trends in Immunology*, *28*(3), 138–145. doi:10.1016/j.it.2007.01.005 PMID:17276138

Farizatto, K. L., & Baldwin, K. T. (2023). Astrocyte-synapse interactions during brain development. *Current Opinion in Neurobiology*, *80*, 102704. doi:10.1016/j.conb.2023.102704 PMID:36913751

Fatemi, S. H., Emamian, E. S., Sidwell, R. W., Kist, D. A., Stary, J. M., Earle, J. A., & Thuras, P. (2002). Human influenza viral infection in utero alters glial fibrillary acidic protein immunoreactivity in the developing brains of neonatal mice. *Molecular Psychiatry*, *7*(6), 633–640. doi:10.1038j.mp.4001046 PMID:12140787

Fatemi, S. H., Laurence, J. A., Araghi-Niknam, M., Stary, J. M., Schulz, S. C., Lee, S., & Gottesman, I. I. (2004). Glial fibrillary acidic protein is reduced in cerebellum of subjects with major depression, but not schizophrenia. *Schizophrenia Research*, *69*(2–3), 317–323. doi:10.1016/j.schres.2003.08.014 PMID:15469203

Faulkner, J. R., Herrmann, J. E., Woo, M. J., Tansey, K. E., Doan, N. B., & Sofroniew, M. V. (2004). Reactive astrocytes protect tissue and preserve function after spinal cord injury. *The Journal of Neuroscience : The Official Journal of the Society for Neuroscience*, *24*(9), 2143–2155. doi:10.1523/JNEUROSCI.3547-03.2004 PMID:14999065

Fawcett, J. W., & Asher, R. A. (1999). The glial scar and central nervous system repair. *Brain Research Bulletin*, *49*(6), 377–391. doi:10.1016/S0361-9230(99)00072-6 PMID:10483914

Fazil, M., Md, S., Haque, S., Kumar, M., Baboota, S., Sahni, J., & Ali, J. (2012). Development and evaluation of rivastigmine loaded chitosan nanoparticles for brain targeting. *European Journal of Pharmaceutical Sciences*, *47*(1), 6–15. doi:10.1016/j.ejps.2012.04.013 PMID:22561106

Fehlings, M. G., & Tighe, A. (2008). Spinal cord injury: The promise of translational research. *Neurosurgical Focus*, *25*(5), E1. doi:10.3171/FOC.2008.25.11.E1 PMID:18980469

Feizerfan, A., & Sheh, G. (2015). Transition from acute to chronic pain. *Continuing Education in Anaesthesia, Critical Care & Pain*, *15*(2), 98–102. doi:10.1093/bjaceaccp/mku044

Fellin, T. (2009). Communication between neurons and astrocytes: Relevance to the modulation of synaptic and network activity. *Journal of Neurochemistry*, *108*(3), 533–544. doi:10.1111/j.1471-4159.2008.05830.x PMID:19187090

Fellner, L., Irschick, R., Schanda, K., Reindl, M., Klimaschewski, L., Poewe, W., Wenning, G. K., & Stefanova, N. (2013). *Toll-like receptor 4 is required for α-synuclein dependent activation of microglia and astroglia.* Wiley. doi:10.1002/glia.22437

Felsky, D., Roostaei, T., Nho, K., Risacher, S. L., Bradshaw, E. M., Petyuk, V., Schneider, J. A., Saykin, A., Bennett, D. A., & De Jager, P. L. (2019). Neuropathological correlates and genetic architecture of microglial activation in elderly human brain. *Nature Communications, 10*(1), 409. doi:10.103841467-018-08279-3 PMID:30679421

Fenn, K. M., Evans, S. B., McCorkle, R., DiGiovanna, M. P., Pusztai, L., Sanft, T., Hofstatter, E. W., Killelea, B. K., Knobf, M. T., Lannin, D. R., Abu-Khalaf, M., Horowitz, N. R., & Chagpar, A. B. (2014). Impact of financial burden of cancer on survivors' quality of life. *Journal of Oncology Practice / American Society of Clinical Oncology, 10*(5), 332–338. doi:10.1200/JOP.2013.001322 PMID:24865220

Fernández-Arjona, M. del M., León-Rodríguez, A., López-Ávalos, M. D., & Grondona, J. M. (2021). Microglia activated by microbial neuraminidase contributes to ependymal cell death. *Fluids and Barriers of the CNS, 18*(1), 15. doi:10.118612987-021-00249-0 PMID:33757539

Ferraiuolo, L., Meyer, K., Sherwood, T. W., Vick, J., Likhite, S., Frakes, A., Miranda, C. J., Braun, L., Heath, P. R., Pineda, R., Beattie, C. E., Shaw, P. J., Askwith, C. C., McTigue, D., & Kaspar, B. K. (2016). Oligodendrocytes contribute to motor neuron death in ALS via SOD1-dependent mechanism. *Proceedings of the National Academy of Sciences of the United States of America, 113*(42), E6496–E6505. doi:10.1073/pnas.1607496113 PMID:27688759

Ferrari, C. C., Pott Godoy, M. C., Tarelli, R., Chertoff, M., Depino, A. M., & Pitossi, F. J. (2006). Progressive neurodegeneration and motor disabilities induced by chronic expression of IL-1β in the substantia nigra. *Neurobiology of Disease, 24*(1), 183–193. doi:10.1016/j.nbd.2006.06.013 PMID:16901708

Ferreira, H. B., Melo, T., Monteiro, A., Paiva, A., Domingues, P., & Domingues, M. R. (2021). Serum phospholipidomics reveals altered lipid profile and promising biomarkers in multiple sclerosis. *Archives of Biochemistry and Biophysics, 697*, 108672. doi:10.1016/j.abb.2020.108672

Ferreira, H. B., Neves, B., Guerra, I. M., Moreira, A., Melo, T., Paiva, A., & Domingues, M. R. (2020). An overview of lipidomic analysis in different human matrices of multiple sclerosis. *Multiple Sclerosis and Related Disorders, 44*, 102189. doi:10.1016/j.msard.2020.102189

Ferreira, S., Pitman, K. A., Wang, S., Summers, B. S., Bye, N., Young, K. M., & Cullen, C. L. (2020). Amyloidosis is associated with thicker myelin and increased oligodendrogenesis in the adult mouse brain. *CullenJournal of Neuroscience Research. Journal of Neuroscience Research, 98*(10), 1905–1932. doi:10.1002/jnr.24672 PMID:32557778

Ferri, C. P., Prince, M., Brayne, C., Brodaty, H., Fratiglioni, L., Ganguli, M., Hall, K., Hasegawa, K., Hendrie, H., Huang, Y., Jorm, A., Mathers, C., Menezes, P. R., Rimmer, E., & Scazufca, M. (2005). Global prevalence of dementia: A Delphi consensus study. *Lancet, 366*(9503), 2112–2117. doi:10.1016/S0140-6736(05)67889-0 PMID:16360788

Feuerstein, G. Z., Wang, X., & Barone, F. C. (1998). The Role of Cytokines in the Neuropathology of Stroke and Neurotrauma. *Neuroimmunomodulation, 5*(3–4), 143–159. doi:10.1159/000026331 PMID:9730680

Feyissa, A. M., Chandran, A., Stockmeier, C. A., & Karolewicz, B. (2009). Reduced levels of NR2A and NR2B subunits of NMDA receptor and PSD-95 in the prefrontal cortex in major depression. *Progress in Neuro-Psychopharmacology & Biological Psychiatry, 33*(1), 70–75. doi:10.1016/j.pnpbp.2008.10.005 PMID:18992785

Fields, R. D. (2008). 'White Matter in Learning, Cognition and Psychiatric Disorders'. *Trends in Neurosciences, 31*(7), 361–370. doi:10.1016/j.tins.2008.04.001 PMID:18538868

Fields, R. D. (2015). 'A New Mechanism of Nervous System Plasticity: Activity-Dependent Myelination'. *Nature Reviews. Neuroscience, 16*(12), 756–767. doi:10.1038/nrn4023 PMID:26585800

Fields, R. D., Araque, A., Johansen-Berg, H., Lim, S. S., Lynch, G., Nave, K. A., Nedergaard, M., Perez, R., Sejnowski, T., & Wake, H. (2014). Glial biology in learning and cognition. *The Neuroscientist, 20*(5), 426–431. doi:10.1177/1073858413504465 PMID:24122821

Fields, R. D., & Stevens-Graham, B. (2002). New Insights into Neuron-Glia Communication. *Science, 298*(5593), 556–562. doi:10.1126cience.298.5593.556 PMID:12386325

Filipi, T., Hermanova, Z., Tureckova, J., Vanatko, O., & Anderova, A. M. (2020). Glial Cells-The Strategic Targets in Amyotrophic Lateral Sclerosis Treatment. *Journal of Clinical Medicine, 9*(1), 261. doi:10.3390/jcm9010261 PMID:31963681

Filippi, M., Bozzali, M., Rovaris, M., Gonen, O., Kesavadas, C., & Ghezzi, A. (2003). Evidence for widespread axonal damage at the earliest clinical stage of multiple sclerosis. *Brain, 126*(2), 433–437. doi:10.1093/brain/awg038 PMID:12538409

Filippo, M. D., Sarchielli, P., Picconi, B., & Calabresi, P. (2008). 'Neuroinflammation and Synaptic Plasticity: Theoretical Basis for a Novel, Immune-Centred, Therapeutic Approach to Neurological Disorders'. *Trends in Pharmacological Sciences, 29*(8), 402–412. doi:10.1016/j.tips.2008.06.005 PMID:18617277

Filosa, J. A., & Iddings, J. A. (2013). 'Astrocyte Regulation of Cerebral Vascular Tone'. *American Journal of Physiology. Heart and Circulatory Physiology, 305*(5), H609–H619. doi:10.1152/ajpheart.00359.2013 PMID:23792684

Filosa, J. A., Morrison, H. W., Iddings, J. A., Du, W., & Kim, K. J. (2015). Beyond neurovascular coupling, the role of astrocytes in the regulation of vascular tone. *Neuroscience, 323*, 96–109. doi:10.1016/j.neuroscience.2015.03.064 PMID:25843438

Findley, C. A., Bartke, A., Hascup, K. N., & Hascup, E. R. (2019). Amyloid Beta-Related Alterations to Glutamate Signaling Dynamics During Alzheimer's Disease Progression. *ASN Neuro, 11*, 175909141985554. doi:10.1177/1759091419855541 PMID:31213067

Fink, T., Davey, D. F., & Ansselin, A. D. (1999). Glutaminergic and adrenergic receptors expressed on adult guinea pig Schwann cells in vitro. *Canadian Journal of Physiology and Pharmacology, 77*(3), 204–210. doi:10.1139/y99-008 PMID:10535694

Finnie, P. S. B., & Nader, K. (2012). The role of metaplasticity mechanisms in regulating memory destabilization and reconsolidation. *Neuroscience and Biobehavioral Reviews, 36*(7), 1667–1707. doi:10.1016/j.neubiorev.2012.03.008 PMID:22484475

Fischer, M. T., Sharma, R., Lim, J. L., Haider, L., Frischer, J. M., Drexhage, J., Don Mahad, M. B., van Horssen, J., & Lassmann, H. (2012). 'NADPH Oxidase Expression in Active Multiple Sclerosis Lesions in Relation to Oxidative Tissue Damage and Mitochondrial Injury'. *Brain, 135*(3), 886–899. doi:10.1093/brain/aws012 PMID:22366799

Fog, C. K., & Kirkegaard, T. (2019). Animal models for Niemann-Pick type C: Implications for drug discovery & development. *Expert Opinion on Drug Discovery, 14*(5), 499–509. doi:10.1080/17460441.2019.1588882 PMID:30887840

Fonalledas-Perelló, M. A., Valero-Politi, J., Lizarraga-Dallo, M. A., & Segura-Cardona, R. (2008). The cerebrospinal fluid lactate is decreased in early stages of multiple sclerosis. *Puerto Rico Health Sciences Journal, 27*(2).

Fonollosa, A., & Tourlet, C. (2023).Pathogenèse et évolution de la sclérose latérale amyotrophique, *Kinésithérapie, la Revue, 23*. doi:10.1016/j.kine.2023.01.008

Fontenas, L., Welsh, T. G., Piller, M., Coughenour, P., Gandhi, A. V., Prober, D. A., & Kucenas, S. (2019). The neuromodulator adenosine regulates oligodendrocyte migration at motor exit point transition zones. *Cell Reports*, *27*(1), 115–128. doi:10.1016/j.celrep.2019.03.013 PMID:30943395

Fornaro, M., Rocchi, G., Escelsior, A., Contini, P., Ghio, M., Colicchio, S., De Berardis, D., Amore, M., Fornaro, P., & Martino, M. (2013). VEGF plasma level variations in duloxetine-treated patients with major depression. *Journal of Affective Disorders*, *151*(2), 590–595. doi:10.1016/j.jad.2013.06.055 PMID:23871390

Fossati, G., Pozzi, D., Canzi, A., Mirabella, F., Valentino, S., Morini, R., Ghirardini, E., Filipello, F., Moretti, M., Gotti, C., Annis, D. S., Mosher, D. F., Garlanda, C., Bottazzi, B., Taraboletti, G., Mantovani, A., Matteoli, M., & Menna, E. (2019). Pentraxin 3 regulates synaptic function by inducing AMPA receptor clustering via ECM remodeling and β1-integrin. *The EMBO Journal*, *38*(1), e99529. doi:10.15252/embj.201899529 PMID:30396995

Frakes, A. E., Ferraiuolo, L., Haidet-Phillips, A. M., Schmelzer, L., Braun, L., Miranda, C. J., Ladner, K. J., Bevan, A. K., Foust, K. D., Godbout, J. P., Popovich, P. G., Guttridge, D. C., & Kaspar, B. K. (2014). Microglia induce motor neuron death via the classical NF-κB pathway in amyotrophic lateral sclerosis. *Neuron*, *81*(5), 1009–1023. doi:10.1016/j.neuron.2014.01.013 PMID:24607225

Franceschi, C., & Campisi, J. (2014). 'Chronic Inflammation (Inflammaging) and Its Potential Contribution to Age-Associated Diseases'. *The Journals of Gerontology. Series A, Biological Sciences and Medical Sciences*, *69*(Suppl_1), S4–S9. doi:10.1093/gerona/glu057 PMID:24833586

Franceschi, E., Cavallo, G., Lonardi, S., Magrini, E., Tosoni, A., Grosso, D., Scopece, L., Blatt, V., Urbini, B., Pession, A., Tallini, G., Crinò, L., & Brandes, A. A. (2007). Gefitinib in patients with progressive high-grade gliomas: A multicentre phase II study by Gruppo Italiano Cooperativo di Neuro-Oncologia (GICNO). *British Journal of Cancer*, *96*(7), 1047–1051. doi:10.1038j.bjc.6603669 PMID:17353924

Franklin, R. J. M., & ffrench-Constant, C. (2008). 'Remyelination in the CNS: From Biology to Therapy'. *Nature Reviews. Neuroscience*, *9*(11), 839–855. doi:10.1038/nrn2480 PMID:18931697

Fraser, C. L., & Arieff, A. I. (1997). Epidemiology, Pathophysiology, and Management of Hyponatremic Encephalopathy. *The American Journal of Medicine*, *102*(1), 67–77. doi:10.1016/S0002-9343(96)00274-4 PMID:9209203

Fray, A. E., Ince, P. G., Banner, S. J., Milton, I. D., Usher, P. A., Cookson, M. R., & Shaw, P. J. (1998). The expression of the glial glutamate transporter protein EAAT2 in motor neuron disease: An immunohistochemical study. *The European Journal of Neuroscience*, *10*(8), 2481–2489. doi:10.1046/j.1460-9568.1998.00273.x PMID:9767379

Frederick, R. T. (2011). Current concepts in the pathophysiology and management of hepatic encephalopathy. *Gastroenterology & Hepatology*, *7*(4), 222. PMID:21857820

Freeman, L. C., & Ting, J. P. (2016). The pathogenic role of the inflammasome in neurodegenerative diseases. *Journal of Neurochemistry*, *136*(S1, Suppl 1), 29–38. doi:10.1111/jnc.13217 PMID:26119245

Friedman, A. J., Phan, J., Schairer, D. O., Champer, J., Qin, M., Pirouz, A., Blecher-Paz, K., Oren, A., Liu, P. T., Modlin, R. L., & Kim, J. (2013). Antimicrobial and Anti-Inflammatory Activity of Chitosan–Alginate Nanoparticles: A Targeted Therapy for Cutaneous Pathogens. *The Journal of Investigative Dermatology*, *133*(5), 1231–1239. doi:10.1038/jid.2012.399 PMID:23190896

Friess, W. (1998). Collagen–biomaterial for drug delivery. *European Journal of Pharmaceutics and Biopharmaceutics*, *45*(2), 113–136. doi:10.1016/S0939-6411(98)00017-4 PMID:9704909

Frigerio, C. S., Wolfs, L., Fattorelli, N., Thrupp, N., Voytyuk, I., Schmidt, I., Mancuso, R., Chen, W.-T., Woodbury, M. E., & Srivastava, G. (2019). The major risk factors for Alzheimer's disease: Age, sex, and genes modulate the microglia response to Aβ plaques. *Cell Reports*, *27*(4), 1293–1306. doi:10.1016/j.celrep.2019.03.099 PMID:31018141

Frisardi, V., Santamato, A., & Cheeran, B. (2016). Parkinson's disease: new insights into pathophysiology and rehabilitative approaches. Park. Dis.

Frisoni, G. B., Altomare, D., Thal, D. R., Ribaldi, F., van der Kant, R., Ossenkoppele, R., Blennow, K., Cummings, J., van Duijn, C., Nilsson, P. M., Dietrich, P.-Y., Scheltens, P., & Dubois, B. (2022). The probabilistic model of Alzheimer disease : The amyloid hypothesis revised. *Nature Reviews. Neuroscience*, *23*(1), 53–66. doi:10.103841583-021-00533-w PMID:34815562

Fritsche, S., Wang, X., Antioxidants, C. J. (2017). Recent advances in our understanding of tocopherol biosynthesis in plants: an overview of key genes, functions, and breeding of vitamin E improved crops. *Antioxidants, mdpi.Com*. doi:10.3390/antiox6040099

Frost, G. R., & Li, Y. M. (2017). The role of astrocytes in amyloid production and Alzheimer's disease. *Open Biology*, *7*(12), 170228. doi:10.1098/rsob.170228 PMID:29237809

Fu, H., Li, B., Hertz, L., & Peng, L. (2012). Contributions in Astrocytes of SMIT1/2 and HMIT to Myo-Inositol Uptake at Different Concentrations and PH. *Neurochemistry International*, *61*(2), 187–194. doi:10.1016/j.neuint.2012.04.010 PMID:22564531

Führmann, T., & Shoichet, M. S. (2018). The role of biomaterials in overcoming barriers to regeneration in the central nervous system. *Biomedical Materials (Bristol, England)*, *13*(5), 050201. doi:10.1088/1748-605X/aac2f6 PMID:29864020

Fujii, S., Tanaka, K. F., Ikenaka, K., & Yamazaki, Y. (2014). Increased adenosine levels in mice expressing mutant glial fibrillary acidic protein in astrocytes fail induction of LTP reversal (depotentiation) in hippocampal CA1 neurons. *Brain Research*, *1578*, 1–13. doi:10.1016/j.brainres.2014.07.005 PMID:25017946

Fultz, M. J., Barber, S. A., Dieffenbach, C. W., & Vogel, S. N. (1993). Induction of IFN-γ in macrophages by lipopolysaccharide. *International Immunology*, *5*(11), 1383–1392. doi:10.1093/intimm/5.11.1383 PMID:8260452

Fünfschilling, U., Supplie, L. M., Mahad, D., Boretius, S., Saab, A. S., Edgar, J., Brinkmann, B. G., Kassmann, C. M., Tzvetanova, I. D., Möbius, W., Diaz, F., Meijer, D., Suter, U., Hamprecht, B., Sereda, M. W., Moraes, C. T., Frahm, J., Goebbels, S., & Nave, K.-A. (2012). Glycolytic oligodendrocytes maintain myelin and long-term axonal integrity. *Nature*, *485*(7399), 517–521. doi:10.1038/nature11007 PMID:22622581

Furman, J. L., Sama, D. M., Gant, J. C., Beckett, T. L., Murphy, M. P., Bachstetter, A. D., Van Eldik, L. J., & Norris, C. M. (2012). Targeting Astrocytes Ameliorates Neurologic Changes in a Mouse Model of Alzheimer's Disease. *The Journal of Neuroscience : The Official Journal of the Society for Neuroscience*, *32*(46), 16129–16140. doi:10.1523/JNEUROSCI.2323-12.2012 PMID:23152597

Furnari, F. B., Cloughesy, T. F., Cavenee, W. K., & Mischel, P. S. (2015). Heterogeneity of epidermal growth factor receptor signaling networks in glioblastoma. *Nature Reviews. Cancer*, *15*(5), 302–310. doi:10.1038/nrc3918 PMID:25855404

Furnari, F. B., Fenton, T., Bachoo, R. M., Mukasa, A., Stommel, J. M., Stegh, A., Hahn, W. C., Ligon, K. L., Louis, D. N., Brennan, C., Chin, L., DePinho, R. A., & Cavenee, W. K. (2007). Malignant astrocytic glioma: Genetics, biology, and paths to treatment. *Genes & Development*, *21*(21), 2683–2710. doi:10.1101/gad.1596707 PMID:17974913

Gabriel, S., Njunting, M., Pomper, J. K., Merschhemke, M., Sanabria, E. R. G., Eilers, A., Kivi, A., Zeller, M., Meencke, H.-J., Cavalheiro, E. A., Heinemann, U., & Lehmann, T.-N. (2004). Stimulus and potassium-induced epileptiform activity in the human dentate gyrus from patients with and without hippocampal sclerosis. *The Journal of Neuroscience : The Official Journal of the Society for Neuroscience*, *24*(46), 10416–10430. doi:10.1523/JNEUROSCI.2074-04.2004 PMID:15548657

Gage, F. H. (2000). Mammalian neural stem cells. *Science*, *287*(5457), 1433–1438. doi:10.1126cience.287.5457.1433 PMID:10688783

Gagne, J. J., & Power, M. C. (2010). *Anti-inflammatory drugs and risk of Parkinson disease A meta-analysis*. Neurology. www.neurology.org

Gale, S. C., Gao, L., Mikacenic, C., Coyle, S. M., Rafaels, N., Murray Dudenkov, T., Madenspacher, J. H., Draper, D. W., Ge, W., Aloor, J. J., Azzam, K. M., Lai, L., Blackshear, P. J., Calvano, S. E., Barnes, K. C., Lowry, S. F., Corbett, S., Wurfel, M. M., & Fessler, M. B. (2014). APOepsilon4 is associated with enhanced *in vivo* innate immune responses in human subjects. *The Journal of Allergy and Clinical Immunology*, *134*(1), 127–134. doi:10.1016/j.jaci.2014.01.032 PMID:24655576

Gallo, J., Raska, M., Kriegova, E., & Goodman, S. B. (2017). Inflammation and its resolution and the musculoskeletal system. *Journal of Orthopaedic Translation*, *10*, 52–67. doi:10.1016/j.jot.2017.05.007 PMID:28781962

Gans, H., Mori, K., Lindsey, E., Kaster, B., Richter, D., Quinlan, R., Dineen, P. A., & Tan, B. H. (1971). Septicemia as a Manifestation of Acute Liver Failure. *Surgery, Gynecology & Obstetrics*, *132*(5), 783–790. PMID:5555405

Gao, HM, Liu, B., Zhang, W. et Hong, JS (2003). Nouvelle thérapie anti-inflammatoire pour la maladie de Parkinson. *Tendances des sciences pharmacologiques, 24*(8), 395-401.

Gao, H. (2017). Perspectives on dual targeting delivery systems for brain tumors. *Journal of Neuroimmune Pharmacology*, *12*(1), 6–16. doi:10.100711481-016-9687-4 PMID:27270720

Gao, H. M., Liu, B., Zhang, W., & Hong, J. S. (2003). Novel anti-inflammatory therapy for Parkinson's disease. *Trends in Pharmacological Sciences*, *24*(8), 395–401. doi:10.1016/S0165-6147(03)00176-7 PMID:12915048

Gao, Y.-J., & Ji, R.-R. (2010). Targeting astrocyte signaling for chronic pain. *Neurotherapeutics; the Journal of the American Society for Experimental NeuroTherapeutics*, *7*(4), 482–493. doi:10.1016/j.nurt.2010.05.016 PMID:20880510

Gao, Y., Ma, J., Tang, J., Liang, X., Huang, C., Wang, S., Chen, L., Wang, F., Tan, C., Chao, F., Zhang, L., Qiu, X., Luo, Y.-M., Xiao, Q., Du, L., Xiao, Q., & Tang, Y. (2017). White matter atrophy and myelinated fiber disruption in a rat model of depression. *The Journal of Comparative Neurology*, *525*(8), 1922–1933. doi:10.1002/cne.24178 PMID:28118485

Garbes, L., Riessland, M., & Wirth, B. (n.d.). *Histone Acetylation as a Potential Therapeutic Target in Motor Neuron Degenerative Diseases*.

Garbuzova-Davis, S., Hernandez-Ontiveros, D. G., Rodrigues, M. C. O., Haller, E., Frisina-Deyo, A., Mirtyl, S., Sallot, S., Saporta, S., Borlongan, C. V., & Sanberg, P. R. (2012). Impaired blood-brain/spinal cord barrier in ALS patients. *Brain Research*, *1469*, 114–128. doi:10.1016/j.brainres.2012.05.056 PMID:22750125

Garcia-Larrea, L., & Peyron, R. (2013). Pain matrices and neuropathic pain matrices. *Reviews in Pain*, *154*(Suppl 1), S29–S43. doi:10.1016/j.pain.2013.09.001 PMID:24021862

Garden, G. A., & Möller, T. (2006). Microglia Biology in Health and Disease. *Journal of Neuroimmune Pharmacology*, *1*(2), 127–137. doi:10.100711481-006-9015-5 PMID:18040779

Gardner, J., & Ghorpade, A. (2003). Tissue inhibitor of metalloproteinase (TIMP)-1: The TIMPed balance of matrix metalloproteinases in the central nervous system. *Journal of Neuroscience Research, 74*(6), 801–806. doi:10.1002/jnr.10835 PMID:14648584

Garg, N., & Smith, T. W. (2015). An update on immunopathogenesis, diagnosis, and treatment of multiple sclerosis. *Brain and Behavior, 5*(9), e00362. doi:10.1002/brb3.362

Garrison, C. J., Dougherty, P. M., Kajander, K. C., & Carlton, S. M. (1991). Staining of glial fibrillary acidic protein (GFAP) in lumbar spinal cord increases following a sciatic nerve constriction injury. *Brain Research, 565*(1), 1–7. doi:10.1016/0006-8993(91)91729-K PMID:1723019

Garros-Regulez, L., Aldaz, P., Arrizabalaga, O., Moncho-Amor, V., Carrasco-Garcia, E., Manterola, L., Moreno-Cugnon, L., Barrena, C., Villanua, J., Ruiz, I., Pollard, S., Lovell-Badge, R., Sampron, N., Garcia, I., & Matheu, A. (2016). mTOR inhibition decreases SOX2-SOX9 mediated glioma stem cell activity and temozolomide resistance. *Expert Opinion on Therapeutic Targets, 20*(4), 393–405. doi:10.1517/14728222.2016.1151002 PMID:26878385

Gartziandia, O., Herrán, E., Ruiz-Ortega, J. A., Miguelez, C., Igartua, M., Lafuente, J. V., Pedraz, J. L., Ugedo, L., & Hernández, R. M. (2016). Intranasal administration of chitosan-coated nanostructured lipid carriers loaded with GDNF improves behavioral and histological recovery in a partial lesion model of Parkinson's disease. *Journal of Biomedical Nanotechnology, 12*(12), 2220–2280. doi:10.1166/jbn.2016.2313 PMID:29372975

Gauthier, L. R., Charrin, B. C., Borrell-Pagès, M., Dompierre, J. P., Rangone, H., Cordelières, F. P., De Mey, J., MacDonald, M. E., Leßmann, V., Humbert, S., & Saudou, F. (2004). Huntingtin controls neurotrophic support and survival of neurons by enhancing BDNF vesicular transport along microtubules. *Cell, 118*(1), 127–138. doi:10.1016/j.cell.2004.06.018 PMID:15242649

Ge, H., Mu, L., Jin, L., Yang, C., Chang, Y., Long, Y., DeLeon, G., Deleyrolle, L., Mitchell, D. A., Kubilis, P. S., Lu, D., Qi, J., Gu, Y., Lin, Z., & Huang, J. (2017). Tumor associated CD70 expression is involved in promoting tumor migration and macrophage infiltration in GBM. *International Journal of Cancer, 141*(7), 1434–1444. doi:10.1002/ijc.30830 PMID:28612394

Geloso, M. C., Corvino, V., Marchese, E., Serrano, A., Michetti, F., & D'Ambrosi, N. (2017). The Dual Role of Microglia in ALS: Mechanisms and Therapeutic Approaches. *Frontiers in Aging Neuroscience, 9*, 242. doi:10.3389/fnagi.2017.00242 PMID:28790913

Gelpi, E., Rahimi, J., Klotz, S., Schmid, S., Ricken, G., Forcen-vega, S., Budka, H., & Kovacs, G. G. (2020). The autophagic marker p62 highlights Alzheimer type II astrocytes in metabolic / hepatic encephalopathy. *Neuropathology, 40*(4), 358–366. doi:10.1111/neup.12660 PMID:32483828

Gemma, C., & Bachstetter, A. D. (2013). The role of microglia in adult hippocampal neurogenesis. *Frontiers in Cellular Neuroscience, 7*(NOV), 1–5. doi:10.3389/fncel.2013.00229 PMID:24319411

Genoud, C., Quairiaux, C., Steiner, P., Hirling, H., Welker, E., & Knott, G. W. (2006). Plasticity of astrocytic coverage and glutamate transporter expression in adult mouse cortex. *PLoS Biology, 4*(11), e343. doi:10.1371/journal.pbio.0040343 PMID:17048987

Gensert, J. M., Baranova, O. V., Weinstein, D. E., & Ratan, R. R. (2007). CD81, a cell cycle regulator, is a novel target for histone deacetylase inhibition in glioma cells. *Neurobiology of Disease, 26*(3), 671–680. doi:10.1016/j.nbd.2007.03.008 PMID:17481908

George, J., Gonçalves, F. Q., Cristóvão, G., Rodrigues, L., Meyer Fernandes, J. R., Gonçalves, T., Cunha, R. A., & Gomes, C. A. (2015). Different danger signals differently impact on microglial proliferation through alterations of ATP release and extracellular metabolism. *Glia*, *63*(9), 1636–1645. doi:10.1002/glia.22833 PMID:25847308

Gerhard, A., Pavese, N., Hotton, G., Turkheimer, F., Es, M., Hammers, A., Eggert, K., Oertel, W., Banati, R. B., & Brooks, D. J. (2006). In vivo imaging of microglial activation with PET in idiopathic Parkinson's disease. *Neurobiology of Disease*, *21*(2), 404–412. doi:10.1016/j.nbd.2005.08.002 PMID:16182554

Ghézali, G., Calvo, C., Pillet, L., Llense, F., Ezan, P., Pannasch, U., Bemelmans, A., Etienne Manneville, S., & Rouach, N. (2018). Connexin 30 controls astroglial polarization during postnatal brain development. *Development (Cambridge, England)*, *145*(4), dev155275. doi:10.1242/dev.155275 PMID:29475972

Ghézali, G., Vasile, F., Curry, N., Fantham, M., Cheung, G., Ezan, P., Cohen-Salmon, M., Kaminski, C., & Rouach, N. (2020). Neuronal Activity Drives Astroglial Connexin 30 in Perisynaptic Processes and Shapes Its Functions. *Cerebral Cortex (New York, N.Y.)*, *30*(2), 753–766. doi:10.1093/cercor/bhz123 PMID:31271200

Giacci, M. K., Bartlett, C. A., Smith, N. M., Iyer, K. S., Toomey, L. M., Jiang, H., Guagliardo, P., Kilburn, M. R., & Fitzgerald, M. (2018). Oligodendroglia Are Particularly Vulnerable to Oxidative Damage after Neurotrauma In Vivo. *The Journal of Neuroscience : The Official Journal of the Society for Neuroscience*, *38*(29), 6491–6504. doi:10.1523/JNEUROSCI.1898-17.2018 PMID:29915135

Gibson, G. E., Zimber, A., Krook, L., Richardson, E. P. Jr, & Visek, W. J. (1974). Brain histology and behavior of mice injected with urease. *Journal of Neuropathology and Experimental Neurology*, *33*(2), 201–211. doi:10.1097/00005072-197404000-00001 PMID:4822569

Giehrl-Schwab, J., Giesert, F., Rauser, B., Lao, C. L., Hembach, S., Lefort, S., Ibarra, I. L., Koupourtidou, C., Luecken, M. D., Truong, D. J., Fischer-Sternjak, J., Masserdotti, G., Prakash, N., Ninkovic, J., Hölter, S. M., Vogt Weisenhorn, D. M., Theis, F. J., Götz, M., & Wurst, W. (2022). Parkinson's disease motor symptoms rescue by CRISPRa-reprogramming astrocytes into GABAergic neurons. *EMBO Molecular Medicine*, *14*(5), e14797. doi:10.15252/emmm.202114797 PMID:35373464

Gilani, A. I., Chohan, M. O., Inan, M., Schobel, S. A., Chaudhury, N. H., Samuel, P., Chuhma, N., Glickstein, S., Merker, R. J., Xu, Q., Small, S. A., Anderson, S. A., Ross, M. E., & Moore, H. (2014). Interneuron precursor transplants in adult hippocampus reverse psychosis-relevant features in a mouse model of hippocampal disinhibition. *Proceedings of the National Academy of Sciences of the United States of America*, *111*(20), 7450–7455. doi:10.1073/pnas.1316488111 PMID:24794528

Gilkes, J. A., & Heldermon, C. D. (2014). Mucopolysaccharidosis III (Sanfilippo Syndrome)-disease presentation and experimental therapies. *Pediatric Endocrinology Reviews; PER*, *12*, 133–140. PMID:25345095

Gilmore, S. A. (1975). Proliferation of non-neuronal cells in spinal cords of irradiated, immature rats following transection of the sciatic nerve. *The Anatomical Record*, *181*(4), 799–811. doi:10.1002/ar.1091810411 PMID:47232

Ginhoux, F., Greter, M., Leboeuf, M., Nandi, S., See, P., Gokhan, S., Mehler, M. F., Conway, S. J., Ng, L. G., Stanley, E. R., Samokhvalov, I. M., & Merad, M. (2010). Fate mapping analysis reveals that adult microglia derive from primitive macrophages. *Science*, *330*(6005), 841–845. doi:10.1126cience.1194637 PMID:20966214

Ginhoux, F., & Merad, M. (2011). Les cellules de la microglie - Leurs origines extra-embryonnaires enfin révélées [Microglia arise from extra-embryonic yolk sac primitive progenitors]. *Medecine sciences. Ms*, *27*(8-9), 719–724. doi:10.1051/medsci/2011278013 PMID:21880259

Ginhoux, F., & Prinz, M. (2015). 'Origin of Microglia: Current Concepts and Past Controversies'. *Cold Spring Harbor Perspectives in Biology*, *7*(8), a020537. doi:10.1101/cshperspect.a020537 PMID:26134003

Gipson, C. D., & Olive, M. F. (2017). Structural and functional plasticity of dendritic spines–root or result of behavior? *Genes Brain & Behavior*, *16*(1), 101–117. doi:10.1111/gbb.12324 PMID:27561549

Gittins, R. A., & Harrison, P. J. (2011). A morphometric study of glia and neurons in the anterior cingulate cortex in mood disorder. *Journal of Affective Disorders*, *133*(1–2), 328–332. doi:10.1016/j.jad.2011.03.042 PMID:21497910

Giussani, P., Prinetti, A., & Tringali, C. (2021). The role of Sphingolipids in myelination and myelin stability and their involvement in childhood and adult demyelinating disorders. *Journal of Neurochemistry*, *156*(4), 403–414. doi:10.1111/jnc.15133

Glaser, T., Resende, R. R., & Ulrich, H. (2013). Implications of purinergic receptor-mediated intracellular calcium transients in neural differentiation. *Cell Communication and Signaling*, *11*(1), 12. doi:10.1186/1478-811X-11-12 PMID:23414261

Glass, C., Saijo, K., Winner, B., & Marchetto. (2010). *Mechanisms underlying inflammation in neurodegeneration.* Cell.com. https://www.cell.com/fulltext/S0092-8674(10)00168-6

Glezer, I., Zekki, H., Scavone, C., & Rivest, S. (2003). Modulation of the innate immune response by NMDA receptors has neuropathological consequences. *The Journal of Neuroscience : The Official Journal of the Society for Neuroscience*, *23*(35), 11094–11103. doi:10.1523/JNEUROSCI.23-35-11094.2003 PMID:14657167

Gnavi, S., Barwig, C., Freier, T., Haastert-Talini, K., Grothe, C., & Geuna, S. (2013). The use of chitosan-based scaffolds to enhance regeneration in the nervous system. *International Review of Neurobiology*, *109*, 1–62. doi:10.1016/B978-0-12-420045-6.00001-8 PMID:24093605

Gober, R., Ardalan, M., Shiadeh, S. M. J., Duque, L., Garamszegi, S. P., Ascona, M., Barreda, A., Sun, X., Mallard, C., & Vontell, R. T. (2022). Microglia activation in postmortem brains with schizophrenia demonstrates distinct morphological changes between brain regions. *Brain Pathology (Zurich, Switzerland)*, *32*(1), e13003. doi:10.1111/bpa.13003 PMID:34297453

Goenaga, J., Araque, A., Kofuji, P., & Herrera Moro Chao, D. (2023). Calcium signaling in astrocytes and gliotransmitter release. *Frontiers in Synaptic Neuroscience*, *15*, 1138577. doi:10.3389/fnsyn.2023.1138577 PMID:36937570

Golarai, G., Greenwood, A., Feeney, D., & Connor, J. (2008). Physiological and structural evidence for hippocampal involvement in persistent seizure susceptibility after traumatic brain injury. *The Journal of Neuroscience : The Official Journal of the Society for Neuroscience*, *21*(21), 8523–8537. doi:10.1523/JNEUROSCI.21-21-08523.2001 PMID:11606641

Goldman, S. M. (2014). Environmental toxins and Parkinson's disease. *Annual Review of Pharmacology and Toxicology*, *54*(1), 141–164. doi:10.1146/annurev-pharmtox-011613-135937 PMID:24050700

Gomes, C., Ferreira, R., George, J., Sanches, R., Rodrigues, D. I., Gonçalves, N., & Cunha, R. A. (2013). Activation of microglial cells triggers a release of brain-derived neurotrophic factor (BDNF) inducing their proliferation in an adenosine A2A receptor-dependent manner: A2A receptor blockade prevents BDNF release and proliferation of microglia. *Journal of Neuroinflammation*, *10*(1), 1–13. doi:10.1186/1742-2094-10-16 PMID:23363775

Gomes-Leal, W. (2012). Microglial physiopathology: How to explain the dual role of microglia after acute neural disorders? *Brain and Behavior*, *2*(3), 345–356. doi:10.1002/brb3.51 PMID:22741103

Gómez-Benito, M., Granado, N., García-Sanz, P., Michel, A., Dumoulin, M., & Moratalla, R. (2020). 'Modeling Parkinson's Disease With the Alpha-Synuclein Protein'. *Frontiers in Pharmacology*, *11*, 356. https://www.frontiersin.org/articles/10.3389/fphar.2020.00356. doi:10.3389/fphar.2020.00356 PMID:32390826

Gomez-Manzano, C., Fueyo, J., Kyritsis, A. P., Steck, P. A., Roth, J. A., McDonnell, T. J, & Alfred Yung, W. K. (1996). Adenovirus-mediated transfer of the p53 gene produces rapid and generalized death of human glioma cells via apoptosis. *Cancer Research*, *56*(4), 694–699. PMID:8630997

Gómez-Nicola, D., Fransen, N. L., Suzzi, S., & Perry, V. H. (2013). Regulation of Microglial Proliferation during Chronic Neurodegeneration. *The Journal of Neuroscience : The Official Journal of the Society for Neuroscience*, *33*(6), 2481–2493. doi:10.1523/JNEUROSCI.4440-12.2013 PMID:23392676

Gong, W., Zhang, S., Zong, Y., Halim, M., Ren, Z., Wang, Y., Ma, Y., Li, B., Ma, L., Zhou, G., Yu, J., Zhang, J., & Liu, Q. (2019). Involvement of the microglial NLRP3 inflammasome in the anti-inflammatory effect of the antidepressant clomipramine. *Journal of Affective Disorders*, *254*, 15–25. doi:10.1016/j.jad.2019.05.009 PMID:31082627

Gong, Z., Ba, L., & Zhang, M. (2022). Dysfunction of the oligodendrocytes in amyotrophic lateral sclerosis. *Journal of Biomedical Research*, *36*(5), 336–342. doi:10.7555/JBR.36.20220009 PMID:36106501

González-Reyes, R. E., Nava-Mesa, M. O., Vargas-Sánchez, K., Ariza-Salamanca, D., & Mora-Muñoz, L. (2017). Involvement of Astrocytes in Alzheimer's Disease from a Neuroinflammatory and Oxidative Stress Perspective. *Frontiers in Molecular Neuroscience*, *10*, 427. doi:10.3389/fnmol.2017.00427 PMID:29311817

Goodarzi, P., & Aghayan, H. (2015). Stem cell-based approach for the treatment of Parkinson's disease. *Journal of the Islamic Republic of Iran*. https://www.ncbi.nlm.nih.gov/pmc/articles/PMC4431356/

Gopalakrishnan, A., Shankarappa, S. A., & Rajanikant, G. K. (2019). Hydrogel scaffolds: Towards restitution of ischemic stroke-injured brain. *Translational Stroke Research*, *10*(1), 1–18. doi:10.100712975-018-0655-6 PMID:30151667

Gopal, K., Gupta, N., Zhang, H., Alshareef, A., Alqahtani, H., Bigras, G., Lewis, J., Douglas, D., Kneteman, N., Lavasanifar, A., & Lai, R. (2016). Oxidative stress induces the acquisition of cancer stem-like phenotype in breast cancer detectable by using a Sox2 regulatory region-2 (SRR2) reporter. *Oncotarget*, *7*(3), 3111–3127. doi:10.18632/oncotarget.6630 PMID:26683522

Görg, B., Bidmon, H.-J., & Häussinger, D. (2013). Gene Expression Profiling in the Cerebral Cortex of Patients with Cirrhosis with and without Hepatic Encephalopathy. *Hepatology (Baltimore, Md.)*, *57*(6), 2436–2447. doi:10.1002/hep.26265 PMID:23325665

Görg, B., Karababa, A., Shafigullina, A., Bidmon, H. J., & Häussinger, D. (2015). Ammonia-induced Senescence in Cultured Rat Astrocytes and in Human Cerebral Cortex in Hepatic Encephalopathy. *Glia*, *63*(1), 37–50. doi:10.1002/glia.22731 PMID:25092802

Göritz, C., Dias, D. O., Tomilin, N., Barbacid, M., Shupliakov, O., & Frisén, J. (2011). A pericyte origin of spinal cord scar tissue. *Science*, *333*(6039), 238–242. doi:10.1126cience.1203165 PMID:21737741

Gosselin, R.-D., Suter, M. R., Ji, R.-R., & Decosterd, I. (2010). Glial cells and chronic pain. *The Neuroscientist*, *16*(5), 519–531. doi:10.1177/1073858409360822 PMID:20581331

Gouw, A. A., Seewann, A., Vrenken, H., Van Der Flier, W. M., Rozemuller, J. M., Barkhof, F., Scheltens, P., & Geurts, J. J. G. (2008). Heterogeneity of white matter hyperintensities in Alzheimer's disease : Post-mortem quantitative MRI and neuropathology. *Brain*, *131*(12), 3286–3298. doi:10.1093/brain/awn265 PMID:18927145

Gowing, G., Philips, T., Van Wijmeersch, B., Audet, J. N., Dewil, M., Van Den Bosch, L., Billiau, A. D., Robberecht, W., & Julien, J. P. (2008). Ablation of proliferating microglia does not affect motor neuron degeneration in amyotrophic lateral sclerosis caused by mutant superoxide dismutase. *The Journal of Neuroscience : The Official Journal of the Society for Neuroscience*, *28*(41), 10234–10244. doi:10.1523/JNEUROSCI.3494-08.2008 PMID:18842883

Grabe, H. J., Ahrens, N., Rose, H.-J., Kessler, C., & Freyberger, H. J. (2001). Neurotrophic factor S100beta in major depression. *Neuropsychobiology*, *44*(2), 88–90. doi:10.1159/000054922 PMID:11490177

Graeber, M. B. (2010). Changing face of microglia. *Science*, *330*(6005), 783–788. doi:10.1126cience.1190929 PMID:21051630

Grassi, S., Giussani, P., Mauri, L., Prioni, S., Sonnino, S., & Prinetti, A. (2020). Lipid rafts and neurodegeneration: Structural and functional roles in physiologic aging and neurodegenerative diseases: Thematic Review Series: Biology of Lipid Rafts. *Journal of Lipid Research*, *61*(5), 636–654. doi:10.1194/jlr.TR119000427

Griciuc, A., Serrano-Pozo, A., Parrado, A. R., Lesinski, A. N., Asselin, C. N., Mullin, K., Hooli, B., Choi, S. H., Hyman, B. T., & Tanzi, R. E. (2013). Alzheimer's disease risk gene CD33 inhibits microglial uptake of amyloid beta. *Neuron*, *78*(4), 631–643. doi:10.1016/j.neuron.2013.04.014 PMID:23623698

Griffin, W. S., Stanley, L. C., Ling, C., White, L., MacLeod, V., Perrot, L. J., White, C. L. III, & Araoz, C. (1989). Brain interleukin 1 and S-100 immunoreactivity are elevated in Down syndrome and Alzheimer disease. *Proceedings of the National Academy of Sciences of the United States of America*, *86*(19), 7611–7615. doi:10.1073/pnas.86.19.7611 PMID:2529544

Griffiths, B. B., Bhutani, A., & Stary, C. M. (2020). Adult neurogenesis from reprogrammed astrocytes. *Neural Regeneration Research*, *15*(6), 973. doi:10.4103/1673-5374.270292 PMID:31823866

Griffiths, I., Klugmann, M., Anderson, T., Yool, D., Thomson, C., Schwab, M. H., Schneider, A., Zimmermann, F., McCulloch, M., Nadon, N., & Nave, K. A. (1998). Axonal swellings and degeneration in mice lacking the major proteolipid of myelin. *Science*, *280*(5369), 1610–1613. doi:10.1126cience.280.5369.1610 PMID:9616125

Griggs, R. & Yermakov, L. (2017). *Formation and disruption of functional domains in myelinated CNS axons*. Elsevier.

Gritsch, S., Lu, J., Thilemann, S., Wörtge, S., Möbius, W., Bruttger, J., Karram, K., Ruhwedel, T., Blanfeld, M., Vardeh, D., Waisman, A., Nave, K.-A., & Kuner, R. (2014). Oligodendrocyte ablation triggers central pain independently of innate or adaptive immune responses in mice. *Nature Communications*, *5*(1), 5472. doi:10.1038/ncomms6472 PMID:25434649

Gritti, A., Parati, E., Cova, L., Frolichsthal, P., Galli, R., Wanke, E., Faravelli, L., Morassutti, D. J., Roisen, F., Nickel, D. D., & Vescovi, A. L. (1996). 'Multipotential Stem Cells from the Adult Mouse Brain Proliferate and Self-Renew in Response to Basic Fibroblast Growth Factor'. *The Journal of Neuroscience : The Official Journal of the Society for Neuroscience*, *16*(3), 1091–1100. doi:10.1523/JNEUROSCI.16-03-01091.1996 PMID:8558238

Grossmann, R., Stence, N., Carr, J., Fuller, L., Waite, M., & Dailey, M. E. (2002). Juxtavascular microglia migrate along brain microvessels following activation during early postnatal development. *Glia*, *37*(3), 229–240. doi:10.1002/glia.10031 PMID:11857681

Grutzendler, J., Kasthuri, N., & Gan, W.-B. (2002). Long-term dendritic spine stability in the adult cortex. *Nature*, *420*(6917), 812–816. doi:10.1038/nature01276 PMID:12490949

Guan, Y.-H., Zhang, L.-J., Wang, S.-Y., Deng, Y.-D., Zhou, H.-S., Chen, D.-Q., & Zhang, L.-C. (2022). The role of microglia in Alzheimer's disease and progress of treatment. *Ibrain*, *8*(1), 37–47. doi:10.1002/ibra.12023 PMID:37786418

Guan, Z., Kuhn, J. A., Wang, X., Colquitt, B., Solorzano, C., Vaman, S., Guan, A. K., Evans-Reinsch, Z., Braz, J., Devor, M., Abboud-Werner, S. L., Lanier, L. L., Lomvardas, S., & Basbaum, A. I. (2016). Injured sensory neuron-derived CSF1 induces microglial proliferation and DAP12-dependent pain. *Nature Neuroscience*, *19*(1), 94–101. doi:10.1038/nn.4189 PMID:26642091

Guerrero, B. L., & Sicotte, N. L. (2020). Microglia in Multiple Sclerosis: Friend or Foe? *Frontiers in Immunology, 11*, 374. doi:10.3389/fimmu.2020.00374 PMID:32265902

Guevara, M., Baccaro, M., Torre, A., Gómez-Ansón, B., Ríos, J., Torres, F., Rami, L., Monté-Rubio, G., Martín-Llahí, M., Arroyo, V. (2009). Hyponatremia Is a Risk Factor of Hepatic Encephalopathy in Patients with Cirrhosis: A Prospective Study with Time-Dependent Analysis. *Official Journal of the American College of Gastroenterology, 104*(6), 1382–89.

Guevara, M., Baccaro, M. E., Ríos, J., Martín-Llahí, M., Uriz, J., Ruiz del Arbol, L., Planas, R., Monescillo, A., Guarner, C., Crespo, J., Bañares, R., Arroyo, V., & Ginès, P. (2010). Risk Factors for Hepatic Encephalopathy in Patients with Cirrhosis and Refractory Ascites: Relevance of Serum Sodium Concentration. *Liver International, 30*(8), 1137–1142. doi:10.1111/j.1478-3231.2010.02293.x PMID:20602681

Gu, L., Wu, D., Tang, X., Qi, X., Li, X., Bai, F., Chen, X., Ren, Q., & Zhang, Z. (2018). Myelin changes at the early stage of 5XFAD mice. *Brain Research Bulletin, 137*, 285–293. doi:10.1016/j.brainresbull.2017.12.013 PMID:29288735

Guo, C., Sun, L., Chen, X., & Zhang, D. (2013). 'Oxidative Stress, Mitochondrial Damage and Neurodegenerative Diseases'. *Neural Regeneration Research, 8*(21), 2003–2014. doi:10.3969/j.issn.1673-5374.2013.21.009 PMID:25206509

Guo, C., Yang, L., Wan, C. X., Xia, Y. Z., Zhang, C., Chen, M. H., Wang, Z. D., Li, Z. R., Li, X. M., Geng, Y. D., & Kong, L. Y. (2016). Anti-neuroinflammatory effect of Sophoraflavanone G from Sophora alopecuroides in LPS-activated BV2 microglia by MAPK, JAK/STAT and Nrf2/HO-1 signaling pathways. *Phytomedicine, 23*(13), 1629–1637. doi:10.1016/j.phymed.2016.10.007 PMID:27823627

Guo, H., Lai, L., Butchbach, M. E., Stockinger, M. P., Shan, X., Bishop, G. A., & Lin, C. L. (2003). Increased expression of the glial glutamate transporter EAAT2 modulates excitotoxicity and delays the onset but not the outcome of ALS in mice. *Human Molecular Genetics, 12*(19), 2519–2532. doi:10.1093/hmg/ddg267 PMID:12915461

Guo, L., Choi, S., Bikkannavar, P., & Cordeiro, M. F. (2022, March 17). Microglia: Key Players in Retinal Ageing and Neurodegeneration. *Frontiers in Cellular Neuroscience, 16*, 804782. doi:10.3389/fncel.2022.804782 PMID:35370560

Guo, S., Song, Z., He, J., Yin, G., Zhu, J., Liu, H., Yang, L., Ji, X., Xu, X., Liu, Z., & Liu, J. (2022). Akt/Aquaporin-4 Signaling Aggravates Neuropathic Pain by Activating Astrocytes after Spinal Nerve Ligation in Rats. *Neuroscience, 482*, 116–131. doi:10.1016/j.neuroscience.2021.12.015 PMID:34942314

Guo, S., Wang, H., & Yin, Y. (2022). Microglia Polarization From M1 to M2 in Neurodegenerative Diseases. *Frontiers in Aging Neuroscience, 14*, 815347. doi:10.3389/fnagi.2022.815347 PMID:35250543

Guo, W., Wang, H., Watanabe, M., Shimizu, K., Zou, S., LaGraize, S. C., Wei, F., Dubner, R., & Ren, K. (2007). Glial-cytokine-neuronal interactions underlying the mechanisms of persistent pain. *The Journal of Neuroscience : The Official Journal of the Society for Neuroscience, 27*(22), 6006–6018. doi:10.1523/JNEUROSCI.0176-07.2007 PMID:17537972

Guo, Z., Zhang, L., Wu, Z., Chen, Y., Wang, F., & Chen, G. (2014). In vivo direct reprogramming of reactive glial cells into functional neurons after brain injury and in an Alzheimer's disease model. *Cell Stem Cell, 14*(2), 188–202. doi:10.1016/j.stem.2013.12.001 PMID:24360883

Gupta, K., Patani, R., Baxter, P., Serio, A., Story, D., Tsujita, T., Hayes, J. D., Pedersen, R. A., Hardingham, G. E., & Chandran, S. (2011). Human embryonic stem cell derived astrocytes mediate non-cell-autonomous neuroprotection through endogenous and drug-induced mechanisms. *Cell Death & Differentiation, 5, 19*(5), 779–787. doi:10.1038/cdd.2011.154

Gustafsson, C. M., Falkenberg, M., & Larsson, N.-G. (2016). 'Maintenance and Expression of Mammalian Mitochondrial DNA'. *Annual Review of Biochemistry, 85*(1), 133–160. doi:10.1146/annurev-biochem-060815-014402 PMID:27023847

Gu, X. L., Long, C. X., Sun, L., Xie, C., Lin, X., & Cai, H. (2010). Astrocytic expression of Parkinson's disease-related A53T -synuclein causes neurodegeneration in mice. *Molecular Brain, 3*(1), 12. doi:10.1186/1756-6606-3-12 PMID:20409326

Guzauskas, G. F., Salzberg, M., & Wang, B. C. (2018). Estimated lifetime survival benefit of tumor treating fields and temozolomide for newly diagnosed glioblastoma patients. *CNS Oncology, 7*(3), CNS23. doi:10.2217/cns-2018-0010 PMID:30124334

Haas-Kogan, D. A., Prados, M. D., Tihan, T., Eberhard, D. A., Jelluma, N., Arvold, N. D., Baumber, R., Lamborn, K. R., Kapadia, A., Malec, M., Berger, M. S., & Stokoe, D. (2005). Epidermal growth factor receptor, protein kinase B/Akt, and glioma response to erlotinib. *Journal of the National Cancer Institute, 97*(12), 880–887. doi:10.1093/jnci/dji161 PMID:15956649

Haaxma, C. A., Bloem, B. R., Borm, G. F., Oyen, W. J. G., Leenders, K. L., Eshuis, S., Booij, J., Dluzen, D. E., & Horstink, M. W. I. M. (2007). Gender differences in Parkinson's disease. *Journal of Neurology, Neurosurgery, and Psychiatry, 78*(8), 819–824. doi:10.1136/jnnp.2006.103788 PMID:17098842

Häberle, J., Görg, B., Toutain, A., Rutsch, F., Benoist, J.-F., Gelot, A., Suc, A.-L., Koch, H. G., Schliess, F., & Häussinger, D. (2006). Inborn error of amino acid synthesis : Human glutamine synthetase deficiency. *Journal of Inherited Metabolic Disease, 29*(2-3), 352–358. doi:10.100710545-006-0256-5 PMID:16763901

Habib, N., McCabe, C., Medina, S., Varshavsky, M., Kitsberg, D., Dvir-Szternfeld, R., Green, G., Dionne, D., Nguyen, L., Marshall, J. L., Chen, F., Zhang, F., Kaplan, T., Regev, A., & Schwartz, M. (2020). Disease-associated astrocytes in Alzheimer's disease and aging. *Nature Neuroscience, 23*(6), 701–706. doi:10.103841593-020-0624-8 PMID:32341542

Hains, B. C., & Waxman, S. G. (2006). Activated microglia contribute to the maintenance of chronic pain after spinal cord injury. *The Journal of Neuroscience : The Official Journal of the Society for Neuroscience, 26*(16), 4308–4317. doi:10.1523/JNEUROSCI.0003-06.2006 PMID:16624951

Halassa, M. M., Fellin, T., & Haydon, P. G. (2007). The tripartite synapse: Roles for gliotransmission in health and disease. *Trends in Molecular Medicine, 13*(2), 54–63. doi:10.1016/j.molmed.2006.12.005 PMID:17207662

Halassa, M. M., Fellin, T., Takano, H., Dong, J.-H., & Haydon, P. G. (2007). 'Synaptic Islands Defined by the Territory of a Single Astrocyte'. *The Journal of Neuroscience : The Official Journal of the Society for Neuroscience, 27*(24), 6473–6477. doi:10.1523/JNEUROSCI.1419-07.2007 PMID:17567808

Hald, A., Nedergaard, S., Hansen, R. R., Ding, M., & Heegaard, A.-M. (2009). Differential activation of spinal cord glial cells in murine models of neuropathic and cancer pain. *European Journal of Pain (London, England), 13*(2), 138–145. doi:10.1016/j.ejpain.2008.03.014 PMID:18499488

Halleskog, C., Mulder, J., Dahlström, J., Mackie, K., Hortobágyi, T., Tanila, H., Kumar Puli, L., Färber, K., Harkany, T., & Schulte, G. (2011). WNT signaling in activated microglia is proinflammatory. *Glia, 59*(1), 119–131. doi:10.1002/glia.21081 PMID:20967887

Halliday, M. R., Rege, S. V., Ma, Q., Zhao, Z., Miller, C. A., Winkler, E. A., & Zlokovic, B. V. (2016). Accelerated pericyte degeneration and blood-brain barrier breakdown in apolipoprotein E4 carriers with Alzheimer's disease. *Journal of Cerebral Blood Flow and Metabolism, 36*(1), 216–227. doi:10.1038/jcbfm.2015.44 PMID:25757756

Hallmann, A. L., Araúzo-Bravo, M. J., Mavrommatis, L., Ehrlich, M., Röpke, A., Brockhaus, J., Missler, M., Sterneckert, J., Schöler, H. R., Kuhlmann, T., Zaehres, H., & Hargus, G. (2017). Astrocyte pathology in a human neural stem cell model of frontotemporal dementia caused by mutant TAU protein. *Scientific Reports, 7*(1), 42991. doi:10.1038rep42991 PMID:28256506

Hamdan, Y. A., Elouali, S., Maloui, A. B., El-Mansoury, B., Oudadesse, H., & Rhazi, M. (2023b). Chitosan and Its Derivatives as Potential Neuro-Protective Agents for the Treatment of Parkinson's Disease. In Experimental and Clinical Evidence of the Neuropathology of Parkinson's Disease (pp. 253-274). IGI Global. doi:10.4018/978-1-6684-5156-4.ch015

Hamdan, Y. A., Elouali, S., Eladlani, N., Lefeuvre, B., Oudadesse, H., & Rhazi, M. (2023a). Investigation on Akis granulifera (Coleoptera, Sahlberg, 1823) as a potential source of chitin and chitosan: Extraction, characterization and hydrogel formation. *International Journal of Biological Macromolecules, 252*, 126292. doi:10.1016/j.ijbiomac.2023.126292 PMID:37573901

Hamd-Ghadareh, S., Salimi, A., Parsa, S., & Mowla, S. J. (2022). Development of three-dimensional semi-solid hydrogel matrices for ratiometric fluorescence sensing of Amyloid β peptide and imaging in SH-SY5 cells: Improvement of point of care diagnosis of Alzheimer's disease biomarker. *Biosensors & Bioelectronics, 199*, 113895. doi:10.1016/j.bios.2021.113895 PMID:34968953

Hamilton, N. B., Clarke, L. E., Lorena Arancibia-Carcamo, I., Kougioumtzidou, E., Matthey, M., Arad, R. K., Whiteley, L., Bergersen, L. H., Richardson, W. D., & Attwell, D. (2017). Endogenous GABA controls oligodendrocyte lineage cell number, myelination, and CNS internode length. *Glia, 65*(2), 309–321. doi:10.1002/glia.23093 PMID:27796063

Hammond, T. R., Dufort, C., Dissing-Olesen, L., Giera, S., Young, A., Wysoker, A., Walker, A. J., Gergits, F., Segel, M., Nemesh, J., Marsh, S. E., Saunders, A., Macosko, E., Ginhoux, F., Chen, J., Franklin, R. J. M., Piao, X., McCarroll, S. A., & Stevens, B. (2019). Single-Cell RNA Sequencing of Microglia throughout the Mouse Lifespan and in the Injured Brain Reveals Complex Cell-State Changes. *Immunity, 50*(1), 253–271.e6. doi:10.1016/j.immuni.2018.11.004 PMID:30471926

Hanani, M. (2005). Satellite glial cells in sensory ganglia: From form to function. *Brain Research. Brain Research Reviews, 48*(3), 457–476. doi:10.1016/j.brainresrev.2004.09.001 PMID:15914252

Hanani, M., Huang, T. Y., Cherkas, P. S., Ledda, M., & Pannese, E. (2002). Glial cell plasticity in sensory ganglia induced by nerve damage. *Neuroscience, 114*(2), 279–283. doi:10.1016/S0306-4522(02)00279-8 PMID:12204197

Handa, A. K., Fatima, T., & Mattoo, A. K. (2018). Polyamines: Bio-Molecules with Diverse Functions in Plant and Human Health and Disease. *Frontiers in Chemistry, 6*, 10. doi:10.3389/fchem.2018.00010 PMID:29468148

Hanisch, U.-K., & Kettenmann, H. (2007). 'Microglia: Active Sensor and Versatile Effector Cells in the Normal and Pathologic Brain'. *Nature Neuroscience, 10*(11), 1387–1394. doi:10.1038/nn1997 PMID:17965659

Hansen, C., Angot, E., Bergström, A.-L., Steiner, J. A., Pieri, L., Paul, G., Outeiro, T. F., Melki, R., Kallunki, P., Fog, K., Li, J.-Y., & Brundin, P. (2011). 'α-Synuclein Propagates from Mouse Brain to Grafted Dopaminergic Neurons and Seeds Aggregation in Cultured Human Cells'. *The Journal of Clinical Investigation, 121*(2), 715–725. doi:10.1172/JCI43366 PMID:21245577

Hansen, D. V., Hanson, J. E., & Sheng, M. (2017). 'Microglia in Alzheimer's Disease'. *The Journal of Cell Biology, 217*(2), 459–472. doi:10.1083/jcb.201709069 PMID:29196460

Hanslik, K. L., Marino, K. M., & Ulland, T. K. (2021). Modulation of Glial Function in Health, Aging, and Neurodegenerative Disease. *Frontiers in Cellular Neuroscience, 15*, 718324. doi:10.3389/fncel.2021.718324 PMID:34531726

Hao, C., Gao, L., Zhang, Y., Wang, W., Yu, G., Guan, H., Zhang, L., & Li, C. (2015). Acetylated chitosan oligosaccharides act as antagonists against glutamate-induced PC12 cell death via Bcl-2/Bax signal pathway. *Marine Drugs, 13*(3), 1267–1289. doi:10.3390/md13031267 PMID:25775423

Hao, C., Wang, W., Wang, S., Zhang, L., & Guo, Y. (2017). An overview of the protective effects of chitosan and acetylated chitosan oligosaccharides against neuronal disorders. *Marine Drugs, 15*(4), 89. doi:10.3390/md15040089 PMID:28333077

Harbo, H. F., Gold, R., & Tintoré, M. (2013). Sex and gender issues in multiple sclerosis. *Therapeutic Advances in Neurological Disorders*, *6*(4), 237–248. doi:10.1177/1756285613488434

Hardingham, G. E., & Bading, H. (2010). Synaptic versus extrasynaptic NMDA receptor signalling: Implications for neurodegenerative disorders. *Nature Reviews. Neuroscience*, *11*(10), 682–696. doi:10.1038/nrn2911 PMID:20842175

Hardy, E., Moulard, J., Walter, A., Ezan, P., Bemelmans, A., Mouthon, F., Charvériat, M., Rouach, N., & Rancillac, A. (2023). Upregulation of astroglial connexin 30 impairs hippocampal synaptic activity and recognition memory. *PLoS Biology*, *21*(4), e3002075. doi:10.1371/journal.pbio.3002075 PMID:37040348

Hardy, J. A., & Higgins, G. A. (1992). Alzheimer's disease : The amyloid cascade hypothesis. *Science*, *256*(5054), 184–185. doi:10.1126cience.1566067 PMID:1566067

Haruwaka, K., Ikegami, A., Tachibana, Y., Ohno, N., Konishi, H., Hashimoto, A., Matsumoto, M., Kato, D., Ono, R., Kiyama, H., Moorhouse, A. J., Nabekura, J., & Wake, H. (2019). Dual microglia effects on blood brain barrier permeability induced by systemic inflammation. *Nature Communications*, *10*(1), 5816. doi:10.103841467-019-13812-z PMID:31862977

Ha, S. K., Moon, E., Lee, P., Ryu, J. H., Oh, M. S., & Kim, S. Y. (2012). Acacetin attenuates neuroinflammation via regulation the response to LPS stimuli in Vitro and in Vivo. *Neurochemical Research*, *37*(7), 1560–1567. doi:10.100711064-012-0751-z PMID:22447574

Hashimoto, K., Sawa, A., & Iyo, M. (2007). Increased levels of glutamate in brains from patients with mood disorders. *Biological Psychiatry*, *62*(11), 1310–1316. doi:10.1016/j.biopsych.2007.03.017 PMID:17574216

Hashimoto, M. Y. S. F. T. Y., Ishikawa, Y., Yokota, S., Goto, F., Bando, T., Sakakibara, Y., & Iriki, M. (1991). Action Site of Circulating Interleukin-1 on the Rabbit Brain. *Brain Research*, *540*(1–2), 217–223. doi:10.1016/0006-8993(91)90510-3 PMID:2054614

Hassanein, T. I., Hilsabeck, R. C., & Perry, W. (2008). Introduction to the hepatic encephalopathy scoring algorithm (HESA). *Digestive Diseases and Sciences*, *53*(2), 529–538. doi:10.100710620-007-9895-0 PMID:17710551

Hassel, B., Tessler, S., Faull, R. L. M., & Emson, P. C. (2008). Glutamate uptake is reduced in prefrontal cortex in Huntington's disease. *Neurochemical Research*, *33*(2), 232–237. doi:10.100711064-007-9463-1 PMID:17726644

Hathway, G. J., Vega-Avelaira, D., Moss, A., Ingram, R., & Fitzgerald, M. (2009). Brief, low frequency stimulation of rat peripheral C-fibres evokes prolonged microglial-induced central sensitization in adults but not in neonates. *Pain*, *144*(1–2), 110–118. doi:10.1016/j.pain.2009.03.022 PMID:19410369

Hattori, Y. (2022). The Multiple Roles of Pericytes in Vascular Formation and Microglial Functions in the Brain. *Life (Basel, Switzerland)*, *12*(11), 1835. doi:10.3390/life12111835 PMID:36362989

Häussinger, D., Kircheis, G., Fischer, R., Schliess, F., & vom Dahl, S. (2000). Hepatic encephalopathy in chronic liver disease: A clinical manifestation of astrocyte swelling and low-grade cerebral edema? *Journal of Hepatology*, *32*(6), 1035–1038. doi:10.1016/S0168-8278(00)80110-5 PMID:10898326

Haydon, P. G. (2001). GLIA: Listening and talking to the synapse. *Nature Reviews. Neuroscience*, *2*(3), 185–193. doi:10.1038/35058528 PMID:11256079

Hayes, M. T. (2019). Parkinson's disease and parkinsonism. *The American Journal of Medicine*, *132*(7), 802–807. doi:10.1016/j.amjmed.2019.03.001 PMID:30890425

Hazell, A. S., Desjardins, P., & Butterworth, R. F. (1999). Chronic exposure of rat primary astrocyte cultures to manganese results in increased binding sites for the 'peripheral-type'benzodiazepine receptor ligand 3H-PK 11195. *Neuroscience Letters*, *271*(1), 5–8. doi:10.1016/S0304-3940(99)00489-9 PMID:10471200

Hazell, A. S., Normandin, L., Nguyen, B., & Kennedy, G. (2003). Upregulation of 'peripheral-type' benzodiazepine receptors in the globus pallidus in a sub-acute rat model of manganese neurotoxicity. *Neuroscience Letters, 349*(1), 13–16. doi:10.1016/S0304-3940(03)00649-9 PMID:12946575

Hazell, A. S., Normandin, L., Norenberg, M. D., Kennedy, G., & Yi, J.-H. (2006). Alzheimer type II astrocytic changes following sub-acute exposure to manganese and its prevention by antioxidant treatment. *Neuroscience Letters, 396*(3), 167–171. doi:10.1016/j.neulet.2005.11.064 PMID:16384640

He, H., Zhou, Y., Zhou, Y., Zhuang, J., He, X., Wang, S., & Lin, W. (2018). *Dexmedetomidine Mitigates Microglia-Mediated Neuroinflammation through Upregulation of Programmed Cell Death Protein 1 in a Rat Spinal Cord Injury Model.* Mary Ann Libert Publications. doi:10.1089/neu.2017.5625

He, F., & Sun, Y. E. (2007). Glial cells more than support cells? *The International Journal of Biochemistry & Cell Biology, 39*(4), 661–665. doi:10.1016/j.biocel.2006.10.022 PMID:17141551

Hefendehl, J., LeDue, J., Ko, R. W. Y., Mahler, J., Murphy, T. H., & MacVicar, B. A. (2016). Mapping synaptic glutamate transporter dysfunction in vivo to regions surrounding Aβ plaques by iGluSnFR two-photon imaging. *Nature Communications, 7*(1), 13441. doi:10.1038/ncomms13441 PMID:27834383

Heine, V. M., Maslam, S., Zareno, J., Joëls, M., & Lucassen, P. J. (2004). Suppressed proliferation and apoptotic changes in the rat dentate gyrus after acute and chronic stress are reversible. *The European Journal of Neuroscience, 19*(1), 131–144. doi:10.1046/j.1460-9568.2003.03100.x PMID:14750971

Heinricher, M. M., Tavares, I., Leith, J. L., & Lumb, B. M. (2009). Descending control of nociception: Specificity, recruitment and plasticity. *Brain Research. Brain Research Reviews, 60*(1), 214–225. doi:10.1016/j.brainresrev.2008.12.009 PMID:19146877

Heins, J., & Zwingmann, C. (2010). Organic Osmolytes in Hyponatremia and Ammonia Toxicity. *Metabolic Brain Disease, 25*(1), 81–89. doi:10.100711011-010-9170-5 PMID:20195726

Hench, L. L. (1998). Biomaterials: A forecast for the future. *Biomaterials, 19*(16), 1419–1423. doi:10.1016/S0142-9612(98)00133-1 PMID:9794512

Heneka, M. T., Carson, M. J., El Khoury, J., Landreth, G. E., Brosseron, F., & Feinstein, D. L. (2015). Neuroinflammation in Alzheimer's disease. *Lancet Neurology, 14*(4), 388–405. doi:10.1016/S1474-4422(15)70016-5 PMID:25792098

Henkel, J. S., Beers, D. R., Siklós, L., & Appel, S. H. (2006). The chemokine MCP-1 and the dendritic and myeloid cells it attracts are increased in the mSOD1 mouse model of ALS. *Molecular and Cellular Neurosciences, 31*(3), 427–437. doi:10.1016/j.mcn.2005.10.016 PMID:16337133

Henneberger, C., Papouin, T., Oliet, S. H. R., & Rusakov, D. A. (2010). Long-term potentiation depends on release of d-serine from astrocytes. *Nature, 463*(7278), 232–236. doi:10.1038/nature08673

Henneberger, C., Bard, L., Panatier, A., Reynolds, J. P., Kopach, O., Medvedev, N. I., Minge, D., Herde, M. K., Anders, S., Kraev, I., Heller, J. P., Rama, S., Zheng, K., Jensen, T. P., Sanchez-Romero, I., Jackson, C. J., Janovjak, H., Ottersen, O. P., Nagelhus, E. A., & Rusakov, D. A. (2020). 'LTP Induction Boosts Glutamate Spillover by Driving Withdrawal of Perisynaptic Astroglia'. *Neuron, 108*(5), 919–936.e11. doi:10.1016/j.neuron.2020.08.030 PMID:32976770

Henneman, W. J. P., Sluimer, J. D., Barnes, J., van der Flier, W. M., Sluimer, I. C., Fox, N. C., Scheltens, P., Vrenken, H., & Barkhof, F. (2009). Hippocampal atrophy rates in Alzheimer disease : Added value over whole brain volume measures. *Neurology, 72*(11), 999–1007. doi:10.1212/01.wnl.0000344568.09360.31 PMID:19289740

Henry, C. J., Huang, Y., Wynne, A. M., & Godbout, J. P. (2009). 'Peripheral Lipopolysaccharide (LPS) Challenge Promotes Microglial Hyperactivity in Aged Mice That Is Associated with Exaggerated Induction of Both pro-Inflammatory IL-1β and Anti-Inflammatory IL-10 Cytokines'. *Brain, Behavior, and Immunity*, *23*(3), 309–317. doi:10.1016/j.bbi.2008.09.002 PMID:18814846

Hensley, K., Fedynyshyn, J., Ferrell, S., Floyd, R. A., Gordon, B., Grammas, P., Hamdheydari, L., Mhatre, M., Mou, S., Pye, Q. N., Stewart, C., West, M., West, S., & Williamson, K. S. (2003). Message and protein-level elevation of tumor necrosis factor alpha (TNF alpha) and TNF alpha-modulating cytokines in spinal cords of the G93A-SOD1 mouse model for amyotrophic lateral sclerosis. *Neurobiology of Disease*, *14*(1), 74–80. doi:10.1016/S0969-9961(03)00087-1 PMID:13678668

Hensley, K., Floyd, R. A., Gordon, B., Mou, S., Pye, Q. N., Stewart, C., West, M., & Williamson, K. (2002). Temporal patterns of cytokine and apoptosis-related gene expression in spinal cords of the G93A-SOD1 mouse model of amyotrophic lateral sclerosis. *Journal of Neurochemistry*, *82*(2), 365–374. doi:10.1046/j.1471-4159.2002.00968.x PMID:12124437

Henstridge, C. M., Tzioras, M., & Paolicelli, R. C. (2019). Glial contribution to excitatory and inhibitory synapse Loss in Neurodegeneration. *Frontiers in Neuroscience*, *13*, 63. doi:10.3389/fncel.2019.00063 PMID:30863284

Herculano-Houzel, S. (2014). The glia/neuron ratio: How it varies uniformly across brain structures and species and what that means for brain physiology and evolution. *Glia*, *62*(9), 1377–1391. doi:10.1002/glia.22683 PMID:24807023

Hernando, S., Herran, E., Figueiro-Silva, J., Pedraz, J. L., Igartua, M., Carro, E., & Hernandez, R. M. (2020). In vitro investigations on dopamine loaded solid lipid nanoparticles. *Journal of Pharmaceutical and Biomedical Analysis*, *185*, 113257. doi:10.1016/j.jpba.2020.113257 PMID:32199326

Héron, B., Mikaeloff, Y., Froissart, R., Caridade, G., Maire, I., Caillaud, C., Levade, T., Chabrol, B., Feillet, F., Ogier, H., Valayannopoulos, V., Michelakakis, H., Zafeiriou, D., Lavery, L., Wraith, E., Danos, O., Heard, J.-M., & Tardieu, M. (2011). Incidence and natural history of mucopolysaccharidosis type III in France and comparison with United Kingdom and Greece. *American Journal of Medical Genetics. Part A*, *155*(1), 58–68. doi:10.1002/ajmg.a.33779 PMID:21204211

Hertz, L., Rothman, D. L., Li, B., & Peng, L. (2015). Chronic SSRI stimulation of astrocytic 5-HT2B receptors change multiple gene expressions/editings and metabolism of glutamate, glucose and glycogen: A potential paradigm shift. *Frontiers in Behavioral Neuroscience*, *9*, 25. PMID:25750618

He, Y., Hakvoort, T. B., Vermeulen, J. L., Lamers, W. H., & Van Roon, M. A. (2007). Glutamine synthetase is essential in early mouse embryogenesis. *Developmental Dynamics*, *236*(7), 1865–1875. doi:10.1002/dvdy.21185 PMID:17557305

He, Y., Ruganzu, J. B., Jin, H., Peng, X., Ji, S., Ma, Y., Zheng, L., & Yang, W. (2020). LRP1 knockdown aggravates Aβ1–42-stimulated microglial and astrocytic neuroinflammatory responses by modulating TLR4/NF-κB/MAPKs signaling pathways. *Experimental Cell Research*, *394*(2), 112166. doi:10.1016/j.yexcr.2020.112166 PMID:32645395

Hickman, S., Izzy, S., Sen, P., Morsett, L., & El Khoury, J. (2018). Microglia in neurodegeneration. In Nature Neuroscience (Vol. 21, Issue 10, pp. 1359–1369). Nature Publishing Group. doi:10.103841593-018-0242-x

Hickman, S. E., Kingery, N. D., Ohsumi, T. K., Borowsky, M. L., Wang, L., Means, T. K., & El Khoury, J. (2013). 'The Microglial Sensome Revealed by Direct RNA Sequencing'. *Nature Neuroscience*, *16*(12), 1896–1905. doi:10.1038/nn.3554 PMID:24162652

Hirase, H., Qian, L., Barthó, P., & Buzsáki, G. (2004). Calcium Dynamics of Cortical Astrocytic Networks In Vivo. *PLoS Biology*, *2*(4), e96. doi:10.1371/journal.pbio.0020096 PMID:15094801

Hirsch, E. C., & Hunot, S. (2009). Neuroinflammation in Parkinson's Disease: A Target for Neuroprotection? *Lancet Neurology*, *8*(4), 382–397. doi:10.1016/S1474-4422(09)70062-6 PMID:19296921

Hirsch, T., Rothoeft, T., Teig, N., Bauer, J. W., Pellegrini, G., De Rosa, L., Scaglione, D., Reichelt, J., Klausegger, A., Kneisz, D., Romano, O., Secone Seconetti, A., Contin, R., Enzo, E., Jurman, I., Carulli, S., Jacobsen, F., Luecke, T., Lehnhardt, M., & De Luca, M. (2017). Regeneration of the entire human epidermis using transgenic stem cells. *Nature*, *551*(7680), 327–332. doi:10.1038/nature24487 PMID:29144448

Hisahara, S., & Shimohama, S. (2011). Dopamine receptors and Parkinson's disease. *International Journal of Medicinal Chemistry*, *2011*, 403039. doi:10.1155/2011/403039 PMID:25954517

Hochstrasser, T., Ullrich, C., Sperner-Unterweger, B., & Humpel, C. (2011). Inflammatory stimuli reduce survival of serotonergic neurons and induce neuronal expression of indoleamine 2, 3-dioxygenase in rat dorsal raphe nucleus organotypic brain slices. *Neuroscience*, *184*, 128–138. doi:10.1016/j.neuroscience.2011.03.070 PMID:21501664

Hoffmann, S., & Beyer, C. (2020). A Fatal Alliance between Microglia, Inflammasomes, and Central Pain. *International Journal of Molecular Sciences*, *21*(11), 3764. doi:10.3390/ijms21113764 PMID:32466593

Hofstetter, S., Tavor, I., Moryosef, S. T., & Assaf, Y. (2013). Short-term learning induces white matter plasticity in the fornix. *The Journal of Neuroscience : The Official Journal of the Society for Neuroscience*, *33*(31), 12844–12850. doi:10.1523/JNEUROSCI.4520-12.2013 PMID:23904619

Hogan, D. B., Jetté, N., Fiest, K. M., Roberts, J. I., Pearson, D., Smith, E. E., Roach, P., Kirk, A., Pringsheim, T., & Maxwell, C. J. (2016). The prevalence and incidence of frontotemporal dementia: A systematic review. *The Canadian Journal of Neurological Sciences*, *43*(S1), S96–S109. doi:10.1017/cjn.2016.25 PMID:27307130

Holm, T. H., Draeby, D., & Owens, T. (2012). Microglia are required for astroglial Toll-like receptor 4 response and for optimal TLR2 and TLR3 response. *Glia*, *60*(4), 630–638. doi:10.1002/glia.22296 PMID:22271465

Holtmaat, A. J. G. D., Trachtenberg, J. T., Wilbrecht, L., Shepherd, G. M., Zhang, X., Knott, G. W., & Svoboda, K. (2005). Transient and persistent dendritic spines in the neocortex in vivo. *Neuron*, *45*(2), 279–291. doi:10.1016/j.neuron.2005.01.003 PMID:15664179

Ho, M. S. (2019). Microglia in Parkinson's Disease. *Advances in Experimental Medicine and Biology*, *1175*, 335–353. doi:10.1007/978-981-13-9913-8_13 PMID:31583594

Hong, S., Dissing-Olesen, L., & Stevens, B. (2016b). New insights on the role of microglia in synaptic pruning in health and disease. *Current Opinion in Neurobiology*, *36*, 128–134. doi:10.1016/j.conb.2015.12.004 PMID:26745839

Horiguchi, S., Fujita, Y., Watanabe, S., Xue, Y., & Nagamatsu, K. (2020). End-to-End Speaker Diarization for an Unknown Number of Speakers with Encoder-Decoder Based Attractors (arXiv:2005.09921). arXiv. /arXiv.2005.09921 doi:10.21437/Interspeech.2020-1022

Horiuchi, M., Maezawa, I., Itoh, A., Wakayama, K., & L. J.-N. (2012). *Amyloid β1–42 oligomer inhibits myelin sheet formation in vitro*. Elsevier. https://www.sciencedirect.com/science/article/pii/S0197458010002216?casa_token=ej2OJHtzdogAAAAA:IgCWwtmXoIHp6Z-

Hoshino, K., Hayakawa, M., & Morimoto, Y. (2017). Minocycline prevents the impairment of hippocampal long-term potentiation in the septic mouse. *Shock (Augusta, Ga.)*, *48*(2), 209–214. doi:10.1097/SHK.0000000000000847 PMID:28187038

Hösli, L., Binini, N., Ferrari, K. D., Thieren, L., Looser, Z. J., Zuend, M., Zanker, H. S., Berry, S., Holub, M., Möbius, W., Ruhwedel, T., Nave, K.-A., Giaume, C., Weber, B., & Saab, A. S. (2022). Decoupling astrocytes in adult mice impairs synaptic plasticity and spatial learning. *Cell Reports*, *38*(10), 110484. doi:10.1016/j.celrep.2022.110484 PMID:35263595

Hou, L., Liu, Y., Wang, X., Ma, H., He, J., Zhang, Y., Yu, C., Guan, W., & Ma, Y. (2011). The effects of amyloid-β42 oligomer on the proliferation and activation of astrocytes in vitro. *In Vitro Cellular & Developmental Biology. Animal*, *47*(8), 573–580. doi:10.100711626-011-9439-y PMID:21858610

Hoveizi, E., Tavakol, S., & Ebrahimi-Barough, S. (2015). Neuroprotective effect of transplanted neural precursors embedded on PLA/CS scaffold in an animal model of multiple sclerosis. *Molecular Neurobiology*, *51*(3), 1334–1342. doi:10.100712035-014-8812-8 PMID:25081586

Howland, D. S., Liu, J., She, Y., Goad, B., Maragakis, N. J., Kim, B., Erickson, J., Kulik, J., DeVito, L., Psaltis, G., DeGennaro, L. J., Cleveland, D. W., & Rothstein, J. D. (2002). Focal loss of the glutamate transporter EAAT2 in a transgenic rat model of SOD1 mutant-mediated amyotrophic lateral sclerosis (ALS). *Proceedings of the National Academy of Sciences of the United States of America*, *99*(3), 1604–1609. doi:10.1073/pnas.032539299 PMID:11818550

Howren, M. B., Lamkin, D. M., & Suls, J. (2009). Associations of depression with C-reactive protein, IL-1, and IL-6: A meta-analysis. *Psychosomatic Medicine*, *71*(2), 171–186. doi:10.1097/PSY.0b013e3181907c1b PMID:19188531

Hsiao, H.-Y., Chen, Y.-C., Chen, H.-M., Tu, P.-H., & Chern, Y. (2013). A critical role of astrocyte-mediated nuclear factor-κB-dependent inflammation in Huntington's disease. *Human Molecular Genetics*, *22*(9), 1826–1842. doi:10.1093/hmg/ddt036 PMID:23372043

Huang, K. F., Hsu, W. C., Chiu, W. T., & Wang, J. Y. (2012). Functional improvement and neurogenesis after collagen-GAG matrix implantation into surgical brain trauma. *Biomaterials*, *33*(7), 2067–2075. doi:10.1016/j.biomaterials.2011.11.040 PMID:22169139

Huang, T.-Y., & Hanani, M. (2005). Morphological and electrophysiological changes in mouse dorsal root ganglia after partial colonic obstruction. *American Journal of Physiology. Gastrointestinal and Liver Physiology*, *289*(4), G670–G678. doi:10.1152/ajpgi.00028.2005 PMID:15920014

Huberman, M., Shalit, F., Roth-Deri, I., Gutman, B., Brodie, C., Kott, E., & Sredni, B. (1994). Correlation of cytokine secretion by mononuclear cells of Alzheimer patients and their disease stage. *Journal of Neuroimmunology*, *52*(2), 147–152. doi:10.1016/0165-5728(94)90108-2 PMID:8034754

Huentelman, M., De Both, M., Jepsen, W., Piras, I. S., Talboom, J. S., Willeman, M., Reiman, E. M., Hardy, J., & Myers, A. J. (2019). Common BACE2 polymorphisms are associated with altered risk for Alzheimer's disease and CSF amyloid biomarkers in APOE epsilon4 non-carriers. *Scientific Reports*, *9*(1), 9640. doi:10.103841598-019-45896-4 PMID:31270419

Hughes, C. E., & Nibbs, R. J. B. (2018). A guide to chemokines and their receptors. *The FEBS Journal*, *285*(16), 2944–2971. doi:10.1111/febs.14466 PMID:29637711

Huh, Y., Ji, R.-R., & Chen, G. (2017). Neuroinflammation, Bone Marrow Stem Cells, and Chronic Pain. *Frontiers in Immunology*, *8*, 1014. https://www.frontiersin.org/articles/10.3389/fimmu.2017.01014. doi:10.3389/fimmu.2017.01014 PMID:28871264

Huizinga, R., & Van, B. J. (2012). 'Phagocytosis of Neuronal Debris by Microglia Is Associated with Neuronal Damage in Multiple Sclerosis'. *Glia*, *60*(3), 422–431. doi:10.1002/glia.22276 PMID:22161990

Hu, J.-G., Wu, X.-J., Feng, Y.-F., Xi, G.-M., Wang, Z.-H., Zhou, J.-S., & Lü, H.-Z. (2012). PDGF-AA and bFGF mediate B104CM-induced proliferation of oligodendrocyte precursor cells. *International Journal of Molecular Medicine*, *30*(5), 1113–1118. doi:10.3892/ijmm.2012.1110 PMID:22922759

Hunot, S., Boissière, F., Faucheux, B., Brugg, B., Mouatt-Prigent, A., Agid, Y., & Hirsch, E. C. (1996). Nitric oxide synthase and neuronal vulnerability in parkinson's disease. *Neuroscience*, *72*(2), 355–363. doi:10.1016/0306-4522(95)00578-1 PMID:8737406

Hu, S., Yang, C., Li, Y., Luo, Q., & Luo, H. (2022). Nanozyme sensor array based on manganese dioxide for the distinction between multiple amyloid β peptides and their dynamic aggregation process. *Biosensors & Bioelectronics, 199*, 113881. doi:10.1016/j.bios.2021.113881 PMID:34915216

Huse, J. T., & Aldape, K. D. (2014). The evolving role of molecular markers in the diagnosis and management of diffuse glioma. *Clinical Cancer Research, 20*(22), 5601–5611. doi:10.1158/1078-0432.CCR-14-0831 PMID:25398843

Hu, X., Li, P., Guo, Y., Wang, H., Leak, R. K., Chen, S., Gao, Y., & Chen, J. (2012). Microglia/macrophage polarization dynamics reveal novel mechanism of injury expansion after focal cerebral ischemia. *Stroke, 43*(11), 3063–3070. doi:10.1161/STROKEAHA.112.659656 PMID:22933588

Iadecola, C. (2017). The Neurovascular Unit Coming of Age: A Journey through Neurovascular Coupling in Health and Disease. *Neuron, 96*(1), 17–42. doi:10.1016/j.neuron.2017.07.030 PMID:28957666

Iadecola, C., & Nedergaard, M. (2007). 'Glial Regulation of the Cerebral Microvasculature'. *Nature Neuroscience, 10*(11), 1369–1376. doi:10.1038/nn2003 PMID:17965657

Igarashi, Y., Utsumi, H., Chiba, H., Yamada-Sasamori, Y., Tobioka, H., Kamimura, Y., Furuuchi, K., Kokai, Y., Nakagawa, T., Mori, M., & Sawada, N. (1999). Glial cell line-derived neurotrophic factor induces barrier function of endothelial cells forming the blood-brain barrier. *Biochemical and Biophysical Research Communications, 261*(1), 108–112. doi:10.1006/bbrc.1999.0992 PMID:10405331

Imrie, J., Dasgupta, S., Besley, G. T. N., Harris, C., Heptinstall, L., Knight, S., Vanier, M. T., Fensom, A. H., Ward, C., Jacklin, E., Whitehouse, C., & Wraith, J. E. (2007). The natural history of Niemann–Pick disease type C in the UK. *Journal of Inherited Metabolic Disease, 30*(1), 51–59. doi:10.100710545-006-0384-7 PMID:17160617

Inazu, M., Takeda, H., & Matsumiya, T. (2003). Functional expression of the norepinephrine transporter in cultured rat astrocytes. *Journal of Neurochemistry, 84*(1), 136–144. doi:10.1046/j.1471-4159.2003.01514.x PMID:12485410

Inoue, K., & Tsuda, M. (2018). Microglia in neuropathic pain: Cellular and molecular mechanisms and therapeutic potential. *Nature Reviews. Neuroscience, 19*(3), 138–152. doi:10.1038/nrn.2018.2 PMID:29416128

Inserm. (2017). Sclérose latérale amyotrophique (SLA) / maladie de Charcot. *Une dégénérescence des motoneurones encore incurable*. https://www.inserm.fr/dossier/sclerose-laterale-amyotrophique-sla-maladie-charcot/.

Iordanishvili, E., Schall, M., Loução, R., & Neuroimage, M. Z. (2019). *Quantitative MRI of cerebral white matter hyperintensities: a new approach towards understanding the underlying pathology*. Elsevier.

Irala, D., Bonafina, A., Fontanet, P. A., Alsina, F. C., Paratcha, G., & Ledda, F. (2016). The GDNF-GFRα1 complex promotes the development of hippocampal dendritic arbors and spines via NCAM. *Development (Cambridge, England), 143*(22), 4224–4235. doi:10.1242/dev.140350 PMID:27707798

Ishibashi, T., Dakin, K. A., Stevens, B., Lee, P. R., Kozlov, S. V., Stewart, C. L., & Fields, R. D. (2006). Astrocytes promote myelination in response to electrical impulses. *Neuron, 49*(6), 823–832. doi:10.1016/j.neuron.2006.02.006 PMID:16543131

Ishikawa, T., Eto, K., Kim, S. K., Wake, H., Takeda, I., Horiuchi, H., Moorhouse, A. J., Ishibashi, H., & Nabekura, J. (2018). Cortical astrocytes prime the induction of spine plasticity and mirror image pain. *Pain, 159*(8), 1592–1606. doi:10.1097/j.pain.0000000000001248 PMID:29672449

Itabashi, S., Arai, H., Matsui, T., Higuchi, S., & Sasaki, H. (1997). Herpes simplex virus and risk of Alzheimer's disease. *Lancet, 349*(9058), 1102. doi:10.1016/S0140-6736(05)62325-2 PMID:9107270

Iwasa, M., & Takei, Y. (2015). Pathophysiology and Management of Hepatic Encephalopathy 2014 Update: Ammonia Toxicity and Hyponatremia. *Hepatology Research*, *45*(12), 1155–1162. doi:10.1111/hepr.12495 PMID:25604571

Iwata, M., Ishida, H., Kaneko, K., & Shirayama, Y. (2016). Learned helplessness activates hippocampal microglia in rats: A potential target for the antidepressant imipramine. *Pharmacology, Biochemistry, and Behavior*, *150*, 138–146. doi:10.1016/j.pbb.2016.10.005 PMID:27769904

Iwata, M., Shirayama, Y., Ishida, H., Hazama, G., & Nakagome, K. (2011). Hippocampal astrocytes are necessary for antidepressant treatment of learned helplessness rats. *Hippocampus*, *21*(8), 877–884. doi:10.1002/hipo.20803 PMID:20572198

Izrael, M., Slutsky, S. G., & Revel, M. (2020). Rising Stars: Astrocytes as a Therapeutic Target for ALS Disease. *Frontiers in Neuroscience*, *14*, 824. doi:10.3389/fnins.2020.00824 PMID:32848579

Izumi, S., Hughes, R. D., Langley, P. G., Pernambuco, J. R. B., & Williams, R. (1995). Acute Phase Response after Liver Transplantation for Fulminant Hepatic Failure and Cirrhosis. *Transplant International*, *8*(5), 340–345. doi:10.1111/j.1432-2277.1995.tb01533.x PMID:7576014

Jack, C. R. Jr, Knopman, D. S., Jagust, W. J., Shaw, L. M., Aisen, P. S., Weiner, M. W., Petersen, R. C., & Trojanowski, J. Q. (2010). Hypothetical model of dynamic biomarkers of the Alzheimer's pathological cascade. *Lancet Neurology*, *9*(1), 119–128. doi:10.1016/S1474-4422(09)70299-6 PMID:20083042

Jacob, C. P., Koutsilieri, E., Bartl, J., Neuen-Jacob, E., Arzberger, T., Zander, N., Ravid, R., Roggendorf, W., Riederer, P., & Grünblatt, E. (2007). Alterations in expression of glutamatergic transporters and receptors in sporadic Alzheimer's disease. *Journal of Alzheimer's Disease*, *11*(1), 97–116. doi:10.3233/JAD-2007-11113 PMID:17361039

Jaeger, V., Demorrow, S., & Mcmillin, M. (2019). The Direct Contribution of Astrocytes and Microglia to the Pathogenesis of Hepatic Encephalopathy. *Journal of Clinical and Translational Hepatology*, *7*(X), 352–361. doi:10.14218/JCTH.2019.00025 PMID:31915605

Jahn, O., Siems, S. B., Kusch, K., Hesse, D., Jung, R. B., Liepold, T., Uecker, M., Sun, T., & Werner, H. B. (2020). The CNS myelin proteome: Deep profile and persistence after post-mortem delay. *Frontiers in Cellular Neuroscience*, *14*, 239. doi:10.3389/fncel.2020.00239

Jäkel, S., & Dimou, L. (2017). Glial Cells and Their Function in the Adult Brain: A Journey through the History of Their Ablation. *Frontiers in Cellular Neuroscience*, *11*, 24. doi:10.3389/fncel.2017.00024 PMID:28243193

Jakobs, M., Lee, D. J., & Lozano, A. M. (2020). Modifying the progression of Alzheimer's and Parkinson's disease with deep brain stimulation. *Neuropharmacology*, *171*, 107860. doi:10.1016/j.neuropharm.2019.107860 PMID:31765650

Jakobsson, J., Bjerke, M., Sahebi, S., Isgren, A., Ekman, C. J., Sellgren, C., Olsson, B., Zetterberg, H., Blennow, K., Pålsson, E., & Landén, M. (2015). Monocyte and microglial activation in patients with mood-stabilized bipolar disorder. Journal of psychiatry & neuroscience. *Journal of Psychiatry & Neuroscience*, *40*(4), 250–258. doi:10.1503/jpn.140183 PMID:25768030

James, G., & Butt, A. M. (2002). P2Y and P2X purinoceptor mediated Ca2+ signaling in glial cell pathology in the central nervous system. *European Journal of Pharmacology*, *447*(2-3), 247–260. doi:10.1016/S0014-2999(02)01756-9 PMID:12151016

Jankovic, J. (2008). Parkinson's disease: Clinical features and diagnosis. *Journal of Neurology, Neurosurgery, and Psychiatry*, *79*(4), 368–376. doi:10.1136/jnnp.2007.131045 PMID:18344392

Jankovic, J., & Aguilar, L. G. (2008). Current approaches to the treatment of Parkinson’s disease. *Neuropsychiatric Disease and Treatment*, *4*(4), 743–757. doi:10.2147/NDT.S2006 PMID:19043519

Jasmin, L., Vit, J.-P., Bhargava, A., & Ohara, P. T. (2010). Can satellite glial cells be therapeutic targets for pain control? *Neuron Glia Biology*, *6*(1), 63–71. doi:10.1017/S1740925X10000098 PMID:20566001

Jaumotte, J. D., & Zigmond, M. J. (2014). Comparison of GDF5 and GDNF as neuroprotective factors for postnatal dopamine neurons in ventral mesencephalic cultures. *Journal of Neuroscience Research*, *92*(11), 1425–1433. doi:10.1002/jnr.23425 PMID:24916473

Jayakumar, A. R. & Norenberg, M. (2013b). *Endothelial-Astrocytic Interactions in Acute Liver Failure*. NIH.

Jayakumar, A. R., & Norenberg, M. D. (2013a). Endothelial-Astrocytic Interactions in Acute Liver Failure. *Metabolic Brain Disease*, *28*(2), 183–186. doi:10.100711011-012-9344-4 PMID:23099995

Jayakumar, A. R., Panickar, K. S., & Ch, R. K. (2006). Oxidative Stress and Mitogen-Activated Protein Kinase Phosphorylation Mediate Ammonia-Induced Cell Swelling and Glutamate Uptake Inhibition in Cultured Astrocytes. *The Journal of Neuroscience : The Official Journal of the Society for Neuroscience*, *26*(18), 4774–4784. doi:10.1523/JNEUROSCI.0120-06.2006 PMID:16672650

Jayakumar, A. R., Rama Rao, K. V., & Norenberg, M. D. (2015). Neuroinflammation in Hepatic Encephalopathy: Mechanistic Aspects. *Journal of Clinical and Experimental Hepatology*, *5*(March), S21–S28. doi:10.1016/j.jceh.2014.07.006 PMID:26041953

Jayakumar, A. R., Rama Rao, K. V., Tong, X. Y., & Norenberg, M. D. (2009). Calcium in the Mechanism of Ammonia-induced Astrocyte Swelling. *Journal of Neurochemistry*, *109*(s1), 252–257. doi:10.1111/j.1471-4159.2009.05842.x PMID:19393035

Jayakumar, A. R., Tong, X. Y., Curtis, K. M., Ruiz-Cordero, R., Abreu, M. T., & Norenberg, M. D. (2014). Increased Toll-like Receptor 4 in Cerebral Endothelial Cells Contributes to the Astrocyte Swelling and Brain Edema in Acute Hepatic Encephalopathy. *Journal of Neurochemistry*, *128*(6), 890–903. doi:10.1111/jnc.12516 PMID:24261962

Jayakumar, A. R., Tong, X. Y., Ospel, J., & Norenberg, M. D. (2012). Role of Cerebral Endothelial Cells in the Astrocyte Swelling and Brain Edema Associated with Acute Hepatic Encephalopathy. *Neuroscience*, *218*, 305–316. doi:10.1016/j.neuroscience.2012.05.006 PMID:22609932

Jayakumar, A. R., Valdes, V., & Norenberg, M. D. (2011). The Na–K–Cl Cotransporter in the Brain Edema of Acute Liver Failure. *Journal of Hepatology*, *54*(2), 272–278. doi:10.1016/j.jhep.2010.06.041 PMID:21056502

Jayakumar, A. R., Valdes, V., Tong, X. Y., Shamaladevi, N., Gonzalez, W., & Norenberg, M. D. (2014). Sulfonylurea Receptor 1 Contributes to the Astrocyte Swelling and Brain Edema in Acute Liver Failure. *Translational Stroke Research*, *5*(1), 28–37. doi:10.100712975-014-0328-z PMID:24443056

Je, J. Y., & Kim, S. K. (2005). Water-soluble chitosan derivatives as a BACE1 inhibitor. *Bioorganic & Medicinal Chemistry*, *13*(23), 6551–6555. doi:10.1016/j.bmc.2005.07.004 PMID:16084727

Jellinger, K. A. (2020). Neuropathological assessment of the Alzheimer spectrum. *Journal of Neural Transmission (Vienna, Austria)*, *127*(9), 1229–1256. doi:10.100700702-020-02232-9 PMID:32740684

Jeon, Y. J., Park, P. J., & Kim, S. K. (2001). Antimicrobial effect of chitooligosaccharides produced by bioreactor. *Carbohydrate Polymers*, *44*(1), 71–76. doi:10.1016/S0144-8617(00)00200-9

Jeon, Y. J., Shahidi, F., & Kim, S. K. (2000). Preparation of chitin and chitosan oligomers and their applications in physiological functional foods. *Food Reviews International*, *16*(2), 159–176. doi:10.1081/FRI-100100286

Jessen, K. R. (2004). Glial cells. *The International Journal of Biochemistry & Cell Biology*, *36*(10), 1861–1867. doi:10.1016/j.biocel.2004.02.023 PMID:15203098

Jessen, K. R., & Arthur-Farraj, P. (2019). 'Repair Schwann Cell Update: Adaptive Reprogramming, EMT, and Stemness in Regenerating Nerves'. *Glia*, *67*(3), 421–437. doi:10.1002/glia.23532 PMID:30632639

Jessen, K. R., & Mirsky, R. (1999). 'Schwann Cells and Their Precursors Emerge as Major Regulators of Nerve Development'. *Trends in Neurosciences*, *22*(9), 402–410. doi:10.1016/S0166-2236(98)01391-5 PMID:10441301

Jessen, K. R., & Mirsky, R. (2005). The origin and development of glial cells in peripheral nerves. *Nature Reviews. Neuroscience*, *6*(9), 9. Advance online publication. doi:10.1038/nrn1746 PMID:16136171

Jessen, K. R., & Mirsky, R. (2016). 'The Repair Schwann Cell and Its Function in Regenerating Nerves'. *The Journal of Physiology*, *594*(13), 3521–3531. doi:10.1113/JP270874 PMID:26864683

Jha, M. K., Jo, M., Kim, J., & Suk, K. (2018). Microglia-Astrocyte Crosstalk : An Intimate Molecular Conversation. *The Neuroscientist*. Advance online publication. doi:10.1177/1073858418783959 PMID:29931997

Jha, M. K., Lee, W.-H., & Suk, K. (2016). Functional polarization of neuroglia: Implications in neuroinflammation and neurological disorders. *Biochemical Pharmacology*, *103*, 1–16. doi:10.1016/j.bcp.2015.11.003 PMID:26556658

Jha, M. K., Seo, M., Kim, J. H., Kim, B. G., Cho, J. Y., & Suk, K. (2013). The secretome signature of reactive glial cells and its pathological implications. *Biochimica et Biophysica Acta (BBA) -. Biochimica et Biophysica Acta. Proteins and Proteomics*, *1834*(11), 2418–2428. doi:10.1016/j.bbapap.2012.12.006 PMID:23269363

Jiang, K., Wang, J., Wu, F., Xue, Q., Yao, Q., Zhang, J., Chen, Y., Zhang, G., Zhu, Z., Yan, H., Zhu, L., & Yip, H. (2020). Dopant-Free Organic Hole-Transporting Material for Efficient and Stable Inverted All-Inorganic and Hybrid Perovskite Solar Cells. *Advanced Materials*, *32*(16), 1908011. doi:10.1002/adma.201908011 PMID:32115824

Jiang, M., Zhuge, X., Yang, Y., Gu, X., & Ding, F. (2009). The promotion of peripheral nerve regeneration by chitooligosaccharides in the rat nerve crush injury model. *Neuroscience Letters*, *454*(3), 239–243. doi:10.1016/j.neulet.2009.03.042 PMID:19429091

Jiang, N., Lv, J., Wang, H., Huang, H., Wang, Q., Lu, C., Zeng, G., & Liu, X. (2020). Ginsenoside Rg1 ameliorates chronic social defeat stress-induced depressive-like behaviors and hippocampal neuroinflammation. *Life Sciences*, *252*, 117669. doi:10.1016/j.lfs.2020.117669 PMID:32298740

Jiang, W., Desjardins, P., & Butterworth, R. F. (2009). Direct Evidence for Central Proinflammatory Mechanisms in Rats with Experimental Acute Liver Failure: Protective Effect of Hypothermia. *Journal of Cerebral Blood Flow and Metabolism*, *29*(5), 944–952. doi:10.1038/jcbfm.2009.18 PMID:19259110

Jiang, X., Zhang, H., Duan, F., & Quan, X. (2017). Identify Huntington's disease associated genes based on restricted Boltzmann machine with RNA-seq data. *BMC Bioinformatics*, *18*(1), 447. doi:10.118612859-017-1859-6 PMID:29020921

Jia, X., Gao, Z., & Hu, H. (2021). Microglia in depression: Current perspectives. *Science China. Life Sciences*, *64*(6), 911–925. doi:10.100711427-020-1815-6 PMID:33068286

Jimenez-Ferrer, I., Jewett, M., Tontanahal, A., Romero-Ramos, M., & Swanberg, M. (2017). Allelic difference in Mhc2ta confers altered microglial activation and susceptibility to α-synuclein-induced dopaminergic neurodegeneration. *Neurobiology of Disease*, *106*, 279–290. doi:10.1016/j.nbd.2017.07.016 PMID:28736195

Jimenez-Sanchez, M., Licitra, F., Underwood, B. R., & Rubinsztein, D. C. (2017). Huntington's disease: Mechanisms of pathogenesis and therapeutic strategies. *Cold Spring Harbor Perspectives in Medicine*, *7*(7), 7. doi:10.1101/cshperspect.a024240 PMID:27940602

Jimenez-Shahed, J., Telkes, I., Viswanathan, A., & Ince, N. F. (2016). GPi oscillatory activity differentiates tics from the resting state, voluntary movements, and the unmedicated Parkinsonian state. *Frontiers in Neuroscience, 10*(SEP), 204266. doi:10.3389/fnins.2016.00436 PMID:27733815

Jin, L., Ge, H., Long, Y., Yang, C., Chang, Y., Mu, L., Sayour, E. J., De Leon, G., Wang, Q. J., Yang, J. C., Kubilis, P. S., Bao, H., Xia, S., Lu, D., Kong, Y., Hu, L., Shang, Y., Jiang, C., Nie, J., & Huang, J. (2018). CD70, a novel target of CAR T-cell therapy for gliomas. *Neuro-Oncology, 20*(1), 55–65. doi:10.1093/neuonc/nox116 PMID:28651374

Jin, R., Aidan, K. Y. C., Wu, J., & Tatia, M. C. L. (2022). 'Relationships between Inflammation and Age-Related Neurocognitive Changes'. *International Journal of Molecular Sciences, 23*(20), 12573. doi:10.3390/ijms232012573 PMID:36293430

Jin, S.-X., Zhuang, Z.-Y., Woolf, C. J., & Ji, R.-R. (2003). P38 mitogen-activated protein kinase is activated after a spinal nerve ligation in spinal cord microglia and dorsal root ganglion neurons and contributes to the generation of neuropathic pain. *The Journal of Neuroscience : The Official Journal of the Society for Neuroscience, 23*(10), 4017–4022. doi:10.1523/JNEUROSCI.23-10-04017.2003 PMID:12764087

Ji, R.-R., Berta, T., & Nedergaard, M. (2013). Glia and pain: Is chronic pain a gliopathy? *Pain, 154*(Supplement 1), S10–S28. doi:10.1016/j.pain.2013.06.022 PMID:23792284

Ji, R.-R., Chamessian, A., & Zhang, Y.-Q. (2016). Pain regulation by non-neuronal cells and inflammation. *Science, 354*(6312), 572–577. doi:10.1126cience.aaf8924 PMID:27811267

Ji, R.-R., Donnelly, C. R., & Nedergaard, M. (2019). Astrocytes in chronic pain and itch. *Nature Reviews. Neuroscience, 20*(11), 667–685. doi:10.103841583-019-0218-1 PMID:31537912

Ji, R.-R., Gereau, R. W. IV, Malcangio, M., & Strichartz, G. R. (2009). MAP kinase and pain. *Brain Research. Brain Research Reviews, 60*(1), 135–148. doi:10.1016/j.brainresrev.2008.12.011 PMID:19150373

Ji, R.-R., & Suter, M. R. (2007). P38 MAPK, microglial signaling, and neuropathic pain. *Molecular Pain, 3*, 33. doi:10.1186/1744-8069-3-33 PMID:17974036

Jiwaji, Z., Tiwari, S. S., Avilés-Reyes, R. X., Hooley, M., Hampton, D., Torvell, M., Johnson, D. A., McQueen, J., Baxter, P., Sabari-Sankar, K., Qiu, J., He, X., Fowler, J., Febery, J., Gregory, J., Rose, J., Tulloch, J., Loan, J., Story, D., & Hardingham, G. E. (2022). Reactive astrocytes acquire neuroprotective as well as deleterious signatures in response to Tau and Aß pathology. *Nature Communications, 13*(1), 1. doi:10.103841467-021-27702-w PMID:35013236

Joe, E., Ringman, J.M., 2019. Cognitive symptoms of Alzheimer's disease: clinical management and prevention. bmj 367.

Johanns, T. M., Miller, C. A., Dorward, I. G., Tsien, C., Chang, E., Perry, A., Uppaluri, R., Ferguson, C., Schmidt, R. E., Dahiya, S., Ansstas, G., Mardis, E. R., & Dunn, G. P. (2016). Immunogenomics of hypermutated glioblastoma: A patient with germline POLE deficiency treated with checkpoint blockade immunotherapy. *Cancer Discovery, 6*(11), 1230–1236. doi:10.1158/2159-8290.CD-16-0575 PMID:27683556

Johanson, C. E., Duncan, J. A. III, Klinge, P. M., Brinker, T., Stopa, E. G., & Silverberg, G. D. (2008). 'Multiplicity of Cerebrospinal Fluid Functions: New Challenges in Health and Disease'. *Cerebrospinal Fluid Research, 5*(1), 10. doi:10.1186/1743-8454-5-10 PMID:18479516

John, C. S., Smith, K. L., Van't Veer, A., Gompf, H. S., Carlezon, W. A. Jr, Cohen, B. M., Öngür, D., & Bechtholt-Gompf, A. J. (2012). Blockade of astrocytic glutamate uptake in the prefrontal cortex induces anhedonia. *Neuropsychopharmacology, 37*(11), 2467–2475. doi:10.1038/npp.2012.105 PMID:22739467

Johnson, D. R., & O'Neill, B. P. (2012). Glioblastoma survival in the United States before and during the temozolomide era. *Journal of Neuro-Oncology, 107*(2), 359–364. doi:10.100711060-011-0749-4 PMID:22045118

Johnson, L. A. (2020). APOE and metabolic dysfunction in Alzheimer's disease. *International Review of Neurobiology, 154*, 131–151. doi:10.1016/bs.irn.2020.02.002 PMID:32739002

Jolly, R. D., Johnstone, A. C., Norman, E. J., Hopwood, J. J., & Walkley, S. U. (2007). Pathology of mucopolysaccharidosis IIIA in Huntaway dogs. *Veterinary Pathology, 44*(5), 569–578. doi:10.1354/vp.44-5-569 PMID:17846229

Jones, V. C., Atkinson-Dell, R., Verkhratsky, A., & Mohamet, L. (2017). Aberrant iPSC-derived human astrocytes in Alzheimer's disease. *Cell Death & Disease, 8*(3), e2696–e2696. doi:10.1038/cddis.2017.89

Jones, E. V., Bernardinelli, Y., Tse, Y. C., Chierzi, S., Wong, T. P., & Murai, K. K. (2011). Astrocytes control glutamate receptor levels at developing synapses through SPARC–β-integrin interactions. *The Journal of Neuroscience : The Official Journal of the Society for Neuroscience, 31*(11), 4154–4165. doi:10.1523/JNEUROSCI.4757-10.2011 PMID:21411656

Jo, S., Yarishkin, O., Hwang, Y. J., Chun, Y. E., Park, M., Woo, D. H., Bae, J. Y., Kim, T., Lee, J., Chun, H., Park, H. J., Lee, D. Y., Hong, J., Kim, H. Y., Oh, S. J., Park, S. J., Lee, H., Yoon, B. E., Kim, Y., ... Lee, C. J. (2014). GABA from reactive astrocytes impairs memory in mouse models of Alzheimer's disease. *Nature Medicine, 20*(8), 886–896. doi:10.1038/nm.3639 PMID:24973918

Jo, W. K., Zhang, Y., Emrich, H. M., & Dietrich, D. E. (2015). Glia in the cytokine-mediated onset of depression: Fine tuning the immune response. *Frontiers in Cellular Neuroscience, 9*(JULY), 1–13. doi:10.3389/fncel.2015.00268 PMID:26217190

Juopperi, T. A., Kim, W. R., Chiang, C.-H., Yu, H., Margolis, R. L., Ross, C. A., Ming, G., & Song, H. (2012). Astrocytes generated from patient induced pluripotent stem cells recapitulate features of Huntington's disease patient cells. *Molecular Brain, 5*(1), 1–14. doi:10.1186/1756-6606-5-17 PMID:22613578

Jurga, A. M., Paleczna, M., Kadluczka, J., & Kuter, K. Z. (2021). Beyond the GFAP-Astrocyte Protein Markers in the Brain. *Biomolecules, 11*(9), 1361. doi:10.3390/biom11091361 PMID:34572572

Jurga, A. M., Paleczna, M., & Kuter, K. Z. (2020). 'Overview of General and Discriminating Markers of Differential Microglia Phenotypes'. *Frontiers in Cellular Neuroscience, 14*, 198. https://www.frontiersin.org/articles/10.3389/fncel.2020.00198. doi:10.3389/fncel.2020.00198 PMID:32848611

Juza, R. M., & Pauli, E. M. (2014). Clinical and surgical anatomy of the liver: A review for clinicians. *Clinical Anatomy (New York, N.Y.), 27*(5), 764–769. doi:10.1002/ca.22350 PMID:24453062

Kabaria, S., Dalal, I., Gupta, K., & Bhurwal, A. (2003, July). Carlos, Minacapelli, D., Catalano, C., & Rustgi, V. (2021). Hepatic encephalopathy: A review. *Annals of Hepatology, 2*(3), 122–130. Advance online publication. doi:10.1016/S1665-2681(19)32137-4

Kadry, H., Noorani, B., & Cucullo, L. (2020). A blood–brain barrier overview on structure, function, impairment, and biomarkers of integrity. *Fluids and Barriers of the CNS, 17*(1), 69. doi:10.118612987-020-00230-3 PMID:33208141

Kaindl, A. M., Degos, V., Peineau, S., Gouadon, E., Chhor, V., Loron, G., Le Charpentier, T., Josserand, J., Ali, C., Vivien, D., Collingridge, G. L., Lombet, A., Issa, L., Rene, F., Loeffler, J.-P., Kavelaars, A., Verney, C., Mantz, J., & Gressens, P. (2012). Activation of microglial N-methyl-D-aspartate receptors triggers inflammation and neuronal cell death in the developing and mature brain. *Annals of Neurology, 72*(4), 536–549. doi:10.1002/ana.23626 PMID:23109148

Kaindl, A. M., Degos, V., Peineau, S., Gouadon, E., Loron, G., Lombet, A., Verney, C., Mantz, J., & Gressens, P. (2008). Microglia express functional NMDA receptors: A novel finding and a promise for innovative treatment of excitotoxic and inflammatory brain disease. *Neuropediatrics*, *39*(05), V23. doi:10.1055-0029-1215736

Kaiser, M., Maletzki, I., Hülsmann, S., Holtmann, B., Schulz-Schaeffer, W., Kirchhoff, F., Bähr, M., & Neusch, C. (2006). 'Progressive Loss of a Glial Potassium Channel (KCNJ10) in the Spinal Cord of the SOD1 (G93A) Transgenic Mouse Model of Amyotrophic Lateral Sclerosis'. *Journal of Neurochemistry*, *99*(3), 900–912. doi:10.1111/j.1471-4159.2006.04131.x PMID:16925593

Kalla, R., Liu, Z., Xu, S., Koppius, A., Imai, Y., Kloss, C. U. A., Kohsaka, S., Gschwendtner, A., Möller, J. C., Werner, A., & Raivich, G. (2001). Microglia and the early phase of immune surveillance in the axotomized facial motor nucleus: Impaired microglial activation and lymphocyte recruitment but no effect on neuronal survival or axonal regeneration in macrophage-colony stimulating factor-defici. *The Journal of Comparative Neurology*, *436*(2), 182–201. doi:10.1002/cne.1060 PMID:11438923

Kalmar, B., & Greensmith, L. (2009). Induction of heat shock proteins for protection against oxidative stress. *Advanced Drug Delivery Reviews*, *61*(4), 310–318. doi:10.1016/j.addr.2009.02.003 PMID:19248813

Kamins, J., & Giza, C. C. (2016). Concussion—mild traumatic brain injury: Recoverable injury with potential for serious sequelae. *Neurosurgery Clinics of North America*, *27*(4), 441–452. doi:10.1016/j.nec.2016.05.005 PMID:27637394

Kaminsky, N., Bihari, O., Kanner, S., & Barzilai, A. (2016). Connecting malfunctioning glial cells and brain degenerative disorders. *Genomics, Proteomics & Bioinformatics*, *14*(3), 155–165. doi:10.1016/j.gpb.2016.04.003 PMID:27245308

Kamphuis, W., Kooijman, L., Orre, M., Stassen, O., Pekny, M., & Hol, E. M. (2015). GFAP and vimentin deficiency alters gene expression in astrocytes and microglia in wild-type mice and changes the transcriptional response of reactive glia in mouse model for Alzheimer's disease. *Glia*, *63*(6), 1036–1056. doi:10.1002/glia.22800 PMID:25731615

Kamphuis, W., Middeldorp, J., Kooijman, L., Sluijs, J. A., Kooi, E. J., Moeton, M., Freriks, M., Mizee, M. R., & Hol, E. M. (2014). Glial fibrillary acidic protein isoform expression in plaque related astrogliosis in Alzheimer's disease. *Neurobiology of Aging*, *35*(3), 492–510. doi:10.1016/j.neurobiolaging.2013.09.035 PMID:24269023

Kang, J., Jiang, L., Goldman, S. A., & Nedergaard, M. (1998). Astrocyte-mediated potentiation of inhibitory synaptic transmission. *Nature Neuroscience*, *1*(8), 683–692. doi:10.1038/3684 PMID:10196584

Kang, J., & Rivest, S. (2007). MyD88-deficient bone marrow cells accelerate onset and reduce survival in a mouse model of amyotrophic lateral sclerosis. *The Journal of Cell Biology*, *179*(6), 1219–1230. doi:10.1083/jcb.200705046 PMID:18086918

Kang, S. H., Li, Y., Fukaya, M., Lorenzini, I., Cleveland, D. W., Ostrow, L. W., Rothstein, J. D., & Bergles, D. E. (2013). Degeneration and impaired regeneration of gray matter oligodendrocytes in amyotrophic lateral sclerosis. *Nature Neuroscience*, *16*(5), 571–579. doi:10.1038/nn.3357 PMID:23542689

Kang, T. C., Kim, D. S., Kwak, S. E., Kim, J. E., Won, M. H., Kim, D. W., Choi, S.-Y., & Kwon, O.-S. (2006). Epileptogenic roles of astroglial death and regeneration in the dentate gyrus of experimental temporal lobe epilepsy. *Glia*, *54*(4), 258–271. doi:10.1002/glia.20380 PMID:16845674

Kaplan, L., Chow, B. W., & Gu, C. (2020). Neuronal regulation of the blood-brain barrier and neurovascular coupling. *Nature Reviews. Neuroscience*, *21*(8), 416–432. doi:10.103841583-020-0322-2 PMID:32636528

Karimi-Abdolrezaee, S., & Billakanti, R. (2012). Reactive astrogliosis after spinal cord injury-beneficial and detrimental effects. *Molecular Neurobiology*, *46*(2), 251–264. doi:10.100712035-012-8287-4 PMID:22684804

Kaslin, J., Ganz, J., & Brand, M. (2008). Proliferation, neurogenesis and regeneration in the non-mammalian vertebrate brain. *Philosophical Transactions of the Royal Society of London. Series B, Biological Sciences*, *363*(1489), 101–122. doi:10.1098/rstb.2006.2015 PMID:17282988

Kassiotis, G., & Kollias, G. (2001). Uncoupling the proinflammatory from the immunosuppressive properties of tumor necrosis factor (TNF) at the p55 TNF receptor level: Implications for pathogenesis and therapy of autoimmune demyelination. *The Journal of Experimental Medicine*, *193*(4), 427–434. doi:10.1084/jem.193.4.427 PMID:11181695

Kassmann, C. M. (2014). Myelin peroxisomes–essential organelles for the maintenance of white matter in the nervous system. *Biochimie*, *98*, 111–118. doi:10.1016/j.biochi.2013.09.020

Kassmann, C. M., Lappe-Siefke, C., Baes, M., Brügger, B., Mildner, A., Werner, H. B., Natt, O., Michaelis, T., Prinz, M., Frahm, J., & Nave, K. A. (2007). Axonal loss and neuroinflammation caused by peroxisome-deficient oligodendrocytes. *Nature Genetics*, *39*(8), 969–976. doi:10.1038/ng2070

Kato, G., Inada, H., Wake, H., Akiyoshi, R., Miyamoto, A., Eto, K., Ishikawa, T., Moorhouse, A. J., Strassman, A. M., & Nabekura, J. (2016). Microglial Contact Prevents Excess Depolarization. *eNeuro*, *3*(June), 1–9. PMID:27390772

Kato, M., Hughes, R. D., Keays, R. T., & Williams, R. (1992). Electron Microscopic Study of Brain Capillaries in Cerebral Edema from Fulminant Hepatic Failure. *Hepatology (Baltimore, Md.)*, *15*(6), 1060–1066. doi:10.1002/hep.1840150615 PMID:1592344

Kato, T. A., & Kanba, S. (2013). Are microglia minding us? Digging up the unconscious mind-brain relationship from a neuropsychoanalytic approach. *Frontiers in Psychology*, *7*, 13. doi:10.3389/fnhum.2013.00013 PMID:23443737

Kaul, R., Gao, G. P., Aloya, M., Balamurugan, K., Petrosky, A., Michals, K., & Matalon, R. (1994). Canavan disease: Mutations among Jewish and non-Jewish patients. *American Journal of Human Genetics*, *55*(1), 34. PMID:8023850

Kaur, D., Sharma, V., & Deshmukh, R. (2019). Activation of microglia and astrocytes: a roadway to neuroinflammation and Alzheimer's disease. In Inflammopharmacology, 27(4), 663–677. Birkhauser Verlag AG. doi:10.100710787-019-00580-x

Kavroulakis, E., Simos, P. G., Kalaitzakis, G., Maris, T. G., Karageorgou, D., Zaganas, I., Panagiotakis, S., Basta, M., Vgontzas, A., & Papadaki, E. (2018). Myelin content changes in probable Alzheimer's disease and mild cognitive impairment : Associations with age and severity of neuropsychiatric impairment. *Journal of Magnetic Resonance Imaging*, *47*(5), 1359–1372. doi:10.1002/jmri.25849 PMID:28861929

Kawabori, M., Kacimi, R., Kauppinen, T., Calosing, C., Kim, J. Y., Hsieh, C. L., Nakamura, M. C., & Yenari, M. A. (2015). Triggering Receptor Expressed on Myeloid Cells 2 (TREM2) Deficiency Attenuates Phagocytic Activities of Microglia and Exacerbates Ischemic Damage in Experimental Stroke. *The Journal of Neuroscience : The Official Journal of the Society for Neuroscience*, *35*(8), 3384–3396. doi:10.1523/JNEUROSCI.2620-14.2015 PMID:25716838

Kawamata, T., Akiyama, H., Yamada, T., & McGeer, P. L. (1992). Immunologic reactions in amyotrophic lateral sclerosis brain and spinal cord tissue. *American Journal of Pathology*, *140*(3), 691–707. PMID:1347673

Kawasaki, Y., Xu, Z.-Z., Wang, X., Park, J. Y., Zhuang, Z.-Y., Tan, P.-H., Gao, Y.-J., Roy, K., Corfas, G., Lo, E. H., & Ji, R.-R. (2008). Distinct roles of matrix metalloproteases in the early- and late-phase development of neuropathic pain. *Nature Medicine*, *14*(3), 331–336. doi:10.1038/nm1723 PMID:18264108

Kazama, S., Kishikawa, J., Kiyomatsu, T., Kawai, K., Nozawa, H., Ishihara, S., & Watanabe, T. (2018). Expression of the stem cell marker CD133 is related to tumor development in colorectal carcinogenesis. *Asian Journal of Surgery*, *41*(3), 274–278. doi:10.1016/j.asjsur.2016.12.002 PMID:28190751

Kehtari, T. (2022). *In Vivo Direct Reprogramming of Reactive Astrocytes into Neurons in a Rat Model of Alzheimer's Disease*.

Keles, G. E., Anderson, B., & Berger, M. S. (1999). The effect of extent of resection on time to tumor progression and survival in patients with glioblastoma multiforme of the cerebral hemisphere. *Surgical Neurology*, *52*(4), 371–379. doi:10.1016/S0090-3019(99)00103-2 PMID:10555843

Keller, A. F., Beggs, S., Salter, M. W., & De Koninck, Y. (2007). Transformation of the output of spinal lamina I neurons after nerve injury and microglia stimulation underlying neuropathic pain. *Molecular Pain*, *3*, 27. doi:10.1186/1744-8069-3-27 PMID:17900333

Keller, A. F., Gravel, M., & Kriz, J. (2009). Live imaging of amyotrophic lateral sclerosis pathogenesis: Disease onset is characterized by marked induction of GFAP in Schwann cells. *Glia*, *57*(10), 1130–1142. doi:10.1002/glia.20836 PMID:19115383

Kellner, V., Kersbergen, C. J., Li, S., Babola, T. A., Saher, G., & Bergles, D. E. (2021). Dual metabotropic glutamate receptor signaling enables the coordination of astrocyte and neuron activity in developing sensory domains. *Neuron*, *109*(16), 2545–2555.e7. doi:10.1016/j.neuron.2021.06.010 PMID:34245686

Kelly, D., & Rothwell, P. M. (2020). Disentangling the multiple links between renal dysfunction and cerebrovascular disease. *Journal of Neurology, Neurosurgery, and Psychiatry*, *91*(1), 88–97. doi:10.1136/jnnp-2019-320526 PMID:31511306

Kelly, R. D., & Cowley, S. M. (2013). The physiological roles of histone deacetylase (HDAC) 1 and 2: Complex co-stars with multiple leading parts. *Biochemical Society Transactions*, *41*(3), 741–749. doi:10.1042/BST20130010 PMID:23697933

Kempuraj, D., Ahmed, M., Selvakumar, G., Thangavel, R., Dhaliwal, A., Dubova, I., Mentor, S., Premkumar, K., Saeed, D., Zahoor, H., Raikwar, S., Zaheer, S., Iyer, S., & Zaheer, A. (2020). Brain injury-Mediated Neuroinflammatory response and Alzheimer's disease. *The Neuroscientist*, *26*(2), 134–155. doi:10.1177/1073858419848293 PMID:31092147

Kengne, F. G., & Decaux, G. (2017). Hyponatremia and the Brain. *Kidney International Reports*, *3*(1), 24–35. doi:10.1016/j.ekir.2017.08.015 PMID:29340311

Kettenmann, H., Hanisch, U.-K., Noda, M., & Verkhratsky, A. (2011). 'Physiology of Microglia'. *Physiological Reviews*, *91*(2), 461–553. doi:10.1152/physrev.00011.2010 PMID:21527731

Kettenmann, H., Kirchhoff, F., & Verkhratsky, A. (2013). Microglia: New roles for the synaptic stripper. *Neuron*, *77*(1), 10–18. doi:10.1016/j.neuron.2012.12.023 PMID:23312512

Kettenmann, H., & Verkhratsky, A. (2008). Neuroglia: The 150 years after. *Trends in Neurosciences*, *31*(12), 653–659. doi:10.1016/j.tins.2008.09.003 PMID:18945498

Keum, J. W., Shin, A., Gillis, T., Mysore, J. S., Elneel, K. A., Lucente, D., Hadzi, T., Holmans, P., Jones, L., & Orth, M. (2016). The HTT CAG-expansion mutation determines age at death but not disease duration in Huntington disease. *American Journal of Human Genetics*, *98*(2), 287–298. doi:10.1016/j.ajhg.2015.12.018 PMID:26849111

Khaing, Z. Z., & Seidlits, S. K. (2015). Hyaluronic acid and neural stem cells: Implications for biomaterial design. *Journal of Materials Chemistry. B*, *3*(40), 7850–7866. doi:10.1039/C5TB00974J PMID:32262899

Khakh, B. S., Beaumont, V., Cachope, R., Munoz-Sanjuan, I., Goldman, S. A., & Grantyn, R. (2017). Unravelling and exploiting astrocyte dysfunction in Huntington's disease. *Trends in Neurosciences*, *40*(7), 422–437. doi:10.1016/j.tins.2017.05.002 PMID:28578789

Khakh, B. S., & Sofroniew, M. V. (2015). 'Diversity of Astrocyte Functions and Phenotypes in Neural Circuits'. *Nature Neuroscience*, *18*(7), 942–952. doi:10.1038/nn.4043 PMID:26108722

Khan, S. A., Peracha, H., Ballhausen, D., Wiesbauer, A., Rohrbach, M., Gautschi, M., Mason, R. W., Giugliani, R., Suzuki, Y., Orii, K. E., Orii, T., & Tomatsu, S. (2017). Epidemiology of mucopolysaccharidoses. *Molecular Genetics and Metabolism, 121*(3), 227–240. doi:10.1016/j.ymgme.2017.05.016 PMID:28595941

Khodagholi, F., Eftekharzadeh, B., Maghsoudi, N., & Rezaei, P. F. (2010). Chitosan prevents oxidative stress-induced amyloid β formation and cytotoxicity in NT2 neurons: Involvement of transcription factors Nrf2 and NF-κB. *Molecular and Cellular Biochemistry, 337*(1-2), 39–51. doi:10.100711010-009-0284-1 PMID:19844776

Kiecolt-Glaser, J. K., Derry, H. M., & Fagundes, C. P. (2015). Inflammation: Depression fans the flames and feasts on the heat. *The American Journal of Psychiatry, 172*(11), 1075–1091. doi:10.1176/appi.ajp.2015.15020152 PMID:26357876

Kierdorf, K., & Prinz, M. (2013). Factors regulating microglia activation. *Frontiers in Cellular Neuroscience, 7*, 44. doi:10.3389/fncel.2013.00044 PMID:23630462

Kierdorf, K., & Prinz, M. (2017). Microglia in steady state. *The Journal of Clinical Investigation, 127*(9), 3201–3209. doi:10.1172/JCI90602 PMID:28714861

Kigerl, K. A., de Rivero Vaccari, J. P., Dietrich, W. D., Popovich, P. G., & Keane, R. W. (2014). Pattern recognition receptors and central nervous system repair. *Experimental Neurology, 258*, 5–16. doi:10.1016/j.expneurol.2014.01.001 PMID:25017883

Kilb, W., & Kirischuk, S. (2022). GABA Release from Astrocytes in Health and Disease. *International Journal of Molecular Sciences, 23*(24), 15859. doi:10.3390/ijms232415859 PMID:36555501

Kim, H., Tator, C. H., & Shoichet, M. S. (2011). Chitosan implants in the rat spinal cord: Biocompatibility and biodegradation. *Journal of Biomedical Materials Research - Part A, 97* A(4), 395–404. doi:10.1002/jbm.a.33070

Kim, J., Choi, D., Jeong, H., Kim, J., & Kim, D. (n.d.). *DJ-1 facilitates the interaction between STAT1 and its phosphatase, SHP-1, in brain microglia and astrocytes: a novel anti-inflammatory function of DJ-1.* Elsevier. https://www.sciencedirect.com/science/article/pii/S0969996113002283

Kimelberg, H. K. (2005). Astrocytic swelling in cerebral ischemia as a possible cause of injury and target for therapy. *Glia, 50*(4), 389–397. doi:10.1002/glia.20174 PMID:15846797

Kimelberg, H. K., & Nedergaard, M. (2010). 'Functions of Astrocytes and Their Potential as Therapeutic Targets'. *Neurotherapeutics; the Journal of the American Society for Experimental NeuroTherapeutics*, 7(4), 338–353. doi:10.1016/j.nurt.2010.07.006 PMID:20880499

Kim, H. G., Ju, M. S., Ha, S. K., Lee, H., Lee, H., Kim, S. Y., & Oh, M. S. (2012). Acacetin Protects Dopaminergic Cells against 1-Methyl-4-phenyl-1,2,3,6-tetrahydropyridine-Induced Neuroinflammation in Vitro and in Vivo. *Biological & Pharmaceutical Bulletin, 35*(8), 1287–1294. doi:10.1248/bpb.b12-00127 PMID:22863927

Kim, H. J., Kim, H. J., Lee, J. Y., Yun, J. Y., Kim, S. Y., Park, S. S., & Jeon, B. S. (2011). Phenotype analysis in patients with early onset Parkinson's disease with and without parkin mutations. *Journal of Neurology, 258*(12), 2260–2267. doi:10.100700415-011-6110-1 PMID:21625934

Kim, K., & Park, C. B. (2020). Femtomolar sensing of Alzheimer's tau proteins by water oxidation-coupled photoelectrochemical platform. *Biosensors & Bioelectronics, 154*, 112075. doi:10.1016/j.bios.2020.112075 PMID:32056970

Kim, M. S., Sung, M. J., Seo, S. B., Yoo, S. J., Lim, W. K., & Kim, H. M. (2002). Water-soluble chitosan inhibits the production of pro-inflammatory cytokine in human astrocytoma cells activated by amyloid β peptide and interleukin-1β. *Neuroscience Letters, 321*(1-2), 105–109. doi:10.1016/S0304-3940(02)00066-6 PMID:11872267

Kim, S. (2018). Competitive Biological Activities of Chitosan and Its Derivatives: Antimicrobial, Antioxidant, Anticancer, and Anti-Inflammatory Activities. *International Journal of Polymer Science*, *2018*, 1–13. doi:10.1155/2018/1708172

Kim, S. H., Smith, C. J., & Van Eldik, L. J. (2004). Importance of MAPK pathways for microglial pro-inflammatory cytokine IL-1 beta production. *Neurobiology of Aging*, *25*(4), 431–439. doi:10.1016/S0197-4580(03)00126-X PMID:15013563

Kim, S. K., Hayashi, H., Ishikawa, T., Shibata, K., Shigetomi, E., Shinozaki, Y., Inada, H., Roh, S. E., Kim, S. J., Lee, G., Bae, H., Moorhouse, A. J., Mikoshiba, K., Fukazawa, Y., Koizumi, S., & Nabekura, J. (2016). Cortical astrocytes rewire somatosensory cortical circuits for peripheral neuropathic pain. *The Journal of Clinical Investigation*, *126*(5), 1983–1997. doi:10.1172/JCI82859 PMID:27064281

Kim, S. K., & Rajapakse, N. (2005). Enzymatic production and biological activities of chitosan oligosaccharides (COS): A review. *Carbohydrate Polymers*, *62*(4), 357–368. doi:10.1016/j.carbpol.2005.08.012

Kim, S. Y., Porter, B. E., Friedman, A., & Kaufer, D. (2016). A potential role for glia-derived extracellular matrix remodeling in postinjury epilepsy. *Journal of Neuroscience Research*, *94*(9), 794–803. doi:10.1002/jnr.23758 PMID:27265805

Kimura, A., Namekata, K., Guo, X., Harada, C., & Harada, T. (2016). Neuroprotection, Growth Factors and BDNF-TrkB Signalling in Retinal Degeneration. *International Journal of Molecular Sciences*, *17*(9), 1584. doi:10.3390/ijms17091584 PMID:27657046

Kimura, T., Kimura, A. K., Ren, M., Monteiro, V., Xu, Y., Berno, B., Schlame, M., & Epand, R. M. (2019). 'Plasmalogen Loss Caused by Remodeling Deficiency in Mitochondria'. *Life Science Alliance*, *2*(4), e201900348. doi:10.26508/lsa.201900348 PMID:31434794

Kim, Y. C., Kim, Y. H., Kim, J. W., & Ha, K. Y. (2016). Transplantation of mesenchymal stem cells for acute spinal cord injury in rats: Comparative study between Intralesional injection and scaffold based transplantation. *Journal of Korean Medical Science*, *31*(9), 1373–1382. doi:10.3346/jkms.2016.31.9.1373 PMID:27510379

Kim, Y. S., Choi, D. H., Block, M. L., Lorenzl, S., Yang, L., Kim, Y. J., Sugama, S., Cho, B. P., Hwang, O., Browne, S. E., Kim, S. Y., Hong, J., Flint Beal, M., & Joh, T. H. (2007). A pivotal role of matrix metalloproteinase-3 activity in dopaminergic neuronal degeneration via microglial activation. *The FASEB Journal*, *21*(1), 179–187. doi:10.1096/fj.06-5865com PMID:17116747

Kim, Y. S., & Joh, T. H. (2006). Microglia, major player in the brain inflammation: Their roles in the pathogenesis of Parkinson's disease. *Experimental & Molecular Medicine*, *38*(4), 333–347. doi:10.1038/emm.2006.40 PMID:16953112

Kim, Y., Park, J., & Choi, Y. K. (2019). 'The Role of Astrocytes in the Central Nervous System Focused on BK Channel and Heme Oxygenase Metabolites: A Review'. *Antioxidants*, *8*(5), 121. doi:10.3390/antiox8050121 PMID:31060341

Kim, Y., Zharkinbekov, Z., Raziyeva, K., Tabyldiyeva, L., Berikova, K., Zhumagul, D., Temirkhanova, K., & Saparov, A. (2023). Chitosan-Based Biomaterials for Tissue Regeneration. *Pharmaceutics*, *15*(3), 3. doi:10.3390/pharmaceutics15030807 PMID:36986668

Kingma, S. D. K., Bodamer, O. A., & Wijburg, F. A. (2015). Epidemiology and diagnosis of lysosomal storage disorders; challenges of screening. *Best Practice & Research. Clinical Endocrinology & Metabolism*, *29*(2), 145–157. doi:10.1016/j.beem.2014.08.004 PMID:25987169

Kinney, G. A., & Spain, W. J. (2002). Synaptically evoked GABA transporter currents in neocortical glia. *Journal of Neurophysiology*, *88*(6), 2899–2908. doi:10.1152/jn.00037.2002 PMID:12466416

Kisler, K., Nelson, A. R., Rege, S. V., Ramanathan, A., Wang, Y., Ahuja, A., Lazic, D., Tsai, P. S., Zhao, Z., Zhou, Y., Boas, D. A., Sakadžić, S., & Zlokovic, B. V. (2017). Pericyte degeneration leads to neurovascular uncoupling and limits oxygen supply to the brain. *Nature Neuroscience*, *20*(3), 406–416. doi:10.1038/nn.4489 PMID:28135240

Kitamura, T., Tsuchihashi, Y., & Fujita, S. (1978). Initial response of silver-impregnated "resting microglia" to stab wounding in rabbit hippocampus. *Acta Neuropathologica*, *44*(1), 31–39. doi:10.1007/BF00691636 PMID:707043

Kitamura, Y., Watanabe, S., Taguchi, M., Takagi, K., Kawata, T., Takahashi-Niki, K., Yasui, H., Maita, H., Iguchi-Ariga, S. M. M., & Ariga, H. (2011). Neuroprotective effect of a new DJ-1-binding compound against neurodegeneration in Parkinson's disease and stroke model rats. *Molecular Neurodegeneration*, *6*(1), 48. doi:10.1186/1750-1326-6-48 PMID:21740546

Klein, U., Kresse, H., & Von Figura, K. (1978). Sanfilippo syndrome type C: deficiency of acetyl-CoA: alpha-glucosaminide N-acetyltransferase in skin fibroblasts. *Proceedings of the National Academy of Sciences of the United States of America*, *75*(10), 5185–5189. doi:10.1073/pnas.75.10.5185 PMID:33384

Klempan, T. A., Ernst, C., Deleva, V., Labonte, B., & Turecki, G. (2009). Characterization of QKI gene expression, genetics, and epigenetics in suicide victims with major depressive disorder. *Biological Psychiatry*, *66*(9), 824–831. doi:10.1016/j.biopsych.2009.05.010 PMID:19545858

Kobayashi, H., Chattopadhyay, S., Kato, K., Dolkas, J., Kikuchi, S.-I., Myers, R. R., & Shubayev, V. I. (2008). MMPs initiate Schwann cell-mediated MBP degradation and mechanical nociception after nerve damage. *Molecular and Cellular Neurosciences*, *39*(4), 619–627. doi:10.1016/j.mcn.2008.08.008 PMID:18817874

Kodachi, T., Matsumoto, S., Mizuguchi, M., Osaka, H., Kanai, N., Nanba, E., Ohno, K., & Yamagata, T. (2017). Severe demyelination in a patient with a late infantile form of Niemann-Pick disease type C. *Neuropathology*, *37*(5), 426–430. doi:10.1111/neup.12380 PMID:28387450

Koerber, J. T., Klimczak, R., Jang, J.-H., Dalkara, D., Flannery, J. G., & Schaffer, D. V. (2009). Molecular evolution of adeno-associated virus for enhanced glial gene delivery. *Molecular Therapy*, *17*(12), 2088–2095. doi:10.1038/mt.2009.184 PMID:19672246

Kofuji, P., & Newman, E. A. (2004). 'Potassium Buffering in the Central Nervous System'. *Neuroscience*, *129*(4), 1043–1054. doi:10.1016/j.neuroscience.2004.06.008 PMID:15561419

Kohno, K., Kitano, J., Kohro, Y., Tozaki-Saitoh, H., Inoue, K., & Tsuda, M. (2018). Temporal Kinetics of Microgliosis in the Spinal Dorsal Horn after Peripheral Nerve Injury in Rodents. *Biological & Pharmaceutical Bulletin*, *41*(7), 1096–1102. doi:10.1248/bpb.b18-00278 PMID:29962405

Kohno, T., Moore, K. A., Baba, H., & Woolf, C. J. (2003). Peripheral nerve injury alters excitatory synaptic transmission in lamina II of the rat dorsal horn. *The Journal of Physiology*, *548*(Pt 1), 131–138. doi:10.1113/jphysiol.2002.036186 PMID:12576493

Koizumi, S., Ohsawa, K., Inoue, K., & Glia, S. K. (2013). *Purinergic receptors in microglia: functional modal shifts of microglia mediated by P2 and P1 receptors*. Wiley doi:10.1002/glia.22358

Koizumi, S., Shigemoto-Mogami, Y., Nasu-Tada, K., Shinozaki, Y., Ohsawa, K., Tsuda, M., Joshi, B. V., Jacobson, K. A., Kohsaka, S., & Inoue, K. (2007). 'UDP Acting at P2Y6 Receptors Is a Mediator of Microglial Phagocytosis'. *Nature*, *446*(7139), 1091–1095. doi:10.1038/nature05704 PMID:17410128

Kölliker-Frers, R., Udovin, L., Otero-Losada, M., Kobiec, T., Herrera, M. I., Palacios, J., Razzitte, G., & Capani, F. (2021). Neuroinflammation: An Integrating Overview of Reactive-Neuroimmune Cell Interactions in Health and Disease. In Mediators of Inflammation. Hindawi Limited. doi:10.1155/2021/9999146

Kondo, S., & Okabe, S. (2013). In vivo two-photon microscopy of microglia. *Methods in Molecular Biology (Clifton, N.J.)*, *1041*, 319–335. doi:10.1007/978-1-62703-520-0_28 PMID:23813390

Kong, H., Sha, L., Fan, Y., Xiao, M., Ding, J., Wu, J., & Hu, G. (2009). Requirement of AQP4 for antidepressive efficiency of fluoxetine: Implication in adult hippocampal neurogenesis. *Neuropsychopharmacology*, *34*(5), 1263–1276. doi:10.1038/npp.2008.185 PMID:18923397

Kong, X., Fu, J., Shao, K., Wang, L., Lan, X., & Shi, J. (2019). Biomimetic hydrogel for rapid and scar-free healing of skin wounds inspired by the healing process of oral mucosa. *Acta Biomaterialia*, *100*, 255–269. doi:10.1016/j.actbio.2019.10.011 PMID:31606531

Koning, N., Swaab, D. F., Hoek, R. M., & Huitinga, I. (2009). 'Distribution of the Immune Inhibitory Molecules CD200 and CD200R in the Normal Central Nervous System and Multiple Sclerosis Lesions Suggests Neuron-Glia and Glia-Glia Interactions'. *Journal of Neuropathology and Experimental Neurology*, *68*(2), 159–167. doi:10.1097/NEN.0b013e3181964113 PMID:19151626

Konishi, H., Kiyama, H., & Ueno, M. (2019). Dual functions of microglia in the formation and refinement of neural circuits during development. *International Journal of Developmental Neuroscience*, *77*(1), 18–25. doi:10.1016/j.ijdevneu.2018.09.009 PMID:30292872

Koo, H. N., Jeong, H. J., Hong, S. H., Choi, J. H., An, N. H., & Kim, H. M. (2002). High molecular weight water-soluble chitosan protects against apoptosis induced by serum starvation in human astrocytes. *The Journal of Nutritional Biochemistry*, *13*(4), 245–249. doi:10.1016/S0955-2863(01)00218-2 PMID:11988407

Kopec, C. D., Li, B., Wei, W., Boehm, J., & Malinow, R. (2006). Glutamate receptor exocytosis and spine enlargement during chemically induced long-term potentiation. *The Journal of Neuroscience : The Official Journal of the Society for Neuroscience*, *26*(7), 2000–2009. doi:10.1523/JNEUROSCI.3918-05.2006 PMID:16481433

Kordower, J. H., Olanow, C. W., Dodiya, H. B., Chu, Y., Beach, T. G., Adler, C. H., Halliday, G. M., & Bartus, R. T. (2013). Disease duration and the integrity of the nigrostriatal system in Parkinson's disease. *Brain*, *136*(8), 2419–2431. doi:10.1093/brain/awt192 PMID:23884810

Kornev, V. A., Grebenik, E. A., Solovieva, A. B., Dmitriev, R. I., & Timashev, P. S. (2018). Hydrogel-assisted neuroregeneration approaches towards brain injury therapy: A state-of-the-art review. *Computational and Structural Biotechnology Journal*, *16*, 488–502. doi:10.1016/j.csbj.2018.10.011 PMID:30455858

Kostrzewa, R. M., & Segura-Aguilar, J. (2003). Novel mechanisms and approaches in the study of neurodegeneration and neuroprotection. A review. *Neurotoxicity Research*, *5*(6), 375–383. doi:10.1007/BF03033166 PMID:14715440

Kotter, M. R., Li, W.-W., Zhao, C., & Franklin, R. J. M. (2006). 'Myelin Impairs CNS Remyelination by Inhibiting Oligodendrocyte Precursor Cell Differentiation'. *The Journal of Neuroscience : The Official Journal of the Society for Neuroscience*, *26*(1), 328–332. doi:10.1523/JNEUROSCI.2615-05.2006 PMID:16399703

Kövari, E., Gold, G., Herrmann, F. R., Canuto, A., Hof, P. R., Bouras, C., & Giannakopoulos, P. (2007). Cortical microinfarcts and demyelination affect cognition in cases at high risk for dementia. *Neurology*, *68*(12), 927–931. doi:10.1212/01.wnl.0000257094.10655.9a PMID:17372128

Koyama, R., & Ikegaya, Y. (2015). Microglia in the pathogenesis of autism spectrum disorders. *Neuroscience Research*, *100*, 1–5. doi:10.1016/j.neures.2015.06.005 PMID:26116891

Koyama, Y. (2014). Signaling molecules regulating phenotypic conversions of astrocytes and glial scar formation in damaged nerve tissues. *Neurochemistry International*, *78*, 35–42. doi:10.1016/j.neuint.2014.08.005 PMID:25180676

Krabbe, G., Matyash, V., Pannasch, U., Mamer, L., Boddeke, H. W. G. M., & Kettenmann, H. (2012). Activation of serotonin receptors promotes microglial injury-induced motility but attenuates phagocytic activity. *Brain, Behavior, and Immunity, 26*(3), 419–428. doi:10.1016/j.bbi.2011.12.002 PMID:22198120

Kraft, A. W., Hu, X., Yoon, H., Yan, P., Xiao, Q., Wang, Y., Gil, S. C., Brown, J., Wilhelmsson, U., Restivo, J. L., Cirrito, J. R., Holtzman, D. M., Kim, J., Pekny, M., & Lee, J. (2013). Attenuating astrocyte activation accelerates plaque pathogenesis in APP/PS1 mice. *The FASEB Journal, 27*(1), 187–198. doi:10.1096/fj.12-208660 PMID:23038755

Krakora, D., Mulcrone, P., Meyer, M., Lewis, C., Bernau, K., Gowing, G., Zimprich, C., Aebischer, P., Svendsen, C. N., & Suzuki, M. (2013). Synergistic effects of GDNF and VEGF on lifespan and disease progression in a familial ALS rat model. *Molecular therapy: the journal of the American Society of Gene Therapy, 21*(8), 1602–1610. doi:10.1038/mt.2013.108

Krauss, M., Weiss, T., Langnaese, K., Richter, K., Kowski, A., Veh, R. W., & Laube, G. (2007). Cellular and subcellular rat brain spermidine synthase expression patterns suggest region-specific roles for polyamines, including cerebellar presynaptic function. *Journal of Neurochemistry, 103*(2), 679–693. doi:10.1111/j.1471-4159.2007.04770.x PMID:17635671

Kreisel, T., Frank, M. G., Licht, T., Reshef, R., Ben-Menachem-Zidon, O., Baratta, M. V., Maier, S. F., & Yirmiya, R. (2014). Dynamic microglial alterations underlie stress-induced depressive-like behavior and suppressed neurogenesis. *Molecular Psychiatry, 19*(6), 699–709. doi:10.1038/mp.2013.155 PMID:24342992

Kremer, D., Göttle, P., Flores-Rivera, J., Hartung, H. P., & Küry, P. (2019). Remyelination in multiple sclerosis: From concept to clinical trials. *Current Opinion in Neurology, 32*(3), 378–384. doi:10.1097/WCO.0000000000000692

Krencik, R., & Zhang, S. C. (2011). Directed differentiation of functional astroglial subtypes from human pluripotent stem cells. *Nature Protocols, 6*(11), 1710–1717. doi:10.1038/nprot.2011.405

Kresse, H., & Neufeld, E. F. (1972). The Sanfilippo A corrective factor: Purification and mode of action. *The Journal of Biological Chemistry, 247*(7), 2164–2170. doi:10.1016/S0021-9258(19)45506-3 PMID:4259567

Kresse, H., Paschke, E., Von Figura, K., Gilberg, W., & Fuchs, W. (1980). Sanfilippo disease type D: Deficiency of N-acetylglucosamine-6-sulfate sulfatase required for heparan sulfate degradation. *Proceedings of the National Academy of Sciences of the United States of America, 77*(11), 6822–6826. doi:10.1073/pnas.77.11.6822 PMID:6450420

Kubota, Y., Takubo, K., Shimizu, T., Ohno, H., Kishi, K., Shibuya, M., Saya, H., & Suda, T. (2009). M-CSF inhibition selectively targets pathological angiogenesis and lymphangiogenesis. *The Journal of Experimental Medicine, 206*(5), 1089–1102. doi:10.1084/jem.20081605 PMID:19398755

Kucheryavykh, L. Y., Benedikt, J., Cubano, L. A., Skatchkov, S. N., Bukauskas, F. F., & Kucheryavykh, Y. V. (2017). Polyamines preserve connexin43-mediated gap junctional communication during intracellular hypercalcemia and acidosis. *Neuroreport, 28*(4), 208–213. doi:10.1097/WNR.0000000000000746 PMID:28134630

Kuchibhotla, K. V., Lattarulo, C. R., Hyman, B. T., & Bacskai, B. J. (2009). Synchronous hyperactivity and intercellular calcium waves in astrocytes in Alzheimer's mice. *Science, 323*(5918), 1211–1215. doi:10.1126cience.1169096 PMID:19251629

Kucukdereli, H., Allen, N. J., Lee, A. T., Feng, A., Ozlu, M. I., Conatser, L. M., Chakraborty, C., Workman, G., Weaver, M., Sage, E. H., Barres, B. A., & Eroglu, C. (2011). Control of excitatory CNS synaptogenesis by astrocyte-secreted proteins Hevin and SPARC. *Proceedings of the National Academy of Sciences of the United States of America, 108*(32), E440–E449. doi:10.1073/pnas.1104977108 PMID:21788491

Kugaya, A., & Sanacora, G. (2005). Beyond monoamines: Glutamatergic function in mood disorders. *CNS Spectrums, 10*(10), 808–819. doi:10.1017/S1092852900010403 PMID:16400244

Kuhn, S., Gritti, L., Crooks, D., & Dombrowski, Y. (2019). Oligodendrocytes in development, myelin generation and beyond. *Cells*, *8*(11), 1424. doi:10.3390/cells8111424

Kuhnt, D., Becker, A., Ganslandt, O., Bauer, M., Buchfelder, M., & Nimsky, C. (2011). Correlation of the extent of tumor volume resection and patient survival in surgery of glioblastoma multiforme with high-field intraoperative MRI guidance. *Neuro-Oncology*, *13*(12), 1339–1348. doi:10.1093/neuonc/nor133 PMID:21914639

Kulijewicz-Nawrot, M., Verkhratsky, A., Chvátal, A., Syková, E., Rodríguez, J.J. (2012). Astrocytic cytoskeletal atrophy in the medial prefrontal cortex of a triple transgenic mouse model of Alzheimer's disease. *Journal of Anatomy, 221*(3), 252-62. . doi:10.1111/j.1469-7580.2012.01536.x

Kulijewicz-Nawrot, M., Syková, E., Chvátal, A., Verkhratsky, A., & Rodríguez, J. J. (2013). Astrocytes and glutamate homoeostasis in Alzheimer's disease: A decrease in glutamine synthetase, but not in glutamate transporter-1, in the prefrontal cortex. *ASN Neuro*, *5*(4), 273–282. doi:10.1042/AN20130017 PMID:24059854

Kumar, M. N. R. (2000). A review of chitin and chitosan applications. *Reactive & Functional Polymers*, *46*(1), 1–27. doi:10.1016/S1381-5148(00)00038-9

Kumirska, J., Weinhold, M. X., Czerwicka, M., & Stepnowski, P. (2011). Influence of the chemical structure and physicochemical properties of chitin-and chitosan-based materials on their biomedical activity. *Biomedical Engineering, Trends in Materials Science*, 25-64.

Kuner, R., & Kuner, T. (2021). Cellular Circuits in the Brain and Their Modulation in Acute and Chronic Pain. *Physiological Reviews*, *101*(1), 213–258. doi:10.1152/physrev.00040.2019 PMID:32525759

Kuo, Y. C., & Chiu, K. H. (2011). Inverted colloidal crystal scaffolds with laminin-derived peptides for neuronal differentiation of bone marrow stromal cells. *Biomaterials*, *32*(3), 819–831. doi:10.1016/j.biomaterials.2010.09.057 PMID:20974492

Kurz, H., & Christ, B. (1998). Embryonic CNS macrophages and microglia do not stem from circulating, but from extravascular precursors. *Glia*, *22*(1), 98–102. doi:10.1002/(SICI)1098-1136(199801)22:1<98::AID-GLIA10>3.0.CO;2-V PMID:9436792

Kushner, P. D., Stephenson, D. T., & Wright, S. (1991). Reactive astrogliosis is widespread in the subcortical white matter of amyotrophic lateral sclerosis brain. *Journal of Neuropathology and Experimental Neurology*, *50*(3), 263–277. doi:10.1097/00005072-199105000-00008 PMID:2022968

Kuzumaki, N., Narita, M., Narita, M., Hareyama, N., Niikura, K., Nagumo, Y., Nozaki, H., Amano, T., & Suzuki, T. (2007). Chronic pain-induced astrocyte activation in the cingulate cortex with no change in neural or glial differentiation from neural stem cells in mice. *Neuroscience Letters*, *415*(1), 22–27. doi:10.1016/j.neulet.2006.12.057 PMID:17316991

Kwon, H. S., & Koh, S.-H. (2020). Neuroinflammation in neurodegenerative disorders : The roles of microglia and astrocytes. *Translational Neurodegeneration*, *9*(1), 42. doi:10.118640035-020-00221-2 PMID:33239064

Lacroix, M., Abi-Said, D., Fourney, D. R., Gokaslan, Z. L., Shi, W., DeMonte, F., Lang, F. F., McCutcheon, I. E., Hassenbusch, S. J., Holland, E., Hess, K., Michael, C., Miller, D., & Sawaya, R. (2001). A multivariate analysis of 416 patients with glioblastoma multiforme: Prognosis, extent of resection, and survival. *Journal of Neurosurgery*, *95*(2), 190–198. doi:10.3171/jns.2001.95.2.0190 PMID:11780887

Laffer, B., Bauer, D., Wasmuth, S., Busch, M., Jalilvand, T. V., Thanos, S., Meyer Zu Hörste, G., Loser, K., Langmann, T., Heiligenhaus, A., & Kasper, M. (2019). Loss of IL-10 Promotes Differentiation of Microglia to a M1 Phenotype. *Frontiers in Cellular Neuroscience*, *13*, 430. doi:10.3389/fncel.2019.00430 PMID:31649508

Lakhani, B., Borich, M. R., Jackson, J. N., Wadden, K. P., Peters, S., Villamayor, A., MacKay, A. L., Vavasour, I. M., Rauscher, A., & Boyd, L. A. (2016). Motor Skill Acquisition Promotes Human Brain Myelin Plasticity. *Neural Plasticity, 2016*, 1–7. doi:10.1155/2016/7526135 PMID:27293906

Lanctôt, K. L., Amatniek, J., Ancoli-Israel, S., Arnold, S. E., Ballard, C., Cohen-Mansfield, J., Ismail, Z., Lyketsos, C., Miller, D. S., Musiek, E., Osorio, R. S., Rosenberg, P. B., Satlin, A., Steffens, D., Tariot, P., Bain, L. J., Carrillo, M. C., Hendrix, J. A., Jurgens, H., & Boot, B. (2017). Neuropsychiatric signs and symptoms of Alzheimer's disease: New treatment paradigms. *Alzheimer's & Dementia: Translational Research & Clinical Interventions, 3*(3), 440–449. doi:10.1016/j.trci.2017.07.001 PMID:29067350

Lang, C., Barco, A., Zablow, L., Kandel, E. R., Siegelbaum, S. A., & Zakharenko, S. S. (2004). Transient expansion of synaptically connected dendritic spines upon induction of hippocampal long-term potentiation. *Proceedings of the National Academy of Sciences of the United States of America, 101*(47), 16665–16670. doi:10.1073/pnas.0407581101 PMID:15542587

Langston, J. W. (2006). The parkinson's complex: Parkinsonism is just the tip of the iceberg. *Annals of Neurology, 59*(4), 591–596. doi:10.1002/ana.20834 PMID:16566021

Lanjhiyana, S. K., Bajpayee, P., Kesavan, K., Lanjhiyana, S., & Muthu, M. S. (2013). Chitosan–sodium alginate blended polyelectrolyte complexes as potential multiparticulate carrier system: Colon-targeted delivery and gamma scintigraphic imaging. *Expert Opinion on Drug Delivery, 10*(1), 5–15. doi:10.1517/17425247.2013.734805 PMID:23106236

Lanquillon, S., Krieg, J.-C., Bening-Abu-Shach, U., & Vedder, H. (2000). Cytokine production and treatment response in major depressive disorder. *Neuropsychopharmacology, 22*(4), 370–379. doi:10.1016/S0893-133X(99)00134-7 PMID:10700656

Laplaud, D. A., Ruiz, C., Wiertlewski, S., Brouard, S., Berthelot, L., & Guillet, M. (2004). Blood T-cell receptor β chain transcriptome in multiple sclerosis. Characterization of the T cells with altered CDR3 length distribution. *Brain, 127*(5), 981–995. doi:10.1093/brain/awh119 PMID:14985265

Lappe-Siefke, C., Goebbels, S., Gravel, M., Nicksch, E., Lee, J., Braun, P. E., Griffiths, I. R., & Nave, K. A. (2003). Disruption of Cnp1 uncouples oligodendroglial functions in axonal support and myelination. *Nature Genetics, 33*(3), 366–374. doi:10.1038/ng1095 PMID:12590258

Largo, C., Cuevas, P., Somjen, G. G., Del Rio, R. M., & Herreras, O. (1996). The effect of depressing glial function in rat brain in situ on ion homeostasis, synaptic transmission, and neuron survival. *The Journal of Neuroscience : The Official Journal of the Society for Neuroscience, 16*(3), 1219–1229. doi:10.1523/JNEUROSCI.16-03-01219.1996 PMID:8558250

Larsen, F. S. (1996). Cerebral Circulation in Liver Failure: Ohm's Law in Force. Seminars in liver disease, 16. Thieme Medical Publishers, Inc.

Larsen, F. S., Hansen, B. A., Pott, F., Ejlersen, E., Secher, N. H., Paulson, O. B., & Knudsen, G. M. (1996). Dissociated Cerebral Vasoparalysis in Acute Liver Failure: A Hypothesis of Gradual Cerebral Hyperaemia. *Journal of Hepatology, 25*(2), 145–151. doi:10.1016/S0168-8278(96)80066-3 PMID:8878774

Larsen, F. S., Strauss, G., Møller, K., & Hansen, B. A. (2000). Regional Cerebral Blood Flow Autoregulation in Patients with Fulminant Hepatic Failure. *Liver Transplantation, 6*(6), 795–800. doi:10.1053/jlts.2000.18705 PMID:11084071

Lassmann, H. (2018). Multiple sclerosis pathology. *Cold Spring Harbor Perspectives in Medicine, 8*(3), a028936. doi:10.1101/cshperspect.a028936

Lassmann, H. (2020). Pathology of inflammatory diseases of the nervous system: Human disease versus animal models. *Glia, 68*(4), 830–844. doi:10.1002/glia.23726 PMID:31605512

Latremoliere, A., & Woolf, C. J. (2009). Central sensitization: A generator of pain hypersensitivity by central neural plasticity. *The Journal of Pain, 10*(9), 895–926. doi:10.1016/j.jpain.2009.06.012 PMID:19712899

Laube, G., Ru¨, R., & Veh, R. W. (1997). Astrocytes, not neurons, show most prominent staining for spermidine/spermine-like immunoreactivity in adult rat brain. *Glia, 19*(2), 171–179. doi:10.1002/(SICI)1098-1136(199702)19:2<171::AID-GLIA8>3.0.CO;2-3 PMID:9034833

Laurence, J. A., & Fatemi, S. H. (2005). Glial fibrillary acidic protein is elevated in superior frontal, parietal and cerebellar cortices of autistic subjects. *Cerebellum (London, England), 4*(3), 206–210. doi:10.1080/14734220500208846 PMID:16147953

Lavoie, S., Chen, Y., Dalton, T. P., Gysin, R., Cuénod, M., Steullet, P., & Do, K. Q. (2009). Curcumin, quercetin, and tBHQ modulate glutathione levels in astrocytes and neurons : Importance of the glutamate cysteine ligase modifier subunit. *Journal of Neurochemistry, 108*(6), 1410–1422. doi:10.1111/j.1471-4159.2009.05908.x PMID:19183254

Lawrence, J., Schardien, K., Wigdahl, B., & Nonnemacher, M. (2023). Roles of neuropathology-associated reactive astrocytes: A systematic review. *Acta Neuropathologica Communications, 11*(1), 42. doi:10.118640478-023-01526-9 PMID:36915214

Lawson, L. J., & Perry, S. (1990). *Heterogeneity In The Distribution And Morphology Of M~Croglia In The Normal Adult Mouse Ibrain.* Wiley.

Lawson, L. J., Perry, V. H., Dri, P., & Gordon, S. (1990). Heterogeneity in the distribution and morphology of microglia in the normal adult mouse brain. *Neuroscience, 39*(1), 151–170. doi:10.1016/0306-4522(90)90229-W PMID:2089275

Lazic, A., Balint, V., Stanisavljevic Ninkovic, D., Peric, M., & Stevanovic, M. (2022). Reactive and Senescent Astroglial Phenotypes as Hallmarks of Brain Pathologies. *International Journal of Molecular Sciences, 23*(9), 4995. doi:10.3390/ijms23094995 PMID:35563385

Leal, M. C., Casabona, J. C., Puntel, M., & Pitossi, F. (2013). Interleukin-1beta and TNF-alpha: Reliable targets for protective therapies in Parkinson's disease? *Frontiers in Cellular Neuroscience, 7*(APR), 45874. doi:10.3389/FNCEL.2013.00053/BIBTEX

Lechtreck, K.-F., Delmotte, P., Robinson, M. L., Sanderson, M. J., & Witman, G. B. (2008). Mutations in Hydin impair ciliary motility in mice. *The Journal of Cell Biology, 180*(3), 633–643. doi:10.1083/jcb.200710162 PMID:18250199

Lécuyer, M. A., Kebir, H., & Prat, A. (2016). Glial influences on BBB functions and molecular players in immune cell trafficking. *Biochimica et Biophysica Acta, 1862*(3), 472–482. doi:10.1016/j.bbadis.2015.10.004 PMID:26454208

Ledeboer, A., Sloane, E. M., Milligan, E. D., Frank, M. G., Mahony, J. H., Maier, S. F., & Watkins, L. R. (2005). Minocycline attenuates mechanical allodynia and proinflammatory cytokine expression in rat models of pain facilitation. *Pain, 115*(1–2), 71–83. doi:10.1016/j.pain.2005.02.009 PMID:15836971

Lee, W. (2022). Acute Liver Failure. *Yamada's Textbook of Gastroenterology,* 1889–1905.

Lee, E., & Chung, W.-S. (2019). Glial control of synapse number in healthy and diseased brain. *Frontiers in Cellular Neuroscience, 13*, 42. doi:10.3389/fncel.2019.00042 PMID:30814931

Lee, H. J., Suk, J. E., Patrick, C., Bae, E. J., Cho, J. H., Rho, S., Hwang, D., Masliah, E., & Lee, S.-J. (2010). Direct transfer of α-synuclein from neuron to astroglia causes inflammatory responses in synucleinopathies. *The Journal of Biological Chemistry, 285*(12), 9262–9272. doi:10.1074/jbc.M109.081125 PMID:20071342

Lee, H. S., Ghetti, A., Pinto-Duarte, A., Wang, X., Dziewczapolski, G., Galimi, F., Huitron-Resendiz, S., Piña-Crespo, J. C., Roberts, A. J., Verma, I. M., Sejnowski, T. J., & Heinemann, S. F. (2014). Astrocytes contribute to gamma oscillations and recognition memory. *Proceedings of the National Academy of Sciences of the United States of America, 111*(32), E3343–E3352. doi:10.1073/pnas.1410893111 PMID:25071179

Lee, H. U., Yamazaki, Y., Tanaka, K. F., Tanaka, K. F., Furuya, K., Sokabe, M., Hida, H., Takao, K., Miyakawa, T., Miyakawa, T., Fujii, S., & Ikenaka, K. (2013). Increased astrocytic ATP release results in enhanced excitability of the hippocampus. *Glia, 61*(2), 210–224. doi:10.1002/glia.22427 PMID:23018918

Lee, H.-G., Wheeler, M. A., & Quintana, F. J. (2022). Function and therapeutic value of astrocytes in neurological diseases. *Nature Reviews. Drug Discovery, 21*(5), 339–358. doi:10.103841573-022-00390-x PMID:35173313

Lee, J., Kim, S. W., & Kim, K. T. (2022). Region-Specific Characteristics of Astrocytes and Microglia: A Possible Involvement in Aging and Diseases. *Cells, 11*(12), 1902. doi:10.3390/cells11121902 PMID:35741031

Lee, M.-C., Yu, W.-C., Shih, Y.-H., Chen, C.-Y., Guo, Z.-H., Huang, S.-J., Chan, J. C. C., & Chen, Y.-R. (2018). Zinc ion rapidly induces toxic, off-pathway amyloid-β oligomers distinct from amyloid-β derived diffusible ligands in Alzheimer's disease. *Scientific Reports, 8*(1), 1–16. doi:10.103841598-018-23122-x PMID:29555950

Lee, M., Schwab, C., & Mcgeer, P. L. (2011). Astrocytes are GABAergic cells that modulate microglial activity. *Glia, 59*(1), 152–165. doi:10.1002/glia.21087 PMID:21046567

Leenders, K. L., Salmon, E. P., Tyrrell, P., Perani, D., Brooks, D. J., Sager, H., Jones, T., Marsden, C. D., & Frackowiak, R. S. J. (1990). The nigrostriatal dopaminergic system assessed in vivo by positron emission tomography in healthy volunteer subjects and patients with Parkinson's disease. *Archives of Neurology, 47*(12), 1290–1298. doi:10.1001/archneur.1990.00530120034007 PMID:2123623

Lee, S. H., Liu, L., Wang, Y. T., & Sheng, M. (2002). Clathrin adaptor AP2 and NSF interact with overlapping sites of GluR2 and play distinct roles in AMPA receptor trafficking and hippocampal LTD. *Neuron, 36*(4), 661–674. doi:10.1016/S0896-6273(02)01024-3 PMID:12441055

Lees, A. J., Hardy, J., & Revesz, T. (2009). Parkinson's disease. *Lancet, 373*(9680), 2055–2066. doi:10.1016/S0140-6736(09)60492-X PMID:19524782

Lee, Y., Morrison, B. M., Li, Y., Lengacher, S., Farah, M. H., Hoffman, P. N., Liu, Y., Tsingalia, A., Jin, L., Zhang, P. W., Pellerin, L., Magistretti, P. J., & Rothstein, J. D. (2012). Oligodendroglia metabolically support axons and contribute to neurodegeneration. *Nature, 487*(7408), 443–448. doi:10.1038/nature11314 PMID:22801498

Lenz, K. M., Nugent, B. M., Haliyur, R., & McCarthy, M. M. (2013). 'Microglia Are Essential to Masculinization of Brain and Behavior'. *The Journal of Neuroscience : The Official Journal of the Society for Neuroscience, 33*(7), 2761–2772. doi:10.1523/JNEUROSCI.1268-12.2013 PMID:23407936

Lepore, A. C., Dejea, C., Carmen, J., Rauck, B., Kerr, D. A., Sofroniew, M. V., & Maragakis, N. J. (2008a). Selective ablation of proliferating astrocytes does not affect disease outcome in either acute or chronic models of motor neuron degeneration. *Experimental Neurology, 211*(2), 423–432. doi:10.1016/j.expneurol.2008.02.020 PMID:18410928

Lepore, A. C., Rauck, B., Dejea, C., Pardo, A. C., Rao, M. S., Rothstein, J. D., & Maragakis, N. J. (2008b). Focal transplantation-based astrocyte replacement is neuroprotective in a model of motor neuron disease. *Nature Neuroscience, 11*(11), 1294–1301. doi:10.1038/nn.2210 PMID:18931666

Lesné, S., Docagne, F., Gabriel, C., Liot, G., Lahiri, D.K., Buée, L., Plawinski, L., Delacourte, A., MacKenzie, E.T., Buisson, A., & Vivien, D. (2003). Transforming growth factor-beta 1 potentiates amyloid-beta generation in astrocytes and in transgenic mice. *JBC, 278*(20), 18408-18. . doi:10.1074/jbc.M300819200

Lessard, C. B., Malnik, S. L., Zhou, Y., Ladd, T. B., Cruz, P. E., Ran, Y., Mahan, T. E., Chakrabaty, P., Holtzman, D. M., Ulrich, J. D., Colonna, M., & Golde, T. E. (2018). High-affinity interactions and signal transduction between Aβ oligomers and TREM 2. *EMBO Molecular Medicine*, *10*(11), e9027. doi:10.15252/emmm.201809027 PMID:30341064

Lev, N., Barhum, Y., Ben-Zur, T., Melamed, E., Steiner, I., & Offen, D. (2013). Knocking out DJ-1 attenuates astrocytes neuroprotection against 6-hydroxydopamine toxicity. *Journal of Molecular Neuroscience*, *50*(3), 542–550. doi:10.100712031-013-9984-9 PMID:23536331

LeWitt, P. A. (2008). Levodopa for the treatment of Parkinson's disease. *The New England Journal of Medicine*, *359*(23), 2468–2476. doi:10.1056/NEJMct0800326 PMID:19052127

Leyns, C. E. G., Gratuze, M., Narasimhan, S., Jain, N., Koscal, L. J., Jiang, H., Manis, M., Colonna, M., Lee, V. M. Y., Ulrich, J. D., & Holtzman, D. M. (2019). TREM2 function impedes tau seeding in neuritic plaques. *Nature Neuroscience*, *22*(8), 1217–1222. doi:10.103841593-019-0433-0 PMID:31235932

Liang, K., Bae, K. H., & Kurisawa, M. (2019). Recent advances in the design of injectable hydrogels for stem cell-based therapy. *Journal of Materials Chemistry. B*, *7*(24), 3775–3791. doi:10.1039/C9TB00485H

Liang, S., Wu, X., & Jin, F. (2018). Gut-Brain Psychology: Rethinking Psychology From the Microbiota-Gut-Brain Axis. *Frontiers in Integrative Neuroscience*, *12*, 33. doi:10.3389/fnint.2018.00033 PMID:30271330

Lian, H., Litvinchuk, A., Chiang, A. C.-A., Aithmitti, N., Jankowsky, J. L., & Zheng, H. (2016). Astrocyte-Microglia Cross Talk through Complement Activation Modulates Amyloid Pathology in Mouse Models of Alzheimer's Disease. *The Journal of Neuroscience : The Official Journal of the Society for Neuroscience*, *36*(2), 577–589. doi:10.1523/JNEUROSCI.2117-15.2016 PMID:26758846

Li, B., McCrudden, C. M., Yuen, H. F., Xi, X., Lyu, P., Chan, K. W., Zhang, S. D., & Kwok, H. F. (2017). CD133 in brain tumor: The prognostic factor. *Oncotarget*, *8*(7), 11144–11159. doi:10.18632/oncotarget.14406 PMID:28055976

Liddelow, S. A., & Barres, B. A. (2017). Reactive Astrocytes: Production, Function, and Therapeutic Potential. *Immunity, 46*(6), 957-967. doi:10.1016/j.immuni.2017.06.006

Liddelow, S. A., Guttenplan, K. A., Clarke, L. E., Bennett, F. C., Bohlen, C. J., Schirmer, L., Bennett, M. L., Münch, A. E., Chung, W.-S., Peterson, T. C., Wilton, D. K., Frouin, A., Napier, B. A., Panicker, N., Kumar, M., Buckwalter, M. S., Rowitch, D. H., Dawson, V. L., Dawson, T. M., & Barres, B. A. (2017). Neurotoxic reactive astrocytes are induced by activated microglia. *Nature*, *541*(7638), 481–487. doi:10.1038/nature21029 PMID:28099414

Li, J., Ramenaden, E. R., Peng, J., Koito, H., Volpe, J. J., & Rosenberg, P. A. (2008). Tumor necrosis factor α mediates lipopolysaccharide-induced microglial toxicity to developing oligodendrocytes when astrocytes are present. *The Journal of Neuroscience : The Official Journal of the Society for Neuroscience*, *28*(20), 5321–5330. doi:10.1523/JNEUROSCI.3995-07.2008 PMID:18480288

Lim, D. A., & Alvarez-Buylla, A. (2016). 'The Adult Ventricular–Subventricular Zone (V-SVZ) and Olfactory Bulb (OB) Neurogenesis'. *Cold Spring Harbor Perspectives in Biology*, *8*(5), a018820. doi:10.1101/cshperspect.a018820 PMID:27048191

Lin, H., Li, G., Peng, X., Deng, A., Ye, L., Shi, L., Wang, T., & He, J. (2021). The use of CRISPR/Cas9 as a tool to study human infectious viruses. *Frontiers in Cellular and Infection Microbiology*, *11*, 590989. doi:10.3389/fcimb.2021.590989 PMID:34513721

Lin, J., Jo, S. B., Kim, T. H., Kim, H. W., & Chew, S. Y. (2020). RNA interference in glial cells for nerve injury treatment. *Journal of Tissue Engineering*, *11*, 2041731420939224. doi:10.1177/2041731420939224 PMID:32670539

Lin, L.-F. H., Doherty, D. H., Lile, J. D., Bektesh, S., & Collins, F. (1993). GDNF: A glial cell line-derived neurotrophic factor for midbrain dopaminergic neurons. *Science*, *260*(5111), 1130–1132. doi:10.1126cience.8493557 PMID:8493557

Liotta, E. M., Romanova, A. L., Lizza, B. D., Rasmussen-torvik, L. J., Kim, M., Francis, B., Sangha, R. S., Carroll, T. J., Ganger, D., Ladner, D. P., Naidech, A. M., Paparello, J. J., Prabhakaran, S., Sorond, F. A., & Maas, M. B. (2017). Osmotic Shifts, Cerebral Edema, and Neurologic Deterioration in Severe Hepatic Encephalopathy. *Critical Care Medicine*, 1–10. PMID:29341965

Li, Q., & Barres, B. A. (2018). 'Microglia and Macrophages in Brain Homeostasis and Disease'. *Nature Reviews. Immunology*, *18*(4), 225–242. doi:10.1038/nri.2017.125 PMID:29151590

Li, S., & Selkoe, D. J. (2020). A mechanistic hypothesis for the impairment of synaptic plasticity by soluble Aβ oligomers from Alzheimer's brain. *Journal of Neurochemistry*. *Journal of Neurochemistry*, *154*(6), 583–597. doi:10.1111/jnc.15007 PMID:32180217

Li, S., & Sheng, Z. H. (2023). Oligodendrocyte-derived transcellular signaling regulates axonal energy metabolism. *Current Opinion in Neurobiology*, *80*, 102722. doi:10.1016/j.conb.2023.102722

Li, T., Chen, X., Zhang, C., Zhang, Y., & Yao, W. (2019). An update on reactive astrocytes in chronic pain. *Journal of Neuroinflammation*, *16*(1), 140. doi:10.118612974-019-1524-2 PMID:31288837

Liu, H., Yang, Y., Xia, Y., Zhu, W., Leak, R., Wei, Z., Wang, J., & Xiaoming, H. (2017). 'Aging of Cerebral White Matter'. *Ageing Research Reviews*, *34*, 64–76. . doi:10.1016/j.arr.2016.11.006

Liu, Y., Zhou, Q., Tang, M., Fu, N., & Shao, W. (2015). *Upregulation of alphaB-crystallin expression in the substantia nigra of patients with Parkinson's disease*. Elsevier. https://www.sciencedirect.com/science/article/pii/S0197458015000482

Liu, C. C., Kanekiyo, T., Xu, H., & Bu, G. (2013). Apolipoprotein E and Alzheimer disease: Risk, mechanisms and therapy. *Nature Reviews. Neurology*, *9*(2), 106–118. doi:10.1038/nrneurol.2012.263 PMID:23296339

Liu, C.-Y., Yang, Y., Ju, W.-N., Wang, X., & Zhang, H.-L. (2018). Emerging Roles of Astrocytes in Neuro-Vascular Unit and the Tripartite Synapse With Emphasis on Reactive Gliosis in the Context of Alzheimer's Disease. *Frontiers in Cellular Neuroscience*, *12*, 193. https://www.frontiersin.org/articles/10.3389/fncel.2018.00193. doi:10.3389/fncel.2018.00193 PMID:30042661

Liu, J., Chang, L., Song, Y., Li, H., & Wu, Y. (2019). The Role of NMDA Receptors in Alzheimer's Disease. *Frontiers in Neuroscience*, *13*, 43. doi:10.3389/fnins.2019.00043 PMID:30800052

Liu, J., Dietz, K., DeLoyht, J. M., Pedre, X., Kelkar, D., Kaur, J., Vialou, V., Lobo, M. K., Dietz, D. M., Nestler, E. J., Dupree, J., & Casaccia, P. (2012). Impaired adult myelination in the prefrontal cortex of socially isolated mice. *Nature Neuroscience*, *15*(12), 1621–1623. doi:10.1038/nn.3263 PMID:23143512

Liu, J.-H., Wang, W., Wu, H., Gong, X., & Moriguchi, T. (2015). Polyamines function in stress tolerance : From synthesis to regulation. *Frontiers in Plant Science*, *6*, 827. doi:10.3389/fpls.2015.00827 PMID:26528300

Liu, J., Wang, L., Tan, J., Ji, P., Gauthier, S., Zhang, Y., Ma, T., & Liu, S. (2012). Burden, anxiety and depression in caregivers of veterans with dementia in Beijing. *Archives of Gerontology and Geriatrics*, *55*(3), 560–563. doi:10.1016/j.archger.2012.05.014 PMID:22749403

Liu, L.-R., Liu, J.-C., Bao, J.-S., Bai, Q.-Q., & Wang, G.-Q. (2020). Interaction of Microglia and Astrocytes in the Neurovascular Unit. *Frontiers in Immunology*, *11*, 1024. doi:10.3389/fimmu.2020.01024 PMID:32733433

Liu, N., Li, Y., Jiang, Y., Shi, S., Niamnud, A., Vodovoz, S. J., Katakam, P. V. G., Vidoudez, C., Dumont, A. S., & Wang, X. (2023). Establishment and Application of a Novel In Vitro Model of Microglial Activation in Traumatic Brain Injury. *The Journal of Neuroscience : The Official Journal of the Society for Neuroscience*, *43*(2), 319–332. doi:10.1523/JNEUROSCI.1539-22.2022 PMID:36446585

Liu, Q. S., Xu, Q., Kang, J., & Nedergaard, M. (2004). Astrocyte activation of presynaptic metabotropic glutamate receptors modulates hippocampal inhibitory synaptic transmission. *Neuron Glia Biology*, *1*(4), 307–316. doi:10.1017/S1740925X05000190 PMID:16755304

Liu, W.-X., Wang, J., Xie, Z.-M., Xu, N., Zhang, G.-F., Jia, M., Zhou, Z.-Q., Hashimoto, K., & Yang, J.-J. (2016). Regulation of glutamate transporter 1 via BDNF-TrkB signaling plays a role in the anti-apoptotic and antidepressant effects of ketamine in chronic unpredictable stress model of depression. *Psychopharmacology*, *233*(3), 405–415. doi:10.100700213-015-4128-2 PMID:26514555

Liu, Y. J., Chen, J., Li, X., Zhou, X., Hu, Y. M., Chu, S. F., Peng, Y., & Chen, N. H. (2019). Research progress on adenosine in central nervous system diseases. *CNS Neuroscience & Therapeutics*, *25*(9), 899–910. doi:10.1111/cns.13190 PMID:31334608

Liu, Z.-Y., Song, Z.-W., Guo, S.-W., He, J.-S., Wang, S.-Y., Zhu, J.-G., Yang, H.-L., & Liu, J.-B. (2019). CXCL12/CXCR4 signaling contributes to neuropathic pain via central sensitization mechanisms in a rat spinal nerve ligation model. *CNS Neuroscience & Therapeutics*, *25*(9), 922–936. doi:10.1111/cns.13128 PMID:30955244

Li, W.-W., Setzu, A., Zhao, C., & Franklin, R. J. M. (2005). Minocycline-mediated inhibition of microglia activation impairs oligodendrocyte progenitor cell responses and remyelination in a non-immune model of demyelination. *Journal of Neuroimmunology*, *158*(1–2), 58–66. doi:10.1016/j.jneuroim.2004.08.011 PMID:15589038

Li, X., Li, M., Tian, L., Chen, J., Liu, R., & Ning, B. (2020). Reactive Astrogliosis: Implications in Spinal Cord Injury Progression and Therapy. *Oxidative Medicine and Cellular Longevity*, *9494352*, 1–14. doi:10.1155/2020/9494352 PMID:32884625

Li, X., Tao, Y., Bradley, R., Du, Z., Tao, Y., Kong, L., Dong, Y., Jones, J., Yan, Y., Harder, C. R. K., Friedman, L. M., Bilal, M., Hoffmann, B., & Zhang, S. C. (2018). Fast Generation of Functional Subtype Astrocytes from Human Pluripotent Stem Cells. *Stem Cell Reports*, *11*(4), 998–1008. doi:10.1016/j.stemcr.2018.08.019 PMID:30269954

Li, Y., Chen, L., Liu, Y., Zhang, Y., Liang, Y., & Mei, Y. (2018). Anti-inflammatory effects in a mouse osteoarthritis model of a mixture of glucosamine and chitooligosaccharides produced by bi-enzyme single-step hydrolysis. *Scientific Reports*, *8*(1), 5624. doi:10.103841598-018-24050-6 PMID:29618773

Li, Y., Liu, H., Xu, Q. S., Du, Y. G., & Xu, J. (2014). Chitosan oligosaccharides block LPS-induced O-GlcNAcylation of NF-κB and endothelial inflammatory response. *Carbohydrate Polymers*, *99*, 568–578. doi:10.1016/j.carbpol.2013.08.082 PMID:24274545

Li, Y., Xia, Y., Yin, S., Wan, F., Hu, J., Kou, L., Sun, Y., Wu, J., Zhou, Q., Huang, J., Xiong, N., & Wang, T. (2021). Targeting Microglial α-Synuclein/TLRs/NF-kappaB/NLRP3 Inflammasome Axis in Parkinson's Disease. *Frontiers in Immunology*, *12*, 719807. doi:10.3389/fimmu.2021.719807 PMID:34691027

Lobsiger, C. S., Boillee, S., McAlonis-Downes, M., Khan, A. M., Feltri, M. L., Yamanaka, K., & Cleveland, D. W. (2009). Schwann cells expressing dismutase active mutant SOD1 unexpectedly slow disease progression in ALS mice. *Proceedings of the National Academy of Sciences of the United States of America*, *106*(11), 4465–4470. doi:10.1073/pnas.0813339106 PMID:19251638

Lobsiger, C. S., & Cleveland, D. W. (2007). Glial cells as intrinsic components of non-cell-autonomous neurodegenerative disease. *Nature Neuroscience*, *10*(11), 1355–1360. doi:10.1038/nn1988 PMID:17965655

Loera-Valencia, R., Cedazo-Minguez, A., Kenigsberg, P. A., Page, G., Duarte, A. I., Giusti, P., Zusso, M., Robert, P., Frisoni, G. B., Cattaneo, A., Zille, M., Boltze, J., Cartier, N., Buee, L., Johansson, G., & Winblad, B. (2019). Current and emerging avenues for Alzheimer's disease drug targets. *Journal of Internal Medicine*, *286*(4), 398–437. doi:10.1111/joim.12959 PMID:31286586

Loera-Valencia, R., Goikolea, J., Parrado-Fernandez, C., Merino-Serrais, P., & Maioli, S. (2019). Alterations in cholesterol metabolism as a risk factor for developing Alzheimer's disease: Potential novel targets for treatment. *The Journal of Steroid Biochemistry and Molecular Biology*, *190*, 104–114. doi:10.1016/j.jsbmb.2019.03.003 PMID:30878503

Loggia, M. L., Chonde, D. B., Akeju, O., Arabasz, G., Catana, C., Edwards, R. R., Hill, E., Hsu, S., Izquierdo-Garcia, D., Ji, R.-R., Riley, M., Wasan, A. D., Zürcher, N. R., Albrecht, D. S., Vangel, M. G., Rosen, B. R., Napadow, V., & Hooker, J. M. (2015). Evidence for brain glial activation in chronic pain patients. *Brain*, *138*(Pt 3), 604–615. doi:10.1093/brain/awu377 PMID:25582579

Lomen-Hoerth, C., Anderson, T., & Miller, B. (2002). The overlap of amyotrophic lateral sclerosis and frontotemporal dementia. *Neurology*, *59*(7), 1077–1079. doi:10.1212/WNL.59.7.1077 PMID:12370467

Long, P., & Corfas, G. (2014). To learn is to myelinate. *Science*, *346*(6207), 298–299. doi:10.1126cience.1261127 PMID:25324370

López-Muguruza, E., Villar-Gómez, N., Matias-Guiu, J. A., Selma-Calvo, B., Moreno-Jiménez, L., Sancho-Bielsa, F., Lopez-Carbonero, J., Benito-Martín, M. S., García-Flores, S., Bonel-García, N., Kamal, O. M.-F., Ojeda-Hernández, D., Matías-Guiu, J., & Gómez-Pinedo, U. (2022). The integration of cell therapy and biomaterials as treatment strategies for remyelination. *Life (Basel, Switzerland)*, *12*(4), 474. doi:10.3390/life12040474

López-Murcia, F. J., Terni, B., & Llobet, A. (2015). SPARC triggers a cell-autonomous program of synapse elimination. *Proceedings of the National Academy of Sciences of the United States of America*, *112*(43), 13366–13371. doi:10.1073/pnas.1512202112 PMID:26420865

Lorenzini, L., Fernandez, M., Baldassarro, V. A., Bighinati, A., Giuliani, A., Calzà, L., & Giardino, L. (2020). White matter and neuroprotection in Alzheimer's dementia. *Molecules (Basel, Switzerland)*, *25*(3), 503. doi:10.3390/molecules25030503 PMID:31979414

Louw, A. M., Kolar, M. K., Novikova, L. N., Kingham, P. J., Wiberg, M., Kjems, J., & Novikov, L. N. (2016). Chitosan polyplex mediated delivery of miRNA-124 reduces activation of microglial cells in vitro and in rat models of spinal cord injury. *Nanomedicine; Nanotechnology, Biology, and Medicine*, *12*(3), 643–653. doi:10.1016/j.nano.2015.10.011 PMID:26582736

Lucassen, P. J., Fuchs, E., & Czéh, B. (2004). Antidepressant treatment with tianeptine reduces apoptosis in the hippocampal dentate gyrus and temporal cortex. *Biological Psychiatry*, *55*(8), 789–796. doi:10.1016/j.biopsych.2003.12.014 PMID:15050859

Lucassen, P. J., Müller, M. B., Holsboer, F., Bauer, J., Holtrop, A., Wouda, J., Hoogendijk, W. J. G., De Kloet, E. R., & Swaab, D. F. (2001). Hippocampal apoptosis in major depression is a minor event and absent from subareas at risk for glucocorticoid overexposure. *American Journal of Pathology*, *158*(2), 453–468. doi:10.1016/S0002-9440(10)63988-0 PMID:11159183

Lucassen, P. J., Vollmann-Honsdorf, G. K., Gleisberg, M., Czéh, B., De Kloet, E. R., & Fuchs, E. (2001). Chronic psychosocial stress differentially affects apoptosis in hippocampal subregions and cortex of the adult tree shrew. *The European Journal of Neuroscience*, *14*(1), 161–166. doi:10.1046/j.0953-816x.2001.01629.x PMID:11488960

Luchena, C., Zuazo-Ibarra, J., Alberdi, E., Matute, C., & Capetillo-Zarate, E. (2018). Contribution of Neurons and Glial Cells to Complement-Mediated Synapse Removal during Development, Aging and in Alzheimer's Disease. *Mediators of Inflammation*, *2018*, 1–12. doi:10.1155/2018/2530414 PMID:30533998

Luciani, M., Gritti, A., & Meneghini, V. (2020). Human iPSC-based models for the development of therapeutics targeting neurodegenerative lysosomal storage diseases. *Frontiers in Molecular Biosciences*, *7*, 224. doi:10.3389/fmolb.2020.00224 PMID:33062642

Lu, H.-J., & Gao, Y.-J. (2023). Astrocytes in Chronic Pain: Cellular and Molecular Mechanisms. *Neuroscience Bulletin*, *39*(3), 425–439. doi:10.100712264-022-00961-3 PMID:36376699

Luissint, A. C., Artus, C., Glacial, F., Ganeshamoorthy, K., & Couraud, P. O. (2012). Tight junctions at the blood-brain barrier: Physiological architecture and disease-associated dysregulation. *Fluids and Barriers of the CNS*, *9*(1), 23. doi:10.1186/2045-8118-9-23 PMID:23140302

Lu, K. (2023). Cellular Pathogenesis of Hepatic Encephalopathy: An Update. *Biomolecules*, *13*(2), 396. doi:10.3390/biom13020396 PMID:36830765

Luo, J., Wärmländer, S., & A. G.-J. (2016). Cross-interactions between the Alzheimer disease amyloid-β peptide and other amyloid proteins: a further aspect of the amyloid cascade hypothesis. *ASBMB*.

Luo, C., Jian, C., Liao, Y., Huang, Q., Wu, Y., Liu, X., Zou, D., & Wu, Y. (2017). The role of microglia in multiple sclerosis. *Neuropsychiatric Disease and Treatment*, *13*, 1661–1667. doi:10.2147/NDT.S140634 PMID:28721047

Luo, H., Liu, H.-Z., Zhang, W.-W., Matsuda, M., Lv, N., Chen, G., Xu, Z.-Z., & Zhang, Y.-Q. (2019). Interleukin-17 Regulates Neuron-Glial Communications, Synaptic Transmission, and Neuropathic Pain after Chemotherapy. *Cell Reports*, *29*(8), 2384–2397.e5. doi:10.1016/j.celrep.2019.10.085 PMID:31747607

Lüscher, C., & Malenka, R. C. (2012). NMDA receptor-dependent long-term potentiation and long-term depression (LTP/LTD). *Cold Spring Harbor Perspectives in Biology*, *4*(6), a005710. doi:10.1101/cshperspect.a005710 PMID:22510460

Luskin, M. B., Pearlman, A. L., & Sanes, J. R. (1988). 'Cell Lineage in the Cerebral Cortex of the Mouse Studied in Vivo and in Vitro with a Recombinant Retrovirus'. *Neuron*, *1*(8), 635–647. doi:10.1016/0896-6273(88)90163-8 PMID:3272182

Lynch, G. S., Dunwiddie, T., & Gribkoff, V. (1977). Heterosynaptic depression: A postsynaptic correlate of long-term potentiation. *Nature*, *266*(5604), 737–739. doi:10.1038/266737a0 PMID:195211

MacDonald, A., Lu, B., Caron, M., Caporicci-Dinucci, N., Hatrock, D., Petrecca, K., Bourque, G., & Stratton, J. A. (2021). 'Single Cell Transcriptomics of Ependymal Cells Across Age, Region and Species Reveals Cilia-Related and Metal Ion Regulatory Roles as Major Conserved Ependymal Cell Functions'. *Frontiers in Cellular Neuroscience*, *15*, 703951. https://www.frontiersin.org/articles/10.3389/fncel.2021.703951. doi:10.3389/fncel.2021.703951 PMID:34335193

MacVicar, B. A., & Choi, H. B. (2017). Astrocytes Provide Metabolic Support for Neuronal Synaptic Function in Response to Extracellular K+. *Neurochemical Research*, *42*(9), 2588–2594. doi:10.100711064-017-2315-8 PMID:28664400

Madeira, C., Lourenco, M. V., Vargas-Lopes, C., Suemoto, C. K., Brandão, C. O., Reis, T., Leite, R. E. P., Laks, J., Jacob-Filho, W., Pasqualucci, C. A., Grinberg, L. T., Ferreira, S. T., & Panizzutti, R. (2015). D-serine levels in Alzheimer's disease: Implications for novel biomarker development. *Translational Psychiatry*, *5*(5), e561. doi:10.1038/tp.2015.52 PMID:25942042

Madrigal, J. L. M., Leza, J. C., Polak, P., Kalinin, S., & Feinstein, D. L. (2009). Astrocyte-derived MCP-1 mediates neuroprotective effects of noradrenaline. *The Journal of Neuroscience : The Official Journal of the Society for Neuroscience, 29*(1), 263–267. doi:10.1523/JNEUROSCI.4926-08.2009 PMID:19129402

Mahmoud, S., Gharagozloo, M., Simard, C., & Gris, D. (2019). Astrocytes Maintain Glutamate Homeostasis in the CNS by Controlling the Balance between Glutamate Uptake and Release. *Cells, 8*(2), 2. doi:10.3390/cells8020184 PMID:30791579

Majewska, A. K., Newton, J. R., & Sur, M. (2006). Remodeling of synaptic structure in sensory cortical areas in vivo. *The Journal of Neuroscience : The Official Journal of the Society for Neuroscience, 26*(11), 3021–3029. doi:10.1523/JNEUROSCI.4454-05.2006 PMID:16540580

Ma, K.-C., Nie, X. J., Höög, A., Olsson, Y., & Zhang, W. W. (1994). Reactive astrocytes in viral infections of the human brain express endothelin-like immunoreactivity. *Journal of the Neurological Sciences, 126*(2), 184–192. doi:10.1016/0022-510X(94)90271-2 PMID:7531760

Makarava, N., Mychko, O., Molesworth, K., Chang, J. C.-Y., Henry, R. J., Tsymbalyuk, N., Gerzanich, V., Simard, J. M., Loane, D. J., & Baskakov, I. V. (2023). Region-Specific Homeostatic Identity of Astrocytes Is Essential for Defining Their Response to Pathological Insults. *Cells, 12*(17), 2172. doi:10.3390/cells12172172 PMID:37681904

Makinodan, M., Rosen, K. M., Ito, S., & Corfas, G. (2012). A critical period for social experience–dependent oligodendrocyte maturation and myelination. *Science, 337*(6100), 1357–1360. doi:10.1126cience.1220845 PMID:22984073

Malatesta, P., Hack, M. A., Hartfuss, E., Kettenmann, H., Klinkert, W., Kirchhoff, F., & Götz, M. (2003). 'Neuronal or Glial Progeny: Regional Differences in Radial Glia Fate'. *Neuron, 37*(5), 751–764. doi:10.1016/S0896-6273(03)00116-8 PMID:12628166

Malberg, J. E., Eisch, A. J., Nestler, E. J., & Duman, R. S. (2000). Chronic antidepressant treatment increases neurogenesis in adult rat hippocampus. *The Journal of Neuroscience : The Official Journal of the Society for Neuroscience, 20*(24), 9104–9110. doi:10.1523/JNEUROSCI.20-24-09104.2000 PMID:11124987

Maletic-Savatic, M., Malinow, R., & Svoboda, K. (1999). Rapid dendritic morphogenesis in CA1 hippocampal dendrites induced by synaptic activity. *Science, 283*(5409), 1923–1927. doi:10.1126cience.283.5409.1923 PMID:10082466

Malmsten, L., Vijayaraghavan, S., Hovatta, O., Marutle, A., & Darreh-Shori, T. (2014). Fibrillar β-amyloid 1-42 alters cytokine secretion, cholinergic signalling and neuronal differentiation. *Journal of Cellular and Molecular Medicine, 18*(9), 1874–1888. doi:10.1111/jcmm.12343 PMID:25109373

Malm, T. M., Iivonen, H., Goldsteins, G., Keksa-Goldsteine, V., Ahtoniemi, T., Kanninen, K., Salminen, A., Auriola, S., Van Groen, T., Tanila, H., & Koistinaho, J. (2007). Pyrrolidine Dithiocarbamate Activates Akt and Improves Spatial Learning in APP/PS1 Mice without Affecting β-Amyloid Burden. *The Journal of Neuroscience : The Official Journal of the Society for Neuroscience, 27*(14), 3712–3721. doi:10.1523/JNEUROSCI.0059-07.2007 PMID:17409235

Ma, M., Li, Y., Dai, S., Chu, M., Sun, L., Liu, L., & Zhou, J.-C. (2023). 'A Meta-Analysis on the Prevalence of Charcot–Marie–Tooth Disease and Related Inherited Peripheral Neuropathies'. *Journal of Neurology, 270*(5), 2468–2482. doi:10.100700415-023-11559-8 PMID:36631678

Man, H.-Y., Lin, J. W., Ju, W. H., Ahmadian, G., Liu, L., Becker, L. E., Sheng, M., & Wang, Y. T. (2000). Regulation of AMPA receptor–mediated synaptic transmission by clathrin-dependent receptor internalization. *Neuron, 25*(3), 649–662. doi:10.1016/S0896-6273(00)81067-3 PMID:10774732

Manji, H. K., Drevets, W. C., & Charney, D. S.HK. (2001). Drevets WC., Charney DS. The cellular neurobiology of depression. *Nature Medicine, 7*(5), 541–547. doi:10.1038/87865 PMID:11329053

Männistö, P. T., & Kaakkola, S. (1999). Catechol-O-methyltransferase (COMT): Biochemistry, molecular biology, pharmacology, and clinical efficacy of the new selective COMT inhibitors. *Pharmacological Reviews*, *51*(4), 593–628. PMID:10581325

Manocha, G. D., Ghatak, A., Puig, K. L., Kraner, S. D., Norris, C. M., & Combs, C. K. (2017). NFATc2 Modulates Microglial Activation in the AβPP/PS1 Mouse Model of Alzheimer's Disease. *Journal of Alzheimer's Disease*, *58*(3), 775–787. doi:10.3233/JAD-151203 PMID:28505967

Manthripragada, A. D., Schernhammer, E. S., Qiu, J., Friis, S., Wermuth, L., Olsen, J. H., & Ritz, B. (2011). *Non-steroidal anti-inflammatory drug use and the risk of Parkinson's disease*. Karger. doi:10.1159/000325653

Mao, X., Ou, M. T., Karuppagounder, S. S., Kam, T.-I., Yin, X., Xiong, Y., Ge, P., Umanah, G. E., Brahmachari, S., Shin, J.-H., Kang, H. C., Zhang, J., Xu, J., Chen, R., Park, H., Andrabi, S. A., Kang, S. U., Gonçalves, R. A., Liang, Y., & Dawson, T. M. (2016). 'Pathological α-Synuclein Transmission Initiated by Binding Lymphocyte-Activation Gene 3'. *Science*, *353*(6307), aah3374. doi:10.1126cience.aah3374 PMID:27708076

Ma, P., Liu, H. T., Wei, P., Xu, Q. S., Bai, X. F., Du, Y. G., & Yu, C. (2011). Chitosan oligosaccharides inhibit LPS-induced over-expression of IL-6 and TNF-α in RAW264.7 macrophage cells through blockade of mitogen-activated protein kinase (MAPK) and PI3K/Akt signaling pathways. *Carbohydrate Polymers*, *84*(4), 1391–1398. doi:10.1016/j.carbpol.2011.01.045

Maphis, N., Xu, G., Kokiko-Cochran, O. N., Jiang, S., Cardona, A., Ransohoff, R. M., Lamb, B. T., & Bhaskar, K. (2015). 'Reactive Microglia Drive Tau Pathology and Contribute to the Spreading of Pathological Tau in the Brain'. *Brain*, *138*(6), 1738–1755. doi:10.1093/brain/awv081 PMID:25833819

Maragakis, N. J., & Rothstein, J. D. (2006). Mechanisms of disease: Astrocytes in neurodegenerative disease. *Nature Clinical Practice. Neurology*, *2*(12), 679–689. doi:10.1038/ncpneuro0355 PMID:17117171

Marangon, D., Boccazzi, M., Lecca, D., & Fumagalli, M. (2020). Regulation of oligodendrocyte functions: Targeting lipid metabolism and extracellular matrix for myelin repair. *Journal of Clinical Medicine*, *9*(2), 470. doi:10.3390/jcm9020470 PMID:32046349

Marcó, S., Pujol, A., Roca, C., Motas, S., Ribera, A., Garcia, M., Molas, M., Villacampa, P., Melia, C. S., & Sanchez, V. (2016). Progressive neurologic and somatic disease in a novel mouse model of human mucopolysaccharidosis type IIIC. *Disease Models & Mechanisms*, *9*(9), 999–1013. PMID:27491071

Marisca, R., Hoche, T., Agirre, E., Hoodless, L. J., Barkey, W., Auer, F., Castelo-Branco, G., & Czopka, T. (2020). Functionally distinct subgroups of oligodendrocyte precursor cells integrate neural activity and execute myelin formation. *Nature Neuroscience*, *23*(3), 363–374. doi:10.103841593-019-0581-2

Markiewicz, I., & Lukomska, B. (2006). The role of astrocytes in the physiology and pathology of the central nervous system. *Acta Neurobiologiae Experimentalis*, *66*(4), 343–358. PMID:17265695

Marques, F., Sousa, J. C., Sousa, N., & Palha, J. A. (2013). Blood-brain-barriers in aging and in Alzheimer's disease. *Molecular Neurodegeneration*, *8*(1), 38. doi:10.1186/1750-1326-8-38 PMID:24148264

Martinac, J. A., Craft, D. K., Su, J. H., Kim, R. C., & Cotman, C. W. (2001). Astrocytes degenerate in frontotemporal dementia: Possible relation to hypoperfusion. *Neurobiology of Aging*, *22*(2), 195–207. doi:10.1016/S0197-4580(00)00231-1 PMID:11182469

Martinez-Banaclocha, M. (2020). Astroglial Isopotentiality and Calcium-Associated Biomagnetic Field Effects on Cortical Neuronal Coupling. *Cells*, *9*(2), 2. doi:10.3390/cells9020439 PMID:32069981

Martinez-Hernandez, A., Bell, K. P., & Norenberg, M. D. (1977). Glutamine synthetase: Glial localization in brain. *Science*, *195*(4284), 1356–1358. doi:10.1126cience.14400 PMID:14400

Martini, R., Fischer, S., López-Vales, R., & David, S. (2008). Interactions between Schwann cells and macrophages in injury and inherited demyelinating disease. *Glia*, *56*(14), 1566–1577. doi:10.1002/glia.20766 PMID:18803324

Martin, S., Al Khleifat, A., & Al-Chalabi, A. (2017). What causes amyotrophic lateral sclerosis? *F1000 Research*, *6*, 371. doi:10.12688/f1000research.10476.1 PMID:28408982

Marzesco, A.-M., Janich, P., Wilsch-Bräuninger, M., Dubreuil, V., Langenfeld, K., Corbeil, D., & Huttner, W. B. (2005). 'Release of Extracellular Membrane Particles Carrying the Stem Cell Marker Prominin-1 (CD133) from Neural Progenitors and Other Epithelial Cells'. *Journal of Cell Science*, *118*(13), 2849–2858. doi:10.1242/jcs.02439 PMID:15976444

Massey, P. V., & Bashir, Z. I. (2007). Long-term depression: Multiple forms and implications for brain function. *Trends in Neurosciences*, *30*(4), 176–184. doi:10.1016/j.tins.2007.02.005 PMID:17335914

Masters, C. L., Bateman, R., Blennow, K., Rowe, C. C., Sperling, R. A., & Cummings, J. L. (2015). Alzheimer's disease. *Nature Reviews. Disease Primers*, *1*(1), 15056. doi:10.1038/nrdp.2015.56 PMID:27188934

Masuch, A., Shieh, C., van Rooijen, N., van Calker, D., & Biber, K. (2016). Mechanism of microglia neuroprotection: Involvement of P2X 7, TNF α, and valproic acid. *Glia*, *64*(1), 76–89. doi:10.1002/glia.22904 PMID:26295445

Matejuk, A., & Ransohoff, R. M. (2020). Crosstalk Between Astrocytes and Microglia : An Overview. *Frontiers in Immunology*, *11*(July), 1–11. doi:10.3389/fimmu.2020.01416 PMID:32765501

Matheus, M. G., Castillo, M., Smith, J. K., Armao, D., Towle, D., & Muenzer, J. (2004). Brain MRI findings in patients with mucopolysaccharidosis types I and II and mild clinical presentation. *Neuroradiology*, *46*(8), 666–672. doi:10.100700234-004-1215-1 PMID:15205860

Mathewson, A. J., & Berry, M. (1985). Observations on the astrocyte response to a cerebral stab wound in adult rats. *Brain Research*, *327*(1-2), 61–69. doi:10.1016/0006-8993(85)91499-4 PMID:3986520

Matias, I., Morgado, J., & Flávia, C. A. G. (2019). 'Astrocyte Heterogeneity: Impact to Brain Aging and Disease'. *Frontiers in Aging Neuroscience*, *11*, 59. https://www.frontiersin.org/articles/10.3389/fnagi.2019.00059. doi:10.3389/fnagi.2019.00059 PMID:30941031

Matos, M., Augusto, E., Oliveira, C. R., & Agostinho, P. (2008). Amyloid-beta peptide decreases glutamate uptake in cultured astrocytes: Involvement of oxidative stress and mitogen-activated protein kinase cascades. *Neuroscience*, *156*(4), 898–910. doi:10.1016/j.neuroscience.2008.08.022 PMID:18790019

Matsumura, K., & Kobayashi, S. (2004). Signaling the Brain in Inflammation: The Role of Endothelial Cells. *Frontiers in Bioscience*, *9*(5), 2819–2826. doi:10.2741/1439 PMID:15353317

Matsuo, R., Yamagishi, M., Wakiya, K., Tanaka, Y., & Ito, E. (2013). Target innervation is necessary for neuronal polyploidization in the terrestrial slug Limax. *Developmental Neurobiology*. doi:10.1002/dneu PMID:23723165

Matsuzaki, M., Honkura, N., Ellis-Davies, G. C. R., & Kasai, H. (2004). Structural basis of long-term potentiation in single dendritic spines. *Nature*, *429*(6993), 761–766. doi:10.1038/nature02617 PMID:15190253

Mattotti, M., Alvarez, Z., Delgado, L., Mateos-Timoneda, M. A., Aparicio, C., Planell, J. A., Alcántara, S., & Engel, E. (2017). Differential neuronal and glial behavior on flat and micro patterned chitosan films. *Colloids and Surfaces. B, Biointerfaces*, *158*, 569–577. doi:10.1016/j.colsurfb.2017.07.045 PMID:28750339

Mauch, D. H., Nagler, K., Schumacher, S., Goritz, C., Müller, E. C., Otto, A., & Pfrieger, F. W. (2001). CNS synaptogenesis promoted by glia-derived cholesterol. *Science*, *294*(5545), 1354–1357. doi:10.1126cience.294.5545.1354 PMID:11701931

Ma, W., & Quirion, R. (2002). Partial sciatic nerve ligation induces increase in the phosphorylation of extracellular signal-regulated kinase (ERK) and c-Jun N-terminal kinase (JNK) in astrocytes in the lumbar spinal dorsal horn and the gracile nucleus. *Pain*, *99*(1–2), 175–184. doi:10.1016/S0304-3959(02)00097-0 PMID:12237195

Mawuenyega, K. G., Sigurdson, W., Ovod, V., Munsell, L., Kasten, T., Morris, J. C., Yarasheski, K. E., & Bateman, R. J. (2010). 'Decreased Clearance of CNS β-Amyloid in Alzheimer's Disease'. *Science*, *330*(6012), 1774–1774. doi:10.1126cience.1197623 PMID:21148344

Mazur, A., Bills, E. H., DeSchepper, K. M., Williamson, J. C., Henderson, B. J., & Risher, W. C. (2021). Astrocyte-derived thrombospondin induces cortical synaptogenesis in a sex-specific manner. *eNeuro*, *8*(4), ENEURO.0014-21.2021. doi:10.1523/ENEURO.0014-21.2021 PMID:34266964

McConnell, J. R., Antonson, D. L., Ong, C. S., Chu, W.-K., Fox, I. J., Heffron, T. G., Langnas, A. N., & Shaw, B. W. Jr. (1995). Proton Spectroscopy of Brain Glutamine in Acute Liver Failure. *Hepatology (Baltimore, Md.)*, *22*(1), 69–74. doi:10.1002/hep.1840220111 PMID:7601435

Mccoy, M. K., Martinez, T. N., Ruhn, K. A., Szymkowski, D. E., Smith, C. G., Botterman, B. R., Tansey, K. E., & Tansey, M. G. (2006). Blocking soluble tumor necrosis factor signaling with dominant-negative tumor necrosis factor inhibitor attenuates loss of dopaminergic neurons. *Social Neuroscience*. doi:10.1523/JNEUROSCI.1504-06.2006 PMID:16971520

Mcdonald, J. W., Althomsons, S. P., Hyrc, K. L., Choi, D. W., & Goldberg, M. P. (1998). Oligodendrocytes from forebrain are highly vulnerable to AMPA/kainate receptor-mediated excitotoxicity. *Nature Medicine*, *4*(3), 291–297. doi:10.1038/nm0398-291 PMID:9500601

McGeer, P., Itagaki, S., & Boyes, B. (1988). *Reactive microglia are positive for HLA-DR in the substantia nigra of Parkinson's and Alzheimer's disease brains*. AAN Enterprises. https://n.neurology.org/content/38/8/1285.short

McGeer, P. L., Itagaki, S., Tago, H., & McGeer, E. G. (1987). Reactive microglia in patients with senile dementia of the Alzheimer type are positive for the histocompatibility glycoprotein HLA-DR. *Neuroscience Letters*, *79*(1-2), 195–200. doi:10.1016/0304-3940(87)90696-3 PMID:3670729

McKee, C. G., Hoffos, M., Vecchiarelli, H. A., & Tremblay, M. È. (2023). Microglia: A pharmacological target for the treatment of age-related cognitive decline and Alzheimer's disease. *Frontiers in Pharmacology*, *14*, 1125982. doi:10.3389/fphar.2023.1125982 PMID:36969855

McKenzie, I. A., Ohayon, D., Li, H., De Faria, J. P., Emery, B., Tohyama, K., & Richardson, W. D. (2014). Motor skill learning requires active central myelination. *Science*, *346*(6207), 318–322. doi:10.1126cience.1254960 PMID:25324381

McKercher, S. R., Torbett, B. E., Anderson, K. L., Henkel, G. W., Vestal, D. J., Baribault, H., Klemsz, M., Feeney, A. J., Wu, G. E., Paige, C. J., & Maki, R. A. (1996). Targeted disruption of the PU. 1 gene results in multiple hematopoietic abnormalities. *The EMBO Journal*, *15*(20), 5647–5658. doi:10.1002/j.1460-2075.1996.tb00949.x PMID:8896458

McKusick, V. A., & Neufeld, E. F. (1983). The mucopolysaccharide storage diseases. In J. B. Stanbury, J. B. Wyngaarden, D. S. Fredrickson, J. L. Goldstein, & M. S. Brown (Eds.), *The Metabolic Basis of Inherited Disease*. McGraw-Hill.

McMullen, E., Hertenstein, H., Strassburger, K., Deharde, L., Brankatschk, M., & Schirmeier, S. (2023). Glycolytically impaired Drosophila glial cells fuel neural metabolism via β-oxidation. *Nature Communications*, *14*(1), 2996. doi:10.103841467-023-38813-x

McNeill, J., Rudyk, C., Hildebrand, M. E., & Salmaso, N. (2021). Ion Channels and Electrophysiological Properties of Astrocytes : Implications for Emergent Stimulation Technologies. *Frontiers in Cellular Neuroscience*, *15*, 644126. doi:10.3389/fncel.2021.644126 PMID:34093129

Medina, A., Watson, S. J., Bunney, W. Jr, Myers, R. M., Schatzberg, A., Barchas, J., Akil, H., & Thompson, R. C. (2016). Evidence for alterations of the glial syncytial function in major depressive disorder. *Journal of Psychiatric Research*, *72*, 15–21. doi:10.1016/j.jpsychires.2015.10.010 PMID:26519765

Meikle, P. J., Hopwood, J. J., Clague, A. E., & Carey, W. F. (1999). Prevalence of lysosomal storage disorders. *Journal of the American Medical Association*, *281*(3), 249–254. doi:10.1001/jama.281.3.249 PMID:9918480

Mei, X., Ezan, P., Giaume, C., & Koulakoff, A. (2010). Astroglial connexin immunoreactivity is specifically altered at β-amyloid plaques in β-amyloid precursor protein/presenilin1 mice. *Neuroscience Letters*, *474*(2), 71–75. doi:10.1016/j.neuroscience.2010.08.001 PMID:20227462

Meletis, K., Barnabé-Heider, F., Carlén, M., Evergren, E., Tomilin, N., Shupliakov, O., & Frisén, J. (2008). Spinal cord injury reveals multilineage differentiation of ependymal cells. *PLoS Biology*, *6*(7), 1494–1507. doi:10.1371/journal.pbio.0060182 PMID:18651793

Ménard, C., Hodes, G. E., & Russo, S. J. (2016). Pathogenesis of depression: Insights from human and rodent studies. *Neuroscience*, *321*, 138–162. doi:10.1016/j.neuroscience.2015.05.053 PMID:26037806

Meneghini, V., Peviani, M., Luciani, M., Zambonini, G., & Gritti, A. (2021). Delivery Platforms for CRISPR/Cas9 Genome Editing of Glial Cells in the Central Nervous System. *Frontiers in Genome Editing*, *3*, 644319. doi:10.3389/fgeed.2021.644319 PMID:34713256

Meng, Q., Lin, M.-S., & Tzeng, I.-S. (2020). Relationship between exercise and Alzheimer's disease: A narrative literature review. *Frontiers in Neuroscience*, *14*, 131. doi:10.3389/fnins.2020.00131 PMID:32273835

Mergenthaler, P., Lindauer, U., Dienel, G. A., & Meisel, A. (2013). Sugar for the brain: The role of glucose in physiological and pathological brain function. *Trends in Neurosciences*, *36*(10), 587–597. doi:10.1016/j.tins.2013.07.001

Michell-Robinson, M. A., Touil, H., Healy, L. M., Owen, D. R., Durafourt, B. A., Bar-Or, A., Antel, J. P., & Moore, C. S. (2015). Roles of microglia in brain development, tissue maintenance and repair. *Brain*, *138*(5), 1138–1159. doi:10.1093/brain/awv066 PMID:25823474

Michels, M., Danielski, L. G., Dal-Pizzol, F., & Petronilho, F. (2014). Neuroinflammation: Microglial activation during sepsis. *Current Neurovascular Research*, *11*(3), 262–270. doi:10.2174/1567202611666140520122744 PMID:24845857

Michinaga, S., & Koyama, Y. (2019). Dual roles of astrocyte-derived factors in regulation of blood-brain barrier function after brain damage. *International Journal of Molecular Sciences*, *20*(3), 571. doi:10.3390/ijms20030571 PMID:30699952

Miguel-Hidalgo, J. J., Baucom, C., Dilley, G., Overholser, J. C., Meltzer, H. Y., Stockmeier, C. A., & Rajkowska, G. (2000). Glial fibrillary acidic protein immunoreactivity in the prefrontal cortex distinguishes younger from older adults in major depressive disorder. *Biological Psychiatry*, *48*(8), 861–873. doi:10.1016/S0006-3223(00)00999-9 PMID:11063981

Miguel-Hidalgo, J. J., Wilson, B. A., Hussain, S., Meshram, A., Rajkowska, G., & Stockmeier, C. A. (2014). Reduced connexin 43 immunolabeling in the orbitofrontal cortex in alcohol dependence and depression. *Journal of Psychiatric Research*, *55*, 101–109. doi:10.1016/j.jpsychires.2014.04.007 PMID:24774648

Miklossy, J., Doudet, D., & Schwab, C. (2006). *Role of ICAM-1 in persisting inflammation in Parkinson disease and MPTP monkeys.* Elsevier. https://www.sciencedirect.com/science/article/pii/S0014488605003730

Milior, G., Lecours, C., Samson, L., Bisht, K., Poggini, S., Pagani, F., Deflorio, C., Lauro, C., Alboni, S., Limatola, C., Branchi, I., Tremblay, M.-E., & Maggi, L. (2016). Fractalkine receptor deficiency impairs microglial and neuronal responsiveness to chronic stress. *Brain, Behavior, and Immunity, 55*, 114–125. doi:10.1016/j.bbi.2015.07.024 PMID:26231972

Miller, A. H., & Raison, C. L. (2016). The role of inflammation in depression: From evolutionary imperative to modern treatment target. *Nature Reviews. Immunology, 16*(1), 22–34. doi:10.1038/nri.2015.5 PMID:26711676

Miller, A. M., Shah, R. H., Pentsova, E. I., Pourmaleki, M., Briggs, S., Distefano, N., Zheng, Y., Skakodub, A., Mehta, S. A., Campos, C., Hsieh, W.-Y., Selcuklu, S. D., Ling, L., Meng, F., Jing, X., Samoila, A., Bale, T. A., Tsui, D. W. Y., Grommes, C., & Mellinghoff, I. K. (2019). Tracking tumour evolution in glioma through liquid biopsies of cerebrospinal fluid. *Nature, 565*(7741), 654–658. doi:10.103841586-019-0882-3 PMID:30675060

Miller, R., & Raff, M. (1984). Fibrous and protoplasmic astrocytes are biochemically and developmentally distinct. *The Journal of Neuroscience : The Official Journal of the Society for Neuroscience, 4*(2), 585–592. doi:10.1523/JNEUROSCI.04-02-00585.1984 PMID:6366155

Miller, S. J. (2018). Astrocyte Heterogeneity in the Adult Central Nervous System. *Frontiers in Cellular Neuroscience, 12*, 401. doi:10.3389/fncel.2018.00401 PMID:30524236

Milligan, E. D., & Watkins, L. R. (2009). Pathological and protective roles of glia in chronic pain. *Nature Reviews. Neuroscience, 10*(1), 23–36. doi:10.1038/nrn2533 PMID:19096368

Mindell, J. A. (2012). Lysosomal acidification mechanisms. *Annual Review of Physiology, 74*(1), 69–86. doi:10.1146/annurev-physiol-012110-142317 PMID:22335796

Miners, J. S., Baig, S., Palmer, J., Palmer, L. E., Kehoe, P. G., & Love, S. (2008). 'SYMPOSIUM: Clearance of Aβ from the Brain in Alzheimer's Disease: Aβ-Degrading Enzymes in Alzheimer's Disease'. *Brain Pathology (Zurich, Switzerland), 18*(2), 240–252. doi:10.1111/j.1750-3639.2008.00132.x PMID:18363935

Min, K.-J., Yang, M., Kim, S.-U., Jou, I., & Joe, E. (2006). Astrocytes induce hemeoxygenase-1 expression in microglia: A feasible mechanism for preventing excessive brain inflammation. *The Journal of Neuroscience : The Official Journal of the Society for Neuroscience, 26*(6), 1880–1887. doi:10.1523/JNEUROSCI.3696-05.2006 PMID:16467537

Minogue, A. M., Jones, R. S., Kelly, R. J., McDonald, C. L., Connor, T. J., & Lynch, M. A. (2014). Age-associated dysregulation of microglial activation is coupled with enhanced blood-brain barrier permeability and pathology in APP/PS1 mice. *Neurobiology of Aging, 35*(6), 1442–1452. doi:10.1016/j.neurobiolaging.2013.12.026 PMID:24439957

Minois, N. (2014). Molecular basis of the 'anti-aging'effect of spermidine and other natural polyamines-a mini-review. *Gerontology, 60*(4), 319–326. doi:10.1159/000356748 PMID:24481223

Miron, V. E., Boyd, A., Zhao, J.-W., Yuen, T. J., Ruckh, J. M., Shadrach, J. L., van Wijngaarden, P., Wagers, A. J., Williams, A., Franklin, R. J. M., & ffrench-Constant, C. (2013). 'M2 Microglia and Macrophages Drive Oligodendrocyte Differentiation during CNS Remyelination'. *Nature Neuroscience, 16*(9), 1211–1218. doi:10.1038/nn.3469 PMID:23872599

Mirzadeh, Z., Doetsch, F., Sawamoto, K., Wichterle, H., & Alvarez-Buylla, A. (2010, May). 'The Subventricular Zone En-Face: Wholemount Staining and Ependymal Flow'. *Journal of Visualized Experiments, 1938*(39). doi:10.3791/1938-v PMID:20461052

Mirzadeh, Z., Merkle, F. T., Soriano-Navarro, M., Garcia-Verdugo, J. M., & Alvarez-Buylla, A. (2008). 'Neural Stem Cells Confer Unique Pinwheel Architecture to the Ventricular Surface in Neurogenic Regions of the Adult Brain'. *Cell Stem Cell, 3*(3), 265–278. doi:10.1016/j.stem.2008.07.004 PMID:18786414

Mishra, A., Bandopadhyay, R., Singh, P. K., Mishra, P. S., Sharma, N., & Khurana, N. (2021a). Neuroinflammation in neurological disorders: pharmacotherapeutic targets from bench to bedside. In Metabolic Brain Disease. Springer. doi:10.100711011-021-00806-4

Mittal, S., Klinger, N. V., Michelhaugh, S. K., Barger, G. R., Pannullo, S. C., & Juhász, C. (2017). Alternating electric tumor treating fields for treatment of glioblastoma: Rationale, preclinical, and clinical studies. *Journal of Neurosurgery, 128*(2), 414–421. doi:10.3171/2016.9.JNS16452 PMID:28298023

Miyagi, M., Ohtori, S., Ishikawa, T., Aoki, Y., Ozawa, T., Doya, H., Saito, T., Moriya, H., & Takahashi, K. (2006). Up-regulation of TNFalpha in DRG satellite cells following lumbar facet joint injury in rats. *European Spine Journal: Official Publication of the European Spine Society, the European Spinal Deformity Society, and the European Section of the Cervical Spine Research Society, 15*(6), 953–958. doi:10.1007/s00586-005-1031-2

Miyoshi, K., Obata, K., Kondo, T., Okamura, H., & Noguchi, K. (2008). Interleukin-18-mediated microglia/astrocyte interaction in the spinal cord enhances neuropathic pain processing after nerve injury. *The Journal of Neuroscience : The Official Journal of the Society for Neuroscience, 28*(48), 12775–12787. doi:10.1523/JNEUROSCI.3512-08.2008 PMID:19036970

Moens, U., Kostenko, S., & Sveinbjørnsson, B. (2013). The Role of Mitogen-Activated Protein Kinase-Activated Protein Kinases (MAPKAPKs) in Inflammation. *Genes, 4*(2), 101–133. doi:10.3390/genes4020101 PMID:24705157

Mogi, M., Kondo, A., Kinpara, K., & Togari, A. (2000). Anti-apoptotic action of nerve growth factor in mouse osteoblastic cell line. *Life Sciences, 67*(10), 1197–1206. doi:10.1016/S0024-3205(00)00705-0 PMID:10954053

Mohammed, M. A., Syeda, J. T., Wasan, K. M., & Wasan, E. K. (2017). An overview of chitosan nanoparticles and its application in non-parenteral drug delivery. *Pharmaceutics, 9*(4), 53. doi:10.3390/pharmaceutics9040053 PMID:29156634

Mohan, V., Edamakanti, C. R., & Sharma, V. (2021). Role of Neuroinflammation in Neurodegenerative Disorders. *The Molecular Immunology of Neurological Diseases*, 41–49. doi:10.1016/B978-0-12-821974-4.00001-7

Mohandas, E., Rajmohan, V., & Raghunath, B. (2009). Neurobiology of Alzheimer's disease. *Indian Journal of Psychiatry, 51*(1), 55. doi:10.4103/0019-5545.44908 PMID:19742193

Molendijk, M. L., Spinhoven, P., Polak, M., Bus, B. A. A., Penninx, B., & Elzinga, B. M. (2014). Serum BDNF concentrations as peripheral manifestations of depression: Evidence from a systematic review and meta-analyses on 179 associations (N= 9484). *Molecular Psychiatry, 19*(7), 791–800. doi:10.1038/mp.2013.105 PMID:23958957

Molofsky, A. V., & Deneen, B. (2015). Astrocyte development : A Guide for the Perplexed. *Glia, 63*(8), 1320–1329. doi:10.1002/glia.22836 PMID:25963996

Molofsky, A. V., Krenick, R., Ullian, E., Tsai, H., Deneen, B., Richardson, W. D., Barres, B. A., & Rowitch, D. H. (2012). Astrocytes and disease: A neurodevelopmental perspective. *Genes & Development, 26*(9), 891–907. doi:10.1101/gad.188326.112 PMID:22549954

Monkul, E. S., Hatch, J. P., Nicoletti, M. A., Spence, S., Brambilla, P., de Lacerda, A. L. T., Sassi, R. B., Mallinger, A. G., Keshavan, M. S., & Soares, J. C. (2007). Fronto-limbic brain structures in suicidal and non-suicidal female patients with major depressive disorder. *Molecular Psychiatry, 12*(4), 360–366. doi:10.1038j.mp.4001919 PMID:17389903

Montagne, A., Zhao, Z., & Zlokovic, B. V. (2017). Alzheimer's disease: A matter of blood-brain barrier dysfunction? *The Journal of Experimental Medicine, 214*(11), 3151–3169. doi:10.1084/jem.20171406 PMID:29061693

Monterey, M. D., Wei, H., Wu, X., & Wu, J. Q. (2021). The Many Faces of Astrocytes in Alzheimer's Disease. *Frontiers in Neurology, 12*, 619626. doi:10.3389/fneur.2021.619626 PMID:34531807

Montgomery, D. L. (1994). Astrocytes: Form, functions, and roles in disease. *Veterinary Pathology*, *31*(2), 145–167. doi:10.1177/030098589403100201 PMID:8203078

Mookherjee, P., Green, P. S., Watson, G. S., Marques, M. A., Tanaka, K., Meeker, K. D., Meabon, J. S., Li, N., Zhu, P., Olson, V. G., & Cook, D. G. (2011). GLT-1 loss accelerates cognitive deficit onset in an Alzheimer's disease animal model. *Journal of Alzheimer's Disease*, *26*(3), 447–455. doi:10.3233/JAD-2011-110503 PMID:21677376

Moore, L., Skop, N. B., Rothbard, D. E., Corrubia, L. R., & Levison, S. W. (2018). Tethered growth factors on biocompatible scaffolds improve stemness of cultured rat and human neural stem cells and growth of oligodendrocyte progenitors. *Methods (San Diego, Calif.)*, *133*, 54–64. doi:10.1016/j.ymeth.2017.08.015 PMID:28887084

Morley, J. E., Farr, S. A., & Nguyen, A. D. (2018). Alzheimer disease. *Clinics in Geriatric Medicine*, *34*(4), 591–601. doi:10.1016/j.cger.2018.06.006 PMID:30336989

Moro, E., Lozano, A. M., Pollak, P., Agid, Y., Rehncrona, S., Volkmann, J., Kulisevsky, J., Obeso, J. A., Albanese, A., Hariz, M. I., Quinn, N. P., Speelman, J. D., Benabid, A. L., Fraix, V., Mendes, A., Welter, M.-L., Houeto, J.-L., Cornu, P., Dormont, D., & Lang, A. E. (2010). Long-term results of a multicenter study on subthalamic and pallidal stimulation in Parkinson's disease. *Movement Disorders*, *25*(5), 578–586. doi:10.1002/mds.22735 PMID:20213817

Morris, G. P., Clark, I. A., Zinn, R., & Vissel, B. (2013). Neurobiology of Learning and Memory Microglia : A new frontier for synaptic plasticity, learning and memory, and neurodegenerative disease research. *Neurobiology of Learning and Memory*, *105*, 40–53. doi:10.1016/j.nlm.2013.07.002 PMID:23850597

Morrison, L. D., Cao, X.-C., & Kish, S. J. (1998). Ornithine decarboxylase in human brain : Influence of aging, regional distribution, and Alzheimer's disease. *Journal of Neurochemistry*, *71*(1), 288–294. doi:10.1046/j.1471-4159.1998.71010288.x PMID:9648877

Moser, H. W., Mahmood, A., & Raymond, G. V. (2007). X-linked adrenoleukodystrophy. *Nature Clinical Practice. Neurology*, *3*(3), 140–151. doi:10.1038/ncpneuro0421 PMID:17342190

Mosser, C. A., Baptista, S., Arnoux, I., & Audinat, E. (2017). Microglia in CNS development: Shaping the brain for the future. *Progress in Neurobiology*, *149–150*, 1–20. doi:10.1016/j.pneurobio.2017.01.002 PMID:28143732

Mosser, J., Douar, A.-M., Sarde, C.-O., Kioschis, P., Feil, R., Moser, H., Poustka, A.-M., Mandel, J.-L., & Aubourg, P. (1993). Putative X-linked adrenoleukodystrophy gene shares unexpected homology with ABC transporters. *Nature*, *361*(6414), 726–730. doi:10.1038/361726a0 PMID:8441467

Moss, J., Gebara, E., Bushong, E. A., Sánchez-Pascual, I., O'Laoi, R., El M'Gharia, I., Kocher-Braissant, J., Ellisman, M. H., & Toni, N. (2016). Fine processes of Nestin-GFP-positive radial glia-like stem cells in the adult dentate gyrus ensheathe local synapses and vasculature. *Proceedings of the National Academy of Sciences of the United States of America*, *113*(18), E2536–E2545. doi:10.1073/pnas.1514652113 PMID:27091993

Mot, A. I., Depp, C., & Nave, K. A. (2022). An emerging role of dysfunctional axon-oligodendrocyte coupling in neurodegenerative diseases. *Dialogues in Clinical Neuroscience*.

Mozzachiodi, R., & Byrne, J. H. (2010). More than synaptic plasticity: Role of nonsynaptic plasticity in learning and memory. *Trends in Neurosciences*, *33*(1), 17–26. doi:10.1016/j.tins.2009.10.001 PMID:19889466

Mueller, S. G., & Weiner, M. W. (2009). Selective effect of age, Apo e4, and Alzheimer's disease on hippocampal subfields. *Hippocampus*, *19*(6), 558–564. doi:10.1002/hipo.20614 PMID:19405132

Muenzer, J. (2004). The mucopolysaccharidoses: A heterogeneous group of disorders with variable pediatric presentations. *The Journal of Pediatrics*, *144*(5), S27–S34. doi:10.1016/j.jpeds.2004.01.052 PMID:15126981

Muenzer, J., Gucsavas-Calikoglu, M., McCandless, S. E., Schuetz, T. J., & Kimura, A. (2007). A phase I/II clinical trial of enzyme replacement therapy in mucopolysaccharidosis II (Hunter syndrome). *Molecular Genetics and Metabolism*, *90*(3), 329–337. doi:10.1016/j.ymgme.2006.09.001 PMID:17185020

Mukhopadhyay, S., Fellows, K., Browne, R. W., Khare, P., Krishnan Radhakrishnan, S., Hagemeier, J., Weinstock-Guttman, B., Zivadinov, R., & Ramanathan, M. (2017). Interdependence of oxysterols with cholesterol profiles in multiple sclerosis. *Multiple Sclerosis*, *23*(6), 792–801. doi:10.1177/1352458516666187

Mulaikal, T. A., & Emond, J. C. (2012). *Liver Anesthesiology and Critical Care Medicine*. Liver Anesthesiology and Critical Care Medicine., doi:10.1007/978-1-4614-5167-9

Müller, M. B., Lucassen, P. J., Yassouridis, A., Hoogendijk, W. J. G., Holsboer, F., & Swaab, D. F. (2001). Neither major depression nor glucocorticoid treatment affects the cellular integrity of the human hippocampus. *The European Journal of Neuroscience*, *14*(10), 1603–1612. doi:10.1046/j.0953-816x.2001.01784.x PMID:11860455

Müller, N., & Schwarz, M. J. (2007). The immune-mediated alteration of serotonin and glutamate: Towards an integrated view of depression. *Molecular Psychiatry*, *12*(11), 988–1000. doi:10.1038j.mp.4002006 PMID:17457312

Mumtaz, K., Ahmed, U. S., Abid, S., Baig, N., Hamid, S., & Jafri, W. (2010). Precipitating factors and the outcome of hepatic encephalopathy in liver cirrhosis. *Journal of the College of Physicians and Surgeons—Pakistan*, *20*(8), 514. PMID:20688015

Muralidar, S., Ambi, S. V., Sekaran, S., Thirumalai, D., & Palaniappan, B. (2020). Role of tau protein in Alzheimer's disease: The prime pathological player. *International Journal of Biological Macromolecules*, *163*, 1599–1617. doi:10.1016/j.ijbiomac.2020.07.327 PMID:32784025

Murthy, Ch. R. K., Rama Rao, K. V., Bai, G., & Norenberg, M. D. (2001). Ammonia-induced Production of Free Radicals in Primary Cultures of Rat Astrocytes. *Journal of Neuroscience Research*, *66*(2), 282–288. doi:10.1002/jnr.1222 PMID:11592125

Muthusamy, N., Sommerville, L. J., Moeser, A. J., Stumpo, D. J., Sannes, P., Adler, K., Blackshear, P. J., Weimer, J. M., & Ghashghaei, H. T. (2015). MARCKS-dependent mucin clearance and lipid metabolism in ependymal cells are required for maintenance of forebrain homeostasis during aging. *Aging Cell*, *14*(5), 764–773. doi:10.1111/acel.12354 PMID:26010231

Muzio, L., Viotti, A., & Martino, G. (2021). 'Microglia in Neuroinflammation and Neurodegeneration: From Understanding to Therapy'. *Frontiers in Neuroscience*, *15*(September), 742065. doi:10.3389/fnins.2021.742065 PMID:34630027

Nagai, J., Rajbhandari, A. K., Gangwani, M. R., Hachisuka, A., Coppola, G., Masmanidis, S. C., Fanselow, M. S., & Khakh, B. S. (2019). Hyperactivity with Disrupted Attention by Activation of an Astrocyte Synaptogenic Cue. *Cell*, *177*(5), 1280–1292 e1220. doi:10.1016/j.cell.2019.03.019 PMID:31031006

Nagai, J., Yu, X., Papouin, T., Cheong, E., Freeman, M. R., Monk, K. R., Hastings, M. H., Haydon, P. G., Rowitch, D., Shaham, S., & Khakh, B. S. (2021). Behaviorally consequential astrocytic regulation of neural circuits. *Neuron*, *109*(4), 576–596. doi:10.1016/j.neuron.2020.12.008 PMID:33385325

Nagelhus, E. A., & Ottersen, O. P. (2013). 'Physiological Roles of Aquaporin-4 in Brain'. *Physiological Reviews*, *93*(4), 1543–1562. doi:10.1152/physrev.00011.2013 PMID:24137016

Nägler, K., Mauch, D. H., & Pfrieger, F. W. (2001). Glia-derived signals induce synapse formation in neurones of the rat central nervous system. *The Journal of Physiology*, *533*(Pt 3), 665–679. doi:10.1111/j.1469-7793.2001.00665.x PMID:11410625

Nagy, C., Suderman, M., Yang, J., Szyf, M., Mechawar, N., Ernst, C., & Turecki, G. (2015). Astrocytic abnormalities and global DNA methylation patterns in depression and suicide. *Molecular Psychiatry*, *20*(3), 320–328. doi:10.1038/mp.2014.21 PMID:24662927

Nagy, J. I., Patel, D., Ochalski, P. A., & Stelmack, G. L. (1996). Connexin30 in rodent, cat and human brain: Selective expression in gray matter astrocytes, co-localization with connexin43 at gap junctions and late developmental appearance. *Neuroscience*, *76*(4), 931–953. doi:10.1016/S0306-4522(98)00191-2 PMID:10197766

Nakada, M., Kita, D., Watanabe, T., Hayashi, Y., Teng, L., Pyko, I. V., & Hamada, J. I. (2011). Aberrant signaling pathways in glioma. *Cancers (Basel)*, *3*(3), 3242–3278. doi:10.3390/cancers3033242 PMID:24212955

Nakahara, J., Maeda, M., Aiso, S., & Suzuki, N. (2012). Current Concepts in Multiple Sclerosis : Autoimmunity Versus Oligodendrogliopathy. *Clinical Reviews in Allergy & Immunology*, *42*(1), 26–34. doi:10.100712016-011-8287-6 PMID:22189514

Nakajima, K., & Kohsaka, S. (2001). Microglia: Activation and their significance in the central nervous system. *Journal of Biochemistry*, *130*(2), 169–175. doi:10.1093/oxfordjournals.jbchem.a002969 PMID:11481032

Nakayama, Y., & Kohno, K. (1974). Number and polarity of the ependymal cilia in the central canal of some vertebrates. *Journal of Neurocytology*, *3*(4), 449–458. doi:10.1007/BF01098732 PMID:4436691

Nalivaeva, N. N., Beckett, C., Belyaev, N. D., & Turner, A. J. (2012). Are amyloid-degrading enzymes viable therapeutic targets in Alzheimer's disease? *Journal of Neurochemistry*, *120*(s1), 167–185. doi:10.1111/j.1471-4159.2011.07510.x PMID:22122230

Nanclares, C., Baraibar, A. M., Araque, A., & Kofuji, P. (2021). egulation of astrocytes-neuronal communication in Alzheimer's disease. *International Journal of Molecular Sciences*, *22*(15), 7887. doi:10.3390/ijms22157887 PMID:34360652

Narita, M., Narita, M., Kaneko, C., Hareyama, N., Miyatake, M., Shindo, K., Miyoshi, K., Nakajima, M., Nagumo, Y., Sato, F., Wachi, H., Seyama, Y., & Suzuki, T. (2006). Chronic pain-induced emotional dysfunction is associated with astrogliosis due to cortical delta-opioid receptor dysfunction. *Journal of Neurochemistry*, *97*(5), 1369–1378. doi:10.1111/j.1471-4159.2006.03824.x PMID:16696849

Narita, M., Yoshida, T., Nakajima, M., Narita, M., Miyatake, M., Takagi, T., Yajima, Y., & Suzuki, T. (2006). Direct evidence for spinal cord microglia in the development of a neuropathic pain-like state in mice. *Journal of Neurochemistry*, *97*(5), 1337–1348. doi:10.1111/j.1471-4159.2006.03808.x PMID:16606373

Nascimento, D. M., Nunes, Y. L., Figueirêdo, M. C. B., de Azeredo, H. M. C., Aouada, F. A., Feitosa, J. P. A., Rosa, M. F., & Dufresne, A. (2018). Nanocellulose nanocomposite hydrogels : Technological and environmental issues. *Green Chemistry*, *20*(11), 2428–2448. doi:10.1039/C8GC00205C

Nasiri, N., Ceramidas, A., Mukherjee, S., Panneerselvan, A., Nisbet, D. R., & Tricoli, A. (2016). Ultra-porous nanoparticle networks: A biomimetic coating morphology for enhanced cellular response and infiltration. *Scientific Reports*, *6*(1), 24305. doi:10.1038rep24305 PMID:27076035

Naskar, S., Datta, S., & Chattarji, S. (2022). Riluzole prevents stress-induced spine plasticity in the hippocampus but mimics it in the amygdala. *Neurobiology of Stress*, *18*, 100442. doi:10.1016/j.ynstr.2022.100442 PMID:35330860

Nasrabady, S. E., Rizvi, B., Goldman, J. E., & Brickman, A. M. (2018a). White matter changes in Alzheimer's disease : A focus on myelin and oligodendrocytes. *Acta Neuropathologica Communications*, *6*(1), 1–10. doi:10.118640478-018-0515-3 PMID:29499767

Navari, R. M. (2016). Olanzapine for the prevention of chemotherapy-induced nausea and vomiting. *Management of Chemotherapy-Induced Nausea and Vomiting: New Agents and New Uses of Current Agents*, 107-120. Springer.

Nave, K. A. (2010). Myelination and support of axonal integrity by glia. *Nature*, *468*(7321), 244–252. doi:10.1038/nature09614

Nave, K.-A., & Werner, H. B. (2014). 'Myelination of the Nervous System: Mechanisms and Functions'. *Annual Review of Cell and Developmental Biology*, *30*(1), 503–533. doi:10.1146/annurev-cellbio-100913-013101 PMID:25288117

Nawrotek, K., Marqueste, T., Modrzejewska, Z., Zarzycki, R., Rusak, A., & Decherchi, P. (2017). T hermogelling chitosan lactate hydrogel improves functional recovery after a C2 spinal cord hemisection in rat. *Journal of Biomedical Materials Research. Part A*, *105*(7), 2004–2019. doi:10.1002/jbm.a.36067 PMID:28324618

Nayak, D., Roth, T. L., & McGavern, D. B. (2014). Microglia development and function. *Annual Review of Immunology*, *32*(1), 367–402. doi:10.1146/annurev-immunol-032713-120240 PMID:24471431

Nedergaard, M., Ransom, B., & Goldman, S. A. (2003). New roles for astrocytes: Redefining the functional architecture of the brain. *Trends in Neurosciences*, *26*(10), 523–530. doi:10.1016/j.tins.2003.08.008 PMID:14522144

Nedergaard, M., & Verkhratsky, A. (2012). 'Artifact versus Reality-How Astrocytes Contribute to Synaptic Events'. *Glia*, *60*(7), 1013–1023. doi:10.1002/glia.22288 PMID:22228580

Nehal, N., Nabi, B., Rehman, S., Pathak, A., Iqubal, A., Khan, S. A., Yar, M. S., Parvez, S., Baboota, S., & Ali, J. (2021). Chitosan coated synergistically engineered nanoemulsion of Ropinirole and nigella oil in the management of Parkinson's disease: Formulation perspective and In vitro and In vivo assessment. *International Journal of Biological Macromolecules*, *167*, 605–619. doi:10.1016/j.ijbiomac.2020.11.207 PMID:33278450

Neher, J. J., Neniskyte, U., & Brown, G. C. (2012). Primary phagocytosis of neurons by inflamed microglia: Potential roles in neurodegeneration. *Frontiers in Pharmacology*, *3*(FEB). doi:10.3389/fphar.2012.00027 PMID:22403545

Nestler, E. J., Gould, E., & Manji, H. (2002). Preclinical models: Status of basic research in depression. *Biological Psychiatry*, *52*(6), 503–528. doi:10.1016/S0006-3223(02)01405-1 PMID:12361666

Neufeld, E. F., & Muenzer, J. (2001). The mucopolysaccharidoses. The Metabolic and Molecular Bases of Inherited Disease. New York: McGraw-Hill.

Neumann, H., Kotter, M. R., & Franklin, R. J. M. (2008). 'Debris Clearance by Microglia: An Essential Link between Degeneration and Regeneration'. *Brain*, *132*(2), 288–295. doi:10.1093/brain/awn109 PMID:18567623

Neu, S. C., Pa, J., Kukull, W., Beekly, D., Kuzma, A., Gangadharan, P., Wang, L.-S., Romero, K., Arneric, S. P., Redolfi, A., Orlandi, D., Frisoni, G. B., Au, R., Devine, S., Auerbach, S., Espinosa, A., Boada, M., Ruiz, A., Johnson, S. C., & Toga, A. W. (2017). Apolipoprotein E genotype and sex risk factors for Alzheimer disease: A meta-analysis. *JAMA Neurology*, *74*(10), 1178–1189. doi:10.1001/jamaneurol.2017.2188 PMID:28846757

Newton, S. S., Collier, E. F., Hunsberger, J., Adams, D., Terwilliger, R., Selvanayagam, E., & Duman, R. S. (2003). Gene profile of electroconvulsive seizures: Induction of neurotrophic and angiogenic factors. *The Journal of Neuroscience : The Official Journal of the Society for Neuroscience*, *23*(34), 10841–10851. doi:10.1523/JNEUROSCI.23-34-10841.2003 PMID:14645477

Newton, S. S., Girgenti, M. J., Collier, E. F., & Duman, R. S. (2006). Electroconvulsive seizure increases adult hippocampal angiogenesis in rats. *The European Journal of Neuroscience*, *24*(3), 819–828. doi:10.1111/j.1460-9568.2006.04958.x PMID:16930411

Ngo, D.-H., & Kim, S.-K. (2014). Chapter Two—Antioxidant Effects of Chitin, Chitosan, and Their Derivatives. In S.-K. Kim (Ed.), *Advances in Food and Nutrition Research* (Vol. 73, pp. 15–31). Academic Press. doi:10.1016/B978-0-12-800268-1.00002-0

Nguyen, P. T., Dorman, L. C., Pan, S., Vainchtein, I. D., Han, R. T., Nakao-Inoue, H., Taloma, S. E., Barron, J. J., Molofsky, A. B., Kheirbek, M. A., & Molofsky, A. V. (2020). Microglial remodeling of the extracellular matrix promotes synapse plasticity. *Cell, 182*(2), 388–403. doi:10.1016/j.cell.2020.05.050 PMID:32615087

Nicaise, A. M., D'Angelo, A., Ionescu, R.-B., Krzak, G., Willis, C. M., & Pluchino, S. (2022). The role of neural stem cells in regulating glial scar formation and repair. *Cell and Tissue Research, 387*(3), 399–414. doi:10.100700441-021-03554-0 PMID:34820704

Nichols, N. R., Day, J. R., Laping, N. J., Johnson, S. A., & Finch, C. E. (1993). GFAP mRNA increases with age in rat and human brain. *Neurobiology of Aging, 14*(5), 421–429. doi:10.1016/0197-4580(93)90100-P PMID:8247224

Nickel, M., & Gu, C. (2018). Regulation of central nervous system myelination in higher brain functions. *Neural Plasticity, 2018*, 1–12. doi:10.1155/2018/6436453 PMID:29692804

Nieto-Quero, A., Chaves-Pena, P., Santín, L. J., Pérez-Martín, M., & Pedraza, C. (2021). Do changes in microglial status underlie neurogenesis impairments and depressive-like behaviours induced by psychological stress? A systematic review in animal models. *Neurobiology of Stress, 15*, 100356. doi:10.1016/j.ynstr.2021.100356 PMID:34355047

Ni, H.-D., Xu, L. S., Wang, Y., Li, H., An, K., Liu, M., Liu, Q., Deng, H., He, Q., Huang, B., Fang, J., & Yao, M. (2019). Astrocyte activation in the periaqueductal gray promotes descending facilitation to cancer-induced bone pain through the JNK MAPK signaling pathway. *Molecular Pain, 15*, 1744806919831909. doi:10.1177/1744806919831909 PMID:30700204

Ni, H., Wang, Y., An, K., Liu, Q., Xu, L., Zhu, C., Deng, H., He, Q., Wang, T., Xu, M., Zheng, Y., Huang, B., Fang, J., & Yao, M. (2019). Crosstalk between NFκB-dependent astrocytic CXCL1 and neuron CXCR2 plays a role in descending pain facilitation. *Journal of Neuroinflammation, 16*(1), 1. doi:10.118612974-018-1391-2 PMID:30606213

Nimmerjahn, A., Kirchhoff, F., & Helmchen, F. (2005). 'Resting Microglial Cells Are Highly Dynamic Surveillants of Brain Parenchyma in Vivo'. *Science, 308*(5726), 1314–1318. doi:10.1126cience.1110647 PMID:15831717

Niraula, A. & Sheridan, J. (n.d.). Microglia priming with aging and stress. *Nature.* https://www.nature.com/articles/npp2016185

Nisbet, D. R., Crompton, K. E., Horne, M. K., Finkelstein, D. I., & Forsythe, J. S. (2008). Neural tissue engineering of the CNS using hydrogels. In Journal of Biomedical Materials Research - Part B Applied Biomaterials, 87(1). doi:10.1002/jbm.b.31000

Nisbet, D. R., Rodda, A. E., Horne, M. K., Forsythe, J. S., & Finkelstein, D. I. (2009). Neurite infiltration and cellular response to electrospun polycaprolactone scaffolds implanted into the brain. *Biomaterials, 30*(27), 4573–4580. doi:10.1016/j.biomaterials.2009.05.011 PMID:19500836

Nisbet, D. R., Yu, L. M. Y., Zahir, T., Forsythe, J. S., & Shoichet, M. S. (2008). Characterization of neural stem cells on electrospun poly (ε-caprolactone) submicron scaffolds: Evaluating their potential in neural tissue engineering. *Journal of Biomaterials Science. Polymer Edition, 19*(5), 623–634. doi:10.1163/156856208784089652 PMID:18419941

Nishida, H., & Okabe, S. (2007). Direct astrocytic contacts regulate the local maturation of dendritic spines. *The Journal of Neuroscience : The Official Journal of the Society for Neuroscience, 27*(2), 331–340. doi:10.1523/JNEUROSCI.4466-06.2007 PMID:17215394

Nishiyama, A., Shimizu, T., Sherafat, A., & Richardson, W. D. (2021, August). Life-long oligodendrocyte development and plasticity. [). Academic Press.]. *Seminars in Cell & Developmental Biology, 116*, 25–37. doi:10.1016/j.semcdb.2021.02.004

Nolte, C., Moller, T., Walter, T., & Kettenmann, H. (1996). Complement 5a controls motility of murine microglial cells in vitro via activation of an inhibitory G-protein and the rearrangement of the actin cytoskeleton. *Neuroscience, 73*(4), 1091–1107. doi:10.1016/0306-4522(96)00106-6 PMID:8809827

Norden, D. M., & Godbout, J. P. (2013). 'Review: Microglia of the Aged Brain: Primed to Be Activated and Resistant to Regulation: Increased Microglial Reactivity with Age'. *Neuropathology and Applied Neurobiology, 39*(1), 19–34. doi:10.1111/j.1365-2990.2012.01306.x PMID:23039106

Norenberg, Michael D, Itzhak, Y., & Bender, A. S. (1997). The peripheral benzodiazepine receptor and neurosteroids in hepatic encephalopathy. *Advances in Cirrhosis, Hyperammonemia, and Hepatic Encephalopathy*, 95–111.

Norenberg, Michael D. (1987b). The Role of Astrocytes in Hepatic Encephalopathy. *Neurochernical Pathology, 6.*

Norenberg, M. D. (1977). A light and electron microscopic study of experimental portal-systemic (ammonia) encephalopathy. Progression and reversal of the disorder. *Laboratory Investigation, 36*(6), 618–627. https://europepmc.org/abstract/MED/559221 PMID:559221

Norenberg, M. D. (1987a). The role of astrocytes in hepatic encephalopathy. *Neurochemical Pathology, 6*(1-2), 13–33. doi:10.1007/BF02833599 PMID:3306480

Norenberg, M. D. (2003). Oxidative and Nitrosative Stress in Ammonia Neurotoxicity. *Hepatology (Baltimore, Md.), 37*(2), 245–248. doi:10.1053/jhep.2003.50087 PMID:12540772

Norenberg, M. D., Jayakumar, A. R., Rama Rao, K. V., & Panickar, K. S. (2007). New Concepts in the Mechanism of Ammonia-Induced Astrocyte Swelling. *Metabolic Brain Disease, 22*(3-4), 219–234. doi:10.100711011-007-9062-5 PMID:17823859

Norenberg, M. D., & Martinez-Hernandez, A. (1979). Fine structural localization of glutamine synthetase in astrocytes of rat brain. *Brain Research, 161*(2), 303–310. doi:10.1016/0006-8993(79)90071-4 PMID:31966

Nosi, D., Lana, D., Giovannini, M. G., Delfino, G., & Zecchi-Orlandini, S. (2021). Neuroinflammation: Integrated nervous tissue response through intercellular interactions at the "whole system" scale. In Cells, 10(5). MDPI. doi:10.3390/cells10051195

Nwe, N., Furuike, T., & Tamura, H. (2009). The mechanical and biological properties of chitosan scaffolds for tissue regeneration templates are significantly enhanced by chitosan from Gongronella butleri. *Materials (Basel), 2*(2), 374–398. doi:10.3390/ma2020374

O'Brown, N. M., Pfau, S. J., & Gu, C. (2018). Bridging barriers: A comparative look at the blood-brain barrier across organisms. *Genes & Development, 32*(7-8), 466–478. doi:10.1101/gad.309823.117 PMID:29692355

O'Regan, G. C., Farag, S. H., Casey, C. S., Wood-Kaczmar, A., Pocock, J. M., Tabrizi, S. J., & Andre, R. (2021). Human Huntington's disease pluripotent stem cell-derived microglia develop normally but are abnormally hyper-reactive and release elevated levels of reactive oxygen species. *Journal of Neuroinflammation, 18*(1), 94. doi:10.118612974-021-02147-6 PMID:33874957

Oberheim, N. A., Goldman, S. A., & Nedergaard, M. (2012). Heterogeneity of astrocytic form and function. *Astrocytes: Methods and Protocols*, 23–45.

Oberheim, N. A., Goldman, S. A., & Nedergaard, M. 2012. 'Heterogeneity of Astrocytic Form and Function'. In Astrocytes. Totowa, NJ: Humana Press. doi:10.1007/978-1-61779-452-0_3

Oberheim, N. A., Takano, T., Han, X., He, W., Lin, J. H., Wang, F., Xu, Q., Wyatt, J. D., Pilcher, W., Ojemann, J. G., Ransom, B. R., Goldman, S. A., & Nedergaard, M. (2009). Uniquely hominid features of adult human astrocytes. *The Journal of Neuroscience : The Official Journal of the Society for Neuroscience*, *29*(10), 3276–3287. doi:10.1523/JNEUROSCI.4707-08.2009 PMID:19279265

Oberheim, N. A., Wang, X., Goldman, S., & Nedergaard, M. (2006). Astrocytic complexity distinguishes the human brain. *Trends in Neurosciences*, *29*(10), 547–553. doi:10.1016/j.tins.2006.08.004 PMID:16938356

Octeau, J. C., Chai, H., Jiang, R., Bonanno, S. L., Martin, K. C., & Khakh, B. S. (2018). An optical neuron-astrocyte proximity assay at synaptic distance scales. *Neuron*, *98*(1), 49.e9–66.e9. doi:10.1016/j.neuron.2018.03.003 PMID:29621490

Oda, S., & Funato, H. (2023). D1- and D2-type dopamine receptors are immunolocalized in pial and layer I astrocytes in the rat cerebral cortex. *Frontiers in Neuroanatomy*, *17*, 1111008. doi:10.3389/fnana.2023.1111008 PMID:36865631

Oenarto, J., Görg, B., Moos, M., Bidmon, H.-J., & Häussinger, D. (2014). Expression of Organic Osmolyte Transporters in Cultured Rat Astrocytes and Rat and Human Cerebral Cortex. *Archives of Biochemistry and Biophysics*, *560*, 59–72. doi:10.1016/j.abb.2014.06.024 PMID:25004465

Ohara, P. T., Vit, J.-P., Bhargava, A., Romero, M., Sundberg, C., Charles, A. C., & Jasmin, L. (2009). Gliopathic Pain: When Satellite Glial Cells Go Bad. *The Neuroscientist*, *15*(5), 450–463. doi:10.1177/1073858409336094 PMID:19826169

Ohgaki, H., Dessen, P., Jourde, B., Horstmann, S., Nishikawa, T., Di Patre, P. L., Burkhard, C., Schüler, D., Probst-Hensch, N. M., Maiorka, P. C., Baeza, N., Pisani, P., Yonekawa, Y., Yasargil, M. G., Lütolf, U. M., & Kleihues, P. (2004). Genetic pathways to glioblastoma: A population-based study. *Cancer Research*, *64*(19), 6892–6899. doi:10.1158/0008-5472.CAN-04-1337 PMID:15466178

Ohgaki, H., & Kleihues, P. (2007). Genetic pathways to primary and secondary glioblastoma. *American Journal of Pathology*, *170*(5), 1445–1453. doi:10.2353/ajpath.2007.070011 PMID:17456751

Ohgaki, H., & Kleihues, P. (2009). Genetic alterations and signaling pathways in the evolution of gliomas. *Cancer Science*, *100*(12), 2235–2241. doi:10.1111/j.1349-7006.2009.01308.x PMID:19737147

Ohmichi, M., Ohmichi, Y., Ohishi, H., Yoshimoto, T., Morimoto, A., Li, Y., Sakurai, H., Nakano, T., & Sato, J. (2014). Activated spinal astrocytes are involved in the maintenance of chronic widespread mechanical hyperalgesia after cast immobilization. *Molecular Pain*, *10*, 6. doi:10.1186/1744-8069-10-6 PMID:24456903

Ojala, J., Alafuzoff, I., Herukka, S.-K., van Groen, T., Tanila, H., & Pirttilä, T. (2009). Expression of interleukin-18 is increased in the brains of Alzheimer's disease patients. *Neurobiology of Aging*, *30*(2), 198–209. doi:10.1016/j.neurobiolaging.2007.06.006 PMID:17658666

Ojeda-Hernández, D. D., Canales-Aguirre, A. A., Matias-Guiu, J., Gomez-Pinedo, U., & Mateos-Díaz, J. C. (2020). Potential of chitosan and its derivatives for biomedical applications in the central nervous system. *Frontiers in Bioengineering and Biotechnology*, *8*, 389. doi:10.3389/fbioe.2020.00389 PMID:32432095

Okada-Ogawa, A., Suzuki, I., Sessle, B. J., Chiang, C.-Y., Salter, M. W., Dostrovsky, J. O., Tsuboi, Y., Kondo, M., Kitagawa, J., Kobayashi, A., Noma, N., Imamura, Y., & Iwata, K. (2009). Astroglia in medullary dorsal horn (trigeminal spinal subnucleus caudalis) are involved in trigeminal neuropathic pain mechanisms. *The Journal of Neuroscience : The Official Journal of the Society for Neuroscience*, *29*(36), 11161–11171. doi:10.1523/JNEUROSCI.3365-09.2009 PMID:19741123

Okada, S., Hara, M., Kobayakawa, K., Matsumoto, Y., & Nakashima, Y. (2018). Astrocyte reactivity and astrogliosis after spinal cord injury. *Neuroscience Research*, *126*, 39–43. doi:10.1016/j.neures.2017.10.004 PMID:29054466

Okamura, N., Suemoto, T., Shiomitsu, T., Suzuki, M., Shimadzu, H., Akatsu, H., Yamamoto, T., Arai, H., Sasaki, H., Yanai, K., Staufenbiel, M., Kudo, Y., & Sawada, T. (2004). A novel imaging probe for in vivo detection of neuritic and diffuse amyloid plaques in the brain. *Journal of Molecular Neuroscience*, *24*(2), 247–255. doi:10.1385/JMN:24:2:247 PMID:15456938

Oksanen, M., Lehtonen, S., Jaronen, M., Goldsteins, G., Hämäläinen, R. H., & Koistinaho, J. (2019). Astrocyte alterations in neurodegenerative pathologies and their modeling in human induced pluripotent stem cell platforms. *Cellular and Molecular Life Sciences*, *76*(14), 2739–2760. doi:10.100700018-019-03111-7 PMID:31016348

Okun, M. S. (2012). Deep-Brain Stimulation for Parkinson's Disease. *NEJM, 367*(16), 1529–1538. doi:10.1056/NEJMct1208070

Okun, E., Griffioen, K. J., Lathia, J. D., Tang, S.-C., Mattson, M. P., & Arumugam, T. V. (2009). Toll-like receptors in neurodegeneration. *Brain Research. Brain Research Reviews*, *59*(2), 278–292. doi:10.1016/j.brainresrev.2008.09.001 PMID:18822314

Olabarria, M., Noristani, H. N., Verkhratsky, A., & Rodríguez, J. J. (2010). 'Concomitant Astroglial Atrophy and Astrogliosis in a Triple Transgenic Animal Model of Alzheimer's Disease'. *Glia*, *58*(7), 831–838. doi:10.1002/glia.20967 PMID:20140958

Oliveira, E. M. L. D., Montani, D. A., Oliveira-Silva, D., Rodrigues-Oliveira, A. F., Matas, S. L. D. A., Fernandes, G. B. P., Silva, I. D. C. G., & Lo Turco, E. G. (2019). Multiple sclerosis has a distinct lipid signature in plasma and cerebrospinal fluid. *Arquivos de Neuro-Psiquiatria*, *77*(10), 696–704. doi:10.1590/0004-282x20190122

Oliveira, J. F., Gomes, C. A., Vaz, S. H., Sousa, N., & Pinto, L. (2016). Editorial: Glial plasticity in depression. *Frontiers in Cellular Neuroscience*, *10*(JUN), 1–2. doi:10.3389/fncel.2016.00163 PMID:27378858

Oliveira, M. I., Santos, S. G., Oliveira, M. J., Torres, A. L., & Barbosa, M. A. (2012). Chitosan drives anti-inflammatory macrophage polarisation and pro-inflammatory dendritic cell stimulation. *European Cells & Materials*, *24*, 136–153. doi:10.22203/eCM.v024a10 PMID:22828991

Olsen, M. L., Khakh, B. S., Skatchkov, S. N., Zhou, M., Lee, C. J., & Rouach, N. (2015). New insights on astrocyte ion channels : Critical for homeostasis and neuron-glia signaling. *The Journal of Neuroscience : The Official Journal of the Society for Neuroscience*, *35*(41), 13827–13835. doi:10.1523/JNEUROSCI.2603-15.2015 PMID:26468182

Olsen, M. L., & Sontheimer, H. (2008). Functional implications for Kir4.1 channels in glial biology: From K+ buffering to cell differentiation. *Journal of Neurochemistry*, *107*(3), 589–601. doi:10.1111/j.1471-4159.2008.05615.x PMID:18691387

Omuro, A. M., Faivre, S., & Raymond, E. (2007). Lessons learned in the development of targeted therapy for malignant gliomas. *Molecular Cancer Therapeutics*, *6*(7), 1909–1919. doi:10.1158/1535-7163.MCT-07-0047 PMID:17620423

Ong, J. P., Aggarwal, A., Krieger, D., Easley, K. A., Karafa, M. T., Van Lente, F., Arroliga, A. C., & Mullen, K. D. (2003). Correlation between Ammonia Levels and the Severity of Hepatic Encephalopathy. *The American Journal of Medicine*, *114*(3), 188–193. doi:10.1016/S0002-9343(02)01477-8 PMID:12637132

Öngür, D., Drevets, W. C., & Price, J. L. (1998). Glial reduction in the subgenual prefrontal cortex in mood disorders. *Proceedings of the National Academy of Sciences of the United States of America*, *95*(22), 13290–13295. doi:10.1073/pnas.95.22.13290 PMID:9789081

Oppenlander, M. E., Wolf, A. B., Snyder, L. A., Bina, R., Wilson, J. R., Coons, S. W., Ashby, L. S., Brachman, D., Nakaji, P., Porter, R. W., Smith, K. A., Spetzler, R. F., & Sanai, N. (2014). An extent of resection threshold for recurrent glioblastoma and its risk for neurological morbidity. *Journal of Neurosurgery*, *120*(4), 846–853. doi:10.3171/2013.12.JNS13184 PMID:24484232

Orellana, J. A., Froger, N., Ezan, P., Jiang, J. X., Bennett, M. V. L., Naus, C. C. G., & Sáez, J. C. (2011a). ATP and glutamate released via astroglial connexin 43 hemichannels mediate neuronal death through activation of pannexin 1 hemichannels. *The Journal of Neuroscience : The Official Journal of the Society for Neuroscience*, *31*(49), 17690–17701. doi:10.1111/j.1471-4159.2011.07210.x PMID:21294731

Orellana, J. A., Sáez, P. J., Shoji, K. F., Schalper, K. A., Palacios-Prado, N., Velarde, V., & Sáez, J. C. (2011b). Modulation of brain hemichannels and gap junction channels by pro-inflammatory agents and their possible role in neurodegeneration. *Antioxidants & Redox Signaling*, *11*(2), 369–399. doi:10.1089/ars.2008.2130 PMID:18816186

Orihuela, R., McPherson, C. A., & Harry, G. J. (2016). Microglial M1/M2 polarization and metabolic states. *British Journal of Pharmacology*, *173*(4), 649–665. doi:10.1111/bph.13139 PMID:25800044

Orkand, R. K., Nicholls, J. G., & Kuffler, S. W. (1966). Effect of nerve impulses on the membrane potential of glial cells in the central nervous system of amphibia. *Journal of Neurophysiology*, *29*(4), 788–806. doi:10.1152/jn.1966.29.4.788 PMID:5966435

Orlovskaya, D. D., Vostrikov, V. M., Rachmanova, V. I., & Uranova, N. A. (2000). Decreased numerical density of oligodendroglial cells in postmortem prefrontal cortex in schizophrenia, bipolar affective disorder and major depression. *Schizophrenia Research*, *1*(41), 105–106. doi:10.1016/S0920-9964(00)90551-6

Orsini, F., De Blasio, D., Zangari, R., Zanier, E. R., & De Simoni, M. G. (2014). The versatility of the complement system in neuroinflammation, neurodegeneration, and brain homeostasis. *Frontiers in Cellular Neuroscience*, *8*, 380. doi:10.3389/fncel.2014.00380 PMID:25426028

Ortinski, P. I., Dong, J., Mungenast, A., Yue, C., Takano, H., Watson, D. J., Haydon, P. G., & Coulter, D. A. (2010). Selective induction of astrocytic gliosis generates deficits in neuronal inhibition. *Nature Neuroscience*, *13*(5), 584–591. doi:10.1038/nn.2535 PMID:20418874

Ortiz, M., Cordoba, J., Doval, E., Jacas, C., Pujadas, F., Esteban, R., & Guardia, J. (2007). Development of a clinical hepatic encephalopathy staging scale. *Alimentary Pharmacology & Therapeutics*, *26*(6), 859–867. doi:10.1111/j.1365-2036.2007.03394.x PMID:17767470

Ortiz, M., Jacas, C., & Córdoba, J. (2005). Minimal hepatic encephalopathy: Diagnosis, clinical significance and recommendations. *Journal of Hepatology*, *42*(1), S45–S53. doi:10.1016/j.jhep.2004.11.028 PMID:15777572

Ortiz-Ortiz, M. A., Morán, J. M., Ruiz-Mesa, L. M., Bonmatty, R. G., & Fuentes, J. M. (2011). Protective effect of the glial cell line-derived neurotrophic factor (GDNF) on human mesencephalic neuron-derived cells against neurotoxicity induced by paraquat. *Environmental Toxicology and Pharmacology*, *31*(1), 129–136. doi:10.1016/j.etap.2010.09.013 PMID:21787677

Osborn, L. M., Kamphuis, W., Wadman, W. J., & Hol, E. M. (2016). Astrogliosis: An integral player in the pathogenesis of Alzheimer's disease. *Progress in Neurobiology*, *144*, 121–141. doi:10.1016/j.pneurobio.2016.01.001 PMID:26797041

Oshio, K., Binder, D. K., Liang, Y., Bollen, A., Feuerstein, B., Berger, M. S., & Manley, G. T. (2005). 'Expression of the Aquaporin-1 Water Channel in Human Glial Tumors'. *Neurosurgery*, *56*(2), 375–381. doi:10.1227/01.NEU.0000148904.57841.6B PMID:15670385

Oshita, A., Kishida, S., Kobayashi, H., Michiue, T., Asahara, T., Asashima, M., & Kikuchi, A. (2003). Identification and characterization of a novel Dvl-binding protein that suppresses Wnt signalling pathway. *Genes to Cells*, *8*(12), 1005–1017. doi:10.1111/j.1365-2443.2003.00692.x PMID:14750955

Osorio, M. J., & Goldman, S. A. (2016). Glial progenitor cell-based treatment of the childhood leukodystrophies. *Experimental Neurology*, *283*, 476–488. doi:10.1016/j.expneurol.2016.05.010 PMID:27170209

Ostrom, Q. T., Bauchet, L., Davis, F. G., Deltour, I., Fisher, J. L., Langer, C. E., Pekmezci, M., Schwartzbaum, J. A., Turner, M. C., Walsh, K. M., Wrensch, M. R., & Barnholtz-Sloan, J. S. (2014). The epidemiology of glioma in adults: A "state of the science" review. *Neuro-Oncology*, *16*(7), 896–913. doi:10.1093/neuonc/nou087 PMID:24842956

Ostrom, Q. T., Gittleman, H., Fulop, J., Liu, M., Blanda, R., Kromer, C., Wolinsky, Y., Kruchko, C., & Barnholtz-Sloan, J. S. (2015). CBTRUS statistical report: Primary brain and central nervous system tumors diagnosed in the United States in 2008-2012. *Neuro-Oncology*, *17*(suppl_4), iv1–iv62. doi:10.1093/neuonc/nov189 PMID:26511214

Ota, M., Sato, N., & Kimura, Y. (2019). Changes of myelin organization in patients with Alzheimer's disease shown by q-space myelin map imaging. *Matsuda Dementia and Geriatric Cognitive Disorders Extra.* Karger. https://karger.com/dee/article/9/1/24/102937

Ota, Y., Zanetti, A. T., & Hallock, R. M. (2013). The role of astrocytes in the regulation of synaptic plasticity and memory formation. *Neural Plasticity*, *2013*, 185463. doi:10.1155/2013/185463 PMID:24369508

Otsuki, K., Uchida, S., Watanuki, T., Wakabayashi, Y., Fujimoto, M., Matsubara, T., Funato, H., & Watanabe, Y. (2008). Altered expression of neurotrophic factors in patients with major depression. *Journal of Psychiatric Research*, *42*(14), 1145–1153. doi:10.1016/j.jpsychires.2008.01.010 PMID:18313696

Ouchi, Y., Yoshikawa, E., Sekine, Y., Futatsubashi, M., Kanno, T., Ogusu, T., & Torizuka, T. (2005). Microglial activation and dopamine terminal loss in early Parkinson's disease. *Annals of Neurology*, *57*(2), 168–175. doi:10.1002/ana.20338 PMID:15668962

Ouyang, Q. Q., Zhao, S., Li, S. D., & Song, C. (2017). Application of chitosan, chitooligosaccharide, and their derivatives in the treatment of Alzheimer's disease. In Marine Drugs. MDPI AG. doi:10.3390/md15110322

Overmyer, M., Helisalmi, S., Soininen, H., Laakso, M., Riekkinen, P. Sr, & Alafuzoff, I. (1999). Reactive microglia in aging and dementia : An immunohistochemical study of postmortem human brain tissue. *Acta Neuropathologica*, *97*(4), 383–392. doi:10.1007004010051002 PMID:10208278

Owen, O. E., Morgan, A. P., Kemp, H. G., Sullivan, J. M., Herrera, M. G., & Cahill, G. J. Jr. (1967). Brain metabolism during fasting. *The Journal of Clinical Investigation*, *46*(10), 1589–1595. doi:10.1172/JCI105650

Ozben, T., & Ozben, S. (2019). Neuro-inflammation and anti-inflammatory treatment options for Alzheimer's disease. *Clinical Biochemistry*, *72*, 87–89. doi:10.1016/j.clinbiochem.2019.04.001 PMID:30954437

Pabon, M. M., Bachstetter, A. D., Hudson, C. E., Gemma, C., & Bickford, P. C. (2011). CX3CL1 reduces neurotoxicity and microglial activation in a rat model of Parkinson's disease. *Journal of Neuroinflammation*, *8*(1), 9. doi:10.1186/1742-2094-8-9 PMID:21266082

Padfield, E., Ellis, H. P., & Kurian, K. M. (2015). Current therapeutic advances targeting EGFR and EGFRvIII in glioblastoma. *Frontiers in Oncology*, *5*, 5. doi:10.3389/fonc.2015.00005 PMID:25688333

Padovani, A., Agosti, C., Premi, E., Bellelli, G., & Borroni, B. (2007). Extrapyramidal symptoms in frontotemporal dementia: Prevalence and clinical correlations. *Neuroscience Letters*, *422*(1), 39–42. doi:10.1016/j.neulet.2007.05.049 PMID:17574750

Pae, C.-U., Marks, D. M., Han, C., & Patkar, A. A. (2008). Does minocycline have antidepressant effect? *Biomedicine and Pharmacotherapy*, *62*(5), 308–311. doi:10.1016/j.biopha.2007.12.005 PMID:18267354

Pallavi, P., Sagar, R., Mehta, M., Sharma, S., Subramanium, A., Shamshi, F., Sengupta, U., Pandey, R. M., & Mukhopadhyay, A. K. (2015). Serum cytokines and anxiety in adolescent depression patients: Gender effect. *Psychiatry Research*, *229*(1–2), 374–380. doi:10.1016/j.psychres.2015.06.036 PMID:26163725

Pallavi, P., Sagar, R., Mehta, M., Sharma, S., Subramanium, A., Shamshi, F., Sengupta, U., Qadri, R., Pandey, R. M., & Mukhopadhyay, A. K. (2013). Serum neurotrophic factors in adolescent depression: Gender difference and correlation with clinical severity. *Journal of Affective Disorders*, *150*(2), 415–423. doi:10.1016/j.jad.2013.04.033 PMID:23769609

Palmer, A. L., & Ousman, S. S. (2018). 'Astrocytes and Aging'. *Frontiers in Aging Neuroscience*, *10*, 337. https://www.frontiersin.org/articles/10.3389/fnagi.2018.00337. doi:10.3389/fnagi.2018.00337 PMID:30416441

Palpagama, T. H., Waldvogel, H. J., Faull, R. L. M., & Kwakowsky, A. (2019). The Role of Microglia and Astrocytes in Huntington's Disease. *Frontiers in Molecular Neuroscience*, *12*, 258. doi:10.3389/fnmol.2019.00258 PMID:31708741

Pangestuti, R., & Kim, SK (2010). Neuroprotective properties of chitosan and its derivatives. *marine drugs, 8*(7), 2117-2128.

Pang, S., Ho, P., Liu, H., Leung, C., Li, L., Chang, E., Ramsden, D., & Ho, S. (2019). The interplay of aging, genetics, and environmental factors in the pathogenesis of Parkinson's disease. *Translational Neurodegeneration*, *8*(1), 23. doi:10.118640035-019-0165-9 PMID:31428316

Pang, Y., Cai, Z., & Rhodes, P. G. (2003). Disturbance of oligodendrocyte development, hypomyelination and white matter injury in the neonatal rat brain after intracerebral injection of lipopolysaccharide. *Brain Research. Developmental Brain Research*, *140*(2), 205–214. doi:10.1016/S0165-3806(02)00606-5 PMID:12586426

Panickar, K. S., & Norenberg, M. D. (2005). Astrocytes in cerebral ischemic injury: Morphological and general considerations. *Glia*, *50*(4), 287–298. doi:10.1002/glia.20181 PMID:15846806

Pankratova, S., Klingelhofer, J., Dmytriyeva, O., Owczarek, S., Renziehausen, A., Syed, N., Porter, A. E., Dexter, D. T., & Kiryushko, D. (2018). The S100A4 protein signals through the ErbB4 receptor to promote neuronal survival. *Theranostics*, *8*(14), 3977–3990. doi:10.7150/thno.22274 PMID:30083275

Pankratov, Y., & Lalo, U. (2015). Role for astroglial α1-adrenoreceptors in gliotransmission and control of synaptic plasticity in the neocortex. *Frontiers in Cellular Neuroscience*, *9*, 230. doi:10.3389/fncel.2015.00230 PMID:26136663

Pannasch, U., Derangeon, M., Chever, O., & Rouach, N. (2012). Astroglial gap junctions shape neuronal network activity. *Communicative & Integrative Biology*, *5*(3), 248–254. doi:10.4161/cib.19410 PMID:22896785

Pannasch, U., & Rouach, N. (2013). An emerging role for astroglial networks in information processing: From synapse to behavior. *Trends in Neurosciences*, *36*(7), 405–417. doi:10.1016/j.tins.2013.04.004 PMID:23659852

Pannasch, U., Vargová, L., Reingruber, J., Ezan, P., Holcman, D., Giaume, C., Syková, E., & Rouach, N. (2011). Astroglial networks scale synaptic activity and plasticity. *Proceedings of the National Academy of Sciences of the United States of America*, *108*(20), 8467–8472. doi:10.1073/pnas.1016650108 PMID:21536893

Pantazatos, S. P., Huang, Y. Y., Rosoklija, G. B., Dwork, A. J., Arango, V., & Mann, J. J. (2017). Whole-transcriptome brain expression and exon-usage profiling in major depression and suicide: Evidence for altered glial, endothelial and ATPase activity. *Molecular Psychiatry*, *22*(5), 760–773. doi:10.1038/mp.2016.130 PMID:27528462

Panza, G. A., Taylor, B. A., MacDonald, H. V., Johnson, B. T., Zaleski, A. L., Livingston, J., Thompson, P. D., & Pescatello, L. S. (2018). Can exercise improve cognitive symptoms of Alzheimer's disease? *Journal of the American Geriatrics Society*, *66*(3), 487–495. doi:10.1111/jgs.15241 PMID:29363108

Paolicelli, R. C., Bolasco, G., Pagani, F., Maggi, L., Scianni, M., Panzanelli, P., Giustetto, M., Ferreira, T. A., Guiducci, E., Dumas, L., Ragozzino, D., & Gross, C. T. (2011). Synaptic pruning by microglia is necessary for normal brain development. *Science*, *333*(6048), 1456–1458. doi:10.1126cience.1202529 PMID:21778362

Papadeas, S. T., Kraig, S. E., O'Banion, C., Lepore, A. C., & Maragakis, N. J. (2011). Astrocytes carrying the superoxide dismutase 1 (SOD1 G93A) mutation induce wild-type motor neuron degeneration in vivo. *Proceedings of the National Academy of Sciences of the United States of America, 108*(43), 17803–17808. doi:10.1073/pnas.1103141108 PMID:21969586

Papuć, E., Science, K. R.-A. (2020). The role of myelin damage in Alzheimer's disease pathology. *Termedia. Archives of Medical Science, 16*(2), 345–351. doi:10.5114/aoms.2018.76863

Pardo, A. C., Wong, V., Benson, L. M., Dykes, M., Tanaka, K., Rothstein, J. D., & Maragakis, N. J. (2006). Loss of the astrocyte glutamate transporter GLT1 modifies disease in SOD1(G93A) mice. *Experimental Neurology, 201*(1), 120–130. doi:10.1016/j.expneurol.2006.03.028 PMID:16753145

Pardo-Moreno, T., González-Acedo, A., Rivas-Domínguez, A., García-Morales, V., García-Cozar, F. J., Ramos-Rodríguez, J. J., & Melguizo-Rodríguez, L. (2022). Therapeutic approach to Alzheimer's disease: Current treatments and new perspectives. *Pharmaceutics, 14*(6), 1117. doi:10.3390/pharmaceutics14061117 PMID:35745693

Parenti, G., Andria, G., & Ballabio, A. (2015). Lysosomal storage diseases: From pathophysiology to therapy. *Annual Review of Medicine, 66*(1), 471–486. doi:10.1146/annurev-med-122313-085916 PMID:25587658

Parent, M., & Parent, A. (2010). Substantia nigra and Parkinson's disease : A brief history of their long and intimate relationship. *The Canadian Journal of Neurological Sciences, 37*(3), 313–319. doi:10.1017/S0317167100010209 PMID:20481265

Parisi, S., Corsa, P., Raguso, A., Perrone, A., Cossa, S., Munafò, T., Sanpaolo, G., Donno, E., Clemente, M. A., Piombino, M., Parisi, F., & Valle, G. (2015). Temozolomide and radiotherapy versus radiotherapy alone in high grade gliomas: A very long-term comparative study and literature review. *BioMed Research International, 2015*, 2015. doi:10.1155/2015/620643 PMID:25815327

Park, H., Han, K. S., Seo, J., Lee, J., Dravid, S. M., Woo, J., Chun, H., Cho, S., Bae, J. Y., An, H., An, H., Koh, W., Yoon, B.-E., Yoon, B.-E., Berlinguer-Palmini, R., Mannaioni, G., Traynelis, S. F., Bae, Y. C., Choi, S.-Y., Lee, C. J. (2015). Channel-mediated astrocytic glutamate modulates hippocampal synaptic plasticity by activating postsynaptic NMDA receptors. *Molecular Brain, 8*(1), pp. 8:7. doi:10.1186/s13041-015-0097-y

Parkhurst, C. N., Yang, G., Ninan, I., Savas, J. N., Yates, J. R. III, Lafaille, J. J., Hempstead, B. L., Littman, D. R., & Gan, W.-B. (2013). 'Microglia Promote Learning-Dependent Synapse Formation through Brain-Derived Neurotrophic Factor'. *Cell, 155*(7), 1596–1609. doi:10.1016/j.cell.2013.11.030 PMID:24360280

Parkinson, J. (2002). An essay on the shaking palsy. *The Journal of Neuropsychiatry and Clinical Neurosciences, 14*(2), 223–236. doi:10.1176/jnp.14.2.223 PMID:11983801

Park, S. H., Kim, J. H., Choi, K. H., Jang, Y. J., Bae, S. S., Choi, B. T., & Shin, H. K. (2013). Hypercholesterolemia accelerates amyloid β-induced cognitive deficits. *International Journal of Molecular Medicine, 31*(3), 577–582. doi:10.3892/ijmm.2013.1233 PMID:23314909

Parpura, V., Heneka, MT, Montana, V., Oliet, SH, Schousboe, A., Haydon, PG, & Verkhratsky, A. (2012). Cellules gliales en (patho)physiologie. *Tourillon de neurochimie, 121*(1), 4-27.

Parpura, V., & Verkhratsky, A. (2012). The astrocyte excitability brief: From receptors to gliotransmission. *Neurochemistry International, 61*(4), 610–621. doi:10.1016/j.neuint.2011.12.001 PMID:22178457

Parsons, D. W., Jones, S., Zhang, X., Lin, J. C. H., Leary, R. J., Angenendt, P., & Kinzler, K. W. (2008). An integrated genomic analysis of human glioblastoma multiforme. *science, 321*(5897), 1807-1812.

Pasandideh, S., & Arasteh, A. (2021). Evaluation of antioxidant and inhibitory properties of Citrus aurantium L. on the acetylcholinesterase activity and the production of amyloid nano–bio fibrils. *International Journal of Biological Macromolecules, 182*, 366–372. doi:10.1016/j.ijbiomac.2021.04.043 PMID:33848544

Pascual, A., Hidalgo-Figueroa, M., Piruat, J. I., Pintado, C. O., Gomez-Diaz, R., & López-Barneo, J. (2008). Absolute requirement of GDNF for adult catecholaminergic neuron survival. *Nature Neuroscience, 11*(7), 755–761. doi:10.1038/nn.2136 PMID:18536709

Pascual, M., Calvo-Rodriguez, M., Núñez, L., Villalobos, C., Ureña, J., & Guerri, C. (2021). Toll-like receptors in neuroinflammation, neurodegeneration, and alcohol-induced brain damage. *IUBMB Life, 73*(7), 900–915. doi:10.1002/iub.2510 PMID:34033211

Pascual, O., Ben Achour, S., Rostaing, P., Triller, A., & Bessis, A. (2012). Microglia activation triggers astrocyte-mediated modulation of excitatory neurotransmission. *Proceedings of the National Academy of Sciences of the United States of America, 109*(4), E197–E205. doi:10.1073/pnas.1111098109 PMID:22167804

Pascual, O., Casper, K. B., Kubera, C., Zhang, J., Revilla-Sanchez, R., Sul, J. Y., & Haydon, P. G. (2005). Astrocytic purinergic signaling coordinates synaptic networks. *Science, 310*(5745), 113–116. doi:10.1126cience.1116916 PMID:16210541

Pasinelli, P., & Brown, R. H. (2006). Molecular biology of amyotrophic lateral sclerosis: Insights from genetics. *Nature Reviews. Neuroscience, 7*(9), 710–723. doi:10.1038/nrn1971 PMID:16924260

Patanow, C. M., Day, J. R., & Billingsley, M. L. (1996). Alterations in hippocampal expression of SNAP-25, GAP-43, stannin and glial fibrillary acidic protein following mechanical and trimethyltin-induced injury in the rat. *Neuroscience, 76*(1), 187–202. doi:10.1016/S0306-4522(96)00335-1 PMID:8971771

Patel, N., Paris, D., Mathura, V., & A. Q.-J. (2005). Inflammatory cytokine levels correlate with amyloid load in transgenic mouse models of Alzheimer's disease. *Journal of Neuroinflammation*. https://link.springer.com/article/10.1186/1742-2094-2-9

Patel, D. C., Tewari, B. P., Chaunsali, L., & Sontheimer, H. (2019). Neuron-glia interactions in the pathophysiology of epilepsy. *Nature Reviews. Neuroscience, 20*(5), 282–297. doi:10.103841583-019-0126-4 PMID:30792501

Paul, I. A., & Skolnick, P. (2003). Glutamate and depression: Clinical and preclinical studies. *Annals of the New York Academy of Sciences, 1003*(1), 250–272. doi:10.1196/annals.1300.016 PMID:14684451

Pedersen, W. A., Fu, W., Keller, J. N., Markesbery, W. R., Appel, S., Smith, R. G., Kasarskis, E., & Mattson, M. P. (1998). Protein modification by the lipid peroxidation product 4-hydroxynonenal in the spinal cords of amyotrophic lateral sclerosis patients. *Annals of Neurology, 44*(5), 819–824. doi:10.1002/ana.410440518 PMID:9818940

Peferoen, L., Kipp, M., van der Valk, P., van Noort, J. M., & Amor, S. (2014a). Oligodendrocyte-microglia cross-talk in the central nervous system. *Immunology, 141*(3), 302–313. doi:10.1111/imm.12163 PMID:23981039

Pekny, M., & Pekna, M. (2014). Astrocyte Reactivity and Reactive Astrogliosis : Costs and Benefits. *Physiological Reviews, 94*(4), 1077–1098. doi:10.1152/physrev.00041.2013 PMID:25287860

Pellerin, B. A., Wollheim, W. M., Feng, X., & Vörösmarty, C. J. (2008). The application of electrical conductivity as a tracer for hydrograph separation in urban catchments. *Hydrological Processes, 22*(12), 1810–1818. doi:10.1002/hyp.6786

Pellicciari, R., Costantino, G., Marinozzi, M., & Natalini, B. (1998). Modulation of glutamate receptor pathways in the search for new neuroprotective agents. *Il Farmaco, 53*(4), 255–261. doi:10.1016/S0014-827X(98)00018-4 PMID:9658582

Peng, L., Verkhratsky, A., Gu, L., & Li, B. (2015). Targeting astrocytes in major depression. *Expert Review of Neurotherapeutics, 15*(11), 1299–1306. doi:10.1586/14737175.2015.1095094 PMID:26471936

Peng, X., Lin, Q., Liu, Y., Jin, Y., Druso, J. E., Antonyak, M. A., Guan, J.-L., & Cerione, R. A. (2013). Inactivation of Cdc42 in embryonic brain results in hydrocephalus with ependymal cell defects in mice. *Protein & Cell*, *4*(3), 231–242. doi:10.100713238-012-2098-2 PMID:23150167

Peng, Z., Zhang, C., Yan, L., Zhang, Y., Yang, Z., Wang, J., & Song, C. (2020). EPA is more effective than DHA to improve depression-like behavior, glia cell dysfunction and hippcampal apoptosis signaling in a chronic stress-induced rat model of depression. *International Journal of Molecular Sciences*, *21*(5), 1769. doi:10.3390/ijms21051769 PMID:32150824

Pentschew, A., Ebner, F. F., & Kovatch, R. M. (1963). Experimental manganese encephalopathy in monkeys: A preliminary report. *Journal of Neuropathology and Experimental Neurology*, *22*(3), 488–499. doi:10.1097/00005072-196307000-00010 PMID:14045007

Pepe, G., Calderazzi, G., De Maglie, M., Villa, A. M., & Vegeto, E. (2014). Heterogeneous induction of microglia M2a phenotype by central administration of interleukin-4. *Journal of Neuroinflammation*, *11*(1), 211. doi:10.118612974-014-0211-6 PMID:25551794

Perea, G., Gomez, R., Mederos, S., Covelo, A., Ballesteros, J. J., Schlosser, L., Hernández-Vivanco, A., Martín-Fernández, M., Quintana, R., Rayan, A., Díez, A., Fuenzalida, M., Agarwal, A., Bergles, D. E., Bettler, B., Manahan-Vaughan, D., Martín, E. D., Kirchhoff, F., & Araque, A. (2016). Activity-dependent switch of GABAergic inhibition into glutamatergic excitation in astrocyte-neuron networks. *eLife*, *5*, e20362. doi:10.7554/eLife.20362 PMID:28012274

Perea, G., Navarrete, M., & Araque, A. (2009). Tripartite synapses: Astrocytes process and control synaptic information. *Trends in Neurosciences*, *32*(8), 421–431. doi:10.1016/j.tins.2009.05.001 PMID:19615761

Perea, J., Bolós, M., & Biomolecules, J. A. (2020). Microglia in Alzheimer's disease in the context of tau pathology. *Mdpi.ComJR Perea, M Bolós, J AvilaBiomolecules •mdpi. Biomolecules*, *10*(10), 1–26. doi:10.3390/biom10101439 PMID:33066368

Pereira, V. H., Salgado, A. J., Oliveira, J. M., Cerqueira, S. R., Frias, A. M., Fraga, J. S., Roque, S., Falcão, A. M., Marques, F., Neves, N. M., Mano, J. F., Reis, R. L., & Sousa, N. (2011). In vivo biodistribution of carboxymethylchitosan/poly(amidoamine) dendrimer nanoparticles in rats. *Journal of Bioactive and Compatible Polymers*, *26*(6), 619–627. doi:10.1177/0883911511425567

Pérez-Cerdá, F., Sánchez-Gómez, M. V., & Matute, C. (2015). Pío del Río Hortega and the discovery of the oligodendrocytes. *Frontiers in Neuroanatomy*, *9*, 92. doi:10.3389/fnana.2015.00092

Perez-Nievas, B. G., & Serrano-Pozo, A. (2018). Deciphering the astrocyte reaction in Alzheimer's disease. *Frontiers in Aging Neuroscience*, *10*, 114. doi:10.3389/fnagi.2018.00114 PMID:29922147

Perrie, W. T., Lee, G. T., Curtis, E. M., Sparke, J., Buller, J. R., & Rossi, M. L. (1993). Changes in the myelinated axons of femoral nerve in amyotrophic lateral sclerosis. *Journal of Neural Transmission. Supplementum*, *39*, 223–233. PMID:8360662

Perry, J. R., Laperriere, N., O'Callaghan, C. J., Brandes, A. A., Menten, J., Phillips, C., Fay, M., Nishikawa, R., Cairncross, J. G., Roa, W., Osoba, D., Rossiter, J. P., Sahgal, A., Hirte, H., Laigle-Donadey, F., Franceschi, E., Chinot, O., Golfinopoulos, V., Fariselli, L., & Mason, W. P. (2017). Short-course radiation plus temozolomide in elderly patients with glioblastoma. *The New England Journal of Medicine*, *376*(11), 1027–1037. doi:10.1056/NEJMoa1611977 PMID:28296618

Perry, V. H., Nicoll, J. A. R., & Holmes, C. (2010). Microglia in neurodegenerative disease. *Nature Reviews. Neurology*, *6*(4), 193–201. doi:10.1038/nrneurol.2010.17 PMID:20234358

Pertin, M., Allchorne, A. J., Beggah, A. T., Woolf, C. J., & Decosterd, I. (2007). Delayed sympathetic dependence in the spared nerve injury (SNI) model of neuropathic pain. *Molecular Pain*, *3*, 21. doi:10.1186/1744-8069-3-21 PMID:17672895

Peters, A. (2004). A fourth type of neuroglial cell in the adult central nervous system. *Journal of Neurocytology*, *33*(3), 345–357. doi:10.1023/B:NEUR.0000044195.64009.27 PMID:15475689

Peters, A., & Feldman, M. L. (1976). The projection of the lateral geniculate nucleus to area 17 of the rat cerebral cortex. I. General description. *Journal of Neurocytology*, *5*(1), 63–84. doi:10.1007/BF01176183 PMID:1249593

Petersen, M. A., & Dailey, M. E. (2004). Diverse microglial motility behaviors during clearance of dead cells in hippocampal slices. *Glia*, *46*(2), 195–206. doi:10.1002/glia.10362 PMID:15042586

Petravicz, J., Fiacco, T. A., & McCarthy, K. D. (2008, May 7). Loss of IP3 receptor-dependent Ca2+ increases in hippocampal astrocytes does not affect baseline CA1 pyramidal neuron synaptic activity. *The Journal of Neuroscience : The Official Journal of the Society for Neuroscience*, *28*(19), 4967–4973. doi:10.1523/JNEUROSCI.5572-07.2008 PMID:18463250

Petroff, O. A. C., Errante, L. D., Rothman, D. L., Kim, J. H., & Spencer, D. D. (2002). Glutamate-glutamine cycling in the epileptic human hippocampus. *Epilepsia*, *43*(7), 703–710. doi:10.1046/j.1528-1157.2002.38901.x PMID:12102672

Pettas, S., Karagianni, K., Kanata, E., Chatziefstathiou, A., Christoudia, N., Xanthopoulos, K., Sklaviadis, T., & Dafou, D. (2022). Profiling Microglia through Single-Cell RNA Sequencing over the Course of Development, Aging, and Disease. *Cells*, *11*(15), 2383. doi:10.3390/cells11152383 PMID:35954228

Phatnani, H., & Maniatis, T. (2015). Astrocytes in Neurodegenerative Disease. *Cold Spring Harbor Perspectives in Biology*, *7*(6), a020628. doi:10.1101/cshperspect.a020628 PMID:25877220

Philips, T., & Rothstein, J. D. (2014). Glial cells in amyotrophic lateral sclerosis. *Experimental neurology, 262m* 111–120. doi:10.1016/j.expneurol.2014.05.015

Philips, T., Bento-Abreu, A., Nonneman, A., Haeck, W., Staats, K., Geelen, V., Hersmus, N., Küsters, B., Van Den Bosch, L., Van Damme, P., Richardson, W. D., & Robberecht, W. (2013). Oligodendrocyte dysfunction in the pathogenesis of amyotrophic lateral sclerosis. *Brain*, *136*(Pt 2), 471–482. doi:10.1093/brain/aws339 PMID:23378219

Philips, T., & Robberecht, W. (2011). Neuroinflammation in amyotrophic lateral sclerosis: Role of glial activation in motor neuron disease. *Lancet Neurology*, *10*(3), 253–263. doi:10.1016/S1474-4422(11)70015-1 PMID:21349440

Phillips, H. S., Kharbanda, S., Chen, R., Forrest, W. F., Soriano, R. H., Wu, T. D., Misra, A., Nigro, J. M., Colman, H., Soroceanu, L., Williams, P. M., Modrusan, Z., Feuerstein, B. G., & Aldape, K. (2006). Molecular subclasses of high-grade glioma predict prognosis, delineate a pattern of disease progression, and resemble stages in neurogenesis. *Cancer Cell*, *9*(3), 157–173. doi:10.1016/j.ccr.2006.02.019 PMID:16530701

Piacentini, R., Li Puma, D. D., Ripoli, C., Elena Marcocci, M., De Chiara, G., Garaci, E., Teresa Palamara, A., & Grassi, C. (2015). Herpes Simplex Virus type-1 infection induces synaptic dysfunction in cultured cortical neurons via GSK-3 activation and intraneuronal amyloid-β protein accumulation. *Scientific Reports*, *5*(1), 15444. doi:10.1038rep15444 PMID:26487282

Piccioni, G., Mango, D., Saidi, A., Corbo, M., & Nisticò, R. (2021). Targeting Microglia-synapse interaction in Alzheimer's disease. *International Journal of Molecular Sciences*, *22*(5), 2342. doi:10.3390/ijms22052342 PMID:33652870

Piehl, F. (2021). Current and emerging disease-modulatory therapies and treatment targets for multiple sclerosis. *Journal of Internal Medicine*, *289*(6), 771–791. doi:10.1111/joim.13215

Pinzani, M. (2015). *Pathophysiology of Liver Fibrosis*. 492–497. doi:10.1159/000374096

Pirttimaki, T.M., Codadu, N.K., Awni, A., Pratik, P., Nagel, D.A., Hill, E.J., Dineley, K.T., Parri, H.R.(2013). α7 Nicotinic receptor-mediated astrocytic gliotransmitter release: Aβ effects in a preclinical Alzheimer's mouse model. *PLoS One, 28*(11), e81828. . doi:10.1371/journal.pone.0081828

Pisanu, A., Lecca, D., Mulas, G., & Wardas, J. (2014). *Dynamic changes in pro-and anti-inflammatory cytokines in microglia after PPAR-γ agonist neuroprotective treatment in the MPTPp mouse model of.* Elsevier. https://www.sciencedirect.com/science/article/pii/S0969996114002423

Pizzino, G., Irrera, N., Cucinotta, M., Pallio, G., Mannino, F., Arcoraci, V., Squadrito, F., Altavilla, D., & Bitto, A. (2017). 'Oxidative Stress: Harms and Benefits for Human Health'. *Oxidative Medicine and Cellular Longevity, 8416763*, 1–13. doi:10.1155/2017/8416763 PMID:28819546

Planas, A. M., Justicia, C., Soriano, M. A., & Ferrer, I. (1998). Epidermal growth factor receptor in proliferating reactive glia following transient focal ischemia in the rat brain. *Glia, 23*(2), 120–129. doi:10.1002/(SICI)1098-1136(199806)23:2<120::AID-GLIA3>3.0.CO;2-A PMID:9600380

Plant, L. D., Boyle, J. P., Smith, I. F., Peers, C., & Pearson, H. A. (2003). The production of amyloid β peptide is a critical requirement for the viability of central neurons. *The Journal of Neuroscience : The Official Journal of the Society for Neuroscience, 23*(13), 5531–5535. doi:10.1523/JNEUROSCI.23-13-05531.2003 PMID:12843253

Platt, F. M., d'Azzo, A., Davidson, B. L., Neufeld, E. F., & Tifft, C. J. (2018). Lysosomal storage diseases. *Nature Reviews. Disease Primers, 4*(1), 27. doi:10.103841572-018-0025-4 PMID:30275469

Plemel, J. R., Liu, W. Q., & Yong, V. W. (2017). Remyelination therapies: A new direction and challenge in multiple sclerosis. *Nature Reviews. Drug Discovery, 16*(9), 617–634. doi:10.1038/nrd.2017.115

Podbielska, M., O'Keeffe, J., & Pokryszko-Dragan, A. (2021). New insights into multiple sclerosis mechanisms: Lipids on the track to control inflammation and neurodegeneration. *International Journal of Molecular Sciences, 22*(14), 7319. doi:10.3390/ijms22147319

Polten, A., Fluharty, A. L., Fluharty, C. B., Kappler, J., von Figura, K., & Gieselmann, V. (1991). Molecular basis of different forms of metachromatic leukodystrophy. *The New England Journal of Medicine, 324*(1), 18–22. doi:10.1056/NEJM199101033240104 PMID:1670590

Ponomarev, E. D., Maresz, K., Tan, Y., & Dittel, B. N. (2007). CNS-Derived Interleukin-4 Is Essential for the Regulation of Autoimmune Inflammation and Induces a State of Alternative Activation in Microglial Cells. *The Journal of Neuroscience : The Official Journal of the Society for Neuroscience, 27*(40), 10714–10721. doi:10.1523/JNEUROSCI.1922-07.2007 PMID:17913905

Poon, C. H., Wang, Y., Fung, M.-L., Zhang, C., & Lim, L. W. (2020). Rodent models of amyloid-beta feature of Alzheimer's disease : Development and potential treatment implications. *Aging and Disease, 11*(5), 1235. doi:10.14336/AD.2019.1026 PMID:33014535

Popiolek-Barczyk, K., & Mika, J. (2016). Targeting the Microglial Signaling Pathways: New Insights in the Modulation of Neuropathic Pain. *Current Medicinal Chemistry, 23*(26), 2908–2928. doi:10.2174/0929867323666160607120124 PMID:27281131

Popov, A., Brazhe, A., Denisov, P., Sutyagina, O., Li, L., Lazareva, N., Verkhratsky, A., & Semyanov, A. (2021). 'Astrocyte Dystrophy in Ageing Brain Parallels Impaired Synaptic Plasticity'. *Aging Cell, 20*(3), e13334. doi:10.1111/acel.13334 PMID:33675569

Poskanzer, K. E., & Yuste, R. (2016). Astrocytes regulate cortical state switching in vivo. *Proceedings of the National Academy of Sciences of the United States of America, 113*(19), E2675–E2684. doi:10.1073/pnas.1520759113 PMID:27122314

Postuma, R. B., Iranzo, A., Hogl, B., Arnulf, I., Ferini-Strambi, L., Manni, R., Miyamoto, T., Oertel, W., Dauvilliers, Y., El Ju, Y., Puligheddu, M., Sonka, K., Pelletier, A., Santamaria, J., Frauscher, B., Leu-Semenescu, S., Zucconi, M., Terzaghi, M., Miyamoto, M., & Montplaisir, J. Y. (2015). Risk factors for neurodegeneration in idiopathic rapid eye movement sleep behavior disorder: A multicenter study. *Annals of Neurology*, *77*(5), 830–839. doi:10.1002/ana.24385 PMID:25767079

Poupětová, H., Ledvinová, J., Berná, L., Dvořáková, L., Kožich, V., & Elleder, M. (2010). The birth prevalence of lysosomal storage disorders in the Czech Republic: Comparison with data in different populations. *Journal of Inherited Metabolic Disease*, *33*(4), 387–396. doi:10.100710545-010-9093-7 PMID:20490927

Poveda, M.-J., Bernabeu, Á., Concepción, L., Roa, E., de Madaria, E., Zapater, P., Pérez-Mateo, M., & Jover, R. (2010). Brain edema dynamics in patients with overt hepatic encephalopathy: A magnetic resonance imaging study. *NeuroImage*, *52*(2), 481–487. doi:10.1016/j.neuroimage.2010.04.260 PMID:20451628

Prabaharan, M. (2008). Review paper: Chitosan derivatives as promising materials for controlled drug delivery. In Journal of Biomaterials Applications, 23(1). doi:10.1177/0885328208091562

Prabaharan, M. (2008). Review Paper: Chitosan derivatives as promising materials for controlled. *Drug Deliv J Biomat Applicat*, *23*(1), 5–36. PMID:18593819

Pramatarova, A., Laganière, J., Roussel, J., Brisebois, K., & Rouleau, G. A. (2001). Neuron-specific expression of mutant superoxide dismutase 1 in transgenic mice does not lead to motor impairment. *The Journal of Neuroscience : The Official Journal of the Society for Neuroscience*, *21*(10), 3369–3374. doi:10.1523/JNEUROSCI.21-10-03369.2001 PMID:11331366

Prasad, S., Dhiman, R. K., Duseja, A., Chawla, Y. K., Sharma, A., & Agarwal, R. (2007). Lactulose improves cognitive functions and health-related quality of life in patients with cirrhosis who have minimal hepatic encephalopathy. *Hepatology (Baltimore, Md.)*, *45*(3), 549–559. doi:10.1002/hep.21533 PMID:17326150

Pratt, D., Pittaluga, S., Palisoc, M., Fetsch, P., Xi, L., Raffeld, M., Gilbert, M. R., & Quezado, M. (2017). Expression of CD70 (CD27L) is associated with epithelioid and sarcomatous features in IDH-Wild-Type glioblastoma. *Journal of Neuropathology and Experimental Neurology*, *76*(8), 697–708. doi:10.1093/jnen/nlx051 PMID:28789475

Prevot, V., Dehouck, B., Sharif, A., Ciofi, P., Giacobini, P., & Clasadonte, J. (2018). The versatile tanycyte: A hypothalamic integrator of reproduction and energy metabolism. *Endocrine Reviews*, *39*(3), 333–368. doi:10.1210/er.2017-00235 PMID:29351662

Price, D. A., Martinez, A. A., Seillier, A., Koek, W., Acosta, Y., Fernandez, E., Strong, R., Lutz, B., Marsicano, G., Roberts, J. L., & Giuffrida, A. (2009). WIN55,212-2, a cannabinoid receptor agonist, protects against nigrostriatal cell loss in the 1-methyl-4-phenyl-1,2,3,6-tetrahydropyridine mouse model of Parkinson's disease. *The European Journal of Neuroscience*, *29*(11), 2177–2186. doi:10.1111/j.1460-9568.2009.06764.x PMID:19490092

Prigent, A., Lionnet, A., Corbillé, A.-G., & Derkinderen, P. (2017). Neuropathologie et physiopathologie de la maladie de Parkinson: Focus sur l'α-synucléine. *Presse Medicale (Paris, France)*, *46*(2), 182–186. doi:10.1016/j.lpm.2016.11.010

Pringsheim, T., Jette, N., Frolkis, A., & Steeves, T. D. L. (2014). The prevalence of Parkinson's disease: A systematic review and meta-analysis. *Movement Disorders*, *29*(13), 1583–1590. doi:10.1002/mds.25945 PMID:24976103

Prinz, M., & Priller, J. (2017). 'The Role of Peripheral Immune Cells in the CNS in Steady State and Disease'. *Nature Neuroscience*, *20*(2), 136–144. doi:10.1038/nn.4475 PMID:28092660

Prunell, G., & Olivera-Bravo, S. (2022). A Focus on Astrocyte Contribution to Parkinson's Disease Etiology. *Biomolecules*, *12*(12), 1745. doi:10.3390/biom12121745 PMID:36551173

Przedborski, S., Jackson-Lewis, V., Djaldetti, R., Liberatore, G., Vila, M., & Vukosavic, S. (2000). The parkinsonian toxin MPTP: Action and mechanism. *Restorative Neurology and Neuroscience*, *16*, 135–142. PMID:12671216

Puchałowicz, K., Tarnowski, M., Baranowska-Bosiacka, I., Chlubek, D., & Dziedziejko, V. (2014). P2X and P2Y receptors—Role in the pathophysiology of the nervous system. *International Journal of Molecular Sciences*, *15*(12), 23672–23704. doi:10.3390/ijms151223672 PMID:25530618

Qian, L., & Flood, P. M. (2008). Microglial cells and Parkinson's disease. *Immunologic Research*, *41*(3), 155–164. doi:10.100712026-008-8018-0 PMID:18512160

Qian, L., Winfree, E., & Bruck, J. (2011). Neural network computation with DNA strand displacement cascades. *Nature*, *475*(7356), 7356. doi:10.1038/nature10262 PMID:21776082

Qin, C., Zhou, L. Q., Ma, X. T., Hu, Z. W., Yang, S., Chen, M., Bosco, D. B., Wu, L. J., & Tian, D. S. (2019). Dual Functions of Microglia in Ischemic Stroke. *Neuroscience Bulletin*, *35*(5), 921–933. doi:10.100712264-019-00388-3 PMID:31062335

Qin, L., Wu, X., Block, M. L., Liu, Y., Breese, G. R., Hong, J.-S., Knapp, D. J., & Crews, F. T. (2007). 'Systemic LPS Causes Chronic Neuroinflammation and Progressive Neurodegeneration'. *Glia*, *55*(5), 453–462. doi:10.1002/glia.20467 PMID:17203472

Qiu, J., Yan, Z., Tao, K., Li, Y., Li, J., Dong, Y., Feng, D., & Chen, H. (2016). Sinomenine activates astrocytic dopamine D2 receptors and alleviates neuroinflammatory injury via the CRYAB/STAT3 pathway after ischemic stroke in mice. *Journal of Neuroinflammation*, *13*(1), 263. doi:10.118612974-016-0739-8 PMID:27724964

Rafiki, A., Boulland, J. L., Halestrap, A. P., Ottersen, O. P., & Bergersen, L. (2003). Highly differential expression of the monocarboxylate transporters MCT2 and MCT4 in the developing rat brain. *Neuroscience*, *122*(3), 677–688. doi:10.1016/j.neuroscience.2003.08.040 PMID:14622911

Raghavendra, V., Tanga, F., & DeLeo, J. A. (2003). Inhibition of microglial activation attenuates the development but not existing hypersensitivity in a rat model of neuropathy. *The Journal of Pharmacology and Experimental Therapeutics*, *306*(2), 624–630. doi:10.1124/jpet.103.052407 PMID:12734393

Rahmani, F., Saghazadeh, A., Rahmani, M., Teixeira, A. L., Rezaei, N., Aghamollaii, V., & Ardebili, H. E. (2019). 'Plasma Levels of Brain-Derived Neurotrophic Factor in Patients with Parkinson Disease: A Systematic Review and Meta-Analysis'. *Brain Research*, *1704*(February), 127–136. doi:10.1016/j.brainres.2018.10.006 PMID:30296429

Raison, C. L., Capuron, L., & Miller, A. H. (2006). Cytokines sing the blues: Inflammation and the pathogenesis of depression. *Trends in Immunology*, *27*(1), 24–31. doi:10.1016/j.it.2005.11.006 PMID:16316783

Raj, D., Yin, Z., Breur, M., Doorduin, J., Holtman, I. R., Olah, M., Mantingh-Otter, I. J., Van Dam, D., De Deyn, P. P., den Dunnen, W., Eggen, B. J. L., Amor, S., & Boddeke, E. (2017). 'Increased White Matter Inflammation in Aging- and Alzheimer's Disease Brain'. *Frontiers in Molecular Neuroscience*, *10*, 206. https://www.frontiersin.org/articles/10.3389/fnmol.2017.00206. doi:10.3389/fnmol.2017.00206 PMID:28713239

Rajkowska, G., & Miguel-Hidalgo, J. (2007). Gliogenesis and glial pathology in depression. *CNS & Neurological Disorders-Drug Targets (Formerly Current Drug Targets-CNS & Neurological Disorders)*, *6*(3), 219–233.

Rajkowska, G., Hughes, J., Stockmeier, C. A., Miguel-Hidalgo, J. J., & Maciag, D. (2013). Coverage of blood vessels by astrocytic endfeet is reduced in major depressive disorder. *Biological Psychiatry*, *73*(7), 613–621. doi:10.1016/j.biopsych.2012.09.024 PMID:23146357

Rajkowska, G., & Miguel-Hidalgo, J. J. (2007). Gliogenesis and glial pathology in depression. *CNS & Neurological Disorders - Drug Targets*, *6*(3), 219–233. doi:10.2174/187152707780619326 PMID:17511618

Rajkowska, G., Miguel-Hidalgo, J. J., Wei, J., Dilley, G., Pittman, S. D., Meltzer, H. Y., Overholser, J. C., Roth, B. L., & Stockmeier, C. A. (1999). Morphometric evidence for neuronal and glial prefrontal cell pathology in major depression. *Biological Psychiatry*, *45*(9), 1085–1098. doi:10.1016/S0006-3223(99)00041-4 PMID:10331101

Rajkowska, G., & Stockmeier, A. (2013). Astrocyte pathology in major depressive disorder: Insights from human postmortem brain tissue. *Current Drug Targets*, *14*(11), 1225–1236. doi:10.2174/13894501113149990156 PMID:23469922

Ramanathan, A., Nelson, A. R., Sagare, A. P., & Zlokovic, B. V. (2015). Impaired vascular-mediated clearance of brain amyloid beta in Alzheimer's disease: The role, regulation and restoration of LRP1. *Frontiers in Aging Neuroscience*, *7*, 136. doi:10.3389/fnagi.2015.00136 PMID:26236233

Rannikko, E. H., Weber, S. S., & Kahle, P. J. (2015). Exogenous α-Synuclein Induces Toll-like Receptor 4 Dependent Inflammatory Responses in Astrocytes. *BMC Neuroscience*, *16*(1), 57. doi:10.118612868-015-0192-0 PMID:26346361

Ransohoff, R. M. (2016). A polarizing question: do M1 and M2 microglia exist? *Nature Neuroscience 2, 19*(8), 987–991. doi:10.1038/nn.4338

Ransohoff, R. M., & Brown, M. A. (2012). Innate immunity in the central nervous system. *The Journal of Clinical Investigation*, *122*(4), 1164–1171. doi:10.1172/JCI58644 PMID:22466658

Ransohoff, R. M., & Cardona, A. E. (2010). 'The Myeloid Cells of the Central Nervous System Parenchyma'. *Nature*, *468*(7321), 253–262. doi:10.1038/nature09615 PMID:21068834

Ransohoff, R. M., & Hugh Perry, V. (2009). Microglial Physiology: Unique Stimuli, Specialized Responses. *Annual Review of Immunology*, *27*(1), 119–145. doi:10.1146/annurev.immunol.021908.132528 PMID:19302036

Rao, Kakulavarapu V Rama, & Norenberg, M. D. (2004). Manganese induces the mitochondrial permeability transition in cultured astrocytes. *Journal of Biological Chemistry, 279*(31), 32333–32338.

Rao, J. S., Zhao, C., Zhang, A., Duan, H., Hao, P., Wei, R. H., Shang, J., Zhao, W., Liu, Z., Yu, J., Fan, K. S., Tian, Z., He, Q., Song, W., Yang, Z., Sun, Y. E., & Li, X. (2018). NT3-chitosan enables de novo regeneration and functional recovery in monkeys after spinal cord injury. *Proceedings of the National Academy of Sciences of the United States of America*, *115*(24), E5595–E5604. doi:10.1073/pnas.1804735115 PMID:29844162

Rao, K. V. (2003). Increased Aquaporin-4 Expression in Ammonia-Treated Cultured Astrocytes. *Neuroreport*, *14*(18), 2379–2382. doi:10.1097/00001756-200312190-00018 PMID:14663195

Rao, K. V. (2014). Aquaporin-4 Deletion in Mice Reduces Encephalopathy and Brain Edema in Experimental Acute Liver Failure. *Neurobiology of Disease*, *63*, 222–228. doi:10.1016/j.nbd.2013.11.018 PMID:24321433

Rao, K. V., Brahmbhatt, M., & Norenberg, M. D. (2013). Microglia Contribute to Ammonia-Induced Astrocyte Swelling in Culture. *Metabolic Brain Disease*, *28*(2), 139–143. doi:10.100711011-012-9339-1 PMID:23065046

Rao, K. V., Panickar, K. S., Jayakumar, A. R., & Norenberg, M. D. (2005). Astrocytes protect neurons from ammonia toxicity. *Neurochemical Research*, *30*(10), 1311–1318. doi:10.100711064-005-8803-2 PMID:16341593

Rao, R. (2010). Brain Aquaporin-4 in Experimental Acute Liver Failure. *Journal of Neuropathology and Experimental Neurology*, *69*(9), 869–879. doi:10.1097/NEN.0b013e3181ebe581 PMID:20720509

Rao, Rama, R., Kakulavarapu, V., Arumugam, R., Jayakumar, X., Tong, V., Alvarez, M., & Michael D. (2010). Marked Potentiation of Cell Swelling by Cytokines in Ammonia-Sensitized Cultured Astrocytes. *Journal of Neuroinflammation*, *7*(1), 1–8. PMID:20942959

Rascol, O., Payoux, P., Ory, F., Ferreira, J. J., Brefel-Courbon, C., & Montastruc, J. L. (2003). Limitations of current Parkinson's disease therapy. *Annals of Neurology*, *53*(S3), S3–S15. doi:10.1002/ana.10513 PMID:12666094

Rasheed, M., Liang, J., Wang, C., Deng, Y., & Chen, Z. (2021). 'Epigenetic Regulation of Neuroinflammation in Parkinson's Disease'. *International Journal of Molecular Sciences*, *22*(9), 4956. doi:10.3390/ijms22094956 PMID:34066949

Rashid, K., Wolf, A., & Langmann, T. (2018). Microglia Activation and Immunomodulatory Therapies for Retinal Degenerations. *Frontiers in Cellular Neuroscience*, *12*, 176. doi:10.3389/fncel.2018.00176 PMID:29977192

Rasmussen, J., & Langerman, H. (2019). Alzheimer's disease–why we need early diagnosis. *Degenerative Neurological and Neuromuscular Disease*, *9*, 123–130. doi:10.2147/DNND.S228939 PMID:31920420

Rassu, G., Soddu, E., Posadino, A. M., Pintus, G., Sarmento, B., Giunchedi, P., & Gavini, E. (2017). Nose-to-brain delivery of BACE1 siRNA loaded in solid lipid nanoparticles for Alzheimer's therapy. *Colloids and Surfaces. B, Biointerfaces*, *152*, 296–301. doi:10.1016/j.colsurfb.2017.01.031 PMID:28126681

Rauch, J. N., Luna, G., Guzman, E., Audouard, M., Challis, C., Sibih, Y. E., Leshuk, C., Hernandez, I., Wegmann, S., Hyman, B. T., Gradinaru, V., Kampmann, M., & Kosik, K. S. (2020). LRP1 is a master regulator of tau uptake and spread. *Nature*, *580*(7803), 381–385. doi:10.103841586-020-2156-5 PMID:32296178

Rawlins, M. D., Wexler, N. S., Wexler, A. R., Tabrizi, S. J., Douglas, I., Evans, S. J. W., & Smeeth, L. (2016). The prevalence of Huntington's disease. *Neuroepidemiology*, *46*(2), 144–153. doi:10.1159/000443738 PMID:26824438

Reale, M., Costantini, E., Jagarlapoodi, S., Khan, H., Belwal, T., & Cichelli, A. (2020). Relationship of wine consumption with Alzheimer's disease. *Nutrients*, *12*(1), 206. doi:10.3390/nu12010206 PMID:31941117

Reardon, D. A., Dresemann, G., Taillibert, S., Campone, M., van den Bent, M., Clement, P., Blomquist, E., Gordower, L., Schultz, H., Raizer, J., Hau, P., Easaw, J., Gil, M., Tonn, J., Gijtenbeek, A., Schlegel, U., Bergstrom, P., Green, S., Weir, A., & Nikolova, Z. (2009). Multicentre phase II studies evaluating imatinib plus hydroxyurea in patients with progressive glioblastoma. *British Journal of Cancer*, *101*(12), 1995–2004. doi:10.1038j.bjc.6605411 PMID:19904263

Redondo, R. L., & Morris, R. G. M. (2011). Making memories last: The synaptic tagging and capture hypothesis. *Nature Reviews. Neuroscience*, *12*(1), 17–30. doi:10.1038/nrn2963 PMID:21170072

Regenold, W. T., Phatak, P., Makley, M. J., Stone, R. D., & Kling, M. A. (2008). Cerebrospinal fluid evidence of increased extra-mitochondrial glucose metabolism implicates mitochondrial dysfunction in multiple sclerosis disease progression. *Journal of the Neurological Sciences*, *275*(1-2), 106–112. doi:10.1016/j.jns.2008.07.032

Reiner, K. S. & C. S. (2021). *Diffuse Liver Disease: Cirrhosis, Focal Lesions in Cirrhosis, and Vascular Liver Disease.* Springer. doi:10.1007/978-3-319-75019-4

Reiner, A., Albin, R. L., Anderson, K. D., D'Amato, C. J., Penney, J. B., & Young, A. B. (1988). Differential loss of striatal projection neurons in Huntington disease. *Proceedings of the National Academy of Sciences of the United States of America*, *85*(15), 5733–5737. doi:10.1073/pnas.85.15.5733 PMID:2456581

Revilla, S., Ursulet, S., Álvarez-López, M. J., Castro-Freire, M., Perpiñá, U., García-Mesa, Y., Bortolozzi, A., Giménez-Llort, L., Kaliman, P., Cristòfol, R., Sarkis, C., & Sanfeliu, C. (2014). Lenti-GDNF gene therapy protects against Alzheimer's disease-like neuropathology in 3xTg-AD mice and MC65 cells. *CNS Neuroscience & Therapeutics*, *20*(11), 961–972. doi:10.1111/cns.12312 PMID:25119316

Rezaianzadeh, A., Talei, A., Rajaeefard, A., Hasanzadeh, J., Tabatabai, H., Tahmasebi, S., & Mousavizadeh, A. (2012). Vascular invasion as an independent prognostic factor in lymph node negative invasive breast cancer. *Asian Pacific Journal of Cancer Prevention*, *13*(11), 5767–5772. doi:10.7314/APJCP.2012.13.11.5767 PMID:23317254

Rhazi, M., Desbriè, J., Tolaimate, A., Alagui, A., & Vottero, P. (2000). *Investigation of different natural sources of chitin: influence of the source and deacetylation process on the physicochemical characteristics of chitosan.*

Rhodes, A., Inker, J., Richardson, J., & Zanjani, F. (2022). Alzheimer's Disease Prevention Health Coaching. J. *Prev. Alzheimer's Dis.*, *9*, 277–285. PMID:35543001

Rial, D., Lemos, C., Pinheiro, H., Duarte, J. M., Gonçalves, F. Q., Real, J. I., Prediger, R. D., Gonçalves, N., Gomes, C. A., Canas, P. M., Agostinho, P., & Cunha, R. A. (2016). Depression as a glial-based synaptic dysfunction. *Frontiers in Cellular Neuroscience*, *9*(JAN2016), 1–11. doi:10.3389/fncel.2015.00521

Richter, N., Beckers, N., Onur, O. A., Dietlein, M., Tittgemeyer, M., Kracht, L., Neumaier, B., Fink, G. R., & Kukolja, J. (2018). Effect of cholinergic treatment depends on cholinergic integrity in early Alzheimer's disease. *Brain*, *141*(3), 903–915. doi:10.1093/brain/awx356 PMID:29309600

Ricks, C. B., Shin, S. S., Becker, C., & Grandhi, R. (2014). Extracellular matrices, artificial neural scaffolds and the promise of neural regeneration. *Neural Regeneration Research*, *9*(17), 1573. doi:10.4103/1673-5374.141778 PMID:25368641

Rikani, A. A., Choudhry, Z., Choudhry, A. M., Rizvi, N., Ikram, H., Mobassarah, N. J., & Tuli, S. (2014). The mechanism of degeneration of striatal neuronal subtypes in Huntington disease. *Annals of Neurosciences*, *21*, 112. doi:10.5214/ans.0972.7531.210308 PMID:25206077

Rinaudo, M. (2006). Chitine et chitosane: Propriétés et applications. *Progrès de la science des polymères*, *31*(7), 603-632.

Rinaudo, M. P. S. (2019). *From Chitin to Chitosan.* Glycopedia.Eu. http://www.glycopedia.eu/IMG/pdf/from_chitin_to_chitosan.pdf.

Rinholm, J. E., Hamilton, N. B., Kessaris, N., Richardson, W. D., Bergersen, L. H., & Attwell, D. (2011). Regulation of oligodendrocyte development and myelination by glucose and lactate. *The Journal of Neuroscience : The Official Journal of the Society for Neuroscience*, *31*(2), 538–548. doi:10.1523/JNEUROSCI.3516-10.2011

Risher, W. C., Patel, S., Kim, I. H., Uezu, A., Bhagat, S., Wilton, D. K., Pilaz, L.-J., Singh Alvarado, J., Calhan, O. Y., Silver, D. L., Stevens, B., Calakos, N., Soderling, S. H., & Eroglu, C. (2014). Astrocytes refine cortical connectivity at dendritic spines. *eLife*, *3*, e04047. doi:10.7554/eLife.04047 PMID:25517933

Rivetti Di Val Cervo, P., Romanov, R. A., Spigolon, G., Masini, D., Martín-Montañez, E., Toledo, E. M., La Manno, G., Feyder, M., Pifl, C., Ng, Y.-H., Sánchez, S. P., Linnarsson, S., Wernig, M., Harkany, T., Fisone, G., & Arenas, E. (2017). Induction of functional dopamine neurons from human astrocytes in vitro and mouse astrocytes in a Parkinson's disease model. *Nature Biotechnology*, *35*(5), 444–452. doi:10.1038/nbt.3835 PMID:28398344

Robel, S., Bardehle, S., Lepier, A., Brakebusch, C., & Götz, M. (2011). Genetic deletion of cdc42 reveals a crucial role for astrocyte recruitment to the injury site in vitro and in vivo. *The Journal of Neuroscience : The Official Journal of the Society for Neuroscience*, *31*(35), 12471–12482. doi:10.1523/JNEUROSCI.2696-11.2011 PMID:21880909

Robel, S., Buckingham, S. C., Boni, J. L., Campbell, S. L., Danbolt, N. C., Riedemann, T., Sutor, B., & Sontheimer, H. (2015). Reactive astrogliosis causes the development of spontaneous seizures. *The Journal of Neuroscience : The Official Journal of the Society for Neuroscience*, *35*(8), 3330–3345. doi:10.1523/JNEUROSCI.1574-14.2015 PMID:25716834

Robel, S., Mori, T., Zoubaa, S., Schlegel, J., Sirko, S., Faissner, A., Goebbels, S., Dimou, L., & Götz, M. (2009). Conditional deletion of β1-integrin in astroglia causes partial reactive gliosis. *Glia*, *57*(15), 1630–1647. doi:10.1002/glia.20876 PMID:19373938

Robillard, K. N., Lee, K. M., Chiu, K. B., & MacLean, A. G. (2016). 'Glial Cell Morphological and Density Changes through the Lifespan of Rhesus Macaques'. *Brain, Behavior, and Immunity*, Microglia. *Physiology & Behavior*, *55*(July), 60–69. doi:10.1016/j.bbi.2016.01.006 PMID:26851132

Rocha, N. P., De Miranda, A. S., & Teixeira, A. L. (2015). Insights into neuroinflammation in Parkinson's disease: From biomarkers to anti-inflammatory based therapies. *BioMed Research International*, *2015*, 1–12. doi:10.1155/2015/628192 PMID:26295044

Rodnight, R. B., & Gottfried, C. (2013). Morphological plasticity of rodent astroglia. *Journal of Neurochemistry*, *124*(3), 263–275. doi:10.1111/jnc.12087 PMID:23278277

Rodrigo, R., Cauli, O., Gomez–Pinedo, U., Agusti, A., Hernandez–Rabaza, V., Garcia–Verdugo, J.-M., & Felipo, V. (2010). Hyperammonemia Induces Neuroinflammation That Contributes to Cognitive Impairment in Rats with Hepatic Encephalopathy. *Gastroenterology*, *139*(2), 675–684. doi:10.1053/j.gastro.2010.03.040 PMID:20303348

Rodriguez, J. I., & Kern, J. K. (2011). Evidence of microglial activation in autism and its possible role in brain underconnectivity. *Neuron Glia Biology*, *7*(2-4), 205–213. doi:10.1017/S1740925X12000142 PMID:22874006

Rodríguez, J. J., Noristani, H. N., & Verkhratsky, A. (2012). 'The Serotonergic System in Ageing and Alzheimer's Disease'. *Progress in Neurobiology*, *99*(1), 15–41. doi:10.1016/j.pneurobio.2012.06.010 PMID:22766041

Rodríguez, J. J., Olabarria, M., Chvatal, A., & Verkhratsky, A. (2009). 'Astroglia in Dementia and Alzheimer's Disease'. *Cell Death and Differentiation*, *16*(3), 378–385. doi:10.1038/cdd.2008.172 PMID:19057621

Rodríguez, J. J., Yeh, C.-Y., Terzieva, S., Olabarria, M., Kulijewicz-Nawrot, M., & Verkhratsky, A. (2014). Complex and region-specific changes in astroglial markers in the aging brain. *Neurobiology of Aging*, *35*(1), 15–23. doi:10.1016/j.neurobiolaging.2013.07.002 PMID:23969179

Röhl, C., Lucius, R., & Sievers, J. (2007). The effect of activated microglia on astrogliosis parameters in astrocyte cultures. *Brain Research*, *1129*(1), 43–52. doi:10.1016/j.brainres.2006.10.057 PMID:17169340

Rojanathammanee, L., Murphy, E. J., & Combs, C. K. (2011). Expression of mutant alpha-synuclein modulates microglial phenotype in vitro. *Journal of Neuroinflammation*, *8*(1), 44. doi:10.1186/1742-2094-8-44 PMID:21554732

Rolyan, H., Scheffold, A., Heinrich, A., & Brain, Y. B.-N. (2011). Telomere shortening reduces Alzheimer's disease amyloid pathology in mice. *Academic*. https://academic.oup.com/brain/article-abstract/134/7/2044/293996

Rone, M. B., Cui, Q. L., Fang, J., Wang, L. C., Zhang, J., Khan, D., Bedard, M., Almazan, G., Ludwin, S. K., Jones, R., Kennedy, T. E., & Antel, J. P. (2016). Oligodendrogliopathy in multiple sclerosis: Low glycolytic metabolic rate promotes oligodendrocyte survival. *The Journal of Neuroscience : The Official Journal of the Society for Neuroscience*, *36*(17), 4698–4707. doi:10.1523/JNEUROSCI.4077-15.2016

Rosas, H. D., Salat, D. H., Lee, S. Y., Zaleta, A. K., Pappu, V., Fischl, B., Greve, D., Hevelone, N., & Hersch, S. M. (2008). Cerebral cortex and the clinical expression of Huntington's disease: Complexity and heterogeneity. *Brain*, *131*(4), 1057–1068. doi:10.1093/brain/awn025 PMID:18337273

Ros-Bernal, F., Hunot, S., Herrero, M. T., Parnadeau, S., Corvol, J.-C., Lu, L., Alvarez-Fischer, D., Carrillo-de Sauvage, M. A., Saurini, F., Coussieu, C., Kinugawa, K., Prigent, A., Höglinger, G., Hamon, M., Tronche, F., Hirsch, E. C., & Vyas, S. (2011). Microglial glucocorticoid receptors play a pivotal role in regulating dopaminergic neurodegeneration in parkinsonism. *Proceedings of the National Academy of Sciences of the United States of America*, *108*(16), 6632–6637. doi:10.1073/pnas.1017820108 PMID:21467220

Rose, C. F., Verkhratsky, A., & Parpura, V. (2013). Astrocyte glutamine synthetase : Pivotal in health and disease. *Biochemical Society Transactions*, *41*(6), 1518–1524. doi:10.1042/BST20130237 PMID:24256247

Rose, C. R., Felix, L., Zeug, A., Dietrich, D., Reiner, A., & Henneberger, C. (2018). Astroglial Glutamate Signaling and Uptake in the Hippocampus. *Frontiers in Molecular Neuroscience*, *10*, 451. doi:10.3389/fnmol.2017.00451 PMID:29386994

Rosenblatt, A. (2022). Neuropsychiatry of Huntington's disease. *Dialogues in Clinical Neuroscience*. PMID:17726917

Rosen, D. R., Siddique, T., Patterson, D., Figlewicz, D. A., Sapp, P., Hentati, A., Donaldson, D., Goto, J., O'Regan, J. P., Deng, H. X., Rahmani, Z., Krizus, A., McKenna-Yasek, D., Cayabyab, A., Gaston, S. M., Berger, R., Tanzi, R. E., Halperin, J. J., Herzfeldt, B., & Brown, R. H. Jr. (1993). Mutations in Cu/Zn superoxide dismutase gene are associated with familial amyotrophic lateral sclerosis. *Nature*, *362*(6415), 59–62. doi:10.1038/362059a0 PMID:8446170

Rosko, L., Smith, V. N., Yamazaki, R., & Huang, J. K. (2019). Oligodendrocyte bioenergetics in health and disease. *The Neuroscientist*, *25*(4), 334–343. doi:10.1177/1073858418793077

Rosoklija, G., Toomayan, G., Ellis, S. P., Keilp, J., Mann, J. J., Latov, N., Hays, A. P., & Dwork, A. J. (2000). Structural abnormalities of subicular dendrites in subjects with schizophrenia and mood disorders: Preliminary findings. *Archives of General Psychiatry*, *57*(4), 349–356. doi:10.1001/archpsyc.57.4.349 PMID:10768696

Ross, C. A., & Tabrizi, S. J. (2011). Huntington ' s disease : From molecular pathogenesis to clinical treatment. *Lancet Neurology*, *10*(1), 83–98. doi:10.1016/S1474-4422(10)70245-3 PMID:21163446

Rothhammer, V., Borucki, D. M., Tjon, E. C., Takenaka, M. C., Chao, C.-C., Ardura-Fabregat, A., De Lima, K. A., Gutiérrez-Vázquez, C., Hewson, P., Staszewski, O., Blain, M., Healy, L., Neziraj, T., Borio, M., Wheeler, M., Dragin, L. L., Laplaud, D. A., Antel, J., Alvarez, J. I., & Quintana, F. J. (2018). Microglial control of astrocytes in response to microbial metabolites. *Nature*, *557*(7707), 724–728. doi:10.103841586-018-0119-x PMID:29769726

Rothstein, J. D., Martin, L. J., & Kuncl, R. W. (1992). Decreased glutamate transport by the brain and spinal cord in amyotrophic lateral sclerosis. *The New England Journal of Medicine*, *326*(22), 1464–1468. doi:10.1056/NEJM199205283262204 PMID:1349424

Rothstein, J. D., Tsai, G., Kuncl, R. W., Clawson, L., Cornblath, D. R., Drachman, D. B., Pestronk, A., Stauch, B. L., & Coyle, J. T. (1990). Abnormal excitatory amino acid metabolism in amyotrophic lateral sclerosis. *Annals of Neurology*, *28*(1), 18–25. doi:10.1002/ana.410280106 PMID:2375630

Rothstein, J. D., Van Kammen, M., Levey, A. I., Martin, L. J., & Kuncl, R. W. (1995). Selective loss of glial glutamate transporter GLT-1 in amyotrophic lateral sclerosis. *Annals of Neurology*, *38*(1), 73–84. doi:10.1002/ana.410380114 PMID:7611729

Roux, L., Madar, A., Lacroix, M. M., Yi, C., Benchenane, K., & Giaume, C. (2015). Astroglial Connexin 43 Hemichannels Modulate Olfactory Bulb Slow Oscillations. *The Journal of Neuroscience : The Official Journal of the Society for Neuroscience*, *35*(46), 15339–15352. doi:10.1523/JNEUROSCI.0861-15.2015 PMID:26586821

Roy, N. S., Wang, S., Jiang, L., Kang, J., Benraiss, A., Harrison-Restelli, C., Fraser, R. A. R., Couldwell, W. T., Kawaguchi, A., Okano, H., Nedergaard, M., & Goldman, S. A. (2000). In vitro neurogenesis by progenitor cells isolated from the adult human hippocampus. *Nature Medicine*, *6*(3), 271–277. doi:10.1038/73119 PMID:10700228

Rueda-CarrascoJ.Inés MateoM.BorrotoA.Jesús Martin-BermejoM.KummerM. P.SchwartzS.López-AtalayaJ. P.AlarconB. HenekaM. T.EsteveP.BovolentaP. (2020). SFRP1 shapes astrocyte to microglia cross-talk in acute and chronic neuroinflammation. BiorXiv. doi:10.1101/2020.03.10.982579

Ruivo, R., Anne, C., Sagné, C., & Gasnier, B. (2009). Molecular and cellular basis of lysosomal transmembrane protein dysfunction. *Biochimica et Biophysica Acta (BBA)-. Molecular Cell Research*, *1793*(4), 636–649. PMID:19146888

Rusakov, D. A. (2015). Disentangling calcium-driven astrocyte physiology. *Nature Reviews. Neuroscience*, *16*(4), 226–233. doi:10.1038/nrn3878 PMID:25757560

Ruskamo, S., Raasakka, A., Pedersen, J. S., Martel, A., Škubník, K., Darwish, T., Porcar, L., & Kursula, P. (2022). Human myelin proteolipid protein structure and lipid bilayer stacking. *Cellular and Molecular Life Sciences*, *79*(8), 419. doi:10.100700018-022-04428-6

Rusznák, Z., Sengul, G., Paxinos, G., Kim, W. S., & Fu, Y. (2018). Odor Enrichment Increases Hippocampal Neuron Numbers in Mouse. *Experimental Neurobiology*, *27*(2), 94–102. doi:10.5607/en.2018.27.2.94 PMID:29731675

Saab, A. S., Tzvetavona, I. D., Trevisiol, A., Baltan, S., Dibaj, P., Kusch, K., Möbius, W., Goetze, B., Jahn, H. M., Huang, W., Steffens, H., Schomburg, E. D., Pérez-Samartín, A., Pérez-Cerdá, F., Bakhtiari, D., Matute, C., Löwel, S., Griesinger, C., Hirrlinger, J., & Nave, K.-A. (2016). 'Oligodendroglial NMDA Receptors Regulate Glucose Import and Axonal Energy Metabolism'. *Neuron*, *91*(1), 119–132. doi:10.1016/j.neuron.2016.05.016 PMID:27292539

Saba, E. S., Karout, M., Nasrallah, L., Kobeissy, F., Darwish, H., & Khoury, S. J. (2021). Long-term cognitive deficits after traumatic brain injury associated with microglia activation. *Clinical Immunology (Orlando, Fla.)*, *230*, 108815. doi:10.1016/j.clim.2021.108815 PMID:34339843

Sabelström, H., Stenudd, M., Réu, P., Dias, D. O., Elfineh, M., Zdunek, S., Damberg, P., Göritz, C., & Frisén, J. (2013). Resident neural stem cells restrict tissue damage and neuronal loss after spinal cord injury in mice. *Science*, *342*(6158), 637–640. doi:10.1126cience.1242576 PMID:24179227

Saboktakin, M. R., Tabatabaie, R. M., Maharramov, A., & Ramazanov, M. A. (2011). Synthesis and characterization of pH-dependent glycol chitosan and dextran sulfate nanoparticles for effective brain cancer treatment. *International Journal of Biological Macromolecules*, *49*(4), 747–751. doi:10.1016/j.ijbiomac.2011.07.006 PMID:21782844

Saftig, P., & Klumperman, J. (2009). Lysosome biogenesis and lysosomal membrane proteins: Trafficking meets function. *Nature Reviews. Molecular Cell Biology*, *10*(9), 623–635. doi:10.1038/nrm2745 PMID:19672277

Sagar, D. R., Burston, J. J., Hathway, G. J., Woodhams, S. G., Pearson, R. G., Bennett, A. J., Kendall, D. A., Scammell, B. E., & Chapman, V. (2011). The contribution of spinal glial cells to chronic pain behaviour in the monosodium iodoacetate model of osteoarthritic pain. *Molecular Pain*, *7*, 88. doi:10.1186/1744-8069-7-88 PMID:22093915

Sahariah, P., & Másson, M. (2017). Antimicrobial chitosan and chitosan derivatives: A review of the structure–activity relationship. *Biomacromolecules*, *18*(11), 3846–3868. doi:10.1021/acs.biomac.7b01058 PMID:28933147

Saijo, K., Crotti, A., & Glass, C. K. (2013). Regulation of microglia activation and deactivation by nuclear receptors. *Glia*, *61*(1), 104–111. doi:10.1002/glia.22423 PMID:22987512

Saijo, K., & Glass, C. K. (2011). 'Microglial Cell Origin and Phenotypes in Health and Disease'. *Nature Reviews. Immunology*, *11*(11), 775–787. doi:10.1038/nri3086 PMID:22025055

Saijo, K., Winner, B., Carson, C. T., Collier, J. G., Boyer, L., Rosenfeld, M. G., Gage, F. H., & Glass, C. K. (2009). A Nurr1/CoREST Pathway in Microglia and Astrocytes Protects Dopaminergic Neurons from Inflammation-Induced Death. *Cell*, *137*(1), 47–59. doi:10.1016/j.cell.2009.01.038 PMID:19345186

Saitgareeva, A. R., Bulygin, K. V., Gareev, I. F., Beylerli, O. A., & Akhmadeeva, L. R. (2020). The role of microglia in the development of neurodegeneration. *Neurological Sciences*, *41*(12), 3609–3615. doi:10.100710072-020-04468-5 PMID:32458252

Salarian, A., Russmann, H., Wider, C., Burkhard, P. R., Vingerhoets, F. J. G., & Aminian, K. (2007). Quantification of tremor and bradykinesia in Parkinson's disease using a novel ambulatory monitoring system. *IEEE Transactions on Biomedical Engineering*, *54*(2), 313–322. doi:10.1109/TBME.2006.886670 PMID:17278588

Salter, M. W., & Stevens, B. (2017). 'Microglia Emerge as Central Players in Brain Disease'. *Nature Medicine*, *23*(9), 1018–1027. doi:10.1038/nm.4397 PMID:28886007

Salzer, J. L. (2015). 'Schwann Cell Myelination'. *Cold Spring Harbor Perspectives in Biology*, *7*(8), a020529. doi:10.1101/cshperspect.a020529 PMID:26054742

Sama, M. A., Mathis, D. M., Furman, J. L., Abdul, H. M., Artiushin, I. A., Kraner, S. D., & Norris, C. M. (2008). Interleukin-1β-dependent Signaling between Astrocytes and Neurons Depends Critically on Astrocytic Calcineurin/NFAT Activity. *The Journal of Biological Chemistry*, *283*(32), 21953–21964. doi:10.1074/jbc.M800148200 PMID:18541537

Samie, M. A., & Xu, H. (2014). Lysosomal exocytosis and lipid storage disorders. *Journal of Lipid Research*, *55*(6), 995–1009. doi:10.1194/jlr.R046896 PMID:24668941

Samii, A., Etminan, M., Wiens, M. O., & Jafari, S. (2009). NSAID use and the risk of parkinsons disease: Systematic review and meta-analysis of observational studies. *Drugs & Aging*, *26*(9), 769–779. doi:10.2165/11316780-000000000-00000 PMID:19728750

Sams, E. C. (2021). 'Oligodendrocytes in the Aging Brain'. *Neuronal Signaling*, *5*(3), NS20210008. doi:10.1042/NS20210008 PMID:34290887

Sanacora, G., & Banasr, M. (2013). From pathophysiology to novel antidepressant drugs: Glial contributions to the pathology and treatment of mood disorders. *Biological Psychiatry*, *73*(12), 1172–1179. doi:10.1016/j.biopsych.2013.03.032 PMID:23726152

Sanacora, G., Rothman, D. L., Mason, G., & Krystal, J. H. (2003). Clinical studies implementing glutamate neurotransmission in mood disorders. *Annals of the New York Academy of Sciences*, *1003*(1), 292–308. doi:10.1196/annals.1300.018 PMID:14684453

Santello, M., Calì, C., & Bezzi, P. (2012). 'Gliotransmission and the Tripartite Synapse'. In Synaptic Plasticity. Springer Vienna. doi:10.1007/978-3-7091-0932-8_14

Santello, M., & Volterra, A. (2009). Astrocytes as major integrators in neurovascular coupling. *The Neuroscientist*, *15*(3), 262–273.

Santos, R., Vadodaria, K. C., Jaeger, B. N., Mei, A., Lefcochilos-Fogelquist, S., Mendes, A. P. D., Erikson, G., Shokhirev, M., Randolph-Moore, L., Fredlender, C., Dave, S., Oefner, R., Fitzpatrick, C., Pena, M., Barron, J. J., Ku, M., Denli, A. M., Kerman, B. E., Charnay, P., & Gage, F. H. (2017). Differentiation of Inflammation-Responsive Astrocytes from Glial Progenitors Generated from Human Induced Pluripotent Stem Cells. *Stem Cell Reports*, *8*(6), 1757–1769. doi:10.1016/j.stemcr.2017.05.011 PMID:28591655

Sapena, R., Morin, D., Zini, R., Morin, C., & Tillement, J.-P. (1996). Desipramine treatment differently down-regulates β-adrenoceptors of freshly isolated neurons and astrocytes. *European Journal of Pharmacology*, *300*(1–2), 159–162. doi:10.1016/0014-2999(96)00060-X PMID:8741184

Sarlus, H., & Heneka, M. T. (2017). Microglia in Alzheimer's disease. *The Journal of Clinical Investigation*, *127*(9), 3240–3249. doi:10.1172/JCI90606 PMID:28862638

Sarris, J., Marx, W., Ashton, M. M., Ng, C. H., Galvao-Coelho, N., Ayati, Z., Zhang, Z.-J., Kasper, S., Ravindran, A., Harvey, B. H., Lopresti, A., Mischoulon, D., Amsterdam, J., Yatham, L. N., & Berk, M. (2021). Plant-based Medicines (Phytoceuticals) in the Treatment of Psychiatric Disorders : A Meta-review of Meta-analyses of Randomized Controlled Trials: Les médicaments à base de plantes (phytoceutiques) dans le traitement des troubles psychiatriques: une méta-revue des méta-analyses d'essais randomisés contrôlés. *Canadian Journal of Psychiatry*, *66*(10), 849–862. doi:10.1177/0706743720979917 PMID:33596697

Sasabe, J., Chiba, T., Yamada, M., Okamoto, K., Nishimoto, I., Matsuoka, M., & Aiso, S. (2007). D-serine is a key determinant of glutamate toxicity in amyotrophic lateral sclerosis. *The EMBO Journal*, *26*(18), 4149–4159. doi:10.1038j.emboj.7601840 PMID:17762863

Sasaki, S., Warita, H., Abe, K., & Iwata, M. (2001). Inducible nitric oxide synthase (iNOS) and nitrotyrosine immunoreactivity in the spinal cords of transgenic mice with a G93A mutant SOD1 gene. *Journal of Neuropathology and Experimental Neurology*, *60*(9), 839–846. doi:10.1093/jnen/60.9.839 PMID:11556540

Sato, K., Takahashi, K., Shigemoto-Mogami, Y., Chujo, K., & Sekino, Y. (2016). Glypican 6 enhances N-methyl-D-aspartate receptor function in human-induced pluripotent stem cell-derived neurons. *Frontiers in Cellular Neuroscience*, *10*, 259. doi:10.3389/fncel.2016.00259 PMID:27895553

Saudou, F., & Humbert, S. (2016). The biology of huntingtin. *Neuron*, *89*(5), 910–926. doi:10.1016/j.neuron.2016.02.003 PMID:26938440

Savchenko, E., Marote, A., Russ, K., Collin, A., Goldwurm, S., Roybon, L., & Pomeshchik, Y. (2018). Generation of a human induced pluripotent stem cell line (CSC-42) from a patient with sporadic form of Parkinson's disease. *Stem Cell Research*, *27*, 78–81. doi:10.1016/j.scr.2018.01.002 PMID:29334629

Sawa, H., Murakami, H., Ohshima, Y., Murakami, M., Yamazaki, I., Tamura, Y., Mima, T., Satone, A., Ide, W., Hashimoto, I., & Kamada, H. (2002). Histone deacetylase inhibitors such as sodium butyrate and trichostatin A inhibit vascular endothelial growth factor (VEGF) secretion from human glioblastoma cells. *Brain Tumor Pathology*, *19*(2), 77–81. doi:10.1007/BF02478931 PMID:12622137

Sawamoto, K., Wichterle, H., Gonzalez-Perez, O., Cholfin, J. A., Yamada, M., Spassky, N., Murcia, N. S., Garcia-Verdugo, J. M., Marin, O., Rubenstein, J. L. R., Tessier-Lavigne, M., Okano, H., & Alvarez-Buylla, A. (2006). 'New Neurons Follow the Flow of Cerebrospinal Fluid in the Adult Brain'. *Science*, *311*(5761), 629–632. doi:10.1126cience.1119133 PMID:16410488

Scanga, V. I., Goraltchouk, A., Nussaiba, N., Shoichet, M. S., & Morshead, C. M. (2010). Biomaterials for neural-tissue engineering — Chitosan supports the survival, migration, and differentiation of adult-derived neural stem and progenitor cells. *Canadian Journal of Chemistry*, *88*(3), 277–287. doi:10.1139/v09-171

Schafer, D. P., Lehrman, E. K., Kautzman, A. G., Koyama, R., Mardinly, A. R., Yamasaki, R., Ransohoff, R. M., Greenberg, M. E., Barres, B. A., & Stevens, B. (2012). Microglia sculpt postnatal neural circuits in an activity and complement-dependent manner. *Neuron*, *74*(4), 691–705. doi:10.1016/j.neuron.2012.03.026 PMID:22632727

Schafer, D. P., Lehrman, E. K., & Stevens, B. (2013). 'The "Quad-partite" Synapse: Microglia-synapse Interactions in the Developing and Mature CNS'. *Glia*, *61*(1), 24–36. doi:10.1002/glia.22389 PMID:22829357

Schafer, D. P., & Stevens, B. (2015). Microglia Function in Central Nervous System Development and Plasticity. *Cold Spring Harbor Perspectives in Biology*, *7*(10), a020545. doi:10.1101/cshperspect.a020545 PMID:26187728

Schiffmann, R., & Elroy-Stein, O. (2006). Childhood ataxia with CNS hypomyelination/vanishing white matter disease—A common leukodystrophy caused by abnormal control of protein synthesis. *Molecular Genetics and Metabolism*, *88*(1), 7–15. doi:10.1016/j.ymgme.2005.10.019 PMID:16378743

Schildkraut, J. J., & Mooney, J. J. (2004). Toward a rapidly acting antidepressant: The normetanephrine and extraneuronal monoamine transporter (uptake 2) hypothesis. *The American Journal of Psychiatry*, *161*(5), 909–911. doi:10.1176/appi.ajp.161.5.909 PMID:15121658

Schilling, L. P., Balthazar, M. L. F., Radanovic, M., Forlenza, O. V., Silagi, M. L., Smid, J., Barbosa, B. J. A. P., Frota, N. A. F., de Souza, L. C., & Vale, F. A. C. (2022). Diagnosis of Alzheimer's disease: Recommendations of the Scientific Department of Cognitive Neurology and Aging of the Brazilian Academy of Neurology. *Dementia & Neuropsychologia*, *16*, 25–39. doi:10.1590/1980-5764-dn-2022-s102pt PMID:36533157

Schipke, C. G., Heuser, I., & Peters, O. (2011). Antidepressants act on glial cells: SSRIs and serotonin elicit astrocyte calcium signaling in the mouse prefrontal cortex. *Journal of Psychiatric Research*, *45*(2), 242–248. doi:10.1016/j.jpsychires.2010.06.005 PMID:20619420

Schipper, H. M. (1996). Astrocytes, brain aging, and neurodegeneration. *Neurobiology of Aging*, *17*(3), 467–480. doi:10.1016/0197-4580(96)00014-0 PMID:8725909

Schlicht, K., Büttner, A., Siedler, F., Scheffer, B., Zill, P., Eisenmenger, W., Ackenheil, M., & Bondy, B. (2007). Comparative proteomic analysis with postmortem prefrontal cortex tissues of suicide victims versus controls. *Journal of Psychiatric Research*, *41*(6), 493–501. doi:10.1016/j.jpsychires.2006.04.006 PMID:16750834

Schmid, C. D., Sautkulis, L. N., Danielson, P. E., Cooper, J., Hasel, K. W., Hilbush, B. S., Sutcliffe, J. G., & Carson, M. J. (2002). Heterogeneous expression of the triggering receptor expressed on myeloid cells-2 on adult murine microglia : Microglial expression of TREM-2. *Journal of Neurochemistry*, *83*(6), 1309–1320. doi:10.1046/j.1471-4159.2002.01243.x PMID:12472885

Schneider, J., Karpf, J., & Beckervordersandforth, R. (2019). Role of astrocytes in the neurogenic niches. *Astrocytes: Methods and Protocols*, 19-33.

Schneider, M. E., Guillade, L., Correa-Duarte, M. A., & Moreira, F. T. C. (2022). Tau proteinsTau 181 Protein detection in Early-Stage Alzheimer's Disease. *Bioelectrochemistry (Amsterdam, Netherlands)*, 108057. doi:10.1016/j.bioelechem.2022.108057

Schnieder, T. P., Trencevska, I., Rosoklija, G., Stankov, A., Mann, J. J., Smiley, J., & Dwork, A. J. (2014). Microglia of prefrontal white matter in suicide. *Journal of Neuropathology and Experimental Neurology*, *73*(9), 880–890. doi:10.1097/NEN.0000000000000107 PMID:25101704

Schofield, E., Kersaitis, C., Shepherd, C. E., Kril, J. J., & Halliday, G. M. (2003). Severity of gliosis in Pick's disease and frontotemporal lobar degeneration: Tau-positive glia differentiate these disorders. *Brain*, *126*(4), 827–840. doi:10.1093/brain/awg085 PMID:12615642

Schools, G. P., & Kimelberg, H. K. (1999). mGluR3 and mGluR5 are the predominant metabotropic glutamate receptor mRNAs expressed in hippocampal astrocytes acutely isolated from young rats. *Journal of Neuroscience Research*, *58*(4), 533–543. doi:10.1002/(SICI)1097-4547(19991115)58:4<533::AID-JNR6>3.0.CO;2-G PMID:10533045

Schrander-Stumpel, C., & Fryns, J.-P. (1998). Congenital hydrocephalus : Nosology and guidelines for clinical approach and genetic counselling. *European Journal of Pediatrics*, *157*(5), 355–362. doi:10.1007004310050830 PMID:9625330

Schreiner, B., Romanelli, E., Liberski, P., Ingold-Heppner, B., Sobottka-Brillout, B., Hartwig, T., Chandrasekar, V., Johannssen, H., Zeilhofer, H. U., Aguzzi, A., Heppner, F., Kerschensteiner, M., & Becher, B. (2015). Astrocyte Depletion Impairs Redox Homeostasis and Triggers Neuronal Loss in the Adult CNS. *Cell Reports*, *12*(9), 1377–1384. doi:10.1016/j.celrep.2015.07.051 PMID:26299968

Schroeter, M. L., Abdul-Khaliq, H., Krebs, M., Diefenbacher, A., & Blasig, I. E. (2008). Serum markers support disease-specific glial pathology in major depression. *Journal of Affective Disorders*, *111*(2–3), 271–280. doi:10.1016/j.jad.2008.03.005 PMID:18430474

Schuchman, E. H., & Wasserstein, M. P. (2015). Types A and B niemann-pick disease. *Best Practice & Research. Clinical Endocrinology & Metabolism*, *29*(2), 237–247. doi:10.1016/j.beem.2014.10.002 PMID:25987176

Schuurman, P. R., Bosch, D. A., Merkus, M. P., & Speelman, J. D. (2008). Long-term follow-up of thalamic stimulation versus thalamotomy for tremor suppression. *Movement Disorders*, *23*(8), 1146–1153. doi:10.1002/mds.22059 PMID:18442104

Schwabe, T., & Srinivasan, K. (2020). *Shifting paradigms: The central role of microglia in Alzheimer's disease*. Elsevier. https://www.sciencedirect.com/science/article/pii/S0969996120302370

Scimemi, A. (2014). 'Structure, Function, and Plasticity of GABA Transporters'. *Frontiers in Cellular Neuroscience*, *8*. https://www.frontiersin.org/articles/10.3389/fncel.2014.00161. doi:10.3389/fncel.2014.00161 PMID:24987330

Scimemi, A., Meabon, J. S., Woltjer, R. L., Sullivan, J. M., Diamond, J. S., & Cook, D. G. (2013). Amyloid-β $_{1-42}$ Slows Clearance of Synaptically Released Glutamate by Mislocalizing Astrocytic GLT-1. *The Journal of Neuroscience : The Official Journal of the Society for Neuroscience*, *33*(12), 5312–5318. doi:10.1523/JNEUROSCI.5274-12.2013 PMID:23516295

Sciotti, V. M., & Van Wylen, D. G. L. (1993). 'Increases in Interstitial Adenosine and Cerebral Blood Flow with Inhibition of Adenosine Kinase and Adenosine Deaminase'. *Journal of Cerebral Blood Flow and Metabolism*, *13*(2), 201–207. doi:10.1038/jcbfm.1993.24 PMID:8436611

Scott, E. W., Simon, M. C., Anastasi, J., & Singh, H. (1994). Requirement of transcription factor PU. 1 in the development of multiple hematopoietic lineages. *Science*, *265*(5178), 1573–1577. doi:10.1126cience.8079170 PMID:8079170

Seifert, G., Schilling, K., & Steinhäuser, C. (2006). Astrocyte dysfunction in neurological disorders : A molecular perspective. *Nature Reviews. Neuroscience*, *7*(3), 194–206. doi:10.1038/nrn1870 PMID:16495941

Seifert, G., & Steinhäuser, C. (2013). 'Neuron–Astrocyte Signaling and Epilepsy'. *Experimental Neurology*, *244*(June), 4–10. doi:10.1016/j.expneurol.2011.08.024 PMID:21925173

Selkoe, D. J. (2015). Alzheimer disease. In *Rosenberg's Molecular and Genetic Basis of Neurological and Psychiatric Disease* (pp. 753–768). Elsevier. doi:10.1016/B978-0-12-410529-4.00067-X

Selkoe, D. J., & Hardy, J. (2016). The amyloid hypothesis of Alzheimer's disease at 25 years. *EMBO Molecular Medicine*, *8*(6), 595–608. doi:10.15252/emmm.201606210 PMID:27025652

Selvakumar, G. P., Ahmed, M. E., Raikwar, S. P., Thangavel, R., Kempuraj, D., Dubova, I., Saeed, D., Zahoor, H., Premkumar, K., Zaheer, S., Iyer, S., & Zaheer, A. (2019). CRISPR/Cas9 Editing of Glia Maturation Factor Regulates Mitochondrial Dynamics by Attenuation of the NRF2/HO-1 Dependent Ferritin Activation in Glial Cells. *Journal of Neuroimmune Pharmacology*, *14*(4), 537–550. doi:10.100711481-019-09833-6 PMID:30810907

Seri, B., Manuel García-Verdugo, J., Mcewen, B. S., & Alvarez-Buylla, A. (2001). Astrocytes give rise to new neurons in the adult mammalian hippocampus. *Journal of Neuroscience*. https://www.jneurosci.org/content/21/18/7153.short

Setiawan, E., Wilson, A. A., Mizrahi, R., Rusjan, P. M., Miler, L., Rajkowska, G., Suridjan, I., Kennedy, J. L., Rekkas, P. V., Houle, S., & Meyer, J. H. (2015). Role of translocator protein density, a marker of neuroinflammation, in the brain during major depressive episodes. *JAMA Psychiatry*, *72*(3), 268–275. doi:10.1001/jamapsychiatry.2014.2427 PMID:25629589

Seto, T., Kono, K., Morimoto, K., Inoue, Y., Shintaku, H., Hattori, H., Matsuoka, O., Yamano, T., & Tanaka, A. (2001). Brain magnetic resonance imaging in 23 patients with mucopolysaccharidoses and the effect of bone marrow transplantation. *Annals of Neurology*, *50*(1), 79–92. doi:10.1002/ana.1098 PMID:11456314

Sevick, J., Kanda, F., Mintorovitch, J., Arieff, A. I., Kucharczyk, J., Tsuruda, J. S., Norman, D., & Moseley, M. E. (1992). Cytotoxic Brain Edema : Assessment with Diffusion-Weighted MR Imaging. *Radiology*, *185*(3), 687–690. doi:10.1148/radiology.185.3.1438745 PMID:1438745

Shabab, T., Khanabdali, R., Moghadamtousi, S. Z., Kadir, H. A., & Mohan, G. (2017). Neuroinflammation pathways: a general review. In International Journal of Neuroscience (Vol. 127, Issue 7, pp. 624–633). Taylor and Francis Ltd. do i:10.1080/00207454.2016.1212854

Shabason, J. E., Tofilon, P. J., & Camphausen, K. (2011). Grand rounds at the National Institutes of Health: HDAC inhibitors as radiation modifiers, from bench to clinic. *Journal of Cellular and Molecular Medicine*, *15*(12), 2735–2744. doi:10.1111/j.1582-4934.2011.01296.x PMID:21362133

Shao, W., Zhang, S.Z., Tang, M., Zhang, X.H., Zhou, Z., Yin, Y.Q., Zhou, Q.B., Huang, Y.Y., Liu, Y.J., Wawrousek, E., Chen, T., Li, S.B., Xu, M., Zhou, J.N., Hu, G., Zhou, J.W. (2013). Suppression of neuroinflammation by astrocytic dopamine D2 receptors via αB-crystallin. *Nature, 7*(7435), 90-4. . doi:10.1038/nature11748

Sharma, A., & Nagalli, S. (2022). Chronic liver disease. In *StatPearls* [Internet]. StatPearls Publishing.

Sharma, K., Zhang, G., & Li, S. (2015). Astrogliosis and Axonal Regeneration. In K.-F. So & X.-M. Xu (Eds.), *Neural Regeneration* (pp. 181–196). Academic Press. doi:10.1016/B978-0-12-801732-6.00011-2

Sharma, N. L., Groselj, B., Hamdy, F. C., & Kiltie, A. E. (2013). The emerging role of histone deacetylase (HDAC) inhibitors in urological cancers. *BJU International*, *111*(4), 537–542. doi:10.1111/j.1464-410X.2012.11647.x PMID:23551441

Sharp, A. H., Loev, S. J., Schilling, G., Li, S., Li, X., Bao, J., Wagster, M. V., Kotzuk, J. A., Steiner, J. P., Lo, A., Hedreen, J., Sisodia, S., Snyder, S. H., Dawson, T. M., Ryugo, D. K., & Ross, C. A. (1995). Widespread Expression of Huntington ' s Disease Gene (IT15). *Protein Product*, *14*, 1065–1074. PMID:7748554

Shelton, R. C., Claiborne, J., Sidoryk-Wegrzynowicz, M., Reddy, R., Aschner, M., Lewis, D. A., & Mirnics, K. (2011). Altered expression of genes involved in inflammation and apoptosis in frontal cortex in major depression. *Molecular Psychiatry*, *16*(7), 751–762. doi:10.1038/mp.2010.52 PMID:20479761

Sheng, J., Li, Q., Liu, T., & Wang, X. (2022). 'Cerebrospinal Fluid Dynamics along the Optic Nerve'. *Frontiers in Neurology*, *13*(August), 931523. doi:10.3389/fneur.2022.931523 PMID:36046631

Shen, S., Sandoval, J., Swiss, V. A., Li, J., Dupree, J., Franklin, R. J. M., & Casaccia-Bonnefil, P. (2008). 'Age-Dependent Epigenetic Control of Differentiation Inhibitors Is Critical for Remyelination Efficiency'. *Nature Neuroscience*, *11*(9), 1024–1034. doi:10.1038/nn.2172 PMID:19160500

Sheridan, G. K., & Murphy, K. J. (2013). Neuron–glia crosstalk in health and disease: fractalkine and CX3CR1 take centre stage. *Open Biology, 3*(1 DEC). doi:10.1098/rsob.130181

Sherman, D. L., & Brophy, P. J. (2005). 'Mechanisms of Axon Ensheathment and Myelin Growth'. *Nature Reviews. Neuroscience*, *6*(9), 683–690. doi:10.1038/nrn1743 PMID:16136172

Sherr, C. J., & Roberts, J. M. (1999). CDK inhibitors: Positive and negative regulators of G1-phase progression. *Genes & Development, 13*(12), 1501–1512. doi:10.1101/gad.13.12.1501 PMID:10385618

Sherwin, C., & Fern, R. (2005). Acute lipopolysaccharide-mediated injury in neonatal white matter glia: Role of TNF-α, IL-1β, and calcium. *The Journal of Immunology : Official Journal of the American Association of Immunologists, 175*(1), 155–161. doi:10.4049/jimmunol.175.1.155 PMID:15972642

Sheth, S., Brito, R., Mukherjee, D., Rybak, L. P., & Ramkumar, V. (2014). Adenosine receptors: Expression, function and regulation. *International Journal of Molecular Sciences, 15*(2), 2024–2052. doi:10.3390/ijms15022024 PMID:24477263

Shi, S., Liang, D., Chen, Y., Xie, Y., Wang, Y., Wang, L., Wang, Z., & Qiao, Z. (2016). Gx-50 reduces β-amyloid-induced TNF-α, IL-1β, NO, and PGE2 expression and inhibits NF-κB signaling in a mouse model of Alzheimer's disease. *European Journal of Immunology, 2016•Wiley Online Library, 46*(3), 665–676. doi:10.1002/eji.201545855

Shi, Y., Zhang, X., Zhang, H., Ge, L., & Zhao, L. (2021). Research progress in pathogenesis and treatment of neuroinflammation. *E3S Web of Conferences, 292*. doi:10.1051/e3sconf/202129203100

Shibata, N., Ohnuma, T., Takahashi, T., Baba, H., Ishizuka, T., Ohtsuka, M., Ueki, A., Nagao, M., & Arai, H. (2002). Effect of IL-6 polymorphism on risk of Alzheimer disease : Genotype-phenotype association study in Japanese cases. *American Journal of Medical Genetics, 114*(4), 436–439. doi:10.1002/ajmg.10417 PMID:11992567

Shi, D. D., Savani, M. R., Levitt, M. M., Wang, A. C., Endress, J. E., Bird, C. E., Buehler, J., Stopka, S. A., Regan, M. S., Lin, Y.-F., Puliyappadamba, V. T., Gao, W., Khanal, J., Evans, L., Lee, J. H., Guo, L., Xiao, Y., Xu, M., Huang, B., & McBrayer, S. K. (2022). De novo pyrimidine synthesis is a targetable vulnerability in IDH mutant glioma. *Cancer Cell, 40*(9), 939–956. doi:10.1016/j.ccell.2022.07.011 PMID:35985343

Shields, S. A., Gilson, J. M., Blakemore, W. F., & Franklin, R. J. M. (1999). 'Remyelination Occurs as Extensively but More Slowly in Old Rats Compared to Young Rats Following Gliotoxin-Induced CNS Demyelination'. *Glia, 28*(1), 77–83. doi:10.1002/(SICI)1098-1136(199910)28:1<77::AID-GLIA9>3.0.CO;2-F PMID:10498825

Shigemoto-Mogami, Y., Hoshikawa, K., Goldman, J. E., Sekino, Y., & Sato, K. (2014). Microglia Enhance Neurogenesis and Oligodendrogenesis in the Early Postnatal Subventricular Zone. *The Journal of Neuroscience : The Official Journal of the Society for Neuroscience, 34*(6), 2231–2243. doi:10.1523/JNEUROSCI.1619-13.2014 PMID:24501362

Shigemoto-Mogami, Y., Hoshikawa, K., & Sato, K. (2018). Activated microglia disrupt the blood-brain barrier and induce chemokines and cytokines in a rat in vitro model. *Frontiers in Cellular Neuroscience, 12*, 494. doi:10.3389/fncel.2018.00494 PMID:30618641

Shin, J. Y., Fang, Z. H., Yu, Z. X., Wang, C. E., Li, S. H., & Li, X. J. (2005). Expression of mutant huntingtin in glial cells contributes to neuronal excitotoxicity. *The Journal of Cell Biology, 171*(6), 1001–1012. doi:10.1083/jcb.200508072 PMID:16365166

Shinojima, N., Tada, K., Shiraishi, S., Kamiryo, T., Kochi, M., Nakamura, H., & Ushio, Y. (2003). Prognostic value of epidermal growth factor receptor in patients with glioblastoma multiforme. *Cancer Research, 63*(20), 6962–6970. PMID:14583498

Shi, Y., Shu, J., Liang, Z., Yuan, S., & Tang, S.-J. (2016). EXPRESS: Oligodendrocytes in HIV-associated pain pathogenesis. *Molecular Pain, 12*, 1744806916656845. doi:10.1177/1744806916656845 PMID:27306410

Shruthi, S., Sumitha, R., Varghese, A. M., Ashok, S., Chandrasekhar Sagar, B. K., Sathyaprabha, T. N., Nalini, A., Kramer, B. W., Raju, T. R., Vijayalakshmi, K., & Alladi, P. A. (2017). Brain-Derived Neurotrophic Factor Facilitates Functional Recovery from ALS-Cerebral Spinal Fluid-Induced Neurodegenerative Changes in the NSC-34 Motor Neuron Cell Line. *Neurodegenerative Diseases, 17*(1), 44–58. doi:10.1159/000447559 PMID:27617773

Shulyatnikova, T. V. A., & Tumanskyi, V. O. (2023). Ammonia level and Alzheimer type 2 astrocytes in the brain of deceased patients with liver cirrhosis of the varying degree. *Pathologia*, *20*(1), 36–44. doi:10.14739/2310-1237.2023.1.276453

Shulyatnikova, T. V., & Shavrin, V. A. (2017). Modern view on hepatic encephalopathy: Basic terms and concepts of pathogenesis. *Патологія*, *14*(3), 371–380. doi:10.14739/2310-1237.2017.3.118773

Shulyatnikova, T. V., & Tumanskiy, V. O. (2022). Immunohistochemical Study of the Brain Aquaporin-4 in the Rat Acute Liver Failure Model. *The Art of Medication*, *1455*(iii), 103–108. doi:10.21802/artm.2022.1.21.103

Shwartz, Y., Gonzalez-Celeiro, M., Chen, C. L., Pasolli, H. A., Sheu, S. H., Fan, S. M. Y., Shamsi, F., Assaad, S., Lin, E. T.-Y., Zhang, B., Tsai, P.-C., He, M., Tseng, Y.-H., Lin, S.-J., & Hsu, Y. C. (2020). Cell types promoting goosebumps form a niche to regulate hair follicle stem cells. *Cell*, *182*(3), 578–593. doi:10.1016/j.cell.2020.06.031 PMID:32679029

Sibille, J., Sibille, J., Pannasch, U., & Rouach, N. (2014). Astroglial potassium clearance contributes to the short-term plasticity of synaptically evoked currents at the tripartite synapse. *Journal of Physiology*, *592*(1), pp. 87-102. doi:10.1113/jphysiol.2013.261735

Sibille, J., Zapata, J., Teillon, J., & Rouach, N. (2015). Astroglial calcium signaling displays short-term plasticity and adjusts synaptic efficacy. *Frontiers in Cellular Neuroscience*, *9*, 189. doi:10.3389/fncel.2015.00189 PMID:26074766

Sidoryk-Wegrzynowicz, M., Wegrzynowicz, M., Lee, E., Bowman, A. B., & Aschner, M. (2011). Role of Astrocytes in Brain Function and Disease. *Toxicologic pathology*, *39*(1), 115-123. doi:10.1177/0192623310385254

Sieber, M. W., Jaenisch, N., Brehm, M., Guenther, M., Linnartz-Gerlach, B., Neumann, H., Witte, O. W., & Frahm, C. (2013). Attenuated Inflammatory Response in Triggering Receptor Expressed on Myeloid Cells 2 (TREM2) Knock-Out Mice following Stroke. *PLoS One*, *8*(1), e52982. doi:10.1371/journal.pone.0052982 PMID:23301011

Sierra, A., Abiega, O., Shahraz, A., & Neumann, H. (2013). Janus-faced microglia : Beneficial and detrimental consequences of microglial phagocytosis. *Frontiers in Cellular Neuroscience*, *7*. https://www.frontiersin.org/articles/10.3389/fncel.2013.00006. doi:10.3389/fncel.2013.00006 PMID:23386811

Sierra, A., Encinas, J. M., Deudero, J. J. P., Chancey, J. H., Enikolopov, G., Overstreet-Wadiche, L. S., Tsirka, S. E., & Maletic-Savatic, M. (2010). Microglia shape adult hippocampal neurogenesis through apoptosis-coupled phagocytosis. *Cell Stem Cell*, *7*(4), 483–495. doi:10.1016/j.stem.2010.08.014 PMID:20887954

Sievers, C., Auer, M. K., Klotsche, J., Athanasoulia, A. P., Schneider, H. J., Nauck, M., Völzke, H., John, U., Schulz, A., Freyberger, H. J., Friedrich, N., Biffar, R., Stalla, G. K., Wallaschofski, H., & Grabe, H. J. (2014). IGF-I levels and depressive disorders: Results from the Study of Health in Pomerania (SHIP). *European Neuropsychopharmacology*, *24*(6), 890–896. doi:10.1016/j.euroneuro.2014.01.008 PMID:24507017

Silva, B., Mantha, O. L., Schor, J., Pascual, A., Plaçais, P. Y., Pavlowsky, A., & Preat, T. (2022). Glia fuel neurons with locally synthesized ketone bodies to sustain memory under starvation. *Nature Metabolism*, *4*(2), 213–224. doi:10.103842255-022-00528-6

Silver, J., & Miller, J. H. (2004). Regeneration beyond the glial scar. *Nature Reviews. Neuroscience*, *5*(2), 146–156. doi:10.1038/nrn1326 PMID:14735117

Simard, J., Kent, T. A., Chen, M., Tarasov, K. V., & Gerzanich, V. (2007). Brain Oedema in Focal Ischaemia: Molecular Pathophysiology and Theoretical Implications. *Lancet Neurology*, *6*(3), 258–268. doi:10.1016/S1474-4422(07)70055-8 PMID:17303532

Simon, D. K., Tanner, C. M., & Brundin, P. (2020). Parkinson Disease Epidemiology, Pathology, Genetics, and Pathophysiology. *Clinics in Geriatric Medicine*, *36*(1), 1–12. doi:10.1016/j.cger.2019.08.002 PMID:31733690

Simoni, E., Caporaso, R., Bergamini, C., Fiori, J., Fato, R., Miszta, P., Filipek, S., Caraci, F., Giuffrida, M. L., Andrisano, V., Minarini, A., Bartolini, M., & Rosini, M. (2016). Polyamine conjugation as a promising strategy to target amyloid aggregation in the framework of Alzheimer's disease. *ACS Medicinal Chemistry Letters*, *7*(12), 1145–1150. doi:10.1021/acsmedchemlett.6b00339 PMID:27994754

Simons, M., & Nave, K.-A. (2016). 'Oligodendrocytes: Myelination and Axonal Support'. *Cold Spring Harbor Perspectives in Biology*, *8*(1), a020479. doi:10.1101/cshperspect.a020479 PMID:26101081

Simpson, J. E., Ince, P. G., Haynes, L. J., Theaker, R., Gelsthorpe, C., Baxter, L., Forster, G., Lace, G. L., Shaw, P. J., Matthews, F. E., Savva, G. M., Brayne, C., & Wharton, S. B. (2010). 'Population Variation in Oxidative Stress and Astrocyte DNA Damage in Relation to Alzheimer-Type Pathology in the Ageing Brain'. *Neuropathology and Applied Neurobiology*, *36*(1), 25–40. doi:10.1111/j.1365-2990.2009.01030.x PMID:19422529

Singh, D. (2022). 'Astrocytic and Microglial Cells as the Modulators of Neuroinflammation in Alzheimer's Disease'. *Journal of Neuroinflammation*, *19*(1), 206. doi:10.118612974-022-02565-0 PMID:35978311

Singh, S. K., Hidau, M. K., Gautam, S., Gupta, K., Singh, K. P., Singh, S. K., & Singh, S. (2018). Glycol chitosan functionalized asenapine nanostructured lipid carriers for targeted brain delivery: Pharmacokinetic and teratogenic assessment. *International Journal of Biological Macromolecules*, *108*, 1092–1100. doi:10.1016/j.ijbiomac.2017.11.031 PMID:29126941

Singh, S. K., Stogsdill, J. A., Pulimood, N. S., Dingsdale, H., Kim, Y. H., Pilaz, L. J., Kim, I. H., Manhaes, A. C., Rodrigues, W. S. Jr, Pamukcu, A., Enustun, E., Ertuz, Z., Scheiffele, P., Soderling, S. H., Silver, D. L., Ji, R.-R., Medina, A. E., & Eroglu, C. (2016). Astrocytes assemble thalamocortical synapses by bridging NRX1α and NL1 via hevin. *Cell*, *164*(1), 183–196. doi:10.1016/j.cell.2015.11.034 PMID:26771491

Si-Tayeb, K., Lemaigre, F. P., & Duncan, S. A. (2010). Organogenesis and development of the liver. *Developmental Cell*, *18*(2), 175–189. doi:10.1016/j.devcel.2010.01.011 PMID:20159590

Si, X., Miguel-Hidalgo, J. J., O'Dwyer, G., Stockmeier, C. A., & Rajkowska, G. (2004). Age-dependent reductions in the level of glial fibrillary acidic protein in the prefrontal cortex in major depression. *Neuropsychopharmacology*, *29*(11), 2088–2096. doi:10.1038j.npp.1300525 PMID:15238995

Si, Z.-Z., Zou, C.-J., Mei, X., Li, X.-F., Luo, H., Shen, Y., Hu, J., Li, X.-X., Wu, L., & Liu, Y. (2023). 'Targeting Neuroinflammation in Alzheimer's Disease: From Mechanisms to Clinical Applications'. *Neural Regeneration Research*, *18*(4), 708. doi:10.4103/1673-5374.353484 PMID:36204826

Skaper, S. D. (2019). Oligodendrocyte precursor cells as a therapeutic target for demyelinating diseases. *Progress in Brain Research*, *245*, 119–144. doi:10.1016/bs.pbr.2019.03.013

Skatchkov, S. N., Bukauskas, F. F., Benedikt, J., Inyushin, M., & Kucheryavykh, Y. V. (2015). Intracellular spermine prevents acid-induced uncoupling of Cx43 gap junction channels. *Neuroreport*, *26*(9), 528–532. doi:10.1097/WNR.0000000000000385 PMID:26011388

Skatchkov, S. N., Woodbury-Fariña, M. A., & Eaton, M. (2014). The role of glia in stress : Polyamines and brain disorders. *Psychiatria Clinica*, *37*(4), 653–678. PMID:25455070

Skop, N. B., Calderon, F., Cho, C. H., Gandhi, C. D., & Levison, S. W. (2016). Optimizing a multifunctional microsphere scaffold to improve neural precursor cell transplantation for traumatic brain injury repair. *Journal of Tissue Engineering and Regenerative Medicine*, *10*(10), E419–E432. doi:10.1002/term.1832 PMID:27730762

Skop, N. B., Singh, S., Antikainen, H., Saqcena, C., Calderon, F., Rothbard, D. E., Cho, C. H., Gandhi, C. D., Levison, S. W., & Dobrowolski, R. (2019). Subacute transplantation of native and genetically engineered neural progenitors seeded on microsphere scaffolds promote repair and functional recovery after traumatic brain injury. *ASN Neuro, 11*, 1759091419830186. doi:10.1177/1759091419830186 PMID:30818968

Skowrońska, M., Zielińska, M., Wójcik-Stanaszek, L., Ruszkiewicz, J., Milatovic, D., Aschner, M., & Albrecht, J. (2012). Ammonia Increases Paracellular Permeability of Rat Brain Endothelial Cells by a Mechanism Encompassing Oxidative/Nitrosative Stress and Activation of Matrix Metalloproteinases. *Journal of Neurochemistry, 121*(1), 125–134. doi:10.1111/j.1471-4159.2012.07669.x PMID:22260250

Ślusarczyk, J., Trojan, E., Głombik, K., Budziszewska, B., Kubera, M., Lasoń, W., Popiołek-Barczyk, K., Mika, J., Wędzony, K., & Basta-Kaim, A. (2015). Prenatal stress is a vulnerability factor for altered morphology and biological activity of microglia cells. *Frontiers in Cellular Neuroscience, 9*, 82. doi:10.3389/fncel.2015.00082 PMID:25814933

Śmiałowska, M., Szewczyk, B., Woźniak, M., Wawrzak-Wleciał, A., & Domin, H. (2013). Glial degeneration as a model of depression. *Pharmacological Reports, 65*(6), 1572–1579. doi:10.1016/S1734-1140(13)71518-4 PMID:24553005

Smith, C. M., Cooksey, E., & Duncan, I. D. (2013). Myelin loss does not lead to axonal degeneration in a long-lived model of chronic demyelination. *The Journal of Neuroscience : The Official Journal of the Society for Neuroscience, 33*(6), 2718–2727. doi:10.1523/JNEUROSCI.4627-12.2013 PMID:23392698

Smith, G. M., & Strunz, C. (2005). Growth factor and cytokine regulation of chondroitin sulfate proteoglycans by astrocytes. *Glia, 52*(3), 209–218. doi:10.1002/glia.20236 PMID:15968632

Smith, M. J., Wallace, A. J., Bennett, C., Hasselblatt, M., Elert-Dobkowska, E., Evans, L. T., Hickey, W. F., van Hoff, J., Bauer, D., Lee, A., Hevner, R. F., Beetz, C., du Plessis, D., Kilday, J.-P., Newman, W. G., & Evans, D. G. (2014). 'Germline *SMARCE1* Mutations Predispose to Both Spinal and Cranial Clear Cell Meningiomas'. *The Journal of Pathology, 234*(4), 436–440. doi:10.1002/path.4427 PMID:25143307

Sneeboer, M. A. M., Snijders, G. J. L. J., Berdowski, W. M., Fernández-Andreu, A., van Mierlo, H. C., Berdenis van Berlekom, A., Litjens, M., Kahn, R. S., Hol, E. M., & de Witte, L. D.Psychiatric Donor Program of the Netherlands Brain Bank. (2019). Microglia in post-mortem brain tissue of patients with bipolar disorder are not immune activated. *Translational Psychiatry, 9*(1), 153. doi:10.103841398-019-0490-x PMID:31127084

Snijders, G. J. L. J., van Zuiden, W., Sneeboer, M. A. M., Berdenis van Berlekom, A., van der Geest, A. T., Schnieder, T., MacIntyre, D. J., Hol, E. M., Kahn, R. S., & de Witte, L. D. (2021). A loss of mature microglial markers without immune activation in schizophrenia. *Glia, 69*(5), 1251–1267. doi:10.1002/glia.23962 PMID:33410555

Snyder, E. M., Nong, Y., Almeida, C. G., Paul, S., Moran, T., Choi, E. Y., Nairn, A. C., Salter, M. W., Lombroso, P. J., Gouras, G. K., & Greengard, P. (2005). Regulation of NMDA receptor trafficking by amyloid-β. *Nature Neuroscience, 8*(8), 1051–1058. doi:10.1038/nn1503 PMID:16025111

Sobolczyk, M., & Boczek, T. (2022). Astrocytic Calcium and cAMP in Neurodegenerative Diseases. *Frontiers in Cellular Neuroscience, 16*, 889939. doi:10.3389/fncel.2022.889939 PMID:35663426

Sock, E., & Wegner, M. (2021). Using the lineage determinants Olig2 and Sox10 to explore transcriptional regulation of oligodendrocyte development. *Developmental Neurobiology, 81*(7), 892–901. doi:10.1002/dneu.22849

Sofroniew, M. V. (2015). Astrogliosis. *Cold Spring Harbor Perspectives in Biology, 7*(2), a020420. doi:10.1101/cshperspect.a020420 PMID:25380660

Sofroniew, M. V., & Vinters, H. V. (2010). 'Astrocytes: Biology and Pathology'. *Acta Neuropathologica, 119*(1), 7–35. doi:10.100700401-009-0619-8 PMID:20012068

Solano, R. M., Casarejos, M. J., Menéndez-Cuervo, J., Rodriguez-Navarro, J. A., De Yébenes, J. G., & Mena, M. A. (2008). Glial dysfunction in parkin null mice: Effects of aging. *The Journal of Neuroscience : The Official Journal of the Society for Neuroscience, 28*(3), 598–611. doi:10.1523/JNEUROSCI.4609-07.2008 PMID:18199761

Song, I., & Dityatev, A. (2018). Crosstalk between glia, extracellular matrix and neurons. *Brain Research Bulletin, 136*, 101–108. doi:10.1016/j.brainresbull.2017.03.003 PMID:28284900

Song, S., Miranda, C. J., Braun, L., Meyer, K., Frakes, A. E., Ferraiuolo, L., Likhite, S., Bevan, A. K., Foust, K. D., McConnell, M. J., Walker, C. M., & Kaspar, B. K. (2016). Major histocompatibility complex class I molecules protect motor neurons from astrocyte-induced toxicity in amyotrophic lateral sclerosis. *Nature Medicine, 22*(4), 397–403. doi:10.1038/nm.4052 PMID:26928464

Sorrentino, Z. (2019). α-Synuclein and Astrocytes: Tracing the Pathways from Homeostasis to Neurodegeneration in Lewy Body Disease. *Acta Neuropathologica, 138*(1), 1–21. . doi:10.1007/s00401-019-01977-2

Spaas, J., Van Veggel, L., Schepers, M., Tiane, A., Van Horssen, J., Wilson, D. M. III, Moya, P. R., Piccart, E., Hellings, N., Eijnde, B. O., Derave, W., Schreiber, R., & Vanmierlo, T. (2021). Oxidative stress and impaired oligodendrocyte precursor cell differentiation in neurological disorders. *Cellular and Molecular Life Sciences, 78*(10), 4615–4637. doi:10.100700018-021-03802-0

Spassky, N., Merkle, F. T., Flames, N., Tramontin, A. D., García-Verdugo, J. M., & Alvarez-Buylla, A. (2005). Adult Ependymal Cells Are Postmitotic and Are Derived from Radial Glial Cells during Embryogenesis. *The Journal of Neuroscience : The Official Journal of the Society for Neuroscience, 25*(1), 10–18. doi:10.1523/JNEUROSCI.1108-04.2005 PMID:15634762

Spector, R., & Johanson, C. E. (1989). 'The Mammalian Choroid Plexus'. *Scientific American, 261*(5), 68–74. doi:10.1038cientificamerican1189-68 PMID:2479097

Spiegel, A. M., Sewal, A. S., & Rapp, P. R. (2014). 'Epigenetic Contributions to Cognitive Aging: Disentangling Mindspan and Lifespan'. *Learning & Memory (Cold Spring Harbor, N.Y.), 21*(10), 569–574. doi:10.1101/lm.033506.113 PMID:25227252

Squarzoni, P., Oller, G., Hoeffel, G., Pont-Lezica, L., Rostaing, P., Low, D., Bessis, A., Ginhoux, F., & Garel, S. (2014). Microglia Modulate Wiring of the Embryonic Forebrain. *Cell Reports, 8*(5), 1271–1279. doi:10.1016/j.celrep.2014.07.042 PMID:25159150

Sridhar, V., Gaud, R., Bajaj, A., & Wairkar, S. (2018). Pharmacocinétique et pharmacodynamique des nanoparticules de sélégiline administrées par voie intranasale avec une administration cérébrale améliorée dans la maladie de Parkinson. Nanomédecine: Nanotechnologie. *Biology and Medicine (Aligarh), 14*(8), 2609–2618.

Srinivasan, R., Huang, B. S., Venugopal, S., Johnston, A. D., Chai, H., Zeng, H., Golshani, P., & Khakh, B. S. (2015). Ca2+ signaling in astrocytes from Ip3r2−/− mice in brain slices and during startle responses in vivo. *Nature Neuroscience, 18*(5), 708–717. doi:10.1038/nn.4001

Sriram, K., Benkovic, S. A., Hebert, M. A., Miller, D. B., & O'Callaghan, J. P. (2004). Induction of gp130-related cytokines and activation of JAK2/STAT3 pathway in astrocytes precedes up-regulation of glial fibrillary acidic protein in the 1-methyl-4-phenyl-1, 2, 3, 6-tetrahydropyridine model of neurodegeneration : Key signaling pathway for astrogliosis in vivo? *The Journal of Biological Chemistry, 279*(19), 19936–19947. doi:10.1074/jbc.M309304200 PMID:14996842

Srivastava, S. (2017). 'The Mitochondrial Basis of Aging and Age-Related Disorders'. *Genes, 8*(12), 398. doi:10.3390/genes8120398 PMID:29257072

Stadelmann, C., Timmler, S., Barrantes-Freer, A., & Simons, M. (2019). Myelin in the central nervous system : Structure, function, and pathology. *Physiological Reviews*, *99*(3), 1381–1431. doi:10.1152/physrev.00031.2018 PMID:31066630

Stancu, I. C., Vasconcelos, B., Terwel, D., & Dewachter, I. (2014). Models of β-amyloid induced Tau-pathology: The long and "folded" road to understand the mechanism. *Molecular Neurodegeneration*, *9*(1), 51. doi:10.1186/1750-1326-9-51 PMID:25407337

Stangel, M., Kuhlmann, T., Matthews, P. M., & Kilpatrick, T. J. (2017). Achievements and obstacles of remyelinating therapies in multiple sclerosis. *Nature Reviews. Neurology*, *13*(12), 742–754. doi:10.1038/nrneurol.2017.139

Steiner, J., Bielau, H., Brisch, R., Danos, P., Ullrich, O., Mawrin, C., Bernstein, H.-G., & Bogerts, B. (2008). Immunological aspects in the neurobiology of suicide: Elevated microglial density in schizophrenia and depression is associated with suicide. *Journal of Psychiatric Research*, *42*(2), 151–157. doi:10.1016/j.jpsychires.2006.10.013 PMID:17174336

Stellwagen, D., Beattie, E. C., Seo, J. Y., & Malenka, R. C. (2005). Differential regulation of AMPA receptor and GABA receptor trafficking by tumor necrosis factor-α. *The Journal of Neuroscience : The Official Journal of the Society for Neuroscience*, *25*(12), 3219–3228. doi:10.1523/JNEUROSCI.4486-04.2005 PMID:15788779

Stence, N., Waite, M., & Dailey, M. E. (2001). Dynamics of microglial activation: A confocal time-lapse analysis in hippocampal slices. *Glia*, *33*(3), 256–266. doi:10.1002/1098-1136(200103)33:3<256::AID-GLIA1024>3.0.CO;2-J PMID:11241743

Stenovec, M., Trkov, S., Lasič, E., Terzieva, S., Kreft, M., Rodríguez Arellano, J. J., Parpura, V., Verkhratsky, A., & Zorec, R. (2016). Expression of familial Alzheimer disease presenilin 1 gene attenuates vesicle traffic and reduces peptide secretion in cultured astrocytes devoid of pathologic tissue environment. *Glia*, *64*(2), 317–329. doi:10.1002/glia.22931 PMID:26462451

Stephan, A. H., Madison, D. V., Mateos, J. M., Fraser, D. A., Lovelett, E. A., Coutellier, L., Kim, L., Tsai, H.-H., Huang, E. J., Rowitch, D. H., Berns, D. S., Tenner, A. J., Shamloo, M., & Barres, B. A. (2013). A dramatic increase of C1q protein in the CNS during normal aging. *Journal of Neuroscience. Social Neuroscience*, *33*(33), 13460–13474. doi:10.1523/JNEUROSCI.1333-13.2013 PMID:23946404

Stephenson, J., Nutma, E., van der Valk, P., & Amor, S. (2018). Inflammation in CNS neurodegenerative diseases. *Immunology*, *154*(2), 204–219. doi:10.1111/imm.12922 PMID:29513402

Stevens, B., Allen, N. J., Vazquez, L. E., Howell, G. R., Christopherson, K. S., Nouri, N., Micheva, K. D., Mehalow, A. K., Huberman, A. D., Stafford, B., Sher, A., Litke, A. M., Lambris, J. D., Smith, S. J., John, S. W. M., & Barres, B. A. (2007). The classical complement cascade mediates CNS synapse elimination. *Cell*, *131*(6), 1164–1178. doi:10.1016/j.cell.2007.10.036 PMID:18083105

Stewart, V. C., Sharpe, M. A., Clark, J. B., & Heales, S. J. (2000). Astrocyte-derived nitric oxide causes both reversible and irreversible damage to the neuronal mitochondrial respiratory chain. *Journal of Neurochemistry*, *75*(2), 694–700. doi:10.1046/j.1471-4159.2000.0750694.x PMID:10899944

Stobart, J. L., & Anderson, C. M. (2013). Multifunctional role of astrocytes as gatekeepers of neuronal energy supply. *Frontiers in Cellular Neuroscience*, *7*, 38. doi:10.3389/fncel.2013.00038 PMID:23596393

Stockmeier, C. A., Mahajan, G., Konick, L. C., Overholser, J. C., Jurjus, G. J., Meltzer, H. Y., Friedman, L., & Rajkowska, G. (2003). Neuronal and glial density is increased and neuronal soma size is decreased in hippocampus in major depressive disorder (MDD). *Biological Psychiatry*, *53*(8), 70S–70S.

Stockmeier, C. A., & Rajkowska, G. (2004). Cellular abnormalities in depression: Evidence from postmortem brain tissue. *Dialogues in Clinical Neuroscience*, *6*(2), 185–197. doi:10.31887/DCNS.2004.6.2/cstockmeier PMID:22033633

Stogsdill, J. A., Ramirez, J., Liu, D., Kim, Y. H., Baldwin, K. T., Enustun, E., Ejikeme, T., Ji, R.-R., & Eroglu, C. (2017). Astrocytic neuroligins control astrocyte morphogenesis and synaptogenesis. *Nature*, *551*(7679), 192–197. doi:10.1038/nature24638 PMID:29120426

Strange, K. (2004). Cellular Volume Homeostasis. *Advances in Physiology Education*, *28*(4), 155–159. doi:10.1152/advan.00034.2004 PMID:15545344

Strauss, G., Hansen, B. A., Knudsen, G. M., & Larsen, F. S. (1998). Hyperventilation Restores Cerebral Blood Flow Autoregulation in Patients with Acute Liver Failure. *Journal of Hepatology*, *28*(2), 199–203. doi:10.1016/0168-8278(88)80006-0 PMID:9514532

Streit, W. J. (2005). Microglia and neuroprotection: Implications for Alzheimer's disease. *Brain Research. Brain Research Reviews*, *48*(2), 234–239. doi:10.1016/j.brainresrev.2004.12.013 PMID:15850662

Streit, W. J., Sammons, N. W., Kuhns, A. J., & Sparks, D. L. (2004). Dystrophic microglia in the aging human brain. *Glia*, *45*(2), 208–212. doi:10.1002/glia.10319 PMID:14730714

Streit, W. J., Walter, S. A., & Pennell, N. A. (1999). Reactive microgliosis. *Progress in Neurobiology*, *57*(6), 563–581. doi:10.1016/S0301-0082(98)00069-0 PMID:10221782

Stupp, R., Mason, W. P., van den Bent, M. J., Weller, M., Fisher, B., Taphoorn, M. J. B., Belanger, K., Brandes, A. A., Marosi, C., Bogdahn, U., Curschmann, J., Janzer, R. C., Ludwin, S. K., Gorlia, T., Allgeier, A., Lacombe, D., Cairncross, J. G., Eisenhauer, E., & Mirimanoff, R. O. (2005). Radiotherapy plus concomitant and aduuvant temozolomide for globlastoma. *The New England Journal of Medicine*, *352*(10), 987–996. doi:10.1056/NEJMoa043330 PMID:15758009

Sturm, D., Bender, S., Jones, D. T., Lichter, P., Grill, J., Becher, O., Hawkins, C., Majewski, J., Jones, C., Costello, J. F., Iavarone, A., Aldape, K., Brennan, C. W., Jabado, N., & Pfister, S. M. (2014). Paediatric and adult glioblastoma: Multiform (epi) genomic culprits emerge. *Nature Reviews. Cancer*, *14*(2), 92–107. doi:10.1038/nrc3655 PMID:24457416

Stys, P. K. (2005). General mechanisms of axonal damage and its prevention. *Journal of the Neurological Sciences*, *233*(1-2), 3–13. doi:10.1016/j.jns.2005.03.031

Stys, P. K., & Tsutsui, S. (2019). Recent advances in understanding multiple sclerosis. *F1000 Research*, 8.

Suárez-Calvet, M., Capell, A., Araque Caballero, M. Á., Morenas-Rodríguez, E., Fellerer, K., Franzmeier, N., Kleinberger, G., Eren, E., Deming, Y., Piccio, L., Karch, C. M., Cruchaga, C., Paumier, K., Bateman, R. J., Fagan, A. M., Morris, J. C., Levin, J., Danek, A., Jucker, M., ... Haass, C. (2018). CSF progranulin increases in the course of Alzheimer's disease and is associated with sTREM 2, neurodegeneration and cognitive decline. *EMBO Molecular Medicine*, *10*(12), e9712. doi:10.15252/emmm.201809712 PMID:30482868

Subhramanyam, C. S., Wang, C., Hu, Q., & Dheen, S. T. (2019). Microglia-mediated neuroinflammation in neurodegenerative diseases. In *Seminars in Cell and Developmental Biology* (Vol. 94, pp. 112–120). Elsevier Ltd., doi:10.1016/j.semcdb.2019.05.004

Su, F., Yi, H., Xu, L., & Zhang, Z. (2015). Fluoxetine and S-citalopram inhibit M1 activation and promote M2 activation of microglia in vitro. *Neuroscience*, *294*, 60–68. doi:10.1016/j.neuroscience.2015.02.028 PMID:25711936

Su, J. H., Nichol, K. E., Sitch, T., Sheu, P., Chubb, C., Miller, B. L., Tomaselli, K. J., Kim, R. C., & Cotman, C. W. (2000). DNA damage and activated caspase-3 expression in neurons and astrocytes: Evidence for apoptosis in frontotemporal dementia. *Experimental Neurology*, *163*(1), 9–19. doi:10.1006/exnr.2000.7340 PMID:10785439

Suk, K. (2014). Combined analysis of the glia secretome and the CSF proteome: neuroinflammation and novel biomarkers. *EPR, 7*(2), 263–274. doi:10.1586/epr.10.6

Suksuphew, S., & Noisa, P. (2015). 'Neural Stem Cells Could Serve as a Therapeutic Material for Age-Related Neurodegenerative Diseases'. *World Journal of Stem Cells*, *7*(2), 502–511. doi:10.4252/wjsc.v7.i2.502 PMID:25815135

Sulehria, T., Corbett, A. M., Sharma, N., Nagarajan, D., Abushamma, A., Gagle, S., & Johnson, A. (n.d.). Increasing Progenitor Cell Proliferation in the Sub-Ventricular Zone: a therapeutic treatment for progressive multiple sclerosis? *Ingentaconnect*. doi:10.2174/1872211314999201117130123

Sultan, S., Li, L., Moss, J., Petrelli, F., Cassé, F., Gebara, E., Lopatar, J., Pfrieger, F. W., Bezzi, P., Bischofberger, J., & Toni, N. (2015). Synaptic Integration of Adult-Born Hippocampal Neurons Is Locally Controlled by Astrocytes. *Neuron*, *88*(5), 957–972. doi:10.1016/j.neuron.2015.10.037 PMID:26606999

Suminaite, D., Lyons, D. A., & Livesey, M. R. (2019). Myelinated axon physiology and regulation of neural circuit function. *Glia*, *67*(11), 2050–2062. doi:10.1002/glia.23665

Sung, C. Y., Yang, C. Y., Chen, W. S., Wang, Y. K., Yeh, J. A., & Cheng, C. M. (2015). Probing neural cell behaviors through micro-/nano-patterned chitosan substrates. *Biofabrication*, *7*(4), 045007. doi:10.1088/1758-5090/7/4/045007 PMID:26685015

Sun, J.-D., Liu, Y., Yuan, Y.-H., Li, J., & Chen, N.-H. (2012). Gap junction dysfunction in the prefrontal cortex induces depressive-like behaviors in rats. *Neuropsychopharmacology*, *37*(5), 1305–1320. doi:10.1038/npp.2011.319 PMID:22189291

Sun, S., Sun, Y., Ling, S. C., Ferraiuolo, L., McAlonis-Downes, M., Zou, Y., Drenner, K., Wang, Y., Ditsworth, D., Tokunaga, S., Kopelevich, A., Kaspar, B. K., Lagier-Tourenne, C., & Cleveland, D. W. (2015). Translational profiling identifies a cascade of damage initiated in motor neurons and spreading to glia in mutant SOD1-mediated ALS. *Proceedings of the National Academy of Sciences of the United States of America*, *112*(50), E6993–E7002. doi:10.1073/pnas.1520639112 PMID:26621731

Sun, Y.-N., Luo, J.-Y., Rao, Z.-R., Lan, L., & Duan, L. (2005). GFAP and Fos immunoreactivity in lumbo-sacral spinal cord and medulla oblongata after chronic colonic inflammation in rats. *World Journal of Gastroenterology*, *11*(31), 4827–4832. doi:10.3748/wjg.v11.i31.4827 PMID:16097052

Sun, Y., Yang, C., Zhu, X., Wang, J. J., Liu, X. Y., Yang, X. P., An, X.-W., Liang, J., Dong, H.-J., Jiang, W., Chen, C., Wang, Z.-G., Sun, H.-T., Tu, Y., Zhang, S., Chen, F., & Li, X. H. (2019). 3D printing collagen/chitosan scaffold ameliorated axon regeneration and neurological recovery after spinal cord injury. *Journal of Biomedical Materials Research. Part A*, *107*(9), 1898–1908. doi:10.1002/jbm.a.36675 PMID:30903675

Suter, M. R., Berta, T., Gao, Y.-J., Decosterd, I., & Ji, R.-R. (2009). Large A-fiber activity is required for microglial proliferation and p38 MAPK activation in the spinal cord: Different effects of resiniferatoxin and bupivacaine on spinal microglial changes after spared nerve injury. *Molecular Pain*, *5*, 53. doi:10.1186/1744-8069-5-53 PMID:19772627

Swain, M., Butterworth, R. F., & Blei, A. T. (1992). Ammonia and Related Amino Acids in the Pathogenesis of Brain Edema in Acute Ischemic Liver Failure in Rats. *Hepatology (Baltimore, Md.)*, *15*(3), 449–453. doi:10.1002/hep.1840150316 PMID:1544626

Swaminathan, M., Ellul, M. A., & Cross, T. J. S. (2018). Hepatic encephalopathy: Current challenges and future prospects. *Hepatic Medicine : Evidence and Research*, *10*, 1–11. doi:10.2147/HMER.S118964 PMID:29606895

Swanson, M. E. V., Scotter, E. L., Smyth, L. C. D., Murray, H. C., Ryan, B., Turner, C., Faull, R. L. M., Dragunow, M., & Curtis, M. A. (2020). Identification of a dysfunctional microglial population in human Alzheimer's disease cortex using novel single-cell histology image analysis. *Acta Neuropathologica Communications*, *8*(1), 170. doi:10.118640478-020-01047-9 PMID:33081847

Szabolcsi, V., & Celio, M. R. (2015). De novo expression of parvalbumin in ependymal cells in response to brain injury promotes ependymal remodeling and wound repair. *Glia*, *63*(4), 567–594. doi:10.1002/glia.22768 PMID:25421913

Szepesi, Z., Manouchehrian, O., Bachiller, S., & Deierborg, T. (2018). Bidirectional Microglia–Neuron Communication in Health and Disease. *Frontiers in Cellular Neuroscience*, *12*(September), 1–26. doi:10.3389/fncel.2018.00323 PMID:30319362

Szopa, W., Burley, T. A., Kramer-Marek, G., & Kaspera, W. (2017). Diagnostic and therapeutic biomarkers in glioblastoma: Current status and future perspectives. *BioMed Research International*, *2017*, 2017. doi:10.1155/2017/8013575 PMID:28316990

Tabata, H. (2015). Diverse subtypes of astrocytes and their development during corticogenesis. *Frontiers in Neuroscience*, *9*, 114. doi:10.3389/fnins.2015.00114 PMID:25904839

Takahashi, H., Koehler, R. C., Brusilow, S. W., & Traystman, R. J. (1991). Inhibition of Brain Glutamine Accumulation Prevents Cerebral Edema in Hyperammonemic Rats. *American Journal of Physiology. Heart and Circulatory Physiology*, *261*(3), H825–H829. doi:10.1152/ajpheart.1991.261.3.H825 PMID:1679605

Takahashi, Y., Sukegawa, K., Aoki, M., Ito, A., Suzuki, K., Sakaguchi, H., Watanabe, M., Isogai, K., Mizuno, S., Hoshi, H., Kuwata, K., Tomatsu, S., Kato, S., Ito, T., Kondo, N., & Orii, T. (2001). Evaluation of accumulated mucopolysaccharides in the brain of patients with mucopolysaccharidoses by 1H-magnetic resonance spectroscopy before and after bone marrow transplantation. *Pediatric Research*, *49*(3), 349–355. doi:10.1203/00006450-200103000-00008 PMID:11228260

Takaku, S., Yanagisawa, H., Watabe, K., Horie, H., Kadoya, T., Sakumi, K., Nakabeppu, Y., Poirier, F., & Sango, K. (2013). GDNF promotes neurite outgrowth and upregulates galectin-1 through the RET/PI3K signaling in cultured adult rat dorsal root ganglion neurons. *Neurochemistry International*, *62*(3), 330–339. doi:10.1016/j.neuint.2013.01.008 PMID:23340048

Takata, F., Nakagawa, S., Matsumoto, J., & Dohgu, S. (2021). Blood-Brain Barrier Dysfunction Amplifies the Development of Neuroinflammation: Understanding of Cellular Events in Brain Microvascular Endothelial Cells for Prevention and Treatment of BBB Dysfunction. In Frontiers in Cellular Neuroscience (Vol. 15). Frontiers Media S.A. doi:10.3389/fncel.2021.661838

Takebayashi, M., Hisaoka, K., Nishida, A., Tsuchioka, M., Miyoshi, I., Kozuru, T., Hikasa, S., Okamoto, Y., Shinno, H., Morinobu, S., & Yamawaki, S. (2006). Decreased levels of whole blood glial cell line-derived neurotrophic factor (GDNF) in remitted patients with mood disorders. *The International Journal of Neuropsychopharmacology*, *9*(5), 607–612. doi:10.1017/S1461145705006085 PMID:16191208

Takeuchi, H., Mizoguchi, H., Doi, Y., Jin, S., Noda, M., Liang, J., & Suzumura, A. (2011). Blockade of gap junction hemichannel suppresses disease progression in mouse models of amyotrophic lateral sclerosis and Alzheimer's disease. *PLoS One*, *6*(6), e21108. doi:10.1371/journal.pone.0021108 PMID:21712989

Talantova, M., Sanz-Blasco, S., Zhang, X., Xia, P., Akhtar, M. W., Okamoto, S. I., Dziewczapolski, G., Nakamura, T., Cao, G., Pratt, A. E., Kang, Y.-J., Tu, S., Molokanova, E., McKercher, S. R., Hires, S. A., Sason, H., Stouffer, D. G., Buczynski, M. W., Solomon, J. P., & Lipton, S. A. (2013). Aβ induces astrocytic glutamate release, extrasynaptic NMDA receptor activation and synaptic loss. *Proceedings of the National Academy of Sciences of the United States of America*, *110*(27), E2518–E2527. doi:10.1073/pnas.1306832110 PMID:23776240

Talifu, Z., Liu, J.-Y., Pan, Y.-Z., Ke, H., Zhang, C.-J., Xu, X., Gao, F., Yu, Y., Du, L.-J., & Li, J.-J. (2023). In vivo astrocyte-to-neuron reprogramming for central nervous system regeneration: A narrative review. *Neural Regeneration Research*, *18*(4), 750. doi:10.4103/1673-5374.353482 PMID:36204831

Tamagawa, K., Morimatsu, Y., Fujisawa, K., Hara, A., & Taketomi, T. (1985). Neuropathological study and chemico-pathoiogical correlation in sibling cases of Sanfilippo syndrome type B. *Brain & Development*, *7*(6), 599–609. doi:10.1016/S0387-7604(85)80008-5 PMID:3938624

Tan, Z., Liu, Y., Xi, W., Lou, H. F., Zhu, L., Guo, Z., Mei, L., & Duan, S. (2017). Glia-derived ATP inversely regulates excitability of pyramidal and CCK-positive neurons. *Nature Communications, 8*(1), 1–14. doi:10.1038/ncomms13772

Tanaka, M., Shih, P. Y., Gomi, H., Yoshida, T., Nakai, J., Ando, R., Furuichi, T., Mikoshiba, K., Semyanov, A., & Itohara, S. (2013). Astrocytic Ca^{2+} signals are required for the functional integrity of tripartite synapses. *Molecular Brain*, *6*(1), 6. doi:10.1186/1756-6606-6-6 PMID:23356992

Tanaka, T., Kai, S., Matsuyama, T., Adachi, T., Fukuda, K., & Hirota, K. (2013). General anesthetics inhibit LPS-induced IL-1β expression in glial cells. *PLoS One*, *8*(12), e82930. doi:10.1371/journal.pone.0082930 PMID:24349401

Tanaka, T., Murakami, K., Bando, Y., & Yoshida, S. (2013). Minocycline reduces remyelination by suppressing ciliary neurotrophic factor expression after cuprizone-induced demyelination. *Journal of Neurochemistry*, *127*(2), 259–270. doi:10.1111/jnc.12289 PMID:23647102

Tan, C. X., Lane, C. J. B., & Eroglu, C. (2021). Role of astrocytes in synapse formation and maturation. *Current Topics in Developmental Biology*, *142*, 371–407. doi:10.1016/bs.ctdb.2020.12.010 PMID:33706922

Tang, Y., Li, T., Li, J., Yang, J., Liu, H., Zhang, X. J., & Le, W. (2013). Jmjd3 is essential for the epigenetic modulation of microglia phenotypes in the immune pathogenesis of Parkinson's disease. *Cell Death & Differentiation, 21*(3), 369–380. doi:10.1038/cdd.2013.159

Tanga, F. Y., Raghavendra, V., & DeLeo, J. A. (2004). Quantitative real-time RT-PCR assessment of spinal microglial and astrocytic activation markers in a rat model of neuropathic pain. *Neurochemistry International*, *45*(2–3), 397–407. doi:10.1016/j.neuint.2003.06.002 PMID:15145554

Tang, L., Ma, Y., Liu, X. L., Chen, L., & Fan, D. S. (2019). Better survival in female SOD1-mutant patients with ALS: A study of SOD1-related natural history. *Translational Neurodegeneration*, *8*(1), 2. doi:10.118640035-018-0142-8 PMID:30637102

Tang, Yu., & Le, W. (2016). 'Differential Roles of M1 and M2 Microglia in Neurodegenerative Diseases'. *Molecular Neurobiology*, *53*(2), 1181–1194. doi:10.100712035-014-9070-5 PMID:25598354

Tan, J., Town, T., Crawford, F., Mori, T., DelleDonne, A., Crescentini, R., Obregon, D., Flavell, R. A., & Mullan, M. J. (2002). Role of CD40 ligand in amyloidosis in transgenic Alzheimer's mice. *Nature Neuroscience*, *5*(12), 1288–1293. doi:10.1038/nn968 PMID:12402041

Tanner, C. M., Goldman, S. M., Ross, G. W., & Grate, S. J. (2014). The disease intersection of susceptibility and exposure: Chemical exposures and neurodegenerative disease risk. *Alzheimer's & Dementia*, *10*(3S), S213–S225. doi:10.1016/j.jalz.2014.04.014 PMID:24924672

Tan, S., Wang, Y., Chen, K., Long, Z., & Zou, J. (2017). Ketamine alleviates depressive-like behaviors via down-regulating inflammatory cytokines induced by chronic restraint stress in mice. *Biological & Pharmaceutical Bulletin*, *40*(8), 1260–1267. doi:10.1248/bpb.b17-00131 PMID:28769008

Tansey, M., & McCoy, M. (n.d.). *Neuroinflammatory mechanisms in Parkinson's disease: potential environmental triggers, pathways, and targets for early therapeutic intervention.* Elsevier. https://www.sciencedirect.com/science/article/pii/S0014488607002701

Tan, Y., Delvaux, E., Nolz, J., Coleman, P. D., Chen, S., & Mastroeni, D. (2018). Upregulation of histone deacetylase 2 in laser capture nigral microglia in Parkinson's disease. *Neurobiology of Aging*, *68*, 134–141. doi:10.1016/j.neurobiolaging.2018.02.018 PMID:29803514

Tanzi, R. E., & Bertram, L. (2005). Twenty years of the Alzheimer's disease amyloid hypothesis: A genetic perspective. *Cell*, *120*(4), 545–555. doi:10.1016/j.cell.2005.02.008 PMID:15734686

Tarkowski, E., Blennow, K., Wallin, A., & Tarkowski, A. (1999). Intracerebral production of tumor necrosis factor-α, a local neuroprotective agent, in Alzheimer disease and vascular dementia. *Journal of Clinical Immunology*, *19*(4), 223–230. doi:10.1023/A:1020568013953 PMID:10471976

Taylor, R. A., & Sansing, L. H. (2013). Microglial responses after ischemic stroke and intracerebral hemorrhage. *Clinical & Developmental Immunology*, *746068*, 1–10. doi:10.1155/2013/746068 PMID:24223607

Tay, T. L., Savage, J. C., Chin, W. H., Bisht, K., & Tremblay, M.-È. (2017). 'Microglia across the Lifespan: From Origin to Function in Brain Development, Plasticity and Cognition: Microglia across the Lifespan'. *The Journal of Physiology*, *595*(6), 1929–1945. doi:10.1113/JP272134 PMID:27104646

Teravskis, P., Covelo, A., Miller, E., Singh, B., Martell-Martínez, H., Benneyworth, M., Gallardo, C., Oxnard, B., Araque, A., Lee, M., & Liao, D. (2018). A53T Mutant Alpha-Synuclein Induces Tau-Dependent Postsynaptic Impairment Independently of Neurodegenerative Changes. *The Journal of Neuroscience : The Official Journal of the Society for Neuroscience*, *38*(45), 9754–9767. doi:10.1523/JNEUROSCI.0344-18.2018 PMID:30249789

Terrillion, C. E., Abazyan, B., Yang, Z., Crawford, J., Shevelkin, A. V., Jouroukhin, Y., Yoo, K. H., Cho, C. H., Roychaudhuri, R., Snyder, S. H., Jang, M. H., & Pletnikov, M. V. (2017). DISC1 in Astrocytes Influences Adult Neurogenesis and Hippocampus-Dependent Behaviors in Mice. *Neuropsychopharmacology*, *42*(11), 2242–2251. doi:10.1038/npp.2017.129

The NINDS NET-PD Investigators. (2006). A randomized, double-blind, futility clinical trial of creatine and minocycline in early Parkinson disease. *Neurology*. PMID:16481597

Thomas, W. E. (1992). Brain macrophages: Evaluation of microglia and their functions. *Brain Research. Brain Research Reviews*, *17*(1), 61–74. doi:10.1016/0165-0173(92)90007-9 PMID:1638276

Thorne, N., Malik, N., Shah, S., Zhao, J., Class, B., Aguisanda, F., Southall, N., Xia, M., McKew, J. C., Rao, M., & Zheng, W. (2016). High-Throughput Phenotypic Screening of Human Astrocytes to Identify Compounds That Protect Against Oxidative Stress. *Stem Cells Translational Medicine*, *5*(5), 613–627. doi:10.5966ctm.2015-0170 PMID:27034412

Thrane, A. S., Thrane, V. R., & Nedergaard, M. (2014). Drowning Stars: Reassessing the Role of Astrocytes in Brain Edema. *Trends in Neurosciences*, *37*(11), 620–628. doi:10.1016/j.tins.2014.08.010 PMID:25236348

Tiane, A., Schepers, M., Rombaut, B., Hupperts, R., Prickaerts, J., Hellings, N., van den Hove, D., & Vanmierlo, T. (2019). From OPC to oligodendrocyte: An epigenetic journey. *Cells*, *8*(10), 1236. doi:10.3390/cells8101236

Tian, L., Ma, L., Kaarela, T., & Li, Z. (2012). Neuroimmune crosstalk in the central nervous system and its significance for neurological diseases. *Journal of Neuroinflammation*, *9*(1), 1–10. doi:10.1186/1742-2094-9-155 PMID:22747919

Tian, Y., Tian, Y., Yuan, Z., Zeng, Y., Wang, S., Fan, X., Yang, D., & Yang, M. (2022). Iron Metabolism in Aging and Age-Related Diseases. *International Journal of Molecular Sciences*, *23*(7), 3612. doi:10.3390/ijms23073612 PMID:35408967

Timmler, S., & Simons, M. (2019). Grey matter myelination. *Glia*, *67*(11), 2063–2070. doi:10.1002/glia.23614 PMID:30860619

Tini, G., Scagliola, R., Monacelli, F., La Malfa, G., Porto, I., Brunelli, C., & Rosa, G. M. (2020). Alzheimer's disease and cardiovascular disease: A particular association. *Cardiology Research and Practice, 2020*, 2020. doi:10.1155/2020/2617970 PMID:32454996

Todd, B. P., Chimenti, M. S., Luo, Z., Ferguson, P. J., Bassuk, A. G., & Newell, E. A. (2021). Traumatic brain injury results in unique microglial and astrocyte transcriptomes enriched for type I interferon response. *Journal of Neuroinflammation, 18*(1), 151. doi:10.118612974-021-02197-w PMID:34225752

Tolaimate, A., Desbrieres, J., Rhazi, M., & Alagui, A. (2003). Contribution to the preparation of chitins and chitosans with controlled physico-chemical properties. *Polymer, 44*(26), 7939–7952. doi:10.1016/j.polymer.2003.10.025

Toledano-Zaragoza, A., & Ledesma, M. D. (2020). Addressing neurodegeneration in lysosomal storage disorders: Advances in Niemann Pick diseases. *Neuropharmacology, 171*, 107851. doi:10.1016/j.neuropharm.2019.107851 PMID:31734384

Toledo, E. M., & Inestrosa, N. C. (2010). Activation of Wnt signaling by lithium and rosiglitazone reduced spatial memory impairment and neurodegeneration in brains of an APPswe/PSEN1ΔE9 mouse model of Alzheimer's disease. *Molecular Psychiatry, 15*(3), 272–285. doi:10.1038/mp.2009.72 PMID:19621015

Tong, H., Zhang, X., Meng, X., Xu, P., Zou, X., & Qu, S. (2017). Amyloid-beta peptide decreases expression and function of glutamate transporters in nervous system cells. *The International Journal of Biochemistry & Cell Biology, 85*, 75–84. doi:10.1016/j.biocel.2017.01.017 PMID:28189841

Tong, X., Ao, Y., Faas, G. C., Nwaobi, S. E., Xu, J., Haustein, M. D., Anderson, M. A., Mody, I., Olsen, M. L., Sofroniew, M. V., & Khakh, B. S. (2014). Astrocyte Kir4.1 ion channel deficits contribute to neuronal dysfunction in Huntington's disease model mice. *Nature Neuroscience, 17*(5), 694–703. doi:10.1038/nn.3691 PMID:24686787

Torres-Ceja, B., & Olsen, M. L. (2022). A closer look at astrocyte morphology: Development, heterogeneity, and plasticity at astrocyte leaflets. *Current Opinion in Neurobiology, 74*, 102550. doi:10.1016/j.conb.2022.102550 PMID:35544965

Torres-Platas, S. G., Cruceanu, C., Chen, G. G., Turecki, G., & Mechawar, N. (2014). Evidence for increased microglial priming and macrophage recruitment in the dorsal anterior cingulate white matter of depressed suicides. *Brain, Behavior, and Immunity, 42*, 50–59. doi:10.1016/j.bbi.2014.05.007 PMID:24858659

Torres-Platas, S. G., Hercher, C., Davoli, M. A., Maussion, G., Labonté, B., Turecki, G., & Mechawar, N. (2011). Astrocytic hypertrophy in anterior cingulate white matter of depressed suicides. *Neuropsychopharmacology, 36*(13), 2650–2658. doi:10.1038/npp.2011.154 PMID:21814185

Torres-Platas, S. G., Nagy, C., Wakid, M., Turecki, G., & Mechawar, N. (2016). Glial fibrillary acidic protein is differentially expressed across cortical and subcortical regions in healthy brains and downregulated in the thalamus and caudate nucleus of depressed suicides. *Molecular Psychiatry, 21*(4), 509–515. doi:10.1038/mp.2015.65 PMID:26033239

Toth, E., Gersner, R., Wilf-Yarkoni, A., Raizel, H., Dar, D. E., Richter-Levin, G., Levit, O., & Zangen, A. (2008). 'Age-Dependent Effects of Chronic Stress on Brain Plasticity and Depressive Behavior: Coping with Chronic Stress'. *Journal of Neurochemistry, 107*(2), 522–532. doi:10.1111/j.1471-4159.2008.05642.x PMID:18752645

Trang, T., Beggs, S., Wan, X., & Salter, M. W. (2009). P2X4-receptor-mediated synthesis and release of brain-derived neurotrophic factor in microglia is dependent on calcium and p38-mitogen-activated protein kinase activation. *The Journal of Neuroscience : The Official Journal of the Society for Neuroscience, 29*(11), 3518–3528. doi:10.1523/JNEUROSCI.5714-08.2009 PMID:19295157

Trapp, B. D., & Nave, K.-A. (2008). Multiple Sclerosis: An Immune or Neurodegenerative Disorder? *Annual Review of Neuroscience, 31*(1), 247–269. doi:10.1146/annurev.neuro.30.051606.094313 PMID:18558855

Trapp, B. D., Peterson, J., Ransohoff, R. M., Rudick, R., Mörk, S., & Bö, L. (1998). Axonal transection in the lesions of multiple sclerosis. *The New England Journal of Medicine*, *338*(5), 278–285. doi:10.1056/NEJM199801293380502 PMID:9445407

Travers, B. S., Tsang, B. K., & Barton, J. L. (2022). Multiple sclerosis: Diagnosis, disease-modifying therapy and prognosis. *Australian Journal of General Practice*, *51*(4), 199–206. doi:10.31128/AJGP-07-21-6103

Tremblay, M.-È., Lowery, R. L., & Majewska, A. K. (2010). Microglial interactions with synapses are modulated by visual experience. *PLoS Biology*, *8*(11), e1000527. doi:10.1371/journal.pbio.1000527 PMID:21072242

Trendelenburg, G., & Dirnagl, U. (2005). Neuroprotective role of astrocytes in cerebral ischemia: Focus on ischemic preconditioning. *Glia*, *50*(4), 307–320. doi:10.1002/glia.20204 PMID:15846804

Tricoire, L., Pelkey, K. A., Erkkila, B. E., Jeffries, B. W., Yuan, X., & McBain, C. J. (2011). A blueprint for the spatiotemporal origins of mouse hippocampal interneuron diversity. *The Journal of Neuroscience : The Official Journal of the Society for Neuroscience*, *31*(30), 10948–10970. doi:10.1523/JNEUROSCI.0323-11.2011 PMID:21795545

Tronel, C., Largeau, B., Santiago Ribeiro, M. J., Guilloteau, D., Dupont, A. C., & Arlicot, N. (2017). Molecular Targets for PET Imaging of Activated Microglia: The Current Situation and Future Expectations. *International Journal of Molecular Sciences*, *18*(4), 802. doi:10.3390/ijms18040802 PMID:28398245

Trullas, R., & Skolnick, P. (1990). Functional antagonists at the NMDA receptor complex exhibit antidepressant actions. *European Journal of Pharmacology*, *185*(1), 1–10. doi:10.1016/0014-2999(90)90204-J PMID:2171955

Tryc, A. B., Goldbecker, A., Berding, G., Rümke, S., Afshar, K., Shahrezaei, G. H., Pflugrad, H., Barg-Hock, H., Strassburg, C. P., Hecker, H., & Weissenborn, K. (2013). Cirrhosis-related Parkinsonism: Prevalence, mechanisms and response to treatments. *Journal of Hepatology*, *58*(4), 698–705. doi:10.1016/j.jhep.2012.11.043 PMID:23220368

Tseng, P.-T., Lee, Y., & Lin, P.-Y. (2013). Age-associated decrease in serum glial cell line-derived neurotrophic factor levels in patients with major depressive disorder. *Progress in Neuro-Psychopharmacology & Biological Psychiatry*, *40*, 334–339. doi:10.1016/j.pnpbp.2012.09.009 PMID:23022671

Tsuda, M., Kohro, Y., Yano, T., Tsujikawa, T., Kitano, J., Tozaki-Saitoh, H., Koyanagi, S., Ohdo, S., Ji, R.-R., Salter, M. W., & Inoue, K. (2011). JAK-STAT3 pathway regulates spinal astrocyte proliferation and neuropathic pain maintenance in rats. *Brain*, *134*(Pt 4), 1127–1139. doi:10.1093/brain/awr025 PMID:21371995

Tsuda, M., Mizokoshi, A., Shigemoto-Mogami, Y., Koizumi, S., & Inoue, K. (2004). Activation of p38 mitogen-activated protein kinase in spinal hyperactive microglia contributes to pain hypersensitivity following peripheral nerve injury. *Glia*, *45*(1), 89–95. doi:10.1002/glia.10308 PMID:14648549

Tsuda, M., Shigemoto-Mogami, Y., Koizumi, S., Mizokoshi, A., Kohsaka, S., Salter, M. W., & Inoue, K. (2003). P2X4 receptors induced in spinal microglia gate tactile allodynia after nerve injury. *Nature*, *424*(6950), 778–783. doi:10.1038/nature01786 PMID:12917686

Tsui, C., Koss, K., Churchward, M. A., & Todd, K. G. (2019). Biomaterials and glia: Progress on designs to modulate neuroinflammation. *Acta Biomaterialia*, *83*, 13–28. doi:10.1016/j.actbio.2018.11.008 PMID:30414483

Tsukada, S., Parsons, C. J., & Rippe, R. A. (2006). *Mechanisms of liver fibrosis B*. Elsevier. doi:10.1016/j.cca.2005.06.014

Tucci, P., & Bagetta, G. (2008). How to study neuroprotection? *Cell Death and Differentiation*, *15*(6), 1084–1085. doi:10.1038/cdd.2008.32

Turan, K., & Nagata, K. (2006). Chitosan-DNA nanoparticles: The effect of cell type and hydrolysis of chitosan on in vitro DNA transfection. *Pharmaceutical Development and Technology, 11*(4), 503–512. doi:10.1080/10837450600940873 PMID:17101521

Turner, B. J., Ackerley, S., Davies, K. E., & Talbot, K. (2010). Dismutase-competent SOD1 mutant accumulation in myelinating Schwann cells is not detrimental to normal or transgenic ALS model mice. *Human Molecular Genetics, 19*(5), 815–824. doi:10.1093/hmg/ddp550 PMID:20008901

Turner, D. A., & Adamson, D. C. (2011). Neuronal-Astrocyte Metabolic Interactions : Understanding the Transition into Abnormal Astrocytoma Metabolism. *Journal of Neuropathology and Experimental Neurology, 70*(3), 167–176. doi:10.1097/NEN.0b013e31820e1152 PMID:21293295

Turner, M. R., Cagnin, A., Turkheimer, F. E., Miller, C. C., Shaw, C. E., Brooks, D. J., Leigh, P. N., & Banati, R. B. (2004). Evidence of widespread cerebral microglial activation in amyotrophic lateral sclerosis: An [11C](R)-PK11195 positron emission tomography study. *Neurobiology of Disease, 15*(3), 601–609. doi:10.1016/j.nbd.2003.12.012 PMID:15056468

Turrigiano, G. (2011). 'Too Many Cooks? Intrinsic and Synaptic Homeostatic Mechanisms in Cortical Circuit Refinement'. *Annual Review of Neuroscience, 34*(1), 89–103. doi:10.1146/annurev-neuro-060909-153238 PMID:21438687

Turrigiano, G. G. (2008). The self-tuning neuron: Synaptic scaling of excitatory synapses. *Cell, 135*(3), 422–435. doi:10.1016/j.cell.2008.10.008 PMID:18984155

Tvrdik, P., & Kalani, M. Y. S. (2017). In Vivo Imaging of Microglial Calcium Signaling in Brain Inflammation and Injury. *International Journal of Molecular Sciences, 18*(11), 2366. doi:10.3390/ijms18112366 PMID:29117112

Twohig, D., & Nielsen, H. M. (2019). α-synuclein in the pathophysiology of Alzheimer's disease. *Molecular Neurodegeneration, 14*(1), 1–19. doi:10.118613024-019-0320-x PMID:31186026

Twomey, C., & McCarthy, J. V. (2005). Pathways of apoptosis and importance in developement. *Journal of Cellular and Molecular Medicine, 9*(2), 345–359. doi:10.1111/j.1582-4934.2005.tb00360.x PMID:15963254

Ucar, B., & Humpel, C. (2018). Collagen for brain repair: Therapeutic perspectives. *Neural Regeneration Research, 13*(4), 595. doi:10.4103/1673-5374.230273 PMID:29722301

Uddin, M. S., & Lim, L. W. (2022). Glial cells in Alzheimer's disease: From neuropathological changes to therapeutic implications. *Ageing Research Reviews, 78*, 101622. doi:10.1016/j.arr.2022.101622 PMID:35427810

Ueno, M., Fujita, Y., Tanaka, T., Nakamura, Y., Kikuta, J., Ishii, M., & Yamashita, T. (2013). Layer V cortical neurons require microglial support for survival during postnatal development. *Nature Neuroscience, 16*(5), 543–551. doi:10.1038/nn.3358

Ulland, T. K., Song, W. M., Huang, S. C.-C., Ulrich, J. D., Sergushichev, A., Beatty, W. L., Loboda, A. A., Zhou, Y., Cairns, N. J., Kambal, A., Loginicheva, E., Gilfillan, S., Cella, M., Virgin, H. W., Unanue, E. R., Wang, Y., Artyomov, M. N., Holtzman, D. M., & Colonna, M. (2017). TREM2 Maintains Microglial Metabolic Fitness in Alzheimer's Disease. *Cell, 170*(4), 649–663.e13. doi:10.1016/j.cell.2017.07.023 PMID:28802038

Ullian, E. M., Sapperstein, S. K., Christopherson, K. S., & Barres, B. A. (2001). Control of synapse number by glia. *science, 291*(5504), 657-661.

Ullian, E.M., Sapperstein, S.K., Christopherson, K.S., Barres, B.A. (2001). Control of synapse number by glia. *Science, 291*(5504), 657–61. doi:10.1126/science.291.5504.657

Ullian, E. M., Christopherson, K. S., & Barres, B. A. (2004). Role for glia in synaptogenesis. *Glia, 47*(3), 209–216. doi:10.1002/glia.20082 PMID:15252809

Ulrich, J., Ulland, T., Colonna, M., & Neuron, D. H. (2017). Elucidating the role of TREM2 in Alzheimer's disease. *Cell*. doi:10.1016/j.neuron.2017.02.042

Umoh, M. E., Dammer, E. B., Dai, J., Duong, D. M., Lah, J. J., Levey, A. I., Gearing, M., Glass, J. D., & Seyfried, N. T. (2018). A proteomic network approach across the ALS - FTD disease spectrum resolves clinical phenotypes and genetic vulnerability in human brain. *EMBO Molecular Medicine*, *10*(1), 48–62. doi:10.15252/emmm.201708202 PMID:29191947

Umpierre, A. D., West, P. J., White, J. A., & Wilcox, K. S. (2019). Conditional Knock-out of mGluR5 from Astrocytes during Epilepsy Development Impairs High-Frequency Glutamate Uptake. *The Journal of Neuroscience : The Official Journal of the Society for Neuroscience*, *39*(4), 727–742. doi:10.1523/JNEUROSCI.1148-18.2018 PMID:30504280

Uranova, N., Orlovskaya, D., Vikhreva, O., Zimina, I., Kolomeets, N., Vostrikov, V., & Rachmanova, V. (2001). Electron microscopy of oligodendroglia in severe mental illness. *Brain Research Bulletin*, *55*(5), 597–610. doi:10.1016/S0361-9230(01)00528-7 PMID:11576756

Urits, I., Adamian, L., Fiocchi, J., Hoyt, D., Ernst, C., Kaye, A. D., & Viswanath, O. (2019). Advances in the Understanding and Management of Chronic Pain in Multiple Sclerosis: A Comprehensive Review. *Current Pain and Headache Reports*, *23*(8), 59. doi:10.100711916-019-0800-2 PMID:31342191

Urushitani, M., Sik, A., Sakurai, T., Nukina, N., Takahashi, R., & Julien, J. P. (2006). Chromogranin-mediated secretion of mutant superoxide dismutase proteins linked to amyotrophic lateral sclerosis. *Nature Neuroscience*, *9*(1), 108–118. doi:10.1038/nn1603 PMID:16369483

Uzdensky, A., Komandirov, M., Fedorenko, G., & Lobanov, A. (2013). Protection effect of GDNF and neurturin on photosensitized crayfish neurons and glial cells. *Journal of Molecular Neuroscience*, *49*(3), 480–490. doi:10.100712031-012-9858-6 PMID:22847529

Vainchtein, I. D., Chin, G., Cho, F. S., Kelley, K. W., Miller, J. G., Chien, E. C., Liddelow, S. A., Nguyen, P. T., Nakao-Inoue, H., Dorman, L. C., Akil, O., Joshita, S., Barres, B. A., Paz, J. T., Molofsky, A. B., & Molofsky, A. V. (2018). Astrocyte-derived interleukin-33 promotes microglial synapse engulfment and neural circuit development. *Science*, *359*(6381), 1269–1273. doi:10.1126cience.aal3589 PMID:29420261

Vainchtein, I. D., & Molofsky, A. V. (2020). Astrocytes and microglia: In sickness and in health. *Trends in Neurosciences*, *43*(3), 144–154. doi:10.1016/j.tins.2020.01.003 PMID:32044129

Valente, M. F., Romariz, S., Calcagnotto, M. E., Ruiz, L., Mello, L. E., Frussa-Filho, R., & Longo, B. M. (2013). Postnatal transplantation of interneuronal precursor cells decreases anxiety-like behavior in adult mice. *Cell Transplantation*, *22*(7), 1237–1247. doi:10.3727/096368912X657422 PMID:23031356

Valstar, M. J., Ruijter, G. J. G., Van Diggelen, O. P., Poorthuis, B. J., & Wijburg, F. A. (2008). Sanfilippo syndrome: A mini-review. *Journal of Inherited Metabolic Disease*, *31*(2), 240–252. doi:10.100710545-008-0838-5 PMID:18392742

van Deijk, A. L. F., Camargo, N., Timmerman, J., Heistek, T., Brouwers, J. F., Mogavero, F., Mansvelder, H. D., Smit, A. B., & Verheijen, M. H. G. (2017). Astrocyte lipid metabolism is critical for synapse development and function in vivo. *Glia*, *65*(4), 670–682. doi:10.1002/glia.23120 PMID:28168742

van Horssen, J., van Schaik, P., & Witte, M. (2019). Inflammation and mitochondrial dysfunction: A vicious circle in neurodegenerative disorders? *Neuroscience Letters*, *710*, 132931. doi:10.1016/j.neulet.2017.06.050

Van Vlierberghe, S., Dubruel, P., & Schacht, E. (2011). Hydrogels à base de biopolymères comme échafaudages pour les applications d'ingénierie tissulaire: Une revue. *Biomacromolecules*, *12*(5), 1387–1408. doi:10.1021/bm200083n PMID:21388145 van Vliet, E. A., Ndode-Ekane, X. E., Lehto, L. J., Gorter, J. A., Andrade, P., Aronica, E., Gröhn, O., & Pitkänen, A. (2020). Long-lasting blood-brain barrier dysfunction and neuroinflammation after traumatic brain injury. *Neurobiology of Disease*, *145*, 105080. doi:10.1016/j.nbd.2020.105080 PMID:32919030

VanGuilder, H. D., Yan, H., Farley, J. A., Sonntag, W. E., & Freeman, W. M. (2010). 'Aging Alters the Expression of Neurotransmission-Regulating Proteins in the Hippocampal Synaptoproteome'. *Journal of Neurochemistry*, *113*(6), 1577–1588. doi:10.1111/j.1471-4159.2010.06719.x PMID:20374424

Vanier, M. T. (2010). Niemann-Pick disease type C. *Orphanet Journal of Rare Diseases*, *5*(1), 1–18. doi:10.1186/1750-1172-5-16 PMID:20525256

Vaquero, J., Chung, C., & Blei, A. T. (2003). Brain Edema in Acute Liver Failure. A Window to the Pathogenesis of Hepatic Encephalopathy. *Annals of Hepatology*, *2*(1), 12–22. doi:10.1016/S1665-2681(19)32153-2 PMID:15094701

Vargas, M. R., Johnson, D. A., Sirkis, D. W., Messing, A., & Johnson, J. A. (2008). Nrf2 activation in astrocytes protects against neurodegeneration in mouse models of familial amyotrophic lateral sclerosis. *The Journal of Neuroscience : The Official Journal of the Society for Neuroscience*, *28*(50), 13574–13581. doi:10.1523/JNEUROSCI.4099-08.2008 PMID:19074031

Vasile, F., Dossi, E., & Rouach, N. (2017). Human astrocytes: Structure and functions in the healthy brain. *Brain Structure & Function*, *222*(5), 2017–2029. doi:10.100700429-017-1383-5 PMID:28280934

Vaz, M., & Silvestre, S. (2020). Alzheimer's disease : Recent treatment strategies. *European Journal of Pharmacology*, *887*, 173554. doi:10.1016/j.ejphar.2020.173554 PMID:32941929

Vedrenne, N. (2013). *Roles of different astrocyte populations in repair mechanisms after central nervous system injury*[Dissertation, Limoges].

Venegas, C., & Heneka, M. T. (2019). Inflammasome-mediated innate immunity in Alzheimer's disease. *The FASEB Journal*, *33*(12), 13075–13084. doi:10.1096/fj.201900439 PMID:31702392

Ventura, R., & Harris, K. M. (1999). Three-dimensional relationships between hippocampal synapses and astrocytes. *The Journal of Neuroscience : The Official Journal of the Society for Neuroscience*, *19*(16), 6897–6906. doi:10.1523/JNEUROSCI.19-16-06897.1999 PMID:10436047

Venturini, A., Passalacqua, M., Pelassa, S., Pastorino, F., Tedesco, M., Cortese, K., Gagliani, M. C., Leo, G., Maura, G., Guidolin, D., Agnati, L. F., Marcoli, M., & Cervetto, C. (2019). Exosomes From Astrocyte Processes: Signaling to Neurons. *Frontiers in Pharmacology*, *10*, 1452. doi:10.3389/fphar.2019.01452 PMID:31849688

Veremeyko, T., Yung, A. W. Y., Dukhinova, M., Strekalova, T., & Ponomarev, E. D. (2019). The Role of Neuronal Factors in the Epigenetic Reprogramming of Microglia in the Normal and Diseased Central Nervous System. *Frontiers in Cellular Neuroscience*, *13*, 453. doi:10.3389/fncel.2019.00453 PMID:31680868

Verkhratsky, A., Parpura, V., Rodriguez-Arellano, J. J., & Zorec, R. (2019). Astroglia in Alzheimer's Disease. In Neuroglia in Neurodegenerative Diseases. Singapore: Springer. doi:10.1007/978-981-13-9913-8_11

Verkhratsky, A., Matteoli, M., Parpura, V., Mothet, J. P., & Zorec, R. (2016). Astrocytes as secretory cells of the central nervous system: Idiosyncrasies of vesicular secretion. *The EMBO Journal*, *35*(3), 239–257. doi:10.15252/embj.201592705 PMID:26758544

Verkhratsky, A., & Nedergaard, M. (2018). 'Physiology of Astroglia'. *Physiological Reviews*, *98*(1), 239–389. doi:10.1152/physrev.00042.2016 PMID:29351512

Verkhratsky, A., Untiet, V., & Rose, C. R. (2020). Ionic signalling in astroglia beyond calcium. *The Journal of Physiology, 598*(9), 1655–1670. doi:10.1113/JP277478 PMID:30734296

Viader, F. (2023, March). La sclérose latérale amyotrophique: une maladie neurodégénérative emblématiqueAmyotrophic lateral sclerosis: An emblematic neurodegenerative disease. *Bulletin de l'Académie Nationale de Médecine, 207*(3), 272–286. doi:10.1016/j.banm.2023.01.002

Vidal-Itriago, A., Radford, R. A. W., Aramideh, J. A., Maurel, C., Scherer, N. M., Don, E. K., Lee, A., Chung, R. S., Graeber, M. B., & Morsch, M. (2022). Microglia Morphophysiological Diversity and Its Implications for the CNS. *Frontiers in Immunology, 13*(October), 997786. doi:10.3389/fimmu.2022.997786 PMID:36341385

Vila, M., & Przedborski, S. (2003). Targeting programmed cell death in neurodegenerative diseases. *Nature Reviews. Neuroscience, 4*(5), 365–375. doi:10.1038/nrn1100 PMID:12728264

Villalba, R. M., & Smith, Y. (2011). Neuroglial plasticity at striatal glutamatergic synapses in Parkinson's disease. *Frontiers in Systems Neuroscience, 5*, 68. doi:10.3389/fnsys.2011.00068 PMID:21897810

Villemagne, V. L., Dore, V., Bourgeat, P., Burnham, S. C., Laws, S., Salvado, O., Masters, C. L., & Rowe, C. C. (2017). Aβ-amyloid and tau imaging in dementia. In *Seminars in Nuclear Medicine* (pp. 75–88). Elsevier. doi:10.1053/j.semnuclmed.2016.09.006

Villemagne, V. L., & Rowe, C. C. (2011). Amyloid imaging. *International Psychogeriatrics, 23*(S2), S41–S49. doi:10.1017/S1041610211000895 PMID:21729418

Vilstrup, H., Amodio, P., Bajaj, J., Cordoba, J., Ferenci, P., Mullen, K. D., Weissenborn, K., & Wong, P. (2014). Hepatic encephalopathy in chronic liver disease: 2014 Practice Guideline by the American Association for the Study of Liver Diseases and the European Association for the Study of the Liver. *Hepatology (Baltimore, Md.), 60*(2), 715–735. doi:10.1002/hep.27210 PMID:25042402

Vincenzi, F., Pasquini, S., Contri, C., Cappello, M., Nigro, M., Travagli, A., Merighi, S., Gessi, S., Borea, P. A., & Varani, K. (2023). Pharmacology of Adenosine Receptors: Recent Advancements. *Biomolecules, 13*(9), 1387. doi:10.3390/biom13091387 PMID:37759787

Vinet, J., van Weering, H. R., Heinrich, A., Kälin, R. E., Wegner, A., Brouwer, N., Heppner, F. L., van Rooijen, N., & Boddeke, W. G. M. H., & Biber, K. (2012). Neuroprotective function for ramified microglia in hippocampal excitotoxicity. *Journal of Neuroinflammation.* http://www.jneuroinflammation.com/content/9/1/27

Virchow, R. L. K. (1858). *Die CELLULARPATHOLOGIE in ihrer Begründung auf physiologische und pathologische Gewebelehre* [Cellular Pathology As Based Upon Physiological and Pathological Histology]. Hirschwald.

Volterra, A., Liaudet, N., & Savtchouk, I. (2014). Astrocyte Ca2+ signalling: an unexpected complexity. *Nature Reviews Neuroscience, 15*(5), 327–335. doi:10.1038/nrn3725

Von Bartheld, C. S., Bahney, J., & Herculano-Houzel, S. (2016). The search for true numbers of neurons and glial cells in the human brain : A review of 150 years of cell counting. *The Journal of Comparative Neurology, 524*(18), 3865–3895. doi:10.1002/cne.24040 PMID:27187682

von Bernhardi, R., Eugenín-von Bernhardi, L., & Eugenín, J. (2015). Microglial Cell Dysregulation in Brain Aging and Neurodegeneration. *Frontiers in Aging Neuroscience, 7*(July), 124. doi:10.3389/fnagi.2015.00124 PMID:26257642

von Figura, K., & Kresse, H. (1972). The Sanfilippo B corrective factor: A N-acetyl-α-D-glucosaminidase. *Biochemical and Biophysical Research Communications, 48*(2), 262–269. doi:10.1016/S0006-291X(72)80044-5 PMID:4261365

Vousden, K. H., & Lane, D. P. (2007). p53 in health and disease. *Nature Reviews. Molecular Cell Biology*, *8*(4), 275–283. doi:10.1038/nrm2147 PMID:17380161

Wagner, K., Unger, L., Salman, M. M., Kitchen, P., Bill, R. M., & Yool, A. J. (2022). Signaling Mechanisms and Pharmacological Modulators Governing Diverse Aquaporin Functions in Human Health and Disease. *International Journal of Molecular Sciences*, *23*(3), 1388. doi:10.3390/ijms23031388 PMID:35163313

Wahis, J., Baudon, A., Althammer, F., Kerspern, D., Goyon, S., Hagiwara, D., Lefevre, A., Barteczko, L., Boury-Jamot, B., Bellanger, B., Abatis, M., Da Silva Gouveia, M., Benusiglio, D., Eliava, M., Rozov, A., Weinsanto, I., Knobloch-Bollmann, H. S., Kirchner, M. K., Roy, R. K., ... Charlet, A. (2021). Astrocytes mediate the effect of oxytocin in the central amygdala on neuronal activity and affective states in rodents. *Nature Neuroscience*, *24*(4), 529–541. doi:10.103841593-021-00800-0 PMID:33589833

Wakabayashi, K., Hayashi, S., Yoshimoto, M., Kudo, H., & Takahash, H. (2000). NACP αsynuclein-positive filamentous inclusions in astrocytes and oligodendrocytes of Parkinson's disease brains. *Acta Neuropathologica*, *99*(1), 14–20. doi:10.1007/PL00007400 PMID:10651022

Wake, H., & Douglas Fields, R. (2011). 'Physiological Function of Microglia'. *Neuron Glia Biology*, *7*(1), 1–3. doi:10.1017/S1740925X12000166 PMID:22857736

Wake, H., Moorhouse, A. J., Jinno, S., Kohsaka, S., & Nabekura, J. (2009). Resting microglia directly monitor the functional state of synapses in vivo and determine the fate of ischemic terminals. *The Journal of Neuroscience : The Official Journal of the Society for Neuroscience*, *29*(13), 3974–3980. doi:10.1523/JNEUROSCI.4363-08.2009 PMID:19339593

Wakeman, D. R., Redmond, D. E. Jr, Dodiya, H. B., Sladek, J. R. Jr, Leranth, C., Teng, Y. D., Samulski, R. J., & Snyder, E. Y. (2014). Human neural stem cells survive long term in the midbrain of dopamine-depleted monkeys after GDNF overexpression and project neurites toward an appropriate target. *Stem Cells Translational Medicine*, *3*(6), 692–701. doi:10.5966ctm.2013-0208 PMID:24744393

Walker, A. K., Budac, D. P., Bisulco, S., Lee, A. W., Smith, R. A., Beenders, B., Kelley, K. W., & Dantzer, R. (2013). NMDA receptor blockade by ketamine abrogates lipopolysaccharide-induced depressive-like behavior in C57BL/6J mice. *Neuropsychopharmacology*, *38*(9), 1609–1616. doi:10.1038/npp.2013.71 PMID:23511700

Wallace, K., Burt, A. D., & Wright, M. C. (2008). Liver fibrosis. *Biochemical Journal*. doi:10.1042/BJ20071570

Wallraff, A., Köhling, R., Heinemann, U., Theis, M., Willecke, K., & Steinhäuser, C. (2006). The impact of astrocytic gap junctional coupling on potassium buffering in the hippocampus. *The Journal of Neuroscience : The Official Journal of the Society for Neuroscience*, *26*(20), 5438–5447. doi:10.1523/JNEUROSCI.0037-06.2006 PMID:16707796

Walrave, L., Vinken, M., Albertini, G., De Bundel, D., Leybaert, L., & Smolders, I. J. (2016). Inhibition of Connexin43 Hemichannels Impairs Spatial Short-Term Memory without Affecting Spatial Working Memory. *Frontiers in Cellular Neuroscience*, *10*, 288. doi:10.3389/fncel.2016.00288 PMID:28066184

Walsh, D. M., & Selkoe, D. J. (2020). Amyloid β-protein and beyond: The path forward in Alzheimer's disease. *Current Opinion in Neurobiology*, *61*, 116–124. doi:10.1016/j.conb.2020.02.003 PMID:32197217

Walterfang, M., Fahey, M., Desmond, P., Wood, A., Seal, M. L., Steward, C., Adamson, C., Kokkinos, C., Fietz, M., & Velakoulis, D. (2010). White and gray matter alterations in adults with Niemann-Pick disease type C: A cross-sectional study. *Neurology*, *75*(1), 49–56. doi:10.1212/WNL.0b013e3181e6210e PMID:20484681

Walton, N. M., Sutter, B. M., Laywell, E. D., Levkoff, L. H., Kearns, S. M., Marshall, G. P. II, Scheffler, B., & Steindler, D. A. (2006). Microglia instruct subventricular zone neurogenesis. *Glia*, *54*(8), 815–825. doi:10.1002/glia.20419 PMID:16977605

Wang, Q., Liu, Y., & Zhou, J. (2015). Neuroinflammation in Parkinson's disease and its potential as therapeutic target. In Translational Neurodegeneration, 4(1). BioMed Central Ltd. doi:10.118640035-015-0042-0

Wang, Y., Ulland, T., & Ulrich, J. (2016). TREM2-mediated early microglial response limits diffusion and toxicity of amyloid plaques. *Journal of Experimental Medicine.* https://rupress.org/jem/article-abstract/213/5/667/42126

Wang, B., Wang, L., Qu, Y., Lu, J., & Xia, W. (2022). Chitosan oligosaccharides exert neuroprotective effects via modulating the PI3K/Akt/Bcl-2 pathway in a Parkinsonian model. *Food & Function, 13*(10), 5838–5853. doi:10.1039/D1FO04374A PMID:35545086

Wang, C., Liu, F., Liu, Y.-Y., Zhao, C.-H., You, Y., Wang, L., Zhang, J., Wei, B., Ma, T., Zhang, Q., Zhang, Y., Chen, R., Song, H., & Yang, Z. (2011). Identification and characterization of neuroblasts in the subventricular zone and rostral migratory stream of the adult human brain. *Cell Research, 21*(11), 1534–1550. doi:10.1038/cr.2011.83 PMID:21577236

Wang, C., Yue, H., Hu, Z., Shen, Y., Ma, J., Li, J., Wang, X.-D., Wang, L., Sun, B., Shi, P., Wang, L., & Gu, Y. (2020). Microglia mediate forgetting via complement-dependent synaptic elimination. *Science, 367*(6478), 688–694. doi:10.1126cience.aaz2288 PMID:32029629

Wang, H., He, Y., Sun, Z., Ren, S., Liu, M., Wang, G., & Yang, J. (2022). Microglia in depression: An overview of microglia in the pathogenesis and treatment of depression. *Journal of Neuroinflammation, 19*(1), 132. doi:10.118612974-022-02492-0 PMID:35668399

Wang, J. Q., Gao, M. Y., Gao, R., Zhao, K. H., Zhang, Y., & Li, X. (2023). Oligodendrocyte lineage cells: Advances in development, disease, and heterogeneity. *Journal of Neurochemistry, 164*(4), 468–480. doi:10.1111/jnc.15728

Wang, J., Song, N., Jiang, H., Wang, J., & Xie, J. (2013). Pro-inflammatory cytokines modulate iron regulatory protein 1 expression and iron transportation through reactive oxygen/nitrogen species production in ventral mesencephalic neurons. *Biochimica et Biophysica Acta. Molecular Basis of Disease, 1832*(5), 618–625. doi:10.1016/j.bbadis.2013.01.021 PMID:23376588

Wang, J., Su, H. K., Zhao, H. F., Chen, Z. P., & To, S. S. T. (2015). Progress in the application of molecular biomarkers in gliomas. *Biochemical and Biophysical Research Communications, 465*(1), 1–4. doi:10.1016/j.bbrc.2015.07.148 PMID:26253473

Wang, L., Pytel, P., Feltri, M. L., Wrabetz, L., & Roos, R. P. (2012). Selective knockdown of mutant SOD1 in Schwann cells ameliorates disease in G85R mutant SOD1 transgenic mice. *Neurobiology of Disease, 48*(1), 52–57. doi:10.1016/j.nbd.2012.05.014 PMID:22668777

Wang, L., Sharma, K., Grisotti, G., & Roos, R. P. (2009). The effect of mutant SOD1 dismutase activity on non-cell autonomous degeneration in familial amyotrophic lateral sclerosis. *Neurobiology of Disease, 35*(2), 234–240. doi:10.1016/j.nbd.2009.05.002 PMID:19442735

Wang, W.-Y., Tan, M.-S., Yu, J.-T., & Tan, L. (2015). 'Role of Pro-Inflammatory Cytokines Released from Microglia in Alzheimer's Disease'. *Annals of Translational Medicine, 3*(10), 136. doi:10.3978/j.issn.2305-5839.2015.03.49 PMID:26207229

Wang, W.-Y., Tan, M.-S., Yu, J.-T., & Tan, L. (2015). Role of pro-inflammatory cytokines released from microglia in Alzheimer's disease. *Annals of Translational Medicine, 3*(10). PMID:26207229

Wang, X. J., Zhang, S., Yan, Z. Q., Zhao, Y. X., Zhou, H. Y., Wang, Y., Lu, G. Q., & Zhang, J. D. (2011). Impaired CD200–CD200R-mediated microglia silencing enhances midbrain dopaminergic neurodegeneration: Roles of aging, superoxide, NADPH oxidase, and p38 MAPK. *Free Radical Biology & Medicine, 50*(9), 1094–1106. doi:10.1016/j.freeradbiomed.2011.01.032 PMID:21295135

Wang, X., Miao, J., Yan, C., Ge, R., Liang, T., Liu, E., & Li, Q. (2016). Chitosan attenuates dibutyltin-induced apoptosis in PC12 cells through inhibition of the mitochondria-dependent pathway. *Carbohydrate Polymers, 151*, 996–1005. doi:10.1016/j.carbpol.2016.06.053 PMID:27474647

Wang, Z., Ma, W., Chabot, J.-G., & Quirion, R. (2009). Cell-type specific activation of p38 and ERK mediates calcitonin gene-related peptide involvement in tolerance to morphine-induced analgesia. *The FASEB Journal, 23*(8), 2576–2586. doi:10.1096/fj.08-128348 PMID:19299480

Ware, A. J., D'Agostino, A. N., & Combes, B. (1971). Cerebral Edema: A Major Complication of Massive Hepatic Necrosis. *Gastroenterology, 61*(6), 877–884. doi:10.1016/S0016-5085(19)33402-X PMID:5125688

Warner-Schmidt, J. L., Vanover, K. E., Chen, E. Y., Marshall, J. J., & Greengard, P. (2011). Antidepressant effects of selective serotonin reuptake inhibitors (SSRIs) are attenuated by antiinflammatory drugs in mice and humans. *Proceedings of the National Academy of Sciences of the United States of America, 108*(22), 9262–9267. doi:10.1073/pnas.1104836108 PMID:21518864

Warren, S. L., Reid, E., Whitfield, P., Helal, A. M., Abo Hamza, E. G., Tindle, R., Moustafa, A. A., & Hamid, M. S. (2023). Cognitive and behavioral abnormalities in individuals with Alzheimer's disease, mild cognitive impairment, and subjective memory complaints. *Current Psychology (New Brunswick, N.J.)*, 1–11. doi:10.100712144-023-04281-1

Wassif, C. A., Cross, J. L., Iben, J., Sanchez-Pulido, L., Cougnoux, A., Platt, F. M., Ory, D. S., Ponting, C. P., Bailey-Wilson, J. E., Biesecker, L. G., & Porter, F. D. (2016). High incidence of unrecognized visceral/neurological late-onset Niemann-Pick disease, type C1, predicted by analysis of massively parallel sequencing data sets. *Genetics in Medicine, 18*(1), 41–48. doi:10.1038/gim.2015.25 PMID:25764212

Weber-Hamann, B., Blum, W. F., Kratzsch, J., Gilles, M., Heuser, I., & Deuschle, M. (2009). Insulin-like growth factor-I (IGF-I) serum concentrations in depressed patients: Relationship to saliva cortisol and changes during antidepressant treatment. *Pharmacopsychiatry, 42*(01), 23–28. doi:10.1055-0028-1085442 PMID:19153943

Webster, M. J., Knable, M. B., Johnston-Wilson, N., Nagata, K., Inagaki, M., & Yolken, R. H. (2001). Immunohistochemical localization of phosphorylated glial fibrillary acidic protein in the prefrontal cortex and hippocampus from patients with schizophrenia, bipolar disorder, and depression. *Brain, Behavior, and Immunity, 15*(4), 388–400. doi:10.1006/brbi.2001.0646 PMID:11782105

Wei, F., Guo, W., Zou, S., Ren, K., & Dubner, R. (2008). Supraspinal glial-neuronal interactions contribute to descending pain facilitation. *The Journal of Neuroscience : The Official Journal of the Society for Neuroscience, 28*(42), 10482–10495. doi:10.1523/JNEUROSCI.3593-08.2008 PMID:18923025

Weinhard, L., Di Bartolomei, G., Bolasco, G., Machado, P., Schieber, N. L., Neniskyte, U., Exiga, M., Vadisiute, A., Raggioli, A., Schertel, A., Schwab, Y., & Gross, C. T. (2018). Microglia remodel synapses by presynaptic trogocytosis and spine head filopodia induction. *Nature Communications, 9*(1), 1–14. doi:10.1038/s41467-018-03566-5

Weintraub, D., & Burn, D. J. (2011). Parkinson's disease: The quintessential neuropsychiatric disorder. *Movement Disorders, 26*(6), 1022–1031. doi:10.1002/mds.23664 PMID:21626547

Weissenborn, K. (1998). Diagnosis of encephalopathy. *Digestion, 59*(Suppl. 2), 22–24. doi:10.1159/000051415 PMID:9718414

Weissenborn, K., Bokemeyer, M., Krause, J., Ennen, J., & Ahl, B. (2005). Neurological and neuropsychiatric syndromes associated with liver disease. *AIDS (London, England), 19*(Suppl 3), S93–S98. doi:10.1097/01.aids.0000192076.03443.6d PMID:16251835

Wei, Z.-Y. D., & Shetty, A. K. (2021). Treating Parkinson's disease by astrocyte reprogramming: Progress and challenges. *Science Advances*, *7*(26), eabg3198. doi:10.1126ciadv.abg3198 PMID:34162545

Wendimu, M. Y., & Hooks, S. B. (2022). 'Microglia Phenotypes in Aging and Neurodegenerative Diseases'. *Cells*, *11*(13), 2091. doi:10.3390/cells11132091 PMID:35805174

Wendon, J., & William Lee, Æ. (2008). Encephalopathy and Cerebral Edema in the Setting of Acute Liver Failure : Pathogenesis and Management. *Neurocritical Care*, *9*(1), 97–102. doi:10.100712028-008-9123-6 PMID:18688582

Wenger, D. A., Rafi, M. A., Luzi, P., Datto, J., & Costantino-Ceccarini, E. (2000). Krabbe disease: Genetic aspects and progress toward therapy. *Molecular Genetics and Metabolism*, *70*(1), 1–9. doi:10.1006/mgme.2000.2990 PMID:10833326

Wen, P. Y., & Kesari, S. (2008). Malignant gliomas in adults. *The New England Journal of Medicine*, *359*(5), 492–507. doi:10.1056/NEJMra0708126 PMID:18669428

White, R. E., & Barry, D. S. (2015). The emerging roles of transplanted radial glial cells in regenerating the central nervous system. *Neural Regeneration Research*, *10*(10), 1548–1551. doi:10.4103/1673-5374.165317 PMID:26692835

Wieghofer, P., & Prinz, M. (2016). Genetic manipulation of microglia during brain development and disease. *Biochimica et Biophysica Acta*, *1862*(3), 299–309. doi:10.1016/j.bbadis.2015.09.019 PMID:26432479

Wierońska, J. M., & Pilc, A. (2009). Metabotropic glutamate receptors in the tripartite synapse as a target for new psychotropic drugs. *Neurochemistry International*, *55*(1–3), 85–97. doi:10.1016/j.neuint.2009.02.019 PMID:19428811

Wigmore, S. J., Walsh, T. S., Lee, A., & Ross, J. A. (1998). Pro-Inflammatory Cytokine Release and Mediation of the Acute Phase Protein Response in Fulminant Hepatic Failure. *Intensive Care Medicine*, *24*(3), 224–229. doi:10.1007001340050554 PMID:9565803

Wiley, J. (2021). Alzheimer's disease facts and figures. *Alzheimer's & Dementia*, *17*(3), 327–406. doi:10.1002/alz.12328 PMID:33756057

Wilhelmsson, U., Li, L., Pekna, M., Berthold, C.-H., Blom, S., Eliasson, C., Renner, O., Bushong, E., Ellisman, M., Morgan, T. E., & Pekny, M. (2004). Absence of Glial Fibrillary Acidic Protein and Vimentin Prevents Hypertrophy of Astrocytic Processes and Improves Post-Traumatic Regeneration. *The Journal of Neuroscience : The Official Journal of the Society for Neuroscience*, *24*(21), 5016–5021. doi:10.1523/JNEUROSCI.0820-04.2004 PMID:15163694

Wilkinson, S. P., Gazzard, B. G., Arroyo, V., Moodie, H., & Williams, R. (1974). Relation of Renal Impairment and Haemorrhagic Diathesis to Endotoxaemia in Fulminant Hepatic Failure. *Lancet*, *303*(7857), 521–524. doi:10.1016/S0140-6736(74)92711-1 PMID:4131970

Willard-Mack, C. L., Koehler, R. C., Hirata, T., Cork, L. C., Takahashi, H., Traystman, R. J., & Brusilow, S. W. (1996). Inhibition of Glutamine Synthetase Reduces Ammonia-Induced Astrocyte Swelling in Rat. *Neuroscience*, *71*(2), 589–599. doi:10.1016/0306-4522(95)00462-9 PMID:9053810

Williamson, J. M., & Lyons, D. A. (2018). Myelin dynamics throughout life: An ever-changing landscape? *Frontiers in Cellular Neuroscience*, *12*, 424. doi:10.3389/fncel.2018.00424

Wiltfang, J., Nolte, W., Weissenborn, K., Kornhuber, J., & Rüther, E. (1998). Psychiatric aspects of portal-systemic encephalopathy. *Metabolic Brain Disease*, *13*(4), 379–389. doi:10.1023/A:1020697127718 PMID:10206828

Winkelmann, A., Loebermann, M., Reisinger, E. C., Hartung, H. P., & Zettl, U. K. (2016). Disease-modifying therapies and infectious risks in multiple sclerosis. *Nature Reviews. Neurology*, *12*(4), 217–233. doi:10.1038/nrneurol.2016.21

Witt, P. A. L., & Fahn, S. (2016). Levodopa therapy for Parkinson disease. *Neurology*, *86*(14, Supplement 1), S3–S12. doi:10.1212/WNL.0000000000002509 PMID:27044648

Wolf, N. I., Ffrench-Constant, C., & van der Knaap, M. S. (2021). Leucodystrophies hypomyélinisantes – Démêler la biologie de la myéline. *Nature Reviews. Neurology*, *17*(2), 88–103. doi:10.103841582-020-00432-1 PMID:33324001

Wong, A. D., Ye, M., Levy, A. F., Rothstein, J. D., Bergles, D. E., & Searson, P. C. (2013). The blood-brain barrier: An engineering perspective. *Frontiers in Neuroengineering*, *6*, 7. doi:10.3389/fneng.2013.00007 PMID:24009582

Wong, K. Y., Roy, J., Fung, M. L., Heng, B. C., Zhang, C., & Lim, L. W. (2020). Relationships between mitochondrial dysfunction and neurotransmission failure in Alzheimer's disease. *Aging and Disease*, *11*(5), 1291. doi:10.14336/AD.2019.1125 PMID:33014538

Wong, P. C., Pardo, C. A., Borchelt, D. R., Lee, M. K., Copeland, N. G., Jenkins, N. A., Sisodia, S. S., Cleveland, D. W., & Price, D. L. (1995). An adverse property of a familial ALS-linked SOD1 mutation causes motor neuron disease characterized by vacuolar degeneration of mitochondria. *Neuron*, *14*(6), 1105–1116. doi:10.1016/0896-6273(95)90259-7 PMID:7605627

Woo, J., Im, S.-K., Chun, H., Jung, S.-Y., Oh, S.-J., Choi, N., Lee, C. J., & Hur, E.-M. (2017). Functional Characterization of Resting and Adenovirus-Induced Reactive Astrocytes in Three-Dimensional Culture. *Experimental Neurobiology*, *26*(3), 158–167. doi:10.5607/en.2017.26.3.158 PMID:28680301

Woolf, C. J. (1983). Evidence for a central component of post-injury pain hypersensitivity. *Nature*, *306*(5944), 686–688. doi:10.1038/306686a0 PMID:6656869

Wright, G., Shawcross, D., Olde Damink, S. W. M., & Jalan, R. (2007). Brain Cytokine Flux in Acute Liver Failure and Its Relationship with Intracranial Hypertension. *Metabolic Brain Disease*, *22*(3-4), 375–388. doi:10.100711011-007-9071-4 PMID:17899343

Wu, Y., Ma, Y., Liu, Z., Geng, Q., & Chen, Z. (2017). *Alterations of myelin morphology and oligodendrocyte development in early stage of Alzheimer's disease mouse model*. Elsevier.

Wu, Z., Sun, L., Hashioka, S., Yu, S., & C. S.-N. (2013). *Differential pathways for interleukin-1β production activated by chromogranin A and amyloid β in microglia*. Elsevier.

Wu, Y., Dissing-Olesen, L., MacVicar, B. A., & Stevens, B. (2015). Microglia: Dynamic mediators of synapse development and plasticity. *Trends in Immunology*, *36*(10), 605–613. doi:10.1016/j.it.2015.08.008 PMID:26431938

Wu, Y., Sriram, G., Fawzy, A. S., Fuh, J. Y., Rosa, V., Cao, T., & Wong, Y. S. (2016). Fabrication and evaluation of electrohydrodynamic jet 3D printed polycaprolactone/chitosan cell carriers using human embryonic stem cell-derived fibroblasts. *Journal of Biomaterials Applications*, *31*(2), 181–192. doi:10.1177/0885328216652537 PMID:27252227

Wu, Z., Parry, M., Hou, X.-Y., Liu, M.-H., Wang, H., Cain, R., Pei, Z.-F., Chen, Y.-C., Guo, Z.-Y., Abhijeet, S., & Chen, G. (2020). Gene therapy conversion of striatal astrocytes into GABAergic neurons in mouse models of Huntington's disease. *Nature Communications*, *11*(1), 1105. doi:10.103841467-020-14855-3 PMID:32107381

Xiao, L., Ohayon, D., McKenzie, I. A., Sinclair-Wilson, A., Wright, J. L., Fudge, A. D., Emery, B., Li, H., & Richardson, W. D. (2016). 'Rapid Production of New Oligodendrocytes Is Required in the Earliest Stages of Motor-Skill Learning'. *Nature Neuroscience*, *19*(9), 1210–1217. doi:10.1038/nn.4351 PMID:27455109

Xiao, M.-Y., Zhou, Q., & Nicoll, R. A. (2001). Metabotropic glutamate receptor activation causes a rapid redistribution of AMPA receptors. *Neuropharmacology*, *41*(6), 664–671. doi:10.1016/S0028-3908(01)00134-4 PMID:11640920

Xie, W., Strong, J. A., & Zhang, J.-M. (2009). Early blockade of injured primary sensory afferents reduces glial cell activation in two rat neuropathic pain models. *Neuroscience*, *160*(4), 847–857. doi:10.1016/j.neuroscience.2009.03.016 PMID:19303429

Xiong, Y., Zeng, Y. S., Zeng, C. G., Du, B. L., He, L. M., Quan, D. P., Zhang, W., Wang, J.-M., Wu, J.-L., Li, Y., & Li, J. (2009). Synaptic transmission of neural stem cells seeded in 3-dimensional PLGA scaffolds. *Biomaterials*, *30*(22), 3711–3722. doi:10.1016/j.biomaterials.2009.03.046 PMID:19375792

Xue, P., Chen, L., Lu, X., Zhang, J., Bao, G., Xu, G., Sun, Y., Guo, X., Jiang, J., Gu, H., & Cui, Z. (2017). Vimentin Promotes Astrocyte Activation After Chronic Constriction Injury. *Journal of Molecular Neuroscience*, *63*(1), 91–99. doi:10.100712031-017-0961-6 PMID:28791619

Xu-Friedman, M. A., Harris, K. M., & Regehr, W. G. (2001). Three-dimensional comparison of ultrastructural characteristics at depressing and facilitating synapses onto cerebellar Purkinje cells. *The Journal of Neuroscience : The Official Journal of the Society for Neuroscience*, *21*(17), 6666–6672. doi:10.1523/JNEUROSCI.21-17-06666.2001 PMID:11517256

Xu, J., Chen, S., Ahmed, S. H., Chen, H., Ku, G., Goldberg, M. P., & Hsu, C. Y. (2001). Amyloid-β peptides are cytotoxic to oligodendrocytes. *The Journal of Neuroscience : The Official Journal of the Society for Neuroscience*, *21*(1), RC118. doi:10.1523/JNEUROSCI.21-01-j0001.2001 PMID:11150354

Xu, J.-P., Zhao, J., & Li, S. (2011). Roles of NG2 glial cells in diseases of the central nervous system. *Neuroscience Bulletin*, *27*(6), 413–421. doi:10.100712264-011-1838-2 PMID:22108818

Xu, W., Huang, H. C., Lin, C. J., & Jiang, Z. F. (2010). Chitooligosaccharides protect rat cortical neurons against copper induced damage by attenuating intracellular level of reactive oxygen species. *Bioorganic & Medicinal Chemistry Letters*, *20*(10), 3084–3088. doi:10.1016/j.bmcl.2010.03.105 PMID:20399653

Xu, Y., Asghar, S., Yang, L., Li, H., Wang, Z., Ping, Q., & Xiao, Y. (2017). Lactoferrin-coated polysaccharide nanoparticles based on chitosan hydrochloride/hyaluronic acid/PEG for treating brain glioma. *Carbohydrate Polymers*, *157*, 419–428. doi:10.1016/j.carbpol.2016.09.085 PMID:27987946

Xu, Z., Xiao, N., Chen, Y., Huang, H., Marshall, C., Gao, J., Cai, Z., Wu, T., Hu, G., & Xiao, M. (2015). Deletion of aquaporin-4 in APP/PS1 mice exacerbates brain Aβ accumulation and memory deficits. *Molecular Neurodegeneration*, *10*(1), 58. doi:10.118613024-015-0056-1 PMID:26526066

Yamada, T., Kawamata, T., Walker, D. G., & McGeer, P. L. (1992). Vimentin immunoreactivity in normal and pathological human brain tissue. *Acta Neuropathologica*, *84*(2), 157–162. doi:10.1007/BF00311389 PMID:1523971

Yamamizu, K., Iwasaki, M., Takakubo, H., Sakamoto, T., Ikuno, T., Miyoshi, M., Kondo, T., Nakao, Y., Nakagawa, M., Inoue, H., & Yamashita, J. K. (2017). RETRACTED: In Vitro Modeling of Blood-Brain Barrier with Human iPSC-Derived Endothelial Cells, Pericytes, Neurons, and Astrocytes via Notch Signaling. *Stem Cell Reports*, *8*(3), 634–647. doi:10.1016/j.stemcr.2017.01.023 PMID:28238797

Yamamuro, T., Hench, L. L., & Wilson, J. (1990). CRC handbook of bioactive ceramics. (No Title). Maclean, F. L., Horne, M. K., Williams, R. J., & Nisbet, D. R. (2018). Biomaterial systems to resolve brain inflammation after traumatic injury. *APL Bioengineering*, *2*(2).

Yamanaka, G., Suzuki, S., Morishita, N., Takeshita, M., Kanou, K., Takamatsu, T., Suzuki, S., Morichi, S., Watanabe, Y., Ishida, Y., Go, S., Oana, S., Kashiwagi, Y., & Kawashima, H. (2021). Role of neuroinflammation and blood-brain barrier permutability on migraine. In International Journal of Molecular Sciences, 22(16). MDPI AG. doi:10.3390/ijms22168929

Yamanaka, K., Chun, S. J., Boillee, S., Fujimori-Tonou, N., Yamashita, H., Gutmann, D. H., Takahashi, R., Misawa, H., & Cleveland, D. W. (2008). Astrocytes as determinants of disease progression in inherited amyotrophic lateral sclerosis. *Nature Neuroscience*, *11*(3), 251–253. doi:10.1038/nn2047 PMID:18246065

Yang, E. J., Kim, J. G., Kim, J. Y., Kim, S. C., Lee, N. H., & Hyun, C. G. (2010). Anti-inflammatory effect of chitosan oligosaccharides in RAW 264.7 cells. *Open Life Sciences*, *5*(1), 95–102. doi:10.247811535-009-0066-5

Yang, F., Trolle Lagerros, Y., Bellocco, R., Adami, H.-O., Fang, F., Pedersen, N. L., & Wirdefeldt, K. (2015). Physical activity and risk of Parkinson's disease in the Swedish National March Cohort. *Brain*, *138*(2), 269–275. doi:10.1093/brain/awu323 PMID:25410713

Yang, R., Zheng, Y., Wang, Q., & Zhao, L. (2018). Curcumin-loaded chitosan–bovine serum albumin nanoparticles potentially enhanced Aβ 42 phagocytosis and modulated macrophage polarization in Alzheimer's disease. *Nanoscale Research Letters*, *13*(1), 1–9. doi:10.118611671-018-2759-z PMID:30350003

Yang, T., Li, S., Xu, H., Walsh, D. M., & Selkoe, D. J. (2017). Large soluble oligomers of amyloid β-protein from Alzheimer brain are far less neuroactive than the smaller oligomers to which they dissociate. *The Journal of Neuroscience : The Official Journal of the Society for Neuroscience*, *37*(1), 152–163. doi:10.1523/JNEUROSCI.1698-16.2016 PMID:28053038

Yao, Z. A., Chen, F. J., Cui, H. L., Lin, T., Guo, N., & Wu, H. G. (2018). Efficacy of chitosan and sodium alginate scaffolds for repair of spinal cord injury in rats. *Neural Regeneration Research*, *13*(3), 502. doi:10.4103/1673-5374.228756 PMID:29623937

Yasuno, F., Kosaka, J., Ota, M., Higuchi, M., Ito, H., Fujimura, Y., Nozaki, S., Takahashi, S., Mizukami, K., Asada, T., & Suhara, T. (2012). Increased binding of peripheral benzodiazepine receptor in mild cognitive impairment–dementia converters measured by positron emission tomography with [11C] DAA1106. *Psychiatry Research: Neuroimaging*, *203*(1), 67–74. doi:10.1016/j.pscychresns.2011.08.013 PMID:22892349

Yeh, C. Y., Verkhratsky, A., Terzieva, S., & Rodríguez, J. J. (2013). Glutamine synthetase in astrocytes from entorhinal cortex of the triple transgenic animal model of Alzheimer's disease is not affected by pathological progression. *Biogerontology*, *14*(6), 777–787. doi:10.100710522-013-9456-1 PMID:23990215

Yeh, C.-Y., Vadhwana, B., Verkhratsky, A., & Rodríguez, J. J. (2011). Early astrocytic atrophy in the entorhinal cortex of a triple transgenic animal model of Alzheimer's disease. *ASN Neuro*, *3*(5), AN20110025. doi:10.1042/AN20110025 PMID:22103264

Yellen, G. (2018). Fueling thought: Management of glycolysis and oxidative phosphorylation in neuronal metabolism. *The Journal of Cell Biology*, *217*(7), 2235–2246. doi:10.1083/jcb.201803152

Ye, Y., Wang, G., Wang, H., & Wang, X. (2011). Brain-derived neurotrophic factor (BDNF) infusion restored astrocytic plasticity in the hippocampus of a rat model of depression. *Neuroscience Letters*, *503*(1), 15–19. doi:10.1016/j.neulet.2011.07.055 PMID:21843601

Yi, C., Verkhratsky, A., & Niu, J. (2023). Pathological potential of oligodendrocyte precursor cells: Terra incognita. *Trends in Neurosciences*, *46*(7), 581–596. doi:10.1016/j.tins.2023.04.003

Yi, H., Wu, L. Q., Bentley, W. E., Ghodssi, R., Rubloff, G. W., Culver, J. N., & Payne, G. F. (2005). Biofabrication with chitosan. *Biomacromolecules*, *6*(6), 2881–2894. doi:10.1021/bm050410l PMID:16283704

Yi, J., Horky, L. L., Friedlich, A. L., Shi, Y., Rogers, J. T., & Huang, X. (2009). L-arginine and Alzheimer's disease. *International Journal of Clinical and Experimental Pathology*, *2*(3), 211. PMID:19079617

Yin, J., Valin, K. L., Dixon, M. L., & Leavenworth, J. W. (2017). The Role of Microglia and Macrophages in CNS Homeostasis, Autoimmunity, and Cancer. *Journal of Immunology Research, 5150678*, 1–12. doi:10.1155/2017/5150678 PMID:29410971

Yin, Y., Zhang, X., Li, Z., Deng, L., Jiao, G., Zhang, B., Xie, P., Mu, H., Qiao, W., & Zou, J. (2013). Glucocorticoid receptor β regulates injury-mediated astrocyte activation and contributes to glioma pathogenesis via modulation of β-catenin/TCF transcriptional activity. *Neurobiology of Disease*, *59*, 165–176. doi:10.1016/j.nbd.2013.07.013 PMID:23906498

Yirmiya, R., Rimmerman, N., & Reshef, R. (2015). Depression as a microglial disease. *Trends in Neurosciences*, *38*(10), 637–658. doi:10.1016/j.tins.2015.08.001 PMID:26442697

Yolas, C., Ozdemir, N. G., Kanat, A., Aydin, M. D., Keles, P., Kepoglu, U., Aydin, N., & Gundogdu, C. (2016). Uncovering a New Cause of Obstructive Hydrocephalus Following Subarachnoid Hemorrhage: Choroidal Artery Vasospasm–Related Ependymal Cell Degeneration and Aqueductal Stenosis—First Experimental Study. *World Neurosurgery*, *90*, 484–491. doi:10.1016/j.wneu.2016.03.049 PMID:27020981

Yong, H. Y. F., Rawji, K. S., Ghorbani, S., Xue, M., & Yong, V. W. (2019). The benefits of neuroinflammation for the repair of the injured central nervous system. *Cellular & Molecular Immunology*, *16*(6), 540–546. doi:10.103841423-019-0223-3 PMID:30874626

Younes, I., & Rinaudo, M. (2015). Chitin and chitosan preparation from marine sources. Structure, properties and applications. *Marine Drugs*, *13*(3), 1133–1174. doi:10.3390/md13031133 PMID:25738328

You, Y., Joseph, C., Wang, C., Gupta, V., Liu, S., Yiannikas, C., Chua, B. E., Chitranshi, N., Shen, T., Dheer, Y., Invernizzi, A., Borotkanics, R., Barnett, M., Graham, S. L., & Klistorner, A. (2019). Demyelination precedes axonal loss in the transneuronal spread of human neurodegenerative disease. *Brain*, *142*(2), 426–442. doi:10.1093/brain/awy338 PMID:30668642

Yuan, Y., Zhang, P., Yang, Y., Wang, X., & Gu, X. (2004). The interaction of Schwann cells with chitosan membranes and fibers in vitro. *Biomaterials*, *25*(18), 4273–4278. doi:10.1016/j.biomaterials.2003.11.029 PMID:15046917

Yue, P., Gao, L., Wang, X., Ding, X., & Teng, J. (2018). Pretreatment of glial cell-derived neurotrophic factor and geranylgeranylacetone ameliorates brain injury in Parkinson's disease by its anti-apoptotic and anti-oxidative property. *Journal of Cellular Biochemistry*, *119*(7), 5491–5502. doi:10.1002/jcb.26712 PMID:29377238

Yu, G., Zhang, Y., & Ning, B. (2021). Reactive Astrocytes in Central Nervous System Injury : Subgroup and Potential Therapy. *Frontiers in Cellular Neuroscience*, *15*, 792764. doi:10.3389/fncel.2021.792764 PMID:35002629

Yun, S., Kam, T., Panicker, N., Kim, S., & Oh, Y. (2018). *Block of A1 astrocyte conversion by microglia is neuroprotective in models of Parkinson's disease.* Nature. https://www.nature.com/articles/s41591-018-0051-5

Yu, S., Xu, X., Feng, J., Liu, M., & Hu, K. (2019). Chitosan and chitosan coating nanoparticles for the treatment of brain disease. *International Journal of Pharmaceutics*, *560*, 282–293. doi:10.1016/j.ijpharm.2019.02.012 PMID:30772458

Yu, T.-W., Lane, H.-Y., & Lin, C.-H. (2021). Novel therapeutic approaches for Alzheimer's disease: An updated review. *International Journal of Molecular Sciences*, *22*(15), 8208. doi:10.3390/ijms22158208 PMID:34360973

Yu, X., Nagai, J., & Khakh, B. S. (2020). Improved tools to study astrocytes. *Nature Reviews. Neuroscience*, *21*(3), 121–138. doi:10.103841583-020-0264-8 PMID:32042146

Zamproni, L. N., Mundim, M. V., Porcionatto, M. A., & des Rieux, A. (2017). Injection of SDF-1 loaded nanoparticles following traumatic brain injury stimulates neural stem cell recruitment. *International Journal of Pharmaceutics*, *519*(1-2), 323–331. doi:10.1016/j.ijpharm.2017.01.036 PMID:28115261

Zarpelon, A. C., Rodrigues, F. C., Lopes, A. H., Souza, G. R., Carvalho, T. T., Pinto, L. G., Xu, D., Ferreira, S. H., Alves-Filho, J. C., McInnes, I. B., Ryffel, B., Quesniaux, V. F. J., Reverchon, F., Mortaud, S., Menuet, A., Liew, F. Y., Cunha, F. Q., Cunha, T. M., & Verri, W. A. Jr. (2016). Spinal cord oligodendrocyte-derived alarmin IL-33 mediates neuropathic pain. *The FASEB Journal, 30*(1), 54–65. doi:10.1096/fj.14-267146 PMID:26310268

Zeisel, A., Hochgerner, H., Lönnerberg, P., Johnsson, A., Memic, F., van der Zwan, J., Häring, M., Braun, E., Borm, L. E., La Manno, G., Codeluppi, S., Furlan, A., Lee, K., Skene, N., Harris, K. D., Hjerling-Leffler, J., Arenas, E., Ernfors, P., Marklund, U., & Linnarsson, S. (2018). Molecular Architecture of the Mouse Nervous System. *Cells, 174*(4), 999–1014. e22. doi:10.1016/j.cell.2018.06.021 PMID:30096314

Zeis, T., Enz, L., & Schaeren-Wiemers, N. (2016). The immunomodulatory oligodendrocyte. *Brain Research, 1641*, 139–148. doi:10.1016/j.brainres.2015.09.021 PMID:26423932

Zemtsova, I., Görg, B., Keitel, V., Bidmon, H. J., Schrör, K., & Häussinger, D. (2011). Microglia Activation in Hepatic Encephalopathy in Rats and Humans. *Hepatology (Baltimore, Md.), 54*(1), 204–215. doi:10.1002/hep.24326 PMID:21452284

Zhan, Y., Paolicelli, R., & Sforazzini, F. (2014). *Deficient neuron-microglia signaling results in impaired functional brain connectivity and social behavior.* Nature. https://www.nature.com/articles/nn.3641

Zhang, Q., Chen, W., Tan, S., & Lin, T. (2017). *Stem Cells for Modeling and Therapy of Parkinson's Disease.* Mary Ann Liebert Publishers. doi:10.1089/hum.2016.116

Zhang, Y., Chen, Y., Wu, J., Manaenko, A., Yang, P., Tang, J., Fu, W., Zhang, J.H. (2015). Activation of Dopamine D2 Receptor Suppresses Neuroinflammation Through αB-Crystalline by Inhibition of NF-κB Nuclear Translocation in Experimental ICH Mice Model. *Stroke, 46*(9), 2637-46. . doi:10.1161/STROKEAHA.115.009792

Zhang, C., Zhang, Y.-P., Li, Y.-Y., Liu, B.-P., Wang, H.-Y., Li, K.-W., Zhao, S., & Song, C. (2019). Minocycline ameliorates depressive behaviors and neuro-immune dysfunction induced by chronic unpredictable mild stress in the rat. *Behavioural Brain Research, 356*, 348–357. doi:10.1016/j.bbr.2018.07.001 PMID:30003978

Zhang, H., Mei, X., Zhang, P., Ma, C., White, F. A., Donnelly, D. F., & Lamotte, R. H. (2009). Altered functional properties of satellite glial cells in compressed spinal ganglia. *Glia, 57*(15), 1588–1599. doi:10.1002/glia.20872 PMID:19330845

Zhang, S., Wang, X. J., Tian, L. P., Pan, J., Lu, G. Q., Zhang, Y. J., Ding, J. Q., & Di Chen, S. (2011). CD200-CD200R dysfunction exacerbates microglial activation and dopaminergic neurodegeneration in a rat model of Parkinson's disease. *Journal of Neuroinflammation, 8*(1), 154. doi:10.1186/1742-2094-8-154 PMID:22053982

Zhang, W., Wang, T., Pei, Z., Miller, D. S., Wu, X., Block, M. L., Wilson, B., Zhang, W., Zhou, Y., Hong, J.-S., & Zhang, J. (2005). Aggregated α-synuclein activates microglia: A process leading to disease progression in Parkinson's disease. *The FASEB Journal, 19*(6), 533–542. doi:10.1096/fj.04-2751com PMID:15791003

Zhang, X., Chen, Y., Wang, C., & Huang, L.-Y. M. (2007). Neuronal somatic ATP release triggers neuron-satellite glial cell communication in dorsal root ganglia. *Proceedings of the National Academy of Sciences of the United States of America, 104*(23), 9864–9869. doi:10.1073/pnas.0611048104 PMID:17525149

Zhang, X., Huang, W. J., & Chen, W. W. (2016). TGF-β1 factor in the cerebrovascular diseases of Alzheimer's disease. *European Review for Medical and Pharmacological Sciences, 20*(24), 5178–5185. PMID:28051272

Zhang, X., Ru, B., Sha, W., Xin, W., Zhou, H., & Zhang, Y. (2014). Performance on the W isconsin card-sorting test and serum levels of glial cell line-derived neurotrophic factor in patients with major depressive disorder. *Asia-Pacific Psychiatry, 6*(3), 302–307. doi:10.1111/appy.12120 PMID:24436216

Zhang, X.-X., Tian, Y., Wang, Z.-T., Ma, Y.-H., Tan, L., & Yu, J.-T. (2021). The epidemiology of Alzheimer's disease modifiable risk factors and prevention. *The Journal of Prevention of Alzheimer's Disease*, *8*, 313–321. doi:10.14283/jpad.2021.15 PMID:34101789

Zhang, X., Zhang, Z., Xie, C., Xi, G., Zhou, H., Zhang, Y., & Sha, W. (2008). Effect of treatment on serum glial cell line-derived neurotrophic factor in depressed patients. *Progress in Neuro-Psychopharmacology & Biological Psychiatry*, *32*(3), 886–890. doi:10.1016/j.pnpbp.2008.01.004 PMID:18262703

Zhang, Y., Chen, K., Sloan, S. A., Bennett, M. L., Scholze, A. R., O'Keeffe, S., Phatnani, H. P., Guarnieri, P., Caneda, C., Ruderisch, N., Deng, S., Liddelow, S. A., Zhang, C., Daneman, R., Maniatis, T., Barres, B. A., & Wu, J. Q. (2014). An RNA-Sequencing Transcriptome and Splicing Database of Glia, Neurons, and Vascular Cells of the Cerebral Cortex. *The Journal of Neuroscience : The Official Journal of the Society for Neuroscience*, *34*(36), 11929–11947. doi:10.1523/JNEUROSCI.1860-14.2014 PMID:25186741

Zhang, Z., Wang, X., & Wang, S. (2008). Isolation and characterization of mesenchymal stem cells derived from bone marrow of patients with Parkinson's disease. *In Vitro Cellular & Developmental Biology. Animal*, *44*(5–6), 169–177. doi:10.100711626-008-9093-1 PMID:18401666

Zhang, Z., Zhang, N., & Ding, S. (2022). Glial cell line-derived neurotrophic factor in brain repair after focal ischemic stroke. *Neural Regeneration Research*, *17*(8), 1735–1736. doi:10.4103/1673-5374.332141 PMID:35017424

Zhan, Y., Paolicelli, R. C., Sforazzini, F., Weinhard, L., Bolasco, G., Pagani, F., Vyssotski, A. L., Bifone, A., Gozzi, A., Ragozzino, D., & Gross, C. T. (2014). Deficient neuron-microglia signaling results in impaired functional brain connectivity and social behavior. *Nature Neuroscience*, *17*(3), 400–406. doi:10.1038/nn.3641 PMID:24487234

Zhao, L., Mandler, M. D., Yi, H., & Feng, Y. (2010). Quaking I controls a unique cytoplasmic pathway that regulates alternative splicing of myelin-associated glycoprotein. *Proceedings of the National Academy of Sciences of the United States of America*, *107*(44), 19061–19066. doi:10.1073/pnas.1007487107 PMID:20956316

Zhao, P., Waxman, S. G., & Hains, B. C. (2007). Extracellular signal-regulated kinase-regulated microglia-neuron signaling by prostaglandin E2 contributes to pain after spinal cord injury. *The Journal of Neuroscience : The Official Journal of the Society for Neuroscience*, *27*(9), 2357–2368. doi:10.1523/JNEUROSCI.0138-07.2007 PMID:17329433

Zhao, W., Beers, D. R., Henkel, J. S., Zhang, W., Urushitani, M., Julien, J. P., & Appel, S. H. (2010). Extracellular mutant SOD1 induces microglial-mediated motoneuron injury. *Glia*, *58*(2), 231–243. doi:10.1002/glia.20919 PMID:19672969

Zhao, Z., Nelson, A. R., Betsholtz, C., & Zlokovic, B. V. (2015). Establishment and Dysfunction of the Blood-Brain Barrier. *Cell*, *163*(5), 1064–1078. doi:10.1016/j.cell.2015.10.067 PMID:26590417

Zhou, B., Zhu, Z., Ransom, B. R., & Tong, X. (2021). Oligodendrocyte lineage cells and depression. *Molecular Psychiatry*, *26*(1), 103–117. doi:10.103841380-020-00930-0 PMID:33144710

Zhou, B., Zuo, Y. X., & Jiang, R. T. (2019). Astrocyte morphology: Diversity, plasticity, and role in neurological diseases. *CNS Neuroscience & Therapeutics*, *25*(6), 665–673. doi:10.1111/cns.13123 PMID:30929313

Zhou, C., Ni, W., Zhu, T., Dong, S., Sun, P., & Hua, F. (2022). Cellular reprogramming and its potential application in Alzheimer's disease. *Frontiers in Neuroscience*, *16*, 884667. doi:10.3389/fnins.2022.884667 PMID:35464309

Zhou, H., Su, J., Hu, X., Zhou, C., Li, H., Chen, Z., Xiao, Q., Wang, B. O., Wu, W., Sun, Y., Zhou, Y., Tang, C., Liu, F., Wang, L., Feng, C., Liu, M., Li, S., Zhang, Y., Xu, H., ... Yang, H. (2020). Glia-to-neuron conversion by CRISPR-CasRx alleviates symptoms of neurological disease in mice. *Cell*, *181*(3), 590–603. doi:10.1016/j.cell.2020.03.024 PMID:32272060

Zhou, Q., Homma, K. J., & Poo, M. M. (2004). Shrinkage of dendritic spines associated with long-term depression of hippocampal synapses. *Neuron*, *44*(5), 749–757. doi:10.1016/j.neuron.2004.11.011 PMID:15572107

Zhou, S., Yang, Y., Gu, X., & Ding, F. (2008). Chitooligosaccharides protect cultured hippocampal neurons against glutamate-induced neurotoxicity. *Neuroscience Letters*, *444*(3), 270–274. doi:10.1016/j.neulet.2008.08.040 PMID:18755243

Zhou, X. E., Suino-Powell, K., Schultz, C. R., Aleiwi, B., Brunzelle, J. S., Lamp, J., Vega, I. E., Ellsworth, E., Bachmann, A. S., & Melcher, K. (2021). Structural basis of binding and inhibition of ornithine decarboxylase by 1-amino-oxy-3-aminopropane. *The Biochemical Journal*, *478*(23), 4137–4149. doi:10.1042/BCJ20210647 PMID:34796899

Zhou, X. F., Deng, Y. S., Chie, E., Xue, Q., Zhong, J. H., McLachlan, E. M., Rush, R. A., & Xian, C. J. (1999). Satellite-cell-derived nerve growth factor and neurotrophin-3 are involved in noradrenergic sprouting in the dorsal root ganglia following peripheral nerve injury in the rat. *The European Journal of Neuroscience*, *11*(5), 1711–1722. doi:10.1046/j.1460-9568.1999.00589.x PMID:10215925

Zhou, X. F., Deng, Y. S., Xian, C. J., & Zhong, J. H. (2000). Neurotrophins from dorsal root ganglia trigger allodynia after spinal nerve injury in rats. *The European Journal of Neuroscience*, *12*(1), 100–105. doi:10.1046/j.1460-9568.2000.00884.x PMID:10651864

Zhou, Y., Borchelt, D., Bauson, J. C., Fazio, S., Miles, J. R., Tavori, H., & Notterpek, L. (2020). Subcellular diversion of cholesterol by gain-and loss-of-function mutations in PMP22. *Glia*, *68*(11), 2300–2315. doi:10.1002/glia.23840 PMID:32511821

Zhou, Z., Chen, S., Huang, Y., Gu, B., Li, J., Wu, C., Yin, P., Zhang, Y., & Li, H. (2022). Simultaneous visualization and quantification of copper (II) ions in Alzheimer's disease by a near-infrared fluorescence probe. *Biosensors & Bioelectronics*, *198*, 113858. doi:10.1016/j.bios.2021.113858 PMID:34871835

Zhuang, Z.-Y., Wen, Y.-R., Zhang, D.-R., Borsello, T., Bonny, C., Strichartz, G. R., Decosterd, I., & Ji, R.-R. (2006). A peptide c-Jun N-terminal kinase (JNK) inhibitor blocks mechanical allodynia after spinal nerve ligation: Respective roles of JNK activation in primary sensory neurons and spinal astrocytes for neuropathic pain development and maintenance. *The Journal of Neuroscience : The Official Journal of the Society for Neuroscience*, *26*(13), 3551–3560. doi:10.1523/JNEUROSCI.5290-05.2006 PMID:16571763

Zhu, Y., Yang, G. Y., Ahlemeyer, B., Pang, L., Che, X. M., Culmsee, C., Klumpp, S., & Krieglstein, J. (2002). Transforming Growth Factor-β1 Increases Bad Phosphorylation and Protects Neurons Against Damage. *The Journal of Neuroscience : The Official Journal of the Society for Neuroscience*, *22*(10), 3898–3909. doi:10.1523/JNEUROSCI.22-10-03898.2002 PMID:12019309

Zia, S., Rawji, K. S., Michaels, N. J., Burr, M., Kerr, B. J., Healy, L. M., & Plemel, J. R. (2020). Microglia Diversity in Health and Multiple Sclerosis. *Frontiers in Immunology*, *11*, 588021. doi:10.3389/fimmu.2020.588021 PMID:33240276

Ziv, Y., Ron, N., Butovsky, O., Landa, G., Sudai, E., Greenberg, N., Cohen, H., Kipnis, J., & Schwartz, M. (2006). Immune cells contribute to the maintenance of neurogenesis and spatial learning abilities in adulthood. *Nature Neuroscience*, *9*(2), 268–275. doi:10.1038/nn1629 PMID:16415867

Zlokovic, B. V. (2005). 'Neurovascular Mechanisms of Alzheimer's Neurodegeneration'. *Trends in Neurosciences*, *28*(4), 202–208. doi:10.1016/j.tins.2005.02.001 PMID:15808355

Zlokovic, B. V. (2011). Neurovascular pathways to neurodegeneration in Alzheimer's disease and other disorders. *Nature Reviews. Neuroscience*, *12*(12), 723–738. doi:10.1038/nrn3114 PMID:22048062

Zonta, M., Sebelin, A., Gobbo, S., Fellin, T., Pozzan, T., & Carmignoto, G. (2003). Glutamate-mediated cytosolic calcium oscillations regulate a pulsatile prostaglandin release from cultured rat astrocytes. *The Journal of Physiology*, *553*(Pt 2), 407–414. doi:10.1113/jphysiol.2003.046706 PMID:14500777

Zotova, E., Holmes, C., Johnston, D., Neal, J. W., Nicoll, J. A. R., & Boche, D. (2011). Microglial alterations in human Alzheimer's disease following Aβ42 immunization. *Neuropathology and Applied Neurobiology*, *37*(5), 513–524. doi:10.1111/j.1365-2990.2010.01156.x PMID:21166690

Zou, L., Zhang, Y., Liu, X., Chen, J., & Zhang, Q. (2019). Biomimetic mineralization on natural and synthetic polymers to prepare hybrid scaffolds for bone tissue engineering. *Colloids and Surfaces. B, Biointerfaces*, *178*, 222–229. doi:10.1016/j.colsurfb.2019.03.004 PMID:30870789

Zuccato, C. & Cattaneo, E. (2014). Huntington's disease. *Neurotrophic factors,* 357–409. Research Gate.

Zuccato, C., Marullo, M., Conforti, P., MacDonald, M. E., Tartari, M., & Cattaneo, E. (2008). Systematic assessment of BDNF and its receptor levels in human cortices affected by Huntington's disease. *Brain Pathology (Zurich, Switzerland)*, *18*(2), 225–238. doi:10.1111/j.1750-3639.2007.00111.x PMID:18093249

Zurolo, E., de Groot, M., Iyer, A., Anink, J., van Vliet, E. A., Heimans, J. J., Reijneveld, J. C., Gorter, J. A., & Aronica, E. (2012). Regulation of Kir4.1 expression in astrocytes and astrocytic tumors: A role for interleukin-1 β. *Journal of Neuroinflammation*, *9*(1), 280. doi:10.1186/1742-2094-9-280 PMID:23270518

About the Contributors

Bilal El-Mansoury, PhD. His research work focuses mainly on glial cells dysfunction in hepatic encephalopathy and neurodegenerative disorders. El-Mansoyry is a research assistant at Frontiers in Neurosciences. He is also a member of the International Society for Hepatic Encephalopathy and Nitrogen Metabolism (ISHEN) and The International Parkinson and Movement Disorder Society (MDS).

El Hiba Omar is professor of physiopathology, neuroscience and histology, Faculty of Sciences, Chouaib Doukkali University, He was born in 12 September 1985 in Marrakech-Safi state (Morocco) and was graduated in 2007 from Cadi Ayyad University (Morocco) with a bachelor degree in Biology. In 2009, hewas graduated from the same university with Master's degree in Life and Health Sciences. He obtained his PhD in 2015 on the "Physiopathology of Hepatic Encephalopathy" from Cadi Ayyad University. He is, since 2014, member of the International Parkinson and Movement Disorders Society (MDS), USA and the vice-President of the Moroccan Society of Ethics and Animal Research (MoSEAR).

Arumugam Jayakumar is an experienced Associate Professor with a demonstrated history of working in the research industry. He is skilled in Protein Chemistry, Molecular Biology, and Biotechnology. He received a PhD in Philosophy from Miami Veterans Hospital. His interests are focused in Neuroscience, Neuropharmacology, Human Physiology, and Neuroimmunology.

***

Ramadane Abakar Ramdane is a research engineer at Nice University hospital.

Hajar Azraida is a Phd student, Laboratory of Pharmacology, Neurobiology, Anthropobiology and Environment, Department of Biology, Faculty of Sciences Semlalia, Cadi Ayyad University, Marrakech, Morocco.

Chaimaa Azzouhri Phd Student in Neuroscience, Faculty of Sciences and Techniques, Sultan Moulay Slimane University

Dharmeswar Barhoi is a regular faculty in the Department of Zoology, The Assam Royal Global University. Dr. Barhoi has experience in teaching as well as in research along with experience in guiding more than 32 Master's students and Ph. D Research Scholars as well. Dr. Barhoi has a specialization in Cell and Molecular Biology He has published more than 15 Research articles and a number of book chapters and contributed to Edited books.

Mohamed Chraa is a professor of neurophysiology.

Ahmed Draoui has a PhD Neuroscience, Pharmacology and Environment Unit (NPEU), Cadi Ayyad UniversityDepartment of Biology.

Hanane El Fatimi is a PhD student in neuroscience, Neurophysiology and oncogenetics laboratory / Neuroscience and Cellular Physiology Team (NPC),Mohammed VI University of Health Sciences, Morocco.

El Ghachi Hafida a PhD student At Cadi Ayyad University, faculty of Sciences Semlalia, Marrakech, Morocco who works on Neurosciences and is part of the Neurosciences, Pharmacology, and Environnement research unit.

Samia Elouali is a PhD student in biology department at Cadi Ayyad University at LIRBEM and member in laboratory LPCM in Mons University in Belgium.

Halima Gamrani is a professor of neuroscience, head of the laboratory of Neuroscience, Pharmacology and Environment, FSSM, UCAM.

Asmaa Haj Khlifa, young motivated third-year PhD student in neurosciences at the Faculty of Sciences Semlalia with a great ability to learn and practice anything related to my knowledge. I completed my bachelor's degree in Life Sciences and master's degree in Physiopathology and Anthropobiology from Cadi Ayyad University

Laadraoui Jawad is a Professor of Neurosciences, Animal Physiology and Biotechnology.

Hamid Kabdy is a Ph.D. Candidate in Biology | Université Cadi Ayyad, Marrakech An avid biology enthusiast driven by the pursuit of scientific discovery, I am currently pursuing my Ph.D. in Biology within the Faculty of Sciences Semlalia at Université Cadi Ayyad in Marrakech.

Hanane Khalki is a professor of physiology and biophysics, Phd on neurosciences and pharmacology

Mohamed Merzouki is a research professor at the Faculty of Science and Technology - Sultan Moulay Slimane University - Beni Mellal (since 2016) and at the Ecole Supérieure de Technologie of Beni Mellal (2013-2016), I am specialized in Ecotoxicology and Biostatistics Modeling, I am also Founding member of the Moroccan Society of Ecotoxicology (SMETox) in 2015. My research is generally concerned with studying the impact of heavy metal pollution on animal physiology. Research on " Heavy metals and modeling of aquatic animal survival" was achieved at the Institute of Marine Substances and Organisms, Department of Ecotoxicology (ISOmer), Nantes, at the Laboratory of Pathology and Marine Invertebrate Genetics (LPGIM), Tremblade Station (IFREMER) and the Institute of Biology and Applied Ecology (IBEA) Angers (France). My current research activities focus on the assessment of the impact of the environment and climate change on health and biodiversity, the analysis of the impact of climate change on animal production and the modeling of the effect of climate change. pollution and climate change on the survival of animal populations.

Hassane Oudadesse is a currently is the Head of the research unit of Biomaterials since 2001, Vice President of University of Rennes 1, Human Resources 2008 - 2012, Director of Master 2, Solid State Chemistry and Materials (CSM) since 2006.

Dipsanu Paul is a master's student studying in the Department of Zoology, The Assam Royal Global University. He has Cell Biology and Genetics as a special paper during his post-graduation. He is working on his special paper project under the supervision of Dr, Dharmeswar Barhoi, Assistant Professor, Department of Zoology, The Assam Royal Global University.

Mohammed Rhazi his research has focused on the valorization of terrestrial and marine fishing wastes by the production of chitinous derivatives and their applications in various fields (pharmacology, biomedical, agroalimentary and water treatment). He has been working since 1996 on the complexation of metal ions by chitosan and its derivatives and their interest in wastewater treatment.

Es-Said Sabir is a Phd sutdent at Metabolic platform, Biochemistry Laboratory, Faculty of Medicine, Cadi Ayyad University, Marrakech, Morocco.

Kamal Smimihis a doctor in neurosciences Faculty of Sciences and Techniques, Sultan Moulay Slimane University, Mococco.

Index

G

H

I

L

M

N

O

P

R

S

T

Premier Reference Source
Digital Therapies in Psychosocial Rehabilitation and Mental Health

Premier Reference Source
Historical and Epidemiological Analyses on the Impact of Infectious Disease on Society

Premier Reference Source
Ethical Implications of Reshaping Healthcare With Emerging Technologies

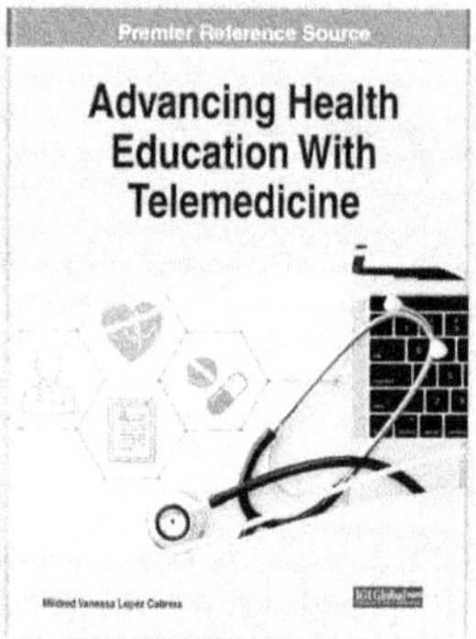

Premier Reference Source
Advancing Health Education With Telemedicine

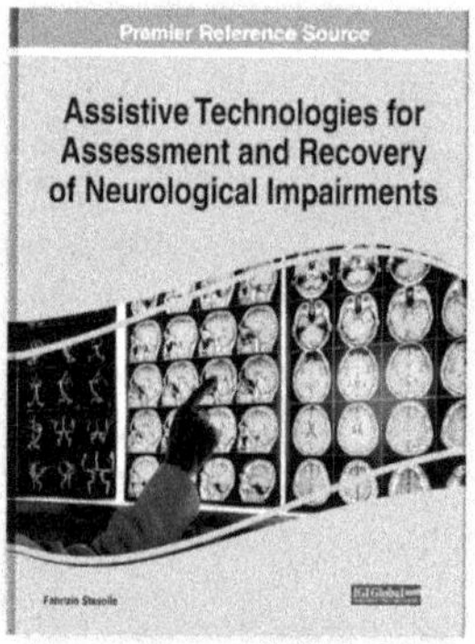

Premier Reference Source
Assistive Technologies for Assessment and Recovery of Neurological Impairments

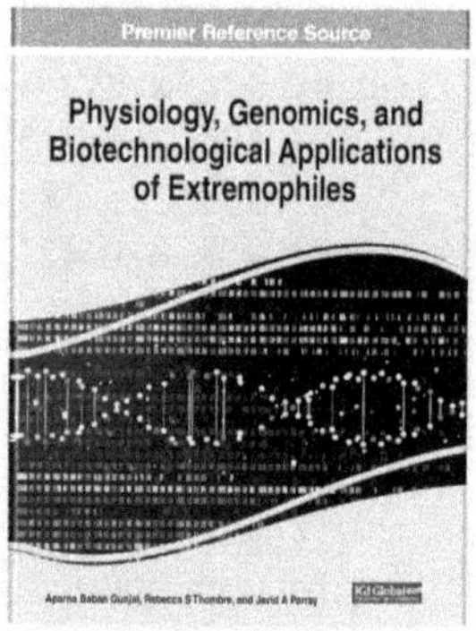

Premier Reference Source
Physiology, Genomics, and Biotechnological Applications of Extremophiles

www.ingramcontent.com/pod-product-compliance
Lightning Source LLC
LaVergne TN
LVHW080618300926
849162LV00005B/1202